Modern Birkhäuser Classics

Many of the original research and survey monographs in pure and applied mathematics published by Birkhäuser in recent decades have been groundbreaking and have come to be regarded as foundational to the subject. Through the MBC Series, a select number of these modern classics, entirely uncorrected, are being re-released in paperback (and as eBooks) to ensure that these treasures remain accessible to new generations of students, scholars, and researchers.

A History of Algebraic and Differential Topology 1900–1960

Jean Dieudonné

Reprint of the 1989 Edition

Birkhäuser
Boston • Basel • Berlin

Jean Dieudonné (deceased)

ISBN 978-0-8176-4906-7 e-ISBN 978-0-8176-4907-4
DOI 10.1007/978-0-8176-4907-4

Library of Congress Control Number: 2009927768

Mathematics Subject Classification (2000): 55-03, 55Pxx, 55Uxx, 55Mxx, 55Rxx

Printed on acid-free paper

Birkhäuser Boston is part of Springer Science+Business Media (www.birkhauser.com)

Jean Dieudonné

A History of Algebraic and Differential Topology 1900–1960

Birkhäuser
Boston · Basel

Jean Dieudonné
120, avenue de Suffren
75015 Paris
France

Library of Congress Cataloging-in-Publication Data
Dieudonné, Jean Alexandre, 1906–
A history of algebraic and differential topology, 1900–1960/Jean Dieudonné.
p. cm.
Bibliography: p.
Includes index.
ISBN 0-8176-3388-X
1. Algebraic topology—History. 2. Differential topology—History. I. Title.
QA612.D54 1988
514'.2—dc19 88-14484

CIP-Titelaufnahme der Deutschen Bibliothek
Dieudonné, Jean:
A history of algebraic and differential topology 1900–1960/
Jean Dieudonné.—Boston; Basel: Birkhäuser, 1989
ISBN 3-7643-3388-X (Basel) Pb.
ISBN 0-8176-3388-X (Boston) Pb.
WG: 27 DBN 88.080022.4 88.06.14
8136 rs

Printed on acid-free paper. Birkhäuser

ISBN 0-8176-3388-X
ISBN 3-7643-3388-X

Typeset by Asco Trade Typesetting Ltd., Hong Kong.
Printed and bound by Braun-Brumfield, Inc., Ann Arbor, MI.
Printed in the U. S. A.

9 8 7 6 5 4 3 2

Preface

Although concepts that we now consider as part of topology had been expressed and used by mathematicians in the nineteenth century (in particular, by Riemann, Klein, and Poincaré), algebraic topology as a part of rigorous mathematics (i.e., with precise definitions and correct proofs) only began in 1900. At first, algebraic topology grew very slowly and did not attract many mathematicians; until 1920 its applications to other parts of mathematics were very scanty (and often shaky). This situation gradually changed with the introduction of more powerful algebraic tools, and Poincaré's vision of the fundamental role topology should play in all mathematical theories began to materialize. Since 1940, the growth of algebraic and differential topology and of its applications has been exponential and shows no sign of slackening.

I have tried in this book to describe the main events in that expansion prior to 1960. The choice of that terminal date does not correspond to any particular occurrence nor to an inflection in the development of the theory. However, on one hand, I wanted to limit the size of this book, which is already a large one; and, on the other hand, it is difficult to have a bird's eye view of an evolution that is still going on around us at an unabated pace. Twenty years from now it will be much easier to describe what happened between 1960 and 1980, and it will probably fill a book as large as this one.

There is one part of the history of algebraic and differential topology that I have not covered at all, namely, that which is called "low-dimensional topology." It was soon realized that some general tools could not give satisfactory results in spaces of dimension 4 at most, and, conversely, methods that were successful for those spaces did not extend to higher dimensions. I feel that a description of the discovery of the properties of these spaces deserves a book by itself, which I hope somebody will write soon.

The literature on algebraic and differential topology is very large, and to analyze each paper would have been unbearably boring. I have tried to focus the history on the emergence of ideas and methods opening new fields of research, and I have gone into some details on the work of the pioneers, even when their methods were later superseded by simpler and more powerful ones. As Hadamard once said, in mathematics simple ideas usually come last.

I assume that the reader is familiar with the elementary part of algebra and "general topology." Whenever I have had to mention striking applications of algebraic topology to other parts of mathematics, I have summarized the notions necessary to understand these applications.

Contents

Part 2. The First Applications of Simplicial Methods and of Homology

Notations

General Notations

N, **Z**, **Q**, **R**, **C**, **H**, $\mathbf{F}_q$: integers $\geqslant 0$, rational integers, rational numbers, real numbers, complex numbers, quaternions, finite field with q elements.
$\mathbf{S}_n$: sphere $\sum_j |\xi_j|^2 = 1$ in $\mathbf{R}^n$.
$\mathbf{D}_n$: ball $\sum_j |\xi_j|^2 \leqslant 1$ in $\mathbf{R}^n$.
$\mathrm{e}_1, \mathrm{e}_2, \ldots, \mathrm{e}_n$: canonical basis of $\mathbf{R}^n$ (also written $\mathrm{e}_0, \mathrm{e}_1, \ldots, \mathrm{e}_{n-1}$).
$\mathbf{T}^n$ n-dimensional torus $\mathbf{R}^n/\mathbf{Z}^n$.
$\mathbf{P}_n(\mathrm{F})$: n-dimensional projective space over a field F.
$\mathrm{GL}(n, \mathrm{F})$: general linear group, i.e., group of automorphisms of the vector space F^n; or group of invertible $n \times n$ matrices with entries in F.
$\mathrm{SL}(n, \mathrm{F})$: for F a commutative field, subgroup of $\mathrm{GL}(n, \mathrm{F})$, consisting of matrices of determinant 1.
$\mathrm{O}(n)$: orthogonal group, subgroup of $\mathrm{GL}(\mathrm{n}, \mathbf{R})$ leaving invariant the euclidean scalar product.
$\mathrm{SO}(n)$: group of rotations $\mathrm{O}(n) \cap \mathrm{SL}(n, \mathbf{R})$.
$\mathrm{U}(n)$: unitary group, subgroup of $\mathrm{GL}(n, \mathbf{C})$ leaving invariant the hermitian scalar product.
$\mathrm{U}(n, \mathbf{H})$, $\mathrm{Sp}(n)$: unitary group over the skew field of quaternions, leaving invariant the hermitian scalar product in $\mathbf{H}^n$.

Categories

$\mathcal{C}$: arbitrary category.
$\mathcal{C}^0$: category dual to $\mathcal{C}$.
Set, ***PSet***: category of sets, of pointed sets.
Gr, ***Ab***: category of groups, of commutative groups.
$\boldsymbol{Mod}_\Lambda$, $\boldsymbol{Vect}_k$, $\boldsymbol{Alg}_k$: categories of modules over a ring Λ, of vector spaces over a field k, of algebras over a field k.
T, ***PT***: category of topological spaces, of pointed spaces.
$\boldsymbol{T}'$: any subcategory of $\boldsymbol{T}$.
$\boldsymbol{T}_1$: category of pairs (X, A), X topological space, A subspace of X.

Notations of Set Theory and General Topology

pt.: set or space with only one element.
Id, Id_E, 1_E: identity map of the set E.
$A \coprod B$: disjoint union of A and B (sets or spaces).
$\bar{A}$, $\mathring{A}$, Fr(A): closure, interior, frontier of a subset A of a topological space.
$\varinjlim X_n$: direct limit of a sequence of sets, spaces, or groups (relative to morphisms $\varphi_{n,n+1}: X_n \to X_{n+1}$).
$\varprojlim X_\alpha$: inverse limit of an ordered family $(X_\alpha)_{\alpha\in I}$ (I ordered set) relative to morphisms $\varphi_{\beta\alpha}: X_\beta \to X_\alpha$ for $\alpha < \beta$.

Quotient Spaces (Part 2, chap. V)

X/A: space obtained by collapsing the subset A of X.
$\tilde{C}X$, CX: cone, reduced cone of X.
$\tilde{S}X$, SX: suspension, reduced suspension of X.
$X \vee Y$, $\bigvee_\alpha X_\alpha$: wedges of pointed spaces.
$X \wedge Y = (X \times Y)/(X \vee Y)$: smash product.
$X * Y$: join of two spaces X, Y.
$X \bigcup_f Y$: attachment of X to Y by means of a continuous map $f: A \to Y$, for a subspace $A \subset X$.
Z_f, $\tilde{Z}_f$: mapping cylinder of $f: X \to Y$, reduced mapping cylinder.
C_f, $\tilde{C}_f$: mapping cone of $f: X \to Y$, reduced mapping cone.

General Notations in Algebra and Homological Algebra

Ker f, Coker f, Im f: kernel, cokernel, image of a homomorphism $f: A \to B$ of modules.
$A \oplus B$: direct sum of two Λ-modules.
$A \otimes_\Lambda B$, $A \otimes B$: tensor product of two Λ-modules.
Hom(A, B): module of homomorphisms $A \to B$.
End(A) = Hom(A, A), ring of endomorphisms of A.
Tor(A, B): Part 1, chap. IV, § 5,B.
Ext(A, B): Part 1, chap. IV, § 5,D.
$H_n(\Pi; G)$, $H^n(\Pi; G)$: homology and cohomology groups of the group Π with coefficients in the group G: Part 3, chap. V, § 1,D.

Homology of Chain Complexes and Cohomology of Cochain Complexes (Part 1, chap. IV, § 5)

$C_\bullet = (C_j)_{j\geq 0}$: chain complex.
$\mathbf{b}_p: C_p \to C_{p-1}$: boundary operator (also written $\mathbf{b}$).
Z_p, $Z_p(C_\bullet) = \ker \mathbf{b}_p$: module of p-cycles.

B_p, $B_p(C_.) = \mathrm{Im}\, \mathbf{b}_{p+1}$: module of p-boundaries.
$H_p(C_.) = Z_p/B_p$: p-th homology module.
$C^{\bullet} = (C^j)_{j \geqslant 0}$: cochain complex.
$\mathbf{d}_p$: $C^p \to C^{p+1}$: coboundary operator (also written $\mathbf{d}$).
Z^p, $Z^p(C^{\bullet}) = \mathrm{Ker}\, \mathbf{d}_p$: module of p-cocycles.
B^p, $B^p(C^{\bullet}) = \mathrm{Im}\, \mathbf{d}_{p-1}$: module of p-coboundaries.
$H^p(C^{\bullet}) = Z^p/B^p$: p-th cohomology module
∂_n: $H_n(C_.) \to H_{n-1}(A_.)$ connecting homomorphism of the homology exact sequence for the exact sequence $0 \to A_. \to B_. \to C_. \to 0$ of chain complexes (also written ∂).
rk(M): rank of a finitely generated $\mathbf{Z}$-module.
$\chi(C_.) = \sum_j (-1)^j \mathrm{rk}(C_j)$: Euler-Poincaré characteristic of a finitely generated chain complex of $\mathbf{Z}$-modules.
$f_. = (f_p)$: $C_. \to C'_.$: chain transformation (Part 1, chap. IV, § 5,F).
$H_p(f_.) = f_{p*}$: $H_p(C_.) \to H_p(C'_.)$: homomorphism in homology corresponding to a chain transformation $f_.$.

Axiomatic Homology and Cohomology (Part 1, chap. IV, § 6,B)

$H_p(X)$, $H_p(X, A)$: homology modules.
$H_p(f)$: $H_p(X, A) \to H_p(Y, B)$: homomorphism corresponding to the morphism f: $(X, A) \to (Y, B)$ in $\boldsymbol{T}_1$ (also written f_*).
$\partial_q(X, A)$: $H_q(X, A) \to H_{q-1}(A)$: connecting homomorphism (written ∂).
$H^p(X)$, $H^p(X, A)$: cohomology modules.
$H^p(f)$: $H^p(Y, B) \to H^p(X, A)$: homomorphism corresponding to the morphism f: $(X, A) \to (Y, B)$ (also written f^*).
∂: $H^{q-1}(A) \to H^q(X, A)$: connecting homomorphism.
$\tilde{H}_0(X)$, $\tilde{H}^0(X)$: reduced 0-homology and 0-cohomology modules.

Singular Homology and Cohomology (Part 1, chap. IV, § 2 and § 3)

Δ_p: standard p-simplex.
$S_.(X; \mathbf{Z}) = (S_p(X; \mathbf{Z}))$: singular complex of X.
$H_.(X; \mathbf{Z}) = (H_p(X; \mathbf{Z}))$: singular homology of X.
$S_p(X, A; \mathbf{Z}) = S_p(X; \mathbf{Z})/S_p(A; \mathbf{Z})$ for a subspace $A \subset X$.
$H_p(X, A; \mathbf{Z})$: relative singular homology.
$u_* = H_p(u)$: $H_p(X; \mathbf{Z}) \to H_p(Y; \mathbf{Z})$: homomorphism in homology deduced from a continuous map u: $X \to Y$.
$b_p(X)$, b_p: p-th Betti number of X.
$S^p(X; G) = \mathrm{Hom}(S_p(X; \mathbf{Z}), G)$: group of singular p-cochains.
$H^p(X; G)$: p-th singular cohomology group of X with coefficients in G.

$H^p(X, A; G)$: p-th relative singular cohomology group, for $A \subset X$.
$u^* = H^p(u)$: $H^p(Y; G) \to H^p(X; G)$: homomorphism in cohomology deduced from a continuous map u: $X \to Y$.
$H^p_c(X; G)$: p-th singular cohomology group of X with compact supports.
$a \smile b$: cup-product of two cohomology classes in $H^\bullet(X; \Lambda)$: Part 1, chap. IV, § 4.
$a \frown u$: cap-product of a homology class a and a cohomology class u: Part 1, chap. IV, § 4.
u/c', $c''\backslash u$: right (left) slant product of a cohomology class u and a homology class $c'(c'')$: Part 1, chap. IV, § 5,H.

Čech and Alexander–Spanier Cohomology (Part 1, chap. IV, § 3)

$\check{H}^p(X; G)$: p-th Čech cohomology group of X with coefficients in G.
$C^p(X; G)$: group of maps of X^{p+1} into G.
$C^p_0(X; G)$: subgroup of $C^p(X; G)$ consisting of maps vanishing in a neighborhood of the diagonal.
$\bar{C}^p(X; G) = C^p(X; G)/C^p_0(X; G)$: group of Alexander–Spanier p-cochains.
$\bar{\delta}_p$: $\bar{C}^p(X; G) \to \bar{C}^{p+1}(X; G)$: p-th coboundary operator.
$\bar{H}^p(X; G)$: p-th Alexander–Spanier cohomology group.

Sheaves, Sheaf Cohomology (Part 1, chap. IV, § 7)

$f_*(\mathscr{F})$: direct image of a sheaf by a continuous map.
$\Gamma(\mathscr{F})$: sections of a sheaf.
$H^\bullet(X; \mathscr{F})$: cohomology of X with coefficients in a sheaf $\mathscr{F}$.
$H^\bullet_\Phi(X; \mathscr{F})$: cohomology of X with supports in Φ and coefficients in $\mathscr{F}$.

Spectral Sequences (Part 1, chap. IV, § 7,D and Part 3, chap. IV, § 3,C)

$M^\bullet = \bigoplus_q M_q$, differential graded module, with $d(M_q) \subset M_{q+1}$, and decreasing filtration F, with $F^p(M^\bullet) = \bigoplus_q M_q \cap F^p(M^\bullet)$.
Z^{pq}_r: set of $z \in F^p(M^\bullet) \cap M_{p+q}$ with $dz \in F^{p+r}(M^\bullet)$, $Z^p_r = \bigoplus_q Z^{pq}_r$, $B^{pq}_{r-1} = M_{p+q} \cap dZ^{p+1-r}_{r-1}$; $E^{pq}_r = Z^{pq}_r/(Z^{p+1,q-1}_{r-1} + B^{pq}_{r-1})$;
d_r: $E^{pq}_r \to E^{p+r,q-r+1}_r$ $d_r \circ d_r = 0$
If $E^\bullet_r = \bigoplus_{p,q} E^{pq}_r$, $H^\bullet(E^\bullet_r) = E^\bullet_{r+1}$ for the coboundary d_r.
$E^{pq} = F^pH^{p+q}(M^\bullet)/F^{p+1}H^{p+q}(M^\bullet)$.
For a graded differential module $M^\bullet = \bigoplus_q M^q$ with $d(M^q) \subset M^{q-1}$ and an increasing filtration F, with $F_p(M^\bullet) = \bigoplus_q M^q \cap F_p(M^\bullet)$, one writes Z^r_{pq}: set of $z \in F_p(M^\bullet) \cap M^{p+q}$ with $dz \in F_{p-r}(M^\bullet)$; $Z^r_p = \bigoplus_q Z^r_{pq}$, $B^r_{pq} = M^{p+q} \cap dZ^r_{p+r}$, $E^r_{pq} = Z^r_{pq}/(Z^{r-1}_{p-1,q+1} + B^{r-1}_{pq})$.

De Rham Cohomology (Part 1, chap. III, § 3)

$\mathscr{E}_p(M)$: smooth p-forms on the manifold M.
$\mathscr{D}_p(M)$: smooth p-forms on M with compact support.
$H^p(M)$: De Rham cohomology, i.e., cohomology of the cochain complex ($\mathscr{E}_p(M)$) for the exterior differential $\omega \mapsto d\omega$.
$H_c^p(M)$: De Rham cohomology with compact support, i.e., cohomology of the subcomplex ($\mathscr{D}_p(M)$).
$\mathscr{E}'_p(M)$: p-currents on M with compact support.
$H'_p(M)$: homology of ($\mathscr{E}'_p(M)$) for the boundary operator $b = {}^t d$.

Fundamental Classes (Part 2, chap. I, § 3,A and chap. IV, § 3,A)

[M]: fundamental homology class of an oriented pseudomanifold M.
e_M^*: cohomology fundamental class, or orientation class of an oriented smooth compact n-dimensional manifold M, class of the n-form ω such that $\int_M \omega = 1$.
s_n: orientation class of the sphere S_n.
$\mu_{M,K}$: for an oriented smooth n-dimensional manifold M and a nonempty compact subset K, fundamental class relative to K, element of $H_n(M, M - K; Z)$.

Degree and Fixed Points (Part 2, chap. I, chap. III, and chap. IV, § 1,B)

$\deg f$: degree of a continuous map $f: M \to M'$, where M, M′ are compact connected oriented pseudomanifolds of the same dimension.
$d(f, M, p)$: degree of f relative to M and p, for a continuous map $f: \overline{M} \to \mathbf{R}^n$ and a point $p \notin f(\mathrm{Fr}(M))$.
$\deg_a f$: local degree at the point $a \in X$, for a C^∞ map: $X \to Y$ of smooth oriented manifolds of the same dimension, a being isolated for f.
lk(A, B): linking number of two chains A, B with no common points, of dimensions k and $n - k$ in a connected oriented n-dimensional combinatorial manifold X, when αA and βB are boundaries, for integers α, β.
Fix(f): set of fixed points of a map $f: X \to X$.
$\Lambda(f)$: Lefschetz number of f, when X is a finite simplicial complex.

Homotopy (Part 3, chap. I and II)

$\mathscr{C}(Y, y_0; X, x_0)$: set of continuous maps $(Y, y_0) \to (X, x_0)$ of pointed spaces, with the compact-open topology.
$[Y, y_0; X, x_0] = \pi_0(\mathscr{C}(Y, y_0; X, x_0))$: set of arcwise connected components of

$\mathscr{C}(Y, y_0; X, x_0)$, or, equivalently, homotopy classes of maps $(Y, y_0) \to (X, x_0)$ of pointed spaces.

$[f]$: homotopy class of a continuous map f of pointed spaces.

$*$: the point $e_1 \in \mathbf{S}_n$

$[0, 1]$, I: the interval $0 \leqslant t \leqslant 1$ in $\mathbf{R}$.

$\Omega(X, x_0) = \mathscr{C}(\mathbf{S}_1, *; X, x_0)$: space of loops in X of origin x_0.

$\pi_n(X, x_0) = \pi_0(\mathscr{C}(\mathbf{S}_n, *; X, x_0)$: n-th homotopy group of the pointed space (X, x_0).

$u_*: \pi_n(X, x_0) \to \pi_n(Y, y_0)$: homomorphism $[f] \mapsto [u \circ f]$ (written $[u] \circ [f]$, for a continuous map u of pointed spaces).

$(\Omega^p(X), x_p)$: pointed iterated loop space of (X, x_0).

E: $[Y, y_0; X, x_0] \to [SY, y_0; SX, x_0]$: homotopy suspension.

$E(f)$ or Sf: natural map $(SY, y_0) \to (SX, x_0)$ deduced from $f: (Y, y_0) \to (X, x_0)$, so that $E([f]) = [Sf]$.

$[u, v] \in \pi_{m+n-1}(X, x_0)$: Whitehead product of $u \in \pi_m(X, x_0)$ and $v \in \pi_n(X, x_0)$.

ε_n: relative homology class in $H_n(\bar{\Delta}_n, \bar{\Delta}_n - \Delta_n; \mathbf{Z})$ of the identity map of $\bar{\Delta}_n$.

h_n: $[f] \to f_*(\varepsilon_n)$: Hurewicz homomorphism $\pi_n(X, x_0) \to H_n(X; \mathbf{Z})$.

$\mathscr{C}(Y, B, y_0; X, A, x_0)$: subspace of $\mathscr{C}(Y, y_0; X, x_0)$ consisting of maps f such that $f(B) \subset A$ (with $y_0 \in B$, $x_0 \in A$).

$[Y, B, y_0; X, A, x_0]$: set of relative homotopy classes in $\mathscr{C}(Y, B, y_0; X, A, x_0)$.

$\Omega^n(X, A, x_0) \simeq \mathscr{C}(\mathbf{D}_n, \mathbf{S}_{n-1}, *; X, A, x_0)$: iterated space of paths.

$\pi_n(X, A, x_0) = \pi_0(\mathscr{C}(\mathbf{D}_n, \mathbf{S}_{n-1}, *; X, A, x_0))$: relative homotopy set for $n \geqslant 1$ (group for $n \geqslant 2$).

$u_*: \pi_n(X, A, x_0) \to \pi_n(Y, B, y_0)$: map $[f] \to [u \circ f]$ for u a map $(X, A) \to (Y, B)$ such that $u(x_0) = y_0$.

$\partial: \pi_n(X, A, x_0) \to \pi_{n-1}(A, x_0)$: connecting map of the homotopy exact sequence.

Fibrations (Part 3, chap. III)

$\lambda = (E, B, F, \pi)$ [or (E, B, F), or (E, B, π)]: fibration with total space E, base space B, typical fiber F, projection π.

$E_b = \pi^{-1}(b)$: fiber at the point $b \in B$.

(f, g): $(E, B, \pi) \to (E', B', \pi')$: morphism of fibrations, with $g: B \to B'$ and $f: E \to E'$ continuous, such that $\pi'(f(x)) = g(\pi(x))$.

$f_b = f | E_b$: continuous map $E_b \to E'_{g(b)}$.

$g^*(\lambda)$: pull-back of a fibration (E, B, π) by a continuous map $g: B' \to B$; $g^*(\lambda) = (E', B', \pi')$ with $E' = E \times_B B' \subset E \times B'$, $\pi' = \mathrm{pr}_2 | E'$.

$\lambda' \times \lambda''$: product fibration $(E' \times E'', B' \times B'', \pi' \times \pi'')$ of two fibrations $\lambda' = (E', B', \pi')$, $\lambda'' = (E'', B'', \pi'')$.

$E' \oplus E''$: direct (or Whitney) sum of two vector bundles E', E'' with base space B.

$E' \otimes E''$: tensor product of two vector bundles E', E'' with base space B.

$P \times^G F$: fiber space associated to a principal fiber space (P, B, G) and to an action of G on F.

$\Omega(X, A, x_0) = \mathscr{C}([-1,1], \{-1,1\}, *; X, A, x_0)$: space of paths in X with origin at x_0 and extremity in $A \subset X$.

$P_{x_0}X = \Omega(X, X, x_0)$: space of paths in X with origin x_0.

$(P_{x_0}X, X, \Omega(X, x_0), \pi)$: fibration with total space the space of paths $P_{x_0}X$, base space X and fiber $\pi^{-1}(x_0) = \Omega(X, x_0)$, space of loops of origin x_0.

$G_{n,p}(F)$: grassmannian of p-dimensional vector subspaces of the vector space F^n over the field (or skew field) F.

$G'_{n,p}(\mathbf{R})$: special real grassmannian of oriented p-dimensional vector spaces in $\mathbf{R}^n$.

$S_{n,p}(F)$: Stiefel manifold of orthonormal frames of p vectors in F^n, for $F = \mathbf{R}$, $\mathbf{C}$ or $\mathbf{H}$.

(E_G, B_G, G): principal fiber space with n-universal total space E_G, n-classifying base space B_G, structural group G.

Characteristic Classes (Part 3, chap. IV)

$w_r(\xi)$ (or $w_r(E)$): Stiefel–Whitney class $w_r(\xi) \in H^r(B; \mathbf{F}_2)$ of a real vector bundle $\xi = (E, B, \mathbf{R}^n)$.

$w(E; t) = \sum_r w_r(E)t^r$ (or $w(E) = \sum_r w_r(E)$), total Stiefel–Whitney class of E.

$p_k(\xi) \in H^{4k}(B; \mathbf{Z})$: Pontrjagin class of the vector bundle ξ.

$e(\xi) \in H^n(B; \mathbf{Z})$: Euler class of an oriented vector bundle $\xi = (E, B, \mathbf{R}^n)$.

$c_k(\xi)$: Chern class of a complex vector bundle $\xi = (E, B, \mathbf{C}^n)$.

$c(E; t) = \sum_r c_r(E)t^r$ (or $c(E) = \sum_r c_r(E)$), total Chern class of E.

$s_k(c(\xi))$: polynomial expressing $t_1^k + t_2^k + \cdots + t_n^k$ in terms of the elementary symmetric polynomials σ_j in the t_j, where σ_j is replaced by $c_k(\xi)$.

$\mathrm{ch}(\xi)$: Chern character of ξ, equal to $\sum_{k=0}^{\infty} \frac{1}{k!} s_k(c(\xi))$.

Part 1 Simplicial Techniques and Homology

Introduction

Ever since the concept of homeomorphism was clearly defined, the "ultimate" problem in topology has been to classify topological spaces "up to homeomorphism". That this was a hopeless undertaking was very soon apparent, the subspaces of the plane $\mathbf{R}^2$ being an obvious example. From this impossibility were born algebraic and differential topology, by a shift of emphasis which consisted in associating "invariant" objects to *some* types of spaces (objects are the same for two homeomorphic spaces). At first these objects were integers, but it was soon realized that much more information could be extracted from invariant algebraic structures such as groups and rings.

The algebraic invariants which have provided most insight into the structure of topological spaces are homology and homotopy groups. In the two first parts of this book I shall describe the evolution of homology. Homotopy theory did not start in earnest until 1930; Part 3 is devoted to its explosive development and its relations with homology, which it soon overwhelmed as a central concept of topology.

A remarkable feature in the history of homology and homotopy is the way in which notions initially introduced in these theories for applications to problems of topology unexpectedly found fruitful applications to other parts of mathematics and have become the starting points of extensive theories: categories and functors, homological algebra, and K-theory are outstanding examples.

In modern, sophisticated terms the central concept in homology is the *category* $\boldsymbol{S}(\boldsymbol{C})$ *of generalized chain complexes* in an abelian category $\boldsymbol{C}$. The objects of that category are infinite sequences

$$\mathrm{A}_\bullet\colon \cdots \leftarrow \mathrm{A}_{n-2} \xleftarrow{d_{n-1}} \mathrm{A}_{n-1} \xleftarrow{d_n} \mathrm{A}_n \xleftarrow{d_{n+1}} \mathrm{A}_{n+1} \leftarrow \cdots. \tag{1}$$

where, for each $n \in \mathbf{Z}$, A_n is an object of $\boldsymbol{C}$ and d_n a morphism of $\boldsymbol{C}$, with the conditions

$$d_{n-1}d_n = 0 \quad \text{for every } n \in \mathbf{Z}. \tag{2}$$

When $\mathrm{A}_n = 0$ for $n < 0$, one speaks of *chain complexes*. The d_n are called *boundary operators*.

J. Dieudonné, *A History of Algebraic and Differential Topology, 1900–1960*,
Modern Birkhäuser Classics, DOI 10.1007/978-0-8176-4907-4_1,

The morphisms are sequences $f_\bullet = (f_n)\colon A_\bullet \to B_\bullet$ where, for each $n \in \mathbf{Z}$, $f_n\colon A_n \to B_n$ is a morphism of $\boldsymbol{C}$, and in the diagram

$$\begin{array}{ccccccccc}
\cdots & \longleftarrow & A_{n-1} & \xleftarrow{d_n} & A_n & \xleftarrow{d_{n+1}} & A_{n+1} & \longleftarrow & \cdots \\
 & & \downarrow{\scriptstyle f_{n-1}} & & \downarrow{\scriptstyle f_n} & & \downarrow{\scriptstyle f_{n+1}} & & \\
\cdots & \longleftarrow & B_{n-1} & \xleftarrow[d'_n]{} & B_n & \xleftarrow[d'_{n+1}]{} & B_{n+1} & \longleftarrow & \cdots
\end{array} \tag{3}$$

all squares are commutative; one says the f_n *commute* with the boundary operators. One has $\operatorname{Im} d_{n+1} \subset \operatorname{Ker} d_n \subset A_n$ for every $n \in \mathbf{Z}$; the quotient $H_n(A_\bullet) = \operatorname{Ker} d_n / \operatorname{Im} d_{n+1}$ is called the *n-th homology object* of $A_\bullet$. From (3) it follows that $f_n(\operatorname{Ker} d_n) \subset \operatorname{Ker}(d'_n)$, $f_n(\operatorname{Im} d_{n+1}) \subset \operatorname{Im}(d'_{n+1})$, so there is a morphism

$$H_n(f_\bullet)\colon H_n(A_\bullet) \to H_n(B_\bullet)$$

deduced canonically from $f_\bullet$, and

$$(A_\bullet, f_\bullet) \mapsto (H_n(A_\bullet), H_n(f_\bullet)) \tag{4}$$

is a *covariant functor* from $\boldsymbol{S}(\boldsymbol{C})$ to $\boldsymbol{C}$. When $\boldsymbol{C}$ is the category $\boldsymbol{Mod}_\Lambda$ of modules over a commutative ring Λ, the elements of A_n (resp. $\operatorname{Ker} d_n$, resp. $\operatorname{Im} d_{n+1}$) are called *n-chains* (resp. *n-cycles*, resp. *n-boundaries*).

It took 50 years to reach this level of abstraction. The first occurrence of a chain complex is to be found in Poincaré's papers of 1900, although he did not use the language of modules and homomorphisms as we do now, but the equivalent one of matrices. After an unsuccessful attempt in 1895 to give a genuine mathematical formulation to the intuitive ideas of Riemann and Betti, probably inspired by earlier work of the nineteenth century on polyhedra and the Euler formula, he restricted himself to compact *triangulated spaces* (or *finite cell complexes*); for a triangulation T consisting of cells homeomorphic to convex polyhedra and in finite number he took *n-chains* consisting in formal linear combinations of oriented cells of T of dimension n with integer coefficients for each integer $n \geqslant 0$; they form a free $\mathbf{Z}$-module $C_n(T)$, and one takes $C_n(T) = 0$ for $n < 0$. To define the boundary operator

$$d_n\colon C_n(T) \to C_{n-1}(T)$$

for $n > 0$ it is enough to define $d_n(\sigma)$ for every n-dimensional cell σ by linearity; Poincaré took $d_n(\sigma)$ equal to the sum of all $(n-1)$-dimensional cells on the frontier of σ, each affected with a coefficient ± 1, chosen according to the orientations taken on the cells of T in such a way that the relation (2) holds. He was obviously guided by the intuitive idea that in $\mathbf{R}^2$ or $\mathbf{R}^3$, "a boundary has no boundary", in the sense that, for instance, the points on the frontier of a closed disk in $\mathbf{R}^2$ look different from the interior points, whereas on the frontier any two points look alike.

From this beginning the evolution of homology went through a series of steps, the description of which constitutes Part 1 of this book. We summarize them below, not necessarily in chronological order.

I) Poincaré had already proved that a subdivision of every cell into smaller cells gives the same homology for the subdivided triangulation. This allows one to only consider *euclidean simplicial complexes*, in which every cell is a rectilinear simplex contained in some $\mathbf{R}^N$ with large N.

II) To define a simplex σ in such a simplicial complex K it is only necessary to know its vertices. Taking a total ordering on the set $\sum$ of all vertices of all simplices in K, and writing $\sigma = (a_0 a_1 \dots a_n)$ for the n-simplex σ with vertices a_j written in the given order, the boundary operator is given by

$$d_n \sigma = \sum_{j=0}^{n} (-1)^j (a_0 a_1 \dots \hat{a}_j \dots a_n). \tag{5}$$

It is clear that the chain complex $C_\bullet(K)$ is completely independent of the topology of K. It can be defined for an *arbitrary* (finite or infinite) set $\sum$ where the "simplices" are finite subsets of $\sum$, subject only to the condition that any subset of a simplex is again a simplex. This defines what is called a *combinatorial complex.*

III) Poincaré does not seem to have thought of what we now call "morphisms" between chain complexes. This definition was made possible by the notion of *simplicial mapping* introduced by Brouwer. If K and L are two euclidean simplicial complexes, a simplicial mapping $f: K \to L$ is characterized by the condition that for each simplex σ of K, $f(\sigma)$ is a simplex (possibly of smaller dimension) of L, and the restriction of f to σ is *affine*. This implies that f is entirely determined by the images $f(a_\alpha)$ of the vertices of K, which are vertices of L. Brouwer was not interested in homology, but the Princeton topologists* deduced from such a map f a morphism $\tilde{f}_\bullet: C_\bullet(K) \to C_\bullet(L)$ in the following way: given total orderings of the vertices of K and of L, if $\sigma = (a_0 a_1 \dots a_n)$ is an n-simplex of K, then $\tilde{f}_n(\sigma) = 0$ if the vertices $f(a_j)$ are not all distinct for $0 \leqslant j \leqslant n$; if they are the vertices of an n-simplex $f(\sigma) = (b_0 b_1 \dots b_n)$ of L (in the chosen ordering), then

$$\tilde{f}_n(\sigma) = \varepsilon(b_0 b_1 \dots b_n)$$

where $\varepsilon = 1$ if the permutation $(f(a_0) f(a_1) \dots f(a_n)) \mapsto (b_0 b_1 \dots b_n)$ is even, $\varepsilon = -1$ if that permutation is odd. The commutativity of (3) for $A_\bullet = C_\bullet(K)$, $B_\bullet = C_\bullet(L)$ is then readily verified.

IV) Poincaré conjectured that if a space X can be triangulated, the homology groups of the chain complex $C_\bullet(T)$ defined by a triangulation T are independent of the chosen triangulation T and of the orientations chosen on each cell. This was first proved by Alexander using *simplicial approximation*, another even more important idea of Brouwer. For two triangulated spaces X, Y and *any* continuous map $f: X \to Y$ it is possible, for each $\varepsilon > 0$, to find subdivisions of the triangulations of X and Y and a simplicial map $g: X \to Y$ relative to these subdivisions such that the distance of f and g is at most ε and f and g are homotopic.

* Chiefly Veblen, Alexander and Lefschetz.

This result accomplished the first goal of homology theory: attach to any triangulable space X a system of "homology groups" $H_n(X)$ (for $n \geqslant 0$) *invariant under homeomorphism.*

V) It is easy to define triangulations on usual compact spaces such as S_n, $\mathbf{P}_n(\mathbf{R}), \mathbf{T}^n$. Poincaré conjectured that C^1 manifolds and algebraic varieties (with singularities) can be triangulated. The second conjecture was proved by van der Waerden and the first one by Cairns around 1930.

VI) Beginning around 1925 several topologists began to try to define homology groups for spaces on which no triangulation is known, or even possible. Two ideas emerged. The first, *singular homology*, introduced in an imperfect form by the Princeton topologists, and defined with maximum generality by Eilenberg, takes as n-chains in any space X the formal linear combinations (with integer coefficients) of the *continuous maps* $\Delta_n \to X$, where Δ_n is the standard n-simplex in $\mathbf{R}^{n+1}$ ("*singular n-chains*"). They thus form a free $\mathbf{Z}$-module $S_n(X)$, generally with infinite basis; the boundary operator d_n: $S_n(X) \to S_{n-1}(X)$ assigns to a continuous map f: $\Delta_n \to X$ the $(n-1)$-chain

$$d_n f = \sum_{j=0}^{n} (-1)^j f \circ s_j$$

where s_j: $\Delta_{n-1} \to \Delta_n$ is the affine map which sends Δ_{n-1} onto the face of Δ_n opposite to the j-th vertex, with conservation of the order of the vertices.

The other method, developed by Vietoris, Alexandroff, Lefschetz and Čech, starts from a (finite or infinite) open covering $\mathscr{U} = (U_\alpha)$ of X; to this covering is associated its *nerve*, a combinatorial complex $N(\mathscr{U})$ whose vertices are the U_α and the n-simplices are the non-empty intersections of n sets U_α. This defines a chain complex $C_{\bullet}(\mathscr{U})$, and from the homology groups $H_n(\mathscr{U})$ of these chain complexes one deduces homology groups $\check{H}_n(X)$ by an inverse limit process on the open coverings $\mathscr{U}$ (or on some family of open coverings, for instance the finite ones).

VII) Earlier, homology groups "with coefficients" were also introduced. Given a commutative group G and a chain complex $A_{\bullet} = (A_n)$ consisting of $\mathbf{Z}$-modules, one can form the chain complex

$$A_{\bullet} \otimes_{\mathbf{Z}} G = (A_n \otimes_{\mathbf{Z}} G)$$

and the homology $H_{\bullet}(A_{\bullet} \otimes_{\mathbf{Z}} G)$, written $H_{\bullet}(A_{\bullet}; G)$, is called the homology of $A_{\bullet}$ "with coefficients in G". Before the general definition of tensor products by Whitney, only the case in which the A_n are free $\mathbf{Z}$-modules was considered. For singular homology, the dimension of $H_p(X; \mathbf{Q})$ is called the p-th *Betti number* $b_p(X)$ of the space X. For an arbitrary group G, $H_p(X; G)$ is entirely determined by G and the groups $H_p(X; \mathbf{Z})$ and $H_{p-1}(X; \mathbf{Z})$ (formula of universal coefficients).

VIII) Lefschetz also defined the new notion of *relative homology* $H_{\bullet}(K, L)$ for a finite simplicial complex K and a subcomplex L. The notion could easily be extended to singular and Čech homology, and its properties later were recognized as the first manifestations of the *homology exact sequence*: if three chain complexes $A_\bullet$, $B_\bullet$, $C_\bullet$ are elements of a short exact sequence of morphisms

$$0 \longrightarrow A_\bullet \xrightarrow{u_\bullet} B_\bullet \xrightarrow{v_\bullet} C_\bullet \longrightarrow 0$$

then there exists an infinite sequence of canonically defined morphisms ∂_n: $H_n(C_\bullet) \to H_{n-1}(A_\bullet)$ such that the sequence

$$\cdots \longrightarrow H_n(A_\bullet) \xrightarrow{H_n(u_\bullet)} H_n(B_\bullet) \xrightarrow{H_n(v_\bullet)} H_n(C_\bullet) \xrightarrow{\partial_n} H_{n-1}(A_\bullet) \longrightarrow \cdots \tag{6}$$

is exact.

For singular homology one considers a space X and a subspace Y and the singular complexes $S_\bullet(X)$, $S_\bullet(Y)$. The natural injection j: $Y \to X$ defines an injective morphism $S_\bullet(j)$: $S_\bullet(Y) \to S_\bullet(X)$, hence a chain complex $S_\bullet(X, Y) = S_\bullet(X)/S_\bullet(j)(S_\bullet(Y))$. The homology $H_\bullet(S_\bullet(X, Y))$ of that chain complex is the *relative singular homology* $H_\bullet(X, Y)$, and one has the *exact sequence of singular homology*

$$\cdots \to H_n(Y) \to H_n(X) \to H_n(X, Y) \to H_{n-1}(Y) \to \cdots \tag{7}$$

IX) From the beginning homology theory had been pervaded by the general (and vague) idea of *duality*. Poincaré clearly considered that the climax of his work in topology was his famous *duality theorem*. For an n-dimensional compact *oriented* C^1 manifold X, where a triangulation T is given, he described a *topological* construction of a "dual" triangulation T^*, reminiscent of the *geometrical* duality of spherical polygons known since the 17th century. It was rigorously shown later that this provides for each oriented p-cell a_p^i of T an oriented "dual" $(n-p)$-cell b_{n-p}^i of T^*, such that the Kronecker intersection index $I(a_p^i, b_{n-p}^j) = \delta_{ij}$. From this it is easy to show that the matrices of the boundary operators $C_p(T) \to C_{p-1}(T)$ and $C_{n-p+1}(T^*) \to C_{n-p}(T^*)$ are *transposes* of each other. Once the invariance of the homology of a triangulation had been established, this gave for the Betti numbers the equality

$$b_{n-p}(X) = b_p(X). \tag{8}$$

X) In conformity with the misconceptions of duality, beginning with Poncelet, for whom it took place in a *single* space instead of two spaces "in duality", Poincaré duality was considered as holding between elements of a single graded group $H_\bullet(X)$; this view was not modified even after De Rham, in 1930, had shown that for a compact C^r manifold X, $H_p(X; \mathbf{R})$ is isomorphic to the quotient of the space of closed differential p-forms on X by the subspace of exact ones.

It was only in 1935, under the influence of Pontrjagin's duality theorem—where a compact commutative group is completely different from its discrete "dual"—that Alexander and Kolmogoroff realized that a p-form α on a smooth compact manifold X can be considered, by integration on a smooth p-chain, as a *homomorphism*

$$c \mapsto \int_c \alpha \tag{9}$$

of the subspace of $C_p(X)$ consisting of those chains, into **R**. This led them to the general definition of *cohomology*.

At the level of *arbitrary* generalized chain complexes (1) this definition is little more than a tautology: change the indexing in (1) by writing $A'_n = A_{-n}$ and $d'_n = d_{-n}$ for every $n \in \mathbf{Z}$, and write (1) as

$$A'^{\bullet}: \cdots \longrightarrow A'_n \xrightarrow{d'_n} A'_{n+1} \xrightarrow{d'_{n+1}} A'_{n+2} \longrightarrow \cdots$$

with $d'_{n+1} d'_n = 0$ for all n; such a sequence takes the name *generalized cochain complex* and d'_n is called the n-th *coboundary operator*; one merely speaks of a *cochain complex* when $A'_n = 0$ for $n < 0$. When $\boldsymbol{C} = \boldsymbol{Mod}_\Lambda$ the elements of A'_n (resp. $\operatorname{Ker} d'_n$, resp. $\operatorname{Im} d'_{n-1}$) are called *n-cochains* (resp. *n-cocycles*, resp. *n-coboundaries*). $H^n(A'^{\bullet}) = \operatorname{Ker} d'_n / \operatorname{Im} d'_{n-1}$ is the *n-th cohomology object* of $A'^{\bullet}$. To each morphism $f^{\bullet}: A'^{\bullet} \to B'^{\bullet}$ corresponds a morphism

$$H^n(f'^{\bullet}): H^n(A'^{\bullet}) \to H^n(B'^{\bullet})$$

so that H^n is a *covariant* functor from the category of generalized cochain complexes to $\boldsymbol{C}$.

But if one takes for $\boldsymbol{C}$ the category $\boldsymbol{Ab}$ of additive commutative groups, there is another, non-trivial, way to obtain a generalized cochain complex from a generalized chain complex $A_{\bullet}$ in $\boldsymbol{Ab}$: for an arbitrary *commutative group* G put

$$A^n_G = \operatorname{Hom}(A_n, G) \tag{10}$$

$$\delta^n = \operatorname{Hom}(d_{n+1}, 1_G): \operatorname{Hom}(A_n, G) \to \operatorname{Hom}(A_{n+1}, G) \tag{11}$$

Then

$$A^{\bullet}_G: \cdots \longrightarrow A^n_G \xrightarrow{\delta^n} A^{n+1}_G \xrightarrow{\delta^{n+1}} A^{n+2}_G \longrightarrow \cdots \tag{12}$$

is a generalized cochain complex. This defines a *contravariant* functor

$$A_{\bullet} \mapsto A^{\bullet}_G$$

which to a morphism $f_{\bullet}: A_{\bullet} \to B_{\bullet}$, associates $f^{\bullet}_G = (f^n_G)$ with

$$f^n_G = \operatorname{Hom}(f_n, 1_G): B^n_G \to A^n_G.$$

Hence for any n there is a *contravariant* functor

$$(A_{\bullet}, f_{\bullet}) \to (H^n(A^{\bullet}_G), H^n(f^{\bullet}_G)).$$

When $A_{\bullet} = C_{\bullet}(K)$, where K is a finite cell complex, one thus defines *cohomology groups* $H^n(K; G) = H^n(A^{\bullet}_G)$ which do not depend on the triangulation of K, but only on its topology. If $f: K \to L$ is a continuous map, one has for each $n \geqslant 0$ a homomorphism

$$f^*: H^n(L; G) \to H^n(K; G) \tag{13}$$

defining a contravariant functor on the category of cell complexes. This is the case which was first considered by Alexander and Kolmogoroff but similar definitions lead to *singular cohomology groups* $H^n(X; G)$ and *Čech cohomology groups* $\check{H}^n(X; G)$ for all spaces X; observe that these are also covariant functors in G.

It is also possible to define cohomology groups of a space X by directly associating a *cochain* complex to X without using a *chain* complex as an intermediate. If X is a smooth manifold, one may consider the *De Rham complex*

$$0 \to \Omega^0 \xrightarrow{d} \Omega^1 \xrightarrow{d} \Omega^2 \to \cdots \xrightarrow{d} \Omega^n \xrightarrow{d} \cdots.$$

where Ω^p is the vector space over $\mathbf{R}$ of all C^∞ differential p-forms (for $p = 0$ the C^∞ functions) and d the exterior differential. One thus obtains the De Rham cohomology spaces $H^n_{DR}(X)$; when X is compact the De Rham theorem gives a canonical isomorphism of $H^n_{DR}(X)$ on the *dual* vector space $(H_n(X; \mathbf{R}))^*$.

There is a similar direct definition of cohomology for an arbitrary space, due to Alexander and Spanier. The n-cochains are equivalence classes of mappings $f: X^{n+1} \to G$ (not necessarily continuous), two mappings being equivalent if they coincide on a neighborhood of the diagonal in X^{n+1}. The coboundary operator δ^n associates to the class of such a map the class of the map

$$(x_0, x_1, \ldots, x_{n+1}) \mapsto \sum_{j=0}^{n+1} (-1)^j f(x_0, \ldots, \hat{x}_j, \ldots, x_{n+1}).$$

XI) Once cohomology had been defined it was possible to better understand duality in algebraic topology. In the first place definition (10) means that a bilinear map $A^n_G \times A_n \to G$ is defined as

$$(f, u) \mapsto f(u) \in G$$

also written $(f, u) \mapsto \langle f, u \rangle$. From the definition (11)

$$\langle \delta^n f, u \rangle = \langle f, d_n u \rangle; \tag{14}$$

it follows that when f is a *cocycle* and u a *cycle*, $\langle f, u \rangle$ only depends on the *classes* $\bar{f} \in H^n(A^\cdot_G)$ and $\bar{u} \in H_n(A_.)$, and can therefore be written $\langle \bar{f}, \bar{u} \rangle$, thus defining a bilinear map

$$H^n(A^\cdot_G) \times H_n(A_.) \to G$$

or equivalently a linear map

$$H^n(A^\cdot_G) \to \mathrm{Hom}(H_n(A_.), G) \tag{15}$$

sending $\bar{f}$ to the linear map $\bar{u} \mapsto \langle \bar{f}, \bar{u} \rangle$. When $A_.$ is a free $\mathbf{Z}$-module the map (15) is surjective, and it is injective if in addition $H_{n-1}(A_.)$ is free.

Various "products" (i.e. bilinear maps) may be defined using duality. If K and L are cell complexes, the products $\sigma \times \tau$ of a cell σ of X and a cell τ of L are the cells of a cell complex $K \times L$. If Λ is a commutative ring, one defines a bilinear map

$$\kappa^{pq}: \mathrm{Hom}(C_p(K), \Lambda) \times \mathrm{Hom}(C_q(L), \Lambda) \to \mathrm{Hom}(C_{p+q}(K \times L), \Lambda)$$

which to $f \times g$ assigns the linear map

$$\kappa^{pq}(f \times g): \sigma \times \tau \mapsto f(\sigma)g(\tau)$$

(product taken in Λ). These maps commute with coboundary operators, hence yield Λ-bilinear maps

$$k^{pq}: \mathrm{H}^p(\mathrm{K};\Lambda) \times \mathrm{H}^q(\mathrm{L};\Lambda) \to \mathrm{H}^{p+q}(\mathrm{K} \times \mathrm{L};\Lambda). \tag{16}$$

When Λ is a field these maps are injective, and $\mathrm{H}^{p+q}(\mathrm{K} \times \mathrm{L};\Lambda)$ is the direct sum of their images (Künneth formulas). Similar maps can be defined for singular or Čech cohomology.

Now consider a space X and the diagonal map

$$\mathrm{D}: x \mapsto (x, x)$$

of X into X × X. By functoriality, this yields linear maps in singular cohomology

$$\mathrm{D}^*: \mathrm{H}^n(\mathrm{X} \times \mathrm{X};\Lambda) \to \mathrm{H}^n(\mathrm{X};\Lambda)$$

for all n, hence for any pair of integers $p \geqslant 0$, $q \geqslant 0$ one has a composed map

$$\mathrm{H}^p(\mathrm{X};\Lambda) \times \mathrm{H}^q(\mathrm{X};\Lambda) \xrightarrow{k^{pq}} \mathrm{H}^{p+q}(\mathrm{X} \times \mathrm{X};\Lambda) \xrightarrow{\mathrm{D}^*} \mathrm{H}^{p+q}(\mathrm{X};\Lambda) \tag{17}$$

which is called the *cup-product* and written

$$(u, v) \mapsto u \smile v. \tag{18}$$

It is easy to see that on the direct sum $\mathrm{H}^{\bullet}(\mathrm{X};\Lambda) = \bigoplus_{n \geqslant 0} \mathrm{H}^n(\mathrm{X};\Lambda)$ the maps (18) define a structure of associative anticommutative graded Λ-algebra, called the *cohomology algebra* of X with coefficients in Λ.

One can also define another Λ-bilinear map

$$\mathrm{H}_{p+q}(\mathrm{X};\Lambda) \times \mathrm{H}^p(\mathrm{X};\Lambda) \to \mathrm{H}_q(\mathrm{X};\Lambda)$$

for singular homology and cohomology, written

$$(c, u) \mapsto c \frown u \tag{19}$$

and called the *cap-product*, such that

$$\langle c \frown u, v \rangle = \langle c, u \smile v \rangle \quad \text{for all } v \in \mathrm{H}^q(\mathrm{X};\Lambda). \tag{20}$$

XII) Using the cap-product, a better version of Poincaré duality can be given. If M is a C^1 compact n-dimensional manifold, connected and *oriented*, there is a privileged class $\mu_n \in \mathrm{H}_n(\mathrm{M};\mathbf{Z})$ which is a generator of that group; if M is equipped with a triangulation T and each n-simplex of T is given the orientation induced by the orientation of M, μ_n is the class of the *sum* of those n-simplices; it is called the *fundamental* (or *orientation*) class of M. Then the homomorphism

$$\mathrm{H}^p(\mathrm{M};\mathbf{Z}) \to \mathrm{H}_{n-p}(\mathrm{M};\mathbf{Z}),$$

defined by

$$u \mapsto \mu_n \frown u,$$

is *bijective*. For nonorientable manifolds the same result holds for homology and cohomology with coefficients in $\mathbf{F}_2$.

A similar definition can be given for another kind of "duality" first discovered in 1922 by Alexander. Suppose Y is a compact subset contained in $\mathbf{S}_n$, homeomorphic to a smooth manifold of dimension $\leqslant n - 1$. Alexander defined (for the first time) homology groups for the *open* set $\mathbf{S}_n - \mathrm{Y}$ (for which it is only possible to construct a triangulation with *infinitely* many cells), and he proved that for $p \leqslant n - 2$, $\mathrm{H}_p(\mathrm{Y}; \mathbf{F}_2)$ is *isomorphic* to $\mathrm{H}_{n-p-1}(\mathbf{S}_n - \mathrm{Y}; \mathbf{F}_2)$, and if $\mathbf{S}_n - \mathrm{Y}$ has q connected components, $\mathrm{H}^{n-1}(\mathrm{Y}; \mathbf{F}_2) \simeq \mathbf{F}_2^{q-1}$.

Using a refined localized version of the fundamental class this can be generalized to a canonical isomorphism for singular homology and cohomology

$$\mathrm{H}^p(\mathrm{Y}; \mathbf{Z}) \xrightarrow{\sim} \mathrm{H}_{n-p}(\mathrm{X}, \mathrm{X} - \mathrm{Y}; \mathbf{Z}) \tag{21}$$

when X is an *oriented* n-dimensional compact manifold and Y a closed submanifold. If it is only supposed that Y is an arbitrary closed subset of the manifold X, the isomorphism (21) still holds, provided one replaces singular cohomology $\mathrm{H}^p(\mathrm{Y}; \mathbf{Z})$ by *Čech cohomology* $\check{\mathrm{H}}^p(\mathrm{Y}; \mathbf{Z})$. Various improvements and similar results can be given: passage to non compact spaces, replacement of $\mathbf{Z}$ by $\mathbf{F}_2$ for a non orientable manifold X, and finally extension of the theorems to C^0 manifolds (not necessarily triangulable).

XIII) Around 1940 topologists began comparing the various definitions of homology and cohomology given in the previous years. It was shown that Čech cohomology and Alexander–Spanier cohomology give isomorphic cohomology groups, whereas the singular cohomology of a space may differ from its Čech cohomology.

Eilenberg and Steenrod inaugurated a new approach by focusing not on the machinery used for the construction of homology or cohomology groups, but on the properties shared by the various theories. They selected a small number of these properties and took them as *axioms* for a theory of homology and cohomology; they showed that many other properties, formerly separately proved for each theory, were in fact consequences of the axioms, and they examined each theory accordingly to see if it satisfied the axioms. Their most interesting result was the proof that on the category of *compact triangulable spaces* all theories verifying the axioms give isomorphic groups; in other words, there is only *one* notion of homology and cohomology in that category.

XIV) The last stage in the evolution of homology theory has been dominated by the concept of *sheaf cohomology*, invented by J. Leray in 1946. It may be considered a general machinery applicable to problems designated by the vague words "passage from local to global properties": when some mathematical object attached to a topological space X can be "restricted" to any open subset U of X, and that restriction is known for sufficiently small sets U, what can be said of that "global" object? Problems of this type had arisen since the 1880's for analytic functions of several complex variables, studied by H. Poincaré, P. Cousin and later H. Cartan and K. Oka. Beginning in 1942 Leray attacked a similar problem for cohomology: given a space X, what can one say on the cohomology algebra $H^{\bullet}(X; \Lambda)$ when the cohomology algebras $H^{\bullet}(U; \Lambda)$ are known for sufficiently small open sets $U \subset X$? The machinery he devised in 1946 in order to tackle that problem was gradually refined and generalized by him and other mathematicians in the ensuing years.

It is first centered on the notion of a *sheaf* $\mathscr{F}$ *of* Λ*-modules on* X. This is defined by assigning to each open subset U of X a Λ-module $\mathscr{F}(U)$, and to each pair of open sets $V \supset U$ a Λ-homomorphism $\rho_{UV}: \mathscr{F}(V) \to F(U)$ satisfying the condition that if $W \supset V \supset U$ are three open sets

$$\rho_{UW} = \rho_{UV} \circ \rho_{VW}. \tag{22}$$

Furthermore, the Λ-modules $\mathscr{F}(U)$ must satisfy two additional axioms:

(F I) If $U = U_1 \cup U_2$ is a union of two open sets, and $s \in \mathscr{F}(U)$ is such that $\rho_{U_1U}(s) = 0$ and $\rho_{U_2U}(s) = 0$, then $s = 0$.

(F II) If U is a union of a family (U_α) of open sets, and a family (s_α) with $s_\alpha \in \mathscr{F}(U_\alpha)$ is such that for $U_\alpha \cap U_\beta \neq \varnothing$,

$$\rho_{U_\alpha \cap U_\beta, U_\alpha}(s_\alpha) = \rho_{U_\alpha \cap U_\beta, U_\beta}(s_\beta)$$

then there exists $s \in \mathscr{F}(U)$ such that $\rho_{U_\alpha U}(s) = s_\alpha$ for every α.

The elements $s \in \mathscr{F}(U)$ are called the *sections of the sheaf* $\mathscr{F}$ *over* U.

When X is connected and locally connected and M is any Λ-module, there exists a sheaf $\mathscr{F}$ on X such that $\mathscr{F}(U) = M$ for all connected open sets U, and ρ_{UV} is the identity if $V \supset U$ are any two connected open sets; it is called the *constant sheaf* defined by M.

One defines a *morphism* $f: \mathscr{F} \to \mathscr{G}$ for two sheaves of Λ-modules on X by assigning to each open set $U \subset X$ a homomorphism $f_U: \mathscr{F}(U) \to \mathscr{G}(U)$ of Λ-modules in such a way that if $V \supset U$ are two open sets, the diagram

$$\begin{array}{ccc} \mathscr{F}(V) & \xrightarrow{\rho_{UV}} & \mathscr{F}(U) \\ {\scriptstyle f_V}\downarrow & & \downarrow{\scriptstyle f_U} \\ \mathscr{G}(V) & \xrightarrow[\rho_{UV}]{} & \mathscr{G}(U) \end{array}$$

is commutative. This defines the *category* $\boldsymbol{Sh}_\Lambda(X)$ of sheaves of Λ-modules on X; it can be shown that it is an *abelian category*.

For each point $x \in X$ the Λ-modules $\mathscr{F}(U)$ for all open neighborhoods U of x form a *direct system* for the homomorphisms ρ_{UV}; they therefore have a direct limit

$$\mathscr{F}(x) = \varinjlim \mathscr{F}(U)$$

called the *stalk of* $\mathscr{F}$ at the point x; its elements are called the *germs* of sections of $\mathscr{F}$ at the point x.

As an example of a sheaf, suppose X is Hausdorff, let $x_0 \in \mathrm{X}$ be any point, and take $\mathscr{F}(\mathrm{U}) = \{0\}$ if $x_0 \notin \mathrm{U}$, $\mathscr{F}(\mathrm{U}) = \mathrm{M}$ if $x_0 \in \mathrm{U}$ (M any Λ-module), with $\rho_{\mathrm{UV}} = 0$ if $x_0 \notin \mathrm{U}$, $\rho_{\mathrm{UV}} = \mathrm{Id_M}$ if $x_0 \in \mathrm{U}$. Then $\mathscr{F}(x) = \{0\}$ if $x \neq x_0$, and $\mathscr{F}(x_0) = \mathrm{M}$ ("skyscraper sheaf").

The second item in Leray's procedure was the definition, for any sheaf $\mathscr{F}$ on X, of groups $\mathrm{H}^n(\mathrm{X}; \mathscr{F})$ for all integers $n \geqslant 0$, called the *cohomology groups of* X *with coefficients in* $\mathscr{F}$. The modern conception of these groups subsumes them under the general idea of "derived functors". If a functor $\mathrm{T}\colon \boldsymbol{Sh}_\Lambda(\mathrm{X}) \to \boldsymbol{Ab}$ transforms monomorphisms into injective homomorphisms (but does not necessarily transform epimorphisms into surjective homomorphisms), a canonical process assigns to T a sequence of *derived functors*

$$\mathrm{R}^n\mathrm{T}\colon \boldsymbol{Sh}_\Lambda(\mathrm{X}) \to \boldsymbol{Ab}$$

which may be considered as characterizing the "failure of exactness" of T: if $0 \to \mathscr{F}' \to \mathscr{F} \to \mathscr{F}'' \to 0$ is an exact sequence, there corresponds to it a *long exact sequence*

$$\begin{aligned} 0 &\to \mathrm{T}(\mathscr{F}') \to \mathrm{T}(\mathscr{F}) \to \mathrm{T}(\mathscr{F}'') \to \mathrm{R}^1\mathrm{T}(\mathscr{F}') \to \mathrm{R}^1\mathrm{T}(\mathscr{F}) \to \mathrm{R}^1\mathrm{T}(\mathscr{F}'') \\ &\to \mathrm{R}^2\mathrm{T}(\mathscr{F}') \to \mathrm{R}^2\mathrm{T}(\mathscr{F}) \to \cdots \end{aligned}$$

Write $\Gamma(\mathscr{F}) = \mathscr{F}(\mathrm{X})$, the group of sections of $\mathscr{F}$ *over* X; this defines a functor Γ with values in $\boldsymbol{Ab}$, and the groups $\mathrm{H}^n(\mathrm{X}; \mathscr{F})$ are the values at $\mathscr{F}$ of the derived functors

$$\mathrm{R}^n\Gamma\colon \boldsymbol{Sh}_\Lambda(\mathrm{X}) \to \boldsymbol{Ab}.$$

When X is compact, connected, and locally connected, if $\mathscr{L}$ is the constant sheaf with stalks equal to the ring Λ, the group $\mathrm{H}^n(\mathrm{X}; \mathscr{L})$ is naturally isomorphic to the Alexander–Spanier cohomology group $\mathrm{H}^n(\mathrm{X}; \Lambda)$.

The third and most remarkable part of Leray's theory attacked the following general problem: given two spaces X, Y and a continuous map $f\colon \mathrm{X} \to \mathrm{Y}$, what are the relations between the cohomology groups of X, of Y, and of the "fibers" $f^{-1}(y)$ for $y \in \mathrm{Y}$? When X and Y are locally compact and f is proper, for any integer q, there is a *sheaf* $\mathscr{H}^q(f)$ on Y whose stalk at each $y \in \mathrm{Y}$ is $\mathrm{H}^q(f^{-1}(y); \Lambda)$. Hence for any $p \geqslant 0$ one may consider the cohomology group

$$\mathrm{H}^p(\mathrm{Y}; \mathscr{H}^q(f)). \tag{23}$$

When X is a product $\mathrm{Y} \times \mathrm{Z}$ and f the first projection, $\mathrm{H}^n(\mathrm{X}; \Lambda)$ is the direct sum of the groups (23) for $p + q = n$ when Λ is a field, by Künneth's formula. In the general case Leray invented a kind of "algebraic approximation", now called a *spectral sequence*, of which (23) is the first term; in the best cases the process gives valuable information on the quotients of a filtration of the group $\mathrm{H}^{p+q}(\mathrm{X}; \Lambda)$. This device has become one of the most useful tools in applications of homological algebra to all branches of mathematics, from logic to operator algebras.

XV) Using sheaf cohomology and an algebraic "dualizing" process, A. Borel and J. Moore have defined a homology theory for locally compact spaces such that the sequence (7) is exact for any closed subset Y of such a space X (whereas it is not in general for Čech homology) and the other Eilenberg–Steenrod axioms are also verified.

CHAPTER I

The Work of Poincaré

§ 1. Introduction

Concepts and results belonging to algebraic and differential topology may already be noted in the eighteenth and nineteenth centuries, but cannot be said to constitute a mathematical discipline in the usual sense. Before Poincaré we should therefore only speak of the *prehistory* of algebraic topology; that period has been described in detail in the recent book by J.C. Pont [373], to which we shall refer when necessary.

It is quite difficult for us to understand the point of view of the mathematicians who undertook to tackle topological problems in the second half of the nineteenth century: When dealing with curves, surfaces, and, later, manifolds of arbitrary dimension, with their intersections or their existence when submitted to various conditions, etc., they relied exclusively on "intuition," and thus followed—with a vengeance—in the footsteps of Riemann, behaving exactly as the analysts of the eighteenth or early nineteenth century in dealing with questions of convergence or continuity! It is certainly incorrect to attribute this attitude (as many authors do) to an unending evolution of a general concept of "rigor" in mathematical arguments, which would be doomed to perpetual change; what history shows us is a *sectorial* evolution of "rigor." Having come long before "abstract" algebra, the proofs in algebra or number theory have never been challenged; around 1880 the canon of "Weierstrassian rigor" in classical analysis gained wide acceptance among analysts, and has *never* been modified. Furthermore, even in manifolds of dimension 2, where the word "intuition" might have had some justification, inaccuracies and inconclusive arguments were pointed out as early as 1873 ([373], pp. 82–83*) in the topological papers of Riemann and Betti.

One is tempted to attribute this schizophrenia of the mathematical community to the compartmentalization still prevalent at the end of the nineteenth

* Beware that the words *linearmente connesso* of Betti, which mean "arcwise connected," are wrongly translated "simply connected" by Pont.

J. Dieudonné, *A History of Algebraic and Differential Topology, 1900–1960*,
Modern Birkhäuser Classics, DOI 10.1007/978-0-8176-4907-4_2,

century, the various mathematical theories having a tendency to ignore each other; however, even a universal mathematician like Hilbert was still occasionally guilty of "intuitive" arguments of a kind that he would indignantly have rejected if they had concerned the convergence of a series ([223], vol. II, p. 327–328); and we shall see that in his topological papers Poincaré himself appeared as one of the worst offenders. It was only after 1910 that uniform standards of what constitutes a correct proof became universally accepted in topology, with Brouwer's work on simplicial approximation ([89], pp. 420–552) and Weyl's treatment of the theory of Riemann surfaces [483]; again, this standard has remained *unchanged* ever since.*

Of course, before Fréchet (1906) and Hausdorff (1914) the general notion of topological space had not been defined; what had become familiar after the work of Weierstrass and Cantor were the elementary topological notions (open sets, closed sets, neighborhoods, continuous mappings, etc.) in the spaces $\mathbf{R}^n$ and their subspaces; these notions had been extended by Riemann (in an "intuitive" way and without any precise definition) to "n-dimensional manifolds" (or rather what we now would call C^r-manifolds with $r \geqslant 1$).

We shall have plenty of opportunity to stress the importance, in algebraic topology, of various *constructions* of new spaces from given ones. Surprisingly enough, the simplest of all, the cartesian *product* of two or more *arbitrary* spaces, does not seem to have been considered before 1908 [456]. The general notion of *quotient space* appeared even later (Part 2, chap. V), but a special case was used (without any precise definition) as early as the 1870s: the "gluing together" of spaces along homeomorphic subspaces. Klein ([273], vol. 3, pp. 36–43) and Poincaré obtained a compact surface by gluing together isometric edges of a "fuchsian" curvilinear polygon, and a little later von Dyck extended that method to define examples of higher-dimensional manifolds: projective space $\mathbf{P}_n(\mathbf{R})$, for example, by identification of symmetric points (with respect to the origin) of the sphere $\mathbf{S}_n$ ([147], pp. 278–279 and 284).

Function spaces did not appear before 1906.†

§2. Poincaré's First Paper: *Analysis Situs*

Toutes les voies diverses où je m'étais engagé successivement me conduisaient à l'*Analysis Situs*. J'avais besoin des données de cette Science pour poursuivre mes études sur les courbes définies par les équations différentielles et pour les étendre aux équations différentielles d'ordre supérieur, et, en particulier, à celles du problème des trois corps. J'en avais besoin pour l'étude des fonctions non uniformes de deux variables. J'en avais besoin pour l'étude des intégrales multiples et pour l'application de cette

* The same situation was repeated in algebraic geometry, where until around 1950 a large part of the arguments were based on "intuition."

† The "continuity methods" used by Klein and Poincaré in their work on the uniformization problem may be regarded as forerunners of the concept of a function as a "point" in some set, of which traces are even to be found in Riemann.

étude au développement de la fonction perturbatrice. Enfin j'entrevoyais dans l'*Analysis Situs* un moyen d'aborder un problème important de théorie des groupes, la recherche des groupes discrets ou des groupes finis contenus dans un groupe continu donné ([364], p. 101).

This is how Poincaré motivated his determination to investigate the basic concepts of algebraic topology, which he called *Analysis Situs*; his first results were announced in a Comptes-Rendus Note of 1892 ([369], pp. 189–192) and developed in a long paper entitled *Analysis Situs* published in 1895 ([369], pp. 193–288); between 1899 and 1905 he returned to the theory in five papers, which he called *Compléments à l'Analysis Situs.* As in so many of his papers, he gave free rein to his imaginative powers and his extraordinary "intuition," which only very seldom led him astray; in almost every section is an original idea. But we should not look for precise definitions, and it is often necessary to guess what he had in mind by interpreting the context. For many results, he simply gave no proof at all, and when he endeavored to write down a proof hardly a single argument does not raise doubts. The paper is really a *blueprint* for future developments of entirely new ideas, each of which demanded the creation of a new technique to put it on a sound basis. We shall devote this chapter to a detailed examination of *Analysis Situs* and of the first two *Compléments.*

Poincaré's starting point was the same as those of his predecessors (he quoted Riemann, Betti, and von Dyck), and his "intuitive" style is very similar to theirs; but immediately novelties appeared that gave rise to algebraic topology as we understand it.

The first and most important one was that, whereas mathematicians before him tried to attach *numbers* invariant under homeomorphism to spaces, Poincaré was the first who introduced the idea of *computing with topological objects*, not only with numbers. He did this in two different ways, by defining the concepts of *homology* and of *fundamental group*; in this chapter we shall concentrate on the first one, postponing to Part 3, chap. I, the theory of the fundamental group, the first step in the homotopy theory of today.

In the first three sections of *Analysis Situs*, Poincaré began, reasonably enough, by trying to define the spaces he considered. They must be subspaces of some $\mathbf{R}^N$, and are what we now call *connected* C^1 differential manifolds. However, most of the examples introduced later in the paper were obtained by the "gluing" process, which we mentioned in § 1, and Poincaré never bothered to show that they satisfied the preceding definition! Among the C^1 manifolds he considered he concentrated on the compact ones without boundary (those he called *closed* (p. 198)), or the ones that are open in a C^1 submanifold W of some $\mathbf{R}^N$, of dimension $N - p$, defined by p *global* equations

$$F_1 = F_2 = \cdots = F_p = 0 \qquad (1)$$

between the N coordinates, the F_j $(1 \leqslant j \leqslant p)$ being defined in a neighborhood of W in $\mathbf{R}^N$, and having in that neighborhood a jacobian matrix of rank p everywhere. He specified that an open set V in W of the kind he considered should be defined by adjoining to the equations (1) a finite set of inequalities $\varphi_\alpha > 0$, where the φ_α are C^1 functions defined in a neighborhood of W. We now know that *any* open subset V of W may be defined in that way; the context shows that Poincaré had a very special kind of open set in mind: he wanted the frontier of V in the space W to be defined as the union of a finite number of submanifolds v_α of dimension $\leqslant N - p - 1$, each of which is the set of points satisfying equations (1) plus *one* equation $\varphi_\alpha = 0$, as well as the remaining inequalities $\varphi_\gamma > 0$ for $\gamma \neq \alpha$. Furthermore, he tacitly assumed that when v_α has dimension $N - p - 1$, every point of v_α is both in the closure of V and of the open subset U_α of W defined by replacing $\varphi_\alpha > 0$ by $\varphi_\alpha < 0$ in the definition of V, and U_α is assumed to be *nonempty*. This seems to be the meaning of his claim that the union of the v_α is the "complete boundary" of V in W (p. 198).*

An unformulated assumption, common to Poincaré and all his predecessors, and repeatedly used by everyone, has to do with the idea of *deformation*, a notion that is never precisely defined but that probably is our present-day concept of *isotopy*; quite often it is asserted that "infinitely near" manifolds may be "deformed" into each other! It is also tacitly assumed that when a finite union of submanifolds v_α in W constitute the "complete boundary" of an open set, then, if one "deforms" in W the v_α, the union of the "deformed" submanifolds is again a "complete boundary."†

To arrive at his conception of *homology*, Poincaré started, as Riemann and Betti had done, from a system of $(q - 1)$-dimensional connected (and apparently compact) submanifolds $v_1, \ldots, v_\lambda$ (without boundary) of a p-dimensional manifold W, whose union constitutes the "complete boundary" of a connected q-dimensional submanifold V of W (with $q \leqslant p$), and to express this fact he elected to write by convention

$$v_1 + v_2 + \cdots + v_\lambda \sim 0, \tag{2}$$

calling this relation a "homology" between the v_j. However, he immediately added a completely new and crucial proviso: *homologies can be combined as ordinary equations*. The context shows that the operations he meant by this are addition and subtraction. This implies that he should define linear relations

$$k_1 v_1 + k_2 v_2 + \cdots + k_\lambda v_\lambda \sim 0 \tag{3}$$

---

* Sometimes (p. 198) he disregarded the v_α of dimension $\leqslant N - p - 2$, and called "complete boundary" of V the union of the v_α of dimension $N - p - 1$, which is not a closed set in general.

† The closest approximation to a definition of "deformation" is probably the one given by Picard and Simart ([362], vol. I, p. 28); they considered a family as given above by a system of equations and inequalities depending on parameters, and assumed that two such manifolds corresponding to "infinitely near" values of the parameters constitute a "complete boundary."

when the k_j are *arbitrary* (positive or negative) integers.* When all k_j are >0, Poincaré said (3) means that there exists a q-dimensional set V in W, the "complete boundary" of which consists in "k_1 manifolds slightly different from v_1, k_2 manifolds slightly different from $v_2, \ldots, k_\lambda$ manifolds slightly different from v_λ" (p. 207). This rather cryptic statement is explained in the exposition of Poincaré's theory given two years later by Picard and Simart ([362], vol. I, pp. 32–34), with examples illustrating the concept: in (3), each $k_j v_j$ should be replaced by the sum of k_j *distinct* manifolds, each one of which is deduced from v_j by "small deformations."

To allow negative integers as coefficients in (3), Poincaré appealed to the concept of *oriented manifold* introduced by Klein for surfaces and generalized by von Dyck for manifolds of arbitrary dimension.† In *Analysis Situs*, Poincaré gave a characterization of orientable manifolds by what is still one of the modern criteria: there exist charts $(U_\lambda, \psi_\lambda)$ such that the transition diffeomorphisms $\psi_\lambda(U_\lambda \cap U_\mu) \to \psi_\mu(U_\lambda \cap U_\mu)$ have positive determinants for all pairs of indices such that $U_\lambda \cap U_\mu \neq \emptyset$. Unfortunately, the relation between the orientation of an open set V in a q-dimensional manifold W of the kind specified above and the orientations of the $(q-1)$-dimensional manifolds v_j on its frontier is quite obscure in *Analysis Situs*. However, the context and the more detailed explanations Poincaré gave in the first *Complément* (p. 294) show that what he had in mind can be expressed in the following way. Let a

* This may well be the first example of a process that has become commonplace in modern mathematics: When one wants to study the objects of some set E, one considers the A-module C(E) of *formal linear combinations* of objects of E with coefficients in a commutative ring A, and then, using properties of the objects of E, one introduces submodules (or quotients of submodules) of C(E); this associates to E algebraic objects invariant under automorphisms of E. Before Poincaré the only similar construction of that type was the formation of "divisors" on an algebraic curve by Dedekind and Weber, although they used a *multiplicative* notation (the submodule consisting here of "principal divisors"). For algebraic surfaces this was later replaced by an additive notation in the Italian school. Could Poincaré have been inspired by that theory? In *Analysis Situs* he does not mention algebraic geometry at all, although in the third and fourth *Compléments* he continued the work of Picard by *applying* homology theory to algebraic surfaces (Part 2, chap. VII, § 1,A).

† Klein's idea was to take a small circuit (the "indicatrix") around a point, and to "orient" it by choosing on it a positive direction. He then conceived this circuit to be moved around continuously on the connected surface, and he distinguished two cases, depending on whether when coming back to the initial point the direction along the circuit is always preserved or may be reversed for some closed paths on the surface (as in the Möbius band). One may replace the circuit by an ordered frame of two vectors in the tangent plane; this is what von Dyck did to extend the notion of "indicatrix" to n-dimensional manifolds, by considering an ordered frame of n independent tangent vectors. The word "orientable" only appeared with Tietze [466] and Alexander; Poincaré used the older denomination of "one-sided" and "two-sided," although Klein and von Dyck had pointed out that this is a notion relative to the embedding of the manifold into a larger one, not an intrinsic concept: $\mathbf{P}_1(\mathbf{R})$ is orientable but one-sided when embedded in $\mathbf{P}_2(\mathbf{R})$ ([373], p. 124; [421], pp. 272–273).

q-dimensional submanifold W of $\mathbf{R}^n$ be defined by $n - q$ equations $\varepsilon_j F_j = 0$ $(1 \leqslant j \leqslant n - q)$, where the rank of the jacobian of the F_j is $n - q$, and the ε_j are ± 1; then W is orientable, and the problem is to define for W an *orientation*: Poincaré did this by defining an orientation as an *ordering* of the equations $\varepsilon_j F_j = 0$, and a *choice of signs* for the ε_j; changing one sign or permuting two equations replaces the orientation by the opposite one.* Then let V be an open subset of W and let v be a $(q - 1)$-dimensional manifold that is a part of the "complete boundary" of V and that is defined by the equations

$$\varepsilon_1 F_1 = \cdots = \varepsilon_{n-q} F_{n-q} = \varphi = 0;$$

as orientation of v take the one defined by *this* ordering; Poincaré called the orientations of V and v *directed related* (resp. *inversely related*) if $\varphi > 0$ (resp. $\varphi < 0$) in a neighborhood of v in V. Poincaré then made the convention that when a "homology" (2) is written the orientations of the v_j must be directly related to the orientation of the submanifold V of which they constitute the "complete boundary"; if the orientation of v_j is inversely related to that of V, v_j must be replaced by $-v_j$ in the relation (2). Nothing is said about the meaning of $k_j v_j$ in (3) when k_j is <0 and different from -1.

Regarding Poincaré's sweeping assertion that one may "add" homologies, we should observe that in doing so he entirely glossed over a nontrivial difficulty: given two q-dimensional manifolds V′, V″, how does one prove there *exists* a q-dimensional manifold having as "complete boundary" the *union* of the "complete boundaries" of V′ and V″? Certainly $V' \cup V''$ will not do in general, for it will *not* be a smooth manifold. (The intersection $V' \cap V''$ is not necessarily empty!) Finally, there is a rule of computation for "homologies" that is not explicitly stated at first, but that will be used in the examples (p. 244): for any integer $c \neq 0$, the homologies $\sum_j k_j v_j \sim 0$ and $\sum_j c k_j v_j \sim 0$ are considered to be *equivalent* (see pp. 244–245).

These definitions determine the frame within which Poincaré intended to develop the ideas introduced by Riemann and Betti. For an n-dimensional manifold U, he defined the *q-dimensional Betti number* P_q for $1 \leqslant q \leqslant n - 1$ by the condition that $P_q - 1$ is the maximum number of distinct compact connected q-dimensional submanifolds (without boundary) contained in U and which are 'independent," i.e., between which there exists no "homology" with coefficients not all 0.† In *Analysis Situs*, Poincaré seemed to be convinced that these numbers are the same as the "orders of connection" defined by Betti; at

* This definition may seem artificial, but it is in fact close to another modern way of defining an orientation by the choice of a q-form that is everywhere $\neq 0$: one has only to consider a q-form ω such that

$$\omega \wedge d(\varepsilon_1 F_1) \wedge \cdots \wedge d(\varepsilon_{n-q} F_{n-q}) = dx_1 \wedge \cdots \wedge dx_n$$

and take its restriction to W.

† Since Lefschetz's *Topology* [304], the term "Betti numbers" designates the numbers $P_q - 1$.

any rate, the first examples (p. 208) coincide with those of Betti (namely, an open ball, the interior of a torus, and the open set between two concentric spheres or between two tori having the same axis). Another example is obtained by considering a ball B: $|x| < \rho$ in $\mathbf{R}^3$, and a finite number of compact connected disjoint surfaces S_j ($1 \leqslant j \leqslant m$) contained in B; the example is the connected component U of $B - \bigcup_j S_j$ having as frontier the union of the S_j and the sphere $|x| = \rho$. The Betti numbers of all these examples are written down with no more proofs than in Betti's paper.

Before turning to different and more interesting examples, Poincaré first paused (section 7) to generalize, for integrals of differential p-forms ($1 \leqslant p \leqslant n - 1$) in a n-dimensional manifold, the relations established by Riemann for $n = 2$, and then by Betti for arbitrary n, between the Betti numbers (for dimensions 1 and $n - 1$ only) and the *periods* of the integrals of 1-forms and ($n - 1$)-forms. Referring to a previous paper of his on double integrals ([367], pp. 440–489), he considered what we now call *closed* p-forms ω (i.e., those for which $d\omega = 0$); he did not introduce the exterior derivative but said that these forms are characterized by "integrability conditions." The periods of such a form are the values of its integral along compact p-dimensional submanifolds without boundary, and Poincaré stated without proof that they are linear combinations with rational coefficients of $P_q - 1$ periods; this of course immediately follows from the Stokes formula, which, surprisingly, is not mentioned explicitly.* Of course we now see this as the first step toward the De Rham theorems (chap. III, §3), but until Cartan's revival of these results, nobody seemed to have thought of them.

In section 9 Poincaré endeavored to prove his central result on homology, the famous *duality theorem* for compact, connected, and orientable n-dimensional manifolds without boundary, which he formulated as $P_p = P_{n-p}$ for $1 \leqslant p \leqslant n - 1$.† Again an important new concept emerged, which only came to full fruition with the work of Lefschetz [300] and of De Rham [389]. Inspired by the results of Kronecker on the number of solutions of a system of equations [288], which he had used himself repeatedly in previous papers (see, for instance, [368], pp. 303–304), Poincaré defined, for two *oriented* submanifolds V_1, V_2 of complementary dimensions p and $n - p$ in an n-dimensional manifold U, and for a common point M where V_1 and V_2 intersect *transversely*,‡ an *intersection number* S(M) equal to ± 1.§ Then, sup-

---

* The formula is written down in [370], vol. 3, p. 10. The first mathematician to have written down Stokes' formula for an arbitrary dimension was probably V. Volterra (*Opere mat.*, vol. I, p. 407).

† Poincaré said the formula was known and used by some mathematicians whom he did not name.

‡ This means that the tangent spaces of V_1 and V_2 at the point M intersect in the single point M.

§ If z (resp. z_1, z_2) is a positive n-vector [resp. a positive p-vector, a positive ($n - p$)-vector] in the tangent space of V (resp. V_1, V_2), one has $z_1 \wedge z_2 = c \cdot z$ for a number $c \neq 0$, and S(M) is the *sign* ± 1 of c.

posing that V_1 and V_2 intersect in a *finite* number of points M_i ($1 \leqslant i \leqslant m$) and that they intersect transversely at each M_i, he put $N(V_1, V_2) = \sum_i S(M_i)$. He then wanted to show that a "homology" (2) between q-dimensional oriented compact connected submanifolds (without boundary) of an n-dimensional manifold U is *equivalent* to the fact that $\sum_j N(V, v_j) = 0$ for *any* oriented $(n - q)$-dimensional submanifold V of U intersecting transversely each v_j in a finite number of points. That theorem easily yields the inequality $P_q \leqslant P_{n-q}$ between Betti numbers of U and, by exchanging q and $n - q$, the duality theorem.

The proof given by Poincaré (probably inspired by Kronecker [288]) for $q = n = 1$ seems correct, at least when the v_j are distinct. (The general case is not considered.) But his attempt to prove the result for $q < n - 1$ is totally unconvincing: it consists in assuming (of course without any justification) that there is a *smooth* $(q + 1)$-dimensional submanifold $W \subset U$ containing all the v_j's, considering the 1-dimensional intersection $V' = V \cap W$, and applying the result obtained for $q = n - 1$ to that curve. Even granted the existence of W, one would still [as Heegaard will observe later (§ 3)] have to prove that each curve $V' \subset W$ has the form $V \cap W$ for an $(n - q)$-dimensional submanifold V.

At the end of section 9 Poincaré considered the case $n = 4k + 2$ and the number $N(V_1, V_2)$ for two oriented $(2k + 1)$-dimensional submanifolds, that only depends on their homology classes. He claimed (without any proof) that it is possible to define $N(V_1, V_2)$ for *any* pair of oriented compact connected manifolds of complementary dimensions by "deforming" them into manifolds for which the conditions of intersection are those described above (the "general position" argument*); the fact that the dimensions of V_1 and V_2 are odd numbers then implies that $N(V_2, V_1) = -N(V_1, V_2)$, and in particular $N(V, V) = 0$ for all $(2k + 1)$-dimensional submanifolds V. Considering the determinant $\det(N(V_i, V_j))$ for a maximal system of "independent" $(2k + 1)$-dimensional submanifolds of U then gave him the result that the Betti number P_{2k+1} of U is odd.

In section 10 of *Analysis Situs* Poincaré introduced a series of very interesting examples of three-dimensional orientable compact connected manifolds, and then devoted several of the next sections to the computation of their Betti numbers. These spaces are obtained by a generalization to three dimensions of the process Klein ([273], vol. 3, pp. 36–43) and Poincaré himself ([366], p. 148) had used to construct surfaces by gluing together edges of a curvilinear polygon. This time Poincaré started from a compact convex polyhedron K and took its quotient by an equivalence relation R. One of his examples is the projective space $\mathbf{P}_3(\mathbf{R})$, obtained (without giving it its name) by identifying symmetric points $\pm z$ of the boundary of a regular octahedron of center O (a

* This had been used by algebraic geometers for many years; the idea is that when the manifolds depend on parameters, they will intersect transversely except for values of the parameters that form a rare set in parameter space. (For a modern version of that idea, see Part 3, chap. VII, § 1,A.)

construction equivalent to the one given by von Dyck [147]). All the other examples are obtained by taking for K the cube

$$0 \leqslant x, y, z \leqslant 1,$$

and the general description of the relation R may be given in the following way: (1) R is a closed relation; (2) no two points in the interior* of an edge (resp. of a face, resp. of K) are equivalent under R; (3) R identifies each closed face (resp. each edge, each vertex) with at least another face (resp. edge, resp. vertex) and the identification of two faces (resp. edges) is given by an affine transformation. (The simplest of these examples is the one where the identifications are given by $x \mapsto x + 1$, $y \mapsto y + 1$, $z \mapsto z + 1$, and then K/R is the torus T^3.) However, Poincaré wanted K/R to be a manifold, and he showed that for that R must satisfy a necessary condition obtained by considering in K/R a small sphere S with center at a point A_0, which is the image of a vertex A of K; the inverse image of S in K is then a union of intersections of K with small spheres having as centers the vertices of K equivalent to A under R. These intersections are spherical triangles, and the union of their images in K/R should be the whole sphere S. In other words, these images should be the faces of a *spherical polyhedron*, and hence satisfy the Euler relation

$$v - e + f = 2, \tag{4}$$

where v is the number of vertices, e is the number of edges, and f is the number of faces. But these numbers are easy to evaluate: f is the number of vertices of K equivalent to A, v the number of equivalence classes of half edges of K whose extremities are the vertices equivalent to A, and finally e is the number of equivalence classes of quarter faces of K adjacent to a vertex equivalent to A. The relation (4) therefore imposes conditions on the relation R that are not always verified, as is shown by one of Poincaré's examples.

Section 11 is devoted to another type of manifold, also inspired by the theory of fuchsian groups, namely, the space of orbits $X = \mathbf{R}^3/G$, where G is a "properly discontinuous" group of diffeomorphisms of $\mathbf{R}^3$ (see Part 3, chap. I, § 1). Poincaré limited himself to the case in which G is generated by three transformations:

$$(x, y, z) \mapsto (x + 1, y, z), \qquad (x, y, z) \mapsto (x, y + 1, z),$$
$$(x, y, z) \mapsto (\alpha x + \beta y, \gamma x + \delta y, z + 1),$$

where α, β, γ, δ are integers such that $\alpha\delta - \beta\gamma = 1$; the cube K is then the closure of a "fundamental domain" for G. (Two of the previous examples turn out to be special cases of this one.)

It is quite remarkable that this example seems to have inspired the general definition of the *fundamental group* $\pi_1(X)$ of a manifold, which follows in

* The interior is taken with respect to the affine linear variety generated by the edge (resp. a face).

section 12 and which we will describe in more detail in Part 3, chap. I, § 1. For $X = \mathbf{R}^3/G$ the group $\pi_1(X)$ is naturally isomorphic to G, since $\mathbf{R}^3$ is simply connected. Bent upon the computation of the Betti numbers $P_1 = P_2$ in his examples, Poincaré singled out the fact that when, in a fundamental group $\pi_1(X)$, there is a relation

$$s_1^{\alpha_1} s_2^{\alpha_2} \cdots s_m^{\alpha_m} = \mathrm{Id} \quad (\alpha_j \text{ positive or negative integers}) \tag{5}$$

between classes s_j of closed curves C_j $(1 \leqslant j \leqslant m)$ with a common point, it gives rise to a "homology"

$$\alpha_1 C_1 + \alpha_2 C_2 + \cdots + \alpha_m C_m \sim 0 \tag{6}$$

between the C_j. (he assumed without proof that a curve which can be "deformed" to a point bounds a smooth surface.) He also assumed without proof that *all* homologies between curves in X are linear combinations of those obtained in that manner for all relations (5) in $\pi_1(X)$.

The determination of the Betti numbers of Poincaré's examples then proceeds as follows. In all these examples opposite faces of the cube K are identified by R; Poincaré stated (without proof) that a system of three generators of the (noncommutative) group $\pi_1(X)$ is obtained by considering for each pair of opposite faces F, F′ an arc joining a point of F to the point of F′ with which it is identified; it becomes a closed curve in X, and Poincaré claimed that the classes in $\pi_1(X)$ of these three curves are the generators of $\pi_1(X)$, provided the three arcs have a common point O inside K. To obtain the relations among these generators, he "deformed" the arcs in such a way that if F_1 and F_2 are two faces of K having a common edge E, the arcs joining O to points of F_1 and F_2 coincide and have a point of E as their extremity (different from a vertex). Each equivalence class of edges of K then yields a relation, and from these relations Poincaré deduced the corresponding "homologies" and finally the value of P_1. Then, in section 14, concentrating on his last type of examples, he launched a long discussion to show it is possible to choose an infinity of matrices $\begin{pmatrix} \alpha & \beta \\ \gamma & \delta \end{pmatrix}$ with integral entries and determinant 1 for which the corresponding groups G are *nonisomorphic*. As the Betti number P_1 is always $\leqslant 4$, he concluded that there are infinitely many three-dimensional manifolds with the *same* Betti numbers, no two of which are homeomorphic.

This led him to raise two interesting questions: (1) Are there two nonhomeomorphic manifolds having the *same* Betti numbers and the *same* fundamental group (cf. [421], p. 279) and (2) given an *arbitrary* group G, does there exist a manifold for which G is the fundamental group (cf. [421], p. 180 and Part 3, chap. II, § 6,F)?

In section 15, Poincaré introduced two more examples. The first example is another way of considering the sphere $\mathbf{S}_2$ as a two-sheeted covering space of the projective plane $\mathbf{P}_2(\mathbf{R})$, by using the Veronese mapping of $\mathbf{S}_2$ into $\mathbf{R}^6$:

$$(x, y, z) \mapsto (x^2, y^2, z^2, xy, yz, zx).$$

The second example is much more interesting, being the *only* example in *Analysis Situs* of the determination of the Betti numbers for a manifold (other than spheres) of *arbitrary* even dimension. Poincaré considered the product $W = \mathbf{S}_{q-1} \times \mathbf{S}_{q-1}$ with $q \geqslant 2$ and the map

$$\pi: (\mathsf{y}, \mathsf{z}) \mapsto ((y_i + z_i)_{1 \leqslant i \leqslant q}, (y_i z_i)_{1 \leqslant i \leqslant q}, (y_i z_k + y_k z_i)_{1 \leqslant i, k \leqslant q})$$

of W into $\mathbf{R}^n$ with $n = q(q + 3)/2$. If H is the diagonal $\mathsf{y} = \mathsf{z}$ in W and $V = \pi(W - H)$, then $W - H$ is a two-sheeted covering space of the $(2q - 2)$-dimensional manifold V; $\pi(W) = \overline{V}$ is its closure in $\mathbf{R}^n$, and $\overline{V} - V = \pi(H)$ is diffeomorphic to $\mathbf{S}_{q-1}$. Poincaré showed that V is orientable when q is odd and nonorientable when q is even. The Betti numbers of W can immediately be obtained from the Künneth theorem (chap. II, §5), but this was not available to Poincaré. He therefore used an *ad hoc* method introducing the two "axes" in the product, $U_1 = \mathbf{S}_{q-1} \times \{Q_0\}$, $U_2 = \{Q_0\} \times \mathbf{S}_{q-1}$, where Q_0 is a point of $\mathbf{S}_{q-1}$; $W - (U_1 \cup U_2) = (\mathbf{S}_{q-1} - \{Q_0\}) \times (\mathbf{S}_{q-1} - \{Q_0\})$ is then diffeomorphic to $\mathbf{R}^{2q-2}$. By "intuitive" deformation arguments Poincaré arrived at the conclusion that any compact submanifold v of W of dimension $h < q - 1$ may be "deformed" into a submanifold contained in $W - (U_1 \cup U_2)$, hence $v \sim 0$, which shows that $P_h = 1$ and, by duality, $P_h = 1$ for $h > q - 1$; if v has dimension $q - 1$, more refined arguments give $v \sim mU_1 + nU_2$ for suitable integers m, n. Finally, to show that U_1 and U_2 are homologically "independent," hence that $P_{q-1} = 3$, he used the following very ingenious argument: let σ be the closed $(q - 1)$-form on $\mathbf{S}_{q-1}$, invariant by rotation and such that $\int_{\mathbf{S}_{q-1}} \sigma = 1$, and let σ_1, σ_2 be the pullbacks of σ on U_1 and U_2; if λ is an irrational number, $\langle \sigma_1 + \lambda\sigma_2, mU_1 + nU_2 \rangle \neq 0$ for any pair of integers m, n not both 0.

Then follows an obscure and totally unconvincing argument (pp. 268–269) leading to the conclusion that for V the Betti number $P_{q-1} = 2$. (An equally obscure passage on pp. 262–263 could be interpreted as meaning that when one deletes from a p-dimensional manifold a q-dimensional submanifold with $q \leqslant p - 2$, the Betti numbers do not change!) Finally it transpires that the purpose of this long discussion was to obtain counterexamples to the result on $P_{n/2}$ mentioned in section 9 for orientable compact connected manifolds of dimension $4k + 2$, since V is orientable and of dimension $4k$, or nonorientable and of dimension $4k + 2$.

In the last three sections of *Analysis Situs* (16–18), Poincaré turned to a new problem with far reaching consequences for his initial outlook in the first and second *Compléments* (§3). Ever since Schläfli ([413] and [373], pp. 29–31) posed the problem, several mathematicians tried to extend to n dimensions the Euler formula (4), limiting themselves most of the time to a *convex* polyhedron (see the bibliography in [138]). What replaced the left-hand side of (4) was the alternating sum

$$\alpha_{n-1} - \alpha_{n-2} + \alpha_{n-3} - \cdots + (-1)^{n-1}\alpha_0, \tag{7}$$

where α_j is the number of j-dimensional faces of the polyhedron. On the other hand, the Euler formula (4) itself had given rise to many uncertainties and errors when mathematicians tried, since the beginning of the nineteenth century, to extend it to nonconvex polyhedra.* The general formula had finally been given by Jordan [268] in the form

$$v - e + f = 3 - \mathrm{P}_1 \tag{8}$$

when the surface of the polyhedron is a manifold having P_1 as "Betti number" in Poincaré's notation. At the end of *Analysis Situs*, Poincaré undertook to prove the corresponding formula giving the value of (7) in $\mathbf{R}^n$ for an arbitrary n; going even beyond that goal, he observed that the theorem is a purely topological one and should therefore be formulated in an arbitrary n-dimensional manifold.

He thus had first to define what is meant by a finite (curvilinear) p-dimensional polyhedron V. His definition is essentially (with some imprecision) the modern one: V is a compact, connected p-dimensional manifold contained in an n-dimensional manifold W; V is the *disjoint* union of a finite family T of "cells"† of various dimensions $\leqslant p$; each j-dimensional cell C is an open subset of a j-dimensional submanifold U of W, such that there exists a homeomorphism (in fact a C^1-diffeomorphism‡ in Poincaré's paper) of an open neighborhood in U of the closure $\bar{\mathrm{C}}$ of C in U, onto an open subset of $\mathbf{R}^j$, mapping $\bar{\mathrm{C}}$ on a closed ball, and C on its interior; the frontier $\bar{\mathrm{C}} - \mathrm{C}$ of C in U must be a union of k-cells of the family T, with $k \leqslant j - 1$. Using a word later introduced by Weyl ([483], p. 21) we shall say that the family T is a *triangulation* of the manifold V. Poincaré observed that any $(p-1)$-cell of T is on the boundary of exactly two p-cells of T. Calling α_j the number of j-cells in T, he wanted to prove that the alternating sum

$$\alpha_p - \alpha_{p-1} + \alpha_{p-2} - \cdots + (-1)^p \alpha_0 \tag{9}$$

depends only on V, and not on the particular triangulation T.

Poincaré made three attempts to give such a proof. The first two rely on the same "natural" idea: if T and T′ are two triangulations of V, the family T″ of intersections of a cell of T and a cell of T′ should be a triangulation of V, and it would be enough to show that passing from T″ to T (or T′) does not change the number (9). However, after sketching a method to yield that result by successive steps leading from T″ to T (or T′) by a sequence of intermediate triangulations, he realized that the intersection of a cell of T and a cell of T′ may have components that are not simply connected (for instance, the intersection of two open caps covering $\mathbf{S}_2$).

---

* See M.B. Brückner, *Vielecke und Vielfläche*, Leipzig, Teubner, 1900. pp. 58–66.

† It seems that the nearest approximation to this notion (for $p \geqslant 3$) before Poincaré was von Dyck's idea of building up a space by successive adjunction or deletion of cells ([373], pp. 147–148); but he does not seem to have thought of a general definition of curvilinear polyhedron.

‡ In applications, Poincaré often met homeomorphisms that were only *piecewise* C^1, so that he should have allowed this possibility in his definition of cells.

To avoid this obstacle, and assuming that V is contained in some $\mathbf{R}^N$, he tried in the second "proof" to compare T with a triangulation T′ of a special type, consisting of the intersections of V with the "triangulation" of $\mathbf{R}^N$ whose N-cells are the connected components of the complement of the union of the hyperplanes $x_j = k\delta$ $(1 \leqslant j \leqslant N, -\infty < k < +\infty)$ with *small enough* δ. This is also doomed to fail, for an intersection of a C^∞-manifold with a linear variety is susceptible of the weirdest pathology, as we mentioned earlier; Poincaré was probably not aware of it. Even if all cells of T are analytic manifolds* their intersections with linear varieties may fail to be smooth, and that fact could not have escaped Poincaré. Nevertheless, if this difficulty is ignored† (for instance, if the cells of T are rectilinear convex polyhedra in the usual sense), Poincaré's argument is remarkable: the fact that the "mesh" δ is arbitrarily small allowed him to assume that the points of the cells of T that are interior to a p-cell of T′ belong to the "star" of a single q-cell v_q of T [a notion which he had introduced earlier (p. 276) under the name of "aster"]. He was then able to use induction on p; if γ_j is the number of j-cells of T that are in the star of v_q $(q \leqslant j \leqslant p)$, by intersecting that star with a small sphere with center belonging to v_q, by induction he obtained the relation

$$\gamma_p - \gamma_{p-1} + \gamma_{p-2} - \cdots \mp \gamma_{q+1} = \begin{cases} 2 & \text{if } p - q \text{ is even} \\ 0 & \text{if } p - q \text{ is odd} \end{cases} \tag{10}$$

and from that he was able to deduce the equality of the numbers (9) for T and T′. One cannot fail to see in this kind of argument a forerunner of the method based on "sufficiently fine subdivisions" of cells, which was later at the root of the correct proofs of Brouwer and Alexander (chap. II and Part 2, chap. I).

In the third "proof" Poincaré aimed not only at proving the independence of (9) from the triangulation, but also at giving its value in terms of the Betti numbers, i.e., the "Euler–Poincaré characteristic"

$$\begin{cases} 3 - P_1 + P_2 - \cdots + P_{p-1} & \text{for even } p \\ P_{p-1} - P_{p-2} + \cdots + P_2 - P_1 & \text{for odd } p. \end{cases} \tag{11}$$

Actually he only considered the cases $p = 2$ and $p = 3$. For $p = 2$, his idea is clear enough: he wanted to show that every 1-cycle is homologous to a 1-cycle consisting only of edges of the polygons of the triangulation, which gave him $\alpha_2 + P_1 - 1$ 1-cycles, between which (by a very obscure argument) he claimed there is only one "homology," so that, in our language, the space of 1-cycles has dimension $\alpha_2 + P_1 - 2$; linear algebra then shows this implies

---

* Poincaré takes for granted that any compact manifold may be submitted to an arbitrarily small "deformation" transforming it into an analytic one (p. 200). In the remainder of *Analysis Situs* and in the *Compléments*, it is never stated explicitly if the manifolds are analytic, and often he submits them to "deformations" that obviously cannot be analytic (e.g., p. 311).

† On p. 311, Poincaré brushed aside similar objections by an argument of "general position."

that the space of 0-boundaries has dimension $\alpha_1 - (\alpha_2 + P_1 - 2)$. But since V is connected, any two vertices of the triangulation may be joined by a path consisting of edges of the triangulation and 0-boundaries form a space of dimension $\alpha_0 - 1$, hence

$$\alpha_0 - 1 = \alpha_1 - (\alpha_2 + P_1 - 2), \tag{12}$$

which is Jordan's formula (8). Of course Poincaré did not use our algebraic language, but it may be said that his argument has a "cohomological" flavor. Indeed, he considered what we now call "0-cochains," of the triangulation T, functions φ assigning an arbitrary real number to each vertex, and to each oriented edge of T he assigned the difference $\varphi(a) - \varphi(b)$ of the values of φ at the origin a and at the extremity b of the edge. This gave him a system of generators of what we now call the "1-coboundaries," and for him the two sides of (12) are just two different evaluations of the number of linearly independent "1-coboundaries."

When Poincaré tried to apply similar methods to the case $n = 3$, he met difficulties for 1-coboundaries, for he wanted to prove that when a 1-cycle K bounds a surface R it also bounds a surface R″ that is the union of some of the 2-cells of T. To try to prove this he "decomposed" R into the union of its intersections with the 3-cells of T, and therefore had to cope again with the inextricable pathologies of the intersections of C^∞-manifolds. Things are much worse for 2-boundaries, and all his assertions regarding them are in fact unsupported by any proof.

Thus ends this fascinating and exasperating paper, which, in spite of its shortcomings, contains the germs of most of the developments of homology during the next 30 years.

§3. Heegaard's Criticisms and Poincaré's First Two *Compléments à l'Analysis Situs*

If we disregard the Picard–Simart chapter of 1897, which is a mere commentary on Poincaré's *Analysis Situs* ([362], vol. I), the first paper to appear on algebraic topology after Poincaré's was the dissertation of the Danish mathematician Heegaard (1871–1948) published in 1898 (and translated into French in 1918 [221]). Heegaard was interested in the topology of complex algebraic surfaces and reduced its study to that of suitably constructed three-dimensional manifolds. His arguments were just as "intuitive" as those of his contemporaries, but this did not prevent him from noting obscurities or flaws in Poincaré's paper. In the definition of a "homology"

$$v_1 + v_2 + \cdots + v_\lambda \sim 0$$

between distinct compact connected manifolds, he questioned the possibility that the v_j have nonempty intersections ([221], p. 214). But what brought his paper to Poincaré's attention was an example of a three-dimensional orientable manifold* for which Heegaard claimed the Betti numbers were $P_1 = 2$, $P_2 = 1$, in contradiction with the duality theorem. When he examined that example, Poincaré realized that, contrary to what he probably had believed in *Analysis Situs*, his definition of "Betti numbers" did not coincide with the Riemann–Betti "orders of connection" (which Heegaard took as definition of *his* "Betti numbers"): the qth "order of connection" is the largest number of *distinct* q-dimensional compact connected manifolds v_j $(1 \leqslant j \leqslant \lambda)$ such that their union is *not* the boundary of a $(q+1)$-dimensional manifold, but this does not preclude the existence of integers m_j $(1 \leqslant j \leqslant \lambda)$ not all equal to ± 1 and for which the "manifold" $m_1 v_1 + m_2 v_2 + \cdots + m_\lambda v_\lambda$ (in the sense of Poincaré) *is* a boundary. The simplest case takes place in projective space $\mathbf{P}_3(\mathbf{R})$, where the projective line $\mathbf{P}_1(\mathbf{R})$ is not a boundary, but $2\mathbf{P}_1(\mathbf{R})$ bounds an open set homeomorphic to $\mathbf{R}^2$.

Poincaré had to admit that his first attempt to prove the duality theorem was completely unsatisfactory, since it seemed to apply to the "orders of connection" as well! It is quite likely that he realized that his arguments were beyond repair, but he rose to the challenge and discovered that the techniques he had used to prove the generalized Euler formula in *Analysis Situs* could lead to a completely new conception of homology; he developed it in the first two *Compléments* published in 1899 and 1900 ([369], pp. 290–370), and it became the *backbone* of "combinatorial topology" ever after.

The numbers attached to a manifold V by Riemann, Betti, and Poincaré himself in *Analysis Situs* were obviously *invariant* under any C^1-diffeomorphism. What Poincaré did in the first *Complément* was to assume that there existed a triangulation of the manifold V and to attach numbers *to that triangulation*. Of course, this immediately raised three questions: (1) Is it possible to define a triangulation on a compact connected C^1-manifold?† (2) Are the numbers defined by Poincaré (which he calls the "reduced Betti numbers" of the triangulation) independent of the chosen triangulation? (3) If so, are they equal to the "Betti numbers" as defined in *Analysis Situs*? We shall see that Poincaré's attempts to give positive answers to these questions could not succeed, and they were only settled much later.

After describing Heegaard's criticisms in the introductory section I of the first *Complément*, Poincaré began, in section II, by making precise the relations (which we have described in §2, and which he had left implicit in *Analysis Situs*) between the orientations of an open subset V of an n-dimensional manifold W and those of the $(n-1)$-dimensional manifolds v_α of its frontier. Then, assuming that a triangulation T of a compact connected manifold V has been given and that orientations have been chosen for each cell of T, he defined for each pair consisting of a q-cell a_i^q and a $(q-1)$-cell a_j^{q-1} of T a

---

* The intersection of the cone $z^2 = xy$ in $\mathbf{C}^3$ with the cylinder $|x|^2 + |y|^2 = 1$.

† Of course, the existence of such a triangulation is obvious on "elementary" manifolds such as $\mathbf{S}_n$, $\mathbf{T}^n$, or $\mathbf{P}_n(\mathbf{R})$.

number ε_{ij}^q equal to 0 if a_j^{q-1} is not contained in the frontier of a_i^q and to $+1$ (resp. -1) if a_j^{q-1} is contained in the frontier of a_i^q and the orientations of a_i^q and of a_j^{q-1} are directly (resp. inversely) related.

Keeping the terminology introduced in *Analysis Situs*, Poincaré called a linear combination of q-cells

$$\sum_i \lambda_i a_i^q \tag{1}$$

with integer coefficients λ_i a "manifold" and the linear combination

$$\sum_{i,j} \lambda_i \varepsilon_{ij}^q a_j^{q-1} \tag{2}$$

the "set of manifolds" constituting its "complete boundary." In fact, once the ε_{ij}^q have been fixed, the handling of these putative "manifolds" by Poincaré was usually purely algebraic. After Poincaré the linear combinations (1) were considered "formal combinations," and more aptly called *chains.** The *boundary* of the *chain* (1) is then the *chain* (2) and should not be confused with a *subset* of V. From now on we shall use this terminology for more clarity and use the word *frontier* of a subset $A \subset V$ and the notation Fr(A) for the set of points of V that are both in the closure of A in V and in the closure of the complement $V - A$. The big step forward in that shift of emphasis is that it will give a *regular* procedure for the computation of the "reduced Betti numbers" of a triangulation, whereas in *Analysis Situs* (and earlier in Betti's paper) the Betti numbers were obtained by unsupported guesswork.

First Poincaré showed that the boundary of the boundary of a chain is 0. Then, mimicking the definitions of *Analysis Situs*, he called the chain $\sum_i \lambda_i a_i^q$ *closed* (our *cycles*) if its boundary is 0, and wrote a "homology"

$$\sum_i \lambda_i a_i^q \sim 0 \tag{3}$$

if the chain $\sum_i \lambda_i a_i^q$ is the boundary of a $(q+1)$-chain. What he called (section III) the "reduced Betti number" P_q' is the largest number such that there exists $P_q' - 1$ closed chains which are *not* linked by a homology with coefficients not all 0. He wrote α_q the number of q-cells of T, and introduced two numbers α_q' and α_q'': in our language $\alpha_q - \alpha_q''$ and $\alpha_q - \alpha_q'$ are, respectively, the dimensions (over the rational field **Q**) of the vector space of closed q-chains and of the vector space of boundaries of $(q+1)$-chains; hence, $P_q' - 1 = \alpha_q' - \alpha_q''$ and $\alpha_q'' = \alpha_{q-1} - \alpha_{q-1}'$ (the space of boundaries of q-chains is isomorphic to a supplementary of the subspace of closed q-chains).† From these relations an "Euler–Poincaré" relation follows at once:

$$\alpha_n - \alpha_{n-1} + \cdots + (-1)^n \alpha_0 = 1 - (P_{n-1}' - 1) + \cdots + (-1)^n (P_1' - 1) + (-1)^n. \tag{4}$$

* That terminology is due to Alexander.

† Lacking a convenient algebraic language, Poincaré's definitions are extremely clumsy; for instance, he says that α_q' is "the number of cells which remain distinct, when cells linked by homologies are not considered as distinct"!

Poincaré's main goal in the first *Complément* was to give a new proof of the duality theorem; but before we examine the fundamental techniques he invented for that purpose, we shall briefly review his unsuccessful efforts to answer the three questions previously mentioned. In sections IV–VI, he wanted to prove that his "reduced Betti numbers" are the same as the "Betti numbers" he had defined in *Analysis Situs*. To this end he had to prove that (A) any compact connected q-manifold v in V is homologous to a q-chain of T (in the sense of *Analysis Situs*) and (B) there are no "homologies" between q-chains of T that are not linear combinations of the α''_q "homologies" defined in section III.

His argument for statement A is given in three enigmatic lines at the end of section V (p. 309); it seems to be based (without being explicitly stated) on what he had done in section 18 of *Analysis Situs*, decomposing the manifold v into a union of its intersections with the cells of a "sufficiently fine" triangulation T′ and replacing each of these intersections by a homologous chain on the boundary of the cell; as usual, this founders on the pathology of intersections of manifolds.

The necessity of passing from a triangulation T to a finer one T′ by suitable subdivisions of the cells of T compelled Poincaré to prove that this operation does not change the "reduced Betti numbers," and this is what he did (this time very carefully) in sections IV and V. The passage from T to T′ is not at first described very clearly, but the fundamental property which emerges is that every q-cell of T′ is a cell of a *triangulation of an h-cell of* T, for a *smallest* dimension $h \geqslant q$, and that this h-cell is uniquely determined. Using this fact, and the fact that the frontier of a q-cell is homeomorphic to S_{q-1}, Poincaré first showed that in a "homology" between q-cells of T′ those for which $h > q$ may be disregarded and then that the "homology" between the remaining q-cells of T′ is equivalent to a "homology" between the uniquely determined q-cells of T, in which everyone of these q-cells of T′ is contained. (There are a number of improperly used symbols in the argument, but that is easily corrected.)

After this clever proof, section VI of the first *Complément*, which follows, shows Poincaré at his worst, intersecting and "deforming" manifolds in the most reckless way without the slightest justification in an attempt to prove statement B in the particular case $n = 4$, $q = 1$. (Even if the argument could be made rigorous, it would not extend to more general cases.)

Finally, in section XI of the first *Complément*, Poincaré tried to prove the existence of a triangulation of an analytic manifold V. He wanted to use induction on the dimension of the manifold.* But he assumed that there is a covering of V by closures of domains of charts such that these domains have no common points and each of them has piecewise analytic frontiers in V, a result that certainly is not obvious; in the course of his argument he once more

* A similar idea leads to a proof of the triangulability of an algebraic variety, smooth or not ([478], [311]).

takes intersections of analytic manifolds without any mention of their possible pathology.

We now come to the three essential innovations that launched "combinatorial topology": *simplicial subdivisions* by the "barycentric" method, the use of "*dual triangulation*",* and, finally, the use of *incidence matrices* and of their "reduction." It will be convenient here to consider simultaneously the first and second *Compléments*, since much of the material introduced in the first was taken up again, improved, and generalized in the second.

The "barycentric" subdivision T′ of a triangulation T of an n-dimensional manifold consists in taking in each cell of T a point (the "barycenter") and defining for each q-cell v of T a *triangulation of v into simplices* by induction on q: consider the "pyramids" having as vertex the barycenter of v, and as bases the simplices in the barycentric subdivisions of the p-cells forming the frontier of v for all $p \leqslant q - 1$;† T′ is then the collection of the triangulations of all the n-cells of T.

To define the "dual" triangulation T* of T Poincaré introduced different groupings of the simplices of T′. He considered the barycenter b_0 of a p-cell a_p of T, and the closures of the $(n - p)$-simplices of T′ having that point as a vertex and intersecting a_p transversely; the union b_{n-p} of these simplices and of those of the simplices in their frontier that have b_0 as a vertex was believed by Poincaré to be homeomorphic to an open ball of $\mathbf{R}^{n-p}$ and was named by him the $(n - p)$-cell *dual* to a_p; T* was then the triangulation of V consisting of all the duals of the cells of T. As a matter of fact Poincaré, in the first *Complément*, only considered in detail the case $n = 4$. This did not prevent him, in the second *Complément*, from freely using the previous construction for all dimensions; the fact that for $n > 4$ the b_{n-p}, for $p > 1$, are not necessarily "cells" in Poincaré's sense was only discovered much later, and it was necessary to modify the definition of triangulation of a manifold accordingly to make sense of the Poincaré construction (chap. II, § 4).

Poincaré then stated without proof (even for $n = 4$) that (apparently for suitable orientations of the cells of T and T*, which he does not define) the coefficients ε_{ij}^{*p} and ε_{ij}^{p} which he introduced for T* and T, respectively, are linked by the relations (p. 340)

$$\varepsilon_{ij}^{n-p+1} = \varepsilon_{ji}^{*p}. \tag{5}$$

As T′ is obtained by subdivision of *both* T and T*, it follows from Poincaré's result in section V of the first *Complément* that the "reduced Betti numbers" P_p and P_p^* of T and T*, respectively, are *the same*. Strangely enough, this only comes as an afterthought at the end of section VII of the first *Complément*, after Poincaré tried to give a direct proof by "deformation" of the relation $\mathrm{P}_1^* = \mathrm{P}_1$ for $n = 4$.

* Poincaré says "polyèdre réciproque."

† This construction (which for $n = 3$ is mentioned by Euler [*Opera Omnia*, (1), t. XXVI, p. 105]) uses the fact that a cell is homeomorphic to a convex polyhedron.

To prove the duality theorem (for "reduced Betti numbers"), it is therefore necessary to show that

$$\mathrm{P}^*_{n-p} = \mathrm{P}_p. \tag{6}$$

This results from the *algorithm* that Poincaré set up for the computation of the "reduced Betti numbers," in an imperfect manner in the first *Complément*, and in its final form in the second; it will be the *decisive* step in the concept of homology.

The algorithm rests on the remark that the numbers α_q, α'_q, and α''_q introduced by Poincaré *do not change* when each system of α_p p-cells (for $0 \leqslant p \leqslant n$) is replaced by an equal number of linear combinations with integral coefficients of those cells, provided that for each p the matrix of these coefficients is invertible. If E_q is the $\alpha_q \times \alpha_{q-1}$ matrix (ε^q_{ij}) (the "incidence matrix"), the corresponding matrix, after these replacements, has the form PE_qQ, where P and Q are, respectively square matrices in $\mathrm{SL}(\alpha_q, \mathbf{Z})$ and $\mathrm{SL}(\alpha_{q-1}, \mathbf{Z})$, and may be chosen arbitrarily. A classical theorem proved by H.S. Smith in 1861* shows that it is possible to choose the unimodular matrices P, Q in such a way that in the matrix $PE_qQ = (\rho_{ij})$, $\rho_{ij} = 0$ for $i \neq j$, the ρ_{ii} (the "invariant factors" of E_q) are integers such that ρ_{ii} divides $\rho_{i+1,i+1}$, and $\rho_{ii} = 0$ if and only if $i > \gamma_q$, where γ_q is the *rank* of E_q (as a matrix over $\mathbf{Q}$). This shows at once that the maximum number of linearly independent closed q-chains [resp. linearly independent boundaries of $(q+1)$-chains] is $\alpha_q - \gamma_q$ (resp. γ_{q+1}), hence

$$\mathrm{P}_q - 1 = \alpha_q - \gamma_q - \gamma_{q+1}. \tag{7}$$

Poincaré also observed that if one of the invariant factors ρ_{ii} is an integer $c > 1$, then there is a $(q-1)$-chain z such that $cz \sim 0$, although z is *not* boundary of a q-chain. It seems likely that the example of the Möbius band led him to interpret this phenomenon as revealing what he calls *torsion intérieure* (interior twisting) in the manifold. To describe it (in section 6 of the second *Complément*) he introduced "closed sequences" $(a^q_1, a^q_2, \ldots, a^q_r)$ of q-cells of T, where $a^q_r = a^q_1$ but all the $r-1$ q-cells $a^q_1, \ldots, a^q_{r-1}$ are distinct, and any two consecutive cells a^q_j, a^q_{j+1} for $1 \leqslant j \leqslant r-1$ have frontiers whose intersection is the closure of a single $(q-1)$-cell b^{q-1}_j of T; the sequence is *orientable* if it is possible to define orientations on the cells a^q_j and b^{q-1}_j $(1 \leqslant j \leqslant r)$ such that b^{q-1}_j is directly related to one of the cells a^q_j, a^q_{j+1} and inversely related to the other. Poincaré then proved that if all the closed sequences of q-cells are orientable, all the corresponding invariant factors ρ_{ii} are 0 or 1. This phenomenon led him to give the name *torsion coefficients* to the ρ_{ii} different from 0 and 1;† he probably thought they were also independent of the triangulation T, but he did not attempt to prove it.

---

* Had Poincaré read Smith's paper, or the version of that paper given by Frobenius in 1879? At any rate he deemed it necessary to reprove the theorem in section 2 of the second *Complément*.

† It is easy to define a closed sequence of three triangles in a triangulation of the Möbius band that is not orientable, and similarly a nonorientable closed sequence of five triangles in a triangulation of the projective plane (see [304], p. 53).

With the help of his algorithm for the computation of the "reduced Betti numbers" it was easy for Poincaré to prove relation (6). By (5), the matrix E^*_{n-p+1} relative to the dual triangulation T* is the *transposed* matrix of E_p, hence its rank γ^*_{n-p+1} is equal to γ_p, and as $\alpha^*_{n-p+1} = \alpha_{p-1}$, then by (7)

$$P^*_{n-p+1} = \alpha^*_{n-p+1} - \gamma^*_{n-p+1} - \gamma^*_{n-p+2} = \alpha_{p-1} - \gamma_p - \gamma_{p-1} = P_{p-1}.$$

Surprisingly, Poincaré does not seem to have been aware of the purely algebraic fact that not only are the ranks of a matrix (with integral entries) and its transposed matrix equal, but that they also have the same invariant factors; at any rate, to show that the torsion coefficients are the same for the triangulation T and its dual T*, he resorted to a partly topological argument that gave him a new proof of (6) at the same time. For that purpose he introduced a new notion in section 5 of the second *Complément*, later known (Part 2, chap. V, § 2,E) as the *join* of two polyhedra and used by many topologists.

In general, if A and B are two nonempty compact subsets of some $\mathbf{R}^N$, it is possible to embed A (resp. B) in a (N + 1)-dimensional vector subspace E (resp. F) of $\mathbf{R}^{2N+2}$, such that E and F are supplementary and such that a point of $\mathbf{R}^{2N+2}$ not in $A \cup B$ cannot be on two distinct straight lines meeting both A and B.* The union of the line segments joining a point of A and a point of B is the *join* of A and B.

Poincaré then considered the dual triangulations T, T* of V, and for each p-cell a_i^p of T and each q-cell b_j^q of T with $p + q = m \leqslant n$, the join of these two cells, which he wrote $a_i^p b_j^q$, and the union of all these joins. His main result was that any m-cycle $\sum_{i,j} \lambda_{ij} a_i^p b_j^q$ for fixed p, q is homologous to a m-cycle $\sum_{h,k} \mu_{hk} a_h^{p-1} b_k^{q+1}$ (a_h^{-1} being deleted when $p = 0$).† By induction on p this enabled him to establish a one-to-one correspondence between homologies of T and homologies of T* from which he finally deduced the equality of the Betti numbers and of the torsion coefficients for T and T*.

We should also mention that Poincaré did not lose sight of his first attempted proof of the duality theorem in *Analysis Situs*. In the second *Complément* he showed that for dual cells a_i^q, b_i^{n-q}, orientations can be chosen such that $N(a_i^q, b_i^{n-q}) = (-1)^q$. In the first *Complément* he proved the following theorem for $n = 3$: the condition for a homology $c\sum_i v_i \sim 0$ between 1-cycles v_i, linear combinations of 1-cells of a triangulation T (c integer $\neq 0$), is that $\sum_i N(v_i, V) = 0$ for all 2-cycles V, linear combinations of 2-cells of the dual triangulation T*. He was probably convinced of the possibility of extending that proof for arbitrary values of n and q, which in fact was done later by Veblen and Weyl [484].

Finally at the end of the second *Complément* Poincaré stated for the first

---

* One has only to embed A (resp. B) in a hyperplane of E (resp. F) that does not contain the origin.

† This generalizes the obvious fact that in a pyramid the base can be continuously deformed into the lateral surface.

time a (wrong) version of the "Poincaré conjecture"; he thought at that time he could prove that when all Betti numbers are equal to 1 and there is no torsion for a compact, connected, orientable n-dimensional manifold V, V is homeomorphic to $\mathbf{S}_n$. It was only in the fifth *Complément* ([369], pp. 475–498) that he obtained for $n = 3$ a counterexample of such a manifold V with $\pi_1(\mathrm{V})$ not trivial (Part 3, chap. I, §4,D); he then formulated the conjecture that for $n = 3$, $\pi_1(\mathrm{V}) = 0$ implies that V is homeomorphic to $\mathbf{S}_3$, a result which to this day is neither proved nor disproved.

With all their shortcomings, the chasm between Poincaré's papers on algebraic topology and what existed before him is so wide that one must agree with the opening sentence of Lefschetz's *Topology*: "Perhaps on no branch of mathematics did Poincaré lay his stamp more indelibly than on topology."

CHAPTER II

The Build-Up of "Classical" Homology

§ 1. The Successors of Poincaré

It took about 30 years to construct a theory of homology applicable to curvilinear "polyhedra," embodying all the ideas of Poincaré and entirely rigorous. In a period in which the number of professional mathematicians was definitely on the increase, it is surprising that this new field of research at first attracted so few people. This is true even if one takes into account topological questions such as the theory of dimension or the theory of fixed points (see Part 2), which until 1920 were not directly linked to homology but attracted much more attention, owing to the spectacular use of simplicial methods by L.E.J. Brouwer (1881–1966). For many years Brouwer himself was completely isolated in Holland; in France, after Poincaré's death and until 1928 only Hadamard and Lebesgue were interested in these questions, but they did not use simplicial methods; Italian mathematicians do not seem to have been attracted at all to topology, nor the English until 1926. The progress in the build-up of homology is entirely due to (1) a handful of mathematicians in Germany, Austria–Hungary, and Denmark: P. Heegaard, M. Dehn (1878–1952), H. Tietze (1880–1964), E. Steinitz (1871–1928), and after 1920 H. Kneser (1898–1973), H. Künneth (1892–1974), W. Mayer (1887–1948), L. Vietoris (1891–), and H. Hopf (1894–1971); and (2) the three members of what may be called the "Princeton school": O. Veblen (1880–1960), J.W. Alexander (1888–1971), and S. Lefschetz (1884–1972).

The first treatise on this "classical" algebraic topology was Veblen's *Analysis Situs*, published in 1922 (but a preliminary version was given as "Colloquium lectures" in 1916); it was followed by the much more complete book *Topology* by Lefschetz (1930), the very popular *Lehrbuch der Topologie* of H. Seifert and W. Threllfall (1934), and the book by P. Alexandroff and H. Hopf entitled *Topologie* I (1935).*

* This was the first example of a projected treatise in several volumes, which stops with the first one; other conspicuous examples are the well-known books by Eilenberg–Steenrod [189] and Godement [208].

J. Dieudonné, *A History of Algebraic and Differential Topology, 1900–1960*,
Modern Birkhäuser Classics, DOI 10.1007/978-0-8176-4907-4_3,

§2. The Evolution of Basic Concepts and Problems

The emphasis Poincaré put on C^1-manifolds (or even analytic ones) was immediately abandoned by his successors. For them the closures of the *cells* of a triangulation are merely deduced by *homeomorphisms* from closures of bounded convex euclidean (rectilinear) polyhedra, so that all notions relative to triangulations are invariant under homeomorphisms. Furthermore, they generalized the notion of (curvilinear) "polyhedron" defined by Poincaré, and until 1925 they only considered the homology of what they called *complexes*. Unfortunately that word is given different meanings by the mathematicians who use it; for the sake of clarity we shall use a terminology that distinguishes these meanings, even if it does not coincide with the one used in the papers we describe. In §§2–5 of this chapter, a *triangulation* will only be defined for a *compact* space X: as with Poincaré, it will mean a *finite partition* T of X in cells of various dimensions, such that the frontier of a cell of T in X is the union of cells of T of strictly lower dimension. Each cell is given an *orientation*; but Poincaré's additional requirement that, for the maximal dimension p of the cells of T, each $(p-1)$-cell should be contained in the frontier of exactly two p-cells of T (see §4) is dropped.* The pair (X, T) (or, by abuse of language, X itself) will be called a *cell complex*; after §5 of this chapter, we shall say *finite cell complex*, since more general "cell complexes" will also be defined. The barycentric subdivision of Poincaré (chap. I, §3) naturally led to the introduction of *simplicial cell complexes*, where the cells of the triangulation T are (curvilinear) simplices and *each face* of a simplex of T *belongs to* T (and is not merely a *union* of simplices of T). This condition still leaves open the possibility that the intersection of the frontiers of two simplices of T of dimension k contains *more than one* simplex of T of dimension $k - 1$.† To get the simplicial complexes obtained by barycentric subdivision that possibility must be excluded; it is easy to see‡ that this is equivalent to the condition that there exists a homeomorphism of X on a compact subset X′ of some $\mathbf{R}^N$ of

* In their first paper [21], Alexander and Veblen impose the condition that in a cell complex where p is the maximal dimension of the cells, every cell of dimension $q < p$ is contained in the frontier of at least one cell of dimension $q + 1$; this was later dropped.

† As an example, consider the usual description of the two-dimensional torus $\mathbf{T}^2$ as obtained by identification of opposite sides of a rectangle, and decompose the rectangle into two triangles by the diagonal.

‡ This is proved in [421], p. 46. If X_n is the union of all simplices of T of dimension $\leqslant n$ (later called the *n-skeleton* of X), the homeomorphism $X \to X'$ is defined by induction on the X_n. The set X_0 consists of the vertices of the simplices of T; each vertex is mapped onto a unit vector of the natural basis of $\mathbf{R}^N$, where N is the number of vertices. The extension of a homeomorphism of X_n onto X'_n to a homeomorphism of X_{n+1} onto X'_{n+1} is then reduced to the case in which X_n and X'_n are the frontiers of two simplices of dimension $n + 1$, in which case the extension is immediate by means of barycentric coordinates.

large dimension such that T is sent by that homeomorphism onto a triangulation T′ of X′ consisting of *rectilinear* simplices such that the faces of each simplex of T′ belong to T′. This is the definition of a *simplicial complex* (X, T) chosen by Lefschetz [304] and which we shall adopt (after § 5 of this chapter we will say *finite simplicial complex*); the simplicial complexes such as (X′, T′) will be called *euclidean simplicial complexes*, and in most questions we may only consider euclidean simplicial complexes; this has the advantage of avoiding all difficulties linked to the intersections of manifolds. Of course, barycentric subdivisions of euclidean simplicial complexes are also taken rectilinear.*

This enlarged concept of triangulation of course changes nothing in Poincaré's definition of the Betti numbers and torsion coefficients of the triangulation, nor in the algorithm for their computation. In fact, that algorithm is so obviously of an algebraic nature and uses so little of topology that, in the very first paper on topology published after Poincaré's *Compléments*, the *Enzyklopädie* article of Dehn and Heegaard [138], there is already an attempt to define "homology" in a purely algebraic context, where the "cells" are elements of finite sets without any topological properties, with an *ad hoc* system of axioms. This axiom system was slightly improved by Steinitz in 1908 [456], but he did not go beyond a notion of "orientation" within this context, and it was only Weyl in 1923 [484] who consistently pursued this idea and built up an algebraic "homology" theory; his axioms, however, like those of Steinitz, were so narrowly tailored to mimic the topological situation that they did not seem applicable to very different topological problems or to algebraic ones.

Weyl had already considered, in addition to Poincaré's incidence matrices, the **Z**-*modules* C_j having as bases the sets of oriented j-cells. In 1925 H. Hopf, at the beginning of his career, spent a year at Göttingen; E. Noether, who then was engaged in the process of liberating linear algebra from matrices and determinants, observed that the boundaries of j-chains defined a *homomorphism* of **Z**-modules

$$\mathbf{b}_j\colon C_j \to C_{j-1} \tag{1}$$

such that

$$\mathbf{b}_{j-1} \circ \mathbf{b}_j = 0, \tag{2}$$

and that the consideration of Betti numbers and torsion coefficients amounted to that of the **Z**-modules

$$H_j = \operatorname{Ker} \mathbf{b}_j / \operatorname{Im} \mathbf{b}_{j+1}; \tag{3}$$

Hopf accordingly used these *homology modules* when writing his 1928 paper

* The orientation of a simplex may be defined by choosing an *order* among its vertices; two orderings give the same orientation if they are deduced from one another by an even permutation.

on the Lefschetz trace formula (Part 2, chap. III, §2). Independently, in 1926, Vietoris also needed to get rid of matrices in order to define homology for more general spaces than simplicial complexes (see below, chap. IV, §2), and he used the definition (3) of homology groups for a simplicial complex, without relating it to general notions of linear algebra [475].

This seemingly innocuous modification was to have important consequences, both for the ulterior development of algebraic topology and later for algebra itself (see chap. IV), since it was clear that the definition of homology modules could at once be extended to *arbitrary* (finite or infinite) sequences $C_{\bullet} = (C_j)_{j \geqslant 0}$ of modules over any ring, and to module homomorphisms $\mathbf{b}_j$ ($j \geqslant 1$) satisfying (2) (when $\mathbf{b}_0$ is taken by convention to be the unique homomorphism $C_0 \to \{0\}$). We shall say that such a system $(C_j, \mathbf{b}_j)$ is a *chain complex*; Mayer, in 1929 [336], was apparently the first to consider such systems, with the additional restriction that the C_j are *free* modules with *finite* bases; he calls them "complexes."*

In particular, he considered the following situation, suggested to him by Vietoris: each C_j has a basis, union of two subsets B_j^1, B_j^2 such that, if C_j^1, C_j^2, and C_j^3 are the $\mathbf{Z}$-modules having as bases B_j^1, B_j^2, and $B_j^1 \cap B_j^2$, respectively, the sequences (C_j^1), (C_j^2), and (C_j^3) are again differential graded modules for the restrictions of the homomorphisms $\mathbf{b}_j$. Mayer looked for a relation between the homology modules H_j^1, H_j^2, H_j^3, and H_j of (C_j^1), (C_j^2), (C_j^3), and (C_j), respectively; he proved that $H_j = E_j \oplus G_{j-1}$, where $G_j \subset H_j^3$ consists of the classes of cycles that are boundaries both in C_j^1 and C_j^2, and E_j consists of the classes of the sums of a cycle of C_j^1 and a cycle of C_j^2. In 1930 Vietoris [476] completed Mayer's result and showed that

$$E_j \simeq (H_j^1 \oplus H_j^2)/(H_j^3/G_j). \tag{4}$$

These results, later incorporated into what became known as the Mayer–Vietoris exact sequence (chap. IV, §6,B), were to have many applications in algebraic topology.†

The first example of a chain complex different from the classical modules of "chains" of a triangulation was linked to a more abstract conception of those chains, which appeared simultaneously around 1926 in papers by Alexander [14], Alexandroff [22], and M.H.A. Newman [356] and was characterized by van der Waerden [477] as "pure combinatorial topology."

---

* We shall also use the notation $C_{\bullet}$ for the *direct sum* $\bigoplus_{j \geqslant 0} C_j$ (in modern terminology, this is a *differential graded module*) when no confusion can arise. For rings of coefficients which are principal ideal rings, it is equivalent to saying that each C_j is free or that their direct sum is free; we will also say in that case that the *chain complex* $C_{\bullet} = (C_j)$ is *free*. More special "abstract" free chain complexes, mimicking the simplicial complexes, were introduced by Tucker, and used by the American school around 1940, in particular for the definition of cohomology.

† Mayer himself gave an application of his results to the usual torus $\mathbf{T}^2$ considered as union of two cylinders, their intersection being also the union of two disjoint cylinders ([336], p. 41).

The underlying idea is that rectilinear euclidean simplices and their orientation are entirely determined by the sequence of their *vertices*. Disregarding anything else, we shall therefore define a *combinatorial complex* as a set V equipped with a (finite or infinite) set $\mathfrak{S}$ of *finite subsets*, the *combinatorial simplices*, submitted only to the restriction that if $S \in \mathfrak{S}$ and $S' \subset S$, then also $S' \in \mathfrak{S}$; the dimension, faces, and orientation of these combinatorial simplices are defined in an obvious way. The module C_j of *alternating j-chains* of the combinatorial complex is then the set of finite linear combinations with integral coefficients

$$\sum_i x^i(a_i^0, a_i^1, \ldots, a_i^j), \tag{5}$$

where $x^i \in \mathbf{Z}$, and $(a_i^0, a_i^1, \ldots, a_i^j)$ is the sequence (in an arbitrary order) of the *distinct* vertices of a j-dimensional simplex, with the identification

$$(a_i^{\pi(0)}, a_i^{\pi(1)}, \ldots, a_i^{\pi(j)}) = \operatorname{sgn}(\pi)(a_i^0, a_i^1, \ldots, a_i^j) \tag{6}$$

for any permutation π of $\{0, 1, \ldots, j\}$. The boundary operator (1) is then defined by*

$$\mathbf{b}_j \cdot (a_i^0, a_i^1, \ldots, a_i^j) = \sum_{k=0}^{j} (-1)^k (a_i^0, \ldots, \widehat{a_i^k}, \ldots, a_i^j) \tag{7}$$

and makes (C_j) into a *chain complex*, the homology of which is, by definition, the homology of the combinatorial complex $(X, \mathfrak{S})$. Another equivalent definition of C_j consists in choosing a total order on X, and considering only in (5) the sequences such that $a_i^0 < a_i^1 < \cdots < a_i^j$ for this order; this shows that C_j is a *free* $\mathbf{Z}$-module.

To each euclidean simplicial complex (X, T) is thus associated a *finite* combinatorial complex $(V, \mathfrak{S})$, where V is the finite set of *all* vertices of *all* simplices of T, and $\mathfrak{S}$ is the subset of $\mathfrak{P}(V)$ consisting of the sets of vertices of all simplices of T. It is clear that there is an isomorphism of the chain complex of (X, T) onto the chain complex of $(V, \mathfrak{S})$, commuting with the boundary operators, and therefore giving a natural isomorphism of the homology of (X, T) onto the homology of $(V, \mathfrak{S})$. Conversely, it is easily shown ([308], p. 97) that for each *finite* combinatorial complex, there exist euclidean simplicial complexes to which it is associated; they are called the *realizations* of the combinatorial complex, and it can be proved that any two realizations of the same combinatorial complex are homeomorphic.

It is possible to define for a *combinatorial complex* $K = (V, \mathfrak{S})$ a notion that reduces to the classical "barycentric subdivision" for simplicial complexes: the *first derived complex* K′ of K is a combinatorial complex, where the set of vertices is the set $\mathfrak{S}$ of *combinatorial simplices* of K: a combinatorial p-simplex of K′ is a set $S_1 \subset S_2 \subset \cdots \subset S_{p+1}$ of $p + 1$ distinct simplices of K, *totally*

* Eilenberg and Mac Lane introduced the convention that a "hat" above a letter means that this letter should be omitted in the sequence in which it is inserted.

ordered by inclusion, the dimensions of which form an arbitrary strictly increasing sequence ([308], p. 164).

Another chain complex emerged with the consideration of "singular simplices," which we will introduce in §3. In a combinatorial complex K, and with the same notations as above, the module C'_j of j-chains of this chain complex consists again of the linear combinations (5), but in which this time a_i^0, a_i^1, ..., a_i^j are vertices of a combinatorial simplex $S \in \mathfrak{S}$ but are *not necessarily distinct*; such sequences $(a_i^0, \ldots, a_i^j)$ with repetitions are called *degenerate simplices* of K. The identification (6) is not applied to degenerate simplices: if the boundary of a degenerate j-simplex is again defined by (7), the right-hand side is a combination of degenerate $(j-1)$-simplices.

It is clear that there is a natural injection $h: C_j \to C'_j$, and a retraction $r: C'_j \to C_j$ obtained by replacing the coefficients of the degenerate simplices by 0; both mappings commute with the boundary operators, and therefore yield homomorphisms $H_j \to H'_j$ and $H'_j \to H_j$ for the homology modules, but it is not immediately obvious that these homomorphisms are *bijective*. This was taken for granted by both Alexander [9] and Lefschetz [304] and the proof was only provided in 1938 by Tucker [471], who showed that if a chain (5) consisting only of *degenerate* simplices is a *cycle*, it is also a *boundary*. The use of the chain complex (C'_j) by these authors was never very explicit; with the work of Eilenberg on singular homology (chap. IV, §2) it gave way to a much less hybrid type, namely, the chain complex (C''_j), where the j-chains are simply the linear combinations of *all* sequences $(a_i^0, a_i^1, \ldots, a_i^j)$ consisting of vertices of the same simplex (distinct or not), but *no* identification is made; the boundary operator is still given by (7). There is a natural surjection $C''_j \to C_j$, the kernel of which is generated by the degenerate simplices and the differences

$$(a_i^{\pi(0)}, a_i^{\pi(1)}, \ldots, a_i^{\pi(j)}) - \operatorname{sgn}(\pi)(a_i^{(0)}, a_i^{(1)}, \ldots, a_i^{(j)}).$$

The elements of C''_j are the *ordered j-chains* of the combinatorial complex; the proof that the homology of (C''_j) is naturally isomorphic to that of (C_j) was initially made by using a homotopy operator, and is an easy consequence of the method of acyclic models (chap. IV, §5,G).

Another novelty in homology was introduced by Tietze [466]* and taken up by Alexander and Veblen [21], the *homology modulo* 2, where the coefficients of the cells in a chain are integers mod 2. This dispenses altogether with any consideration of orientation of the cells, and the "incidence matrices" now have coefficients in the field $\mathbf{F}_2$ of two elements, hence are equivalent to matrices (ρ_{ij}) with $\rho_{ij} = 0$ if $i \neq j$ and $\rho_{ii} = 0$ or 1 (or, equivalently, the homology modules are now vector spaces over $\mathbf{F}_2$). This does not give new topological invariants, since the dimension of that vector space for dimension p is

* This seems to be the first paper that questions the validity of Poincaré's arguments, and points to pathologies in the theory of differential manifolds ([466], pp. 32, 36, and 41)

the sum of the p-dimensional Betti number and of the number of torsion coefficients in dimensions p and $p-1$ which are not divisible by 2; it is now possible to generalize the duality theorem for *nonorientable* n-dimensional compact connected triangulated manifolds, that expresses the isomorphism of the p-dimensional and the $(n-p)$-dimensional homology vector spaces over $\mathbf{F}_2$. Later Alexander considered more generally "homology modulo m" for any integer m [14], and Lefschetz realized that the "homology with division" of Poincaré was simply "rational homology" with coefficients in the field $\mathbf{Q}$ ([302], p. 234), but this still did not yield any invariant not expressible by the known ones.

These attempts testify to the persistence of the search for a system of numerical or algebraic invariants of a topological space that would entirely characterize it up to homeomorphism, on the model of what Jordan and von Dyck had succeeded in doing (with insufficient proofs) for surfaces ([373], p. 139); we saw in chapter I that the introduction by Poincaré of homology and of the fundamental group was certainly motivated in part by this search. But even for dimension 3, where the Poincaré conjecture remained undecided, it was soon realized that the fundamental group was not sufficient to determine an orientable manifold up to homeomorphism. This followed from the study of a remarkable family of three-dimensional, compact, connected orientable manifolds, first defined by Tietze in 1908, and now called the *lens spaces* ([466], § 20). For an odd prime p and an integer q such that $0 \leqslant q \leqslant p-1$, the lens space $\mathrm{L}(p, q)$ is defined by Tietze as the quotient space $\mathbf{D}_3/\mathrm{R}$, where $\mathbf{D}_3$ is the ball $|x| \leqslant 1$ in $\mathbf{R}^3$, and R is the equivalence relation whose classes consist of the one-element sets $\{x\}$ for $|x| < 1$, and of the orbits of the cyclic group $\mathbf{Z}/p\mathbf{Z}$, acting on the sphere $\mathbf{S}_2$: $|x| = 1$ by the action

$$(k,(\varphi,\theta)) \mapsto \left(\varphi + \frac{2kq\pi}{p},(-1)^k\theta\right) \tag{8}$$

φ and θ being the usual longitude and latitude. Later another equivalent definition of $\mathrm{L}(p,q)$ was formulated as the space of orbits of the group $\mathbf{Z}/p\mathbf{Z}$ acting on the sphere $\mathbf{S}_3$: this sphere is considered to be the manifold $|z_1|^2 + |z_2|^2 = 1$ in the space $\mathbf{C}^2$, and the action is

$$(k,(z_1,z_2)) \mapsto (\omega^k z_1, \omega^{kq} z_2) \tag{9}$$

with $\omega = e^{2\pi i/p}$. The fundamental group of $\mathrm{L}(p,q)$ is $\mathbf{Z}/p\mathbf{Z}$, and the homology modules are $\mathrm{H}_1 = \mathbf{Z}/p\mathbf{Z}$, $\mathrm{H}_2 = 0$, so that the value of q is irrelevant; nevertheless, Tietze suspected (but could not prove) that, for instance, $\mathrm{L}(5,1)$ and $\mathrm{L}(5,2)$ are not homeomorphic. This was proved in 1919 by Alexander [10], using a construction of $\mathrm{L}(5,1)$ and $\mathrm{L}(5,2)$ different from that of Tietze, whose paper was not mentioned;* another proof was provided by de Rham in 1931 (see Part 2, chap. VI, § 3,A), using the notion of linking coefficient (Part 2, chap. I, § 3).

* See [421], p. 216.

More urgent than this ultimate and more and more elusive goal* was the immediate necessity to prove conclusively that the homology modules defined by two different triangulations of the same compact, connected space X are isomorphic (the *invariance problem*), which would show that these homology modules only depend on the homeomorphism class of the triangulable space. A "natural" method would have been to show that, for two triangulations T, T′ of X, there existed two suitable subdivisions of T and of T′ that could be deduced from one another by a homeomorphism of X; this was given the name *Hauptvermutung* in algebraic topology by H. Kneser [274], but for a long time it could only be proved for complexes of dimension 2, and remained undecided for higher dimensions. It was finally shown much later [349] that the "Hauptvermutung" is true for dimension 3, but counterexamples exist for dimension $\geqslant 5$. The invariance property must therefore be proved by independent means.

During that period the concepts of deformation, homotopy, and isotopy finally acquired a precise meaning. The words *homotopy* and *isotopy* were coined by Dehn and Heegaard in their *Enzyklopädie* article with a purely combinatorial definition adapted to their "abstract" conception of homology ([138], pp. 205–207), and they were not retained by later workers, with the exception of Steinitz.† Brouwer seems to have been the first to give our present definition of homotopy ([89], p. 462): two continuous mappings $f\colon X \to Y$, $g\colon X \to Y$ are homotopic if there exists a continuous mapping $F\colon X \times [0, 1] \to Y$ such that $F(x, 0) = f(x)$ and $F(x, 1) = g(x)$ in X.

The final touches to the homology theory of cell complexes were brought about by the theory of intersections (§ 4), the introduction of product spaces (§ 5), and, finally, the concept of relative homology (§ 6). Around 1930 algebraic topology was ready for further extensions and new concepts.

§ 3. The Invariance Problem

There are two proofs of the independence of homology from the triangulation of a simplicial complex. Both are essentially due to Alexander ([9] and [14]);‡ they both use the new ideas of *simplicial mapping* and *simplicial approximation*, and the first one is also based on a new concept, the *singular chains*.

Simplicial mappings are a natural extension to n dimensions of the classical

---

* It has finally been proved by A.A. Markov [332] that there *cannot exist* any algorithm (in the sense of the theory of recursive functions) that would allow one to determine if two euclidean simplicial complexes X, Y of dimension $\geqslant 4$ are homeomorphic or not. He considers the fundamental groups $\pi_1(X)$, $\pi_1(Y)$; these groups may be *any* group finitely generated and finitely presented, and the algorithm would enable one to decide if two such groups are isomorphic or not. But it is known that no such algorithm exists.

† Steinitz only uses the Dehn–Heegaard notion of "homotopy" to introduce an abstract notion of "orientation"; Dehn and Heegaard themselves do not seem to have used it at all for questions of homology.

‡ For a third, indirect, proof by Alexander, see § 6.

notion of *piecewise linear* function of a real variable. Let X be a euclidean simplicial complex; a *simplicial mapping* f of X into a euclidean simplicial complex Y is a continuous mapping such that for *all* p, and any p-simplex S of X, f(S) is *contained* in a q-simplex of Y for a number $q \leqslant p$ and the restriction of f to S is *affine*. This restriction is therefore entirely determined by the values of f at the vertices of S, which must be vertices of a q-simplex of Y for a number $q \leqslant p$, *not necessarily distinct*, but otherwise entirely arbitrary.

For *combinatorial complexes* X, Y (§ 2) a *simplicial mapping* will be a map $f\colon X \to Y$ such that for *all* p and any p-simplex S of X (which are finite subsets of X here), f(S) is contained in a q-simplex of Y for some $q \leqslant p$. Thus there is a one-to-one correspondence between the simplicial mappings of a euclidean simplicial complex X into a euclidean simplicial complex Y, and the simplicial mappings of the combinatorial complex associated to X into the combinatorial complex associated to Y. By linearity, if $(C'_j(X))$ and $(C'_j(Y))$ are the chain complexes of ordinary *and degenerate* simplices in X and Y, one deduces for each j, from a simplicial mapping $f\colon X \to Y$, a homomorphism $\tilde{f}_j\colon C'_j(X) \to C'_j(Y)$, by the formula

$$\tilde{f}_j((a_0, a_1, \ldots, a_j)) = (f(a_0), f(a_1), \ldots, f(a_j)) \tag{10}$$

for each (ordinary or degenerate) j-simplex $(a_0, a_1, \ldots, a_j)$. If $g\colon Y \to Z$ is a second simplicial mapping and $h = g \circ f\colon X \to Z$, h is also a simplicial mapping and $\tilde{h} = \tilde{g} \circ \tilde{f}$. Furthermore, it can easily be shown that for the boundary operators

$$\mathbf{b}_j \circ \tilde{f}_j = \tilde{f}_{j-1} \circ \mathbf{b}_j; \tag{11}$$

hence $\tilde{f}_{\bullet} = (\tilde{f}_j)$ is a homomorphism of *chain complexes*, which yields a homomorphism $f_*\colon (H'_j(X)) \to (H'_j(Y))$ of graded homology modules. If it is composed with the natural homomorphisms $(H_j(X)) \to (H'_j(X))$ and $(H'_j(Y)) \to (H_j(Y))$ (§ 2), this also gives a homomorphism $(H_j(X)) \to (H_j(Y))$, which topologists identified with f_* even before it had been proved that the natural maps $H_j \to H'_j$ are isomorphisms. The homomorphism $f_*\colon (H_j(X)) \to (H_j(Y))$ can also be defined directly, if the definition (10) is modified by taking $\tilde{f}_j((a_0, a_1, \ldots, a_j)) = 0$ when the simplex $(f(a_0), f(a_1), \ldots, f(a_j))$ is degenerate. With the same notations as above, $(g \circ f)_* = g_* \circ f_*$.

The idea of simplicial approximation is due to Brouwer ([89], p. 459). He considered two euclidean simplicial complexes X, Y (satisfying some additional conditions that we disregard) and a continuous map $f\colon X \to Y$ such that for *any* simplex σ of X $f(\sigma)$ is contained in a simplex of Y. Then for any $\varepsilon > 0$ there is a triangulation T′ of X obtained by repeated barycentric subdivisions of the given one T, and a map $g\colon X \to Y$ that coincides with f at the vertices of the new triangulation, is such that $|f(x) - g(x)| \leqslant \varepsilon$ for all $x \in X$ and is an *affine* map in every simplex of T′. We shall return in Part 2, chaps. I, II, and III to describe the way he used this result with great virtuosity to prove his famous theorems without linking them to homology.

In his first proof Alexander realized that he could extend Brouwer's method to *all* euclidean simplicial complexes X, Y and to an *arbitrary* continuous map $f: X \to Y$.* The stars of the triangulation of Y form an open covering of Y; replacing the triangulation of X by one obtained by repeated barycentric subdivisions, it can be assumed that for the new triangulation the image of any *star* of X is contained in one of the *stars* of Y. Thus, if $(a_k)_{1 \leqslant k \leqslant N}$ are the vertices of the new triangulation of X, and $(b_j)_{1 \leqslant j \leqslant N'}$ are the vertices of the triangulation of Y, let $b_{\varphi(k)}$ be one of the vertices of Y whose star contains the image of the star of a_k. If $j + 1$ vertices a_k are the distinct vertices of a j-simplex of X, the $b_{\varphi(k)}$ are (not necessarily distinct) vertices of a q-simplex of Y with $q \leqslant j$. If a simplicial mapping g is defined by $g(a_k) = b_{\varphi(k)}$ for all k, g is a simplicial approximation of f, with $|f(x) - g(x)| \leqslant 3\delta$, where δ is the maximum diameter of the simplices of Y. Furthermore, for any $x \in X$, $f(x)$ and $g(x)$ are the extremities of a segment contained in Y, and therefore f and g are *homotopic*.

The notion of *singular chain* also arose from the need to consider continuous maps $f: X \to Y$ between euclidean simplicial complexes, both having arbitrary dimensions. It was first mentioned by Dehn–Heegaard in their *Enzyklopädie* article [138]; they of course realized that phenomena such as the Peano curve implied that the image $f(E)$ of a cell E may exhibit the weirdest pathology, so they included in their conception not only the image $f(E)$, but also the cell E itself in rather vague terms;† they do not seem to have made any use of it to prove anything.

In [9] Alexander had the idea‡ that the singular simplices might be used to define new kinds of chains by linear combination, and be provided with boundary operators with which one could define new homology modules that *ipso facto* would be independent of any triangulation; the invariance problem would then be solved if he could define isomorphisms of these modules on the homology modules of an arbitrary triangulation. At least this is what we may guess from the context of his paper, for his definition of singular cells is simply translated from Dehn–Heegaard. He never said when two images of different p-cells by two continuous mappings should be identified, nor what the boundary of a singular cell should be. This vagueness was only partly improved in the successive versions of Alexander's proof given (this time for cell complexes) by Veblen ([474], p. 102), van der Waerden [477], and Lefschetz ([304], chap. II); it was only in a short note published in 1933 that Lefschetz, "to clear up misconceptions," defined a singular cell on a space X [305]: he considers pairs (e_p, f), where e_p is a p-dimensional oriented convex polyhedron in some $\mathbf{R}^N$, and $f: \bar{e}_p \to X$ is a continuous mapping; singular p-cells are classes of such

* Although Alexander did not mention any paper on algebraic topology with the exceptions of Poincaré's and his own joint paper with Veblen [21], it is quite certain that he knew Brouwer's work, for it is quoted in a 1913 paper by Veblen.

† "Wir nennen C'_n, aufgefasst als das Abbild eines bestimmten C_n, *einen n-dimensionalen Komplex mit Singularitäten*, . . . , und geben ihm die Bezeichnung $C'_n(C_n)$" (p. 164).

‡ Alexander only considered homology mod 2 on manifolds.

pairs for the equivalence relation $(e_p, f) \equiv (e'_p, f')$, where $f' = f \circ u$ and u is an *affine bijection* $u\colon e'_p \to e_p$.* If $\mathbf{b}_p$ denotes the boundary map for euclidean polyhedra, then the equivalence

$$(\mathbf{b}_p e_p, f | \mathbf{b}_p e_p) \equiv (\mathbf{b}_p e'_p, f' | \mathbf{b}_p e'_p)$$

holds when the equivalence relation is extended to "singular chains," linear combinations of singular cells;† this clearly defines a boundary operator for these new "chains," and from these data one deduces by the standard method, applicable to all chain complexes, homology modules that this time obviously only depend on the space X up to homeomorphism. To make things precise, we shall attach the qualification "topological" or the index "top" to the notions entering in the homology of singular chains. For any continuous mapping $g\colon \mathrm{X} \to \mathrm{Y}$ the image by $\tilde{g}$ of a singular cell on X is defined by

$$\tilde{g}(e_p, f) = (e_p, g \circ f) \tag{12}$$

and therefore this can be extended by linearity to a homomorphism $\tilde{g}_p\colon \mathrm{C}_p^{\mathrm{top}}(\mathrm{X}) \to \mathrm{C}_p^{\mathrm{top}}(\mathrm{Y})$ of singular chains, permuting with boundary operators and yielding a homomorphism $g_*\colon (\mathrm{H}_j^{\mathrm{top}}(\mathrm{X})) \to (\mathrm{H}_j^{\mathrm{top}}(\mathrm{Y}))$ of graded homology modules with the relation $(g_1 \circ g_2)_* = g_{1*} \circ g_{2*}$ for two continuous mappings.

Granted this clarification, Alexander's method may be stated as follows: For a triangulation T of a euclidean simplicial complex X, there is a homomorphism

$$\mathrm{H}_j \to \mathrm{H}_j^{\mathrm{top}} \tag{13}$$

from the homology defined by chains of T to the homology of singular chains, defined in a natural way: each p-simplex E_p of the triangulation T is identified to the singular p-simplex $(\mathrm{E}_p, \mathrm{Id.})$, and its boundary with the (singular) boundary of that singular simplex. What has to be shown is that (13) is *bijective*, or equivalently that: (A) every topological p-cycle w_p is topologically homologous to a p-cycle of T; (B) every p-cycle z_p of T that is a topological boundary is also a boundary of T.

Some preliminary results are needed. First is the fact that the homology of T is naturally identified with the homology of any triangulation T′ deduced from T by barycentric subdivision. We have seen in chap. I that Poincaré had already given a substantially correct proof of that result, and others were proposed by Tietze ([466], p. 42), Alexander himself ([9], p. 153), Veblen ([474], p. 90), and Lefschetz ([304], p. 68). This invariance by subdivision is immediately extended to the homology of singular chains,‡ and has as a consequence the fact that in the proof of A (resp. B) the singular chain w_p [resp.

---

* This allows one to take all the e_p equal to the same simplex, which will be done later (chap. IV, § 2).

† In addition, Lefschetz imposed the relation $(-e_p, f) = -(e_p, f)$, where $-e_p$ is the simplex e_p with opposite orientation.

‡ A subdivision of a singular cell (e_p, f) consists of the singular simplices $(e_p^i, f | e_p^i)$, where (e_p^i) is the family of p-simplices of a subdivision of e_p.

the singular $(p + 1)$-chain of which z_p is the topological boundary] may consist of singular simplices whose images in X are *arbitrarily small.*

The second result that emerges from Veblen's invariance proof ([474], p. 102) and more clearly from Lefschetz's, is, at last, a correct statement and proof of the invariance of homology under homotopy (the irrelevance of "deformation," so long taken for granted, as we have seen in chap. I). Writing $S_p(\mathbf{R}^N)$ the **Z**-module of the simplicial p-chains in $\mathbf{R}^N$, linear combinations with integral coefficients of the oriented euclidean simplices in $\mathbf{R}^N$, start with an elementary simplicial subdivision* of a product $\Delta_p \times I$ in $\mathbf{R}^{N+1}$ (a "prism"), where Δ_p is a p-dimensional euclidean simplex in $\mathbf{R}^N$ and $I =]0, 1[$. With suitable orientations, we obtain the relation between p-chains in $\mathbf{R}^{N+1}$

$$\mathbf{b}_{p+1}(\Delta_p \times I) = \Delta_p \times \{1\} - \Delta_p \times \{0\} + (\mathbf{b}_p(\Delta_p) \times I) \tag{14}$$

and by linearity this gives in the **Z**-module $S_p(\mathbf{R}^{N+1})$ the relation

$$\mathbf{b}_{p+1}(P_p(z_p)) = z_p \times \{1\} - z_p \times \{0\} + P_{p-1}(\mathbf{b}_p z_p), \tag{15}$$

where, for each integer q, $z_q \mapsto P_q(z_q)$ is the linear map of $S_q(\mathbf{R}^N)$ into $S_{q+1}(\mathbf{R}^{N+1})$ that coincides with the map $\Delta_q \mapsto \Delta_q \times I$ on each q-simplex Δ_q. From (15), by applying to both sides the homomorphism deduced from a continuous map $F: X \times I \to X$ as shown above, where X is a euclidean simplicial complex, this immediately gives the first example of a *homotopy formula* for singular p-chains z_p in X (cf. chap. IV, §5,F):

$$\tilde{f}(z_p) - \tilde{g}(z_p) = \mathbf{b}_{p+1}(\tilde{F}(z_p \times I)) - \tilde{F}((\mathbf{b}_p z_p \times I)), \tag{16}$$

where $f(x) = F(x, 1)$, $g(x) = F(x, 0)$, from which it follows at once that if z_p is a singular p-*cycle*, $\tilde{f}(z_p)$ and $\tilde{g}(z_p)$ are topologically homologous.

To prove A, after subdividing the singular simplices of w_p in order to be able to apply the Alexander construction of simplicial approximations described above, one shows that there exists a *homotopy* of w_p on another singular chain w'_p, whose singular simplices (e_p, g) are such that g is an *affine* map of e_p into a p-simplex of T sending vertices of e_p into vertices of that p-simplex. This would clinch the matter, except that the affine map g is not necessarily bijective.

This difficulty was ignored by Alexander and van der Waerden; Veblen's proof is very obscure and he does not seem to have distinguished, for cycles of T, between the concepts of "topologically homologous to 0" (i.e., being boundary of a singular chain) and "homologous to 0 in T" (i.e., being boundary of a chain of T). Lefschetz realized that w'_p is not identified with a p-chain of

---

* If $A_0 A_1 \cdots A_p$ is the sequence of vertices of Δ_p, identified with $\Delta_p \times \{0\}$, and $B_0 B_1 \cdots B_p$ is the sequence of the vertices $(A_j, 1)$ of $\Delta_p \times \{1\}$, the subdivision consists of the $(p + 1)$-simplices

$$(-1)^k A_0 A_1 \cdots A_k B_k B_{k+1} \cdots B_p$$

for $0 \leqslant k \leqslant p$; this generalizes the decomposition of a prism into tetrahedra, which goes back to Euclid.

T but with a p-chain in the module C'_p of chains of ordinary *and degenerate* simplices of T; but we have seen that he identified the homology of (C'_j) and of (C_j), a result that was only proved in 1938.

The proof of proposition B is very similar. The singular $(p + 1)$-chain of which z_p is the topological boundary is subdivided in such a way that a homotopy of that $(p + 1)$-chain can be defined as above, with the added proviso that the vertices of z_p remain invariant under the homotopy; z_p is then identified to the boundary of a $(p + 1)$-chain w_{p+1} of C'_{p+1}, which may contain degenerate simplices; but as z_p does not contain degenerate simplices, it is also the boundary of the $(p + 1)$-chain obtained by deleting from w_{p+1} the degenerate simplices, and that is a chain of T.

Alexander's second proof [14] did not use singular simplices any more, and relies exclusively on simplicial approximation. It was enough to show that if two *euclidean* simplicial complexes X, X′ are homeomorphic, and T (resp. T′) is a triangulation of X (resp. X′), then the homology modules $H_\bullet(T)$ and $H_\bullet(T')$ are isomorphic. Let $f: X \to X'$ be a homeomorphism, with inverse $g = f^{-1}: X' \to X$. Let (T_i) [resp. (T'_j)] be the sequence of successive barycentric subdivisions of T (resp. T′); the maximum diameter of the simplices of T_i (resp. T'_j) tends to 0; hence, for each index i, there is an index j and a simplicial approximation g_{ij} of g, from T'_j to T_i; similarly, there is an index $k > i$ and a simplicial approximation f_{jk} of f, from T_k to T'_j. The composite $h_{ik} = g_{ij} \circ f_{jk}$ is then a simplicial map of T_k into T_i; suppose i and k large enough; then, owing to the relation $g \circ f = 1_X$, for every p, every p-simplex σ of T_i, and every p-simplex $\tau \subset \sigma$ of T_k [which is a p-simplex of the $(k - i)$-th barycentric subdivision of σ], h_{ik} sends every vertex of τ to a vertex of σ.

Let $\tilde{f}_{jk}: C'_p(T_k) \to C'_p(T'_j)$ and $\tilde{g}_{ij}: C'_p(T'_j) \to C'_p(T_i)$ be the homomorphisms of modules of ordinary *and degenerate* p-chains corresponding to the simplicial maps f_{jk} and g_{ij}, and $\tilde{h}_{ik} = \tilde{g}_{ij} \circ \tilde{f}_{jk}: C'_p(T_k) \to C'_p(T_i)$ their composite. On the other hand, let sd_{k-i} be the homomorphism of $C'_p(T_i)$ into $C'_p(T_k)$ that associates to every p-simplex of T_i the *sum* of the p-simplices of T_k contained in it, with the same orientation; then $\tilde{h}_{ik}(\mathrm{sd}_{k-i}(\sigma)) = \sigma + \theta_{ik}$, where θ_{ik} is a *degenerate* p-chain. This lemma is proved by induction on p, being obvious by definition for $p = 0$. The assumption on h_{ik} implies that for any p-simplex τ of T_k contained in σ, either $\tilde{h}_{ik}(\tau) = \pm\sigma$ or $\tilde{h}_{ik}(\tau)$ is a degenerate p-simplex; hence $\tilde{h}_{ik}(\mathrm{sd}_{k-i}(\sigma)) = c\,.\,\sigma + \theta_{ik}$, where c is a constant and θ_{ik} is a degenerate chain. But as $\tilde{h}_{ik} \circ \mathrm{sd}_{k-i}$ is a simplicial map,

$$\tilde{h}_{ik}(\mathrm{sd}_{k-i}(\mathbf{b}_p\sigma)) = c\,.\,\mathbf{b}_p\sigma + \mathbf{b}_p\theta_{ik}$$

and $\mathbf{b}_p\theta_{ik}$ is degenerate. On the other hand, the induction hypothesis implies

$$\tilde{h}_{ik}(\mathrm{sd}_{k-i}(\mathbf{b}_p\sigma)) = \mathbf{b}_p\sigma + \theta'_{ik},$$

where θ'_{ik} is degenerate; the comparison of the two formulas gives $c = 1$.*

* This lemma is a special case of the Sperner lemma, proved two years later by the same method ([30], p. 376).

Assuming, as Alexander does, that the homology $H_\bullet(T_i)$ can be identified to the homology $H'_\bullet(T_i)$ of ordinary *and degenerate* chains, $(h_{ik})_* = (g_{ij})_* \circ (f_{jk})_*$ is the *identity* in $H_\bullet(T_i)$; similarly a "right inverse" is obtained for $(f_{jk})_*$, which proves that $(f_{jk})_*$ is an isomorphism; the theorem then results from the fact that $H_\bullet(T)$ (resp. $H_\bullet(T')$) is isomorphic to $H_\bullet(T_k)$ (resp. $H_\bullet(T'_j)$).

§ 4. Duality and Intersection Theory on Manifolds

A. The Notion of "Manifold"

After the invariance problem had been solved, two main items remained in the implementation of the program outlined by Poincaré: a rigorous proof of the duality theorem and a complete theory of intersections, barely begun by Poincaré (chap. I, § 2). Obvious examples show that in neither question can one work with a general cell complex; some restrictions have to be introduced in order to make available the arguments Poincaré used for his "manifolds."

We have seen (chap. I, § 2) that the concept of a C^r-manifold for $r \geqslant 1$ was clear to Poincaré. In what follows we will systematically use the name n-dimensional C^0-*manifold* to designate what is also called a locally euclidean space, namely, a Hausdorff space in which any point has a compact neighborhood homeomorphic to a closed ball in $\mathbf{R}^n$.* The triangulability of C^r-manifolds for $r \geqslant 1$ was only proved in 1930 (chap. III, § 2); but (except for $n \leqslant 3$) the triangulability of C^0-manifolds remained undecided until about 1960, when counterexamples were found for dimensions $\geqslant 5$. In the meantime, in order to use simplicial methods, topologists had to settle for more tractable definitions of "manifolds."

In fact, several definitions were proposed ([308], pp. 342–343); the first one was described by Veblen ([474], pp. 91–95) and it is a definition that is based on a *given triangulation* T into "cells" [in the sense of Poincaré (chap. I, § 2)] of the compact space X, but Veblen did not investigate its invariance under homeomorphism. The definition generalizes Poincaré's condition that for the maximal dimension n of the cells of T, each $(n-1)$-cell should be in the frontier of exactly two n-cells: for any k-cell C $(k \leqslant n-1)$, let $Z^{n-k-1}(C)$ be the union of the j-cells $(j \leqslant n-k-1)$ that are in the frontiers of the n-cells having C in their frontier but the closures of which do not meet the closure of C. Then (X, T) is a manifold (without boundary) in Veblen's sense if, for all $k \leqslant n-1$ and all k-cells C, $Z^{n-k-1}(C)$ is homeomorphic to the sphere S_{n-k-1}.

However, since (as Poincaré had shown in his fifth *Complément*) the homology of a sphere is not enough to characterize it up to homeomorphism, it was not possible to verify Veblen's condition by purely combinatorial means, and, in particular, it was not at all obvious that it would be satisfied by a triangu-

* Of course, to be sure that this definition is meaningful, one has to invoke Brouwer's theorem on the invariance of dimension (Part 2, chap. II, § 1).

lated C^r-manifold, so that the proof of Poincaré's duality theorem for these manifolds, given by Veblen, could not be considered as conclusive. This observation was first made in print by Vietoris in 1928; he therefore proposed to consider only what he called *h-manifolds*, defined by induction on the dimension in the following way: such a manifold is a compact n-dimensional simplicial complex (X, T) for which the frontier of the star of each vertex of T is an $(n-1)$-dimensional h-manifold with the homology of $\mathbf{S}_{n-1}$. A similar (unpublished) observation was made by Alexander, who proposed to weaken Veblen's condition by requiring only that the $Z^{n-k-1}(C)$ be cell complexes with the same homology as $\mathbf{S}_{n-k-1}$. This definition was adopted by Lefschetz in 1929 and by most of the later writers under the name of *combinatorial manifolds*; it is easily shown that they are the same as Vietoris' h-manifolds.

B. Computation of Homology by Blocks

We have seen in § 3 that after Poincaré it was essentially known that the homology of a cell complex is naturally isomorphic to the homology of a simplicial subdivision of the complex. But if one starts with a *simplicial* complex (X, T) and regroups simplices into "blocks" [as in Poincaré's construction of "dual cells" (chap. I, § 3)], it is useful to know conditions that allow the computation of the homology of the complex to be performed by using *only* these "blocks" of simplices.

This question was analyzed by Seifert and Threlfall in their book [421]. They defined a *system of blocks* by giving, for each $p \geqslant 0$, a *basis* of the $\mathbf{Z}$-submodule K_p of the (free) $\mathbf{Z}$-module C_p of p-chains of T, satisfying for each $p \geqslant 0$ the two following conditions (where as usual Z_p and B_p are the submodules of cycles and boundaries in C_p):

1. $\mathbf{b}_p K_p = K_{p-1} \cap B_{p-1}$;
2. $Z_p = (K_p \cap Z_p) + \mathbf{b}_{p+1} K_{p+1}$.

This implies that $Z_p = B_p + (K_p \cap Z_p)$, hence, for the homology groups

$$H_p = Z_p/B_p \simeq (K_p \cap Z_p)/(K_p \cap B_p) = (K_p \cap Z_p)/\mathbf{b}_{p+1} K_{p+1}. \tag{17}$$

In other words, (K_p) is a *chain complex* for the same boundary operator as (C_p), and the homology of (K_p) is isomorphic to the homology of (X, T). This is useful not only for proving Poincaré duality (see below), but also for practical computation of homology modules for explicitly given complexes.

C. Poincaré Duality for Combinatorial Manifolds

The simplest proof of Poincaré duality for an oriented combinatorial manifold X with a *simplicial* triangulation T is the one described by Pontrjagin ([374], p. 186). He considered the barycentric subdivision T′ of T, and "regrouped," as did Poincaré, the simplices of T′ into "dual cells," forming the dual triangulation T* of T. Any such "dual cell" E of dimension k has a frontier F such that the pair $(\bar{E}, F)$ has the same relative homology (§ 6) as the pair $(\mathbf{D}_k, \mathbf{S}_{k-1})$ consisting of the unit ball $\mathbf{D}_k$ in $\mathbf{R}^k$ and its frontier $\mathbf{S}_{k-1}$. This follows from the definition of combinatorial manifolds, and implies that the "dual cells" of T* form a *system of blocks* with which one may compute the homology of X: only condition 2 of B needs a proof, which can be done most simply by descending induction on the dimension k ([421], p. 235). It is still necessary to check the relation between the incidence matrices of T and of T*, but this follows easily from the definitions.

The proofs of Vietoris and Lefschetz ([304], pp. 135–140) are similar: start from a combinatorial manifold X, whose triangulation T into "cells" is related to a simplicial triangulation T′ by the fact that each "cell" is the *star* of a vertex of T′ (the *center* of the star), and T′ is the barycentric subdivision of T. Then assume that the frontier of a k-star is a union of stars of dimension $\leqslant k - 1$, that each k-star, for $k \leqslant n - 1$, is in the frontier of a $(k + 1)$-star, and finally that the homology of the frontier of a k-star is isomorphic to the homology of $\mathbf{S}_{k-1}$. This implies, as in the particular case of a simplicial complex, that the "dual" cells obtained by the Poincaré construction have the same properties, and the Poincaré duality follows as before.

In their 1934 book Seifert and Threlfall showed that it is possible to replace in this proof the definition of combinatorial manifold by a definition independent of the triangulation: it is enough to suppose that the compact space X is *triangulable* and that it is an n-dimensional *generalized manifold* in the sense defined in 1933 by Lefschetz [306] and Čech [122] (Part 2, chap. IV, § 3); here this simply means that for any $x \in \mathrm{X}$, the relative homology (§ 6) $\mathrm{H}_q(\mathrm{X}, \mathrm{X} - \{x\}\mathrm{H}; \mathbf{Z})$ is 0 for $q \neq n$ and isomorphic to $\mathbf{Z}$ for $q = n$ ([421], pp. 236–241).*

The duality theorems proved by Lefschetz and Čech in these papers of 1933 applied to generalized manifolds that were not necessarily triangulable, and therefore had to be proved by other methods (see Part 2, chap. IV, § 3).

D. Intersection Theory for Combinatorial Manifolds

When Poincaré's construction of "dual cells" is possible, it is easy to extend in a "cell complex" X his definition of the "Kronecker index" $\mathrm{N}(\mathrm{V}_1, \mathrm{V}_2)$ (chap. I, § 2) to a "Kronecker index" $\mathrm{N}(a_p, b^*_{n-p})$, where a_p is a p-cell of the complex and b^*_{n-p} its dual cell, both oriented: one transcribes the definition of Poincaré using the oriented vector spaces that are the directions of one of the simplices constituting a_p (resp. b^*_{n-p}) with vertex at the intersection of a_p and b^*_{n-p} ([369], p. 242). This was done in 1923 by Veblen and Weyl [484]. Assuming that the homology of X could be computed by using both the given "cell complex" and its dual, they defined in that way a bilinear form on the product $\mathrm{H}_p \times \mathrm{H}_{n-p}$ of the homology modules, and this form determines a duality between H_p and H_{n-p}. Actually there was very little to add to Poincaré's arguments to reach that conclusion, and it is a bit surprising that he did not do it himself, even taking into account the clumsy character of the linear algebra he had at his disposal.

* These conditions are satisfied by C^r manifolds for $r \geqslant 1$.

The papers by Alexander and Lefschetz on intersections, which date from about the same time as those of Veblen and Weyl, are much more ambitious. Both authors started their mathematical careers in algebraic geometry, which Alexander abandoned almost immediately in favor of topology. Lefschetz, on the contrary, kept a continued and vigorous interest in the topic for more than 30 years, and we shall see later (Part 2, chap. VII, § 1,B) how, by expanding the ideas of Picard and Poincaré, he "planted the harpoon of algebraic topology into the body of the whale of algebraic geometry," to use his own words ([296], p. 13). But it should be emphasized here that if in the hands of Lefschetz algebraic geometry was transformed by this injection of algebraic topology, the latter, as we shall see presently and later, received from him impulses inspired by algebraic geometry just as valuable as the ones it gave in return (see [435]).

In the type of algebraic geometry begun around 1870 by Clebsch, Brill, and M. Noether and followed by Halphen, Picard, Humbert, Zeuthen and the Italian school, algebraic subvarieties of a complex projective space and algebraic families of such varieties were a fundamental tool. Under ill-defined conditions, for two subvarieties V, W of a third variety X, the combinations $V + W$, $V - W$, kV (for an integer k) were considered as subvarieties (or "virtual" subvarieties) when V and W have the same dimension, as well as the "product" $V.W$ when $\dim V + \dim W \geqslant \dim X$; in the best cases $V + W$ would be the set-theoretic union and $V.W$ the set-theoretic intersection, but the complexity of the general definitions ruled out any possibility of dealing with varieties (or classes of "equivalent" varieties in some sense) as elements of a group or a ring. We may wonder if Poincaré was not inspired by these would be algebraic operations when he introduced his "chains" of varieties in algebraic topology. At any rate this analogy was central in Lefschetz's early work, and Poincaré's algebraic manipulations probably appealed to him more than the complicated geometric constructions of the Italians; following ideas of Picard he combined algebraic and topological arguments in an original way and obtained remarkable new results (see Part 2, chap. VII, § 1,B); but since he was as reckless as Poincaré (and the Italians) in his use of "intuition," none of these results could be supported at that time by a convincing proof, which he and others could only supply 10 years later.

When Alexander and Lefschetz shifted their investigations to cell complexes and combinatorial manifolds, they naturally were led to generalize the concept of intersection to arbitrary cycles. But Alexander did not make any effort to clarify nor even to define that concept, which apparently he considered "intuitive" enough; his short notes on the subject [13] were only bent on showing by examples that the formulas giving intersections of cycles on two manifolds X, Y could be essentially different* even if the homology modules of X and Y are isomorphic.

---

* In today's terminology, the intersection *rings* (or cohomology *rings*) of X and Y are not isomorphic. Other examples were given by de Rham ([388], p. 104).

Lefschetz took the matter much more seriously. Although at first he did not express it in this form, what he needed was, for an n-dimensional compact, connected, and oriented combinatorial manifold X, and for any two integers p, q such that $0 \leqslant p, q \leqslant n$, a bilinear mapping

$$(z_p, z_q) \mapsto z_p \,.\, z_q$$

of $H_p \times H_q$ into H_{p+q-n} (replaced by 0 if $p + q < n$), such that

$$z_q \,.\, z_p = (-1)^{(n-p)(n-q)} z_p \,.\, z_q \tag{18}$$

and

$$(z_p \,.\, z_q) \,.\, z_r = z_p \,.\, (z_q \,.\, z_r) \tag{19}$$

for any three integers p, q, r in $[0, n]$. Of course, when z_p and z_q are homology classes of cycles that are submanifolds and whose intersection is a $(p + q - n)$-dimensional submanifold, the homology class of that submanifold should be $z_p \,.\, z_q$ up to sign; furthermore, when $q = n - p$, $z_p \,.\, z_{n-p}$ is a scalar multiple λz_0 of the homology class z_0 of any point of X, and the scalar λ should be the "Kronecker index" defined by Poincaré, which Lefschetz wrote $(z_p \,.\, z_{n-p})$.

We shall see later (chapter IV, §4) that once cohomology was introduced, it was easy to define the products $z_p \,.\, z_q$ for manifolds using the "cup-product" of Whitney; here therefore, we shall only give a sketchy description of the direct methods which Lefschetz initially used in [300] and [301] to define the products $z_p \,.\, z_q$.

His idea was to consider *singular cycles* C_p, C_q having, respectively, z_p, z_q as homology classes, and deduce from them a $(p + q - n)$-singular cycle having $z_p \,.\, z_q$ as homology class. As could be expected, all he could actually do was to define a whole family of singular cycles, all homologous to each other, by a fairly complicated approximation process, of which he published two variants.

Both variants start with the definition of the oriented intersection "product" P . Q of two oriented convex polyhedra of respective dimensions p, q contained in a third one R of dimension n with $p + q \geqslant n$; P, Q, R are open in the respective linear affine varieties V_P, V_Q, and V_R they generate. If $P \cap Q = \varnothing$, take $P \,.\, Q = 0$; otherwise, P . Q is only defined when $V_P \cap V_Q$ has dimension $s = p + q - n$; $P \cap Q$ is then a convex polyhedron open in $V_P \cap V_Q$, and there is a way of assigning to $V_P \cap V_Q$ an orientation canonically dependent on those of V_P, V_Q and V_R;* P . Q is then the convex polyhedron $P \cap Q$ with that orientation, and

$$Q \,.\, P = (-1)^{(n-p)(n-q)} P \,.\, Q. \tag{20}$$

When P and Q satisfy all these conditions, they are said to be "in general position."

* Orienting an n-dimensional vector space means choosing a decomposable n-vector spanning that space. Let u, v, w be decomposable multivectors orienting the directions of V_P, V_Q, V_R; then w defines a "regressive" product $u \vee v$, and that s-vector orients the direction of $V_P \cap V_Q$.

Now let X be an n-dimensional euclidean simplicial complex, which is a compact, connected, oriented *combinatorial manifold* with triangulation T. If C_0 is a p-chain and C'_0 is a q-chain of T, the "intersection product" $C_0 . C'_0$ can be defined by linearity provided that when a p-simplex of C_0 and a q-simplex of C'_0 have a nonempty intersection, they are contained in the closure of the same n-simplex of T and are in general position; $C_0 . C'_0$ is then a $(p + q - n)$-chain.

Now suppose C is a singular p-chain and C′ is a singular q-chain on X. Their *geometric intersection* is by definition the intersection of their images in X; assume that $p + q \geqslant n$, and that the geometric intersections of C with $\mathbf{b}_q C'$ and of $\mathbf{b}_p C$ with C′ are empty. The general idea is, after suitable subdivisions of T, C, and C′, to apply to C and C′ a refined (and somewhat complicated) version of the Alexander approximation process (§ 3), in order to obtain a p-chain C_0 of T and a q-chain C'_0 of T which satisfy the above condition and are such that $\mathbf{b}_p C_0 \cap C'_0 = C_0 \cap \mathbf{b}_q C'_0 = \varnothing$. In his first version [300] Lefschetz had to introduce an additional condition on chains C_0 and C'_0 in "general position" in order to ensure that the relation

$$\mathbf{b}_s(C_0 . C'_0) = \mathbf{b}_p C_0 . C'_0 + (-1)^{n-q} C_0 . \mathbf{b}_q C'_0 \tag{21}$$

holds; when the above condition on the boundaries of C and C′ is added the approximation process yields an *s-cycle* $C_0 . C'_0$. He could then easily show that the homology class of that cycle did not depend on the approximation used as long as that approximation deformed the singular chains by an amount smaller than a fixed quantity depending only on T, C, and C′.

In the second variant [301] he had the idea of using the "intersection product" P . Q of oriented convex polyhedra *only* when P is a p-cell of a (not necessarily simplicial) triangulation T of X, and Q is a q-cell of the *dual* triangulation T*. There is then no need to suppose "general position" for P and Q: automatically, either $P \cap Q = \varnothing$ or $P \cap Q$ is a $(p + q - n)$-dimensional convex polyhedron, and the application of the approximation process is greatly simplified.

However, in both variants, it is still necessary to prove that the homology class $z_p . z_q$ obtained is also independent of the triangulation chosen on X, and in both cases this necessitates a long and complicated argument.

Today the properties of the intersection products $z_p . z_q$ are expressed by saying that they define, by linearity, a structure of (associative and anti-commutative) *ring*, on the direct sum

$$H_{\bullet} = \bigoplus_{0 \leqslant p \leqslant n} H_p \tag{22}$$

of the homology modules, and that this ring is an invariant of the complex X under homeomorphism. It is a curious reflection on the clumsiness of algebra before van der Waerden that this formulation, which seems so obvious to us, was only given by Hopf in 1930 [242] (perhaps again under the influence of E. Noether). Alexander and Lefschetz, in the case of homology over the rationals, picked up bases (z_p^i) in each H_p, wrote out the expressions of the intersection products

$$z_p^i \cdot z_q^j = \sum_k \alpha_k^{ij} z_{p+q-n}^k \tag{23}$$

and then limited themselves to saying that the systems (α_k^{ij}) of rational numbers are "tensors" invariant under homeomorphisms!

§ 5. Homology of Products of Cell Complexes

Even the cartesian product of two arbitrary sets (excepting of course subsets of the $\mathbf{R}^n$) was not a notion in common use at the end of the nineteenth century (although it had been explicitly defined by Cantor). Only among algebraic geometers did it occasionally occur, for instance when Cayley considered the product of two algebraic curves or C. Segre the product of two complex projective spaces of arbitrary dimension; then these products were immediately given a structure of algebraic variety. The first mathematician who introduced the concept of a topological space, product of two given topological spaces, was apparently Steinitz in 1908 [456], but the investigation of the relations between the topology of the two factor spaces and the topology of their product was only begun independently by Künneth ([289], [290]) and Lefschetz [301] in 1923.

Both limited themselves to euclidean (rectilinear) compact connected cell complexes; actually, once the invariance problem had been solved (§ 3), computation of the homology of X × Y for two such cell complexes X, Y was an exercise in elementary linear algebra; for simplicity we shall describe it in the algebraic language of today. From the given triangulations T(X), T(Y) of X and Y into convex polyhedra (not necessarily simplices) a similar triangulation T(X × Y) is derived by taking all products A × B for A ∈ T(X) and B ∈ T(Y), and the **Z**-module $S_p(X \times Y)$ of p-chains of T(X × Y) is just the direct sum

$$S_p(X \times Y) = \bigoplus_{0 \leqslant k \leqslant p} (S_k(X) \otimes S_{p-k}(Y)). \tag{24}$$

Now the "reduction" of Poincaré's incidence matrices amounts to a decomposition

$$S_p(X) = Z_p(X) \oplus F_p(X) \tag{25}$$

into a direct sum of two submodules such that the boundary map $\mathbf{b}_p$ is 0 in $Z_p(X)$ and is an injection $F_p(X) \to Z_{p-1}(X)$ in $F_p(X)$; by the theory of invariant factors there are bases (e_{p-1}^j) of $Z_{p-1}(X)$ and (f_p^i) of $F_p(X)$ such that for those bases the matrix of $\mathbf{b}_p$ considered as an injection of $F_p(X)$ into $Z_{p-1}(X)$ is the matrix (ρ_{ij}) defined in chap. I, § 3, after removal of the zero columns. Now for the boundary map in T(X × Y), if A is a k-cell of T(X) and B is a $(p-k)$-cell of T(Y),

$$\mathbf{b}_p(A \times B) = \mathbf{b}_k A \times B + (-1)^k A \times \mathbf{b}_{p-k} B. \tag{26}$$

As $S_p(X \times Y)$ splits into a direct sum of the **Z**-modules

$$F_k(X) \otimes F_{p-k}(Y), \qquad F_k(X) \otimes Z_{p-k}(Y), \qquad Z_k(X) \otimes F_{p-k}(Y),$$
$$Z_k(X) \otimes Z_{p-k}(Y)$$

for $0 \leqslant k \leqslant p$, the matrix of $\mathbf{b}_p: S_p(X \times Y) \to S_{p-1}(X \times Y)$ splits accordingly into *blocks*, all of which are trivially written down immediately, with the exception of the matrix of

$$\mathbf{b}_p: F_k(X) \otimes F_{p-k}(Y) \to (Z_{k-1}(X) \otimes F_{p-k}(Y)) \oplus (F_k(X) \otimes Z_{p-k-1}(Y)); \quad (27)$$

but the "reduction" of that matrix is at once brought down to the "reduction" of 2×2 matrices. One thus obtains a regular algorithm for computing the homology modules of $X \times Y$ when one knows those of X and Y; in particular, one has for the Betti numbers the "Künneth formula"

$$b_p(X \times Y) = \sum_{0 \leqslant k \leqslant p} b_k(X) b_{p-k}(Y) \quad (28)$$

from which one deduces at once for the Euler–Poincaré characteristics

$$\chi(X \times Y) = \chi(X)\chi(Y). \quad (29)$$

It is just as easy to compute the intersection ring of $X \times Y$ when X and Y are oriented combinatorial manifolds; from formula (26) it follows that the cartesian product of two cycles is a cycle; if we denote by $z_p \times z'_q$ the homology class of the cartesian product of a cycle of class z_p in X and of a cycle of class z'_p in Y, then

$$(z_k \times z'_{p-k}).(u_h \times u'_{q-h}) = (z_k . u_h) \times (z'_{p-k} . u'_{q-h}). \quad (30)$$

§ 6. Alexander Duality and Relative Homology

Until 1920 homology had only been defined for finite cell complexes (connected or not). In a remarkable paper [11] published in 1922 (the first draft of which goes back to 1916) Alexander broke new ground by considering the homology of *open* subsets of an $\mathbf{R}^n$; at the same time he showed how the Brouwer theorems, proved by him without reference to homology (Part 2, chaps. I and II), could be inserted into the theory of homology and extended in that way.

He considered a subspace of a sphere $\mathbf{S}_n$ ($n \geqslant 2$), which is a *compact* (connected or not) *curvilinear cell complex* X of dimension $m < n$ (for instance, a closed Jordan curve, for $n = 2$ and $m = 1$). He first had to define the homology of the open set $\mathbf{S}_n - X$. Alexander did not do this formally, but considered a simplicial subdivision T *of* $\mathbf{S}_n$ and the sequence of triangulations T_j obtained by successive barycentric subdivisions of T; p-chains *of* $\mathbf{S}_n - X$ are then p-chains of *any* T_j that are linear combinations of p-simplices *contained in* $\mathbf{S}_n - X$. In order to add a p-chain C of T_j and a p-chain C′ of T_k for $k > j$ (both combinations of p-simplices contained in $\mathbf{S}_n - X$), he replaced each simplex in C by the *sum* of the simplices of T_k into which it is decomposed. (Actually Alexander worked in homology mod. 2, in which he did not have to bother with signs.) Boundary operators and homology are then defined as usual, but *a priori* the homology modules might not be finitely generated.

Another equivalent method is to define the homology of an *arbitrary* open subset U of $\mathbf{R}^n$ by an extension of the concept of *triangulation* for such a set: this time it means a *locally finite* partition T of U into cells of various dimensions, with the condition that the frontier of any cell of T is a *finite* union of cells of T of strictly lower dimension;* more generally, from now on we shall call a space equipped with such a triangulation a *cell complex*. The p-chains are then defined as linear combinations of a *finite* number of cells of T, and boundaries and homology are defined as for finite cell complexes [30]. Alexander's remarkable result was that if $X \subset \mathbf{S}_n$ is a *finite curvilinear cell complex*, all Betti numbers (mod 2) of $\mathbf{S}_n - X$ are *finite* and satisfy *Alexander duality*

$$\dim H_p(X; \mathbf{F}_2) = \dim H_{n-p-1}(\mathbf{S}_n - X; \mathbf{F}_2) \quad \text{for } 1 \leqslant p \leqslant n-2, \tag{31}$$

$$\dim H_0(\mathbf{S}_n - X; \mathbf{F}_2) = \dim H_{n-1}(X; \mathbf{F}_2) + 1,$$

$$\dim H_{n-1}(\mathbf{S}_n - X; \mathbf{F}_2) = \dim H_0(X; \mathbf{F}_2) - 1 \tag{32}$$

[with $H_p(X; \mathbf{F}_2) = 0$ by convention when p is larger than the dimensions of the simplices of X].

The very ingenious and rather intricate proof relies on splitting X into a union of two (curvilinear) cell complexes Y, Z, and , from the knowledge of the cases in which X is replaced by Y, Z or $Y \cap Z$, to deduce the result for X: a typical "Mayer–Vietoris" procedure (although, as we have seen in §2, the papers of Mayer and Vietoris were only published 7 years later). This is applied in three steps, each one using the results of the preceding one.

The first step concerns the case in which X is homeomorphic to a closed cube of any dimension $m \leqslant n$; then it is shown that $H_k(\mathbf{S}_n - X) = 0$, except for $k = 0$ [a generalization of Brouwer's "no separation" theorem (Part 2, chap. II, §4)]. The procedure consists in an induction on m, splitting X into two half cubes and using contradiction, by an infinite iteration of the splitting into cubes with diameters tending to 0. The second step is devoted to the case in which X is homeomorphic to a sphere $\mathbf{S}_m$; induction on m, splitting X into two closed hemispheres with an intersection homeomorphic to $\mathbf{S}_{m-1}$, to each

* To prove the existence of such a triangulation for an *arbitrary* open subset U of $\mathbf{R}^n$, one may consider the closed n-dimensional cubes of $\mathbf{R}^n$ having as vertices the points of $2^{-k}\mathbf{Z}^n$, and 2^{-k} as lengths of their edges; let C_k be the set of those cubes contained in U, and C'_k the subset of C_k consisting of cubes having no common interior point with the cubes of C_{k-1}; the triangulation T is obtained by decomposing each cube belonging to the union of the C'_k for all integers $k \geqslant 1$ into disjoint open cubes of all dimensions $\leqslant n$. This method was already used by Runge in $\mathbf{R}^2$ ([30], p. 143); it was considered as well known by Brouwer ([89], p. 316).

of which the first case can be applied. This already contains as a particular case the Jordan–Brouwer theorem (Part 2, chap. II, § 3) for $m = n - 1$.

The general case (for which X may be taken as a curvilinear *simplicial* complex) is treated in the third step, by a double induction on $m = \dim X$ and on the number of simplices of maximum dimension m in X. Write $b'_p(X) = \dim H_p(X; \mathbf{F}_2)$ for simplicity. First split X into the union of the closure Y of one m-simplex and the closed complement Z of that simplex and prove that

$$b'_p(X) = b'_p(Z) \quad \text{and} \quad b'_{n-p-1}(\mathbf{S}_n - X) = b'_{n-p-1}(\mathbf{S}_n - Z) \quad \text{for } p \leqslant m - 2. \tag{33}$$

By induction, assume that

$$b'_m(Z) - b'_{m-1}(Z) = b'_{n-m-1}(\mathbf{S}_n - Z) - b'_{n-m}(\mathbf{S}_n - Z) \tag{34}$$

and show that

$$b'_m(X) - b'_{m-1}(X) = b'_{n-m-1}(X) - b'_{n-m}(X) \tag{35}$$

[add 1 on the right-hand sides of (34) and (35) when $m = n - 1$]. Finally, after having split off *all simplices of dimension m*, one gets a cell complex Z_0 of dimension $\leqslant m - 1$; the induction hypothesis shows that $0 = b'_m(Z_0) = b'_{n-m-1}(\mathbf{S}_n - Z_0)$ and $b'_{m-1}(Z_0) = b'_{n-m}(\mathbf{S}_n - Z_0)$. The final step consists in proving that $b'_m(X) = b'_{n-m-1}(\mathbf{S}_n - X)$ (with 1 added on the left-hand side if $m = n - 1$) by looking at the $(m - 1)$-simplices of Z_0 and at the m-simplices of X of which these $(m - 1)$-simplices are faces; finally $b'_{m-1}(X) = b'_{n-m}(\mathbf{S}_n - X)$ by (35).

At the end of the paper Alexander observed that relations (31) and (32) yield a third proof of the independence of homology from the triangulation used to compute it, since the triangulations of X and of $\mathbf{S}_n$ are independent of each other.

Alexander's paper was, on one hand, the starting point of investigations by several mathematicians (Vietoris, Alexandroff, Lefschetz, Pontrjagin, Čech) aiming at generalizing homology modules to spaces other than compact complexes or open subsets of $\mathbf{R}^n$; we shall describe these developments in chap. IV, § 2.

On the other hand, it led Lefschetz to introduce the new and important concept of *relative homology*. In his first publication on that subject [302] he introduced this notion for homology with rational coefficients and for very general spaces, thus linking with the generalizations we just mentioned. In his book [304] Lefschetz separated the homology of finite cell complexes from its generalizations and allowed coefficients in $\mathbf{Z}$, $\mathbf{Z}/m\mathbf{Z}$, or $\mathbf{Q}$. If K is a finite euclidean simplicial complex, L is a union of simplices of K, the boundary $\mathbf{b}_L C$ of a chain C of K *relative to* L (or *mod* L) is the sum of the simplices of K in the expression of $\mathbf{b}C$ that *are not contained in* L, so that a p-chain C of K is a *cycle mod* L (resp. is *homologous to* 0 *mod* L) if its (usual) boundary is a combination of simplices of L [resp. if there is a $(p + 1)$-chain of K whose (usual) boundary is the sum of C and of simplices of L]; hence, the definition of the *homology modules of* K *mod* L, which Lefschetz wrote $H_p(K, L)$. More generally, he also gave the corresponding definitions when K is an "open complex," an open subset of a compact euclidean simplicial complex K′ that is a union of simplices of K′.

With these definitions Lefschetz first considered the case in which K is an n-dimensional euclidean simplicial complex, and L is a subcomplex of K (which is automatically closed in K since the faces of a simplex of L have to be in L). He then proved that the homology of K mod L only depends on the topology of $K - L$ (first appearance of what later will be called "excision"); for that purpose he adapted the homotopy process used by Alexander in his first proof of invariance (§3) in such a way that a singular chain having its frontier in L is deformed into another chain whose frontier remains in L ([304], p. 86). Lefschetz also investigated the relations between the homology of K, the homology of L, and the homology of K mod L ([304], pp. 149–150), which later took their final form in the exact sequence of relative homology (chap. IV, §6,B).

Supposing that $K - L$ is also an orientable combinatorial manifold, Lefschetz first generalized to the homology of K mod L the Poincaré duality. He denoted by K^* the union of the duals of the simplices of K that are not simplices of L; then he showed that K^* is a compact simplicial complex, and that its incidence matrix of $(n - p)$-chains and $(n - p - 1)$-chains is, up to sign, the transpose of the incidence matrix of the $(p + 1)$-chains and p-chains of $K - L$.

From this he was able to deduce, by an entirely different method, Alexander's duality theorem, at least in the special case in which L is a *subcomplex* of S_n [for a suitable (curvilinear) triangulation of S_n] by showing that in this case the relative homology module $H_p(S_n, L)$ is isomorphic to the "absolute" homology module $H_{p-1}(L)$ for $1 < p < n$, by an argument that again is essentially part of the exact sequence of relative homology ([304], pp. 143–144).

Finally, Lefschetz took up the more general situation in which $L_1 \subset L$ are two subcomplexes of K, and by a more refined argument, he can show that if $L_2 = L - L_1$, then for the Betti numbers

$$b_p(K - L_1, L_2) = b_{n-p}(K - L_2, L_1) \tag{36}$$

and a corresponding relation holds for the torsion coefficients ([304], pp. 141–142).

CHAPTER III

The Beginnings of Differential Topology

§ 1. Global Properties of Differential Manifolds

We have seen (chap. I, § 2) that Poincaré already had a conception of n-dimensional manifolds of class C^r (r integer $\geqslant 1$ or ∞) and of analytic manifolds (sometimes called C^ω manifolds) which was essentially the modern one. In papers dealing with such a manifold X, it was usually assumed that in addition X was a submanifold of some $\mathbf{R}^N$, with the C^r (or C^ω) structure induced by the canonical structure of $\mathbf{R}^N$: each point of X has a neighborhood V in $\mathbf{R}^N$ such that $V \cap X$ is defined by the annulation of $N - n$ C^r (resp. analytic) functions defined in V and having a jacobian matrix of rank $N - n$ at each point. Until 1935 nobody seems to have tried to prove the existence of such an "embedding" for a manifold "abstractly" given by a system of charts and transition homeomorphisms; nor, for that matter, had anybody studied global properties of such embedded manifolds, except in relation to special connections defined on X (mostly Riemannian structures).

The proof of the most basic results of the theory of differential manifolds was the work of a single man, Hassler Whitney (1907–). In his papers ([506] and [507]) of 1936 (announced in the note [504] of 1935), he broke entirely new ground. He distinguished *immersions* (called by him "regular maps") as C^r-maps such that the tangent map is injective at every point, and *embeddings*, which are *injective* immersions. His first theorem was that for any connected m-dimensional C^r manifold M, there is a C^r immersion of M into $\mathbf{R}^{2m}$, and a *proper* embedding (i.e., one for which the inverse image of a compact set is compact) of M in $\mathbf{R}^{2m+1}$.* The idea of the proof, for a *compact* manifold M, was to cover M by a finite number of open sets U_j ($1 \leqslant j \leqslant h$) that are domains of charts φ_j: $U_j \to \mathbf{R}^m$. The map $f = (\varphi_1, \varphi_2, \ldots, \varphi_h)$ is then an immersion of M into $\mathbf{R}^{hm}$, going down to $\mathbf{R}^{2m}$ or $\mathbf{R}^{2m+1}$ by suitable projections of $f(\mathrm{M})$ into these spaces. If M is not compact, M is exhausted by an increasing sequence (V_k) of relatively compact open subsets; the immersion is then defined

* Later Whitney was able to show that there is even an *embedding* of M into $\mathbf{R}^{2m}$; the proof is long and difficult [515].

J. Dieudonné, *A History of Algebraic and Differential Topology, 1900–1960*,
Modern Birkhäuser Classics, DOI 10.1007/978-0-8176-4907-4_4,

inductively on each V_k. Whitney added the remarkable (and more difficult) result that the image of M by the C^r embedding may be chosen as an *analytic* submanifold of $\mathbf{R}^{2m+1}$.

In the very ingenious proof of that theorem Whitney introduced what was to become one of the most useful notions in the theory of differentiable manifolds: the concept of *tubular neighborhood*. Given an m-dimensional C^r manifold M embedded as a closed submanifold in an n-dimensional manifold V, there exists an open neighborhood T of M in V (a "tube") and a surjective C^r map π: T → M (a "projection") such that:

1. $\pi(x) = x$ for all $x \in M$;
2. the sets $\pi^{-1}(x)$, when x takes values in M, are disjoint, and each of them is C^r-diffeomorphic to an open ball in $\mathbf{R}^{n-m}$;
3. for any $x \in M$, there is an open neighborhood U of x in M such that there exists a C^r-diffeomorphism $\varphi: U \times B \to \pi^{-1}(U)$ with $\pi(\varphi(y, b)) = y$ for $y \in U$, where B is an open ball in $\mathbf{R}^{n-m}$.

These conditions can be expressed by saying that T may be identified with the *normal vector bundle* of M in V, a notion which Whitney actually defined in a companion note presented to the Academy on the same day (see Part 3, chap. III, § 1).

As an example of the use of tubular neighborhoods the Weierstrass approximation theorem for continuous functions can be generalized in the following way: if K is a compact subset of a C^r manifold X, and f: K → Y is a continuous map of K into a C^r manifold Y, then, for a distance d defining the topology of Y, and for any $\varepsilon > 0$, there is an open neighborhood U of K in X and a C^r map g of U into Y such that $d(f(x), g(x)) \leqslant \varepsilon$ for all $x \in K$. One is immediately (by the embedding theorem) brought back to the case $X = \mathbf{R}^m$, $Y = \mathbf{R}^n$; given a tubular neighborhood T of a relatively compact open neighborhood V of $f(K)$, the Weierstrass theorem shows that there exist n polynomials h_j in m variables such that, for $h = (h_1, \ldots, h_n)$ one has $d(f(x), h(x)) \leqslant \delta$, where $\delta \leqslant \varepsilon/2$ is such that the points at a distance $\leqslant \delta$ of $f(K)$ belong to T; the function $g(x) = \pi(h(x))$ is the required one. Furthermore, all maps g: K → Y such that $d(f(x), g(x)) \leqslant \varepsilon/2$ in V are *homotopic* to f; if the restrictions of f and g to the interior $\mathring{K}$ are of class C^r, the homotopy between f and g may be taken of class C^r in $\mathring{K} \times]0, 1[$.

During the next few years, these useful results were completed, first by the introduction of C^r *partitions of unity*, which, for any open subset U of a C^r manifold X, any closed subset $F \subset U$, and every C^r numerical function f *defined in* U, proved the existence of a C^r function g *defined in* X and such that $g(x) = f(x)$ in F; partitions of unity also made possible a very simple proof of Stokes' formula [52]. The use of partitions of unity easily gives a generalization of the approximation theorem mentioned above, when the set K is any closed subset of X.

Another property that was much used later is *Sard's theorem* [407]: if X, Y are two C^r manifolds, the *critical points* of a C^r mapping f: X → Y are the points $x \in X$ such that the tangent mapping $T_x(f)$: $T_x(X) \to T_{f(x)}(Y)$ at that point *is not surjective*; the theorem states that if $r > n/p$, where $n = \dim X$ and $p = \dim Y$, the image $f(E)$ of the set $E \subset X$ of the critical points of f is a *null set* [which implies that $Y - f(E)$ is *everywhere dense*, which is the most often applied corollary of the theorem].*

§ 2. The Triangulation of C^1 Manifolds

Ever since the publication of Poincaré's papers of 1899–1900 the existence of triangulations on differential manifolds and on algebraic varieties had been open problems. They were solved almost simultaneously for algebraic varieties by van der Waerden [478] and Lefschetz [304]; the latter even extended his result to analytic varieties (see Part 2, chap. VII, § 1).

For C^1 manifolds the first correct proof of the existence of a C^1 triangulation was announced in 1930 by Cairns [96], and published in detail in 1934 [97]; he only considered closed submanifolds M of some space $\mathbf{R}^n$ (which turned out not to be a restriction after Whitney's embedding theorem). Later Brouwer ([88] pp. 453–458), Freudenthal [204], and J.H.C. Whitehead [497] published modifications of Cairns' proof (Brouwer's proof having the peculiarity of having been written in conformity with the principles of intuitionism). The most perspicuous of these variants is probably the one inserted by Whitney in his book on *Geometric Integration Theory* ([516], pp. 124–135). He considered a triangulation T *of* $\mathbf{R}^n$ in sufficiently small euclidean simplices σ_k^i, where σ_k^i has dimension k, with the following properties:

1. $M \cap \sigma_k^i = \varnothing$ if $k < s = n - \dim M$; hence $M \cap \mathrm{Fr}(\sigma_s^i) = \varnothing$.
2. $M \cap \sigma_s^i$ has at most one point.
3. The angles between the tangent space to M at the unique point of $M \cap \sigma_s^i$ (when this set is not empty) and the linear s-dimensional linear variety of $\mathbf{R}^n$ generated by σ_s^i are not too small.

Then, for $k \geqslant s$, Whitney proved that the sets $M \cap \sigma_k^i$ that are not empty are "almost" convex polyhedra, and that they are all contained in a tubular neighborhood U of M in $\mathbf{R}^n$. The projections, by the map π: U → M, of the closures of the simplices meeting M then yield the required C^1 triangulation of M.

§ 3. The Theorems of de Rham

Between Poincaré's brief mention of their "periods" in *Analysis Situs* (chap. I, § 2) and 1920 nobody seems to have investigated the *global* theory of exterior differential forms. E. Cartan used them extensively during that time in his

* Less precise forms of Sard's theorem had earlier been obtained by A.B. Brown [90], A. Morse [352], and Whitney. The latter also gave an example of a C^1 map of $\mathbf{R}^2$ into $\mathbf{R}$, such that the image of the set of critical points of that map is the whole interval [0, 1].

papers on Lie groups and on Pfaff systems but only for *local* theorems; it was only in 1922 that he wrote a book on *Integral Invariants* [104] where global properties began to be considered. In particular, he proved that in a space $\mathbf{R}^n$ any closed C^1 p-form is the exterior differential of a $(p-1)$-form, a fact that was classical for $p = 1$; he called that result "the converse of Poincaré's lemma," but, in fact, unknown to Cartan it had already been proved (in a different language) by Volterra in 1889 (*Opere matematiche*, vol. I, pp. 407–422). In the same book Cartan noted that the result could not be generalized to an arbitrary differential manifold, since on the sphere $\mathbf{S}_2$ every 2-form is closed but it is not always the exterior differential of a 1-form.

Cartan's interest in topology was aroused a little later by the fundamental papers of Weyl in 1925 inaugurating the global theory of Lie groups. One of the key results proved by Weyl was that the fundamental group of a compact Lie group is finite, enabling him to use integration theory on arbitrary compact Lie groups. This was the spark at the origin of the wonderful series of papers that Cartan published between 1927 and 1935 on the global theory of Lie groups and homogeneous spaces, in which the topological properties were in the foreground [98]. It was certainly quite natural for him to again try to use in these researches the exterior differential forms on a C^1 manifold that had become through the years a kind of "trademark" of his papers. It is probable that when he became acquainted with homology he was struck by the analogy between the formula $\mathbf{b}_{p-1} \circ \mathbf{b}_p = 0$ for boundary operators, and the formula $d(d\omega) = 0$ for exterior differential forms. At any rate, pursuing the analogy in order to prove the theorems he had in mind for compact Lie groups, he was led to consider, for each integer $p \geqslant 0$, the maximum number of closed p-forms ω_j such that no linear combination $\sum_j \lambda_j \omega_j$, with real coefficients λ_j not all 0, should be exact; and he conjectured that for an oriented C^1 manifold M of dimension n, that number is equal to the *Betti number* b_p of M.

Now, there is a natural bilinear map

$$(c, \omega) \mapsto \langle c, \omega \rangle = \int_c \omega \tag{1}$$

where c runs through the p-cycles of a C^1 triangulation* of M and ω through the closed C^1 p-forms on M; and Stokes' formula shows that $\langle c, \omega \rangle = 0$ if either ω is exact or c is a boundary. Assuming the Betti numbers of M are finite, the proof of Cartan's conjecture boils down to two results:

I. If the closed form ω is such that $\langle c, \omega \rangle = 0$ for *all p-cycles*, it is exact.
II. If the p-cycle c is such that $\langle c, \omega \rangle = 0$ for *all* closed p-forms, it is a boundary.

In 1928 Cartan published a *Comptes-rendus* note [101] conjecturing (in a slightly different but equivalent formulation) the truth of these two results. At

* Both Cartan and de Rham explicitly assumed that such a triangulation exists.

that time de Rham was working toward his thesis, and after reading Cartan's note he very quickly saw that he could prove theorems I and II. Starting with a C^1 triangulation T of M, his method consisted in considering the dual triangulation T* (chap. I, § 3); letting a_q^i be the q-cells of T, and A_{n-q}^i be the $(n-q)$-cell of T* dual to a_q^i, de Rham first showed that he could construct for each A_q^j an open set $D(A_q^j)$, diffeomorphic to an open ball in $\mathbf{R}^n$, such that:

1. $\overline{D(A_q^j)}$ is contained in the interior of the union of the closures of the n-cells having a_{n-q}^j in their boundary.
2. If A_{q-1}^k is contained in the boundary of A_q^j, $D(A_{q-1}^k)$ is contained in $D(A_q^j)$.

Next he defined for each A_q^j what he called an "elementary form $\omega(A_q^j)$" associated to that cell: an $(n-q)$-form on M, vanishing outside $D(A_q^j)$, and such that:

1. If $\mathbf{b}_q(A_q^j) = \sum_k \lambda_{jk} A_{q-1}^k$, then $d(\omega(A_q^j)) = \sum_k \lambda_{jk}\omega(A_{q-1}^k)$.
2. $\langle a_q^i, \omega(A_{n-q}^j)\rangle = \delta_{ij}$ (Kronecker's delta).

Having done this, de Rham proved that *any* C^1 p-form on M can be written $\omega + d\zeta$, where ω is a linear combination of *elementary p-forms* and ζ is a $(p-1)$-form. From that result it is easy to deduce theorems I and II of Cartan [389].

Of course, when cohomology was defined (chap. IV, § 3), it was quite clear that, on a smooth differential manifold, if $\mathscr{E}_p(M)$ is the vector space (over $\mathbf{R}$) consisting of the C^∞ p-forms on M, the exterior differentials

$$d_p\colon \mathscr{E}_p(M) \to \mathscr{E}_{p+1}(M)$$

define a *cochain complex* $(\mathscr{E}_p(M))_{p\geqslant 0}$, hence cohomology vector spaces $H^p(M)$. Similarly, if $\mathscr{D}_p(M)$ is the subspace of $\mathscr{E}_p(M)$ consisting of *compactly supported* p-forms, the restrictions $d_p|\mathscr{D}_p(M)$ define another cochain complex $(\mathscr{D}_p(M))_{p\geqslant 0}$, hence other cohomology vector spaces $H_c^p(M)$. Finally, if M is *oriented*, the bilinear map

$$(\beta, \alpha) \to \int_M \beta \wedge \alpha \tag{2}$$

defined in $\mathscr{D}_{n-p}(M) \times \mathscr{E}_p(M)$, can easily be shown to establish a *duality* identifying $H^p(M)$ with the *dual* vector space $(H_c^{n-p})^*$. After 1950 the relations between these spaces and other cohomology modules were put in a more general context by the theory of sheaf cohomology (chap. IV, § 7,E).

De Rham himself did not try to translate his results in cohomological language before 1950, but from 1935 onward he tried to define new objects that would include both p-forms and p-chains as special cases; he called them *p-currents* by analogy with electric currents, which can be represented by 1-chains when they consist of electricity flowing through a net of thin wires, and by 2-forms when the current is a flow of electric charges in a three-dimensional body, and the 2-form defines the flux of these charges through an element of surface. At first de Rham limited himself to what he called *elementary p-currents* on an n-dimensional C^∞ manifold M: he considered the pairs (c_{p+k}, ω_k), where c_{p+k} is a C^∞ $(p + k)$-chain and ω_k is a C^∞ k-form (k being any integer such that $0 \leqslant k \leqslant n - p$); the vector space over $\mathbf{R}$ consisting of the formal linear combinations of these pairs is divided by the subspace generated by the elements $(c, 0)$, $(0, \omega)$, $(c' + c'', \omega) - (c', \omega) - (c'', \omega)$, $(c, \omega' + \omega'') - (c, \omega') - (c, \omega'')$, and $\lambda(c, \omega) - (c, \lambda\omega)$ for $\lambda \in \mathbf{R}$; the quotient space C_p is by definition the space of elementary p-currents. Then a linear map $\mathbf{b}_p$: $C_p \to C_{p-1}$ can be defined by

$$\mathbf{b}_p . (c_{p+k}, \omega_k) = (c_{p+k}, d\omega_k) + (-1)^k (\mathbf{b}_{p+k} c_{p+k}, \omega_k) \tag{3}$$

and it is easy to see that this defines $(C_p)_{p \geqslant 0}$ as a chain complex, hence there are homology vector spaces for elementary currents.

In 1950 de Rham gave a much more general definition of currents, using the techniques introduced by Sobolev and L. Schwartz in the theory of distributions [393]. The vector space $\mathscr{E}_p(X)$ over $\mathbf{R}$, consisting of the C^∞ p-forms on a C^∞ manifold X is equipped with a unique topology of Fréchet space having the following property: in the domain U of any chart of X, convergence of a sequence $(\omega^{(v)})$ of p-forms means that, when $\omega^{(v)}$ is expressed in local coordinates, its coefficients as well as their derivatives of any order converge uniformly in U. The *dual* $\mathscr{E}'_p(X)$ is then the space of all linear forms T on $\mathscr{E}_p(X)$ that are continuous; it can be shown that for any such linear form there is a compact set $K \subset X$ such that $T(\omega) = 0$ for all forms ω having a support that does not meet K. The elements of $\mathscr{E}'_p(X)$ are the *compactly supported real p-currents* on X; as the exterior differential d: $\mathscr{E}_p(X) \to \mathscr{E}_{p+1}(X)$ is a linear map, continuous for the Fréchet topologies, it has a *transposed* map $\mathbf{b}_{p+1}$: $\mathscr{E}'_{p+1}(X) \to \mathscr{E}'_p(X)$, and these maps define a *chain complex* $(\mathscr{E}'_p(X))_{p \geqslant 0}$, hence homology vector spaces $H'_p(X)$.

Although de Rham did not use the cohomological language, he also proved in substance that the homology vector space $H'_p(X)$ could be identified naturally with the cohomology vector space $H_c^{n-p}(X)$ defined by compactly supported $(n - p)$-forms when X is an *oriented* n-dimensional manifold. Indeed, in that case, integration of n-forms on X enables one to define a natural injective homomorphism

$$j_p: \mathscr{D}_{n-p}(X) \to \mathscr{E}'_p(X)$$

by the relation

$$\langle j_p(\beta), \alpha \rangle = \int_X \beta \wedge \alpha \tag{4}$$

for each p-form α. This map sends closed forms into cycles and exact forms into boundaries, and defines a linear map

$$j_p^*: H_c^{n-p}(X) \to H'_p(X). \tag{5}$$

De Rham proved that this map is bijective by a process generalizing the *regularization* of distributions used by Sobolev and Schwartz. More precisely, he proved a "homotopy formula" (chap. II, §3 and chap. IV, §5,F) for p-

currents on X: for each p, there exist two sequences of linear maps

$$\mathsf{R}_m\colon \mathscr{E}'_p(\mathrm{X}) \to \mathscr{E}'_p(\mathrm{X}), \qquad \mathsf{A}_m\colon \mathscr{E}'_p(\mathrm{X}) \to \mathscr{E}'_{p+1}(\mathrm{X})$$

such that:

1. For each compactly supported p-current S,

$$\mathsf{R}_m(\mathrm{S}) - \mathrm{S} = \mathsf{A}_m(\mathbf{b}_p\mathrm{S}) + \mathbf{b}_{p+1}(\mathsf{A}_m(\mathrm{S})) \tag{6}$$

   and for $m \geqslant m_0$ (index dependent on S), $\mathsf{R}_m(\mathrm{S}) = j_p(\beta_m)$ for a compactly supported $(n-p)$-form β_m.
2. If $\mathrm{S} = j_p(\alpha)$, $\mathsf{A}_m(\mathrm{S}) = j_{p+1}(\gamma)$ for some compactly supported $(n-p-1)$-form γ.
3. For each neighborhood U of the support of S, the supports of $\mathsf{A}_m(\mathrm{S})$ and $\mathsf{R}_m(\mathrm{S})$ are contained in U for large enough m.
4. R_m and A_m are weakly continuous, and for given S, $\mathsf{A}_m(\mathrm{S})$ tends weakly to 0 when m tends to $+\infty$.

Another way to formulate de Rham's result is to say that the cohomology vector space $\mathrm{H}^p(\mathrm{X})$ is naturally isomorphic to the *dual* vector space $(\mathrm{H}'_p(\mathrm{X}))^*$. In chap. IV, § 6,C, we shall see how de Rham related the vector spaces $\mathrm{H}'_p(\mathrm{X})$ to singular cohomology without using sheaf theory.

After 1935 the continuation of differential topology was closely linked to the development of the ideas of *homotopy* and *fiber bundles*, and these will be taken up in Part 3, chap. IV.

CHAPTER IV

The Various Homology and Cohomology Theories

§ 1. Introduction

After 1925 topologists began to look for an extension of the notion of homology groups, until then restricted to the very special and rather artificial case of cell complexes, that would have meaning for the most general topological spaces possible; several such extensions were proposed. The concept of homology groups was enlarged by taking elements of an *arbitrary* commutative group as coefficients of chains; the duality theorems of topology were thus put in relation with the duality of commutative groups, at first for finite groups only. But the discovery of Pontrjagin duality for locally compact commutative groups yielded a method of dealing with general discrete or compact commutative groups as coefficients. It also gave birth to a trend that was very popular until around 1950 (although all but later abandoned*), namely, to consider homology groups as topological groups for suitably chosen topologies and to use the Pontrjagin theory to express the duality theorems in topology. Another closely related development inspired by duality was the introduction of the concept of cohomology and the discovery of the products connected with it. Algebraic notions thus played an ever increasing part in homology theories, and after 1940 the mutual interaction of algebra and topology became the central topic, leading, on one hand, to the axiomatization of homology theories, and, on the other hand, by a kind of backlash, to the creation of homological algebra and of the concepts of category and functor. Finally, a new idea emerged: the group of values taken by cochains of a space X could *vary* in X instead of being a fixed group; it is best expressed in the notion of *sheaf cohomology*, which, with the attendant algebraic device of spectral sequences, rapidly conquered many mathematical theories far removed at first sight from topology.

Until 1940 the topologists who had built up the homology of complexes,

* Witness the most widely used books, dating from the period 1960–1970 such as [5], [212], [440].

J. Dieudonné, *A History of Algebraic and Differential Topology, 1900–1960*,
Modern Birkhäuser Classics, DOI 10.1007/978-0-8176-4907-4_5,

above all Alexander and Lefschetz, were still active in its extensions. They were joined in the United States by newcomers such as Tucker and Steenrod (1910–1971), who were students of Lefschetz at Princeton, and Whitney, who came from Harvard, and who between 1935 and 1940 displayed an extraordinary production of new ideas in almost any question connected with algebraic topology, as we saw in chap. III and will again observe in Parts 2 and 3. In Zürich Hopf had led a brilliant school since 1931 [Stiefel (1909–1979), Eckmann, Gysin, Samelson], and in the USSR Alexandroff (1896–1982), influenced by E. Noether and Lefschetz since 1925, was joined by Pontrjagin (1908–), and (for a short time) by Kurosh (1908–1971) and Kolmogoroff (1903–1987). In Czechoslovakia there appeared a very original topologist, Čech (1893–1960), but (probably due to the unauspicious political circumstances) he had no followers. In contrast Brouwer, in the Netherlands, at last found pupils worthy of him in Freudenthal (1905–), E.van Kampen and Hurewicz (1904–1956).* The latter originally came from Poland; he and his younger compatriot Eilenberg (1915–) might have started a school of algebraic topology there, but by 1939 both were in the United States, where they brought a new impetus to the American school, especially when Eilenberg joined forces with the algebraist Mac Lane (1908–) and created homological algebra with him. Finally, in England M.H.A. Newman (1897–1984), who had been isolated since he had started working on homology in 1926, acquired a brilliant student and colleague in the person of J.H.C. Whitehead (1904–1960). The French school only began in 1945 with Leray (1906–), Koszul (1921–), and H. Cartan (1904–).

Perhaps again as a consequence of World War II and the political turmoil that preceded it, until 1950 the number of these newcomers to algebraic topology was not much greater than in the preceding period; it was only in the postwar generation that the explosion came. With the exception of Alexander, Lefschetz, and Leray, all topologists named above took an active part in the first steps of homotopy theory until after 1950, when it was no longer possible to dissociate it from homology (see Part 3).

§ 2. Singular Homology versus Čech Homology and the Concept of Duality

As defined by Lefschetz in 1933 [305] the notion of singular simplex in a space X, which itself was a compact euclidean simplicial complex, clearly did not use any of these special properties of X, and actually was meaningful for *any* topological space X. Lefschetz's definitions, however, had some defects stemming from the equivalence relations he introduced between the pairs (e_p, f) (chap. II, § 3); the most troublesome was that the **Z**-module of singular p-chains was not free. It was only in 1943 that Lefschetz's definitions were

* For a spirited account of life around Brouwer in 1925–1930, see [29].

slightly modified by Eilenberg [172] to overcome these defects. His definitions were substantially equivalent to the ones now commonly adopted: if Δ_p is the "standard" euclidean simplex, i.e., the set of points $(x_0, x_1, \ldots, x_p)$ in $\mathbf{R}^{p+1}$ such that $x_j > 0$ for $0 \leqslant j \leqslant p$ and $x_0 + x_1 + \cdots + x_p = 1$, a *singular p-simplex in* X is a continuous map $s: \bar{\Delta}_p \to X$. The *singular p-chains* are the elements of the **Z**-module $S_p(X; \mathbf{Z})$ [also written $S_p(X)$] of formal linear combinations of the singular p-simplices in X, with integer coefficients.

Then let σ_j be the continuous map $\sigma_j: \Delta_{p-1} \to \Delta$ defined by

$$\sigma_j(x_0, x_1, \ldots, x_{p-1}) = (x_0, \ldots, x_{j-1}, 0, x_j, \ldots, x_{p-1}). \tag{1}$$

Define the *boundary* of a singular p-simplex s by

$$\mathbf{b}_p s = \sum_{j=0}^{p} (-1)^j s \circ \sigma_j. \tag{2}$$

By linearity this defines the boundary of any singular p-chain, and it is readily verified that these homomorphisms make $S_\bullet(X) = (S_p(X))_{p \geqslant 0}$ into a *free chain complex* (chap. II, § 2); each $S_p(X)$ will in general have an *infinite* basis. The corresponding homology **Z**-modules are the *singular homology groups* $H_p(X; \mathbf{Z})$ of the space X. In this conception all equivalence relations have disappeared; these homology groups are shown to be isomorphic to those defined by Lefschetz by using the homotopy operators which eliminate the degenerate simplices (chap. II, § 3).

If A is any subspace of X, $S_p(A)$ is a free **Z**-submodule of $S_p(X)$ having a free supplementary [with basis all the continuous maps $s: \bar{\Delta}_p \to X$ for which $s(\bar{\Delta}_p) \not\subset A$]; hence,

$$S_\bullet(X, A; \mathbf{Z}) = (S_p(X; \mathbf{Z})/S_p(A; \mathbf{Z}))_{p \geqslant 0}$$

is again a *free chain complex*, the boundary operator being defined by

$$\bar{\mathbf{b}}_p \bar{s} = \overline{\mathbf{b}_p s} \tag{3}$$

for any class $\bar{s} \in S_p(X; \mathbf{Z})/S_p(A; \mathbf{Z})$. This yields the *relative* singular homology module $H_p(X, A; \mathbf{Z})$. If $u: X \to Y$ is a continuous map with $u(A) \subset B$ for subsets $A \subset X$, $B \subset Y$, a homomorphism of **Z**-modules

$$S_p(u): S_p(X, A; \mathbf{Z}) \to S_p(Y, B; \mathbf{Z})$$

can be deduced from it by putting $S_p(u)(\bar{s}) = \overline{u \circ s}$. Since these commute with the boundary operators, they define homomorphisms

$$u_* = H_p(u): H_p(X, A; \mathbf{Z}) \to H_p(Y, B; \mathbf{Z}) \tag{4}$$

such that, if $v: Y \to Z$ is a continuous map and $v(B) \subset C$, then $H_p(v \circ u) = H_p(v) \circ H_p(u)$.

The fact that singular homology was somewhat neglected in the 1930s apparently arose from the realization of its inadequacy in the attempts at extension of Alexander duality (chap. II, § 6) to an *arbitrary* compact set $X \subset \mathbf{S}_n$ different from $\mathbf{S}_n$. This is seen on a simple example given in 1925 by Alexandroff

[23]: consider in $\mathbf{R}^2$ the open set U consisting of pairs (x, y) such that

$$0 < x < 2/3\pi, \qquad \sin 1/x < y < 2;$$

X is the frontier of U. It immediately follows that U and the complement of $\bar{\mathrm{U}}$ in $\mathbf{R}^2$ are connected open sets; hence, $\mathrm{H}_0(\mathbf{R}^2 - \mathrm{X}; \mathbf{F}_2) \cong \mathbf{F}_2^2$. But any singular 1-simplex in X has an image that does not meet some open subset V of X defined by

$$0 < x < a, \qquad y = \sin 1/x$$

for some $a > 0$ (depending on the 1-simplex); from that it follows that the singular homology $\mathrm{H}_1(\mathrm{X}; \mathbf{F}_2) = 0$, and the first Alexander relation (32) of chap. II, § 6 is not satisfied.

This independently led Alexandroff [22], Vietoris [475], and Lefschetz [304] to define *other* homology modules for compact metric spaces. Their common starting point was the idea of "approximating" a compact metric space X in a certain sense by a sequence of simplicial complexes or combinatorial complexes (chap. II, § 2) a device which may be traced back to Brouwer for compact subsets of $\mathbf{R}^2$ ([89], pp. 523–526; see Part 2, chap. I, § 1).

For each $\varepsilon > 0$, Vietoris defined an *ε-complex* (or *ε-chain*) on X to be a linear combination (with coefficients in $\mathbf{Z}$ or $\mathbf{F}_2$) of a finite set of combinatorial simplices (finite subsets of X), each one of which has a diameter $\leqslant \varepsilon$. (The supremum of the diameters of these simplices is called the *mesh* of the ε-complex.) A *Vietoris p-cycle* is then a sequence $\gamma_p = (\gamma_p^n)_{n \geqslant 0}$ of finite combinatorial p-cycles (chap. II, § 2), such that: (1) the mesh of γ_p^n tends to 0; (2) for each n, one may write $\gamma_p^{n+1} - \gamma_p^n = \mathbf{b}_{p+1}\alpha_{p+1}^n$, where the mesh of the chain α_{p+1}^n also tends to 0. Write $\gamma_p \underset{\varepsilon}{\sim} 0$ if there exists a sequence (β_{p+1}^n) of combinatorial $(p+1)$-chains whose mesh tends to 0, and $\gamma_p^n = \mathbf{b}_{p+1}\beta_{p+1}^n$. Linear combinations of Vietoris cycles with coefficients in $\mathbf{Z}$, or $\mathbf{F}_q$ for prime q, or $\mathbf{Q}$, are defined in an obvious way, giving the Vietoris homology modules by the usual process. (Vietoris himself only considers coefficients in $\mathbf{Z}$ or $\mathbf{F}_2$; he called "Brouwer numbers" the dimension over $\mathbf{F}_2$ of his homology groups; these numbers may be infinite for general compact metric spaces.) In a later paper Vietoris showed that these numbers are 0 for m-cycles if the dimension of X [in the sense of Lebesgue and Brouwer (Part 2, chap. II)] is smaller than m; furthermore, he proved that they coincide with the Betti numbers mod. 2 when X is a finite euclidean cell complex.

The point of view of Lefschetz [304] was very close to that of Vietoris, but he only considered compact subsets of some $\mathbf{R}^{\mathrm{N}}$; the chains he took have coefficients in $\mathbf{Q}$ or in some $\mathbf{F}_q$. Starting from a simplicial triangulation T of $\mathbf{S}_{\mathrm{N}}$, he associated to X and T the neighborhood N(T) consisting of the union of all simplices of T meeting X. Taking a sequence (T_n) of triangulations whose mesh tends to 0, he defined a "p-cycle γ_p on X" as a sequence $(\Gamma_p^n)_{n \geqslant 0}$ of simplicial p-cycles in $\mathbf{S}_{\mathrm{N}}$ such that Γ_p^n is contained in $\mathrm{N}(\mathrm{T}_n)$ and that for each n there is an $m > n$ such that Γ_p^s and Γ_p^t are *homologous* in $\mathrm{N}(\mathrm{T}_n)$ for $s \geqslant m$, $t \geqslant m$; γ_p is homologous to 0 if for each n there is an $m > n$ such that $\Gamma_p^s \sim 0$

in $N(T_n)$ for all $s \geqslant m$. He thus obtained, as did Vietoris, "Brouwer numbers" (which may be infinite) $\beta_p(X)$, and he proved the extension of Alexander duality

$$\beta_p(X) = \dim H_{N-p-1}(S_N - X; \mathbf{Q}) \quad \text{for } 1 \leqslant p \leqslant N-2 \tag{5}$$

$$\beta_{N-1}(X) + 1 = \dim H_0(S_N - X; \mathbf{Q}), \quad \beta_0(X) - 1 = \dim H_{N-1}(S_N - X; \mathbf{Q}). \tag{6}$$

He did not show the Brouwer numbers equal to the Betti numbers when X is a finite (curvilinear) cell complex; he had corresponding results for coefficients in $\mathbf{F}_q$.

Alexandroff's method of approximation [22] is based on a different and new idea, related to the concept of dimension introduced by Lebesgue and Brouwer (Part 2, chap. II). It consists in associating to each *finite* open covering $\mathscr{U} = (U_\alpha)$ of a compact metric space X, a combinatorial complex (finite set) $N(\mathscr{U})$ called the *nerve* of the covering. Its vertices are the U_α, and the p-simplices of $N(\mathscr{U})$ are the subsets $\{U_{\alpha_0}, U_{\alpha_1}, \ldots, U_{\alpha_p}\}$ of $p + 1$ sets of the covering such that the intersection $U_{\alpha_0} \cap U_{\alpha_1} \cap \cdots \cap U_{\alpha_p} \neq \varnothing$. A finite open covering $\mathscr{U}' = (U'_\beta)$ of X is *finer* than $\mathscr{U} = (U_\alpha)$ if each U'_β is contained in at least one U_α; if $\varphi(\beta)$ is defined as *one* of the indices α such that $U'_\beta \subset U_\alpha$, the mapping $U'_\beta \mapsto U_{\varphi(\beta)}$ is a *simplicial mapping* (chap. II, § 3) of the nerve $N(\mathscr{U}')$ into the nerve $N(\mathscr{U})$, which depends on the choice of φ; but for the homology groups, the corresponding $\varphi_*: H_\bullet(N(\mathscr{U}')) \to H_\bullet(N(\mathscr{U}))$ is *independent* of that choice.

Alexandroff then said that a sequence $(\mathscr{U}_n)$ of finite open coverings of X is a *projection spectrum* if the maximum diameter of the sets belonging to $\mathscr{U}_n$ tends to 0 with $1/n$, and if $\mathscr{U}_{n+1}$ is finer than $\mathscr{U}_n$ for all n. He showed that for any integer $p \geqslant 0$, the dimension of the vector space $H_p(N(\mathscr{U}_n); \mathbf{F}_2)$ tends to a (finite or infinite) limit, which is independent of the "projection spectrum" $(\mathscr{U}_n)$ and which is none other than the "Brouwer number" $\beta_p(X)$ of Vietoris. Independently of Lefschetz, he also proved relations (5) and (6) for coefficients in $\mathbf{F}_2$ when X is a subspace of some $\mathbf{R}^N$.

In the ensuing years these results were extended around three new themes: (1) direct and inverse limits, (2) definition of topologies on homology groups, and (3) the concept of duality between commutative groups, culminating around 1935 in what is now called Čech homology.

The third of these themes is most developed in Pontrjagin's paper of 1931 [374]. All the "duality" theorems proved before him for the situation considered by Alexander (i.e., a closed subset $X \subset \mathbf{R}^N$ homeomorphic to a euclidean simplicial complex) only concerned "Betti numbers" for homology with coefficients in a *field* ($\mathbf{Q}$ or an $\mathbf{F}_p$); Lefschetz himself had pointed out in his book ([304], p. 144) that no similar theorem was known for homology with coefficients in a ring $\mathbf{Z}/m\mathbf{Z}$ for an *arbitrary* integer m. The problem attacked by Pontrjagin was to obtain results linking homology *groups* of X and $\mathbf{R}^N - X$ (for any compact subset X of $\mathbf{R}^N$) instead of relations between *numbers*; as he observed ([374], p. 171), even when the coefficients are taken in a field, the equality of Betti numbers only guarantees the isomorphy of the homology groups when these numbers are *finite*.

It is a well-known fact that among the concepts of linear algebra, duality was the one least understood until around 1940, owing to the accidental circumstance that the dual of a vector space of finite dimension over a commutative field is isomorphic to that space. A similar accident occurs in H. Weber's theory of finite commutative groups, but there, fortunately, nobody had identified a group with its group of characters. Although Pontrjagin made no mention of characters before 1932, it is possible that they led him to the concept of duality that he develops in his 1931 paper. He based it on the notion of a **Z**-*bilinear* map

$$\varphi\colon (u, v) \mapsto \langle u, v\rangle$$

of a product $U \times V$ of two *finitely generated* groups into a third commutative group M, which in fact he restricted to $\mathbf{Z}$ or $\mathbf{Z}/m\mathbf{Z}$. He called the pair (U, V) *primitive* (with respect to φ) if φ is nondegenerate (i.e., $\langle u, v\rangle = 0$ for all $v \in V$ implies $u = 0$, and $\langle u, v\rangle = 0$ for all $u \in U$ implies $v = 0$). He showed that this always implies (for his choice of M) that U and V are isomorphic: if $M = \mathbf{Z}$, U and V are isomorphic to some $\mathbf{Z}^r$, and if $M = \mathbf{Z}/m\mathbf{Z}$, U and V are direct sums of modules $\mathbf{Z}/q_j\mathbf{Z}$, where the q_j are divisors of m.

He then applied this notion to the Poincaré duality theorem for a compact orientable combinatorial manifold X (chap. II, § 4) of dimension n and showed that $H_p(X; \mathbf{Z}/m\mathbf{Z})$ and $H_{n-p}(X; \mathbf{Z}/m\mathbf{Z})$ form a *primitive pair* for the bilinear form $\langle a, b\rangle$ equal to the class mod m of the Kronecker index $(z \,.\, z')$, where z (resp. z') is the homology class with integer coefficients of a cycle of homology class a (resp. b) with coefficients in $\mathbf{Z}/m\mathbf{Z}$. Similarly, if X is a *subcomplex* of some triangulation of $\mathbf{S}_n$ contained in $\mathbf{R}^N$, Pontrjagin proved that for $1 \leqslant p \leqslant n - 2$, $H_p(X; \mathbf{Z}/m\mathbf{Z})$ and $H_{n-p-1}(\mathbf{R}^n - X; \mathbf{Z}/m\mathbf{Z})$ are a primitive pair for the bilinear form $\langle a, b\rangle$, which this time is the class mod. m of the linking coefficient"* of a p-cycle of class a and a $(n - p - 1)$-cycle of class b.

Pontrjagin had similar results for homology with coefficients in $\mathbf{Z}$ but he had to restrict himself to what he called the "Betti groups," namely, the quotients of the homology groups by their torsion subgroups. From this he obtained the isomorphism theorems already proved by Lefschetz, but his emphasis had changed, duality being the main result and isomorphism an accidental consequence.

To handle the case in which X is an arbitrary compact subspace of $\mathbf{R}^n$, Pontrjagin first defined the *direct limit* of a sequence (G_n) of (not necessarily commutative) groups, relatively to a sequence of homomorphisms

---

* This notion, which had already been defined by Gauss for closed curves in ordinary space, was defined in general by Brouwer (Part 2, chap. I, § 3): an $(n - p - 1)$-cycle C′ in $\mathbf{S}_n - X$ is the *boundary in* $\mathbf{S}_n$ of an $(n - p)$-chain C″, since $H_{n-p-1}(\mathbf{S}_n) = 0$ for $1 \leqslant n - p - 1 < n$. The linking coefficient of a p-cycle C and of C′ is the Kronecker index $(a \,.\, b'')$, where b'' is the homology class of C″.

$$\varphi_{n+1,n}: \mathrm{G}_n \to \mathrm{G}_{n+1};$$

he took all sequences (x_n) such that for each n, $x_n \in \mathrm{G}_n$ and $x_{n+1} = \varphi_{n+1,n}(x_n)$, and he considered two such sequences (x_n), (y_n) as equivalent if $x_n = y_n$ from a certain index on. If (x_n) is equivalent to (x'_n) and (y_n) is equivalent to (y'_n), then the sequence $(x_n y_n)$ is equivalent to $(x'_n y'_n)$, and this defines on the set of equivalence classes of sequences a structure of commutative group, the direct limit of the sequence (G_n) with respect to the homomorphisms $\varphi_{n+1,n}$, later written $\varinjlim \mathrm{G}_n$ when no ambiguity may result about the $\varphi_{n+1,n}$. Pontrjagin showed that if he replaced the sequence (G_n) by a subsequence (G_{n_k}) and the homomorphisms $\varphi_{n+1,n}$ by

$$\psi_{k+1,k} = \varphi_{n_{k+1},n_{k+1}-1} \circ \varphi_{n_{k+1}-1,n_{k+1}-2} \circ \cdots \circ \varphi_{n_k+1,n_k}$$

there is a natural isomorphism of $\varinjlim \mathrm{G}_{n_k}$ onto $\varinjlim \mathrm{G}_n$.

He also considered, for a sequence (G_n) of groups, homomorphisms in the *opposite* direction $\theta_{n,n+1}: \mathrm{G}_{n+1} \to \mathrm{G}_n$, but, surprisingly, added that "such a sequence has no limit group" ([374], p. 196), thus missing the notion of inverse limit! However, specializing his definitions to commutative groups, he observed that if $(\mathrm{U}_1, \mathrm{V}_1)$, $(\mathrm{U}_2, \mathrm{V}_2)$ are two "primitive pairs," relative to the same group M, to each homomorphism $u: \mathrm{U}_1 \to \mathrm{U}_2$ corresponds a unique (later called "transposed") homomorphism ${}^t u: \mathrm{V}_2 \to \mathrm{V}_1$ such that

$$\langle u(x_1), y_2 \rangle = \langle x_1, {}^t u(y_2) \rangle \tag{7}$$

(in modern notation). Therefore, when $(\mathrm{U}_n, \mathrm{V}_n)$ is a sequence of "primitive pairs," to an "inverse sequence" of homomorphisms $\theta_{n,n+1}: \mathrm{U}_{n+1} \to \mathrm{U}_n$, there corresponds a unique "direct sequence" ${}^t\theta_{n,n+1}: \mathrm{V}_n \to \mathrm{V}_{n+1}$, for which Pontrjagin defined a direct limit $\varinjlim \mathrm{V}_n$; he called that group "the dual to the inverse system (U_n)," and it served as a substitute for the missing inverse limit.

Next, Pontrjagin applied these notions to the "Alexander duality" for a compact subset $\mathrm{X} \subset \mathbf{R}^n$. Like Lefschetz he took a fundamental decreasing sequence of closed neighborhoods $\mathrm{N}(\mathrm{T}_j)$ of X (see above) and considered (for coefficients in $\mathbf{Z}/m\mathbf{Z}$) the maps

$$\mathrm{H}_p(\mathrm{N}(\mathrm{T}_{j+1})) \to \mathrm{H}_p(\mathrm{N}(\mathrm{T}_j)) \tag{8}$$

$$\mathrm{H}_{n-p-1}(\mathbf{R}^n - \mathrm{N}(\mathrm{T}_j)) \to \mathrm{H}_{n-p-1}(\mathbf{R}^n - \mathrm{N}(\mathrm{T}_{j+1})) \tag{9}$$

for $1 \leqslant p \leqslant n-1$. From the definition of "linking coefficients" it easily follows that these maps are transposed of each other. As the direct sequence $(\mathrm{H}_{n-p-1}(\mathbf{R}^n - \mathrm{N}(\mathrm{T}_j)))$ has $\mathrm{H}_{n-p-1}(\mathbf{R}^n - \mathrm{X})$ as a direct limit, the "dual," in Pontrjagin's sense, of the inverse sequence $(\mathrm{H}_p(\mathrm{N}(\mathrm{T}_j)))$ is isomorphic to that group. Finally Pontrjagin showed that this is still true if, instead of the $\mathrm{H}_p(\mathrm{N}(\mathrm{T}_j))$, he takes as inverse sequence $(\mathrm{H}_p(\mathrm{N}(\mathscr{U}_j)))$ based on the "projection spectrum" corresponding to the Alexandroff "nerves" of a decreasing sequence of finite open coverings $(\mathscr{U}_j)$ of X in $\mathbf{R}^n$.

Čech's paper [119], published in 1932, was obviously written without knowledge of Pontrjagin's paper of the previous year. In that paper Čech was not interested in duality theory, but wanted to define homology groups (with coefficients in **Q**) for the most general topological space X. Instead of considering Alexandroff's "projection spectra," which are *sequences* of finite open coverings, he at once took *all finite open coverings* $\mathscr{U}$ of X; they constitute a directed set (for the "refinement" relation), since, for two such coverings $\mathscr{U}'$, $\mathscr{U}''$, the finite open covering $\mathscr{U}$ consisting of the nonempty intersections of a set of $\mathscr{U}'$ and a set of $\mathscr{U}''$ is finer than both $\mathscr{U}'$ and $\mathscr{U}''$. He thus obtained an *inverse system* of groups $(\mathrm{H}_p(\mathrm{N}(\mathscr{U});\mathbf{Q}))$ for each p (thus dispensing with Alexandroff's "equivalence" between projection spectra). He then introduced for that particular system the now familiar concept of *inverse limit*, which Pontrjagin had missed: its elements are the families $(a_{\mathscr{U}})$, where $a_{\mathscr{U}} \in \mathrm{H}_p(\mathrm{N}(\mathscr{U});\mathbf{Q})$, and if $\mathscr{U}$ is finer than $\mathscr{U}'$, $a_{\mathscr{U}'}$ is the image of $a_{\mathscr{U}}$ by the natural homomorphism $\mathrm{H}_p(\mathrm{N}(\mathscr{U});\mathbf{Q}) \rightarrow \mathrm{H}_p(\mathrm{N}(\mathscr{U}');\mathbf{Q})$; he thus attached to X homology vector spaces $\check{\mathrm{H}}_p(\mathrm{X};\mathbf{Q})$ over **Q**. He could also define relative homology vector spaces $\check{\mathrm{H}}_p(\mathrm{X}, \mathrm{A};\mathbf{Q})$ for any subspace A of X: If $\mathrm{N}_\mathrm{A}(\mathscr{U})$ is the subcomplex of $\mathrm{N}(\mathscr{U})$ formed of the simplices meeting A, take the inverse limit of the vector spaces $\mathrm{H}_p(\mathrm{N}(\mathscr{U}), \mathrm{N}_\mathrm{A}(\mathscr{U});\mathbf{Q})$. Instead of considering the set $\mathfrak{F}$ of *all* finite open coverings of X, a subset $\mathfrak{F}'$ of $\mathfrak{F}$ may be taken if it is *cofinal* to $\mathfrak{F}$, i.e., such that for every covering $\mathscr{U} \in \mathfrak{F}$ there is a covering $\mathscr{U} \in \mathfrak{F}'$ finer than $\mathscr{U}$. Finally, Čech showed that his definitions extend to coefficients in $\mathbf{Z}/m\mathbf{Z}$, and that for completely normal spaces X, finite *open* coverings may be replaced by finite *closed* coverings without changing the homology (later it was shown that it is enough to suppose X normal).

The theme of *topological* homology groups first appears in Vietoris's paper [475]: if $\gamma_p = (\gamma_p^n)$ and $\delta_p = (\delta_p^n)$ are two Vietoris p-cycles for a compact metric space X, their distance is taken as the infimum of the numbers $\rho \geqslant 0$ such that $\gamma_p^n - \delta_p^n = \mathbf{b}_{p+1}\alpha_{p+1}^n$ and the mesh of the $(p+1)$-chain α_{p+1}^n is ρ. This gives a distance on the homology group $\mathrm{H}_p(\mathrm{X};\mathbf{F}_2)$, and Vietoris showed that $\mathrm{H}_p(\mathrm{X};\mathbf{F}_2)$ is compact for that distance.

This was one of Pontrjagin's starting points in his 1934 paper [375]; the other came from Alexandroff's extension of homology over the coefficient rings **Z**, $\mathbf{Z}/m\mathbf{Z}$ and **Q** used until then, to homology with coefficients, first in an arbitrary commutative *ring* [23], and next in an *arbitrary commutative group* G: for a finite simplicial euclidean complex K, with α_p simplices of dimension p, the group $\mathrm{C}_p(\mathrm{K};\mathrm{G})$ of p-chains with coefficients in G was simply the product G^{α_p}. Later, when tensor products of arbitrary commutative groups were defined,* $\mathrm{C}_p(\mathrm{K};\mathrm{G})$ could be written $\mathrm{G} \otimes \mathrm{C}_p(\mathrm{K};\mathbf{Z})$, and the boundary homomorphism was $1 \otimes \mathbf{b}_p$; similar definitions could be given for relative homology of a euclidean simplicial complex, for singular homology and, by a passage to inverse limit, to Čech homology.

The decisive tool in Pontrjagin's paper was, of course, his famous discovery of the duality between discrete and compact commutative groups based on

---

* This was done in 1938 by Whitney [511].

Weyl's theory of linear representations of compact Lie groups and on the recent discovery of Haar measure on all compact groups. His general bilinear map $U \times V \to M$ of 1931 was now specialized to the case in which M is the (compact) torus group $\mathbf{T} = \mathbf{R}/\mathbf{Z}$, U is discrete and V is the group $\hat{U} = \mathrm{Hom}(U, \mathbf{T})$ of all characters of U, $\langle u, v\rangle = v(u)$, and V is given the topology of simple convergence, for which it is compact. Pontrjagin's theorem [375] states that the maps $v \mapsto \langle u, v\rangle$ are *all continuous characters* of V, and that *any* compact commutative group V has the form $\hat{U}$, where U is the discrete group of all continuous characters of V.

When K is a finite euclidean simplicial complex and G is an arbitrary, at most denumerable, discrete commutative group, one defines on the group of p-chains with coefficients in the dual group $\hat{G}$ the product topology, which immediately yields on $H_p(K;\hat{G})$ a topology for which that group is compact metric. But to define what we now call the Čech group $\check{H}_p(X;\hat{G})$ for an arbitrary compact metric space X, Pontrjagin used the Vietoris process, not being aware of the notion of inverse limit,* and obtained $\check{H}_p(X;\hat{G})$ as a compact group. His main theorem was the final version of Alexander's duality theorem for the homology of $\mathbf{R}^n - X$ with arbitrary coefficients, X being a *compact* subspace of $\mathbf{R}^n$. If G is an arbitrary denumerable discrete group, the compact group $\check{H}_p(X;\hat{G})$ is the Pontrjagin *dual* of the discrete group $H_{n-p-1}(\mathbf{R}^n - X; G)$ for $1 \leqslant p \leqslant n-1$; the latter is defined as a direct limit, as in Pontrjagin's paper of 1931, with G replacing $\mathbf{Z}/m\mathbf{Z}$; it may also be defined as the homology of finite chains in the infinite cell complex obtained from a triangulation of $\mathbf{R}^n - X$ (chap. II, § 6).

In his thesis of 1936 [443], Steenrod began by rounding off the theory of Čech homology. He gave general definitions of inverse limits for topological spaces as well as for groups, and he proved† that an inverse limit of compact spaces is compact ([443], p. 671), that

$$(\varprojlim G_\alpha)\hat{} = \varinjlim \hat{G}_\alpha \tag{10}$$

for an inverse system (G_α) of compact groups, and that for an inverse system (X_α) of compact spaces the Čech homology is

$$\check{H}_p(\varprojlim X_\alpha; G) = \varprojlim \check{H}_p(X_\alpha; G) \tag{11}$$

for any topological group G (p. 691). Using the Pontrjagin theory of locally compact commutative groups, he finally observed that for a *connected* compact commutative group A the first Čech homology group $\check{H}_1(A;\mathbf{T})$ is isomorphic (as a topological group) to A itself (p. 693).

In the same paper, Steenrod was led to modify the definition of the homo-

* In a short *Comptes Rendus* note [130], Chevalley pointed out that the concept of inverse limit (at least for sequences) had already been defined by Herbrand in his arithmetical study of infinite extensions of number fields.

† As observed by Eilenberg and Mac Lane ([177], p. 790) that proof is not quite complete.

logy groups $H_p(K;G)$ for a euclidean simplicial complex K, when G is a *topological* group (not necessarily discrete or compact). He pointed out that the group $Z_p(K;G)$ of p-cycles is always closed in the group $C_p(K;G)$ of p-chains equipped with the product topology, but the group $B_p(K;G)$ of p-boundaries needs not be closed. (It is closed, however, when G is compact or discrete.) Steenrod therefore took as homology groups

$$H'_p(K;G) = Z_p(K;G)/\overline{B_p(K;G)}, \tag{12}$$

which is thus always a Hausdorff topological group. Using this definition in the Čech inverse limit process, new Čech homology groups $\check{H}'_p(X;G)$ are obtained for a compact space X (coinciding with the usual groups when G is discrete or compact). Furthermore, Steenrod showed that $H'_p(X;G)$ is then the topological direct sum of two closed subgroups,* which only depend on G and on the particular Čech group $\check{H}_p(X;\mathbf{T})$; Čech pointed out to him that one of these subgroups is isomorphic to the group $\mathrm{Hom}((\check{H}_p(X;\mathbf{T}))^{\wedge}, G)$, but the precise structure of the other subgroup (which is the inverse limit of torsion groups) could not at that time be determined, even if G is discrete or compact (see § 5D).

Following earlier ideas of Alexandroff and Lefschetz, Steenrod proposed in 1940, in a very original paper [444], a new definition of homology groups of compact metric spaces. Let K be a denumerable infinite *combinatorial complex* (chap. II, § 2), such that every simplex in K is a component of the boundary of only a *finite* number of simplices of K (what one calls a *locally finite*, or *star finite*, combinatorial complex†). For such a complex, Lefschetz had observed that it was possible to define *infinite* chains and their boundaries: an infinite p-chain of K with coefficients in a group G is an element $z = (\alpha_s)_{s \in K_p}$ of the product G^{K_p}, where K_p is the set of p-simplices of K; the boundary $\mathbf{b}_p s$ of each of these simplices being defined as usual, the boundary $\mathbf{b}_p z$ is the family $(\beta_t)_{t \in K_{p-1}}$, where for each $t \in K_{p-1}$, β_t is the sum of the coefficients of t in the boundaries of the $s \in K_p$, each multiplied by α_s; this is meaningful for locally finite combinatorial complexes. Homology groups $H_p(K;G)$ can therefore be defined in the usual way, as can relative homology groups $H_p(K, L; G)$ for a subcomplex L of K; furthermore, if G is a topological group, the homology groups $H'_p(K;G)$ may also be defined as above for finite simplicial complexes.

Lefschetz, in his definition of homology for compact subspaces of an $\mathbf{R}^N$ described above ([304], p. 327), had attached an infinite, locally finite combinatorial complex to the inverse system of finite simplicial complexes he used, by a process that can in fact be applied to any system (K_n) of finite simplicial complexes with simplicial maps $\pi_n\colon K_{n+1} \to K_n$. If K_n is a euclidean simplicial complex contained in a finite-dimensional vector space E_n and if E is the direct sum of the E_n, the infinite complex K is obtained by joining each vertex x_{n+1}

* Steenrod had thought that his proof extended to noncompact spaces, but his argument was incorrect.

† A typical example of such complexes is provided by a triangulation of an *open* subset of $\mathbf{R}^n$.

of K_{n+1} in E to its image $\pi_n(x_{n+1}) \in K_n$ by a line segment and adding to the simplices of all the K_n's those obtained by a simplicial decomposition of each of the polyhedra "joining" K_{n+1} to K_n.* Steenrod considered this Lefschetz complex K when the K_n and π_n constitute a "projection spectrum" of Alexandroff for a space X (see above), and called it a "fundamental complex" of X; it is always possible for the first complex K_0 to have only one vertex, and then the "geometric" definition of K shows that K is contractible to the point K_0; hence, the usual $H_p(K; G)$ defined by finite cycles are all 0 for $p \geqslant 1$. But Steenrod showed that the homology groups $H''_p(K; G)$ defined by *infinite* cycles as above only depend on the topology of X and not on the "projection spectrum" chosen,† and may therefore be written $H''_p(X; G)$. Furthermore, he considered in $H''_p(K; G)$ the subgroup $\tilde{H}''_p(K; G)$ consisting of the homology classes of the infinite cycles which are infinite sums of finite cycles,‡ and showed that there is a natural isomorphism of the quotient $H''_p(K; G)/\tilde{H}''_p(K; G)$ onto the Čech homology group $\check{H}_{q-1}(X; G)$. When G is compact, $\tilde{H}''_q(K; G) = 0$; hence, only Čech homology groups are obtained in that way. When X is a compact subspace of $\mathbf{R}^n$, there is a duality theorem for these new groups: if K′ is the infinite combinatorial complex defined by a triangulation of $\mathbf{R}^n - X$, $H''_q(X; G)$ is isomorphic to the group $H''_p(K'; G)$ defined by *infinite* chains of the locally finite combinatorial complex K′.§

Finally, Steenrod applied his definition to the *solenoids*: for any infinite sequence $A = (a_1, a_2, \ldots, a_n, \ldots)$ of integers $a_n \geqslant 2$, the solenoid Σ_A is defined as the inverse limit $\varprojlim G_n$, where $G_n = \mathbf{T}$ for every n, and the homomorphism $\theta_{n,n+1}\colon G_{n+1} \to G_n$ is multiplication by a_n in $\mathbf{T}$. The compact group Σ_A is connected but not locally connected, and its Pontrjagin dual is the subgroup of the additive group $\mathbf{Q}$ of the rationals generated by the products $(a_1 a_2 \ldots a_n)^{-1}$. Steenrod then showed that the Čech group $\check{H}_1(\Sigma_A; \mathbf{Z}) = 0$, whereas $H''_1(\Sigma_A; \mathbf{Z})$ is isomorphic to $\mathbf{Z}^{\mathbf{N}}/M$, where the subgroup M consists of the infinite sequences (x_n) such that the system of equations $a_n y_{n+1} - y_n = x_n$ has a solution in $\mathbf{Z}^{\mathbf{N}}$; this group $\mathbf{Z}^{\mathbf{N}}/M$ has the power of the continuum.

---

* For a precise definition, order the vertices of the K_n in such a way that if $x_{n+1} \leqslant y_{n+1}$ in K_{n+1}, then $\pi_n(x_{n+1}) \leqslant \pi_n(y_{n+1})$ in K_n; then for each simplex

$$x_{n+1}^{i_0} < \cdots < x_{n+1}^{i_q}$$

of K_{n+1} introduce the simplices

$$\{x_{n+1}^{i_0}, \ldots, x_{n+1}^{i_k}, \pi_n(x_{n+1}^{i_k}), \ldots, \pi_n(x_{n+1}^{i_q})\}$$

for $0 \leqslant k \leqslant q$.

† To prove this result, he gave another definition of $H''_p(X; G)$ based on a generalization of singular simplices (chap. II, § 3).

‡ For such a sum $\sum_k z_k$ to have a meaning, for each m there must be an integer r such that the z_k of index $k \geqslant r$ do not contain any of the simplices of the K_j for $j \leqslant m$, nor any of the simplices "joining" them.

§ All the definitions of homology proposed in this § 2 are *functorial* (§ 6, E), in the same sense as singular homology; i.e., to a continuous map u: X → Y corresponds a homomorphism u_* of the homology of X into the homology of Y, with the usual properties.

§ 3. Cohomology

In a compact n-dimensional *combinatorial manifold* M (chap. II, § 4), the Poincaré construction of the dual complex T^* of a given triangulation T of M provides in a natural way what we now call a dual basis of a basis (a_p^i) of the vector space $H_p(M; \mathbf{Q})$: if a_p^i is the homology class of a p-cycle of T and b_{n-p}^i is the homology class of the dual of that cycle in T^*, (b_{n-p}^i) is a basis of $H_{n-p}(M; \mathbf{Q})$ such that $(a_p^i . b_{n-p}^j) = \delta_{ij}$ for all i, j. This could not, however, be done on an *arbitrary* finite euclidean simplicial complex. Nevertheless, Hopf [241] had succeeded in extending the fixed point formula that Lefschetz had proved (by using the Poincaré construction) for combinatorial manifolds (see Part 2, chap. III) to such general complexes. In order to try to explain this surprising result, Lefschetz introduced, for an arbitrary finite euclidean simplicial complex X, a substitute for the missing cycles of T^*, which he called *pseudocycles* [304]. He first embedded X as a subcomplex of a triangulation T of some S_N with large enough N, and considered the neighborhood U of X in S_N consisting of the simplices of T whose closures meet X. He then observed that if (a_p^i) is again a basis of $H_p(X; \mathbf{Q})$, which are classes of cycles of the triangulation $T \cap X$, it was possible to find elements b_{N-p}^i of $H_{n-p}(S_N, S_N - U; \mathbf{Q})$ such that $(a_p^i . b_{N-p}^j) = \delta_{ij}$ and that b_{N-p}^i is the class of a relative cycle $\text{mod}(S_N - U)$, whose boundary does not meet X.

The Lefschetz pseudocycles are often considered to be the forerunners of cohomology, but they lack an essential ingredient, the coboundary operator. The motivations for the definition of cohomology as described by Alexander in [17] and [19] are quite different. Since, on one hand, for an arbitrary compact space X the Čech homology groups $\check{H}_p(X; \mathbf{T})$ are compact, they have discrete Pontrjagin duals, and it was quite natural to try to give a direct definition of these groups even when no Poincaré or Alexander duality was available. On the other hand, the de Rham theorems (chap. III, § 3) established between the closed p-forms on a compact C^1 manifold M, modulo exact p-forms, and the homology classes of smooth p-cycles, a duality given by the integral of a form over a chain, considered as a bilinear mapping.* This last motivation is particularly apparent in the note [17] of 1935: starting from a (finite or infinite) combinatorial complex K, and its group of (finite) p-chains $S_p(K; A)$ with coefficients in a discrete commutative group A, Alexander considered the group $S^p(K; \hat{A})$ of *all* functions defined on $S_p(K; A)$ with values in the dual $\hat{A}$ of A; if $S_p(K; A)$ is identified with the direct sum $A^{(K_p)}$, where K_p is the set of p-simplices of K, $S^p(K; \hat{A})$ is the Pontrjagin dual of that direct sum, the product $\hat{A}^{K_p}$; for $z_p = \sum_j \alpha_j \sigma_p^j$ (where the σ_p^j are the p-simplices of K, and $\alpha_j \in A$, with $\alpha_j = 0$ except for a finite number of indices), if $\hat{z}'_p \in S^p(K; \hat{A})$ is such that $\hat{z}'_p(\sigma_p^j) = \beta_j \in \hat{A}$, then

$$\langle z_p, \hat{z}'_p \rangle = \sum_j \langle \alpha_j, \beta_j \rangle \in \mathbf{T};$$

* Alexander does not mention de Rham's thesis.

Alexander called this the "integral of $\hat{z}'_p$ over z_p." The transposed homomorphism of $\mathbf{b}_p$: $S_p(K; A) \to S_{p-1}(K; A)$ is a continuous homomorphism $\mathbf{d}_{p-1}$: $S^{p-1}(K; \hat{A}) \to S^p(K; \hat{A})$, which Alexander called "derivative" and which is now known as the *coboundary* operator, so that the relation

$$\langle \mathbf{b}_p z_p, \hat{z}'_{p-1} \rangle = \langle z_p, \mathbf{d}_{p-1} \hat{z}'_{p-1} \rangle$$

was considered by Alexander to be the analog of Stokes' theorem.

Alexander also pointed out in [17] that the definitions are similar when $A = \mathbf{Z}$ and $\hat{A}$ is replaced by $\mathbf{R}$, the bilinear map being replaced by ordinary multiplication. In [18], which is chiefly concerned with the cup-product (see §4), he dropped the duality properties linking homology and cohomology, and for any commutative ring Λ defined the Λ-module of p-cochains as the set of all group homomorphisms $S_p(K; \Lambda) \to \Lambda$.

The same definitions were given simultaneously and independently by Kolmogoroff [282]. A natural problem was then to extend the concept of cohomology from cell complexes to topological spaces. This was done by Eilenberg [172] for the singular theory, who did not need any new concepts: taking the $\mathbf{Z}$-module $S_p(X; \mathbf{Z})$ of singular p-chains in a space X, he simply defined the group of *singular p-cochains* $S^p(X; G)$ for an arbitrary commutative group G as the group $\mathrm{Hom}(S_p(X; \mathbf{Z}), G)$. The coboundary $\mathbf{d}_p$: $S^p(X; G) \to S^{p+1}(X; G)$ is defined by the formula

$$(\mathbf{d}_p f)(z) = f(\mathbf{b}_{p+1} z) \tag{13}$$

for any $(p + 1)$-chain $z \in S_{p+1}(X; \mathbf{Z})$. The p-cochains f such that $\mathbf{d}_p f = 0$ are called *p-cocycles*, and the images $\mathbf{d}_{p-1} g$ of $(p - 1)$-cochains are called *p-coboundaries*. As $\mathbf{d}_p \circ \mathbf{d}_{p-1} = 0$, the group $B^p(X; G)$ of coboundaries is contained in the group $Z^p(X; G)$ of p-cocycles. The quotient group $Z^p(X; G)/B^p(X; G)$ is the *p-th cohomology group* $H^p(X; G)$ of X with coefficients in G. The *relative singular cohomology groups* $H^p(X, Y; G)$ are similarly defined by replacing $S_p(X; \mathbf{Z})$ by $S_p(X; \mathbf{Z})/S_p(Y; \mathbf{Z})$ (which is still a *free* $\mathbf{Z}$-module). Eilenberg also showed that any continuous map $u: X \to X'$ naturally defines a homomorphism $\tilde{u}$: $S^\bullet(X'; G) \to S^\bullet(X; G)$ of graded groups by $\tilde{u}(f) = f \circ u$ for $f \in S^p(X'; G)$; since this map commutes with coboundaries, he deduced from it a homomorphism $u^* = H^\bullet(u)$: $H^\bullet(X'; G) \to H^\bullet(X; G)$ of graded cohomology groups.

If $\varphi: G \to G'$ is a homomorphism of commutative groups, the map $f \mapsto \varphi \circ f$ of is a homomorphism of $S^p(X; G)$ into $S^p(X; G')$, and one has $\varphi \circ (\mathbf{d}_p f) = \mathbf{d}_p(\varphi \circ f)$; hence, $Z^p(X; G)$ is mapped into $Z^p(X; G')$ and $B^p(X; G)$ into $B^p(X; G')$, so that a homomorphism φ_*: $H^\bullet(X; G) \to H^\bullet(X; G')$ of graded cohomology groups is defined.

Finally, regarding duality, Eilenberg only observed that the relations established by Alexander between homology and cohomology of cell complexes are also applicable to the chain complex $K = S_\bullet(X)$ for any space X.

A p-cochain $f \in S^p(X; G)$ on a Hausdorff space X has a *compact support* if there is a compact set $K \subset X$ such that $f(z) = 0$ for every p-chain $z \in S_p(X; \mathbf{Z})$ *whose support does not meet* K. The set $S_c^p(X; G)$ of those cochains is a subgroup of $S^p(X; G)$, and $\mathbf{d}_p$ maps $S_c^p(X; G)$ into $S_c^{p+1}(X; G)$, so that, as above, p-cocycles and p-coboundaries with compact support can be defined, as can *singular cohomology groups with compact supports*, written $H_c^p(X; G)$.

Another type of cohomology could be based on the Alexandroff–Čech process starting from nerves of *finite* open coverings $\mathscr{U}$ of the space X (§ 2). Instead of the homology groups of the finite combinatorial complexes $N(\mathscr{U})$, consider their cohomology groups $H^p(N(\mathscr{U}); G)$ with coefficients in an arbitrary commutative group G; this time they form a *direct system*, and their direct limit is

$$\check{H}^p(X; G) = \varinjlim H^p(N(\mathscr{U}); G), \tag{14}$$

which by definition is the *Čech cohomology group* based on finite coverings. It follows from Steenrod's thesis [443] that if X is a *compact* space and G a discrete group, there is a natural isomorphism

$$\check{H}_p(X; \hat{G}) \simeq (\check{H}^p(X; G))^{\wedge} \tag{15}$$

of the Pontrjagin dual of $\check{H}^p(X; G)$ onto the homology group of X with coefficients in the compact dual $\hat{G}$ of G.

But Čech cohomology groups can also be defined based on *arbitrary* open coverings $\mathscr{U}$ of X; $N(\mathscr{U})$ is then an infinite combinatorial complex that in general is *not* locally finite. Nevertheless, for *any* infinite combinatorial complex K, cohomology groups can (in contrast with homology) be defined based on *infinite cochains*. Indeed, infinite p-cochains may be identified with elements $z = (\alpha_s)_{s \in K_p}$ of the product G^{K_p}; but here the coboundary $\mathbf{d}_p z$ can be defined without additional assumptions on the infinite combinatorial complex K:

$$\mathbf{d}_p z = (\gamma_t)_{t \in K_{p+1}},$$

where γ_t is defined as follows for any $t \in K_{p+1}$. If $\mathbf{b}_{p+1} t = \sum_s \varepsilon_s s$, with $\varepsilon_s = \pm 1$, then $\gamma_t = \sum_s \varepsilon_s \alpha_s$, and this sum is always *finite*. These cohomology groups are in general distinct from the $\check{H}^p(X; G)$ defined above; their properties were chiefly investigated by Dowker [145] and by Eilenberg and Steenrod [189] in their book, in connection with their axiomatic approach; we shall postpone a more detailed description of those papers to § 6, A.

In his 1940 paper [444] Steenrod related the Čech cohomology group $H^p(X; G)$ of a compact space to the "fundamental complex" K he had attached to X; he proved that $\check{H}^p(X; G)$ is isomorphic to the cohomology group $H^{p+1}(K; G)$ based on *finite* cochains of K.

Finally, in 1935, in a second note [18] published simultaneously with [17], Alexander proposed yet another definition of cohomology in which he considered only compact metric spaces; but later, Spanier, in his thesis [438], generalized and simplified Alexander's definition, following an idea of A. Wallace. In the Alexander–Spanier theory for an arbitrary topological space X, the commutative group $C^p(X; G)$ of *all* mappings of X^{p+1} into G (isomorphic to $G^{X^{p+1}}$) is first considered for any commutative group G; in this group the subgroup $C_0^p(X; G)$ of all mappings f that vanish in a neighborhood depending on f) of the diagonal of X^{p+1} is singled out. On $C^p(X;G)$, a coboundary operator δ_p: $C^p(X;G) \to C^{p+1}(X;G)$, is defined by

$$\delta_p f(x_0, x_1, \ldots, x_{p+1}) = \sum_{k=0}^{p+1} (-1)^k f(x_0, \ldots, \hat{x}_k, \ldots, x_{p+1}), \tag{16}$$

checking that $\delta_{p+1} \circ \delta_p = 0$. Furthermore,

$$\delta_p(C_0^p(X;G)) \subset C_0^{p+1}(X;G);$$

hence, by passage to the quotients $\bar{C}^p(X;G) = C^p(X;G)/C_0^p(X;G)$, a coboundary operator

$$\bar{\delta}_p\colon \bar{C}^p(X;G) \to \bar{C}^{p+1}(X;G)$$

is again obtained. The corresponding cohomology groups $\bar{H}^p(X;G)$ are the *Alexander–Spanier cohomology groups*; relative cohomology groups $\bar{H}^p(X, A; G)$ can also be defined for any subspace A of X in a natural way.

An important property of the Alexander–Spanier cohomology is what is called *tautness* in a paracompact space X. If A is a closed subset of X, the open neighborhoods of A in X form an ordered set for the relation $U \supset V$, and the natural maps $\bar{H}^q(U;G) \to \bar{H}^q(V;G)$ define a direct system $(\bar{H}^q(U;G))$ of groups over that ordered set. There is a natural homomorphism

$$\varinjlim \bar{H}^q(U;G) \to \bar{H}^q(A;G);$$

the tautness property is that this homomorphism is *bijective* (a property which is not true for singular cohomology) ([440], p.316). Again, the results of Spanier on these groups will be better understood in connection with the axiomatic theory of cohomology (§6,B).*

§4. Products in Cohomology

A. The Cup Product

Alexander's and Kolmogoroff's definitions of cohomology were accompanied by a result that came as a total surprise to the topologists assembled for the Moscow conference in September 1935 ([250], p. 14); they showed (again independently) that it was possible, for an *arbitrary* finite euclidean simplicial complex K, to define bilinear mappings

$$H^p(K;\Lambda) \times H^q(K;\Lambda) \to H^{p+q}(K;\Lambda) \tag{17}$$

which, when Λ is a commutative ring, give on

$$H^\bullet(K;\Lambda) = \bigoplus_p H^p(K;\Lambda)$$

* All definitions of cohomology are also *functorial*, but to a continuous map u: $X \to Y$ corresponds this time a homomorphism u^* of the cohomology of Y into the cohomology of X.

a structure of graded *associative ring*. This was expected for the case when Λ is a *field* and K is an orientable compact combinatorial *manifold* of dimension n, for then Poincaré duality and the fact that $H^p(K;\Lambda)$ and $H_p(K;\Lambda)$ are vector spaces in duality over Λ [see below, § 5,D, formula (63)] give a natural isomorphism

$$j_p\colon H^p(K;\Lambda) \xrightarrow{\sim} H_{n-p}(K;\Lambda); \tag{18}$$

the mapping (17) can then be deduced by these isomorphisms from the intersection product of Lefschetz (chap. II, § 4),

$$H_{n-p}(K;\Lambda) \times H_{n-q}(K;\Lambda) \to H_{n-p-q}(K;\Lambda). \tag{19}$$

What was completely unexpected was that the Alexander–Kolmogoroff product (17) could be defined on an *arbitrary* finite simplicial complex K, where Poincaré duality did not apply.

Alexander's [18] initial definition was not correct, and in [19] he adopted a modified one suggested by Čech [123] and found independently by Whitney [510], who also contributed the notation

$$(a,b) \mapsto a \smile b \tag{20}$$

for the map (17), hence the name "cup product." Start with a definition of a bilinear map for *cochains*

$$S^p(K;\Lambda) \times S^q(K;\Lambda) \to S^{p+q}(K;\Lambda), \tag{21}$$

also written $(f,g) \mapsto f \smile g$; put an arbitrary order on the vertices of the triangulation of K; for a $(p+q)$-simplex $(x_0,\ldots,x_p,x_{p+1},\ldots,x_{p+q})$, take

$$(f \smile g)(x_0,\ldots,x_{p+q}) = f(x_0,\ldots,x_p)g(x_p,x_{p+1},\ldots,x_{p+q}). \tag{22}$$

Then

$$\mathbf{d}_{p+q}(f \smile g) = \mathbf{d}_p f \smile g + (-1)^p f \smile \mathbf{d}_q g, \tag{23}$$

which implies that when f and g are *cocycles* the same is true for $f \smile g$ and the cohomology class of $f \smile g$ only depends on the cohomology classes a of f and b of g, hence can be written $a \smile b$. Finally, these bilinear maps define a structure of associative ring on $H^{\bullet}(K;\Lambda)$ and

$$b \smile a = (-1)^{pq} a \smile b. \tag{24}$$

These definitions were immediately extended by Eilenberg [172] to *singular cohomology* of arbitrary spaces; if $u\colon X \to Y$ is a continuous mapping, the corresponding map $u^*\colon H^{\bullet}(Y;\Lambda) \to H^{\bullet}(X;\Lambda)$ is a *ring* homomorphism. Finally, for any subspace A of X the relative cohomology group $H^{\bullet}(X,A;\Lambda)$ is again a ring for the restriction of the cup product.

Note that it is possible to define the cup product for more general coefficient groups: suppose we have a bilinear map

$$\varphi\colon G \times G' \to G''$$

for three commutative groups G, G′, G″; then we define a bilinear map

$$S^p(K;G) \times S^q(K;G') \to S^{p+q}(X;G'')$$

by replacing (22) with

$$(f \smile g)(x_0, \ldots, x_{p+q}) = \varphi(f(x_0, \ldots, x_p), g(x_p, x_{p+1}, \ldots, x_{p+q})).$$

From that we deduce as above a bilinear map

$$H^p(K;G) \times H^q(K;G') \to H^{p+q}(K;G'').$$

When X is an orientable n-dimensional combinatorial manifold (chap. II, §4), the natural isomorphism j_p of (18) gives a new definition of the intersection products (19), which bypasses the difficult "general position" arguments of Lefschetz (chap. II, §4). It is only necessary to reverse the arguments made above and to define the intersection product $z_p . z_q$ for $z_p \in H_p(X;\Lambda)$, $z_q \in H_q(X;\Lambda)$, and $p + q \geqslant n$ by

$$z_p . z_q = j_{2n-p-q}(j_{n-p}^{-1}(z_p) \smile j_{n-q}^{-1}(z_q)). \tag{25}$$

This explained a remarkable (and hitherto mysterious) result obtained by Hopf in 1930 for intersection rings. If X and Y are two combinatorial manifolds of *same* dimension n, and u: X → Y a continuous mapping, simple examples show that the linear map u_*: $H_\bullet(X;\mathbf{Q}) \to H_\bullet(Y;\mathbf{Q})$ is *not* in general a ring homomorphism for the intersection *rings*. But by a clever use of the graph of u in the product X × Y (inspired by Lefschetz's methods, see Part 2, chap. III), Hopf was able to define in the *opposite* direction a ring homomorphism φ: $H_\bullet(Y;\mathbf{Q}) \to H_\bullet(X;\mathbf{Q})$, which he called "Umkehrhomomorphismus" [242]. From the relation (25) this follows immediately by taking

$$\varphi(z_p) = j_{n-p}(u^*(j_{n-p}^{-1}(z_p))). \tag{26}$$

The identification of Hopf's "Umkehrhomomorphismus" with (26) was carried out by Freudenthal [203]. In another paper [202], he also showed how to reduce to the cup product an operation defined for compact euclidean cell complexes by Gordon (a student of Pontrjagin) [209]. Let $X \subset \mathbf{R}^n$ be such a complex; Gordon considered two cycles $u \in Z_p(\mathbf{R}^n - X)$, $v \in Z_q(\mathbf{R}^n - X)$ with empty intersection. They may be considered as cycles *in* $\mathbf{R}^n$, and as such written $u = \mathbf{b}_{p+1}x$, $v = \mathbf{b}_{q+1}y$ for chains x, y in $\mathbf{R}^n$; then the "intersection product" $x . y$ may be taken in the sense of Lefschetz; its boundary in $\mathbf{R}^n$ is $u . y \pm x . v$, hence also a cycle in $\mathbf{R}^n - X$. Gordon showed that its homology class only depends on those of u and v; hence, it might be taken as a "product" of these classes and thus define a bilinear map

$$H_p(\mathbf{R}^n - X) \times H_q(\mathbf{R}^n - X) \to H_{p+q+1-n}(\mathbf{R}^n - X). \tag{27}$$

But the Alexander–Pontrjagin duality

$$H_p(\mathbf{R}^n - X) \simeq H^{n-p-1}(X)$$

implies that (27) can also be considered as a bilinear map

$$\mathrm{H}^{n-p-1}(\mathrm{X}) \times \mathrm{H}^{n-q-1}(\mathrm{X}) \to \mathrm{H}^{2n-p-q-2}(\mathrm{X}) \tag{28}$$

and Freudenthal proved that (28) is indeed the cup product.

B. The Functional Cup Product

In view of applications to homotopy theory Steenrod, in 1948 [448], defined a "functional cup product" associated to a continuous map $f\colon \mathrm{X}' \to \mathrm{X}$. The idea is to associate to cohomology classes $u \in \mathrm{H}^p(\mathrm{X})$, $v \in \mathrm{H}^q(\mathrm{X})$ a cohomology class in $\mathrm{H}^{p+q-1}(\mathrm{X}')$. If a, b are cocycles in the classes u, v, there is a $(p-1)$-cochain c in X′ such that $\mathbf{d}c = \tilde{f}(a)$, such that $c \smile \tilde{f}(b)$ is a $(p+q-1)$-cocycle whose cohomology class in $\mathrm{H}^{p+q-1}(\mathrm{X}')$ would be associated to u and v. This imposes restrictions on u and v. First we must have $f^*(u) = 0$, so that $\tilde{f}(a)$ is a coboundary $\mathbf{d}c$. Then $c \smile \tilde{f}(b)$ should be a cocycle; however,

$$\mathbf{d}(c \smile \tilde{f}(b)) = (\mathbf{d}c) \smile \tilde{f}(b) = \tilde{f}(a) \smile \tilde{f}(b) = \tilde{f}(a \smile b),$$

so one must also have $u \smile v = 0$; if $a \smile b = \mathbf{d}w$, then $(c \smile \tilde{f}(b)) - \tilde{f}(w)$ is a cocycle, but its cohomology class depends on the choice of c and w; altering c changes it by an element of $\mathrm{H}^{p-1}(\mathrm{X}') \smile f^*(v)$, and altering w changes it by an element of $f^*(\mathrm{H}^{p+q-1}(\mathrm{X}))$. So, for each $v \in \mathrm{H}^q(\mathrm{X})$, Steenrod defined a subgroup $\mathrm{K}^p(f, v) \subset \mathrm{H}^p(\mathrm{X})$ consisting of the classes u such that

$$f^*(u) = 0 \quad \text{and} \quad u \smile v = 0.$$

Then he defined a subgroup

$$\mathrm{L}^{p+q-1}(f, v) = f^*(\mathrm{H}^{p+q-1}(\mathrm{X})) + \mathrm{H}^{p-1}(\mathrm{X}') \smile f^*(v)$$

and obtained an element

$$u \smile_f v \in \mathrm{H}^{p+q-1}(\mathrm{X}')/\mathrm{L}^{p+q-1}(f, v),$$

the *functional cup product* of u and v. Its main interest is when $\mathrm{K}^p(f, v) = \mathrm{H}^p(\mathrm{X})$ and $\mathrm{L}^{p+q-1}(f, v) = 0$; we shall meet such cases later.

C. The Cap Product

Another bilinear mapping

$$\mathrm{H}_{p+q}(\mathrm{X}; \Lambda) \times \mathrm{H}^p(\mathrm{X}; \Lambda) \to \mathrm{H}_q(\mathrm{X}; \Lambda) \tag{29}$$

was introduced by Čech [123], and independently by Whitney [510], who wrote it $(a, u) \mapsto a \frown u$, and gave it the name "cap product." For a finite euclidean simplicial complex K, start again with a bilinear map

$$\mathrm{S}_{p+q}(\mathrm{K}; \Lambda) \times \mathrm{S}^p(\mathrm{K}; \Lambda) \to \mathrm{S}_q(\mathrm{K}; \Lambda)$$

defined, for a $(p+q)$-simplex $(x_0, \ldots, x_{p+q})$ and a p-cochain f, by

$$(x_0, \ldots, x_{p+q}) \frown f = f(x_0, \ldots, x_p)(x_p, \ldots, x_{p+q}).$$

Then

$$\mathbf{b}_q(z \frown f) = (-1)^p((\mathbf{b}_{p+q}z \frown f) - (z \frown \mathbf{d}_p f)),$$

from which it follows that, if z is a cycle and f is a cocycle, $z \frown f$ is a cycle and its homology class only depends on the homology class of z and the cohomology class of f, hence the definition of (29). Check that for $v \in \mathrm{H}^q(\mathrm{K}; \Lambda)$,

$$\langle a \frown u, v \rangle = \langle a, u \smile v \rangle \tag{30}$$

(which shows that $a \mapsto a \frown u$ is the *transposed* map of $v \mapsto u \smile v$); if $c \in \mathrm{H}_{p+q+r}(\mathrm{X}; \Lambda)$,

$$c \frown (u \smile v) = (c \frown u) \frown v. \tag{31}$$

For $q = 0$,

$$\varepsilon(a \frown u) = \langle a, u \rangle \tag{32}$$

where $\varepsilon\colon \mathrm{H}_0(\mathrm{K}; \Lambda) \to \Lambda$ is the "augmentation" (cf. § 5, F) which to every 0-chain $\sum_j \lambda_j x_j$ with x_j vertices of K and $\lambda_j \in \Lambda$ associates $\sum_j \lambda_j$. Finally, for a continuous map $f\colon \mathrm{K} \to \mathrm{L}$,

$$f_*(a \frown f^*(u)) = f_*(a) \frown u. \tag{33}$$

Again, these results were extended to singular homology and singular cohomology by Eilenberg [172] in an obvious way.

§ 5. The Growth of Algebraic Machinery and the Forerunners of Homological Algebra

Between 1940 and 1955 there was a gradual recognition that properties of homology or cohomology of a space should be presented in two steps: first, purely algebraic properties of differential graded modules (or "chain complexes," see chap. II, § 2), and then application of these properties to a particular type of such modules related to the space under consideration; in particular, it was only in this second step that differences between the behavior of homology and cohomology (§ 4) would appear. This conscious process of extracting algebraic theorems from the contents of topological theorems was to lead to the birth of homological algebra at the end of this period, accompanied by the very convenient language of categories and functors. We shall review the main algebraic devices which emerged in that way.

A. Exact Sequences

In 1941, in a short announcement without proofs [257] concerned with dimension theory, Hurewicz for the first time combined known homomorphisms in cohomology* in a sequence†

* Hurewicz does not mention explicitly the cohomology theory he is using.

† In [257], $\mathrm{H}^{q+1}(\mathrm{B})$ is replaced erroneously by $\mathrm{H}^{q+1}(\mathrm{A} - \mathrm{B})$.

$$\mathrm{H}^q(\mathrm{A}) \xrightarrow{j^*} \mathrm{H}^q(\mathrm{B}) \xrightarrow{\partial} \mathrm{H}^{q+1}(\mathrm{A}-\mathrm{B}) \xrightarrow{v} \mathrm{H}^{q+1}(\mathrm{A}) \xrightarrow{j^*} \mathrm{H}^{q+1}(\mathrm{B});$$

A is a locally compact space, B is a closed subset of A. The cohomology groups are over $\mathbf{Z}$; j: B → A is the natural injection, v maps the class of a cocycle in A − B to its extension in A (not otherwise defined in the text). Finally, ∂ is defined as follows: if z is a q-cocycle in B, it can be considered as the restriction of a q-cochain x in A, and the coboundary $\mathbf{d}_q x$ has a restriction to B equal to 0, hence can be considered as a $(q+1)$-cocycle in A − B; if $\bar{z}$ is the class of z in $\mathrm{H}^q(\mathrm{B})$, $\partial\bar{z}$ is the class of that $(q+1)$-cocycle in $\mathrm{H}^{q+1}(\mathrm{A}-\mathrm{B})$. Hurewicz's essential remark was that in that sequence the *image* of each homomorphism is exactly the *kernel* of the next one, a property later called *exactness.*

At the same time, independently, Eckmann [148] and Ehresmann and Feldbau [166] each described relations between homotopy groups of a fiber space, of its base and its fiber, which amounted to what later was called the *homotopy exact sequence* of a fiber space (Part 3, chap. III, § 2,C) without using arrows, but at that time nobody noted the relation between these results and the sequence written by Hurewicz.

The next two appearances in homology theory of the notion of exact sequence* occurred independently in 1945. The first was in the announcement by Eilenberg and Steenrod [188] of their axiomatic theory of homology (see § 6,B), under the name of "natural system of groups and homomorphisms"

$$\cdots \to \mathrm{H}_q(\mathrm{X}) \to \mathrm{H}_q(\mathrm{X},\mathrm{A}) \to \mathrm{H}_{q-1}(\mathrm{A}) \to \mathrm{H}_{q-1}(\mathrm{X}) \to \cdots \to \mathrm{H}_0(\mathrm{X},\mathrm{A}) \to 0,$$

where X is an arbitrary topological space and A is a closed subset. The homology was not specified, since this is an axiom that all theories of homology must verify.

The second one was the central result in the first topological paper published by H. Cartan [106]. He considered a locally compact space E, a closed subset F ≠ E in E, and the open complement E − F. To each locally compact space X, he attached homology groups $\Gamma^r(\mathrm{X})$ with coefficients in $\mathbf{T} = \mathbf{R}/\mathbf{Z}$. They are defined by considering the Alexandroff compactification $\hat{\mathrm{X}}$ of X (when X is not compact), which adds to X a single point ω "at infinity," and $\Gamma^r(\mathrm{X})$ is the Čech *relative homology group* $\check{\mathrm{H}}_r(\hat{\mathrm{X}}, \omega; \mathbf{T})$, which is compact; when X is compact, $\Gamma^r(\mathrm{X})$ is just the Čech homology group $\check{\mathrm{H}}_r(\mathrm{X}; \mathbf{T})$. Another way of defining $\Gamma^r(\mathrm{X})$ is to apply the Čech inverse limit process, not to all open coverings of X, but to coverings (U_α) for which each U_α is either relatively compact or is the complement of a compact set. Now, if E is locally compact and F ≠ E is a closed subset of E, to the injection j: F → E corresponds a homomorphism j_*: $\Gamma^r(\mathrm{F}) \to \Gamma^r(\mathrm{E})$ in the usual way. But there is also a homomorphism g_*: $\Gamma^r(\mathrm{E}) \to \Gamma^r(\mathrm{E}-\mathrm{F})$; when neither F nor E − F is compact, g is

* In the 1944 paper [249] where Hopf (independently of Eilenberg and Mac Lane) defined the homology of groups (Part 3, chap. V, § 1,B), he explicitly introduced *free resolutions* for an A-module where A is an arbitrary (not necessarily commutative) ring, but he did not use arrows.

a continuous map of the compactification $E \cup \{\omega\}$ into the compactification $(E - F) \cup \{\omega\}$ such that $g(x) = x$ for $x \in E - F$ and $g(x) = \omega$ for $x \in F$; it is defined similarly in the other cases. Finally Cartan showed that there exists a homomorphism ∂: $\Gamma^r(E - F) \to \Gamma^{r-1}(F)$ such that the sequence

$$\cdots \to \Gamma^r(F) \xrightarrow{j_*} \Gamma^r(E) \xrightarrow{g_*} \Gamma^r(E - F) \xrightarrow{\partial} \Gamma^{r-1}(F) \xrightarrow{j_*} \Gamma^{r-1}(E) \to \cdots \tag{34}$$

is exact (although he neither used that term nor wrote arrows). His proofs were first done for finite euclidean simplicial complexes and then passed to the inverse limit.

Particularly interesting features in his paper are the properties he deduced from the exact sequence (34), as a foretaste of its future power in topology. He first proved two consequences of what was later called the Mayer–Vietoris exact sequence (see §6,B), and from them he deduced the possibility of defining a cycle on E from the knowledge of its images on the open sets forming a base for the topology of E (this may now be regarded as a special case of the Leray spectral sequence of a covering; see §7,B). The principal applications he had in mind were the extensions of theorems known for *combinatorial* manifolds to *arbitrary* C^0-*manifolds* for which triangulations were not available in general; we shall return in Part 2, chap. IV to that problem, for which Cartan obtained the first substantial result in that 1945 paper, namely, the generalization of the Jordan–Brouwer theorem.

The next step was taken in 1947 by Kelley and Pitcher [271]; to them is due the name "exact sequence." They observed that this notion is meaningful for arbitrary commutative groups and homomorphisms of groups and that the previously considered exact sequences in homology and cohomology are special cases of a purely algebraic result applying to *chain complexes* in the sense of Mayer (chap. II, §2), i.e., sequences $C_\bullet = (C_j)_{j \geqslant 0}$ of commutative groups, with boundary homomorphisms $\mathbf{b}_j$: $C_j \to C_{j-1}$ such that $\mathbf{b}_0 = 0$ and $\mathbf{b}_{j-1} \circ \mathbf{b}_j = 0$ for $j \geqslant 1$. They also needed the algebraic concept of homomorphism $f_\bullet$: $C_\bullet \to C'_\bullet$ of chain complexes (also called *chain transformation*): a system (f_j) of homomorphisms f_j: $C_j \to C'_j$ such that $f_{j-1} \circ \mathbf{b}_j = \mathbf{b}_j \circ f_j$ for the boundary operators of $C_\bullet$ and $C'_\bullet$. This definition enabled them to deduce homomorphisms $H_j(f_\bullet) = f_{j*}$: $H_j(C_\bullet) \to H_j(C'_\bullet)$ for the homology groups. We saw in chap. II that such homomorphisms had already been used in homology theory in special contexts. The *homology exact sequence* is relative to a short exact sequence of chain complexes

$$0 \to A_\bullet \underset{f_\bullet}{\to} B_\bullet \underset{g_\bullet}{\to} C_\bullet \to 0$$

where $f_\bullet$ and $g_\bullet$ are chain transformations, and exactness means that each sequence

$$0 \to A_j \underset{f_j}{\to} B_j \underset{g_j}{\to} C_j \to 0$$

is exact. There exist then homomorphisms

$$\partial_n: H_n(C_\bullet) \to H_{n-1}(A_\bullet)$$

such that the (infinite) sequence of groups and homomorphisms

$$\cdots \to H_n(A_.) \xrightarrow[H_n(f_.)]{} H_n(B_.) \xrightarrow[H_n(g_.)]{} H_n(C_.) \xrightarrow[\partial_n]{} H_{n-1}(A_.) \to \cdots \to H_0(A_.) \to 0$$

is exact. The mapping ∂_n is defined as follows. For any cycle $z \in C_n$, there is a chain $y \in B_n$ such that $z = g_n(y)$, and $0 = \mathbf{b}_n z = g_{n-1}(\mathbf{b}_n y)$, hence $\mathbf{b}_n y = f_{n-1}(x)$ where $x \in A_{n-1}$; x is a cycle and its class $\bar{x}$ in $H_{n-1}(A_.)$ only depends on the class $\bar{z}$ of z, hence $\partial_n \bar{z}$ is taken equal to $\bar{x}$.

Kelley and Pitcher also investigated the behavior of exact sequences under direct or inverse limits. They showed that if $(A_\alpha, \varphi_{\beta\alpha})$, $(B_\alpha, \psi_{\beta\alpha})$, $(C_\alpha, \theta_{\beta\alpha})$ are three direct systems of commutative groups such that for each α there are homomorphisms

$$A_\alpha \xrightarrow[u_\alpha]{} B_\alpha \xrightarrow[v_\alpha]{} C_\alpha$$

forming an exact sequence and if for $\alpha \leqslant \beta$ the diagram

$$\begin{array}{ccccc} A_\alpha & \xrightarrow{u_\alpha} & B_\alpha & \xrightarrow{v_\alpha} & C_\alpha \\ \downarrow{\scriptstyle \varphi_{\beta\alpha}} & & \downarrow{\scriptstyle \psi_{\beta\alpha}} & & \downarrow{\scriptstyle \theta_{\beta\alpha}} \\ A_\beta & \xrightarrow[u_\beta]{} & B_\beta & \xrightarrow[v_\beta]{} & C_\beta \end{array}$$

is commutative, then the sequence

$$\varinjlim A_\alpha \xrightarrow[\varinjlim u_\alpha]{} \varinjlim B_\alpha \xrightarrow[\varinjlim v_\alpha]{} \varinjlim C_\alpha$$

is exact, but the corresponding result for inverse limits does not hold in general.

The chain complexes were generalized by Eilenberg and Steenrod in their 1952 book [189]. The limitation of the indices to integers $\geqslant 0$ is completely unessential, and they may run through the set $\mathbf{Z}$ of all rational integers; among these *generalized chain complexes*, the chain complexes are those for which $C_j = 0$ for $j < 0$. A simple change of notation then leads to the notion of *generalized cochain complex* $C^\cdot = (C^j)$, where $C^j = C_{-j}$ and the coboundary $\mathbf{d}_j$: $C^j \to C^{j+1}$ is just $\mathbf{b}_{-j}$. The only thing that changes is the terminology: the elements of C^j are called j-cochains, $\operatorname{Ker} \mathbf{d}_j$ is the group of j-cocycles, and $\operatorname{Im} \mathbf{d}_{j-1}$ is the group of j-coboundaries; finally, $H^j(C^\cdot) = \operatorname{Ker} \mathbf{d}_j / \operatorname{Im} \mathbf{d}_{j-1}$ is the jth cohomology group of the generalized cochain complex; this is simply a *cochain complex* if $C^j = 0$ for $j < 0$. Any notion or result relative to chain complexes is therefore immediately transferred to cochain complexes.

An important special case of generalized chain complexes is the one in which $C_. = (C_j)$ consists of *vector spaces* over a field k, and the boundary operators $\mathbf{b}_j$ are linear maps of vector spaces. Then the *dual* vector spaces $C^j = C_j^* = \operatorname{Hom}(C_j, k)$ form a *generalized cochain complex* for the coboundaries

$$\mathbf{d}_j = {}^t\mathbf{b}_{j+1} : C^j \to C^{j+1}$$

transposed of the boundaries. If $f_. = (f_j)$: $C_. \to C'_.$ is a chain transformation that consists of linear maps, the transposed maps ${}^t f_j = f^j \colon C_j'^* \to C_j^*$ constitute a cochain transformation ${}^t f_\bullet \colon C'^\bullet \to C^\bullet$, and the corresponding homomorphisms

$$H^j({}^t f_\bullet)\colon H^j(C'^\bullet) \to H^j(C^\bullet)$$

are the transposed homomorphisms ${}^t H_j(f_\bullet)$. If

$$0 \to A_\bullet \xrightarrow[f_\bullet]{} B_\bullet \xrightarrow[g_\bullet]{} C_\bullet \to 0$$

is a short exact sequence of chain complexes over k,

$$0 \to C^\bullet \xrightarrow[{}^t g_\bullet]{} B^\bullet \xrightarrow[{}^t f_\bullet]{} A^\bullet \to 0$$

is an exact sequence of cochain complexes, and the corresponding cohomology exact sequence is

$$\cdots \to H^n(C^\bullet) \xrightarrow[{}^t H_n(g_\bullet)]{} H^n(B^\bullet) \xrightarrow[{}^t H_n(f_\bullet)]{} H^n(A^\bullet) \xrightarrow[{}^t \partial_{n+1}]{} H^{n+1}(C^\bullet) \to \cdots.$$

The exact sequences in homology and cohomology naturally led to the study of exact sequences of **Z**-modules (or more generally of R-modules for any commutative ring R) independently of the theory of chain complexes. Some useful lemmas were formulated, such as the famous "five lemma," which assumes that in the commutative diagram of R-modules

$$\begin{array}{ccccccccc}
G_1 & \longrightarrow & G_2 & \longrightarrow & G_3 & \longrightarrow & G_4 & \longrightarrow & G_5 \\
\downarrow{\scriptstyle u_1} & & \downarrow{\scriptstyle u_2} & & \downarrow{\scriptstyle u_3} & & \downarrow{\scriptstyle u_4} & & \downarrow{\scriptstyle u_5} \\
G_1' & \longrightarrow & G_2' & \longrightarrow & G_3' & \longrightarrow & G_4' & \longrightarrow & G_5'
\end{array}$$

the two lines are exact; then (1) if u_1 is surjective, u_2 and u_4 injective, u_3 is injective; (2) if u_5 is injective, u_2 and u_4 surjective, u_3 is surjective. In particular, if u_1, u_2, u_4, u_5 are bijective, so is u_3; in that form it appears for the first time in the Eilenberg–Steenrod book ([189], p. 16). Another one is the "snake lemma": given a commutative diagram of R-modules

$$\begin{array}{ccccccccc}
0 & \longrightarrow & A & \longrightarrow & B & \longrightarrow & C & \longrightarrow & 0 \\
 & & \downarrow{\scriptstyle u} & & \downarrow{\scriptstyle v} & & \downarrow{\scriptstyle w} & & \\
0 & \longrightarrow & A' & \longrightarrow & B' & \longrightarrow & C' & \longrightarrow & 0
\end{array}$$

where the two lines are exact, there exist canonically defined homomorphisms making the sequence

$$0 \to \operatorname{Ker} u \to \operatorname{Ker} v \to \operatorname{Ker} w \to \operatorname{Coker} u \to \operatorname{Coker} v \to \operatorname{Coker} w \to 0$$

exact ([113], p. 40).

Not only exact homology sequences but other parts of homology theory as well were pushed back into the algebraic theory of chain complexes. For instance, if a chain complex $C_\bullet = (C_j)$ is such that $C_j = 0$ except for a finite number of indices, and the C_j are finitely generated $\mathbf{Z}$-modules, the *Euler–Poincaré characteristic* of $C_.$ is*

$$\chi(C_.) = \sum_j (-1)^j \operatorname{rk}(C_j). \tag{35}$$

Kelley and Pitcher observed that if for such chain complexes there is an exact sequence

$$0 \to A_. \to B_. \to C_. \to 0, \tag{36}$$

then

$$\chi(B_.) = \chi(A_.) + \chi(C_.). \tag{37}$$

Earlier [336] Mayer had noted that

$$\chi(H_.(C_.)) = \chi(C_.) \tag{38}$$

for the homology of $C_.$. When $C_.$ is the complex of chains of a finite simplicial complex this gives back the Euler–Poincaré formula (4) of chap. I, § 3.

B. The Functors ⊗ and Tor

In 1935 Čech inaugurated a new direction in homology theory [125], the investigation of the way in which homology groups $H_p(X; G)$ depend on the group of coefficients G. He limited himself to the case in which X is a denumerable locally finite combinatorial complex,† but in fact the treatment is entirely algebraic, and consists of the study of the relations between $H_.(C_.; \mathbf{Z})$ and $H_.(C_.; G)$ for a *free* chain complex $C_.$ and an *arbitrary* discrete commutative group G.

Čech's arguments and results were expressed in terms of generators of modules and of linear relations among them. After tensor products of arbitrary $\mathbf{Z}$-modules had been defined by Whitney in 1938 [511], it became possible to use this notion to obtain an intrinsic expression of Čech's theorems, but the complete statement of those theorems in such terms does not seem to have appeared in print before the Eilenberg–Steenrod book ([189], p. 161). It is this formulation that we shall use to describe Čech's arguments, in order to simplify the exposition.

In the free $\mathbf{Z}$-module C_p of p-chains with coefficients in $\mathbf{Z}$, the submodule Z_p of p-cycles is also free, as well as its submodule B_p of p-boundaries. Furthermore, as the boundary map $\mathbf{b}_p: C_p \to Z_{p-1}$ has kernel Z_p, the fact that its image B_{p-1} is free implies that there is in C_p a free submodule F_p (not unique in general) such that $C_p = Z_p \oplus F_p$. Since C_p is free, the G-module of p-chains with coefficients in G is, by definition (§ 2), $C_p \otimes G$, so

---

* For a finitely generated $\mathbf{Z}$-module M, the quotient M/T of M by its torsion submodule is free, and the *rank* rk(M) is the number of elements of a basis of that $\mathbf{Z}$-module.

† Independently, Alexandroff and Hopf treated the same problem for *finite* complexes in their book ([30], p. 233).

$$C_p \otimes G = (Z_p \otimes G) \oplus (F_p \otimes G), \tag{39}$$

since Z_p and F_p are also free. However, the boundary map

$$\mathbf{b}_p \otimes 1 \colon C_p \otimes G \to Z_{p-1} \otimes G \tag{40}$$

has a kernel (the group of p-cycles with coefficients in G) that contains $Z_p \otimes G$ but may be *larger*, and is therefore written $(Z_p \otimes G) \oplus Z'_p$, where Z'_p is the kernel of the restriction of $\mathbf{b}_p \otimes 1$ to $F_p \otimes G$. The image of $C_{p+1} \otimes G$ by $\mathbf{b}_{p+1} \otimes 1$ (the group of p-boundaries with coefficients in G) is contained in $Z_p \otimes G$. From the exact sequence defining the homology group $H_p = H_p(C_\bullet; G)$

$$C_{p+1} \xrightarrow[\mathbf{b}_{p+1}]{} Z_p \to H_p \to 0 \tag{41}$$

the exactness of

$$C_{p+1} \otimes G \xrightarrow[\mathbf{b}_{p+1} \otimes 1]{} Z_p \otimes G \to H_p \otimes G \to 0 \tag{42}$$

follows, and

$$H_p(C_\bullet; G) \simeq (H_p \otimes G) \oplus Z'_p. \tag{43}$$

Finally, as $F_p \otimes G$ is isomorphic with $B_{p-1} \otimes G$, the exact sequence defining Z'_p is

$$0 \to Z'_p \to B_{p-1} \otimes G \to Z_{p-1} \otimes G \to H_{p-1} \otimes G \to 0. \tag{44}$$

Although expressed in the language of generators and relations, this was historically the first example of a general phenomenon that was noted by Whitney when he defined the tensor product $A \otimes B$ of two arbitrary $\mathbf{Z}$-modules. If a homomorphism $f \colon A \to A''$ is surjective, the same is true of $f \otimes 1 \colon A \otimes B \to A'' \otimes B$, but if the homomorphism $g \colon A' \to A$ is injective, the corresponding homomorphism $g \otimes 1 \colon A' \otimes B \to A \otimes B$ is not necessarily injective. However, it was only much later ([440], p. 225) that it was realized that the kernel of $g \otimes 1$ only depends on B and on $A'' = A/g(A')$ (a fact that Čech had proved in the special case he was studying); this kernel was then written $A'' * B$ or $\mathrm{Tor}(A'', B)$ and called the *torsion product* of A'' and B, because (as Čech also had noted) it only depends on the torsion submodules of A'' and B. Čech's result could then be written as an isomorphism

$$H_p(C_\bullet \otimes G) \simeq (H_p \otimes G) \oplus \mathrm{Tor}(H_{p-1}, G) \tag{45}$$

or as an exact sequence which splits

$$0 \to H_p \otimes G \to H_p(C_\bullet \otimes G) \to \mathrm{Tor}(H_{p-1}, G) \to 0, \tag{46}$$

where the first homomorphism is deduced from the natural homomorphism

$$\mathrm{Ker}(\mathbf{b}_p) \otimes G \to \mathrm{Ker}(\mathbf{b}_p \otimes 1).$$

Another important result concerning the dependence of $H_\bullet(C_\bullet; G)$ on the group G came from the following observation. If $C_\bullet$ is a *free* chain complex and

$$0 \to G' \to G \to G'' \to 0$$

is an exact sequence of commutative groups, the corresponding sequence

$$0 \to C_{\bullet} \otimes G' \to C_{\bullet} \otimes G \to C_{\bullet} \otimes G'' \to 0$$

is also exact and there is a corresponding homology exact sequence

$$\cdots \to H_p(C_{\bullet}; G') \to H_p(C_{\bullet}; G) \to H_p(C_{\bullet}; G'') \xrightarrow{\beta} H_{p-1}(C_{\bullet}; G') \to \cdots.$$

The connecting homomorphisms $\beta\colon H_p(C_{\bullet}; G'') \to H_{p-1}(C_{\bullet}; G')$ are called the *Bockstein homomorphisms* in homology; they were first considered by Bockstein [53] for the exact sequence $0 \to \mathbf{Z} \xrightarrow{\mu} \mathbf{Z} \to \mathbf{Z}/m\mathbf{Z} \to 0$, where μ is multiplication by m.

C. The Künneth Formula for Chain Complexes

We have seen that the homology of the product $K' \times K''$ of two *finite cell complexes* with coefficients in $\mathbf{Z}$ is the homology of the free chain complex $S_{\bullet}(K' \times K'')$ defined by formula (24) of chap. II, § 5 with the boundary operator defined by formula (26), *loc. cit.* Obviously this can again be transferred to a purely algebraic operation. If $C'_{\bullet}$ and $C''_{\bullet}$ are any two chain complexes, consider the chain complex $C_{\bullet} = C'_{\bullet} \otimes C''_{\bullet}$, defined by

$$C_r = \bigoplus_{0 \leqslant p \leqslant r} (C'_p \otimes C''_{r-p}) \tag{47}$$

and by the boundary operator

$$\mathbf{b}_r(z'_p \otimes z''_{r-p}) = \mathbf{b}_p z'_p \otimes z''_{r-p} + (-1)^p z'_p \otimes \mathbf{b}_{r-p} z''_{r-p}; \tag{48}$$

this formula shows first that $\mathbf{b}_{r-1} \circ \mathbf{b}_r = 0$, and then, if z'_p and z''_{r-p} are cycles, so is $z'_p \otimes z''_{r-p}$, and if either z'_p or z''_{r-p} is a boundary, $z'_p \otimes z''_{r-p}$ is also a boundary. This yields a natural homomorphism of graded $\mathbf{Z}$-modules

$$H_{\bullet}(C'_{\bullet}) \otimes H_{\bullet}(C''_{\bullet}) \to H_{\bullet}(C'_{\bullet} \otimes C''_{\bullet}). \tag{49}$$

From general theorems of homological algebra given by H. Cartan and Eilenberg ([113], p. 111) it follows that if $C'_{\bullet}$ and $C''_{\bullet}$ are *free* chain complexes, the preceding homomorphism is inserted in a split exact sequence

$$\begin{aligned} 0 \to \bigoplus_{0 \leqslant p \leqslant r} (H_p(C'_{\bullet}) \otimes H_{r-p}(C''_{\bullet})) &\to H_r(C'_{\bullet} \otimes C''_{\bullet}) \\ \to \bigoplus_{0 \leqslant q \leqslant r-1} \operatorname{Tor}(H_q(C'_{\bullet}), H_{r-1-q}(C''_{\bullet})) &\to 0 \end{aligned} \tag{50}$$

(*Künneth-formula for chain complexes*). A direct proof can be given as follows ([440], pp. 228–229): Consider chain complexes $Z''_{\bullet} = (Z''_j)$ and $B''_{\bullet} = (B''_{j-1})$ of cycles and boundaries of $C''_{\bullet}$ with boundaries equal to 0. There is then a short exact sequence of chain complexes

$$0 \to Z''_{\bullet} \to C''_{\bullet} \xrightarrow{\mathbf{b}''_{\bullet}} B''_{\bullet} \to 0 \tag{51}$$

the first arrow being the natural injection. These chain complexes are free, so this gives the exact sequence

$$0 \to C'_\bullet \otimes Z''_\bullet \to C'_\bullet \otimes C''_\bullet \xrightarrow{1 \otimes \mathbf{b}''_\bullet} C'_\bullet \otimes B''_\bullet \to 0. \tag{52}$$

The corresponding homology exact sequence is therefore

$$\cdots \to H_q(C'_\bullet \otimes Z''_\bullet) \xrightarrow{\mu} H_q(C'_\bullet \otimes C''_\bullet) \to H_q(C'_\bullet \otimes B''_\bullet) \xrightarrow{\partial} H_{q-1}(C'_\bullet \otimes Z''_\bullet) \to \cdots. \tag{53}$$

It is easy to check that μ is the surjective map

$$\bigoplus_j (H_j(C'_\bullet) \otimes Z''_{q-j}) \to \bigoplus_j (H_j(C'_\bullet) \otimes H_{q-j}(C''_\bullet)) \tag{54}$$

and ∂ is the map

$$\bigoplus_{0 \leqslant j \leqslant q-1} (H_{q-1-j}(C'_\bullet) \otimes B''_j) \to \bigoplus_{0 \leqslant j \leqslant q-1} (H_{q-1-j}(C'_\bullet) \otimes Z''_j) \tag{55}$$

coming from the natural injections $B''_j \to Z''_j$; by (46) the kernel of that homomorphism is $\bigoplus_{0 \leqslant j \leqslant q-1} \mathrm{Tor}(H_{q-1-j}(C'_\bullet), H_j(C''_\bullet))$, hence the sequence (50).

D. The Functors Hom and Ext

The discovery of the role played by extensions of groups in homology did not come from such a systematic search for "universal coefficients" as Čech's paper on homology of infinite combinatorial complexes based on *finite* chains. We saw in § 4 that in his thesis of 1936 Steenrod had essentially tried to do the same for Čech homology of compact metric spaces but had only partially succeeded. In the same spirit, in the 1940 paper where he introduced the new homology groups $H''_q(X; G)$, the variation of those groups when the coefficient group G varies played an important part. It is precisely the description he had given of the groups $H''_1(\Sigma_A; \mathbf{Z})$ for the solenoids, which started the collaboration of Eilenberg and Mac Lane and their famous 1942 paper [177].

At that time Mac Lane was working on the group Ext(G, H) of extensions of a commutative group G by another commutative group H: a commutative group E is an *extension* of G by H if there is an exact sequence*

$$0 \to G \xrightarrow{\alpha} E \xrightarrow{\beta} H \to 0. \tag{56}$$

Two such extensions (E, α, β) and (E', α', β') are *equivalent* if there is an isomorphism $\theta: E \xrightarrow{\sim} E'$ such that the diagram

$$\begin{array}{ccccc} & & E & & \\ & \overset{\alpha}{\nearrow} & & \overset{\beta}{\searrow} & \\ 0 \to G & & \downarrow \theta & & H \to 0 \\ & \underset{\alpha'}{\searrow} & & \underset{\beta'}{\nearrow} & \\ & & E' & & \end{array} \tag{57}$$

---

* One may define more generally an extension E of two *not necessarily commutative* G, H, meaning that G is a *normal* subgroup of E, and H is isomorphic to the quotient group E/G; they were studied in a particular case by I. Schur [418] and in general by O. Schreier.

is commutative. Given a *section* $u: \mathrm{H} \to \mathrm{E}$ for β, i.e., a map such that $\beta(u(h)) = h$ for all $h \in \mathrm{H}$, then

$$u(h) + u(h') = u(h + h') + \alpha(f(h, h')), \tag{58}$$

where f is a map of $\mathrm{H} \times \mathrm{H}$ into G (a "factor set"). The associativity and commutativity of the group law in E impose on f the conditions

$$\begin{cases} f(h,k) + f(h+k,l) = f(h,k+l) + f(k,l) \\ f(k,h) = f(h,k) \end{cases} \tag{59}$$

for h, k, l in H. Furthermore, if $u': \mathrm{H} \to \mathrm{E}$ is another section of β, then

$$u'(h) = u(h) + \alpha(g(h)), \tag{60}$$

where g is a map of H into G, and

$$\varphi(h,k) = g(h) + g(k) - g(h+k) \tag{61}$$

belongs to the additive group $\mathrm{Z(G, H)} = \mathrm{Z}$ of factor sets. To say that the factor set f belongs to the subgroup $\mathrm{B(G, H)} = \mathrm{B}$ consisting of the special factor sets (61) means that there is a section $u: \mathrm{H} \to \mathrm{E}$ of β that is a homomorphism, hence, the sequence (56) *splits*, and the extension E, isomorphic to $\mathrm{G} \oplus \mathrm{H}$, is *trivial*.

Any extension $(\mathrm{E}, \alpha, \beta)$ thus determines a unique element of the factor group Z/B. Conversely, for any $f \in \mathrm{Z}$ the set $\mathrm{E}_f = \{(g, h) \in \mathrm{G} \times \mathrm{H}\}$ is a commutative group for the group law $(g_1, h) + (g_2, k) = (g_1 + g_2 + f(h,k), h + k)$, and the triple $(\mathrm{E}_f, \alpha, \beta)$, where $\alpha: \mathrm{G} \to \mathrm{E}_f$ is the natural injection and $\beta: \mathrm{E}_f \to \mathrm{H}$ is the second projection, is an extension of G by H; two factor sets f_1, f_2 give equivalent extensions if and only if $f_1 - f_2 \in \mathrm{B}$. In this way the set $\mathrm{Ext(G, H)} = \mathrm{Z/B}$ was given the structure of a *commutative group*. This group structure on equivalence classes of extensions was discovered by R. Baer in 1934 ([37], [38]).

Mac Lane computed the group $\mathrm{Ext}(\mathbf{Z}, \mathbf{T}_p)$ for the "p-adic solenoid" $\mathbf{T}_p$ [i.e., the solenoid corresponding to the set $\mathrm{A} = (p, p^2, \ldots, p^n, \ldots)$ for a prime number p (§ 2)]; it is the Pontrjagin dual of the discrete additive group of the ring $\mathbf{Z}[1/p]$. Eilenberg noted that from Steenrod's description it followed that $\mathrm{Ext}(\mathbf{Z}, \mathbf{T}_p)$ is isomorphic with the Steenrod group $\mathrm{H}''_1(\mathbf{T}_p; \mathbf{Z})$, and he and Mac Lane set out to investigate thoroughly this unexpected coincidence. They first found another definition of the group Ext(G, H). If $\mathrm{H} = \mathrm{F/R}$, where F is a free commutative group, then $0 \to \mathrm{R} \to \mathrm{F} \to \mathrm{H} \to 0$ is an extension corresponding to a factor set $f_0 \in \mathrm{Z(R, H)}$. To each $\theta \in \mathrm{Hom(H, G)}$ corresponds a factor set $f_\theta = \theta \circ f_0 \in \mathrm{Z(G, H)}$, and if θ can be *extended* to a homomorphism of F into G, then $f_\theta \in \mathrm{B(G, H)}$. It is then easy to see that, if $\mathrm{Hom(F|R, G)}$ designates the subgroup of Hom(R, G) consisting of restrictions to R of homomorphisms of F into G, there is an isomorphism

$$\mathrm{Hom(R, G)/Hom(F|R, G)} \simeq \mathrm{Ext(G, H)}. \tag{62}$$

The 1942 Eilenberg–Mac Lane paper is fairly long, because they considered both homology and cohomology groups for compact metric spaces with all the definitions proposed by Čech and Steenrod for arbitrary groups of coefficients, taking into account the peculiarities linked to the topology assigned to these groups. Their two main results were:

1. For Čech cohomology groups, the complete description for a compact metric space X and a discrete group G ([177], p. 823):
$$\check{H}^q(X; G) \simeq (G \otimes \check{H}^q(X; \mathbf{Z})) \oplus \mathrm{Hom}_{\mathrm{cont}}(\hat{G}, T^q(X; \mathbf{Z})) \qquad (63)$$
where $T^q(X; \mathbf{Z})$ is the torsion subgroup of $\check{H}^q(X; \mathbf{Z})$, and $\mathrm{Hom}_{\mathrm{cont}}$ means the subgroup of Hom consisting of continuous homomorphisms.
2. For the Steenrod homology groups, there is a similar decomposition (for discrete G) ([177], p. 824)
$$H''_q(X; G) \simeq \mathrm{Hom}(\check{H}^{q-1}(X; \mathbf{Z}), G) \oplus \mathrm{Ext}(G, \check{H}^q(X; \mathbf{Z})), \qquad (64)$$
which in particular "explains" Mac Lane's observation for $X = \Sigma_A$; then
$$H''_1(\Sigma_A; \mathbf{Z}) \simeq \mathrm{Ext}(\mathbf{Z}, \Sigma_A)$$
and the right side is isomorphic to $\mathbf{Z}_A/\mathbf{Z}$, where $\mathbf{Z}_A = \varinjlim_n \mathbf{Z}/\mathbf{Z}a_1 a_2 \cdots a_n$.

In that paper, among other things, Eilenberg and Mac Lane observed that, when working with an infinite combinatorial complex K, anything proved for groups of infinite q-cochains $C^q(K; G) = \mathrm{Hom}(C_q(K; \mathbf{Z}), G)$ [where $C_q(K; \mathbf{Z})$ is the group of finite q-chains] applies just as well to groups of infinite q-chains $C'_q(K; G) = \mathrm{Hom}(C'^q(K; \mathbf{Z}), G)$ [where $C'^q(K; \mathbf{Z})$ is the group of finite q-cochains] ([177], p. 813). This meant that (just as Čech with his introduction of Tor), they had to deal with a *purely algebraic property*: given a free chain complex $C_\bullet = (C_q)$ over $\mathbf{Z}$, and the *cochain complex* $C^\bullet = (\mathrm{Hom}(C_n, G))_{n \geqslant 0}$ associated to it, the problem is to describe its cohomology group $H^n(C^\bullet; G)$ in terms of $H_n(C_\bullet; \mathbf{Z}) = H_n$ and G. $H^n(C^\bullet; G) = Z^n/B^n$, where Z^n is the group of n-cocycles and B^n is the group of n-coboundaries, and by definition Z^n is the subgroup of $\mathrm{Hom}(C_n, G)$ consisting of n-cochains that vanish in the group B_n of n-boundaries. As the group Z_n of n-cycles is a direct summand of C_n, it is possible to associate to each n-cocycle $f \in Z^n$ its restriction $\bar{f}$ to Z_n, and as $\bar{f}$ vanishes in B_n, and every n-coboundary $g \in B^n$ vanishes in Z_n, f determines a unique homomorphism of $H_n = Z_n/B_n$ into G. This thus defines a natural homomorphism
$$\beta\colon H^n(C^\bullet; G) \to \mathrm{Hom}(H_n, G). \qquad (65)$$
This homomorphism is *surjective* because any homomorphism of Z_n into G that vanishes in B_n is the restriction $\bar{f}$ of a homomorphism $f\colon C_n \to G$ vanishing in B_n (since Z_n is a direct summand of C_n), i.e., an n-cocycle. To investigate $\mathrm{Ker}\,\beta$, note that if there is an n-cocycle f such that $\bar{f} = 0$, then f vanishes in Z_n and can be written $g \circ \mathbf{b}_n$, where $g \in \mathrm{Hom}(B_{n-1}, G)$. However, two such cocycles f_1, f_2 only give the same class in $H^n(C^\bullet; G)$ if $f_1 - f_2 \in B^n$, and $\mathrm{Ker}\,\beta$ is the quotient of $\mathrm{Hom}(B_{n-1}, G)$ by the subgroup of its elements g such that

$g \circ \mathbf{b}_n$ is a restriction of an n-coboundary. This means that g is the restriction of a homomorphism $g_0\colon C_{n-1} \to G$, or equivalently (since $Z_{n-1} \supset B_{n-1}$ is a direct summand of C_{n-1}), the restriction of a homomorphism $g_1\colon Z_{n-1} \to G$. Finally, by (62), as Z_{n-1} is free,

$$\operatorname{Ker}\beta \cong \operatorname{Hom}(B_{n-1}, G)/\operatorname{Hom}(Z_{n-1} | B_{n-1}, G) \cong \operatorname{Ext}(H_{n-1}, G).$$

We thus obtain an exact sequence

$$0 \to \operatorname{Ext}(H_{n-1}, G) \to H^n(C^{\bullet}, G) \to \operatorname{Hom}(H_n, G) \to 0 \tag{66}$$

and this sequence immediately splits: the choice of a supplement F_n to Z_n in C_n determines a unique map $\operatorname{Hom}(Z_n, G) \to \operatorname{Hom}(C_n, G)$ by assigning to a homomorphism of Z_n into G its extension to C_n which vanishes in F_n.

If $C_{\bullet}$ is a free chain complex and

$$0 \to G' \to G \to G'' \to 0$$

is an exact sequence of commutative groups, the corresponding sequence

$$0 \to \operatorname{Hom}(C_{\bullet}, G') \to \operatorname{Hom}(C_{\bullet}, G) \to \operatorname{Hom}(C_{\bullet}, G'') \to 0$$

is also exact; therefore there is a cohomology exact sequence

$$\cdots \to H^p(C^{\bullet}; G') \to H^p(C^{\bullet}; G) \to H^p(C^{\bullet}; G'') \xrightarrow{\beta} H^{p+1}(C^{\bullet}; G') \to \cdots.$$

The connecting homomorphisms $\beta\colon H^p(C^{\bullet}; G'') \to H^{p+1}(C^{\bullet}; G')$ are again called the *Bockstein homomorphisms* in cohomology.

E. The Birth of Categories and Functors

The work done by Eilenberg and Mac Lane on the role of the groups Ext(G, H) in homology led them almost immediately afterward [178] to very general considerations on various aspects of group theory that would ultimately bring new points of view in many parts of mathematics and exert a deep influence on subsequent work in algebra, algebraic topology, and algebraic geometry in particular.

Perhaps the custom they had adopted of systematically using notations such as Hom(G, H), Ext(G, H), G $\bigcirc$ H (their notation for the tensor product), Ch(G) (dual group of G), Annih(L) [orthogonal in Ch(G) of a subgroup L of G], etc., for the various groups they defined in their 1942 paper, suggested to them that they were defining each time a kind of "function" which assigned a commutative group to an *arbitrary* commutative group (or to a pair of such groups) according to a fixed rule. Perhaps to avoid speaking of the "paradoxical" "set of all commutative groups," they coined the word "functor" for this kind of correspondence; of course their originality did not lie there, but in their crucial observation that a "functor" acted not only on groups, but also on *homomorphisms of groups*. To any homomorphism $u\colon G_1 \to G_2$ (resp. $v\colon H_1 \to H_2$), there corresponds a well-determined homomorphism

$$\operatorname{Hom}(u, 1)\colon \operatorname{Hom}(G_2, H) \to \operatorname{Hom}(G_1, H)$$

$$[\text{resp. Hom}(1, v)\colon \text{Hom}(G, H_1) \to \text{Hom}(G, H_2)]$$

defined by $g \mapsto g \circ u$ (resp. $f \mapsto v \circ f$).

In general, a *covariant functor* $F\colon G \mapsto F(G)$ associates to a homomorphism $u\colon G_1 \to G_2$ a homomorphism $F(u)\colon F(G_1) \to F(G_2)$, whereas a *contravariant functor* associates to u a homomorphism $F(u)\colon F(G_2) \to F(G_1)$. In addition,

$$\begin{cases} F(1_G) = 1_{F(G)}, & F(v \circ u) = F(v) \circ F(u) \text{ for covariant functors,} \\ F(1_G) = 1_{F(G)}, & F(v \circ u) = F(u) \circ F(v) \text{ for contravariant functors.} \end{cases} \tag{67}$$

It is in this new concept that they found the answer to what they described as their principal motivation: to give a precise formulation to the distinction between "natural" and "unnatural" isomorphisms, the typical "unnatural" isomorphisms being those between a finite dimensional vector space (resp. a finite commutative group) and its dual vector space (resp. its Pontrjagin dual), whereas there is a unique "natural" isomorphism of the space (resp. the group) onto its second dual (resp. its second Pontrjagin dual).

As we have noted above (§ 2), that distinction was of very recent origin, since until 1930 almost all mathematicians had been gleefully identifying vectors and linear forms; we must credit Eilenberg and Mac Lane with their foresight in guessing that such distinctions would become very significant in the mathematics of later days. More generally, they defined a *natural transformation* ξ of a functor F into a functor G: for each commutative group X, $\xi(X)$ is a homomorphism $F(X) \to G(X)$, such that for any homomorphism $u\colon X \to Y$, there is a commutative diagram

$$\begin{array}{ccc} F(X) & \xrightarrow{F(u)} & F(Y) \\ {\scriptstyle \xi(X)}\downarrow & & \downarrow{\scriptstyle \xi(Y)} \\ G(X) & \xrightarrow[G(u)]{} & G(Y) \end{array} \tag{68}$$

If in addition $\xi(X)$ is bijective for all X, then ξ is a *natural equivalence*. The typical example is the natural transformation ξ such that

$$\xi(X)\colon X \to \text{Ch}(\text{Ch}(X))$$

is the natural isomorphism of the Pontrjagin theory. They arrived at this definition by observing that when mathematicians spoke vaguely of a "natural" isomorphism, they had in mind the unexpressed idea that the isomorphism could be defined *generally* for any group without using any special property of the group.

In the 1945 paper in which they expanded their 1942 note [180] Eilenberg and Mac Lane introduced the word "category" to designate a type of mathematical object having a common structure, to which were *compulsorily* added what was later called "morphisms" between these objects. They gave a large number of examples of various categories and functors between categories. In particular, they showed that generalized chain complexes (with chain transformations as morphisms) may be considered as forming a category, and $C_\bullet \mapsto H_j(C_\bullet)$ is a covariant functor from that category to the category of commutative groups.

It was precisely for the category of chain complexes that in 1953 [186] Eilenberg and Mac Lane gave the first application of their ideas which was not what they themselves called "abstract nonsense" (purely verbal), namely, the method of *acyclic models*, that later became a useful tool in algebraic topology and homological algebra. (We shall discuss it below in section G.)

F. Chain Homotopies and Chain Equivalences

The homotopy operators of Lefschetz (chap. II, § 3) can be defined in a purely algebraic way for chain complexes; this was done by Eilenberg in his paper on singular homology [172] for *free* chain complexes, a restriction that is irrelevant and was abandoned in the Eilenberg–Steenrod book [189]. Let $C_\bullet = (C_j)$ and $C'_\bullet = (C'_j)$ be two generalized chain complexes, and let $u_\bullet = (u_j)$ and $v_\bullet = (v_j)$ be two chain transformations of $C_\bullet$ into $C'_\bullet$. A *chain homotopy* between $u_\bullet$ and $v_\bullet$ is a sequence $h_\bullet = (h_j)$ of homomorphisms

$$h_j\colon C_j \to C'_{j+1}$$

such that for all indices $j \in \mathbf{Z}$, and all $x_j \in C_j$,

$$v_j(x_j) - u_j(x_j) = \mathbf{b}_{j+1}(h_j(x_j)) + h_{j-1}(\mathbf{b}_j x_j).$$

From this it follows at once that for any cycle z_j, $v_j(z_j) - u_j(z_j)$ is a boundary, hence $H_\bullet(u_\bullet) = H_\bullet(v_\bullet)$ for the homology homomorphisms deduced from $u_\bullet$ and $v_\bullet$.

This leads to the concept of *chain equivalence* between two generalized chain complexes $C_\bullet$, $C'_\bullet$: it consists in two chain transformations $f_\bullet\colon C_\bullet \to C'_\bullet$, $g_\bullet\colon C'_\bullet \to C_\bullet$ such that the composite chain transformations $g_\bullet \circ f_\bullet\colon C_\bullet \to C_\bullet$, $f_\bullet \circ g_\bullet\colon C'_\bullet \to C'_\bullet$ are, respectively, chain homotopic to the *identity* transformations of $C_\bullet$ and $C'_\bullet$. When such transformations exist, the homology graded $\mathbf{Z}$-modules $H_\bullet(C_\bullet)$ and $H_\bullet(C'_\bullet)$ are therefore *isomorphic*.

If $C_\bullet = (C_j)_{j\geqslant 0}$ is a chain complex, by definition $\mathbf{b}_0 = 0$, hence 0-chains are 0-cycles, and the sequence $C_1 \xrightarrow[\mathbf{b}_1]{} C_0 \to H_0(C_\bullet) \to 0$ is exact. More generally, an *augmented chain complex* $\tilde{C}_\bullet = (C_j)$ is a generalized chain complex with $C_j = 0$ for $j \leqslant -2$; it is usually written

$$\cdots \to C_j \xrightarrow[\mathbf{b}_j]{} C_{j-1} \to \cdots \to C_1 \xrightarrow[\mathbf{b}_1]{} C_0 \xrightarrow[\varepsilon]{} A \to 0, \tag{69}$$

where ε is called the *augmentation* of the (unaugmented) chain complex $C_\bullet = (C_j)_{j\geqslant 0}$ (ker ε only *contains* the group of 0-boundaries, but is not necessarily equal to it); we write $\tilde{C}_\bullet = (C_\bullet, \varepsilon)$. The augmented complex $(C_\bullet, \varepsilon)$ (or $C_\bullet$ itself) is called *acyclic* if the sequence (69) is *exact*, which means that $H_j(C_\bullet) = 0$ for $j \geqslant 1$ and $H_0(C_\bullet) = A$.

Augmented *acyclic* complexes $(C_\bullet, \varepsilon)$ with $C_\bullet$ *free* were first introduced by

Hopf in 1945 in connection with his concept of cohomology of groups (see Part 3, chap. V, § 1,D). His main theorem concerned chain transformations between two augmented complexes $(C_\bullet, \varepsilon)$, $(C'_\bullet, \varepsilon')$, where $C_\bullet$ is *free* and $(C'_\bullet, \varepsilon')$ is *acyclic*; then

1. For *any* given homomorphism $\varphi: A \to A'$, there exist chain transformations $f_\bullet: C_\bullet \to C'_\bullet$ such that the diagram

$$\begin{array}{ccccccccccccc} \cdots \longrightarrow & C_j & \xrightarrow{\mathbf{b}_j} & C_{j-1} & \longrightarrow \cdots \longrightarrow & C_1 & \xrightarrow{\mathbf{b}_1} & C_0 & \xrightarrow{\varepsilon} & A & \longrightarrow 0 \\ & \downarrow{\scriptstyle f_j} & & \downarrow{\scriptstyle f_{j-1}} & & \downarrow{\scriptstyle f_1} & & \downarrow{\scriptstyle f_0} & & \downarrow{\scriptstyle \varphi} & \\ \cdots \longrightarrow & C'_j & \xrightarrow[\mathbf{b}_j]{} & C'_{j-1} & \longrightarrow \cdots \longrightarrow & C'_1 & \xrightarrow[\mathbf{b}_1]{} & C'_0 & \xrightarrow[\varepsilon']{} & A' & \longrightarrow 0 \end{array} \tag{70}$$

   is commutative.
2. Any two chain transformations $f_\bullet, g_\bullet$ having property 1 are *chain homotopic*.

To define $f_\bullet = (f_j)$ and the chain homotopy $h_\bullet = (h_j)$ his method (which in fact Hopf only applied when *both* complexes are free and acyclic) consisted in using induction on j to define the f_j and h_j. Since the C_j are *free*, it is enough to define f_j and h_j for each element e_j of a *basis* of C_j.

As ε' is surjective, take $f_0(e_0)$ such that $\varepsilon'(f_0(e_0)) = \varphi(\varepsilon(e_0))$. Then, if the f_i are defined for $i < j$ and

$$\mathbf{b}_{j-1}(f_{j-1}(\mathbf{b}_j e_j)) = f_{j-2}(\mathbf{b}_{j-1}(\mathbf{b}_j e_j)) = 0,$$

the acyclicity of $C'_\bullet$ shows that $f_{j-1}(\mathbf{b}_j e_j)$ is a boundary and it is thus possible to define $f_j(e_j)$ in such a way that $\mathbf{b}_j(f_j(e_j)) = f_{j-1}(\mathbf{b}_j e_j)$.

Similarly, if $g_\bullet$ is a second chain transformation such that $\varepsilon' \circ g_0 = \varphi \circ \varepsilon$, then $\varepsilon'(g_0(e_0) - f_0(e_0)) = 0$; hence, there is $h_0(e_0)$ such that $\mathbf{b}_1(h_0(e_0)) = g_0(e_0) - f_0(e_0)$. Then, if the h_i are defined for $i < j$, such that

$$g_{j-1}(e_{j-1}) - f_{j-1}(e_{j-1}) = \mathbf{b}_j(h_{j-1}(e_{j-1})) + h_{j-2}(\mathbf{b}_{j-1} e_{j-1}),$$

then

$$\mathbf{b}_j(g_j(e_j) - f_j(e_j)) = g_{j-1}(\mathbf{b}_j e_j) - f_{j-1}(\mathbf{b}_j e_j) = \mathbf{b}_j(h_{j-1}(\mathbf{b}_j e_j))$$

and by acyclicity of $C'_\bullet$, $h_j(e_j)$ may be defined such that

$$g_j(e_j) - f_j(e_j) = h_{j-1}(\mathbf{b}_j e_j) + \mathbf{b}_{j+1}(h_j(e_j)).$$

If $C_\bullet$ and $C'_\bullet$ are *both* free and acyclic and $\varphi: H_0(C_\bullet) \to H_0(C'_\bullet)$ is *bijective*, $C_\bullet$ and $C'_\bullet$ are then *chain equivalent* [in that case the isomorphisms $H_p(C_\bullet) \cong H_p(C'_\bullet)$ reduce to $0 = 0$, but Hopf wanted to establish uniqueness (up to chain equivalence) of free acyclic complexes with given initial module A (later called *free resolutions* of A; see Part 3, chap. V, § 1)].

In his 1942 book ([308], p. 158) Lefschetz had already used an inductive process similar to the preceding one in proofs of isomorphism of homology modules defined by different means. He essentially used the fact that in the chain complexes $C_\bullet = (C_j)$ he was considering each C_j has a basis consisting of euclidean simplices or continuous images of such simplices that were *contractible* as topological spaces and therefore had vanishing homology in dimensions $\geqslant 1$ (they were called *acyclic carriers*).

It is this topological method that Eilenberg and Mac Lane were able to translate into purely algebraic terms in their "method of acyclic models," which later had many applications.

G. Acyclic Models and the Eilenberg–Zilber Theorem

It is probably clearer to start with the Eilenberg–Zilber theorem, which was one of the first applications of the method of acyclic models [190]. The problem is the following: let X, Y be arbitrary topological spaces; to X, Y and their product $X \times Y$ are associated the singular chain complexes $S_\bullet(X)$, $S_\bullet(Y)$, and $S_\bullet(X \times Y)$ (§ 2); the tensor product $S_\bullet(X) \otimes S_\bullet(Y)$, which is again a chain complex [formula (47)] can also be formed. The theorem is that there is a *chain equivalence* between $S_\bullet(X \times Y)$ and $S_\bullet(X) \otimes S_\bullet(Y)$, hence their homology modules are isomorphic.

Each **Z**-module $S_n(X \times Y)$ is free, having a basis formed of the pairs (f, g), where $f: \bar{\Delta}_n \to X$ and $g: \bar{\Delta}_n \to Y$ are arbitrary continuous maps. Similarly, $S_p(X) \otimes S_q(Y)$ is a free **Z**-module having as basis the tensor products $u \otimes v$, where $u: \bar{\Delta}_p \to X$ and $v: \bar{\Delta}_q \to Y$ are arbitrary continuous maps. There are canonical augmentations $\varepsilon: S_0(X \times Y) \to \mathbf{Z}$ and $\varepsilon': S_0(X) \otimes S_0(Y) \to \mathbf{Z}$, defined by $\varepsilon(x, y) = 1$ and $\varepsilon'(x \otimes y) = 1$ for every pair $(x, y) \in X \times Y$ [where x (resp. y) is identified with the map $\Delta_0 \to X$ (resp. $\Delta_0 \to Y$) taking the value x (resp. y)].

First define $\varphi: \mathbf{Z} \to \mathbf{Z}$ as the identity. Let $\varphi_0(x, y) = x \otimes y$. Suppose the maps

$$\varphi_{i,X,Y}: S_i(X \times Y) \to (S_\bullet(X) \otimes S_\bullet(Y))_i = \bigoplus_{h+k=i} (S_h(X) \otimes S_k(Y))$$

have been defined, for $i < j$, for *every pair* (X, Y) of spaces; assume that they commute with the boundary homomorphisms, and that if $\alpha: X \to X'$, $\beta: Y \to Y'$ are any two continuous maps,

$$\varphi_{i,X',Y'} \circ (S_\bullet(\alpha, \beta)) = (S_\bullet(\alpha) \otimes S_\bullet(\beta)) \circ \varphi_{i,X,Y}. \tag{71}$$

As any element of $S_j(X \times Y)$ is a sum of *basis elements* written

$$S_\bullet(f, g)(1_{\Delta_j}, 1_{\Delta_j}),$$

it is only necessary to define $\varphi_{j,\Delta_j,\Delta_j}(1_{\Delta_j}, 1_{\Delta_j})$, and $\varphi_{j,X,Y}$ will be defined for all spaces X, Y by (71). But for $X = Y = \Delta_j$, observe that Δ_j is *contractible*, and it follows from the Künneth theorem for chain complexes that $H_q(S_\bullet(\Delta_j) \otimes S_\bullet(\Delta_j)) = 0$ for $q \geqslant 1$. We can therefore apply the Hopf procedure (described in § 5,F) to define $\varphi_{j,\Delta_j,\Delta_j}$, and it is then easy to verify that (71) is still valid when i is replaced by j; the existence of chain transformations of $S_\bullet(X \times Y)$ into $S_\bullet(X) \otimes S_\bullet(Y)$ is thus proved.

Similarly, to prove that any two such chain transformations are chain homotopic, assume that the homomorphisms $h_{i,X,Y}$ have been defined for $i < j$ and for every pair (X, Y) of spaces, and that

$$h_{i,X',Y'} \circ (S_{\bullet}(\alpha, \beta)) = (S_{\bullet}(\alpha) \otimes S_{\bullet}(\beta)) \circ h_{i,X,Y} \tag{72}$$

with the same notations as above. Then the Hopf procedure is applied to the case $X = Y = \Delta_j$, and $h_{j,X,Y}$ is defined and satisfies (72).

In the opposite direction, if $\psi_0(x \otimes y) = (x, y)$, the same arguments show that the restriction of $\psi_{j,X,Y}$ to each $S_p(X) \otimes S_q(Y)$ must be defined for $p + q = j$. The basis elements of $S_p(X) \otimes S_q(Y)$ are written $(S_{\bullet}(u) \otimes S_{\bullet}(v))(1_{\Delta_p} \otimes 1_{\Delta_q})$; it is therefore enough to define the $\psi_{j,X,Y}$ for $X = \Delta_p$, $Y = \Delta_q$, and as $\Delta_p \times \Delta_q$ is contractible, the Hopf procedure applies again for the definitions of the chain transformation and of the chain homotopy.

To understand the theorem of acyclic models, we first express (71) and (72) in functorial language: consider the *category of pairs* (X, Y) of topological spaces, where the *morphisms* are the pairs (α, β) of continuous maps; the maps $(X, Y) \mapsto S_{\bullet}(X \times Y)$, $(\alpha, \beta) \mapsto S_{\bullet}((\alpha, \beta))$ define a *functor* G from that category to the category of chain complexes, and the maps $(X, Y) \mapsto S_{\bullet}(X) \otimes S_{\bullet}(Y)$, $(\alpha, \beta) \mapsto S_{\bullet}(\alpha) \otimes S_{\bullet}(\beta)$ another such functor G′; finally (71) and (72) mean that $\varphi_{X,Y} = (\varphi_{i,X,Y})$ and $h_{X,Y} = (h_{i,X,Y})$ are *natural transformations* of G into G′.

To generalize, consider an *arbitrary* category $\boldsymbol{C}$, and covariant functors G, G′ from $\boldsymbol{C}$ to the category of chain complexes; the role of the pairs (Δ_p, Δ_q) is played by a set $\mathfrak{M}$ of objects of $\boldsymbol{C}$ called *models*; the functor G is called *acyclic* if $H_q(G(M)) = 0$ for $q \geqslant 1$ and for each object $M \in \mathfrak{M}$; G is *free* if, for each $q \geqslant 0$, there is a family of objects $(M_q^{(i)})_{i \in I_q}$ belonging to $\mathfrak{M}$ and such that, for any object X of $\boldsymbol{C}$, the **Z**-module $(G(X))_q$ has a *basis* consisting of elements $G(f)(g_i)$, where g_i is an element of $(G(M_q^{(i)}))_q$ for some $i \in I_q$, and f runs through the set of morphisms $\text{Mor}(M_q^{(i)}, X)$. [For the functor $S_{\bullet}(X \times Y)$ in the Eilenberg–Zilber theorem, the family $(M_q^{(i)})$ is the singular q-simplex $(1_{\Delta_q}, 1_{\Delta_q})$ in the space $\Delta_q \times \Delta_q$. For the functor $S_{\bullet}(X) \otimes S_{\bullet}(Y)$, $(M_q^{(i)})$ consists in all pairs (Δ_h, Δ_k) for $h + k = q$, and g_i is the tensor product $1_{\Delta_h} \otimes 1_{\Delta_k}$.]

The *theorem of acyclic models* then says that if G and G′ are two functors from $\boldsymbol{C}$ to the category of chain complexes, such that G *is free and* G′ *acyclic*, then any natural transformation of functors

$$\zeta: H_0 \circ G \to H_0 \circ G'$$

is such that

$$\zeta(X)(H_0(G(X))) = H_0(\xi(X)(G(X))),$$

where $\xi: G \to G'$ is a natural transformation of functors; furthermore, any two such natural transformations ξ, ξ' for which $\zeta = \zeta'$ are such that $\xi(X)$ and $\xi'(X)$ are chain homotopic for all X. When both G and G′ are free and acyclic (as in the case of the Eilenberg–Zilber theorem), if $H_0(G(X)) \to H_0(G'(X))$ is an isomorphism for all X in $\boldsymbol{C}$, then the chain transformation $G(X) \to G'(X)$ is a chain equivalence.

As an example of substitution of the theorem of acyclic models for earlier similar arguments restricted to particular cases, consider the proof of the isomorphism of the homology $H_\bullet(K; \mathbf{Z})$ of a finite euclidean simplicial complex K and the homology $H_\bullet(K'; \mathbf{Z})$ of its "first derived" complex K'. It is immediate to check that the chain complexes $C_\bullet(K)$ and $C_\bullet(K')$ of *alternating chains* are both *free* and *acyclic*; to establish a chain equivalence between them, one has only to define an isomorphism $H_0(C_\bullet(K)) \cong H_0(C_\bullet(K'))$. First suppose K connected; then any vertex of K may be joined to a particular vertex x_0 by a broken line consisting of edges of simplices of K; this proves that every vertex is homologous to x_0, hence $H_0(C_\bullet(K))$ is the module generated by the class of x_0. The same is true for $H_0(C_\bullet(K'))$, and the isomorphism simply assigns to the class of x_0 in $H_0(C_\bullet(K))$ the class of x_0 in $H_0(C_\bullet(K'))$. If K is not connected, the argument is repeated for each connected component.

The interest of this method is that it can be transferred to arbitrary *combinatorial* complexes, by "mimicking" simplicial complexes. A combinatorial complex K is *connected* if any vertex x can be "joined" to a fixed vertex x_0 by a sequence $(x_0, x_1, \ldots, x_p)$, where $x_p = x$, and any consecutive pair (x_j, x_{j+1}) is a 1-simplex; if K is connected, $H_0(C_\bullet(K))$ is again generated by the class of x_0. It is then necessary to prove that if K is connected, so is K'; a vertex of K' is a simplex S of K, and there is a totally ordered set $(\{x\}, S_1, \ldots, S_p)$ of simplices of K such that x is a vertex of K and $S_p = S$; as this ordered set is a simplex of K', this shows S and $\{x\}$, vertices of K', may be "joined" by 1-simplices of K'; one then similarly "joins" $\{x\}$ and $\{x_0\}$ in K' using a chain of 1-simplices of K joining x and x_0. The end of the argument is then the same as above, $C_\bullet(K)$ and $C_\bullet(K')$ being trivially free and acyclic; the extension to non-connected complexes is easy.

H. Applications to Homology and Cohomology of Spaces: Cross Products and Slant Products

Any result on homology or cohomology of chain complexes immediately yields corresponding results for *singular* homology and cohomology. For instance, if X is any space and A is a subspace of X, the chain complex $S_\bullet(X)/S_\bullet(A)$ is free; the "universal coefficient theorem" for homology of chain complexes [formula (46)] gives, for any commutative group G, the split exact sequence for singular homology

$$0 \to H_p(X, A; \mathbf{Z}) \otimes G \to H_p(X, A; G) \to \operatorname{Tor}(H_{p-1}(X, A; \mathbf{Z}), G) \to 0 \quad (73)$$

and the "universal coefficient theorem" for cohomology of chain complexes [formula (66)] gives the split exact sequence for singular cohomology

$$0 \to \operatorname{Ext}(H_{p-1}(X, A; \mathbf{Z}), G) \to H^p(X, A; G) \to \operatorname{Hom}(H_p(X, A; \mathbf{Z}), G) \to 0. \quad (74)$$

If X and Y are arbitrary spaces, it follows from the Eilenberg–Zilber theorem that there is a natural isomorphism of graded $\mathbf{Z}$-modules

$$H_\bullet(X \times Y; \mathbf{Z}) \cong H_\bullet(S_\bullet(X) \otimes S_\bullet(Y)) \quad (75)$$

and the Künneth formula for chain complexes [formula (50)] gives a split exact sequence

$$0 \to \bigoplus_{0 \leqslant p \leqslant r} (H_p(X; \mathbf{Z}) \otimes H_{r-p}(Y; \mathbf{Z})) \to H_r(X \times Y; \mathbf{Z})$$

$$\to \bigoplus_{0 \leqslant q \leqslant r-1} \mathrm{Tor}(H_q(X; \mathbf{Z}), H_{r-q-1}(Y; \mathbf{Z})) \to 0 \qquad (76)$$

(Künneth formula for singular homology). This gives rise to the definition, for $z \in H_p(X; \mathbf{Z})$ and $z' \in H_q(Y; \mathbf{Z})$, of the *homology cross product* $z \times z' \in H_{p+q}(X \times Y; \mathbf{Z})$, as the image of $z \otimes z'$ by the first homomorphism of (76) (Lefschetz [308], p. 173).

On the other hand, there is a cochain transformation of cochain complexes

$$\mathrm{Hom}(S_\bullet(X), G) \otimes \mathrm{Hom}(S_\bullet(Y), G') \to \mathrm{Hom}(S_\bullet(X) \otimes S_\bullet(Y), G \otimes G') \qquad (77)$$

for any two commutative groups G, G′; the Eilenberg–Zilber chain equivalence

$$S_\bullet(X \times Y) \to S_\bullet(X) \otimes S_\bullet(Y)$$

gives a cochain equivalence

$$\mathrm{Hom}(S_\bullet(X) \otimes S_\bullet(Y), G \otimes G') \to \mathrm{Hom}(S_\bullet(X \times Y), G \otimes G') \qquad (78)$$

and, composing (77) and (78), a cochain transformation is obtained

$$\mathrm{Hom}(S_\bullet(X), G) \otimes \mathrm{Hom}(S_\bullet(Y), G') \to \mathrm{Hom}(S_\bullet(X \times Y), G \otimes G')$$

that yields a functorial homomorphism of singular cohomology groups

$$H^p(X; G) \otimes H^q(Y; G') \to H^{p+q}(X \times Y; G \otimes G'). \qquad (79)$$

In particular, if the additive group of a commutative ring Λ is taken for G and G′, and the natural homomorphism $\Lambda \otimes_{\mathbf{Z}} \Lambda \to \Lambda$ is applied, then there arise homomorphisms

$$H^p(X; \Lambda) \otimes H^q(Y; \Lambda) \to H^{p+q}(X \times Y; \Lambda), \qquad (80)$$

and for $u \in H^p(X; \Lambda)$ and $v \in H^q(Y; \Lambda)$, the image of $u \otimes v$ is written $u \times v$, the *cohomology cross product* of u and v.

For X = Y, this gives a new definition of the cup product, as first observed by Lefschetz ([308], p. 190): the *diagonal mapping* $\delta: x \mapsto (x, x)$ of X into X × X, yields for cohomology a homomorphism

$$\delta^*: H^\bullet(X \times X; \Lambda) \to H^\bullet(X; \Lambda)$$

hence

$$u \smile v = \delta^*(u \times v). \qquad (81)$$

The derivation of this equality is a good example of the method of acyclic models: it can be applied to the free and acyclic functors $S_\bullet(X)$ and $S_\bullet(X) \otimes S_\bullet(X)$, with respective models (Δ_p) and $((\Delta_p, \Delta_q))$; the map $x \mapsto x \otimes x$ of $S_0(X)$ into $S_0(X) \otimes S_0(X)$ can therefore be extended to a chain transformation

$$S_\bullet(X) \to S_\bullet(X) \otimes S_\bullet(X), \tag{82}$$

and any two such chain transformations are chain homotopic, hence yield the same mapping $H_\bullet(X; \mathbf{Z}) \to H_\bullet(S_\bullet(X) \otimes S_\bullet(X))$, from which, as above, functorial homomorphisms can be deduced, *independent* of the choice of the chain transformation (82),

$$H^p(X; \Lambda) \otimes H^q(X; \Lambda) \to H^{p+q}(X; \Lambda). \tag{83}$$

Now we can easily verify that we may take for (82) the map which to each simplex $(x_0, \ldots, x_p)$ associates $\sum_{0 \leqslant j \leqslant p} (x_0, \ldots, x_j) \otimes (x_j, \ldots, x_p)$, and then the corresponding map (83) is just the cup product as defined by Čech and Whitney (§4). But we may also define a chain transformation (82) as the *composite*

$$S_\bullet(X) \xrightarrow{S_\bullet(\delta)} S_\bullet(X \times X) \xrightarrow{\tau} S_\bullet(X) \otimes S_\bullet(X),$$

where τ is one of the Eilenberg–Zilber chain equivalences; then the map (83) is given by the right-hand side of (81), hence our conclusion.

Still another type of "product" was introduced in 1953 by Steenrod [453] in connection with his previous definition of "reduced powers" (Part 3, chap. IV, §2,C). Again, it can first be defined for arbitrary *free chain complexes* $C'_\bullet$, $C''_\bullet$, and arbitrary commutative groups G, H. Consider an element $u \in (\mathrm{Hom}(C'_\bullet \otimes C''_\bullet, G))_n$ and an element $c' \in C'_q \otimes H$, for an index $q \leqslant n$. Then $c' = \sum_i \sigma'_i \otimes h_i$, where $h_i \in H$ and (σ'_i) is a basis of C'_q; the (right) *slant product* u/c' is then an element of $(\mathrm{Hom}(C''_\bullet, G \otimes H))_{n-q}$ defined by

$$(u/c')(\tau'') = \sum_i u(\sigma'_i \otimes \tau'') \otimes h_i. \tag{84}$$

The relation

$$\mathbf{d}_{n-q}(u/c') = (\mathbf{d}_n u/c') - (-1)^{n-q}(u/\mathbf{b}_q c') \tag{85}$$

shows that the slant product u/c' of a cocycle u and a cycle c' is a cocycle, and if in addition u is a coboundary or c' is a boundary, u/c' is a coboundary. Therefore a (right) *slant product*

$$H^n(C'_\bullet \otimes C''_\bullet; G) \times H_q(C'_\bullet; H) \to H^{n-q}(C''_\bullet; G \otimes H) \tag{86}$$

is defined, and written $(w, z) \mapsto w/z$. This procedure favors the second factor in the product $C'_\bullet \otimes C''_\bullet$; similarly, a "left" slant product may be defined. For $c'' \in C''_q \otimes H$, define $c''\backslash u$ by

$$(c''\backslash u)(\tau') = \sum_i u(\tau' \otimes \sigma''_i) \otimes h_i \tag{87}$$

if (σ''_i) is a basis of C''_q and $c'' = \sum_i \sigma''_i \otimes h_i$, thus obtaining a (left) slant product written $(w, z) \mapsto z \backslash w$

$$H^n(C'_\bullet \otimes C''_\bullet; G) \times H_q(C''_\bullet; H) \to H^{n-q}(C'_\bullet; G \otimes H). \tag{88}$$

Using the Eilenberg–Zilber theorem, these definitions apply to the chain complexes $C'_\bullet = S_\bullet(X)$, $C''_\bullet = S_\bullet(Y)$ of singular chains in two spaces X, Y, to give *slant products*

$$H^n(X \times Y; G) \times H_q(Y; H) \to H^{n-q}(X; G \otimes H),$$
$$H^n(X \times Y; G) \times H_q(X; H) \to H^{n-q}(Y; G \otimes H). \qquad (89)$$

§ 6. Identifications and Axiomatizations

The different definitions of homology and cohomology groups soon led to the conviction that at least for some types of spaces they must yield *isomorphic* groups. This was gradually proved for larger and larger classes of spaces, all of them containing the class of finite euclidean simplicial complexes for which homology had been defined in the first place (chap. II). Eilenberg and Steenrod ([188], [189]) discovered that this coincidence of homology theories on this "minimal" class was due to the fact that they all shared common properties. This led them to an *axiomatic definition* of homology and cohomology that was later to acquire more importance with the introduction of "extraordinary" cohomology theories (Part 3, chap. VII). In another direction, Leray [313] was able, for locally compact spaces, to subsume all special processes used in the definition of the cohomology theories under a general method; this was the first germ of sheaf cohomology, which he introduced shortly afterward (§ 7).

A. Comparison of Vietoris, Čech, and Alexander–Spanier Theories

In his 1942 book ([308], p. 273) Lefschetz defined, for a *compact metric space* X, a natural isomorphism of the Čech homology groups on the Vietoris homology groups. His idea was to consider, for each nonempty open set U of X, a point $\varphi(U) \in U$; if $U_0, U_1, \ldots, U_p$ are open sets with nonempty intersection [hence form a Čech p-simplex (§ 2)], the diameter of the set $\{\varphi(U_0), \varphi(U_1), \ldots, \varphi(U_p)\}$ is at most twice the supremum of the diameters of the sets U_j. If $(\mathcal{U}_n)$ is a sequence of finite open coverings of X such that $\mathcal{U}_{n+1}$ is finer than $\mathcal{U}_n$ for each n, and the mesh of $\mathcal{U}_n$ tends to 0, we have seen (§ 2) that the Čech homology groups $\check{H}_p(X; G)$, for any commutative group G, can be defined as the inverse limits of the homology groups $H_p(N(\mathcal{U}_n); G)$. Lefschetz proved, in a rather complicated way,* that for every n, φ defines a simplicial mapping φ_n of $N(\mathcal{U}_n)$ into the complex of Vietoris chains and that the corresponding maps $(\varphi_n)_*$ for homology groups have an inverse limit that is an isomorphism of the Čech homology $\check{H}_\bullet(X; G)$ onto the Vietoris homology with coefficients in G.†

* He does not seem to have taken into account the fact that in general φ is not injective.

† In the same book, Lefschetz, a little later ([308], p. 285), shows that, for compact metric spaces, there is also a natural isomorphism of the Čech cohomology onto the Alexander cohomology, but his definition of the latter is not the same as the one used by Spanier.

In later years this identification was reconsidered in more general contexts. First the definition of Vietoris homology was simultaneously extended to all compact spaces by Spanier [438], and to all topological spaces by Hurewicz, Dugundji, and Dowker [259]. Consider an arbitrary (finite or infinite) open covering $\mathscr{U} = (U_\alpha)$ of a space X; this defines a *combinatorial complex* $K_{\mathscr{U}}$ consisting of the combinatorial simplices (finite sequences of points of X) *contained in one of the sets* U_α. If $\mathscr{U}'$ is an open covering of X, finer than $\mathscr{U}$, then $K_{\mathscr{U}'}$ is a subcomplex of $K_{\mathscr{U}}$ and there is a natural chain transformation

$$C''_\bullet(K_{\mathscr{U}'}) \to C''_\bullet(K_{\mathscr{U}}) \tag{89}$$

between the corresponding chain complexes of *ordered chains* (with coefficients in an arbitrary commutative group G) (chap. II, § 2). The $C''_\bullet(K_{\mathscr{U}})$ therefore form an inverse system indexed by the directed set of all open coverings of X; the homology graded groups $H_\bullet(C''_\bullet(K_{\mathscr{U}}))$ also form an inverse system, and by definition the inverse limit of that system of graded groups is the graded generalized *Vietoris homology* of X with coefficients in G. When X is a compact metric space, these groups (for $G = \mathbf{Z}$ or $G = \mathbf{F}_2$) are naturally isomorphic to those initially defined by Vietoris (§ 2).

The interesting feature of that definition is that it immediately provided a connection with the Alexander–Spanier cohomology (§ 3). Indeed, consider the *cochain complex* $(C^p(K_{\mathscr{U}}; G))$ where $C^p(K_{\mathscr{U}}; G) = \mathrm{Hom}(C''_p(K_{\mathscr{U}}; \mathbf{Z}), G)$, which form a *direct system*; a *cochain equivalence*

$$\bar{C}^\bullet(X; G) \to \varinjlim C^\bullet(K_{\mathscr{U}}; G) \tag{90}$$

can be defined between the Alexander–Spanier cochain complex (§ 3) and the direct limit. First define for each $p \geqslant 0$ a homomorphism

$$\lambda_p\colon C^p(X; G) \to \varinjlim C^p(K_{\mathscr{U}}; G) \tag{91}$$

in the following way: for any $f \in C^p(X; G)$ and any open covering $\mathscr{U} = (U_\alpha)$ of X, consider the restrictions $f \mid U_\alpha^{p+1}$, which are *p*-cochains of $C^p(K_{\mathscr{U}}; G)$; when $\mathscr{U}$ varies, all these cochains have the same image in $\varinjlim C^p(K_{\mathscr{U}}; G)$, which is by definition $\lambda_p(f)$. The homomorphism λ_p is *surjective*: if $\tilde{u} \in \varinjlim C^p(K_{\mathscr{U}}; G)$, there is a covering $\mathscr{U} = (U_\alpha)$ and an element $u \in C^p(K_{\mathscr{U}}; G)$ having $\tilde{u}$ as its image in the direct limit; taking $f \in C^p(X; G)$ such that

$$f(x_0, \ldots, x_p) = \begin{cases} u(x_0, x_1, \ldots, x_p) & \text{if } (x_0, x_1, \ldots, x_p) \in U_\alpha^{p+1} \text{ for some } \alpha, \\ 0 & \text{otherwise,} \end{cases}$$

then $\lambda_p(f) = \tilde{u}$. Finally, the *kernel* of λ_p consists in the $f \in C^p(X; G)$ for which there is a covering $\mathscr{U} = (U_\alpha)$ such that $f(x_0, x_1, \ldots, x_p) = 0$ for all systems $(x_0, x_1, \ldots, x_p)$ the elements of which are all in some U_α. This means (§ 3) that $f \in C_0^p(X; G)$, and passage to the quotient in (91) yields the requested isomorphism; it is easy to check that it commutes with the coboundary operator.

The introduction of the combinatorial complexes $K_{\mathscr{U}}$ enabled Dowker, by a very ingenious argument, to identify for an *arbitrary* topological space X both the *Vietoris and Čech homologies* and the *Alexander–Spanier and Čech cohomologies* (based on arbitrary open coverings) [146]. He first proved a general result concerning two *arbitrary sets* (with no topology) X, Y, and a "relation" between X and Y, i.e., an *arbitrary* nonempty subset R of the product $X \times Y$. He defined *combinatorial complexes* L_X on X, M_Y on Y: the p-simplices of L_X (resp. M_Y) are the sets $\{x_0, x_1, \ldots, x_p\}$ (resp. $\{y_0, y_1, \ldots, y_p\}$) of p elements belonging to some subset $R^{-1}(y)$ for an $y \in Y$ [resp. to some subset $R(x)$ for an $x \in X$]. Now consider the "*first derived*" combinatorial complex M'_Y, and recall (chap. II, §2) that the vertices of M'_Y are the simplices of M_Y; each one therefore is a subset $\sigma = \{y_0, y_1, \ldots, y_p\}$ of *some* $R(x)$, and a mapping of the set of vertices of M'_Y into the set of vertices of L_X is defined by assigning to σ an element x for which $\sigma \subset R(x)$. There is therefore the possibility of defining *several* such mappings, but it may be shown that all these mappings are *simplicial*, and any two of them yield the *same* homomorphism $H_\bullet(M'_Y) \to H_\bullet(L_X)$ in homology, or $H^\bullet(L_X) \to H^\bullet(M'_Y)$ in cohomology (with any commutative group as group of coefficients). As there is a natural isomorphism $H_\bullet(M_Y) \cong H_\bullet(M'_Y)$ [resp. $H^\bullet(M'_Y) \cong H^\bullet(M_Y)$] (§5,G), this thus defines natural homomorphisms $H_\bullet(M_Y) \to H_\bullet(L_X)$ and $H^\bullet(L_X) \to H^\bullet(M_Y)$. If the procedure is repeated, exchanging the roles of X and Y finally establishes natural *isomorphisms* $H_\bullet(L_X) \cong H_\bullet(M_Y)$ and $H^\bullet(L_X) \cong H^\bullet(M_Y)$.

This general result can now be applied to the case in which X is a topological space, and $Y = \mathscr{U} = (U_\alpha)$ is an *open covering* of X. The relation R is simply $x \in U_\alpha$; therefore, M_Y is simply the *nerve* $N(\mathscr{U})$ of $\mathscr{U}$ (§2), whereas L_X is the combinatorial complex $K_{\mathscr{U}}$ defined above. Applying Dowker's general abstract theorem, and then taking inverse limits (on the directed set of open coverings of X) for homology and direct limits for cohomology, one finally obtains the natural isomorphisms announced above.

This result may easily be extended to relative homology and relative cohomology. Instead of taking limits over the directed set of all open coverings of X, one may also consider other directed sets of open coverings (for instance, finite open coverings).

B. The Axiomatic Theory of Homology and Cohomology

The various homology and cohomology theories appear as complicated machines, the end product of which is an assignment of a graded group to a topological space, through a series of processes which look so arbitrary that one wonders why they succeed at all. In a remarkable book [189] (announced in [188]) Eilenberg and Steenrod endeavored to break through this maze of unpleasant mathematics by adopting a totally different viewpoint, concentrating on *properties* of these end products rather than on the various methods devised to get them. This is the *axiomatic theory of homology* (*and cohomology*).

Their first original idea in the selection of their axioms was to think of the assignment of homology groups (with given coefficients) to some kinds of topological spaces as a *covariant functor* $X \mapsto H_\bullet(X)$ from the category $\boldsymbol{T}$ of topological spaces (or a subcategory $\boldsymbol{T}'$ of $\boldsymbol{T}$) to the category of graded commutative groups; in other words, to each morphism $u: X \to Y$ in $\boldsymbol{T}$ or $\boldsymbol{T}'$ (morphisms being arbitrary continuous maps for $\boldsymbol{T}$, sometimes restricted by additional conditions when $\boldsymbol{T}' \neq \boldsymbol{T}$) must correspond a homomorphism $H_{\bullet}(u): H_{\bullet}(X) \to H_{\bullet}(Y)$ of graded commutative groups (§ 2).

Another noteworthy feature of the book is that they included *relative homology* (resp. *cohomology*) by introducing the category $\boldsymbol{T}_1$ of *pairs* (X, A) consisting of a space X and one of its subspaces A (when $\boldsymbol{T}$ is replaced by a subcategory $\boldsymbol{T}'$, some restrictions may be imposed on A). A *morphism* $f: (X, A) \to (Y, B)$ of $\boldsymbol{T}_1$ is then a continuous map $f: X \to Y$ such that $f(A) \subset B$; $\boldsymbol{T}$ is made a subcategory of $\boldsymbol{T}_1$ by identifying X with the pair $(X, \varnothing)$.

These basic notions allowed them to formulate, in addition to the conditions (67) satisfied by *all* functors, their first axiom for homology, the *exactness axiom*: it calls for the existence of a *natural transformation* $\partial_{\bullet}: H_{\bullet} \to H_{\bullet}$ (§ 5,E) such that (a) $\partial_{\bullet}(X, A) = (\partial_q(X, A))_{q \geqslant 0}$, where $\partial_q(X, A)$ is a homomorphism $H_q(X, A) \to H_{q-1}(A)$; (b) for any pair (X, A), the sequence

$$\cdots \to H_q(A) \xrightarrow{H_q(i)} H_q(X) \xrightarrow{H_q(j)} H_q(X, A) \xrightarrow{\partial_q(X, A)} H_{q-1}(A)$$
$$\to \cdots \to H_0(X, A) \to 0 \tag{92}$$

is exact, i being the injection $A \to X$ and j being the injection $(X, \varnothing) \to (X, A)$; one says (92) is the *exact homology sequence* of the pair (X, A).

Three more properties are taken as axioms; two are quite familiar:

1. The *homotopy axiom* (the old idea that two cycles that can be "deformed into one another" are homologous, cf. chap. I): if f and g are two homotopic mappings $(X, A) \to (Y, B)$, then $H_{\bullet}(f) = H_{\bullet}(g)$.
2. The *dimension axiom* stating that if X is reduced to a single point P (sometimes written X = pt.), $H_q(P) = 0$ for $q \geqslant 1$; the group $G = H_0(P)$, which is independent of the choice of P (up to isomorphism), is called the *coefficient group* of the homology theory.

For any space X in $\boldsymbol{T}$, the unique map $f: X \to P$ defines a homomorphism $H_0(f): H_0(X) \to G$; the kernel $\tilde{H}_0(X)$ of that map is called the *reduced* 0-dimensional homology group of X; for any map $g: X \to Y$ in $\boldsymbol{T}$, $H_0(g)$ maps $\tilde{H}_0(X)$ into $\tilde{H}_0(Y)$. In the homology sequence (92), $H_0(X)$ and $H_0(A)$ may be replaced by $\tilde{H}_0(X)$ and $\tilde{H}_0(A)$, respectively, without disturbing the exactness of the sequence.

The last axiom is essentially concerned with *relative* homology, and can be traced back to Lefschetz's idea that for a euclidean simplicial complex K and a subcomplex L, the homology of K modulo L depends only on the space $K - L$ (chap. II, § 6), and therefore would not change if the simplices of L that do not meet $\overline{K - L}$ were *deleted* from K and L. But nobody before Eilenberg and Steenrod had given a general expression to that property, which they named:

3. The *excision axiom*, saying that if an open set U is such that its *closure* $\bar{U}$ in X is contained in the *interior* of a subset A of X, then the inclu-

sion map $j\colon (X - U, A - U) \to (X, A)$ yields an *isomorphism* $H_\bullet(j)\colon H_\bullet(X - U, A - U) \to H_\bullet(X, A)$.

There are corresponding axioms for *cohomology* obtained by the usual process of "reversing arrows"; we only formulate explicitly the exactness axiom, which says that for every pair (X, A) the *exact cohomology sequence* is

$$\cdots \to H^{q-1}(A) \to H^q(X, A) \to H^q(X) \to H^q(A) \to \cdots. \tag{93}$$

The *reduced* 0-dimensional cohomology group $\tilde{H}^0(X)$ is the cokernel of the homomorphism $G \to H^0(X)$; the sequence (93) remains exact when $H^0(X)$ is replaced by $\tilde{H}^0(X)$ and $H^0(A)$ by $\tilde{H}^0(A)$.

By applying these axioms in a convenient way Eilenberg and Steenrod first generalized the exact sequence (92) to the case of a *triple* (X, A, B), where $B \subset A \subset X$ and the maps $i\colon (A, B) \to (X, B)$ and $j\colon (X, B) \to (X, A)$ are morphisms of $\boldsymbol{T}_1$; there is then an exact sequence

$$\cdots \to H_q(A, B) \xrightarrow{H_q(i)} H_q(X, B) \xrightarrow{H_q(j)} H_q(X, A) \xrightarrow{\bar{\partial}_q} H_{q-1}(A, B) \to \cdots \tag{94}$$

where $\bar{\partial}_q = H_{q-1}(j'') \circ \partial_q$, j'' being the injection $A \to (A, B)$ and ∂_q being the map $H_q(X, A) \to H_{q-1}(A)$ of (92).

The homotopy axiom leads to the notion of *homotopically equivalent pairs* (X, A), (Y, B) such that there exist morphisms $f\colon (X, A) \to (Y, B)$ and $g\colon (Y, B) \to (X, A)$ of $\boldsymbol{T}_1$ for which $g \circ f$ is homotopic to the identity of (X, A) and $f \circ g$ is homotopic to the identity of (Y, B); this implies that $H_\bullet(f)$ and $H_\bullet(g)$ are *isomorphisms* inverse to each other.

An example of homotopically equivalent pairs that frequently occurs is when a pair (X′, A′) contained in (X, A) is a *strong deformation retract* of (X, A). This means that there exists a homotopy $h\colon X \times I \to X$ such that $h(x, 0) = x$ for all $x \in X$, $h(x, 1) \in X'$ for $x \in X$, and $h(x, 1) \in A'$ for $x \in A$, and, finally, $h(x, t) = x$ for all $x \in X'$ and all $t \in I$ (cf. Part 3, chap. II, §2,D). Then if $i\colon (X', A') \to (X, A)$ is the natural inclusion and $r\colon (X, A) \to (X', A')$ is defined by $r(x) = h(x, 1)$, the two morphisms i and r define a homotopy equivalence between (X, A) and (X′, A′). For instance, if X is a finite euclidean simplicial complex, then any subcomplex A of X is a strong deformation retract of some neighborhood of A; any compact submanifold Y of a C^0 manifold X is a strong deformation retract of a fundamental sequence of open neighborhoods of Y in X.

The excision axiom introduces in a space X the notion of *excisive couples* (X_1, X_2) of subspaces of X by the condition that the morphism $(X_1, X_1 \cap X_2) \to (X_1 \cup X_2, X_2)$ induces an *isomorphism* of homology groups. The axiom says that $(A, X - U)$ is an excisive couple. For any excisive couple (X_1, X_2) Eilenberg and Steenrod showed that the sequence

$$\begin{aligned} \cdots \to H_q(X_1, X_1 \cap X_2) \to H_q(X, X_2) \to H_q(X, X_1 \cup X_2) \\ \to H_{q-1}(X_1, X_1 \cap X_2) \to \cdots \end{aligned} \tag{95}$$

is exact. The most important case is when $X = X_1 \cup X_2$; there is then an exact sequence

$$\cdots \to H_q(X_1 \cap X_2) \to H_q(X_1) \oplus H_q(X_2) \to H_q(X) \to H_{q-1}(X_1 \cap X_2) \to \cdots, \tag{96}$$

very useful for computations of homology groups, called the *Mayer–Vietoris* exact sequence, since from it one easily recovers the earlier results of Mayer and Vietoris on simplicial complexes (chap. II, § 2).

The remainder of Eilenberg and Steenrod's book is devoted to the examination of the various homology and cohomology theories from the point of view of their relation to the axioms. The result on which they put the greatest emphasis is a *uniqueness* theorem. They considered the subcategory $\boldsymbol{T}'$ of *triangulable pairs* (X, A): this means that there is a finite triangulation T of X in simplices such that A is a *subcomplex* of the finite simplicial complex (X, T) (which of course implies that X is compact and A is closed in X). Then if $H_\bullet$ and $H'_\bullet$ are any two homology theories on $\boldsymbol{T}'$ satisfying the axioms, for each homomorphism $\varphi: G \to G'$ of the coefficient groups there is a natural transformation $\xi: H_\bullet \to H'_\bullet$ (§ 5,E) which coincides with φ in degree 0; in particular, if φ is an isomorphism, the homomorphisms $H_q(X, A) \to H'_q(X, A)$ are bijective. In other words, taking the identity for φ, for the category $\boldsymbol{T}'$ of triangulable pairs there is essentially *only one* homology theory with coefficients in G; it may be called the *simplicial homology*, since the groups $H_q(X, A)$ can be computed by the Lefschetz process (chap. II, § 6).

To prove this result, their method was to consider, for a given homology theory $H_\bullet$ satisfying the axioms, with given coefficient group G, and a given triangulation T of a triangulable pair (X, A), the closed subspaces X^q of X, where X^q (the *q-skeleton*) is the union of the simplices of T of dimension $\leqslant q$*; the spaces $X^q \cup A$ are then subcomplexes of (X, T) and one can consider the groups $\bar{C}_q(T, A) = H_q(X^q \cup A, X^{q-1} \cup A)$. These groups are made into a *chain complex* by the boundary operator $\bar{\partial}$ of the homology sequence (94) of the triple $(X^q \cup A, X^{q-1} \cup A, X^{q-2} \cup A)$. The main point is to show that $\bar{C}_q(T, A)$ may be naturally *identified* with $C_q(T, A) \otimes G$, where $C_q(T, A)$ is the **Z**-module of alternating chains (chap. II, § 2) formed with the combinatorial simplices, the vertices of which are vertices of simplices of T, none of which is in A.

The introduction of the group G in the structure of $\bar{C}_q(T, A)$ is made possible by the fact that, for the unit closed ball $\mathbf{D}_n$ in $\mathbf{R}^n$ and its boundary $\mathbf{S}_{n-1}$ in the homology theory $H_\bullet$, $H_q(\mathbf{D}_n) = 0$ for $q \neq 0$, $H_q(\mathbf{S}_n) = 0$ if $q \neq 0, n$, and $H_0(\mathbf{S}_n) = H_n(\mathbf{S}_n) = G$ up to isomorphism; hence, by the exact sequence (94) of triples,

$$H_q(\mathbf{D}_n, \mathbf{S}_{n-1}) = 0 \text{ for } q \neq n, \qquad H_n(\mathbf{D}_n, \mathbf{S}_{n-1}) = G;$$

this can be proved by the usual method (induction on n and decomposition of $\mathbf{S}_n$ in two closed hemispheres), using the axioms exclusively. It remains

* This notation should not be confused with that of the product of q copies of X.

to show that $H_\bullet(\bar{C}_\bullet(T, A))$ is naturally isomorphic to $H_\bullet(X, A)$, which needs repeated ingenious applications of the homology sequence (94) of well-chosen triples.

Once this identification is made, the proof of the uniqueness theorem is clear, since a homomorphism $\varphi: G \to G'$ obviously defines a chain transformation $\bar{C}_\bullet(T, A) \to \bar{C}'_\bullet(T, A)$.

The next item in Eilenberg and Steenrod's book is the investigation, for *arbitrary pairs* (X, A), of the axioms for *singular* homology and cohomology (note that X is not necessarily a Hausdorff space). The dimension axiom is trivially verified, and the exactness of the homology sequence (92) is a special case of the exact homology sequence for chain complexes (§ 5,A). The homotopy axiom is proved by establishing a chain equivalence

$$S_\bullet(X \times I, A \times I) \rightleftarrows S_\bullet(X, A)$$

where I is the interval [0, 1]; this is done with the help of a homotopy operator. Later it was observed ([440], p. 174) that one could simply apply the method of acyclic models (§ 5,G).

Finally, the proof of the excision axiom has to be done in two steps; the first one establishes the invariance of singular homology under the process that corresponds to barycentric subdivision (chap. I, § 3); it consists in defining an endomorphism sd of $S_\bullet(X; \mathbf{Z})$ in the following way. First, $\mathrm{sd}(1_{\Delta_p})$ is defined for each p as the singular chain $\sum_i \pm\delta_i$, where the δ_i are the linear bijections of Δ_p on the p-simplices of the barycentric subdivision of Δ_p, and the signs are suitably chosen; next, for any singular p-simplex $\sigma = S_p(f) \circ 1_{\Delta_p}$, $\mathrm{sd}(\sigma) = S_p(f) \circ \mathrm{sd}(1_{\Delta_p})$, and sd extends to $S_\bullet(X; \mathbf{Z})$ by linearity. It is then proved that there is a chain homotopy between sd and the identity. The second step considers a more general situation than the one needed for the excision axiom. For any open covering $\mathscr{U} = (U_\alpha)$ of X, define a subchain complex $S_\bullet^{\mathscr{U}}(X)$ of $S_\bullet(X)$, namely the one having as basis the singular simplices whose image is contained in one of the U_α. For any singular chain σ, there is an integer m such that $\mathrm{sd}^m(\sigma) \in S_\bullet^{\mathscr{U}}(X)$; from this result and the first step of the proof, it follows that $S_\bullet(X)$ and $S_\bullet^{\mathscr{U}}(X)$ are chain equivalent. The excision axiom is a particular case of this result, corresponding to a covering by two open sets. Examples show that for singular homology the assumptions made in the excision axiom cannot be improved; the result may be false if it is only assumed that the open set U is contained in A, but $\bar{U}$ is not contained in the interior of A ([189], p. 268).

Again, all these properties are matched by corresponding ones for singular cohomology.*

---

* One can, for a euclidean simplicial complex X with triangulation T, define an isomorphism of its simplicial homology to its singular homology in the following way: one orders the vertices of T, and to each p-simplex of T with vertices $a_0 < a_1 < \cdots < a_p$, one assigns the singular p-simplex, which is an affine map sending each vertex e_j of Δ_p onto a_j. This defines a chain transformation of the chain complex $C_\bullet(T)$ of the triangulation into the singular chain complex $S_\bullet(X)$, and one proves it is a chain *equivalence* (chap. II, § 3).

The end of Eilenberg and Steenrod's book is principally concerned with Čech homology and cohomology, and it gives a systematic treatment of the relations of these theories with the axioms; most of the results had previously been proved by Dowker [145] and Spanier [438]. Following Čech's original definition (§§ 2 and 3), Dowker [145] studied Čech cohomology based on *finite* open coverings, but he found that for noncompact spaces this gave surprisingly pathological cohomology groups: for the real line the first Čech cohomology group based on finite coverings is isomorphic to the quotient $C(\mathbf{R})/BC(\mathbf{R})$, where $C(\mathbf{R})$ is the vector space of all continuous real functions in $\mathbf{R}$ and $BC(\mathbf{R})$ is the subspace of all bounded continuous functions; that group has the cardinal of the continuum, a very unexpected property!* This example shows that the homotopy axiom in particular certainly *cannot* be verified for such a cohomology theory, since every continuous map $\mathbf{R} \to \mathbf{R}$ is homotopic to a constant map. This led Dowker to propose replacing finite open coverings by *arbitrary* open coverings in the definition of Čech homology and cohomology (§ 3), and in [145], he proved that, with this modification, the Eilenberg–Steenrod axioms are verified for Čech cohomology of *arbitrary* spaces and coefficient groups, but *not* for Čech homology.

The procedure Dowker followed was naturally suggested by the definition of the Čech groups by a passage to the (direct or inverse) limit on homology and cohomology of suitably defined pairs (K_α, L_α) of infinite combinatorial complexes. He first investigated whether the homology and cohomology groups of these pairs (K_α, L_α) satisfy the axioms, and then whether the axioms are preserved by direct or inverse limits. This worked out smoothly for the excision axiom, but for the homotopy axiom he had to use special coverings of the product $X \times I$, consisting of products $U_\lambda \times V_i$, where the U_λ are open in X and, for each λ, (V_i) is a conveniently chosen finite covering of I by open intervals (the proof given by Eilenberg and Steenrod is an improvement of Dowker's).

Finally, the exactness axiom is easily verified for each pair (K_α, L_α), and the homology (resp. cohomology) sequence for (X, A) is the inverse (resp. direct) limit of those of (K_α, L_α). Direct limits preserve exactness, but the same is not true for inverse limits. Čech cohomology therefore satisfies all the axioms, but there are examples of compact pairs (X, A) for which the exactness axiom is *not* satisfied by Čech homology with coefficients in $\mathbf{Z}$ ([189], p. 265)†; exactness still holds for compact pairs (X, A) when one takes coefficients in a field, or compact coefficient groups and topologized (compact) homology groups; it also holds for triangulable pairs and arbitrary coefficients.

An interesting property satisfied by Čech homology and cohomology is a stronger form of the excision axiom: if (X, A) and (Y, B) are compact pairs, and if $f: (X, A) \to (Y, B)$ is a morphism such that the restriction of f to $X - A$ is a

* An "explanation" of this result is that the Čech cohomology of a locally compact space, based on finite coverings, is isomorphic to the cohomology of its *Stone–Čech compactification* ([189], p. 282).

† As a consequence, this homology theory has not been much used.

homeomorphism of X − A onto Y − B, then $\check{H}_\bullet(f)\colon \check{H}_\bullet(X, A) \to \check{H}_\bullet(Y, B)$ and $\check{H}^\bullet(f)\colon \check{H}^\bullet(Y, B) \to \check{H}^\bullet(X, A)$ are isomorphisms (the ultimate form of Lefschetz's idea of relative homology).

The isomorphism between Čech and Alexander–Spanier cohomology (§ 6,A) implies that the latter also satisfies the Eilenberg–Steenrod axioms. This had been proved directly for compact pairs by Spanier [438].

For *locally compact* spaces, Eilenberg and Steenrod also generalized the definition of H. Cartan (§ 5,A) to arbitrary coefficients, and characterized it axiomatically ([189], pp. 273–276).

C. Cohomology of Smooth* Manifolds

Let X be a *compact* space, Y be a *closed* subset of X, and suppose $Y \neq X$. Then, for any open neighborhood V of Y in X,

$$Z^p(X, V; G) \subset Z_c^p(X - Y; G)$$

since X − V is compact, and any *p*-cocycle that has value 0 for every singular simplex with support contained in V obviously has compact support. Similarly $B^p(X, V; G) \subset B_c^p(X - Y; G)$; hence, there is a natural homomorphism

$$H^p(X, V; G) \to H_c^p(X - Y; G).$$

Furthermore, $H_c^p(X - Y; G)$ is the union of the images of these homomorphisms, for all open neighborhoods V of Y.

From these remarks, it is easily deduced that if there exists a fundamental system of neighborhoods of Y such that Y is a strong deformation retract (section B) of each of these neighborhoods, then there is a natural isomorphism

$$H^p(X, Y; G) \xrightarrow{\sim} H_c^p(X - Y; G).$$

Using this isomorphism, under the same assumptions, the exact cohomology sequence (93) can be written as

$$\cdots \to H_c^p(X - Y; G) \xrightarrow{a} H^p(X; G) \xrightarrow{j^*} H^p(Y; G) \xrightarrow{\partial} H_c^{p+1}(X - Y; G) \to \cdots. \quad (97)$$

The map *a* associates to the class of a *p*-cocycle in X − Y with compact support the class of that cocycle considered as a cocycle in X (taking the value 0 on the singular simplices whose support does not meet a compact subset of X − Y).

The preceding result is particularly valid when X is a *smooth compact manifold* and Y is a *closed submanifold*, since Y is a strong deformation retract of its tubular neighborhoods (chap. III, § 1). The use of Alexander–Spanier cohomology allows a relaxation on the assumptions on the closed subspace Y of a smooth compact manifold X: the exact sequence (97) is still valid for *any* closed subspace Y of X provided the cohomology groups are taken in the Alexander–Spanier sense. If X is orientable, Alexander duality (chap. II, § 6) generalizes as follows: for any two closed sets A, B of X such that $B \subset A$, there

---

* The word "smooth" is synonymous to C^∞.

is an isomorphism

$$H_q(X - B, X - A; G) \simeq \bar{H}_c^{n-q}(A, B; G),$$

where the left-hand side is singular homology and the right-hand side is Alexander–Spanier cohomology with compact supports ([440], pp. 318 and 342).

With the use of Borel–Moore homology and of concepts related to *local* homological properties of spaces, still much more general results could be obtained (Part 2, chap. IV, § 3).

In his book [393], de Rham showed that on a smooth manifold the cohomology groups he had defined by means of differential forms and the homology groups of currents are naturally isomorphic to singular cohomology and homology groups with real coefficients. He introduced *smooth* singular simplices in a smooth manifold X, which are restrictions to a standard simplex Δ_p of a C^∞ map of a neighborhood of $\bar{\Delta}_p$ in X; they generate a subspace of $S_p(X; \mathbf{R})$, written $S_p^\infty(X; \mathbf{R})$, and the boundary operator $\mathbf{b}_p$ maps $S_p^\infty(X; \mathbf{R})$ into $S_{p-1}^\infty(X; \mathbf{R})$ so that he could define homology vector spaces (over $\mathbf{R}$) $H_p^\infty(X; \mathbf{R})$. Each smooth p-simplex $s: \Delta_p \to X$ defines a p-current $\tilde{s}$ on X in the following way: for every smooth p-form α, if ${}^t s(\alpha)$ is the pull-back of the form α on Δ_p, then $\langle \tilde{s}, \alpha \rangle = \int_{\Delta_p} {}^t s(\alpha)$. From this definition a linear map $S_p^\infty(X; \mathbf{R}) \to \mathscr{E}'_p(X)$ is derived, and next a linear map $H_p^\infty(X; \mathbf{R}) \to H'_p(X)$. De Rham proved that this map is *bijective*, and that the obvious map $H_p^\infty(X; \mathbf{R}) \to H_p(X; \mathbf{R})$ is also *bijective*. The proofs are long and intricate. De Rham bypassed the triangulation of manifolds (chap. III, § 2) by reducing the proofs to the case in which X is a bounded open subset of an $\mathbf{R}^N$. He took a simplicial (rectilinear) triangulation $\mathfrak{S}$ of X (chap. II, § 6). To each p-simplex s of that triangulation is naturally associated a p-current $\tilde{s}$ on X, defined as above; de Rham first proved that any p-current with compact support on X, which is a cycle, is homologous to a linear combination of these currents associated to $\mathfrak{S}$. Next he had to show that if a cycle in $Z_p^\infty(X; \mathbf{R})$ maps onto a boundary in $Z'_p(X)$, then it is already a boundary belonging to $B_p^\infty(X; \mathbf{R})$. He first proved that, by subdivision of $\mathfrak{S}$ and use of the Alexander construction (chap. II, § 3) he could replace $Z_p(X; \mathbf{R})$ by the subspace $L_p(\mathfrak{S}; \mathbf{R})$ generated by the simplicial maps of Δ_p on the p-simplices of $\mathfrak{S}$. Then he showed that if $z \in L_p(\mathfrak{S}; \mathbf{R})$ is such that $\tilde{z}$ is the boundary of a $(p + 1)$-current, $\tilde{z}$ is also the boundary of the image of a $(p + 1)$-chain of $L_{p+1}(\mathfrak{S}; \mathbf{R})$, by an algebraic argument resting on the duality of vector spaces and on lemmas on extension of a smooth differential form into a euclidean simplex when it is only defined on a neighborhood of its frontier ([516], p. 137). Finally, using a homotopy operator, he proved that if $z \in L_p(\mathfrak{S}; \mathbf{R})$ is such that $\tilde{z}$ is the boundary of a $(p + 1)$-current $\tilde{w}$, with $w \in L_{p+1}(\mathfrak{S}; \mathbf{R})$, then there is also a $(p + 1)$-chain $u \in L_{p+1}(\mathfrak{S}; \mathbf{R})$ such that $z = \mathbf{b}_{p+1} u$.

From this result it is easily deduced, by duality, that there is a natural isomorphism

$$H^p(X) \simeq H^p(X; \mathbf{R})$$

mapping the vector space $H^p(X)$ of classes of closed p-forms modulo exact p-forms, onto the singular cohomology group $H^p(X; \mathbf{R})$.

D. Cubical Singular Homology and Cohomology

For some constructions, in particular in the work of Serre on the topology of fibrations (Part 3, chap. IV, § 3,C), it was found convenient to modify the definition of singular homology and cohomology by replacing the standard simplices Δ_n by the standard *cubes* I^n (with $I = [0, 1]$ in $\mathbf{R}$) (I^0 being the origin 0 by convention). More precisely, for any topological space X, a *singular cube* in X is a continuous map $q: I^n \to X$. Let $Q_n(X; \mathbf{Z})$ be the free group generated by the singular n-cubes in X and define $Q_{\bullet}(X; \mathbf{Z}) = (Q_n(X; \mathbf{Z}))_{n \geqslant 0}$ as a *chain complex* in the following way: for any singular n-cube q, take

$$\mathbf{b}_n q(x_1, x_2, \dots, x_n)$$
$$= \sum_{i=1}^{n-1} (-1)^i (q(x_1, \dots, x_{i-1}, 0, x_i, \dots, x_{n-1}) - q(x_1, \dots, x_{i-1}, 1, x_i, \dots, x_{n-1}))$$

(each bracket is the difference between the values of q at corresponding points z, $z + e_i$, on the opposite faces of I^n defined by $x_i = 0$ and $x_i = 1$). It is easy to check that $\mathbf{b}_{n-1} \circ \mathbf{b}_n = 0$. A singular n-cube q is called *degenerate* if $q(x_1, \dots, x_{n-1}, x_n) = q(x_1, \dots, x_{n-1}, y_n)$ for all values of $x_1, \dots, x_{n-1}, x_n, y_n$ in I. The degenerate n-cubes form a basis for a $\mathbf{Z}$-submodule $D_n(X; \mathbf{Z})$ of $Q_n(X; \mathbf{Z})$, and $\mathbf{b}_n D_n(X; \mathbf{Z}) \subset D_{n-1}(X; \mathbf{Z})$, so that the quotients $\bar{Q}_n(X; \mathbf{Z}) = Q_n(X; \mathbf{Z})/D_n(X; \mathbf{Z})$ are *free* $\mathbf{Z}$-modules and form again a chain complex. Its homology (resp. cohomology) is called the *cubical* (*singular*) *homology* (resp. *cohomology*) of X with coefficients in $\mathbf{Z}$.

The relations between singular homology and cubical homology are deduced from the existence of a mapping $\theta: I^n \to \Delta_n$ defined by

$$(x_1, \dots, x_n) \mapsto (1 - x_1, x_1(1 - x_2), \dots, x_1 x_2 \cdots x_{n-1}(1 - x_n), x_1 x_2 \cdots x_n). \quad (98)$$

To each singular n-simplex $s: \Delta_n \to X$ corresponds a singular n-cube $q = s \circ \theta$, and this defines a chain transformation $S_{\bullet}(X; \mathbf{Z}) \to \bar{Q}_{\bullet}(X; \mathbf{Z})$. Eilenberg and Mac Lane proved by the method of acyclic models that this is a *chain equivalence* [186]. There is thus a natural isomorphism of the singular homology of X onto its cubical homology.

E. Leray's 1945 Paper

During World War II Leray was a prisoner of war from 1940 to 1945. He organized a university in his prison camp and himself gave a course on algebraic topology, a field he had become interested in in connection with his collaboration with Schauder on applications of degree theory of functional analysis (Part 2, chap. VII). He became dissatisfied both with the methods using triangulations and with those using inverse or direct limits. In 1942 he published a series of four Notes in the *Comptes rendus* outlining a new and original way of defining and studying cohomology; he elaborated his method in a long paper published in 1945 [313]. All methods used until then to define the cohomology of a space X consisted in making geometric constructions in X (triangulations, singular complexes, or coverings) and then *deducing* cochain complexes from these constructions, the elements of these complexes (the cochains) being *functions* defined in the sets thus constructed. Leray's idea was to *bypass* completely the first leg of this process, and to start right away with an "abstract" cochain complex $C^\bullet = (C^p)_{p \geqslant 0}$ of A-modules (the ring A being equal to $\mathbf{Z}$, or $\mathbf{Z}/m\mathbf{Z}$, or $\mathbf{Q}$), whose elements are not necessarily functions.

The link with the space X (which at first is an arbitrary topological space) is provided by the notion of *support*. This is a map S which to each element $k \in C^\bullet$ assigns a part $S(k)$ of X, subject to the following conditions:

$$S(0) = \varnothing, \qquad S(k - k') \subset S(k) \cup S(k'), \qquad S(\mathbf{d}k) \subset S(k). \tag{99}$$

However, these conditions were not written in that way in Leray's paper of 1945. There he assumed that the C^p were 0 except for a finite number of degrees and that each C^p had a *finite basis* $(e^{p,\alpha})$ over A; then he defined the map S by considering its values $S(e^{p,\alpha}) = |e^{p,\alpha}|$, which he explicitly assumed to be *different from* $\varnothing$, taking for each p,

$$S\left(\sum_\alpha a_\alpha e^{p,\alpha}\right) = \bigcup_\alpha |e^{p,\alpha}| \tag{100}$$

for all linear combinations with coefficients a_α *different from* 0,* Leray called the pair $(C^\bullet, S)$ a *concrete complex*; the definition of S depends on the chosen basis.† Some operations (see below) may lead to a similar definition with some of the sets $|e^{p,\alpha}|$ equal to the empty set; then the $e^{p,\alpha}$ such that $|e^{p,\alpha}| \neq \varnothing$ form the basis of a subcomplex $C'^\bullet$ of $C^\bullet$, and $(C'^\bullet, S|C'^\bullet)$ is what, in 1947, Leray (following H. Cartan's terminology) called the *separated concrete complex* associated to $(C^\bullet, S)$.

To obtain a cohomology theory Leray needed two operations on his "concrete complexes." One was the *intersection* $E . C^\bullet$ with a nonempty subset E of X (corresponding to a passage from global to local properties): consider the concrete complex $(C^\bullet, S_E)$, where $S_E(k) = E \cap S(k)$; $E . C^\bullet$ is the separated complex associated to $(C^\bullet, S_E)$.

The other operation is called the "intersection" of two concrete complexes $(C^\bullet, S)$ and $(C'^\bullet, S')$. Its "abstract" complex is just the usual tensor product $C^\bullet \otimes C'^\bullet$ (as defined in § 5,C), and

$$|e^{p,\alpha} \otimes e'^{q,\beta}| = |e^{p,\alpha}| \cap |e'^{q,\beta}|; \tag{101}$$

---

* To satisfy the condition $S(\mathbf{d}k) \subset S(k)$, Leray imposed on the sets $|e^{p,\alpha}|$ to be such that if $\mathbf{d}e^{p,\alpha} = \sum_\beta c^p_{\alpha\beta} e^{p+1,\beta}$ one must have $|e^{p+1,\beta}| \subset |e^{p,\alpha}|$ for all pairs (α, β) such that $c^p_{\alpha\beta} \neq 0$.

† Suppose C^1 has only two elements $e^{1,1}$, $e^{1,2}$ as basis; then $S(e^{1,1} + e^{1,2}) = |e^{1,1}| \cup |e^{1,2}|$; but $f^{1,1} = e^{1,1} + e^{1,2}$ and $f^{1,2} = e^{1,2}$ also form a basis with $|f^{1,1}| = |e^{1,1}| \cup |e^{1,2}|$ and $|f^{1,2}| = |e^{1,2}|$ and now $S(e^{1,1}) = S(f^{1,1} - f^{1,2}) = |e^{1,1}| \cup |e^{1,2}|$.

the intersection $C^{\bullet}.C'^{\bullet}$ is the associated separated complex. Now $C^{\bullet}.(C'^{\bullet}.C''^{\bullet}) = (C^{\bullet}.C'^{\bullet}).C''^{\bullet}$ and $E.(C^{\bullet}.C'^{\bullet}) = (E.C^{\bullet}).(E.C'^{\bullet})$ for any subset E of X. From the point of view of cohomology, $C'^{\bullet}.C^{\bullet}$ may be identified with $C^{\bullet}.C'^{\bullet}$ by the natural isomorphism

$$x_p \otimes x'_q \simeq (-1)^{pq} x'_q \otimes x_p. \tag{102}$$

The central objects in Leray's theory are special (separated) "concrete complexes" which he calls "couvertures."* They are the concrete complexes satisfying three additional conditions: (1) the sets $|e^{p,\alpha}|$ are closed in X; (2) for each $x \in X$ the cohomology of the intersection $x.C^{\bullet}$ is such that $H^0(x.C^{\bullet}) \simeq A$ and $H^p(x.C^{\bullet}) = 0$ for $p > 0$; (3) the element $u = \sum_\alpha e^{0,\alpha}$ is a *cocycle* called the "unit cocycle" of $C^{\bullet}$.

The ultimate goal of Leray's 1945 paper clearly was to define a *cohomology algebra* $H^{\bullet}(X)$ over A by using "couvertures," such that $H^{\bullet}(X) \simeq H^{\bullet}(C^{\bullet})$ (as graded A-modules) for at least one "couverture" $C^{\bullet}$. The restrictions to finite fixed bases for the C^p led him to a fairly complicated and ineffective way of defining $H^{\bullet}(X)$: he considered *all* "couvertures" $C^{\bullet}_\alpha$ of X, with their bases and "unit cocycles" u_α, and the direct sum $U = \bigoplus_\alpha C^{\bullet}_\alpha$; he singled out the submodule V in that graded A-module generated by all elements $z - u_\alpha.z$ for $z \in U$, α arbitrary; since u_α has degree 0, V is a graded submodule of U, and as $\mathbf{d}u_\alpha = 0$, $\mathbf{d}(u_\alpha.z) = u_\alpha.\mathbf{d}z$, so that $\mathbf{d}(V) \subset V$; hence U/V is a *cochain complex* over A, and by definition $H^{\bullet}(X)$ is its cohomology. It has a structure of *anticommutative graded algebra* over A: indeed, if $z^p \in C^p_\alpha$, $z'^q \in C^q_\beta$, then $u_\gamma.(z^p.z'^q) = (u_\gamma.z^p).z'^q = z^p.(u_\gamma.z'^q)$, and if z^p and z'^q are cocycles (resp. coboundaries), the same is true for $z^p.z'^q$; finally $z'^q.z^p = (-1)^{pq} z^p.z'^q$.

The most original result of Leray's paper was his use of what he called "simple" normal spaces (now usually called "acyclic"), namely, those for which $H^0(X) \simeq A$ and $H^p(X) = 0$ for $p > 0$. He first proved the homotopy axiom, implying that a contractible compact space is simple. The proof relies on a crucial lemma about "couvertures": if $C^{\bullet}$ is a "couverture," $C'^{\bullet}$ is a "concrete complex", and if for each $k \in C'^{\bullet}$, $H^0(S(k).C^{\bullet}) \simeq A$ and $H^p(S(k).C^{\bullet}) = 0$ for $p > 0$, then $H^{\bullet}(C'^{\bullet}.C^{\bullet})$ and $H^{\bullet}(C'^{\bullet})$ are *isomorphic* graded A-modules. This, and the properties of normal spaces, enabled Leray to replace his unwieldy general definition of $H^{\bullet}(X)$, when X is normal, by *restricting* the "couvertures" $C^{\bullet}_\alpha$ to a *subfamily* Φ, such that: (1) the intersection of two "couvertures" of Φ is still in Φ, and (2) there exist "couvertures" in Φ with *arbitrary small supports.* His next step consisted in proving that the cohomology of $X \times [0,1]$ is isomorphic to $H^{\bullet}(X)$ by a convenient use of a family Φ, and from this the homotopy axiom easily follows.

The new *fundamental result* was the consideration of "couvertures" $C^{\bullet}$ of a compact space X such that all supports of elements of $C^{\bullet}$ are *simple*; then there is a natural *isomorphism* $H^{\bullet}(X) \simeq H^{\bullet}(C^{\bullet})$; in applications, the theorem is used when the supports are all *contractible.*

---

* This terminology having rapidly disappeared, there is no need to translate it.

As examples of "couvertures," Leray gave in his 1945 paper those deduced from a *finite closed covering* $\mathscr{R} = (\mathrm{F}_j)_{1 \leqslant j \leqslant \mathrm{N}}$ of a space X. Letting M be the A-module $\bigoplus_{j=1}^{\mathrm{N}} \mathrm{A}e_j$ (isomorphic to A^{N}), he defined the "abstract" complex $\mathrm{C}^{\bullet}$ by $\mathrm{C}^p = \bigwedge^{p+1} \mathrm{M}$, $(p+1)$-st exterior power of M, and took as a basis of C^p the $(p+1)$-vectors $e_{\mathrm{H}} = e_{j_0} \wedge e_{j_1} \wedge \cdots \wedge e_{j_p}$ for all strictly increasing sequences $\mathrm{H}: j_0 < j_1 < \cdots < j_p$ of $p+1$ numbers in $\{1, 2, \ldots, \mathrm{N}\}$. If $u = \sum_{j=1}^{\mathrm{N}} e_j$, the coboundary in $\mathrm{C}^{\bullet}$ is given by $z \mapsto u \wedge z$; finally, if

$$\mathrm{S}(e_{\mathrm{H}}) = \mathrm{F}_{j_0} \cap \mathrm{F}_{j_1} \cap \cdots \cap \mathrm{F}_{j_p},$$

the "concrete complex" associated with the covering $\mathscr{R}$ is the *separated* complex associated to $(\mathrm{C}^{\bullet}, \mathrm{S})$. It is easily seen to be a "couverture" with u for unit cocycle, but in general its cohomology is not 0 in degrees $p > 0$.

Leray also gave examples of "couvertures" that are not obtained from a finite closed covering of X by the preceding construction. The most interesting one gives (by application of the theorem on simple supports) the cohomology of the sphere $\mathbf{S}_n$: each C^p for $0 \leqslant p \leqslant n$ has a basis of two elements $e^{p,1}$, $e^{p,2}$ with the coboundaries $\mathbf{d}e^{p,1} = -\mathbf{d}e^{p,2} = e^{p+1,1} + e^{p+1,2}$ for $p < n$, and $\mathbf{d}e^{n,1} = \mathbf{d}e^{n,2} = 0$; the supports $|e^{p,1}|$ and $|e^{p,2}|$ are, respectively, the hemispheres of $\mathbf{S}_p$ defined by $x_p \geqslant 0$ and $x_p \leqslant 0$.

As we shall see in §7, in the Notes of 1946 in which he defined sheaf cohomology and spectral sequences, Leray was still using the same notion of "couverture" as in his 1945 paper. But in the lectures he gave in 1947 (from which he published a survey without proofs in 1949, as a forerunner of his long paper of 1950 [321]), he had apparently realized that this definition was too restrictive and unduly bound to the choice of bases, and this led him to more conceptual definitions. The main changes with respect to the 1945 paper are: (1) the spaces under consideration are *locally compact*; (2) the "abstract" complex $\mathrm{C}^{\bullet}$ of A-modules is only subject to the condition of being *without torsion*, but otherwise arbitrary; (3) the supports $\mathrm{S}(k)$ are not defined with the help of a basis, but are just closed sets satisfying (99). The "concrete complexes" with arbitrarily small supports are replaced by the new notion of *fine complex*, at which Leray and Cartan independently arrived at the same time; it generalizes the notion of *partition of unity* to complexes in the following way: for any finite open covering $(\mathrm{U}_i)_{1 \leqslant i \leqslant n}$ of the space X, where for each i, either $\bar{\mathrm{U}}_i$ or $\mathrm{X} - \mathrm{U}_i$ is compact (called a "proper" covering), there exist endomorphisms r_i of the A-module $\mathrm{C}^{\bullet}$, such that $r_i(\mathrm{C}^p) \subset \mathrm{C}^p$ for all $p \geqslant 0$, and which satisfy the conditions:

$$r_1(k) + r_2(k) + \cdots + r_n(k) = k \qquad \text{for all } k \in \mathrm{C}^{\bullet}, \tag{103}$$

$$\mathrm{S}(r_i(k)) \subset \bar{\mathrm{U}}_i \cap \mathrm{S}(k) \qquad \text{for } 1 \leqslant i \leqslant n \text{ and all } k \in \mathrm{C}^{\bullet}. \tag{104}$$

In this new context the definitions of the 1945 paper had to be modified. The intersection $\mathrm{E} \,.\, \mathrm{C}^{\bullet}$ of a concrete complex $\mathrm{C}^{\bullet}$ with a subset E of X has as "abstract" complex the quotient $\mathrm{C}^{\bullet}/\mathrm{C}^{\bullet}_{\mathrm{X}-\mathrm{E}}$, where $\mathrm{C}^{\bullet}_{\mathrm{X}-\mathrm{E}}$ is the subcomplex of $\mathrm{C}^{\bullet}$ consisting of cochains with supports in $\mathrm{X} - \mathrm{E}$. The intersection of two concrete complexes $C^\bullet$, $C'^\bullet$, is now written $C^\bullet \bigcirc C'^\bullet$ and is defined as follows: for each $x \in X$, consider the natural homomorphism

$$f_x: C^\bullet \otimes C'^\bullet \to x.C^\bullet \otimes x.C'^\bullet,$$

and for $h \in C^\bullet \otimes C'^\bullet$, define $S(h)$ as the set of $x \in X$ such that $f_x(h) \neq 0$; $C^\bullet \bigcirc C'^\bullet$ is then the separated complex associated to $(C^\bullet \otimes C'^\bullet, S)$; for $k \in C^\bullet$, $k' \in C'^\bullet$, $k \bigcirc k'$ is the element of $C^\bullet \bigcirc C'^\bullet$ image of $k \otimes k'$. "Couvertures" are defined as before, except that the "unit cocycle" u is simply supposed to be an element of C^0, such that $S(u) = X$ and $H^0(x.C^\bullet) = A.(x.u)$ for $x \in X$.

The fundamental lemma now concerns an arbitrary "couverture" $C^\bullet$ with unit u, and a *fine* concrete complex K with *compact supports*. Then the map $k \mapsto u \bigcirc k$ of K into $C^\bullet \bigcirc K^\bullet$ is an injective homomorphism of cochain complexes, and the lemma says that the corresponding map $H^\bullet(C^\bullet \bigcirc K') \to H^\bullet(K^\bullet)$ is *bijective*; it may be proved by the same technique of induction on the degree (see [57]), although in 1947 Leray chose to prove it by an application of spectral sequences. The lemma implies that if $K^\bullet$ and $K'^\bullet$ are *both* fine "couvertures" with *compact supports*, $H^\bullet(K^\bullet)$ and $H^\bullet(K'^\bullet)$ are naturally *isomorphic*, being both naturally isomorphic with $H^\bullet(K^\bullet \bigcirc K'^\bullet)$. This of course defines the cohomology $H^\bullet(X)$ over A, up to isomorphism, as $H^\bullet(K^\bullet)$ for *any* fine "couverture" with compact supports.

In his 1945 paper Leray had not tried to make any connection between his theory and previous cohomology theories. In his 1947 lectures he showed that, for a locally compact space X, the cochain complex $\bar{C}^\bullet(X; A) = (\bar{C}^p(X; A))_{p \geqslant 0}$ defining the Alexander–Spanier cohomology (§3) could be given a structure of "fine couverture." For any function $f \in C^p(X; A)$, a closed set $S(f)$ in X can be defined as consisting of the points $x \in X$ for which there exists a fundamental system of neighborhoods (V_λ) of x, such that f is not identically 0 in V_λ^{p+1} for any λ; then the separated complex associated to $(C^\bullet(X; A), S)$ is precisely $\bar{C}^\bullet(X; A)$, i.e., the cochain complex whose cohomology is the Alexander–Spanier cohomology $\bar{H}^\bullet(X; A)$ (*loc. cit.*). The fact that $\bar{C}^\bullet(X; A)$ is *fine* is seen by exhibiting, for any finite open covering (U_i) of X $(1 \leqslant i \leqslant n)$, at first endomorphisms r_i *of* $C^\bullet(X; A)$ satisfying (103) and (104), and then observing that the r_i map the subcomplex $C_0(X; A)$ into itself: for any $f \in C^p(X; A)$, the $r_i(f)$ are explicitly defined by

$$r_1(f)(x_0, x_1, \ldots, x_p) = \begin{cases} f(x_0, x_1, \ldots, x_p) & \text{if } x_0 \in U_1, \\ 0 & \text{otherwise,} \end{cases}$$

and, for $k \geqslant 1$

$$r_k(f)(x_0, x_1, \ldots, x_p) = \begin{cases} f(x_0, \ldots, x_p) & \text{if } x_0 \in U_k - (U_1 \cup \cdots \cup U_{k-1}), \\ 0 & \text{otherwise.} \end{cases}$$

The unit cocycle u of $\bar{C}^\bullet(X; A)$ is of course the class of the constant function on X equal to 1. To check that $\bar{C}^\bullet(X; A)$ is a "couverture," it is therefore only necessary to compute the cohomology $H^\bullet(x.\bar{C}^\bullet(X; A))$ for any $x \in X$. The definition of the coboundaries δ_p [§3, formula (16)] first shows that if

$\bar{f} \in \bar{C}^0(X; A)$ is such that $x . \overline{\delta_0 f} = 0$, then $\bar{f}$ is the class of a function f: $X \to A$, constant in a neighborhood of x, so that $H^0(x . \bar{C}^{\bullet}(X; A)) = A . (x . u)$. For $p > 0$, and $f \in C^p(X; A)$, define $g_x \in C^{p-1}(X; A)$ by $g_x(y_0, y_1, \ldots, y_{p-1}) = f(x, y_0, \ldots, y_{p-1})$; then it is easy to see that if the class $\bar{f}$ of f in $x . \bar{C}^p$ is a cocycle, there exists a neighborhood V of x such that

$$(f - \delta_p g_x)(y_0, y_1, \ldots, y_{p-1}, y_p) = 0 \quad \text{for } (y_0, \ldots, y_p) \in V^{p+1}$$

hence $\bar{f}$ is a coboundary.

To apply his uniqueness theorem to the Alexander–Spanier cohomology Leray had to limit himself to compact spaces, since only "couvertures" with compact supports are admissible. A little later, however, Fary [192] showed that all of Leray's arguments were still valid if the third property, defining "couvertures," was weakened: it is enough to assume that, for each compact subset Z of X, there exists a $u_Z \in C^0$ such that, for *every* $x \in Z$, $H^0(x . C^{\bullet}) = A . (x . u_Z)$ ("relative unit cocycle"). Then on a locally compact space X the cohomology $H^{\bullet}(C^{\bullet})$, for *all* "fine couvertures" $C^{\bullet}$ with compact supports, is naturally isomorphic to the *Alexander–Spanier cohomology with compact supports* $\bar{H}^{\bullet}_c(X; A)$ [i.e., the cohomology of the subcomplex $\bar{C}^{\bullet}_c(X; A)$ of the cochain complex $\bar{C}^{\bullet}(X; A)$, consisting of the classes of the cochains $f \in C^{\bullet}(X; A)$ with compact supports].

In his seminars on sheaf theory, which started in 1948 (§ 7), H. Cartan proved in a similar way that on a C^∞ differentiable manifold X, the de Rham cochain complex $C^{\bullet} = (\mathscr{D}_p(X))_{p \geqslant 0}$ (chap. III, § 3) is a "fine couverture," if the usual support is taken as support of a differential form; the fact that it is a fine complex is proved by using a C^∞ partition of unity, and the computation of $H^{\bullet}(x . C^{\bullet})$ uses the Poincaré lemma.

Finally, on an arbitrary locally compact space X, the complex of *singular cochains* (§ 3) is also a "fine complex" in a natural way; it is not always a "couverture," but this property is valid under additional hypotheses on X, for instance the HLC condition (Part 2, chap. IV, § 2).

§ 7. Sheaf Cohomology

A. Homology with Local Coefficients

In algebraic topology until 1935 the consideration of chains with coefficients other than the integers had been limited to taking these coefficients in a *fixed* commutative group (chap. IV, § 2). During the next period, this point of view was enlarged in several ways.

In 1935 Reidemeister wanted to study the homology of a *covering space* $\tilde{K}$ (Part 3, chap. I) of a finite euclidean simplicial complex K. If π: $\tilde{K} \to K$ is the projection, then replacing K by a convenient subdivision implies that, above any simplex σ of K, $\pi^{-1}(\sigma)$ is a product $D \times \sigma$, where D is a discrete set independent of σ. This is apparently what led Reidemeister to consider $\tilde{K}$ as a simplicial complex whose simplices are $\{x\} \times \sigma$ where $x \in D$ and σ is a simplex of K; now D is given the structure of a commutative group, so that

$$(\{x\} \times \sigma) + (\{y\} \times \sigma) = \{x + y\} \times \sigma.$$

The boundary operators in $\tilde{K}$ must be defined; simple examples show that if τ is a q-simplex of K contained in the frontier of a p-simplex σ ($q < p$), $\{x\} \times \tau$ will not in general be contained in the frontier of $\{x\} \times \sigma$ (the "sheets" cross!); the q-simplex of $\tilde{K}$ above τ in the frontier of $\{x\} \times \sigma$ will be of the form $\{x'\} \times \tau$ for an $x' \in D$ that may be different from x. So, to each pair (σ, τ), where τ is contained in the frontier of σ, is assigned a *permutation* $\gamma_{\sigma\tau}$ of D, and to preserve the linearity of the boundary map, an *automorphism* of the group D must be taken for $\gamma_{\sigma\tau}$. Then define the boundary operator $\mathbf{b}_p$ in $\tilde{K}$ as follows: if $\mathbf{b}_p\sigma = \sum_{j=0}^{p}(-1)^j\tau_j$,

$$\mathbf{b}_p(\{x\} \times \sigma) = \sum_{j=0}^{p} (\{(-1)^j\gamma_{\sigma\tau_j}(x)\} \times \tau_j). \tag{105}$$

In order that $\mathbf{b}_{p-1} \circ \mathbf{b}_p = 0$, the $\gamma_{\sigma\tau}$ must be such that

$$\gamma_{\sigma\xi} = \gamma_{\sigma\tau} \circ \gamma_{\tau\xi} \tag{106}$$

when τ is contained in the frontier of σ and ξ is contained in the frontier of τ [386].

In 1942 [446] Steenrod, in his work on fibre bundles and obstructions (Part 3, chap. III, §2,E), considered homology (or cohomology) with *local coefficients* from a more general viewpoint; the next year he returned to the subject with more detail in a paper specially devoted to that concept [445]. Unlike Reidemeister, instead of having a single group, he attaches a group G_x to *each point* x of an *arcwise connected* space X; these groups are all *isomorphic**; more precisely, for every *path* α_{yx}: $[0, 1] \to X$ from x to y, there is given an isomorphism $\varphi(\alpha_{yx})$: $G_x \tilde{\to} G_y$, satisfying the following conditions: (1) if α_{yx}^{-1} denotes the path $t \mapsto \alpha_{yx}(1 - t)$ from y to x, then $\varphi(\alpha_{yx}^{-1}) = (\varphi(\alpha_{yx}))^{-1}$; (2) if $\alpha_{zy} \vee \alpha_{yx}$ denotes the path from x to z, juxtaposition of α_{yx} and α_{zy}, then $\varphi(\alpha_{zy} \vee \alpha_{yx}) = \varphi(\alpha_{zy}) \circ \varphi(\alpha_{yx})$; (3) if α_{yx} and β_{yx} are two paths from x to y, which are *homotopic* by a homotopy fixing x and y, then $\varphi(\beta_{yx}) = \varphi(\alpha_{yx})$. Such a family (G_x) is called a *local system of groups* on X.

Now suppose the G_x are *commutative* groups. To any singular p-simplex s: $\bar{\Delta}_p \to X$, associate the group G_x for $x = s(1, 0, \ldots, 0)$, which is simply written G_s. A *singular p-chain with coefficients in the local system* (G_x) is a formal linear combination $z = \sum_s g_s . s$, where the $g_s \in G_s$ are 0 except for a finite number of p-simplices s, and define on the set $C_p(X, (G_x))$ of these p-chains a structure of commutative group by the condition that $g_s . s + g'_s . s = (g_s + g'_s) . s$, addition being in G_s. To make the graded group $C_\bullet(X, (G_x)) = \bigoplus_p C_p(X, (G_x))$ into a chain complex, define the boundary of $g_s . s$ for any singular p-simplex s: As

* This condition may appear to be unduly restrictive, but it is for instance satisfied if G_x is the fundamental group $\pi_1(X, x)$.

usual, let σ_j $(0 \leqslant j \leqslant p)$ be the map

$$(\xi_0, \xi_1, \ldots, \xi_{p-1}) \mapsto (\xi_0, \xi_1, \ldots, \xi_{j-1}, 0, \xi_j, \ldots, \xi_{p-1}) \tag{107}$$

of $\bar{\Delta}_{p-1}$ into $\bar{\Delta}_p$; let $y_j = s(\sigma_j(1, 0, \ldots, 0))$, and let $\alpha_{y_j x}$ be the image by s of a *path in* $\bar{\Delta}_p$ from the vertex $(1, 0, \ldots, 0)$ of $\bar{\Delta}_p$ to the image by σ_j of the vertex $(1, 0, \ldots, 0)$ of $\bar{\Delta}_{p-1}$.* As $\bar{\Delta}_p$ is simply connected, the isomorphism $\varphi(\alpha_{y_j x})\colon G_s \tilde{\to} G_{s \circ \sigma_j}$ does not depend on the choice of $\alpha_{y_j x}$; then as boundary,

$$\mathbf{b}_p(g_s . s) = \sum_{j=0}^{p} (-1)^j (\varphi(\alpha_{y_j x})(g_s)) . (s \circ \sigma_j). \tag{108}$$

Because the $\bar{\Delta}_p$ are simply connected, $\mathbf{b}_{p-1} \circ \mathbf{b}_p = 0$. This therefore defines the *singular homology* $H_{\bullet}(X; (G_x))$ *with coefficients in the local system* (G_x).

The *singular cohomology* $H^{\bullet}(X; (G_x))$ *with coefficients in* (G_x) is defined in a similar way: a *singular p-cochain* is a function f which, assigns *a value* $f(s)$ *in* G_s to every p-simplex s; they obviously form a commutative group for the addition $(f + g)(s) = f(s) + g(s)$ for every singular p-simplex s. Here the coboundary is defined by

$$\mathbf{d}_{p-1} f(s) = \sum_{j=0}^{p} (-1)^j \varphi(\alpha_{y_j x})(f(s \circ \sigma_j)) \tag{109}$$

with the same notations as above, for every singular p-simplex s and every $(p-1)$-cochain f. If the G_x have a structure of *ring*, the $\varphi(\alpha_{yx})$ being homomorphisms of rings, a natural structure of graded anticommutative ring on $H^{\bullet}(X, (G_x))$ arises in the usual way.

Let a be a point of X; for any *loop* α_{aa} of origin and extremity a, $\varphi(\alpha_{aa})$ is an automorphism of G_a that only depends on the class α_a of α_{aa} in the fundamental group $\pi_1(X, a)$, and may thus be written $\varphi_a(\alpha_a)$; $\alpha_a \mapsto \varphi_a(\alpha_a)$ is then a *homomorphism* of $\pi_1(X, a)$ in $\mathrm{Aut}(G_a)$.

Conversely, given a commutative group G_a and a homomorphism

$$\varphi_a\colon \pi_1(X, a) \to \mathrm{Aut}(G_a) \tag{110}$$

there is a *unique* (up to isomorphism) local system (G_x) on X for which φ_a is obtained by the preceding construction. Construct the product $G_a \times \Gamma_x$ for each $x \in X$, where Γ_x is the set of *all* paths from a to x, then define an equivalence relation

$$R_x\colon (g, \alpha_{xa}) \equiv (g', \beta_{xa}) \tag{111}$$

by the condition $g' = \varphi_a(\beta_{xa}^{-1} \vee \alpha_{xa})(g)$ [where the juxtaposition $\beta_{xa}^{-1} \vee \alpha_{xa}$ should really be replaced by its class in $\pi_1(X, a)$]. If $G_x = (G_a \times \Gamma_x)/R_x$, check that the law

$$(g', \beta_{xa})(g, \alpha_{xa}) = (\varphi_a(\alpha_{xa}^{-1} \vee \beta_{xa})(g'g), \alpha_{xa}) \tag{112}$$

* For $j > 0$, the path in $\bar{\Delta}_p$ is reduced to a single point.

is compatible with the relation R_x, and defines on G_x a structure of commutative group. Finally, for any path α_{yx} from x to y, define a map $G_a \times \Gamma_x \to G_a \times \Gamma_y$ by

$$(g, \alpha_{xa}) \mapsto (g, \alpha_{yx} \vee \alpha_{xa}); \tag{113}$$

it is compatible with the equivalence relations R_x and R_y, and, by passage to quotients, it defines an isomorphism $\varphi(\alpha_{yx})$ of G_x onto G_y, which establishes (G_x) as a local system.

For this interpretation of local systems, formula (108) for the boundary is written

$$\mathbf{b}_p((g, \alpha_{xa}).s) = \sum_{j=0}^{p} (-1)^j((\varphi_a(\alpha_{y_ja}^{-1} \vee \alpha_{y_jx} \vee \alpha_{xa})(g), \alpha_{y_ja}).(s \circ \sigma_j)), \tag{114}$$

which has the same form as the boundary formula (105) of Reidemeister's theory which it generalizes to singular homology*; (105) may be recovered by the chain transformation of the chain complex of a triangulation into the singular chain complex (§ 6,B and chap. II, § 3).

At the end of his paper Steenrod considered the possibility of using local coefficients for Čech homology and cohomology. He realized that this is not possible by his method if X is not locally simply connected; his proposal to consider only coverings finer than a given one would have been suitable if he added that this fixed covering should consist of simply connected sets. But a good definition can only be given in the context of sheaf cohomology (§ 7,F).

B. The Concept of Sheaf

In May 1946 Leray published two Notes in the *Comptes Rendus* ([314], [315]), in which he introduced for the first time the notions of *sheaf*, of *sheaf cohomology*, and of *spectral sequence*. In retrospect, it is difficult to exaggerate the importance of these concepts, which very rapidly became not only powerful tools in algebraic topology, but spread to many other parts of mathematics, some of which seem very remote from topology, such as algebraic geometry, number theory, and mathematical logic. These applications certainly went far beyond the wildest dreams of the inventor of these notions, and they undoubtedly rank at the same level in the history of mathematics as the methods invented by Poincaré and Brouwer.

Leray's motivation is clearly expressed in the second Note. If X and Y are topological spaces, and $f: Y \to X$ is a continuous map, the general problem is to relate the homology of X and the homology of Y, using properties of f. The only case in which such a problem had been considered was when X and Y are compact metric spaces: in his paper [475] of 1927 Vietoris had proved

* As a matter of fact, Steenrod assumed in most of his paper that X is triangulated, and it was only at the end that he considered "continuous chains," i.e., singular chains in the sense of Lefschetz (chap. II, § 3).

(with his definition of homology) that if the homology modules $H_q(f^{-1}(x))$ are 0 for $1 \leqslant q \leqslant n$ and for every $x \in X$, then f defines an isomorphism of $H_q(Y)$ onto $H_q(X)$ for the same values of q. Using the fact that any neighborhood of a fiber $f^{-1}(x_0)$ contains a neighborhood that is a union of fibers $f^{-1}(z)$, he showed that any q-cycle in X is homologous to the image by f of a q-cycle of Y.*

It was therefore natural to look for a more general result by considering the cohomology of each fiber $f^{-1}(x)$, and seeing if it was possible to "reconstruct" the cohomology $H^{\bullet}(Y)$ from the knowledge of $H^{\bullet}(X)$ and of the $H^{\bullet}(f^{-1}(x))$ (a sweeping generalization of the "Künneth theorem"). To put it otherwise, the question was how to use the information given by the assignment, to a variable point $x \in X$, of a **Z**-module $\mathscr{F}(x)$ that is not subject to the very stringent restrictions of Steenrod's "local systems." Leray's originality lay in his consideration, not of the "punctual" modules $\mathscr{F}(x)$, but of a family $\mathscr{F}(E)$ of A-modules (or A-algebras) indexed by *all closed sets* E of X (in the 1946 Notes, X is supposed normal; in Leray's later work, X is locally compact). In 1946 he says that such a family is a *normal sheaf* (later, in 1947, a *continuous sheaf*, and we shall merely say a *sheaf*) if:

L1. for each pair of closed sets $E_1 \supset E_2$ in X, there is a homomorphism

$$\rho_{E_2E_1}: \mathscr{F}(E_1) \to \mathscr{F}(E_2) \tag{115}$$

such that ρ_{EE} is the identity of $\mathscr{F}(E)$, and, if $E_1 \supset E_2 \supset E_3$,

$$\rho_{E_3E_1} = \rho_{E_3E_2} \circ \rho_{E_2E_1};$$

L2. for each $z \in \mathscr{F}(E)$, there is a closed neighborhood V of E in X, and a $z' \in \mathscr{F}(V)$ such that $z = \rho_{EV}(z')$;

L3. if $\rho_{E_2E_1}(z) = 0$ for a $z \in \mathscr{F}(E_1)$, there is in the subspace E_1 a closed neighborhood W of E_2 such that $\rho_{E_2W}(z) = 0$.

The conditions L2 and L3 may also be expressed by saying that

$$\mathscr{F}(E) = \varinjlim \mathscr{F}(V)$$

for the direct system of closed neighborhoods V of E, ordered by inclusion. Leray proved in his 1945 paper ([313], lemmas 22 and 23) that the cohomology modules $H^p(E, A) = \mathscr{F}(E)$ satisfy these conditions. In his 1946 Notes he pointed out that sheaves have some analogy with the "local systems" of Steenrod,† but even if he only considered the "punctual" modules $\mathscr{F}(x)$, the gain in generality is tremendous: whereas in a local system (G_x) all the G_x are isomorphic, there are "skyscraper sheaves" $\mathscr{F}(E)$ such that $\mathscr{F}(x) = 0$ except for *one* point x_0.‡

---

* Leray does not mention Vietoris, and gives the result for cohomology instead of homology.

† The analogy is not quite evident with Leray's definition of a sheaf; it is only in the Cartan–Lazard definition that local systems become special cases of sheaves (see below).

‡ One has only to take $\mathscr{F}(E) = 0$ is $x_0 \notin E$, $\mathscr{F}(E) = M$, a fixed A-module if $x_0 \in E$ and $\rho_{E_2E_1}$ is the identity in M if $x_0 \in E_2 \subset E_1$.

Of course this definition only proved its value when Leray showed how, by the use of spectral sequences, he could give in many cases a solution to the general problem mentioned at the beginning of this section. But before we examine it we shall quickly see how Leray's initial definition of a sheaf was transformed between 1950 and 1960.

His announcements, and the subsequent lectures he gave on the subject, created quite a stir in the mathematical circles in Paris, particularly that of H. Cartan and his graduate students of that period (most notably Koszul and Serre). Between 1947 and 1951 these topics certainly were the occasions of many exchanges and discussions but not much of this activity has been preserved in publications.

In 1947 there was a Symposium on algebraic topology in Paris, sponsored by the Centre National de la Recherche Scientifique, in which Leray gave a short survey of his theory [318] and Cartan gave a paper on his own first results on the same subject,* but before the Proceedings of that Symposium were published in 1949 Cartan's conceptions on sheaves had changed, and he withdrew the paper. During the academic year 1948–1949 Cartan started his famous series of Seminars ([423]–[426]), and half of the first one was devoted to sheaf theory; again that half was not reproduced in the later edition (available on a commercial basis) of that Seminar. Then in 1950 Leray's long memoir appeared, with full details on sheaf cohomology and its applications ([321] and [322]). This was followed by the Cartan Seminar of 1950–1951, in which he developed his new conceptions, and in 1951 by A. Borel's expository lectures on Leray's theory [57].

As we shall see in more detail (§ 7,F) below, one of Leray's results was a device for computing the cohomology of a space when the cohomology of each subset of a suitable *covering* of the space is known (generalizing his fundamental result of 1945 on "couvertures" with contractible supports). This kind of result can be thought of as a "passage from local properties to global properties." Now this kind of problem had already been attacked in another part of mathematics, namely, the theory of meromorphic functions of several complex variables. In 1883 Poincaré proved that a function of two complex variables x, y, which is a quotient of two functions holomorphic in a *sufficiently small neighborhood* around each point $(x_0, y_0) \in \mathbf{C}^2$, is in fact a quotient of two functions *holomorphic in the whole space* $\mathbf{C}^2$. In 1895 Cousin generalized Poincaré's theorem to some types of open subsets of $\mathbf{C}^n$, and in so doing he introduced new problems of "passage from local to global." These problems are of a degree of difficulty far greater than the trivial one of defining a function by its restrictions to the sets of a covering, for the objects to be defined globally are *classes* of functions, for instance classes of the type $f + \mathrm{H}$, where f is a meromorphic function and H is the set of *all* holomorphic functions in a given

* In the Proceedings of that Symposium, Leray and Cartan published a joint paper on applications of sheaf cohomology to the actions of groups on topological spaces [114].

open set. After 1934 the work of H. Cartan and Oka on functions of several complex variables was centered on such problems, which had been neglected after Cousin. It is certainly more than a coincidence that in Cartan's paper of 1945 [106] on homology (§ 5,A) he proved typical theorems of "passage from local to global" for homology groups $\mathscr{F}(U) = H_n(U; T)$ of open subsets U of a locally compact space of dimension* $\leqslant n$:

F1. If U is a union of open subsets U_i, and s', s'' are two elements of $\mathscr{F}(U)$ having same images in each U_i, then $s' = s''$.
F2. If U is a union of open subsets U_i, and the $s_i \in \mathscr{F}(U_i)$ are such that, for each pair (i,j) for which $U_i \cap U_j \neq \emptyset$, s_i and s_j have the same image in $\mathscr{F}(U_i \cap U_j)$, then there is an $s \in \mathscr{F}(U)$ having image s_i in each $\mathscr{F}(U_i)$.

However, both in that paper and in his work on ideals of holomorphic functions Cartan needed modules attached to *open* subsets. He was therefore led to modify Leray's definition of a sheaf by considering families $U \mapsto \mathscr{F}(U)$ assigning an A-module to each *open* subset U of a space X together with A-homomorphisms $\rho_{VU}: \mathscr{F}(U) \to \mathscr{F}(V)$ for pairs $U \supset V$ of open sets, such that $\rho_{UU} = \mathrm{Id}.$ and $\rho_{WU} = \rho_{WV}\rho_{VU}$ for $U \supset V \supset W$. These families later became known as *presheaves*. For points $x \in X$, the *stalks* $\mathscr{F}(x)$ are defined as

$$\mathscr{F}(x) = \varinjlim \mathscr{F}(U), \tag{116}$$

where the direct limit is over the directed set (by the relation $\supset$) of open neighborhoods of x in X; if one calls *sections over* U of the presheaf $\mathscr{F}$ the elements of $\mathscr{F}(U)$, the elements of $\mathscr{F}(x)$ are the *germs* of sections of $\mathscr{F}$ at the point x.

For each section $s \in \mathscr{F}(U)$ and each $x \in U$ let $s_x = \rho_{x,U}(s)$ be the germ of s at the point x, $\rho_{x,U}: \mathscr{F}(U) \to \mathscr{F}(x)$ being the canonical map into the direct limit; and associate to s the map

$$\tilde{s}: x \mapsto s_x \tag{117}$$

of U into the disjoint union $\coprod_{x \in U} \mathscr{F}(x)$ of the fibers $\mathscr{F}(x)$ for $x \in U$. Let $\tilde{\mathscr{F}}(U)$ be the subset of $\coprod_{x \in U} \mathscr{F}(x)$, image of $\mathscr{F}(U)$ by the map $s \mapsto \tilde{s}$; it is again an A-module ($\tilde{s} + \tilde{s}'$ being the map $x \mapsto s_x + s'_x$). Furthermore, for an open set $V \subset U$ and a point $x \in V$, $\rho_{x,U} = \rho_{x,V}\rho_{VU}$, hence $s_x = (\rho_{VU}(s))_x$. Let $\tilde{\rho}_{VU}$ be the map of $\tilde{\mathscr{F}}(U)$ into $\tilde{\mathscr{F}}(V)$ defined by† $\tilde{s} \mapsto (\rho_{VU}(s))^{\sim}$; for these maps the family $U \mapsto \tilde{\mathscr{F}}(U)$ is again a presheaf. But in general the maps $s \mapsto \tilde{s}$ of $\mathscr{F}(U)$ into $\tilde{\mathscr{F}}(U)$ are not bijective. Indeed, it is easily verified that $\tilde{\mathscr{F}}$ *satisfies the conditions* F1 *and* F2 above; it is the presheaves having that property that are the *sheaves* in Cartan's terminology,‡ and $\tilde{\mathscr{F}}$ is the sheaf *associated* with

* Recall that, although $X \mapsto \Gamma^r(X)$ is a *covariant* functor, for locally compact spaces X, there is a natural homomorphism $\Gamma^r(X) \to \Gamma^r(U)$ for any *open* subset U of X.

† This definition is meaningful, for if $\rho_{x,U}(s) = \rho_{x,U}(s')$ for two sections, there is a neighborhood $W \subset V \subset U$ of x such that $\rho_{WU}(s) = \rho_{WU}(s')$.

‡ For an example of a presheaf $\mathscr{F}$ that is not a sheaf, take $X = \mathbf{R}$, and $\mathscr{F}(U)$ equal to the $\mathbf{R}$-vector space of all *bounded* continuous functions in U; then $\tilde{\mathscr{F}}(U)$ is the vector space of *all* functions continuous in U, bounded or not ([57], p. V-3).

the presheaf $\mathscr{F}$; the maps $\mathscr{F}(\mathrm{U}) \to \tilde{\mathscr{F}}(\mathrm{U})$ are bijective if and only if $\mathscr{F}$ is a sheaf.

This definition of sheaves is the first one given in Godement's book [208]; it was later universally adopted, chiefly through the influence of Grothendieck. But in his 1950–1951 Seminar, on a suggestion of M. Lazard, Cartan adopted an equivalent definition based on topological considerations. Let $\mathscr{F}$ be a topological space and $p: \mathscr{F} \to \mathrm{X}$ be a surjective continuous map that is a *local homeomorphism*, i.e., each $z \in \mathscr{F}$ has an open neighborhood V in $\mathscr{F}$ such that the restriction $p|\mathrm{V}: \mathrm{V} \to \mathrm{X}$ is a homeomorphism of V onto an *open* subset of X. This is by definition a *sheaf of sets* over X; a *section* of $\mathscr{F}$ over an open subset U of X is a *continuous* map $s: \mathrm{U} \to \mathscr{F}$ such that $p(s(x)) = x$ for $x \in \mathrm{U}$; when, for each $x \in \mathrm{X}$, $p^{-1}(x)$ has a structure of A-module (resp. A-algebra) such that the laws of composition are continuous for the topology induced on $p^{-1}(x)$, $\mathscr{F}$ is a *sheaf of* A-*modules* (resp. A-*algebras*) over X. If, for this definition, for each open subset U of X, $\mathscr{F}(\mathrm{U})$ is defined as $p^{-1}(\mathrm{U})$, it follows immediately that $\mathrm{U} \mapsto \mathscr{F}(\mathrm{U})$ is a sheaf according to the previous definition. Conversely, if $\mathrm{U} \mapsto \mathscr{F}(\mathrm{U})$ is a sheaf over X, the set $\mathscr{F}$ is the disjoint union of the $\mathscr{F}(x)$ for $x \in \mathrm{X}$, and the topology on $\mathscr{F}$ is defined as the coarsest one such that for each open subset U of X the $s \in \mathscr{F}(\mathrm{U})$ are continuous.

From this "topological" definition it follows at once that if two sections s, s' belonging to $\mathscr{F}(\mathrm{U})$ coincide at a point $x \in \mathrm{U}$, they also coincide in an *open neighborhood* of x in U. In particular, if $\mathscr{F}$ is a sheaf of A-modules, the map $x \mapsto 0_x$ is a section of $\mathscr{F}$ over X, written 0; for any section $s \in \mathscr{F}(\mathrm{U})$, the set of points $x \in \mathrm{U}$ such that $s_x \neq 0_x$ is *closed* in U; it is called the *support* of s.

The usual algebraic notions for modules (resp. algebras)—submodule (resp. subalgebra, ideal), quotient module (resp. quotient algebra), sum and intersection of submodules, direct sum of modules, graded module, homomorphism, kernel and image of a homomorphism, exact sequence of modules, module of homomorphisms, tensor product, and direct limit—are immediately extended to *presheaves*. A subpresheaf $\mathscr{G}$ of a presheaf $\mathscr{F}$ is such that, for each open subset U of X, $\mathscr{G}(\mathrm{U})$ is a submodule of $\mathscr{F}(\mathrm{U})$, and for an open set $\mathrm{V} \subset \mathrm{U}$, the diagram

$$\begin{array}{ccc} \mathscr{G}(\mathrm{U}) & \longrightarrow & \mathscr{F}(\mathrm{U}) \\ {\scriptstyle \rho_{\mathrm{VU}}}\Big\downarrow & & \Big\downarrow{\scriptstyle \rho_{\mathrm{VU}}} \\ \mathscr{G}(\mathrm{V}) & \longrightarrow & \mathscr{F}(\mathrm{V}) \end{array}$$

commutes. But when applied to *sheaves*, these algebraic processes may yield presheaves that are *not* sheaves. If $\mathscr{G}$ is a subsheaf of a sheaf of modules $\mathscr{F}$, $\mathrm{U} \mapsto \mathscr{F}(\mathrm{U})/\mathscr{G}(\mathrm{U})$ is not in general a sheaf; the *quotient sheaf* $\mathscr{F}/\mathscr{G}$ is defined as the sheaf *associated* with that presheaf; similarly for the other definitions.

If $\mathrm{U} \mapsto \mathscr{O}_{\mathrm{U}}$ is a presheaf of A-algebras, then $\mathscr{O}$-Modules, their homomorphisms, presheaves of homomorphisms, and tensor products are defined in the same way.

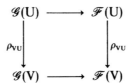

To each A-module M there corresponds a *constant sheaf*: in the "topological" conception it is the product $M \times X$, where M is given the discrete topology [so that M is a trivial covering space (Part 3, chap. I, § 2) of X], and p is the second projection (note that if $U \subset X$ is not connected, sections of that sheaf over U are only *locally constant* maps $U \to M$). More generally, any covering space of X may be taken; if X is arcwise connected and $\mathscr{F}$ is connected, this gives for the system of stalks $(\mathscr{F}(x))_{x \in X}$ a *local system* in the sense of Steenrod (§ 7,A).

Finally, sections of a sheaf $\mathscr{F}$ can be defined over *any* subset S of X; here the topological definition is simplest, a section s over S being a continuous map $s\colon S \to \mathscr{F}$ such that $p(s(x)) = x$ for all $x \in S$. If X is *paracompact* and S is closed in X, or if X is Hausdorff and S has a fundamental system of *paracompact neighborhoods*, then

$$\mathscr{F}(S) = \varinjlim \mathscr{F}(U) \tag{118}$$

for the set $\mathscr{F}(S)$ of sections over S, the direct limit being taken over the directed set of open neighborhoods of S. This shows that $S \mapsto \mathscr{F}(S)$ is then a *sheaf in the sense of Leray*.

C. Sheaf Cohomology

In his 1946 Notes Leray considered, over a space X, a sheaf $\mathscr{F}$ (in his sense) of A-modules or A-algebras (apparently the ring A of scalars is arbitrary). He defined the cohomology of X *with coefficients in* $\mathscr{F}$ by a direct generalization of his definitions in his 1945 paper (§ 6,D); more precisely, he defined the cohomology of a "couverture" $C^\bullet = (C^p)$ of X *with coefficients in* $\mathscr{F}$. As we mentioned in § 6,D, at that time he still assumed that each C^p has a finite A-basis $(e^{p,\alpha})$, and he considered "linear combinations" $\sum_\alpha s_\alpha e^{p,\alpha}$, where the s_α are not scalars but *sections* of $\mathscr{F}$: each s_α is an element of $\mathscr{F}(E)$ for a closed set E containing the support $|e^{p,\alpha}|$; with the condition that if $E' \supset E \supset |e^{p,\alpha}|$ is another closed set, $s_\alpha e^{p,\alpha} = (\rho_{EE'}(s_\alpha))e^{p,\alpha}$ for all $s_\alpha \in \mathscr{F}(E')$. The coboundary is defined by

$$\mathbf{d}_p\left(\sum_\alpha s_\alpha e^{p,\alpha}\right) = \sum_\alpha s_\alpha \mathbf{d}_p e^{p,\alpha}. \tag{119}$$

Leray then stated without any further detail that provided the space X is normal, his results of 1945 would carry over without difficulty.

In his 1947 survey [318] Leray was more explicit; as we have seen (§ 6,D), he limited himself to a *locally compact* space X, and considered a *fine "couverture"* $C^\bullet$ *with compact supports*. First he gave a precise definition of what should be meant in that case by the sums $\sum_\alpha s_\alpha e^{p,\alpha}$ of his 1946 Notes. In order to do this, he defined a "concrete complex" $C^\bullet \bigcirc \mathscr{F}$, the needed generalization of the "intersection" $C^\bullet . C'^\bullet$ of his 1945 paper. In defining "linear combinations" $\sum_\alpha k_\alpha \bigcirc s_\alpha$ for $k_\alpha \in C^p$, $s_\alpha \in \mathscr{F}(E_\alpha)$, and $S(k_\alpha) \subset E_\alpha$ he followed the same pattern Whitney had in his definition of tensor products [511]: he first considered the module of formal linear combinations of the pairs (k, s) for all $k \in C^p$, and all

$s \in \mathscr{F}(\mathrm{E})$ such that $\mathrm{S}(k) \subset \mathrm{E}$, and then took the quotient of that module by the submodule generated by the usual elements

$$(k + k', s) - (k, s) - (k', s), \quad (k, s + s') - (k, s) - (k, s'),$$

$$(\lambda k, s) - \lambda(k, s), \quad \lambda(k, s) - (k, \lambda s)$$

for $\lambda \in \mathrm{A}$ as well as by the elements

$$(k, s) - (k, \rho_{\mathrm{EE'}}(s))$$

for closed sets $\mathrm{E}' \supset \mathrm{E} \supset \mathrm{S}(k)$. This gives an "abstract complex",* and the support of the image of $\sum_\alpha (k_\alpha, s_\alpha)$ in that complex is defined as the set of $x \in \mathrm{X}$ such that $\sum_\alpha (x \,.\, k_\alpha) \otimes (s_\alpha(x)) \neq 0$; finally the corresponding separated complex is by definition $\mathrm{C}^\bullet \bigcirc \mathscr{F}$; it is again a "fine couverture" with compact supports.

The fundamental lemma on fine "couvertures" (§6,D) shows that, for two fine "couvertures" $\mathrm{C}^\bullet$, $\mathrm{C}'^\bullet$, the graded modules $\mathrm{H}^\bullet(\mathrm{C}^\bullet \bigcirc \mathscr{F})$ and $\mathrm{H}^\bullet(\mathrm{C}'^\bullet \bigcirc \mathscr{F})$ are naturally isomorphic, since both are isomorphic to $\mathrm{H}^\bullet((\mathrm{C}^\bullet \bigcirc \mathrm{C}'^\bullet) \bigcirc \mathscr{F})$; this defines (up to isomorphism) the *cohomology* $\mathrm{H}^\bullet(\mathrm{X}; \mathscr{F})$ *of* X *with coefficients in* $\mathscr{F}$.

A. Borel, in the second edition of his lectures [57], showed how this proof could be adopted to the Cartan–Lazard definition of sheaves. To any "concrete complex" $\mathrm{K}^\bullet = (\mathrm{K}^p)$ in Leray's sense is associated a Cartan–Lazard graded sheaf of A-modules $\mathscr{K}^\bullet = (\mathscr{K}^p)$. In the notation of [57] the space of $\mathscr{K}^\bullet$ is the disjoint union of the $x \,.\, \mathrm{K}^\bullet$ for $x \in \mathrm{X}$. Borel defined a neighborhood of u for any $u \in x \,.\, \mathrm{K}^\bullet$ by considering a point $y \in \mathrm{X}$ and the set of elements $y \,.\, k$, where k varies over the set of all elements of $\mathrm{K}^\bullet$ such that $u = x \,.\, k$; the set of all these neighborhoods, when y varies in X, is a fundamental system of neighborhoods of u.

Cartan, on the other hand, defined a *fine* graded sheaf $\mathscr{F}^\bullet = (\mathscr{F}_p)$ over X in the same way as a fine complex (§6,D): for each *open locally finite covering* (U_i) of X, there must be endomorphisms r_i of $\mathscr{F}^\bullet$ such that $r_i(\mathscr{F}_p) \subset \mathscr{F}_p$ for all p, $r_i(s_x) = 0$ if $s_x \in \mathscr{F}^\bullet(x)$ and $x \notin \bar{\mathrm{U}}_i$, and finally $\sum_i r_i = \mathrm{Id}.$. Then, if $\mathrm{K}^\bullet$ is a fine "couverture," the corresponding sheaf $\mathscr{K}^\bullet$ is fine, and for any Cartan–Lazard sheaf $\mathscr{F}$, $\mathrm{K}^\bullet \bigcirc \mathscr{F}$ can be naturally identified with the cochain complex of *sections of* $\mathscr{K}^\bullet \otimes \mathscr{F}$ *over* X *with compact supports.*

Cartan's conceptions on sheaf cohomology in his 1950–1951 Seminar differed from the preceding ones on the following points: (1) the space X is an *arbitrary* topological space; (2) several types of cohomology on X are introduced (to accommodate for instance both de Rham cohomologies $\mathrm{H}^\bullet(\mathrm{X})$ and $\mathrm{H}^\bullet_c(\mathrm{X})$ defined in chap. III, §3, or both the Alexander–Spanier cohomology and the Alexander–Spanier cohomology with compact supports (§6,D)); (3) homological algebra and the functorial language were used to a much greater extent. These conceptions were later presented in a more systematic way, with some simplifications, improvements, and shorter proofs, in Godement's book [208], which we will follow for the sake of simplicity (see also [235]).

---

* The coboundary of the image of (k, s) is the image of $(\mathbf{d}k, s)$.

A *family of supports* in a topological space X is a set Φ of *closed* subsets of X such that:

Φ1. The union of two sets belonging to Φ belongs to Φ.
Φ2. Any closed subset of a set belonging to Φ belongs to Φ.

In what follows we only consider sheaves of commutative groups; for such a sheaf $\mathscr{F}$ over X, we write $\Gamma(\mathscr{F})$ for the group $\mathscr{F}(X)$ of sections of $\mathscr{F}$ over the whole space X, and $\Gamma_\Phi(\mathscr{F})$ for the subgroup of $\Gamma(\mathscr{F})$ consisting of sections with *supports belonging to* Φ.

The main construction by which Cartan defined cohomology consists in considering the presheaf $\mathscr{L}^0 = \mathscr{C}^0(X;\mathscr{F})$ defined by taking, for each open subset U of X

$$\mathscr{L}^0(U) = \prod_{x \in U} \mathscr{F}(x) \tag{120}$$

[i.e. *all* maps $s: x \mapsto s_x$ (continuous or not) for $x \in U$ and $s_x \in \mathscr{F}(x)$]; the map $\rho_{VU}: \mathscr{L}^0(U) \to \mathscr{L}^0(V)$ that defines the presheaf is such that $\rho_{VU}(s)$ is the ordinary *restriction* $s|V$. It is easy to check that $\mathscr{L}^0$ is a *sheaf* of commutative groups, and that $\mathscr{F}(U)$ is a *subgroup* of $\mathscr{L}^0(U)$ for all open subsets U of X, so that this defines an *injective* homomorphism $j: \mathscr{F} \to \mathscr{L}^0$.

Next he *iterated* that construction to form a *resolution* of sheaves in the sense of homological algebra ([119] and [215]):

$$\mathscr{C}^{\bullet}(X;\mathscr{F}): 0 \to \mathscr{F} \xrightarrow{\ j\ } \mathscr{L}^0 \xrightarrow{\ \mathbf{d}_0\ } \mathscr{L}^1 \xrightarrow{\ \mathbf{d}_1\ } \mathscr{L}^2 \to \cdots; \tag{121}$$

$\mathscr{L}^1$ is defined as $\mathscr{C}^0(X; \mathscr{L}^0/j(\mathscr{F}))$, and $\mathbf{d}_0$ is the composite map $\mathscr{L}^0 \to \mathscr{L}^0/j(\mathscr{F}) \to \mathscr{L}^1$; in general $\mathscr{L}^k$ is $\mathscr{C}^0(X; \mathscr{L}^{k-1}/\mathbf{d}_{k-1}(\mathscr{L}^{k-2}))$, and $\mathbf{d}_k$ is the composite map

$$\mathscr{L}^{k-1} \to \mathscr{L}^{k-1}/\mathbf{d}_{k-1}(\mathscr{L}^{k-2}) \to \mathscr{L}^k;$$

(121) is called the *canonical resolution of* $\mathscr{F}$. From (121) is deduced a *cochain complex* of $\mathbf{Z}$-modules

$$C^{\bullet}_\Phi(X;\mathscr{F}): 0 \to \Gamma_\Phi(\mathscr{F}) \to \Gamma_\Phi(\mathscr{L}^0) \to \Gamma_\Phi(\mathscr{L}^1) \to \cdots$$

and the cohomology $H^{\bullet}(C^{\bullet}_\Phi(X;\mathscr{F}))$ is by definition the *cohomology* $H^{\bullet}_\Phi(X;\mathscr{F})$ *of* X *with supports in* Φ *and coefficients in* $\mathscr{F}$. It is a covariant *functor* from the category of sheaves of commutative groups over X, to the category of commutative groups. We will postpone to § 7,E the investigation of the relation of $H^{\bullet}_\Phi(X;\mathscr{F})$ with the usual cohomology theories when $\mathscr{F}$ is a *constant* sheaf. When Φ is the family of *all closed subsets* of X, one writes $H^{\bullet}(X;\mathscr{F})$ instead of $H^{\bullet}_\Phi(X;\mathscr{F})$.

This sheaf cohomology has the three additional properties:

I. From the definition it follows that there is a natural isomorphism

$$\Gamma_\Phi(\mathscr{F}) \cong H^0_\Phi(X;\mathscr{F}). \tag{122}$$

II. For any exact sequence of sheaves

$$0 \to \mathscr{F}' \to \mathscr{F} \to \mathscr{F}'' \to 0 \tag{123}$$

the sequence of cochain complexes

$$0 \to C^{\bullet}_{\Phi}(X; \mathscr{F}') \to C^{\bullet}_{\Phi}(X; \mathscr{F}) \to C^{\bullet}_{\Phi}(X; \mathscr{F}'') \to 0 \tag{124}$$

is also exact; the exact sequence of sheaf cohomology is then

$$\cdots \to H^p_{\Phi}(X; \mathscr{F}') \to H^p_{\Phi}(X; \mathscr{F}) \to H^p_{\Phi}(X; \mathscr{F}'') \xrightarrow{\partial} H^{p+1}_{\Phi}(X; \mathscr{F}') \to \cdots. \tag{125}$$

Furthermore, for a commutative diagram of sheaves

$$\begin{array}{ccccccccc} 0 & \longrightarrow & \mathscr{F}' & \longrightarrow & \mathscr{F} & \longrightarrow & \mathscr{F}'' & \longrightarrow & 0 \\ & & \downarrow & & \downarrow & & \downarrow & & \\ 0 & \longrightarrow & \mathscr{G}' & \longrightarrow & \mathscr{G} & \longrightarrow & \mathscr{G}'' & \longrightarrow & 0 \end{array}$$

with exact lines, the diagrams

$$\begin{array}{ccc} H^p_{\Phi}(X; \mathscr{F}'') & \xrightarrow{\partial} & H^{p+1}_{\Phi}(X; \mathscr{F}') \\ \downarrow & & \downarrow \\ H^p_{\Phi}(X; \mathscr{G}'') & \xrightarrow[\partial]{} & H^{p+1}_{\Phi}(X; \mathscr{G}') \end{array} \tag{126}$$

are commutative.

III. A sheaf $\mathscr{F}$ over X is *flabby* if, for each *open* subset U of X, the map $\rho_{UX}: \mathscr{F}(X) \to \mathscr{F}(U)$ is *surjective* (in other words, each section of $\mathscr{F}$ over U is the image of a section of $\mathscr{F}$ over X). For any flably sheaf $\mathscr{F}$

$$H^n_{\Phi}(X; \mathscr{F}) = 0 \quad \text{for all } n \geqslant 1. \tag{127}$$

This follows from the fact that the sheaves $\mathscr{L}^j$ of (121) are always flabby and that for any exact sequence of *flabby* sheaves

$$0 \to \mathscr{G}^0 \to \mathscr{G}^1 \to \mathscr{G}^2 \to \cdots$$

the sequence of complexes

$$0 \to C^{\bullet}_{\Phi}(\mathscr{G}^0) \to C^{\bullet}_{\Phi}(\mathscr{G}^1) \to C^{\bullet}_{\Phi}(\mathscr{G}^2) \to \cdots$$

is also exact.

It is easy to show that properties I, II, and III *characterize* the functor $\mathscr{F} \mapsto H^{\bullet}_{\Phi}(X; \mathscr{F})$ (for given X and Φ) up to isomorphism. For any sheaf $\mathscr{F}$ over X and any subset E of X, the sheaf $\mathscr{F}|E$ *induced by* $\mathscr{F}$ over E is defined in the simplest way by taking the topological definition, the space $\mathscr{F}|E$ being simply the *subspace* $p^{-1}(E)$ of $\mathscr{F}$.

Suppose E is *locally closed* in X; then, if $\mathscr{G}$ is a sheaf *over* E, there is a unique sheaf $\mathscr{G}^X$ over X such that $\mathscr{G}^X|E = \mathscr{G}$ and $\mathscr{G}^X|(X - E) = 0$; if $\mathscr{F}$ is a sheaf *over* X, there is a unique sheaf $\mathscr{F}_E$ over E such that $(\mathscr{F}_E)^X|E = \mathscr{F}|E$ and $(\mathscr{F}_E)^X|(X - E) = 0$; in general $(\mathscr{F}_E)^X$ *is not a subsheaf of* $\mathscr{F}$, but this property holds if E is *open* in X.

When E is *closed* in X, $(\mathscr{F}_E)^X(U) = \mathscr{F}(E \cap U)$ for all open subsets U of X, and there is an exact sequence of sheaves over X

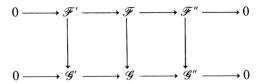

$$0 \to (\mathscr{F}_{X-E})^X \to \mathscr{F} \to (\mathscr{F}_E)^X \to 0 \tag{128}$$

from which, using (125), there follows an exact cohomology sequence

$$\cdots \to H_\Phi^n(X; (\mathscr{F}_{X-E})^X) \to H_\Phi^n(X; \mathscr{F}) \to H_\Phi^n(X; (\mathscr{F}_E)^X) \to H_\Phi^{n+1}(X; (\mathscr{F}_{X-E})^X) \to \cdots. \tag{129}$$

For *any* subset E of X and any sheaf $\mathscr{G}$ over X such that $\mathscr{G}|(X - E) = 0$, there is a natural chain equivalence of complexes of sheaves

$$\mathscr{C}^\bullet(X; \mathscr{G})|E \simeq \mathscr{C}^\bullet(\mathscr{G}|E). \tag{130}$$

When E is *closed* (130) yields an isomorphism*

$$H_\Phi^\bullet(X; \mathscr{G}) \simeq H_{\Phi|E}^\bullet(E; \mathscr{G}|E)$$

where $\Phi|E$ is the set of all subsets $S \in \Phi$ contained in E. In the sequence (129) $H_\Phi^\bullet(X; (\mathscr{F}_E)^X)$ can therefore be replaced by $H_{\Phi|E}^\bullet(E; \mathscr{F}|E)$.

Suppose that, in addition to $\Phi 1$ and $\Phi 2$, Φ satisfies the two conditions:

$\Phi 3$. The sets belonging to Φ are *paracompact*.
$\Phi 4$. For each $S \in \Phi$, there is a closed neighborhood V of S belonging to Φ.

A family Φ of supports satisfying $\Phi 3$ and $\Phi 4$ is called a *paracompactifying* family.

Then, when E is *closed*, there is a natural isomorphism

$$H_{\Phi|(X-E)}^\bullet(X - E; \mathscr{G}|(X - E)) \simeq H_\Phi^\bullet(X; (\mathscr{G}_{X-E})^X)$$

so that, in that case, the exact sequence of cohomology (129) becomes

$$\begin{aligned} \cdots \to H_{\Phi|(X-E)}^n(X - E; \mathscr{F}|(X - E)) \to H_\Phi^n(X; \mathscr{F}) \to H_{\Phi|E}^n(E; \mathscr{F}|E) \\ \to H_{\Phi|(X-E)}^{n+1}(X - E; \mathscr{F}|(X - E)) \to \cdots. \end{aligned} \tag{131}$$

This sequence is in particular valid when X is *locally compact and paracompact* and Φ is the family of *all closed* subsets of X, or of *all compact* subsets of X.

D. Spectral Sequences

The most original and striking part of Leray's 1946 Notes is undoubtedly his investigation of the problem of the cohomological study of a continuous map by means of constructions that led to spectral sequences, and for which I do not think there was the slightest precursor.

In the first Note he observed that for any normal space X and ring A (presumably one of the rings of coefficients of his 1945 paper), he could define the *q-th cohomology sheaf* $\mathscr{H}^q(X; A)$ of X for any integer $q \geqslant 0$ with coefficients in A by the condition that for any closed subset E of X

$$\mathscr{H}^q(X; A)(E) = H^q(E; A).^\dagger \tag{132}$$

---

* There is a similar result of Leray in [321], p. 82.

† The cohomology groups $H^q(E; A)$ are those defined in his 1945 paper [313].

On the other hand, for two normal spaces X, Y, a *closed* continuous map $f: X \to Y$ and any sheaf $\mathscr{F}$ (in his sense) over X, Leray defined the *direct image* $f_*(\mathscr{F})$ as a sheaf over Y such that for any closed subset E of Y

$$f_*(\mathscr{F})(E) = \mathscr{F}(f^{-1}(E)).^* \tag{133}$$

As he had also defined the cohomology of Y with coefficients in a sheaf of A-modules, he put all these notions together and for any pair of integers $p \geqslant 0$, $q \geqslant 0$ he introduced the cohomology A-modules

$$P_1^{p,q} = H^p(Y; f_*(\mathscr{H}^q(X; A))). \tag{134}$$

These were already completely new ideas but what followed (in the second Note) was even more extraordinary. Leray asserted that for each $q \geqslant 0$ there is, in the A-module $H^{q,0} = H^q(X; A)$, a sequence of submodules

$$0 = H^{-1,q+1} \subset H^{0,q} \subset H^{1,q-1} \subset \cdots \subset H^{q-1,1} \subset H^{q,0} \tag{135}$$

and for each pair (p, q), a sequence of submodules of $P_1^{p,q}$

$$0 = Q_0^{p,q} \subset Q_1^{p,q} \subset \cdots \subset Q_{q-1}^{p,q} \subset P_{q+1}^{p,q} \subset P_q^{p,q} \subset \cdots \subset P_1^{p,q} \tag{136}$$

linked by exact sequences of homomorphisms

$$0 \to Q_{q-1}^{p,q} \to P_{q+1}^{p,q} \xrightarrow{\Gamma^{p,q}} H^{p,q}/H^{p+1,q-1} \to 0 \tag{137}$$

and, for each r such that $1 \leqslant r \leqslant q$

$$0 \to P_{r+1}^{p,q} \to P_r^{p,q} \xrightarrow{\Delta_r^{p,q}} Q_r^{p+r+1,q-r}/Q_{r-1}^{p+r+1,q-r} \to 0 \tag{138}$$

He also gave a very cryptic and incomplete sketch of the definitions of these objects in terms of "couvertures" of X and Y, with the definitions and notations of his 1945 paper. The remarkable thing about these results is that once the $P_r^{p,q}$ and $\Delta_r^{p,q}$ are known, the successive quotients of the sequence (135) are also known, which is a major step toward understanding the relations established by f between the cohomologies of X and Y [when A is a field, the vector spaces $H^p(X; A)$ are entirely determined once the quotients $H^{p,q}/H^{p+1,q-1}$ are known].

It was no mean performance by Koszul, one year after Leray's announcements ([284], [285]), to discern the algebraic pattern hidden behind this maze of formulas, and to describe it so clearly that Leray immediately adopted it in his 1947 survey and in his large memoirs of 1950. As he never published proofs of his 1946 Notes in the formulation he had given them, we shall first describe the technique introduced by Koszul and its refinements.

Koszul's starting point was the notion of a *differential* A-*module*, i.e., a pair (M, d) (most of the time simply written M) consisting of an A-module M and an endomorphism d of M (the "differential") such that $d \circ d = 0$.† For such a

---

* The condition that f is closed is needed to ensure that for $y \in Y$, the inverse images of neighborhoods of y form a fundamental system of neighborhoods of $f^{-1}(y)$.

† Following Leray, Koszul supposed that there is in addition on M an algebra structure for which d is a derivation.

module, the submodule Z [or Z(M)] of *cocycles* is the kernel of d, the submodule B [or B(M)] of *coboundaries* is the image of d, and the *cohomology module* H [or H(M)] is the quotient Z/B. The *generalized cochain complexes* of Eilenberg–Mac Lane (§ 5,A) are the *graded* differential modules $E^{\bullet} = \bigoplus_{p \in \mathbf{Z}} E_p$ with the condition $d(E_p) \subset E_{p+1}$. This can be generalized to graded differential modules with the condition $d(E_p) \subset E_{p+r}$ for an integer $r \geqslant 1$ (d is then a *differential of degree* r); the module $H(E^{\bullet})$ is then graded by the submodules

$$H^p(E^{\bullet}) = \mathrm{Ker}(d_p)/\mathrm{Im}(d_{p-r}) \quad \text{with } d_p = d|E_p. \tag{139}$$

The essential notion of Koszul's Notes (following a suggestion of Cartan) is not the graded differential module but the *filtered* differential module M, defined by a sequence $(F^p(M))_{p \in \mathbf{Z}}$ such that $F^{p+1}(M) \subset F^p(M)$ and $d(F^p(M)) \subset F^p(M)$ for all $p \in \mathbf{Z}$. Writing for short $M^p = F^p(M)$, there are corresponding filtrations $Z^p = M^p \cap Z$ and $B^p = M^p \cap B$ on Z and B, with $B^p \subset Z^p$. The image of Z^p in $H = Z/B$ is immediately seen to be

$$F^pH(M) = Z^p/B^p \tag{140}$$

and these submodules define a filtration on H(M).

In most applications, $\bigcap_{p \in \mathbf{Z}} F^p(M) = 0$, so that the elements of $F^p(M)$ may be intuitively thought of as "more and more negligible" as p tends to $+\infty$; an element $z \in M^p$ should thus be considered "almost a cocycle" if $dz \in M^{p+r}$ for some large r. This leads to the idea that the set Z_r^p of the elements $z \in M^p$ such that $dz \in M^{p+r}$ should be considered an "approximation" of the set Z^p of p-cocycles. From the inclusion $M^{p+1} \subset M^p$ it follows that $Z_{r-1}^{p+1} \subset Z_r^p$, and from the relation $d \circ d = 0$ that $B_{r-1}^p = dZ_{r-1}^{p+1-r} \subset Z_r^p$, leading by (139) to the quotients

$$E_r^p = Z_r^p/(Z_{r-1}^{p+1} + B_{r-1}^p). \tag{141}$$

Now, there are the inclusions

$$\begin{aligned} 0 = B_0^p \subset B_1^p \subset \cdots \subset B_s^p \subset B_{s+1}^p \subset \cdots \subset B^p \subset Z^p \subset \cdots \subset Z_{r+1}^p \subset Z_r^p \\ \subset \cdots \subset Z_1^p \subset Z_0^p = M^p, \end{aligned} \tag{142}$$

in particular $Z_r^{p+1} \subset Z_{r-1}^{p+1}$, and from the comparison of formulas (140) and (141) E_r^p can be considered an "approximation" to

$$E_\infty^p = F^pH(M)/F^{p+1}H(M) = Z^p/(Z^{p+1} + B^p).$$

From these definitions, it follows that the differential d of M applies Z_r^p into Z_r^{p+r}, and $B_{r-1}^p + Z_{r-1}^{p+1}$ into B_{r-1}^{p+r}, defining a homomorphism

$$d_r \colon E_r^p \to E_r^{p+r} \quad \text{for all } p, \tag{143}$$

and as $d^2 = 0$, $d_r^2 = 0$, so that d_r is a *differential of degree* r in the graded module

$$E_r^{\bullet} = \bigoplus_{p \in \mathbf{Z}} E_r^p, \quad \text{also written } E_r^{\bullet}(M). \tag{144}$$

An easy computation shows that, for that differential,

$$\mathrm{H}^p(\mathrm{E}_r^{\bullet}) = \mathrm{E}_{r+1}^p \quad \text{for all } p,$$

or equivalently

$$\mathrm{H}^{\bullet}(\mathrm{E}_r^{\bullet}) = \mathrm{E}_{r+1}^{\bullet}. \tag{145}$$

For any filtered module N, the *graded* module

$$\mathrm{gr}^{\bullet}(\mathrm{N}) = \bigoplus_{p \in \mathbf{Z}} \mathrm{F}^p(\mathrm{N})/\mathrm{F}^{p+1}(\mathrm{N})$$

is *associated* with the filtration F. The graded modules $\mathrm{E}_r^{\bullet}$ may thus be considered as "approximations" to the graded module

$$\mathrm{gr}^{\bullet}(\mathrm{H}(\mathrm{M})) = \bigoplus_{p \in \mathbf{Z}} \mathrm{F}^p\mathrm{H}(\mathrm{M})/\mathrm{F}^{p+1}\mathrm{H}(\mathrm{M}). \tag{146}$$

When the filtration $[\mathrm{F}^p(\mathrm{M})]$ is *finite*, i.e., $\mathrm{F}^p(\mathrm{M}) = \mathrm{M}$ for $p < u$ and $\mathrm{F}^p(\mathrm{M}) = 0$ for $p > v$, $\mathrm{E}_r^{\bullet} = \mathrm{gr}^{\bullet}(\mathrm{H}(\mathrm{M}))$ *exactly* for $r \geqslant v - u$, so the determination of the $\mathrm{gr}^{\bullet}(\mathrm{M})\,(= \mathrm{E}_r^{\bullet}$ for $r \leqslant u - v)$ and of the d_r for $u - v \leqslant r \leqslant v - u$ gives the actual computation of $\mathrm{gr}^{\bullet}\mathrm{H}(\mathrm{M})$.

In general, the sequence $(\mathrm{E}_r^{\bullet})$ of graded modules is called the *spectral sequence* of the filtered differential module (M, F), and the graded module $\mathrm{gr}^{\bullet}\mathrm{H}(\mathrm{M})$ is its *abutment*.

In many cases, one starts from a *graded* differential module

$$\mathrm{M}^{\bullet} = \bigoplus_{q \in \mathbf{Z}} \mathrm{M}_q \quad \text{with } d(\mathrm{M}_q) \subset \mathrm{M}_{q+1}$$

and the grading is *compatible* with the filtration F, i.e.,

$$\mathrm{F}^p(\mathrm{M}^{\bullet}) = \bigoplus_{q \in \mathbf{Z}} \mathrm{M}_q \cap \mathrm{F}^p(\mathrm{M}^{\bullet}); \tag{147}$$

then $\mathrm{M}^{\bullet}$ is a *filtered generalized cochain complex*. Write

$$\mathrm{Z}_r^{pq} = \mathrm{Z}_r^p \cap \mathrm{M}_{p+q}, \qquad \mathrm{B}_r^{pq} = \mathrm{B}_r^p \cap \mathrm{M}_{p+q} \tag{148}$$

so that $\mathrm{Z}_r^p = \bigoplus_q \mathrm{Z}_r^{pq}$, $\mathrm{B}_r^p = \bigoplus_q \mathrm{B}_r^{pq}$, and, for each pair (p, q),

$$0 = \mathrm{B}_0^{pq} \subset \mathrm{B}_1^{pq} \subset \cdots \subset \mathrm{B}_s^{pq} \subset \mathrm{B}_{s+1}^{pq} \subset \cdots \subset \mathrm{Z}_{r+1}^{pq} \subset \mathrm{Z}_r^{pq} \subset \cdots \subset \mathrm{Z}_1^{pq} \subset \mathrm{Z}_0^{pq}$$
$$= \mathrm{M}^p \cap \mathrm{M}_{p+q}.$$

Furthermore, $\mathrm{B}_{r-1}^{pq} \subset \mathrm{Z}_r^{pq}$, $\mathrm{Z}_{r-1}^{p+1,q-1} \subset \mathrm{Z}_r^{pq}$, and therefore

$$\mathrm{E}_r^{pq} = \mathrm{Z}_r^{pq}/(\mathrm{Z}_{r-1}^{p+1,q-1} + \mathrm{B}_{r-1}^{pq}) \tag{149}$$

can be defined.* For an element of E_r^{pq}, p is often called the *filtering degree*, q is the *complementary degree*, and $p + q$ is the *total degree*; now

$$d_r(\mathrm{E}_r^{pq}) \subset \mathrm{E}_r^{p+r,q-r+1} \tag{150}$$

---

* In the case considered in Leray's Note [315], the term E_r^{pq} is equal to $\mathrm{P}_r^{p,q}/\mathrm{Q}_{r-2}^{p,q}$ with the notations of (136) [285].

and in $E^{\cdot}_{r+1}$, identified to $H^{\cdot}(E^{\cdot}_r)$, the elements of E^{pq}_{r+1} are the cohomology classes of cocycles of E^{pq}_r. The filtration of the cohomology module $H^{\cdot}(M^{\cdot}) = \bigoplus_q H^q(M^{\cdot})$ is compatible with the grading, each $H^q(M^{\cdot})$ being filtered by

$$F^p H^q(M^{\cdot}) = H^q(M^{\cdot}) \cap F^p H^{\cdot}(M^{\cdot})$$

and the E^{pq}_r are "approximations" of the modules

$$E^{pq}_{\infty} = F^p H^{p+q}(M^{\cdot})/F^{p+1} H^{p+q}(M^{\cdot}). \tag{151}$$

After Koszul's Note, many expositions of the theory were soon given from different points of view ([113], [335], [429]); a particularly thorough study is in the book by Cartan and Eilenberg. Spectral sequences yield important natural homomorphisms in special cases. If $F^p(M^{\cdot}) = M^{\cdot}$ for $p \leqslant 0$ [or equivalently $F^0(M^{\cdot}) = M^{\cdot}$], there are natural homomorphisms

$$H^n(M^{\cdot}) \to E^{0,n}_r \quad \text{for } r \geqslant 1, \tag{152}$$

and if $M_n \cap F^p(M^{\cdot}) = 0$ for $p > n$, there are natural surjective homomorphisms

$$E^{n,0}_r \to E^{n,0}_{r+1} \quad \text{for } r \geqslant 2 \tag{153}$$

and the group $F^p H^n(M^{\cdot})/F^{p+1} H^n(M^{\cdot})$ is then the *direct limit* of the $E^{n,0}_r$ for these homomorphisms; this implies the existence of natural homomorphisms

$$E^{n,0}_2 \to H^n(M^{\cdot}). \tag{154}$$

The homomorphisms (152) and (154) are called *edge homomorphisms*. If *both* conditions $F^0(M^{\cdot}) = M^{\cdot}$ and $M_n \cap F^p(M^{\cdot}) = 0$ for $p > n$ are satisfied, there is also a natural *exact sequence of terms of low degree*

$$0 \to E^{1,0}_2 \to H^1(M^{\cdot}) \to E^{0,1}_2 \xrightarrow{d_3} E^{2,0}_2 \to H^2(M^{\cdot}). \tag{155}$$

An important example of the application of spectral sequences is given by the *double generalized cochain complexes* studied in [113]: they are direct sums

$$M^{\cdot\cdot} = \bigoplus_{(p,q) \in \mathbf{Z} \times \mathbf{Z}} M^{pq}$$

of submodules, equipped with *two* "partial differentials," endomorphisms d_1, d_2 such that $d_1(M^{pq}) \subset M^{p+1,q}$, $d_2(M^{pq}) \subset M^{p,q+1}$, $d_1^2 = d_2^2 = d_1 d_2 + d_2 d_1 = 0$. These relations allow a definition of an ordinary generalized cochain complex $(M^{\cdot}, d)$, where

$$M^r = \bigoplus_{p+q=r} M^{pq} \tag{156}$$

and the "total differential" d is defined by

$$d|M^{pq} = d_1|M^{pq} + d_2|M^{pq}.$$

On the other hand, d_1 and d_2 may also be considered as ordinary differentials in the generalized cochain complexes

$$(M^{\cdot}_{\mathrm{I}}, d_1) \quad \text{with } M^{\cdot}_{\mathrm{I}} = \bigoplus_{p \in \mathbf{Z}} M^p_{\mathrm{I}}, \quad M^p_{\mathrm{I}} = \bigoplus_{q \in \mathbf{Z}} M^{pq}, \quad d_1(M^p_{\mathrm{I}}) \subset M^{p+1}_{\mathrm{I}},$$

$$(M^{\cdot}_{\mathrm{II}}, d_2) \quad \text{with } M^{\cdot}_{\mathrm{II}} = \bigoplus_{q \in \mathbf{Z}} M^q_{\mathrm{II}}, \quad M^q_{\mathrm{II}} = \bigoplus_{p \in \mathbf{Z}} M^{pq}, \quad d_2(M^q_{\mathrm{II}}) \subset M^{q+1}_{\mathrm{II}},$$

to which correspond cohomology modules

$$H_I^{\cdot}(M_I^{\cdot}) = \bigoplus_{p\in\mathbf{Z}} H_I^p(M_I^{\cdot}), \qquad H_{II}^{\cdot}(M_{II}^{\cdot}) = \bigoplus_{q\in\mathbf{Z}} H_{II}^q(M_{II}^{\cdot}).$$

But $M_I^{\cdot}$ may be considered as the direct sum of generalized cochain complexes

$$(M_{II}^{\cdot,q}, d_1) \quad \text{with } M_{II}^{\cdot,q} = \bigoplus_{p\in\mathbf{Z}} M^{pq}$$

and d_2 as a cochain transformation $M_{II}^{\cdot,q} \to M_{II}^{\cdot,q+1}$, which yields homomorphisms d_2^* in cohomology, defining a generalized cochain complex

$$H_I^p(M_{II}^{\cdot})\colon \cdots \to H^p(M_{II}^{\cdot,q}) \xrightarrow{d_2^*} H^p(M_{II}^{\cdot,q+1}) \to \cdots \tag{157}$$

and similarly $M_{II}^{\cdot}$ is the direct sum of generalized cochain complexes

$$H_{II}^q(M_I^{\cdot})\colon \cdots \to H^q(M_I^{p\cdot}) \xrightarrow{d_1^*} H^q(M_I^{p+1,\cdot}) \to \cdots. \tag{158}$$

Finally, on $(M^{\cdot}, d)$, there are *two filtrations* compatible with the grading (156), namely,

$$F_I^p(M^{\cdot}) = \bigoplus_{r\geqslant p} M_I^r, \qquad F_{II}^q(M^{\cdot}) = \bigoplus_{s\geqslant p} M_{II}^s. \tag{159}$$

They give rise to two spectral sequences written, respectively, $(I_r^{p,q})$ and $(II_r^{p,q})$; computations show that

$$I_2^{pq} = H_I^p(H_{II}^q(M_I^{\cdot})), \qquad II_2^{pq} = H_{II}^q(H_I^p(M_{II}^{\cdot})). \tag{160}$$

Another method of definition for spectral sequences has been proposed by Massey [335]. It is based on the notion of *exact couple*: this is a pair of modules (A, C), equipped with three homomorphisms

$$\begin{array}{ccc} A & \xrightarrow{f} & A \\ & {\scriptstyle h}\nwarrow \quad \swarrow{\scriptstyle g} & \\ & C & \end{array} \tag{161}$$

such that

$$\operatorname{Im} f = \operatorname{Ker} g, \qquad \operatorname{Im} g = \operatorname{Ker} h, \qquad \operatorname{Im} h = \operatorname{Ker} f. \tag{162}$$

Then $d = g \circ h\colon C \to C$ is such that $d^2 = 0$, so that (C, d) is a differential module.

Let $C' = H(C)$ its homology, and $A' = \operatorname{Im} f = \operatorname{Ker} g$. Then $h(Z(C)) \subset A'$, $h(B(C)) \subset A'$, so that h induces a homomorphism $h'\colon C' = Z(C)/B(C) \to A'$. If $f'\colon A' \to A'$ is the restriction of f, there exists a well-determined homomorphism $g'\colon A' \to C'$, such that

$$\begin{array}{ccc} A' & \xrightarrow{f'} & A' \\ & {\scriptstyle h'}\nwarrow \quad \swarrow{\scriptstyle g'} & \\ & C' & \end{array} \tag{163}$$

is again an exact couple, called the *derived couple* of (161). The construction may be iterated, and starting from bigraded modules A, C and homogeneous homomorphisms f, g, h, it can be shown that the sequence $(C^{(n)}, d^{(n)})$ of differential modules obtained in this way is the sequence $(E_r^{\bullet})$ of terms of a spectral sequence.

E. Applications of Spectral Sequences to Sheaf Cohomology

We first return to the genesis of spectral sequences in Leray's 1946 Notes; the sequences he defined to study the cohomology of a *closed* continuous map are *not* spectral sequences in the sense of Koszul, but the ideas behind their definition are unmistakably the same, namely computing the quotients of the sequence (135) by a construction of modules "approximating" them. In his 1945 paper (§ 6,D) Leray had already defined the *inverse image* $f^{-1}(L^{\bullet})$ of a "concrete complex" $L^{\bullet}$ on a space Y by a closed continuous map $f: X \to Y$; he considered the *same* "abstract complex" as the one underlying $L^{\bullet}$, but to each $l \in L^{\bullet}$ he associated a *support in* X, the inverse image $f^{-1}(S(l))$. This defines a "concrete complex" on X, and the separated complex associated to it is $f^{-1}(L^{\bullet})$ by definition*; its main property (which Leray only proved in his 1947 survey [318]) is that, if $L^{\bullet}$ is a *fine* "couverture" on Y, and $K^{\bullet}$ a fine "couverture" on X, then $C^{\bullet} = f^{-1}(L^{\bullet}) \bigcirc K^{\bullet}$ is again a fine "couverture" on X, and therefore can be used to compute the cohomology $H^{\bullet}(X)$. By definition,†

$$C^{p+q} = \bigoplus_{m} f^{-1}(L^{q+m}) \bigcirc K^{p-m} \tag{164}$$

and the coboundary of $C^{\bullet}$ is obtained by using *both* coboundaries of $L^{\bullet}$ and $K^{\bullet}$ [§ 5, formula (48)] {in contrast with the coboundary of $L^{\bullet} \bigcirc \mathscr{F}$ when $\mathscr{F}$ is a sheaf on Y [formula (119)]}.

Leray's idea was to *compare* the modules $P_1^{p,q}$ and $H^{p+q}(X)$. If $\mathscr{F}^p = f_*(\mathscr{H}^p(X; A))$, an element of $P_1^{p,q}$ is the cohomology class $\mathrm{cl}(z)$ of a *cocycle* $z \in L^q \bigcirc \mathscr{F}^p$; by definition,

$$z = \sum_{\alpha} u^{p,\alpha} l^{q,\alpha}$$

where $l^{q,\alpha} \in L^q$ and $u^{p,\alpha} = v^{p,\alpha} . f^{-1}(S(l^{q,\alpha}))$ is a cocycle in $K^p . f^{-1}(S(l^{q,\alpha}))$. As there is a natural map $L^{\bullet} \to f^{-1}(L^{\bullet})$, Leray could associate to z the cochain in $f^{-1}(L^{\bullet}) \bigcirc K^{\bullet}$

$$w = \sum_{\alpha} v^{p,\alpha} f^{-1}(l^{q,\alpha}).$$

Since $\sum_{\alpha} u^{p,\alpha} d_p(l^{q,\alpha}) = 0$ and $u^{p,\alpha}$ is a cocycle, the coboundary $d_{p+q} w$ in

---

* This definition obviously was at the origin of Leray's definition of the "direct image" of a sheaf [formula (133)].

† The module $f^{-1}(L^{q+m}) \bigcirc K^{p-m}$ is the image in $f^{-1}(L^{\bullet}) \bigcirc K^{\bullet}$ of the tensor product $f^{-1}(L^{q+m}) \otimes K^{p-m}$.

$f^{-1}(\mathrm{L}^\cdot) \bigcirc \mathrm{K}^\cdot$ is a linear combination of terms of the form

$$v'^{p-t,\beta} f^{-1}(l'^{q+t+1,\beta}) \quad \text{for integers } t \geqslant 0. \tag{165}$$

The same, of course, is true for any cochain in

$$\bigoplus_{s \geqslant 1} f^{-1}(\mathrm{L}^{q+s}) \bigcirc \mathrm{K}^{p-s}. \tag{166}$$

Leray then defined $\mathrm{P}_r^{p,q}$ for $r \geqslant 2$ as the submodule of $\mathrm{P}_1^{p,q}$ consisting of the elements $\mathrm{cl}(z)$ such that, when adding to z a suitable cochain of (166), all elements (165) in the coboundary $d_{p+q}z'$ of the cochain z' thus obtained *are such that* $t \geqslant r$. The submodule $\mathrm{Q}_r^{p-r,q+r+1}$ then consists of the cohomology classes of the elements

$$\sum_\beta v'^{p-r,\beta} l'^{q+r+1,\beta}$$

of $\mathrm{L}^{q+r+1} \bigcirc \mathscr{F}^{p-r}$ corresponding to the elements

$$\sum_\beta v'^{p-r,l} f^{-1}(l'^{q+r+1,\beta})$$

in $d_{p+q}z'$. When $r = p + 1$, $d_{p+q}z' = 0$, and $\mathrm{cl}(z')$ can therefore be identified with an element of $\mathrm{H}^{p+q}(\mathrm{X})$, linking the sequences (139) and (136).

As mentioned above, Leray never published a detailed description of the preceding constructions outlined in his 1946 Notes, but adopted the Koszul presentation; he put on $\mathrm{C}^\cdot$ the *filtration* defined by

$$\mathrm{F}^p\mathrm{C}^\cdot = \bigoplus_{j \geqslant p} f^{-1}(\mathrm{L}^j) \bigcirc \mathrm{K}^\cdot \tag{167}$$

and considered the corresponding spectral sequence. This meant working with cochains instead of cohomology classes, certainly a substantial simplification. Most of the results Leray proved in this manner in his 1950 paper [321] had already been announced in the 1946 Notes, however, showing that he had obtained them by the earlier technique outlined in these Notes.

We shall describe these results of Leray (with the exception of those on fibrations, which we postpone to Part 3, chap. IV) in the generalized form they took within the context of Cartan's theory, using the improvements and simplifications due to Cartan and Godement.

A basic result is a method for computing the cohomology $\mathrm{H}_\Phi^\cdot(\mathrm{X}; \mathscr{F})$ by more easily applicable techniques than the one used in its definition in §7,D). Let

$$\mathscr{L}^\cdot : 0 \to \mathscr{L}^0 \xrightarrow{\delta_0} \mathscr{L}^1 \xrightarrow{\delta_1} \mathscr{L}^2 \cdots \tag{168}$$

be a *cochain complex of sheaves*. Associate to it the *double complex of modules*

$$\mathrm{K}^{\cdot\cdot} = \bigoplus_{p \geqslant 0, q \geqslant 0} \mathrm{C}_\Phi^p(\mathrm{X}; \mathscr{L}^q)$$

where the first differential

$$d_1 : \mathrm{C}_\Phi^p(\mathrm{X}; \mathscr{L}^q) \to \mathrm{C}_\Phi^{p+1}(\mathrm{X}; \mathscr{L}^q)$$

is the differential of the cochain complex $\mathrm{C}_\Phi^\cdot(\mathrm{X}; \mathscr{F})$ for $\mathscr{F} = \mathscr{L}^q$ [formula

(122)]; the second differential

$$d_2 \colon C^p_\Phi(X; \mathscr{L}^q) \to C^p_\Phi(X; \mathscr{L}^{q+1})$$

is deduced by functoriality from the map $(-1)^p \delta_q \colon \mathscr{L}^q \to \mathscr{L}^{q+1}$. The conditions $d_1^2 = 0$, $d_2^2 = 0$, $d_1 d_2 + d_2 d_1 = 0$ are easily checked, and two spectral sequences are obtained, for which the computation of the terms E_2 gives

$$I_2^{pq} = H^p_\Phi(X; \mathscr{H}^q(\mathscr{L}^\bullet)) \tag{169}$$

where $\mathscr{H}^\bullet(\mathscr{L}^\bullet)$ is the cohomology of the complex of sheaves $\mathscr{L}^\bullet$, and

$$II_2^{pq} = H^p(H^q_\Phi(X; \mathscr{L}^\bullet)) \tag{170}$$

where $H^q_\Phi(X; \mathscr{L}^\bullet)$ is the cochain complex of modules $(H^q_\Phi(X; \mathscr{L}^s))_{s \geqslant 0}$, with the differential δ_s^* deduced by functoriality from $\delta_s \colon \mathscr{L}^s \to \mathscr{L}^{s+1}$. The abutment of the first spectral sequence is the graded module associated to a filtration of $H^\bullet(\Gamma_\Phi(\mathscr{L}^\bullet))$.

From these results and the properties of spectral sequences it follows in particular that if $\mathscr{L}^\bullet$ is a *resolution* of the sheaf $\mathscr{F}$, for which

$$H^q(H^q_\Phi(X; \mathscr{L}^\bullet)) = 0 \quad \text{for } q \geqslant 1 \text{ and } p \geqslant 0, \tag{171}$$

there is a natural *isomorphism* of graded modules

$$H^\bullet(\Gamma_\Phi(\mathscr{L}^\bullet)) \xrightarrow{\sim} H^\bullet_\Phi(X; \mathscr{F})$$

generalizing the definition of sheaf cohomology by means of the *canonical* resolution (§ 7,C).

This property can in particular be applied to compare the sheaf cohomology with coefficients in a *constant* sheaf $\mathscr{A}$ associated to the constant presheaf $U \mapsto A$, where A is any ring, with other usual cohomology theories. We have seen (§ 7,C) that to the "concrete complex" of Alexander–Spanier cochains over the ring A (§ 6,D) is canonically associated a cochain complex of sheaves, which is a resolution $(\mathscr{L}^\bullet)$ of the constant sheaf $\mathscr{A}$ and consists of *fine* sheaves; but then $H^q_\Phi(X; \mathscr{L}^s) = 0$ for all $s \geqslant 0$ and $q \geqslant 1$ if the family Φ is *paracompactifying*; therefore, for such a family condition (171) is satisfied and $H^\bullet_\Phi(X; \mathscr{A})$ is naturally isomorphic to the Alexander–Spanier cohomology with coefficients in A and supports in Φ. Similarly, on a C^∞ manifold X, the germs of p-forms constitute a *fine* sheaf Ω^p, hence the cohomology $H^\bullet(X; \Omega^\bullet)$ [resp. $H^\bullet_\Phi(X; \Omega^\bullet)$, where Φ is the family of *compact* subsets] is naturally isomorphic to the de Rham cohomology $H^\bullet(X; \mathbf{R})$ [resp. $H^\bullet_c(X; \mathbf{R})$] (chap. III, § 3).

When $\mathscr{F}$ is a sheaf of *algebras* over a commutative ring A, the cohomology $H^\bullet(X; \mathscr{F})$ inherits a structure of algebra over A, and that structure was always prominent in Leray's papers. More generally, products similar to those defined in cohomology with coefficients in a ring (§ 4) can be defined in sheaf cohomology. If $\mathscr{F}$ and $\mathscr{G}$ are two sheaves of A-modules over X, it is possible to define "cup-products" as linear maps

$$H^p_\Phi(X; \mathscr{F}) \otimes_A H^q_\Psi(X; \mathscr{G}) \to H^{p+q}_\Theta(X; \mathscr{F} \otimes \mathscr{G})$$

where $\Theta = \Phi \cap \Psi$; these constructions were developed in Cartan's seminar and in Godement's book [208].

In Cartan's theory Leray's results on the cohomology of a continuous map $f: X \to Y$ take the following form: if $\mathscr{F}$ is a sheaf over X, the cohomology group $H^q(f^{-1}(U); \mathscr{F})$ may be associated to any open set U in Y, and it is immediately seen that $U \mapsto H^q(f^{-1}(U); \mathscr{F})$ is a presheaf. Let $\mathscr{H}^q(f; \mathscr{F})$ be the associated *sheaf on* Y. There is then a spectral sequence (often called the *Leray spectral sequence of f*) having as E_2 terms

$$E_2^{pq} = H^p(Y; \mathscr{H}^q(f; \mathscr{F})),$$

the abutment of which is the bigraded group derived from a filtration on the graded group $H^\bullet(X; \mathscr{F})$; it can be generalized to the case of cohomology with supports in a family of subsets.

When X and Y are locally compact and f is a *proper* map, $\mathscr{H}^q(f; \mathscr{F})(y)$ is isomorphic to $H^q(f^{-1}(y); \mathscr{F})$, which justifies the notation. In particular, if, for each $y \in Y$ and $1 \leqslant q \leqslant m$, $H^q(f^{-1}(y); A) = 0$ and $H^0(f^{-1}(y); A) \simeq A$, there are natural isomorphisms $H^j(Y; A) \simeq H^j(X; A)$ for $0 \leqslant j \leqslant m$; this generalizes the Vietoris theorem mentioned above (§7,B) for compact metric spaces [475], extended by Begle to all compact spaces [47].

F. Coverings and Sheaf Cohomology

Let $\mathfrak{M} = (M_i)_{i \in I}$ be an arbitrary open covering of a space X; we have seen in §6,B how Dowker associated to any commutative group G the cohomology groups $H^p(N(\mathfrak{M}); G)$ of the *nerve* $N(\mathfrak{M})$ with coefficients in G. This construction can be generalized by substituting for G a *presheaf* $\mathscr{G}$ of commutative groups over X: for each combinatorial simplex $s = (M_{i_0}, M_{i_1}, \ldots, M_{i_p})$ of $N(\mathfrak{M})$, let $M_s = M_{i_0} \cap M_{i_1} \cap \cdots \cap M_{i_p}$, which by definition is not empty. A *p-cochain* of $N(\mathfrak{M})$ with coefficients in $\mathscr{G}$ is a map f which, to each p-simplex $s \in N(\mathfrak{M})$, associates *an element* $f(s)$ *of* $\mathscr{G}(M_s)$; if, for a $(p+1)$-simplex of $N(\mathfrak{M})$, $t = (M_{i_0}, M_{i_1}, \ldots, M_{i_{p+1}})$,

$$t_k = (M_{i_0}, \ldots, M_{i_{k-1}}, M_{i_{k+1}}, \ldots, M_{i_{p+1}}) \quad \text{for } 0 \leqslant k \leqslant p+1$$

the coboundary operator is given by

$$(\mathbf{d}_p f)(t) = \sum_{0 \leqslant k \leqslant p+1} (-1)^k \rho_{M_t M_{t_k}}(f(t_k)) \in \mathscr{G}(M_t) \tag{172}$$

thus defining a cochain complex of modules, written $C^\bullet(\mathfrak{M}; \mathscr{G})$ for short, hence a graded cohomology module $H^\bullet(\mathfrak{M}; \mathscr{G})$; both $C^\bullet(\mathfrak{M}; \mathscr{G})$ and $H^\bullet(\mathfrak{M}; \mathscr{G})$ are *covariant functors* of $\mathscr{G}$.

One can also define a cochain complex $C^\bullet_\Phi(\mathfrak{M}; \mathscr{G})$ for any family of supports Φ: a p-cochain f in that complex is restricted by the condition that there is a set $T \in \Phi$ such that $f(s)_x = 0$ in $\mathscr{G}_x$ for all points $x \notin T \cap M_s$.

The definition of $C^\bullet(\mathfrak{M}; \mathscr{G})$ only uses the groups $\mathscr{G}(M_s)$ and not all groups $\mathscr{G}(U)$ for arbitrary open sets U; therefore, $C^\bullet(\mathfrak{M}; \mathscr{G})$ and $H^\bullet(\mathfrak{M}; \mathscr{G})$ can still be defined when $\mathscr{G}$ is what one calls a *system of coefficients* on $N(\mathfrak{M})$: this means a map $s \mapsto \mathscr{G}(s)$ which associates a module to each simplex s of $\mathrm{N}(\mathfrak{M})$, with restriction homomorphisms $\rho_{ts}: \mathscr{G}(s) \to \mathscr{G}(t)$ when $s \subset t$ (hence $\mathrm{M}_t \subset \mathrm{M}_s$), with the usual condition $\rho_{us} = \rho_{ut} \circ \rho_{ts}$ for $s \subset t \subset u$. Now let U be any open subset of X; $\mathfrak{M} \cap \mathrm{U} = (\mathrm{M}_i \cap \mathrm{U})_{i \in \mathrm{I}}$ is then an open covering of U, and the cochain complex $\mathrm{C}^{\bullet}(\mathfrak{M} \cap \mathrm{U}; \mathscr{G})$ for a presheaf $\mathscr{G}$ over X can be defined; furthermore, if $\mathrm{V} \subset \mathrm{U}$ is another open set, there is a cochain transformation

$$\mathrm{C}^{\bullet}(\mathfrak{M} \cap \mathrm{U}; \mathscr{G}) \to \mathrm{C}^{\bullet}(\mathfrak{M} \cap \mathrm{V}; \mathscr{G})$$

which defines $\mathrm{U} \mapsto \mathrm{C}^{\bullet}(\mathfrak{M} \cap \mathrm{U}; \mathscr{G})$ as a *cochain complex of presheaves* $(\mathscr{C}^p(\mathfrak{M}; \mathscr{G}))_{p \geqslant 0}$. When $\mathscr{G}$ is a *sheaf*, the $\mathscr{C}^p(\mathfrak{M}; \mathscr{G})$ are also *sheaves*, and they constitute a *resolution* of $\mathscr{G}$

$$\mathscr{C}^{\bullet}(\mathfrak{M}; \mathscr{G}): 0 \to \mathscr{G} \to \mathscr{C}^0(\mathfrak{M}; \mathscr{G}) \to \mathscr{C}^1(\mathfrak{M}; \mathscr{G}) \to \cdots.$$

When $\mathscr{G}$ is a *sheaf*, all the preceding constructions and results are still valid when $\mathfrak{M}$ is a *closed locally finite covering* of X.

When $\mathscr{G}$ is a *sheaf*, one can therefore consider either the double complex $\mathrm{C}^{\bullet}(\mathrm{X}; \mathscr{C}^{\bullet}(\mathfrak{M}; \mathscr{G}))$ when $\mathfrak{M}$ is a closed locally finite covering, or the double complex $\mathrm{C}^{\bullet}(\mathfrak{M}; \mathscr{L}^{\bullet})$ when $\mathfrak{M}$ is an open covering and $\mathscr{L}^{\bullet}$ is the canonical resolution (§ 7,C) of G; in both cases, one gets natural homomorphisms

$$\mathrm{H}^n(\mathfrak{M}; \mathscr{G}) \to \mathrm{H}^n(\mathrm{X}; \mathscr{G}) \tag{173}$$

and a spectral sequence abutting to the bigraded module associated to a filtration of $\mathrm{H}^{\bullet}(\mathrm{X}; \mathscr{G})$, and having as E_2 terms

$$\mathrm{E}_2^{pq} = \mathrm{H}^p(\mathfrak{M}; \mathscr{H}^q(\mathscr{G})). \tag{174}$$

For closed locally finite coverings, this is an extension to general spaces of a theorem of Leray: when $\mathrm{H}^q(\mathrm{M}_s; \mathscr{G}) = 0$ for $q \geqslant 1$ and for all simplices $s \in \mathrm{N}(\mathfrak{M})$, the homomorphism (173) is *bijective*; this generalizes the result of Leray's 1945 paper on "couvertures" with simple supports (§ 6,D).

If $\mathfrak{N}$ is an open covering finer than $\mathfrak{M}$, one can define a simplicial mapping of the nerves: $\mathrm{N}(\mathfrak{N}) \to \mathrm{N}(\mathfrak{M})$, from which one can deduce a cochain transformation $\mathrm{C}^{\bullet}_{\Phi}(\mathfrak{M}; \mathscr{G}) \to \mathrm{C}^{\bullet}_{\Phi}(\mathfrak{N}; \mathscr{G})$ for any presheaf $\mathscr{G}$, in the same manner as for a group G (§ 3). The corresponding homomorphism $\mathrm{H}^{\bullet}_{\Phi}(\mathfrak{M}; \mathscr{G}) \to \mathrm{H}^{\bullet}_{\Phi}(\mathfrak{N}; \mathscr{G})$ again does not depend on the choice of the simplicial mapping $\mathrm{N}(\mathfrak{N}) \to \mathrm{N}(\mathfrak{M})$. It is therefore possible to define the *Čech cohomology of* X *with coefficients in the presheaf* $\mathscr{G}$ *and supports in* Φ

$$\check{\mathrm{H}}^{\bullet}_{\Phi}(\mathrm{X}; \mathscr{G}) = \varinjlim \mathrm{H}^{\bullet}_{\Phi}(\mathfrak{M}; \mathscr{G}) \tag{175}$$

the direct limit being taken on the directed set of all open coverings $\mathfrak{M}$ of X. To any exact sequence of *presheaves*

$$0 \to \mathscr{G}' \to \mathscr{G} \to \mathscr{G}'' \to 0$$

corresponds a Čech cohomology exact sequence

$$\cdots \to \check{\mathrm{H}}^n_{\Phi}(\mathrm{X}; \mathscr{G}') \to \check{\mathrm{H}}^n_{\Phi}(\mathrm{X}; \mathscr{G}) \to \check{\mathrm{H}}^n_{\Phi}(\mathrm{X}; \mathscr{G}'') \to \check{\mathrm{H}}^{n+1}_{\Phi}(\mathrm{X}; \mathscr{G}') \to \cdots.$$

For a *sheaf* $\mathscr{G}$, there is a natural homomorphism

$$\check{H}^{\bullet}_{\Phi}(X;\mathscr{G}) \rightarrow H^{\bullet}_{\Phi}(X;\mathscr{G}) \tag{176}$$

and a spectral sequence abutting on the bigraded module associated to a filtration of $H^{\bullet}_{\Phi}(X;\mathscr{G})$, and having as E_2 terms

$$E_2^{pq} = H^p_{\Phi}(X;\mathscr{H}^q(\mathscr{G})). \tag{177}$$

If Φ is also a *paracompactifying* family, then the natural map (176) is *bijective*.

If $\mathscr{F}$ is a presheaf and $\tilde{\mathscr{F}}$ is the associated sheaf (§ 7,B)), then, if Φ is a *paracompactifying* family, the natural homomorphism

$$\check{H}^{\bullet}_{\Phi}(X;\mathscr{F}) \rightarrow \check{H}^{\bullet}_{\Phi}(X;\tilde{\mathscr{F}})$$

is *bijective.*

When $\mathscr{A}$ is the constant presheaf $U \mapsto A$, $\check{H}^{\bullet}_{\Phi}(X;\mathscr{A})$ is identical with the usual Čech cohomology group $\check{H}^{\bullet}_{\Phi}(X;A)$.

G. Borel–Moore Homology

We have seen that Čech cohomology [or equivalently Alexander–Spanier cohomology (§ 6,A)] is well behaved for *any* space and *any* group of coefficients (§ 6,B), whereas Čech homology can only be used in special cases, such as compact spaces with compact coefficient groups (§ 6,B). Cohomology starts with application of a "duality" functor Hom(., L) to *chain complexes*; we may wonder if the application of the functor Hom(., L) to *cochain complexes* might not lead to a better homology theory, a kind of "bidual" of the classical ones.

This was apparently first attempted in 1957 [62] by A. Borel, who limited himself to *locally compact* spaces and coefficients in a *principal ideal ring* L (although his most important results concern the case in which L is a *field*). His main purpose was to build up a homology theory with which he could obtain a "Poincaré duality" for more general spaces than either the C^0 manifolds considered by Cartan around 1948–1950 or the *compact* "generalized manifolds" studied since 1934 by Čech, Lefschetz, Wilder, and Begle, who also limited themselves to coefficients in a field (see Part 2, chap. IV, § 3). In this first paper Borel took as "dual" of a L-module M the L-module Hom(M, L); although he used sheaf cohomology to some extent he did not study in detail the general properties of the homology theory he described, his investigations being mainly limited to "generalized manifolds."

Two years later, in a joint paper with J.C. Moore [66], Borel returned to the problem from a different angle. They still only considered locally compact spaces, but they changed the notion of "duality" for modules, which enabled them to take any *Dedekind ring* as a ring of coefficients (the applications were still mainly concerned with principal ideal rings). Their new approach used sheaves to a much greater extent, but their main innovation was to apply the notion of an *injective module.*

This was first introduced by R. Baer in 1940 [39], and given prominence in the Cartan–Eilenberg book on homological algebra [113]. For any ring Λ, a

Λ-module I is called *injective* if, for *any* Λ-module M and *any* submodule M′ of M, any homomorphism $M' \to I$ is the restriction of a homomorphism $M \to I$; when $\Lambda = \mathbf{Z}$, by transfinite induction the quotient $\mathbf{Q}/\mathbf{Z}$ is injective. The main properties of injective modules are:

1. Any exact sequence $0 \to I \to M \to M'' \to 0$ where I is injective is *split*.
2. Any Λ-module M is isomorphic to a *submodule of an injective module.*

This injective module can be made to depend *functorially* on M. First, if $\Lambda = \mathbf{Z}$, the free $\mathbf{Z}$-module

$$F(M) = \mathbf{Z}^{(M^*)}$$

has as basis the elements $\neq 0$ of M. There is a natural exact sequence of $\mathbf{Z}$-modules

$$0 \to R(M) \to F(M) \to M \to 0$$

and R(M) is naturally embedded in $F(M) \otimes_{\mathbf{Z}} \mathbf{Q}$; if $\overline{M} = (F(M) \otimes_{\mathbf{Z}} \mathbf{Q})/R(M)$, there is a natural injection $M \to \overline{M}$, and $\overline{M}$ is an injective $\mathbf{Z}$-module.

Now if Λ is arbitrary and M is a Λ-module, first consider M as a $\mathbf{Z}$-module, and take the corresponding injective $\mathbf{Z}$-module $\overline{M}$. Then one has the exact sequence of Λ-modules

$$0 \to \mathrm{Hom}_{\mathbf{Z}}(\Lambda, M) \to \mathrm{Hom}_{\mathbf{Z}}(\Lambda, \overline{M});$$

as a Λ-module, M is naturally identified with $\mathrm{Hom}_{\mathbf{Z}}(\Lambda, M)$ and $I(M) = \mathrm{Hom}_{\mathbf{Z}}(\Lambda, \overline{M})$ is an *injective* Λ-module, which depends functorially on M.

This definition of injective modules can be transferred to the definition of *injective sheaves* of Λ-modules: a sheaf $\mathscr{I}$ is injective if for any sheaf $\mathscr{L}$ and any subsheaf $\mathscr{L}'$, any homomorphism of sheaves $\mathscr{L}' \to \mathscr{I}$ is the restriction of a homomorphism $\mathscr{L} \to \mathscr{I}$; an injective sheaf is *flabby* (§ 7,C).

Again, any sheaf $\mathscr{F}$ is functorially isomorphic to a subsheaf of an injective sheaf $\mathscr{I}^0(\mathscr{F})$, which is defined as the presheaf

$$U \mapsto \prod_{x \in U} I(\mathscr{F}_x)$$

where $I(\mathscr{F}_x)$ is the injective module associated to the stalk $\mathscr{F}_x$, as defined above. Then the same method used in § 7,C leading to the canonical resolution $\mathscr{C}^{\bullet}(X; \mathscr{F})$ of a sheaf $\mathscr{F}$ [§ 7,C, formula (121)], can be applied by merely replacing $\mathscr{C}^0(X; \mathscr{F})$ by $\mathscr{I}^0(\mathscr{F})$; this yields a *canonical injective resolution* of $\mathscr{F}$

$$\mathscr{I}^{\bullet}(\mathscr{F}): 0 \to \mathscr{F} \to \mathscr{I}^0(\mathscr{F}) \to \mathscr{I}^1(\mathscr{F}) \to \cdots \qquad (178)$$

which can be used to compute the cohomology $H^{\bullet}_{\Phi}(X; \mathscr{F})$, since the injective sheaves $\mathscr{I}^p(\mathscr{F})$ are flabby (§ 7,E).

After these preliminaries, the first step in the definition of the Borel–Moore homology is purely algebraic. L is a Dedekind ring, K is its field of fractions; then K/L is an *injective* L-module. To a *cochain complex* $M^{\bullet} = (M^r)_{r \geqslant 0}$ of L-modules, is associated a *generalized chain complex* $D_{\bullet}(M^{\bullet}) = (D_r(M^{\bullet}))_{r \geqslant -1}$ defined in the following way: $D_r(M^{\bullet})$ is the set of homomorphisms f of $M^{\bullet}$ into the direct sum $K \oplus (K/L)$ such that

$$f(M^r) \subset K, \qquad f(M^{r+1}) \subset K/L, \qquad f(M^s) = 0 \quad \text{for } s \neq r, r+1.$$

If $\varphi\colon K \to K/L$ is the natural map, the boundary operator $\mathbf{b}_r\colon D_r(M^\bullet) \to D_{r-1}(M^\bullet)$ is given by

$$(\mathbf{b}_r f)(x) = (-1)^{r+1} f(\mathbf{d}_{r-1} x) \quad \text{for } x \in M^{r-1},$$

$$(\mathbf{b}_r f)(x) = \varphi(f(x)) + (-1)^{r+1} f(\mathbf{d}_r x) \quad \text{for } x \in M^r,$$

$$(\mathbf{b}_r f)(x) = 0 \quad \text{for } x \in M^s,\ s \neq r-1, r.$$

The next step is to associate to a *cochain complex of sheaves* (of L-modules) $\mathscr{F}^\bullet$ the cochain complex of L-modules $\Gamma_c(\mathscr{F}^\bullet)$, consisting of sections of $\mathscr{F}^\bullet$ over X with compact supports. Then, using the definition of $D_\bullet(M^\bullet)$ for a cochain complex of L-modules given above, a *generalized chain complex* of sheaves $\mathscr{D}_\bullet(\mathscr{F}^\bullet)$ is defined by taking the sheaves associated to the presheaves

$$U \mapsto D_\bullet(\Gamma_c(\mathscr{F}^\bullet | U)).$$

If in particular the cochain complex $\mathscr{I}^\bullet(L)$, canonical injective resolution of the constant sheaf L [formula (178)], is taken for $\mathscr{F}$, then one defines

$$\mathscr{C}_{H\bullet}(X; L) = \mathscr{D}_\bullet(\mathscr{I}^\bullet(L)).$$

For any family Φ of supports in X,

$$C^\Phi_{H\bullet}(X; L) = \Gamma_\Phi(\mathscr{C}_{H\bullet}(X; L))$$

is therefore a generalized chain complex of L-modules; finally its homology group

$$H^\Phi_n(X; L) = H_n(C^\Phi_{H\bullet}(X; L)) \tag{179}$$

is by definition the *Borel–Moore n-dimensional homology group of* X *with coefficients in* L *and supports in* Φ. When Φ is the family of all closed sets (resp. all compact subsets), one simply writes $H_n(X; L)$ [resp. $H^c_n(X; L)$].

The justification for this involved definition is that Borel–Moore homology has almost all the good properties that can be expected. If $f\colon X \to Y$ is a *proper* map of locally compact spaces, there is for every n a natural homomorphism

$$f_n\colon H_n(X; L) \to H_n(Y; L). \tag{180}$$

Similarly, for any continuous map $f\colon X \to Y$, there is for every n a natural homomorphism

$$f_n\colon H^c_n(X; L) \to H^c_n(Y; L); \tag{181}$$

furthermore, if f and g are homotopic, $f_n = g_n$ for each n.

For a locally closed subspace A of a locally compact space X, *relative Borel–Moore homology* $H_\bullet(X, A; L)$ can be defined, satisfying the usual exact sequence and the excision theorem [87]. If A is closed, $H^\bullet(X, A; L)$ is naturally isomorphic to $H_\bullet(X - A; L)$.

For any closed subspace F of X, the natural injection $i\colon F \to X$ yields a natural homomorphism $i_n\colon H_n(F; L) \to H_n(X; L)$; but, if $U = X - F$, there is also a natural homomorphism

$$j_n\colon H_n(X; L) \to H_n(U; L) \tag{182}$$

which is defined by considering the restrictions to U of the sections of the sheaf $\mathscr{C}_{H^\bullet}(X; L)$. Moreover, there is an *exact homology sequence*

$$\cdots \to H_q(F; L) \xrightarrow{i_q} H_q(X; L) \xrightarrow{j_q} H_q(U; L) \to H_{q-1}(F; L) \to \cdots \tag{183}$$

There is also the expected *split* exact sequence linking homology to sheaf cohomology with compact supports [see § 5H, formula (74)]

$$0 \to \mathrm{Ext}(H_c^{p+1}(X; L), L) \to H_p(X; L) \to \mathrm{Hom}(H_c^p(X; L), L) \to 0. \tag{184}$$

Finally, there is a kind of Künneth exact sequence, in which homology and cohomology are mixed:

$$0 \to \bigoplus_{r+s=p+1} \mathrm{Ext}(H_c^r(X), H_s(Y)) \to H_p(X \times Y) \to \bigoplus_{r+s=p} \mathrm{Hom}(H_c^r(X), H_s(Y)) \to 0 \tag{185}$$

(all coefficients in L). However, there are examples ([87], p. 235) of locally compact spaces X, Y, for which $H_1(X \times Y; \mathbf{Z}) \neq 0$, $H_0(X; \mathbf{Z}) \otimes H_1(Y; \mathbf{Z}) = 0$, and $\mathrm{Tor}(H_0(X; \mathbf{Z}), H_0(Y; \mathbf{Z})) = 0$, so that the usual Künneth sequence [§ 5H, formula (76)] is invalid for Borel–Moore homology.

When X is a *compact* space, and $\mathfrak{m}$ is a maximal ideal in L, $H_\bullet(X; L/\mathfrak{m})$ is the *Čech homology* $\check{H}_\bullet(X; L/\mathfrak{m})$. When X is a *compact metric* space, the groups $H_n(X; L)$ are the Steenrod groups $H_n''(X; L)$ (§ 2).

From the fact that $D_r(M^\bullet)$ is defined for $r = -1$ and $\neq 0$ in general, it is possible that $H_{-1}(X; L) \neq 0$ although no such example is known. It has been proved that if X is locally connected, $H_{-1}(X; L) = 0$ ([66], p. 151); this is also true if X is metric and separable ([87], p. 185).

Another feature of Borel–Moore homology is that in general there is no relation between Borel–Moore homology groups of the same space but with two different coefficient rings.

Finally, for any *sheaf* $\mathscr{F}$ of L-modules over X, it is possible to define Borel–Moore homology with supports in Φ and *coefficients in* $\mathscr{F}$: consider the graded sheaf

$$\mathscr{C}_{H^\bullet}(X; \mathscr{F}) = \mathscr{F} \otimes \mathscr{C}_{H^\bullet}(X; L)$$

and the generalized chain complex of L-modules

$$C_{H^\bullet}^{\Phi}(X; \mathscr{F}) = \Gamma_\Phi(\mathscr{C}_{H^\bullet}(X; \mathscr{F})).$$

The Borel–Moore homology with supports in Φ and coefficients in $\mathscr{F}$ is the homology

$$H_\bullet^\Phi(X; \mathscr{F}) = H_\bullet(C_{H^\bullet}^\Phi(X; \mathscr{F})).$$

If Φ is *paracompactifying* and $0 \to \mathscr{F}' \to \mathscr{F} \to \mathscr{F}'' \to 0$ is an exact sequence of sheaves of L-modules over X, there is an exact homomology sequence

$$\cdots \to H_n^{\Phi}(X; \mathscr{F}') \to H_n^{\Phi}(X; \mathscr{F}) \to H_n^{\Phi}(X; \mathscr{F}'') \to H_{n-1}^{\Phi}(X; \mathscr{F}') \to \cdots .$$

§ 8. Homological Algebra and Category Theory

A. Homological Algebra

Sheaf theory and spectral sequences spurred new developments in homological algebra and the theory of categories after 1950. To the first functors of commutative groups or modules Eilenberg and Mac Lane (§ 5,E) gave as examples, they soon added many more, as did Hopf, Cartan, Hochschild, Chevalley, E. Artin, and Tate, some of whom were motivated by algebraic topology, others by purely algebraic theories (associative algebras, finite groups, Lie algebras).

In their book on homological algebra (a name which they introduced) published in 1956 [113], Cartan and Eilenberg made a thorough study of these methods and results, to which they added many of their own. The most prominent concepts, which dominate the book and allow a unified presentation of all the examples mentioned above, are the projective and injective modules and the *derived functors.*

Projective modules are generalizations of *free* modules. A projective Λ-module* is a submodule of a free Λ-module that is a *direct summand.* For any Λ-module M and any quotient module M″ of M, any homomorphism $f: P \to M''$, where P is a projective Λ-module, can be "lifted" to a homomorphism $g: P \to M$, i.e., one has $f = \pi \circ g$, where $\pi: M \to M''$ is the natural homomorphism. When P is projective, any exact sequence

$$0 \to M' \to M \to P \to 0$$

splits (a trivial result when P is free).

We have seen (§ 5,F) how *free resolutions* of a Λ-module had been introduced by Hopf in 1945 as *exact* chain complexes

$$(C_\bullet, \varepsilon): \cdots \to C_j \to C_{j-1} \to \cdots \to C_1 \to C_0 \xrightarrow{\varepsilon} A \to 0 \qquad (186)$$

where the C_j are *free*; as any Λ-module is a quotient of a free module, such resolutions exist for *any* Λ-module A. The two fundamental properties proved by Hopf for these resolutions (§ 5,F) are still valid when the C_j are merely *projective* Λ-modules, in which case (186) is called a *projective resolution* of A.

Cartan and Eilenberg observed that the way Hopf used the free resolutions to define the *homology of groups* (Part 3, chap. V, § 1,D) could be used for all *additive covariant functors* $T: \boldsymbol{Mod}_\Lambda \to \boldsymbol{Ab}$ from the category $\boldsymbol{Mod}_\Lambda$ of Λ-

* For simplicity's sake, we assume that Λ is commutative, but Cartan and Eilenberg also considered left or right modules over a noncommutative ring, as well as *bimodules* over two such rings.

modules to the category ***Ab*** of commutative groups: for any pair of homomorphisms $u\colon A \to B$, $u'\colon A \to B$ of Λ-modules

$$T(u + u') = T(u) + T(u')$$

for the corresponding homomorphisms of commutative groups $T(A) \to T(B)$. If (186) is any projective resolution of A, a chain complex of commutative groups can be deduced from the complex (186):

$$T(C_\bullet)\colon \cdots \to T(C_j) \to T(C_{j-1}) \to \cdots \to T(C_1) \to T(C_0) \to 0 \qquad (187)$$

and the corresponding homology groups $H_n(T(C_\bullet))$ are *independent* of the chosen projective resolution $(C_\bullet, \varepsilon)$, up to isomorphism; if

$$L_n T(A) = H_n(T(C_\bullet)) \qquad (188)$$

then $L_n T\colon \boldsymbol{Mod}_\Lambda \to \boldsymbol{Ab}$ is a functor that Cartan and Eilenberg call the *n-th left derived functor* of T. When $T(A) = B \otimes_\Lambda A$ for a fixed Λ-module B, $L_n T(A)$ is written $\mathrm{Tor}_n^\Lambda(B, A)$; $\mathrm{Tor}_1^\Lambda(B, A)$ is just the functor Tor(B, A) defined in § 5,B.

For any exact sequence of Λ-modules

$$0 \to A \to B \to C \to 0$$

there is an *exact homology sequence* of left derived functors

$$\cdots \to L_{n+1}T(C) \to L_n T(A) \to L_n T(B) \to L_n T(C) \to \cdots. \qquad (189)$$

The properties of injective modules (§ 7,G) led Cartan and Eilenberg to introduce, in a similar way, *injective resolutions* of a Λ-module A: they are *exact cochain complexes*

$$(C^\bullet, \varepsilon)\colon 0 \to A \xrightarrow{\varepsilon} C^0 \to C^1 \to \cdots \to C^j \to C^{j+1} \to \cdots \qquad (190)$$

where the C^j are injective Λ-modules; again, such resolutions exist for *any* Λ-module A. For any covariant additive functor $T\colon \boldsymbol{Mod}_\Lambda \to \boldsymbol{Ab}$, a cochain complex of commutative groups can be deduced from (190)

$$T(C^\bullet)\colon 0 \to T(C^0) \to T(C^1) \to \cdots \to T(C^j) \to T(C^{j+1}) \to \cdots \qquad (191)$$

and the corresponding cohomology groups $H^n(T(C^\bullet))$ again only depend on A up to isomorphism; writing

$$R^n T(A) = H^n(T(C^\bullet)), \qquad (192)$$

Cartan and Eilenberg called the functor $R^n T\colon \boldsymbol{Mod}_\Lambda \to \boldsymbol{Ab}$ the *n-th right derived functor* of T. When $T(A) = \mathrm{Hom}_\Lambda(B, A)$ for a fixed Λ-module B, $R^n T(A)$ is written $\mathrm{Ext}_\Lambda^n(B, A)$; $\mathrm{Ext}_\Lambda^1(B, A)$ is just the functor Ext(B, A) defined in § 5,D).

For any exact sequence of Λ-modules

$$0 \to A \to B \to C \to 0$$

there is an *exact cohomology sequence* of right derived functors

$$\cdots \to R^n T(A) \to R^n T(B) \to R^n T(C) \to R^{n+1} T(A) \to \cdots. \qquad (193)$$

Although Cartan and Eilenberg only considered the category of Λ-modules over an arbitrary ring Λ, topologists used categorical concepts in other contexts, as we have already seen for homology theories by Eilenberg and Steenrod (§6,B), and as we shall also see later in the emerging theory of homotopy (Part 3, chap. II). At the end of the 1950s categorical notions also became one of the main tools in Grothendieck's recasting of algebraic geometry into the new frame of the theory of schemes. In all these applications it was soon realized that it was necessary to make frequent use of properties and constructions applicable to *all* categories, which had not been mentioned by Eilenberg and Mac Lane. Their proofs are invariably trivial; we shall review those which are linked to three of the most conspicuous themes in the applications of the theory of categories: dual categories, representable functors, and abelian categories.

B. Dual Categories

In a Note of 1948, developed in 1950 [353], Mac Lane explicitly remarked that many notions in group theory naturally come *in pairs*; moreover, when they could be expressed by conditions on diagrams of homomorphisms, they could be deduced from one another by "reversing the arrows." The simplest example is the pair consisting of *injective homomorphisms*, expressed by the exactness of the sequence

$$0 \to M' \to M,$$

and *surjective homomorphisms*, expressed by the exactness of the sequence

$$0 \leftarrow M'' \leftarrow M.$$

Another example given by Mac Lane is the pair of notions *direct product* and *free product* in group theory. The direct product $D = G \times H$ of two groups and its projections $p_1\colon D \to G$, $p_2\colon D \to H$ are such that for any group X and any pair of homomorphisms $u\colon X \to G$, $v\colon X \to H$, there is a unique homomorphism $w\colon X \to D$ for which the diagram

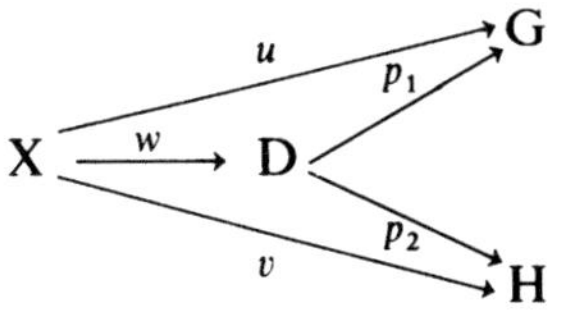

(194)

becomes commutative. Similarly, Mac Lane observed that the free product*

* To define $G * H$, consider the monoid M consisting in *words* of any length whose elements are either in G or in H, the composition law $(w, w') \mapsto ww'$ being juxtaposition. An equivalence relation R is defined in M by taking as equivalent two words w_1, w_2 such that in one of them there are two consecutive elements that belong both to G or both to H; one replaces them by their product in G (resp. H) to obtain the other word. Then $G * H$ is the quotient M/R.

$F = G * H$ and its injections $i_1: G \to F$, $i_2: H \to F$ are such that for any group X and any pair of homomorphisms $u: G \to X$, $v: H \to X$, there is a unique homomorphism $w: F \to X$ for which the diagram

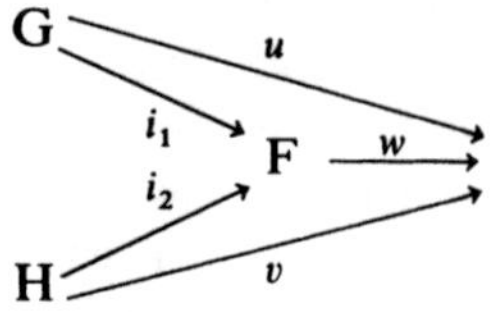

(195)

becomes commutative, and clearly the diagrams (194) and (195) are deduced from one another by "reversal of the arrows."

Mac Lane also thought of the pair of notions consisting in *injective* Λ-modules and *free* Λ-modules. Injective modules I can be defined by the property that, given a diagram of homomorphisms

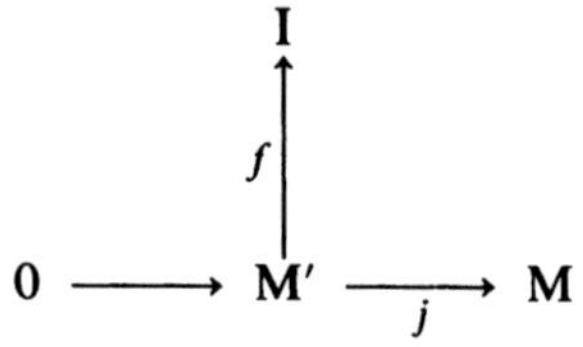

$$\begin{array}{c} I \\ \uparrow f \\ 0 \longrightarrow M' \xrightarrow{j} M \end{array}$$
(196)

where the line is exact, there exists a homomorphism $g: M \to I$ for which the diagram

becomes commutative. For a free Λ-module F, for a diagram (with exact line) obtained from (196) by reversal of arrows

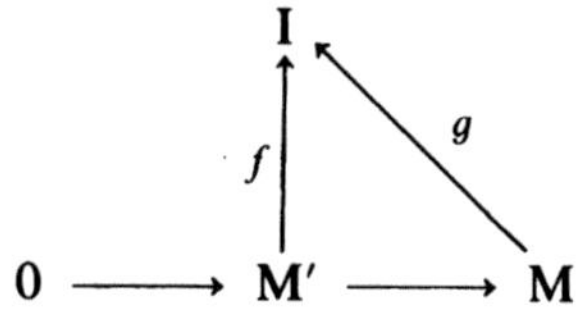

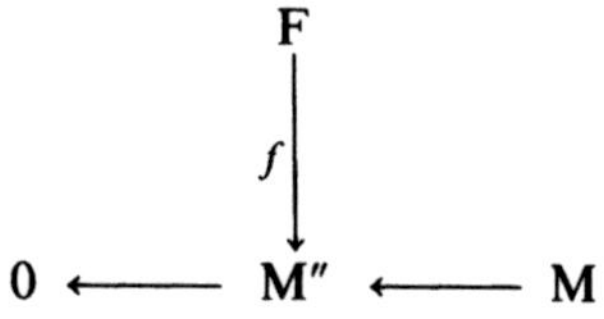

$$\begin{array}{c} F \\ \downarrow f \\ 0 \longleftarrow M'' \longleftarrow M \end{array}$$
(197)

there exists similarly a homomorphism $h: F \to M$ for which the diagram

becomes commutative. However that property *does not characterize* free

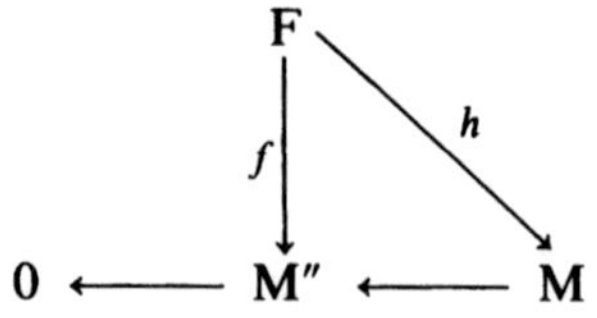

Λ-modules but *projective* Λ-modules, not yet defined in 1948; but Mac Lane limited himself to the case $\Lambda = \mathbf{Z}$, where projective modules are free.

In 1955 [215] Grothendieck found a general principle underlying these observations, the fact that *all* categories can be considered as coming in "*dual pairs*" (or "opposite pairs"). Indeed, for any category $\boldsymbol{C}$, let $\mathrm{Mor}_{\boldsymbol{C}}(\mathrm{X}, \mathrm{Y})$ be the set of morphisms $\mathrm{X} \to \mathrm{Y}$ in $\boldsymbol{C}$, for any two objects X, Y of $\boldsymbol{C}$. The *dual* category $\boldsymbol{C}^0$ has the *same* objects as $\boldsymbol{C}$, but for any two such objects X, Y,

$$\mathrm{Mor}_{\boldsymbol{C}^0}(\mathrm{X}, \mathrm{Y}) = \mathrm{Mor}_{\boldsymbol{C}}(\mathrm{Y}, \mathrm{X}) \tag{198}$$

and the composition law

$$\mathrm{Mor}_{\boldsymbol{C}^0}(\mathrm{X}, \mathrm{Y}) \times \mathrm{Mor}_{\boldsymbol{C}^0}(\mathrm{Y}, \mathrm{Z}) \to \mathrm{Mor}_{\boldsymbol{C}^0}(\mathrm{X}, \mathrm{Z})$$

is $(u, v) \mapsto uv$; it can immediately be seen that $\boldsymbol{C}^0$ is a category and that $(\boldsymbol{C}^0)^0 = \boldsymbol{C}$.

One of the obvious consequences of that definition is that there is no longer any need to distinguish covariant and contravariant functors: a contravariant functor $\boldsymbol{C} \to \boldsymbol{C}'$ is the same thing as a (covariant) functor $\boldsymbol{C}^0 \to \boldsymbol{C}'$ (or $\boldsymbol{C} \to \boldsymbol{C}'^0$). More generally, to *any* notion or property concerning a category $\boldsymbol{C}$ is "paired" another ("dual" to the first) for the *same* category $\boldsymbol{C}$, obtained by applying the primitive one to $\boldsymbol{C}^0$.

In the same paper [215], although it was chiefly oriented toward abelian categories, Grothendieck completed the vocabulary of the theory of categories, already partly laid down by Eilenberg and Mac Lane (§5,E). A *subcategory* $\boldsymbol{C}'$ of a category $\boldsymbol{C}$ is such that objects of $\boldsymbol{C}'$ are some objects of $\boldsymbol{C}$, for any two objects X′, Y′ of $\boldsymbol{C}'$, $\mathrm{Mor}_{\boldsymbol{C}'}(\mathrm{X}', \mathrm{Y}') \subset \mathrm{Mor}_{\boldsymbol{C}}(\mathrm{X}, \mathrm{Y})$, for any three objects X′, Y′, Z′ of $\boldsymbol{C}'$, the composition map

$$\mathrm{Mor}_{\boldsymbol{C}'}(\mathrm{X}', \mathrm{Y}') \times \mathrm{Mor}_{\boldsymbol{C}'}(\mathrm{Y}', \mathrm{Z}') \to \mathrm{Mor}_{\boldsymbol{C}'}(\mathrm{X}', \mathrm{Z}')$$

is the restriction of the composition map in $\boldsymbol{C}$, and finally the identity $1_{\mathrm{X}'}$ for an object of $\boldsymbol{C}'$ is the same in $\boldsymbol{C}'$ and in $\boldsymbol{C}$. A subcategory $\boldsymbol{C}'$ of $\boldsymbol{C}$ is called *full* if, for each pair of objects X′, Y′ of $\boldsymbol{C}'$, $\mathrm{Mor}_{\boldsymbol{C}'}(\mathrm{X}', \mathrm{Y}') = \mathrm{Mor}_{\boldsymbol{C}}(\mathrm{X}', \mathrm{Y}')$; an example is given by the category $\boldsymbol{Ab}$ of commutative groups, which is a full subcategory of the category $\boldsymbol{Gr}$ of groups.

A functor $\mathrm{F}\colon \boldsymbol{C}_1 \to \boldsymbol{C}_2$ is called *faithful* (resp. *fully faithful*) if, for any pair X, Y of objects of $\boldsymbol{C}_1$, the map $u \mapsto \mathrm{F}(u)$ of $\mathrm{Mor}_{\boldsymbol{C}_1}(\mathrm{X}, \mathrm{Y})$ into $\mathrm{Mor}_{\boldsymbol{C}_2}(\mathrm{F}(\mathrm{X}), \mathrm{F}(\mathrm{Y}))$ is *injective* (resp. *bijective*). When F is fully faithful, $\mathrm{F}(\boldsymbol{C}_1)$ is a full subcategory of $\boldsymbol{C}_2$; if, in addition, for any object X_2 of $\boldsymbol{C}_2$ there is an object X_1 of $\boldsymbol{C}_1$ and an *isomorphism* $\mathrm{F}(\mathrm{X}_1) \xrightarrow{\sim} \mathrm{X}_2$, F is called an *equivalence of categories*. In the category $\boldsymbol{Vect}_k$ of finite dimensional vector spaces over a field k, the full subcategory consisting of the vector spaces k^n (for $n \geqslant 0$) is equivalent to $\boldsymbol{Vect}_k$.

C. Representable Functors

Before category theory was invented it had been noted in some parts of mathematics that from some privileged objects, often called "universal," one could construct *all* objects of the same kind. The oldest example is probably the *universal covering manifold* $\tilde{X}$ of a connected C^∞ manifold X, defined by Poincaré when X is a surface (Part 3, chap. I, § 1). Let $p: \tilde{X} \to X$ be the natural projection, $\tilde{x}_0$ be a point of $\tilde{X}$, and $x_0 = p(\tilde{x}_0)$; consider a *simply connected* C^∞ manifold Y, a point $y_0 \in Y$, and a C^∞ map $f: Y \to X$ such that $f(y_0) = x_0$; there is then a *unique* C^∞ map $\tilde{f}: Y \to \tilde{X}$ such that $\tilde{f}(y_0) = \tilde{x}_0$ and $f = p \circ \tilde{f}$. In other words, *all* C^∞ *maps* of simply connected manifolds into connected manifolds are known once the C^∞ maps of simply connected manifolds into *simply connected* manifolds are known.

In his Note of 1948 [333] Mac Lane made a similar observation on a much more elementary concept, the *product* $G \times H$ of two groups: the diagram (194) shows that for each pair (u, v) of homomorphisms of groups $X \to G$, $X \to H$ there is a *unique* homomorphism $w: X \to G \times H$ such that $u = p_1 \circ w$, $v = p_2 \circ w$.

Analogous remarks were made between 1955 and 1965 by several mathematicians (Yoneda, Freyd) and were finally systematized under the general idea of *representable functors*, which became one of the main tools Grothendieck used in his theory of schemes.

The starting point is the consideration, for *any* category $\boldsymbol{C}$ and any object X of $\boldsymbol{C}$, of the contravariant functor

$$\boldsymbol{h}_X: \boldsymbol{C}^0 \to \boldsymbol{Set}$$

into the category of sets, defined as follows: for any object Y of $\boldsymbol{C}$,

$$\boldsymbol{h}_X(Y) = \mathrm{Mor}_{\boldsymbol{C}}(Y, X) \tag{199}$$

and for any morphism $u \in \mathrm{Mor}_{\boldsymbol{C}}(Y, Z)$,

$$\boldsymbol{h}_X(u): \mathrm{Mor}_{\boldsymbol{C}}(Z, X) \to \mathrm{Mor}_{\boldsymbol{C}}(Y, X) \quad \text{is the map } v \mapsto vu. \tag{200}$$

For any morphism $w: X \to X'$ of $\boldsymbol{C}$, there is a *natural transformation* [§ 5,E, formula (68)] of functors $\boldsymbol{h}_w: \boldsymbol{h}_X \to \boldsymbol{h}_{X'}$ defined by taking, for any object Y of $\boldsymbol{C}$, the map

$$\boldsymbol{h}_w(Y): \mathrm{Mor}_{\boldsymbol{C}}(Y, X) \to \mathrm{Mor}_{\boldsymbol{C}}(Y, X') \quad \text{such that } \boldsymbol{h}_w(Y)(v) = wv. \tag{201}$$

When $\boldsymbol{h}_w(Y)$ is *injective* for all objects Y of $\boldsymbol{C}$, the morphism $w: X \to X'$ is called a *monomorphism*.

If $F: \boldsymbol{C}^0 \to \boldsymbol{Set}$ is now *any* contravariant functor and x is any element of the set F(X), a *natural transformation* of functors

$$\beta_x: \boldsymbol{h}_X \to F$$

is defined as follows: for any $v \in \mathrm{Mor}_{\boldsymbol{C}}(Y, X)$, $F(v)$ is a map $F(X) \to F(Y)$, and the map

$$\beta_x(Y): \mathrm{Mor}_{\boldsymbol{C}}(Y, X) \to F(Y) \quad \text{is such that } \beta_x(Y)(v) = (F(v))(x). \tag{202}$$

Functors of type $\boldsymbol{h}_X$ are much easier to handle than general functors; it is therefore very useful to know that for a functor $F: \boldsymbol{C}^0 \to \boldsymbol{Set}$ there is an object X of $\boldsymbol{C}$ and an element $x \in F(X)$ such that β_x is an *isomorphism* of functors.

Then F is called *representable*, and the pair (X, x) (or simply X) represents F. If another pair (X', x') also represents F, there is a *unique isomorphism* $w\colon \mathrm{X} \xrightarrow{\sim} \mathrm{X}'$ such that $x = \mathrm{F}(w)(x')$.

To subsume Mac Lane's remark on the product of two groups under this general concept it is enough to consider, for two objects X, Y of a category $\boldsymbol{C}$, the functor $\mathrm{F}\colon \boldsymbol{C}^0 \to \boldsymbol{Set}$ defined as follows:

$$\mathrm{F(T)} = \mathrm{Mor(T, X)} \times \mathrm{Mor(T, Y)}$$

for any object T of $\boldsymbol{C}$, and for any morphism $v\colon \mathrm{T} \to \mathrm{T}'$, the map

$$\mathrm{F}(v)\colon \mathrm{F(T')} \to \mathrm{F(T)} \quad \text{such that } \mathrm{F}(v)(f', g') = (f'v, g'v).$$

To say that this functor is representable means that there is an object Z of $\boldsymbol{C}$ and a pair $(p_1, p_2) \in \mathrm{Mor(Z, X)} \times \mathrm{Mor(Z, Y)}$ such that $v \mapsto (p_1 v, p_2 v)$ is an isomorphism of functors

$$\mathrm{Mor(T, Z)} \xrightarrow{\sim} \mathrm{Mor(T, X)} \times \mathrm{Mor(T, Y)}.$$

When F is representable, the object Z is called the (*categorical*) *product* of X and Y and is usually written $\mathrm{X} \times \mathrm{Y}$. In the categories $\boldsymbol{Set}$, $\boldsymbol{Gr}$, $\boldsymbol{Ab}$, $\boldsymbol{Mod}_\Lambda$, products exist for all pairs (X, Y); this is also true in the category of commutative rings with unit element (morphisms being ring homomorphisms sending unit element to unit element), but it is *not true* for the full subcategory of *fields*, since the ring $\mathrm{X} \times \mathrm{Y}$, product of two fields, is never a field.

Another even simpler example is given by the "*final*" functor $\mathrm{F}\colon \boldsymbol{C}^0 \to \boldsymbol{Set}$, defined by $\mathrm{F(X)} = \{a\}$, $\mathrm{F}(u) = 1_{\{a\}}$ for any object X of $\boldsymbol{C}$ and any $u \in \mathrm{Mor}_{\boldsymbol{C}}(\mathrm{X}, \mathrm{Y})$ ($\{a\}$ being any singleton). Saying F is representable means that there exists an object e of $\boldsymbol{C}$ such that $\mathrm{Mor}_{\boldsymbol{C}}(\mathrm{Y}, e)$ has *only one* element for all objects Y of $\boldsymbol{C}$; e is then called a *final object* of $\boldsymbol{C}$, and any two of them are uniquely isomorphic. For the category $\boldsymbol{Gr}$ the one element group is a final object, but again the category of fields has *no* final object.

A last example, important in the applications, is the notion of *kernel* of *two* morphisms u_1, u_2 in the same set $\mathrm{Mor}_{\boldsymbol{C}}(\mathrm{X}, \mathrm{Y})$ (Freyd). First consider the category $\boldsymbol{Set}$: if E, F is a pair of sets, and $f_1\colon \mathrm{E} \to \mathrm{F}$, $f_2\colon \mathrm{E} \to \mathrm{F}$ are two arbitrary maps, the *kernel* (or *set of coincidences*) of the pair (f_1, f_2) is the subset $\mathrm{N} = \mathrm{Ker}(f_1, f_2)$ consisting of the elements $x \in \mathrm{E}$ such that $f_1(x) = f_2(x)$; then the diagram of maps

$$\mathrm{M} \xrightarrow[j]{} \mathrm{E} \underset{f_2}{\overset{f_1}{\rightrightarrows}} \mathrm{F}$$

(where j is *injective*) is *exact* if $j(\mathrm{M}) = \mathrm{Ker}(f_1, f_2)$.

Now return to an arbitrary category $\boldsymbol{C}$ and to the pair of morphisms (u_1, u_2). Define a functor $\mathrm{F}\colon \boldsymbol{C}^0 \to \boldsymbol{Set}$ by taking, for any object T of $\boldsymbol{C}$,

$$\mathrm{F(T)} = \mathrm{Ker}(\boldsymbol{h}_{u_1}(\mathrm{T}), \boldsymbol{h}_{u_2}(\mathrm{T}))$$

[formula (201)], and for a morphism $v\colon \mathrm{T} \to \mathrm{T}'$, $\mathrm{F}(v)$ equal to the restriction to $\mathrm{F(T')}$ of the map $\boldsymbol{h}_{\mathrm{X}}(v)$ [formula (200)]. Saying that F is *representable* means there exists a pair (N,j) consisting of an object N of $\boldsymbol{C}$ and a *monomorphism* j: N → X, such that the diagram of *maps*

$$\mathrm{Mor}_{\boldsymbol{C}}(\mathrm{T},\mathrm{N}) \xrightarrow[h_j(\mathrm{T})]{} \mathrm{Mor}_{\boldsymbol{C}}(\mathrm{T},\mathrm{X}) \overset{h_{u_1}(\mathrm{T})}{\underset{h_{u_2}(\mathrm{T})}{\rightrightarrows}} \mathrm{Mor}_{\boldsymbol{C}}(\mathrm{T},\mathrm{Y})$$

is *exact* for all objects T of $\boldsymbol{C}$; N is called the *kernel* of the pair of morphisms (u_1, u_2) and written $\mathrm{Ker}(u_1, u_2)$; the diagram of *morphisms*

$$\mathrm{N} \xrightarrow[j]{} \mathrm{X} \overset{u_1}{\underset{u_2}{\rightrightarrows}} \mathrm{Y}$$

is called *exact*.

The "dual" process (§ 8,B) applied to the preceding notions yields new notions for *any* category $\boldsymbol{C}$:

1. A monomorphism in $\boldsymbol{C}^0$ is called an *epimorphism* in $\boldsymbol{C}$: it is a morphism w: X → X′ such that, for any object Y of $\boldsymbol{C}$, the map $v \mapsto vw$ of $\mathrm{Mor}_{\boldsymbol{C}}(\mathrm{X}', \mathrm{Y})$ into $\mathrm{Mor}_{\boldsymbol{C}}(\mathrm{X}, \mathrm{Y})$ is *injective*.
2. If $\boldsymbol{C}^0$ has a final object e, e is called an *initial object of* $\boldsymbol{C}$; all these objects are uniquely isomorphic to each other, and are such that $\mathrm{Mor}_{\boldsymbol{C}}(e, \mathrm{X})$ has *only one* element for all objects X of $\boldsymbol{C}$. In the category ***Set***, ∅ is an initial object. Let $\boldsymbol{Alg}_k$ be the category of algebras (with unit element) over a field k (morphisms being k-algebra homomorphisms sending unit element on unit element); it has an initial element equal to k.
3. If for two objects X, Y of $\boldsymbol{C}$ their product *in* $\boldsymbol{C}^0$ exists, it is a triplet written $(\mathrm{X} \coprod \mathrm{Y}, j_1, j_2)$. The object $\mathrm{X} \coprod \mathrm{Y}$ is called the (*categorical*) *sum* of X and Y, and $j_1: \mathrm{X} \to \mathrm{X} \coprod \mathrm{Y}$, $j_2: \mathrm{Y} \to \mathrm{X} \coprod \mathrm{Y}$ are morphisms such that $v \mapsto (vj_1, vj_2)$ defines an isomorphism of functors

$$\mathrm{Mor}_{\boldsymbol{C}}(\mathrm{X} \textstyle\coprod \mathrm{Y}, \mathrm{T}) \simeq \mathrm{Mor}_{\boldsymbol{C}}(\mathrm{X}, \mathrm{T}) \times \mathrm{Mor}_{\boldsymbol{C}}(\mathrm{Y}, \mathrm{T}).$$

   Mac Lane's observation concerning the free product of groups can be expressed by saying that in the category ***Gr***, the "sum" of any two objects X, Y exists and is the *free product* X ∗ Y. In the category $\boldsymbol{Alg}_k$ of k-algebras, the "sum" is the *tensor product* $\mathrm{X} \otimes_k \mathrm{Y}$, equipped with the two natural injective homomorphisms $j_1: x \mapsto x \otimes 1$, $j_2: y \mapsto 1 \otimes y$.
4. If kernels exist *in* $\boldsymbol{C}^0$, they are called *cokernels* in $\boldsymbol{C}$. For two morphisms u_1, u_2 in $\mathrm{Mor}_{\boldsymbol{C}}(\mathrm{X}, \mathrm{Y})$, the cokernel $\mathrm{K} = \mathrm{Coker}(u_1, u_2)$ is an object of $\boldsymbol{C}$, equipped with an *epimorphism* p: Y → K, such that the diagram of maps

$$\mathrm{Mor}_{\boldsymbol{C}}(\mathrm{K},\mathrm{T}) \xrightarrow[h'_p(\mathrm{T})]{} \mathrm{Mor}_{\boldsymbol{C}}(\mathrm{Y},\mathrm{T}) \overset{h'_{u_1}(\mathrm{T})}{\underset{h'_{u_2}(\mathrm{T})}{\rightrightarrows}} \mathrm{Mor}_{\boldsymbol{C}}(\mathrm{X},\mathrm{T})$$

   for any object T of $\boldsymbol{C}$ is exact [for $v \in \mathrm{Mor}_{\boldsymbol{C}}(\mathrm{U}, \mathrm{V})$, $h'_v(\mathrm{T})$ is the map $w \mapsto wv$ of $\mathrm{Mor}_{\boldsymbol{C}}(\mathrm{V}, \mathrm{T})$ into $\mathrm{Mor}_{\boldsymbol{C}}(\mathrm{U}, \mathrm{T})$]. The diagram of *morphisms*

$$\mathrm{X} \overset{u_1}{\underset{u_2}{\rightrightarrows}} \mathrm{Y} \xrightarrow[p]{} \mathrm{K}$$

   is called *exact*.

In 1959 Kan published a remarkable observation, similar to the one made by Mac Lane in 1948. Not only categories, but also many *functors*, he said naturally come *in pairs*, by another kind of "duality" that until then was only known in special cases [270]. This is the concept of *adjoint functors*: if $\boldsymbol{C}$, $\boldsymbol{C}'$ are two categories and $\mathrm{F}: \boldsymbol{C} \to \boldsymbol{C}'$ is a functor, then F has a *right adjoint* $\mathrm{F}^{\mathrm{ad}}: \boldsymbol{C}' \to \boldsymbol{C}$ (resp. a *left adjoint* ${}^{\mathrm{ad}}\mathrm{F}: \boldsymbol{C}' \to \boldsymbol{C}$) if there is an *isomorphism of bifunctors**

$$\mathrm{Mor}_{\boldsymbol{C}}(\mathrm{T}, \mathrm{F}^{\mathrm{ad}}(\mathrm{X}')) \xrightarrow{\sim} \mathrm{Mor}_{\boldsymbol{C}'}(\mathrm{F}(\mathrm{T}), \mathrm{X}')$$

[resp.

$$\mathrm{Mor}_{\boldsymbol{C}}({}^{\mathrm{ad}}\mathrm{F}(\mathrm{X}'), \mathrm{T}) \xrightarrow{\sim} \mathrm{Mor}_{\boldsymbol{C}'}(\mathrm{X}', \mathrm{F}(\mathrm{T}))].$$

If F^{ad} (resp. ${}^{\mathrm{ad}}\mathrm{F}$) exists,† ${}^{\mathrm{ad}}(\mathrm{F}^{\mathrm{ad}})$ (resp. $({}^{\mathrm{ad}}\mathrm{F})^{\mathrm{ad}}$) exists and is equal to F.

For $\boldsymbol{C} = \boldsymbol{C}' = \boldsymbol{Mod}_{\Lambda}$, where Λ is a commutative ring, adjoint functors had appeared in linear algebra: for any Λ-module E and the functor $\mathrm{F}: \mathrm{X} \mapsto \mathrm{E} \otimes_{\Lambda} \mathrm{X}$, the functor $\mathrm{Y} \mapsto \mathrm{Hom}_{\Lambda}(\mathrm{E}, \mathrm{Y})$ is equal to a right adjoint F^{ad} of F, as follows from the classical isomorphism

$$\mathrm{Hom}_{\Lambda}(\mathrm{E} \otimes_{\Lambda} \mathrm{X}, \mathrm{Y}) \xrightarrow{\sim} \mathrm{Hom}_{\Lambda}(\mathrm{X}, \mathrm{Hom}_{\Lambda}(\mathrm{E}, \mathrm{Y})).$$

A much less obvious example given by Kan is the *forgetful functor* $\mathrm{F}: \boldsymbol{Gr} \to \boldsymbol{Set}$, which assigns to a group (resp. a homomorphism of groups) that group considered as a *set* (resp. the homomorphism considered as a *map*); it has a *right adjoint* F^{ad} which, to a set T (resp. a map $\mathrm{T} \to \mathrm{T}'$) assigns the *free group* L(T) generated by T [resp. the homomorphism $\mathrm{L}(\mathrm{T}) \to \mathrm{L}(\mathrm{T}')$ which on T coincides with the given map].

Kan's paper showed that there are many other examples of adjoint functors in algebra and in topology.

D. Abelian Categories

A little earlier than Grothendieck, Buchsbaum [95] enlarged Mac Lane's remarks of 1948–1950 and arrived at the idea of abelian categories, at least for some types of categories. He was motivated by the striking way in which Cartan and Eilenberg formulated a large number of their theorems [113]: they came in "double entry," so to speak, a statement for right derived functors being immediately followed by a corresponding one for left derived functors, or vice versa, the proof of the second statement being most often left to the reader. Buchsbaum realized that these pairs of statements were in fact that

---

* The product $\boldsymbol{C} \times \boldsymbol{C}'$ of two categories has as objects the pairs $(\mathrm{X}, \mathrm{X}')$ of an object X of $\boldsymbol{C}$ and an object X′ of $\boldsymbol{C}'$; morphisms are pairs of morphisms (u, u') that are composed componentwise. A *bifunctor* is a functor from a product $\boldsymbol{C} \times \boldsymbol{C}'$ to another category.

† The existence of F^{ad} is equivalent to the fact that for *any* object X′ of $\boldsymbol{C}'$, the functor

$$h_{\mathrm{X}'} \circ \mathrm{F}: \boldsymbol{C}^0 \to \boldsymbol{Set}$$

is *representable*.

same statement applied in succession to the category $\boldsymbol{Mod}_\Lambda$ and to its dual, and he extended the concept of dual categories to more general ones, which he called "exact," and which (as well as the "bicategories" defined earlier by Mac Lane) were special cases of the *abelian categories* defined by Grothendieck in [215].

He first considered a more general kind, the *additive* categories, satisfying the following three conditions:

1. For each pair of objects X, Y of $\boldsymbol{C}$, $\mathrm{Mor}_{\boldsymbol{C}}(X, Y)$ is equipped with a structure of *commutative group*, such that, for all morphisms $u: Z \to X$, $v: Z \to Y$, and all pairs of morphisms (f, g) of $\mathrm{Mor}_{\boldsymbol{C}}(X, Y)$,

$$(f + g)u = fu + gu, \qquad v(f + g) = vf + vg.$$

2. $\boldsymbol{C}$ has a final object A, for which $\mathrm{Mor}_{\boldsymbol{C}}(A, A) = \{0\}$; it is also an initial object, usually written 0.
3. Products $X \times Y$ are defined for every pair X, Y of objects of $\boldsymbol{C}$; this implies that the sum $X \coprod Y$ also exists and is naturally isomorphic to $X \times Y$.

Abelian categories are additive categories satisfying two additional axioms:

4. For each morphism $f: X \to Y$, the pair $(f, 0)$ has both a kernel N and a cokernel K (§ 8,C). In other words, there is an exact diagram of morphisms

$$N \xrightarrow[j]{} X \underset{0}{\overset{f}{\rightrightarrows}} Y \xrightarrow[p]{} K,$$

where j is a monomorphism and p is an epimorphism. One writes $j = \mathrm{Ker}(f)$, $p = \mathrm{Coker}(f)$, and also, if no confusion arises, $N = \mathrm{Ker}(f)$ and $K = \mathrm{Coker}(f)$.

5. Every monomorphism can be written $\mathrm{Ker}(f)$ for some morphism f, and every epimorphism can be written $\mathrm{Coker}(g)$ for some morphism g.

It can then be shown that in an abelian category, every morphism $f: X \to Y$ can be naturally factorized as

$$X \xrightarrow{\mathrm{Coker}(\mathrm{Ker}(f))} I \xrightarrow{\mathrm{Ker}(\mathrm{Coker}(f))} Y$$

(a classical property in $\boldsymbol{Mod}_\Lambda$ which Mac Lane had taken as an axiom in his 1950 paper); I is called the *image* of f and written $\mathrm{Im}(f)$. A sequence of morphisms $X \xrightarrow{f} Y \xrightarrow{g} Z$ is called *exact* when $\mathrm{Im}(f) = \mathrm{Ker}(g)$; for a monomorphism $j: N \to X$ the sequence $0 \to N \xrightarrow{j} X$ is exact, and for an epimorphism $p: X \to K$ the sequence $X \xrightarrow{p} K \to 0$ is exact.

The typical abelian category is $\boldsymbol{Mod}_\Lambda$, and almost all notions and constructions concerning Λ-modules [for instance injective modules, projective modules (§ 8,A)), spectral sequences (§ 7,D)] carry over to all abelian categories. Another example of abelian categories is the category of *sheaves of* Λ-*modules* over a topological space (the study of which was the main motivation of Grothendieck). However in an abelian category $\boldsymbol{C}$ it is not always true that for an object X of *C*, there is a monomorphism of X into an injective object (resp. an epimorphism of a projective object onto X); when that property is true for *any* object X, one says *C* has *enough injective* (resp. *enough projective*) objects. For instance, the category of sheaves of Λ-modules over a space has enough injective objects (§ 7, G), but not enough projective objects in general. When an abelian category has enough injective (resp. projective) objects, one can transfer to that category the definition of right (resp. left) derived functors and all the corresponding results of Cartan–Eilenberg for $\boldsymbol{Mod}_\Lambda$ still hold.

An example of an additive category which is not an abelian category is given by the category whose objects are *topological commutative groups*, and the morphisms *continuous homomorphisms*.

Part 2 The First Applications of Simplicial Methods and of Homology

Introduction

I have gathered in Part 2 various results linked to homology which do not use homotopy theory; most of them were discovered prior to its creation. I think it is better to dissociate them from the foundational material described in Part 1, where they would have digressed from the main themes.

I) We start with the epoch-making results of Brouwer in 1910–1912, which may rightly be called the *first proofs* in algebraic topology, since Poincaré's papers can only be considered as blueprints for theorems to come. However, nobody understands why Brouwer never mentioned these papers, nor tried to apply his fundamental discovery of simplicial approximation to bring to life the theorems guessed by Poincare (as Alexander did a little later).

Instead, he used his discovery to define rigorously the concept of *degree* of a continuous map, and then proceeded, mostly by fantastically complicated constructions, relying *exclusively* on that notion, to prove the celebrated "Brouwer theorems".

Most of these have become fairly simple consequences of homology theory. We first take the degree. Let M, M′ be two smooth connected compact oriented manifolds having the same dimension n; let μ_n, μ'_n be their fundamental classes. Then for any continuous map $f\colon \mathrm{M} \to \mathrm{M}'$, the map

$$f_*\colon \mathrm{H}_n(\mathrm{M}; \mathbf{Z}) \to \mathrm{H}_n(\mathrm{M}'; \mathbf{Z})$$

is entirely determined by $f_*(\mu_n) = c \,.\, \mu'_n$, where $c \in \mathbf{Z}$; the integer c is the *degree* of f, written $\deg(f)$. When f is not surjective, $c = 0$. When f is C^∞ and surjective there exist points $y \in \mathrm{M}'$ such that $f^{-1}(y)$ is a finite set $\{x_1, x_2, \ldots, x_r\}$, and furthermore, for each j, f is a local homeomorphism of an open neighborhood V_j of x_j onto an open neighborhood W_j of y. Let $\varepsilon_j = 1$ if $f \,|\, \mathrm{V}_j$ preserves orientation, $\varepsilon_j = -1$ if it reverses orientation; then $\deg(f) = \sum_{j=1}^{r} \varepsilon_j$. When f is merely continuous however, it may happen that *all* fibers $f^{-1}(y)$ have the power of the continuum, even for $\mathrm{M} = \mathrm{M}' = \mathbf{S}_1$; nobody before Brouwer had seen how to circumvent this obstacle.

The first result Brouwer proved using the degree of a map is the "*invariance of dimension*", the fact (first conjectured by Dedekind) that there cannot exist a homeomorphism $f\colon \mathbf{R}^m \to \mathbf{R}^n$ when $m \neq n$. This is immediate *via* homology

J. Dieudonné, *A History of Algebraic and Differential Topology, 1900–1960*,
Modern Birkhäuser Classics, DOI 10.1007/978-0-8176-4907-4_6,

theory, since f would extend to a homeomorphism $\bar{f}\colon \mathbf{S}_m \to \mathbf{S}_n$, and these two spaces have different homology groups.

The *Jordan–Brouwer theorem* states that if Σ is a subset of $\mathbf{R}^n$ homeomorphic to $\mathbf{S}_{n-1}$, then $\mathbf{R}^n - \Sigma$ has exactly two connected components. If a homeomorphism $f\colon \mathbf{S}_{n-1} \simeq \Sigma$ can be extended to a homeomorphism of the closed ball $\mathbf{D}_n$ onto a subspace E of $\mathbf{R}^n$ (which is not always possible), that subspace is the closure of the bounded component of $\mathbf{R}^n - \Sigma$. These statements easily follow from the computation of $H_\bullet(\mathbf{R}^n - \Sigma)$ and $H_\bullet(\mathbf{R}^n - E)$; this can be done directly or by application of Alexander duality. An elementary corollary of these results is "*invariance of domain*": if $f\colon U \to \mathbf{R}^n$ is a continuous injective map, where U is an open subset of $\mathbf{R}^n$, then $f(U)$ is open in $\mathbf{R}^n$.

II) The number n is thus a topological invariant of the space $\mathbf{R}^n$ and of its open subsets, but its definition involves constructions which have nothing to do with topology. Already in 1903 Poincaré had suggested a purely topological definition of an integer which should be called "the dimension" of X for any space X; in 1911 Lebesgue proposed another definition based on very different constructions. Brouwer (using as usual the notion of degree) was able to prove that both definitions give the expected number n for $\mathbf{R}^n$.

The concept of dimension for larger and larger categories of spaces was much studied between 1920 and 1940, but its links with homology are tenuous: the cube $[0, 1]^n$ has dimension n, but it is contractible for every n, hence, for all homology theories it has the homology of a point. On the other hand, there are compact subspaces X of $\mathbf{R}^n$ which have singular homology groups $H_m(X) \neq 0$ for *infinitely* many integers m.

III) One of the most straightforward applications of the concept of degree found by Brouwer concerns the continuous maps f of the sphere $\mathbf{S}_n$ (resp. the closed ball $\mathbf{D}_n$) into itself; he showed that for $\deg(f) \neq (-1)^{n+1}$ (resp. for *any* f) there is in $\mathbf{S}_n$ (resp. $\mathbf{D}_n$) a *fixed point* x for f, i.e., $f(x) = x$. Both of the theorems are consequences of the invariance of the degree under homotopy. If $f\colon \mathbf{S}_n \to \mathbf{S}_n$ is such that $f(x) \neq x$ for all $x \in \mathbf{S}_n$, f is homotopic to the antipode map $s\colon x \mapsto -x$, and $\deg(s) = (-1)^{n+1}$. Similarly if $f\colon \mathbf{D}_n \to \mathbf{D}_n$ is such that $f(x) \neq x$ for all $x \in \mathbf{D}_n$, the map $g\colon \mathbf{S}_{n-1} \to \mathbf{S}_{n-1}$ given by $g(y) = (f(y) - y)/(|f(y) - y|)$ is everywhere defined and continuous; the assumption shows that on one hand, g is homotopic to the constant map $y \mapsto f(0)/|f(0)|$, and on the other hand it is also homotopic to the antipode map of degree $(-1)^n$; this contradiction proves Brouwer's fixed point theorem.

This result by Brouwer almost immediately found applications to generalizations of classical theorems. For instance, if a merely *continuous* map $f\colon \mathbf{C} \to \mathbf{C}$ is such that $|f(z)| < |z|^k$ for large $|z|$, the equation $z^k + f(z) = 0$ has a solution in $\mathbf{C}$, a generalization of the "fundamental theorem of algebra".

IV) Of course, for an arbitrary compact space X and a continuous map $f\colon X \to X$, the existence of fixed points for f depends on f (the case $X = \mathbf{D}_n$ being exceptional). This fact was given a remarkable expression in the *Lefschetz formula*, first proved for compact smooth manifolds by Lefschetz and generalized by H. Hopf. For any n-dimensional simplicial complex X, define the

Lefschetz number

$$\Lambda(f) = \sum_{0 \leqslant p \leqslant n} (-1)^p \operatorname{Tr}((f_*)_p). \tag{1}$$

Hopf proved that if f has *no* fixed point, $\Lambda(f) = 0$.

If X is a C^0 manifold and f has only a *finite* number of fixed points, it is possible to assign to each of these points a an integer $j(a) \in \mathbf{Z}$ called its *index*, and one has the *Hopf formula*

$$\Lambda(f) = \sum_{a \in \operatorname{Fix}(f)} j(a). \tag{2}$$

The number $j(a)$ is defined by considering a homeomorphism h of an open neighborhood of a onto an open neighborhood of 0 in $\mathbf{R}^n$ with $h(a) = 0$. Then for small enough $\rho > 0$, $g = h \circ f \circ h^{-1}$ is defined for $|x| \leqslant \rho$, and $j(a)$ is the *degree* of the map $x \mapsto g(x)/|g(x)|$ of the sphere $|x| = \rho$ into $\mathbf{S}_{n-1}$.

Closely related to (2) is Hopf's *index theorem for vector fields*. Let M be a compact C^1 manifold and Z a C^1 vector field on M; it is assumed that there are only a *finite* numbers of points $a \in M$ for which $Z(a) = 0$ ("singular points" of the field). Write $t \mapsto F_Z(x, t)$ the solution of the differential equation $v'(t) = Z(v(t))$ defined in a neighborhood of 0 in $\mathbf{R}$ and such that $v(0) = x \in M$; then, for t small enough, the fixed points of the map $x \mapsto F_Z(x, t)$ are exactly the singular points of Z; by definition the *index* of the singular point a of Z is the index $j(a)$ of that map. It follows from (2) that the sum $\sum_a j(a)$ extended to all singular points of Z is the *Euler–Poincaré characteristic* $\chi(M)$. This result is also true for merely continuous vector fields on M, since they may be approximated by C^1 fields. That theorem had been proved in 1911 by Brouwer for $M = \mathbf{S}_n$ using an earlier definition of the index of a vector field which had been introduced for $M = \mathbf{S}_2$ by Poincaré in 1881.

V) It was early realized that *global* topological properties for a space (such as compactness or connectedness) give very little information on the structure of the space; much more is known from the corresponding *localized* properties. Similarly the knowledge of homology groups does not say much about a space; contractible spaces (such as cones) all have the homology of a point. Around 1930 topologists began to study, with the help of homological notions, properties of a space "around a point x", for instance the groups $H_p(X, X - \{x\})$ of relative homology. R. Wilder was the most active protagonist in these studies; he showed conclusively how properties of subsets of $\mathbf{R}^2$ investigated since Schoenflies by "point-set topologists" could be greatly generalized and better understood with the help of "local" homological notions. The best results have been obtained in connection with Borel–Moore homology; in particular, it has provided a purely topological definition of spaces for which the duality theorems of Poincaré and Alexander are valid.

VI) With the beginning in 1925 of the *global* theory of Lie groups the determination of their topological invariants came to the fore. The study of the De Rham cohomology of these groups was inaugurated by Elie Cartan. He proved that a connected Lie group is diffeomorphic to the product of a compact Lie group and a space $\mathbf{R}^n$, so that the problem is reduced to the topology of compact Lie groups; passage to the universal covering group then shows that it is enough to know the cohomology of the *simple** compact Lie groups. Around 1935 the cohomology algebra $H^{\bullet}(G; \mathbf{R})$ was determined by several authors for the four *classical* groups; E. Cartan observed that in all these cases $H^{\bullet}(G; \mathbf{R})$ is isomorphic to the cohomology algebra of a product of *l odd dimensional spheres*

$$\mathbf{S}_{r_1} \times \mathbf{S}_{r_2} \times \cdots \times \mathbf{S}_{r_l} \qquad (l \text{ being the rank of G}). \tag{3}$$

This coincidence was "explained" by Hopf in 1939 using the functoriality of cohomology. For a topological group G, let $m: G \times G \to G$ be the multiplication; it determines a homomorphism of Λ-algebras

$$m^*: H^{\bullet}(G; \Lambda) \to H^{\bullet}(G \times G; \Lambda)$$

preserving the graduation. When Λ is a field, the Künneth formula shows that m^* can be considered as a homomorphism

$$m^*: H^{\bullet}(G; \Lambda) \to H^{\bullet}(G; \Lambda) {}^{g}\!\otimes_{\Lambda} H^{\bullet}(G; \Lambda)$$

and it is easily seen that for $z \in H^n(G; \Lambda)$,

$$m^*(z) = 1 \otimes z + z \otimes 1 + \sum_{\substack{i+j=n \\ i>0, j>0}} x_i \otimes y_j \tag{4}$$

where $x_i \in H^i(G; \Lambda)$, $y_j \in H^j(G; \Lambda)$.

Hopf then more generally considered the purely algebraic problem which consists in describing the structure of an anticommutative graded algebra A over a field Λ, for which there is a homomorphism

$$c: A \to A {}^{g}\!\otimes_{\Lambda} A$$

preserving the graduation and satisfying (4); these are now called *Hopf algebras*. He showed that if the field Λ has characteristic 0, there is a vector subspace $E \subset A$, generated by elements $e_1, \ldots, e_m$ of *odd degrees*, such that A is isomorphic to the *exterior algebra* $\bigwedge E$. In particular this proves that $H^{\bullet}(G; \mathbf{R})$ is isomorphic to the cohomology algebra of a product (3) for all simple compact Lie groups (not only the classical ones). More generally, $H^{\bullet}(M; \Lambda)$ is a Hopf algebra not only for topological groups, but also for all spaces M for which there exists a continuous map $m: M \times M \to M$ and an element $e \in M$ such that both $x \mapsto m(e, x)$ and $x \mapsto m(x, e)$ are homotopic to the identity 1_M; they are now called H-*spaces*, and have been the subject of many studies in recent years.

VII) Beginning with Cayley, Klein, and Poincaré, mathematicians did not hesitate to "glue" together spaces along subspaces, or to interpret as a "point"

---

* A Lie group G is called *simple* (or, better, *quasi-simple*) if it contains no nontrivial *closed* normal subgroup.

somewhere a family of curves or surfaces, considered as "equivalent" under some relation. All this was done in a purely "intuitive" way. It took a long time during the first third of the twentieth century to subsume these constructions and many others under the general notion of *quotient space* of a topological space by an equivalence relation. I thought it might be useful to bring together the very imaginative examples of that construction in algebraic and differential topology, and to describe how the homology of these new spaces is related to the homology of the spaces used in their construction.

The most influential of these constructions have been *mapping cylinders* and *CW-complexes*, both due to J.H.C. Whitehead. His main objectives in introducing them concerned homotopy theory, as we show in Part 3. But the notion of CW-complex has given topologists a generalization of Poincaré's cell complexes which is a far more flexible and versatile tool. In particular it has provided easy computations of the homology of important homogeneous spaces, such as grassmannians and Stiefel manifolds. It also yields a simpler presentation of the remarkable relations discovered by Marston Morse between the Betti numbers of a smooth manifold X and the critical points of a smooth function defined in X.

VIII) The last chapter of Part 2 is devoted to the early applications of homology to other parts of mathematics.

It can be surmised that when Poincaré began in 1895 to form "linear combinations of varieties" in order to define homology he might have been inspired by the "linear systems of curves on a surface" used by the algebraic geometers of his time. At any rate, he never ceased to be keenly interested in the way E. Picard started to apply homology to algebraic surfaces, and he himself wrote several papers on such applications. After 1905 Picard's and Poincaré's ideas were expanded and generalized by Severi and above all by Lefschetz, who obtained remarkable relations between Betti numbers of complex projective algebraic varieties of any dimension and their purely algebraic invariants. After De Rham's theorem was proved, Hodge used it to extend to smooth algebraic varieties of any dimension the analytic methods Riemann had pioneered for algebraic curves; this put Lefschetz's results in a more natural context and provided new topological properties for complex manifolds.

Another part of mathematics where homology found fruitful applications is functional analysis. Many functional equations can be written

$$F(u) = u$$

where u belongs to some function space Ω and F is an *operator*, i.e., a map of Ω into itself. Hence one may prove the existence of a solution of a functional equation by applying general theorems on fixed points of operators in a function space; in fact the Liouville–Picard "method of successive approximations" is just an elementary example of that idea.

The Brouwer fixed point theorem provided new possibilities of such applications, first by G.D. Birkhoff and O. Kellogg in 1922, followed by several papers of J. Schauder. The space Ω is a convex subset of a Banach space, and must satisfy compactness properties for suitable topologies on that space. The proof consists in approximating Ω by a finite dimensional convex simplicial complex K_n and simultaneously approximating F by a simplicial map $F_n: K_n \to K_n$; Brouwer's fixed point theorem is then applied to K_n and F_n, yielding the existence in Ω of a sequence (a_n) for which $F_n(a_n) = a_n$; it has a limit which is a fixed point of F.

For more particular maps $x \mapsto x - u(x)$, where u maps bounded sets into relatively compact ones, it is possible to go further and to generalize Brouwer's degree as well. This was done in 1934 in a famous paper by Leray and Schauder. It yielded, for some partial differential equations, existence theorems which looked beyond reach.

But the most remarkable application of homology theory during the period 1920–1940 is what M. Morse called the "Calculus of variations in the large". On a complete n-dimensional smooth Riemannian manifold M, he considered two points p, q and the set $\Omega = \Omega(M, p, q)$ of piecewise smooth continuous paths joining p and q. The set Ω can be equipped with a distance for which it is a metric space, infinite dimensional if $n \geqslant 2$; the length $L(\gamma)$ of any path $\gamma \in \Omega$ is continuous for that topology. Let Ω_c be the subspace consisting of all $\gamma \in \Omega$ such that $L(\gamma) \leqslant c$; by a skillful combination of differential geometry and topology, Morse was able to prove that Ω_c has the same homology as a compact smooth manifold B with boundary, provided q is not conjugate to p along *any* geodesic $\gamma \in \Omega_c$ (the set of these points q is dense in M). Furthermore, there is only a finite number of geodesics $\gamma \in \Omega_c$, and the knowledge of the points conjugate to p along those geodesics gives information on the Betti numbers of B, *via* the Morse relations involving the critical points of a smooth function.

CHAPTER I

The Concept of Degree

§ 1. The Work of Brouwer

L.E.J. Brouwer (1881–1966) had started his career with papers on geometry and mechanics, but by 1909 he had shifted his interests to parts of mathematics far less popular at that time, and in which he was entirely self-taught. He first tackled Hilbert's famous "5th problem" and showed that all C^0 groups of transformations of the real line are in fact Lie groups. This work, and attempts to extend it to transformation groups of $\mathbf{R}^2$, led him to study what was known at the time about the topology of the plane. This had started with Cantor's theory of sets, and was closely linked to the general study of functions of real variables and of their often surprising properties; the results that had attracted the most attention were the Jordan theorem on the domains limited by a simple closed curve (1893), and the Peano "curve" filling a square (1890), leading to the investigation of the various meanings that could be given to the word "curve." The papers published by Schoenflies around 1900 were devoted to such questions (see chap. IV, § 3), and Brouwer thought he could use their results for his own purposes. He soon discovered that many of them were either incorrect or insufficiently proved, and in 1910 he published a paper ([89], pp. 352–366) containing many counterexamples. The most unexpected of these immediately brought international recognition to Brouwer: it was an example of an "indecomposable" compact connected set in the plane (one which cannot be written as the union of two proper compact connected subsets), which has the extraordinary property that it is the frontier of *three* connected components of the complement of the set (p. 359).

From these early papers it would have been difficult to foresee the breakthrough accomplished by Brouwer in the years 1910–1912, owing to a complete change of outlook and a remarkably skillful use of the new concept *simplicial approximation* that he introduced (see Part 1, chap. II, § 2). In a rapid succession of papers published in less than two years, the "Brouwer theorems" (as they are still called) made him famous overnight. They solved a whole batch of problems on n-dimensional spaces for *arbitrary n* that had looked intractable to the previous generation: invariance of dimension of open sets

J. Dieudonné, *A History of Algebraic and Differential Topology, 1900–1960*,
Modern Birkhäuser Classics, DOI 10.1007/978-0-8176-4907-4_7,

in $\mathbf{R}^n$, invariance of domain, extension of the Jordan curve theorem, existence of fixed points of continuous mappings, singularities of vector fields, and, finally, based on ideas of Poincaré and Lebesgue, a definition of the notion of *dimension* for arbitrary compact metric spaces.

In retrospect, it therefore seems legitimate to consider Brouwer as the cofounder, with Poincaré, of simplicial topology. More precisely, it may be said that Poincaré defined the *objects* of that discipline, but it is Brouwer who imagined *methods* by which theorems about these objects could be *proved*, something Poincaré had been unable to do.

It is all the more surprising then that Brouwer did not attempt to use his techniques in order to put Poincaré's "theorems" in simplicial homology on less shaky foundations [we recall that this was only done by Alexander in 1915–1922, using Brouwer's simplicial approximation (Part 1, chap. II, § 3)]. Brouwer did not publish any important paper on topology after 1913, devoting the major part of his career to an intuitionist reconstruction of mathematics. But even in his 1911–1912 papers his attitude toward the concepts of simplicial homology is puzzling. He freely used the notion of triangulation for what he called "n-dimensional manifolds," meaning a finite or locally finite simplicial complex X, such that for every vertex there is a homeomorphism of its star in X onto the star of the origin of $\mathbf{R}^n$ in a rectilinear triangulation of a compact neighborhood of 0, mapping simplices onto simplices.* Brouwer never referred to anybody for such a definition, nor for the concept of barycentric subdivision; although in [89], pp. 455–456 he gave detailed proofs of the existence of barycentric coordinates, and carefully defined orientable "manifolds," it seems that he thought he was dealing with well-known notions for which he did not need any reference. As he had never used them before 1911, it is of course possible that he reinvented them independently; but even so it is very unlikely that nobody should have pointed out to him that there already existed a sizable body of knowledge in these matters, summarized for instance in the Dehn–Heegaard *Enzyklopädie* article [138].

At any rate, Brouwer never showed any interest for homological concepts in his "n-dimensional manifolds." The only paper of his having some connection with Betti numbers is the last one published in the years 1911–1912 ([89], pp. 523–526), but it is sharply different from the others. It deals with what we now would call a very special case of Lefschetz's extension of Alexander duality (Part 1, chap. IV, § 2): namely, to prove that for any connected compact subspace X of $\mathbf{R}^2$, the number of components of $\mathbf{R}^2 - X$ does not change when X is replaced by another compact subspace of $\mathbf{R}^2$ homeomorphic to X. One would expect to see Brouwer using his trusted simplicial methods (as Lefschetz will do in [301]), but there is nothing of the kind. His topological notions

* These complexes are not as general as the "combinatorial manifolds" defined by Alexander and Lefschetz (Part 1, chap. II, § 4); for an example of a four-dimensional combinatorial manifold which is not a "four-dimensional manifold" in the sense of Brouwer, see [421], p. 241.

seem to be directly lifted from Riemann: it is asserted that in the plane an "$(h + 1)$-fach zusammenhängendes Gebiet" contains h "closed simple cuves" c_j such that any other"closed curve" may be derived by deformation from a finite number of the c_j (p. 523); certainly no such theorem had been proved in 1912.* The whole paper is written in that style, extremely sketchy and unconvincing, with many steps in the proof merely outlined without any detail.† Of course, the paper is not entirely without value, and Vietoris could find the germ of his homology theory (Part 1, chap. IV, § 2) in Brouwer's attempts to characterize the number of connected components of $\mathbf{R}^2 - X$ by sequences of chains of points of X.‡ Unfortunately, it must be admitted that this paper represents the extreme case of Brouwer's gradual tendency to revert to the "intuitive" style of his predecessors, in contrast with the high level of rigor that he had achieved in his first papers of 1911. We shall have opportunities to witness this regression when we examine his proof of the Jordan–Brouwer theorem in chapter II.

§ 2. The Brouwer Degree

It is quite remarkable that Brouwer proved all his big theorems by a skillful (and sometimes quite tortuous) use of a single concept he discovered and studied in the first days of 1910, the *degree* of a map ([89], p. 419). That concept was not completely new, but it had only appeared in very special cases, and not always under the form of Brouwer's definition. The example quoted in [89], p. 462, is the map R of $\mathbf{S}_2 = \mathbf{C} \cup \{\infty\}$ onto itself, defined as the continuous extension of a rational function $z \mapsto P(z)/Q(z)$ of algebraic degree n (the highest degree of the polynomials without common factor P, Q). Then n is the number of solutions p of the equation $R(p) = q$ for a given $q \in \mathbf{S}_2$ distinct from a finite number of "critical values" of R.

Another example is the "winding number" of a path around a point in the Cauchy theory of integration along curves in $\mathbf{C}$: if $\zeta \mapsto \varphi(\zeta)$ is a piecewise C^1 map of the unit circle $\mathbf{U}$: $|\zeta| = 1$ into $\mathbf{C}$, the winding number of φ around a point $a \notin \varphi(\mathbf{U})$ is the integer

$$j(a) = \frac{1}{2\pi i} \int_{\mathrm{U}} \frac{\varphi'(\zeta)\,d\zeta}{\varphi(\zeta) - a}, \tag{1}$$

* A standard proof would consist in taking an infinite simplicial triangulation of $\mathbf{R}^2 - X$ and use Alexander's deformation methods to reduce the "curves" to paths consisting of edges of the simplices of the triangulation ([421], p. 164); nobody except Brouwer himself could have done that at that time, but he did not do it.

† On p. 524, one might interpret what is said in § 2 as implying the fact that when one replaces X by the set X_ε of points at a distance $\leqslant \varepsilon$ from X, the number of connected components of $\mathbf{R}^2 - X_\varepsilon$ is the same as the number of components of $\mathbf{R}^2 - X$ when ε is small enough, which is false.

‡ In 1927, Alexandroff showed how one may give an entirely correct proof of Brouwer's result by using similar arguments [23].

which enters in Cauchy's formula for a function f holomorphic in the disk $|z| \leqslant 1$:

$$j(a)f(a) = \frac{1}{2\pi i}\int_{U}\frac{f(\varphi(\zeta))\varphi'(\zeta)\,d\zeta}{\varphi(\zeta)-a}. \tag{2}$$

Intuitively $j(a)$ is "the algebraic number of times $\varphi(U)$ turns around the point a," and simple cases of maps φ lead to the following evaluation of $j(a)$: take a half line D of origin a, and suppose it cuts $\varphi(U)$ transversely in finitely many points; then $j(a)$ is the difference between the number of these points where the argument of $\varphi(\zeta)$ is increasing and the number of points where it is decreasing.

Brouwer does not mention the winding number, but there is a variant of that notion with which he was familiar, the *index at a point a* of a piecewise C^1 vector field $x \mapsto v(x)$ in the plane having only isolated zeroes; this had been defined by Poincaré in 1881 in a particular case, and Brouwer had used it in a paper in 1910 ([89], p. 316) (see chap. III, § 3). Suppose v does not vanish for $x \neq a$ in a neighborhood of a; then to each point x of a small circle γ of center a, assign the unit vector $v(x)/|v(x)|$; this may be considered as a map of the unit circle U into itself, and its winding number around 0 is constant when the radius of γ tends to 0; by definition that constant value is the index of v at the point a.

Generalizing such notions to arbitrary dimensions must have appeared intractable, because of the pathological properties continuous maps may possess (think of the Peano curve!). Nevertheless, this is what Brouwer achieved with his idea of simplicial approximation. We shall describe in some detail (with slight complements and modificatons for the sake of clarity) the fundamental ideas of his proof, in order to put in evidence the two features that characterize almost all his proofs of 1911–1912: a remarkable originality and a great complexity.

Given are two compact, connected, oriented n-dimensional "manifolds" (in Brouwer's sense) M, M′, with triangulations T, T′. For simplicity we may assume that M and M′ are euclidean simplicial complexes* with no loss of generality (see Part 1, chap. II, § 2), although Brouwer did not make that assumption.

First step. Let ε be smaller than the diameter of any n-simplex of T′; let S_0 be the set of vertices of T, and γ_0 be a map of S_0 into M′ such that for any n-simplex of T, the diameter of the image by γ_0 of the set of its vertices is $< \varepsilon/2$. For any n-simplex s of T such that the images of its vertices by γ_0 are all contained in the interior of an n-simplex E of T′, we define γ as the affine map of s into E coinciding with γ_0 at the vertices of s; we shall say that γ is a *piecewise affine* map. We first suppose that for any n-simplex s of T where γ is defined, $\gamma(s)$ is *nondegenerate*.

* Euclidean simplices for Brouwer are always *closed*.

Let E be any n-simplex of T', and let J be the homothetic of E with respect to the barycenter of E, in such a ratio $1 - \alpha$ that the distance of J to the boundary of E is $> \varepsilon$. Let Ω be the set of interior points of J that do not belong to the image under γ of any p-simplex of T for $p \leqslant n - 2$; it immediately follows that Ω is an open *connected* set.

For any point P in Ω that does not belong to the image by γ of any $(n - 1)$-simplex of T, let $p \geqslant 0$ (resp. $q \geqslant 0$) be the number of n-simplices of T where γ is defined, such that γ preserves (resp. reverses) orientation in it (they are respectively called *positive* and *negative* simplices) and such that P is in the image of that simplex by γ. The crucial property is that the number $p - q$ is *independent of* P in Ω.

To prove this Brouwer considered two points P_1, P_2 in Ω, joined by a polygonal arc L in Ω, and he let P vary on L from P_1 to P_2. The number $p - q$ might only vary when P crosses the image by γ of an $(n - 1)$-simplex t of T. But by assumption t is the face of exactly *two* n-simplices s', s'' of T, and the assumption on γ_0, together with the fact that P is in the interior of J, show that γ is defined in *both* s' and s'', and that $\gamma(s')$ and $\gamma(s'')$ are in the interior of E. Then when L crosses $\gamma(t)$ at the point Q either p and q increase by the same amount, or both decrease by the same amount [Q might of course belong to the images of several $(n - 1)$-simplices of T].

Before going further it is necessary to get rid of the hypothesis that the images $\gamma(s)$ are nondegenerate. Suppose then that $\beta_0: S_0 \to M'$ satisfies the same conditions as γ_0 above, except that the $\beta(s)$ may be degenerate; however if $P \in J$ is in an image $\beta(s)$ without being in the image by β of any $(n - 1)$-simplex of T, then $\beta(s)$ certainly is nondegenerate; the numbers p and q are therefore still defined. To show that $p - q$ is again independent of P in Ω, Brouwer argued by contradiction: β may be arbitrarily approximated by a map γ for which the $\gamma(s)$ are nondegenerate; if for two points P_1, P_2 in J, not in the image by β of an $(n - 1)$-simplex of T, the values of $p_1 - q_1$ and $p_2 - q_2$ were different, P_1 and P_2 would not belong to the image by γ of an $(n - 1)$-simplex of T if γ is close enough to β, and the values of $p_1 - q_1$ (resp. $p_2 - q_2$) for β and γ would be the same, bringing the desired contradiction.

Brouwer added to this a very useful interpretation of the number $c(\beta, E)$, the constant value of $p - q$: take the sum $\Sigma(J)$ of the (euclidean) volumes of all intersections of J with the images by β of the n-simplices contained in E, each volume being affected with the sign + for positive simplices or the sign − for the negative ones; this sum is equal to $c(\beta, E) . \text{vol}(J)$.*

Second step. Prove that the number $c(\beta, E)$ is *independent* of the choice of the simplex E in the triangulation T'. Using the fact that M' is connected and oriented, consider only two such n-simplices E_1, E_2 having as common face a $(n - 1)$-simplex F. The idea is to "straighten" the union $E_1 \cup E_2$, by considering a homeomorphism φ of that union onto a "double pyramid" E' =

* The introduction of the volume of J is not at all artificial; see below, formula (7).

$E'_1 \cup E'_2$, union of two rectilinear, n-simplices E'_1, E'_2 *in* $\mathbf{R}^n$, having a common face F'; φ is such that its restrictions to E_1 and E_2 are *affine* maps respectively onto E'_1 and E'_2. If J' is the homothetic of E' with respect to the barycenter *of* F', with ratio $1 - (1 - 1/n)\alpha$, then J' contains the images J'_1 and J'_2 of J_1 and J_2 by φ. The map $\varphi \circ \beta$ can now be extended to the n-simplices of T such that the images by $\varphi \circ \beta_0$ of its vertices are all contained in *the interior of* E'. The argument of the first step may then be repeated, replacing β by that extension β' of $\varphi \circ \beta$, and J by J'; as β' coincides with $\varphi \circ \beta$ on the n-simplices s whose image by β are in the interior of E_1 or the interior of E_2, this proves that $c(\beta, E_1) = c(\beta, E_2)$; one may thus write $c(\beta)$ instead of $c(\beta, E)$.

Third step. The triangulations T and T′ being given, α: $M \to M'$ is now an *arbitrary* continuous map. For any $\varepsilon > 0$, it is then possible to replace T by a simplicial subdivision T_ε such that the map β_0^ε of the set of vertices of T_ε into M′, which coincides with α on these vertices, satisfies the conditions enumerated in the first step, so that the number $c(\beta^\varepsilon)$ is defined. It remains to prove that this number is *independent* of T_ε when ε tends to 0. If T_1, T_2 are any two of the subdivisions T_ε, β_1, β_2 the corresponding maps β^ε, one first considers the case in which T_2 is a subdivision of T_1; there is then a *homotopy* β_t between the maps β_1 and β_2 in each n-simplex of T_1 in which β_1 is defined, and the corresponding number $\Sigma_t(J)$ varies continuously with t, hence $c(\beta_t)$ also, which shows that $c(\beta_1) = c(\beta_2)$. Finally, if T_1 and T_2 are arbitrary subdivisions of T, one considers a common subdivision T_3 of T_1 and T_2.

This therefore defines unambiguously the *degree* deg(α) as the common value of the $c(\beta^\varepsilon)$. Its fundamental property, which Brouwer proved immediately after the definition, is its *invariance under homotopy*: for this, one considers two homotopic maps α_1, α_2, and the corresponding approximations β_1, β_2 for which the subdivision T_ε is the same: it is then possible to interpolate between β_1 and β_2 a finite number of similar piecewise affine maps, such that any two consecutive ones differ only at *one* vertex of T_ε, and are arbitrarily close to one another; this implies that they have the *same* degree.

Brouwer also observed that, by the same arguments, when M is compact, connected, and oriented, but M′ is compact, connected, and *not orientable*, or when M′ is *not compact*, the degree is always 0.

An examination of the preceding proof shows that, whereas one has to assume that M′ is connected this is not necessary for M: if M_j $(1 \leqslant j \leqslant k)$ are the connected components of M, deg(α) is the *sum* of the degrees $\deg(\alpha|M_j)$ of the restrictions of α to the M_j. This *additivity property* of the degree even holds when the M_j are n-dimensional manifolds (resp. pseudomanifolds, see §3,A) having nonempty intersections that are unions of n-simplices of each M_j in which they are contained, for instance if the M_j are pseudomanifolds that are parts of boundaries of $(n + 1)$-dimensional pseudomanifolds with boundaries. Brouwer did not mention this, but he uses it implicitly later.

Finally Brouwer conjectured in a footnote ([89], p. 463) that the degree is independent of the triangulations chosen on M and M′. A little later ([89], pp. 504–506) he was able to prove that conjecture as a consequence of the *multiplicative property* of the degree: if M, M′, M″ are three compact, connected, oriented n-manifolds (in Brouwer's sense), with triangulations T, T′, T″, and α: M → M′, α': M′ → M″ two continuous maps, then

$$\deg(\alpha' \circ \alpha) = \deg(\alpha')\deg(\alpha). \tag{3}$$

His sketched proof consists, after suitable subdivisions of T and T′, in considering piecewise affine approximations β and β' of α and α', for which $\deg(\alpha) = c(\beta)$ and $\deg(\alpha') = c(\beta')$; the approximation $\beta' \circ \beta$ (which is piecewise affine) can be used to define $\deg(\alpha' \circ \alpha)$, and it is therefore enough to prove the relation

$$c(\beta' \circ \beta) = c(\beta')c(\beta). \tag{4}$$

To compute $c(\beta')$, consider an n-simplex E″ of T″, and the corresponding homothetic n-simplex J″; let P″ be a point of J″, not in the image of any $(n-1)$-simplex of T′; let $E'_1, \ldots, E'_{p'}$ (resp. $E'_{p'+1}, \ldots, E'_{p'+q'}$) be the n-simplices of T′ for which P″ is the image by β' of one of their interior points, and for which β' preserves (resp. reverses) orientation. It is then possible to subdivide T and to choose the homothetic simplices J'_i of the E'_i in such a way that P″ does not belong to the images by β' of any of the $E'_i - J'_i$. Then, for each i, P″ is the image by β' of a point P'_i of J'_i, and there is a neighborhood of P'_i in J'_i that contains a point which is in the interior of the images by β of p_i (resp. q_i) n-simplices of T for which β preserves (resp. reverses) orientation. The recipe for the computation of $c(\beta' \circ \beta)$ therefore gives the number

$$\sum_{i=1}^{p'} p_i + \sum_{i=p'+1}^{p'+q'} q_i - \sum_{i=1}^{p'} q_i - \sum_{i=p'+1}^{p'+q'} p_i; \tag{5}$$

since all the $p_i - q_i$ are equal to $c(\beta)$ and $c(\beta') = p' - q'$, this proves (4).

Once the multiplicative property (3) is proved, consider two different triangulations T'_1, T'_2 of M′ and a continuous map α: M → M′, and write $\alpha = \varepsilon \circ \alpha$, where ε is the identity map of M′. Let $\deg_1(\alpha)$, $\deg_2(\alpha)$ be the degrees of α computed with respect to T and T'_1 and with respect to T and T'_2; then $\deg_2(\alpha) = \deg_{12}(\varepsilon)\deg_1(\alpha)$, where $\deg_{12}(\varepsilon)$ is the degree of ε computed with respect to the triangulations T'_1 and T'_2; it follows at once from the definition that $\deg_{12}(\varepsilon) = 1$. A similar argument proves the independence on the triangulation of M.

Another multiplicative property, later used by topologists but not mentioned by Brouwer, is that if f: M → N, $f' =$ M′ → N′ are two continuous maps, where M and N (resp. M′ and N′) are "manifolds" of the same dimension, then for the map $f \times f'$: M × M′ → N × N′,

$$\deg(f \times f') = \deg(f).\deg(f').$$

§ 3. Later Improvements and Variations

A. Homological Interpretation of the Degree

As soon as homology was conceived as a theory of homology *groups*, and the invariance theorem for finite simplicial complexes was proved (Part 1, chap.

II, § 3), it was possible to define the Brouwer degree in a much simpler way. First an examination of Brouwer's proofs shows that they are still valid for more general simplicial complexes, which Brouwer himself defined a little later ([89], p. 477), the *closed connected pseudomanifolds* (or simply *pseudomanifolds*). These are defined as being homeomorphic to a finite euclidean simplicial complex X, such that:

1. X is the union of its n-dimensional (closed) simplices;
2. X is connected;
3. any $(n-1)$-dimensional simplex of X is exactly a face of *two* n-dimensional simplices.

[Replacing "two" by "one or two" in point 3 gives the more general notion of "pseudomanifold-with-boundary" (cf. Part 3, chap. VII, § 1,H), the "boundary" being the union of the $(n-1)$-simplices that are a face of only *one* n-simplex.] The pseudomanifolds of course include the Brouwer "n-manifolds" as well as the "combinatorial n-manifolds" of Alexander and Lefschetz as special cases (Part 1, chap. II, § 4).* Orientation being defined as usual, it follows at once from the definitions that for an n-dimensional compact pseudomanifold M there are no n-cycles other than 0 if M is not orientable; if M is oriented the n-cycles are exactly the multiples, by an arbitrary (positive or negative) integer, of the *sum* of all positively oriented n-simplices, called the *fundamental n-cycle*. Hence, $H_n(M; Z) = 0$ if M is not orientable and $H_n(M; Z) \simeq Z$ if M is oriented, the class [M] of the fundamental n-cycle being a basis of the Z-module $H_n(M; Z)$; it is called the homology *fundamental class* of M (cf. chap. IV, § 3,A).

If M and M′ are now compact connected n-dimensional oriented pseudomanifolds, and $f: M \to M'$ a continuous map, then

$$f_*([M]) = c[M']$$

for the corresponding homomorphism f_* (Part 1, chap. II, § 3), and the integer c is the *Brouwer degree* of f. This follows from the Alexander construction (Part 1, chap. II, § 3) of a simplicial approximation g of f which is homotopic to f, and from Brouwer's definition of $\deg(g)$ given in § 2; that interpretation of the degree seems to have been given for the first time by Hopf in 1930 [242]. All this extends to the case in which M is not connected, but its connected components are compact oriented pseudomanifolds; M is still called a *pseudomanifold* if no confusion arises.

One can of course replace Z by F_2 for the coefficients of homology modules, and then one has the definition of the *degree modulo* 2, which also applies to nonorientable compact pseudomanifolds; in the definition of § 2, one has to replace the difference $p - q$ by the class mod. 2 of the number of n-simplices whose images contain a point P in Ω.

* It is easy to give examples of pseudomanifolds that are not "manifolds" in the sense of Brouwer, for instance the unreduced suspension (chap. V, § 2,C) of an $(n-1)$-dimensional smooth manifold that is not a homology sphere.

We note in passing that for an oriented compact combinatorial manifold (Part 1, chap. II, §4), Poincaré duality can be expressed in the following way ([440], p. 305): the cap product (Part 1, chap. IV, §4)

$$z \mapsto z \frown [\mathrm{M}]$$

is an *isomorphism* of the **Z**-module $\mathrm{H}^q(\mathrm{M}; \mathbf{Z})$ onto $\mathrm{H}_{n-q}(\mathrm{M}; \mathbf{Z})$ for $0 \leqslant q \leqslant n$. There is a similar result for nonorientable combinatorial manifolds, [M] being then the class of the sum of all n-simplices, in $\mathrm{H}_n(\mathrm{M}; \mathbf{F}_2)$. When M is a compact oriented C^∞ *manifold*, the class [M] can be identified to the homology class, in the n-th homology vector space $\mathrm{H}'_n(\mathrm{M})$ over **R** based on the currents, of the current equal to the *constant function* 1 on M (Part 1, chap. III, §3). Then if M is connected, the dual basis of [M] in the de Rham cohomology vector space $\mathrm{H}^n(\mathrm{M})$ is the class e^*_{M} (called cohomology *fundamental class*, or *orientation class* of M) of an n-form ω on M such that $\int_{\mathrm{M}} \omega = 1$. If M, M′ are compact, connected, oriented C^∞ n-manifolds, and $f\colon \mathrm{M} \to \mathrm{M}'$ a C^∞ map,

$$f^*(e^*_{\mathrm{M}'}) = \deg(f) . e^*_{\mathrm{M}}; \tag{6}$$

equivalently, for any n-form ω' on M′, its pullback ${}^t f(\omega')$ on M is such that

$$\int_{\mathrm{M}} {}^t f(\omega') = \deg(f) . \int_{\mathrm{M}'} \omega'. \tag{7}$$

A simple application of Sard's theorem (Part 1, chap. III, §1) in that case yields a definition of $\deg(f)$ similar to Brouwer's: there exist points $y \in \mathrm{M}'$ such that:

(P) $f^{-1}(y)$ is a finite set $\{x_1, x_2, \ldots, x_r\}$ in M, and for each index k there is an open neighborhood V_k of x_k such that the V_k are disjoint and the restriction of f to each V_k is a *homeomorphism* of V_k onto an open neighborhood of y.

Then

$$\deg(f) = \sum_{k=1}^{r} \varepsilon(x_k), \tag{8}$$

where $\varepsilon(x_k) = 1$ if the restriction of f to V_k preserves orientation, and $\varepsilon(x_k) = -1$ if it reverses orientation.

All this, of course, can be proved without any triangulation of M and M′ (using, for instance, de Rham's cohomology); if one uses C^1 triangulations (Part 1, chap. III, §2), formula (8) shows again that the definition coincides with Brouwer's.

B. Order of a Point with Respect to a Hypersurface; the Kronecker Integral and the Index of a Vector Field

Many applications of the degree are relative to the cases in which M or M′ is equal to the sphere $\mathbf{S}_n$. If $\mathrm{M} = \mathrm{M}' = \mathbf{S}_1$, a continuous map $f\colon \mathbf{S}_1 \to \mathbf{S}_1$ can be written $e^{it} \mapsto e^{i\psi(t)}$, where $t \in \mathbf{R}$, and $\psi\colon \mathbf{R} \to \mathbf{R}$ is continuous and $\psi(t + 2\pi) = \psi(t) + 2n\pi$ for some $n \in \mathbf{Z}$. Then $\deg(f) = n$, the "winding number" of f (§2).

More generally, let X be a compact, connected, oriented $(n-1)$-dimensional C^∞ manifold, and let $f: X \to \mathbf{R}^n - \{0\}$ be a C^∞ map. Let $p: \mathbf{R}^n - \{0\} \to \mathbf{S}_{n-1}$ be the projection $x \mapsto x/|x|$; then $p \circ f$ is a C^∞ map of X into $\mathbf{S}_{n-1}$, and $\deg(p \circ f)$ (for $\mathbf{S}_{n-1}$ oriented toward the exterior) is called the *order* of 0 with respect to the "hypersurface" $f(X)$ (or with respect to f). It can be computed, using (7), by the *Kronecker integral*

$$\deg(p \circ f) = \frac{1}{\Omega_n} \int_X {}^t f(\tau^{(n)}), \tag{9}$$

where Ω_n is the measure of $\mathbf{S}_{n-1}$, and, for $z = (\xi^1, \ldots, \xi^n)$ in $\mathbf{R}^n - \{0\}$ and $r(z) = ((\xi^1)^2 + \cdots + (\xi^n)^2)^{1/2}$, $\tau^{(n)}$ is the $(n-1)$-form on $\mathbf{R}^n - \{0\}$, invariant by rotations:

$$\frac{1}{r^n(z)} \sum_{j=0}^{n} (-1)^j \xi^j \, d\xi^1 \wedge \cdots \wedge \widehat{d\xi_j} \wedge \cdots \wedge d\xi^n. \tag{10}$$

For $n = 1$, this of course gives back the Cauchy integral (1) for $j(0)$. For arbitrary n, Hadamard proposed in (1910) [217] the integral (9) as a *definition* of the order. The integral (9) had been introduced for $X \subset \mathbf{R}^n$, by Kronecker [288] in order to study the roots of an equation $f(x) = 0$ in an open bounded subset U of $\mathbf{R}^n$, f being a C^∞ map of $\bar{U}$ into $\mathbf{R}^n$. The frontier X of $\bar{U}$ is supposed to be a smooth hypersurface, and it is assumed that $f(x) \neq 0$ in X; finally 0 is supposed to be a regular value for f, so that 0 has the property (P) (§ 3,A). Then Stokes' formula shows that if $f^{-1}(0) = \{x_1, x_2, \ldots, x_r\}$ in U,

$$\deg(p \circ (f|X)) = (-1)^n \sum_{k=1}^{r} \varepsilon(x_k). \tag{11}$$

If this degree is $\neq 0$, this proves the existence of *at least* one root of $f(x) = 0$ in U; if in addition all the $\varepsilon(x_k)$ have the same sign, the absolute value of $\deg(p \circ (f|X))$ gives the *number* of these roots.

As defined above (§ 2), the Poincaré index at a point a of a vector field v in $\mathbf{R}^2$, having only isolated singularities, is the *order* of 0 with respect to the restriction of v to a small circle of center a; its generalization to a notion of *index* for a continuous vector field with only isolated singularities in $\mathbf{R}^n$ (or in a C^∞ manifold) is therefore obvious.

C. Linking Coefficients

Another intuitive geometric notion in $\mathbf{R}^3$ was, for two Jordan curves C_1, C_2 without common points, the "number of times C_1 turns around C_2," and Gauss had already introduced an integral that he considered expressed that number [206]. In 1912 ([89], pp. 511–520) Brouwer generalized that notion to the situation of two compact, connected, oriented pseudomanifolds M, N in $\mathbf{R}^n$, with empty intersection, M having dimension $d \leqslant n-1$ and N dimension $n-d-1$; $M \times N$ is then a compact connected pseudomanifold (for a simplicial subdivision of the product of two simplices), on which the orientation is the product of the orientation of M by the orientation of N. Then $f: (x, y) \mapsto x - y$ is a continuous map of M × N into $\mathbf{R}^n - \{0\}$, and the *order* of 0 with respect to f is defined; that integer lk(M, N) is what Brouwer called the "looping coefficient" of M and N (the name has now been changed to *linking coefficient*); $\mathrm{lk}(\mathrm{N}, \mathrm{M}) = (-1)^{(d+1)(n-d)} \mathrm{lk}(\mathrm{M}, \mathrm{N})$, and when M and N are C^∞ manifolds, lk(M, N) can be computed by a Kronecker integral (9), which reduces to Gauss' integral for $n = 3$ and $d = 1$ [328]. Here again M may be replaced by a disjoint union of pseudomanifolds M_j; then $\mathrm{lk}(\mathrm{M}, \mathrm{N}) = \sum_j \mathrm{lk}(\mathrm{M}_j, \mathrm{N})$.

It is likely that Brouwer had already thought of that definition when he defined the degree of a map. But in the Comptes Rendus Note that Lebesgue published about the invariance of dimension in March 1911 (chap. II, §3) he used another concept of "linked" manifolds ("variétés enlacées"). Approximating M and N by "polygonal varieties" M′, N′, he said that M and N were "enlacées" if, when N′ is deformed continuously to a single point not on M′, the moving variety "crosses" M′ an *odd* number of times. In his 1912 paper Brouwer took up this idea and generalized it to the linking number of *oriented* manifolds: for nonoriented ones, if the "linking number mod. 2" is defined as above but starting from the degree mod. 2, Lebesgue's definition amounts to saying that the linking number mod. 2 is $\neq 0$.

Brouwer first observed that if the injections $\mathrm{M} \to \mathbf{R}^n$, $\mathrm{N} \to \mathbf{R}^n$ are replaced by sufficiently close piecewise affine maps g, h (not necessarily injective), lk(M, N) is still equal to the degree of $(x, y) \mapsto g(x) - h(y)$. Next he supposed that there exists an $(n - d)$-dimensional connected orientable pseudomanifold S with boundary, h(N) being its boundary. Taking arbitrarily small modifications of g, assume that no p-simplex of g(M) intersects a q-simplex of S if $p + q \leqslant n - 1$ ("general position"); then the intersection of S and g(M) consists of a *finite* number of points, each of which is the intersection of a d-simplex of g(M) and a $(n - d)$-simplex of S, affected with the number ± 1 obtained by the rule formulated by Poincaré (Part 1, chap. I, §2). What is to be proved is that the *sum* of these numbers is equal to lk(M, N) for suitable orientations of M and N.

The proof consists in first *translating* g(M) by a large vector v such that S and $g(\mathrm{M}) + v$ are contained in two balls with no common point, in which case both numbers to be compared are zero. Then consider the sets $g(\mathrm{M}) + \lambda v$ where λ varies from 0 to 1; by a "general" choice of v, these two numbers may only change when λ crosses a value λ_0 such that a $(d - 1)$-simplex of $g(\mathrm{M}) + \lambda_0 v$ meets S, or a d-simplex of $g(\mathrm{M}) + \lambda_0 v$ meets an $(n - d - 1)$-simplex of S. Because g(M) is a pseudomanifold and S is a pseudomanifold-with-boundary, there is no change at all in the first case, nor in the second unless the $(n - d - 1)$-simplex of S is contained in h(N); but then, on crossing the value λ_0, both $\mathrm{lk}(g(\mathrm{M}) + \lambda v, h(\mathrm{N}))$ and the sum of the signed intersection numbers of S and $g(\mathrm{M}) + \lambda v$ change by the *same* amount.

This second definition of linking coefficients only uses intersection numbers, and it can be generalized to define lk(A, B) when A and B are two *chains* with no common points, of dimensions k and $n - k$ in a connected oriented combinatorial manifold X of dimension n, which are both *boundaries* in the complex X. As A is the boundary of a $(k + 1)$-chain C, and the simplices of B do not meet the simplices in the frontier of C, the intersection number (C . B) is defined (Part 1, chap. II, § 4) and by definition it is the linking number lk(A, B). As B is a boundary, this number does not depend on the choice of the $(k + 1)$-chain C with boundary A.

If αA and βB are boundaries for two integers α, β, a linking number lk(A, B) can still be defined as equal to $1/\alpha\beta\,\mathrm{lk}(\alpha A, \beta B)$. This rational number, taken *modulo* 1, then only depends on the homology classes of A and B in $H_\cdot(X; \mathbf{Z})$; in particular the self-linking number lk(A, A) can be defined in $\mathbf{Q}/\mathbf{Z}$ when $n = 2k + 1$ ([421], pp. 277–279). It is by means of this invariant, applied to a 1-cycle, that Alexander proved that the lens spaces L(5, 1) and L(5, 2) are not homeomorphic, for he showed that this self-linking number has the form $\pm v^2 q/p \bmod . 1$, and therefore $L(p, q)$ and $L(p, q')$ can only be homeomorphic if $q' \equiv \pm v^2 q \pmod{p}$ for some integer v ([421], p. 279).

D. Localization of the Degree

The Brouwer concept of the degree is a *global* notion; it may be refined by *localizing* it on M or on M′. We postpone to chap. IV, § 1,B, the localization on M. Regarding localization on M′, in its simplest form, take $M' = \mathbf{R}^n$, M being a *bounded open subset* of $\mathbf{R}^n$; for a continuous map $f: \bar{M} \to \mathbf{R}^n$ and a point $p \in \mathbf{R}^n - f(\mathrm{Fr}(M))$, we want to define an integer $d(f, M, p) \in \mathbf{Z}$ such that:

1. if a homotopy $F: \bar{M} \times [0, 1] \to \mathbf{R}^n$ is such that p does not belong to the image of the frontier Fr(M) by *any* partial map $x \mapsto F(x, t)$, then the number $d(F(.\,, t), M, p)$ is *constant*;
2. if property (P) (see § 3,A) holds for f at the point p, then

$$d(f, M, p) = \sum_{k=1}^{r} \varepsilon(x_k). \tag{12}$$

The proof of existence of this "degree of f relative to M and p" can be done by elementary analysis, considering first C^1 functions and using Sard's theorem, and then approximating any continuous map f by C^1 maps (see, e.g., [328]). In [239] Hopf tackled continuous maps directly, but he had to use his rather deep theorem characterizing homotopy classes of maps in an S_n by their degree (Part 3, chap. II, § 1). Much later [324], Leray observed that $d(f, M, p)$ can be defined by much more elementary means. $\mathbf{R}^n$ is an open set in S_n; using the Tietze–Urysohn theorem, it is possible to extend $f: \bar{M} \to \mathbf{R}^n$ to a continuous map $F: S_n \to S_n$ such that $p \notin F(S_n - M)$; then one *defines* $d(f, M, p) = \deg(F)$, this number being independent of the choice of the extension F.

It is easy to deduce most of the properties of $d(f, M, p)$ from the properties of the degree; the most useful one is that the relation $d(f, M, p) \neq 0$ implies the existence of *at least* one point $q \in M$ such that $f(q) = p$. It can also be shown that the function $p \mapsto d(f, M, p)$ is *constant* in every connected component Δ of the (open) complement in $\mathbf{R}^n$ of $f(\mathrm{Fr}(\mathrm{M}))$, and so $d(f, \mathrm{M}, \Delta)$ is the common value of $d(f, \mathrm{M}, p)$ in Δ. Now, one has the remarkable *multiplication theorem* of Leray [324]: let V be a bounded open set in $\mathbf{R}^n$ containing $f(\bar{\mathrm{M}})$, and let Δ_i be the connected components of the open set $\mathrm{V} - f(\mathrm{Fr}(\mathrm{M}))$ (there may be infinitely many); then, for any continuous map $g\colon \bar{\mathrm{V}} \to \mathbf{R}^n$ such that $p \notin g(f(\mathrm{Fr}(\mathrm{M})) \cup g(\mathrm{Fr}(\mathrm{V}))$,

$$d(g \circ f, \mathrm{M}, p) = \sum_i d(g, \Delta_i, p)\, d(f, \mathrm{M}, \Delta_i) \tag{13}$$

[if $\mathrm{V} - f(\mathrm{Fr}(\mathrm{M}))$ has infinitely many connected components Δ_i, it may be shown that the right-hand side of (13) has only finitely many terms not 0]. In particular, if M is connected, and f is a *homeomorphism* of M onto an open subset M′ of $\mathbf{R}^n$, then $d(f, \mathrm{M}, \mathrm{M}') = \pm 1$.*

Finally this reduction of the localized degree to a suitable global degree enables one to apply to the computation of $d(f, \mathrm{M}, p)$ the simplicial method of §2, using triangulations of $\mathbf{S}_n$. If U is an open connected subset contained in $\mathbf{S}_n - f(\mathrm{Fr}(\mathrm{M}))$, then

$$d(f, \mathrm{M}, \mathrm{U}) = d(f, f^{-1}(\mathrm{U}), \mathrm{U}) \tag{14}$$

by computing the corresponding degree deg(F) using a simplex of the triangulation contained in U.†

In 1928 Hopf was able to generalize the definition of $d(f, \mathrm{M}, p)$ when M is an open relatively compact subset of any C^0 manifold and f is a continuous map of $\bar{\mathrm{M}}$ into a C^0 manifold M′. As it is not possible to use triangulations here, Hopf's idea was to mimick Brouwer's method of approximation by "simpler" maps; piecewise affine maps were unavailable and Hopf replaced them by maps g satisfying property (P) above (§3,A). In the statement of that property the neighborhoods V_k may be chosen homeomorphic to open balls in $\mathbf{R}^n$, and the properties of the localized degree relative to such an open set also give the definition $d(g, \mathrm{M}, p)$ in that situation; $d(g, \mathrm{M}, p)$ does not depend on the choice of the approximation g of f, hence the definition of $d(f, \mathrm{M}, p)$, with properties similar to when M is an open bounded subset of $\mathbf{R}^n$ and $\mathrm{M}' = \mathbf{R}^n$. In particular, if M and M′ are any two compact, connected, oriented C^0 *manifolds*, all the numbers $d(f, \mathrm{M}, p)$ are equal, extending the definition of $\deg(f)$ without postulating the existence of triangulations. In these definitions, however, Hopf again had to use his theorem on homotopy classes. Another more elementary proof of the existence of the degree for C^0 manifolds was published in the same number of *Mathematische Annalen* by W. Wilson, a student of Brouwer [519].

Even *before* defining the degree in general, Brouwer, in his very first paper

---

* This is mentioned by Brouwer in a footnote of his second paper on the Jordan–Brouwer theorem ([89], p. 502).

† This is an interpretation of Brouwer's argument in his second proof of invariance of domain ([89], pp. 509–510).

of the 1911–1912 series ([89] pp. 430–434), in which he proved the invariance of dimension, used, as his main lemma, a property of the localized degree $d(f, \mathrm{K}, p)$ in a particular case: K is the cube $[-1, 1]^n$, f is a continuous map of K into $\mathbf{R}^n$ such that $|f(x) - x| < \frac{1}{2}$, and p belongs to $[-\frac{1}{2}, \frac{1}{2}]^n$. By exactly the same argument of approximation by piecewise affine maps developed in ([89] pp. 454–472), and using the fact that f is homotopic to the identity in K, he showed that $d(f, \mathrm{K}, p) \neq 0$, hence $f(\mathrm{K})$ contains $[-\frac{1}{2}, \frac{1}{2}]^n$.

§4. Applications of the Degree

As was said above, almost all properties used by Brouwer in his papers of 1911–1913 are proved by using the degree, as we shall see in the next chapters. The notion of degree was also found to yield simple proofs of classical theorems and establishing new ones in differential geometry. We shall give here some simple examples.

Suppose f and g are two continuous maps of $\mathbf{S}_{n-1}$ into $\mathbf{R}^n - \{0\}$, such that for any point $z \in \mathbf{S}_{n-1}$, the point 0 is not on the segment joining $f(z)$ and $g(z)$ in $\mathbf{R}^n$. Then $(z, t) \mapsto tf(z) + (1 - t)g(z)$, mapping $\mathbf{S}_{n-1} \times [0, 1]$ into $\mathbf{R}^n - \{0\}$, is a homotopy between f and g, hence the *orders* of 0 with respect to $f(\mathbf{S}_{n-1})$ and $g(\mathbf{S}_{n-1})$ are the same (Poincaré–Bohl theorem). In particular, this is the case if, in $\mathbf{S}_{n-1}$, $f(z) \neq 0$ and $|g(z) - f(z)| < |f(z)|$ (a very useful property for $n = 2$, called Rouché's theorem in the theory of analytic functions of a complex variable). This, for instance, shows that in $\mathbf{C}$ an equation

$$z^k + f(z) = 0 \quad (k \text{ integer} > 0), \tag{15}$$

where f is continuous in $\mathbf{C}$, and $|f(z)| < |z|^k$ for large $|z|$, has at least one root, since the order of 0 with respect to the images of a large circle $|z| = \mathrm{R}$ by the maps $z \mapsto z^k$ and $z \mapsto z^k + f(z)$ are equal. This contains the "fundamental theorem of algebra" as a special case; the result can easily be extended when $\mathbf{C}$ is replaced by the algebra of quaternions or of octonions ([189], pp. 308–310).

In 1933 Borsuk proved interesting results on *odd* continuous maps f: $\mathbf{S}_n \to \mathbf{S}_n$, i.e., such that $f(-z) = -f(z)$ for all $z \in \mathbf{S}_n$; he showed that $\deg(f)$ is then an *odd integer*. Suppose for simplicity that f is C^∞; there is then an elementary proof of that theorem by induction on n (see [30], pp. 484–485). Using Sard's theorem and applying a rotation, in $\mathbf{R}^{n+1}$, one may assume $\mathbf{e}_{n+1} \notin f(\mathbf{S}_{n-1})$ and $f^{-1}(\mathbf{e}_{n+1})$ is a finite set of $\mathbf{S}_n$; it is necessary to prove that it has an *odd* number of points. Let q: $\mathbf{S}_n \to \mathbf{R}^n$ be the orthogonal projection on a diametral hyperplane, and p: $\mathbf{R}^n - \{0\} \to \mathbf{S}_{n-1}$ be the map $\mathbf{x} \mapsto \mathbf{x}/|\mathbf{x}|$. The composite map $p \circ q \circ (f|\mathbf{S}_{n-1})$ is then defined in $\mathbf{S}_{n-1}$; it is an odd map, and by the induction hypothesis it has an odd degree. If D_+ is the hemisphere defined in $\mathbf{S}_n$ by the inequality $x_{n+1} \geqslant 0$, having $\mathbf{S}_{n-1}$ as boundary, this implies that the sum of the number of points of $\mathrm{D}_+ \cap f^{-1}(\mathbf{e}_{n+1})$ and of the number of points of $\mathrm{D}_+ \cap f^{-1}(-\mathbf{e}_{n+1})$ is odd. But since f is odd, $\mathrm{D}_+ \cap f^{-1}(-\mathbf{e}_{n+1})$

has as many points as $(-\mathrm{D}_+) \cap f^{-1}(\mathrm{e}_{n+1})$, and as $\mathbf{S}_n = \mathrm{D}_+ \cup (-\mathrm{D}_+)$ and $f^{-1}(\mathrm{e}_{n+1}) \cap \mathbf{S}_{n-1} = \varnothing$, this ends the proof.

An obvious consequence is that, when n continuous real functions f_j $(1 \leqslant j \leqslant n)$ are odd, they have at least one common zero on $\mathbf{S}_n$. Applying this to $f_j(z) = g_j(z) - g_j(-z)$, where now the g_j are *arbitrary* real continuous functions in $\mathbf{S}_n$, shows that there is at least one point $z \in \mathbf{S}_n$ such that $g_j(-z) = g_j(z)$ for $1 \leqslant j \leqslant n$ (theorem of Borsuk–Ulam).

CHAPTER II

Dimension Theory and Separation Theorems

§ 1. The Invariance of Dimension

Before 1870 mathematicians only dealt with those subsets X of an $\mathbf{R}^N$ that could (at least locally) be "parametrized" by (usually C^1) injective maps into X of open subsets of some $\mathbf{R}^n$. It was tacitly assumed that the position of a point in $\mathbf{R}^n$ could only be completely determined by a system of n real numbers. The discovery by Cantor in 1877 of a *bijection* of $\mathbf{R}$ onto $\mathbf{R}^n$, for any n, came as a complete surprise and seemed to threaten the bases of analysis. Cantor's map was wildly discontinuous, but the discovery of the Peano curve (1890) showed that there existed *continuous* (although not injective) maps of $\mathbf{R}$ *onto* $\mathbf{R}^n$. The only hope that remained of salvaging the classical notion of dimension was the one expressed by Dedekind as soon as Cantor had communicated his theorem to him: there should not exist *bicontinuous bijections* of $\mathbf{R}^m$ onto $\mathbf{R}^n$ for $m \neq n$. This was elementary for $m = 1, n > 1$, since a point disconnects $\mathbf{R}$ but not $\mathbf{R}^n$; several mathematicians before 1910 were also able to tackle the cases $m = 2$ and $m = 3, n > m$. But the general proof of Dedekind's conjecture was only obtained by Brouwer in the first of the series of papers which he started in 1911 ([89], pp. 430–434).

Brouwer's proof is based on the key lemma showing that if a continuous map f of $[-1, 1]^n$ into $\mathbf{R}^n$ is such that $|f(x) - x| < \frac{1}{2}$ for all x, then $f([-1, 1]^n)$ contains the cube $[-\frac{1}{2}, \frac{1}{2}]^n$ (chap. I, § 3,D). He used that lemma to show that there may not exist an *injective* continuous map g of $[-1, 1]^n$ onto a *rare* subset C of $\mathbf{R}^n$. The proof is by contradiction: Brouwer showed that if such a map existed, it would be possible to define a continuous map h: $C \rightarrow [-1, 1]^n$, such that $h(C)$ would be rare and $|h(g(x)) - x| < \frac{1}{2}$ for all $x \in [-1, 1]^n$, in contradiction with the lemma.

To construct h, start from a sufficiently fine triangulation T of a cube $K \supset C$ and consider the union F of the n-simplices of T that meet C. Define $h_0(a)$ for each vertex a of an n-simplex $\sigma \subset F$ as one of the points of $[-1, 1]^n$ such that $g(h_0(a)) \in \sigma$, then extend h_0 to a piecewise affine map h_0 of F into $[-1, 1]^n$. If h is the restriction of h_0 to C, then $h(\sigma \cap C)$ is rare for each n-simplex $\sigma \subset F$, hence h has the required properties provided T has been chosen fine enough.

J. Dieudonné, *A History of Algebraic and Differential Topology, 1900–1960*,
Modern Birkhäuser Classics, DOI 10.1007/978-0-8176-4907-4_8,

From this theorem, Brouwer obtained the invariance of dimension in two steps:

1. Suppose $m > n$; a cube K of $\mathbf{R}^m$ contains a rare image K′ of $[-1, 1]^n$ by a continuous injection j. If there existed a continuous injective map f: $K \to [-1, 1]^n$, the map $f \circ j$ would contradict the theorem.
2. If a cube K′ of $\mathbf{R}^n$ contained the image of $[-1, 1]^m$ by a homeomorphism g, as there exists a continuous injection h: $K' \to [-1, 1]^m$ such that $h(K')$ is rare, $h \circ g$ would again contradict the theorem.

These two corollaries imply the nonexistence of a homeomorphism of $\mathbf{R}^m$ onto $\mathbf{R}^n$ for $m \neq n$.

§ 2. The Invariance of Domain

A result closely related to the invariance of dimension was called the "invariance of domain": if A is a compact subset of an $\mathbf{R}^n$ and f: $A \to \mathbf{R}^n$ is a continuous *injective* map, f sends interior points of A to interior points of $f(A)$ [which implies that it maps the interior of A homeomorphically onto the interior of $f(A)$]. This property implies invariance of dimension: suppose there existed a homeomorphism f of an open set $U \neq \emptyset$ in $\mathbf{R}^m$ onto an open subset of $\mathbf{R}^n$ with $n < m$; one may consider $\mathbf{R}^n$ as a rare subset of $\mathbf{R}^m$, and for V open nonempty and relatively compact in $\mathbf{R}^m$ and $\bar{V} \subset U$, $f(\bar{V})$, considered as a subset of $\mathbf{R}^m$, would have no interior point.

This is essentially the argument by which Baire, in 1907, wanted to prove the invariance of dimension ([40], [41]). He then endeavored to reduce the invariance of domain to a weak* generalization of the Jordan curve theorem to n dimensions: if f is a homeomorphism of the closed ball $\mathbf{D}_n$: $|x| \leqslant 1$ of $\mathbf{R}^n$ onto a subset of $\mathbf{R}^n$, the complement of $f(\mathbf{S}_{n-1})$ in $\mathbf{R}^n$ has two connected components [traditionally called the "interior" and "exterior" of $f(\mathbf{S}_{n-1})$, the "exterior" being the unbounded one].

In assuming this result, Baire also had to assume that $f(\mathbf{D}_n)$ was not contained in the "exterior" of $f(\mathbf{S}_{n-1})$.† He considered the concentric open balls $B(\rho)$: $|x| < \rho$ for $0 < \rho \leqslant 1$ [$B(1) = \mathbf{D}_n$], and their boundaries $S(\rho)$: $|x| = \rho$. Then $f(B(1))$ is contained by assumption in the "interior" A of $f(S(1))$, and by contradiction $f(B(1)) = A$. Indeed, if that were not the case, there would be a point $y \in A$ not in the closed set $f(\overline{B(1)}) = f(B(1)) \cup f(S(1))$ and hence a ball

* That this is not the real generalization of the Jordan theorem is due to the fact that a continuous injection of $\mathbf{S}_{n-1}$ into $\mathbf{R}^n$ cannot in general be extended to a continuous injection of $\mathbf{D}_n$ into $\mathbf{R}^n$.

† If one knows that the order of an "interior" point A with respect to $f(\mathbf{S}_{n-1})$ is ± 1 (see § 3), it implies the fact that $f(\mathbf{D}_n)$ is not contained in the "exterior" of $f(\mathbf{S}_{n-1})$, for as $S(\rho)$ tends to a point when ρ tends to 0, the order of A with respect to $f(S(\rho))$ would tend to 0, although it must be constant.

γ of center y and radius r that does not meet $f(\overline{B(1)})$. It is impossible for γ to be contained in the "interior" of $f(S(\rho))$ for all ρ, since the diameter of $f(S(\rho))$ tends to 0 with ρ. Let ρ_0 be the g.l.b. of the $\rho > 0$ such that γ is contained in the "interior" of $f(S(\rho))$. Then, for a sequence (ε_k) tending to 0, y would be at a distance $\geqslant r$ of the "interior" of $f(S(\rho_0 - \varepsilon_k))$ and at a distance $\geqslant r$ of the "exterior" of $f(S(\rho_0 + \varepsilon_k))$, which is impossible by continuity.

Baire, however, could not prove the weak generalization of the Jordan theorem which he needed.* It was again Brouwer who gave two different proofs of the invariance of domain. The first one ([89], p. 485) does not use the Jordan–Brouwer theorem, but what we may call for short the *no separation theorem* (NS), for which Brouwer gave a proof in the same paper (see § 4):

(NS) If U is a connected open subset of $\mathbf{R}^n$, and $F \subset U$ is a homeomorphic image of a compact subset A of $\mathbf{S}_{n-1}$, *distinct* from $\mathbf{S}_{n-1}$, then $U - F$ is connected.

To deduce the invariance of domain from this, Brouwer argued by contradiction: let f be an injective continuous map of $\bar{U}$ into $\mathbf{R}^n$, where U is a nonempty bounded open set in $\mathbf{R}^n$, and suppose there exists a point $P \in U$ such that $f(P)$ is *not* interior to $f(U)$. Let $Q \neq P$ be another point of U; by assumption, there are spheres Σ of center $f(P)$ and arbitrary small radius that are not contained in $f(\bar{U})$; take the radius of such a sphere Σ smaller than the distance of $f(P)$ to $f(Q)$ and such that $F = f^{-1}(\Sigma \cap f(\bar{U}))$ is contained in a closed ball $B \subset U$ of center P that does not contain Q. By (NS), P and Q may be joined by a polygonal line $L \subset U$ that does not meet F; then $f(L) \subset f(U)$ would join $f(P)$ and $f(Q)$ without meeting $\Sigma \cap f(U)$, which is the desired contradiction.

Brouwer's second proof ([89], pp. 509–510) is simpler and only uses properties of the degree [or rather of its localization (chap. I, § 3,D)]. With the same notations, let $P \in U$ and let I be a small open ball of center P such that $\bar{I}$, union of I and its boundary, the sphere K, is contained in U. Let H be the connected component of the open set $\mathbf{R}^n - f(K)$ that contains $f(P)$, hence also $f(I)$ since $f(I) \cap f(K) = \varnothing$; the proof consists in showing that $f(I) = H$. Brouwer's argument, which is only sketched, is clearer if we use the localized degree $d(f, I, p)$; if $f(I) \neq H$ it would imply $d(f, I, H) = 0$ since $\mathrm{Fr}(I) = K$ and $H \cap f(K) = \varnothing$. In his proof of invariance of dimension (§ 1), however, Brouwer had shown that there exists a nonempty open ball $\gamma' \subset f(I)$; then $\gamma = f^{-1}(\gamma') \cap I$ is open in $\mathbf{R}^n$, and the restriction of f to γ is a homeomorphism onto γ'; hence $d(f, \gamma, \gamma') = \pm 1$ (chap. I, § 3,D). If $p \in \gamma'$, then $d(f, I, p) = d(f, I, H) = 0$; but $d(f, I, p) = d(f, \gamma, \gamma')$ [*loc. cit.* formula (14)] and therefore the assumption $f(I) \neq H$ implies a contradiction.

* He complained in a letter to Brouwer that his bad health prevented him from mustering the energy needed to elaborate his ideas.

§ 3. The Jordan–Brouwer Theorem

The full generalization of the Jordan curve theorem (now called the Jordan–Brouwer theorem) was first tackled by Lebesgue and Brouwer in 1911. We can split the problem into three parts. Given a subset J of $\mathbf{R}^n$ homeomorphic to $\mathbf{S}_{n-1}$,

(i) The complement $\mathbf{R}^n - \mathrm{J}$ has at least two connected components.
(ii) J is the boundary of every connected component of $\mathbf{R}^n - \mathrm{J}$.
(iii) $\mathbf{R}^n - \mathrm{J}$ has at most two connected components.

A. Lebesgue's Note

Part (i) is independent of the other two and also of (NS). In March 1911 Lebesgue published a sketch of a proof in a *Comptes rendus* note ([294], pp. 173–175). At first Brouwer had doubts that this sketch could be elaborated into a correct proof ([89], p. 452); because of Lebesgue's imprecise language, he thought J was any $(n-1)$-dimensional compact connected manifold in $\mathbf{R}^n$. Later he admitted that (i) could indeed be proved by Lebesgue's method, but [probably owing to his contemporary controversy with Lebesgue on the definition of dimension (see § 5)] he did not wish to write out a complete proof himself. Lebesgue did not write anything on the matter after his *Comptes rendus* note, so no complete proof of (i) was available until 1922.

Lebesgue's method relies on an ingenious interpretation of part (i): for $0 \leqslant k \leqslant n-1$, let L_k be a subset of $\mathbf{R}^n$ homeomorphic to $\mathbf{S}_k$; then there exists a subset L'_{n-k-1} of $\mathbf{R}^n$, homeomorphic to $\mathbf{S}_{n-k-1}$, and such that L_k and L'_{n-k-1} are "enlacées" (i.e., their linking number mod. 2 is $\neq 0$). For $k = n-1$, $\mathbf{S}_{n-k-1} = \mathbf{S}_0$ consists of two points, and the statement is thus equivalent to (i). For $k = 0$, the theorem is trivial, and Lebesgue's proof is by *induction on k*. He considered a piecewise affine approximation g to a homeomorphism $f: \mathrm{S}_k \rightarrow \mathrm{L}_k$; let A_+, A_-, and L_{k-1} be the images by f of the hemispheres D_+, D_- and of their common boundary S_{k-1}.* By the inductive assumption L_{k-1} is linked by a homeomorphic image L'_{n-k} of $\mathbf{S}_{n-k}$. Replacing L'_{n-k} by an arbitrarily close piecewise affine approximation $h(\mathbf{S}_{n-k})$, makes the intersection $g(\mathbf{D}_+) \cap h(\mathbf{S}_{n-k})$ finite, and it has an odd number of points (if not, replace D_+ by D_-). If P is one of these points, by a slight change of g, it may be taken to be the intersection of a k-simplex of $g(\mathbf{S}_k)$ and an $(n-k)$-simplex σ of $h(\mathbf{S}_{n-k})$ and belongs to the interior of these simplices; then the boundary of σ in $h(\mathbf{S}_{n-k})$ links $g(\mathbf{S}_k)$.

B. Brouwer's First Paper on the Jordan–Brouwer Theorem

Brouwer published two papers on the Jordan–Brouwer theorem. The first one ([89], pp. 489–494), exclusively deals with part (iii) of the problem. Part

* This seems to be the first occurrence of this splitting of the sphere, which will be used again and again later in many contexts.

(ii) is dismissed with the remark that it follows from the (NS) theorem (§ 2), for which he had written a proof in an earlier paper (see § 4), without giving any detail. For any point $x_0 \in J$, it is enough to delete from J the interior of an arbitrarily small $(n-1)$-simplex σ of a sufficiently fine triangulation of J containing x_0. If G_1 and G_2 are two connected components of $\mathbf{R}^n - J$, $y_1 \in G_1$ and $y_2 \in G_2$, there is, by the (NS) theorem, a polygonal arc joining y_1 and y_2 in $\mathbf{R}^n - (J - \sigma)$; on that arc there are points of G_1 and points of G_2 at a distance from x_0 smaller than the diameter of σ; this proves (ii).

The proof of (iii) occupies four pages; it is quite involved and, in spite of its length, full of cryptic statements that make it very hard to follow in detail. What follows is my own interpretation and simplification of what I think are the main points of Brouwer's arguments. He repeatedly uses a lemma first stated in the paper on the (NS) theorem ([89], p. 478):

(L) The boundary F of a pseudomanifold-with-boundary P (chap. I, § 3,A), of dimension n, is a disjoint union of closed $(n-1)$-dimensional pseudomanifolds F_j.

Simple examples show that, if taken literally, this is not correct, for an $(n-2)$-simplex of F may be contained in more than two $(n-1)$-simplices of F. Brouwer acknowledges this but dismisses the matter by saying that p-simplices of F, for $p \leqslant n-2$, that appear to contradict the fact that the F_j are pseudomanifolds and are pairwise disjoint, should be "demultiplied" ("als verschieden zu betrachten sind") so to speak. It would have been clearer if he had bothered to give a proof, and said that one can do away with those occurrences by slightly moving the vertices of F!

The proof of (iii) is essentially based on the idea of linking number, which Brouwer only defined in a general way six months later; here it is used in the particular case of a polygonal Jordan curve L and the frontier j of an $(n-1)$-simplex σ of a (curvilinear) triangulation T of J; his arguments can be simplified by using the definition of linking numbers as degrees of mappings (chap. I, § 3,C). Let E be the unbounded component of $\mathbf{R}^n - J$, G another (bounded) component, P a point of G; the bulk of Brouwer's proof consists in constructing a polygonal Jordan curve L containing P and such that $\mathrm{lk}(L,j) = \pm 1$.

He first constructed in $\mathbf{R}^n$ an infinite locally finite $(n-1)$-dimensional simplicial complex $g \subset G$ whose closure in $\mathbf{R}^n$ is $g \cup j$. Starting with the cubical subdivisions A_ν of $\mathbf{R}^n$ whose vertices are the points of $2^{-\nu}\mathbf{Z}^n$, for each integer τ, let μ_τ be the union of the closed cubes of A_ν that meet the interior of σ and have a distance at least $\sqrt{n}/2^{\tau-1}$ from $J'' = \overline{J - \sigma}$; if τ is taken large enough, P does not belong to any $\mu_{\tau+k}$ for $k \geqslant 0$. The union V_τ of the $\mu_{\tau+k}$ for $k \geqslant 0$ is a kind of "thickening" of σ in $\mathbf{R}^n$ with a "decent" boundary; $V_\tau \cup J''$ is closed and connected, and $\overline{V}_\tau \cap J'' = j$. Define I_τ as the intersection of G and of the open component in $\mathbf{R}^n$ of the complement of $V_\tau \cup J''$ that contains P; g is the part of the boundary of I_τ contained in V_τ, the union of the g_ν, where g_ν is a finite rectilinear cell complex, the cells of which are cells in the frontiers of some of the cubes whose union is $\mu_{\tau+k}$ for $\tau + k \leqslant \nu$. After subdividing of the cubes into simplices and using lemma (L), one sees that g_ν is the disjoint union of finitely many $(n-1)$-dimensional pseudomanifolds.

To construct L, one first joins P to a point R on one of the rectilinear simplices of g, by a polygonal arc L_1 contained* in I_τ. On the other hand, one can join P, by a polygonal arc L_2 contained in I_τ, to a point B′ arbitrarily close to a point of $J - \sigma$ [using (ii)]; then [again using (ii)], one can join B′ to a point B″ in E by a line segment of arbitrarily small length s_2. Similarly, one can join R to a point A′ of V_τ arbitrarily close to a point in the interior of σ, by a polygonal arc L_3 in V_τ; then, again using (ii), a line segment s_1 of arbitrarily small length joins A′ to a point A″ in E; finally, one may join A″ and B″ by a polygonal arc L_4 contained in E. The polygonal Jordan curve L is the union of L_1, L_3, s_1, L_4, s_2, and L_2.

If g were a closed pseudomanifold with boundary j, one would have $\mathrm{lk}(L,j) = \pm 1$, since L meets g in the single point R. But the argument by which Brouwer proved the equivalence of the definition of the linking number as a degree and its definition by counting intersection points does not apply to "open" complexes such as g. To circumvent this difficulty, Brouwer apparently considered the connected component γ_ν of g_ν containing R, which is a rectilinear $(n-1)$-dimensional pseudomanifold with boundary η_ν, and he takes for granted that η_ν tends to j when ν tends to $+\infty$, but gives no proof for this statement. Taking ν large enough and a sufficiently fine triangulation of γ_ν, a simplicial mapping φ of η_ν into j can be defined, homotopic to the identity,† so that $\mathrm{lk}(L, \eta_\nu)$ is equal to the degree of the map $(x, y) \mapsto (\varphi(x) - y)/|\varphi(x) - y|$ of $\eta_\nu \times L$ into S_{n-1}; by the multiplicative property of the degree, this implies that

$$\mathrm{lk}(L, \eta_\nu) = \deg(\varphi) . \mathrm{lk}(L, j); \tag{1}$$

but for γ_ν and η_ν, the equivalence of the two definitions of the linking number applies, so that $\mathrm{lk}(L, \eta_\nu) = \pm 1$, and from (1) it follows that $\mathrm{lk}(L,j) = \pm 1$.

Now assume there exists a third (bounded) component G′ of $\mathbf{R}^n - J$, and construct the corresponding intersection I'_τ of G′ and an open component in $\mathbf{R}^n$ of the complement of $V_\tau \cap J''$. If g' is the part of the boundary of I'_τ contained in V_τ, the construction of the polygonal Jordan curve L shows that $L \cap g' = \emptyset$, if s_1 and s_2 are small enough. But then $\mathrm{lk}(L,j) = 0$ by the argument made above where g is replaced by g'; this brings the required contradiction.

C. Brouwer's Second Paper on the Jordan–Brouwer Theorem

This paper immediately follows the first one in *Mathematische Annalen* ([89], pp. 498–505). In it Brouwer capitalized on the hard work he did in the first

* To simplify the language, we say that a polygonal arc is "contained" in an open set I_τ if the complement of its extremities is a subset of I_τ.

† Note that φ need not be injective.

paper to obtain additional properties of the "Jordan hypersurfaces" in $\mathbf{R}^n$, generalizing results Schoenflies had proved for Jordan curves in $\mathbf{R}^2$.

(I) J is *accessible* from both components I and E (the "interior" and "exterior" of J) of $\mathbf{R}^n - \mathrm{J}$. This means that for any point A of J, there is a Jordan arc having A as one extremity and contained in I (resp. in E). The idea is to consider a sequence (T_k) of triangulations of J obtained by repeated subdivisions of T, and a decreasing sequence (σ_k) of $(n-1)$-simplices of the triangulation T_k, whose diameter tends to 0, such that A is interior to each σ_k. For each k, Brouwer constructed a "thickening" $\mathrm{V}_{\tau_k}^{(k)}$ of σ_k as in the first paper, for a sufficiently large τ_k, in such a way that $\mathrm{V}_{\tau_{k+1}}^{(k+1)}$ is contained in the interior of $\mathrm{V}_{\tau_k}^{(k)}$. Then, starting from a point $\mathrm{P}_0 \in \mathrm{I}$ not in $\mathrm{V}_{\tau_1}^{(1)}$, he constructed a sequence of polygonal arcs L_k, joining a point $\mathrm{P}_k \in \mathrm{V}_{\tau_k}^{(k)} - \mathrm{V}_{\tau_{k+1}}^{(k+1)}$ to a point $\mathrm{P}_{k+1} \in \mathrm{V}_{\tau_{k+1}}^{(k+1)} - \mathrm{V}_{\tau_{k+2}}^{(k+2)}$ and contained in $\mathrm{V}_{\tau_k}^{(k)} \cap \mathrm{I}$. The union of the L_k and of the point A is the required Jordan arc. The same argument applies for a point in E.

(II) A similar argument proves the property called "Unbewaltheit" by Schoenflies: if Q and Q′ are two points of J, and $m(\mathrm{Q},\mathrm{Q}')$ is the infimum of the *diameters* of Jordan arcs joining Q and Q′ in I (resp. E), then $m(\mathrm{Q},\mathrm{Q}')$ tends to 0 with the distance $d(\mathrm{Q},\mathrm{Q}')$ of the two points in $\mathbf{R}^n$. This time one considers a sequence $(\mathrm{Q}_k, \mathrm{Q}'_k)$ of pairs of points of J with $d(\mathrm{Q}_k, \mathrm{Q}'_k)$ tending to 0, and a sequence (σ_k) of $(n-1)$-simplices of triangulations of J, whose diameter tends to 0, and are such that both Q_k and Q'_k are in the interior of σ_k. The construction in I) shows that $m(\mathrm{Q}_k, \mathrm{Q}'_k)$ is at most the diameter of a "thickening" $\mathrm{V}_{\tau_k}^{(k)}$, which obviously tends to 0 when the diameter of σ_k tends to 0 and τ_k tends to $+\infty$.

(III) Finally, Brouwer sketched a proof that the *order* of a point $\mathrm{P} \in \mathrm{I}$ with respect to J (chap. I, § 3,B) is ± 1 (that order is of course constant in I). With the notations of the first paper, he took for granted that there exists an $(n-1)$-simplex σ of the triangulation T of J, and a half-line D of origin a suitable point P of I, such that $\mathrm{D} \cap \mathrm{J}'' = \varnothing$. To show this, take the first point of intersection Q of J and of an oriented line D_0 that meets I and does not meet the $(n-2)$-simplices of T; then if σ is the $(n-1)$-simplex of T containing Q, the distance of Q to $\mathrm{J}'' = \overline{\mathrm{J} - \sigma}$ is > 0. There is therefore a point $\mathrm{P} \in \mathrm{D}_0 \cap \mathrm{I}$ close enough to Q that the half-line D of origin P and containing Q satisfies the requirement.

Next he took a subdivision T_1 of T, and considered the piecewise affine map h of J into $\mathbf{R}^n$ coinciding with the identity on the vertices of T_1; h is homotopic with the identity by a homotopy F whose image does not contain P if T_1 is fine enough. Hence the order of P with respect to J is, up to sign, the sum of the intersection numbers of D and of the rectilinear complex $\mathrm{J}_1 = h(\mathrm{J})$ [one may always suppose that D does not meet the $(n-2)$-simplices of T_1]. Brouwer stated without proof that this number m is ± 1. It is possible to supply a simple argument justifying this claim by using the construction of the first paper: first take a polygonal arc L′ joining P to a point P′ of $\mathrm{D} \cap \mathrm{E}$, which does not meet $\mathrm{J}''_1 = h(\mathrm{J}'')$, and next a polygonal arc L″ joining P′ to P and which does not meet $\sigma_1 = h(\sigma)$. If $\mathrm{L} = \mathrm{L}' \cup \mathrm{L}''$, the construction gives

$\mathrm{lk}(L, j_1) = \pm 1$, where $j_1 = h(j)$. Now, if L_0 is the segment of D joining P and P′, then by definition $\mathrm{lk}(L_0 \cup L'', j_1) = \pm m$. If L'_0 is the loop $L_0 \cup L'$, it is only necessary to show that $\mathrm{lk}(L'_0, j_1) = 0$, and as L'_0 does not meet J''_1, and j_1 is homotopic to a point *in the complex* J''_1, this is obvious.

There is also in this second paper a curious section in which Brouwer claimed to have proved (by a fairly intricate construction) the orientability of J. Did he forget that by definition J is homeomorphic to S_{n-1}, and that S_{n-1} is orientable as a "manifold" in his sense, for any triangulation, according to his own definition of orientability ([89], p. 458)?

§ 4. The No Separation Theorem

In *Mathematische Annalen*, this paper precedes the one on the Jordan–Brouwer theorem, and is entitled "Proof of the invariance of domain," although invariance of domain is only mentioned in the last section; the bulk of the paper (six pages) consists in the proof of what we have called in § 2 the "no separation theorem." It is certainly the most intricate proof of all Brouwer's theorems and the most difficult to follow; the details are so sketchy that I find it impossible to give more than a summary of the main arguments as I understand them.

A preliminary result is a generalization of a theorem of Janiszewski on sets of the plane [267]: let P, Q be two points of S_{n-1}, X and Y be two disjoint relatively closed subsets of the open ball B_n: $|x| < 1$. Suppose P and Q are *not separated* by X nor by Y in B_n, a statement which Brouwer interpreted as meaning that there are Jordan arcs L, M, joining P and Q in B_n such that $L \cap X = M \cap Y = \varnothing$; then P and Q are not separated by $X \cup Y$, i.e., there is a Jordan arc N joining P and Q in B_n such that $N \cap (X \cup Y) = \varnothing$. Brouwer's proof consists in approximating X and Y by neighborhoods that are subcomplexes of a sufficiently fine triangulation T of $\overline{B}_n$, and showing that the theorem may be proved when X and Y are replaced by these neighborhoods. In that simpler case Brouwer used, in addition to lemma (L) of § 3, the following unproved assertion:

(L′) A subcomplex K of T separates P and Q if and only if any polygonal arc joining P and Q in B_n, and which does not meet any $(n-2)$-simplex of T, meets K in an *odd* number of points.

He then simply observed that if a polygonal arc joining P and Q in B_n and having empty intersections with the $(n-2)$-simplices of T meets each of the subcomplexes X, Y in an even number of points, it also meets $X \cup Y$ in an even number of points.

Brouwer then used this theorem to show that if the points P, Q in S_{n-1} are separated in B_n by a relatively closed subset X of B_n, then they also are separated by a suitably chosen *connected component* of X.

The proof of (NS) proper is by contradiction, and can be divided into three steps.

First step. Let J be a "Jordan hypersurface" in $\mathbf{R}^n$, M be a closed subset of J *distinct from* J. By arguments that are not at all clear, Brouwer claimed that the assumption that $\mathbf{R}^n - \mathrm{M}$ has more than one connected component leads to the following situation: P is a *frontier point* of M in J, D is an open ball of $\mathbf{R}^n$ with center at P, H is the $(n-1)$-dimensional sphere, boundary of D in $\mathbf{R}^n$, A, B are two points of H separated by $\mathrm{M} \cap \mathrm{D}$ in D. From the preliminary result he deduced that there is a connected set t contained in $\mathrm{M} \cap \mathrm{D}$, relatively closed in D, containing P and separating A and B in D. Let u be the intersection $\bar{t} \cap \mathrm{H}$, contained in $\mathrm{H} \cap \mathrm{M}$, and G the connected component of $\mathrm{M} - u$ in J containing t. The first step in Brouwer's argument was to show that $t \neq \mathrm{G}$; otherwise P would be an interior point of $t = \mathrm{G}$, contrary to the assumption that P is a frontier point of M in J. For a sufficiently fine triangulation T of J there is therefore an $(n-1)$-simplex of T contained in $\mathrm{G} - t$.

Second step. For the second and third step Brouwer found it easier to transform $\mathbf{R}^n - \{\mathrm{B}\}$ by an inversion of pole B, bringing about the following situation (where we use the *same* notation for elements of the former situation and for their transforms by inversion): D is now an open half space of $\mathbf{R}^n$, having a hyperplane H as its frontier, and one has $\mathrm{A} \in \mathrm{H}$; u is a closed subset of H that does not contain A; $\mathrm{G} \subset \mathrm{D}$ is a homeomorphic image of a subset of J, open in J; $u = \bar{\mathrm{G}} \cap \mathrm{H}$; finally t is a subset of G, relatively closed in G, $u = \bar{t} \cap \mathrm{H}$, and $\mathrm{G} - t$ contains an $(n-1)$-simplex σ of a triangulation T of J. If $\pi: \mathrm{H} - \{\mathrm{A}\} \to \mathrm{S}$ is the projection from A of $\mathrm{H} - \{\mathrm{A}\}$ onto an $(n-2)$-dimensional sphere $\mathrm{S} \subset \mathrm{H}$ of center A, then, as $\mathrm{A} \notin u$, the restriction $p: u \to \mathrm{S}$ of π to u is defined. The second step of the proof consists in *extending* p to a continuous map $\bar{p}: t \cup u \to \mathrm{S}$. As nothing is known of the connected set t, p is in fact extended to a continuous map $p_0: (\mathrm{G} - \sigma) \cup u \to \mathrm{S}$, and then $\bar{p}$ is the restriction of p_0 to $t \cup u$.

Begin by triangulating the open subset G of J by the usual method, taking a sequence (T_ν) of successive subdivisions of T, whose mesh tends to 0. G_ν is the union of the simplices of T_ν contained in G, and $\mathrm{G}_\nu \subset \mathrm{G}_{\nu+1}$; $g_\nu = \mathrm{G}_{\nu+1} - \mathrm{G}_\nu$ converges uniformly to u, and G is the union of the g_ν. Next define p_0 in two steps:

First take a sufficiently large number r, and define p_0 on the union G'_r of all the g_ν for $\nu \geqslant r$ by projecting each vertex of all g_ν for $\nu \geqslant r$ on H by the orthogonal projection $f: \mathbf{R}^n \to \mathrm{H}$; p_0 is then defined on the vertices of G'_r as the map $\pi \circ f$. Extend it to a piecewise affine map $\mathrm{G}'_r \to \mathrm{S}$ (using barycentric coordinates in the simplices of each T_ν and in simplices of S). This defines p_0 on the *frontier* E_r *in* G of the union G''_r of the G_ν for $\nu < r$.

Next define p_0 on $\mathrm{G} - \sigma$ by extending it "backward," so to speak, from E_r; for each $(n-1)$-simplex σ_r in G''_r with one of its faces τ_r in E_r, assign an arbitrary value in S to the only vertex of σ_r not in τ_r; then p_0 can be extended from τ_r to the whole of σ_r as a piecewise affine map (again using barycentric coordinates in curvilinear simplices). Then p_0 is defined on the union of these simplices σ_r, hence it is known on the frontier E_{r-1} of the union of the remaining simplices of G_r'', and the procedure can be repeated. This would not work if one wanted to define p_0 in the *whole* set G (there would be an "obstruction" in σ); but it does work for $G - \sigma$.

Third step. Let N be the connected component of the open complement of t in the half space D, such that $A \in \bar{N}$. Let T′ be a triangulation of the half space $\bar{D}$, such that the n-simplices of T′ that meet H have as an intersection with H a p-simplex of their frontier ($p \leqslant n - 1$); these intersections form a triangulation T″ of H. Then construct a triangulation in the usual way for the set $N \cup (H - u)$ (open in $\bar{D}$) by taking successive subdivisions T'_ν of T′ with mesh tending to 0, and defining N_ν as the union of the simplices of T'_ν contained in $N \cup (H - u)$; A may always be supposed interior to an $(n - 1)$-simplex σ_0 of that triangulation. By lemma (L), the frontier of N_ν in $\mathbf{R}^n$ is the union of $N_\nu \cap H$ (which contains A) and a union L_ν of pseudomanifolds-with-boundary, and $F_\nu = L_\nu \cap H$ is a union of closed $(n - 2)$-dimensional pseudomanifolds.

For each vertex C of L not in H let C_1 be a point of t at a distance $d(C, t)$ of C, and let $q(C) = \bar{p}(C_1) \in S$; if $C \in F_\nu$, let $q(C) = \pi(C)$. Then extend q to L_ν as a piecewise affine map in S (using barycentric coordinates as above); q is then a continuous map of L_ν into S.

The contradiction needed to end the proof consists in computing, for sufficiently large values of ν, the degree of the restriction $q|F_\nu$ (as a mapping into S) in two different ways, using the fact that F_ν is *both* the intersection $L_\nu \cap H$ and the frontier of $\bar{N}_\nu \cap H$ in H. For the first computation take ν large enough; for any $(n - 1)$-simplex σ_1 in L_ν, $q(\sigma_1)$ is then contained in a half sphere of S (depending on σ_1). The degree of the restriction of q to the boundary of σ_1 is then 0. By the additivity of the degree and the fact that any $(n - 2)$-simplex of L_ν is the face of two $(n - 1)$-simplices except those in F_ν the degree of $q|F_\nu$ is 0.

For the second computation, consider the $(n - 1)$-simplices of $\bar{N}_\nu \cap H$; it may be assumed that they are so small that, with the exception of σ_0 (which contains A), their images by π each belong to a half sphere of S; the degree of the restriction of π to the boundary of each such simplex is therefore 0. The additivity of the degree then shows that the degree of $q|F_\nu = \pi|F_\nu$ is the same as the degree of the restriction of π to the boundary of σ_0, and it is clear that the latter is ± 1.

§ 5. The Notion of Dimension for Separable Metric Spaces

The theorem on the invariance of dimension (§ 1) did not give a definition of the word "dimension" as a number attached to a topological space and invariant under homeomorphisms except for spaces locally homeomorphic to $\mathbf{R}^n$ ("pure" C^0 manifolds), and even for these spaces the introduction of the auxiliary space $\mathbf{R}^n$ was not satisfactory for a notion that should have been an intrinsic one. This incongruity was stressed by Poincaré in 1903 [371] and again in 1912, the last year of his life [372], in articles written for a nonmathematical public. He pointed out that, just as in classical geometry, one thought of a surface as "limiting" a solid, a curve as "limiting" a surface, and a point as "limiting" a curve, it should be possible to define the "dimension" of a connected space by an *inductive process*: the dimension should be *one* if the space may be disconnected by points, *two* if it may be disconnected by sets of dimension 1, *three* if it may be disconnected by sets of dimenion 2, "and so on."

Meanwhile, in October 1910, Lebesgue, who had heard from Blumenthal of Brouwer's proof of the invariance of dimension (then in the process of being published in *Mathematische Annalen*, of which Blumenthal was one of the editors) sketched, in a letter to Blumenthal (which the latter published immediately after Brouwer's proof) another proof, based on a completely new and remarkable idea ([293], pp. 170–171). Observing that for a covering of a plane domain by sufficiently small closed "bricks" there always are points of the domain belonging to at least three bricks, he stated as a theorem that for any finite covering (E_j) of an open bounded connected set D in $\mathbf{R}^n$ by sufficiently small closed sets there always are points in D belonging to at least $n + 1$ sets. He added that for a cube D it is always possible to find a finite covering by arbitrary small parallelotopes for which no point of D belongs to more than $n + 1$ sets of the covering (both statements of course together imply the invariance of dimension).

This last part was easy enough to show by a simple arrangement of "bricks" in the cube D; but although Lebesgue's sketch of a proof for the first statement was later seen to be capable of yielding a correct argument, the way in which he tried to apply it led to incorrect statements, as Brouwer almost immediately observed. The proof is easily reduced to the case in which D is the cube $[0, 1]^n$, and the E_j are unions of closed cubes of side $1/2^\nu$, having as vertices points of $2^{-\nu}\mathbf{Z}^n$ for sufficiently large ν; it is only necessary to suppose that no E_j meets *both* opposite faces C_i, C'_i of D (defined, respectively, by $x_i = 0$ and $x_i = 1$) for $1 \leqslant i \leqslant n$. Lebesgue's idea was to inductively construct nonempty closed sets $K_1 \supset K_2 \supset \cdots \supset K_n$, for which it could be proved that each K_h contains points belonging to at least $h + 1$ sets E_j (cf. [261], p. 43). He thought he could define the K_h by taking the union G_1 of those E_j that meet C_1, and letting K_1 be a connected component of the frontier of G_1 in $\mathbf{R}^n$ contained in D, different from C_1 and meeting *both* C_1 and C'_i for $2 \leqslant i \leqslant n$. He could then take the union G_2 of those E_j not contained in G_1 and meeting both K_2 and C_2, and let K_2 be a connected component of the frontier of $G_2 \cap K_1$, not contained in C_2 and meeting *both* C_i and C'_i for $3 \leqslant i \leqslant n$. Lebesgue claimed he could proceed inductively in this way (without giving any detail) to define the K_h; however Brouwer found a counterexample (for $n = 3$) where Lebesgue's procedure does not yield any set K_3 having the properties he claimed ([89], p. 545). It was only in 1921 that Lebesgue published a correct proof of his theorem ([295], pp. 177–206).

In the meantime Brouwer had taken up Poincaré's idea in 1913, and had given it mathematical content ([89], pp. 540–546). He first observed that

Poincaré's tentative definition had to be slightly modified to really conform to intuition*: if one deletes the vertex of a cone with two sheets in $\mathbf{R}^3$, the cone is disconnected although no one would consider its dimension to be 1! For a space E,† Brouwer said that two disjoint closed sets F, F′ are *separated* by a set C if any connected subset of E that meets F and F′ also meets C‡; he then defined a space of *dimension* 0 as one containing no connected set with more than one point, and a space E of *dimension* $n > 0$ by the property that n is the smallest integer > 0 such that *any two* disjoint closed subsets of E are separated by a subset of dimension $\leqslant n - 1$. That definition can immediately be localized: a space E has *dimension n at a point* P if P has a fundamental system of neighborhoods of dimension n.

The bulk of Brouwer's paper is devoted to proving that, with his definition of dimension, $\mathbf{R}^n$ has dimension n at every point. By induction on n, it is easy to show that this dimension is $\leqslant n$. To prove that it is $\geqslant n$, an argument similar to Lebesgue's reduced the proof to a simplicial version of Lebesgue's theorem:

(S) Let $\sigma = A_1 A_2 \cdots A_{n+1}$ be an n-simplex in $\mathbf{R}^n$, and consider a triangulation T of σ in rectilinear simplices, none of which meets *both* $A_1 A_2 \cdots A_\nu$ and $A_{\nu+1} A_{\nu+2} \cdots A_{n+1}$, for any $\nu \leqslant n$. Define σ_j inductively for $1 \leqslant j \leqslant n$ by letting γ be the subcomplex of T, union of all the n-simplices of T having A_1 as one of their vertices; lemma (L) of §3 shows that γ is a pseudomanifold-with-boundary; the part σ_1 of that boundary, the union of the $(n-1)$-simplices that does not contain A_1, is a union of closed pseudomanifolds, and $A_1 \notin \sigma_1$. In general, σ_ν is defined by induction on $\nu \leqslant n$: let γ_ν be the subcomplex of σ_ν, union of the $(n-\nu)$-simplices of σ_ν that meet $A_1 A_2 \cdots A_{\nu+1}$, but do not meet $A_1 A_2 \cdots A_\nu A_{\nu+2} \cdots A_{n+1}$; this is again a pseudomanifold-with-boundary; the part $\sigma_{\nu+1}$ of that boundary which is the union of the $(n-\nu-1)$-simplices of σ_ν that do not meet $A_1 A_2 \cdots A_\nu$ is a union of closed pseudomanifolds. Then the σ_ν, which form a decreasing sequence of sets, are all *nonempty*.

The proof uses the properties of the degree of a map, and, as usual, is very sketchy and has to be interpreted to make sense. Let π_ν be the projection of $\sigma - (A_1 A_2 \cdots A_\nu)$ onto $A_{\nu+1} A_{\nu+2} \cdots A_{n+1}$ [$\pi_\nu(M)$ being the intersection of $A_{\nu+1} \cdots A_{n+1}$ with the ν-dimensional linear affine variety generated by M and $A_1 A_2 \cdots A_\nu$]. Let p_ν be the restriction of π_ν to σ_ν.

---

* A similar observation had already been made by Riesz [396].

† This paper is the only one of the period 1911–1913 in which Brouwer considers general topological spaces. He says his spaces must be "Normalmenge in Fréchetschen Sinne" (?) but does not use any property beyond the definition of a topological space.

‡ In the paper as it was first published, he had written "closed connected subset" instead of "connected subset"; after Urysohn had pointed out to him that this definition was incompatible with the proof of the main theorem of Brouwer's paper, the latter published in 1923 a corrected version ([89], p. 547), which he elaborated in a 1924 paper ([89], pp. 554–557).

The induction starts with the obvious remark that the degree of p_1 is equal to 1.* The main point of the proof is to show that if the degree of p_ν is 1, so is the degree of $p_{\nu+1}$; this of course implies that $\sigma_\nu \neq \emptyset$ for all ν.

The passage from ν to $\nu + 1$ is done by considering each $(n - \nu - 1)$-simplex of $\sigma_\nu \cap (A_1 \cdots A_\nu A_{\nu+2} \cdots A_{n+1})$, which is the face of a unique $(n - \nu)$-simplex of σ_ν; it follows easily, by a continuity argument, that the restriction of $p_{\nu+1}$ to $\sigma_\nu \cap (A_1 \cdots A_\nu A_{\nu+2} \cdots A_{n+1})$, considered as a mapping into $A_{\nu+2} \cdots A_{n+1}$, has degree 1. On the other hand, the restriction of $p_{\nu+1}$ to the frontier of each $(n - \nu)$-simplex meeting $A_1 \cdots A_\nu A_{\nu+2} \cdots A_{n+1}$ has degree 0. By additivity of the degree, it follows that, deleting all these simplices from σ_ν, which by definition gives as remnant $\gamma_{\nu+1}$, the restriction of $p_{\nu+1}$ to $\sigma_{\nu+1}$ has degree 1.

§ 6. Later Developments

The first complete proofs of the "no separation" (§ 4) and Jordan–Brouwer (§ 3) theorems entirely devoid of the obscurities linked to the fantastic complexity of Brouwer's constructions were given by Alexander in 1922. They constitute the first and second steps, respectively, in the proof of his duality theorem (Part 1, chap. II, § 6). As we have seen, these proofs, based on convenient splittings of a cube or a sphere, are reminiscent of the (later) Mayer–Vietoris theorems. Indeed the use of the general Mayer–Vietoris exact sequence in cohomology (Part 1, chap. IV, § 6) very easily determines the whole de Rham cohomology $H^\cdot(\mathbf{R}^n - X)$ (Part 1, chap. III, § 3) when X is homeomorphic to a cube or to a sphere, and the "no separation" and Jordan–Brouwer theorems are just consequences of the computation of $H^0(\mathbf{R}^n - X)$.

Another way of obtaining these theorems was used by Leray [324] who proved a general result containing both as special cases†: if K and K′ are two homeomorphic compact subsets of $\mathbf{R}^n$, then $\mathbf{R}^n - K$ and $\mathbf{R}^n - K'$ have the same cardinal number (finite or infinite) of connected components. This follows from the multiplicative property of the localized degree [chap. I, § 3, formula (13)] and the purely algebraic property of invariance of (linear) dimension of a vector space over $\mathbf{Q}$.

Although Brouwer gave a definition of the notion of dimension applying to arbitrary spaces, he was obviously chiefly interested in proving that for $\mathbf{R}^n$ that definition gives the number n. This is probably the reason why his paper was considered merely another way of proving the invariance of dimension, and the fact that he had given a general definition of dimension was neglected. At any rate, when in 1922 Urysohn and Menger proposed (independently of

* As a simplex is not a "manifold" in Brouwer's sense, it is in fact the localized degree $d(p_1 . I, M)$ which is equal to 1, where I is the interior of the simplex $A_2 A_3 \cdots A_{n+1}$ and M is a point of I. Similarly for the p_ν, $\nu \geqslant 2$.

† It also contains the "invariance of closed curves" that Brouwer had attempted to prove ([89], pp. 523–526).

each other) a definition that is equivalent to Brouwer's for locally connected or compact separable metric spaces, they were at first unaware of Brouwer's priority.

The Urysohn–Menger definition applies to *all separable metric spaces.* For them the empty set has dimension -1, and the dimension of a nonempty space is the least integer $n \geqslant 0$ for which every point has a fundamental system of neighborhoods whose boundaries have dimension $< n$ (the dimension is taken to be $+\infty$ if there is no such integer n).*

This definition's consequences were studied in the period, extending to about 1940, during which dimension theory became a very active branch of mathematics. But apart from the Brouwer theorem on the dimension of $\mathbf{R}^n$ the methods of proof in that theory belonged to general (also called "set-theoretic") topology and made no use of triangulations or homology.† We will therefore not describe all the results of that theory, but refer the reader to [261]. Some of results, however have interesting connections with algebraic topology.

First, Lebesgue's theorem furnishes (for separable metric spaces) an alternative definition of dimension. The *order* of a finite open covering $\mathfrak{R}$ of a space E is the largest integer p such that there exists $p + 1$ distinct sets of $\mathfrak{R}$ with nonempty intersection. If $m(\mathfrak{R})$ is the g.l.b. of the orders of the finite open coverings of E *finer* than $\mathfrak{R}$, Lebesgue's theorem says that for $\mathbf{R}^n$ the l.u.b. of the $m(\mathfrak{R})$ for all finite open coverings $\mathfrak{R}$ of $\mathbf{R}^n$ is equal to n. For a general space E this l.u.b. is the dimension of E as defined by Urysohn and Menger.

From this it follows at once that for a separable metric space E of dimension n, the *Čech homology groups* $\check{\mathrm{H}}_p(\mathrm{E}; \mathrm{G})$ based on *finite* open coverings (Part 1, chap. IV, § 2) are all 0 for $p > n$. Surprisingly enough this is not true for *singular* homology groups: there exist compact metric spaces of finite dimension for which *infinitely many* singular homology groups are $\neq 0$ [44]. On the other hand, there are obviously contractible compact spaces of any finite dimension, so that there are no very strong links between dimension and homology of a space. In Part 3, chap. II, we shall see that homotopy theory is much closer to the notion of dimension.

With the arrival of sheaf cohomology (Part 1, chap. IV, § 7,C), another notion of "dimension" of a space could be defined. A space X, on which is given a family Φ of supports (Part 1, chap. IV, § 7,C), has *finite Φ-dimension* if there is an integer $n \geqslant 0$ such that

$$\mathrm{H}^i_\Phi(\mathrm{X}; \mathscr{F}) = 0 \quad \text{for every } i > n \text{ and } \textit{every sheaf } \mathscr{F} \text{ over X;} \tag{2}$$

the smallest integer n having that property is called the *Φ-dimension* of X and

* Brouwer's definition differs from that of Urysohn–Menger because he takes totally disconnected spaces to have dimension 0, whereas for the Urysohn–Menger definition, there are totally disconnected spaces of arbitrary finite dimension ([261], p. 23).

† Brouwer's proof was later replaced by a purely combinatorial lemma of Sperner ([30], p. 376).

written $\dim_\Phi X$; when Φ is the family of all closed sets in X, n is called the *cohomological dimension* of X (or simply *dimension*) if no confusion arises. If $\Phi_1 \supset \Phi_2$ are two families of supports on X, and if X has Φ_1-dimension $\leqslant n$, then it has Φ_2-dimension $\leqslant n$. If Y is a subset of X that is *locally closed*, and if X has Φ-dimension $\leqslant n$, then Y has Φ'-dimension $\leqslant n$, where $\Phi' = \Phi \cap Y$. When X is *metrizable* and has cohomological dimension $\leqslant n$, the same is true for *every subset* of X. For a *paracompact* space X to have cohomological dimension $\leqslant n$ it is necessary and sufficient that each point of X have a neighborhood of cohomological dimension $\leqslant n$. If a compact metric space has dimension $\leqslant n$ in the sense of Urysohn–Menger, it also has a cohomological dimension $\leqslant n$ ([66], [87]).

The condition (2) may be restricted by considering only sheaves $\mathscr{F}$ of modules over a fixed Dedekind ring Λ; if (2) holds for all such sheaves $\mathscr{F}$ and *all* paracompactifying families Φ, one says the dimension of X *over* Λ is $\leqslant n$, and the smallest integer n having that property is the *dimension of* X *over* Λ, written $\dim_\Lambda X$; it is also the smallest integer n for which the cohomology with compact supports $H_c^{n+1}(U;\Lambda) = 0$ for all *open* subsets U of X [208]. When $\dim_\Lambda X \leqslant n$, the *Borel–Moore homology* (Part 1, chap. IV, § 7,F) satisfies

$$H_q^\Phi(X;\Lambda) = 0 \quad \text{for } q \geqslant n+1 \text{ and any family } \Phi \text{ of supports;} \tag{3}$$

$$H_q^x(X;\Lambda) = 0 \quad \text{for } q \geqslant n+1 \text{ and all } x \in X. \tag{4}$$

If $\mathscr{F}$ is the constant sheaf Λ, or if Φ is paracompactifying, there is a canonical isomorphism

$$H_n^\Phi(X;\mathscr{F}) \simeq \Gamma_\Phi(\mathscr{H}_n(X;\Lambda) \otimes \mathscr{F}) \quad \text{([66], pp. 151–152).} \tag{5}$$

CHAPTER III

Fixed Points

§ 1. The Theorems of Brouwer

Brouwer had been considering continuous maps of the sphere S_2 into itself as early as 1909; he first studied the particular case of a *bijection** f (which is therefore bicontinuous) preserving orientation, and he gave a proof that in that case there exists at least one *fixed point* x for f, i.e., such that $f(x) = x$; the proof is very long (nine pages) and involved, using intricate arguments on deformations of curves on S_2 ([89], pp. 195–205). In 1910 he gave another proof of the same result as a corollary to the existence of at least one singular point for a continuous vector field on S_2 (§ 3) by another intricate argument ([89], pp. 303–318).

It was only in 1911, in the paper in which he gave the definition of the degree of a map (chap. I, § 2), that he realized that this notion could be used to prove that a continuous map f of S_n into itself, satisfying *the only condition that* $\deg(f) \neq (-1)^{n+1}$, has at least one fixed point. Equivalently, he showed that if f has *no* fixed point, then $\deg(f) = (-1)^{n+1}$; but his first proof is far from simple, and uses the computation (done earlier in that paper) of the sum of the indices of a continuous vector field on S_n having only isolated singularities (see § 3). Fixing a point O on S_n, he considered, for every point P $\neq$ O for which $f(\mathrm{P}) \neq$ O, the unit vector tangent at P to the arc of the circle through O, P and $f(\mathrm{P})$ having extremities at P and $f(\mathrm{P})$ and not containing O.† To apply his theorem on vector fields, he had to define the vector field in the neighborhood of O and of the points of $f^{-1}(\mathrm{O})$ where the previous definition is meaningless.

Finally, in the next paper he published in 1911 ([89], pp. 454–472), Brouwer arrived at a very simple proof without using vector fields: if f has no fixed point, the consideration of the great circle joining x and $f(x)$ at once provides a *homotopy* of f to the antipodal map $x \mapsto -x$ for which the degree is obviously $(-1)^{n+1}$.

* Brouwer only assumed that f is injective, but by degree theory (which he had not invented at that time) it follows that f is necessarily bijective.
† He had already used that device in 1910 for $n = 2$ ([89], p. 315).

J. Dieudonné, *A History of Algebraic and Differential Topology, 1900–1960*,
Modern Birkhäuser Classics, DOI 10.1007/978-0-8176-4907-4_9,

Being linked to the as yet unfamiliar notion of degree, this result did not attract much attention from the mathematicians of that time. Things were quite different for the corollary Brouwer added concerning a continuous map g of a *cube* I^n into itself. He showed that such a map *always* has at least one fixed point. His argument consisted in replacing I^n by the homeomorphic upper hemisphere D_+ of the sphere S_n and extending g to a continuous map $f: S_n \to D_+$ by taking $f(x) = g(s(x))$ in the lower hemisphere D_-, s being the symmetry with respect to the equator; then $\deg(f) = 0$ since f is not surjective, and a fixed point of f must of necessity be a fixed point of g. The interest aroused by this result was due to its unexpected generality, which made possible its application to *existence* proofs in analysis, using much weaker assumptions than had been customary in earlier existence theorems; later it was realized that Brouwer's fixed point theorem could even be used in infinite-dimensional spaces, under assumptions allowing suitable approximations by finite dimensional compact sets (see chap. VII).

§ 2. The Lefschetz Formula

It is clear that for a continuous map f of a compact space X into itself the existence of fixed points will in general depend not only on the space X, but on f itself (the Brouwer case $X = I^n$ being an exception). This fact was given precise expression in a remarkable formula discovered by Lefschetz in 1926 [300].

Lefschetz limited himself to a *combinatorial manifold* X (Part 1, chap. II, § 4), but considerably enlarged the concept of "fixed point." He first observed that it was a special case of "coincidences" for two continuous maps f, g of X into itself, namely, the points $x \in X$ such that $f(x) = g(x)$. As he was at that time working on the topology of product spaces, he translated that notion in terms of the *graphs* $\Gamma(f)$ and $\Gamma(g)$ of f and g in the product space $X \times X$ which is also a combinatorial manifold: a "coincidence" is the first projection in X of a *common point* of $\Gamma(f)$ and $\Gamma(g)$. Lefschetz was thus led back to a problem of *intersection*, a question on which we have seen he was also working (Part 1, chap. II, §§ 4 and 5).

It is quite obvious that he was strongly influenced by the similar problems in algebraic geometry, and in particular by the theory of *correspondences*, studied since the middle of the nineteenth century by Chasles and the school of "enumerative geometry" (de Jonquières, Zeuthen, Schubert), then by Hurwitz in the theory of Riemann surfaces, and which had been thoroughly investigated by Severi in the first years of the twentieth century; this influence explains the rather unusual frame within which Lefschetz developed his theory.

Let X be a compact, connected, orientable combinatorial manifold without boundary, of dimension n, Lefschetz studied that he calls a "transformation" T in X, by which he means an *n-cycle Γ_T in the product space* $X \times X$. If T′ is a second "transformation" in X, the algebraic intersection number* $(\Gamma_T . \Gamma_{T'})$ is defined (Part 1, chap. IV, §4). Once homology bases (distinct or not), (γ_p^i), (δ_p^j), are known for $H_\bullet(X; \mathbf{Q})$, as well as their multiplication table in the "intersection ring" $H_\bullet(X \times X; \mathbf{Q})$, the $\gamma_p^i \times \delta_q^j$ for $p + q = r$ form a base of $H_r(X \times X; \mathbf{Q})$ by Künneth's theorem, and the intersection products of these elements in $H_\bullet(X \times X; \mathbf{Q})$ are given by formula (30) of Part 1, chap. II, §5. The number $(\Gamma_T . \Gamma_{T'})$ could therefore be computed at once from the expressions

$$\Gamma_T = \sum_{0 \leqslant p \leqslant n} \varepsilon_p^{ij}(\gamma_p^i \times \delta_{n-p}^j), \qquad \Gamma_{T'} = \sum_{0 \leqslant p \leqslant n} \varepsilon_p'^{ij}(\gamma_p^i \times \delta_{n-p}^j). \tag{1}$$

But Lefschetz's original idea was to look for another computation of that number by introducing *actions* of T and T′ on the homology groups $H_p(X; \mathbf{Q})$. Even before singular homology had been defined, it was possible to associate to every continuous map $f: X \to Y$ of finite cell complexes, a homomorphism

$$f_*: H_\bullet(X; \mathbf{Q}) \to H_\bullet(Y; \mathbf{Q})$$

of graded vector spaces, by simplicial approximation (Part 1, chap. II, §3). Lefschetz [probably inspired by similar processes in algebraic geometry, the images of divisors by correspondences (see [299])], extended this idea to his "transformations." He considered a homology class $\alpha_p \in H_p(X; \mathbf{Q})$ and its product $\alpha_p \times [X]$ by the fundamental class of X (chap. I, §3,A) in $H_{p+n}(X \times X; \mathbf{Q})$; its intersection $\Gamma_T . (\alpha_p \times [X])$ with Γ_T is a class in $H_p(X \times X; \mathbf{Q})$, and the image of that class by the homomorphism $(\mathrm{pr}_2)_*$ in $H_p(X; \mathbf{Q})$ is, by definition, the image $T_*(\alpha_p)$ by the action of T on $H_p(X; \mathbf{Q})$.

From his intersection theory (Part 1, chap. II, §4), Lefschetz deduced the fundamental result

$$(\Gamma_T . (\gamma_p^i \times \delta_{n-p}^j)) = (-1)^p (T_*(\gamma_p^i) . \delta_{n-p}^j) \tag{2}$$

which gave him the expressions of the ε_p^{ij} as linear forms in the coefficients of the matrix (α_p^{ij}) of the homomorphism $(T_*)_p$, the restriction of T_* to $H_p(X; \mathbf{Q})$. From these expressions he derived the expression of $(\Gamma_T . \Gamma_{T'})$ as function of the matrices of the $(T_*)_p$ and $(T'_*)_q$. He did not at first express this formula in terms of traces of matrices, but in a second paper [301] he obtained such an expression, and in particular when T′ is the identity (so that $\Gamma_{T'}$ is the *diagonal* Δ of $X \times X$, which is an n-cycle), he arrived at the famous *Lefschetz formula*

$$(\Gamma_T . \Delta) = \sum_{0 \leqslant p \leqslant n} (-1)^p \operatorname{Tr}((T_*)_p). \tag{3}$$

When the cycle Γ_T and the diagonal Δ intersect "transversally" in a finite number of points, the left-hand side of (3) could be interpreted as the "algebraic number of fixed points" of the "transformation" T.

---

* We abuse language by writing an intersection number for *cycles* instead of writing it for their *homology classes.*

In 1928 Hopf returned to the initial problem of the existence of fixed points for an arbitrary continuous *map* $f: X \to X$, but this time he considered not merely a combinatorial manifold X, but an *arbitrary* finite euclidean simplicial complex of dimension *n*. He associated to such a map, according to (3), what came to be called the *Lefschetz number* of f

$$\Lambda(f) = \sum_{0 \leqslant p \leqslant n} (-1)^p \operatorname{Tr}((f_*)_p) \tag{4}$$

and he proved first that if f has *no* fixed point, then $\Lambda(f) = 0$.

As X is compact, the assumption implies that $|f(x) - x| \geqslant \delta > 0$ for all $x \in X$. There is therefore a subdivision K of the triangulation of X and a simplicial approximation g of f for that triangulation, homotopic to f and such that $|g(x) - x| \geqslant \delta/2 > 0$ for $x \in X$; since $g_* = f_*$, it is enough to prove the theorem for g instead of f. If $(\sigma_j)_{1 \leqslant j \leqslant \alpha_p}$ is the canonical basis of the **Q**-vector space $C_p(K)$ of the *p*-chains of K, and if the diameters of the simplices of K are small enough, the endomorphism $\tilde{g}_p$ of $C_p(T)$ corresponding to g (Part 1, chap. II, § 3) is such that

$$\tilde{g}_p(\sigma_j) = \pm\sigma_k \quad \textit{for an index } k \neq j \text{ if } g(\sigma_j) \text{ is not degenerate,}$$
$$\tilde{g}_p(\sigma_j) = 0 \qquad \text{otherwise;}$$

this implies that $\operatorname{Tr}(\tilde{g}_p) = 0$. From this Hopf concluded that all he had to do was prove the formula that he rightly considered the natural generalization of the Euler–Poincaré formula [Part 1, chap. I, § 3, formula (4)]:

$$\sum_{p=0}^{n} (-1)^p \operatorname{Tr}(\tilde{g}_p) = \sum_{p=0}^{n} (-1)^p \operatorname{Tr}((g_*)_p) \tag{5}$$

for every simplicial map $g: X \to X$; it reduces to the Euler–Poincaré formula when g is the identity. The proof is similar, using the fact that $\tilde{g}_p(Z_p) \subset Z_p$, $\tilde{g}_p(B_p) \subset B_p$ for cycles and boundaries and that

$$\operatorname{Tr}(\tilde{g}_p) = \operatorname{Tr}(\tilde{g}_p | Z_p) + \operatorname{Tr}(\tilde{g}_{p-1} | B_{p-1}),$$
$$\operatorname{Tr}((g_*)_p) = \operatorname{Tr}(\tilde{g}_p | Z_p) - \operatorname{Tr}(\tilde{g}_p | B_p).$$

When $\Lambda(f) \neq 0$ and X is again a *combinatorial manifold*, so that (3) is applicable for $T = f$, Hopf gave an interpretation of the left-hand member when f has only a finite number of fixed points, by defining for each fixed point a of f an *index* j_a, the definition of which is meaningful for any C^0 manifold, triangulable or not. Consider a homeomorphism h of an open neighborhood of a in X [with $h(a) = 0$], onto an open neighborhood of 0 in $\mathbf{R}^n$; then, for sufficiently small $\rho > 0$, $g = hfh^{-1}$ is defined in the ball B: $|x| \leqslant \rho$ and is a continuous map $B \to \mathbf{R}^n$, with only one fixed point 0. The map $x \mapsto g(x)/|g(x)|$ is defined on S: $|x| = \rho$ and maps S into S_{n-1}, so that its *degree* is defined; it is independent of ρ and of the choice of the homeomorphism h, and its value is by definition the index j_a. Hopf's interpretation of (3) for $T = f$ is then

$$\sum_{a \in \mathrm{Fix}(f)} j_a = \Lambda(f), \tag{6}$$

Fix(f) being the finite set of fixed points of f.

Hopf's first proof of (6) ([241a], p. 153) is particularly interesting. In the neighborhood of a fixed point a, he modified both the cell complex X and the map f. One may assume that a is contained in an n-simplex σ, of frontier τ, and (with the preceding notation) the homeomorphism h maps $\bar{\sigma}$ onto B and τ onto S; a homotopy can modify f in a neighborhood of a in such a way that $f(\tau)$ does not meet $\bar{\sigma}$. Then Hopf added a new n-simplex σ' to X by gluing it to σ along τ in such a way that $\bar{\sigma} \cup \sigma'$ becomes homeomorphic to $\mathbf{S}_n$, τ being mapped on the "equator" $\mathbf{S}_{n-1}$. Transferring to $\bar{\sigma} \cup \sigma'$ the symmetry with respect to the equator gives an automorphism s of $\bar{\sigma} \cup \sigma'$, exchanging σ and σ' and leaving the points of τ invariant. Next Hopf changed f in $\bar{\sigma}$, replacing it by $\bar{f} = s \circ f$, and defined $\bar{f}$ in σ' equal to $f \circ s$.

Doing this for every fixed point of f yields a cell complex X′ and a continuous map $\bar{f}$ of X′ into itself with *no fixed point*; $\Lambda(\bar{f}) = 0$; but the construction gives the relation

$$\Lambda(\bar{f}) = \Lambda(f) - \sum_{a \in \mathrm{Fix}(f)} j_a$$

hence the result. This is one of the first examples of the use of *attachment* of new cells to a cell complex that later became an important tool (see chap. V, § 3).

Hopf's second proof [241b] starts from a triangulation T of X such that all the fixed points of f belong to n-simplices. He refined T to a sufficiently iterated subdivision T′, for which he constructed a simplicial approximation g homotopic to f, such that there are *no fixed r-simplices* of T′ for g when $r < n$; then $\mathrm{Tr}(\tilde{g}_r) = 0$ for $r < n$ and $\mathrm{Tr}(\tilde{g}_n) = \sum_{a \in \mathrm{Fix}(f)} j_a$, so that formula (6) becomes a consequence of (5).

Lefschetz endeavored to generalize his formula to compact metric spaces using Vietoris homology, but Hopf provided him with an example of a compact subset X of $\mathbf{R}^2$ and a continuous map without fixed point for which $\Lambda(f) \neq 0$ both for singular and Vietoris homology: X is the union of two concentric circles and a spiral curve winding between both and asymptotic to each of them, whereas f is just a rotation of a fixed angle ω for points on each circle and on the spiral ([304], p. 347). Later Lefschetz realized that the validity of the formula could be recovered by making assumptions on the "local connectedness in the sense of homology" on X (cf. chap. IV) [46].

§ 3. The Index Formula

We have already mentioned (chap. I, § 2) that in 1881 Poincaré, in his work on global theory of differential equations, had introduced the notion of *index* for a vector field on the sphere $\mathbf{S}_2$. He was studying in $\mathbf{R}^2$ the integral curves of a differential equation

$$\frac{dx}{X} = \frac{dy}{Y}$$

where X and Y are *polynomials*. He took a point O in $\mathbf{R}^3$ not in the plane, and projected from O the vector field (X, Y) on a sphere S of center O, extending it by continuity on the "equator" of S (section by the plane parallel to $\mathbf{R}^2$). This gave him a vector field on S, symmetrical with respect to O. He showed that there were always at least *two* (*symmetrical*) *singular points* of the field (i.e., points where the field vanishes). Then he restricted himself to "general" such fields in the following sense: (1) X and Y have the same degree m; (2) if X_m, Y_m are their homogeneous parts of degree m, $xY_m - yX_m$ is not identically 0; (3) the curves $X = 0$ and $Y = 0$ intersect transversally in points not on the equator; (4) the roots of the homogeneous equation $xY_m - yX_m = 0$ are simple.

Next Poincaré introduced the notion of *index* of *any* closed curve on an hemisphere of S containing no singular point: if h (resp. k) is the number of points where Y/X passed from $-\infty$ to $+\infty$ (resp. from $+\infty$ to $-\infty$) when moving on the (positively oriented) curve, the index is defined as $i = (h - k)/2$. He showed that $i = \pm 1$ for a small enough curve around a singular point, and took that value as the *index of the singular point*; he then proved the remarkable result that the *sum* of the indices of the singular points is *equal to* 2 ([365], p. 29).

In 1909 Brouwer, who at that time probably was not aware of Poincaré's paper, considered a vector field on $\mathbf{S}_2$ that he only supposed *continuous* (whereas in Poincaré's case, the field is *analytic* at nonsingular points); he wanted to prove that there exists *at least one* singular point. He argued by contradiction, using the detailed study of the trajectories of the vector field (he could not use local uniqueness since the field is not supposed to be C^1) ([89], p. 279).

In his 1911 paper on the definition of the degree ([89], pp. 454–472) Brouwer considered, for any n, a vector field on $\mathbf{S}_n$ that he merely supposed continuous, with at most finitely many singular points; he proceeded to prove that the sum of the indices of the singular points is 2 for even n, 0 for odd n. To apply his definition of the degree to that problem he used a very complicated and obscure process, starting from a simplicial triangulation T of $\mathbf{S}_n$ obtained by intersections of $\mathbf{S}_n$ with hyperplanes, among which is the equator; T is supposed symmetrical with respect to the equator and the singular points of the vector field are all contained in the interior of n-simplices of T. If T is fine enough, the sum of the indices of the singular points of the vector field is given by a sum of degrees of maps, written $c_{1\alpha}$ and $c_{2\alpha}$. To define $c_{1\alpha}$, project each n-simplex $s_{1\alpha}$ of T in the northern hemisphere stereographically on the tangent hyperplane H_1 at the north pole, consider the map of the frontier of the projected simplex in $\mathbf{S}_{n-1}$ given by the (stereographically projected) vector field in H_1, and take its degree $c_{1\alpha}$; do the same for the southern hemisphere, stereographically projected on the tangent hyperplane H_2 at the south pole, to get the degrees $c_{2\alpha}$. Brouwer showed that, owing to the symmetry of T with respect to the equator, the sum of the degrees $c_{1\alpha}$ and $c_{2\alpha}$ (for all n-simplices of T) reduces to the sum of the degrees of two maps of the equator $\mathbf{S}_{n-1}$ into itself. He then claimed that the computation of that sum could be reduced to the case of a *constant* vector field on $\mathbf{S}_{n-1}$, but his description of what he does to reach that result is so sketchy and intricate that it is hard to decide if his procedure really constitutes a proof.

In 1925 ([238], p. 2) Hopf announced that Brouwer's theorem for vector fields on $\mathbf{S}_n$ would generalize to arbitrary compact "manifolds" X: for a continuous vector field on X, with finitely many singular points, the sum of the indices of these points is equal to the *Euler–Poincaré characteristic*. Hopf indicated that this result could be derived from the theory of fixed points of continuous maps. Alexandroff and Hopf showed in their book ([30], p. 549) how this can be done very simply for a C^1 *manifold* X and a C^1 *vector field* Z on X by considering the *flow* $(x, t) \mapsto F_Z(x, t)$ of Z. Recall that this is defined for all $x \in X$ and all $t \in \mathbf{R}$; if $v(t) = F_Z(x, t)$, $t \mapsto v(t)$ is the integral curve of the field Z starting from $x = v(0)$, i.e., $v'(t) = Z(v(t))$. A compactness argument shows that there is an interval $|t| \leqslant \varepsilon$ such that the fixed points of the map $x \mapsto F_Z(x, t)$ are exactly the singular points of Z for any t in that interval, with the same indices. Since that map is also obviously homotopic to the identity, the result follows from formula (6). It can be generalized to a vector field Z on a C^1 manifold X that is merely supposed continuous, for such a field is homotopic to a C^1 vector field with the same singular points.

The notion of vector field on X is not clearly defined for a combinatorial manifold X, since there may be several distinct differential structures on X (or none at all) compatible with the topology. In 1928 [240] Hopf considered vectors attached to each point of X and satisfying conditions depending not only on the topology of X but on its triangulation, and he proved that they still satisfy the index formula.

CHAPTER IV

Local Homological Properties

§ 1. Local Invariants

Local properties of topological spaces were considered at the beginning of the twentieth century, chiefly by Schoenflies, who was a pioneer in that matter. They were mainly studied for subsets of $\mathbf{R}^2$, and without any intervention of homological notions. Examples of these properties are accessibility and "Unbewaltheit," which we saw developed by Brouwer using simplicial methods but still no homology (chap. II, § 3,C). After 1910 the concept of *local connectedness** was also the theme of many papers in "point-set" (or "analytic") topology (see [517] and [518], chap. I).

The fact that all contractible spaces have the same homology showed that homology is a very coarse notion to use for the description of properties of a space invariant under homeomorphism. At the end of the 1920s the idea emerged that, just as global connectedness of a space is a property that gives very little information, and "localizing" it gives much more, so one could perhaps "localize" *homology groups* of any dimension in order to make a deeper study of the topology of a space.

In this chapter, it shall always be understood that "homology group" means reduced homology group (Part 1, chap. IV, §6,E).

The first instance of such ideas probably occurs in print in a footnote of a 1928 paper by Alexandroff ([27], p. 181, note 63), in which he introduces the notion of "r-local connectedness" for any $r \geqslant 0$; we shall examine it in § 2; he mentions that Alexander had considered the same definition but did not publish it.

A. Local Homology Groups and Local Betti Numbers

It was only in 1934 that Alexandroff [28], Čech [122], and Seifert and Threlfall in their book ([421], chap. VIII) independently gave definitions of "local" homology groups or Betti numbers.

* For the many uncertainties and even priority claims to which the notion of connectedness gave rise in the early 1900s, see [89], p. 486.

J. Dieudonné, *A History of Algebraic and Differential Topology, 1900–1960*,
Modern Birkhäuser Classics, DOI 10.1007/978-0-8176-4907-4_10,

Alexandroff only considered compact subspaces of $\mathbf{R}^n$ and Vietoris homology (Part 1, chap. IV, §2) with rational coefficients; Seifert and Threlfall limited themselves to locally finite simplicial complexes and simplicial homology; Čech gave definitions for arbitrary topological spaces and used Čech homology based on finite open coverings (Part 1, chap. IV, §2) with coefficients in $\mathbf{Q}$ or in a finite field .

Both Alexandroff and Čech referred to Lefschetz's "relative homology" (Part 1, chap. II, §6), whereas Seifert and Threlfall gave direct definitions and only mentioned relative homology in a footnote. The natural procedure stemming from relative homology would be to take the relative homology groups $H_p(X, X - \{x\}; G)$ as local invariants at a point $x \in X$ for *some* homology theory (Part 1, chap. IV, §6,B), and if that theory satisfies the excision axiom (*loc.cit.*) these groups may be replaced by $H_p(V, V - \{x\}; G)$ where V is an arbitrary open neighborhood of x; however, this is not the way the authors mentioned above proceeded.

They attached to any point $x \in X$ an "r-dimensional Betti number $p_r(x)$ at x" for every $r \geqslant 0$, in the following way (reformulated for convenience in the present language, and for *any* homology theory with coefficients in a field). Consider two open neighborhoods $U \supset V$ of x, and the natural map

$$H_r(X, X - U) \to H_r(X, X - V);$$

write $p_{r,U,V}$ the rank of that homomorphism [dimension of the image of $H_r(X, X - U)$], which decreases when V decreases and hence has a limit $p_{r,U}$ (finite integer or $+\infty$) for the directed set $\mathfrak{U}(x)$ of open neighborhoods of x. Furthermore, when U decreases $p_{r,U}$ increases and hence has a limit $p_r(x)$ for the directed set $\mathfrak{U}(x)$. Observe that instead of the dimension $p_{r,U,V}$, the homology groups $H_r(X, X - U)$ themselves may be considered, and one can take direct limits over the directed set $\mathfrak{U}(x)$. The group obtained in that manner is not necessarily isomorphic to $H_r(X, X - \{x\})$.

Suppose x has a fundamental decreasing system of neighborhoods (U_m), such that $X - U_m$ is a strong deformation retract (Part 1, chap. IV, §6,B) of $X - \{x\}$ and of $X - U_n$ for $n > m$. It then follows from the exact sequence of relative homology [Part 1, chap. IV, §6,B, formula (94)] that the maps

$$H_r(X, X - U_m) \to H_r(X, X - U_n) \to H_r(X, X - \{x\})$$

are all *bijective*; hence the groups obtained by the preceding limit processes actually are the $H_r(X, X - \{x\})$, which then deserve to be called the *local homology groups* at x.

This is particularly the case when X is a C^0-manifold or a locally finite simplicial complex. In the first case, if the dimension of X at the point x is $n > 0$,

$$\begin{aligned} &H_j(X, X - \{x\}) = 0 && \text{for } j \neq n \\ &H_n(X, X - \{x\}; \Lambda) \simeq \Lambda && \text{for any ring } \Lambda. \end{aligned} \tag{1}$$

In the second case (the only one considered by Seifert and Threlfall), if $\mathrm{St}(x)$ is the star of x for a triangulation of X, $X - \mathrm{St}(x)$ is a strong deformation retract of $X - \{x\}$. As $H_p(\overline{St(x)}) = 0$ for all $p \geqslant 0$, since $\overline{St(x)}$ is contractible,

$$H_p(X, X - \{x\}) \simeq H_{p-1}(K_x) \tag{2}$$

where K_x is the subcomplex $\overline{St(x)} - St(x)$ of the triangulation of X. This is actually the *definition* given by Seifert and Threlfall for the local homology groups, and of course they had to prove it independent of the triangulation of X ([421], pp. 120–125). They used these groups to show that some properties, defined *a priori* with respect to some triangulation, are in fact independent of the choice of that triangulation; for instance, this is the case for the union of the j-simplices that are not on the frontier of a $(j+1)$-simplex $[0 \leqslant j \leqslant \dim(X)]$.

B. Application to the Local Degree

Let u be a C^∞ map of $\mathbf{R}^n$ into $\mathbf{R}^n$ such that $u(0) = 0$, $u(\mathbf{R}^n - \{0\}) \subset \mathbf{R}^n - \{0\}$, so that u defines an endomorphism u^* of $H^{n-1}(\mathbf{R}^n - \{0\}; \mathbf{Z})$, which is isomorphic to $\mathbf{Z}$. Then $u^*(\zeta) = c\zeta$ for any cohomology class ζ, and $c \in \mathbf{Z}$; the integer c is called the *local degree* of u at 0, and written $\deg_0(u)$. If the jacobian J of u at 0 is $\neq 0$, then $\deg_0(u) = 1$ if $J > 0$ and $\deg_0(u) = -1$ if $J < 0$.

Now consider two smooth manifolds X, Y, both oriented and having the same dimension $n \geqslant 2$, and let $f: X \to Y$ be a C^∞ map. A point $a \in X$ is *isolated* for f if there is an open neighborhood U of a such that $f(x) \neq f(a)$ for $x \in \bar{U} - \{a\}$. One may assume that U is the domain of a chart $\varphi: U \to \mathbf{R}^n$ of X such that $\varphi(U) = \mathbf{R}^n$ and $\varphi(a) = 0$ and there is a chart $\psi: V \to \mathbf{R}^n$ of Y such that $f(U) \subset V$, $\psi(V) = \mathbf{R}^n$ and $\psi(f(a)) = 0$. Then define the *local degree* $\deg_a f$ at the point a as $\deg_a f = \deg_0(\psi \circ f \circ \varphi^{-1})$; it does not depend on the choices of U, V, φ, and ψ. If the tangent mapping $T_a(f)$ is a bijection of $T_a(X)$ onto $T_{f(a)}(Y)$, then $\deg_a f = 1$ if $T_a(f)$ preserves orientations, $\deg_a f = -1$ if not.

Let Z be another smooth oriented manifold of dimension n, and $g: Y \to Z$ be a C^∞ map such that $f(a)$ is isolated for g; then a is isolated for $g \circ f$, and

$$\deg_a(g \circ f) = \deg_{f(a)} g \cdot \deg_a f.$$

Finally, suppose X and Y are compact and connected and that there is a point $y_0 \in Y$ such that $f^{-1}(y_0) = \{x_1, x_2, \ldots, x_r\}$, a finite set. The x_j are isolated for f, and

$$\deg f = \sum_{j=1}^{r} \deg_{x_j} f.$$

C. Later Developments

The papers of Alexandroff and Čech defined Betti numbers $p_r(x)$ but not groups attached to a point x. Alexandroff proposed definitions of other groups at a point x, the dimension of which may be different from $p_r(x)$. One of his definitions is similar to one that is better understood within the context of Borel–Moore homology: the definitions and notations of Part 1, chap. IV,

§ 7,G give the *homology graded sheaf* $\mathscr{H}_\bullet(\mathscr{C}_{H\bullet}(X; L))$ for the generalized chain complex of sheaves $\mathscr{C}_{H\bullet}(X; L)$, written $\mathscr{H}_\bullet(X; L)$. The *stalk* $(\mathscr{H}_j(X; L))_x$ at a point x can be called the j-th *local homology group at* x; the exact sequence of relative homology shows that it is isomorphic to the Borel–Moore relative homology group $H_j(X, X - \{x\}; L)$.

The work of Alexandroff and Čech was considerably enlarged and diversified by Wilder between 1935 and 1955. He conclusively showed how all the results (mainly relative to plane sets) obtained by the "point-set topologists" of the Polish and American schools who shunned simplicial methods could be enormously generalized and put in their proper perspective by the use of homological notions [518]. He not only used Čech homology, but also Čech cohomology with compact supports and coefficients in a field (which did not yet exist when Alexandroff and Čech wrote their papers): for two open neighborhoods $U \supset V$ of x in a locally compact space X, there is a natural homomorphism $H_c^r(V) \to H_c^r(U)$ (Part 1, chap. IV, § 7,G). If $p_{U,V}^r$ is the dimension of the image of that homomorphism, the numbers $p_{U,V}^r$ behave exactly as the numbers $p_{r,U,V}$ of Alexandroff; hence, by the same limit processes a number $p^r(x)$ can be attached to each $x \in X$, called the *local co-Betti number* at x, which is an integer or $+\infty$; Wilder showed that in fact $p_r(x) = p^r(x)$ for all $x \in X$ ([518], p. 191).

Wilder's book contains a large number of local properties linked to homology and cohomology. Since it was written when modern algebraic techniques (Part 1, chap. IV, § 5) had not yet been introduced into algebraic topology, it would be worthwhile rewriting it with the help of these techniques, which very likely would make it shorter and more perspicuous.

In the remainder of this chapter, we shall restrict our description to the notions and results of [518] that have proved most striking and useful in other directions in algebraic topology (see [385]).

D. Phragmén–Brouwer Theorems and Unicoherence

As an illustration of Wilder's ideas, I think it worthwhile to insert as a small digression an example of topological properties that are put into a better light when they are connected with notions of algebraic topology.

In 1885 Phragmén published a short note on topology of plane sets [361] in which he proved the following property: if A is a compact connected subset of $\mathbf{R}^2$, and U is the unbounded connected component of the open set $\mathbf{R}^2 - A$, then the frontier of U is connected. His method consisted in decomposing $\mathbf{R}^2$ into squares with sides of length 2^{-m}, considering the union of those squares that met the frontier of U, and letting m tend to infinity.

In one of his first topological papers, in which he gave a new proof of the Jordan theorem for plane curves, Brouwer extended Phragmén's result by showing that the frontier of *any* connected component of $\mathbf{R}^2 - A$ is connected ([89], p. 378). Later it was discovered that this property is linked to several others, and "point-set topologists" were able to extend them when $\mathbf{R}^2$ is replaced by much more general spaces X. But apparently it was only in the

Alexandroff–Hopf ([30], p. 292) book that these properties were shown to depend on the fact that $H_1(X; \mathbf{Z}) = 0$. The key property is:

If X is a Hausdorff arcwise connected space, *such that* $H_1(X; \mathbf{Z}) = 0$, and if A, B are two nonempty *disjoint* closed sets such that $X - A$ and $X - B$ are arcwise connected (neither A nor B "cuts" the space), then $X - (A \cup B)$ also is arcwise connected ($A \cup B$ does not "cut" the space). This is an immediate consequence of the Mayer—Vietoris homology exact sequence.

Elementary arguments of "point-set topology" easily produce from that property the following so-called "Phragmén–Brouwer theorems," under the additional assumption that X is *locally arcwise connected.*

(i) If A, B are two closed nonempty sets in X such that $A \cap B = \varnothing$, and if x, y belong both to the same connected component of $X - A$ and to the same connected component of $X - B$, then they also belong to the same connected component of $X - (A \cup B)$.
(ii) If A is a closed, connected, nonempty subset of X, each connected component of $X - A$ has a connected frontier.
(iii) If A, B are two closed connected subsets of X such that $X = A \cup B$, then $A \cap B$ is connected (a property that was much studied under the name of "unicoherence").
(iv) If A is a closed subset of X, and C_1, C_2 two nonempty connected components of $X - A$ having the same frontier B, then B is connected.

§2. Homological and Cohomological Local Connectedness

In a *locally connected* space X each $x \in X$ has a fundamental system of open neighborhoods that are connected. It follows from the definitions (Part 1, chap. IV, §3) that for Alexander–Spanier cohomology, 0-cocycles are just locally constant functions; hence for a connected space X the reduced cohomology $\tilde{H}^0(X) = 0$. Conversely, if a compact space K is the union of two nonempty open and closed sets U_1, U_2, then a function constant in U_1 and constant in U_2 but with different values is locally constant; hence $\tilde{H}^0(K) \neq 0$. From this it follows at once that for locally compact spaces X, saying that X is locally connected is equivalent to saying that $p^0(x) = 0$ for all $x \in X$.

This leads to the generalization of local connectedness formulated by Alexandroff in 1929 and mentioned in §1. He said that X is *homologically locally connected in dimension* $q \geqslant 0$ (later abbreviated into $q - \text{lc}$) at a point x, if for every open neighborhood U of x there is an open neighborhood $V \subset U$ of x such that every q-cycle in V bounds in U. There is, however, no direct relation between that property and the fact that $p_q(x) = 0$, as Alexandroff himself showed by examples ([28], p. 9). What $p_q(x) = 0$ [or equivalently $p^q(x) = 0$] means is the corresponding notion for Čech–Alexander cohomology with coefficients in a field: X is *cohomologically locally connected in dimension* q (abbreviated to q − clc) at the point x if for any open neighborhood U of x there is an open neighborhood $V \subset U$ of x such that the image of the homomorphism $H^q(U) \to H^q(V)$ is 0.

Examples show that at a point x of a locally compact space X, X may be q − lc for all integers q in an arbitrary finite set, but *not* q − lc for the other values of q ([304], p. 92). In 1935 Lefschetz [307] and Wilder defined the property of being lc^n at a point as meaning that the space is q − lc at that point for *all* values $q \leqslant n$. They needed this for their definition of generalized manifolds (see § 3); the notion was studied in detail by Begle for compact spaces [46]. There is a corresponding notion (clc^n) for cohomology.

Results concerning these notions are now best expressed in the context of Borel–Moore homology. In their notation (L being a Dedekind ring) the locally compact space X is homologically (resp. cohomologically) locally connected in dimension q [abbreviated to $q - hlc_L$ (resp. $q - clc_L$)] at the point x if, for any neighborhood U of x, there is a neighborhood $V \subset U$ of x such that the image of the homomorphism

$$H_q^c(V; L) \to H_q^c(U; L) \quad [\text{resp. } H^q(U; L) \to H^q(V; L)] \tag{3}$$

is 0. The space is $q - hlc_L$ (resp. $q - clc_L$) if it has that property at every point, and hlc_L^r (resp. clc_L^r) if it is $q - hlc_L$ (resp. $q - clc_L$) for all integers $q \leqslant r$. Finally, X is hlc_L (resp. clc_L) if for any neighborhood U of any point x it is possible to choose the neighborhood $V \subset U$ *independently of* q such that the image of the map (3) is 0 for every q.

For a hlc_L^r space X and an L-module B, there is for every $q \leqslant r$ a split exact sequence

$$0 \to \operatorname{Ext}(H_{q-1}^c(X; L), B) \to H^q(X; B) \to \operatorname{Hom}(H_q^c(X; L), B) \to 0 \tag{4}$$

corresponding to the exact sequence for $H_q(X; B)$ applicable to all locally compact spaces [Part 1, chap. IV, § 7,G), formula (184)].

Property hlc_L^r implies clc_L^r, but clc_L^r only implies hlc_L^{r-1}. When L is a field, however, hlc_L^r and clc_L^r are equivalent, and hlc_L^r and clc_L^r are always equivalent.

If X is *compact* and hlc_L^r, then the L-modules $H_q(X; L)$ and $H^q(X; L)$ are finitely generated for $q \leqslant r$; $\operatorname{Ext}(H^{q+1}(X; L), L)$ is then the torsion submodule of $H_q(X; L)$ and $\operatorname{Ext}(H_{q-1}(X; L), L)$ the torsion submodule of $H^q(X; L)$.

We conclude this section with the remark that *singular homology* can be used for the definition of local properties instead of Čech homology or Borel–Moore homology. This was done in 1935 by Lefschetz,* who defined properties q − HLC, HLC^r, and HLC by replacing Čech homology by singular homology in the definitions of q − lc, lc^r, and hlc. The important property

* Do not confuse these notions with other concepts of "local connectedness" based on homotopy rather than on homology, which we shall consider in Part 3, chap. II, § 2,B. They were also introduced by Lefschetz, who used the symbol LC (with indices or exponents) to designate them (the H in HLC stands for "homology").

of HLC spaces is that for them Alexander–Spanier cohomology is naturally *isomorphic* to singular cohomology.

§ 3. Duality in Manifolds and Generalized Manifolds

A. Fundamental Classes and Duality

Local properties of a C^∞ manifold M are used to extend the concept of "fundamental class" in the homology of a compact manifold (chap. I, § 3,A) to "relative fundamental classes" for a noncompact one.

Suppose M is an *oriented* smooth n-dimensional manifold (connected or not). Choose an orientation on $\mathbf{R}^n$ and on $\mathbf{S}_{n-1}$ and let γ_n be the generator of the group $H_n(\mathbf{R}^n, \mathbf{R}^n - \{0\}; \mathbf{Z}) \simeq \mathbf{Z}$ that is mapped on $[\mathbf{S}_{n-1}]$ by the isomorphism $H_n(\mathbf{R}^n, \mathbf{R}^n - \{0\}; \mathbf{Z}) \xrightarrow{\partial} H_{n-1}(\mathbf{S}_{n-1}; \mathbf{Z})$. For any chart φ: $V \to \mathbf{R}^n$ preserving orientation, and $x \in V$ such that $\varphi(x) = 0$, there is an isomorphism φ_*: $H_.(V, V - \{x\}; \mathbf{Z}) \xrightarrow{\sim} H_.(\mathbf{R}^n, \mathbf{R}^n - \{0\}; \mathbf{Z})$. Thus $H_p(V, V - \{x\}; \mathbf{Z}) = 0$ for $p \neq n$ and $H_n(V, V - \{x\}; \mathbf{Z})$ is isomorphic to $H_n(\mathbf{R}^n, \mathbf{R}^n - \{0\}; \mathbf{Z})$. By excision, this gives a composite isomorphism

$$H_n(M, M - \{x\}; \mathbf{Z}) \xrightarrow{\sim} H_n(\mathbf{R}^n, \mathbf{R}^n - \{0\}; \mathbf{Z})$$

which is independent of the chart φ; let μ_x be the element of $H_n(M, M - \{x\}; \mathbf{Z})$ mapped onto γ_n by that isomorphism. Now let $K \subset M$ be any compact subset. Then there exists a unique class $\mu_{M,K} \in H_n(M, M - K; \mathbf{Z})$, called the *fundamental class* relative to K, such that for any $x \in K$ the image of $\mu_{M,K}$ by the homomorphism

$$j_*: H_n(M, M - K; \mathbf{Z}) \to H_n(M, M - \{x\}; \mathbf{Z})$$

deduced from the natural injection is the class μ_x. The proof uses a technique similar to the one in H. Cartan's paper of 1945 [106]. Consider first the case $M = \mathbf{R}^n$ and then the case in which K is small enough, then apply the Mayer–Vietoris exact sequence to treat the union of finitely many such compact sets by induction on their number. Poincaré duality for homology and cohomology of M with *integer* coefficients can then be obtained by considering M as union of an increasing sequence (K_m) such that each K_m is a compact neighborhood of K_{m-1}. Let z_m be a relative n-cycle whose homology class is $\mu_{M,K} \in H_n(M, M - K_m; \mathbf{Z})$. Then, for each p-cocycle f on M with compact support, the class of the cap product $z_m \frown f$ is the same for all sufficiently large m and only depends on the class c of f in $H_c^p(M; \mathbf{Z})$. Call $D_M c$ that class in $H_{n-p}(M; \mathbf{Z})$; then the homomorphism

$$D_M: H_c^p(M; \mathbf{Z}) \to H_{n-p}(M; \mathbf{Z})$$

is bijective (*Poincaré duality*).

In a similar way for a closed subset A of M, there is an isomorphism

$$D_{M,A}: \bar{H}_c^p(A; \mathbf{Z}) \to H_{n-p}(M, M - A; \mathbf{Z})$$

for Alexander–Spanier cohomology with compact supports and singular homology (*Alexander duality*).

There are analogous results for nonorientable manifolds and coefficients in $\mathbf{F}_2$.

B. Duality in Generalized Manifolds

Until 1930 Poincaré and Alexander duality theorems for integer coefficients had only been proved for orientable compact *triangulable* C^0-manifolds. This was soon felt to be an unsatisfactory situation, since the notion of triangulation depends on auxiliary subspaces $\mathbf{R}^n$, whereas the duality theorems only deal with homology and cohomology; even an extension to all C^0-manifolds (for which triangulability was unknown) would have suffered from the same defect. Starting with Čech [121] and Lefschetz [306] in 1933 topologists endeavored to define classes of spaces by *purely homological conditions* which would include both combinatorial manifolds and C^0-manifolds, and for which the duality theorems would hold.

The general idea was to impose homological properties known to hold for C^0-manifolds on these spaces, particularly *local* homological conditions (§ 2). Several definitions were proposed in succession by Wilder, Alexandroff and Pontrjagin [31], P. Smith [437] and Begle [46]. Here again the introduction of Borel–Moore homology, with substantial improvements by Bredon [87], brought a more satisfactory state of the theory.

If L is a Dedekind ring, a locally compact space X is a *homology n-manifold* over L (abbreviated $n - \mathrm{hm}_L$) if:

1. The cohomological dimension $\dim_L X$ of X over L (chap. II, § 6) is finite.
2. The relative Borel–Moore homology

$$H_q(X, X - \{x\}; L) = \begin{cases} L & \text{for } q = n \\ 0 & \text{for } q \neq n \end{cases} \tag{5}$$

for any $x \in X$.

These conditions imply that the cohomological dimension $\dim_L X \leqslant n + 1$ and that the sheaves $\mathscr{H}_q(X; L)$ are 0 for $q \neq n$. Bredon has also proved that $\mathcal{O} = \mathscr{H}_n(X; L)$ is *locally isomorphic to the constant sheaf* L. One says $\mathcal{O}$ is the *orientation sheaf* and X is *orientable over* L if $\mathcal{O}$ is isomorphic to L; an isomorphism of $\mathcal{O}$ onto L is called an *orientation* of X over L.

We have seen that in 1945 H. Cartan had already started to drop assumptions of differentiability or triangulability in the theory of "manifolds" (Part 1, chap. IV, § 5,A). In 1947 he realized that sheaf theory (which he still used at that time in Leray's formulation) provided a way to "localize" the concept of orientation. In his 1950–1951 Seminar he defined a generalized cochain complex (with indices $\leqslant 0$) of sheaves of singular chains and introduced an orientation sheaf in that context, with the help of which he could prove Poincaré and Alexander duality theorems for C^0-manifolds.

In the context of Borel–Moore homology the duality theorems are derived from a spectral sequence applicable to all locally compact spaces X with *finite* cohomological dimension. Suppose $\dim_L X \leqslant n$, and let $\mathscr{B}^{\bullet}$ be the generalized cochain complex of sheaves defined by

$$\mathscr{B}^q = \mathscr{C}_{H,n-q}(X; L) \tag{6}$$

so that $\mathscr{H}^q(\mathscr{B}^{\bullet}) = 0$ for $q < 0$, and $\mathscr{H}^0(\mathscr{B}^{\bullet}) = \mathscr{H}_n(X; L)$. Then ([66], p. 152) for any *paracompactifying* family of supports Φ there is a spectral sequence having as E_2 terms

$$E_2^{pq} = H_\Phi^p(X; \mathscr{H}^q(\mathscr{B}^{\bullet})) \tag{7}$$

and $H^0(\Gamma_\Phi(\mathscr{B}^{\bullet}))$ for abutment with a suitable filtration.

If X is now a *homology n-manifold* over L and $\dim_\Phi X < +\infty$, there is a natural isomorphism

$$H_\Phi^p(X; \mathscr{O} \otimes L) \simeq H_{n-p}^\Phi(X; L) \tag{8}$$

("Poincaré duality"). In addition, if A is a closed set in X, and $\dim_{\Phi|A} X < +\infty$, there are natural isomorphisms

$$H_\Phi^p(X, X - A; \mathscr{O} \otimes L) \simeq H_{n-p}^{\Phi|A}(A; L) \tag{9}$$

$$H_{\Phi\cap(X-A)}^p(X - A; \mathscr{O} \otimes L) \simeq H_{n-p}^\Phi(X, A; L) \tag{10}$$

("Alexander duality").

In the Borel–Moore theory a *generalized n-manifold* X over L (abbreviated $n - \mathrm{gm}_L$), also called *cohomology n-manifold* ($n - \mathrm{cm}_L$), is an $n - \mathrm{hm}_L$ which is also clc_L (§ 2), and $\dim_L X \leqslant n$. If L is a field, a metric $n - \mathrm{hm}_L$ space is also a $n - \mathrm{cm}_L$.

Using excision and the Künneth theorem, it is easy to see that combinatorial manifolds of dimension n in the sense of Alexander (Part 1, chap. II, § 4) are generalized n-manifolds over **Z**.

C^0-manifolds are trivially generalized manifolds, but generalized manifolds are genuine generalizations of C^0-manifolds. There are generalized manifolds of dimension 4 in which for some points x there is an open neighborhood U of x such that for *no* open neighborhood $V \subset U$ of x is $V - \{x\}$ simply connected ([421], p. 241).

The main interest of generalized manifolds is that they are much easier to work with than C^0-manifolds. For instance, if a product $A \times B$ of locally compact spaces is a generalized manifold, both A and B are generalized manifolds. In the theory of transformation groups, fixed point sets and "slices" in a generalized manifold are generalized manifolds.

Wilder's general program was to find conditions under which the Schoenflies results for $\mathbf{R}^2$ could be extended to generalized manifolds. A whole chapter of his book ([518], chap. 12) is devoted to the notion of *accessibility*. He generalized Schoenflies' "Unbewaltheit" (chap. II, §) to the notion of *uniform local q-connectedness*: in a compact space X, an open subset D is *uniformly locally connected in dimension* q (abbreviated to q – ulc) if for every finite open covering $\mathfrak{U} = (U_\alpha)$ of X there exists a finite open covering $\mathfrak{V} = (V_\beta)$ of X finer than U and such that, for any V_β, there exists a $U_\alpha \supset V_\beta$ for which the image of the map $H_q(V_\beta \cap D) \to H_q(U_\alpha \cap D)$ is 0; D is ulc^r if it is q – ulc for $0 \leqslant q \leqslant r$.

We only mention here a few of the numerous properties proved by Wilder.

1. If X is an orientable n – gm which is a homology sphere [$H_q(X) = 0$ for $q \neq n$] and M is a compact $(n - 1)$ – gm contained in X, then the components of X – M are ulc^{n-1}.
2. If X is as in 1 and $M \subset X$ is the common frontier in X of at least two connected open sets, one of which is ulc^{n-2}, then M is an orientable $(n - 1)$ – gm.
3. If X is an orientable n – gm such that $H_1(X) = 0$ and $U \subset X$ is an open connected set which is ulc^{n-2} and has a connected frontier B in X, then B is an orientable $(n - 1)$ – gm.
4. Finally, if X is an orientable generalized manifold and $f: X \to Y$ a surjective continuous map of X onto a Hausdorff space Y, such that the reduced homology of each fiber $f^{-1}(y)$ is 0, then Y is an *orientable generalized manifold* and $f_*: H_\bullet(X) \to H_\bullet(Y)$ is an isomorphism [a remarkable refinement of the Vietoris-Begle theorem (Part 1, chap. IV, §§ 7,B and 7,E)].

CHAPTER V

Quotient Spaces and Their Homology

In this chapter I gather miscellaneous results and techniques related to homology, in all of which the concept of *quotient space* plays a more or less important part. Many of the constructions described were invented for their use in homotopy theory, leading to major results in that theory (see Part 3).

§ 1. The Notion of Quotient Space

We saw in Part 1, chap. I that since the 1870s mathematicians had been freely using "identifications" of points (or "gluing" of subspaces) on an "intuitive" basis without any attempt at precise definitions, which in fact could not possibly be given before the fundamental notions of topology such as limit and neighborhood had been completely clarified.

The general problem, as we now state it in our language, is to assign a suitable topology to the set X/R of equivalence classes for an equivalence relation R on a topological space X. If p: X → X/R is the natural map, what is now known as the *quotient topology* of the topology of X by R is defined on X/R by taking as open sets the sets U such that $p^{-1}(U)$ is open is X. It is the finest topology on X/R for which the map p is continuous, and X/R, with that topology, is called the *quotient space* of X by R.

It was only in the 1930s that this topology was considered for the most general spaces, at the end of somewhat tortuous attempts. Since equivalence classes for R are subsets of X, it is not surprising that the problem of defining a topology on X/R did not emerge until some work had been done on topologies on the *set* 𝔓(X) *of all subsets* of X, or more precisely on concepts of *limits of sequences of subsets* (since for most mathematicians in the early years of the twentieth century, the idea of *limit of a sequence* was the fundamental topological notion). Such notions of "limit", which apparently were first considered in a Note of Painlevé in 1909 [359], appear in the books on topology by Hausdorff [220] and Kuratowski [291], but without any mention of the special case in which only the parts of X belonging to a given *partition* of the space are considered.

J. Dieudonné, *A History of Algebraic and Differential Topology, 1900–1960*,
Modern Birkhäuser Classics, DOI 10.1007/978-0-8176-4907-4_11,

The first mention of a topology on X/R seems to be in the work of R.L. Moore on special partitions of a metric space X, which he calls "upper semi-continuous collections" [351]. They are characterized by the fact that when A_0 is a fixed set and A is a variable set of the partition, if the distance $d(A_0, A)$ tends to 0, then the Hausdorff "Abweichung"

$$\sup_{x \in A_0, y \in A} (d(x, A), d(y, A_0))$$

also tends to 0. When spaces other than metric spaces were considered, this condition meant that for any neighborhood U of A_0, there is a neighborhood $V \subset U$ such that if A meets V it is contained in U. Moore said that a point $z_0 \in X/R$ is in the closure of a set $H \subset X/R$ if for each neighborhood W of $p^{-1}(z_0)$ there is a point $z \in H$ such that $p^{-1}(z) \subset W$ (but he does not use the map p to express that property). If for each $z \in X/R$ the fiber $p^{-1}(z)$ is compact, it is easy, using the "upper semicontinuity," to see that, for the topology thus defined on X/R, p is continuous. If the relation R is also *closed*,* that topology is indeed the quotient topology.

In 1934 Seifert and Threlfall devoted a whole section of their book [421] (pp. 31–35) to what they call "Identifizieren," defining a topology on the set X/R; the relation R is arbitrary and X is any topological space (in fact the spaces they consider are more general than what we now call topological spaces, since the only axiom they impose on open sets is that any union of open sets is open). They take as neighborhoods of a point $z \in X/R$ the images $p(U)$ of *all* neighborhoods U of $p^{-1}(z)$ in X; the map p is then continuous for that topology. However, an open neighborhood of $p^{-1}(z)$ in X does not always contain an open neighborhood of the form $p^{-1}(V)$ for an open neighborhood V of z, so that their topology may be strictly finer than the quotient topology. It is again identical to the quotient topology when the relation R is *closed.* This is proved by Seifert and Threlfall when X is a compact subset of some $\mathbf{R}^n$; they also showed that in that case the topology of X/R is characterized by the fact that if a map $g: X/R \to Y$ into a topological space Y is such that $g \circ p: X \to Y$ is continuous, then g itself is continuous for the quotient topology. When categorical notions were later introduced (Part 1, chap. IV, § 8), this could be expressed in more pedantic terms by saying that if $\mathrm{Cont}_{\mathbf{R}}(X, Y)$ is the set of continuous maps of X into Y that are *constant* on each equivalence class of R, the functor $Y \mapsto \mathrm{Cont}_{\mathbf{R}}(X, Y)$ is *represented* by the pair $(X/R, p)$ with the quotient topology on X/R (in the category $\boldsymbol{T}$ of all topological spaces). Alexandroff and Hopf in their 1935 book independently gave the definition of the quotient topology for a T_1-space X, assuming that the equivalence classes are closed in X ([30], p. 61). They also mention its "categorical" characterization.

The first completely general definition of the quotient topology is ap-

---

* In a topological space X, an equivalence relation R is *closed* if for any closed subset F of X, the union of the equivalence classes of the points of F is also closed.

parently given in Bourbaki's treatise [85] published in 1940. This book also contains results concerning subspaces and products of quotient spaces. In particular, it is not always true that the product $(X/R) \times (Y/S)$ is naturally homeomorphic to $(X \times Y)/(R \times S)$ * but it is so when Y and Y/S are both compact. Quotient spaces intervene naturally in the study of continuous maps $f: X \to Y$; such a map is always naturally factorized into three continuous maps

$$X \xrightarrow{p} X/R \xrightarrow{h} f(X) \xrightarrow{i} Y$$

where p is the natural map, h is bijective, and i is injective. When f is open (resp. closed) [i.e. maps open (resp. closed) subsets of X onto open (resp. closed) subsets of Y], h is a homeomorphism and i is open (resp. closed); an important case is when X is *compact* and Y is *Hausdorff*, for then f is closed.

§ 2. Collapsing and Identifications

A. Collapsing

If X is a topological space and A is a subset of X, consider the partition of X consisting of A and of the singletons $\{x\}$ for $x \in X - A$. The corresponding quotient space is written X/A and it is obtained by *collapsing* (or *shrinking*) A to a point; it was used by J.H.C. Whitehead as early as 1938 ([492], p. 107). If A is closed in X, the restriction of the *collapsing map* $p: X \to X/A$ to $X - A$ is a homeomorphism of $X - A$ onto $X/A - \{p(A)\}$. For any cohomology theory such that $H^{\bullet}(X, A) \cong H^{\bullet}_c(X - A)$ and $H^{\bullet}(X/A, \{p(A)\}) \simeq H^{\bullet}_c(X/A - \{p(A)\})$ it also follows from the exact cohomology sequence that $H^p(X/A) \simeq H^p(X, A)$ for $p \geqslant 1$ and $\tilde{H}^0(X/A) \simeq \tilde{H}^0(X, A)$. This is applicable when the closed set A is a strong deformation retract (Part 1, chap. IV, § 6,B) of a fundamental sequence of open neighborhoods of A. If in addition A is contractible, $H^p(X/A) \simeq H^p(X)$ for $p \geqslant 1$ and $\tilde{H}^0(X/A) \simeq \tilde{H}^0(X)$. If $A \subset B$ are two closed subsets of X, the image $p(B)$ into X/A is homeomorphic to B/A, and there is a natural homeomorphism $(X/A)/(B/A) \xrightarrow{\sim} X/B$.

B. Cones

Let X be any topological space, and consider the space $Y = X \times [0, 1]$ and its closed subspace $A_0 = X \times \{0\}$. The space Y/A_0 is called the *unreduced cone over* X and written $\tilde{C}X$; it is clearly contractible, and X is identified to the closed subspace image of $X \times \{1\}$ in $\tilde{C}X$. From the exact sequence of relative homology,

$$H_q(\tilde{C}X, \tilde{C}X - X) \simeq H_{q-1}(X) \quad \text{for } q > 1,$$
$$H_1(\tilde{C}X, \tilde{C}X - X) \simeq \tilde{H}_0(X). \tag{1}$$

* This was wrongly asserted in the first edition of that book.

If $x_0 \in X$, the *reduced cone* (or simply *cone*) *over* (X, x_0) is the quotient space

$$CX = Y/((X \times \{0\}) \cup (\{x_0\} \times [0, 1])). \tag{2}$$

It is also contractible to the point x_0.

C. Suspension

Keeping the same notations as in B, let $A_1 = X \times \{1\}$, and consider the space $Y/(A_0 \cup A_1) \simeq \tilde{C}X/X$, written $\tilde{S}X$ and called the *unreduced suspension* of X: all points of $X \times \{0\}$ are identified to a single point a_0 and all points of $X \times \{1\}$ to a single point a_1. That construction was introduced by Freudenthal in 1937 [201], and its properties were a landmark in homotopy theory (Part 3, chap. II, § 6,E). The space $\tilde{S}X$ is covered by two open subsets $U_0 = \tilde{S}X - \{a_0\}$, $U_1 = \tilde{S}X - \{a_1\}$ which are contractible, and X is a strong deformation retract of $U_0 \cap U_1$. This implies, by the Mayer–Vietoris exact sequence, the existence of a natural isomorphism

$$\partial: H^p(X) \xrightarrow{\sim} H^{p+1}(\tilde{S}X) \quad \text{for } p \geqslant 1; \tag{3}$$

for $p = 0$, $H^0(X)$ has to be replaced in (3) by the reduced cohomology $\tilde{H}^0(X)$, and $\tilde{H}^0(\tilde{S}X) = 0$.

It can immediately be verified that $\tilde{S}(\mathbf{S}_n)$ is homeomorphic to $\mathbf{S}_{n+1}$; the isomorphism (3) thus gives another way of computing $H^\bullet(\mathbf{S}_n)$ by induction on n.

Write $[x, t]$ the image of the point $(x, t) \in X \times [0, 1]$ in $\tilde{S}X$, so that $[x, 0] = a_0$, $[x, 1] = a_1$. If x_0 is a point of X, the *reduced suspension* (or simply *suspension*) SX of X is obtained by collapsing the set of all $[x_0, t]$ to a single point; it can also be written CX/X. For a CW-complex (§ 3) the natural map $\tilde{S}X \to SX$ gives isomorphisms in homology and cohomology.

D. Wedge and Smash Product

Let X, Y be any two spaces, $x_0 \in X$, $y_0 \in Y$. The *wedge* $X \vee Y$ of the "pointed" spaces (X, x_0) and (Y, y_0) is defined by considering the disjoint sum $X \coprod Y$ and identifying x_0 and y_0 in that space. It is homeomorphic to the subspace

$$(\{x_0\} \times Y) \cup (X \times \{y_0\}) \tag{4}$$

of the product $X \times Y$. If $\{x_0\}$ (resp. $\{y_0\}$) is a strong deformation retract of one of its neighborhoods, then

$$\tilde{H}^0(X \vee Y) \simeq \tilde{H}^0(X) \oplus \tilde{H}^0(Y), \qquad H^p(X \vee Y) \simeq H^p(X) \oplus H^p(Y) \text{ for } p \geqslant 1. \tag{5}$$

The space obtained by shrinking to a point the subspace (4) of $X \times Y$ is called the *smash product* of (X, x_0) and (Y, y_0) and is usually written $X \wedge Y$; moreover,

$$[0, 1] \wedge X \simeq CX, \qquad \mathbf{S}_1 \wedge X \simeq SX. \tag{6}$$

For an arbitrary family (X_α, x_α) of pointed spaces, the *wedge* $\bigvee_\alpha X_\alpha$ is similarly defined by taking the topological sum $\coprod_\alpha X_\alpha$, and collapsing the set of all x_α to a single point.

E. Join of Two Spaces

We have seen that Poincaré had already considered the "join" of two polyhedra (Part 1, chap. I, § 3). This notion can be defined for two arbitrary spaces X, Y. Consider the product $X \times Y \times [0, 1]$ and in that space the partition formed of:

$$\begin{array}{ll} \text{the singletons } \{(x, y, t)\} & \text{for } 0 < t < 1, x \in X, y \in Y; \\ \text{the sets } \{x\} \times Y \times \{0\} & \text{for all } x \in X; \\ \text{the sets } X \times \{y\} \times \{1\} & \text{for all } y \in Y. \end{array}$$

The *join* $X * Y$ is the quotient space of $X \times Y \times [0, 1]$ by the corresponding equivalence relation. The natural map $p\colon X \times Y \times [0, 1] \to X * Y$ sends $X \times Y \times \{0\}$ onto a closed subspace homeomorphic to X, $X \times Y \times \{1\}$ onto a closed subspace homeomorphic to Y.

Write $[x, y, t] = p(x, y, t)$; the complement U_0 (resp. U_1) of $p(X \times Y \times \{0\})$ (resp. $p(X \times Y \times \{1\})$) is retracted on X (resp. Y) by the homotopy

$$([x, y, t], u) \mapsto [x, y, tu] \quad (\text{resp. } ([x, y, t], u) \mapsto [x, y, (1 - u)t + u]).$$

On the other hand, $X \times Y$ may be identified with a deformation retract of $U_0 \cap U_1$ by the homotopy

$$([x, y, t], u) \mapsto [x, y, \tfrac{1}{2}u + t(1 - u)].$$

The Mayer–Vietoris sequence then shows that

$$H_n(X * Y) \simeq H_n(X \times Y)/H_n(X \vee Y) \quad \text{for } n \geqslant 1. \tag{7}$$

The unreduced cone $\tilde{C}X$ is the join of X and a single point, the unreduced suspension $\tilde{S}X$ the join of X and a pair of points.

When X and Y are CW-complexes (§ 3), $X * Y$ has the same homology as $X \wedge Y$.

F. Doubling

Let M be a pseudomanifold-with-boundary of dimension n, and let S be its boundary (chap. I, § 3,A). Consider a homeomorphism $h\colon M \to M'$ of M on another pseudomanifold-with-boundary M′ with boundary S′, so that $h(S) = S'$ ([421], p. 129). In the disjoint topological sum $Y = M \coprod M'$ let R be the equivalence relation whose classes are the pairs $\{x, h(x)\}$ for $x \in S$ and the singletons $\{y\}$ for the other points. In $X = Y/R$ the points of S and $h(S)$ are identified by h. X is called the *double* of M; it is a pseudomanifold of dimension n without boundary, and it was studied by Lefschetz ([301], p. 258–270) and van Kampen. The map $f\colon M \coprod M' \to M$, such that $f(x) = x$ for $x \in M$, $f(x') = h^{-1}(x')$ for $x' \in M'$ is factorized into

$$M \coprod M' \xrightarrow{p} Y \xrightarrow{q} M$$

where M is made into a quotient space of Y by the map q.

The same thing can be done when M is a "C^0-manifold-with-boundary" of dimension n, with boundary L: L is closed in M, M − L is an n-dimensional C^0-manifold, and every point $x \in L$ has an open neighborhood V in M such that there is a homeomorphism of V onto $\mathbf{R}^{n-1} \times [0, 1[$ sending $V \cap L$ onto $\mathbf{R}^{n-1} \times \{0\}$; L is then a C^0-manifold of dimension $n - 1$ (cf. Part 3, chap. VII, § 1).

It was proved in 1962 [94] that if M is *compact*, there is a homeomorphism of $L \times [0, 1[$ on a neighborhood U of L in M keeping the points of L fixed. This implies that L has a fundamental system of neighborhoods of which it is a strong deformation retract; it follows that $H_p(M, L; \mathbf{Z})$ is isomorphic to $H_p(X, X - M; \mathbf{Z})$, M being identified with a closed subset of its double X. Alexander duality then shows that if M is orientable, the *Lefschetz duality* holds for singular homology and cohomology ([440], p. 298):

$$H^{n-q}(M, L; \mathbf{Z}) \simeq H_q(M; \mathbf{Z}), \qquad H_q(M, L; \mathbf{Z}) \simeq H^{n-q}(M; \mathbf{Z}). \tag{8}$$

There is also the exact sequence

$$\cdots \to H^p(X) \to H^p(M) \oplus H^p(M) \to H^p(L) \to H^{p+1}(X) \to \cdots \tag{9}$$

which in particular implies, for the Euler–Poincaré characteristics,

$$\chi(X) = 2\chi(M) - \chi(L); \tag{10}$$

therefore, if X is oriented, as $\chi(X)$ is even (and 0 if n is odd), $\chi(L)$ is *even*.

For nonorientable C^0-manifolds-with-boundary, there are similar results for homology and cohomology with coefficients in $\mathbf{F}_2$.

G. Connected Sums

A construction similar to the "double" is the process of forming "connected sums" of two connected C^0-manifolds X, Y of same dimension n ([421], p. 218). Consider a point $x_0 \in X$, a point $y_0 \in Y$, and two charts $\varphi: U \to B$, $\psi: V \to B$, where U (resp. V) is an open neighborhood of x_0 (resp. y_0) and B is the open ball $|z| < 1$ in $\mathbf{R}^n$. Let B′ be the closed ball $|z| \leqslant \frac{1}{2}$ in $\mathbf{R}^n$, S its boundary [an $(n - 1)$-sphere] and h a homeomorphism of S onto itself. Take the disjoint sum

$$(X - \varphi^{-1}(B')) \coprod (Y - \psi^{-1}(B'))$$

and identify the points $x \in \varphi^{-1}(S)$ and $y \in \psi^{-1}(S)$ in that space if $\psi(y) = h(\varphi(x))$. It is easy to see that the quotient space Z is again a connected C^0-manifold, called a *connected sum* of X and Y and written X # Y. It can be proved that, except perhaps for $n = 3$, if X and Y are oriented and h reverses orientation, Z is oriented and does not depend on the choices made, up to homeomorphism ([400], pp. 42–45). If X and Y are C^r-manifolds with $r \geqslant 1$,

$X \# Y$ is also a C^r-manifold (and the intrinsic character of $X \# Y$ is then much easier to prove). Under the same assumptions,

$$H^p(X \# Y) \simeq H^p(X) \oplus H^p(Y) \quad \text{for } 1 \leqslant p \leqslant n-1. \tag{11}$$

§ 3. Attachments and CW-Complexes

The idea of *adjunction space* or *attachment of a space* to another one was introduced in 1938 by J.H.C. Whitehead for the study of homotopy ([492], p. 115). Let X be a topological space, A be a subspace of X, and $f: A \to Y$ be a continuous map; consider the disjoint sum $X \coprod Y$ and in it the equivalence relation R for which the classes are the singletons $\{x\}$ for $x \in X - A$ and the sets $\{y\} \cup f^{-1}(y)$ for $y \in Y$. The quotient space $(X \coprod Y)/R$ is said to be obtained by *attachment of* X *to* Y *along* A *by means of* f and is sometimes written $X \cup_f Y$. If $p: X \coprod Y \to X \cup_f Y$ is the natural map, it can immediately be verified that the restriction of p to Y is a homeomorphism onto $p(Y)$, so that Y can be *identified* with a subspace of $X \cup_f Y$. Similarly if A is *closed* in X, the restriction of p to $X - A$ is a homeomorphism of $X - A$ onto the open subset $(X \cup_f Y) - p(Y)$. If X and Y are normal spaces and A is closed in X, then $X \cup_f Y$ is normal.

A. The Mapping Cylinder

Let $f: X \to Y$ be a continuous map. The *mapping cylinder* Z_f of f, defined in 1939 by J.H.C. Whitehead for use in his work on homotopy ([493], p. 115), is obtained by attaching the closed subspace $X \times [0, 1]$ to Y along the closed subspace $X \times \{1\}$ by means of the map $(x, 1) \mapsto f(x)$; X and Y are thus identified to closed subspaces of Z_f by the maps $x \mapsto (x, 0) \mapsto p(x, 0)$ and $y \mapsto p(y)$. Y is also a *strong deformation retract* of Z_f by the homotopy $(z, u) \mapsto r(z, u)$ of $Z_f \times [0, 1]$ into Z_f, defined by

$$\begin{cases} r(p(x,t),u) = p(x,(1-u)t+u) & \text{for } x \in X,\ 0 \leqslant t, u \leqslant 1, \\ r(p(y),u) = p(y) & \text{for } y \in Y. \end{cases} \tag{12}$$

The *reduced mapping cylinder* $\tilde{Z}_f$ is obtained by collapsing the set $p(\{x_0\} \times [0, 1])$ in Z_f to a point; Y is still a strong deformation retract of $\tilde{Z}_f$.

B. The Mapping Cone

A construction similar to the preceding one gives the *mapping cone* C_f of the map f, defined by Barratt in 1955 in a particular case [43] and generalized by Puppe in 1958 [384]. Instead of $X \times [0, 1]$, consider the reduced cone CX (§ 2,B), in which X is naturally embedded, and attach CX to Y along X by means of f, so that $C_f = Y \cup_f CX$. Another definition is $C_f \simeq \tilde{Z}_f/X$.

If A is a closed subspace of X and $i: A \to X$ is the natural injection, then X/A is naturally homeomorphic to $(X \cup_i CA)/CA$.

C. The CW-Complexes

In his study of the homology of grassmannians (§4) begun in 1933 Ehresmann was led to consider a partition of a space X into "cells" of a more general type than those in use up to that time. For each dimension p a cell of dimension p is still homeomorphic to an open ball in $\mathbf{R}^p$, but its boundary in X is not necessarily homeomorphic to $\mathbf{S}_{p-1}$ as for usual cells. This idea was developed in more general contexts by J.H.C. Whitehead in 1941 ([492], p. 273) under the name of "membrane complex," which he changed in 1950 to CW-*complex* (abbreviation of "closure finite complex with weak topology"), in a systematic exposition of the properties of these spaces ([493], pp. 95–105).

By definition, a CW-complex (also called CW-*space* in some texts) is a Hausdorff space K equipped with a *partition* $(E_0, E_1, \dots, E_n, \dots)$ into a finite or denumerable family of subsets. Each E_n is itself a *disjoint* union of a family $(e_j^n)_{j \in J_n}$ indexed by a set of *arbitrary* cardinal. The e_j^n are called the *n-cells* of K. For each cell e_j^n, there is a continuous map f_j^n of the unit closed ball $\mathbf{D}_n$: $|z| \leqslant 1$ in $\mathbf{R}^n$ into K, such that

a. $f_j^n(\mathbf{D}_n) = \overline{e_j^n}$, closure of e_j^n in K (which therefore is *compact*),
b. the restriction $f_j^n|\mathring{\mathbf{D}}_n$ to the *open* ball $|z| < 1$ is a *homeomorphism* of that ball onto e_j^n, and
c. if K^{n-1} is the $(n-1)$-*skeleton* of K, union of the E_m for $m \leqslant n-1$, $f_j^n(\mathbf{S}_{n-1})$ is contained in K^{n-1}, and only meets a *finite* family of m-cells for $m \leqslant n-1$.

A *finite* CW-complex, by definition, is a CW-complex having only a finite number of cells (of all dimensions); it is *compact*. For infinite CW-complexes Whitehead also imposed the condition that

d. K has the "weak" topology, i.e., a subset A of K is closed in K if and only if its intersection $A \cap \overline{e_j^n}$ is closed in $\overline{e_j^n}$ for every cell.

Each e_j^n is then open in E_n. For that topology, K is always *paracompact* [348]. The *dimension* of K is the largest integer n such that $E_n \neq \varnothing$ if such an integer exists and $+\infty$ if not.

The usual cell complexes are CW-complexes. A simple example of CW-complex is given by the sphere $\mathbf{S}_n$ with two cells, one e^0 consisting of a single point, to which is attached a single n-cell. In order to visualize the map $p\colon \mathbf{D}_n \to \mathbf{S}_n$, identify $\mathbf{D}_n$ with the lower hemisphere of $\mathbf{S}_n$ (defined by $0 \leqslant \theta \leqslant \pi/2$, where θ is the angle of the vector x with the vector $-\mathrm{e}_{n+1}$ in $\mathbf{R}^{n+1}$); $p(\mathrm{x})$ is then the point on the same meridian as x but the angle of $p(\mathrm{x})$ and $-\mathrm{e}_{n+1}$ is 2θ; the attaching map $g\colon \mathbf{S}_{n-1} \to \{\mathrm{e}_{n+1}\}$ is constant.

A CW-*subcomplex* of a CW-complex K is a union L of a set of cells of K such that if $e_j^n \subset L$, then $\overline{e_j^n} \subset L$. It is a closed subspace of K, and every neighborhood of L in K contains a neighborhood of which L is a strong neighborhood retract; furthermore, on the space K/L obtained by collapsing L to a point (§2,A) there is a natural structure of CW-complex [459]. In general the closure $\overline{e_j^n}$ of a cell in K is not necessarily a CW-subcomplex. The n-skeleton K^n of K is a CW-subcomplex of dimension n, and E_n is open in K^n.

Any compact subset of a CW-complex K meets only a finite number of cells and is contained in a finite subcomplex of K.

The product K × L of two CW-complexes is not always a CW-complex but it has such a structure if K or L is locally compact.

One of the main properties of CW-complexes is that it is possible to define a continuous map $f: K \to X$ of a CW-complex into a space X step by step by defining them in succession on the n-skeletons K^n of K. If f is already defined on K^n, consider on the closure of each $(n+1)$-cell e_α^{n+1} a continuous map $f^{n+1}: \overline{e_\alpha^{n+1}} \to X$, and check that f^{n+1} coincides with the already defined f on $K^n \cap \overline{e_\alpha^{n+1}}$.

The process of attaching cells allows one to construct CW-complexes for which the sets of indices J_n have arbitrary cardinality. Start with an arbitrary discrete space J_0 and attach n-cells step by step. If K^{n-1} is an $(n-1)$-dimensional CW-complex and if n-cells $(e_\alpha^n)_{\alpha \in J_n}$ are attached to K^{n-1} by maps f_α^n, then the resulting space K^n is a CW-complex of dimension n. If the process is repeated for all values of n, and K is the set, union of the increasing sequence of the spaces K^n, and if the *fine* topology is taken on K (i.e., a subset F of K is closed for that topology if and only if its intersection with each K^n is closed in K^n *), then the space K thus obtained is a CW-complex ([459], p. 71).

Note that this construction gives the definition of a *simplicial complex* (as a space with a triangulation) without *any* restriction on the cardinal number of the simplices of each dimension nor on the cardinal number of the simplices whose closure contains a given simplex. *Any* combinatorial complex (Part 1, chap. II. § 2) thus gives rise to a triangulation of a space to which this combinatorial complex corresponds.

The construction of a CW-complex by successive attachments of cells yields information on its homology. If K^n is the n-skeleton, the singular homology $H_\bullet(K^n, K^{n-1}; \mathbf{Z})$ is entirely determined by the cardinal number of the set of n-cells. Let M be a subset of $K^n - K^{n-1}$ whose intersection with each n-cell reduces to one point, then $H_\bullet(K^n, K^{n-1}) \simeq H_\bullet(K^n, K^n - M)$, because K^{n-1} is a strong deformation retract of $K^n - M$. If $U = K^n - K^{n-1}$, $H_\bullet(K^n, K^n - M) \simeq H_\bullet(U, U - M)$ by excision. Finally, as U is the disjoint union of all n-cells, $H_p(U, U - M) = 0$ for $p \neq n$ and $H_n(U, U - M) \simeq \mathbf{Z}^{(J_n)}$. In a similar way, $H^n(K^n, K^{n-1}; \mathbf{Z}) \simeq \mathbf{Z}^{(J_n)}$. All this extends to arbitrary rings of coefficients.

Suppose that K is a *finite* CW-complex of dimension n, and let m_k be the number of k-cells. Elementary computations of dimensions in the homology exact sequence of a triple (Part 1, chap. IV, § 6,B) give the inequalities

$$\dim_F H_p(K; F) \leqslant m_p \qquad (13)$$

for any field F, and in particular for the Euler–Poincaré characteristic

* Milnor ([347], p. 63) rightly protests against the use of the name "weak topology" for that topology, since it is contrary to the use of the term "weak" in functional analysis.

$$\chi(\mathrm{K}) = \sum_{p=0}^{n} (-1)^p m_p. \tag{14}$$

The notion of CW-complex may be generalized to that of *relative CW-complex* (X, A): this is the space obtained by successive attachments of n-cells for increasing values of $n \geqslant 1$, but starting with an *arbitrary* Hausdorff space A instead of a discrete space. The *n-skeleton* $(\mathrm{X}, \mathrm{A})^n$ is defined by induction, taking $(\mathrm{X}, \mathrm{A})^0 = \mathrm{A}$, and $(\mathrm{X}, \mathrm{A})^n$ is the subspace obtained by attachment of all the n-cells of (X, A) to $(\mathrm{X}, \mathrm{A})^{n-1}$. The *relative dimension* of (X, A) is the smallest integer n for which $(\mathrm{X}, \mathrm{A}) = (\mathrm{X}, \mathrm{A})^n$, and $+\infty$ if $(\mathrm{X}, \mathrm{A}) \neq (\mathrm{X}, \mathrm{A})^n$ for all n.

§ 4. Applications: I. Homology of Grassmannians, Quadrics, and Stiefel Manifolds

A. Homology of Projective Spaces

Among the "usual" spaces whose homology with coefficients in **Z** is not computed in a trivial way, the projective spaces $\mathbf{P}_n(\mathrm{F})$, where F is one of the fields **R**, **C**, or **H**, were some of the first to be considered. The homology of $\mathbf{P}_n(\mathbf{C})$ was determined in papers by Hopf ([242], p. 30) and van der Waerden [478]; their elementary method also gives the "intersection ring" of $\mathbf{P}_n(\mathbf{C})$, and can immediately be extended to $\mathbf{P}_n(\mathbf{H})$. For $\mathbf{P}_n(\mathbf{R})$ a similar method is described in the book by Seifert and Threlfall ([421], p. 118).

The principle of these proofs is an induction on n. If $(x_j)_{0 \leqslant j \leqslant n}$ are homogeneous coordinates in $\mathbf{P}_n(\mathrm{F})$, the subspace defined by the equations $x_{k+1} = \cdots = x_n = 0$ is identified with $\mathbf{P}_k(\mathrm{F})$. If $\mathrm{F} = \mathbf{R}$ and s_m is a singular m-cycle with $m < n$, it can be assumed, by simplicial approximation, that its image does not contain the point $(0, 0, \ldots, 0, 1)$. The homotopy

$$((x_0, x_1, \ldots, x_n), t) \mapsto (x_0, x_1, \ldots, x_{n-1}, tx_n)$$

shows that s_m is homologous to a singular m-cycle in $\mathbf{P}_{n-1}(\mathbf{R})$, and the process may be continued until one arrives at an m-cycle in the space $\mathbf{P}_m(\mathbf{R})$. For $\mathrm{F} = \mathbf{C}$ (resp. $\mathrm{F} = \mathbf{H}$), the same result obtains for $m < 2n$ (resp. $m < 4n$), with $\mathbf{P}_{[m/2]}(\mathbf{C})$ [resp. $\mathbf{P}_{[m/4]}(\mathbf{H})$] replacing $\mathbf{P}_m(\mathbf{R})$. For $\mathrm{F} = \mathbf{C}$ (resp. $\mathrm{F} = \mathbf{H}$) it is then easy to see by induction on n that

$$\mathrm{H}_k(\mathbf{P}_n(\mathbf{C}); \mathbf{Z}) = 0 \quad \text{if } k \text{ is odd}, \qquad \mathrm{H}_k(\mathbf{P}_n(\mathbf{C}); \mathbf{Z}) \simeq \mathbf{Z} \text{ if } k \text{ is even}, \tag{15}$$

$$\mathrm{H}_k(\mathbf{P}_n(\mathbf{H}); \mathbf{Z}) = 0 \quad \text{if } k \not\equiv 0 \ (\mathrm{mod}.\,4), \qquad \mathrm{H}_k(\mathbf{P}_n(\mathbf{H}); \mathbf{Z}) \simeq \mathbf{Z} \text{ if } k \equiv 0 \ (\mathrm{mod}.\,4). \tag{16}$$

For $\mathrm{F} = \mathbf{R}$, $\mathbf{P}_n(\mathbf{R})$ is orientable if n is odd, nonorientable if n is even. The induction then uses the general result that for a connected compact pseudomanifold M of dimension n, $\mathrm{H}_n(\mathrm{M}; \mathbf{Z}) \simeq \mathbf{Z}$ is M is orientable, but if M is nonorientable, $\mathrm{H}_n(\mathrm{M}; \mathbf{Z}) = 0$ and the matrix of the boundary map $\mathbf{b}_n\colon \mathrm{C}_n \to \mathrm{C}_{n-1}$ from n-chains to $(n-1)$-chains has a single invariant factor $\neq 1$, equal to 2 ([421], p. 89). The final result is

$$H_k(\mathbf{P}_n(\mathbf{R}); \mathbf{Z}) \simeq \begin{cases} 0 & \text{for } k \text{ even and } 2 \leqslant k \leqslant n, \\ \mathbf{Z}/2\mathbf{Z} & \text{if } k \text{ is odd and } 1 \leqslant k \leqslant n-1, \\ \mathbf{Z} & \text{if } k = 0 \text{ and } k = n \text{ when } n \text{ is odd.} \end{cases} \tag{17}$$

More sophisticated proofs use CW-complexes (see below) or fiber bundles (Part 3, chap. IV, § 2).

B. Homology of Grassmannians

If F is one of the fields **R**, **C**, or **H**, the *grassmannian* $G_{n,p}(F)$ is the set of all p-dimensional vector subspaces of the vector space F^n over F; $G_{n+1,1}(F)$ is therefore the projective space $\mathbf{P}_n(F)$. When $F = \mathbf{R}$ (resp. $F = \mathbf{C}$, $F = \mathbf{H}$), $G_{n,p}(\mathbf{R})$ is naturally a homogeneous space for the orthogonal group $O(n, \mathbf{R})$ [resp. the unitary group $U(n, \mathbf{C})$, resp. $U(n, \mathbf{H})$], hence has a natural structure of C^∞ manifold.

In 1933–1934 Ehresmann, in his thesis ([154], [155]), determined a basis for the singular homology $H_\bullet(G_{n,p}(\mathbf{C}); \mathbf{Z})$ of all complex grassmannians by the introduction of a special structure of CW-complex on these spaces. For that purpose he used algebraic subvarieties of $G_{n,p}(\mathbf{C})$ introduced in algebraic geometry by H. Schubert in 1879 [417]. For every $k < n$, a *Schubert symbol* σ *of order* k is a sequence of k integers σ_j such that

$$1 \leqslant \sigma_1 < \sigma_2 < \cdots < \sigma_k \leqslant n. \tag{18}$$

Denote by $\mathbf{C}^k$ the vector subspace of $\mathbf{C}^n$ spanned by the k first vectors of the canonical basis. To each Schubert symbol σ can be associated the subset $e(\sigma)$ of $G_{n,p}(\mathbf{C})$ consisting of the p-dimensional vector subspaces X such that

$$\dim(X \cap \mathbf{C}^{\sigma_i}) = i, \qquad \dim(X \cap \mathbf{C}^{\sigma_i - 1}) = i - 1 \quad \text{for } 1 \leqslant i \leqslant k. \tag{19}$$

The closure $\overline{e(\sigma)}$ in $G_{n,p}(\mathbf{C})$ is called a *fundamental Schubert variety*; it is an algebraic variety (with singularities).

In his thesis Ehresmann proved that $e(\sigma)$ is homeomorphic to an open ball in the space $\mathbf{R}^{2m(\sigma)}$, where

$$m(\sigma) = (\sigma_1 - 1) + (\sigma_2 - 2) + \cdots + (\sigma_k - k). \tag{20}$$

Furthermore, he showed that the boundary $\overline{e(\sigma)} - e(\sigma)$ is the union of the closed cells $\overline{e(\tau)}$, where τ is any one of the Schubert symbols obtained by replacing σ_i by $\sigma_i - 1$ $(1 \leqslant i \leqslant k)$, provided the sequence thus obtained still satisfies (18).

The fact that in the CW-complex $G_{n,p}(\mathbf{C})$ thus defined, there is no cell of *odd dimension* immediately implies that $H_k(G_{n,p}(\mathbf{C}); \mathbf{Z})$ is a *free* **Z**-module, having as basis the homology classes of the Schubert varieties $\overline{e(\sigma)}$ such that $2m(\sigma) = k$. Ehresmann also determined the intersection numbers of the $\overline{e(\sigma)}$ of complementary dimensions, obtaining by topological means the formulas given by Schubert and Severi.

In a later paper [156] he studied the homology of the real grassmannians

$G_{n,p}(\mathbf{R})$ in the same way. All definitions are the same, the field $\mathbf{C}$ being everywhere replaced by $\mathbf{R}$, and $e(\sigma)$ is now homeomorphic to an open ball in the space $\mathbf{R}^{m(\sigma)}$, so that there are now cells of every dimension $\leqslant (p+1)(n-p)$. Ehresmann explicitly computed (for suitable orientations) the boundary operator

$$e(\sigma_1, \sigma_2, \ldots, \sigma_k) \mapsto \sum_{i=1}^{k} \eta_i e(\sigma_1, \ldots, \sigma_{i-1}, \sigma_i - 1, \sigma_{i+1}, \ldots, \sigma_k) \tag{21}$$

where on the right-hand side only the meaningful Schubert symbols must be kept, and

$$\begin{cases} \eta_i = 0 & \text{if } \sigma_i + k + i \text{ is odd,} \\ \eta_i = 2(-1)^{\sigma_1+\sigma_2+\cdots+\sigma_i} & \text{if } \sigma_i + k + i \text{ is even.} \end{cases} \tag{22}$$

In 1947 Chern [126] completed that result by determining the ring structure of the singular cohomology $H^{\bullet}(G_{n,p}(\mathbf{R}); \mathbf{F}_2)$. Pontrjagin [381] computed by similar means the homology classes with coefficients in $\mathbf{Z}$ or in $\mathbf{F}_2$ of the *special grassmannian* $G'_{n,p}(\mathbf{R})$, a two-sheeted covering space of $G_{n,p}(\mathbf{R})$ whose points are the *oriented* p-dimensional vector subspaces of $\mathbf{R}^n$ (see Part 3, chap. IV, § 1,B). Ehresmann also showed how similar methods can determine the homology of flag manifolds.

C. Homology of Quadrics and Stiefel Manifolds

For F denoting one of the fields $\mathbf{R}$, $\mathbf{C}$, or $\mathbf{H}$, the *Stiefel manifolds* $\mathbf{S}_{n,p}(\mathrm{F})$ were defined by Stiefel in 1936 during his investigations of vector fields on spheres [457] (see Part 3, chap. IV, § 1,A). Let $(\mathsf{x}|\mathsf{y})$ be the hermitian scalar product

$$(\mathsf{x}|\mathsf{y}) = x_1\bar{y}_1 + x_2\bar{y}_2 + \cdots + x_n\bar{y}_n$$

on the vector space F^n over F. For $1 \leqslant p \leqslant n-1$, $\mathbf{S}_{n,p}(\mathrm{F})$ is the subspace of F^{np} consisting of the *p-tuples* $(\mathsf{x}_1, \ldots, \mathsf{x}_p)$ of *orthonormal* vectors of F^n [i.e., $(\mathsf{x}_i|\mathsf{x}_j) = \delta_{ij}$].

Stiefel (and independently Whitney in [505]) only considered the space $\mathbf{S}_{n,p}(\mathbf{R}) = \mathbf{S}_{n,p}$, and identified it with the space of pairs

$$(\mathsf{x}_1, (\mathsf{y}_2, \ldots, \mathsf{y}_p))$$

formed by a point $\mathsf{x}_1 \in \mathbf{S}_{n-1}$ and a system $(\mathsf{y}_2, \ldots, \mathsf{y}_p)$ of unitary vectors of origin x_1 in the tangent hyperplane to $\mathbf{S}_{n-1}$ at the point x_1. Let s' (resp. s'') be the stereographic projection of $\mathbf{S}_{n-1} - \{-\mathsf{e}_n\}$ (resp. $\mathbf{S}_{n-1} - \{\mathsf{e}_n\}$) onto $\mathbf{R}^{n-1}$, and let $\mathrm{S}'_{n,p}$ (resp. $\mathrm{S}''_{n,p}$) be the part of $\mathbf{S}_{n,p}$ corresponding to the points x_1 in the closed hemisphere of $\mathbf{S}_{n-1}$ with n-th coordinate $\xi_n \geqslant 0$ (resp. $\xi_n \leqslant 0$), so that

$$\mathbf{S}_{n,p} = \mathrm{S}'_{n,p} \cup \mathrm{S}''_{n,p}.$$

Furthermore, the fact that s' and s'' preserve orthogonality of tangent vectors easily implies that $\mathrm{S}'_{n,p}$ and $\mathrm{S}''_{n,p}$ are both homeomorphic to the product $\mathbf{D}_{n-1} \times \mathbf{S}_{n-1,p-1}$, and $\mathrm{S}'_{n,p} \cap \mathrm{S}''_{n,p}$ homeomorphic to $\mathbf{S}_{n-2} \times \mathbf{S}_{n-1,p-1}$. Using this decomposition of $\mathbf{S}_{n,p}$ as the union of two product spaces, to which he applied induction and the Künneth formula, Stiefel showed that $\mathbf{S}_{n,p}$ is a compact orientable C^∞ manifold of dimension $p(2n-p-1)/2$ and constructed a triangulation of that manifold that enabled him to partially compute the simplicial homology of $\mathbf{S}_{n,p}$ by reduction to the case $p = 2$. His results are

$$H_r(\mathbf{S}_{n,p}) = 0 \qquad \text{for } 1 \leqslant r < n-p,$$

$$H_{n-p}(\mathbf{S}_{n,p}) \cong \begin{cases} \mathbf{Z} & \text{if } n-p \text{ is even,} \\ \mathbf{Z}/2\mathbf{Z} & \text{if } n-p \text{ is odd} \end{cases}$$

[Whitney only computed (without proof) $H_{n-p}(\mathbf{S}_{n,p})$].

In [155] and [156], Ehresmann had studied the homology of the space of maximal linear subvarieties (of dimension n) in a projective quadric over **R** or **C**

$$x_0^2 + x_1^2 + \cdots + x_n^2 - x_{n+1}^2 - \cdots - x_{2n+1}^2 = 0 \tag{23}$$

[or equivalently, the subspace of $G_{2n+2,n+1}(F)$ (for $F = \mathbf{R}$ or $\mathbf{C}$) consisting of the maximal isotropic subspaces for the quadratic form (23)]. He defined a structure of CW-complex for that space by the same method (using Schubert symbols) he applied to grassmannians. In 1939 [157] he considered more generally the space of maximal linear subvarieties (of dimension p) in the projective real quadric $Q_{n,p}$

$$x_0^2 + x_1^2 + \cdots + x_n^2 - x_{n+1}^2 - \cdots - x_{n+p+1}^2 = 0 \quad \text{for } p < n, \tag{24}$$

and observed, by a simple geometric construction, that this space is homeomorphic to the Stiefel manifold $\mathbf{S}_{n+1,p+1}(\mathbf{R})$. He extended this homeomorphism to $\mathbf{S}_{n+1,p+1}(F)$ for $F = \mathbf{C}$ or $F = \mathbf{H}$ by considering the complex or quaternionic "hyperquadric"

$$x_0\bar{x}_0 + x_1\bar{x}_1 + \cdots + x_n\bar{x}_n - x_{n+1}\bar{x}_{n+1} - \cdots - x_{n+p+1}\bar{x}_{n+p+1} = 0 \tag{25}$$

and the space of maximal complex (resp. quaternionic) linear subvarieties that they contain, and showed that his general method could again apply to that space. As the Stiefel manifold $\mathbf{S}_{n+1,n+1}(F)$ is the unitary group in the space F^{n+1}, he had thus described another method of computation for the homology of these groups.

In the theory of fiber bundles (Part 3, chap. III, §2,E) it was found that the computation of the homotopy groups of the Stiefel manifolds $\mathbf{S}_{n,p}(\mathbf{R})$ was a useful tool. J.H.C. Whitehead described another decomposition of $\mathbf{S}_{n,p}(\mathbf{R})$ as a CW-complex for that purpose ([492], pp. 303–355). This was further studied by C. Miller in 1951 [338], who used it to describe the cohomology of $\mathbf{S}_{n,p}(\mathbf{R})$. The cells are again in one-to-one correspondence with some Schubert symbols σ. Consider first the group of rotations $SO(n,\mathbf{R}) = \mathbf{S}_{n,n-1}(\mathbf{R})$; for each component σ_j of σ, let H_{σ_j} be the half-space of $\mathbf{R}^{\sigma_j}$ whose points have their σ_j-th coordinate >0. On the other hand, for any $\mathsf{x} \in \mathbf{S}_{n-1}$, let $h(\mathsf{x})$ be the symmetry with respect to the hyperplane orthogonal to x, and let $f(\mathsf{x})$ be the rotation

$h(\mathsf{x})h(\mathsf{e}_1)$. Then consider in $\mathrm{SO}(n, \mathbf{R})$ the set $\mathrm{R}(\sigma)$ consisting of all the rotations of the form

$$f(\mathsf{x}_k)f(\mathsf{x}_{k-1})\cdots f(\mathsf{x}_1)$$

for $\mathsf{x}_j \in \mathbf{S}_{n-1} \cap \mathrm{H}_{\sigma_j}$. For all Schubert symbols satisfying the inequalities

$$n - p \leqslant \sigma_1 < \sigma_2 < \cdots < \sigma_k < n$$

the set $\mathrm{B}(\sigma) \subset \mathbf{S}_{n,p}(\mathbf{R})$ then consists of the elements

$$u.(\mathsf{e}_{n-p+1}, \mathsf{e}_{n-p+2}, \ldots, \mathsf{e}_n)$$

for all $u \in \mathrm{R}(\sigma)$. J.H.C. Whitehead proved that $\mathrm{B}(\sigma)$ is homeomorphic to an open ball of dimension $m = \sigma_1 + \sigma_2 + \cdots + \sigma_k$; they constitute the cells of a CW-complex structure on $\mathbf{S}_{n,p}(\mathbf{R})$, and the closure of $\mathrm{B}(\sigma)$ in $\mathbf{S}_{n,p}(\mathbf{R})$ is homeomorphic to the "stunted projective space" $\mathbf{P}_m(\mathbf{R})/\mathbf{P}_{m-1}(\mathbf{R})$. Miller also determined the boundary formulas

$$\mathrm{B}(\sigma_1, \sigma_2, \ldots, \sigma_k) \mapsto \sum_j \alpha_j \mathrm{B}(\sigma_1, \ldots, \sigma_j - 1, \ldots, \sigma_k)$$

where the sum is extended only to the Schubert symbols which are meaningful, and

$$\alpha_j = ((-1)^j + 1)(-1)^{\sigma_k + \sigma_{k-1} + \cdots + \sigma_{j+1}}$$

from which he could compute explicitly the homology with coefficients in $\mathbf{Z}$ as well as the cohomology algebra. The same method applies to complex and quaternionic Stiefel manifolds.

The homology of projective complex quadrics was determined by E. Cartan [98]. J. Nordon, a student of Ehresmann, completely described the homology of projective real quadrics, using a CW-complex structure and the fact that for quadrics not homeomorphic to spheres a product of two spheres is a covering space [357].

Further progress in the theory of Lie groups and homogeneous spaces showed that Schubert varieties can be generalized in other Lie groups [210].

§5. Applications: II. The Morse Inequalities

If M is a pure C^∞ manifold of dimension n, f is a real valued C^∞ function defined on M, and x_0 is a *critical point* of f (Part 1, chap. III, §1), then, in a sufficiently small neighborhood of x_0, the Taylor expansion of $f(x) - f(x_0)$, expressed in terms of local coordinates $u_1, \ldots, u_n$, begins with terms of degree $\geqslant 2$; x_0 is called a *nondegenerate* critical point if the sum of the terms of degree 2 in that expansion is a *nondegenerate quadratic form*. Nondegenerate critical points are *isolated*. The *index* of a quadratic form (degenerate or not) is the maximal dimension of a vector subspace on which the form takes value <0 except at the origin. The *index of a critical point* x_0 is the index of the quadratic form consisting of the terms of degree 2 in the Taylor expansion of $f(x) - f(x_0)$; it is a number independent of the choice of local coordinates.

As a nondegenerate critical point x_0 is characterized by an *inequality* between the derivatives of the function f at x_0 (with respect to the local coordinates), it is to be expected that "in general" critical points are nondegenerate. This was made precise by Marston Morse in 1931 [354] when M is *compact*: for the C^r topology (r integer $\geqslant 2$), the functions on M with nondegenerate critical points form a *dense* set in the space of C^r functions. The elegant proof consists in embedding M into a space $\mathbf{R}^N$ in such a way that the function f is equal to the first projection pr_1 in M, and choosing a point $z = (-c + \varepsilon_1, \varepsilon_2, \ldots, \varepsilon_N)$ in $\mathbf{R}^N$ with large $c > 0$ and small ε_j ($1 \leqslant j \leqslant N$); the function

$$g(x) = (|x - z|^2 - c^2)/2c$$

approaches f on M as closely as one wishes in the C^r topology, and it is easy to see that its critical points are nondegenerate.

For the purposes of his researches in the calculus of variations (chap. VII, § 3), this approximation property allowed Morse to restrict his investigations, for C^∞ functions on a *compact* manifold M, to functions having *only* nondegenerate critical points; the number of these points is then *finite*. Letting C_j be the number of these points of *index* j ($0 \leqslant j \leqslant n$), Morse discovered relations between these numbers and the homology (mod. 2) of M.

Some special cases of these relations had been met before Morse. In 1885 Poincaré, in his work on differential equations on surfaces, had shown that "in general" a differential equation

$$\frac{dx}{X} = \frac{dy}{Y} \tag{26}$$

on a compact orientable surface $M \subset \mathbf{R}^3$ (where X and Y are polynomials) has singular points (those where X and Y vanish simultaneously) classified as nodes, saddle points, and foci, and that the respective numbers N, S, and F of these points satisfy the relation

$$N - S + F = 2 - 2g \tag{27}$$

where g is the genus of M ([365], p. 125) The "level curves" $f(x, y) = \alpha$ of a C^∞ function f defined on M may be considered as integral curves of a differential equation (26) (with functions X, Y more general than in Poincaré's paper). In general the numbers N, S, F are exactly the numbers C_2, C_1, and C_0 of critical points of f.

In 1917, in his work on dynamical systems ([48], p. 42), G.D. Birkhoff had to study the critical points of a function f defined on an n-dimensional manifold M. In addition to the usual relative extrema of f he considered what he called a *minimax* critical point, which he defined by the property that if x_0 is such a point and $f(x_0) = c$, the set M_c of points $x \in M$ such that $f(x) < c$ is not connected around x_0. Introducing the index of a critical point, he showed that the minimax points are those of index 1. In order to study those points when M is *compact* and orientable, he had the idea of considering the open subsets M_α of M defined by $f(x) < \alpha$ when α *varies*, and arguments in the Poincaré "intuitive" style gave him the inequality

$$C_1 - C_0 \geqslant R_1 - R_0 \quad \text{where } R_j = \dim H_j(M; \mathbf{F}_2). \tag{28}$$

In 1925 M. Morse showed that these results are particular cases of the inequalities valid for any C^∞ function f on a *compact* C^∞ manifold with nondegenerate critical points:

$$\begin{gathered} C_0 \geqslant R_0, \\ C_1 - C_0 \geqslant R_1 - R_0, \\ \cdots\cdots\cdots\cdots\cdots\cdots\cdots \\ C_{n-1} - C_{n-2} + \cdots + (-1)^{n-2}C_0 \geqslant R_{n-1} - R_{n-2} + \cdots + (-1)^{n-2}R_0 \end{gathered} \tag{29}$$

to which must be added the equality

$$C_n - C_{n-1} + \cdots + (-1)^{n-1}C_0 = R_n - R_{n-1} + \cdots + (-1)^{n-1}R_0. \tag{30}$$

In order to prove these results he took up Birkhoff's idea of looking at the variation with α of the open M_α, and in particular (using this time the rigorous simplicial homology theory developed by Veblen, Alexander, and Lefschetz) seeing what happens to their "Betti numbers in homology mod. 2" (which at that time were still called "connectivities") when α crosses a value of f at a critical point. His intricate arguments are better understood by using the concept of "relative homology" of Lefschetz, as Seifert and Threlfall did in their book on Morse theory ([422], pp. 85–92).

Let x_0 be a nondegenerate critical point of f, and $f(x_0) = \gamma$. Consider a relative k-cycle Z_k *modulo* M_γ on the subspace $M_\gamma \cup \{x_0\}$. If k is not equal to the index of x_0, Z_k is a relative boundary *modulo* M_γ; if k is the index of x_0, there is exactly one relative k-cycle Z_k (over $\mathbf{F}_2$), up to relative boundaries, which is *not* a relative boundary *modulo* M_γ. Then there are two possibilities for that k-cycle: it may be an *absolute boundary in* M_γ or not; in the second case, Z_k becomes an *absolute boundary in* $M_\gamma \cup \{x_0\}$. The "k-th connectivity" therefore *increases by* 1 in the first case and *decreases by* 1 in the second.

If m_k^+ (resp. m_k^-) is the number of critical points of the first (resp. second) type, then

$$C_k = m_k^+ + m_k^- \qquad \text{and} \qquad R_k = m_k^+ - m_{k+1}^-$$

hence

$$C_k - R_k = m_k^- + m_{k+1}^-$$

and as $m_0^- = m_{n+1}^- = 0$, this immediately implies (29) and (30).

The modern presentation of the Morse inequalities ([345], [461], [363]) makes essential use of the idea of *attaching cells* to a space, and has found other quite startling applications to differential topology. The central result is that the existence of a C^∞ function f on a compact C^∞ manifold M having only nondegenerate critical points implies that M has the same homology as a CW-complex defined by attachment of suitable cells corresponding to the critical points. The Morse inequalities (29) and the relation (30) then follow from the general properties (13) and (14) of the homology of a CW-complex (§ 3), and are in fact valid for homology over an *arbitrary* field.

In more detail, the first step consists in considering, for two numbers $a < b$, the closed subset $f^{-1}([a,b])$ when it contains *no* critical point. Then, using a Riemannian structure on M, the assumption implies that the vector field $\operatorname{grad}(f)$ is $\neq 0$ everywhere in $f^{-1}([a,b])$. For a suitably chosen C^∞ vector field X defined on the whole manifold, vanishing outside of a neighborhood of $f^{-1}([a,b])$, and proportional to $\operatorname{grad}(f)$ in $f^{-1}([a,b])$, the flow* $F_X(x,t)$ of X is defined in the whole space $M \times \mathbf{R}$, and each map $h_t\colon x \mapsto F_X(x,t)$ is a homeomorphism of M onto itself. For $a < c < d < b$, h_{d-c} transforms M_c into M_d, and M_c is thus a strong deformation retract of M_d.

The crucial point is what happens in the vicinity of a critical point x_0 of f, of index k. Using the Taylor formula and the Gram–Schmidt orthonormalization process, Morse ([354], p. 172) showed that there are in a neighborhood U of x_0 local coordinates $u_1, u_2, \ldots, u_n$, vanishing at the point x_0 and such that the local expression of f in U is

$$f(x) = c - Q_1(u) + Q_2(u) \tag{31}$$

with

$$Q_1(u) = u_1^2 + u_2^2 + \cdots + u_k^2, \qquad Q_2(u) = u_{k+1}^2 + \cdots + u_n^2. \tag{32}$$

The next step is to modify f in U in such a way that for the modified function g the sets N_t defined by $g(x) < t$ are easier to study than the M_t. More precisely, g has to satisfy the following properties:

1. g coincides with f in $M - U$.
2. If $\varepsilon > 0$ is small enough, so that the ball B_ε defined by

$$\sum_{j=1}^{n} u_j^2 \leqslant 2\varepsilon$$

is contained in U, then $M_{c+\varepsilon} = N_{c+\varepsilon}$, and g has no critical point in $g^{-1}([c-\varepsilon, c+\varepsilon])$.
3. g has the same critical points as f, with the same index.

Property 2 and the preliminary result then imply that $N_{c-\varepsilon}$ is a *strong deformation retract* of $N_{c+\varepsilon} = M_{c+\varepsilon}$, the critical point x_0 (for both f and g) being in $N_{c-\varepsilon} - M_{c-\varepsilon}$.

---

* Recall that the flow of a C^∞ vector field X on a C^∞ manifold M is a function $(x,t) \mapsto F_X(x,t)$ defined in the subset $\operatorname{dom}(F_X) \subset M \times \mathbf{R}$ consisting of the points (x,t) where $x \in M$ and $t \in J_x$, where J_x is the largest open interval containing 0 in $\mathbf{R}$, where the solution $t \mapsto v(t)$ of the differential equation

$$v'(t) = X(v(t))$$

taking the value x for $t = 0$, is defined, and $F_X(x,t)$ is the value of that solution.

Such a function g can be constructed in the following way:

$$g(x) = f(x) - h(x) \text{ in } \mathrm{M} \tag{33}$$

where $h(x) = 0$ for $x \notin \mathrm{B}_\varepsilon$, and for $x \in \mathrm{B}_\varepsilon$,

$$h(x) = \varphi(\mathrm{Q}_1(\mathrm{u}) + 2\mathrm{Q}_2(\mathrm{u})) \tag{34}$$

where $\varphi\colon \mathbf{R} \to \mathbf{R}$ is a C^∞ function such that

$$\varphi(0) > \varepsilon, \qquad \varphi(t) = 0 \text{ for } t \geqslant 2\varepsilon, \qquad \text{and} \qquad -1 < \varphi'(t) \leqslant 0 \quad \text{in } \mathbf{R}. \tag{35}$$

It can also be shown that there is a function r defined and C^∞ in $[0, 1] \times \mathrm{M}$ with values in M and such that

$$\begin{cases} r(t, x) = x & \text{for } x \in \mathrm{M} - \mathrm{U}, \\ r(t, x) = (u_1, \ldots, u_k, \psi(t, \mathrm{u})u_{k+1}, \ldots, \psi(t, \mathrm{u})u_n) & \text{for } x \in \mathrm{U}, \end{cases} \tag{36}$$

where $\psi(t, \mathrm{u})$ is a real C^∞ function such that $\psi(0, \mathrm{u}) = 0$ and $\psi(1, \mathrm{u}) = 1$. Then r is a homotopy that retracts $\mathrm{N}_{c-\varepsilon}$ onto $\mathrm{M}_{c-\varepsilon} \cup \mathrm{D}$, where $\mathrm{D} \subset \mathrm{U}$ is defined by the relations $\mathrm{Q}_1(\mathrm{u}) \leqslant \varepsilon$ and $\mathrm{Q}_2(\mathrm{u}) = 0$, so that D is homeomorphic to a closed ball in $\mathbf{R}^k$, and its intersection with $\mathrm{M}_{c-\varepsilon}$ is the sphere $\mathrm{Q}_1(\mathrm{u}) = \varepsilon$. Therefore, $\mathrm{M}_{c-\varepsilon} \cup \mathrm{D}$ is a *strong deformation retract* of $\mathrm{M}_{c+\varepsilon}$ and is obtained by *attaching a k-cell* (§ 3) to $\mathrm{M}_{c-\varepsilon}$.

In the same way, for each critical point of f, it can be seen that M has the same homology as a finite *CW-complex* having C_k k-cells for each k such that $0 \leqslant k \leqslant n$. From this, and the inequalities (13), the Morse inequalities (29) follow.

CHAPTER VI

Homology of Groups and Homogeneous Spaces

With a few exceptions concerning "abstract" groups defined "by generators and relations" or commutative groups of special types, for mathematicians of the nineteenth century a "group" is a *group of transformations*, that is, a set G of bijections of a set X onto itself, the composition of two elements of G and the inverse of an element of G being in G. This was already obvious when the first groups were defined by Cauchy and Galois as groups of permutations of a finite set. That properties of such a group G would imply properties of X is the substance of the famous "Erlangen Program" of Klein. Conversely, the fact that an "abstract" group is isomorphic to a group of transformations having special geometric properties, such as linear transformations of a vector space, may yield useful information on the group itself, an idea basic in the theory of linear representations created by Frobenius.

These ideas acquired even more significance and power with the introduction of continuity and differentiability considerations both in the space X and in the group G in the work of Lie and his followers, although it took a long time for most mathematicians to realize the wealth of applications opened by those richer structures.

§ 1. The Homology of Lie Groups

Until 1925 the bulk of papers on Lie groups were concerned with *local* properties reduced by Lie theory to the algebraic study of their Lie algebras. Global Lie groups were not unknown; the "classical" groups had been considered by Lie himself, and E. Cartan had given descriptions of global groups corresponding to the five exceptional simple complex Lie algebras. It was known that on a global Lie group there sits a measure invariant by left (or right) translations. Hurwitz, in 1897, was the first to use that fact in order to introduce the "mean value" of a continuous function on the orthogonal group by analogy with the "mean value" of a function on a finite set, which device enabled him to construct invariants for the actions of the orthogonal group. When in 1924 I. Schur took up that idea and showed that by its use the Frobenius theory of characters could be generalized for the orthogonal group, H. Weyl, who was aware of E. Cartan's theory of linear representations of

J. Dieudonné, *A History of Algebraic and Differential Topology, 1900–1960*,
Modern Birkhäuser Classics, DOI 10.1007/978-0-8176-4907-4_12,

semisimple Lie algebras, realized that for *compact* Lie groups Cartan's results could be obtained by *global* arguments of topology and measure theory, generalizing Schur's method.

This was a challenge to E. Cartan, who during the next 10 years produced a series of wonderful papers on the topological properties of global Lie groups and their homogeneous spaces [98]. His main tools, as always, were the exterior differential forms; we have seen (Part 1, chap. III, § 3) how, in order to use them in topology he took up Poincaré's remark on their relation with homology and how de Rham was able to base the real cohomology of a C^∞ manifold on simple computations with these forms without any "combinatorial" technique.

One of Cartan's main theorems is that a connected Lie group G is diffeomorphic to the product of a maximal *compact* connected subgroup K and a subvariety diffeomorphic to some $\mathbf{R}^n$. This implies that K is a strong deformation retract of G, hence the singular homology (resp. cohomology) of G is isomorphic to the singular homology (resp. cohomology) of K for *any* ring of coefficients, thus reducing the homological theory of Lie groups to the case of compact groups.

Cartan only studied *real* cohomology, using de Rham's theorems. Let G be a compact Lie group, operating on the right on a C^∞ manifold M by a C^∞ operation $(s, x) \mapsto x . s$. Since there exists a measure ds on G, invariant by left and right translations and of total mass 1, Hurwitz's idea of taking *mean values* on G of an arbitrary exterior p-form α on M may be applied: for any point $x \in \mathrm{M}$, the mean value $m(\alpha)$ of a p-form α takes the value

$$m(\alpha)(x) = \int_{\mathrm{G}} \alpha(x . s)\, ds. \tag{1}$$

Now $m(\alpha)$ is *invariant* under the action of G on M, and if α is closed (resp. exact), then $m(\alpha)$ is also closed (resp. exact). Furthermore, α and $m(\alpha)$ are *cohomologous* on M; if $\mathrm{H}^{\bullet}_{\mathrm{G}}(\mathrm{M})$ is the graded subspace of the real cohomology space $\mathrm{H}^{\bullet}(\mathrm{M})$, consisting of the classes of the differential forms invariant under the action of G, this defines an *isomorphism*

$$m^*\colon \mathrm{H}^{\bullet}(\mathrm{M}) \simeq \mathrm{H}^{\bullet}_{\mathrm{G}}(\mathrm{M}). \tag{2}$$

Cartan's interpretation of the real cohomology $\mathrm{H}^{\bullet}(\mathrm{G})$ of a compact Lie group G is obtained as a corollary by consideration of the action $((s, t), x) \mapsto s^{-1}xt$ of $\mathrm{G} \times \mathrm{G}$ on G. A p-form is invariant under that action if it is *bi-invariant*, that is, invariant under both left and right translations in G. The Lie–Cartan theory implies that for such a form α, $d\alpha = 0$, so that when one computes $\mathrm{H}^{\bullet}_{\mathrm{G}\times\mathrm{G}}(\mathrm{G})$, all cochains are cocycles and all coboundaries are 0. Hence the fundamental result that $\mathrm{H}^{\bullet}(\mathrm{G})$ *is isomorphic to the graded algebra* $\mathfrak{b}^*(\mathrm{G})$ *of all bi-invariant differential forms.*

The explicit determination of $\mathrm{H}^{\bullet}(\mathrm{G})$ is thus reduced to an algebraic problem. The group G operates on the dual $\mathfrak{g}^*$ of the Lie algebra $\mathfrak{g}$ by the coadjoint representation $s \mapsto {}^t\mathrm{Ad}(s)$; $\mathfrak{b}^*(\mathrm{G})$ is the sum of the *one-dimensional* subspaces of $\mathfrak{g}^*$ stable for that representation; for a compact group, they can in principle be determined by Cartan's method of highest weights. This method can also be applied to homogeneous spaces of a compact Lie group, and examples were given by Cartan himself and by Ehresmann in his thesis and later papers.

The *explicit* determination of the real cohomology $H^\bullet(G)$ was first made along those lines by R. Brauer [86] for the *classical* compact groups of the four classes A, B, C, D. Pontrjagin obtained the same results simultaneously and independently by a direct construction of homology bases [379]. In those cases their results confirmed the guess made earlier by Cartan: the real homology of such a group G is the same as the homology of a product

$$\mathbf{S}_{m_1} \times \mathbf{S}_{m_2} \times \cdots \times \mathbf{S}_{m_l} \tag{3}$$

of l *spheres of odd dimension*, l being the *rank* of G (dimension of a maximal torus in G). Cartan had also proved earlier that for *all* connected compact Lie groups G the Euler–Poincaré characteristic $\chi(G)$ is equal to 2^l. He had shown that the Poincaré polynomial can be expressed by the integral formula

$$P_G(T) = \int_G \det(\mathrm{Ad}(s) + T \,.\, I)\, ds. \tag{4}$$

Then, using Weyl's integration formula in a compact group, he could deduce from (4) that $P_G(T)$ is divisible by $(T + 1)^l$, and $P_G(1) = 2^l$.

We recall that the homology of the *classical* groups can be deduced as special case from the homology of Stiefel manifolds, obtained by decomposition of these manifolds into CW-complexes (chap. V, § 4).

§ 2. H-Spaces and Hopf Algebras

A. Hopf's Theorem

In his survey of what was known in 1936 on the homology of compact Lie groups [103], E. Cartan conjectured that there should be a general result implying that the homology of the classical groups is the same as the homology of a product of odd-dimensional spheres. In 1939 Hopf discovered that general result [246], showing that it held for more general spaces than compact Lie groups, and was in fact a consequence of two properties:

1. The *functorial* character of homology and cohomology;
2. The existence of a law of composition of a very general type on the space.

We shall first consider cohomology, although Hopf elected to express his results for homology using his "Umkehrhomomorphismus" (Part 1, chap. IV, § 4), which is somewhat more awkward.* Lefschetz had observed that for any space X the natural "diagonal" map $\delta: x \mapsto (x, x)$ of X into X × X gives a homomorphism of graded Λ-modules

* This was a deliberate choice, for Hopf mentioned Freudenthal's paper in which he connects the "Umkehrhomomorphismus" with cohomology [203].

$$\delta^*\colon \mathrm{H}^{\bullet}(\mathrm{X} \times \mathrm{X}; \Lambda) \to \mathrm{H}^{\bullet}(\mathrm{X}; \Lambda) \tag{5}$$

for any cohomology theory applicable to all spaces and all rings of coefficients. But if Λ is a *field*, the Künneth formula yields a natural isomorphism of graded Λ-modules

$$\mathrm{H}^{\bullet}(\mathrm{X}; \Lambda) \otimes_{\Lambda} \mathrm{H}^{\bullet}(\mathrm{X}; \Lambda) \simeq \mathrm{H}^{\bullet}(\mathrm{X} \times \mathrm{X}; \Lambda). \tag{6}$$

Hence δ^* may be considered as a Λ-linear map

$$\mathrm{H}^{\bullet}(\mathrm{X}; \Lambda) \otimes_{\Lambda} \mathrm{H}^{\bullet}(\mathrm{X}; \Lambda) \to \mathrm{H}^{\bullet}(\mathrm{X}; \Lambda) \tag{7}$$

respecting the graduation. The properties of associativity and anticommutativity of the graded Λ-algebra $\mathrm{H}^{\bullet}(\mathrm{X}; \Lambda)$ thus defined follow from the consideration of the commutative diagrams

$$\begin{array}{ccc} \mathrm{X} & \xrightarrow{\delta} & \mathrm{X} \times \mathrm{X} \\ {\scriptstyle\delta}\downarrow & & \downarrow{\scriptstyle 1 \times \delta} \\ \mathrm{X} \times \mathrm{X} & \xrightarrow[\delta \times 1]{} & \mathrm{X} \times \mathrm{X} \times \mathrm{X} \end{array} \qquad \begin{array}{ccc} \mathrm{X} \times \mathrm{X} & \xrightarrow{\sigma} & \mathrm{X} \times \mathrm{X} \\ {\scriptstyle\delta}\nwarrow & & \nearrow{\scriptstyle\delta} \\ & \mathrm{X} & \end{array}$$

where σ is the exchange of factors $(x, y) \mapsto (y, x)$.*

This yields the cohomology *algebra* $\mathrm{H}^{\bullet}(\mathrm{X}; \Lambda)$ over the field Λ, and its functorial character follows from the commutativity of the diagram

$$\begin{array}{ccc} \mathrm{X} & \xrightarrow{\delta_{\mathrm{X}}} & \mathrm{X} \times \mathrm{X} \\ {\scriptstyle f}\downarrow & & \downarrow{\scriptstyle f \times f} \\ \mathrm{Y} & \xrightarrow[\delta_{\mathrm{Y}}]{} & \mathrm{Y} \times \mathrm{Y} \end{array}$$

for any continuous map $f\colon \mathrm{X} \to \mathrm{Y}$.

Now suppose that on X there is given a continuous law of composition $m\colon \mathrm{X} \times \mathrm{X} \to \mathrm{X}$, usually written $m(x, y) = xy$. This defines a homomorphism of *graded algebras*

$$m^*\colon \mathrm{H}^{\bullet}(\mathrm{X}; \Lambda) \to \mathrm{H}^{\bullet}(\mathrm{X} \times \mathrm{X}; \Lambda) \tag{8}$$

and, for a field Λ, $\mathrm{H}^{\bullet}(\mathrm{X} \times \mathrm{X}; \Lambda)$ is naturally isomorphic to the skew tensor product†

---

* In $\mathrm{H}^{\bullet}(\mathrm{X} \times \mathrm{X}; \Lambda)$,

$$\sigma^*(u_p \times v_q) = (-1)^{pq} v_q \times u_p$$

if u_p is homogeneous of degree p and v homogeneous of degree q ([440], p. 233).

† For two graded algebras A, B over a field Λ, the underlying vector space of the skew product $\mathrm{A}^g \otimes_{\Lambda} \mathrm{B}$ is the tensor product $\mathrm{A} \otimes_{\Lambda} \mathrm{B}$, and the multiplication is defined by

$$(u_p \otimes v_q)(u_r \otimes v_s) = (-1)^{qr}(u_p u_r) \otimes (v_q v_s)$$

for homogeneous elements u_p, v_q, u_r, v_s of degrees p, q, r, s.

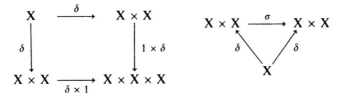

$$H^{\bullet}(X; \Lambda)^{g} \otimes_{\Lambda} H^{\bullet}(X; \Lambda). \tag{9}$$

Of course this applies to topological groups, but Hopf discovered that to prove Cartan's conjecture for Lie groups he needed much less. Suppose there is in X a "homotopy neutral element" e such that $e.e = e$, and for any $x \in X$ the maps $x \mapsto x.e$ and $x \mapsto e.x$ are homotopic to the identity. Such spaces are now called *H-spaces* (no other property of the "multiplication" m is assumed). Hopf showed that for such a space any homogeneous element $z \in H^{\bullet}(X; \Lambda)$ one may write

$$m^*(z) = 1 \otimes z + z \otimes 1 + \sum_{i,j} x_i \otimes y_j \tag{10}$$

where, in the sum, x_i and y_j are homogeneous elements (not uniquely determined) for which $\deg x_i > 0$, $\deg y_j > 0$, and $\deg z = \deg x_i + \deg y_j$. He then realized that *all* he needed for the proof of Cartan's conjecture was formula (10) ([246], p. 142). In fact he supposed even less on the "multiplication" in X, namely, that (for X compact) the *degrees* of the translations $y \mapsto x_0 y$ and $x \mapsto x y_0$ are both $\neq 0$ for all elements x_0, y_0. He proved that in that case (10) is replaced by

$$m^*(z) = 1 \otimes \lambda(z) + \mu(z) \otimes 1 + \sum_{i,j} x_i \otimes y_j \tag{11}$$

where λ and μ are now *automorphisms* of the graded algebra $H^{\bullet}(X; \Lambda)$. But if m^* is replaced by $m'^* = (\lambda^{-1} \otimes \mu^{-1}) \circ m^*$, this homomorphism satisfies (10). It is not hard to trace these generalized assumptions to Hopf's work of 1935 on the continuous maps $\mathbf{S}_{2k-1} \to \mathbf{S}_k$ (Part 3, chap. II, § 1,C), where he defined on $\mathbf{S}_k$ (for k odd) a "multiplication" for which the degrees of $x \mapsto x y_0$ and $y \mapsto x_0 y$ were ± 1 and ± 2. It follows from the Künneth formula that if two connected compact oriented manifolds M_1, M_2 have "multiplications" m_1, m_2 satisfying (11) for suitable automorphisms, then the same is true for the product $M_1 \times M_2$ for the multiplication

$$((x_1, x_2), (y_1, y_2)) \mapsto (m_1(x_1, y_1), m_2(x_2, y_2)).$$

The analogy Cartan observed between compact Lie groups and products of spheres thus appeared in a new light and became more understandable; it remained to be proved that the relation (10) for elements of $H^{\bullet}(X; \Lambda)$ and a field Λ are enough to determine the structure of the graded algebra $H^{\bullet}(X; \Lambda)$.

This is in fact a *purely algebraic* problem. A graded, associative algebra A with unit element 1 over a field Λ is now called a *Hopf algebra* if there exists a *homomorphism of graded algebras*

$$\Delta: A \to A^{g} \otimes_{\Lambda} A \tag{12}$$

(called a *comultiplication*) which, for every homogeneous element $z \in A$ of degree >0, satisfies a relation

$$\Delta(z) = 1 \otimes z + z \otimes 1 + \sum_{i,j} x_i \otimes y_j \tag{13}$$

with homogeneous elements x_i, y_j such that $\deg x_i > 0$, $\deg y_j > 0$, and $\deg z = \deg x_i + \deg y_j$. The theorem which Hopf proved in his 1939 paper can be formulated as follows:

If the field Λ *has characteristic* 0 *and* A *is a finite dimensional anticommutative Hopf algebra over* Λ, *with* $A^0 = \Lambda$, *then there is a vector subspace* E *of* A *having a basis* $(x_1, \ldots, x_l)$ *consisting of homogeneous elements of* odd *degrees, such that* A *is isomorphic to the exterior algebra* $\bigwedge(E)$.

His proof can be presented in the following way.

First observe that in a graded anticommutative algebra every (left or right) graded ideal is two-sided; if the algebra has finite dimension, every homogeneous element of degree >0 is nilpotent.

Let $\mathfrak{m}$ be the maximal ideal of A generated by the elements of degree >0, and for each $p \geqslant 1$, let E_p be a vector subspace of A^p *supplementary to* $A^p \cap \mathfrak{m}^2$. Let E be the direct sum of the E_p for $p \geqslant 1$, and consider a basis $\{x_1, \ldots, x_m\}$ of E, union of the bases of the E_p, with an ordering such that $\deg x_j \leqslant \deg x_{j+1}$ for all $j \geqslant 1$. It is enough to prove for each $k \leqslant m$ that:

(P_k) If $x_k \in E_p$, p is *odd*, the products $x_{j_1} x_{j_2} \cdots x_{j_h}$ for all sequences $j_1 < j_2 < \cdots < j_h \leqslant k$ are linearly independent, and the vector space generated by 1 and by these products contains A^q for $q < p$.

Then (P_m) will imply Hopf's theorem, because if Λ has characteristic 0, A is anticommutative and the x_j have odd degrees, then $x_j^2 = 0$, and $x_j x_k = -x_k x_j$ for $j \neq k$.

The proof is by induction on k (P_0 is void). Let B_{k-1} be the graded subalgebra of A having as *basis* over Λ the unit element 1 and the products $x_{j_1} x_{j_2} \cdots x_{j_h}$ for $j_1 < j_2 < \cdots < j_h \leqslant k-1$ (if $k = 1$, $B_0 = \Lambda . 1$); let $\mathfrak{a}_{k-1}$ be the graded ideal of A generated by $B_{k-1} \cap \mathfrak{m}$. Observe first that $x_k \notin \mathfrak{a}_{k-1}$, otherwise one would have

$$x_k = \sum_{j=0}^{r} c_j x_k^j$$

where each c_j is a sum of products of an element of $B_{k-1} \cap \mathfrak{m}$ and an element of A (which may be a scalar); equality of degrees implies $x_k = c_0$. But then x_k would be sum of elements λx_j ($\lambda \in \Lambda$, $j \leqslant k-1$) of B_{k-1} and of products of at least two elements of $\mathfrak{m}$; however, as $x_k \in E_p$, this would contradict the definition of E_p.

Next, let $\mathfrak{n}_{k-1}$ be the ideal $A \otimes \mathfrak{a}_{k-1}$ of $A \otimes A$. From (13) and the assumption (P_{k-1}),

$$\Delta(x_k) \equiv 1 \otimes x_k + x_k \otimes 1 \qquad \text{mod}\,.\,\mathfrak{n}_{k-1}, \tag{14}$$

$$\Delta(x_j) \equiv x_j \otimes 1 \qquad \text{mod}\,.\,\mathfrak{n}_{k-1} \text{ for } j \leqslant k-1. \tag{15}$$

The relations (15) imply that for any $b \in B_{k-1}$

$$\Delta(b) \equiv b \otimes 1 \qquad \text{mod}\,.\,\mathfrak{n}_{k-1}. \tag{16}$$

If p were *even*, a relation $x_k^r = 0$ for an $r \geqslant 2$ would be impossible, which is absurd. Indeed, suppose r is the *smallest* integer $\geqslant 2$ such that $x_k^r = 0$; as p is even, x_k is in the *center* of A, hence it follows from (14) that

$$0 = \Delta(x_k^r) \equiv (1 \otimes x_k + x_k \otimes 1)^r \qquad \text{mod}\,.\,\mathfrak{n}_{k-1}.$$

Taking into account that for the degrees of the second factor of $\mathrm{A} \otimes \mathrm{A}$, $\mathfrak{n}_{k-1}$ is a graded ideal, it follows that

$$0 \equiv r x_k^{r-1} \otimes x_k \qquad \text{mod}\,.\,\mathfrak{n}_{k-1}.$$

As we have seen that $x_k \notin \mathfrak{a}_{k-1}$, this is only possible if $r x_x^{r-1} = 0$, hence $x_k^{r-1} = 0$ since Λ has characteristic 0, and this contradicts the definition of r.

From anticommutativity of A and from the fact that p is odd, it follows that $x_k^2 = 0$; to prove that the $x_{j_1} \cdots x_{j_h}$ are linearly independent for $j_1 < \cdots < j_h \leqslant k$ it is enough to show that there is no relation

$$b_0 + b_1 x_k = 0 \quad \text{with } b_0, b_1 \in \mathrm{B}_{k-1}, \text{ and } b_1 \neq 0.$$

Again, from (14) and (16), it would follow that

$$0 = \Delta(b_0 + b_1 x_k) \equiv b_1 \otimes x_k \qquad \text{mod}\,.\,\mathfrak{n}_{k-1},$$

but the result $x_k \notin \mathfrak{a}_{k-1}$ shows that this is impossible if $b_1 \neq 0$.

Finally, if $x_{k-1} \in \mathrm{A}^p$, property (P_{k-1}) implies that $\mathrm{B}_{k-1} \supset \mathrm{A}^q$ for $q < p$. If not, $\mathrm{B}_{k-1} \supset \mathrm{A}^q$ for $q < p - 1$ by (P_{k-1}), and $\mathrm{E}_{p-1} \subset \mathrm{B}_{k-1}$, hence $\mathrm{A}^{p-1} \subset \mathrm{B}_k$ by definition.

Hopf's theorem was later extended by Leray [313] and A. Borel [58] to Hopf algebras for which it is only supposed that each A^n is finitely generated and the field of scalars Λ may be of arbitrary characteristic p, but is *perfect*. They established that there is again a system $\{x_1, x_2, \ldots, x_m\}$ of homogeneous linearly independent generators of A such that a basis of the vector space A consists of the unit element and of some of the products $x_1^{r_1} x_2^{r_2} \cdots x_m^{r_m}$. For each x_j let s_j, the *height* of x_j, be the smallest integer s such that $x_j^s = 0$, and $s_j = +\infty$ if no such integers exist; then in the preceding products each exponent r_j may take all values such that $0 \leqslant r_j \leqslant s_j$. Characteristic 2 is somewhat exceptional. If $p \neq 2$ and $\deg x_j$ is even, then $s_j = +\infty$ or s_j must be a power of p; if $p \neq 2$ and $\deg x_j$ is odd, then $s_j = 2$. If $p = 2$, s_j is always $+\infty$ or a power of 2.

Hopf algebras may also be defined over any ring, in particular over **Z**. Even when the algebra has no torsion and is defined over **Z** its structure may be very complicated ([55], p. 405). However, when a Hopf algebra A over **Z** has *finite rank* and no torsion Hopf's theorem still holds: A is isomorphic to an exterior algebra $\bigwedge(\mathrm{E})$, where E is a free **Z**-module having a finite basis of homogeneous elements of odd degrees.

B. Samelson's Theorem and Pontrjagin Product

In Hopf's theorem the subspaces $\mathrm{E}_p \subset \mathrm{A}^p$ are not uniquely determined. At the end of his paper Hopf put forward the conjecture that, in the particular case in which the H-space under consideration is a *group* G, there is a *canonical* system of generators of the algebra $H^{\bullet}(G; \Lambda)$ consisting of the *primitive* elements of that algebra, the homogeneous elements z for which

$$m^*(z) = 1 \otimes z + z \otimes 1. \tag{17}$$

Hopf's conjecture was proved by Samelson in 1941 [402]. The theorem can also be given a purely algebraic formulation. Suppose A is a finite dimensional anticommutative Hopf algebra over a field Λ, having the properties:

1. the comultiplication (12) is *coassociative*, that is, the diagram

$$\begin{array}{ccc} A & \xrightarrow{\Delta} & A \otimes A \\ {\scriptstyle\Delta}\downarrow & & \downarrow{\scriptstyle \Delta \otimes 1} \\ A \otimes A & \xrightarrow[1 \otimes \Delta]{} & A \otimes A \otimes A \end{array} \tag{18}$$

   commutes;
2. there are linearly independent homogeneous elements $x_1, x_2, \ldots, x_r$ with $\deg x_j \leqslant \deg x_{j+1}$, having odd degrees, and such that A is isomorphic to the exterior algebra $\bigwedge(E)$ over the vector subspace E generated by the x_j.

Then if E_0 is the vector subspace generated by the homogeneous *primitive* elements, those satisfying

$$\Delta(u) = 1 \otimes u + u \otimes 1, \tag{19}$$

the graded algebra A is also isomorphic to the exterior algebra $\bigwedge(E_0)$, and E_0 has a basis of r primitive elements u_j such that $\deg u_j = \deg x_j$ for $1 \leqslant j \leqslant r$.

Samelson assumed in his proof that Λ has characteristic 0, but Leray gave a proof valid for fields Λ of *arbitrary* characteristic [313]. The primitive elements u_j are constructed by induction on their degree. Suppose $u_1, \ldots, u_k$ have already been determined and that they generate the same subalgebra as $x_1, \ldots, x_k$; look for u_{k+1} of the form

$$u_{k+1} = x_{k+1} + P(u_1, \ldots, u_k) \tag{20}$$

where P is a polynomial, and determine P in such a way that u_{k+1} is primitive by writing

$$\Delta(x_{k+1}) = 1 \otimes x_{k+1} + x_{k+1} \otimes 1 + \sum_{H,K} c_{HK} u_H \otimes u_K \tag{21}$$

where, for any subset H of p elements $j_1 < j_2 < \cdots < j_p$ of $\{1, 2, \ldots, k\}$, $u_H = u_{j_1} \wedge u_{j_2} \wedge \cdots \wedge u_{j_p}$, and the summation is extended to all pairs of nonempty complementary subsets H, K of $\{1, 2, \ldots, k\}$. Suppose there is a coefficient $c_{H_0K_0} = 0$; writing

$$x'_{k+1} = x_{k+1} \pm c_{H_0K_0} u_{H_0} \wedge u_{K_0}$$

with a suitable sign, it is possible to show that the number of terms in

$$\Delta(x'_{k+1}) - 1 \otimes x'_{k+1} - x'_{k+1} \otimes 1$$

is *smaller* than the number of terms in

$$\Delta(x_{k+1}) - 1 \otimes x_{k+1} - x_{k+1} \otimes 1,$$

from which the theorem follows.

Samelson's theorem implies that the original form of Hopf's conjecture is true for all H-spaces X such that the "multiplication" m is associative; that condition may be expressed by saying that the diagram

$$\begin{array}{ccc} X \times X \times X & \xrightarrow{m \times 1} & X \times X \\ {\scriptstyle 1 \times m}\downarrow & & \downarrow{\scriptstyle m} \\ X \times X & \xrightarrow[m]{} & X \end{array} \tag{22}$$

is commutative, and therefore the corresponding diagram for the cohomology algebras

$$\begin{array}{ccc} H^{\bullet}(X;\Lambda) & \xrightarrow{m^*} & H^{\bullet}(X;\Lambda) \otimes H^{\bullet}(X;\Lambda) \\ {\scriptstyle m^*}\downarrow & & \downarrow{\scriptstyle m^* \otimes 1} \\ H^{\bullet}(X;\Lambda) \otimes H^{\bullet}(X;\Lambda) & \xrightarrow[1 \otimes m^*]{} & H^{\bullet}(X;\Lambda) \otimes H^{\bullet}(X;\Lambda) \otimes H^{\bullet}(X;\Lambda) \end{array} \tag{23}$$

is also commutative.

Samelson's proof used *duality*. If A is a finite-dimensional Hopf algebra over a field Λ and A^* is the dual *vector space*, the dual vector space of $A \otimes_{\Lambda} A$ is naturally identified with $A^* \otimes_{\Lambda} A^*$, so that

$$\langle x \otimes y, u \otimes v \rangle = \langle x, u \rangle \langle y, v \rangle. \tag{24}$$

To the comultiplication $\Delta\colon A \to A \otimes A$ there corresponds by duality its transpose

$${}^t\Delta\colon A^* \otimes A^* \to A^* \tag{25}$$

so that

$$\langle x, {}^t\Delta(u \otimes v) \rangle = \langle \Delta(x), u \otimes v \rangle. \tag{26}$$

The element ${}^t\Delta(u \otimes v)$ is called the *Pontrjagin product* of u and v in A^*, and can be denoted by $u \vee v$. The unit element of Λ is a unit element for that product, and $u \vee v$ is distributive on both sides with respect to addition. By duality the diagram (19) yields a commutative diagram

$$\begin{array}{ccc} A^* \otimes A^* \otimes A^* & \xrightarrow{{}^t\Delta \otimes 1} & A^* \otimes A^* \\ {\scriptstyle 1 \otimes {}^t\Delta}\downarrow & & \downarrow{\scriptstyle {}^t\Delta} \\ A^* \otimes A^* & \xrightarrow[{}^t\Delta]{} & A^* \end{array}$$

and therefore the product $u \vee v$ is associative if and only if the comultiplication Δ is coassociative. In that case the *primitive* elements of A are exactly those that are *orthogonal to the decomposable elements of* A^*, that is, those that are products of two or more homogeneous elements of degree $\geqslant 1$. If A has finite dimension and is isomorphic to $\bigwedge(E_0)$, A^* is also an exterior algebra over a vector subspace generated by homogeneous elements u_j^* such that $\deg(u_j^*) = \deg(u_j)$.

If X is an H-*space*, the multiplication $m\colon X \times X \to X$ defines in the *homology graded vector space* $H_\bullet(X;\Lambda)$ a multiplication

$$m_*\colon H_\bullet(X;\Lambda) \otimes_\Lambda H_\bullet(X;\Lambda) \to H_\bullet(X;\Lambda)$$

which is called the *Pontrjagin product in homology*, first considered by Pontrjagin in 1939 [377]; its transpose is the comultiplication m^* in the Hopf algebra $H^\bullet(X;\Lambda)$. If multiplication in X is associative, $H_\bullet(X;\Lambda)$ is an associative algebra; when $H^\bullet(X;\Lambda)$ is an exterior algebra over the space of primitive elements of odd degree, the same is true for $H_\bullet(X;\Lambda)$. There are compact Lie groups G for which $H_\bullet(G;\mathbf{F}_2)$ is not commutative ([55], p. 405).

C. Interpretation of the Rank in Cohomology

In 1940 Hopf completed his work on the homology of groups by giving a purely topological interpretation of the number l of spheres in the product (3), which has the homology of an H-space X that is a connected, compact, oriented manifold and has an *associative* multiplication. For such a space the k-th power map $p_k\colon x \mapsto x^k$ of X into itself is well defined for all integers $k \geqslant 1$. Hopf proved that the *degree* of that map is exactly k^l for all values of $k \geqslant 1$ ([247], pp. 152–174).

The proof relies on the fact that if u, v are two continuous maps of X into itself, the map $w\colon x \mapsto u(x)v(x)$ may be written as the composite

$$X \xrightarrow{\delta} X \times X \xrightarrow{u \times v} X \times X \xrightarrow{m} X \tag{27}$$

where δ is the diagonal and m is the multiplication. If u^*, v^* are the endomorphisms of $H^\bullet(X;\Lambda)$ corresponding to u and v, the map $z \mapsto w^*(z)$ can be factorized as

$$H^\bullet(X;\Lambda) \xrightarrow{m^*} H^\bullet(X;\Lambda) \otimes H^\bullet(X;\Lambda) \xrightarrow{u^* \otimes v^*} H^\bullet(X;\Lambda) \otimes H^\bullet(X;\Lambda) \xrightarrow{\delta^*} H^\bullet(X;\Lambda)$$

and this implies that for a *primitive* element z

$$w^*(z) = u^*(z) + v^*(z);$$

hence for every integer $k \geqslant 1$

$$p_k^*(z) = k.z.$$

But since $H^\bullet(X;\Lambda)$ can be identified with the exterior algebra $\bigwedge(x_1, x_2, \ldots, x_l)$ where the x_j are primitive elements, the cohomology space of maximum dimension has a basis consisting of the unique element $x_1 \wedge x_2 \wedge \cdots \wedge x_l$, hence the restriction of p_k^* to that space is multiplication by k^l, which establishes the result [chap. I, §3,A, formula (6)].

D. Remarks

We shall see in Part 3, chap. IV, § 4 how the use of fiber spaces and spectral sequences considerably increased the knowledge of homology of Lie groups, H-spaces, and homogeneous spaces ([346], [426]).

The concept of "Hopf algebra" now appears as a special case of a module A over a commutative ring Λ also equipped with a Λ-linear map $\Delta: A \to A \otimes_\Lambda A$, called a comultiplication, or with *both* a comultiplication *and* a multiplication (not necessarily associative) $A \otimes_\Lambda A \to A$ (the latter, when considered alone, giving on A a structure of *algebra*). Bourbaki has given the name of *cogebra* to the structure defined by a comultiplication, and the name of *bigebra* to the structure defined by both a multiplication and a comultiplication on the same Λ-module A; that module is *not necessarily graded* any more, and the multiplication and comultiplication of a bigebra are usually submitted to restrictive axioms linking them in various ways. It has been found that such structures abound in many parts of algebra.

§ 3. Action of Transformation Groups on Homology

A. Complexes with Automorphisms

In his thesis ([389], pp. 106–110) de Rham generalized Tietze's lens spaces (Part 1, chap. II, § 2) to higher dimensions. He considered a cyclic group J of arbitrary order h acting without fixed points on the sphere S_{2n+1} (identified with the subspace $\sum_{j=0}^{n} |z_j|^2 = 1$ of $\mathbf{C}^{n+1}$) by the rotations R^k $(0 \leqslant k \leqslant h-1)$, where R is given by

$$(z_0, z_1, \ldots, z_n) \mapsto (\zeta_0 z_0, \zeta_1 z_1, \ldots, \zeta_n z_n) \tag{28}$$

with ζ_k a primitive h-th root of unity such that $\zeta_k^{l_k} = e^{2\pi i/h}$, the l_k $(0 \leqslant k \leqslant n)$ being integers (mod. h) prime to h. The space of orbits $X = S_{2n+1}/J$ is written $L(l_0, l_1, \ldots, l_n)$ and called a *generalized lens space*; S_{2n+1} is its universal covering space, with $\pi_1(X) = J$. To compute the homology of X, de Rham considered a triangulation of S_{2n+1} consisting, for each dimension q with $0 \leqslant q \leqslant 2n+1$, of h q-cells $R^k a^q$ $(0 \leqslant k \leqslant h-1)$ deduced from a single one a^q by the action of J. The a^q are defined by the following relations between the coordinates of their points in S_{2n+1}:

$$\begin{cases} \text{for } q = 2r+1, & \sum_{j=0}^{r} |z_j|^2 = 1, \quad 0 < \arg z_r < 2\pi/h, \\ \text{for } q = 2r, & \sum_{j=0}^{r} |z_j|^2 = 1, \quad 0 = \arg z_r. \end{cases} \tag{29}$$

The boundary operator is then defined on the a^q by

$$\begin{cases} \mathbf{b}(a^{2r+1}) = (R^{l_r} - 1).a^{2r}, \\ \mathbf{b}(a^{2r}) = (1 + R + \cdots + R^{h-1}).a^{2r-1} \end{cases} \tag{30}$$

and on the other cells by

$$\mathbf{b}(\mathrm{R}^j . a^q) = \mathrm{R}^j . (\mathbf{b}(a^q)). \tag{31}$$

The definition of this triangulation shows that the projections of its cells on X form a triangulation of X. The chain complex $C_\bullet(X; \mathbf{Z})$ of that triangulation can therefore be identified with the submodule of $C_\bullet(\mathbf{S}_{2n+1}; \mathbf{Z})$ having as basis the sums

$$s_q = \sum_{j=0}^{h-1} \mathrm{R}^j . a^q \quad (0 \leqslant q \leqslant 2n+1) \tag{32}$$

with boundary operator

$$\mathbf{b}(s_q) = \sum_{j=0}^{h-1} \mathrm{R}^j . (\mathbf{b}(a^q)). \tag{33}$$

From these formulas de Rham also very simply derived the intersection numbers and the linking coefficients (when defined). of the elements of $H_\bullet(X; \mathbf{Z})$. For $n = 1$ he could thus recover in a much more natural and perspicuous manner Alexander's result showing that the lens spaces L(5, 1) and L(5, 2) are not homeomorphic. He generalized this result to the spaces $L(l_0, l_1, \ldots, l_n)$, proving that a necessary condition for $L(l_0, l_1, \ldots, l_n)$ and $L(l'_0, l'_1, \ldots, l'_n)$ to be homeomorphic is the existence of an integer m, prime to h and such that

$$l_0 l_1 \cdots l_n \equiv \pm m^{n+1} l'_0 l'_1 \cdots l'_n \pmod{h} \tag{34}$$

(see Part 3, chap. II, § 2,C).

A further step forward was taken by Reidemeister in 1935 [387]. His goal was to find other necessary conditions for the existence of a homeomorphism between two lens spaces $L(p,q), L(p,q')$. He considered, more generally, a finite cell complex K and a Galois covering $\tilde{\mathrm{K}}$ of K (Part 3, chap. I, § 2,VIII) with group $G = \mathrm{Aut}_{\mathrm{K}}(\tilde{\mathrm{K}})$, $\tilde{\mathrm{K}}$ being given the structure of cell complex described in Part 3, chap. I, § 3,B. The group G acts linearly on the graded $\mathbf{Z}$-module of chains $C_\bullet(\tilde{\mathrm{K}})$ [resp. on $C_\bullet(\tilde{\mathrm{K}}) \otimes_{\mathbf{Z}} \mathbf{C}$] preserving graduation, and commuting with the boundary operator. The chain complex $C_\bullet(\mathrm{K})$ [resp. $C_\bullet(\mathrm{K}) \otimes_{\mathbf{Z}} \mathbf{C}$] may then be identified with the graded submodule of $C_\bullet(\tilde{\mathrm{K}})$ [resp. $C_\bullet(\tilde{\mathrm{K}}) \otimes_{\mathbf{Z}} \mathbf{C}$] consisting of chains *invariant* under G. Since $C_\bullet(\tilde{\mathrm{K}})$ [resp. $C_\bullet(\tilde{\mathrm{K}}) \otimes_{\mathbf{Z}} \mathbf{C}$] can be considered as a graded module over the *group algebra* $\mathbf{Z}[G]$ [resp. $\mathbf{C}[G]$], that module structure completely determines $C_\bullet(\mathrm{K})$ [resp. $C_\bullet(\mathrm{K}) \otimes_{\mathbf{Z}} \mathbf{C}$] as the $\mathbf{Z}$-module (resp. $\mathbf{C}$-vector space) of $C_\bullet(\tilde{\mathrm{K}})$ [resp. $C_\bullet(\tilde{\mathrm{K}}) \otimes_{\mathbf{Z}} \mathbf{C}$] annihilated by all the elements $g - 1 \in \mathbf{Z}[G] \subset \mathbf{C}[G]$ for $g \in G$.

If the "Hauptvermutung" (Part 1, chap. II, § 2) holds for cell complexes of dimension $\leqslant n$, the condition for homeomorphy of two such complexes K, K′ would be the existence of two *subdivisions* K_1, K'_1 of these complexes that would be isomorphic as *combinatorial complexes*, a property that may be called *combinatorial homeomorphy*. In 1935 it was not known if the "Hauptvermutung" was true for dimensions >2; nevertheless, Reidemeister attacked the problem of combinatorial homeomorphy, and his solution was the starting point for further studies of the topology of generalized lens spaces.

Since a subdivision of K lifts to a subdivision of the covering complex $\tilde{K}$, *stable* under G, Reidemeister reformulated the problem: if $\tilde{K}$, $\tilde{K}'$ are two cell complexes on which acts the same group G, does there exist two subdivisions $\tilde{K}_1$, $\tilde{K}'_1$ of $\tilde{K}$, $\tilde{K}'$, stable under G, and such that the **Z**[G]-*modules* $C_\bullet(\tilde{K}_1)$ and $C_\bullet(\tilde{K}'_1)$ are isomorphic? He only treated the case $n = 3$, but his student W. Franz [199] and independently de Rham [391] extended his method to arbitrary finite cell complexes and arbitrary groups of automorphisms.

Their starting point is a *complex with automorphisms*, a pair (K, G) consisting of a finite cell complex K and a finite group G of homeomorphisms of K onto itself, which transform every cell of K into a cell of K, and commute with the boundary operator. If τ is a cell of K contained in the frontier of a cell σ, then for every $g \in G$, $g(\tau)$ is contained in the frontier of $g(\sigma)$. To explain Reidemeister's method we may assume, for simplicity, that G operates *freely*, that is, *no* cell of K is globally invariant by an element $g \neq 1_K$ of G; then every $C_p(K)$, considered as a **Z**[G]-*module*, is *free*. A basis of that module is obtained by taking a representative a_i^p ($1 \leqslant i \leqslant m_p$) in each orbit of G acting on the set of p-cells of K and the closure of the union of these orbits is then the p-skeleton K_p of K. From such a basis $(a_i^p)_{1 \leqslant i \leqslant m_p}$ of $C_p(K)$ a whole set $S_p(K)$ of bases of $C_p(K)$, called *distinguished bases*, is deduced by application to (a_i^p) of the group of matrices in $\mathbf{GL}(m_p, \mathbf{Z}[G])$ generated by: (1) the permutation matrices, (2) the diagonal matrices $\operatorname{diag}(g_1, \ldots, g_{m_p})$, where the g_j are elements of G, and (3) the "transvection" matrices $B_{ij}(\lambda)$ with diagonal elements equal to 1 and all other elements 0, except the element at the (i, j)-th place, which is equal to any $\lambda \neq 0$ in **Z**[G]. Clearly $S_p(K)$ does not depend on the particular set (a_i^p) of representatives of the orbits.

Two complexes (K, G), (K′, G) with actions of the same abstract group G are considered *isomorphic* if there is a **Z**[G]-*isomorphism* $C_\bullet(K) \xrightarrow{\sim} C_\bullet(K')$ that maps $S_p(K)$ onto $S_p(K')$ for each p.

The method consists in studying the passage from the **Z**[G]-module $C_\bullet(K)$ to $C_\bullet(K_1)$ when K_1 is a subdivision of K, stable under G; this is done by performing the subdivision one cell at a time. A cell complex L with action of G is called *trivial* if $C_k(L) = 0$ except for two values p, $p - 1$ of k and if the boundary operator is such that $\mathbf{b}(C_p(L)) = C_{p-1}(L)$ and $\mathbf{b}(C_{p-1}(L)) = 0$; then $C_\bullet(K_1)$ is isomorphic to a direct sum

$$C_\bullet(K) \oplus C_\bullet(L_1) \oplus \cdots \oplus C_\bullet(L_r)$$

where the L_j are *trivial*.

B. The Franz–Reidemeister Torsion

Reidemeister was chiefly interested in the problem of combinatorial homeomorphy of two lens spaces $L(p, q)$, $L(p, q')$ for prime p. Applying the preceding criterion directly to the triangulations defined by (29), he obtained the equations

$$(1 - \zeta^q)(1 - \zeta^{-q}) = (1 - \zeta^{q'})(1 - \zeta^{-q'}) \tag{35}$$

which must be satisfied for every p-th root of unity $\zeta \neq 1$, implying $q' = \pm q$.

Then Franz [199] and de Rham ([391], pp. 174–184) independently tackled the general problem and succeeded in defining, for some finite cell complexes K on which a finite group G acts *freely* and for each homomorphism θ: $\mathbf{C}[G] \to \mathbf{C}$ of the group ring $\mathbf{C}[G]$, an element of $\mathbf{C}$ which, up to multiplication by $\theta(\pm g)$ for some $g \in G$, is *invariant* by subdivisions of K and stable under G.

The idea is to split the chain complex $C_\bullet(K; \mathbf{C})$ with coefficients in $\mathbf{C}$, considered as a $\mathbf{C}[G]$-module, into the direct sum $C'_\bullet(K; \mathbf{C}) \oplus C''_\bullet(K; \mathbf{C})$, where $C'_\bullet(K; \mathbf{C})$ [resp. $C''_\bullet(K; \mathbf{C})$] is the direct sum of the $C_p(K; \mathbf{C})$ for all odd p (resp. even p). The boundary operator in $C_\bullet(K; \mathbf{C})$ can then be considered as a pair of homomorphisms of graded modules

$$\mathbf{b}': C'_\bullet(K; \mathbf{C}) \to C''_\bullet(K; \mathbf{C}), \qquad \mathbf{b}'': C''_\bullet(K; \mathbf{C}) \to C'_\bullet(K; \mathbf{C}). \tag{36}$$

Assume in addition that

$$\mathrm{Ker}(\mathbf{b}'') = \mathrm{Im}(\mathbf{b}'), \qquad \mathrm{Ker}(\mathbf{b}') = \mathrm{Im}(\mathbf{b}''). \tag{37}$$

{De Rham then called the $\mathbf{C}[G]$-complex K *acyclic*, but this does not mean that all Betti numbers of K of dimension $\geqslant 1$ are 0; for instance, the $\mathbf{C}[G]$-complex defined by (30) and (31), with underlying space $\mathbf{S}_{2n+1}$, is "acyclic" in De Rham's sense.}

Then use the following (trivial) lemma on finite-dimensional vector spaces over an arbitrary commutative field k:

Let $E' = M' \oplus N'$, $E'' = M'' \oplus N''$ be two finite-dimensional vector spaces over k such that

$$\dim M' = \dim N'' = m', \qquad \dim M'' = \dim N' = m'' \tag{38}$$

and let u': $N'' \xrightarrow{\sim} M'$, u'': $N' \xrightarrow{\sim} M''$ be two isomorphisms. Let r', r'', s', s'' be one-element bases of the respective exterior products

$$\bigwedge^{m'} M', \quad \bigwedge^{m''} M'', \quad \bigwedge^{m''} N', \quad \bigwedge^{m'} N'';$$

then

$$\left(\bigwedge^{m''} u'\right)(s'') = c'r', \qquad \left(\bigwedge^{m'} u''\right)(s') = c''r'' \tag{39}$$

for scalars c', c'' in k, and the *quotient* c'/c'' only depends on u', u'' and on the one-element bases

$$r' \wedge s' \in \bigwedge^{m'+m''} E', \qquad r'' \wedge s'' \in \bigwedge^{m'+m''} E''.$$

The lemma cannot be directly applied to $E' = C'_\bullet(K)$ and $E'' = C''_\bullet(K)$, since $A = \mathbf{C}[G]$ is not a field in general. However, if θ is a homomorphism of A onto $\mathbf{C}$, it extends to homomorphisms of graded modules

$$C'_\bullet(K) \to C'_\bullet(\theta), \qquad C''_\bullet(K) \to C''_\bullet(\theta)$$

where $C'_\bullet(\theta)$ and $C''_\bullet(\theta)$ are graded $\mathbf{C}$-vector spaces having as bases the a_i^p for p odd (resp. even), and yields corresponding homomorphisms

$$\mathbf{b}'_\theta;\ C'_\bullet(\theta) \to C''_\bullet(\theta), \qquad \mathbf{b}''_\theta\colon C''_\bullet(\theta) \to C'_\bullet(\theta).$$

Suppose the relations (37) are still satisfied when $\mathbf{b}'$, $\mathbf{b}''$ are replaced by $\mathbf{b}'_\theta$, $\mathbf{b}''_\theta$. Then the lemma applies to

$$E' = C'_\bullet(\theta), \qquad E'' = C''_\bullet(\theta),$$

$$M' = \operatorname{Ker} \mathbf{b}'_\theta, \qquad M'' = \operatorname{Ker} \mathbf{b}''_\theta, \qquad u' = \mathbf{b}'_\theta | N', \qquad u'' = \mathbf{b}''_\theta | N'',$$

and $r' \wedge s'$ (resp. $r'' \wedge s''$) is the exterior product of the vectors a_i^p for p odd (resp. even); this gives an element $c'_\theta / c''_\theta \in \mathbf{C}$. When the basis (a_i^p) is replaced by another distinguished basis of the set $S_p(K)$ for each p, c'_θ / c''_θ is multiplied by an element $\theta(\pm g)$, where $g \in G$ is *independent of* θ. The set of all elements c'_θ / c''_θ for all such homomorphisms θ is what de Rham called the *torsion* of the complex K with automorphism group G. Its fundamental property is that *it does not change when* K *is replaced by an arbitrary subdivision* stable under G.

In the case of a cyclic group G of order h with generator γ, $\mathbf{C}[G]$ is a direct sum of h fields isomorphic to $\mathbf{C}$, corresponding to the homomorphisms $\theta_\zeta\colon \mathbf{C}[G] \to \mathbf{C}$, such that $\theta_\zeta(\gamma) = \zeta$ for all h-th roots of unity ζ. If the action of G on K is given by (28), and conditions (37) are satisfied, then when the root of unity ζ is primitive, the value of c'_θ / c''_θ for $\theta = \theta_\zeta$ is

$$(\pm\zeta)^a \prod_{k=0}^{n} (\zeta^{l_k} - 1)$$

where a is some integer independent of the choice of ζ. If the complexes $L(l_0, l_1, \ldots, l_n)$ and $L(l'_0, l'_1, \ldots, l'_n)$ have the same torsion, the relation

$$\prod_{k=0}^{n} (\zeta^{l_k} - 1) = (\pm\zeta)^d \prod_{k=0}^{n} (\zeta^{l'_k} - 1)$$

holds for some integer d, hence, taking norms

$$\prod_{k=0}^{n} (\zeta^{l_k} - 1)(\zeta^{-l_k} - 1) = \prod_{k=0}^{n} (\zeta^{l'_k} - 1)(\zeta^{-l'_k} - 1) \tag{40}$$

for *all* primitive h-th roots of unity ζ. If h is prime, the theory of cyclotomic fields shows this implies the condition

$$\{l_1, l_2, \ldots, l_n\} = \{\pm l'_1, \pm l'_2, \ldots, \pm l'_n\} \text{ for some choice of signs.} \tag{41}$$

Using Dirichlet's results on L functions, Franz was able to prove this condition holds for *any* integer h [200].

This theory was later extended by J.H.C. Whitehead to the notions of simple homotopy type and what is now called Whitehead torsion (Part 3, chap. II, § 7).

C. Fixed Points of Periodic Automorphisms

Similar ideas were used by Paul Smith in his work on periodic automorphisms of a topological space ([395], [437]). The starting point is again a cyclic group G of order h consisting of homeomorphisms of a space X onto itself, and its action on the homology of X is studied, but the emphasis is different. The focus of interest is not the space of orbits X/G but the subset L of points of X *fixed under every* $g \in G$ (or equivalently the fixed points of a generator γ of the cyclic group G); furthermore, the goal is to deal with the most general possible spaces and with homeomorphisms that are not linked to triangulations.

P. Smith used Čech homology, and thus had to start with preliminary results on a simplicial complex with the group G acting as in the papers of Reidemeister and de Rham; he then took limits over the nerves of "special" open finite coverings of X invariant under G. He was not interested in the whole group $C_{\bullet}(K)$ of chains, considered as a $\mathbf{Z}[G]$-module, but only in the chains *annihilated* by one of the two elements of $\mathbf{Z}[G]$:

$$\tau = 1 - \gamma, \qquad \sigma = 1 + \gamma + \gamma^2 + \cdots + \gamma^{h-1}, \tag{42}$$

following some earlier ideas of M. Richardson [394]. After passage to the limit on "special" coverings of X this gives homology modules $\check{H}_k(\tau; \Lambda)$ and $\check{H}_k(\sigma; \Lambda)$ with an arbitrary ring of coefficients Λ. His efforts were especially directed to the case in which h is a *prime* number p and Λ is the field $\mathbf{F}_p$; by a subtle analysis of the action of σ and τ on chains he showed that these homology modules contain $\check{H}_k(L; \mathbf{F}_p)$ when $L \neq \emptyset$ and obtained many remarkable relations between them, enabling him to prove two famous theorems:

(i) Suppose X is locally compact and finite dimensional and that $\check{H}_k(X; \mathbf{F}_p) = 0$ for $k \geqslant 1$ and $\check{H}_0(X; \mathbf{F}_p) \simeq \mathbf{F}_p$; then there is *at least* one fixed point for any cyclic group G of order p^m.

(ii) Suppose X is a "homology n-sphere over $\mathbf{F}_p$," that is, X has the same homology as the sphere $\mathbf{S}_n$ with coefficients in $\mathbf{F}_p$. Then, if the set L of fixed points under a cyclic group G of order p^m is not empty, it is *also* a homology r-sphere over $\mathbf{F}_p$ for some r such that $0 \leqslant r \leqslant n$.

P. Smith's proofs are quite intricate. In 1952 Floyd showed how similar results could be formulated for sheaf cohomology in a simpler way [195]. The space X is locally compact and a finite group G acts on X. If Φ is a paracompactifying family on X (Part 1, chap. IV, §7,C) and $\pi: X \to X/G$ is the natural projection on the space of orbits X/G, that space is locally compact, π is open and proper, and there is the Leray spectral sequence (Part 1, chap. IV, §7,E) with E_2 terms

$$E_2^{pq} = H_{\Phi}^p(X/G; \mathscr{H}^q(\pi; \Lambda)) \tag{43}$$

for any ring of coefficients Λ; $\mathscr{H}^q(\pi; \Lambda)$ is the sheaf on X/G associated to the presheaf $U \mapsto H^q(\pi^{-1}(U); \Lambda)$, but as every fiber $\pi^{-1}(y)$ is finite for $y \in X/G$, $E_2^{pq} = 0$ for $q \neq 0$, and it follows from the spectral sequence that there is a natural isomorphism

$$H_{\Phi/G}^p(X/G; \mathscr{A}) \xrightarrow{\sim} H_{\Phi}^p(X; \Lambda) \tag{44}$$

where Φ/G is the family of sets $N \subset X/G$ such that $\pi^{-1}(N) \in \Phi$ (it is a paracompactifying family), and $\mathscr{A}$ is the sheaf on X/G associated to the presheaf $U \mapsto H^0(\pi^{-1}(U); \Lambda)$. As the spaces $\pi^{-1}(y)$ are the orbits of G, the group G acts as a group of automorphisms of the sheaf $\mathscr{A}$. In the situation considered by

P. Smith G is a cyclic group of *prime* order p and Λ is the field $\mathbf{F}_p$; Floyd introduced the elements (42) as operating on $\mathscr{A}$.

(I) He first considered the case in which the set L of fixed points of G in X is *empty*. By an argument similar to those used by Richardson and P. Smith [395], he showed that in the sequence of subsheaves

$$\mathscr{A} \supset \tau\mathscr{A} \supset \tau^2\mathscr{A} \supset \cdots \supset \tau^{p-1}\mathscr{A} \supset 0 \tag{45}$$

all quotients $\tau^k\mathscr{A}/\tau^{k+1}\mathscr{A}$ are isomorphic to the constant sheaf $\mathbf{F}_p$ on X/G. From the spectral sequence derived from the filtration (45), it follows that if $\dim_{\mathbf{F}_p} H^{\bullet}(X; \mathbf{F}_p) < +\infty$, then $\dim_{\mathbf{F}_p} H^{\bullet}(X/G; \mathbf{F}_p) < +\infty$ and

$$\chi(X; \mathbf{F}_p) = p \,.\, \chi(X/G; \mathbf{F}_p) \tag{46}$$

where, for any locally compact space Y, the "Euler characteristic" of Y over $\mathbf{F}_p$ is defined as $\chi(Y; \mathbf{F}_p) = \sum_n (-1)^n \dim_{\mathbf{F}_p} H^n(Y; \mathbf{F}_p)$.

(II) Suppose now that $L \neq \varnothing$, and write $\mathscr{A}_L$ the sheaf on X/G whose stalks are $\mathscr{A}_y$ for $y \in \pi(L)$ and 0 outside $\pi(L)$; there is then a surjective homomorphism $\eta: \mathscr{A} \to \mathscr{A}_L$ [Part 1, chap. IV, § 8,C, formula (128)]. It can immediately be seen that the sequences

$$0 \to \tau\mathscr{A} \to \mathscr{A} \xrightarrow{\sigma \oplus \eta} \sigma\mathscr{A} \oplus \mathscr{A}_L \to 0,$$
$$0 \to \sigma\mathscr{A} \to \mathscr{A} \xrightarrow{\tau \oplus \eta} \tau\mathscr{A} \oplus \mathscr{A}_L \to 0$$

are exact by looking at the stalks at each point of X/G. The corresponding cohomology sequences for the "*special*" *cohomology groups* $H^{\bullet}_{\Phi/G}(X/G; \tau\mathscr{A})$ and $H^{\bullet}_{\Phi/G}(X/G; \sigma\mathscr{A})$ [written for short $H^{\bullet}(\tau)$ and $H^{\bullet}(\sigma)$]:

$$\cdots \to H^n(\tau) \to H^n(X/G; \mathscr{A}) \to H^n(\sigma) \oplus H^n(X/G; \mathscr{A}_L) \to H^{n+1}(\tau) \to \cdots \tag{47}$$

$$\cdots \to H^n(\sigma) \to H^n(X/G; \mathscr{A}) \to H^n(\tau) \oplus H^n(X/G; \mathscr{A}_L) \to H^{n+1}(\sigma) \to \cdots \tag{48}$$

are called the *P. Smith sequences.*

From these exact sequences applied when $\dim_{\mathbf{F}_p} X < +\infty$ (chap. II, § 6), relations between dimensions of cohomology spaces are deduced:

$$\dim_{\mathbf{F}_p} H^n(X/G - \pi(L); \mathbf{F}_p) + \sum_{i \geqslant n} \dim_{\mathbf{F}_p} H^i(L; \mathbf{F}_p) \leqslant \sum_{i \geqslant n} \dim_{\mathbf{F}_p} H^i(X; \mathbf{F}_p), \tag{49}$$

$$\chi(X; \mathbf{F}_p) \equiv \chi(L; \mathbf{F}_p) \qquad \text{mod}\,.\, p, \tag{50}$$

this easily gives P. Smith's theorems as corollaries in the formulation of Floyd:

(i′) If $\dim_{\mathbf{F}_p} X < +\infty$, $H^i_{\Phi}(X; \mathbf{F}_p) = 0$ for $i \geqslant 1$, and $H^0_{\Phi}(X; \mathbf{F}_p) \simeq \mathbf{F}_p$, then L is not empty, $H^i_{\Phi}(L; \mathbf{F}_p) = 0$ for $i \geqslant 1$, and $H^0_{\Phi}(L; \mathbf{F}_p) \simeq \mathbf{F}_p$.

(ii′) If $\dim_{\mathbf{F}_p} X < +\infty$ and $H^{\bullet}_{\Phi}(X; \mathbf{F}_p) \simeq H^{\bullet}(\mathbf{S}_n; \mathbf{F}_p)$ for some n, then, when $L \neq \varnothing$, there exists an r with $0 \leqslant r \leqslant n$ and $n - r$ even, such that $H^{\bullet}_{\Phi}(L; \mathbf{F}_p) \simeq H^{\bullet}(\mathbf{S}_r; \mathbf{F}_p)$.

After 1945 the study of transformation groups was continued and greatly expanded with more sophisticated tools such as cohomology of groups, fiber spaces, and spectral sequences.

CHAPTER VII

Applications of Homology to Geometry and Analysis

§ 1. Applications to Algebraic Geometry

A. Early Applications

Algebraic geometry and algebraic topology have been linked from the beginning, since the first topological invariant, the genus of a compact Riemann surface, was introduced by Riemann in his path-breaking study of abelian integrals. His successors extended the use of similar notions to the theory of modular and automorphic functions.

During the years he devoted to the study of algebraic surfaces (1883–1906) E. Picard used all the means at his disposal, such as abelian integrals on curves, double integrals, linear systems, in a completely unsystematic way. We shall limit the description of his results and of those of his successors to purely topological properties of algebraic varieties.

As early as 1888 Picard was using Betti's "orders of connectivity" and "deforming two-dimensional cycles" on an algebraic surface considered as a four-dimensional real "variety." His most interesting results can be deduced from a method he invented, which will later be called the study of a "pencil" of algebraic curves on an irreducible surface S. Let

$$f(x, y, z) = 0 \tag{1}$$

be the equation of S in nonhomogeneous complex coordinates, and consider for every $y \in \mathbf{C}$ the plane curve C_y of equation $f(x, y, z) = 0$ between x and z. After a suitable linear change of coordinates it may be assumed that:

1. C_y is irreducible for all $y \in \mathbf{C}$.
2. C_y has the same genus p for all $y \in \mathbf{C}$, with the exception of a finite number of points a_k $(1 \leqslant k \leqslant N)$.
3. The plane $y = a_k$ has a unique point of contact A_k with S and A_k is a double point of C_{a_k} with distinct tangents, so that the genus of C_{a_k} is $p - 1$.

This tool is first used by Picard to study the 1-cycles on S. He showed that it is possible to transform by "deformation" any 1-cycle into a 1-cycle carried

J. Dieudonné, *A History of Algebraic and Differential Topology, 1900–1960*,
Modern Birkhäuser Classics, DOI 10.1007/978-0-8176-4907-4_13,

by a curve C_y for $y \neq a_k$ ([362], vol. I, p. 86). He always considered two 1-cycles deduced from one another by "deformation" to be "equivalent," and could thus restrict himself to those contained in the curves C_y. In the plane C of the variable y he took a fixed point a distinct from the a_k. Drawing loops l_k of origin a around each a_k, he investigated the problem of how a 1-cycle γ on C_a "varies continuously" on C_y when y varies around the loop l_k and returns to a. At first he limited himself to a surface of equation

$$z^2 = g(x, y) \tag{2}$$

where g is irreducible, and claimed that when γ is a 1-cycle around the two roots z_1, z_2 of (2) it returns to an equivalent cycle when y varies along all loops l_k, and when y tends to an a_k the cycle shrinks to a point, from which he concluded that the first Betti number of S is $R_1 = 0$ ([362], vol. I, p. 88). Next he claimed the same conclusion for a surface of equation

$$z^m = g(x, y) \tag{3}$$

for any integer $m \geqslant 2$, and finally any "general" surface S can be "deformed" into a surface of type (3) and therefore must also have $R_1 = 0$ ([362], vol. I, p. 91). However, in ([362], vol. I. p. 93) he said in substance that for any surface which is the product $C_1 \times C_2$ of two smooth projective irreducible curves of genus $\geqslant 1$, $R_1 \geqslant 2$.

He then returned to his original problem. Probably to feel on more secure ground, he at first substituted for the variation of a 1-cycle on C_y the variation of the period $\omega(y)$ along that cycle of an abelian integral of the second kind on C_y

$$I(y) = \int \frac{P(x, y, z)\,dx}{f'_z} \tag{4}$$

where the polynomial P is independent of the cycle γ. He could then avail himself of an earlier result of Fuchs, who had shown that, as functions of y, all periods of $I(y)$ are integrals of a linear differential equation (E) of order $2p$ with polynomial coefficients and for which the a_k are "regular" singular points in the sense of Fuchs. The variation of the 1-cycles is thus mirrored in the well-understood (since Riemann and Fuchs) variation of the integrals of (E) around the singular points a_k. In modern terms, the fundamental group $\pi_1(C - \{a_1, a_2, \ldots, a_N\})$ acts on the homology module $H_1(C_a; Z)$ that the Fuchs–Picard equation identifies with the module of periods of $I(a)$. The important fact recognized by Picard is that the image of $\pi_1(C - \{a_1, a_2, \ldots, a_N\})$ in the group of automorphisms $GL(2p, Z)$ may be a proper subgroup P of $GL(2p, Z)$. In other words, there is a submodule L of the module of periods of $I(a)$, consisting of elements *invariant* under the action if $\pi_1(C - \{a_1, a_2, \ldots, a_N\})$; Picard showed that the rank of that submodule is the *Betti number* R_1 of the surface S, and later he proved that R_1 is an *even* number ([362], vol. II, p. 423).

After reading Poincaré's first paper on *Analysis situs*, Picard was emboldened to study the variation of the 1-cycles themselves. He showed that when y varies in the loop l_k, a 1-cycle γ is transformed into a 1-cycle homologous to

$$\tau_k(\gamma) = \gamma + m\delta_k \tag{5}$$

where m is an integer and δ_k is a 1-cycle independent of γ ([362], vol. II, p. 334). He also considered the 2-cycles generated by the variation of a 1-cycle in C_y when y varies in a loop l_k ([362], vol. I, p. 108).

In 1902 Poincaré, who followed Picard's researches in algebraic geometry with great interest, entered the field himself with his third and fourth *Compléments*, using both Picard's method and his newly introduced concepts of combinatorial topology. With the help of the group P (which he called the "Picard group"), he gave, in the third *Complément*, a regular method to determine the fundamental group $\pi_1(S)$ for a surface S of equation (2) and applied it to explicit examples; in particular, he showed that $\pi_1(S) = 0$ if f is irreducible.

In the fourth *Complément*, Poincaré undertook to define a triangulation on a surface of equation (1) with ordinary singularities. Using the above notations he considered in the plane **C** the segments of origin a and extremity a_k for $1 \leqslant k \leqslant N$; the complement Q, in the Riemann sphere, of the union of these segments, can therefore be considered a polygon with 2N sides. When $y \in Q$, C_y has genus p, and is therefore always homeomorphic to a fixed Riemann surface R. Although Poincaré did not speak of products, he did take the product $\bar{Q} \times R$ and the cells that are the products of the cells of a triangulation of R with the vertex, the 2N sides and the interior Q of $\bar{Q}$. He then described at length how the "incidence matrices" of that triangulation of S can be determined for dimensions 0 to 4. In his description of Picard's method, he also observed that in formula (5) the cycles δ_k tend to a single point when a tends to a_k. For that reason he called them *evanescent* cycles.

B. The Work of Lefschetz

The preceding papers by Picard and Poincaré were written in the same style as Poincaré's first two *Compléments*, and were open to the same criticisms. The same is true of the first papers by Lefschetz on algebraic geometry written between 1915 and 1924 [299], as he himself realized somewhat later when he turned his attention to rigorous methods in algebraic topology. He then claimed that all his previous results in algebraic geometry could be given rigorous proofs, but he never bothered to write down these proofs himself. This job actually was done by A. Wallace in 1958 [480]; in that book the details of the technique are far from obvious, being quite long and intricate.

The novelty in Lefschetz's approach was first that in the study of the topology of an algebraic surface S, he could dispense with the Fuchs–Picard equation and work directly with topological notions. He took for granted the existence in some projective space $\mathbf{P}_N(\mathbf{C})$ with $N > 3$, of a smooth model S* of the function field of an algebraic irreducible surface S of equation (1) from which S could be recovered by a suitable projection on $\mathbf{P}_3(\mathbf{C})$. In 1932 [310] he pointed out, on simple examples, that the smooth models of the same surface S needed not be homeomorphic, contrary to what (according to him) was believed by many algebraic geometers. He explicitly observed that on S* there is a naturally defined orientation and that an algebraic curve on S* is not only a topological 2-chain but a topological *2-cycle*. Finally he made essential use of the theory of *intersections* outlined by Poincaré and of the elementary (but capital) remark that the topological intersection number $(C_1 . C_2)$ of two *algebraic curves* on S (considered as 2-cycles) (Part I, chap. II, §4) is in fact equal to the *number of common points* (taking into account multiplicities) of C_1 and C_2, as defined in algebraic geometry, when C_1 and C_2 are given their natural orientation.

This first enabled him to give the precise form of Picard's formula (5) for the variation of a 1-cycle γ:

$$\tau_k(\gamma) = \gamma + (\gamma . \delta_k)\delta_k. \tag{6}$$

Using this formula and Poincaré's construction of cells in S, Lefschetz showed that if s_k is the segment joining a and a_k, and Σ_k is the set of points of S that project on s_k, then there is a 2-chain Δ_k on Σ_k having as boundary the "evanescent" 1-cycle $\delta_k = \Delta_k \cap C_a$; the 1-cycle $\Delta_k \cap C_y$ is deduced from δ_k by continuity when y varies along the segment s_k, and $\Delta_k \cap C_{a_k}$ is the double point of C_{a_k}. He then described the homology groups $H_j(S; \mathbf{Z})$ for $j = 1, 2, 3$ with the help of the chains δ_k, Δ_k and a basis of the homology group $H_1(C_a; \mathbf{Z})$; in particular he gave a purely topological proof of the fact that homology classes of the δ_k form a submodule of rank $2p - R_1$, and of the fact that R_1 is an even number.

He then extended his methods to algebraic varieties V of arbitrary dimension n, contained in a complex projective space $\mathbf{P}_N(\mathbf{C})$. The pencil of curves introduced by Picard is now replaced by the sections $V \cap H_y$ of V by complex hyperplanes H_y passing through a well-chosen linear variety of complex dimension $N - 2$ in $\mathbf{P}_N(\mathbf{C})$. His fundamental result, now called the *weak Lefschetz theorem*, is that for general values of the complex parameter y the natural map

$$H_j(V \cap H_y; \mathbf{Z}) \to H_j(V; \mathbf{Z}) \tag{7}$$

is bijective for $0 \leqslant j \leqslant n - 2$ and surjective for $j = n - 1$. This implies for the Betti numbers the inequality $R_p \geqslant R_{p-2}$ for $p \leqslant n$. Lefschetz also showed that R_{2p+1} is always an even number (which may be 0), and finally that $R_{2p} > 0$ for even dimensional Betti numbers. This last result follows from the simple remark that an algebraic subvariety M of V, of complex dimension p, is a $2p$-cycle that cannot be homologous to zero, because it is easy to define an algebraic subvariety L of complex dimension $n - p$ that has a finite and nonempty intersection with M; this implies that the topological intersection number $(L . M) \geqslant 1$.

C. The Triangulation of Algebraic Varieties

In his papers on complex projective algebraic varieties of arbitrary dimension Lefschetz assumed that they could be triangulated. The first mathematician to have undertaken a proof was van der Waerden in 1930; he described in a single page ([478], p. 360) a bare outline of a method that was proved correct when it was elaborated later.

The idea is to prove first the existence of a triangulation on a *real* algebraic projective variety W, that is, the subset of a real projective space $\mathbf{P}_N(\mathbf{R})$ defined by a finite number of equations

$$F_k(x_0, x_1, \ldots, x_N) = 0$$

where the F_k are homogeneous polynomials with real coefficients. Using a stereographic projection, it is always possible to assume that W is a compact subset of $\mathbf{R}^N$. The proof is by induction on N; we describe it in a little more detail than the author, for the convenience of the reader.

The statement to be proved is:

(T_N) Given in $\mathbf{R}^N$ a finite set of compact real algebraic varieties V_α and a finite set of bounded open connected sets U_β such that the frontier in $\mathbf{R}^N$ of each U_β is a finite union of closed subsets of some V_α, then there is a finite curvilinear triangulation T_N of a cube C containing the V_α and the U_β such that each V_α, each U_β, and each frontier of a U_β is a union of cells of T_N.

Statement (T_1) is trivial. Assuming (T_{N-1}), it is always possible, by a linear change of nonhomogeneous coordinates in $\mathbf{R}^N$, to suppose that in the equations

$$H_k\colon f_k(x_1, x_2, \ldots, x_N) = 0$$

of the hypersurfaces used to define each V_α as an intersection of some of them there is a term of highest degree m_k that is a monomial $c \,.\, x_N^{m_k}$. For each point $x' \in \mathbf{R}^{N-1}$ the total number of points of intersection of the H_k with $\{x'\} \times \mathbf{R}$ is bounded. Let $\Omega \subset \mathbf{R}^{N-1}$ be the set of points x' for which no point in $\{x'\} \times \mathbf{R}$ belongs to two distinct H_k; then, for each $x \in H_k \cap (\{x'\} \times \mathbf{R})$, x is nonsingular on H_k and the projection $H_k \to \mathbf{R}^{N-1}$ is étale at x. The set Ω is open and its frontier in $\mathbf{R}^{N-1}$ is the union of a finite number of algebraic varieties V'_λ. Using (T_{N-1}), there is a triangulation T_{N-1} of a cube C′ containing the projections of the V_α and the U_β, each cell of which is contained in some V'_λ or some connected component U'_μ of Ω.

Van der Waerden assumed that for each cell σ of T_{N-1} the number of points of intersection of the union of the H_k with $\{x'\} \times \mathbf{R}$ is constant for all $x' \in \sigma$, that the locus of each of these points is a cell of same dimension p as σ, and that the union of the open segments on $\{x'\} \times \mathbf{R}$ having as extremities the points on the union of the H_k is a cell of dimension $p + 1$. These cells are those of the triangulation T_N.

To pass to complex projective algebraic varieties it is only necessary to embed the complex projective space $\mathbf{P}_N(\mathbf{C})$ as a real compact algebraic variety in some $\mathbf{R}^m$.

At the end of the proof van der Waerden claimed without any detail that it could be extended to analytic varieties, ignoring obvious difficulties such as the existence of infinitely many points forming a discrete set in an intersection of two analytic varieties of complementary dimensions. In his 1930 book ([304], p. 364) Lefschetz sketched a method essentially following the same strategy as van der Waerden but applicable to analytic varieties. Some details were still lacking, and were supplied in a long paper [91] by A.B. Brown and B. Koopman with considerable care in the handling of analytic functions of several complex variables, using extensively Osgood's classical treatise. Finally, still keeping the same general method of induction on the dimension, Lefschetz and J.H.C. Whitehead simplified the proof of Brown and Koopman in a quasidefinitive version of the triangulation theorem [311].

D. The Hodge Theory

In a series of papers beginning in 1930 W.V.D. Hodge inaugurated a new and powerful method of study of the homology of analytic and algebraic manifolds based on a combination of de Rham's theorems (Part 1, chap. III, § 3) and a generalization of the way Riemann had used harmonic functions in his theory of abelian integrals [237].

Any differential manifold M of dimension n can be equipped with many Riemannian metrics

$$ds^2 = \sum_{i,j} g_{ij}(u_1, u_2, \ldots, u_n) du_i du_j$$

(in local coordinates). Beltrami had shown that it is always possible for such a metric to define an operator (depending on the metric) that generalizes the usual laplacian on $\mathbf{R}^n$ and therefore gives rise to the notion of *harmonic functions* on the Riemannian manifold. Hodge showed that it is also possible to define a notion of *harmonic exterior differential form*: the metric on M canonically defines a metric on the tangent bundle T(M), hence also, by standard multilinear algebra, a metric on any bundle of tensors on M. In particular, let $(\alpha, \beta) \mapsto g_p(\alpha, \beta)$ be the positive nondegenerate symmetric bilinear form defined on the vector space of p-forms on M. When M is orientable, this defines a *duality* between p-forms and $(n - p)$-forms: to each p-form α is associated a $(n - p)$-form $* \alpha$, characterized by the relation

$$\beta \wedge (* \alpha) = g_p(\alpha, \beta) v \tag{8}$$

for all p-forms α, β, where v is the volume form on the Riemannian manifold M. If d is the exterior derivative, it has a *transposed* operator δ for that duality, defined by

$$\delta = -(*) \circ d \circ (*) \tag{9}$$

which maps p-forms onto $(p - 1)$-forms. The *Hodge laplacian*

$$\Delta = d \circ \delta + \delta \circ d \tag{10}$$

transforms p-forms into p-forms and generalizes Beltrami's laplacian which is its special case for $p = 0$ (up to sign). This defines *harmonic* (real or complex valued) *p-forms* as those for which $\Delta\alpha = 0$.

Hodge's fundamental theorem (in which he uses the theory of elliptic partial differential equations) is that when the Riemannian manifold M is *compact and orientable* there is *exactly one* harmonic p-form in every *de Rham cohomology class* in $\mathrm{H}^p(\mathrm{M}; \mathbf{C})$. In other words, $\mathrm{H}^p(\mathrm{M}; \mathbf{C})$ is naturally isomorphic to the vector space of complex valued harmonic p-forms, and the same is true for $\mathrm{H}^p(\mathrm{M}; \mathbf{R})$ and real valued harmonic p-forms. From this follows a very simple proof of Poincaré's duality theorem for compact orientable C^∞ manifolds, because if α is a harmonic p-form, $* \alpha$ is a harmonic $(n - p)$-form.

The importance of the Hodge theorem stems from its application to complex analytic manifolds. Such a manifold M of complex dimension n is a C^∞ real differential manifold of dimension $2n$. The existence of local complex coordinates with holomorphic transition maps between charts implies a natural *decomposition* of the complex differential p-forms on M: such a p-form is called a form of *type* (r, s) with $0 \leqslant r \leqslant n$, $0 \leqslant s \leqslant n$ and $r + s = p$ if for local complex coordinates $z_1, z_2, \ldots, z_n$ it can be written

$$\sum_{j_1, \ldots, j_r, k_1, \ldots, k_s} \mathrm{A}_{j_1 \cdots j_r k_1 \cdots k_s}(z_1, \ldots, z_n)\, dz_{j_1} \wedge \cdots \wedge dz_{j_r} \wedge d\bar{z}_{k_1} \wedge \cdots \wedge d\bar{z}_{k_s}.$$

This definition does not depend on the choice of local complex coordinates, and every p-form (for $0 \leqslant p \leqslant 2n$) can be uniquely written

$$\omega = \sum_{r+s=p} \omega^{r,s} \tag{11}$$

where $\omega^{r,s}$ has type (r, s) (with $\omega^{r,s} = 0$ if $r > n$ or $s > n$).

On each complex manifold M of complex dimension n there are *hermitian* metrics, that is, Riemannian metrics that can be expressed in complex local coordinates as

$$ds^2 = \sum_{j,k} h_{jk}\, dz_j\, d\bar{z}_k \quad \text{with } h_{kj} = \bar{h}_{jk}. \tag{12}$$

A sesquilinear hermitian form has an imaginary part that is a real alternating form; to the metric (12) is therefore naturally attached a *differential real 2-form*

$$\Theta = \frac{i}{2} \sum_{j,k} h_{jk}\, d\bar{z}_k \wedge dz_j. \tag{13}$$

In 1933 E. Kähler observed that for the complex projective space $\mathbf{P}_n(\mathbf{C})$, $d\Theta = 0$ for the corresponding 2-form of the classical hermitian metric of Fubini–Study, and the same is true for complex submanifolds of $\mathbf{P}_n(\mathbf{C})$ with the induced metric (they are smooth algebraic varieties by a theorem of Chow).

The complex manifolds on which there is a hermitian metric such that $d\Theta = 0$ are now called *kählerian*. Hodge showed that complex harmonic differential forms have remarkable special properties on these manifolds.

In the first place, the Hodge laplacian Δ permutes with the natural projections $\omega \mapsto \omega^{r,s}$. If ω is a complex harmonic p-form, so is $\omega^{r,s}$, and the space H^p of complex harmonic p-forms splits into a direct sum

$$\mathrm{H}^p = \bigoplus_{r+s=p} \mathrm{H}^{r,s} \tag{14}$$

where $\mathrm{H}^{r,s}$ is the subspace of harmonic p-forms of type (r, s). If M is *compact* and $h^{r,s} = \dim_{\mathbf{R}} \mathrm{H}^{r,s}$, the Betti number R_p is given by

$$\mathrm{R}_p = \sum_{r+s=p} h^{r,s} \quad \text{for } 1 \leqslant p \leqslant 2n \tag{15}$$

by the Hodge theorem. On the other hand, if ω has type (r, s) its complex conjugate $\bar{\omega}$ has type (s, r), so that

$$h^{s,r} = h^{r,s}, \tag{16}$$

giving Lefschetz's theorem that for *odd* p, R_p is an *even* number (which may be 0).

If Θ^k denotes the $2k$-form, the exterior product of k factors equal to the 2-form Θ, then Θ^k is harmonic and $\neq 0$ for $k \leqslant n$; this gives another of Lefschetz's results for compact kählerian manifolds:

$$\mathrm{R}_{2k} > 0 \quad \text{for } 1 \leqslant k \leqslant n. \tag{17}$$

Finally, on a compact kählerian manifold, Hodge showed that the linear map

$$\mathrm{L}\colon \omega \mapsto \Theta \wedge \omega \tag{18}$$

permutes with Δ, sending harmonic forms to harmonic forms. A harmonic p-form ω is called *primitive* if $p \leqslant n$ and

$$\mathrm{L}^{n-p+1}\omega = 0. \tag{19}$$

Let $\mathrm{P}^{r,s}$ be the subspace of $\mathrm{H}^{r,s}$ consisting of primitive forms of type (r, s) with $r + s \leqslant n$, and $\rho^{r,s}$ its dimension. If $r + s < n$,

$$\mathrm{H}^{r+1,s+1} = \mathrm{L}(\mathrm{H}^{r,s}) \oplus \mathrm{P}^{r+1,s+1} \tag{20}$$

and L is injective in $\mathrm{H}^{r,s}$; hence

$$h^{r+1,s+1} = h^{r,s} + \rho^{r+1,s+1}, \tag{21}$$

giving another of Lefschetz's theorems

$$\mathrm{R}_p - \mathrm{R}_{p-2} = \sum_{r+s=p} \rho^{r,s} \geqslant 0 \quad \text{for } p \leqslant n. \tag{22}$$

Hodge also showed that the linear map

$$\mathrm{L}^{n-p}\colon \mathrm{H}^p \to \mathrm{H}^{2n-p} \tag{23}$$

is *bijective* for $p \leqslant n$. For smooth algebraic subvarieties M of $\mathbf{P}_N(\mathbf{C})$ this can be interpreted in purely cohomological terms: if n is the complex dimension of M, the class of Θ in the de Rham group $H^2(M; \mathbf{C})$ is the Poincaré dual (chap. I, § 3,A) of the homology class in $H_{2n-2}(M; \mathbf{C})$ of a section of M by a hyperplane and L may be considered as the cup-product by that cohomology class. The fact that the map

$$\smile^{n-p}: H^p(M; \mathbf{C}) \to H^{2n-p}(M; \mathbf{C}) \tag{24}$$

is bijective is called the *strong Lefschetz theorem.*

By 1950 algebraic geometry had accumulated a large number of results, obtained by two very different techniques. One of these used "transcendental" means, such as triangulations and exterior differential forms and their integrals, and most results proved in this way were still valid for some types of nonalgebraic compact complex manifolds (e.g., kählerian manifolds). The other was based on algebraic notions, from which it was gradually realized that it was possible to expel all use of analysis so that the results could be extended to "abstract" algebraic geometry, the study of algebraic varieties defined over an arbitrary field. There were of course relations between both kinds of notions, going back to Picard, such as the fact that the first Betti number of a surface is given by $R_1 = 2q$, where q is the "irregularity," defined by the consideration of "adjoint" surfaces. Substantial parts of both techniques were unified by the introduction of the concepts of fiber bundles and sheaf cohomology, as we shall see in Part 3, chap. VII, §§ 2 and 3.

§ 2. Applications to Analysis

A. Fixed Point Theorems

In analysis it has been traditional since the nineteenth century to reduce existence problems to particular types of equations to which iterative processes can be applied. The best known example, which goes back to Cauchy and Liouville, is the problem of existence of a solution of a differential equation

$$y' = f(x, y)$$

taking a given value y_0 for $x = x_0$. This is equivalent to the existence of a solution of

$$y(x) = F(x, y(x)), \quad \text{with } F(x, y(x)) = y_0 + \int_{x_0}^{x} f(t, y(t))\, dt; \tag{25}$$

the well-known method of "successive approximations" consists in proving that under suitable assumptions the functions $y_n(x)$ defined recursively by $y_0(x) = y_0$ and $y_{n+1}(x) = F(x, y_n(x))$ converge to a solution of (25).

Banach, in his thesis, subsumed all applications of that method to a general theorem applicable to any Banach space, often called the *contraction principle*: let E be a Banach space, Ω be a *convex bounded open* subset of E, and F be a map of Ω into itself such that

$$\|F(x) - F(y)\| \leqslant k\|x - y\| \tag{26}$$

for x, y in Ω, k being a fixed number *such that* $0 \leqslant k < 1$; then there is a *unique* point $z \in \Omega$ such that $F(z) = z$, and it is obtained as the limit of the sequence $(x_n)_{n \geqslant 0}$ where $x_0 \in \Omega$ is arbitrary and $x_{n+1} = F(x_n)$ [42].

Applications of that theorem were therefore restricted to cases in which *both* existence and uniqueness must be proved. Brouwer's fixed point theorem (chap. III, § 1) was quite different, since the mapping had to satisfy much less stringent assumptions than (26) and only *existence* of a fixed point was proved, but its validity was limited to finite-dimensional spaces.

The first mathematicians who sought to prove fixed point theorems in *function spaces* were G.D. Birkhoff and O. Kellogg in 1922 [50]. They were apparently unaware of Brouwer's general theorem of 1911,* so they first proved it by approximating a continuous map $\mathbf{D}_n \to \mathbf{D}_n$ by a polynomial map. They then showed that the theorem could be extended to a continuous map $F: K \to K$, where K is a *convex compact* subset of the space $C(0, 1)$ of continuous functions in $[0, 1]$, or of the Hilbert space l^2. A little later [408] Schauder generalized their result to a convex compact subset of a Banach space E having a "basis" in the sense of Banach, then to a convex compact set in *any* Banach space [409], and finally to a *weakly compact* convex set K in a separable Banach space, F being *weakly continuous* [410].

The method common to all these papers consists in "approximating" K by a sequence of *finite-dimensional* compact convex subsets K_n and showing that Brouwer's theorem gives the existence in K_n of a point z_n such that $\|F(z_n) - z_n\|$ tends to 0 with $1/n$; using compactness of K, a subsequence of (z_n) tends to a fixed point of F. The set K_n is constructed by covering K with N balls of centers x_j $(1 \leqslant j \leqslant N)$ and radius $1/n$; K_n is the finite-dimensional convex hull of the set of the x_j, which is compact. Consider a sufficiently fine triangulation T of K_n such that for every simplex σ of T the diameter $\delta(F(\sigma)) \leqslant 1/n$. For each vertex y_i of T let $x_{j(i)}$ be one of the points such that $\|F(y_i) - x_{j(i)}\| \leqslant 1/n$, and define F_n as the simplicial map $(K_n, T) \to (K_n, T)$ such that $F_n(y_i) = x_{j(i)}$ for every vertex y_i. Then, for any simplex σ of T, $\delta(F_n(\sigma)) \leqslant 3/n$. By Brouwer's fixed point theorem there is a point $z_n \in K_n$ such that $F_n(z_n) = z_n$. If σ is the simplex of T containing z_n, there is a vertex y_j of σ such that $\|z_n - F_n(y_j)\| \leqslant 3/n$, hence $\|z_n - F(y_j)\| \leqslant 4/n$, and $\|z_n - F(z_n)\| \leqslant 5/n$.†

The interest of this method of course lies in the kind of applications it allows. We cannot enter into the details of the various analytic devices necessary for

* They only mentioned Brouwer's paper of 1910 on homeomorphisms of S_2 onto itself ([89], pp. 244–249).

† In [409] Schauder believed his proof valid for all metric complete topological vector spaces, not only for Banach spaces. But for these more general spaces his arguments are not correct, because it is not possible in general to give an upper bound of the diameter of the convex hull of the set $\{x_{j(i)}\}$ corresponding to the vertices y_i of a simplex σ of T, which would be independent of the dimension of σ.

each of these applications to define the compact set K and the map F to which the general theorem can be applied, and only mention two of these applications given by Schauder. The first one proves the existence of at least one solution of an equation

$$\Delta z = f\left(x, y, z, \frac{\partial z}{\partial x}, \frac{\partial z}{\partial y}\right) \tag{27}$$

in a domain $\Omega \subset \mathbf{R}^2$ with smooth boundary, vanishing on that boundary, under the only assumption that f is bounded and continuous for bounded values of the five variables on which it depends; the method consists in transforming (27) into an integrodifferential equation

$$z(x, y) = \iint_\Omega \mathrm{G}(x, y, \xi, \eta) f\left(\xi, \eta, z(\xi, \eta), \frac{\partial z}{\partial \xi}, \frac{\partial z}{\partial \eta}\right) d\xi\, d\eta \tag{28}$$

where G is the Green function for Ω [411].

The other application is to the local Cauchy problem for quasilinear hyperbolic equations

$$\sum_{i,k} \mathrm{A}_{ik}\left(x_1, \ldots, x_n, z, \frac{\partial z}{\partial x_1}, \ldots, \frac{\partial z}{\partial x_n}\right) \frac{\partial^2 z}{\partial x_i \partial x_k} = \mathrm{A}\left(x_1, \ldots, x_n, z, \frac{\partial z}{\partial x_1}, \ldots, \frac{\partial z}{\partial x_n}\right). \tag{29}$$

Here the idea is, for a *given* function $z(x_1, \ldots, x_n)$, to solve the local Cauchy problem for the *linear* hyperbolic equation in the unknown function Z

$$\sum_{i,k} \mathrm{A}_{ik}\left(x_1, \ldots, x_n, z, \frac{\partial z}{\partial x_1}, \ldots, \frac{\partial z}{\partial x_n}\right) \frac{\partial^2 \mathrm{Z}}{\partial x_i \partial x_k} = \mathrm{A}\left(x_1, \ldots, x_n, z, \frac{\partial z}{\partial x_1}, \ldots, \frac{\partial z}{\partial x_n}\right). \tag{30}$$

This solution $\mathrm{Z}(z)$ depends on the chosen function z, and the problem is to determine a convex compact set K in a suitable Banach space of functions E such that $z \mapsto \mathrm{Z}(z)$ maps K into itself and the fixed point theorem is applicable, so that there exists a function $u \in \mathrm{K}$ for which $\mathrm{Z}(u) = u$ [412].

In 1935 A. Tychonoff was able to extend Schauder's fixed point theorem to convex compact subsets K in *any Hausdorff locally convex space* E, using the Brouwer fixed point theorem in a new and ingenious way [472]. Although at that time uniform structures had not yet been defined, Tychonoff made use of one of the properties of the unique uniform structure on the compact space K: that for every entourage V of that structure there is an entourage V′ such that $\mathrm{V}'^2 \subset \mathrm{V}$. In terms of finite open coverings of K this means that for any such covering (W_α) of K there is a finer finite open covering (U_β) [the "half" of (W_α)] such that, for every U_β, there is a $\mathrm{W}_\alpha \supset \mathrm{U}_\beta$ such that any other U_γ which *meets* U_β is also contained in W_α (this is of course what Tychonoff proved directly).

The proof is by contradiction: assuming the continuous map $f\colon \mathrm{K} \to \mathrm{K}$ has no fixed point, Tychonoff easily showed, by compactness, that there is a finite open covering (W_α) of K such that the W_α are convex sets and $f(\mathrm{W}_\alpha) \cap \mathrm{W}_\alpha = \varnothing$

for each α. Let (U_β) be a corresponding "half" covering, take in each U_β a point x_β, and consider the finite-dimensional convex hull L of the set of the x_β. There is a simplicial triangulation T of L such that for every simplex σ of T there is an index β such that $f(\sigma) \subset U_\beta$. If y_j $(1 \leqslant j \leqslant n)$ are the vertices of a simplex σ of T, define $\varphi(y_j)$ equal to x_β, where U_β is *one* of the open sets containing $f(y_j)$. Then φ can be extended to a map of (L, T) in itself that is affine in every simplex of T; by Brouwer's fixed point theorem, there is a point $a \in L$ such that $\varphi(a) = a$. Let σ be the simplex of T such that $a \in \sigma$, and let y_j $(1 \leqslant j \leqslant p)$ be its vertices; there is a U_{β_0} such that $f(a)$ and *all* the $f(y_j)$ are in U_{β_0}. On the other hand, for each j, $f(y_j)$ and $\varphi(y_j)$ belong both to one U_{β_j}; since $U_{\beta_0} \cap U_{\beta_j} \neq \varnothing$ for $1 \leqslant j \leqslant p$, U_{β_0} and *all* U_{β_j} for $1 \leqslant j \leqslant p$ are contained in some W_α. However, since W_α is convex, $a = \varphi(a) \in W_\alpha$, and as $f(a) \in W_\alpha$, a contradiction has been reached.

B. The Leray–Schauder Degree

In 1930 Schauder was also interested in the Fredholm–Riesz theory of linear compact maps in normed spaces. This may have led him to generalize the concept of compact map to *nonlinear* ones, and to study, in the spirit of Riesz, "perturbations" of the identity of type $f: x \mapsto x - u(x)$, where u is a *nonlinear compact map*, a continuous map of an open subset Ω of a normed space E into a normed space F such that the image by u of any *bounded* subset of Ω is a *relatively compact* subset of F.

In 1932 [411] he showed that for a map of the preceding type f, E and F being Banach spaces, if in addition f is *injective*, then it maps open subsets of Ω onto open subsets of F. He applied this result to the proof of the following theorem. Consider the equation

$$\Delta z - f\left(x, y, z, \frac{\partial z}{\partial x}, \frac{\partial z}{\partial y}\right) = \psi(x, y) \tag{31}$$

and suppose that for an open subset U of $\mathbf{R}^2$ it has *at most* one solution taking given values $\varphi(s)$ on the boundary of U; then, if it is known that for particular functions φ_0, ψ_0 there *exists* such a solution, the same is true for functions φ, ψ sufficiently close to φ_0, ψ_0.

Then, in a famous paper [325] in 1934, Leray and Schauder discovered that for a map $f: x \mapsto x - u(x)$ with u a *compact* map, the Brouwer concept of *degree* could be defined, which gave new proofs of Schauder's previous theorems and had many other applications to the theory of partial differential equations.

A compact map $u: \bar{\omega} \to F$, with ω a *bounded* open subset of E, can be *approximated* arbitrarily by a continuous map of $\bar{\omega}$ into a *finite-dimensional* subspace of F:* By assumption $K = \overline{u(\omega)}$ is a compact subset of F, hence, for

* For a long time it was not known if it was possible to approximate a *linear* compact map by *linear* maps of finite rank. This approximation property holds in all "usual" function spaces, such as C(0, 1) or l^2; only recently have Banach spaces been found where it is not valid.

every $\varepsilon > 0$, there exist a finite number of points $y_k \in \mathrm{K}$ such that the balls of center y_k and radius ε cover K. Define in K the continuous functions

$$\mu_k(y) = \begin{cases} \varepsilon - \|y - y_k\| & \text{if } \|y - y_k\| \leqslant \varepsilon, \\ 0 & \text{if } \|y - y_k\| \geqslant \varepsilon. \end{cases} \tag{32}$$

Then for every $x \in \bar{\omega}$ the $\mu_k(u(x))$ are not all 0, hence the map

$$u_\varepsilon \colon x \mapsto \left(\sum_k \mu_k(u(x)) y_k\right) \Big/ \left(\sum_k \mu_k(u(x))\right) \tag{33}$$

takes its value in the vector subspace F_ε generated by the y_k, is defined and continuous in $\bar{\omega}$, and its definition implies that for $x \in \bar{\omega}$

$$\|u(x) - u_\varepsilon(x)\| \leqslant \varepsilon. \tag{34}$$

It is then possible to define the (local) *degree* $d(f, \omega, p)$ for any Banach space E, any map $x \mapsto f(x) = x - u(x)$ of a bounded open subset $\omega \subset \mathrm{E}$ into E, with u a compact map of $\bar{\omega}$ into E, and any point $p \in \mathrm{E}$ that does not belong to $f(\mathrm{Fr}(\omega))$ (by compactness of u, the image by f of any closed subset of $\bar{\omega}$ is closed in E). Approach u by maps u_ε as above, and let $\mathrm{E}_{n_\varepsilon}$ be the finite-dimensional subspace of E, of dimension n_ε, generated by p, the y_k and a point of ω. The map $f_\varepsilon \colon x \mapsto x - u_\varepsilon(x)$ is then defined in the bounded open subset $\omega_{n_\varepsilon} = \mathrm{E}_{n_\varepsilon} \cap \omega$ of the finite-dimensional vector space $\mathrm{E}_{n_\varepsilon}$ and maps ω_{n_ε} into $\mathrm{E}_{n_\varepsilon}$, so that the degree $d(f_\varepsilon, \omega_{n_\varepsilon}, p)$ is defined (chap. I, §3,D). Leray and Schauder defined

$$d(f, \omega, p) = d(f_\varepsilon, \omega_{n_\varepsilon}, p)$$

and had to prove the definition independent of the choices of ε and u_ε, provided ε is small enough.

This is a consequence of their second main lemma, which deals with finite-dimensional spaces: let ω_{n+1} be an open bounded subset of $\mathbf{R}^{n+1}$, such that $\omega_n = \omega_{n+1} \cap \mathbf{R}^n$ is not empty; let $\mathrm{F} \colon x \mapsto x + g_n(x)$ be a continuous map of $\bar{\omega}_{n+1}$ into $\mathbf{R}^{n+1}$, such that $g_n(x) \in \mathbf{R}^n$ for $x \in \bar{\omega}_{n+1}$, and let $b \in \mathbf{R}^n$ be a point that does not belong to $\mathrm{F}(\mathrm{Fr}(\omega_{n+1}))$; then, if f is the restriction of F to $\bar{\omega}_n$, it is a map of $\bar{\omega}_n$ into $\mathbf{R}^n$. Since $b \notin f(\mathrm{Fr}(\omega_n))$, the local degree $d(f, \omega_n, b)$ is defined and one shows

$$d(\mathrm{F}, \omega_{n+1}, b) = d(f, \omega_n, b). \tag{35}$$

The proof uses simplicial approximation and Brouwer's definition (chap. I, §3,D). Care must be taken in the choice of the triangulation T of ω_{n+1}: the $(n + 1)$-simplices must have no vertex in $\mathbf{R}^n$, and one of their n-dimensional faces must be in a hyperplane parallel to $\mathbf{R}^n$. The intersections of $\mathbf{R}^n$ and of the simplices of T that meet $\mathbf{R}^n$ then constitute a triangulation of ω_n, and the restriction to $\mathbf{R}^n$ of a simplicial approximation of F is a simplicial approximation of f; equation (35) follows easily.

The Leray–Schauder degree of course has all the properties of the Brouwer degree; in particular, if $d(f, \omega, p) \neq 0$, the equation $f(x) = p$ has at least one solution in ω. The properties of continuity of the degree show that, if $(x, \lambda) \mapsto u(x, \lambda)$ is a map of $\omega \times \mathrm{I}$ into E (where I is an interval in $\mathbf{R}$) that is uniformly continuous and such that each partial map $x \mapsto u(x, \lambda)$ is compact, then, if for *one* value $\lambda_0 \in I$, the equation $x = u(x, \lambda_0)$ has *exactly one* solution in ω, the equation $x = u(x, \lambda)$ has *at least one* solution in ω for every $\lambda \in I$. For further details, see [436a].

§ 3. The Calculus of Variations in the Large (Morse Theory)

The notion of geodesic curve on a surface goes back to Johann Bernoulli, who defined it as providing the minimum length of a curve on the surface between any two of its points. Their *local* theory was well understood after the work of the nineteenth century geometers on surfaces (Gauss, Jacobi, O. Bonnet), and later on Riemannian manifolds (Riemann, Christoffel, Levi-Civita). But two main *global* problems on geodesics on a Riemannian manifold M of dimension n can be formulated as follows: (1) Does an arc of geodesic with extremities p, q actually have minimum length among all rectifiable curves joining p and q (one then says it is a *minimal geodesic* between p and q)? (2) How many geodesic arcs are there joining two points of M?

Locally these problems have a complete answer: each point of M has an open neighborhood V such that for any two distinct points p, q of V there is exactly *one* arc of a geodesic *contained in* V and joining p and q, and it is the *unique minimal geodesic* between p and q.

Until 1920 the only general results on the global problem came from Jacobi's deep investigations of problem 1: he had shown that on a geodesic curve C of origin x_0, there exists in general a sequence of points $x_1, x_2, \ldots,$ the *conjugate* points of x_0 on C, such that any arc of C that does not contain any of the x_j for $j \geqslant 1$ is a minimal geodesic; but if it does contain an x_j, then in every neighborhood of C there exist piecewise smooth arcs joining two points p, q of C, the length of which is strictly smaller than the length of the arc of C between p and q.

In a series of papers beginning in 1928 Marston Morse attacked the preceding problems by a bold combination of differential geometry and algebraic topology applied to suitable function spaces, which he called "calculus of variations in the large." He considered the set

$$\Omega = \Omega(\mathrm{M}; p, q) \tag{36}$$

of *piecewise smooth paths* on M having fixed extremities p, q defined as continuous (*not* necessarily injective) maps $\gamma\colon [0, 1] \to \mathrm{M}$ such that $\gamma(0) = p$, $\gamma(1) = q$, and there are a finite number of points

$$t_0 = 0 < t_1 < t_2 < \cdots < t_{m-1} < t_m = 1 \tag{37}$$

such that in every *closed* interval $[t_i, t_{i+1}]$, γ is a C^∞ function. The parametrization is *always* chosen such that

$$\text{for } t_j \leqslant t \leqslant t_{j+1}, \quad t - t_j = \frac{t_{j+1} - t_j}{l_j} \int_{t_j}^{t} \left\| \frac{d\gamma}{du} \right\| du \quad \text{with } l_j = \int_{t_j}^{t_{j+1}} \left\| \frac{d\gamma}{du} \right\| du; \tag{38}$$

in other words, $t - t_j$ is proportional to the *length* of the image of $[t_j, t]$ by γ. Then

$$L(\gamma) = \sum_{j=0}^{m} l_j, \tag{39}$$

the *length* of γ, is a function of γ in Ω. A minimal arc from p to q should be a path γ for which $L(\gamma)$ is *minimum* in Ω, and a geodesic arc from p to q should be a path that is a "critical point" for the function L. This at first has no meaning, since Ω is not a differential manifold; the whole of Morse's theory consists in showing that it is possible to substitute for Ω genuine differential manifolds to which his results on critical points (chap. V, §4) can be applied.

Almost all of the ideas introduced by Morse in this process were new [including his investigation of critical points of a function defined on a smooth manifold (chap. V, §4)]. He applied his method not only to geodesics, but also to extremals of more general problems of the calculus of variations (in one variable), and to extremals joining both two points or two submanifolds. His proofs have gradually been simplified: in [422], Seifert and Threlfall replaced absolute cycles by relative ones and dropped the index, which was reintroduced by Bott, and finally J. Milnor improved on Bott's presentation by eliminating the construction of auxiliary hypersurfaces. We shall follow his streamlined description [345].

To study the geodesics joining two points p, q it is convenient, instead of working with the length $L(\gamma)$, to work with the *energy* of a path $\gamma\colon [a, b] \to M$, defined by

$$E_a^b(\gamma) = \int_a^b \left\| \frac{d\gamma}{du} \right\|^2 du \tag{40}$$

With the chosen parametrization (38) $E(\gamma) = (b - a)L(\gamma)^2$ and the extremals of E are again the geodesics, but the computations are easier with E. Milnor divided his presentation of the Morse theory into several steps.

Step 1 is essentially a presentation of the classical Lagrange method that brings to light the analogy with the critical points of a C^∞ function on M. No topology is put on Ω; a *variation* of a path $\gamma \in \Omega$ is a continuous map α into M, defined in a product $]-\varepsilon, \varepsilon[\times [0, 1]$ with the following properties

1. $\alpha(0, t) = \gamma(t)$,
2. $\alpha(u, 0) = p$, $\alpha(u, 1) = q$ for $-\varepsilon < u < \varepsilon$, and
3. there is a decomposition (37) such that α is C^∞ in each set

$$]-\varepsilon, \varepsilon[\times [t_i, t_{i+1}].$$

A *variation vector field* $t \mapsto W(t)$ is associated to each variation α, where $W(t)$ is a tangent vector in the tangent space $T_{\gamma(t)}M$ to M, defined by

$$W(t) = \frac{\partial \alpha}{\partial u}(0, t); \tag{41}$$

it is a continuous map of $[0, 1]$ into the tangent bundle T(M), *smooth* in each interval $[t_i, t_{i+1}]$. These maps [for all decompositions (37)] are the substitute for the "tangent vectors" at the "point" γ; they form an infinite-dimensional vector space written $T\Omega(\gamma)$.

More generally the interval $]-\varepsilon, \varepsilon[$ can be replaced in the definition of a variation by a neighborhood of 0 in some $\mathbf{R}^n$, defining an *n-parameter variation.*

A *critical path* $\gamma_0 \in \Omega$ for a function $F: \Omega \to \mathbf{R}$ is defined by the condition that for *every* variation α of γ_0 the function

$$u \mapsto F(\alpha(u, .))$$

is derivable for $u = 0$ and its derivative is 0.

Step 2 is again a modern presentation of the formulas of Riemannian geometry, giving the "first variation" and "second variation" of the energy $E(\gamma) = E_0^1(\gamma)$ of a path $\gamma \in \Omega$, which form the basis of Jacobi's results.

First consider an arbitrary path $\omega \in \Omega$, its "velocity" $V(t) = d\omega/dt$, and its "acceleration" in the Riemannian sense

$$A(t) = \nabla_t V(t) \tag{42}$$

(where ∇_t is the covariant derivative). They belong to $T_{\omega(t)}M$ for each $t \in [0, 1]$, are defined and continuous in each interval $[t_i, t_{i+1}]$ in which ω is smooth, and have limits at both extremities. Now let α be a variation of ω and $t \mapsto W(t)$ be the corresponding variation vector field (41). The "first variation formula" gives the first derivative

$$\frac{1}{2}\frac{d}{du}E(\alpha(u, .))|_{u=0} = -\sum_i (W(t_i)|V(t_i+) - V(t_i-)) - \int_0^1 (W(t)|A(t))\,dt \tag{43}$$

where $(x|y)$ is the scalar product of two vectors in a tangent space. It follows from this formula that $\gamma \in \Omega$ is a critical path for E if and only if γ is a (smooth) *geodesic.*

Next fix such a geodesic γ and consider a two-parameter variation:

$$\alpha: U \times [0, 1] \to M$$

where U is a neighborhood of 0 in $\mathbf{R}^2$, so that

$$\alpha(0, 0, t) = \gamma(t), \qquad \frac{\partial \alpha}{\partial u_1}(0, 0, t) = W_1(t). \qquad \frac{\partial \alpha}{\partial u_2}(0, 0, t) = W_2(t)$$

where W_1 and W_2 are in $T\Omega(\gamma)$. The "second variation" formula gives the mixed second derivative

$$\begin{aligned} &\frac{1}{2}\frac{\partial^2}{\partial u_1 \partial u_2}E(\alpha(u_1, u_2, .))|_{(0,0)} \\ &= -\sum_i (W_2(t_i)|\nabla_t W_1(t_i+) - \nabla_t W_1(t_i-)) \\ &\quad - \int_0^1 (W_2(t)|\nabla_t^2 W_1(t) + R(V(t) \wedge W_1(t)).V(t))\,dt \end{aligned} \tag{44}$$

where $Z \mapsto R(X \wedge Y).Z$ is the curvature of the Levi-Civita connection. The left-hand side of (44) is thus a *bilinear symmetric form*

$$(W_1, W_2) \mapsto E_{**}(W_1, W_2) \tag{45}$$

on the product $T\Omega(\gamma) \times T\Omega(\gamma)$. For a one-parameter variation α

$$E_{**}(W, W) = \frac{1}{2}\frac{d^2}{du^2} E(\alpha(u, .))|_{u=0}, \tag{46}$$

from which it follows that if γ is a *minimal* geodesic in Ω, $E_{**}(W, W) \geqslant 0$ in $T\Omega(\gamma)$. As usual, we shall speak of E_{**} indifferently as a symmetric bilinear form or as a quadratic form $W \mapsto E_{**}(W, W)$.

Formula (44) naturally leads to the junction with Jacobi's work: consider the smooth vector fields $t \mapsto J(t)$ along γ satisfying the equation

$$\nabla_t^2 J(t) + R(V(t) \wedge J(t)).V(t) = 0 \quad \text{for } 0 \leqslant t \leqslant 1. \tag{47}$$

With respect to a frame along γ moving by parallel translation this relation is equivalent to a system of n linear homogeneous differential equations of order 2 with C^∞ coefficients; the solutions J of (47) are called the *Jacobi fields* along γ and form a vector space of dimension $2n$. If for a value $a \in]0, 1]$ of the parameter t there exists a Jacobi field along γ that is not identically 0 but *vanishes for* $t = 0$ *and* $t = a$, then the points $p = \gamma(0)$ and $r = \gamma(a)$ are *conjugate* along γ with a *multiplicity* equal to the dimension of the vector space of Jacobi fields vanishing for $t = 0$ and $t = a$.

Jacobi fields may also be defined as variation vector fields for "geodesic variations" of the path γ: they are C^∞ maps

$$\alpha:]-\varepsilon, \varepsilon[\times [0, 1] \to M$$

such that for any $u \in]-\varepsilon, \varepsilon[$, $t \mapsto \alpha(u, t)$ is a geodesic and $\alpha(0, t) = \gamma(t)$ [no conditions are imposed on $\alpha(u, 0)$ and $\alpha(u, 1)$].

It can be proved that the Jacobi fields along γ that vanish at p and q [hence belong to $T\Omega(\gamma)$] are exactly the vector fields $J \in T\Omega(\gamma)$ such that

$$E_{**}(J, W) = 0 \quad \text{for } every\ W \in T\Omega(\gamma). \tag{48}$$

Although $T\Omega(\gamma)$ is infinite dimensional, the form E_{**} is again called *degenerate* if the vector space of the Jacobi fields vanishing at p and q is not reduced to 0, and the dimension of that vector space is called the *nullity* of E_{**}; E_{**} is thus degenerate if and only if p and q are conjugate along γ and the nullity of E_{**} is the multiplicity of q.

Step 3 is the beginning of Morse's contributions. He first considered a *fixed* geodesic $\gamma: [0, 1] \to M$ with extremities $p = \gamma(0)$, $q = \gamma(1)$ and the bilinear symmetric form $E_{**}: T\Omega(\gamma) \times T\Omega(\gamma) \to \mathbf{R}$. By analogy with the finite-dimensional quadratic form, the *index* of E_{**} is defined as the maximum dimension of a vector subspace of $T\Omega(\gamma)$ in which E_{**} is *strictly negative* [i.e., nondegenerate and taking values $E_{**}(W, W) < 0$ except for $W = 0$]. Morse's central result gives the value of the index of E_{**} and is known as the *index theorem.*

Suppose a subdivision (37) is chosen such that each arc $\gamma([t_{i-1}, t_i])$ is contained in an open set $U_i \subset M$ such that any two points of U_i are joined by a unique geodesic arc contained in U_i that is *minimal*; $\gamma([t_{i-1}, t_i])$ is such an arc. In the infinite-dimensional vector space $T\Omega(\gamma)$, consider the two vector subspaces:

(i) $T\Omega(\gamma; t_0, t_1, \ldots, t_m)$ consisting of all continuous vector fields $t \mapsto W(t)$ along γ, vanishing for $t = 0$ and $t = 1$, and such that each restriction $W|[t_{i-1}, t_i]$ is a *Jacobi field* (hence smooth) along $\gamma|[t_{i-1}, t_i]$; that subspace is finite dimensional;
(ii) T' consisting of the vector fields $t \mapsto W(t)$ along γ, such that $W(t_0) = 0$, $W(t_1) = 0, \ldots, W(t_m) = 0$.

$T\Omega(\gamma)$ is then the *direct sum* $T\Omega(\gamma; t_0, t_1, \ldots, t_m) \oplus T'$; these two subspaces are orthogonal for the bilinear form E_{**}, and E_{**} is *strictly positive* in T', so that the index of E_{**} is equal to the index of its restriction to the subspace $T\Omega(\gamma; t_0, t_1, \ldots, t_m)$. This follows easily from a suitable construction of "variations" along γ and the fact that a "broken Jacobi field" in $T\Omega(\gamma; t_0, t_1, \ldots, t_m)$ is uniquely defined by its values at $t_0, t_1, \ldots, t_m$ (in particular T' contains no such field except 0).

To compute the nullity and index of E_{**}, due to this decomposition, apply their definitions either to vector subspaces of $T\Omega(\gamma)$ or to vector subspaces of $T\Omega(\gamma; t_0, t_1, \ldots, t_m)$. The computation of the index of E_{**} is done by considering the geodesic arc γ_τ: $[0, \tau] \to M$, the restriction of γ to $[0, \tau]$, and its energy

$$E(\gamma_\tau) = \tau \int_0^\tau \left\| \frac{d\gamma}{du} \right\|^2 du.$$

E_{**}^τ is the corresponding quadratic form on $T\Omega(\gamma_\tau)$, and $\lambda(\tau)$ its index; one studies the variation of $\lambda(\tau)$ when τ varies from 0 to 1, and $\lambda(1)$ is the index of E_{**}.

From the fact that γ_τ is a minimal geodesic for small enough τ it follows that $\lambda(\tau) = 0$ in a neighborhood of 0. The space $T\Omega(\gamma_\tau)$ can be identified to a subspace of $T\Omega(\gamma_{\tau'})$ for $0 < \tau < \tau'$ by extending any $W \in T\Omega(\gamma_\tau)$ to a vector field $W' \in T\Omega(\gamma_{\tau'})$ such that $W'(t) = 0$ for $\tau \leqslant t \leqslant \tau'$. This easily implies that λ is *increasing* in $[0, 1]$.

The crucial part of the proof is the study of the continuity of λ at a point $\tau \in [0, 1]$, first on the left and then on the right. There is continuity on the left; more precisely

$$\lambda(\tau - \varepsilon) = \lambda(\tau) \tag{49}$$

for sufficiently small $\varepsilon > 0$. Here it is convenient to consider only broken Jacobi fields. First suppose $\tau < 1$, and choose a subdivision (37) such that $t_i < \tau < t_{i+1}$. Then, for ε small enough and $t_i < \tau - \varepsilon \leqslant \tau' \leqslant \tau + \varepsilon < t_{i+1}$, the spaces $T\Omega(\gamma_{\tau'}; t_0, t_1, \ldots, t_i, \tau')$ are identified to subspaces of same dimension ni of $T\Omega(\gamma; t_0, \ldots, t_m)$ that vary continuously with τ' (in the corresponding Grassmannian), and $E_{**}^{\tau'}$ is a quadratic form on the space $T\Omega(\gamma_{\tau'}; t_0, \ldots, t_i, \tau')$

that also "varies continuously" with τ' in a sense easy to make precise by choice of bases. It is then clear that if E^{τ}_{**} is strictly negative on a subspace of $T\Omega(\gamma_\tau; t_0, t_1, \ldots, t_i, \tau)$ of dimension $\lambda(\tau)$, $E^{\tau'}_{**}$ is strictly negative on a subspace of $T\Omega(\gamma_{\tau'}; t_0, t_1, \ldots, t_i, \tau')$ of same dimension, so that $\lambda(\tau') \geqslant \lambda(\tau)$ for $\tau - \varepsilon \leqslant \tau' \leqslant \tau + \varepsilon$. As λ is increasing, this proves (49). The proof for $\tau = 1$ is done by a similar argument.

The last part of the proof consists in showing that for $\varepsilon > 0$ sufficiently small

$$\lambda(\tau + \varepsilon) = \lambda(\tau) + \nu, \tag{50}$$

where ν is the *nullity* of E^{τ}_{**}. From the preceding study of E^{τ}_{**} on broken Jacobi fields it follows that there is a subspace of $T\Omega(\gamma_\tau; t_0, \ldots, t_i, \tau)$ of dimension $ni - \lambda(\tau) - \nu$ in which E^{τ}_{**} is strictly positive; the same argument of continuity then proves that for $\tau - \varepsilon \leqslant \tau' \leqslant \tau + \varepsilon$, $\lambda(\tau') \leqslant \lambda(\tau) + \nu$; it remains only to prove that

$$\lambda(\tau + \varepsilon) \geqslant \lambda(\tau) + \nu. \tag{51}$$

This is done by returning to the consideration of *arbitrary* subspaces of $T\Omega(\gamma_\tau)$. There are $\lambda(\tau)$ linearly independent vector fields $W_1, W_2, \ldots, W_{\lambda(\tau)}$ in that space, spanning a subspace in which E^{τ}_{**} is strictly negative, and these vector fields can be identified as above to vector fields of $T\Omega(\gamma_{\tau+\varepsilon})$. On the other hand, there are ν linearly independently (smooth) Jacobi fields $J_1, \ldots, J_\nu$ defined in $[0, 1]$ and vanishing for $t = 0$ and $t = \tau$; let $J_h^{\tau+\varepsilon}$ $(1 \leqslant h \leqslant \nu)$ be their restrictions to $[0, \tau + \varepsilon]$. By the second variation formula (44)

$$\begin{cases} E^{\tau+\varepsilon}_{**}(J_h^{\tau+\varepsilon}, W_i) = 0 & \text{for } 1 \leqslant h \leqslant \nu,\ 1 \leqslant i \leqslant \lambda(\tau), \\ E^{\tau+\varepsilon}_{**}(J_h^{\tau+\varepsilon}, X_k) = 2\delta_{hk} & \text{for } 1 \leqslant h, k \leqslant \nu, \end{cases} \tag{52}$$

where the X_k are suitably chosen vector fields in $T\Omega(\gamma_{\tau+\varepsilon})$. It is then easy to prove that $E^{\tau+\varepsilon}_{**}$ is strictly negative in the vector space having as basis

$$W_1, \ldots, W_{\lambda(\tau)}, c^{-1}J_1 - cX_1, \ldots, c^{-1}J_\nu - cX_\nu$$

for sufficiently small $c > 0$, proving (51).

From these properties the index theorem immediately follows: *the index of* E_{**} *is the sum of the multiplicities of the points conjugate to* p *along* γ *and distinct from* q.

We have seen that the dimension of $T\Omega(\gamma; t_0, t_1, \ldots, t_m)$ is finite; it follows that the index of E_{**} is always *finite*, and therefore the number of points conjugate to p along γ is also *finite*.

Step 4 of Morse theory introduces a *topology* on the set $\Omega = \Omega(M; p, q)$. On every connected Riemannian manifold M the usual topology can be defined by a *distance* $\rho(x, y)$, the g.l.b. of the lengths of all piecewise smooth paths joining x and y. For any pair of paths ω_1, ω_2 in $\Omega(M; p, q)$, consider the function

$$d(\omega_1, \omega_2) = \sup_{0 \leqslant t \leqslant 1} \rho(\omega_1(t), \omega_2(t)) + \left(\int_0^1 \left(\frac{ds_1}{dt} - \frac{ds_2}{dt}\right)^2 dt\right)^{1/2} \tag{53}$$

where $s_1(t)$ [resp. $s_2(t)$] is the length of the path $\tau \mapsto \omega_1(\tau)$ [resp. $\tau \mapsto \omega_2(\tau)$]

defined in $[0, t]$. This is a *distance* on Ω such that the function

$$\omega \mapsto \mathrm{E}_a^b(\omega)$$

is *continuous* for that distance.

Now suppose that the Riemannian manifold M is *complete*. Let $c > 0$ be a number such that the subset Ω^c of Ω consisting of the piecewise smooth paths for which $\mathrm{E}(\omega) < c$ is not empty. Consider a subdivision (37) and the set $\mathrm{B} \subset \Omega^c$ consisting of the piecewise smooth paths ω such that each restriction $\omega|[t_{i-1}, t_i]$ is a (smooth) geodesic for $1 \leqslant i \leqslant m$. It is possible to take the subdivision (37) such that for each broken geodesic $\omega \in \mathrm{B}$ each restriction $\omega|[t_{i-1}, t_i]$ is the unique minimal arc between $\omega(t_{i-1})$ and $\omega(t_i)$ and depends continuously on these two points. A broken geodesic $\omega \in \mathrm{B}$ is then uniquely determined by the $(m - 1)$-tuple

$$(\omega(t_1), \omega(t_2), \ldots, \omega(t_{m-1})) \in \mathrm{M}^{m-1} \tag{54}$$

and the map $\omega \mapsto (\omega(t_1), \omega(t_2), \ldots, \omega(t_{m-1}))$ is a *homeomorphism* of the subset B of the space Ω^c onto an open subset of M^{m-1}; pulling back the C^∞ structure of that open subset defines B as a C^∞ *manifold.*

Let E′ be the restriction of E to B. Then:

(i) E′ is a C^∞ map $\mathrm{B} \to \mathbf{R}$, and for each $a < c$ the inverse image $\mathrm{E}'^{-1}([0, a])$ is compact.

(ii) The set $\mathrm{E}'^{-1}([0, a])$ is a *strong deformation retract* (Part 1, chap. IV, § 6B) of the set $\mathrm{E}^{-1}([0, a])$. This is proved by a construction of Morse: for $t_{i-1} \leqslant u \leqslant t_i$ and any $\omega \in \mathrm{E}^{-1}([0, a])$, $r_u(\omega)$ is the piecewise smooth path from p to q such that
   a. $r_u(\omega)|[t_{j-1}, t_j]$ coincides with the unique minimal geodesic between $\omega(t_{j-1})$ and $\omega(t_j)$ for all $j \leqslant i - 1$;
   b. $r_u(\omega)|[t_{i-1}, u]$ is the minimal geodesic from $\omega(t_{i-1})$ to $\omega(u)$;
   c. $r_u(\omega)|[u, 1]$ is the restriction $\omega|[u, 1]$.
   It is then easy to see that r_u is a homotopy defining $\mathrm{E}'^{-1}([0, a])$ as strong deformation retract of $\mathrm{E}^{-1}([0, a])$.

(iii) The critical points of E′ in B are the same as those of E in $\mathrm{E}^{-1}([0, c])$, namely, the smooth geodesics from p to q of length $< \sqrt{c}$. Along each of these geodesics the index (resp. nullity) of E'_{**} is the same as the index (resp. nullity) of E_*.

This ends the reduction made by Morse to the finite-dimensional situation:

Suppose that on the complete Riemannian manifold M *the points p, q are not conjugate along* any *geodesic of length* $\leqslant \sqrt{a}$. *Then the homology of the space* $\mathrm{E}^{-1}([0, a])$ *is the same as that of a compact* C^∞ *manifold on which is defined a* C^∞ *function* E′ *with a finite number of critical points. Each of these points corresponds to a geodesic joining p and q, and the index of each critical point of* E′ *is equal to the index of* E_{**} *along the corresponding geodesic.*

In particular only a finite number of geodesics of length $\leqslant \sqrt{a}$ can join p and q; the example of antipodal points on a sphere $\mathbf{S}_n$ shows the necessity of the condition that p and q should not be conjugate along a geodesic joining them.

In order to apply this theorem, it is necessary to know that for every point $p \in \mathrm{M}$ there are points $q \in \mathrm{M}$ such that q is not conjugate to p along any geodesic. The "exponential" map $\exp_p$: $\mathrm{T}_p(\mathrm{M}) \to \mathrm{M}$ associates to each tangent vector $v \in \mathrm{T}_p(\mathrm{M})$ the extremity of the geodesic arc γ: $[0, 1] \to \mathrm{M}$ such that $\gamma(0) = p$ and $d\gamma/dt(0) = v$. As M is complete, this map is defined everywhere in $\mathrm{T}_p(\mathrm{M})$, and it is easy to show that $q = \exp_p(v)$ is conjugate to p along γ if and only if the point v is critical for the map $\exp_p$. It then follows from Sard's theorem (Part 1, chap. III, § 1) that the points $q \in \mathrm{M}$ which are not conjugate to p along *any* geodesic form a *dense* subset of M.

After 1950 it was belatedly realized that the results of Morse had unexplored potentialities that brought substantial process in homotopy theory (Part 3, chap. V, § 6B).

Part 3 Homotopy and its Relation to Homology

Introduction

In this part I retrace the history (until around 1960) of two fundamental concepts, *homotopy groups* and *fibrations*, and of their interactions both between themselves and with homology and other structures.

I) The concept of homotopy may be traced back to Lagrange's method in the Calculus of variations. The vague intuitive idea of "deformation" is found (without any definition) in many mathematical papers during the nineteenth century and even later. Brouwer, in 1911, is the first to have given the general definition of a homotopy between two continuous maps $f: X \to Y$, $g: X \to Y$; it is a continuous map

$$F: X \times [0,1] \to Y$$

such that $F(x,0) = f(x)$ and $F(x,1) = g(x)$ for all $x \in X$. It is clear that the existence of a homotopy is an equivalence relation in the space $\mathscr{C}(X,Y)$ of continuous maps of X into Y.

Most results in homotopy theory are concerned with *pointed spaces* (X, x_0) with a privileged point $x_0 \in X$; they form a category when morphisms are defined as continuous maps $f: (X, x_0) \to (Y, y_0)$ subject to the condition $f(x_0) = y_0$. A homotopy between two such maps f, g is then a continuous map

$$F: (X, x_0) \times [0,1] \to (Y, y_0)$$

such that $F(x,0) = f(x)$, $F(x,1) = g(x)$, and $F(x_0, t) = y_0$ for all $t \in [0,1]$. The existence of such a map again defines an equivalence relation; the equivalence class of f is written $[f]$, and the set of equivalence classes $[X, x_0; Y, y_0]$ or merely $[X; Y]$. In what follows only pointed spaces are considered.

II) The idea of homotopy groups appeared in Poincaré's paper of 1895. He defined $\pi_1(X, x_0)$ as the set $[\mathbf{S}_1, *; X, x_0]$*; a representative of a class of that set, called a *loop*, can be considered as a continuous map $\gamma: [a,b] \to X$ such

---

* $\mathbf{S}_n$ is defined as the subset of $\mathbf{R}^{n+1}$ satisfying the equation

$$x_0^2 + x_1^2 + \cdots + x_n^2 = 1$$

and the chosen privileged point is $* = (1, 0, 0, \ldots, 0)$.

J. Dieudonné, *A History of Algebraic and Differential Topology, 1900–1960*,
Modern Birkhäuser Classics, DOI 10.1007/978-0-8176-4907-4_14,

that $\gamma(a) = \gamma(b) = x_0$. Poincaré defined a law of composition on $\pi_1(X, x_0)$ by first defining the *juxtaposition* $\gamma = \gamma_1 \vee \gamma_2$ of two loops: if γ_1 and γ_2 are defined in $[0, 1]$, then γ: $[0, 1] \to X$ is such that $\gamma(t) = \gamma_1(2t)$ for $0 \leqslant t \leqslant \frac{1}{2}$ and $\gamma(t) = \gamma_2(2t - 1)$ for $\frac{1}{2} \leqslant t \leqslant 1$. It is easy to see that the class $[\gamma_1 \vee \gamma_2]$ only depends on the classes $[\gamma_1]$ and $[\gamma_2]$; and if $[\gamma_1 \vee \gamma_2] = [\gamma_1]\cdot[\gamma_2]$, the law

$$([\gamma_1], [\gamma_2]) \mapsto [\gamma_1]\cdot[\gamma_2]$$

defines a structure of *group* (noncommutative in general).

III) It was natural to consider the sets $[S_n; X]$ for $n > 1$, and the generalization of the law of composition defined by Poincaré on $[S_1; X]$ was not very difficult; in fact, this was done by Čech in 1932, and he mentioned that Dehn had thought of that definition earlier but did not publish it.

In 1935 Hurewicz found a more useful definition for those groups. The set $\Omega(X, x_0)$ of the loops $S_1 \to X$ considered by Poincaré is a topological space for the compact-open topology, and it has a natural privileged point, the constant loop x_1: $S_1 \to \{x_0\}$. Hurewicz therefore considered the group $\pi_1(\Omega(X, x_0), x_1)$ and found that the process could be repeated inductively. The successive spaces of loops are

$$\Omega^p(X, x_0) = \Omega(\Omega^{p-1}(X, x_0), x_{p-1}) \tag{1}$$

where x_{p-1} is the constant map $S_1 \to \{x_{p-2}\}$; the corresponding group

$$\pi_1(\Omega^{p-1}(X, x_0), x_{p-1}) \tag{2}$$

is the *p-th homotopy group* $\pi_p(X, x_0)$; it is a covariant functor from pointed spaces to groups. It is easily seen that for an H-space X, $\pi_1(X, x_0)$ is commutative; since $\Omega(X, x_0)$ is an H-space for juxtaposition, the groups $\pi_n(X, x_0)$ are *commutative for* $n \geqslant 2$. As a set, $\pi_n(X, x_0)$ is also identified to the set $\pi_0(\Omega^n(X, x_0))$ of arcwise connected components of the space $\Omega^n(X, x_0)$ having as privileged point the arcwise connected component of x_0.

A path α: $[0, 1] \to X$ with extremities a, b in X naturally defines an isomorphism $s(\alpha)$: $\pi_n(X, a) \simeq \pi_n(X, b)$; in particular the fundamental group $\pi_1(X, x_0)$ naturally operates by automorphisms on every $\pi_n(X, x_0)$.

IV) A little later Hurewicz defined the *relative homotopy groups* $\pi_n(X, A, x_0)$ (with $n \geqslant 2$) for any subset A of X with $x_0 \in A$, in the same manner. Here he considered the set $\Omega(X, A, x_0)$ consisting of the continuous *paths* α: $[0, 1] \to X$ such that $\alpha(0) = x_0$ and $\alpha(1) \in A$; this is also a topological space for the compact-open topology. A homotopy between two such paths α_1, α_2 is a continuous map F: $[0, 1] \times [0, 1] \to X$ such that $F(s, 0) = \alpha_1(s)$, $F(s, 1) = \alpha_2(s)$, $F(0, t) = x_0$, and $F(1, t) \in A$ for all (s, t); $\pi_1(X, A, x_0)$ is the set of homotopy classes of paths in $\Omega(X, A, x_0)$.

Then define by induction for $n \geqslant 2$

$$\Omega^n(X, A, x_0) = \Omega(\Omega^{n-1}(X, A, x_0), x_{n-1}) \tag{3}$$

where x_{n-1} is the constant loop $S_1 \to \{x_{n-2}\}$; the relative homotopy group is defined by

$$\pi_n(X, A, x_0) = \pi_1(\Omega^n(X, A, x_0), x_{n-1}); \tag{4}$$

it is commutative for $n \geqslant 3$ and is a covariant functor in the category of pairs (X, A). A natural homeomorphism

$$\Omega^n(X, A, x_0) \simeq \Omega(\Omega^{n-1}(X, x_0), \Omega^{n-1}(A, x_0), x_{n-1}) \tag{5}$$

is defined so that

$$\pi_n(X, A, x_0) \simeq \pi_1(\Omega^{n-1}(X, x_0), \Omega^{n-1}(A, x_0), x_{n-1}). \tag{6}$$

Furthermore, there is a natural map

$$\partial\colon \pi_1(X, A, x_0) \to \pi_0(A)$$

which associates the arcwise component of A containing $\alpha(1)$ to the class of a path $\alpha \in \Omega(X, A, x_0)$. Using (6), this defines for each $n \geqslant 2$ a natural homomorphism of *groups* $\partial\colon \pi_n(X, A, x_0) \to \pi_{n-1}(A, x_0)$.

In 1945 J.H.C. Whitehead proved that the sequence

$$\cdots \to \pi_{n+1}(X, A, x_0) \xrightarrow{\partial} \pi_n(A, x_0) \xrightarrow{j_*} \pi_n(X, x_0) \xrightarrow{i_*} \pi_n(X, A, x_0) \xrightarrow{\partial} \pi_{n-1}(A, x_0)$$
$$\cdots \to \pi_1(X, x_0) \xrightarrow{i_*} \pi_1(X, A, x_0) \xrightarrow{\partial} \pi_0(A) \to \pi_0(X) \tag{7}$$

is *exact*, where $j\colon (X, x_0) \to (X, A)$, $i\colon (A, x_0) \to (X, x_0)$ are the natural injections of pairs of pointed spaces.*

V) In 1935 Hurewicz defined the concept of *homotopy equivalence* between two pointed spaces (X, x_0), (Y, y_0). It is a continuous map $f\colon (X, x_0) \to (Y, y_0)$ such that there exists another continuous map $g\colon (Y, y_0) \to (X, x_0)$ for which the composed maps $g \circ f$ and $f \circ g$ are respectively homotopic to the identity in X and in Y; g is then a homotopy equivalence, called a *homotopy inverse* of f. When there exists a homotopy equivalence of X into Y, they have the *same homotopy type*. The homology of a space only depends on its homotopy type, and the set [X; Y] only depends on the homotopy types of X and Y; in particular the homotopy groups of X also only depend on the homotopy type of X.

In 1939 J.H.C. Whitehead introduced other equivalence relations; f: $(X, x_0) \to (Y, y_0)$ is an *n-equivalence* if the corresponding homomorphism f_*: $\pi_r(X, x_0) \to \pi_r(Y, y_0)$ is bijective for $1 \leqslant r < n$ and surjective for $r = n$; f is a *weak homotopy equivalence* if it is an n-equivalence for every $n > 0$. He then proved the remarkable result that when X and Y are CW-complexes, any weak homotopy equivalence $f\colon X \to Y$ is in fact a *homotopy equivalence.*

VI) J.H.C. Whitehead defined still another notion of a combinatorial nature for CW-complexes, which he called *simple homotopy equivalence.* It is related to what he called the *torsion* $\tau(f)$, defined only for *homotopy equivalences* $f\colon X \to Y$ between two CW-complexes. He began by defining, by purely group-

* The exactness of the last map ∂ means that the image $i_*(\pi_1(X, x_0))$ is mapped by ∂ on the arcwise connected component of x_0 in A.

theoretical means, a commutative group Wh(G) for *any* group G, now called the *Whitehead group* of G; the torsion $\tau(f)$ takes its values in $\mathrm{Wh}(\pi_1(X))$. Homotopy equivalences between X and Y are thus classified according to their torsion, and those for which $\tau(f) = 0$ are called simple homotopy equivalences. The spaces X and Y have the same *simple homotopy type* if there exists a simple homotopy equivalence $f: X \to Y$. But it was only in 1972 that it was proved that two homeomorphic CW-complexes have the same simple homotopy type, a nontrivial result since that notion is attached to a particular decomposition of a CW-complex into cells.

The Whitehead torsion generalized earlier combinatorial invariants defined by Reidemeister, Franz, and De Rham in the 1930s in order to classify *lens spaces*, defined as spaces of orbits for free action of cyclic groups on a sphere S_n. Using his notion of torsion, in 1940 Whitehead was able to give an example of two lens spaces having the same homotopy type but which are not homeomorphic.

VII) The definition of homotopy groups by Hurewicz had been preceded in 1925–1930 by the pioneering work of H. Hopf on the maps of spheres into spheres, which may rightly be called the starting point of homotopy theory. Brouwer had conjectured that the homotopy class of a continuous map $f: S_n \to S_n$ is determined by $\deg(f)$, and had sketched a proof for $n = 2$. In 1925 Hopf completely proved Brouwer's conjecture, which now can be expressed by saying that $[f] \mapsto \deg(f)$ is an *isomorphism*

$$\pi_n(S_n) \simeq \mathbf{Z}. \tag{8}$$

It is easy to see that for $n > 1$ and $m < n$ all continuous maps $f: S_m \to S_n$ are homotopic to a constant, in other words $\pi_m(S_n) = 0$. But until 1930 nobody had any idea of what the set $[S_m; S_n]$ was for $m > n$. The breakthrough was due to Hopf, who proved what we now write

$$\pi_3(S_2) \simeq \mathbf{Z}. \tag{9}$$

His method was to attach to any continuous map $f: S_3 \to S_2$ an integer, the *Hopf invariant* $H(f)$; for a simplicial map, this is the linking number of $f^{-1}(x)$ and $f^{-1}(y)$ for two points x, y of S_2 in general position. Hopf proved that $H(f)$ only depends on the class $[f]$, and showed that the "Hopf fibration", a restriction to S_3 of the natural map

$$\mathbf{C}^2 - \{0\} \to \mathbf{P}_1(\mathbf{C}) \simeq S_2,$$

has an invariant equal to ± 1.

From that time on the determination of the groups $\pi_m(S_n)$ for $m > n$ has been one of the outstanding open problems in topology. Hurewicz's definition of the homotopy groups enabled him to give simpler proofs for (8) and (9), and he also proved that

$$\pi_m(S_1) = 0 \quad \text{for } m \geqslant 2. \tag{10}$$

In 1935 Hopf extended his definition of the Hopf invariant to continuous maps $f: \mathbf{S}_{2k-1} \to \mathbf{S}_k$; if k is odd one always has $\mathrm{H}(f) = 0$; but for k even, Hopf constructed maps for which $\mathrm{H}(f) = \pm 2$, so that $\pi_{2k-1}(\mathbf{S}_k)$ contains a subgroup isomorphic to $\mathbf{Z}$; he even exhibited maps f with $\mathrm{H}(f) = \pm 1$ for $k = 2, 4$ and 8.

VIII) After Hurewicz's papers appeared in 1935, and before 1950, the only noticeable advance in the investigations of the $\pi_m(\mathbf{S}_n)$ was the discovery of the properties of *homotopy suspension* by H. Freudenthal in 1937. For any pointed space (X, x_0), he considered the product $\mathrm{X} \times [-1, 1]$ and the quotient space (SX, x_0) obtained by collapsing to x_0 the subspace

$$(\mathrm{X} \times \{-1\} \cup (\mathrm{X} \times \{1\}) \cup (\{x_0\} \times [-1, 1]).$$

This defines a covariant functor $(\mathrm{X}, x_0) \mapsto (\mathrm{SX}, x_0)$, the "suspension", in an obvious way in the category of pointed spaces; it is immediately seen that $\mathrm{S}(\mathbf{S}_n)$ is homeomorphic to $\mathbf{S}_{n+1}$.

For any pointed space (Y, y_0) the definitions yield a natural bijection

$$[\mathrm{X}, x_0; \Omega\mathrm{Y}, y_1] \cong [\mathrm{SX}, x_0; \mathrm{Y}, y_0] \tag{11}$$

(later expressed by saying that S and Ω are adjoint functors). On the other hand, if $p: \mathrm{X} \times [-1, 1] \to \mathrm{SX}$ is the collapsing map, there is a natural map

$$s: (\mathrm{Y}, y_0) \to (\Omega\mathrm{SY}, y_1) \tag{12}$$

such that $s(y)$ is the loop $t \mapsto p(y, t)$ in SY. This defines a map

$$s_*: [\mathrm{X}, x_0; \mathrm{Y}, y_0] \to [\mathrm{X}, x_0; \Omega\mathrm{SY}, y_1]$$

and by composition with (11) we get the *homotopy suspension*

$$\mathrm{E}: [\mathrm{X}, x_0; \mathrm{Y}, y_0] \to [\mathrm{SX}, x_0; \mathrm{SY}, y_0] \tag{13}$$

and in particular a group homomorphism

$$\mathrm{E}: \pi_r(\mathbf{S}_n) \to \pi_{r+1}(\mathbf{S}_{n+1}) \quad \text{for } r \geqslant 1, n \geqslant 1. \tag{14}$$

Freudenthal proved the striking result that (14) *is bijective for* $1 \leqslant r < 2n - 1$ *and surjective for* $r = 2n - 1$. In addition he showed that the kernel of the Hopf invariant $\mathrm{H}: \pi_{2r+1}(\mathbf{S}_{r+1}) \to \mathbf{Z}$ is the image of $\mathrm{E}: \pi_{2r}(\mathbf{S}_r) \to \pi_{2r+1}(\mathbf{S}_{r+1})$. Finally he proved that the kernel of $\mathrm{E}: \pi_3(\mathbf{S}_2) \to \pi_4(\mathbf{S}_3)$ is the subgroup of $\pi_3(\mathbf{S}_2)$ formed of the classes of maps with even Hopf invariant, which gave him the result

$$\pi_4(\mathbf{S}_3) \simeq \mathbf{Z}/2\mathbf{Z}. \tag{15}$$

From the Freudenthal theorem it follows that the sequence

$$\pi_r(\mathbf{S}_n) \xrightarrow{\mathrm{E}} \pi_{r+1}(\mathbf{S}_{n+1}) \xrightarrow{\mathrm{E}} \cdots \xrightarrow{\mathrm{E}} \pi_{r+k}(\mathbf{S}_{n+k}) \xrightarrow{\mathrm{E}} \cdots$$

is *stationary*: all maps E are isomorphisms for $k > r - 2n + 1$. This was the first appearance of *stability* in homotopy theory; the group $\pi_{r+k}(\mathbf{S}_k)$ for $k > r + 1$ is independent of k, up to isomorphism; it is called the *r-stem* or the *r-th stable homotopy group*; in particular

$$\pi_{n+1}(S_n) \simeq \mathbf{Z}/2\mathbf{Z} \quad \text{for } n \geqslant 3. \tag{16}$$

Hopf also generalized the Brouwer conjecture and determined the set $[X; S_n]$ when X is an n-dimensional finite simplicial complex; his result was put by Whitney in the simpler form

$$[X; S_n] \simeq H^n(X; \mathbf{Z}).$$

IX) Poincaré had noticed (without proof) that to any relation between the classes of *loops* in $\pi_1(X, x_0)$ there corresponds a "homology" between their classes as *cycles*. This later was formalized as defining a *surjective homomorphism* $h_1: \pi_1(X, x_0) \to H_1(X; \mathbf{Z})$, the kernel of which is the group of commutators of $\pi_1(X, x_0)$.

Similarly, for any $n > 1$, to any map $f: (S_n, *) \to (X, x_0)$ can be associated the singular chain which is the difference of the restrictions of f to the hemispheres D_+ and D_-, naturally identified with the standard simplex Δ_n. Hurewicz showed that by passage to quotients one obtains a homomorphism

$$h_n: \pi_n(X, x_0) \to H_n(X; \mathbf{Z}),$$

but in general h_n is neither injective nor surjective. However, Hurewicz could prove that if X is $(n-1)$*-connected*, i.e. $\pi_j(X, x_0) = 0$ for $1 \leqslant j \leqslant n-1$, then h_n is *bijective* (the *absolute Hurewicz isomorphism theorem*).

In the same way, Hurewicz defined a natural homomorphism

$$h_n: \pi_n(X, A, x_0) \to H_n(X, A; \mathbf{Z})$$

and stated that if $\pi_j(X, A, x_0) = 0$ for $1 \leqslant j \leqslant n-1$, h_n is bijective (the *relative Hurewicz isomorphism theorem*), but he did not publish a proof; later several were given.

Combining that theorem and the homotopy exact sequence, J.H.C. Whitehead proved that if $f: (X, x_0) \to (Y, y_0)$ is an *n-equivalence*, then the homomorphisms in singular homology

$$f_*: H_r(X; \mathbf{Z}) \to H_r(Y; \mathbf{Z})$$

are bijective for $r < n$, surjective for $r = n$; if X and Y are simply connected, the converse is true.

X) The origin of the concept of fibration can be found in notions like vector fields or moving frames on a smooth manifold X; they deal with what may be thought of as a kind of function defined on X, but which, for every $x \in X$, takes its value $f(x)$ in a set E_x *variable with* x. A fibration is a space E which, as a set, is the disjoint union of "fibers" $E_x (x \in X)$ with a suitable topology; a *section* of the fibration E is then a continuous map $s: X \to E$ such that $s(x) \in E_x$ for each $x \in X$.

The first general definition was given by Whitney in 1935, and his ideas were developed a little later by himself and by Ehresmann, Feldbau, and Steenrod. The kind of fibration they studied is what is now called (locally trivial) *fiber space*. It is best defined as a quadruplet $\xi = (E, B, F, p)$ where E is the "total space," B the "base space," F the "typical fiber," and p, the "projection," is a continuous surjective map p: E → B satisfying the condition that each point $b \in$ B has an open neighborhood U for which there is a homeomorphism

$$\varphi\colon \mathrm{U} \times \mathrm{F} \cong p^{-1}(\mathrm{U}) \tag{17}$$

satisfying

$$p(\varphi(y, t)) = y \quad \text{for } y \in \mathrm{U} \text{ and } t \in \mathrm{F}. \tag{18}$$

A fiber space can thus be regarded as a product "locally," but in general not "globally." All fibers $\mathrm{E}_b = p^{-1}(b)$ are homeomorphic to F.

XI) Special examples of fiber spaces had in fact been considered (in an intuitive way) by Poincaré in 1883, the *covering spaces* of a space X (in Poincaré's case, X is an open subset of **C**); they are fiber spaces with base space X and *discrete fibers*. They became the subject of a rigorous general theory between 1913 and 1934. It was shown that they possess a remarkable "lifting" property. If Y is a covering space of (X, x_0) with projection p: Y → X, and $y_0 \in p^{-1}(x_0)$, then for any path α: [0, 1] → X with $\alpha(0) = x_0$ there is a unique path $\tilde{\alpha}$: [0, 1] → Y with $\tilde{\alpha}(0) = y_0$ such that $\alpha = p \circ \tilde{\alpha}$. Furthermore, if φ: [0, 1] × [0, 1] → X is a homotopy between two paths α_1, α_2 (with $\varphi(0, t) = x_0$ for all $t \in [0, 1]$) there exists a unique homotopy $\tilde{\varphi}$ between $\tilde{\alpha}_1$ and $\tilde{\alpha}_2$ such that $\varphi = p \circ \tilde{\varphi}$.

For a general fiber space $(\mathrm{E}, \mathrm{B}, \mathrm{F}, p)$ the existence of a lifting of a path is no longer guaranteed. But a very useful result remains: if $\mathrm{H}\colon (\mathrm{Z}, z_0) \times [0, 1] \to (\mathrm{B}, b_0)$ is a homotopy between two continuous maps f, g with $\mathrm{H}(z_0, t) = b_0$ for all $t \in [0, 1]$, and if f can be lifted to a continuous map $\tilde{f}\colon (\mathrm{Z}, z_0) \to (\mathrm{E}, a_0)$ with $p(a_0) = b_0$, then H can also be lifted to a homotopy $\tilde{\mathrm{H}}\colon (\mathrm{Z}, z_0) \times [0, 1] \to (\mathrm{E}, a_0)$ such that $\tilde{\mathrm{H}}(z, 0) = \tilde{f}(z)$ and $p . \tilde{\mathrm{H}} = \mathrm{H}$ (in general one cannot hope that $\tilde{\mathrm{H}}$ is unique). This is called the *covering homotopy property* for the system $(\mathrm{Z}, \mathrm{E}, \mathrm{B}, p)$.

Around 1940 it was realized that the covering homotopy property for a fiber space implies that the projection p: E → B defines an *isomorphism* p_*: $\pi_n(\mathrm{E}, \mathrm{F}, a_0) \cong \pi_n(\mathrm{B}, b_0)$ for all $n > 0$. In conjunction with the exact homotopy sequence (7) this gives the *exact homotopy sequence of fiber spaces*

$$\begin{aligned} \cdots \to \pi_n(\mathrm{F}, a_0) &\xrightarrow{j_*} \pi_n(\mathrm{E}, a_0) \xrightarrow{p_*} \pi_n(\mathrm{B}, b_0) \xrightarrow{\partial} \pi_{n-1}(\mathrm{F}, a_0) \\ &\to \cdots \to \pi_0(\mathrm{F}) \xrightarrow{j_*} \pi_0(\mathrm{E}) \xrightarrow{p_*} \pi_0(\mathrm{B}) \to 0 \end{aligned} \tag{19}$$

linking the homotopy groups of the three spaces E, B, F.

XII) In 1940 Hurewicz and Steenrod had already shown that the covering homotopy property is valid for maps p: E → B more general than the projections of fiber spaces. An important example of that phenomenon surfaces as one of the main ingredients in Serre's work on homotopy groups: for any arcwise connected space (X, x_0), consider the space $\mathrm{P} = \Omega(\mathrm{X}, \mathrm{X}, x_0)$ of *all* continuous paths α: [0, 1] → X such that $\alpha(0) = x_0$, and let p: P → X be the map $\alpha \mapsto \alpha(1)$. Then for *any* space (Z, z_0) the system $(\mathrm{Z}, \mathrm{P}, \mathrm{X}, p)$ has the covering homotopy property; this follows from a more general theorem proved by

Borsuk in 1937. It has become customary to call *fibrations* the systems (E, B, p) such that (Z, E, B, p) has the covering homotopy property for all spaces (Z, z_0) (or merely for finite simplicial complexes). The fibers $p^{-1}(b)$ are not in general homeomorphic to the "typical fiber" $F = p^{-1}(b_0)$; if B is arcwise connected, they only have the same homotopy type. The exact homotopy sequence (19) is valid for all fibrations.

XIII) The most important fiber spaces are the *principal fiber spaces* and the *vector bundles*. A principal fiber space (P, B, G, p) has a typical fiber G which is a topological group, called the *structural group*. G acts continuously (on the right) on P in such a way that for the homeomorphisms (18) defining P

$$\varphi(y, t) \cdot s = \varphi(y, ts) \quad \text{for } y \in U \text{ and } t, s \text{ in } G. \tag{20}$$

If $\sigma\colon U \mapsto p^{-1}(U)$ is a section, then $\varphi(y, t) = \sigma(y) \cdot t$. The fibers $p^{-1}(b)$ are then the *orbits* of the action of G on P; on each fiber the action of G is *simply transitive*; B is identified with the space of orbits $G \backslash P$. A typical example of principal fiber space is a Lie group G with structural group a closed subgroup $H \subset G$; the action of H on G is right translation $(x, s) \mapsto xs$, the fibers are the right classes xH, and B is identified with the homogeneous space G/H.

A real (resp. complex) *vector bundle* is a fiber space (E, B, F, p) where F is a vector space $\mathbf{R}^m$ (resp. $\mathbf{C}^m$) and the homeomorphisms (18) are such that the map $t \mapsto \varphi(y, t)$ of F onto $p^{-1}(y)$ is **R**-*linear* (resp. **C**-*linear*) for every $y \in U$. The most important vector bundle is the *tangent bundle* $T(M)$ to a C^1 manifold M, where the fiber $T_x(M)$ at $x \in M$ is the tangent space at that point. If N is a submanifold of M, the tangent bundle $T(N)$ is a vector subbundle of $T(M)$, and the quotient $T(M)/T(N)$ is called the *normal bundle* of N in M.

When Ehresmann defined principal fiber spaces in 1940 he observed that when the structural group G of a locally compact principal fiber space (P, B, G, p) *acts* continuously on a locally compact space F, this canonically defines a fiber space with base space B and typical fiber F *associated* to P. This is done by letting G operate on the right on $P \times F$ by

$$(x, y) \cdot s = (x . s, s^{-1} . y);$$

the space of orbits for that action is written $P \times^G F$; if $\pi\colon P \times F \to P \times^G F$ is the natural projection, $x . y = \pi(x, y) \in P \times^G F$. Then $p_F(x . y) = p(x)$ does not depend on $y \in F$, and $(P \times^G F, B, F, p_F)$ is the fiber space associated to P and to the action of G on F. In particular, every real (resp. complex) vector bundle of rank m is associated to a principal fiber space with structural group the orthogonal group $O(m)$ (resp. the unitary group $U(m)$).

XIV) As soon as fiber spaces had been defined there came the problem of classifying, for a given base space B and typical fiber F, all fiber spaces (E, B, F, p) up to an *isomorphism*; for two fiber spaces E, E′ with same base space B and projections p, p' this means a homeomorphism $f\colon E \to E'$ with $p(y) = p'(f(y))$, so that f is a homeomorphism of each fiber E_b onto E'_b. Around 1940 Whitney and Steenrod attacked the problem for vector bundles and principal fiber spaces with structural group $O(m)$.

For any fiber space $\xi = (E, B, F, p)$ and continuous map g: $B' \to B$, Whitney defined in 1935 the *pull-back* $g^*(\xi) = (E', B', F, p')$ as the fiber product $E \times_B B'$ (the subspace of the product $E \times B'$ consisting of the pairs (y, b') for which $p(y) = g(b')$); the projection p': $E' \to B'$ is then the restriction of the second projection pr_2; the map $(y, b') \mapsto y$ of E' into E is a homeomorphism of each fiber $p'^{-1}(b')$ onto the fiber $p^{-1}(g(b'))$. If E is a vector bundle (resp. a principal fiber space), so is E'.

In 1937 Whitney observed that if ξ is a real vector bundle of rank r with base space a finite simplicial complex B, there exists a continuous map g: $B \to G_{n,r}$ into the *grassmannian* of r-planes in $\mathbf{R}^n$ for large enough n, such that ξ is isomorphic to the *pull-back* $g^*(\xi')$, where $\xi' = (U_{n,r}, G_{n,r}, \mathbf{R}^r, p')$ is the tautological bundle* with base $G_{n,r}$. Then Steenrod took up the idea and proved that any two such pull-backs of ξ' by maps g_1, g_2 of B into $G_{n,r}$ are isomorphic if and only if g_1 and g_2 are *homotopic*, so that the set of isomorphism classes of vector bundles of rank r and base space B can be identified with $[B; G_{n,r}]$ for large n. That same set can also be identified with the set of isomorphism classes of principal fiber spaces with base space B and structural group $O(r)$; one has only to replace the tautological bundle $U_{n,r}$ by the Stiefel manifold $S_{n,r}$.

The same results hold for complex vector bundles and principal fiber spaces with structural group $U(r)$: replace $G_{n,r}$ by the complex grassmannian $G_{n,r}(\mathbf{C})$. Later these results were extended to all principal fiber spaces (P, B, G, p) where B is a CW-complex of dimension $\leqslant n$ and G a topological group.

The important fact for the classification is that to obtain (P, B, G, p) as a pull-back of (P', B', G, p') for *all* CW-complexes B of dimension $\leqslant n$, one must assume that $\pi_i(P') = 0$ for $1 \leqslant i \leqslant n$. Principal fiber spaces with group G satisfying that condition are called *n-universal*, and their base space B' *n-classifying* for the group G. The limitation on the dimension of B can be lifted: it is enough that there exists a sequence of continuous maps u_n: $B \to [0, 1]$ such that P is trivializable over each open set $u_n^{-1}(]0, 1])$ and these open sets form a covering of B. Milnor then showed how to construct a *universal* principal fiber space P_G which is contractible; its base space $B_G = P_G/G$ is called *classifying* for the group G. For any base space B satisfying the above condition the set of isomorphism classes of principal fiber spaces with base space B and structural group G is identified with $[B; B_G]$.

XV) In their pioneering work of 1935 Stiefel and Whitney were looking for conditions implying the existence of a section s of a vector bundle (E, B, F, p) such that $s(b) \neq 0$ for all $b \in B$, B being a simplicial complex. They found that a sufficient condition is the vanishing of a certain cohomology class of B with values in $\mathbf{F}_2$.† Pontrjagin was apparently the first to observe, in 1942, that the Whitney-Steenrod classification determines for each real vector bundle $\xi =$

* $U_{n,r}$ is defined as the set of pairs (V, y) in $G_{n,r} \times \mathbf{R}^n$ such that $y \in V$.

† Stiefel only used homology and did not speak of fiber spaces.

$(E, B, \mathbf{R}^m, p)$ a subalgebra of the cohomology algebra $H^{\bullet}(B; \Lambda)$, namely the image

$$g^*(H^{\bullet}(G_{m+n,m}; \Lambda)$$

for the map $g\colon B \to G_{m+n,m}$ which defines ξ as a pull-back and is determined up to homotopy.* Ehresmann's cell decomposition of the grassmannian enabled him to define m well-determined classes which are *generators* of its cohomology algebra over $\mathbf{Z}$; their images by g^* are called *characteristic classes* of the vector bundle ξ.

Pontrjagin dealt with oriented vector bundles, associated to principal fiber spaces with structural group $SO(m)$ instead of $O(m)$ and base space $G'_{m+n,m}$, a double cover of $G_{m+n,m}$. He also limited himself to the tangent bundle of a manifold, and his results were complicated by the fact that $H^{\bullet}(G'_{m+n,m}; \mathbf{Z})$ has torsion. In 1945 Chern similarly used the Whitney-Steenrod theory for complex vector bundles, where this time the cohomology of the complex grassmannian over $\mathbf{Z}$ was *free*; the classes defined by Pontrjagin are more easily obtained by considering the Chern classes of the *complexified* bundle of a real vector bundle.

Pontrjagin and Chern independently found that their characteristic classes for the tangent bundle of a Riemannian or Hermitian manifold could be expressed (in the De Rham cohomology) by well-determined classes of differential forms. This was later generalized to arbitrary vector bundles on a smooth manifold by A. Weil, using the theory of connections.

Finally, the Stiefel-Whitney classes can also be defined as characteristic classes pulled back from the cohomology of $G_{m+n,m}$ with coefficients in $\mathbf{F}_2$.

XVI) As soon as Leray had invented sheaf cohomology and spectral sequences in 1946 he applied them to fiber spaces (X, B, F, p) where X, B and F are locally compact and arcwise connected and B is locally arcwise connected. He used the Alexander–Spanier cohomology with compact supports and with coefficients in a Λ-module M. When $\pi_1(B)$ acts trivially on $H^{\bullet}(X_b; M)$ the spectral sequences starts with the terms

$$H^p(B; H^q(F; M))$$

and when M is a field, or $M = \mathbf{Z}$ and B or F has no torsion, that term is isomorphic to $H^p(B; M) \otimes H^q(F; M)$. These expressions enabled him to prove many results on the relations between the cohomology of B, F and X. If two of the cohomology modules $H^{\bullet}(B; M)$, $H^{\bullet}(F; M)$ and $H^{\bullet}(X; M)$ are finitely generated, so is the third one. If $M = K$ is a field, the corresponding Betti numbers satisfy

$$b_k(X) \leqslant b_k(B \times F) \quad \text{and} \quad \chi(X) = \chi(B)X(F). \tag{21}$$

If the injection $i\colon F \to X$ is such that $i^*\colon H^{\bullet}(X; M) \to H^{\bullet}(F; M)$ is surjective, then

---

* Pontrjagin only used homology; Wu Wen Tsün simplified his technique by using cohomology.

when X is compact and M = K is a field,

$$H^{\bullet}(X; K) \simeq H^{\bullet}(B : K) \otimes H^{\bullet}(F; K), \tag{22}$$

as in the case of a product. This result was independently proved by G. Hirsch* and is known as the *Leray–Hirsch theorem.*

In his work on homotopy groups Serre needed spectral sequences for a general fibration, where the fibers are not locally compact in general. Since he could not use the Alexander–Spanier cohomology, he had to extend Leray's arguments to singular cohomology; he also showed that there is a similar spectral sequence for singular homology.

From the Leray or Serre spectral sequences it is possible to deduce important exact sequences earlier described by more direct methods. Let $\xi = (E, B, \mathbf{R}^m, p)$ be an oriented vector bundle of rank m, and E^0 the complement of the zero section (identified with B); then one has the *Gysin exact sequence*:

$$\cdots \to H^{r-m}(B; Z) \xrightarrow{g} H^r(B; Z) \xrightarrow{p_0^*} H^r(E^0; Z) \to H^{r-m+1}(B; Z) \to \cdots \tag{23}$$

where p_0 is the restriction of p to E^0 and g is the cup-product $c \mapsto c \smile e(\xi)$ with the Euler class of ξ. There are similar sequences for homology, and for unoriented vector bundles (with $\mathbf{F}_2$ as ring of coefficients).

When $\xi = (E, \mathbf{S}_n, F, p)$ is a fiber space with base space a sphere $\mathbf{S}_n$ and a fiber F which is a finite simplicial complex, one has the *Wang exact sequence* in singular homology:

$$\cdots \to H_{r-n+1}(F) \to H_r(F) \to H_r(E) \to H_{r-n}(F) \to \cdots \tag{24}$$

There is a similar sequence in singular cohomology.

XVII) As an application of spectral sequences Leray was able to compute in 1946 the cohomology of the homogeneous space G/T, where G is a compact classical Lie group and T a maximal torus; in 1949 he extended his results to homogeneous spaces G/U with T ⊂ U ⊂ G and proved a formula, conjectured by G. Hirsch, which gives the Betti numbers of G/U when those of G and of U are known.

Several mathematicians then investigated the cohomology of principal fiber spaces more generally. The most complete results were obtained by A. Borel in 1953, using a deep algebraic study of spectral sequences. Its main application concerns the relation between the cohomology of a compact Lie group G and the cohomology of a classifying space B_G. The general philosophy is that when K is a field of arbitrary characteristic, and $H^{\bullet}(G; K) \simeq \bigwedge(x_1, \ldots, x_m)$ for homogeneous elements of odd degree (a condition verified when K has characteristic 0, by Hopf's theorem), then $H^{\bullet}(B_G; K) \simeq K[y_1, \ldots, y_m]$, an algebra of polynomials, where each y_j is homogeneous of degree $\deg y_j = 1 + \deg x_j$. Theproof relies on the fact that the spectral sequence of the universal principal fiber space (P_G, B_G, G) starts with terms

* Hirsch's method also enabled him to prove (21) independently of Leray.

$$K[y_1,\ldots,y_m]\otimes\bigwedge(x_1,\ldots,x_m);$$

in fact this is an algebra already introduced a little earlier by A. Weil when $K = \mathbf{R}$. Many of A. Borel's other results concerned the cohomology of G/U for closed subgroups U of G. These advances not only gave the Betti numbers of all Lie groups, but for the first time they gave access to their *torsion*.

XVIII) Beginning in 1941 more sophisticated relations were described between homotopy and homology. First Hopf discovered that for a locally finite arcwise connected simplicial complex K the quotient $H_2(K)/h_2(\pi_2(K))$ (h_2 being the Hurewicz homomorphism) is entirely *determined* by the fundamental group $\pi_1(K)$: there is a purely group-theoretic construction applicable to any group Π, and it gives the above quotient when it is performed on $\Pi = \pi_1(K)$. No similar results exist for $H_n(K)/h_n(\pi_n(K))$ when $n \geqslant 3$, but Hopf and Eilenberg–Mac Lane independently found that *if* $\pi_i(X) = 0$ *for* $2 \leqslant i \leqslant n-1$, then all homology groups $H_r(X;G)$ for $r < n$ are again determined by group-theoretic constructions performed on $\pi_1(K)$. The two procedures they described looked different but were later shown to be equivalent. Applied to *any* group Π and any commutative group G they produce commutative groups $H_r(\Pi;G)$ for all $r \geqslant 0$ now called the *homology groups of the (discrete) group* Π *with coefficients in* G; similar constructions give the *cohomology groups* $H^r(\Pi;G)$.

In this way those groups made their entrance and inaugurated what is now called *homological algebra*. One of the first applications of these groups was made in 1947 by Leray and H. Cartan to the spectral sequence of a connected covering space X of a space B with automorphism group Π. The spectral sequence now starts with the groups

$$H^p(\Pi;H^q(X;\Lambda))$$

and gives information on the groups $H^{p+q}(B;\Lambda)$.

XIX) The method used by Eilenberg and Mac Lane in the definition of the homology groups of a group Π consisted in constructing by purely algebraical means a chain complex $K_\bullet(\Pi,1)$, the homology of which is $H_\bullet(\Pi;G)$. If $\pi_i(X) = 0$ for $2 \leqslant i \leqslant n-1$, then

$$H_r(X;G) \simeq H_r(\pi_1(X);G) \quad \text{for } r < n.$$

They also considered spaces X for which $\pi_j(X) = 0$ for *all* $j > m$* and their method led to construction of other chain complexes $K_\bullet(\Pi,m)$ generalizing $K_\bullet(\Pi,1)$. But in general this did not give them the homology $H_\bullet(X;G)$ for $\Pi = \pi_m(G)$, *except* in the case where *also* $\pi_j(X) = 0$ for $1 \leqslant j < m$; in other words, for such spaces $\pi_m(X)$ is the *only* non trivial homomopy group; then, for *all* $n \geqslant 1$

* Such spaces already had been considered by Hurewicz for $m = 1$, under the name of *aspherical* spaces; he had shown that if a finite simplicial complex X is aspherical, its homotopy type is determined by $\pi_1(X)$.

$$\begin{cases} H_n(X;G) \simeq H_n(K_\bullet(\pi_m(X),m);G) \\ H^n(X;G) \simeq H^n(K_\bullet(\pi_m(X),m);G) \end{cases} \tag{25}$$

These spaces are now called *Eilenberg–Mac Lane spaces*, and have acquired a central position in algebraic and differential topology.

Their existence for each $m > 0$ follows from a general result of J.H.C. Whitehead: Given an *arbitrary* sequence $(G_n)_{n \geqslant 1}$ of groups, commutative except possibly for $n = 1$, there exists a CW-complex Z for which $\pi_n(Z) = G_n$ for *all* $n \geqslant 1$, and such a space is determined up to a homotopy equivalence.

It has become customary to write $K(\Pi, n)$ an Eilenberg–Mac Lane space which is a CW-complex with $\pi_n(K(\Pi, n)) \simeq \Pi$. For any commutative group G there is a natural isomorphism

$$H^n(K(\Pi, n); G) \simeq \text{Hom}(\Pi, G). \tag{26}$$

In particular, if $G = \Pi$, there is an element

$$c \in H^n(K(\Pi, n); \Pi)$$

which corresponds by (26) to the identity in End(Π); it is called the *canonical cohomology class* of $K(\Pi, n)$. For any CW-complex X the map

$$[f] \mapsto f^*(c)$$

is a bijection

$$[X; K(\Pi, n)] \simeq H^n(X; \Pi). \tag{27}$$

The many applications of Eilenberg–Mac Lane spaces require the knowledge of *all* their cohomology groups (not only (26)). Preliminary work based on (25) was done by Eilenberg and Mac Lane themselves, but the chain complex $K_\bullet(\Pi, n)$ is unwieldy. They started to modify it by an algebraic device which they called "bar construction," but this did not give them explicit expressions for $H_{n+k}(K(\Pi, n); \mathbf{Z})$ beyond $k = 10$. Only by deep and lengthy manipulations of more general "constructions" did H. Cartan finally succeed in describing $H^\bullet(K(\Pi, n); G)$ for all finitely generated groups Π and G. Earlier Serre had elucidated the case $G = \mathbf{F}_2$ by topological methods.

XX) Hopf's result showing that $H_2(X)/h_2(\pi_2(X))$ only depends on the group $\pi_1 = \pi_1(X)$ raised the problem: does $h_2(\pi_2(X))$ also only depend on π_1, $\pi_2 = \pi_2(X)$ and on the natural action of π_1 on π_2? In 1946 Eilenberg and Mac Lane gave a counterexample to that conjecture, and going farther, they defined an explicit element $k^3 \in H^3(\pi_1; \pi_2)$ which brought the missing information needed to completely determine $H_2(X; \mathbf{Z})$.

A little later that result was greatly generalized by Postnikov; his construction can be presented in several ways. The most suggestive relies on a method independently discovered by G. Whitehead and Cartan–Serre: it "kills" in succession the homotopy groups of a space by passage to suitable fiber spaces, just as the passage from a CW-complex to its universal covering space replaces a space by another one which has the same homotopy groups π_i for $i \geqslant 2$, but for which π_1 has vanished.

In that presentation, the "Postnikov tower" of X is a sequence of fibrations

$$Z_0 \overset{p_1}{\leftarrow} Z_1 \overset{p_2}{\leftarrow} Z_2 \leftarrow \cdots \overset{p_n}{\leftarrow} Z_n \leftarrow \cdots$$

where $p_{n+1}: Z_{n+1} \to Z_n$ is the projection of a fibration with typical fiber an *Eilenberg-Mac Lane space* $K(\pi_n(X), n)$; p_{n+1} is completely defined by an *invariant class* $k^{n+2} \in H^{n+2}(Z_n; \pi_{n+1}(X))$ generalizing the Eilenberg–Mac Lane invariant. There is an inverse system of maps $f_n: X \to Z_n$ such that the inverse limit

$$f = \varprojlim f_n : X \to Z = \varprojlim Z_n$$

is a *weak homotopy equivalence*. By Whitehead's theorem it follows that the homology

$$H_\bullet(X; \mathbf{Z}) \simeq H_\bullet(Z; \mathbf{Z})$$

is completely *determined* by the homotopy groups $\pi_n(X)$ and the Postnikov invariants k^n.

Postnikov towers have proved useful in several parts of algebraic topology, in particular for a generalization of the theory of *obstructions*, first developed by Eilenberg in 1940.

XXI) The Eilenberg–Mac Lane spaces also provide a description (independently due to Serre and to Eilenberg–Mac Lane) of all cohomology operations. Given two integers $n \geqslant 0$, $q \geqslant 0$, and two commutative groups A, B, a *cohomology operation* of *type* (q, n, A, B) is the assignment to each CW-complex X of a map $\xi(X): H^q(X; A) \to H^n(X; B)$ satisfying the following condition: for any CW-complex Y and any continuous map $v: X \to Y$, the diagram

$$\begin{array}{ccc} H^q(Y; A) & \xrightarrow{v^*} & H^q(X; A) \\ \downarrow{\scriptstyle \xi(Y)} & & \downarrow{\scriptstyle \xi(X)} \\ H^n(Y; B) & \xrightarrow[v^*]{} & H^n(X; B) \end{array}$$

is commutative. An example is the cup-square: $u \mapsto u \smile u$, which is a map $H^q(X; A) \to H^{2q}(X; A)$.

If $\iota \in H^q(K(A, q); A)$ is the canonical cohomology class of $K(A, q)$, it follows from (27) that, for any $x \in H^q(X; A)$, there is a continuous map

$$g_{x,A,q}: X \to K(A, q)$$

such that $g^*_{x,A,q}(\iota) = x$. Now let α be any element of $H^n(K(A, q); B)$, and consider the map $g^*_{x,A,q}: H^n(K(A, q); B) \to H^n(X; B)$. Then

$$x \mapsto \xi(X)(x) = g^*_{x,A,q}(\alpha)$$

defines a cohomology operation of type (q, n, A, B) and *any* cohomology operation of that type corresponds in that way to a unique element

$$\alpha \in H^n(K(A, q); B).$$

This general theorem was motivated by the discovery, due to N. Steenrod, of new, hitherto unsuspected, cohomology operations. In 1947, in order to determine the set $[K; S_n]$ when K is a finite simplicial complex of dimension $n + 1$, he described, by a fairly intricate construction, generalizations of the cup-square now known as the *Steenrod squares*; they are functorial homomorphisms, now written

$$Sq^i: H^m(X; \mathbf{F}_2) \to H^{m+i}(X; \mathbf{F}_2).$$

In 1950 he described similar homomorphisms for all primes $p > 2$, the *Steenrod reduced powers*

$$\mathscr{P}_p^k: H^m(X; \mathbf{F}_p) \to H^{m+2k(p-1)}(X; \mathbf{F}_p).$$

These operations and the algebras of operators they generate (called the *Steenrod algebras*) have found many applications in algebraic and differential topology. The most spectacular was the proof by J.F. Adams that the *only* cases in which maps $f: S_{2k+1} \to S_{k+1}$ may have Hopf invariant ± 1 are the ones discovered by Hopf, $k = 1, 3$ or 7.

XXII) The 1950s saw significant advances in homotopy theory. The first was the remarkable finiteness theorems proved by Serre in his 1951 thesis: for $m > n$, $\pi_m(S_n)$ is *finite* if n is *odd* and also if n is *even and* $m \neq 2n - 1$; $\pi_{2n-1}(S_n)$ for n even is the direct sum of $\mathbf{Z}$ and a finite group. For n odd, the result follows from two statements:

1. $\pi_m(S_n)$ is finitely generated;
2. for a field K of characteristic 0, $\pi_m(S_n) \otimes K = 0$.

Both are proved by the following strategy, using loop spaces as in Hurewicz's definition. If X is arcwise connected and simply connected, consider the sequence of spaces

$$X_0 = X, T_1, X_1, T_2, X_2, \ldots$$

where T_j is the universal covering space of X_{j-1}* and $X_j = \Omega(T_j)$ the space of loops of T_j. Then $\pi_{n+1}(X) = H_1(X_n; \mathbf{Z})$. Under the single assumption that all homology groups $H_j(X; \mathbf{Z})$ are finitely generated, it follows from the spectral sequence of the space of paths and the spectral sequence of covering spaces that all $H_j(X_m; \mathbf{Z})$ are also finitely generated for *all* m; this takes care of the first statement for $X = S_n$.

The second statement is proved by first showing that for $X = S_n$ and $m \leqslant n - 1$, $H^\bullet(X_m; K)$ is alternately a K-algebra of polynomials generated by a single element of degree $n - m$ if m is odd and an exterior algebra generated by a single element of degree $n - m$ if m is even. Then the spactral sequence

* One must assume all the universal covering spaces T_j do exist; this is the case when X is a finite CW-complex.

of covering spaces gives $H_i(T_n; K) = 0$ for $i > 0$. Serre proved that this implies $\pi_i(T_n) \otimes K = 0$ for all $i > 0$; since $\pi_i(T_n) = \pi_{i+n-1}(S_n)$ this ended the proof. The details, involving many spectral sequences, are fairly intricate.

Serre proceeded in a similar way to get the $\pi_m(S_n)$ for n even, using Stiefel manifolds instead of spheres.

The *explicit* computation of the $\pi_m(S_n)$ did not fare so well. The fact that $\pi_{n+2}(S_n) \simeq \mathbf{Z}/2\mathbf{Z}$ for $n \geqslant 2$ was only proved in 1950 by Pontrjagin and G. Whitehead independently, and it took the combined efforts of several topologists to obtain $\pi_6(S_3) \simeq \mathbf{Z}/12\mathbf{Z}$. Using various methods, Serre and other mathematicians determined $\pi_{n+k}(S_n)$ for $3 \leqslant k \leqslant 8$, but the complexity of the proofs increased with k. At present, due to the work of H. Toda and his school, these groups are known for $k \leqslant 30$, but no general pattern has yet emerged; only for p-components and values of n related to the prime p do some general results exist.

XXIII) Another open problem attracted topologists in the 1950's, the computation of the homotopy groups $\pi_i(G)$ for a compact Lie group G. For the classical groups $U(n, F)$ with $F = \mathbf{R}$, $\mathbf{C}$ or $\mathbf{H}$, several methods reduced the problem to the computation of homotopy groups of spheres, and tables were obtained for $i \leqslant 12$. Here again there were *stability* phenomena: for the unitary group $U(n) = U(n, \mathbf{C})$, all $\pi_i(U(n))$ are isomorphic for $n > i + 1$. A remarkable *periodicity* also appeared: for $2 \leqslant i < 2n$, the groups $\pi_i(U(n))$ are alternately 0 and $\mathbf{Z}$, whereas they become very irregular for $i \geqslant 2n$. In 1956 R. Bott showed by entirely new methods that this periodicity holds for *all* pairs (n, i) with $i < 2n$; similar periodicities (with period 8 instead of 2) take place for $SO(n)$ and $U(n, \mathbf{H})$ (also written $Sp(n)$).

His idea was to use Morse theory. By stability, $\pi_{i-1}(U(m)) \simeq \pi_{i-1}(U(2m))$ for $i \leqslant 2m$. Using the homotopy exact sequence for the fibrations $(U(2m), S_{2m,m}(\mathbf{C}), U(m))$ and $(S_{2m,m}(\mathbf{C}), G_{2m,m}(\mathbf{C}), U(m))$, one gets $\pi_{i-1}(U(m)) \simeq \pi_i(G_{2m,m}(\mathbf{C}))$ for $i \leqslant 2m$; finally $\pi_j(SU(n)) \simeq \pi_j(U(n))$ for $j \geqslant 2$. The crux of the proof was therefore to show that

$$\pi_{i+1}(SU(2m)) \simeq \pi_i(G_{2m,m}(\mathbf{C})) \quad \text{for } i \leqslant 2m;$$

from the definition of homotopy groups, this is equivalent to

$$\pi_i(\Omega(SU(2m), I_{2m}) \simeq \pi_i(G_{2m,m}(\mathbf{C})).$$

The space of loops $\Omega(SU(2m), I_{2m})$ has the same homotopy type as the space P of paths joining I_{2m} and $-I_{2m}$. Bott first showed that for $i \leqslant 2m$

$$\pi_i(P) = \pi_i(P_{\min})$$

where $P_{\min}$ is the subspace of P consisting of *minimal geodesics*; then he explicitly computed the groups $\pi_i(P_{\min})$ using the knowledge of the geodesics of $SU(2m)$ derived from Lie theory. Both steps relied heavily on Morse's theorems.

It is convenient to express Bott's periodicity by introducing the "infinite unitary group" $U = \varinjlim U(m)$; Bott's theorem has then the simpler form that there exist *weak homotopy equivalences*

$$\mathbf{U} \to \Omega(\mathbf{B_U}), \qquad \mathbf{B_U} \times \mathbf{Z} \to \Omega(\mathbf{U}).$$

Other proofs for that formulation, and for the similar results concerning the "infinite" groups $\mathbf{O} = \varinjlim \mathrm{O}(m)$ and $\mathbf{Sp} = \varinjlim \mathrm{Sp}(m)$, do not use Morse theory.

XXIV) In the 1950s there also were new departures in algebraic and differential topology, with the introduction of *new groups and rings* attached to topological spaces or to smooth manifolds.

Cobordism appeared as a revival of Poincaré's unsuccessful 1895 attempts to define homology using only manifolds. Smooth manifolds (without boundary) are again considered as "negligible" when they are *boundaries* of smooth manifolds-with-boundary. But there is a big difference, which keeps definition of "addition" of manifolds from running into the difficulties encountered by Poincaré; it is now the disjoint union. The (unoriented) *cobordism relation* between two compact smooth manifolds M_1, M_2 of same dimension n simply means that their disjoint union is the boundary of an $(n+1)$-dimensional smooth manifold-with-boundary. This is an equivalence relation, and the classes for that relation of n-dimensional manifolds form a *commutative group* $\mathfrak{N}_n$ in which every element has order 2. The direct sum $\mathfrak{N}_\bullet = \bigoplus_{n\geqslant 0} \mathfrak{N}_n$ is a *ring* for the multiplication of classes deduced from the cartesian product of manifolds.

This idea is not very deep, but R. Thom used it with a remarkable originality; mustering all the resources algebraic topology had accumulated in a half century, he was able to determine the structure of the graded ring $\mathfrak{N}_\bullet$, namely

$$\mathfrak{N}_\bullet \simeq \mathbf{F}_2[v_2, v_4, v_5, v_6, v_8, v_9, \ldots] \tag{28}$$

an $\mathbf{F}_2$-algebra of polynomials in infinitely many indeterminates v_j of degree j, with $j \neq 2^k - 1$. He used the following ideas:

1. For every vector bundle $\xi = (\mathrm{E}, \mathrm{B}, \mathbf{R}^k)$ with B a finite CW-complex, he considered the finite CW-complex $\mathrm{T}(\xi)$ (now called the pointed *Thom space* of ξ) obtained by compactifying E with a single point at infinity, which is the privileged point of that space.
2. When B is a smooth manifold, for every $m > k$, he defined a natural homomorphism

$$\lambda_\xi\colon \pi_m(\mathrm{T}(\xi)) \to \mathfrak{N}_{m-k} \tag{29}$$

by associating to each continuous map of pointed spaces $f\colon \mathbf{S}_m \to \mathrm{T}(\xi)$ a C^∞ map g homotopic to f, for which $g^{-1}(\mathrm{B})$ is an $(m-k)$-dimensional smooth submanifold of $\mathbf{S}_m$; he proved that any two such maps are cobordant.
3. Specializing to the vector bundle $\gamma_{m,k} = (\mathrm{U}_{m,k}, \mathrm{G}_{m,k}, \mathbf{R}^k)$, he proved that

$$\lambda_{\gamma_{m,k}}\colon \pi_{n+k}(\mathrm{T}(\gamma_{m,k})) \to \mathfrak{N}_n$$

is bijective for m and k large (another case of stability) by direct geometric constructions inspired by Steenrod's proof of the classifying theorem for vector bundles.

4. Using the Steenrod squares as Serre had done for the study of the cohomology of the Eilenberg–Mac Lane spaces, Thom finally proved that for $h < k$ and m large, the homotopy groups $\pi_{k+n}(\mathrm{T}(\gamma_{m,k}))$ are the same as those of a certain space, the product of Eilenberg–Mac Lane spaces. This also enabled him to determine the multiplication in $\mathfrak{N}_{\bullet}$.

Thom also defined *oriented cobordism*, where only oriented smooth manifolds (both with and without boundary) are considered; there is a corresponding graded ring $\Omega_{\bullet}$ of oriented cobordism classes. But he could only determine Ω_n explicitly for $n \leqslant 7$; he showed that $\Omega_{\bullet} \otimes_{\mathbf{Z}} \mathbf{Q}$ is a graded polynomial algebra

$$\mathbf{Q}[w_4, w_8, \ldots, w_{4j}, \ldots]$$

where w_{4j} is the class of the projective complex space $\mathbf{P}_{2j}(\mathbf{C})$. Later work by Milnor, Averbuch, Rokhlin, and C.T.C. Wall completely elucidated the rather complicated structure of the ring $\Omega_{\bullet}$ itself.

XXV) The starting point of K-*theory* was even simpler that the initial idea of cobordism, and it is almost unbelievable that nobody thought of it before Grothendieck did in 1957. Since the beginning of the 20th century it had been well known that any commutative monoid* without zero divisor could be canonically imbedded in a commutative group which it generates; it is in that way that negative integers are defined from the additive monoid $\mathbf{N}$, and the positive rational numbers from the multiplicative monoid $\mathbf{N}^* = \mathbf{N} - \{0\}$. Grothendieck applied to monoids the idea of "universal" object, already known at that time, and which he a little later elaborated in the now familiar concept of "representable functor" and used with great power. For any commutative monoid M there is a commutative group K(M), defined up to a unique isomorphism, and a canonically defined homomorphism of monoids $\varphi\colon \mathrm{M} \to \mathrm{K}(\mathrm{M})$ such that for any commutative group G, any homomorphism of monoids $u\colon \mathrm{M} \to \mathrm{G}$ uniquely factorizes into

$$u\colon \mathrm{M} \xrightarrow{\varphi} \mathrm{K}(\mathrm{M}) \xrightarrow{v} \mathrm{G}$$

where v is a homomorphism of groups. The proof is the (now commonplace) method of "generators and relations" reminiscent of Whitney's definition of the tensor product in 1938: $\mathbf{Z}[\mathrm{M}]$ is the $\mathbf{Z}$-module of all formal linear combinations $\sum_j \lambda_j \{m_j\}$ of one-element subsets of M with coefficients in $\mathbf{Z}$, K(M) is the quotient of $\mathbf{Z}[\mathrm{M}]$ by the submodule generated by the elements $\{m + m'\} - \{m\} - \{m'\}$, and $\varphi(m)$ is the image of $\{m\}$ in that quotient.

Grothendieck, and a little later Atiyah and Hirzebruch, applied that construction to the additive monoid of *isomorphism classes of complex vector bundles* with a base space of a finite CW-complex X. Addition is deduced from

* A monoid is a set equipped with an associative law of composition having a neutral element.

the usual Whitney sum, and the Grothendieck group is written K(X); $X \mapsto K(X)$ is a contravariant functor in the category of finite CW-complexes. If $\gamma(E)$ is the image in K(X) of a vector bundle E with base space X, the tensor product of vector bundles defines on K(X) a structure of commutative ring by $\gamma(E)\gamma(F) = \gamma(E \otimes F)$.

There is a natural homomorphism of groups

$$\mathrm{ch}_X: K(X) \to H^{\bullet}(X; \mathbf{Q})$$

defined by combinations of Chern classes. It enabled Grothendieck to obtain a "relative" version of the Riemann–Roch–Hirzebruch theorem for algebraic varieties (later Atiyah and Hirzebruch did the same for smooth manifolds). For a morphism $f: X \to Y$, it relates the maps ch_X, ch_Y and maps $K(X) \to K(Y)$ and $H^{\bullet}(X; \mathbf{Q}) \to H^{\bullet}(Y; \mathbf{Q})$ deduced from f.*

XXVI) The central result of Atiyah and Hirzebruch in their work on K(X) was the proof that the natural map

$$K(X) \otimes K(\mathbf{S}_2) \to K(X \times \mathbf{S}_2)$$

deduced from the product of vector bundles on X and on $\mathbf{S}_2$ is an *isomorphism* of rings; their proof used Bott's periodicity theorem, but several more elementary proofs were soon found. This led them to use the iterated suspension to define

$$K^{-n}(X) = K(S^n X) \quad \text{for } n \geqslant 0;$$

for a subspace $Y \subset X$ which is a sub-CW-complex, they introduced a "relative" K-group

$$K^{-n}(X, Y) = K^{-n}(X/Y)$$

where X/Y is the CW-complex obtained by collapsing Y to a point. With these definitions they proved that there exists a natural exact sequence

$$\cdots \to K^{-n-1}(Y) \to K^{-n}(X, Y) \to K^{-n}(X) \to K^{-n}(Y) \to \cdots \to K^0(X) \to K^0(Y).$$

(where in fact there are only 6 different groups, since $K^{-n-2}(X) \simeq K^{-n}(X)$). This of course reminds one of the homology exact sequence, and in fact Atiyah and Hirzebruch showed that all Eilenberg–Steenrod axioms are satisfied by the functors $K^{-n}(X, Y)$, with the obvious exception of the "dimension" axiom, since for X = pt. (one point space) the $K^{-n}(\mathrm{pt.})$ are not all 0 for $n \neq 0$.

This was the first example of functors having these axiomatic properties, defining what was called *generalized homology* and *generalized cohomology*. Other ones soon followed; a general construction is based on the concept of *spectrum* of spaces, defined by Lima in 1958 and linked to the notion of suspension: it is a sequence of spaces

* The functors K and H being contravariant, such maps can only exist under special assumptions on the spaces X and Y.

$$E: E_1, E_2, \ldots, E_n, \ldots$$

with given continuous maps

$$\rho_n: SE_n \to E_{n+1}.$$

It is used to define generalized cohomology groups as direct limits

$$H_E^q(X) = \varinjlim [S^n X; E_{q+n}]$$

and there is a similar definition of generalized homology groups.

Singular cohomology corresponds to $E_n = K(\Pi, n)$, and the Atiyah–Hirzebruch functors to $E_{2n} = B_U$, $E_{2n+1} = U$. The Thom cobordism groups can also be obtained by this process. It is therefore not surprising that after 1960 generalized homology and cohomology have become central concepts in algebraic topology.

CHAPTER I

Fundamental Group and Covering Spaces

§ 1. Covering Spaces

In modern mathematics three notions are interconnected in such a way that each one essentially determines the other two: *fundamental group*, *covering space*, and *properly discontinuous group*. Historically, they appeared in the reverse order.

In modern terminology a properly discontinuous group G is a group of *homeomorphisms* $s: x \mapsto s.x$ of a topological space X onto itself having the property that for any $x \in X$ there is an open neighborhood U of x such that for any two distinct elements s, s' of G, the intersection $s.U \cap s'.U \neq \varnothing$. This implies that for any $x \in X$, the *orbit* $G.x$ of x is a *discrete* subset of X, and that the map $s \mapsto s.x$ of G into $G.x$ is *bijective* (a property also expressed by saying that there are *no fixed points* in X for homeomorphisms $s \in G$ distinct from the identity e: for any $x \in X$, the relation $s.x = x$ implies $s = e$).

Trivial examples of properly discontinuous groups are the groups of integral translations $x \mapsto x + na$ ($n \in \mathbf{Z}, a \neq 0$) in $\mathbf{R}$, $z \mapsto z + m\omega_1 + n\omega_2$ in $\mathbf{C}$ [$(m, n) \in \mathbf{Z}^2$, ω_1/ω_2 not real]. The latter appeared at the beginning of the theory of elliptic functions in unpublished papers by Gauss. Much less obvious is the group $G = \Gamma(2)$ of transformations

$$z \mapsto (\alpha z - \beta i)/(\gamma + \delta i z)$$

of the half-plane P: $\operatorname{Re} z > 0$, where α, β, γ, δ are integers such that α, δ are odd, β, γ are even and $\alpha\delta - \beta\gamma = 1$; it was studied by Gauss in connection with his discovery of the first example of *modular functions*, which are holomorphic in P and invariant under G.

In this context Gauss had already introduced the idea of *fundamental domain*: in general it is a subset D of X such that two distinct points of D do not belong to the same orbit under G, and the disjoint sets $s.D$ for all $s \in G$ form a *partition* of X.* For the group G of translations $z \mapsto z + m\omega_1 + n\omega_2$, D can be taken as the "parallelogram of periods", i.e., the set of points $z = u\omega_1 + v\omega_2$ such that $0 \leqslant u < 1$, $0 \leqslant v < 1$. For $G = \Gamma(2)$ Gauss showed that D can be taken as the set of points $z \in P$ such that

* In general, such a set is not unique.

J. Dieudonné, *A History of Algebraic and Differential Topology, 1900–1960*,
Modern Birkhäuser Classics, DOI 10.1007/978-0-8176-4907-4_15,

$$-1 < \operatorname{Im} z \leqslant 1 \qquad \text{and} \qquad -1 < \operatorname{Im}(1/z) \leqslant 1.$$

When, after Riemann, modular functions were rediscovered and actively studied (Dedekind, Klein), as well as quotients of solutions of linear differential equations of the second order (Schwarz, Fuchs), many more properly discontinuous groups of transformations of P and fundamental domains for these groups were defined and investigated. They all were included as particular cases in the general theory of fuchsian and kleinian groups created by Poincaré [366].

With Riemann a second member of the triad we mentioned at the beginning made its emergence. There is no indication that anybody before Riemann had thought of a surface consisting of "many sheets, superposed on each other, and covering many times the same part of the plane." The applications of this concept made by Riemann to the theory of analytic functions of a complex variable show that he had in mind the modern concept of a "ramified covering surface" of an open subset X of the sphere S_2: a surface T equipped with a continuous "projection" p: $T \rightarrow X$ such that, except for a finite set $R \subset X$ (the "ramification points") there is an open neighborhood V of x in $X - R$ for any $x \in X - R$ such that $p^{-1}(V)$ is a disjoint union of open sets W_j in T, for which each restriction $p|W_j$ is a *homeomorphism* onto V. The surface $p^{-1}(X - R)$ is an *unramified* covering surface of $X - R$.

The relation between an unramified covering surface and the fundamental domain of a properly discontinuous group operating on an open subset of **C** is probably simplest for the group G of all translations $z \mapsto z + n$ in **C** for $n \in \mathbf{Z}$; a fundamental domain B is defined by the conditions $0 \leqslant \operatorname{Re} z < 1$. Consider the "cylinder" K obtained by "gluing" together the lines $\operatorname{Re} z = 0$ and $\operatorname{Re} z = 1$ forming the boundary of B [i.e., identifying the points $(0, y)$ and $(1, y)$ for all $y \in \mathbf{R}$ (Part 2, chap. V, § 1)]. For $z \in B$, let $\tilde{z}$ be the natural image of z in K. Then the plane **C** is an unramified covering of K with infinitely many "sheets," the projection $p(z)$ of a point $z \in \mathbf{C}$ being equal to $\tilde{z}'$, where z' is the unique point in B for which $z = z' + n$ for an integer $n \in \mathbf{Z}$.

In the last third of the nineteenth century this construction of a surface K by "gluing" together parts of the boundary of a fundamental domain F of a properly discontinuous group G of homeomorphisms of an open subset U of **C** was commonly used in the works of Schwarz, Klein, and Poincaré. "Suppose," Poincaré said, ([366], p. 147) "that we cut the [fuchsian] polygon R_0 [along its boundary], and bend it in a continuous way so that corresponding points [for the fuchsian group] of the boundary are brought together; after this operation, R_0 has become a closed surface." The set U, union of the transforms by G of F, has then become an unramified covering of that surface, and the transforms of F by G are the "sheets" of that covering space. The same process is used to define ramified coverings, which correspond to groups G for which some points of U (the projections of the "ramification points") are invariant by at least one transformation of G distinct from the identity. However, the modern definition of K as the *space of orbits* U/G (quotient space of U, the equivalence classes being the orbits) was not introduced until 1930 (Part 2, chap. V, § 1).

The detailed study by Klein of the subgroups of the modular group SL(2, Z) and of their fundamental domains gave examples of covering spaces $T \to K$, where T is not simply connected, being a surface obtained by the same gluing process as K but for a subgroup of G. In 1883 Poincaré described a process that in a sense "reverses" the association of a "surface of orbits" K to a group G operating in a *simply connected* open set $U \subset C$. He did not speak of surfaces, but starting from "nonuniform" analytic functions $y_1, y_2, \ldots, y_m$, holomorphic in an open arcwise connected open set $V \subset C$, he constructed a covering surface U of V that is *simply connected* and by which the y_j are "uniformized": there are *single-valued* genuine holomorphic functions $\tilde{y}_j$ *defined in* U such that the (generally different) values that $\tilde{y}_j$ takes at the points of U which are above the *same* point $z \in V$ are the "values of the multiform function" y_j at the point z ([368], pp. 57–69). When the y_j are "algebraic functions," this construction amounts to finding a common *universal covering space* (not connected in general) for a finite number of Riemann surfaces from which the ramification points have been deleted.

Poincaré's construction does not substantially differ from the modern one: starting from a fixed point $a \in V$, he considered the paths joining a to z in V for each $z \in V$; two such paths γ, γ' are said to be *equivalent* if the analytic continuations of each y_j along γ and γ' take the same value at z and if γ and γ' are homotopic in V (the extremities a, z remaining fixed). The equivalence classes for this relation are the *points of* U *above* z; Poincaré took it for granted that "intuitively" the set U of all these points could be considered as a surface (topology did not exist in 1883!).

Finally, we saw in Part 1, chap I, § 2 that in his paper *Analysis Situs* Poincaré defined the third member of our triad, the *fundamental group* of a connected manifold X. His definition was essentially the same as ours, described in an "intuitive" language: *loops* of origin $a \in X$ (Poincaré calls them "chemins") are paths starting at a and coming back to the same point; the composition $\alpha\beta$ of two loops of origin a is the loop obtained by *juxtaposition*, first moving along α and then along β; the *inverse* α^{-1} is obtained by moving along α in the opposite direction.

Poincaré considered two loops that can be "deformed" into one another "equivalent", their common origin being fixed (remember homotopy had not yet been precisely defined)*; he then called the group defined by composition of the equivalence classes the *fundamental group* of X at the point a, which we now write $\pi_1(X, a)$. Poincaré showed that if b is another point of X and ω is a path joining b to a, by associating to a loop β of origin b the loop $\omega\beta\omega^{-1}$ of origin a, an isomorphism of $\pi_1(X, b)$ onto $\pi_1(X, a)$ could be defined.†

Poincaré did not mention the fundamental group again in the two first

* In *Analysis Situs*, he apparently thought that a loop "equivalent to 0" was one that bounds a connected surface ([369], p. 241); he corrected the error in the fifth "*Complément*" (*ibid.*, p. 469).

† This justifies the notation $\pi_1(X)$ for a group $\pi_1(X, x_0)$ with unspecified x_0.

Compléments, but he devoted the third one to the determination of $\pi_1(X)$ when X is a special algebraic surface (Part 2, chap. VII, § 1,A). In the fifth *Complément* ([369], pp. 435–498) he again used it in an essential way to construct a three-dimensional compact connected manifold X without boundary having the same homology as the sphere S_3 but a *nontrivial* fundamental group. Written in the usual "intuitive" style of Poincaré, this extremely long and involved construction defining X as generated by "moving" surfaces depending on a real parameter is impossible to describe in detail here; for another interpretation of that space, see § 4D. In defining generators of the fundamental group of X in his constructions, Poincaré found it necessary to summarize the (then classical) description of a compact Riemann surface Y of genus g by gluing edges of a "fuchsian" polygon P with $4g$ edges. Apparently for the only time in all his papers he explicitly mentioned (without proof) the isomorphism between $\pi_1(Y)$ and the "fuchsian group" having P as fundamental domain.

After Poincaré the stage was set for the development of the complete theory of the three interrelated notions mentioned at the beginning of this section.

§ 2. The Theory of Covering Spaces

The definition of an (unramified) covering surface (§ 1) immediately extends to arbitrary topological spaces but is only interesting for *arcwise-connected* spaces. The essential property linking a space X and a covering space Y with projection $p: Y \to X$, are two statements that we may call the *path lifting theorem*:

1. If $t \mapsto \beta(t)$ is a path in X defined in $[0, 1]$ with $\beta(0) = a$, then for any point $b \in p^{-1}(a)$ there is a unique path $\gamma: [0, 1] \to Y$ with $\gamma(0) = b$ and $\beta = p \circ \gamma$ (a "lifting" of β).
2. If $\beta: [0, 1] \to X$ and $\beta': [0, 1] \to X$ are two paths in X with $\beta(0) = \beta'(0) = a$ and if $\varphi: [0, 1] \times [0, 1] \to X$ is a homotopy in X between β and β' leaving a fixed, there exists a homotopy $\psi: [0, 1] \times [0, 1] \to Y$ between the path γ lifting β and the path γ' lifting β' with the same origin such that $\varphi = p \circ \psi$ ("lifting" of the homotopy φ).

It is quite likely that Poincaré and his immediate successors, such as Tietze and Dehn, were aware of that theorem, even if they did not formulate it explicitly (still less try to prove it!). It was stated and proved for surfaces by H. Weyl in 1913 [483]. In 1928 Reidemeister published a short paper on fundamental groups and covering spaces [386] in which he only considered combinatorial manifolds. He deplored the fact that, except for dimension 2, no previous treatment of these questions was available. His own treatment is in fact limited to dimension 3, although he claims it might be generalized to any dimension; it is written in the Poincaré style, without genuine proofs, and the path lifting theorem is not mentioned.

It was only in 1934 that, in their book [421], Seifert and Threlfall gave an admirable and thorough elaboration of the relations between fundamental groups and covering spaces based on the path lifting theorem: although limited to locally finite simplicial complexes (of any dimension), it is essentially definitive, and can be extended to more general spaces with only minor modifications. We shall follow their description, keeping in mind that many of their results had already been formulated in imperfect forms before them and probably did not seem very new to their contemporaries (but certainly their proofs did!).

I. For a covering space Y of X with projection $p\colon \mathrm{Y} \to \mathrm{X}$ and any point $a \in \mathrm{X}$ the group $\pi_1(\mathrm{X}, a)$ *operates* naturally on $p^{-1}(a)$. For any loop γ in X of origin a and any point $b \in p^{-1}(a)$ there is a *unique* path γ_b in Y of origin b such that $p \circ \gamma_b = \gamma$. Furthermore, if γ' is a loop of origin a in X homotopic to γ, the corresponding path γ'_b is homotopic to γ_b, hence has the same extremity since $p^{-1}(a)$ is discrete; this extremity is a point that therefore only depends on b and on the class $u \in \pi_1(\mathrm{X}, a)$ of γ, and that may be written $b\,.\,u$. It is readily verified that for the neutral element e of $\pi_1(\mathrm{X}, a)$, $b\,.\,e = b$, and that $(b\,.\,u)\,.\,v = b\,.\,(uv)$ for any two elements u, v of $\pi_1(\mathrm{X}, a)$, hence $\pi_1(\mathrm{X}, a)$ operates *on the right* on $p^{-1}(a)$.

II. The fundamental group $\pi_1(\mathrm{X}, a)$ is a covariant *functor* (Part 1, chap. IV, §8) in the category $\boldsymbol{T}$ of all topological spaces. For any continuous map $f\colon \mathrm{X} \to \mathrm{X}'$ and any loop γ in X of origin a, $f \circ \gamma$ is a loop in X' of origin $f(a)$. If γ_1, γ_2 are two loops in X of origin a homotopic in X, $f \circ \gamma_1$ and $f \circ \gamma_2$ are homotopic in X'. Hence there is a map $f_*\colon \pi_1(\mathrm{X}, a) \to \pi_1(\mathrm{X}', f(a))$, which to each class $u \in \pi_1(\mathrm{X}, a)$ associates the class in $\pi_1(\mathrm{X}', f(a))$ of the loops $f \circ \gamma$ for $\gamma \in u$; it is immediately verified that f_* is a *homomorphism* of groups.

III. In their book Seifert and Threlfall only considered covering spaces that were arcwise-connected. For an arcwise-connected space X and an arbitrary covering space Y of X the question arises whether the arcwise-connected *components* of Y are again covering spaces of X. This is easily seen to be the case if X is *locally arcwise-connected*, which means that the arcwise-connected components of any open subset of X are open. But for spaces X which do not fulfil that condition counterexamples are known of covering spaces Y whose arcwise-connected components are not covering spaces of X ([440], p. 64).

IV. Let Y be a covering space of an arcwise-connected space X and $p\colon \mathrm{Y} \to \mathrm{X}$ be the projection. In order that Y be arcwise-connected it is necessary and sufficient that for one point $a \in \mathrm{X}$, $\pi_1(\mathrm{X}, a)$ operates *transitively* on $p^{-1}(a)$ (and then the same property holds for all points of X). Then for any point $b \in p^{-1}(a)$ the homomorphism

$$p_*\colon \pi_1(\mathrm{Y}, b) \to \pi_1(\mathrm{X}, a) \tag{1}$$

is *injective*, and its image is the *stabilizer* S_b of b for the action of $\pi_1(\mathrm{X}, a)$ on $p^{-1}(a)$. If $b' = b\,.\,u$ is another point of $p^{-1}(a)$, with $u \in \pi_1(\mathrm{X}, a)$, the stabilizer

$$\mathrm{S}_{b'} = u^{-1}\mathrm{S}_b u. \tag{2}$$

V. The question of "lifting" a map $f\colon \mathrm{Z} \to \mathrm{X}$ to a map $\tilde{f}\colon \mathrm{Z} \to \mathrm{Y}$ into a covering space Y of X with projection $p\colon \mathrm{Y} \to \mathrm{X}$ may be considered for spaces

Z other than an interval of **R**. Suppose X, Y, Z are arcwise-connected; let $c \in Z$ and $f(c) = a$. Then the existence of $\tilde{f}$ such that $p \circ \tilde{f} = f$ immediately implies that if $\tilde{f}(c) = b$ and $a = p(b)$, then

$$f_*(\pi_1(Z, c)) \subset p_*(\pi_1(Y, b)). \tag{3}$$

Conversely, that condition implies that if two paths γ, γ' join c to a point $z \in Z$, the paths in Y with origin b lifting $f \circ \gamma$ and $f \circ \gamma'$ have the same extremity, by IV. This extremity thus only depends on z and may be written $\tilde{f}(z)$; if Z is locally arcwise-connected, $\tilde{f}$ is continuous. Condition (3) is always satisfied if Z is *simply connected*; the existence of the "lifting" $\tilde{f}$ is then sometimes called the *monodromy principle*.

VI. Let X be an arcwise-connected and locally arcwise-connected space. There is then a relation of *domination* between the arcwise-connected covering spaces of X: such a covering space (Y', p') dominates (Y, p) if there is a continuous map $q: Y' \to Y$ for which Y′ is a covering space of Y and $p' = p \circ q$. It follows from IV and V that this condition is equivalent to the following: There exist $b \in Y$ and $b' \in Y'$ such that $p(b) = p'(b')$ and the group $p'_*(\pi_1(Y', b'))$ is *conjugate* in $\pi_1(X, p(b))$ to a subgroup of $p_*(\pi_1(Y, b))$. In particular, if $p_*(\pi_1(Y, b))$ and $p'_*(\pi_1(Y', b'))$ are conjugate, there is a unique *homeomorphism* $g: Y' \to Y$ such that $p' = p \circ g$; such a homeomorphism is also called an X-*isomorphism* of Y′ onto Y.

VII. From this last result it follows that if, to a class of X-isomorphic arcwise-connected covering spaces of X is associated the corresponding class of conjugate subgroups of $\pi_1(X, a)$, the map thus defined is *injective*. It is natural to ask if it is *bijective*; in other words, for any subgroup H of $\pi_1(X, a)$, are there arcwise-connected covering spaces Y of X such that $\pi_1(Y, b)$ is isomorphic to H for a point $b \in Y$ above a?

An obvious approach is to generalize Poincaré's construction of 1883 (§ 1): for a point $x \in X$ two paths β, β' joining a to x are H-equivalent if the class of $\beta'\beta^{-1}$ in $\pi_1(X, a)$ *belongs to* H; Y is the set of these H-*equivalence classes*, the map $p: Y \to X$ being obvious. Next define a topology on Y, assuming X is *locally arcwise-connected*; for any arcwise-connected open set U in X containing x, and any path α joining a to x, $\langle \alpha, U \rangle$ is the set of H-equivalence classes of paths $\alpha\gamma$, where γ is a path *in* U of origin x; these sets form a base for a topology on Y for which p is continuous and open.

It remains to be seen that Y is a covering space of X, and that, for a point $b \in p^{-1}(a)$, $p_*(\pi_1(Y, b)) = H$. This, however, is only true if X satisfies an additional property: Consider an open subset $U \subset X$ such that $p^{-1}(U)$ is trivial, and all loops "attached to U" of origin a. These are the loops $\alpha^{-1}\gamma\alpha$, where α is a path from a to a point $x_0 \in U$, and γ a *loop in* U of origin x_0; all such loops lift to *loops in* Y. Let $\mathscr{U}$ be a covering of X by open sets U_α such that $p^{-1}(U_\alpha)$ is trivial, and let $\pi_1(\mathscr{U}, a)$ be the subgroup of $\pi_1(X, a)$ generated by the classes of all loops attached to a $U_\alpha \in \mathscr{U}$; then, as these loops all lift to loops in Y,

$$\pi_1(\mathscr{U}, a) \subset p_*(\pi_1(Y, b)) \tag{4}$$

for all $b \in p^{-1}(a)$. The necessary and sufficient condition for the existence of the connected covering space Y such that $p_*(\pi_1(Y, b)) = H$ is the existence of a covering $\mathscr{U}$ of X by open sets such that

$$\pi_1(\mathscr{U}, a) \subset H. \tag{5}$$

This condition is always satisfied if there is an open covering $\mathscr{U}$ of X consisting of arcwise-connected and *simply connected* open sets; this is always the case if X is a C^0-manifold or a locally finite simplicial complex. Seifert and Threlfall could therefore prove the existence of a covering space of such a complex with fundamental group an *arbitrary* subgroup of $\pi_1(X, a)$ ([421], pp. 189–192).

VIII. Let X be an arcwise-connected space and Y be an arcwise-connected covering space of X, $p: Y \to X$ the natural map. An X-*automorphism* (or "decktransformation") of Y is a homeomorphism f of Y onto itself such that $p(f(y)) = p(y)$ for all $y \in Y$; restricted to any fiber $p^{-1}(x)$, f is a *permutation* of that (discrete) space. In fact such an X-automorphism is entirely determined by its restriction to *one* fiber $p^{-1}(a)$, for if $y \in Y$ and β is a path joining a to $x = p(y)$, β^{-1} lifts to a unique path β_y^{-1} joining y to a point $b \in p^{-1}(a)$, such that $p \circ \beta_y = \beta$. The group $\mathrm{Aut}_X(Y)$ of all X-automorphisms is thus isomorphic to its restriction to any fiber.

To determine all X-automorphisms of Y, it is necessary to see what condition a point $b' \in p^{-1}(a)$ must satisfy in order for there to exist an X-automorphism f of Y such that $f(b) = b'$. Since f_* transforms $\pi_1(Y, b)$ into $\pi_1(Y, b')$, $p_*(\pi_1(Y, b')) = p_*(\pi_1(Y, b))$; this means that the stabilizers S_b and $S_{b'}$ must be the same, and if $b' = b \,.\, u$, $u^{-1} S_b u = S_b$ by (2). In other words u must belong to the *normalizer* $N(S_b)$ in $\pi_1(X, a)$; but by VI that condition is sufficient, and $\mathrm{Aut}_X(Y)$ is therefore *isomorphic to* $N(S_b)/S_b$.

In particular, b' can be *any* point of $p^{-1}(a)$ if and only if S_b is a *normal* subgroup of $\pi_1(X, a)$, and then $S_{b'} = S_b$ for all points $b' \in p^{-1}(a)$; the group $\mathrm{Aut}_X(Y)$ then acts *transitively* on each fiber $p^{-1}(a)$ and is isomorphic to $\pi_1(X, a)/S_b$. These covering spaces were already considered by the immediate successors of Poincaré, who called them "regular"; later they also were named "Galois coverings" of X, and for any such space Y, $\mathrm{Aut}_X(Y)$ was called the *Galois group* of Y. These coverings were also characterized as those for which either *every* lifting in Y of a loop in X of origin a is a loop in Y or *none* is a loop.

IX. From VII it follows that for an arcwise- and locally arcwise-connected space X to have a *simply connected* arcwise-connected covering space $\tilde{X}$, a necessary and sufficient condition is that there exist an open covering $\mathscr{U}$ of X by arcwise-connected sets, such that $\pi_1(\mathscr{U}, a) = 0$. By VI such a space *dominates* all other arcwise-connected covering spaces of X and is therefore called a *universal covering space* of X; it is unique up to X-isomorphism. The covering space $\tilde{X}$ is a Galois covering space, and $\mathrm{Aut}_X(\tilde{X})$ is isomorphic to $\pi_1(X, a)$. If $f: X \to Y$ is a continuous map, and the universal covering spaces $\tilde{X}$, $\tilde{Y}$ exist, then f can be "lifted" to a continuous map $\tilde{f}: \tilde{X} \to \tilde{Y}$ such that the diagram

$$\begin{array}{ccc} \tilde{X} & \xrightarrow{\tilde{f}} & \tilde{Y} \\ \downarrow & & \downarrow \\ X & \xrightarrow[f]{} & Y \end{array}$$

commutes; this follows at once from V.

X. Let X be an arcwise-connected space, Y an arcwise-connected covering space of X, $p: Y \to X$ the natural map. For any $x \in X$ there is an open neighborhood U of x such that $p^{-1}(U)$ is the disjoint union of open subsets V_α of Y such that $p|V_\alpha$ is a homeomorphism onto U, hence $\mathrm{Aut}_X(Y)$ is a *properly discontinuous* group of homeomorphisms of Y. Conversely, suppose G is a properly discontinuous group of homeomorphisms of a space Z and let p be the natural map of Z onto the space of orbits Z/G. If V is an open set of Z such that all open sets $s.V$ for $s \in G$ are distinct, the image $p(V)$ in Z/G is open and $p^{-1}(p(V))$ is the disjoint union of the open sets $s.V$, the restriction of p to each $s.V$ being a homeomorphism on $p(V)$. If Z is arcwise-connected and locally arcwise-connected, so is Z/G. Since G operates transitively in each orbit, Z is a *Galois* covering of Z/G and $\mathrm{Aut}_{Z/G}(Z)$ is isomorphic to G.

In particular, if Z is *simply connected*, it is the universal covering of Z/G, and $\pi_1(Z/G, x_0)$ is isomorphic to G.

XI. Another approach to covering spaces was developed by Chevalley [131], following a suggestion of H. Cartan. It is no longer based on the use of paths and deals with *connected* (not necessarily arcwise-connected) spaces. The definition of a *simply connected* space X is then that any connected covering space of X is trivial. If every point of a space X has a simply connected neighborhood, one says X is *locally simply connected.* Chevalley proved that any connected, locally connected and locally simply connected space X has a simply connected covering space $\tilde{X}$ ([131], pp. 54–56) and that $\tilde{X}$ is unique up to X-isomorphism. The fundamental group $\pi_1(X)$ is then defined as the group of X-automorphisms of $\tilde{X}$. There are arcwise-connected and locally arcwise-connected spaces X that are simply connected in the sense of Chevalley but for which the fundamental group $\pi_1(X, a)$ (by Poincaré's definition) is not reduced to the identity ([440], p. 84).

Finally, mention must be made of the notion that generalizes *ramified* Riemann surfaces. A *branched covering* Y of a space X is a space with a continuous map $p: Y \to X$ such that there is a dense open subset $X - B$ for which $p^{-1}(X - B)$ is a covering space of $X - B$; B is the *branch set* (or *branch locus*) of the covering. One of the main topics is the possible extension to a branched covering of a covering space of $X - B$ ([400], pp. 292–296).

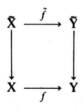

§ 3. Computation of Fundamental Groups

A. Elementary Properties

It follows at once from the definition that for two "pointed spaces" (X, x_0), (Y, y_0) there is a natural isomorphism

$$\pi_1(X \times Y, (x_0, y_0)) \cong \pi_1(X, x_0) \times \pi_1(Y, y_0) \tag{6}$$

deduced from the map $z \mapsto (\mathrm{pr}_1 z, \mathrm{pr}_2 z)$.

If $A \subset X$ is a strong deformation retract of X (Part 1, chap. IV, § 6,B), the homomorphism $i_*: \pi_1(A, x_0) \to \pi_1(X, x_0)$ corresponding to the natural injection $i: A \to X$ is an isomorphism for every $x_0 \in A$, since any loop in X of origin x_0 is homotopic to a loop in A by a homotopy fixing the points of A.

If X is obtained by adjoining to a space A a family of n-cells with $n \geqslant 2$ (Part 2, chap. V, § 3), and $i: A \to X$ is the natural injection, the homomorphism $i_*: \pi_1(A, x_0) \to \pi_1(X, x_0)$ is surjective for $x_0 \in A$ ([440], p. 146).

B. Fundamental Groups of Simplicial Complexes

For a space X the computation of $\pi_1(X, a)$ starting directly from the definition is usually unwieldy, and the examples given (without proof) by Poincaré are not very convincing. As early as 1908 Tietze [466] gave an algorithm for that computation which can be applied whenever a (finite or not) *triangulation* T of a space X is available. Instead of considering arbitrary paths $[\alpha, \beta] \to X$, only *edge paths* are used: juxtapositions of paths $\psi: [\alpha, \beta] \to X$ such that $\psi(\alpha) = v$ and $\psi(\beta) = v'$ are vertices of T and $\psi(]\alpha, \beta[)$ is an 0-simplex (when $v = v'$) or a 1-simplex. The equivalence class of such paths with fixed values v, v' is written (v, v') for the relation $\psi' = \psi \circ \rho$ where ρ is a homomorphism of an interval onto an interval.

Starting from an 0-simplex $\{x_0\}$ of T, the Poincaré definition is applied only to *edge-loops*, i.e., edge-paths having origin and extremity at x_0. Tietze assumed without proof that *any* loop of origin x_0 is homotopic to an edge-loop.* Then, instead of the topological concept of homotopy between edge-loops, he used a *combinatorial* notion mimicking it, earlier defined by Dehn and Heegaard [138]: in a juxtaposition of edge-paths $\alpha_1 \alpha_2 \cdots \alpha_r$, suppressing (or intercalating) consecutive pairs $(v, v')(v', v)$, or consecutive triples $(v, v')(v', v'')(v'', v)$ when there is a *2-simplex* of T having v, v', v'' as vertices, yields "combinatorially homotopic" edge-paths. Again, the fact that an edge-loop that is topologically homotopic to a point is also "combinatorially homotopic" to that point was taken for granted by Tietze.

The justification of Tietze's assumptions is provided by the simplicial approximation method of Alexander (Part 1, chap. II, § 3). For the first one the

* Remember that at that time the general notion of homotopic maps had not been explicitly defined.

method is applied to a map of [0, 1] in X by subdividing [0, 1] into small intervals; for the second the method is applied to a map of $[0, 1] \times [0, 1]$ into X by subdividing the square into small triangles. This is done in detail in Seifert and Threlfall's book ([421], pp. 158–162).

The possibility of considering only edge-loops instead of general loops leads to a description of $\pi_1(X, x_0)$ "by generators and relations." For each vertex v of T choose an edge-path β_v of origin x_0 and extremity v. Then, for all 1-simplices of T of extremities v, v', consider the edge-loop $\beta_v(v, v')\beta_{v'}^{-1}$; their classes are generators of $\pi_1(X, x_0)$. The relations between these generators correspond to the 1-chains of T that are boundaries of 2-chains.

This method was later improved. A *tree* in T is a one-dimensional connected and simply-connected subcomplex of T. Let A be a *maximal* tree in T, that necessarily contains *all* vertices of T, and consider an abstract group G having generators written $[v_1, v_2]$ in one-to-one correspondence with *all* 1-simplices $\{v_1, v_2\}$ of T. Between these generators the relations are: $[v_1, v_2]\cdot[v_2, v_1] = e$, $[v_1, v_2] = e$ if the 1-simplex $\{v_1, v_2\}$ is contained in A, and finally $[v_1, v_2]\cdot[v_2, v_3]\cdot[v_3, v_1] = e$ if v_1, v_2, v_3 are the vertices of a 2-simplex of T.

This abstract group G is isomorphic to $\pi_1(X, x_0)$ ([440], p. 140). For instance, this construction shows that a closed n-simplex $\bar{\Delta}_n$ is simply connected and that $\pi_1(X, x_0)$ only depends on the 2-*skeleton* T^2 of T.

C. Covering Spaces of Complexes

When X is a connected simplicial complex (finite or not), any covering space Y of X may be given a natural structure of simplicial complex. It is best to consider X and Y combinatorial complexes (Part 1, chap. II, § 2). If H is a subgroup of $\pi_1(X, a)$ corresponding to Y, the vertices of Y are the H-equivalence classes (§ 2,VII) c_v corresponding to the vertices v of X. The 1-simplices are the pairs $\{c_v, c_{v'}\}$ such that $\{v, v'\}$ is a 1-simplex of X and, if β_v is a path of the class c_v, the class of the loop $\beta_v(v, v')\beta_{v'}^{-1}$ belongs to H. Finally for any $p \geqslant 2$ a p-simplex of Y is a set $\{c_{v_0}, c_{v_1}, \ldots, c_{v_p}\}$ such that all pairs $\{c_{v_i}, c_{v_j}\}$ are 1-simplices of Y and $\{v_0, v_1, \ldots, v_p\}$ is a p-simplex of X ([421], p. 190).

It can also be proved that any covering space of a CW-complex (Part 2, chap. V, § 3) is similarly equipped in a natural way with a structure of CW-complex ([493], pp. 103–105). If Y is such a connected covering space of a CW-complex X corresponding to a subgroup of finite order d of $\pi_1(X)$, then the Euler–Poincaré characteristics are related by

$$\chi(Y) = d \, . \, \chi(X). \tag{7}$$

D. The Seifert–van Kampen Theorem

Let X be an arcwise-connected space and X_1, X_2 be two arcwise-connected *open* subspaces of X such that $X = X_1 \cup X_2$. Then the open subspace $X_0 = X_1 \cap X_2$ is not empty; if it is arcwise-connected and $i_1 \colon X_0 \to X_1$, $i_2 \colon X_0 \to X_2$

are the natural injections, then van Kampen proved in 1932 [473] that for a point $x_0 \in X_0$, $\pi_1(X, x_0)$ is naturally isomorphic to the *quotient of the free product*

$$\pi_1(X_1, x_0) * \pi_1(X_2, x_0)$$

by the normal subgroup generated by the elements

$$i_{1*}(u) i_{2*}(u^{-1}) \tag{8}$$

where u takes all values in $\pi_1(X_0, x_0)$.

The method is to decompose any loop γ: $[a, b] \to X$ of origin x_0 into small paths by a subdivision of $[a, b]$ into small intervals I_k such that $\gamma(I_k)$ is contained in $X_{v(k)}$, where $v(k)$ is equal to 0, 1, or 2. Using for each k an auxiliary path α_k joining x_0 to the origin of $\gamma(I_k)$ and entirely contained in $X_{v(k)}$, a decomposition $\gamma = \gamma_1 \gamma_2 \cdots \gamma_k \cdots \gamma_r$ is obtained, where $\gamma_k = \alpha_k(\gamma | I_k)\alpha_{k+1}^{-1}$ is a loop in $X_{v(k)}$. This proves that the natural map

$$\pi_1(X_1, x_0) * \pi_1(X_2, x_0) \to \pi_1(X, x_0)$$

is surjective.

To show that the kernel of that map is generated by the products (8), the following result must be proved: if in the previous decomposition the class of γ_k in $\pi_1(X_{v(k)}, x_0)$ is denoted by c_k, and if ψ_1: $\pi_1(X_1, x_0) \to H$, ψ_2: $\pi_1(X_2, x_0) \to H$ are two homomorphisms such that

$$\psi_1 \circ i_{1*} = \psi_2 \circ i_{2*}$$

then, if γ is homotopic to the constant loop $[a, b] \to \{x_0\}$,

$$\psi_{v(1)}(c_1)\psi_{v(2)}(c_2) \cdots \psi_{v(r)}(c_r) = e \tag{9}$$

where $\psi_0 = \psi_1 \circ i_{1*} = \psi_2 \circ i_{2*}$.

Let h: $[a, b] \times [0, 1] \to X$ be a homotopy such that

$$h(t, 0) = \gamma(t), \qquad h(t, 1) = x_0, \qquad h(a, s) = h(b, s) = x_0.$$

By decomposing each I_k into smaller intervals, and $[0, 1]$ into small intervals $[s_j, s_{j+1}]$, it may be assumed that

$$h(I_k \times [s_j, s_{j+1}]) \subset Y_{\mu(k, j)}$$

where $\mu(k, j)$ is equal to 0, 1, or 2 [it needs not be equal to $v(k)$ for $s_j \neq 0$] and

$$Y_{\mu(k, j)} = X_{\mu(k, j)} \cap X_{\mu(k+1, j)} \cap X_{\mu(k, j-1)} \cap X_{\mu(k, j+1)}.$$

For each (k, j), let β_{kj} be a path joining x_0 to $h(t_k, s_j)$ in $Y_{\mu(k, j)}$, where t_k is the origin of I_k, and consider the loops in $Y_{\mu(k, j)}$

$$\gamma_{kj} = \beta_{kj}(h|(I_k \times \{j\}))\beta_{k+1, j}^{-1},$$
$$\delta_{kj} = \beta_{kj}(h|(\{t_k\} \times [s_j, s_{j+1}]))\beta_{k, j+1}^{-1},$$

and their classes c_{kj} and d_{kj} in $\pi_1(Y_{\mu(k, j)}, x_0)$; from the definitions it follows at once that

$$\psi_{\mu(k,j)}(d_{kj}c_{k,j+1}d_{k+1,j}^{-1}c_{kj}^{-1}) = e$$

and

$$\psi_{\mu(k,j)}(d_{k+1,j}) = \psi_{\mu(k+1,j)}(d_{k+1,j}).$$

These relations imply that the two elements

$$\psi_{\mu(1,j)}(c_{1j})\psi_{\mu(2,j)}(c_{2j})\cdots\psi_{\mu(r,j)}(c_{rj}),$$

$$\psi_{\mu(1,j+1)}(c_{1,j+1})\psi_{\mu(2,j+1)}(c_{2,j+1})\cdots\psi_{\mu(r,j+1)}(c_{r,j+1})$$

are equal; but for $s_j = 0$ they are equal to (9) and for $s_{j+1} = 1$, to e, hence the conclusion.

Seifert proved an identical statement at the the same time for a locally finite cell complex X and two subcomplexes X_1, X_2 of X, but the van Kampen theorem cannot be extended to the case of *closed* arcwise-connected subspaces X_1, X_2 of an arcwise-connected space X without additional assumptions; there is an example in which X_1 and X_2 are closed and simply connected, $X_1 \cap X_2$ is reduced to a single point, but $X = X_1 \cup X_2$ is not simply connected ([213]; see [358]).

E. Fundamental Group and One-Dimensional Homology Group

Although Poincaré did not use homology *groups*, we have seen (Part 1, chap. I, § 2) that he observed that relations

$$s_1^{\alpha_1}s_2^{\alpha_2}\cdots s_k^{\alpha_k} = e$$

in the fundamental group $\pi_1(X, a)$ gave rise to "homologies"

$$\alpha_1 S_1 + \alpha_2 S_2 + \cdots + \alpha_k S_k \sim 0$$

where S_j is a one-dimensional cycle containing the point a, and which, for a convenient orientation, can be considered a loop of class s_j. Poincaré also stated without proof that all "homologies" in dimension 1 could be obtained in that way. It was not very difficult to translate this idea into a correct proof that the singular homology group $H_1(X; \mathbf{Z})$ is isomorphic to the "abelianized" fundamental group, i.e., the quotient of $\pi_1(X, a)$ by its group of commutators. This can be found in Seifert and Threlfall's book ([421], pp. 171–174) for a locally fnite simplicial complex, but it can be done in the same way for any arcwise-connected space X. A loop γ: $[0, 1] \rightarrow X$ with origin a can be considered a continuous map $[\gamma]$: $\bar{\Delta}_1 \rightarrow X$ such that $[\gamma](e_0) = [\gamma](e_1) = a$, a singular 1-cycle in $Z_1(X)$. Furthermore, if $(t, u) \mapsto F(t, u)$ is a homotopy between two loops γ, γ' of origin a, i.e., a continuous map F: $[0, 1] \times [0, 1] \rightarrow X$ such that

$$F(t, 0) = F(t, 1) = a \quad \text{for } t \in [0, 1],$$

$$F(0, u) = \gamma(u), \qquad F(1, u) = \gamma'(u) \quad \text{for } u \in [0, 1],$$

then there is a singular 2-simplex $s: \bar{\Delta}_2 \to X$ such that

$$\mathbf{b}_2(s - c) = [\gamma'] - [\gamma]$$

where c is the singular 2-simplex equal to the constant a in $\bar{\Delta}_2$. Just take

$$s(t(u\mathrm{e}_2 + (1-u)\mathrm{e}_0) + (1-t)(u\mathrm{e}_1 + (1-u)\mathrm{e}_0)) = F(t,u) \quad \text{for } 0 \leqslant t, u \leqslant 1.$$

This defines a homomorphism

$$h: \pi_1(X, a) \to H_1(X; \mathbf{Z}). \tag{10}$$

It must be shown that

1. h is surjective. For each $x \in X$, choose a path $\alpha(x)$ from a to x and to each singular simplex $s: \bar{\Delta}_1 \to X$, associate the loop

   $$l(s) = \alpha(s(\mathrm{e}_0))s(\alpha(s(\mathrm{e}_0)))^{-1};$$

   it is easy to prove that if $z = \sum_{j=1}^{r} \lambda_j s_j$ is a singular 1-cycle, the class of z in $H_1(X; \mathbf{Z})$ is the image by h of

   $$l(s_1)^{\lambda_1} l(s_2)^{\lambda_2} \cdots l(s_r)^{\lambda_r}.$$

2. The kernel of h is the commutator subgroup of $\pi_1(X, a)$. It is enough to show that for any singular 1-boundary $s_0 - s_1 + s_2$, where s_0, s_1, s_2 are singular 1-simplices such that

   $$s_1(\mathrm{e}_0) = s_0(\mathrm{e}_1), \qquad s_2(\mathrm{e}_0) = s_1(\mathrm{e}_1), \qquad s_0(\mathrm{e}_0) = s_2(\mathrm{e}_1),$$

   then the loop $l(s_0)l(s_1)^{-1}l(s_2)$ is null homotopic; this is a simple consequence of the fact that the identity map on the boundary $\bar{\Delta}_2 - \Delta_2$ is homotopic to a constant map *in* $\bar{\Delta}_2$.

§ 4. Examples and Applications

A. Fundamental Groups of Graphs

In algebraic topology a simplicial complex of *dimension* 1 is often called a *graph*. If K is a connected graph, its fundamental group $\pi_1(K, x_0)$ is *free*. This follows from the Tietze construction, improved by the use of a maximal tree T (§ 3,B): since there is no 2-simplex in K, the generators $[v_1, v_2]$ for which the 1-simplex $\{v_1, v_2\}$ is not contained in T satisfy *no* relation, and the others are equal to the neutral element.

Conversely, *any* free group F with an arbitrary family $(x_j)_{j \in J}$ of generators is isomorphic to the fundamental group of a connected graph K. Define K as a combinatorial complex: the vertices consist of two families $(v_j)_{j \in J}$, $(w_j)_{j \in J}$ and an additional vertex z; the 1-simplices are the pairs $\{z, v_j\}$, $\{z, w_j\}$, and $\{v_j, w_j\}$ for $j \in J$; then $\pi_1(K, z)$ is the free group on the generators $[v_j, w_j]$. This correspondence between homeomorphism classes of graphs and fundamental groups led to the first example of an application of algebraic topology to group theory. In 1927 O. Schreier showed by purely algebraic methods that *any subgroup of a free group is free* [416]. This result has a very simple proof using the theory of fundamental groups: suppose a connected graph X has a fundamental group $\pi_1(X, x_0)$ isomorphic to a given free group F, and let H be a subgroup of F; then there is a connected covering space Y of X such that, for a point y_0 above x_0, $\pi_1(Y, y_0)$ is isomorphic to H (§ 2,VII), and since Y has a natural structure of graph (§ 3,C), H is free. If X is a connected graph with n_0 vertices and n_1 1-simplices, a maximal tree in X has n_0 vertices and $n_0 - 1$ 1-simplices, hence $\pi_1(X, x_0)$ has $n_1 - n_0 + 1$ generators. If H is a subgroup of F of index $(F : H) = m$, the corresponding covering graph Y of X has mn_0 vertices and mn_1 1-simplices, hence $\pi_1(Y, y_0)$ is a free group with $m(n_1 - n_0) + 1$ generators [134]; this is another of Schreier's results for free groups with a *finite* number of generators.

B. The "Gruppenbild"

There are many other applications of fundamental groups to group theory. By using attachment of 2-cells to a graph (Part 2, chap. V, § 3) it is possible to construct for any group G a complex of dimension 2 having G as fundamental group ([474], p. 145). (This "realizability" property is a particular case of a much more general result for higher homotopy groups; see chap. II, § 6,F.)

A useful construction for the study of groups defined by generators and relations is Dehn's "Gruppenbild" ([421], p. 328). Suppose a group G is generated by a finite number n of elements g_ρ $(1 \leqslant \rho \leqslant n)$; take a set $(M_s)_{s \in G}$ indexed by G, and define a graph $\Gamma(G)$ having as vertices the M_s: Assign $2n$ 1-simplices having as common vertex M_s to any $s \in G$:

$$S_\rho(s) = \{M_s, M_{sg_\rho}\}, \qquad S_\rho^{-1}(s) = \{M_s, M_{sg_\rho^{-1}}\}.$$

Any product

$$S_{\rho_1}^{\varepsilon_1} S_{\rho_2}^{\varepsilon_2} \cdots S_{\rho_k}^{\varepsilon_k}$$

with $\varepsilon_j = \pm 1$ corresponds to the paths of origin M_s (for the various $s \in G$) consisting of the juxtaposition of the successive 1-simplices

$$S_{\rho_1}^{\varepsilon_1}(s), S_{\rho_2}^{\varepsilon_2}(sg_{\rho_1}^{\varepsilon_1}), \ldots, S_{\rho_k}^{\varepsilon_k}(sg_{\rho_1}^{\varepsilon_1} g_{\rho_2}^{\varepsilon_2} \cdots g_{\rho_{k-1}}^{\varepsilon_{k-1}}).$$

The relations

$$g_{\rho_1}^{\varepsilon_1} g_{\rho_2}^{\varepsilon_2} \cdots g_{\rho_p}^{\varepsilon_p} = e$$

between the generators g_ρ exactly correspond to the *loops* in $\Gamma(G)$. But if G is the fundamental group $\pi_1(X, a)$ of a simplicial complex X, Reidemeister observed [386] that one can "concretize" the graph $\Gamma(G)$. Consider the universal covering space $\tilde{X}$ of X, and take for M_s all the points of $\tilde{X}$ above a; if the g_ρ are loops of origin a whose classes generate $\pi_1(X, a)$, then $S_\rho(s)$ is the path of origin M_s which projects on g_ρ.

C. Fundamental Group of an H-Space

Let $(x, y) \mapsto xy$ be the product in an H-space X (Part 2, chap. VI, § 2) and e be the "homotopy identity." For two paths $t \mapsto \gamma_1(t)$, $t \mapsto \gamma_2(t)$, maps of $[0, 1]$ into X, write $\gamma_1 * \gamma_2$ the path $t \mapsto \gamma_1(t)\gamma_2(t)$; then, if γ_1' (resp. γ_2') is homotopic to γ_1 (resp. γ_2), $\gamma_1' * \gamma_2'$ is homotopic to $\gamma_1 * \gamma_2$. If points of X are identified with constant paths, γ_1 is a path from e to a, and γ_2 a path from e to b, then $\gamma_1 * \gamma_2$ is homotopic to each of the juxtapositions $(\gamma_1 * e)(a * \gamma_2)$ and $(e * \gamma_2)(\gamma_1 * b)$. When $a = b = e$, a loop γ of origin e can be identified with $e * \gamma$ and with $\gamma * e$; this proves that the law of composition in the fundamental group $\pi_1(\mathrm{X}, e)$ is the law of composition of classes of loops, corresponding to $(\gamma_1, \gamma_2) \mapsto \gamma_1 * \gamma_2$, and that this law is *commutative*.

This applies in particular to a *topological group* G when it is assumed that G is arcwise-connected and locally arcwise-connected. If G′ is a connected covering space of G and e' is a point of G′ above the neutral element e, there is a unique structure of topological group on G′ such that the projection p: G′ → G is a homomorphism and e' the neutral element; $p^{-1}(e)$ is isomorphic to a quotient group of $\pi_1(\mathrm{G}, e)$ and is contained in the center of G′. If there exists a universal covering space $\tilde{\mathrm{G}}$ of G, the corresponding structure of group on $\tilde{\mathrm{G}}$ (defined up to isomorphism) defines $\tilde{\mathrm{G}}$ as the *universal covering group* of G, having a center that contains a group isomorphic to $\pi_1(\mathrm{G}, e)$. Such a group always exists if G is a Lie group.

There are corresponding statements for the Chevalley conception of fundamental group (§ 2,XI).

D. Poincaré Manifolds

The three-dimensional manifold constructed by Poincaré in his fifth *Complément* has been the topic of many investigations ([400], pp. 224, 290, 308–311; [421], pp. 216–221, 227) in which it is defined in many different ways. The simplest way is to obtain it as a space of orbits $\mathrm{P} = \mathbf{S}_3/\mathrm{I}$, where I is a *finite* subgroup of the group $\mathbf{S}_3$ of quaternions of norm one, chosen in such a way that P is orientable and I is equal to its group of commutators; then $\pi_1(\mathrm{P}, e) \simeq$ I, but $\mathrm{H}_1(\mathrm{P}; \mathbf{Z}) = \mathrm{H}_2(\mathrm{P}; \mathbf{Z}) = 0$ by § 3,E and Poincaré duality. It turns out that there is such a group; $\mathbf{S}_3$ may be identified to the double covering Spin(3) of $\mathbf{SO}(3)$ and I is the inverse image in $\mathbf{S}_3$ of the finite group of rotations of the regular icosahedron (or dodecahedron), which is a simple group of order 60. Furthermore the symmetry $x \mapsto -x$, which belongs to I, is a commutator in I, hence I is a group of order 120, equal to its group of commutators. The manifold P is also called "icosahedral space" or "dodecahedral space."

E. Knots and Links

The basic problem of topology, classifying spaces up to homeomorphism, can be extended to pairs (X, A) where $\mathrm{A} \subset \mathrm{X}$, a homeomorphism

$h: (X, A) \rightrightarrows (X', A')$ being a homeomorphism of X onto X′ mapping A onto A′ [in the language of categories, they are the isomorphisms in the category $\boldsymbol{T}_1$ of pairs (Part 1, chap. IV, §5E)]. The restriction of h to $X - A$ is then a homeomorphism onto $X' - A'$ so that h is a relative homeomorphism in the sense of Lefschetz (Part 1, chap. II, §6), but the converse is not true. When $X' = X$ and (X, A) is homeomorphic to (X, A′), one says A and A′ are *equivalent* subspaces of X.*

Most of the work done in that direction has been concentrated on the case in which $X = \mathbf{S}_n$ (or $X = \mathbf{R}^n$) and A is homeomorphic to a sphere $\mathbf{S}_k$ with $k < n$ or to a disjoint union of finitely many such spheres. Already in the nineteeth century the case $n = 3$, $k = 1$ [(X, A) being called a *classical knot* when A is homeomorphic to a circle $\mathbf{S}_1$, a *classical link* if A is homeomorphic to a disjoint union of finitely many sets homeomorphic to $\mathbf{S}_1$] had a peculiar fascination (not only for professional mathematicians), due to its immediate "physical" interpretation [460].

Concerning that problem, Tietze had already observed [466] that an embedding of $\mathbf{S}_1$ in $\mathbf{R}^3$ may be very "wild," with infinitely many "knotted" portions accumulating in the vicinity of a point ([400], p. 224). A distinction therefore had to be made between such "wild" embeddings and "tame" ones: the latter are defined by the condition that for each point x_0 of the image A of $\mathbf{S}_1$ there is a neighborhood U of x_0 in $\mathbf{R}^3$ and a homeomorphism h of U onto an open ball of $\mathbf{R}^3$ such that $h(U \cap A)$ is a diameter in that ball. This is immediately generalized to any pairs (n, k) with $k < n$; an equivalent definition consists in considering a triangulation T of $\mathbf{S}_n$ and taking for A a k-dimensional subcomplex of T homeomorphic to $\mathbf{S}_k$.

As soon as topological tools became available topologists such as Dehn [137] and, a little later, Alexander [15], applied them to the study of classical knots and links and to their higher-dimensional generalizations (also called knots and links). These early works have grown into a very extensive theory, which at present engages the efforts of many mathematicians (see [400]); we shall mostly consider *tame* embeddings.

A first distinction has to be made according to the value of $n - k$; if $n - k \geqslant 3$, *all tame embeddings* of $\mathbf{S}_k$ into $\mathbf{S}_n$ are *equivalent* [400] (this, however, does not extend to embeddings of disjoint unions of at least two spaces homeomorphic to $\mathbf{S}_k$).†

For $n = 2$, $k = 1$, it is still true that any two embeddings (tame or not) of $\mathbf{S}_1$ into $\mathbf{S}_2$ are equivalent (Schönflies theorem [415]). But for $n = 3$, $k = 2$, there are "wild" embeddings of $\mathbf{S}_2$ into $\mathbf{S}_3$ (the "Alexander horned sphere," see [400], pp. 76–81) that are not equivalent to the standard embedding. To ensure an equivalent embedding, a kind of "global tameness" condition must be added, namely, the existence, for the image A of $\mathbf{S}_2$ in $\mathbf{S}_3$, of a "bicollar,"

* There are other definitions of equivalence ([400], p. 3).

† If one only considers C^∞ embeddings, the condition $n - k \geqslant 3$ is no longer sufficient to ensure equivalence of all C^∞ embeddings of $\mathbf{S}_k$ into $\mathbf{S}_n$ [218].

i.e., a neighborhood U of A such that there is a homeomorphism of A $\times$ $]-1, 1[$ onto U, reducing to the map $(x, 0) \mapsto x$ in A $\times$ $\{0\}$. That same condition also implies equivalence for embeddings of $\mathbf{S}_{n-1}$ into $\mathbf{S}_n$ for any $n > 3$.

Let A and B be two closed sets in $\mathbf{R}^n$, both homeomorphic to $\mathbf{S}_{n-1}$, such that B is contained in the bounded component of $\mathbf{R}^n - \mathrm{A}$. If U is the intersection of the bounded component of $\mathbf{R}^n - \mathrm{A}$ and the unbounded component of $\mathbf{R}^n - \mathrm{B}$ and both A and B are "bicollared", $\bar{\mathrm{U}}$ is homeomorphic to $\mathbf{S}_{n-1} \times [-1, 1]$. This remained for a long time known as the "annulus conjecture"; it was only proved for $n \neq 3$ in 1969 [400].

For the remainder of this section we will only consider the case $n - k = 2$, which is the one usually meant when one speaks of knots and links.

By definition, if two tame knots (resp. links) A, A′ in $\mathbf{S}_n$ are equivalent, $\mathbf{S}_n - \mathrm{A}$ and $\mathbf{S}_n - \mathrm{A}'$ are homeomorphic; it is not known if that condition implies conversely that A and A′ are equivalent. At any rate the *fundamental group* $\pi_1(\mathbf{S}_n - \mathrm{A})$ (called the "knot group" or the "link group") attaches one of the most important algebraic invariants of knot type or link type to any tame knot or link, but there are nonequivalent knots having isomorphic knot groups.

For $n = 3$ there is a regular way to describe the knot group by generators and relations. A tame knot K in $\mathbf{R}^3$ can be considered a Jordan polygon; it is possible to project it on a plane (taken as the xy-plane in $\mathbf{R}^3$) in such a way that the projected polygon K′ has only double points that are not on its vertices. Once K is oriented it is decomposed into a juxtaposition of simple arcs $a_i = \mathrm{P}_i\mathrm{P}_{i+1}$ $(0 \leqslant i \leqslant m,\ \mathrm{P}_{m+1} = \mathrm{P}_0)$ by the points P_i projected on the double points of K′, and such that the other point Q_i having the same projection as P_i is *above* P_i; let $j(i)$ be the index such that Q_i is contained in the arc $a_{j(i)}$. Taking a base point $b \in \mathbf{R}^3 - \mathrm{K}$ above K, assign a loop of origin b going around a_i to each arc a_i. The classes c_i of these loops generate $\pi_1(\mathbf{R}^3 - \mathrm{K})$ and the relations between them are

$$c_{j(i)}c_i = c_{i+1}c_{j(i)} \quad \text{or} \quad c_i c_{j(i)} = c_{j(i)}c_{i+1}$$

depending on the respective orientations in $\mathbf{R}^2$ of the oriented projections of a_i and $a_{j(i)}$.

Even for $n = 3$ it has not yet been possible to find a finite system of algebraic invariants that would characterize the tvpe of a classical knot. We shall only consider the invariants related to homology. For any dimension n it follows from Alexander duality (Part 1, chap. II, § 6) that if $\mathrm{K} \subset \mathbf{S}_n$ is homeomorphic to $\mathbf{S}_{n-2}$, then $\mathrm{H}_1(\mathbf{S}_n - \mathrm{K}; \mathbf{Z}) \simeq \mathbf{Z}$. In other words (§ 3,E), the space $\mathrm{X} = \mathbf{S}_n - \mathrm{K}$ is such that the quotient of $\pi_1(\mathrm{X})$ by its commutator subgroup is isomorphic to $\mathbf{Z}$. It follows (§ 3,VII) that to the group $\mathbf{Z}$ (resp. $\mathbf{Z}/m\mathbf{Z}$) there corresponds a covering space $\tilde{\mathrm{X}}_\infty$ (resp. $\tilde{\mathrm{X}}_m$) of X such that $\mathrm{Aut}_\mathrm{X}(\tilde{\mathrm{X}}_\infty) \simeq \mathbf{Z}$ [resp. $\mathrm{Aut}_\mathrm{X}(\tilde{\mathrm{X}}_m) \simeq \mathbf{Z}/m\mathbf{Z}$]. The invariant factors $\neq 1$ of the finitely generated commutative group $\mathrm{H}_1(\tilde{\mathrm{X}}_m)$ are called the *m-th torsion numbers* of the knot K and obviously are invariant under equivalence.

An even more interesting invariant comes from the consideration of the homology groups $H_i(\tilde{X}_\infty; Z)$ for $1 \leqslant i \leqslant n$. Let t be a generator of the group $G = \mathrm{Aut}_X(\tilde{X}_\infty)$ isomorphic to Z. Since G acts on every Z-module $H_i(\tilde{X}_\infty; Z)$, that action can be extended to the *group algebra* Z[G], which is the algebra of "Laurent polynomials" $Z[t, t^{-1}]$ with integer coefficients; if

$$P(t) = c_{-h}t^{-h} + \cdots + c_{-1}t^{-1} + c_0 + c_1 t + \cdots + c_k t^k$$

the action is given by

$$(P, u) \mapsto \sum_{j=-h}^{k} c_j(t^j . u) \quad \text{for } u \in H_i(\tilde{X}_\infty; Z).$$

The Z-module $H_i(\tilde{X}_\infty; Z)$ is thus given a structure of $Z[t, t^{-1}]$-module known as the *i-th Alexander invariant* of K. When that module is generated by a single element u_i, the ideal $\mathfrak{a}_i$ of $Z[t, t^{-1}]$, annihilator of u_i, is called the *Alexander ideal*; if it is a principal ideal, it is generated by a unique polynomial of $Z[t]$, called the *Alexander polynomial* [15]. There are several ways to compute it [400]; it can often distinguish between knots for which all other invariants are the same.

A similar theory may be done for links, but for a link L with r components, $H_1(L; Z)$ is isomorphic to Z^r; if $X = S_n - L$, the groups $H_i(\tilde{X}_\infty)$ are modules over "Laurent polynomials" in r variables.

CHAPTER II

Elementary Notions and Early Results in Homotopy Theory

Until around 1930 the concept of homotopy (as defined by Brouwer in 1911, see Part 1, chap. II, § 2) essentially appeared as an auxiliary notion, chiefly used as a tool in the proofs of the theorems on homology. The theory of the fundamental group stood apart, as a kind of refinement of one-dimensional homology. It was H. Hopf who, by his pioneering study of maps into spheres between 1926 and 1935, inaugurated homotopy theory.

§ 1. The Work of H. Hopf

A. Brouwer's Conjecture

One of the main results of Brouwer's theory of the degree of a map (Part 2, chap. I) was its invariance under homotopy. In a talk at the International Congress of Mathematicians in 1912 he expressed his belief that, conversely, if M is a connected, compact, orientable n-dimensional manifold (in the sense of Brouwer) and f and g are two continuous maps of M into $\mathbf{S}_n$ which have the same degree, then they are homotopic. In that talk he sketched a proof of that statement for $n = 2$ and $\mathrm{M} = \mathbf{S}_2$, and he detailed that proof a year later ([89], pp. 527–537). It is one of these fantastically complicated and obscure proofs in which he so often indulged; he went through four successive reductions, in order to finally apply a result of Klein on compact Riemann surfaces of genus 0; although at one point he uses simplicial approximation, the validity of the arguments certainly is questionable.

Hopf proved the Brouwer conjecture in 1926 [239]. The proof is long, ingenious, and intricate, and very much inspired by Brouwer's techniques, but fortunately much more precise. I think it is instructive to compare that proof "starting from scratch" with the much simpler later ones, when new topological and algebraic tools had been invented.

The main tool in the proof is a generalization obtained by Hopf in 1925 of Brouwer's criterion for the existence of fixed points for a continuous map of $\mathbf{S}_n$ into itself (Part 2, chap. III, § 1). He considered more generally a compact, connected, orientable n-dimensional combinatorial manifold M, and two

J. Dieudonné, *A History of Algebraic and Differential Topology, 1900–1960*,
Modern Birkhäuser Classics, DOI 10.1007/978-0-8176-4907-4_16,

continuous maps f_1, f_2 of M into $\mathbf{S}_n$; the *points of coincidence* of f_1 and f_2 in M are those for which $f_1(x) = f_2(x)$; it is assumed that the number of these points is finite. For each point of coincidence x an *index of coincidence* is defined: identifying a neighborhood of x and a neighborhood of $f_1(x) = f_2(x)$ with a ball of $\mathbf{R}^n$, consider a small $(n-1)$-dimensional sphere S of center x, and the vector

$$y \mapsto \frac{f_1(y) - f_2(y)}{|f_1(y) - f_2(y)|} \tag{1}$$

on S; the index of coincidence is the degree of the map $\mathrm{S} \to \mathbf{S}_{n-1}$ thus defined. Hopf's result was that the sum of the coincidence indices at all coincidence points of f_1 and f_2 is

$$(-1)^n \deg f_1 + \deg f_2. \tag{2}$$

The proof of Brouwer's conjecture is by induction on n; for $n = 1$ the result is obvious and of course was well known. We describe the proof by breaking it up into four parts.

1. An obvious generalization of Brouwer's last proof of his fixed point theorem (Part 2, chap. III, § 1) is: if g_1, g_2 are two continuous maps of M into $\mathbf{S}_n$ *without* coincidence points, then it is possible to replace g_2 by a homotopic map g_2' such that $g_2'(x) = -g_1(x)$ for all $x \in \mathrm{M}$ by considering the great circle joining $g_1(x)$ and $g_2(x)$ for all points $x \in \mathrm{M}$ such that $g_1(x) \neq -g_2(x)$.

2. Next suppose that g_1 and g_2 have only *one* point of coincidence x_0 and that $\deg(g_2) = (-1)^{n+1} \deg(g_1)$; Hopf's 1925 result showed that the index of coincidence at x_0 is 0. Let p be the stereographic projection onto the hyperplane H tangent to $\mathbf{S}_n$ at the point $g_1(x_0) = g_2(x_0)$ and let B a neighborhood of x_0 in M identified with a small closed ball of $\mathbf{R}^n$ with boundary S; then

$$x \mapsto v(x) = p(g_1(x)) - p(g_2(x))$$

is a vector field in B with values in H vanishing only at x_0. If B_1 is a closed ball of smaller radius concentric to B and S_1 is its boundary, the assumption implies that on S_1 the map $x \mapsto v(x)/|v(x)|$ has degree 0, and by the induction hypothesis there is a homotopy of the restriction $v|\mathrm{S}_1$ to a constant map. An elementary lemma then shows that there is a homotopy of v itself to a vector field v_1 defined in B_1, coinciding with v on S_1, which does not vanish in B_1. From this deduce a homotopy of g_2 to a map g_2' such that

$$v_1(x) = p(g_1(x)) - p(g_2'(x))$$

in B_1; now g_1 and g_2' have *no* coincidence point.

3. The third step deals with two *arbitrary* continuous maps h_1, h_2 of M into $\mathbf{S}_n$ with a finite number of coincidence points, and proves that by a homotopy on h_2 it is always possible to reach the situation in which h_1 and h_2 have only *one* coincidence point. By simplicial approximation it can always be supposed that h_1 and h_2 are simplicial maps and that there is a point $y_0 \in \mathbf{S}_n$ and a small ball $\mathrm{B} \subset \mathbf{S}_n$ of center y_0 such that $h_1^{-1}(\mathrm{B})$ [resp. $h_2^{-1}(\mathrm{B})$] is a disjoint union of

closed balls U_j $(1 \leqslant j \leqslant l)$ [resp. V_k $(1 \leqslant k \leqslant m)$] in M; the U_j have no common point with the V_k, and $h_1|U_j$ (resp. $h_2|V_k$) is a homeomorphism onto B.

Now let p be the stereographic projection onto the hyperplane H tangent to $\mathbf{S}_n$ at $-y_0$. If t is a translation in H, $h_2' = p^{-1}tph_2$ is homotopic to h_2; it is possible to choose t such that the points of coincidence of h_1 and h_2' belong to the disjoint union W of the interiors of the U_j and the V_k. Using a triangulation of M and the inequality $n \geqslant 2$, it is possible to construct a set $E \subset M$ homeomorphic to a closed ball in $\mathbf{R}^n$, containing W and such that $h_1(E) \cup h_2'(E) \neq \mathbf{S}_n$.

Then take a point z_0 in $\mathbf{S}_n$ that is not in $h_1(E) \cup h_2'(E)$ and let q be the stereographic projection onto the hyperplane H′ tangent to $\mathbf{S}_n$ at $-z_0$. Associate in H′ the vector $w(x) = q(h_1(x)) - q(h_2'(x))$ to each point $x \in E$; $w(x) \neq 0$ on the boundary S′ of E and an elementary lemma shows there is a homotopy transforming w into a vector field w_1 defined in E, coinciding with w on S′ and vanishing at only *one* point $u_0 \in E$. Then there is a homotopy of h_2' to h_2'' such that $q(h_1(x)) - q(h_2''(x)) = w_1(x)$ in E, and therefore u_0 is the only point of coincidence of h_1 and h_2''.

4. The end of the proof is now very short. Suppose f_1 and f_2 have equal degrees; then if s: $y \mapsto -y$ is the symmetry, $g_1 = f_1$ and $g_2 = s \circ f_2$ are such that $\deg(g_2) = (-1)^{n+1}\deg(g_1)$. A first homotopy reduces to the case in which g_1 and g_2 have only one point of coincidence, by 3; a second to the case in which g_1 and g_2 have no point of coincidence, by 2; and finally a third to the case in which $g_2 = s \circ g_1$ by 1, so that after a homotopy, $f_2 = f_1$.

Hopf added to the Brouwer conjecture the fact that for *any* integer $d \in \mathbf{Z}$ there is a continuous map f: $M \to \mathbf{S}_n$ such that $\deg(f) = d$. He first considered the case $M = \mathbf{S}_n$; if $n = 1$, the map $z \mapsto z^d$ of $\mathbf{S}_1$ into itself has degree d. For $n > 1$ he used induction on n; considering $\mathbf{S}_{n-1}$ as defined by the equation $x_{n+1} = 0$ in $\mathbf{S}_n \subset \mathbf{R}^{n+1}$, he let g: $\mathbf{S}_{n-1} \to \mathbf{S}_{n-1}$ have degree d; any point x of $\mathbf{S}_n$ other than $\pm \mathsf{e}_{n+1}$ can be written (θ, y), where $\mathsf{y} \in \mathbf{S}_{n-1}$ and $\theta \in \,]0, \pi[$ is the angle of the vectors x and e_{n+1}. He then took $f(\mathsf{x}) = (\theta, g(\mathsf{y}))$ for $\mathsf{x} \neq \pm \mathsf{e}_{n+1}$ and $f(\pm \mathsf{e}_{n+1}) = \pm \mathsf{e}_{n+1}$. The remaining problem is to define a continuous map f: $M \to \mathbf{S}_n$ of degree ± 1. Hopf took a (rectilinear) triangulation T of M with vertices $v_1, v_2, \ldots, v_k$; let $v_1', v_2', \ldots, v_k'$ be k points in $\mathbf{R}^{n+1}$ such that no set of $n + 2$ of these points are in a linear subvariety of $\mathbf{R}^{n+1}$ of dimension n. He then defined a piecewise affine map h: $M \to \mathbf{R}^{n+1}$ by the condition that $h(v_j) = v_j'$ for $1 \leqslant j \leqslant k$ and that h be affine in each simplex of T. Finally, he considered a straight line D in $\mathbf{R}^{n+1}$ intersecting $h(M)$ in a finite set, no point of which belongs to the image by h of a p-simplex of T for $p \leqslant n - 1$. He let $a_1 < a_2 < \cdots < a_r$ be the successive intersections of D with $h(M)$ ordered by an orientation of D, and he took a point b between a_1 and a_2. Then, if p is the projection from b of $h(M)$ onto a sphere S of center b, the map $x \mapsto p(h(x))$ of M onto S has degree ± 1, by Brouwer's definition.

In 1933 [244] Hopf generalized and simplified his solution of Brouwer's conjecture; we shall see in §4,B how that paper was incorporated in the nascent general theory of homotopy.

B. The Hopf Invariant

Ever since his first paper Hopf's work in algebraic topology had been centered on topological properties of *mappings* and of their connection with algebraic notions; his papers on the degree of a map (Part 2, chap. I, § 3,D), on the "Umkehrhomomorphismus" (Part 1, chap. IV, § 4), and on the Brouwer conjecture were all typical of that orientation. They were all concerned with continuous maps $X \to Y$ for combinatorial manifolds of *same* dimension; it was natural that Hopf should turn his attention to maps $f: X \to \mathbf{S}_n$ when the compact combinatorial manifold X has a dimension $m \neq n$. For $m < n$ these maps are of a trivial kind, since by a suitable simplicial approximation it can be assumed that $f(X)$ is not the whole sphere $\mathbf{S}_n$, and therefore a homotopy reduces f to a *constant* map. Aside from the fairly trivial case $n = 1$, however, before 1930 nobody knew anything about continuous maps $f: X \to \mathbf{S}_n$ when $\dim X > n$, even maps $f: \mathbf{S}_m \to \mathbf{S}_n$ with $m > n$. It was clear that the consideration of the corresponding homomorphism $f_*: H_.(\mathbf{S}_m) \to H_.(\mathbf{S}_n)$ could give no information, since for $p > 0$ either $H_p(\mathbf{S}_m) = 0$ or $H_p(\mathbf{S}_n) = 0$. There remained, however, a topological principle of classification, the "homotopy class" of Brouwer. In particular, since f_*, acting on reduced homology, is always 0, was it true that f is always homotopic to a constant map?

The breakthrough came in 1930 when, by a brilliant combination of the Brouwer technique and of his own idea of "Umkehrhomomorphismus" (Part 1, chap. IV, § 4), Hopf succeeded in proving that there are *infinitely* many "homotopy classes" of maps of $\mathbf{S}_3$ into $\mathbf{S}_2$ [243]. Again the proof, starting from scratch, is long (14 pages) and burdened with many technical details, but it is always quite clear.

I. The bulk of the proof concerns euclidean (rectilinear) simplicial complexes X, Y with triangulations T, T′ and *simplicial* maps; chain complexes are used not only for T and T′, but for subdivisions of these triangulations. For simplicity we suppose Y to be of dimension 2, X of dimension $n > 2$.

For each 2-simplex σ of T and every point ξ *interior* to a 2-simplex τ of T′, $\varphi_\sigma(\xi)$ is defined as 0 if $\xi \notin f(\sigma)$; but if $\xi \in f(\sigma)$, the restriction of f to σ is an affine bijection of σ onto τ, and if x is the unique point of σ such that $f(x) = \xi$, then $\varphi_\sigma(\xi)$ is taken equal to x if $f|\sigma$ preserves orientation, $-x$ if it reverses orientation. For any 2-chain $Z = \sum_j a_j \sigma_j$ (with the $a_j \in \mathbf{Z}$), $\varphi_Z(\xi)$ is then defined as the 0-chain $\sum_j a_j \varphi_{\sigma_j}(\xi)$, so that after subdivisions of T and T′ such that ξ and its inverse images by f are vertices of the subdivided complexes, the map $Z \mapsto \varphi_Z(\xi)$ is a homomorphism of the $\mathbf{Z}$-module $C_2(T)$ of 2-chains into the $\mathbf{Z}$-module $C_0(T)$ of 0-chains.

Starting from that definition, Hopf showed by induction on p that it was possible to define, for any p-chain Z of T, a $(p-2)$-chain $\varphi_Z(\xi)$ of some subdivision of T, in such a way that for the boundary opertors

$$\mathbf{b}_{p-2}\varphi_Z(\xi) = \varphi_{\mathbf{b}_p Z}(\xi), \tag{3}$$

and $Z \mapsto \varphi_Z(\xi)$ is a linear map of $C_p(T)$ into $C_{p-2}(T)$ [in particular $\varphi_0(\xi) = 0$].

In his 1931 paper Hopf only used this construction for $n = 3$ and $n = 4$. For $n = 3$ he assumed X to be oriented combinatorial manifold; if [X] is its fundamental 3-cycle (Part 2, chap. I, § 3) and $\xi \in f(\mathrm{X})$, he showed that for a 2-chain Z in T the intersection number $(\mathrm{Z}\,.\,\varphi_{[\mathrm{X}]}(\xi))$ is defined and equal to the degree of f at the point ξ. By this he meant that if $\mathrm{Z} = \sum_j a_j\sigma_j$, where the σ_j are 2-simplices of T, that degree is $\sum_j a_j d(f, \sigma_j, \xi)$ (Part 2, chap. I, § 3,D). To prove this equality, Hopf used the fact that any 2-simplex σ of T that meets $f^{-1}(\xi)$ has only one point in $f^{-1}(\xi)$. Since every 2-simplex of T is a face of *exactly two* 3-simplices, $\varphi_{[\mathrm{X}]}(\xi)$ is a 1-*cycle*, and the intersection number $(\sigma\,.\,\varphi_{[\mathrm{X}]}(\xi))$ is $+1$ (resp. -1) if $f|\sigma$ preserves (resp. reverses) orientation.

After these generalities Hopf only considered the case in which X (resp. Y) is a euclidean simplicial complex homeomorphic to $\mathbf{S}_3$ (resp. $\mathbf{S}_2$). Since for any $\xi \in \mathrm{Y}$ such that $\varphi_{[\mathrm{X}]}(\xi)$ is defined it is a 1-cycle, it is also a *boundary* of a 2-chain K in X (provided the triangulations T and T′ have been conveniently subdivided). If $\tilde{f}$ is the homomorphism $\mathrm{C}'_2(\mathrm{T}) \to \mathrm{C}'_2(\mathrm{T}')$ deduced from f, the boundary of $\tilde{f}(\mathrm{K})$ is a multiple of the *degenerate* 1-chain $\{\xi\}$, hence in $\mathrm{C}_2(\mathrm{T}')$ (Part 1, chap. II, § 2) it is 0, so that $\tilde{f}(\mathrm{K})$ is a 2-*cycle* in Y and as Y has dimension 2, $\tilde{f}(\mathrm{K})$ is homologous to a cycle $\gamma_\xi(f)\,.\,[\mathrm{Y}]$, the multiple of the fundamental cycle [Y] by an integer $\gamma_\xi(f)$, now called the *Hopf invariant* of f. That number can also be defined as the local degree $d(f, \mathrm{K}, \eta)$ for all points $\eta \neq \xi$ in Y (Part 2, chap. I, § 2).

The remainder of the proof first establishes the properties of $\gamma_\xi(f)$ (always for *simplicial* maps f).

1. The number $\gamma_\xi(f)$ is independent of the choice of the 2-chain K with boundary $\varphi_{[\mathrm{X}]}(\xi)$; this is due to the interpretation of $\gamma_\xi(f)$ as the intersection number $(\mathrm{K}\,.\,\varphi_{[\mathrm{X}]}(\eta))$ for any point $\eta \neq \xi$ interior to a 2-simplex of Y, hence it is equal to the *linking coefficient*

$$\mathrm{lk}(\varphi_{[\mathrm{X}]}(\xi), \varphi_{[\mathrm{X}]}(\eta)) \tag{4}$$

and the assertion results from the relation

$$\mathrm{lk}(\varphi_{[\mathrm{X}]}(\xi), \varphi_{[\mathrm{X}]}(\eta)) = \mathrm{lk}(\varphi_{[\mathrm{X}]}(\eta), \varphi_{[\mathrm{X}]}(\xi)) \tag{5}$$

(Part 2, chap. I, § 3,C). One can therefore write $\gamma(f)$ instead of $\gamma_\xi(f)$, or $\gamma(f, \mathrm{T})$, or $\gamma(f, \mathrm{T}, \mathrm{T}')$ to emphasize dependence on the triangulations T, T′.

2. The second step consists in proving that if $\bar{f}$ is a simplicial approximation of the *simplicial* map f constructed by Alexander's process (Part 1, chap. II, § 3) for two suitable subdivisions $\bar{\mathrm{T}}$, $\bar{\mathrm{T}}'$ of T and T′, such that $f(\mathrm{Star}(v)) \subset \mathrm{Star}(\bar{f}(v))$ for any vertex v of $\bar{\mathrm{T}}$, then

$$\gamma(\bar{f}, \bar{\mathrm{T}}, \bar{\mathrm{T}}') = \gamma(f, \mathrm{T}, \mathrm{T}'). \tag{6}$$

Hopf had to prove that if $\bar{\tau}'$ and $\bar{\tau}'_1$ are two distinct 2-simplices of $\bar{\mathrm{T}}'$, and if ξ is interior to $\bar{\tau}'$ and η is interior to $\bar{\tau}'_1$, then

$$\mathrm{lk}(\bar{\varphi}_{[\mathrm{X}]}(\xi), \bar{\varphi}_{[\mathrm{X}]}(\eta)) = \mathrm{lk}(\varphi_{[\mathrm{X}]}(\xi), \varphi_{[\mathrm{X}]}(\eta)), \tag{7}$$

where $\bar{\varphi}_{[\mathrm{X}]}$ is defined for the triangulations $\bar{\mathrm{T}}$, $\bar{\mathrm{T}}'$ and the map $\bar{f}$. He reduced the problem to showing the existence of a 2-chain K of $\bar{T}$, whose support meets neither $f^{-1}(\xi)$ nor $\bar{f}^{-1}(\eta)$, and whose boundary is $\varphi_{[X]}(\xi) - \bar{\varphi}_{[X]}(\xi)$, which implies

$$\text{lk}(\bar{\varphi}_{[X]}(\xi), \varphi_{[X]}(\eta)) = \text{lk}(\varphi_{[X]}(\xi), \varphi_{[X]}(\eta))$$

and a similar argument then proves that

$$\text{lk}(\bar{\varphi}_{[X]}(\xi), \varphi_{[X]}(\eta)) = \text{lk}(\bar{\varphi}_{[X]}(\xi), \bar{\varphi}_{[X]}(\eta)).$$

The existence of K follows from the fact that the 1-cycle

$$Z = \bar{\varphi}_{[X]}(\xi) - \varphi_{[X]}(\xi)$$

has an intersection number equal to 0 with *every* 2-simplex of T; by successive reduction of the total number of points of intersection of the support of Z with the 2-simplices of T, Hopf finally arrived at the situation in which a 1-cycle homologous to Z is a sum of a finite number of 1-cycles, each of which has a support *contained* in a 3-simplex of T, hence is a boundary.

3. Next Hopf proved that if f_0, f_1 are two *simplicial* maps relative to the triangulations T, T′, which are *homotopic*, then

$$\gamma(f_0, T, T') = \gamma(f_1, T, T'). \tag{8}$$

He first extended the triangulation T of X to a triangulation T_0 of $E = X \times [0, 1]$, and replaces the homotopy F of f_0 to f_1, defined in E, by a suitable simplicial approximation G, for suitable subdivisions of T_0 and T′. Let C_0 and C_1 be the fundamental 3-cycles on $X \times \{0\}$ and $X \times \{1\}$; there is a 4-chain B on E such that $\mathbf{b}_4 B = C_0 - C_1$. Then from (3) it follows that

$$\mathbf{b}_2 \varphi_B(\xi) = \varphi_{C_0}(\xi) - \varphi_{C_1}(\xi).$$

Let K_0, K_1 be 2-chains on $X \times \{0\}$ and $X \times \{1\}$ such that

$$\mathbf{b}_2 K_0 = \varphi_{C_0}(\xi), \qquad \mathbf{b}_2 K_1 = \varphi_{C_1}(\xi);$$

$Z = K_1 - K_0 - \varphi_B(\xi)$ is then a 2-cycle on E. By projection on $X \times \{0\}$, Z is homologous in E to a 2-cycle Z_0 on $X \times \{0\}$, hence $\tilde{G}(Z) \sim \tilde{G}(Z_0)$ on Y; but on $X \times \{0\}$, the 2-cycle Z_0 is homologous to 0 (since X is homeomorphic with S_3), hence $\tilde{G}(Z_0) = \tilde{f}_0(Z_0) = 0$ on Y (since on Y, homeomorphic to S_2, the only 2-cycle homologous to 0 is 0). Therefore $\tilde{G}(Z) = 0$, which means that

$$\tilde{f}_1(K_1) - \tilde{f}_0(K_0) - \tilde{G}(\varphi_B(\xi)) = 0.$$

Since the 2-cycle $\tilde{G}(\varphi_B(\xi))$ has support $\{\xi\}$, $\tilde{G}(\varphi_B(\xi)) = 0$, hence $\tilde{f}_0(K_0) = \tilde{f}_1(K_1)$, and by definition this means that $\gamma(f_0) = \gamma(f_1)$.

At this stage we can define $\gamma(f, T, T')$ for *any continuous map* $f: X \to Y$. Any two simplicial approximations f_1, f_2 of f obtained by the Alexander process for subdivisions of T and T′ are homotopic to f, hence $\gamma(f_1, T, T') = \gamma(f_2, T, T')$, so that $\gamma(f, T, T')$ may be defined as the common value of $\gamma(f_1, T, T')$ for *all* these simplicial approximations f_1 of f.

4. Before showing that $\gamma(f, T, T')$ is independent of the triangulations T, T′,

Hopf investigated its behavior relative to continuous maps $g: \mathrm{X}_1 \to \mathrm{X}$, $h: \mathrm{Y} \to \mathrm{Y}_1$, where X_1 (resp. Y_1) is another euclidean simplicial complex homeomorphic to $\mathbf{S}_3$ (resp. $\mathbf{S}_2$) with triangulation T_1 (resp. T'_1). Using simplicial approximations of f, g, h for suitable subdivisions of T, T', T_1, T'_1 and the definition of the degree given by Brouwer he proved that

$$\gamma(f \circ g, \mathrm{T}_1, \mathrm{T}') = \deg(g) \,.\, \gamma(f, \mathrm{T}, \mathrm{T}'), \tag{9}$$

$$\gamma(h \circ f, \mathrm{T}, \mathrm{T}'_1) = (\deg(h))^2 \,.\, \gamma(f, \mathrm{T}, \mathrm{T}'). \tag{10}$$

5. The last step in the definition of $\gamma(f)$ uses the Brouwer device (Part 2, chap. I, § 1): if T_1, T_2 are two triangulations of $\mathbf{S}_3$ and T'_1, T'_2 are two triangulations of $\mathbf{S}_2$, it follows from (9) and (10) that for any continuous map $f: \mathbf{S}_3 \to \mathbf{S}_2$,

$$\gamma(f, \mathrm{T}_1, \mathrm{T}'_1) = \deg(1_{\mathbf{S}_3}) \deg(1_{\mathbf{S}_2})^2 \gamma(f, \mathrm{T}_2, \mathrm{T}'_2) = \gamma(f, \mathrm{T}_2, \mathrm{T}'_2).$$

6. From the relation $\gamma_0(\xi) = 0$ it is clear that when f is homotopic to a constant map, $\gamma(f) = 0$; thus, to prove the significance of his invariant, Hopf still had to exhibit an example of a map $f: \mathbf{S}_3 \to \mathbf{S}_2$ with $\gamma(f) \neq 0$. He found a beautiful example, now known as the *Hopf fibration* over $\mathbf{S}_2$ (Part 3, chap. III, § 1), for which $\gamma(f) = 1$ [by (9), this implies that there are maps $f: \mathbf{S}_3 \to \mathbf{S}_2$ for which $\gamma(f)$ is an *arbitrary* integer $n \in \mathbf{Z}$]. He used the fact that the complex projective line $\mathbf{P}_1(\mathbf{C})$ is homeomorphic to $\mathbf{S}_2$ (the "Riemann sphere"): let $p: \mathbf{C}^2 - \{0\} \to \mathbf{P}_1(\mathbf{C})$ be the natural map sending the point (z_1, z_2) to the line joining 0 to that point; $\mathbf{S}_3$ is the subspace of $\mathbf{C}^2 - \{0\}$ defined by $|z_1|^2 + |z_2|^2 = 1$; f is just the restriction of p to $\mathbf{S}_3$. If $f(z_1, z_2) = \xi$ for a point $(z_1, z_2) \in \mathbf{S}_3$, the set $f^{-1}(\xi)$ is the great circle $t \mapsto (z_1 e^{it}, z_2 e^{it})$; the relation $\gamma(f) = \pm 1$ follows from the fact that if $\xi \neq \eta$, the great circle $f^{-1}(\xi)$ cuts a 2-sphere Σ of $\mathbf{S}_3$, having $f^{-1}(\eta)$ as great circle, in exactly two points; hence it cuts one of the hemispheres of Σ having $f^{-1}(\eta)$ as boundary in exactly *one* point.

C. Generalizations to Maps from $\mathbf{S}_{2k-1}$ into $\mathbf{S}_k$

In 1935 Hopf, perhaps prodded by the first Hurewicz notes on homotopy groups (§ 3) just published, returned to the study of continuous maps of spheres into spheres [245]. He realized that his technique of 1931 could easily be extended to continuous maps $f: \mathbf{S}_m \to \mathbf{S}_k$, provided that for simplicial maps the linking coefficient $\mathrm{lk}(f^{-1}(\xi), f^{-1}(\eta))$ could be defined for two distinct points ξ, η of $\mathbf{S}_k$; this is only the case if $2(m - k) + 1 = m$, i.e., $m = 2k - 1$. Without bothering to repeat the long sequence of arguments leading to the definition of $\gamma(f)$, Hopf focused his paper on the search for maps f for which $\gamma(f) \neq 0$. He first observed that if k is *odd*, then $\gamma(f) = 0$ for *all* continuous maps $f: \mathbf{S}_{2k-1} \to \mathbf{S}_k$. This follows from the relation

$$\mathrm{lk}(f^{-1}(\xi_1), f^{-1}(\xi_2)) = (-1)^k \mathrm{lk}(f^{-1}(\xi_2), f^{-1}(\xi_1))$$

for two distinct points ξ_1, ξ_2 of $\mathbf{S}_k$ (Part 2, chap. I, § 3,C); therefore, $\gamma_{\xi_1}(f) =$

$-\gamma_{\xi_2}(f)$ if k is odd (with the notations of B), but if ξ_3 is a third point of $\mathbf{S}_k$, $\gamma_{\xi_1}(f) = -\gamma_{\xi_3}(f) = \gamma_{\xi_2}(f)$, hence $\gamma_{\xi_1}(f) = 0$.

To show that $\gamma(f)$ may be $\neq 0$ when k is even Hopf invented, for all integers $r \geqslant 1$, a remarkable family of continuous maps $\mathbf{S}_{2r+1} \to \mathbf{S}_{r+1}$. He starts from a decomposition of $\mathbf{S}_{2r+1}$ as a union of two closed sets V^+, V^-, both homeomorphic to $\mathbf{S}_r \times \mathbf{D}_{r+1}$, such that $\mathrm{V}^+ \cap \mathrm{V}^-$ is homeomorphic to $\mathbf{S}_r \times \mathbf{S}_r$. He used a generalization of a device introduced by Heegaard for $r = 1$; in $\mathbf{S}_{2r+1}$, considered as the subspace of $\mathbf{R}^{2r+2}$ defined by

$$x_1^2 + x_2^2 + \cdots + x_{2r+2}^2 = 1,$$

the subsets V^+ and V^- are defined by

$$\mathrm{V}^+\colon x_1^2 + x_2^2 + \cdots + x_{r+1}^2 \leqslant 1/2$$

$$\text{(equivalent to } x_{r+2}^2 + x_{r+3}^2 + \cdots + x_{2r+2}^2 \geqslant 1/2\text{)}, \tag{11}$$

$$\mathrm{V}^-\colon x_1^2 + x_2^2 + \cdots + x_{r+1}^2 \geqslant 1/2$$

$$\text{(equivalent to } x_{r+2}^2 + x_{r+3}^2 + \cdots + x_{2r+2}^2 \leqslant 1/2\text{)}. \tag{12}$$

Their common boundary P_{2r} in $\mathbf{S}_{2r+1}$ is their intersection, equal to $((1/\sqrt{2})\mathbf{S}_r) \times ((1/\sqrt{2})\mathbf{S}_r)$. The map $\mathbf{S}_r \times ((1/\sqrt{2})\mathbf{D}_{r+1}) \to \mathrm{V}^+$ defined by

$$(\mathsf{y}, \mathsf{z}) \mapsto (z_1, \ldots, z_{r+1}, y_1(1 - |\mathsf{z}|^2)^{1/2}, \ldots, y_{r+1}(1 - |\mathsf{z}|^2)^{1/2}) \tag{13}$$

is a homeomorphism, whose inverse will be written

$$\mathsf{x} \mapsto (v^+(\mathsf{x}), w^+(\mathsf{x})). \tag{14}$$

Similarly the map $\mathbf{S}_r \times ((1/\sqrt{2})\mathbf{D}_{r+1}) \to \mathrm{V}^-$ defined by

$$(\mathsf{y}, \mathsf{z}) \mapsto (y_1(1 - |\mathsf{z}|^2)^{1/2}, \ldots, y_{r+1}(1 - |\mathsf{z}|^2)^{1/2}, z_1, \ldots, z_{r+1}) \tag{15}$$

is a homeomorphism, whose inverse will be written

$$\mathsf{x} \mapsto (v^-(\mathsf{x}), w^-(\mathsf{x})). \tag{16}$$

Suppose now that a continuous map

$$g\colon \mathrm{P}_{2r} \to \mathbf{S}_r \tag{17}$$

has been defined; it is then possible to extend g to a continuous map

$$f\colon \mathbf{S}_{2r+1} \to \mathbf{S}_{r+1}$$

in the following way. Let p_0 be the point $(1/\sqrt{2}, 0, \ldots, 0)$ in $(1/\sqrt{2})\mathbf{S}_r$, and define g_1 (resp. g_2) as the restriction of g to $((1/\sqrt{2})\mathbf{S}_r) \times \{p_0\}$ [resp. $\{p_0\} \times ((1/\sqrt{2})\mathbf{S}_r)$]. Let E_{r+1}^+ and E_{r+1}^- be the two hemispheres $x_{r+2} \geqslant 0$ and $x_{r+2} \leqslant 0$ in the sphere $\mathbf{S}_{r+1}$, and for any point $\mathsf{z} \in \mathbf{D}_{r+1}$ let $h^+(\mathsf{z})$ and $h^-(\mathsf{z})$ be the points in E_{r+1}^+ and E_{r+1}^- that project orthogonally to z. Now let $\psi\colon (1/\sqrt{2})\mathbf{D}_{r+1} \to [\frac{1}{2}, 1]$ be a continuous function that is not constant and is equal to 1 in $(1/\sqrt{2})\mathbf{S}_r$. Then take

$$\begin{aligned} f(\mathbf{x}) &= h^{+}\left(\psi(w^{+}(\mathbf{x})).g_1\left(\frac{1}{\sqrt{2}}v^{+}(\mathbf{x})\right)\right) \quad \text{in } \mathrm{V}^{+}_{2r+1}, \\ f(\mathbf{x}) &= h^{-}\left(\psi(w^{-}(\mathbf{x})).g_2\left(\frac{1}{\sqrt{2}}v^{-}(\mathbf{x})\right)\right) \quad \text{in } \mathrm{V}^{-}_{2r+1}. \end{aligned} \tag{18}$$

The idea behind this construction is to start from a map g such that the degrees of g_1 and g_2 are both $\neq 0$; then Hopf showed that for the corresponding function f defined by (18),

$$\gamma(f) = \pm\deg(g_1)\deg(g_2) \neq 0 \tag{19}$$

by the following argument.

A *basis* for the **Z**-module $\mathrm{H}_r(\mathrm{P}_{2r};\mathbf{Z})$ is given by the classes of the two r-cycles

$$\mathrm{Z}_1 = \{p_0\} \times \left[\frac{1}{\sqrt{2}}\mathbf{S}_r\right], \qquad \mathrm{Z}_2 = \left[\frac{1}{\sqrt{2}}\mathbf{S}_r\right] \times \{p_0\}. \tag{20}$$

In the space V^{-}_{2r+1}, Z_1 is the boundary of an $(r+1)$-chain having as support the set U defined by

$$\mathrm{U}\colon x_1 \geqslant \frac{1}{\sqrt{2}}, x_2 = \cdots = x_{r+1} = 0, \qquad x^2_{r+2} + \cdots + x^2_{2r+2} \leqslant 1/2. \tag{21}$$

This set has only one common point with the sphere Y_c defined, for $c < (1/\sqrt{2})$ by

$$x_1^2 + \cdots + x_{r+1}^2 = 1 - c^2, \qquad x_{r+2} = c, \qquad x_{r+3} = \cdots = x_{2r+2} = 0. \tag{22}$$

In the sphere $\mathbf{S}_{2r+1}$, Z_1 is homologous to 0, hence U and Y_c are the supports of two *cycles* $[\mathrm{U}]$, $[\mathrm{Y}_c]$, and the intersection number of these cycles is ± 1. But it is clear that $[\mathrm{Y}_c]$ is homologous to Z_2 in $\mathbf{S}_{2r+1}$, hence

$$\mathrm{lk}(\mathrm{Z}_1, \mathrm{Z}_2) = \pm 1 \tag{23}$$

(Part 2, chap. I, § 3, C). Now f may be taken simplicial, and if ξ_2 is an interior point of an $(r+1)$-simplex in E^{-}_{r+1}, $\varphi_{[\mathbf{S}_{2r+1}]}(\xi_2)$ is defined as an r-cycle in V^{-}_{2r+1}. As the restriction of f to the support of Z_1 is g_1, its restriction to U has degree $\deg(g_1)$, hence the intersection number of $[\mathrm{U}]$ and of $\varphi_{[\mathbf{S}_{2r+1}]}(\xi_2)$ is $\pm\deg(g_1)$; by (23), this implies that $\varphi_{[\mathbf{S}_{2r+1}]}(\xi_2)$ is homologous to $\pm\deg(g_1).\mathrm{Z}_2$. Similarly, if ξ_1 is an interior point of an $(r+1)$-simplex in E^{+}_{r+1}, then $\varphi_{[\mathbf{S}_{2r+1}]}(\xi_1)$ is defined as an r-cycle in V^{+}_{2r+1} and is homologous to $\pm\deg(g_2).\mathrm{Z}_1$; finally

$$\gamma(f) = \mathrm{lk}(\varphi_{[\mathbf{S}_{2r+1}]}(\xi_1), \varphi_{[\mathbf{S}_{2r+1}]}(\xi_2)) = \pm\deg(g_1)\deg(g_2). \tag{24}$$

The last part of the proof consists in finding a map g such that $\deg(g_1) = \pm 1$, $\deg(g_2) = \pm 2$. To show that this is possible when r is *odd*, Hopf defined (for any r) $(1/\sqrt{2})g(x_1, x_2)$ as the point of $(1/\sqrt{2})\mathbf{S}_r$ that is the symmetrical of x_1 with respect to the hyperplane orthogonal to x_2. It is then clear that $\deg(g_1) = -1$; but $g(x_1, -x_2) = g(x_1, x_2)$ and g_2 maps the closed hemisphere D of pole

p_0 onto $\mathbf{S}_r$ with degree ± 1, and *if r is odd*, the symmetry $s\colon x \mapsto -x$ has degree $+1$; g_2 has degree ± 2, since the restriction of g_2 to $-\mathrm{D}$ is identical with the restriction of $g_2 \circ s$ to D.

This theorem immediately raises the question of the existence of continuous maps $g\colon \mathbf{S}_r \times \mathbf{S}_r \to \mathbf{S}_r$ for which *both* $\deg(g_1)$ and $\deg(g_2)$ are ± 1, which would imply that the Hopf invariant $\gamma(f) = \pm 1$. Hopf showed that such maps exist for $r = 1$, 3, or 7. He derived this result from the existence of a *norm* on $\mathbf{R}^{r+1}$, that is, a continuous function $\mathrm{N}\colon \mathbf{R}^{r+1} \to \mathbf{R}_+$ such that for *some* continuous map $(\mathsf{x}, \mathsf{y}) \mapsto \mathsf{xy}$ of $\mathbf{R}^{r+1} \times \mathbf{R}^{r+1}$ into $\mathbf{R}^{r+1}$,

$$\mathrm{N}(\mathsf{xy}) = \mathrm{N}(\mathsf{x})\mathrm{N}(\mathsf{y}) \tag{25}$$

and furthermore $\mathrm{N}(\mathsf{x}) > 0$ for $\mathsf{x} \neq 0$. For $r + 1 = 2$, 4, or 8 such maps exist for the "product" $(\mathsf{x}, \mathsf{y}) \mapsto \mathsf{xy}$ in the algebra of complex numbers for $r + 1 = 2$, the algebra of quaternions for $r + 1 = 4$ and the (nonassociative) Cayley algebra of "octonions"* for $r + 1 = 8$. It was suspected for a long time that these values of r were the *only ones* for which maps $f\colon \mathbf{S}_{2r+1} \to \mathbf{S}_{r+1}$ exist with $\gamma(f) = 1$, and gradually partial results accumulated, restricting the possible values of r by various methods, until J.F. Adams in 1958 succeeded in proving the conjecture, using very sophisticated tools (chap. VI, § 5,D).

§ 2. Basic Notions in Homotopy Theory

A. Homotopy and Extensions

Let X, Y be topological spaces, A be a subspace of X, and $f\colon \mathrm{A} \to \mathrm{Y}$ be a continuous map. An *extension* of f to X is a continuous map $g\colon \mathrm{X} \to \mathrm{Y}$ such that $g|\mathrm{A} = f$; if $j\colon \mathrm{A} \to \mathrm{X}$ is the natural injection, this also means that

$$g \circ j = f.$$

* In the algebra of quaternions, the *conjugate* of a quaternion $q = s + xi + yj + zk$ is the quaternion $\bar{q} = s - xi - yj - zk$; one has $\mathrm{N}(q) = q\bar{q} = \bar{q}q = s^2 + x^2 + y^2 + z^2$. Octonions are pairs of quaternions $p = (q_1, q_2)$, with multiplication defined by

$$(q_1, q_2)(q_1', q_2') = (q_1 q_1' - \bar{q}_2' q_2, q_2 \bar{q}_1' + q_2' q_1);$$

it is neither associative nor commutative, but satisfies the weaker relations

$$p_1^2 p_2 = p_1(p_1 p_2), \qquad p_1 p_2^2 = (p_1 p_2) p_2.$$

The *conjugate* of an octonion $p = (q_1, q_2)$ is the octonion $\bar{p} = (\bar{q}_1, -q_2)$; the product $\mathrm{N}(p) = p\bar{p} = \bar{p}p$ is again a scalar satisfying (25), and

$$\mathrm{N}(p) = \mathrm{N}(q_1) + \mathrm{N}(q_2) > 0 \text{ for } p \neq 0.$$

In every one of the three cases, the map $f\colon \mathbf{S}_{2r+1} \to \mathbf{S}_{r+1}$ can be defined directly: $\mathbf{S}_{2r+1}$ is the subspace of $\mathbf{R}^{2r+2}$ defined as the set of pairs (z_1, z_2) of complex numbers (resp. quaternions, resp. octonions) such that $\mathrm{N}(z_1) + \mathrm{N}(z_2) = 1$, and

$$f(z_1, z_2) = (2z_1\bar{z}_2, \mathrm{N}(z_1) - \mathrm{N}(z_2)) \in \mathbf{R}^{r+2}.$$

It is easy to check that $f(z_1, z_2) \in \mathbf{S}_{r+1}$, and one can also prove that f is a *submersion*.

This last relation implies

$$g_* \circ j_* = f_* \quad (\text{resp. } f^* = j^* \circ g^*)$$

for the corresponding homomorphisms in homology (resp. cohomology). In particular $\operatorname{Ker}(j_*) \subset \operatorname{Ker}(f_*)$ is a necessary condition for the existence of an extension of f to X. For instance, if $H_p(X) = 0$ and $f_*: H_p(A) \to H_p(Y)$ is not 0, there is no extension of f to X.

A homotopy between two maps $f: X \to Y$ and $g: X \to Y$ can be considered as a special case of extension: consider in the space $X \times [0, 1]$ the subspace

$$A = (X \times \{0\}) \cup (X \times \{1\})$$

and the continuous map $G: A \to Y$ defined by

$$G(x, 0) = f(x) \quad \text{and} \quad G(x, 1) = g(x);$$

a homotopy F from f to g is then an *extension* of G to $X \times [0, 1]$.

In his paper of 1933 in which he determined the "homotopy classes" of maps into $\mathbf{S}_n$ of any n-dimensional combinatorial manifold M, Hopf pointed out that (after reduction to simplicial maps) the problem of existence of a homotopy between two continuous maps of M into $\mathbf{S}_n$ amounts to finding, for an n-dimensional subcomplex L of an $(n + 1)$-dimensional simplicial complex K, conditions for a simplicial map $L \to \mathbf{S}_n$ to have an extension to K ([244], p. 86).

B. Retracts and Extensions

At about the same time two other mathematicians, Borsuk and Lefschetz, were interested in extension problems for other reasons.

A *retraction* $r: X \to A$ of a space X onto a subspace A is a continuous map such that $r(X) = A$ and $r(x) = x$ for $x \in A$. If $j: A \to X$ is the natural injection, this definition also means that $r \circ j$ is the *identity* 1_A, or that r is a "left inverse" of j. A subspace A of X for which there is a retraction onto A is called a *retract* of X [70]. The main (obvious) property of a retract A of X is that *any* continuous map $f: A \to Y$ has at least one *extension* $g: X \to Y$, namely, $g = f \circ r$.

A retract A of a Hausdorff space X is closed in X, since it is the set of points $x \in X$ for which $r(x) = x$. From the relation $r \circ j = 1_A$ it follows that for homology (resp. cohomology)

$$r_* \circ j_* = 1_{H_\cdot(A)} \quad (\text{resp. } j^* \circ r^* = 1_{H^\cdot(A)})$$

which impose restrictions on A to be a retract; $\mathbf{S}_{n-1}$ is not a retract of $\mathbf{R}^m$ for $m \geqslant n$. A less trivial example (Steenrod) can be given by using the ring structure of $H^\cdot(A)$ and $H^\cdot(X)$ and the fact that j^* and r^* are ring homomorphisms: $\operatorname{Im} r^*$ is a subring and $\operatorname{Ker} j^*$ an ideal of $H^\cdot(X)$, and from the relation $j^* \circ r^* = 1_{H^\cdot(A)}$, it follows that $H^\cdot(X) = \operatorname{Im} r^* \oplus \operatorname{Ker} j^*$ and $\operatorname{Im} r^*$ is isomorphic to $H^\cdot(A)$. Take $X = \mathbf{P}_2(\mathbf{C})$ and $A = \mathbf{P}_1(\mathbf{C})$; there is then a unique decomposition $H^\cdot(X) = M \oplus N$ into two graded submodules, with M isomorphic to the module $H^{\bullet}(A)$: in dimensions 0 and 2, M is the whole module and N is 0, and in dimension 4, M is 0 and N is the whole module. However M is not a subring, because from the structure of $H^{\bullet}(X)$, it follows that if $u \neq 0$ in $H^2(X)$, $u \smile u \neq 0$. So $\mathbf{P}_1(\mathbf{C})$ is not a retract of $\mathbf{P}_2(\mathbf{C})$.

The notion of retract suffers from a defect when one wants to study a subspace A of a space X: it implies the consideration of the *whole* space X, so that points of X that may be "very far" from A, still, by their presence, may prevent the existence of a retraction $X \to A$. To palliate this defect, Borsuk introduced two other notions [71]. *Restrict* the concept of retract by the condition that for *every* space X and *every* homeomorphism j of A onto a subspace $j(A)$ of X, $j(A)$ be a retract of X; A is then called an *absolute retract.* The other notion, on the contrary, *expands* the concept by only requiring that A be a retract of *some* unspecified neighborhood of A in X. Then A is called a *neighborhood retract* of X; for instance, $\mathbf{S}_{n-1}$ is now a neighborhood retract of $\mathbf{R}^n$, although not a retract.

The most useful notion introduced by Borsuk is intermediate, so to speak, between the two last ones: a separable metrizable space Y is an *absolute neighborhood retract* (abbreviated ANR) if for any homeomorphism of Y onto a subspace Z of an arbitrary space X, Z is a neighborhood retract of X. An *equivalent* definition is that, for every closed subset A of a metrizable space X, every continuous map $f: A \to Y$ has an extension to a neighborhood of A in X.

A retract of an ANR is an ANR; an open subspace of an ANR is an ANR; a product of two ANR's is an ANR. If Y_1, Y_2 are two closed subsets of a metrizable space Y, and if Y_1, Y_2 and $Y_1 \cap Y_2$ are ANR's, then $Y_1 \cup Y_2$ is an ANR.

Compact ANR's have been characterized by Borsuk as being homeomorphic to neighborhood retracts of the Hilbert cube.*

It turns out that a little earlier Lefschetz had considered a notion closely related to the notion of ANR, namely the *locally contractible* (abbreviated LC) spaces: they are defined by the condition that any point x has a neighborhood which is a subspace contractible to the point x [309]. Since the Hilbert cube is locally contractible, any compact metrizable ANR is locally contractible. In [71] Borsuk showed conversely that, any locally contractible compact metrizable space of *finite dimension* is an ANR, but the condition of finite dimension cannot be dropped ([69a], pp. 124–127).

The CW-complexes are locally contractible ([493], p. 102), but compact ANR's may exhibit very "pathological" features ([69a], chap. VI).

Lefschetz also introduced concepts weaker than local contractibility. He said that a space X is $p-\text{LC}$ at a point x if every neighborhood U of x contains a neighborhood such that every injective continuous map $\mathbf{S}_p \to V$ is homotopic in U to a point. The space X is LC^p (resp. LC^{ω}) at x if it is $q-\text{LC}$ at x for every $q \leqslant p$ (resp. for every q). To put that notion in relation with the

---

* The Hilbert cube is the subspace of l^2 consisting of the sequences $\mathsf{x} = (x_n)$ such that $|x_n| \leqslant 1/n$ for every n; it is homeomorphic to the compact space $[0, 1]^{\mathbf{N}}$.

idea of neighborhood retract he also weakened the notion of ANR: a space X is an ANR^p if, for every space Y such that X is a subspace of Y and $\dim(Y - X) \leqslant p$, then X is a neighborhood retract of Y. He then showed that, for compact metric spaces, the property of being ANR^{p+1} is equivalent to being LC^p at every point [309].

C. Homotopy Type

If X, Y are two topological spaces, the relation "f is homotopic to g," written $f \sim g$, between two elements of the set $\mathscr{C}(X; Y)$ of continuous maps of X into Y is an *equivalence relation*, as it results immediately from the definition. If $u: X_1 \to X$ and $v: Y \to Y_1$ are continuous, the relation $f \sim g$ implies $v \circ f \circ u \sim v \circ g \circ u$. The set of equivalence classes in $\mathscr{C}(X; Y)$ for the relation $f \sim g$ is often written $[X; Y]$ or $\pi(X; Y)$. The continuous maps $u: X_1 \to X$ and $v: Y \to Y_1$ therefore define maps

$$u^*: [X; Y] \to [X_1; Y] \quad \text{and} \quad v_*: [X; Y] \to [X; Y_1]$$

such that $(u_1 \circ u_2)^* = u_2^* \circ u_1^*$ and $(v_1 \circ v_2)_* = v_{1*} \circ v_{2*}$. More generally, if f_1, f_2 are two elements of $\mathscr{C}(X; Y)$ and g_1, g_2 are two elements of $\mathscr{C}(Y; Z)$, and if $f_1 \sim f_2$, $g_1 \sim g_2$, then $g_1 \circ f_1 \sim g_2 \circ f_2$. If $\boldsymbol{T}$ is the category of topological spaces (Part 1, chap. IV, § 8), then $X \mapsto [X; Y]$ (resp. $Y \mapsto [X; Y]$) is a *contravariant* (resp. *covariant*) *functor* $\boldsymbol{T} \to \boldsymbol{Set}$.

In his 1935 notes on homotopy [256] Hurewicz introduced the concept of *homotopy equivalence* for a continuous map $f: X \to Y$. There exists a continuous map $g: Y \to X$ such that

$$f \circ g \sim 1_Y \quad \text{and} \quad g \circ f \sim 1_X;$$

g is called a *homotopy inverse* of f; it is also a homotopy equivalence, determined up to homotopy. Two spaces X, X′ have the same *homotopy type* if there exists a homotopy equivalence $f: X \to X'$. This is an equivalence relation between topological spaces, and the set $[X; Y]$ only depends on the homotopy types of X and Y. All contractible spaces have the homotopy type of a space reduced to a single point.

The concept of "homotopy inverse" may be "decomposed" [198], that is, $f: X \to Y$ has a *right* (resp. *left*) *homotopy inverse* if there exists a continuous map $g: Y \to X$ such that $f \circ g \sim 1_Y$ (resp. $g \circ f \sim 1_X$). That property implies that the homomorphism $f_*: H_\bullet(X) \to H_\bullet(Y)$ is *surjective* (resp. *injective*); the existence of *both* right and left homotopy inverses for f implies that f is a homotopy equivalence.

When there is a continuous map $f: X \to Y$ that has a *left* homotopy inverse, the space X is said to be *dominated* by Y. If A is a subspace of X, the fact that the injection $j: A \to X$ has a *right* homotopy inverse means that there exist a continuous map $g: X \to A$ and a continuous map $F: X \times [0, 1] \to X$ such that $F(x, 0) = j(g(x))$ and $F(x, 1) = x$ for all $x \in X$; F is called a *deformation* of X into A.

When the use of "categorical" language became widespread in the 1950s, the preceding notions could be expressed in that language by the introduction of the *homotopy category* $\boldsymbol{H}$. It has the same objects as the category $\boldsymbol{T}$ of topological spaces, but the morphisms are the *homotopy classes* of continuous maps, so that $\mathrm{Mor}_{\boldsymbol{H}}(X, Y) = [X; Y]$; classes of "homotopy inverses" then become "inverse morphisms" (left, right, or two-sided).

Example: Homotopy Types of Lens Spaces

In 1940 J.H.C. Whitehead achieved substantial progress in the theory of lens spaces. For two lens spaces $L(p, q)$, $L(p, q')$ where p is prime, he proved that the condition $q' \equiv \pm v^2 q (\mathrm{mod}.\, p)$ for some integer v, which Alexander had shown to be *necessary* for homeomorphy of the two spaces (Part 2, chap. I, § 3,C), is *necessary and sufficient* for $L(p, q)$ and $L(p, q')$ to have the *same homotopy type* [496]. His proof was based on his theory of *simple homotopy type* (§ 7). In 1943 Franz, unaware of Whitehead's result, gave another proof [200], which was extended and simplified in 1960 by de Rham ([388], pp. 575–580), who proved the corresponding result for $(2n + 1)$-dimensional generalized lens spaces for any $n \geqslant 2$ (Part 2, chap. VI, § 3,A). The two generalized lens spaces L, L′ are respectively defined as the spaces of orbits $\mathbf{S}_{2n+1}/G$, $\mathbf{S}_{2n+1}/G'$, where G and G′ are the cyclic groups of order h generated by the respective rotations of $\mathbf{S}_{2n+1}$:

$$\begin{aligned} R&: (z_0, z_1, \ldots, z_n) \mapsto (\zeta_0 z_0, \zeta_1 z_1, \ldots, \zeta_n z_n) \quad \text{with } \zeta_k = e^{2\pi i m_k/h}, \\ R'&: (z_0, z_1, \ldots, z_n) \mapsto (\zeta'_0 z_0, \zeta'_1 z_1, \ldots, \zeta'_n z_n) \quad \text{with } \zeta'_k = e^{2\pi i m'_k/h}, \end{aligned} \tag{26}$$

where the m_k and m'_k are prime to h. Let c (resp. c') be a generator of the homology group $H_{2n+1}(L; \mathbf{Z})$ [resp. $H_{2n+1}(L'; \mathbf{Z})$] that is isomorphic to $\mathbf{Z}$, and g (resp. g') be a generator of the fundamental group $\pi_1(L)$ [resp. $\pi_1(L')$] that is isomorphic to $\mathbf{Z}/h\mathbf{Z}$. The necessary and sufficient condition for L and L′ to have the same homotopy type is obtained in several steps:

1. Let $f: L \to L'$ be any continuous map; $f_*(c) = dc'$ and $f_*(g) = g'^a$, where $d = \deg(f)$ and a is an integer mod. h. The map f lifts to a map $F: \mathbf{S}_{2n+1} \to \mathbf{S}_{2n+1}$ of degree d such that

$$F \circ R = R'^a \circ F \tag{27}$$

(chap. I, § 2,V). Two maps f_0, f_1 of L into L′ are homotopic if and only if there is a homotopy $E: \mathbf{S}_{2n+1} \times [0, 1] \to \mathbf{S}_{2n+1}$ such that, for the liftings F_0, F_1,

$$\begin{aligned} &E(z, 0) = F_0(z), \qquad E(z, 1) = F_1(z), \qquad \text{and} \\ &E(R(z), t) = R'^a(E(z, t)) \quad \text{for } 0 \leqslant t \leqslant 1 \end{aligned} \tag{28}$$

(chap. I, § 2).

2. If one writes that both sides of (27) have the same degree the relation

$$d m_0 m_1 \cdots m_n \equiv a^n m'_0 m'_1 \cdots m'_n \quad (\mathrm{mod}.\, h) \tag{29}$$

is obtained at once.

3. For any integer $d \in \mathbf{Z}$ satisfying (29) there is a map $F: \mathbf{S}_{2n+1} \to \mathbf{S}_{2n+1}$ of degree d, satisfying (27) [239].

First exhibit a map with degree d_0 for which $d \equiv d_0 \pmod{h}$. Let p_k be an integer such that $p_k m_k \equiv a \pmod{h}$ and define F by the relations

$$F(z_0, z_1, \ldots, z_n) = (F_0(z_0), F_1(z_1), \ldots, F_n(z_n))$$

with

$$F_k(r_k e^{2\pi i \varphi_k}) = r_k e^{2\pi i p_k m'_k \varphi_k} \quad \text{for each } k.$$

F satisfies (27) and has a degree d_0 such that

$$|d_0| = p_0 m'_0 p_1 m'_1 \cdots p_n m'_n.$$

To pass from that map to one with degree d, the following general *lemma* is used. Suppose $g: \mathbf{S}_{2n+1} \to \mathbf{S}_{2n+1}$ is a continuous map which is C^∞ in a dense open subset U; then, for any integer $r \in \mathbf{Z}$ and any open set $V \subset U$, there exists a continuous map $g_1: \mathbf{S}_{2n+1} \to \mathbf{S}_{2n+1}$, equal to g outside V and of degree $\deg(g) + r$.

The application of that lemma is done in the following way: U is the set where all the $z_k \neq 0$, V is a small open subset such that the images $R^j(V)$ are all distinct; F is modified in V according to the lemma, and in each of the $R^j(V)$ for $j \geqslant 1$ in such a way that the modified function F_1 still satisfies (27); then $\deg(F_1) = \deg(F) + hr = d_0 + hr$.

To prove the lemma, consider a point x_0 in $\mathbf{S}_{2n+1}$ such that $g^{-1}(x_0)$ is discrete and finite; take distinct points y_k in V with $1 \leqslant k \leqslant r$, which do not belong to $g^{-1}(x_0)$, and for each y_k take a neighborhood $V_k \subset U$ such that the V_k are disjoint and do not meet $g^{-1}(x_0)$. Let C be a small ball of center x_0 such that $g^{-1}(C)$ does not meet any V_k, and for each k consider a diffeomorphism u_k of a small ball $B_k \subset V_k$ of center y_k onto C such that $u_k(y_k) = x_0$, with a jacobian having the sign of r. Finally, let λ be a map of $\mathbf{S}_{2n+1}$ into $[0,1]$, equal to 1 in $\frac{1}{2}C$, to 0 outside C; define

$$v_k(x) = \lambda(u_k(x))u_k(x) + (1 - \lambda(u_k(x)))g(x) \quad \text{for } x \in u_k^{-1}(\tfrac{1}{2}C).$$

Finally, take $g_1(x) = v_k(x)/|v_k(x)|$ for $x \in u_k^{-1}(\frac{1}{2}C)$, and $g_1(x) = g(x)$ elsewhere.

4. Conversely, suppose F_0 and F_1 are two continuous maps of $\mathbf{S}_{2n+1}$ into itself, both satisfying (27) and having the same degree d which verifies (29); then there exists a homotopy E satisfying the relations (28).

The space $\mathbf{S}_{2n+1}$ is triangulated in cells $a_j^q = R^j . a^q$ $(0 \leqslant q \leqslant 2n+1, 0 \leqslant j \leqslant h-1)$, where the a^q are defined by formulas (29) of Part 2, chap. VI, §3,A. Then, on $\mathbf{S}_{2n+1} \times [0,1]$ a triangulation by the cells $a_j^q \times [0,1]$, $a_j^q \times \{0\}$, and $a_j^q \times \{1\}$ is obtained. Using the third condition (28), it is enough to define E on the cells $a^q \times [0,1]$; this is done by induction on the dimension q. The values of E on the vertices of the $a_j^q \times [0,1]$ are given by the first two relations (28); using obstruction theory (see below, §3D) and the fact that $\pi_i(\mathbf{S}_{2n+1}) = 0$ for $i \leqslant 2n$, the extension is possible for $q \leqslant 2n$. Let E_j be the restriction of E to the boundary of $a^{2n+1} \times [0,1]$; to extend it to the interior of that cell it is sufficient that the degree of E_j be 0 by the Brouwer–Hopf theorem. The degrees of all E_j have the same value e, hence the sum of these degrees is he, but the sum of the fundamental cycles of the $a_j^{2n} \times [0, 1]$ is equal to the difference $[\mathbf{S}_{2n+1}] \times \{0\} - [\mathbf{S}_{2n+1}] \times \{1\}$, hence he is the difference of the degrees of F_0 and F_1. If this difference is 0, then $e = 0$ and this proves the existence of the homotopy E.

5. Suppose now that for some integer a prime to h, the relation (29) is satisfied for $d = 1$ or $d = -1$. Let b be an integer such that $ab \equiv 1 \pmod{h}$; then

$$dm_0'm_1'\cdots m_n' \equiv b^n m_0 m_1 \cdots m_n \pmod{h}$$

and from 4 it follows that there exist two continuous maps $f: L \to L'$, f': $L' \to L$ of same degree $d = \pm 1$ such that $f_*(g) = g'^a$, $f'_*(g') = g^b$. Then the maps $f' \circ f$ and $f \circ f'$ have degree $+1$, $f'_* \circ f_*(g) = g$ and $f_* \circ f'_*(g') = g'$; by the Brouwer–Hopf theorem applied to their liftings to $\mathbf{S}_{2n+1}$ they are homotopic to the identity.

D. Retracts and Homotopy

We already met the notion of *strong deformation retract* (Part 1, chap. IV, § 6,B); it was first introduced by Borsuk in 1933 [72]. A subspace A of a space X that is a strong deformation retract of X not only has the same homology and cohomology as X, but also the *same homotopy type*. For instance, the existence on a compact C^∞ manifold X of a C^∞ function f having only nondegenerate critical points implies that X has the homotopy type of a CW-complex *explicitly described* in terms of the indices of the critical points of f (Part 2, chap. V, § 4).

If $A \subset X$ is a strong deformation retract of X, for the retraction $r: X \to A$, then any continuous map $f: Y \to X$ is homotopic in X to the map $r \circ f: Y \to A$. A useful example of strong deformation retract is given by the *box lemma*: consider the product $X = \mathbf{D}_n \times [0, 1]$ and the subspace

$$A = (\mathbf{S}_{n-1} \times [0, 1]) \cup (\mathbf{D}_n \times \{0\}).$$

Then, if c is the point $(0, 2)$ in $\mathbf{R}^n \times \mathbf{R}$, a retraction $r: X \to A$ is defined by taking for $r(x)$ the point where the line joining c and x meets A; the map $(x, t) \mapsto F(x, t) = t.r(x) + (1 - t)x$ is then such that $F(x, 0) = x$, $F(x, 1) = r(x)$.

When X is an ANR, in order to prove that a closed subset $A \subset X$ is a strong deformation retract of X, it is enough to show that there exists a homotopy $F: X \times [0, 1] \to X$ such that $F(x, 0) = x$ in X and $x \mapsto F(x, 1)$ is a *retraction* of X onto A, but it is not necessary to assume that $F(x, t) = x$ for all $x \in A$ and all $t \in [0, 1]$ ([403], p. 448).

If A is a closed subspace of a space X, one says the pair (X, A) has the *homotopy extension property* relatively to a space Y if for every continuous map $f_0: X \to Y$, every homotopy $F: A \times [0, 1] \to Y$ between the restriction $g_0 = f_0|A$ and a map $g_1: A \to Y$ has an *extension* $H: X \times [0, 1] \to Y$ which is a homotopy between f_0 and a map $f_1: X \to Y$ *extending* g_1.

This is equivalent to an ordinary problem of extension: consider the product $X \times [0,1]$ and in it the closed subspace

$$W = (A \times [0,1]) \cup (X \times \{0\}),$$

and the continuous map $F_1\colon W \to Y$ equal to $(x,0) \mapsto f_0(x)$ in $X \times \{0\}$ and to F in $A \times [0,1]$; the requested homotopies H are the *extensions* of F_1 to $X \times [0,1]$. If the extension f_1 of g_1 is *prescribed*, then the homotopies H between f_0 and f_1 extending F are the extensions to $X \times [0,1]$ of the map F'_1 of

$$W' = W \cup (X \times \{1\})$$

into Y, equal to F_1 in W and to $(x,1) \mapsto f_1(x)$ in $X \times \{1\}$. The existence of H in these two problems will be established if W (resp. W′) is a *retract* of $X \times [0,1]$.

It is in that problem that the ANR spaces are useful because of the following *lemma*. Suppose *both* X and A are separable metrizable ANR's; then (§2,B) W and W′ are ANR's, and therefore every neighborhood of W (resp. W′) contains a neighborhood T (resp. T′) such that F_1 (resp. F'_1) has an extension $G_1\colon T \to Y$ (resp. $G'_1\colon T' \to Y$). By compactness, any "fiber" $\{x\} \times [0,1]$ in $A \times [0,1]$ has a neighborhood of the form $U \times [0,1]$ contained in T (resp. T′), hence there is a neighborhood of $A \times [0,1]$ contained in T (resp. T′) *of the form* $V \times [0,1]$, where V is a neighborhood of A.

This lemma has the following consequences:

1. If X and A are both ANR's, there is a fundamental system of open neighborhoods V_λ of A in X such that A is a *strong deformation retract* of each V_λ: observe that each open neighborhood U of A in X is an ANR, and replace in the lemma X by a neighborhood $U' \subset U$ of A for which there is a retraction $r\colon U' \to A$, Y being replaced by A and f_0 and f_1 by r.

2. If *both* X and A are ANR's, then (X, A) has the *homotopy extension property* with respect to *any* space Y. Indeed, with the notations of the lemma, W is an ANR, hence there is a neighborhood T of W in $X \times [0,1]$ and a retraction $r\colon T \to W$. This shows that $(x,t) \mapsto G(x,t) = F(r(x,t))$ is a map of T into Y such that $G(x,0) = F(x,0)$. Furthermore, T contains a set $U \times [0,1]$, where U is a neighborhood of A in X, hence there is a continuous real function $u\colon X \to [0,1]$ such that $u(x) = 0$ in $X - U$ and $u(x) = 1$ in A. Then

$$(x,t) \mapsto H(x,t) = G(x, t\,.\,u(x)) \tag{30}$$

is continuous in $X \times [0,1]$ and such that $H(x,0) = G(x,0) = F(x,0)$ for $x \in A$.

This result was proved by J.H.C. Whitehead for compact spaces and extended by S. Hu for metrizable spaces [253].

3. Suppose now that *the space* Y *is an* ANR. Borsuk proved [71] that for *any* paracompact space X and any closed subspace A of X the pair (X, A) has the homotopy extension property *with respect to* Y. Indeed, with the notations introduced in the above lemma, there exists an extension G of F_1 to a neighborhood T of W since Y is an ANR, and T contains a neighborhood $U \times [0,1]$

of $A \times [0, 1]$ for a neighborhood U of A. Then the definition (30) is applicable and $H(x, 0) = F(x, 0)$ for $x \in A$.

After the theory of fiber spaces had been developed (chap. III) their properties appeared as "dual" in some sense to the extension problem, and when a pair (X, A) has the homotopy extension property for *all* spaces Y, the natural injection $A \to X$ is called a *cofibration*. If $A \subset B \subset X$ and the injections $A \to B$ and $B \to X$ are cofibrations, the injection $A \to X$ is a cofibration.

An important property of a cofibration $i: A \to X$ is that if A is *contractible* (in itself), then the *collapsing map* $p: (X, x_0) \to (X/A, \bar{x}_0)$ with $x_0 \in A$, $\bar{x}_0 = p(x_0) = p(A)$ (Part 2, chap. V, § 2,A) is a *homotopy equivalence*. Indeed the assumptions imply the existence of a continuous map $h: X \times [0, 1] \to X$ such that $h(x, 0) = x$ for $x \in X$, $h(x, t) \in A$ and $h(x, 1) = x_0$ for $x \in A$, so that one may write $h(x, 1) = k(p(x))$ where $k: (X/A, \bar{x}_0) \to (X, x_0)$ is continuous, hence $k \circ p \sim 1_X$. On the other hand, $p(h(x, t)) = h_1(p(x), t)$, where $h_1: (X/A) \times [0, 1] \to X/A$ is continuous, and $h_1(y, 0) = y$, $h_1(y, 1) = p(k(y))$ for $y \in X/A$, so $p \circ k \sim 1_{X/A}$.

Examples of Cofibrations: I. Mapping Cylinder

Let $f: X \to Y$ be a continuous map, Z_f be its mapping cylinder (Part 2, chap. V, § 3,A), and $p: Y \coprod (X \times [0, 1]) \to Z_f$ be the natural projection. There are two injections $i: x \mapsto p(x, 0)$ of X into Z_f and $j: y \mapsto p(y)$ of Y into Z_f; both i and j are *cofibrations*. To see that i is a cofibration it is enough to define a retraction R′ of the space $Z_f \times [0, 1]$ on its subspace

$$(Z_f \times \{0\}) \cup (X \times [0, 1])$$

[where X is identified with $i(X)$]. This is done by a variation of the "box lemma" (see above) applied to each product $p(\{x\} \times [0, 1]) \times [0, 1]$. Explicitly,

$$R'(p(y), s) = (p(y), 0) \quad \text{for } y \in Y,$$

$$R'(p(x, t), s) = \begin{cases} (p(x, 0), (s - 2t)/(1 - t)) & \text{if } 0 \leqslant t \leqslant s/2 \\ (p(x, (2t - s)/(2 - s)), 0) & \text{if } s/2 \leqslant t \leqslant 1 \end{cases} \quad \text{for } x \in X.$$

The method is the same for j; this time define a retraction R″ of the space $Z_f \times [0, 1]$ on its subspace

$$(Z_f \times \{0\}) \cup (Y \times [0, 1]) \quad [Y \text{ identified with } j(Y)]$$

by

$$R''(p(y), s) = (p(y), s) \quad \text{for } y \in Y,$$

$$R''(p(x, t), s) = \begin{cases} (p(x, 2t/(2 - s)), 0) & \text{if } 0 \leqslant t \leqslant (2 - s)/2 \\ (p(x, 1), (s + 2t - 2)/t) & \text{if } (2 - s)/2 \leqslant t \leqslant 1 \end{cases} \quad \text{for } x \in X.$$

Recall (Part 2, chap. V, § 3,A) that Y is a strong deformation retract of Z_f; if $r: Z_f \to Y$ is the corresponding retraction, $f = r \circ i$, so that f is *homotopic to* i in Z_f. This can be expressed by saying that *any* continuous map may be multiplied on the left by a homotopy equivalence so that the product is homotopic to a *cofibration*.

II. Mapping Cone

With the same assumptions, consider the mapping cone C_f (Part 2, chap. V, §3,B) and the injection $j: Y \to C_f$; j is a *cofibration*. The space C_f is equal to Z_f/A, where A is a subspace of Z_f (*loc. cit.*), then the retraction R″ defined above is such that $R''(A \times [0,1]) \subset A \times [0,1]$; passing to the quotient, R″ yields a retraction R''_0 of $C_f \times [0,1]$ onto the subspace $(C_f \times \{0\}) \cup (Y \times [0,1])$.

III. CW-Complexes

Let (X, A) be a *relative CW-complex* (Part 2, chap. V, §3,C). Then the natural injection $j: A \to X$ is a *cofibration*. It is enough to prove that when X is obtained by attaching n-cells e_α^n to A by maps $f_\alpha: \mathbf{D}_n \to X$. It is necessary to show that $W = (X \times \{0\}) \cup (A \times [0,1])$ is a retract of $X \times [0,1]$. By the box lemma, there is a retraction

$$r: \mathbf{D}_n \times [0,1] \to (\mathbf{D}_n \times \{0\}) \cup (\mathbf{S}_{n-1} \times [0,1]).$$

Then the retraction R of $X \times [0,1]$ on W is defined by taking

$$R(x,t) = (x,t) \quad \text{for } x \in A,$$

$$R(f_\alpha(z),t) = f_\alpha(\varphi(z,t)) \quad \text{if } r(z,t) = (\varphi(z,t),t') \in \mathbf{D}_n \times [0,1].$$

E. Fixed Points and Retracts

The interest expressed by Lefschetz for locally contractible spaces clearly stemmed from his desire to extend as much as possible his fixed point formula (Part 2, chap. III, §2); in ([304], p. 347) he proved its validity for *compact* locally contractible spaces. Later he realized that any compact ANR, X, is *dominated* (§2,C) by a finite euclidean simplicial complex Y; for any $\varepsilon > 0$, Y and the continuous maps $f: X \to Y$ and $g: Y \to X$ can be chosen such that there is a homotopy $F: X \times [0,1] \to X$ between $g \circ f$ and 1_X having the additional property that the diameter of $F(\{x\} \times [0,1])$ is $\leqslant \varepsilon$ for any $x \in X$. From this result it is easy to deduce the fixed point formula for X from the fact that it holds for Y.

Much work was done later to refine the conclusions of the Lefschetz fixed point formula, for which we refer to [94a].

F. The Lusternik–Schnirelmann Category

Around 1930 Lusternik and Schnirelmann attached a new numerical *invariant* of the homotopy type ([329], [414]) to a pair (X, A) of topological spaces. A nonempty subspace Y of X is *deformable to a point* in X if the natural injection $Y \to X$ is homotopic *in* X to a constant map $Y \to \{a\}$ for some $a \in X$; any subspace Z of Y is then also deformable to a point. If $A \subset X$ is the union of a finite number of subspaces of X deformable to a point, the *smallest* number of subspaces of such a union is called the *Lusternik–Schnirelmann category of* A *in* X and written $\mathrm{cat}_X(A)$; if no such union exists, $\mathrm{cat}_X(A) = +\infty$. The number $\mathrm{cat}_X(X)$ is simply written cat(X) and called the *category of the space* X. The relation $\mathrm{cat}(X) = 1$ means that X is contractible. Elementary properties are, for subspaces A, B of X, that if $A \subset B$, then $\mathrm{cat}_X(A) \leqslant \mathrm{cat}_X(B)$ and for any two subspaces $\mathrm{cat}_X(A \cup B) \leqslant \mathrm{cat}_X(A) + \mathrm{cat}_X(B)$. If two arcwise-connected spaces X_1, X_2 have finite category

$$\sup(\mathrm{cat}(X_1), \mathrm{cat}(X_2)) \leqslant \mathrm{cat}(X_1 \times X_2) \leqslant \mathrm{cat}(X_1) + \mathrm{cat}(X_2) - 1$$

Even if X is arcwise-connected and simply connected, there may exist subspaces A of X such that $\mathrm{cat}_X(A) > 1$ [196].

The Lusternik–Schnirelmann category of a space is not determined by its cohomology: clearly $\mathrm{cat}(\mathbf{S}_n) = 2$ for all $n \geqslant 1$, but if P is the Poincaré space (chap. I, § 4,D), $\mathrm{cat}(P) > 2$ although P and $\mathbf{S}_3$ have isomorphic cohomology algebras [73]; this can be proved using the general fact that if $\mathrm{cat}(X) \leqslant 2$, then $\pi_1(X)$ is a free group [196].

There are, however, relations between category and cohomology. If X is a smooth compact manifold such that $\mathrm{cat}(X) = n$, any cup-product $u_1 \smile u_2 \smile \cdots \smile u_n = 0$ in $H^{\bullet}(X; \Lambda)$ if the (homogeneous) classes u_j are not scalars.

Using that property and construction of suitable coverings, one proves $\mathrm{cat}(\mathbf{T}^n) = n + 1$, $\mathrm{cat}(\mathbf{P}_n(\mathbf{R})) = n + 1$, and $\mathrm{cat}(\mathbf{S}_{n,m}(\mathbf{C})) = m$ for complex Stiefel manifolds.

§ 3. Homotopy Groups

A. The Hurewicz Definition

We have seen (Part 2, chap. VII) how the study of the homology of suitable *spaces of functions* can yield results on problems of geometry or analysis in finite-dimensional spaces. A similar idea led Hurewicz, in a series of four Notes published in 1934–1936 [256],* to define the first *algebraic objects* in homotopy theory. His central tool was the introduction of a *topology* on the set $\mathscr{C}(Y; X)$ of continuous maps of a space Y into a space X. When Y is compact and X is a metric space with distance d, a *distance* can be defined in $\mathscr{C}(Y; X)$ by

$$d(f, g) = \sup_{y \in Y} d(f(y), g(y)), \tag{31}$$

generalizing the familiar normed space $\mathscr{C}(I)$ of real continuous functions in a compact interval, so that convergence for the distance (31) may be called *uniform convergence* in Y. Around 1940 this idea was extended to provide a

---

* In these Notes he announced a paper that would contain detailed proofs of his theorems, but that paper was never published.

topology on the set $\mathscr{C}(Y; X)$ for *any* pair of topological spaces X, Y. For any quasicompact set $K \subset Y$ and any open set $U \subset X$, let $T(K, U)$ be the set of all maps $f \in \mathscr{C}(Y; X)$ such that $f(K) \subset U$. Since

$$T(K_1 \cup K_2, U_1 \cap U_2) \subset T(K_1, U_1) \cap T(K_2, U_2),$$

the sets $T(K, U)$ form a *basis* for a topology on $\mathscr{C}(Y; X)$ called the *compact-open* topology, or the topology of *compact convergence*. Clearly

$$(Y, X) \mapsto \mathscr{C}(Y; X)$$

is a *bifunctor* $\boldsymbol{T} \times \boldsymbol{T} \to \boldsymbol{T}$, covariant in X and contravariant in Y. It has a very useful *fundamental property*:

When Y is a *locally compact* space, X is an arbitrary space, and Z is a Hausdorff space, continuity of a map $f: Z \times Y \to X$ is *equivalent* to continuity of the map $\tilde{f}: Z \to \mathscr{C}(Y; X)$, where

$$\tilde{f}(z) = (y \mapsto f(z, y)); \tag{32}$$

there is a natural bijection $f \mapsto \tilde{f}$ of sets

$$\mathscr{C}(Z \times Y; X) \rightleftarrows \mathscr{C}(Z; \mathscr{C}(Y; X)), \tag{33}$$

but this is also a *homeomorphism* for the compact-open topologies.

This property was the point of departure of Hurewicz. Let Y be a locally compact space and X be any topological space; a homotopy F between two maps f_0, f_1 in $\mathscr{C}(Y; X)$ is a continuous map $(t, y) \mapsto F(t, y)$ of $[0, 1] \times Y$ into X with $F(0, y) = f_0(y)$ and $F(1, y) = f_1(y)$; the corresponding map

$$t \mapsto \tilde{F}(t) \in \mathscr{C}(Y; X)$$

is a *path* in $\mathscr{C}(Y; X)$ of origin f_0 and extremity f_1. There is thus a natural *bijection**

$$[Y; X] \rightleftarrows \pi_0(\mathscr{C}(Y; X)) \tag{34}$$

of the set $[Y; X]$ of *classes of homotopic maps* of Y into X onto the set of *arcwise connected components* of the space $\mathscr{C}(Y; X)$. In more sophisticated language, it was later said that (34) is a *bifunctor* (covariant in X, contravariant in Y)

$$\boldsymbol{LC} \times \boldsymbol{T} \to \boldsymbol{Set} \tag{35}$$

where $\boldsymbol{LC}$ is the full subcategory of $\boldsymbol{T}$ consisting of locally compact spaces.

To work with loops and fundamental groups it is necessary to consider the subspaces $\mathscr{C}(Y, y_0; X, x_0)$ of $\mathscr{C}(Y; X)$ consisting of the continuous maps f such that $f(y_0) = x_0$ for $y_0 \in Y$, $x_0 \in X$; one says f is a map of *pointed spaces*

$$f: (Y, y_0) \to (X, x_0). \tag{36}$$

In categorical language, these maps are the morphisms in the category $\boldsymbol{PT}$ of

---

* For any space Z, $\pi_0(Z)$ is the *set* of arcwise-connected components of Z.

pointed spaces, in which the objects are the pairs (X, x_0); for a homotopy F between two such maps, it is *always* required that $F(t, y_0) = x_0$ for all $t \in [0, 1]$. One also uses the category ***PH***, where the objects are the pointed spaces, and the morphisms the homotopy classes of the maps (36).

The set of homotopy classes $[f]$ of maps (36), in the preceding sense, is written

$$[Y, y_0; X, x_0].$$

The *loops* in a pointed space (X, x_0) are then the maps of pointed spaces*

$$(\mathbf{S}_1, *) \to (X, x_0),$$

so that the set of these loops

$$\Omega(X, x_0) = \mathscr{C}(\mathbf{S}_1, *; X, x_0) \tag{37}$$

is now given the *topology of a subspace of* $\mathscr{C}(\mathbf{S}_1; X)$,† and, just as above, there is a natural *bijection*

$$\pi_1(X, x_0) \xrightarrow{\sim} \pi_0(\mathscr{C}(\mathbf{S}_1, *; X, x_0)). \tag{38}$$

The natural generalization that Hurewicz introduced was to consider for *any* integer $n > 1$ *the arcwise-connected components of the space* $\mathscr{C}(\mathbf{S}_n, *; X, x_0)$.

The key property is that if a_n designates the *constant* map $\mathbf{S}_{n-1} \to \{x_0\}$ for $n \geqslant 2$, there is a *natural bijection*

$$\pi_0(\mathscr{C}(\mathbf{S}_n, *; X, x_0)) \xrightarrow{\sim} \pi_1(\mathscr{C}(\mathbf{S}_{n-1}, *; X, x_0), a_n). \tag{39}$$

Identify the space $\mathscr{C}(\mathbf{S}_n, *; X, x_0)$ with the subspace of $\mathscr{C}([0, 1]^n; X)$ consisting of the maps f such that $f(\mathrm{Fr}([0, 1]^n)) = \{x_0\}$; that subspace is naturally homeomorphic to the space $\mathscr{C}(Y, y_0; X, x_0)$, where Y is the collapsed space $[0, 1]^n/\mathrm{Fr}([0, 1]^n)$, y_0 being the image of $\mathrm{Fr}([0, 1]^n)$ (Part 2, chap. IV, § 2,A). Furthermore, we defined in Part 2, chap. IV, § 2,C a canonical homeomorphism of (Y, y_0) onto $(\mathbf{S}_n, *)$ when $\mathbf{S}_n$ is considered as a CW-complex with only two cells.‡ Using the fact that $[0, 1]^n = [0, 1] \times [0, 1]^{n-1}$, the fundamental property of the compact-open topology yields a natural identification $f \mapsto \tilde{f}$ of $\mathscr{C}([0, 1]^n; X)$ with $\mathscr{C}([0, 1]; \mathscr{C}([0, 1]^{n-1}; X))$, where for $(t, t_1, \ldots, t_{n-1}) \in [0, 1]^n$,

$$\tilde{f}(t)(t_1, \ldots, t_{n-1}) = f(t, t_1, \ldots, t_{n-1}). \tag{40}$$

Hence the natural homeomorphism

$$\mathscr{C}(\mathbf{S}_n, *; X, x_0) \xrightarrow{\sim} \mathscr{C}(\mathbf{S}_1, *; \mathscr{C}(\mathbf{S}_{n-1}, *; X, x_0), a_n) = \Omega(\mathscr{C}(\mathbf{S}_{n-1}, *; X, x_0), a_n), \tag{41}$$

which yields the natural bijection (39) by (38).

---

* The sphere $\mathbf{S}_n$ is always considered as the subspace of $\mathbf{R}^{n+1}$ defined by $\sum_{j=1}^{n+1} \xi_j^2 = 1$, and the vector $\mathbf{e}_1$ is noted as $*$.

† If x_1 is the constant loop $\mathbf{S}_1 \to \{x_0\}$, $(X, x_0) \mapsto (\Omega(X, x_0), x_1)$ is a contravariant functor in the category ***PH***.

‡ We identify the closed unit ball $\mathbf{D}_n \subset \mathbf{R}^n$ with $[0, 1]^n$, to which it is homeomorphic.

By chap. I, § 1 it follows from (39) that there is on $\pi_0(\mathscr{C}(\mathrm{S}_n, *; \mathrm{X}, x_0))$ a natural structure of *group*; this group is denoted by $\pi_n(\mathrm{X}, x_0)$ and called the *n-th homotopy group* of the pointed space (X, x_0). Applying the definitions of Poincaré and using (40), the group law in $\pi_n(\mathrm{X}, x_0)$ is defined as follows: $\pi_n(\mathrm{X}, x_0)$ is the set of classes of continuous maps $f\colon [0,1]^n \to \mathrm{X}$ such that $f(\mathrm{Fr}([0,1]^n)) = \{x_0\}$; two such maps f, g are in the same class if there is a homotopy between f and g leaving invariant all points of $\mathrm{Fr}([0,1]^n)$; if $[f_1]$, $[f_2]$ are two classes in $\pi_n(\mathrm{X}, x_0)$, the sum $[f_1] + [f_2]$ is the class of the map

$$(t, t_1, \ldots, t_{n-1}) \mapsto \begin{cases} f_1(2t, t_1, \ldots, t_{n-1}) & \text{if } 0 \leqslant t \leqslant 1/2, \\ f_2(2t-1, t_1, \ldots, t_{n-1}) & \text{if } 1/2 \leqslant t \leqslant 1. \end{cases} \tag{42}$$

The neutral element 0 is the class of the constant map $[0,1]^n \to \{x_0\}$ and the inverse element $-[f]$ of $[f]$ is the class $[f \circ s]$, where

$$s(t, t_1, \ldots, t_{n-1}) = (1-t, t_1, \ldots, t_{n-1}). \tag{43}$$

This definition includes the definition of the fundamental group $\pi_1(\mathrm{X}, x_0)$; but for $n \geqslant 2$ the group $\pi_n(\mathrm{X}, x_0)$ is *commutative* (hence the additive notation). Indeed, the *space* $\Omega(\mathrm{X}, x_0)$ for any pointed space (X, x_0) is an *H-space* for the juxtaposition of loops (Part 2, chap. VI, § 2,A), since the constant loop $[0,1] \to \{x_0\}$ is a "homotopy neutral element" for that law; but for $n \geqslant 2$, by (39) and (41) there is a natural group isomorphism

$$\pi_n(\mathrm{X}, x_0) \simeq \pi_1(\Omega(\mathscr{C}(\mathrm{S}_{n-2}, *; \mathrm{X}, x_0), a_{n-1})); \tag{44}$$

the assertion follows from chap. I, § 4,C.

If X_0 is the arcwise-connected component of the point x_0 in X, it is clear that any map $(\mathrm{S}_n, *) \to (\mathrm{X}, x_0)$ takes its values in X_0 for $n \geqslant 1$; hence $\pi_n(\mathrm{X}, x_0) = \pi_n(\mathrm{X}_0, x_0)$.

B. Elementary Properties of Homotopy Groups

The preceding definition and the properties of the spaces $\mathscr{C}(\mathrm{Y}; \mathrm{X})$ imply the following elementary properties of $\pi_n(\mathrm{X}, x_0)$:

1. $\pi_n\colon (\mathrm{X}, x_0) \mapsto \pi_n(\mathrm{X}, x_0)$ is a *covariant functor*

$$\boldsymbol{PH} \to \boldsymbol{Gr}. \tag{45}$$

Indeed, for any map $u\colon (\mathrm{X}, x_0) \to (\mathrm{Y}, y_0)$, the homomorphism

$$u_*\colon \pi_n(\mathrm{X}, x_0) \to \pi_n(\mathrm{Y}, y_0)$$

maps each class $[f]$ to the class $[u \circ f]$ that only depends on the homotopy class $[u]$ and can be written $[u] \circ [f]$.

2. The map

$$\pi_n(\mathrm{X} \times \mathrm{Y}, (x_0, y_0)) \to \pi_n(\mathrm{X}, x_0) \times \pi_n(\mathrm{Y}, y_0)$$

deduced by 1 from $z \mapsto (\mathrm{pr}_1(z), \mathrm{pr}_2(z))$ is an isomorphism; this can be generalized to arbitrary products:

$$\pi_n\left(\prod_\alpha X_\alpha, (x_\alpha)\right) \simeq \prod_\alpha \pi_n(X_\alpha, x_\alpha).$$

3. If Z is a *covering space* of X, p: Z → X is the natural projection, and $z_0 \in Z$ is any point of $p^{-1}(x_0)$, then *for* $n \geqslant 2$ the map

$$p_*: \pi_n(Z, z_0) \to \pi_n(X, x_0) \tag{46}$$

is an *isomorphism*. Indeed, S_n is then simply connected, therefore any map $(S_n, *) \to (X, x_0)$ lifts uniquely to a map $(S_n, *) \to (Z, z_0)$, and the homotopy class of the latter only depends on the homotopy class of the former (chap. I, § 2).

4. $\pi_n(X, x_0) = 0$ for all n for a contractible space X, since any map $(S_n, *) \to (X, x_0)$ is homotopic to the constant map $S_n \to \{x_0\}$.

From 3 and 4 it follows in particular that

$$\pi_n(S_1) = 0 \quad \text{for } n \geqslant 2 \tag{47}$$

since the covering space **R** of S_1 is contractible.

C. Suspensions and Loop Spaces

For a pointed space (X, x_0), *iterated loop spaces* $(\Omega^p(X), x_p)$ are defined by induction: x_1 is the constant loop $S_1 \to \{x_0\}$ and in general x_p is the constant loop $S_1 \to \{x_{p-1}\}$ in $(\Omega^{p-1}(X), x_{p-1})$. Then one defines $\Omega^0(X) = X$ and

$$\Omega^p(X) = \Omega(\Omega^{p-1}(X), x_{p-1}). \tag{48}$$

From this definition, (37), and (41)

$$\Omega^{m+n}(X) \cong \mathscr{C}(S_m, *; \Omega^n(X), x_n) \tag{49}$$

by induction on m and n; hence there are natural isomorphisms

$$\pi_n(X, x_0) \rightleftarrows \pi_{n-1}(\Omega(X), x_1) \rightleftarrows \cdots \rightleftarrows \pi_{n-p}(\Omega^p(X), x_p) \rightleftarrows \cdots \rightleftarrows \pi_0(\Omega^n(X)). \tag{50}$$

Later [440] these properties, which concern maps of spheres into arbitrary spaces, were generalized for any pair of pointed spaces (X, x_0), (Y, y_0), where Y is a Hausdorff space. There is a natural homeomorphism

$$\mathscr{C}(Y, y_0; \Omega X, x_1) \rightleftarrows \mathscr{C}(SY, y_0; X, x_0) \tag{51}$$

where y_0 is identified with its image in the *reduced suspension* SY (Part 2, chap. V, § 2,C). By (33) $\mathscr{C}(Y, y_0; \Omega(X), x_1)$ is identified with the subspace of $\mathscr{C}([0,1] \times Y; X)$ consisting of the maps $(t, y) \mapsto f(t, y)$ such that $f(0, y) = f(1, y) = x_0$ and $f(t, y_0) = x_0$ for $0 \leqslant t \leqslant 1$; but these are exactly the maps

$$[0,1] \times Y \xrightarrow{p} SY \xrightarrow{g} X,$$

where p is the natural map on the quotient space and $g \in \mathscr{C}(SY, y_0; X, x_0)$. From (51) is deduced a natural bijection

$$[Y, y_0; \Omega X, x_1] \rightleftarrows [SY, y_0; X, x_0]. \tag{52}$$

In categorical language this is expressed by saying that in the subcategory of ***PH*** consisting of pointed Hausdorff spaces,

$$(\mathrm{Y}, y_0) \mapsto (\mathrm{SY}, y_0) \quad \text{and} \quad (\mathrm{X}, x_0) \mapsto (\Omega \mathrm{X}, x_1)$$

are *adjoint functors* (Part 1, chap. IV, § 8,C).

Now $(\Omega\mathrm{X}, x_1)$ is not only a H-space, but an *H-group*; in general, this is a pointed H-space (Z, z_0) in which: (1) the multiplication m is such that the diagram

$$\begin{CD} \mathrm{Z}\times\mathrm{Z}\times\mathrm{Z} @>{1\times m}>> \mathrm{Z}\times\mathrm{Z} \\ @V{m\times 1}VV @VV{m}V \\ \mathrm{Z}\times\mathrm{Z} @>>{m}> \mathrm{Z} \end{CD}$$

commutes up to homotopy (i.e., commutes in the category ***PH***); (2) z_0 is a "homotopy neutral element" (Part 2, chap. VI, § 2,A) and there is a "homotopy inverse" $i: (\mathrm{Z}, z_0) \to (\mathrm{Z}, z_0)$, meaning that the diagrams

$$\begin{array}{ccccc} \mathrm{Z} & \xrightarrow{(1,z_0)} & \mathrm{Z}\times\mathrm{Z} & \xleftarrow{(z_0,1)} & \mathrm{Z} \\ & \nwarrow^{\mathrm{Id}} & \downarrow m & \nearrow_{\mathrm{Id}} & \\ & & \mathrm{Z} & & \end{array} \qquad \begin{array}{ccccc} \mathrm{Z} & \xrightarrow{(i,1)} & \mathrm{Z}\times\mathrm{Z} & \xleftarrow{(1,i)} & \mathrm{Z} \\ & \nwarrow^{z_0} & \downarrow m & \nearrow_{z_0} & \\ & & \mathrm{Z} & & \end{array}$$

(where z_0 is identified with the map $\mathrm{Z} \to \{z_0\}$) also commute in ***PH***. For such a H-group (Z, z_0) and any pointed space (Y, y_0), the set $[\mathrm{Y}, y_0; \mathrm{Z}, z_0]$ is given a law of composition

$$([f],[g]) \mapsto [f].[g] \tag{53}$$

for which $[f].[g]$ is the homotopy class of the map

$$\mathrm{Y} \xrightarrow{\ \delta\ } \mathrm{Y}\times\mathrm{Y} \xrightarrow{f\times g} \mathrm{Z}\times\mathrm{Z} \xrightarrow{\ m\ } \mathrm{Z} \tag{54}$$

[where δ is the diagonal map: $\delta(y) = (y, y)$] and it is readily seen that this makes $[\mathrm{Y}, y_0; \mathrm{Z}, z_0]$ into a *group*; if the diagram

$$\begin{array}{ccc} \mathrm{Z}\times\mathrm{Z} & \xrightarrow{\ \sigma\ } & \mathrm{Z}\times\mathrm{Z} \\ & {}_{m}\searrow \quad \swarrow_{m} & \\ & \mathrm{Z} & \end{array}$$

[where σ is the reversal map: $\sigma(z_1, z_2) = (z_2, z_1)$] commutes in ***PH***, one says the H-group (Z, z_0) is commutative, and the group $[\mathrm{Y}, y_0; \mathrm{Z}, z_0]$ is *commutative*.

In particular, $(\Omega(\mathrm{X}), x_1)$ is an H-group for any pointed space (X, x_0), and if (X, x_0) is itself an H-space, then $(\Omega(\mathrm{X}), x_1)$ is commutative, hence $[\mathrm{Y}, y_0; \Omega\mathrm{X}, x_1]$ {and by (52), $[\mathrm{SY}, y_0; \mathrm{X}, x_0]$} are *groups*, and for any $p \geqslant 2$, $[\mathrm{Y}, y_0; \Omega^p\mathrm{X}, x_p]$ is a *commutative group*; by (52), the same is true of

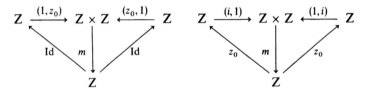

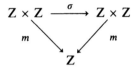

$$[S^kY, y_0; \Omega^{p-k}X, x_{p-k}]$$

for all $k \leqslant p$, all these groups being naturally *isomorphic*.

In particular the homotopy group $\pi_n(X, x_0)$ is naturally identified with the group $[S_n, *; X, x_0]$, generalizing the definition of the fundamental group. In fact, such a definition had been proposed by Čech at the International Congress of Mathematicians in 1932 [121], and he reported that the same idea had earlier been considered by Dehn, who had not published it. Čech himself did not pursue the study of these groups, because at that time it was believed that significant generalizations of the fundamental group should be noncommutative in general. The group law on the set $[S_n, *; X, x_0]$ can be presented in the following way [254]. Identify S_n with the sphere $\sum_{j=1}^{n+1} \xi_j^2 = 1$ in $\mathbf{R}^{n+1}$, let D_+ and D_- be the two closed hemispheres in S_n defined by $\xi_{n+1} \geqslant 0$ and $\xi_{n+1} \leqslant 0$, and let f, g be two continuous maps of S_n into X taking the value x_0 at the point $*$. Since D_+ and D_- are contractible, the restrictions $f|D_-$ and $g|D_+$ are homotopic to the constant maps $D_- \to \{x_0\}$ and $D_+ \to \{x_0\}$, respectively. From the homotopy extension property (§ 2,D) it follows that there exists a map $f_1: (S_n, *) \to (X, x_0)$ homotopic to f and such that $f_1(D_-) = f_1(S_{n-1}) = \{x_0\}$ and a map $g_1: (S_n, *) \to (X, x_0)$ homotopic to g and such that $g_1(D_+) = g_1(S_{n-1}) = \{x_0\}$. Then if h is the map equal to f_1 in D_+ and to g_1 in D_-, the sum $[f] + [g]$ of the homotopy classes of f and g in $\pi_n(X, x_0)$ is the homotopy class $[h]$.

The continuous maps $f: (S_n, *) \to (X, x_0)$ such that $[f] = 0$ are by definition those for which there is a continuous map $F: S_n \times [0,1] \to X$ such that $F(y, 0) = f(y)$, $F(y, 1) = x_0$, and $F(a, t) = x_0$ for $0 \leqslant t \leqslant 1$ where $a = *$ (for convenience); but any point $z \in \mathbf{D}_{n+1}$ other than a can be written $z = ta + (1-t)y$ for $y \in S_n$ and $0 \leqslant t < 1$; the function $z \mapsto G(z) = F(y, t)$ for $z \neq a$, $G(a) = x_0$, is then continuous in $\mathbf{D}_{n+1}$ and extends f. The converse is immediate, and the maps $f: (S_n, *) \to (X, x_0)$ such that $[f] = 0$ are therefore identified to those which can be *continuously extended* to the ball $\mathbf{D}_{n+1}$.

Finally, from this description it follows that for any map $f: (S_n, *) \to (X, x)$ the class $-[f]$ is the class of $f \circ s$, where s is the symmetry $\xi_{n+1} \mapsto -\xi_{n+1}$ with respect to the hyperplane $\xi_{n+1} = 0$ (generally identified with $\mathbf{R}^n$): if f_1 is the map homotopic to f and such that $f_1(D_-) = \{x_0\}$, then $g_1 = f_1 \circ s$ is homotopic to $f \circ s$, and if h is equal to f_1 in D_+ and to g_1 in D_-, it can be extended to $\mathbf{D}_{n+1}$ by $h(ty + (1-t)s(y)) = f_1(y)$ for all $y \in D_+$.

D. The Homotopy Suspension

If in (51) $X = SY$ and $x_0 = y_0$, to the identity 1_{SY} is associated a canonical map

$$s: (Y, y_0) \to (\Omega SY, y_1) \tag{55}$$

such that $s(y)$ is the loop $t \mapsto p(t, y)$ in SY. By functoriality this determines a map $s_*: [X, x_0; Y, y_0] \to [X, x_0; \Omega SY, y_1]$, and using the bijection (52) a natural map is obtained:

$$E: [Y, y_0; X, x_0] \to [SY, y_0; SX, x_0] \qquad \text{(also written S)} \tag{56}$$

called the *homotopy suspension*, which was introduced in 1937 by Freudenthal [201], and which has become one of the most important concepts in homotopy theory. For any map $f: (Y, y_0) \to (X, x_0)$, $E([f])$ can be defined directly as the homotopy class of the map $E(f): (SY, y_0) \to (SX, x_0)$ (also written Sf), which sends $p(y, t)$ to $p(f(y), t)$.

In particular, this defines a natural transformation (Part 1, chap. IV, § 5,E) of the functor π_n to the functor $\pi_{n+1} \circ S$:

$$E: \pi_n(X, x_0) \to \pi_{n+1}(SX, x_0). \tag{57}$$

In contrast with (50), this map is not necessarily injective nor surjective: $\pi_2(S_1) = 0$ but $\pi_3(S_2) \neq 0$, and we shall see later that $\pi_4(S_3)$ is finite although $\pi_3(S_2) \simeq \mathbf{Z}$.

E. Whitehead Products

In 1941, in order to study the homotopy of CW-complexes, J.H.C. Whitehead defined a map ([492], p. 238)

$$\pi_m(X, x_0) \times \pi_n(X, x_0) \to \pi_{m+n-1}(X, x_0) \quad \text{for } m \geqslant 1, n \geqslant 1 \tag{58}$$

in the following way. Write $I = [0, 1]$ and start from two continuous maps

$$f: I^m \to X, \qquad g: I^n \to X,$$

such that $f(\mathrm{Fr}(I^m)) = \{x_0\}$ and $g(\mathrm{Fr}(I^n)) = \{x_0\}$. The frontier $\mathrm{Fr}(I^{m+n})$ can be written as the union $(I^m \times \mathrm{Fr}(I^n)) \cup (\mathrm{Fr}(I^m) \times I^n)$, the intersection of these two sets being $\mathrm{Fr}(I^m) \times \mathrm{Fr}(I^n)$. Define a continuous map $h: \mathrm{Fr}(I^{m+n}) \to X$ by

$$h(s, t) = \begin{cases} f(s) & \text{for } (s, t) \in I^m \times \mathrm{Fr}(I^n), \\ g(t) & \text{for } (s, t) \in \mathrm{Fr}(I^m) \times I^n, \end{cases}$$

since $f(s) = g(t) = x_0$ for $(s, t) \in \mathrm{Fr}(I^m) \times \mathrm{Fr}(I^n)$. The homotopy class w of h only depends on the homotopy classes u, v of f and g, hence one may write

$$w = [u, v] \tag{59}$$

defining the map (58). For $m = n = 1$, $[u, v]$ is the commutator $uvu^{-1}v^{-1}$, which justifies the notation (59). Whitehead proved that for fixed u, $v \mapsto [u, v]$ is a homomorphism of groups, so that for $n \geqslant 2$

$$[u, v_1 + v_2] = [u, v_1] + [u, v_2]. \tag{60}$$

Furthermore, for $m + n \geqslant 3$,

$$[v, u] = (-1)^{mn}[u, v] \tag{61}$$

and, for $m + n + p \geqslant 4$, there is a *Jacobi identity* [355]

$$(-1)^{mp}[[u, v], w] + (-1)^{nm}[[v, w], u] + (-1)^{pn}[[w, u], v] = 0. \tag{62}$$

Finally, for any continuous map $\varphi: (X, x_0) \to (Y, y_0)$

$$\varphi_*([u, v]) = [\varphi_*(u), \varphi_*(v)]. \tag{63}$$

Steenrod has shown that the Whitehead product can be related to his "functional cup-product" (Part 1, chap. IV, §4,B). Let $\alpha \in \pi_m(X, x_0)$, $\beta \in \pi_n(X, x_0)$ be the homotopy classes of $f: (S_m, *) \to (X, x_0)$ and $g: (S_n, *) \to (X, x_0)$, respectively. These maps define a map

$$h: S_m \vee S_n \to X$$

such that $h(x, *) = f(x)$ and $h(*, x) = g(x)$. Now let $k: S_{m+n-1} \to S_m \vee S_n$ be such that $[k] = [\iota_m, \iota_n]$, the Whitehead product of the classes ι_m, ι_n of the natural injections $S_m \to S_m \vee S_n$, $S_n \to S_m \vee S_n$. Then the Whitehead product $[\alpha, \beta]$ is the homotopy class of the composite map

$$F: S_{m+n-1} \xrightarrow{k} S_m \vee S_n \xrightarrow{h} X.$$

Since $k^*(H^{m+n-1}(S_m \vee S_n)) = 0$, $F^*(H^{m+n-1}(X)) = 0$. This implies that for cohomology classes $u \in H^m(X)$, $v \in H^n(X)$ such that $u \smile v = 0$, the functional cup-product $u \smile_F v$ is defined and is an element of $H^{m+n-1}(S_{m+n-1})$ equal to $h^*(u) \smile_k h^*(v)$.

F. Change of Base Points

We have seen that in an arcwise-connected space X, a path α of origin x_0 and extremity x_1 determines an isomorphism of $\pi_1(X, x_0)$ onto $\pi_1(X, x_1)$ (chap. I, § 1). This extends to all homotopy groups.

To see this, use the "box lemma" (§ 2,D): for a map $f: I^n \to X$ equal to x_0 in $Fr(I^n)$, there is an extension F to $I^n \times I$ of the map equal to $(x, 0) \mapsto f(x)$ in $I^n \times \{0\}$, and to $(x, t) \mapsto \alpha(t)$ on $Fr(I^n) \times I$. The homotopy extension property for ANR's (§ 2,D) shows that the homotopy class of $x \mapsto F(x, 1)$ in $\pi_n(X, x_1)$ only depends on the class $[f]$ in $\pi_n(X, x_0)$, and therefore we have defined a map

$$\sigma_n(\alpha): \pi_n(X, x_0) \to \pi_n(X, x_1). \tag{64}$$

Using the definition of addition in $\pi_n(X, x_0)$ and $\pi_n(X, x_1)$ given in § 3,C for the Čech definition of homotopy groups, it is easy to verify that $\sigma_n(\alpha)$ is a homomorphism of groups. The box lemma and the homotopy extension property also show that if α' is a path from x_0 to x_1 homotopic to α (with fixed extremities), $\sigma_n(\alpha') = \sigma_n(\alpha)$. Furthermore, $\sigma_n(\alpha\beta) = \sigma_n(\alpha) \circ \sigma_n(\beta)$ for the juxtaposition of two paths α, β, thus $\sigma_n(\alpha) \circ \sigma_n(\alpha^{-1})$ is the identity, and $\sigma_n(\alpha)$ is an isomorphism.

In particular, if γ is a loop in X of origin x_0, $\sigma_n(\gamma)$ is an automorphism of $\pi_n(X, x_0)$ that only depends on the class u of α in $\pi_1(X, x_0)$, and it therefore can be written $\sigma_n(u)$. The map $u \mapsto \sigma_n(u)$ is a homomorphism of $\pi_1(X, x_0)$ into the group $\mathrm{Aut}(\pi_n(X, x_0))$, and for any $v \in \pi_n(X, x_0)$,

$$\sigma_n(u) . v - v = [v, u]. \tag{65}$$

A space X is called *n-simple* if $\sigma_n(u)$ is the identity for all $u \in \pi_1(X, x_0)$; for any point $x_1 \in X$, the isomorphism $\sigma_n(\alpha)$ is the same for *all* paths α of origin

x_0 and extremity x_1. In general, the groups $\pi_n(X, x)$ for $x \in X$ form a *local system* of groups (Part 1, chap. IV, § 7,A).

When X is n-simple and f, g are two maps $S_n \to X$ in the same homotopy class, but with different values $f(*) = x_0$, $g(*) = x_1$, there is by assumption a path α: I $\to$ X of origin x_0 and extremity x_1 such that $\sigma_n(\alpha)$ sends $[f]$ to $[g]$, so that the class of f *in the set* $[S_n; X]$ can be naturally identified with its class in $\pi_n(X; x_0)$. This defines on the set $[S_n; X]$ a group structure independent of x_0; that group is often written $\pi_n(X)$, and by abuse of language that notation is also used for $\pi_n(X; x_0)$ even when X is not n-simple.

To say that a space X is 1-*simple* means that the fundamental group $\pi_1(X)$ is *commutative*. An arcwise-connected space X for which $\pi_1(X) = 0$ or $\pi_n(X) = 0$ is n-simple.

Let X be an arcwise-connected *H-space*, and let f, g be two maps defining homotopy classes u, v in $\pi_m(X, x_0)$ and $\pi_n(X, x_0)$; if p: $I^{m+n} \to X$ is the map defined by

$$p(s, t) = f(s) . g(t)$$

(product defined by the multiplication in X), the homotopy class of $p | \mathrm{Fr}(I^{m+n})$ is the Whitehead product $[u, v]$; this implies $[u, v] = 0$; hence, by (65), X is n-*simple for every* n ([429], p. 477).

§ 4. First Relations between Homotopy and Homology

A. The Hurewicz Homomorphism

Hurewicz's second Note of 1935 was devoted to relations between the homotopy groups $\pi_n(X, x_0)$ of a pointed space (X, x_0) and the singular homology groups $H_n(X; \mathbf{Z})$ with integer coefficients; he only spoke of metric spaces and used Vietoris homology, but in a footnote ([256], II, p. 525) he mentions that one can use singular homology as well, and for that homology theory his arguments apply to an arbitrary space.

Generalizing the well-known homomorphism $\pi_1(X, x_0) \to H_1(X; \mathbf{Z})$ (chap. I, § 3,B), he defined for each n a natural homomorphism

$$h_n \colon \pi_n(X, x_0) \to H_n(X; \mathbf{Z}), \tag{66}$$

which is now called the *Hurewicz homomorphism.* As the closure $\bar{\Delta}_n$ of the standard n-simplex is homeomorphic to the closed cube $[0, 1]^n$, elements of $\pi_n(X, x_0)$ may be identified with the connected components of the space of maps $f \colon \bar{\Delta}_n \to X$ such that $f(\bar{\Delta}_n - \Delta_n) = \{x_0\}$. For such a map, consider the homomorphism in relative singular homology [Part 1, chap. IV, § 2, formula (4)]:

$$f_* \colon H_n(\bar{\Delta}_n, \bar{\Delta}_n - \Delta_n; \mathbf{Z}) \to H_n(X, \{x_0\}; \mathbf{Z}).$$

The exact sequence of homology for $n \geqslant 2$ (Part 1, chap. IV, § 6,B) yields natural isomorphisms

$$H_n(X; \mathbf{Z}) \simeq H_n(X, \{x_0\}; \mathbf{Z}),$$

$$H_n(\bar{\Delta}_n, \bar{\Delta}_n - \Delta_n; \mathbf{Z}) \xrightarrow{\sim} H_{n-1}(\mathbf{S}_{n-1}; \mathbf{Z}).$$

The inverse image ε_n by the second isomorphism of the fundamental class $[\mathbf{S}_{n-1}]$ can also be defined as the class of the relative singular n-cycle defined by the identity map of $\bar{\Delta}_n$. From the homotopy axiom of homology theory it follows that the element $f_*(\varepsilon_n) \in H_n(X; \mathbf{Z})$ only depends on the homotopy class $[f]$ of f in $\pi_n(X, x_0)$, so that a map (66) is defined by

$$h_n: [f] \to f_*(\varepsilon_n). \tag{67}$$

Using the homotopy operators between a singular simplex and its subdivisions (Part 1, chap. II, § 3), as well as the definition of $[f] + [g]$ in the Čech approach to homotopy groups (§ 3,C), it is easy to see that h_n is a homomorphism of groups. In general it is neither injective nor surjective: the Hopf fibration implies that $\pi_3(\mathbf{S}_2) \neq 0$, whereas $H_3(\mathbf{S}_2; \mathbf{Z}) = 0$. On the other hand, as $\mathbf{R}^2$ is the universal covering space of $\mathbf{T}^2$, $\pi_2(\mathbf{T}^2) = 0$, whereas $H_2(\mathbf{T}^2; \mathbf{Z}) \neq 0$. But in his second Note of 1935 Hurewicz discovered a remarkable case in which h_n is bijective, namely, when $\pi_j(X) = 0$ for $1 \leqslant j \leqslant n - 1$; he called that result (which he proved for finite simplicial complexes) the "equivalence theorem"; it is now referred to as the *absolute Hurewicz isomorphism theorem*, and has become one of the key results in homotopy theory.

Hurewicz's proof relies on the fact that for a subcomplex L of a finite simplicial complex K, the pair (K, L) has the homotopy extension property, since both K and L are ANR's (§§ 2,B and 2,C). For an n-dimensional simplicial complex K and for $0 \leqslant q \leqslant n$, he considered the *$q$-skeleton* K_q, the subcomplex union of all simplices of K of dimension $\leqslant q$. Using induction on q, and the homotopy extension property, he proved the *preliminary lemma*:

If X is an arcwise-connected space such that $\pi_i(X, x_0) = 0$ for $1 \leqslant i \leqslant n - 1$, then any continuous map $f: K \to X$ is homotopic to a map $g: K \to X$ such that $g(K_{n-1}) = \{x_0\}$.

To prove surjectivity of h_n, Hurewicz considered an arbitrary n-cycle $z = \sum_j \lambda_j s_j^n$, where the s_j^n are continuous maps $\bar{\Delta}_n \to X$ such that $\sum_j \lambda_j \mathbf{b} s_j^n = 0$. It is easy to define a finite euclidean simplicial complex K (depending on z) and a continuous map $f: K \to X$ such that $z = f(z')$ where $z' = \sum_j \lambda_j t_j^n$ is an n-cycle in K. By the preliminary lemma, f is homotopic to a map $g: K \to X$ such that $g(K_{n-1}) = \{x_0\}$. By the Čech definition of addition in $\pi_n(X, x_0)$ the homology class $\bar{z}'$ of z' is

$$\bar{z}' = h_n\left(\sum_j \lambda_j [f \circ t_j^n]\right)$$

and as the class $\bar{z}$ of z is equal to $\bar{z}'$, this shows that $\bar{z}$ is in the image of h_n.

To prove injectivity, suppose f is identified with a map $\bar{\Delta}_{n+1} - \Delta_{n+1} \to X$, and that $h_n([f]) = 0$ in $H_n(X; \mathbf{Z})$. Using the preliminary lemma, assume that if $\sigma_i: \bar{\Delta}_n \to \bar{\Delta}_{n+1}$ is the canonical map onto the i-th face of $\bar{\Delta}_{n+1}$, $f(\sigma_i(\bar{\Delta}_n - \Delta_n)) =$

$\{x_0\}$ for each i. The condition $h_n([f]) = 0$ means by definition that the singular cycle $\sum_{i=0}^{n} \pm(f \circ \sigma_i)$ is a boundary in $Z_n(X; \mathbf{Z})$; there is therefore a finite family of singular $(n+1)$-simplices $s_j^{n+1}: \bar{\Delta}_{n+1} \to X$ such that for $0 \leqslant i \leqslant n+1$, $f \circ \sigma_i$ is equal to one of the $s_j^{n+1} \circ \sigma_i$. But the map

$$(z, t) \mapsto s_j^{n+1}(t\mathbf{e}_i + (1-t)z)$$

[where $\mathbf{e}_i$ is the vertex of Δ_{n+1} opposite to $\sigma_i(\Delta_n)$] is a homotopy between $s_j^{n+1} \circ \sigma_i$ and the constant map $\bar{\Delta}_n \to \{x_0\}$. Applying this for $0 \leqslant i \leqslant n+1$, it follows that f is homotopic to the constant map $\bar{\Delta}_{n+1} - \Delta_{n+1} \to \{x_0\}$, hence $[f] = 0$.

A similar proof can be given for *cubical* singular homology (Part 1, chap IV, §6,D).

Since $\pi_1(\mathbf{S}_n) = 0$ for $n \geqslant 2$, the Hurewicz theorem immediately implies

$$\pi_i(\mathbf{S}_n) = 0 \quad \text{for } i < n, \tag{68}$$

$$\pi_n(\mathbf{S}_n) \simeq \mathbf{Z}, \tag{69}$$

the last relation being the Brouwer conjecture for spheres (§1).

A space X is called *n-connected** if $\pi_i(X) = 0$ for $1 \leqslant i \leqslant n$.

B. Application to the Hopf Classification Problem

In his 1933 paper ([244], p. 40) Hopf characterized the elements of the set $[X; \mathbf{S}_n]$ of homotopy classes when X is an n-dimensional finite simplicial complex. Two maps f, g belong to the same class if and only if they define the same homomorphisms

$$H_n(X; \mathbf{Z}) \to H_n(\mathbf{S}_n; \mathbf{Z}) \quad \text{and} \quad H_n(X; \mathbf{Z}/m\mathbf{Z}) \to H_n(\mathbf{S}_n; \mathbf{Z}/m\mathbf{Z}) \quad \text{for } all\ m \geqslant 2.$$

Hopf's method for proving sufficiency of that condition consists in considering the *k-skeletons* X_k (§4,A) for $0 \leqslant k \leqslant n$, and defining, by induction on k, the homotopy between $f|X_k$ and $g|X_k$; he showed that, as in Borsuk's work, the problem reduces to extending a continuous map from $X_k \times [0,1]$ to $X_{k+1} \times [0,1]$. In his third Note of 1935 [256] Hurewicz, using his "equivalence theorem" (§4,A), showed how Hopf's result could be extended to the case in which $\mathbf{S}_n$ is replaced by any $(n-1)$*-connected* space Y (§4,A); his method did not substantially differ from Hopf's, except that he used his "preliminary lemma." Then in 1937 Whitney observed that an equivalent but simpler formulation of the Hopf–Hurewicz criterion was that the maps in *cohomology*

$$H^n(Y; \mathbf{Z}) \to H^n(X; \mathbf{Z})$$

be the same for f and g [509] [see formula (88) in C below].

---

* For $n = 1$, this agrees with the traditional expression "simply connected" which goes back to Riemann. But in classical Analysis, the term "n times connected" applied to open sets in $\mathbf{R}^2$, meant something quite different, namely, that the Betti number b_1 is equal to $n - 1$.

C. Obstruction Theory

This use of cohomology was pushed one step further by Eilenberg in 1939 in a paper which we shall examine in more detail [168]. He gave their natural extension to the Hopf–Hurewicz theorems and to the technique of "climbing up" along the skeletons in his theory of *obstructions*. His new idea was to attach to maps $X \to Y$ elements of the *singular cohomology groups of* X with coefficient groups equal to (commutative) *homotopy groups of* Y.*

Eilenberg's theory dealt with continuous maps of a *locally finite cell complex* X into an *n-simple* space Y (§ 3,F), so that the elements of $\pi_n(Y)$ are the homotopy classes of *all* continuous maps of $\mathbf{S}_n$ into Y; such classes are therefore attached to maps of the n-skeleton X_n of X into Y. The method can be presented in the following way ([254], pp. 176–180).

I. Let f be a continuous map of X_n into Y. Each $(n+1)$-cell τ of X has a definite orientation; let φ_τ be a homeomorphism of the unit ball $\mathbf{D}_{n+1}$ onto $\bar{\tau}$ preserving orientation, and consider the restriction $\varphi_\tau|\mathbf{S}_n$, which is a homeomorphism of $\mathbf{S}_n$ onto the frontier $\mathrm{Fr}(\tau)$; then the class of $f \circ (\varphi_\tau|\mathbf{S}_n)$ does not depend on the choice of φ_τ, owing to the Brouwer–Hopf theorem (§ 1). By linearity, this defines a **Z**-module homomorphism of the chain complex $C_{n+1}(X;\mathbf{Z})$:

$$b_{n+1}(f)\colon C_{n+1}(X;\mathbf{Z}) \to \pi_n(Y), \tag{70}$$

in other words, an $(n+1)$-*cochain* on X with values in $\pi_n(Y)$, such that

$$\langle b_{n+1}(f), \tau\rangle = [f \circ (\varphi_\tau|\mathbf{S}_n)]. \tag{71}$$

This definition shows that $\langle b_{n+1}(f), \tau\rangle = 0$ if and only if $f|\mathrm{Fr}(\tau)$ has an *extension* to τ.

The main property of $b_{n+1}(f)$ is that it is a *cocycle*, in other words

$$\langle b_{n+1}(f), \mathbf{b}\sigma\rangle = 0 \tag{72}$$

for any $(n+2)$-cell σ of X. Consider the $(n+1)$-dimensional subcomplex L of X consisting of all the cells of X having support in $\mathrm{Fr}(\sigma)$; the **Z**-module $Z_n(L)$ of n-cycles of L is identical to the submodule $B_n(L)$ of n-boundaries, since $H_n(L) = 0$, L being homeomorphic to $\mathbf{S}_{n+1}$. But $Z_n(L)$ is also the **Z**-module of n-cycles of the n-skeleton L_n, and as there are no boundaries in the group $C_n(L_n;\mathbf{Z})$, $Z_n(L)$ is identified with the homology **Z**-module $H_n(L_n)$. However, $H_j(L_n) = H_j(L) = 0$ for $1 \leqslant j \leqslant n-1$, and $\pi_1(L_n) = \pi_1(L) = 0$ (chap. I, § 3,B); by Hurewicz's theorem, the Hurewicz map is an isomorphism $h_n\colon \pi_n(L_n) \simeq H_n(L_n)$. For any $(n+1)$-cell τ of L, it follows from the definitions that

$$\langle b_{n+1}(f), \tau\rangle = (f|L_n) \circ (h_n^{-1}(\mathbf{b}\tau)),$$

and as $\mathbf{b}(\mathbf{b}\sigma) = 0$, (72) is proved.

---

* "Measuring" by a cohomology class the fact that a continuous map defined in a subspace A of a space X cannot be extended to a map continuous in X is an idea which probably appeared for the first time in the work of Stiefel and Whitney of 1935 on the existence of continuous sections of a fiber space (chap. IV, § 1).

II. Note that if A is a (closed) subcomplex of X and f is a continuous map $X_n \cup A \to Y$, then $b_{n+1}(f)$ is a *relative* $(n+1)$-*cocycle* modulo A, because if the $(n+1)$-cell τ is contained in A, $f|\mathrm{Fr}(\tau)$ has an extension to τ, hence $\langle b_{n+1}(f), \tau\rangle = 0$.

Consider now *two* continuous maps f_0, f_1 of $X_n \cup A$ into Y such that their restrictions to $X_{n-1} \cup A$ are *homotopic*. In the cell complex $X_n \times I$ with $I = [0, 1]$,

$$W_n = ((X_{n-1} \cup A) \times I) \cup ((X_n \cup A) \times \{0\}) \cup ((X_n \cup A) \times \{1\}) \tag{73}$$

is a subcomplex, and there is therefore a continuous map $F\colon W_n \to Y$ such that

$$F(x, 0) = f_0(x), \qquad F(x, 1) = f_1(x) \quad \text{for } x \in X_n \cup A. \tag{74}$$

The $(n+1)$-cells of $X \times I$ are either of the form $\rho \times I$, where ρ is an n-cell of X, or $\tau \times \{0\}$, or $\tau \times \{1\}$, where τ is an $(n+1)$-cell of X. Define $(n+1)$-cochains $b'_{n+1}(f_0)$ and $b'_{n+1}(f_1)$ on $X \times I$ by

$$\langle b'_{n+1}(f_0), \tau \times \{0\}\rangle = \langle b_{n+1}(f_0), \tau\rangle, \quad b'_{n+1}(f_0) \text{ has value 0 on other cells,} \tag{75}$$

$$\langle b'_{n+1}(f_1), \tau \times \{1\}\rangle = \langle b_{n+1}(f_1), \tau\rangle, \quad b'_{n+1}(f_1) \text{ is 0 on other cells.} \tag{76}$$

On the other hand, F is defined on the n-cells of $X \times I$, hence the $(n+1)$-cocycle $b_{n+1}(F)$ is defined, and

$$\begin{aligned} \langle b_{n+1}(F), \tau \times \{0\}\rangle &= \langle b'_{n+1}(f_0), \tau \times \{0\}\rangle, \\ \langle b_{n+1}(F), \tau \times \{1\}\rangle &= \langle b'_{n+1}(f_1), \tau \times \{1\}\rangle. \end{aligned} \tag{77}$$

From these relations, for the $(n+1)$-cochain on $X \times I$

$$u = b_{n+1}(F) - b'_{n+1}(f_0) - b'_{n+1}(f_1) \tag{78}$$

one has

$$\mathbf{d}u = -\mathbf{d}b'_{n+1}(f_0) - \mathbf{d}b'_{n+1}(f_1); \tag{79}$$

but as $b_{n+1}(f_0)$ and $b_{n+1}(f_1)$ are cocycles,

$$\langle b'_{n+1}(f_0), \mathbf{b}(\tau \times I)\rangle = -\langle b_{n+1}(f_0), \tau\rangle,$$

$$\langle b'_{n+1}(f_1), \mathbf{b}(\tau \times I)\rangle = \langle b_{n+1}(f_1), \tau\rangle,$$

for any $(n+1)$-cell τ of X. Hence

$$\langle \mathbf{d}u, \tau \times I\rangle = \langle b_{n+1}(f_0), \tau\rangle - \langle b_{n+1}(f_1), \tau\rangle \qquad \text{for an } (n+1)\text{-cell } \tau \text{ in X;} \tag{80}$$

$$\langle \mathbf{d}u, \sigma \times \{0\}\rangle = 0, \qquad \langle \mathbf{d}u, \sigma \times \{1\}\rangle = 0 \quad \text{for an } (n+2)\text{-cell } \sigma \text{ in X.} \tag{81}$$

This shows that there is an *n-cochain* $c(f_0, f_1, F)$ *on* X such that

$$\langle c_n(f_0, f_1, F), \rho\rangle = \langle u, \rho \times I\rangle \quad \text{for any } n\text{-cell } \rho \text{ in X,} \tag{82}$$

$$\mathbf{d}c_n(f_0, f_1, F) = b_{n+1}(f_0) - b_{n+1}(f_1). \tag{83}$$

Furthermore, $c_n(f_0, f_1, F)$ is 0 on the n-cells ρ contained in A, since u is 0 for the corresponding cells $\rho \times I$. In other words, $c_n(f_0, f_1, F)$ is an element of $C^n(X, A; \pi_n(Y))$, a *relative n-cochain* on X with respect to A, with values in $\pi_n(Y)$.

III. Suppose given a continuous map $f_0: X_n \cup A \to Y$, a continuous map $g_1: X_{n-1} \cup A \to Y$, and a homotopy

$$G: (X_{n-1} \cup A) \times I \to Y$$

such that $G(x, 0) = f_0(x)$ and $G(x, 1) = g_1(x)$ for $x \in X_{n-1} \cup A$. Then, for *any* relative n-cochain $c \in C^n(X, A; \pi_n(Y))$, there is a map $f_1: X_n \cup A \to Y$ *extending* g_1, and a homotopy $F: W_n \to Y$ *extending* G, such that

$$c_n(f_0, f_1, F) = c. \tag{84}$$

The proof consists in defining F in $\sigma \times I$, for every n-cell σ of X. The closure of $\sigma \times I$ in $X \times I$ is homeomorphic to the ball $\mathbf{D}_{n+1}$, and there is therefore a homeomorphism

$$\psi_\sigma: \mathbf{S}_n \simeq \operatorname{Fr}(\sigma \times I) = (\sigma \times \{0\}) \cup (\operatorname{Fr}(\sigma) \times I) \cup (\sigma \times \{1\})$$

which preserves orientation. Let $h_\sigma: \operatorname{Fr}(\sigma \times I) \to Y$ be a continuous map such that the class $[h_\sigma \circ \psi_\sigma] = c(\sigma)$. As $T = (\sigma \times \{0\}) \cup (\operatorname{Fr}(\sigma) \times I)$ is homeomorphic to $\mathbf{D}_n$, any two continuous maps of T into Y are homotopic; using the homotopy extension property, one may assume that

$$h_\sigma(x, t) = G(x, t) \quad \text{for } (x, t) \in \operatorname{Fr}(\sigma) \times I,$$

$$h_\sigma(x, 0) = f_0(x) \quad \text{for } x \in \sigma.$$

Then define F by $F(x, t) = h_\sigma(x, t)$ in $\operatorname{Fr}(\sigma \times I)$, and $f_1(x) = F(x, 1)$ for $x \in \sigma$. This definition on each n-cell σ of X implies that F is continuous in W_n and f_1 in $X_n \cup A$.

IV. With the help of these properties Eilenberg solved the problems of extension and homotopy *stepwise on the skeletons of* X by showing that they depend on the vanishing of *relative cohomology classes*; the space Y is always assumed *n-simple* for *every* $n \geqslant 1$.

First suppose given a continuous map $f: X_n \cup A \to Y$; then the cocycle $b_{n+1}(f) \in Z^{n+1}(X, A; \pi_n(Y))$ is defined, as well as its cohomology class

$$\beta_{n+1}(f) \in H^{n+1}(X, A; \pi_n(Y)). \tag{85}$$

The condition $\beta_{n+1}(f) = 0$ is then *necessary and sufficient* for the existence of a continuous map $h: X_{n+1} \cup A \to Y$ that is *not* necessarily an extension of f, *but* an extension of the *restriction* $f|(X_{n-1} \cup A)$.

To prove necessity, note that if h exists and if $g = h|(X_n \cup A)$, then $b_{n+1}(g) = 0$ by I. As f and g have the same restriction to $X_{n-1} \cup A$, the n-cochain $c_n(f, g, G_0)$ is defined for $G_0(x, t) = f(x)$ in $(X_{n-1} \cup A) \times I$, by II. From (83) it follows that $b_{n+1}(f) = \mathbf{d}c_n(f, g, G_0)$, hence $\beta_{n+1}(f) = 0$.

Conversely, suppose $b_{n+1}(f)$ is the relative coboundary of an n-cochain c; by III there is a homotopy $F: W_n \to Y$ extending G_0 such that if $h(x) = F(x; 1)$ in $X_n \cup A$, $c_n(f, h, F) = c$; then by (83) $b_{n+1}(h) = 0$, hence h (which coincides with f in $X_{n-1} \cup A$) can be extended to any $(n + 1)$-cell of X.

V. Now suppose that f, g are two continuous maps of X into Y such that

$f|A = g|A$ and there exists a "relative" homotopy between the restrictions of f and g to $X_{n-1} \cup A$, with respect to A (see § 5), i.e., a continuous map

$$G: (X_{n-1} \cup A) \times I \to Y$$

such that $G(x, 0) = f(x)$, $G(x, 1) = g(x)$ for $x \in X_{n-1} \cup A$, and $G(x, t) = f(x) = g(x)$ for $x \in A$ and $t \in I$. Since f and g are defined in X, the map $F: W_n \to Y$ such that $F(x, 0) = f(x)$ and $F(x, 1) = g(x)$ for $x \in X_n \cup A$, and $F(x, t) = G(x, t)$ in $(X_{n-1} \cup A) \times I$ is continuous; if $f_n = f|(X_n \cup A)$ and $g_n = g|(X_n \cup A)$, then $c_n(f_n, g_n, F)$ is defined. Furthermore, since f_n and g_n are extended to f_{n+1} and g_{n+1} in $X_{n+1} \cup A$, $b_{n+1}(f_n) = 0$ and $b_{n+1}(g_n) = 0$ by I; relation (83) then shows that $c_n(f_n, g_n, F)$ is a *relative cocycle* in $Z^n(X, A; \pi_n(Y))$, so that its cohomology class

$$\gamma_n(f_n, g_n, F) \in H^n(X, A; \pi_n(Y)) \tag{86}$$

is defined. The condition $\gamma_n(f_n, g_n, F) = 0$ is then *necessary and sufficient* for the existence of an extension $F': W_{n+1} \to Y$, *not* of F, but of the *restriction* of F to W_{n-1} such that $F'(x, 0) = f(x)$ and $F'(x, 1) = g(x)$ for $x \in X_n \cup A$. From the description of the cells in $X \times I$, $W_n = (X \times I)_n \cup (A \times I)$; since $c_n(f_n, g_n, F)$ is a coboundary, by (78), $b_{n+1}(F)$ is a coboundary in $Z^{n+1}(X \times I, A \times I; \pi_n(Y))$. The theorem is thus a consequence of IV applied to the pair $(X \times I, A \times I)$.

VI. When X is a locally finite cell complex and A is a (closed) subcomplex of X, it is possible to give necessary and sufficient conditions of a cohomological nature for a continuous map $f: A \to Y$ to have an *extension to* X under the following assumptions (with $n \geqslant 2$):

1. $\pi_i(Y) = 0$ for $1 \leqslant i \leqslant n-1$ (Y is therefore m-simple for any $m \geqslant 1$);
2. $H^{j+1}(X, A; \pi_j(Y)) = 0$ for $j \geqslant n+1$; observe that this is always the case if $X - A$ has dimension $\leqslant n+1$.

Then it is immediate that f can be extended to a map $f_n: X_n \cup A \to Y$, and any two such extensions are relatively homotopic with respect to A; therefore, the class $\beta_{n+1}(f_n) \in H^{n+1}(X, A; \pi_n(Y))$ is defined and only depends on f, so that it may be written $\beta_{n+1}(f)$. The condition $\beta_{n+1}(f) = 0$ is then necessary and sufficient (by IV) for the existence of an extension f_{n+1} of f_{n-1} (hence of f) to $X_{n+1} \cup A$. When that condition is satisfied, $\beta_{n+2}(f_{n+1})$ is defined, but is 0 by condition 2, so that there is an extension f_{n+2} of f_n (hence of f) to $X_{n+2} \cup A$. The argument can be repeated, giving a succession of extensions f_{n+2r} of f to $X_{n+2r} \cup A$, such that $f_{n+2(r+1)}$ is an extension of f_{n+2r}. As X is locally finite, there is an extension g of f to X coinciding with f_{n+2r} on each $X_{n+2r} \cup A$. Therefore the condition $\beta_{n+1}(f) = 0$ is *necessary and sufficient* for the existence of an extension of f to X.

VII. Similarly, suppose f, g are two continuous maps of X into Y, with the assumptions (for $n \geqslant 2$):

1. $\pi_i(Y) = 0$ for $1 \leqslant i \leqslant n-1$;
2. $H^j(X, A; \pi_j(Y)) = 0$ for $j \geqslant n+1$, if $X - A$ has a dimension $\leqslant n$.

Then, if $f|A = g|A$, it is possible to give a necessary and sufficient condition of a cohomological nature for f and g to be *relatively homotopic* with respect to A. By V, if f_j and g_j are the restrictions of f and g to $X_j \cup A$, there is a relative homotopy G_{n-1} between f_{n-1} and g_{n-1} with respect to A. By the homotopy extension property, there exist maps f', g' of X into Y, respectively, relatively homotopic to f and g with respect to A, that coincide in $X_{n-1} \cup A$; if $J(x, t) = f'(x) = g'(x)$ in $(X_{n-1} \cup A) \times I$, the element $c_n(f', g', J)$ is defined and is a relative cocycle in $Z^n(X, A; \pi_n(Y))$; its cohomology class $\gamma_n(f', g', J) \in H^n(X, A; \pi_n(Y))$ does not depend on the choice of f' and g', hence can be written $\gamma_n(f, g)$. The condition $\gamma_n(f, g) = 0$ is then necessary and sufficient (by V) for the existence of a homotopy G_n between f_n and g_n that coincides with G_{n-2} in $(X_{n-2} \cup A) \times I$. When that condition is satisfied, $\gamma_{n+1}(f, g)$ is defined and is 0 by assumption 2, hence there is a homotopy G_{n+1} between f_{n+1} and g_{n+1} that coincides with G_{n-1} in $(X_{n-1} \cup A) \times I$. Finally, repeating the argument as in VI, the condition $\gamma_n(f, g) = 0$ is *necessary and sufficient* for the existence of a relative homotopy with respect to A between f and g.

VIII. Suppose the assumptions of VI are verified, and let $f: X \to Y$ be a continuous map and $\gamma \in H^n(X, A; \pi_n(Y))$ be a cohomology class. Then there is a continuous map $g: X \to Y$ such that $f|A = g|A$ and $\gamma_n(f, g) = \gamma$.

Let c be a relative cocycle mod. A whose class is γ; by III, if $f_j = f|(X_j \cup A)$, there is a map $g_n: X_n \cup A \to Y$ such that $g_n|(X_{n-1} \cup A) = f_{n-1}$ and $c_n(f_n, g_n, F) = c$ for a relative homotopy F between f_n and g_n. Since $\mathbf{d}c = 0$, it follows from (83) that $b_{n+1}(g_n) = b_{n+1}(f_n) = 0$ because f is defined in $X_{n+1} \cup A$. Therefore, by I there is an extension g_{n+1} of g to $X_{n+1} \cup A$. By IV, since $\beta_{n+1}(g_n) = 0$, there is an extension g of g_n to X by VII, and $\gamma_n(f, g) = \gamma$.

Eilenberg's theory is easily extended to the case in which (X, A) is only supposed to be a *relative* CW-*complex* ([490], pp. 228–235).

IX. The relative cocycles $\beta_{n+1}(f)$ and $\gamma_n(f, g)$ may be defined for an $(n - 1)$-*connected* space Y and an *arbitrary* pair (X, A). By Hurewicz's theorem the singular homology group $H_n(Y)$ may be naturally identified with $\pi_n(Y)$; $H_j(Y) = 0$ for $1 \leqslant j \leqslant n - 1$, therefore [Part 1, chap. IV, § 5D), formula (66)], $H^n(Y; H_n(Y))$ is naturally isomorphic to $\mathrm{End}(H_n(Y))$. Let κ be the element in $H^n(Y; H_n(Y))$ which corresponds to the identity by that isomorphism. For any continuous map $f: X \to Y$, the map

$$f^*: H^n(Y; H_n(Y)) \to H^n(X; H_n(Y))$$

associates an element $\alpha(f) = f^*(\kappa) \in H^n(X; H_n(Y))$ to f. If A is now a subspace of X and $g: A \to Y$ is a continuous map, the image of $\alpha(g)$ by the homomorphism of the cohomology exact sequence

$$\partial: H^n(A; H_n(Y)) \to H^{n+1}(X, A; H_n(Y))$$

is called the *obstruction* $\beta(g)$ of g (relatively to X). When X is a locally finite complex, $\beta(g) = \beta_{n+1}(g)$ as defined in VI.

Similarly, if f and g are two continuous maps of X into Y that coincide in A, then for any commutative group G, $f^* - g^*$ is a homomorphism of

$H^n(Y; G)$ into $H^n(X, A; G)$ when Y is $(n-1)$-*connected*, so that the pair (f, g) defines an element $\gamma(f, g) = f^*(\kappa) - g^*(\kappa)$ of $H^n(X, A; H_n(Y))$, called the *deviation* between f and g. When X is a locally finite complex, $\gamma(f, g) = \gamma_n(f, g)$ as defined in VII.

With the later discovery of more sophisticated relations between homotopy and homology, obstruction theory could be greatly generalized (chap. V, § 2,C).

X. If we return to the classification problem of Hopf–Hurewicz (§ 4,B), we see that if Y is $(n-1)$-connected and X is a locally finite cell complex, then under the condition

$$H^j(X; \pi_j(Y)) = 0 \quad \text{and} \quad H^{j+1}(X; \pi_j(Y)) = 0 \quad \text{for } j \geqslant n+1 \tag{87}$$

there is a natural bijection

$$[X, x_0; Y, y_0] \stackrel{\sim}{\rightarrow} H^n(X; \pi_n(Y)); \tag{88}$$

it is given by associating the element $\gamma_n(f, e)$ to a homotopy class $[f]$, where e is the constant map $X \rightarrow \{y_0\}$. This is the *Hopf–Hurewicz–Whitney theorem.* In particular, for $X = Y = \mathbf{S}_n$ the first Hopf result is recovered:

$$\pi_n(\mathbf{S}_n) \simeq \mathbf{Z} \quad \text{for } n \geqslant 1. \tag{89}$$

There have been generalizations of the classification problem when X is a more general space and $Y = \mathbf{S}_n$, using Čech cohomology theory. The most interesting one was given by Dowker [144]: if X is locally compact, metrizable, and separable, then there is a bijection

$$[X; \mathbf{S}_n] \stackrel{\sim}{\rightarrow} \check{H}_f^n(X), \tag{90}$$

where $\check{H}_f^n(X)$ is the Čech cohomology group with integer coefficients, based on *finite* open coverings of X (Part 1, chap. IV, § 6,B); this implies the rather surprising result that

$$[\mathbf{R}; \mathbf{S}_n] \simeq C(\mathbf{R})/BC(\mathbf{R}) \tag{91}$$

(*loc. cit.*).

§ 5. Relative Homotopy and Exact Sequences

A. Relative Homotopy Groups

In his second Note of 1935 [256] Hurewicz considered a topological group G and a closed subgroup H, and investigated the relations among the spaces of continuous functions $\mathscr{C}(Y; G)$, $\mathscr{C}(Y; H)$, and $\mathscr{C}(Y; G/H)$. In chapter III we shall describe in some detail the results he announced on this question and see how they were precursors of the homotopy theory of fibrations, begun in a Note he published in 1940 together with Steenrod [260]. It was in that Note that the notion of *relative homotopy* appeared in print for the first time, without a complete definition.* But in 1938, in unpublished lectures, Hurewicz had outlined a "relative" theory of homotopy following the same pattern as in his definition of the groups $\pi_n(X, x_0)$ [253]. Let A be a nonempty subspace of a space X, and $x_0 \in A$. Instead of the space $\mathscr{C}(\mathbf{S}_n, *; X, x_0)$, consider in the space $\mathscr{C}(\mathbf{D}_n; X)$ the subspace

$$\mathscr{C}(\mathbf{D}_n, \mathbf{S}_{n-1}, *; X, A, x_0) \tag{92}$$

whose elements are the continuous maps $f\colon \mathbf{D}_n \to X$ such that

$$f(\mathbf{S}_{n-1}) \subset A, \qquad f(*) = x_0.$$

The *relative homotopy set* $\pi_n(X, A, x_0)$ is by definition the set

$$\pi_n(X, A, x_0) = \pi_0(\mathscr{C}(\mathbf{D}_n, \mathbf{S}_{n-1}, *; X, A, x_0))$$

of arcwise-connected components of the space (92); for $A = \{x_0\}$, we recover $\pi_n(X, x_0)$, and it is clear that if X_0 and A_0 are the arcwise-connected components of x_0 in X and A, $\pi_n(X, A, x_0) = \pi_n(X_0, A_0, x_0)$.

To interpret the arcwise-connected components of (92), we need the general definition of *relative homotopy* between two maps f, g of a pair (X, A) into a pair (Y, B). It is a continuous map $F\colon X \times [0, 1] \to Y$ such that $F(x, 0) = f(x)$ and $F(x, 1) = g(x)$, and $F(x, t) \in B$ for all $x \in A$ and $t \in [0, 1]$. This again defines an equivalence relation in the space $\mathscr{C}(X, A; Y, B)$ of maps of pairs, and the set of equivalence classes is written $[X, A; Y, B]$. In the category of pointed spaces maps $(X, A, x_0) \to (Y, B, y_0)$ (with $x_0 \in A$ and $y_0 \in B$) always map x_0 to y_0, and relative homotopies F are always such that $F(x_0, t) = y_0$ for all $t \in [0, 1]$; the set of relative homotopy classes is then written $[X, A, x_0; Y, B, y_0]$.

With these definitions it is clear that two maps f, g in $\mathscr{C}(\mathbf{D}_n, \mathbf{S}_{n-1}, *; X, A, x_0)$ belong to the same arcwise component if and only if there exists a relative homotopy between f and g considered as maps $(\mathbf{D}_n, \mathbf{S}_{n-1}, *) \to (X, A, x_0)$. The natural homeomorphism (41) of § 3,A generalizes to a natural homeomorphism

$$\mathscr{C}(\mathbf{D}_n, \mathbf{S}_{n-1}, *; X, A, x_0) \simeq \Omega(\mathscr{C}(\mathbf{D}_{n-1}, \mathbf{S}_{n-2}, *; X, A, x_0), a_{n-1}) \tag{93}$$

for $n \geqslant 2$, a_{n-1} being the constant map $\mathbf{D}_{n-1} \to \{x_0\}$.

The proof is patterned after the proof of (41). The ball $\mathbf{D}_n$ is identified with $[0, 1]^n$, and $\mathbf{D}_{n-1}$ with the "lid" $L_{n-1} = [0, 1]^{n-1} \times \{1\}$; the maps $(\mathbf{D}_n, \mathbf{S}_{n-1}, *) \to (X, A, x_0)$ are identified with the continuous maps $f\colon [0, 1]^n \to X$ such that $f(L_{n-1}) \subset A$ and $f(t_1, t_2, \ldots, t_n) = x_0$ in $[0, 1]^n - L_{n-1}$. The homeomorphism (93) again results from the identification (40) of § 3A. From (93) is deduced a natural bijection for $n \geqslant 2$

$$\pi_n(X, A, x_0) \simeq \pi_1(\mathscr{C}(\mathbf{D}_{n-1}, \mathbf{S}_{n-2}, *; X, A, x_0), a_{n-1}) \tag{94}$$

---

* In that Note there is no definition of relative homotopy; it is merely stated that the elements of the relative homotopy group $\pi_n(X, A, x_0)$ are "represented" by maps $f\colon \mathbf{D}_n \to X$ such that $f(\mathbf{S}_{n-1}) \subset A$ and $f(*) = x_0$. A general reference to Hurewicz's Notes of 1935 does not seem to be justified, since there is no mention of relative homotopy in these Notes.

which gives *for* $n \geqslant 2$ a natural *group* structure on $\pi_n(X, A, x_0)$. When $\mathbf{D}_n$ is identified to $[0, 1]^n$ as above, the group law in $\pi_n(X, A, x_0)$ is still given by formula (42) of § 3,A. In order that the image $[f]$ in $\pi_n(X, A, x_0)$ of a map $f \in \mathscr{C}(\mathbf{D}_n, \mathbf{S}_{n-1}, *; X, A, x_0)$ be the neutral element of $\pi_n(X, A, x_0)$ (for $n \geqslant 2$), it is necessary and sufficient that there exists a relative homotopy with respect to $\mathbf{S}_{n-1}$ between f and a continuous map $g\colon \mathbf{D}_n \to X$ such that $g(\mathbf{D}_n) \subset A$.

The relative homotopy sets may also be considered as "absolute" homotopy sets: write

$$\Omega(X, A, x_0) = \mathscr{C}(\mathbf{D}_1, \mathbf{S}_0, *; X, A, x_0) \tag{95}$$

the space of paths in X with origin at x_0 and *arbitrary* extremity in A, and define by induction on n

$$\Omega^n(X, A, x_0) = \Omega(\Omega^{n-1}(X, A, x_0), x_{n-1}), \tag{96}$$

x_{n-1} being the constant map $\mathbf{D}_1 \to \{x_{n-2}\}$. From (93) follows by induction on n a natural homeomorphism

$$\mathscr{C}(\mathbf{D}_n, \mathbf{S}_{n-1}, *; X, A, x_0) \simeq \Omega^n(X, A, x_0) \tag{97}$$

hence natural bijections for $n \geqslant 2$

$$\begin{aligned} \pi_n(X, A, x_0) \simeq \pi_{n-1}(\Omega(X, A, x_0), x_1) \simeq \cdots \simeq \pi_{n-k}(\Omega^k(X, A, x_0), x_k) \simeq \cdots \\ \simeq \pi_0(\Omega^n(X, A, x_0)). \end{aligned} \tag{98}$$

This proves in particular that *for* $n \geqslant 3$ the group $\pi_n(X, A, x_0)$ is *commutative.* For $n = 1$, $\pi_1(X, A, x_0) = \pi_0(\Omega(X, A, x_0))$ is a set with "privileged" element x_1, but no natural group structure.

A useful natural homeomorphism is

$$\Omega^n(X, A, x_0) \simeq \Omega(\Omega^{n-1}(X, x_0), \Omega^{n-1}(A, x_0), x_{n-1}). \tag{99}$$

To define it, consider the identification of $\mathbf{D}_n$ with $[0, 1]^n$ introduced above and to each $f \in \Omega^n(X, A, x_0)$, associate the map

$$f'\colon [0, 1] \to \mathscr{C}([0, 1]^{n-1}; \Omega^{n-1}(X, x_0)) \tag{100}$$

defined by

$$f'(t_n)(t_1, t_2, \ldots, t_{n-1}) = f(t_1, t_2, \ldots, t_{n-1}, t_n). \tag{101}$$

It is clear that $f'(0) = x_{n-1}$, and $f'(1)$ belongs to $\Omega^{n-1}(A, x_0)$, hence the homeomorphism (99) is derived from the fundamental property of the compact-open topology (§ 3,A).

If $\boldsymbol{PH}_1$ is the homotopy category of pairs (X, A, x_0) of pointed spaces (where morphisms are homotopy classes for relative homotopy), then $(X, A, x_0) \mapsto \pi_n(X, A, x_0)$ is a covariant functor $\boldsymbol{PH}_1 \to \boldsymbol{Gr}$ for $n \geqslant 2$ (and a covariant functor $\boldsymbol{PH}_1 \to \boldsymbol{Set}$ for $n = 1$).

Let x_1 be a point of A in the arcwise-connected component of x_0 in A. Let α be any *path in* A from x_0 to x_1 and $f\colon [0, 1]^n \to X$ be a continuous map such that $f(L_{n-1}) \subset A$ and $f([0, 1]^n - L_{n-1}) = \{x_0\}$; a double application of the box lemma (§ 2D) shows that there is an extension $F: [0,1]^n \times [0,1] \to X$ of the map equal to $(x,0) \mapsto f(x)$ in $[0,1]^n \times \{0\}$ and to $(x,t) \mapsto \alpha(t)$ in $([0,1]^n - L_{n-1}) \times [0,1]$, such that $F(L_{n-1} \times [0,1]) \subset A$. The homotopy class of $x \mapsto F(x,1)$ in $\pi_n(X, A, x_1)$ then only depends on the class of f in $\pi_n(X, A, x_0)$, hence this defines a map $\sigma_n(\alpha): \pi_n(X, A, x_0) \to \pi_n(X, A, x_1)$; the properties of that map are similar to those of $\sigma_n(\alpha)$ for the homotopy groups $\pi_n(X, x_0)$ and $\pi_n(X, x_1)$ (§ 3,D). In particular, the fundamental group $\pi_1(X, x_0)$ acts naturally on $\pi_n(X, A, x_0)$ for $n \geqslant 2$; the pair (X, A, x_0) is called *n-simple* if that action is trivial.

B. The Exact Homotopy Sequence

Let (X, A) be a pair of spaces, with $x_0 \in A$. The injections $j: A \to X$, and $i: \{x_0\} \to A$ may be considered respectively as maps of pairs

$$j: (A, x_0) \to (X, x_0), \qquad i: (X, x_0) \to (X, A)$$

leaving x_0 invariant. Therefore, they give rise to group homomorphisms for $n \geqslant 2$

$$\pi_n(A, x_0) \xrightarrow{j_*} \pi_n(X, x_0) \xrightarrow{i_*} \pi_n(X, A, x_0)$$

(for $n = 1, j_*$ is still a group homomorphism but i_* is only a map of sets). There is a natural map

$$\partial: \pi_n(X, A, x_0) \to \pi_{n-1}(A, x_0) \tag{102}$$

for $n \geqslant 2$; it associates to the homotopy class of a map $f: [0,1]^n \to X$ such that $f(L_{n-1}) \subset A$ and $f([0,1]^n - L_{n-1}) = \{x_0\}$, the homotopy class of the map

$$[0,1]^{n-1} \simeq L_{n-1} \xrightarrow{f|L_{n-1}} A,$$

and it is clear that this map is a homomorphism of groups. There is also such a map (102) for $n = 1$: the elements of $\pi_1(X, A, x_0)$ are the homotopy classes of the paths $\gamma: [0,1] \to X$ such that $\gamma(0) = x_0$ and $\gamma(1) \in A$, and it is clear that if two such paths γ, γ' are relatively homotopic (with x_0 fixed), then $\gamma(1)$ and $\gamma'(1)$ are in the same arcwise connected component of A; hence there is a map of *sets*

$$\partial: \pi_1(X, A, x_0) \to \pi_0(A).$$

Finally, the injection j defines trivially a map $j_*: \pi_0(A) \to \pi_0(X)$. The infinite sequence

$$\cdots \to \pi_{n+1}(X, A, x_0) \xrightarrow{\partial} \pi_n(A, x_0) \xrightarrow{j_*} \pi_n(X, x_0) \xrightarrow{i_*} \pi_n(X, A, x_0)$$
$$\xrightarrow{\partial} \pi_{n-1}(A, x_0) \to \cdots \xrightarrow{\partial} \pi_1(A, x_0) \xrightarrow{j_*} \pi_1(X, x_0) \xrightarrow{i_*} \pi_1(X, A, x_0) \xrightarrow{\partial} \pi_0(A) \xrightarrow{j_*} \pi_0(X)$$

was first considered (without using arrows) by J.H.C. Whitehead in 1945 ([498], p. 345). He proved that the sequence is *exact*: for the last four terms the image of $\pi_1(X, x_0)$ by i_* is the inverse image $\partial^{-1}(A_0)$ of the arcwise-connected component A_0 of x_0 in A, and A_0 is the inverse image by j_* of the arcwise-connected component of x_0 in X.

The proof can be reduced to proving the exactness for the last four maps, since by (99), there is a commutative diagram

$$\begin{array}{ccccccccc} \pi_n(A) & \longrightarrow & \pi_n(X) & \longrightarrow & \pi_n(X, A) & \longrightarrow & \pi_{n-1}(A) & \longrightarrow & \pi_{n-1}(X) \\ \big\downarrow \wr & & \big\downarrow \wr & & \big\downarrow \wr & & \big\downarrow \wr & & \big\downarrow \wr \\ \pi_1(A') & \longrightarrow & \pi_1(X') & \longrightarrow & \pi_1(X', A') & \longrightarrow & \pi_0(A') & \longrightarrow & \pi_0(X') \end{array}$$

where $A' = \Omega^{n-1}(A)$, $X' = \Omega^{n-1}(X')$, and the base points have been suppressed for short.

However, in the sequence (103):

1. Exactness at $\pi_0(A)$ is a trivial consequence of the definitions.
2. Exactness at $\pi_1(X, A, x_0)$ means that if there is *in* A a path β from x_0 to x, then there is a relative homotopy F with respect to A between any path α *in* X from x_0 to x and a loop in X of origin x_0. To define it, take

$$F(t, s) = \alpha(2t/(2 - s)) \quad \text{for } 0 \leqslant t \leqslant (2 - s)/2,$$

$$F(t, s) = \beta(3 - s - 2t) \quad \text{for } (2 - s)/2 \leqslant t \leqslant 1.$$

3. Exactness at $\pi_1(X, x_0)$ means that if α is a loop of origin x_0 in X, β is a path of origin x_0 in A, and there is a relative homotopy F with respect to A, deforming α into β such that

$$F(t, 0) = \alpha(t), \qquad F(t, 1) \in A, \qquad F(0, s) = x_0, \qquad F(1, s) = \beta(s),$$

then there is a homotopy $\gamma(t, s)$ such that

$$\gamma(0, s) = \gamma(1, s) = x_0, \qquad \gamma(t, 0) = \alpha(t), \qquad \gamma(t, 1) \in A.$$

Take

$$\gamma(t, s) = F(2t, s) \quad \text{for } 0 \leqslant t \leqslant 1/2,$$

$$\gamma(t, s) = F(1, 2s(1 - t)) \quad \text{for } 1/2 \leqslant t \leqslant 1.$$

If $f: X \to X'$ is a continuous map, A' is a subspace of X' such that $f(A) \subset A'$, and $x'_0 \in A'$ such that $f(x_0) = x'_0$, the diagram

$$\begin{array}{ccccccccccc} \cdots & \longrightarrow & \pi_n(A, x_0) & \longrightarrow & \pi_n(X, x_0) & \longrightarrow & \pi_n(X, A, x_0) & \longrightarrow & \pi_{n-1}(A, x_0) & \longrightarrow & \cdots \\ & & \big\downarrow f_* & & \big\downarrow f_* & & \big\downarrow f_* & & \big\downarrow f_* & & \\ \cdots & \longrightarrow & \pi_n(A', x'_0) & \longrightarrow & \pi_n(X', x'_0) & \longrightarrow & \pi_n(X', A', x'_0) & \longrightarrow & \pi_{n-1}(A', x'_0) & \longrightarrow & \cdots \end{array}$$

is commutative.

Easy consequences follow from the homotopy exact sequence (103). If A is contractible (*in itself*) to x_0, then $\pi_n(X, A, x_0) = \pi_n(X, x_0)$; if X is contractible to x_0, $\pi_n(X, A, x_0) = \pi_{n-1}(A, x_0)$. If A is a *retract* of X, $\pi_n(X, A, x_0)$ is commutative and

$$\pi_n(X, x_0) \simeq \pi_n(A, x_0) \oplus \pi_n(X, A, x_0) \quad \text{for } n \geqslant 2. \tag{104}$$

If X is *deformable* into A with $x_0 \in A$ fixed, then

$$\pi_n(A, x_0) \simeq \pi_n(X, x_0) \oplus \pi_{n+1}(X, A, x_0) \quad \text{for } n \geqslant 2. \tag{105}$$

If A is contractible *in* X to $x_0 \in A$, so that there is a homotopy fixing x_0 between the injection $A \to X$ and the constant map $A \to \{x_0\}$, then

$$\pi_n(X, A, x_0) \simeq \pi_n(X, x_0) \oplus \pi_{n-1}(A, x_0) \quad \text{for } n \geqslant 3. \tag{106}$$

For instance, if $X = \mathbf{S}_n$ and A is a proper closed arcwise-connected subspace of $\mathbf{S}_n$,

$$\pi_i(\mathbf{S}_n, A, *) \simeq \pi_i(\mathbf{S}_n, *) \oplus \pi_{i-1}(A, *) \quad \text{for } i \geqslant 3, \tag{107}$$

a result proved in 1940 by Hurewicz and Steenrod [260].

Finally, for the wedge $(X \vee Y, (x_0, y_0))$ of two Hausdorff pointed spaces considered as a subspace of $(X \times Y, (x_0, y_0))$,

$$\pi_n(X \vee Y, (x_0, y_0)) \simeq \pi_n(X, x_0) \oplus \pi_n(Y, y_0) \oplus \pi_{n+1}(X \times Y, X \vee Y, (x_0, y_0)) \tag{108}$$

for $n \geqslant 2$.

C. Triples and Triads

Consider a triple (X, A, B) (Part 1, chap. IV, § 6,B), that is, a space X and two subspaces $B \subset A \subset X$, with $x_0 \in B$. Then the injections of pairs

$$i\colon (A, B) \to (X, B), \qquad j\colon (X, B) \to (X, A)$$

define maps of sets (group homomorphisms for $n \geqslant 2$)

$$i_*\colon \pi_n(A, B, x_0) \to \pi_n(X, B, x_0), \qquad j_*\colon \pi_n(X, B, x_0) \to \pi_n(X, A, x_0).$$

On the other hand, consider the injection $h\colon (A, \{x_0\}) \to (A, B)$ and the map $\partial\colon \pi_n(X, A, x_0) \to \pi_{n-1}(A, x_0)$ of the homotopy exact sequence (103); by composition, this gives a map

$$\bar{\partial}\colon \pi_n(X, A, x_0) \xrightarrow{\partial} \pi_{n-1}(A, x_0) \xrightarrow{h_*} \pi_{n-1}(A, B, x_0)$$

and we thus get an infinite sequence

$$\cdots \to \pi_{n+1}(X, A, x_0) \xrightarrow{\bar{\partial}} \pi_n(A, B, x_0) \xrightarrow{i_*} \pi_n(X, B, x_0) \xrightarrow{j_*} \pi_n(X, A, x_0)$$
$$\to \cdots \to \pi_2(X, B, x_0) \xrightarrow{j_*} \pi_2(X, A, x_0) \xrightarrow{\bar{\partial}} \pi_1(A, B, x_0) \xrightarrow{i_*} \pi_1(X, B, x_0) \xrightarrow{j_*} \pi_1(X, A, x_0). \tag{109}$$

This sequence is *exact* [with the same meaning as in (103) for the last five terms] and is called the *homotopy exact sequence of the triple* (X, A, B); it reduces to (103) for $B = \{x_0\}$. A proof may be given by following the

Eilenberg–Steenrod procedure for the homology exact sequence of the triple (X, A, B) (Part 1, chap. IV, §6B).

A *triad* is a system (X, A, B) where A and B are again two subspaces of X, but *only* submitted to the condition that $A \cap B$ is *not empty*. If we introduce abbreviated notations (with $x_0 \in A \cap B$):

$$P = \Omega(X, B, x_0), \qquad Q = \Omega(A, A \cap B, x_0),$$

it is clear that $Q \subset P$; since the constant map $x_1 \colon [0, 1] \to \{x_0\}$ belongs to Q, we can write the homotopy exact sequence (103)

$$\cdots \to \pi_{n+1}(P, Q, x_1) \to \pi_n(Q, x_1) \to \pi_n(P, x_1) \to \pi_n(P, Q, x_1) \to \cdots \to \pi_0(Q) \to \pi_0(P). \tag{110}$$

Now, by (98) there are natural isomorphisms

$$\pi_n(Q, x_1) \cong \pi_{n+1}(A, A \cap B, x_0), \qquad \pi_n(P, x_1) \cong \pi_{n+1}(X, B, x_0).$$

If we define the *n-th homotopy group of the triad* (X, A, B) by

$$\pi_n(X, A, B, x_0) = \pi_{n-1}(P, Q, x_1) \quad \text{for } n \geqslant 2; \tag{111}$$

from the exactness of (110) follows the exactness of the sequence

$$\begin{aligned} \cdots \to \pi_{n+1}(X, A, B, x_0) \xrightarrow{\bar{\partial}} \pi_n(A, A \cap B, x_0) \xrightarrow{i_*} \pi_n(X, B, x_0) \xrightarrow{j_*} \pi_n(X, A, B, x_0) \\ \to \cdots \to \pi_2(X, A, B, x_0) \xrightarrow{\bar{\partial}} \pi_1(A, A \cap B, x_0) \xrightarrow{i_*} \pi_1(X, B, x_0), \end{aligned} \tag{112}$$

which is called the *homotopy exact sequence of the triad* (X, A, B); it reduces to (109) for $B \subset A$.

Equivalent descriptions of $\pi_n(X, A, B, x_0)$ are the following ones [459]:

1. Consider the two hemispheres E^+_{n-1}, E^-_{n-1} in $\mathbf{S}_{n-1}$, respectively defined by $\xi_n \geqslant 0$ and $\xi_n \leqslant 0$, and the subspace

$$\mathscr{C}(\mathbf{D}_n, E^+_{n-1}, E^-_{n-1}, *; X, A, B, x_0) \tag{113}$$

of $\mathscr{C}(\mathbf{D}_n; X)$, consisting of the maps f such that

$$f(E^+_{n-1}) \subset A, \qquad f(E^-_{n-1}) \subset B, \qquad f(*) = x_0. \tag{114}$$

Then $\pi_n(X, A, B, x_0)$ is the set of arcwise-connected components of the space (113).

2. A useful variant is to replace $\mathbf{D}_n$, E^+_{n-1}, E^-_{n-1}, $*$ respectively by

$$\mathbf{D}_{n-1} \times [0, 1], \quad \mathbf{S}_{n-2} \times [0, 1], \quad \mathbf{D}_{n-1} \times \{1\}, \quad (\mathbf{D}_{n-1} \times \{0\}) \cup (\{*\} \times [0, 1]).$$

D. The Barratt–Puppe Sequence

The category ***PSet*** of pointed sets has objects that are pairs (X, x_0) of an arbitrary set X and an element $x_0 \in X$; morphisms $(X, x_0) \to (Y, y_0)$ are the maps $f \colon X \to Y$ such that $f(x_0) = y_0$. In that category, an *exact sequence* of morphisms

$$(A, a) \xrightarrow{g} (B, b) \xrightarrow{f} (C, c)$$

is defined by the condition that $\mathrm{Im}(g) = f^{-1}(c)$ (which may be called the "kernel" of f).

Now consider in the category ***PT*** of pointed *topological spaces*, a sequence of morphisms, i.e., *continuous* maps

$$(X, x_0) \xrightarrow{f} (Y, y_0) \xrightarrow{g} (Z, z_0).$$

It is called *coexact* if, for *every* pointed space (W, w_0), the sequence of morphisms of pointed sets

$$[X, x_0; W, w_0] \xleftarrow{f^*} [Y, y_0; W, w_0] \xleftarrow{g^*} [Z, z_0; W, w_0]$$

is *exact*. It was first observed by Barratt [43], and generalized by Puppe [384], that *any* continuous map $f\colon (X, x_0) \to (Y, y_0)$ is the first term of an infinite *coexact* sequence

$$\begin{aligned}(X, x_0) &\xrightarrow{f} (Y, y_0) \xrightarrow{Jf} (C_f, y_0) \xrightarrow{Pf} (SX, x_0) \xrightarrow{Sf} (SY, y_0)\\ &\xrightarrow{SJf} (SC_f, y_0) \xrightarrow{SPf} (S^2X, x_0) \xrightarrow{S^2f} (S^2Y, y_0) \xrightarrow{S^2Jf} (S^2C_f, y_0) \longrightarrow \cdots\end{aligned} \tag{115}$$

usually called the *Barratt–Puppe sequence.* $C_f = Y \cup_f CX$ is the (reduced) mapping cone of f and Jf the natural injection $Y \to C_f$ (Part 2, chap. V, § 3,B); SX is identified with the space C_f/Y and the collapsing map Pf associates to the image of (x, t) in CX the image of the same point in SX. Finally, Sf is the suspension of the map f sending the image of (x, t) to the image of $(f(x), t)$.

To prove (115) is coexact, first check it for f and Jf. The fact that $(Jf) \circ f$ is homotopic to the constant map follows from the definitions and the fact that CX is contractible. Conversely, let $u\colon Y \to W$ be a map such that $u \circ f$ is homotopic to the constant map $Y \to \{w_0\}$; if $[x, 1]$ is the point of CX image of $(x, 1) \in X \times [0, 1]$, there is a map $v\colon CX \to W$ such that $v([x, 1]) = u(f(x))$, and therefore there is a continuous map $g\colon C_f \to W$ such that $g([x, 1]) = v([x, 1])$ and $g(y) = u(y)$; then $(Jf^*)([g]) = [u]$. Iterating that result, the infinite sequence

$$(Y, y_0) \xrightarrow{Jf} (C_f, y_0) \xrightarrow{J^2f} (C_{Jf}, y_0) \xrightarrow{J^3f} (C_{J^2f}, y_0) \longrightarrow \cdots \tag{116}$$

is coexact. On the other hand [using the fact that $(X \cup_i CA)/CA \simeq X/A$, see Part 2, chap. V, § 3,B], there is a homotopy commutative diagram

$$\begin{array}{ccccc}(C_f, y_0) & \xrightarrow{J^2f} & (C_{Jf}, y_0) & \xrightarrow{J^3f} & (C_{J^2f}, y_0)\\ \| & & \downarrow & & \downarrow\\ (C_f, y_0) & \xrightarrow[Pf]{} & (SX, x_0) & \xrightarrow[Sf]{} & (SY, y_0)\end{array} \tag{117}$$

where the vertical arrows are homotopy equivalences. The end of the proof uses the bijection (52) of § 3,C; for any space (W, w_0), the commutativity of the diagram (where the base points have been omitted)

$$\begin{array}{ccccccccc}
[S^nX;W] & \xleftarrow{S^nf^*} & [S^nY;W] & \xleftarrow{S^nJf^*} & [S^nC_f;W] & \xleftarrow{S^nPf^*} & [S^{n+1}X;W] & \xleftarrow{S^{n+1}f^*} & [S^{n+1}Y;W] \\
\wr\! \| & & \wr\! \| & & \wr\! \| & & \wr\! \| & & \wr\! \| \\
[X;\Omega^nW] & \xleftarrow[f^*]{} & [Y;\Omega^nW] & \xleftarrow[Jf^*]{} & [C_f;\Omega^nW] & \xleftarrow[Pf^*]{} & [X;\Omega^{n+1}W] & \xleftarrow[Sf^*]{} & [Y;\Omega^{n+1}W]
\end{array}$$

implies that the upper line is exact since the lower line is exact.

There is a similar sequence for maps of pairs $(X, A) \to (Y, B)$ ([440], p. 369).

E. The Relative Hurewicz Homomorphism

The definition of the "absolute" Hurewicz homomorphism given in §4,A generalizes to any pair (X, A, x_0): there is a natural homomorphism of groups

$$h_n\colon \pi_n(X, A, x_0) \to H_n(X, A; \mathbf{Z}) \quad \text{for } n \geqslant 2, \tag{118}$$

where the right-hand side is the relative singular homology group. Each map $f\colon \bar{\Delta}_n \to X$ such that $f(\bar{\Delta}_n - \Delta_n) \subset A$ and $f(*) = x_0$ defines a homomorphism depending only on the homotopy class $[f]$ in $\pi_n(X, A, x_0)$

$$f_*\colon H_n(\bar{\Delta}_n, \bar{\Delta}_n - \Delta_n; \mathbf{Z}) \to H_n(X, A; \mathbf{Z}). \tag{119}$$

If the class $\varepsilon_n \in H_n(\bar{\Delta}_n, \bar{\Delta}_n - \Delta_n; \mathbf{Z})$ is defined as in §4,A the element $f_*(\varepsilon_n)$ can be associated to $[f]$, which defines the map (118), even for $n = 1$; the proof that it is a homomorphism of groups for $n \geqslant 2$ is done as in §4,A.

The definitions imply that the diagram of exact sequences

$$\begin{array}{ccccccccccc}
\cdots \to & \pi_{n+1}(X, A, x_0) & \to & \pi_n(A, x_0) & \to & \pi_n(X, x_0) & \to & \pi_n(X, A, x_0) & \to & \pi_{n-1}(A, x_0) & \to \cdots \\
& \downarrow{\scriptstyle h_{n+1}} & & \downarrow{\scriptstyle h_n} & & \downarrow{\scriptstyle h_n} & & \downarrow{\scriptstyle h_n} & & \downarrow{\scriptstyle h_{n-1}} & \\
\cdots \to & H_{n+1}(X, A; \mathbf{Z}) & \to & H_n(A; \mathbf{Z}) & \to & H_n(X; \mathbf{Z}) & \to & H_n(X, A; \mathbf{Z}) & \to & H_{n-1}(A; \mathbf{Z}) & \to \cdots
\end{array} \tag{120}$$

is commutative.

A pair (X, A) of arcwise-connected spaces is called *n-connected* if the groups $\pi_k(X, A, x_0) = 0$ for all $x_0 \in A$ and $1 \leqslant k \leqslant n$.

The *relative Hurewicz isomorphism theorem* asserts that if the pair (X, A) is $(n-1)$*-connected* and *n-simple*, then h_n is an *isomorphism*. Hurewicz did not publish a proof of that theorem; very likely his intended proof was built along the lines of the method he had used for the "absolute" isomorphism theorem (§4,A). Both isomorphisms theorems were completed by Fox in 1943 [198], who showed that under the same assumptions on (X, A) the Hurewicz map h_{n+1} is *surjective*. Since then many proofs have been published; we shall meet some sophisticated ones later (chap. V, §4).

F. The First Whitehead Theorem

From the relative Hurewicz isomorphism theorem and the homotopy exact sequence can be deduced a remarkable relation between the action of a continuous map on homotopy groups and homology groups, discovered by J.H.C. Whitehead ([493], pp. 9–21).

A continuous map $f: (X, x_0) \to (Y, y_0)$ of arcwise-connected spaces is called an *n-equivalence* if the homomorphism $f_*: \pi_r(X, x_0) \to \pi_r(Y, y_0)$ is *bijective* for $1 \leqslant r < n$, *surjective* for $r = n$. Whitehead's theorem is:

(i) If f is an n-equivalence, the homomorphisms of singular homology groups $f_*: H_r(X; Z) \to H_r(Y; Z)$ are *bijective* for $r < n$, *surjective* for $r = n$.
(ii) If X and Y are also *simply connected*, the *converse* is true.

When X and Y are not simply connected, assertion (ii) is false; for a counterexample, see ([440], p. 420).

The proof is a good illustration of the use of the mapping cylinder (Part 2, chap. V, § 3,A).

A. *Special case*: X is a subspace of Y with $x_0 = y_0$ and f is the injection.

The homotopy exact sequence (103) and the assumptions of (i) imply that $\pi_q(Y, X, x_0) = 0$ for $1 \leqslant q \leqslant n - 1$; by the relative Hurewicz isomorphism theorem, $H_q(Y, X; Z) = 0$ for $1 \leqslant q \leqslant n - 1$. The conclusion of (i) follows from the homology exact sequence. If in addition $\pi_1(X, x_0) = \pi_1(Y, y_0) = 0$, then $H_1(X; Z) = 0$ and $H_1(Y; Z) = 0$, so the assumptions of (ii) imply $H_q(Y, X; Z) = 0$ by the homology exact sequence for $1 \leqslant q \leqslant n - 1$. The relative Hurewicz isomorphism theorem then yields (ii) by the homotopy exact sequence, using the fact that the pair (Y, X) is $(n - 1)$-connected and n-simple.

B. *General case*: Consider the mapping cylinder Z_f of f, the natural embedding $i: X \to Z_f$, and the retraction $r: Z_f \to Y$, for which Y is a strong deformation retract of Z_f; the maps

$$r_*: \pi_q(Z_f, y_0) \to \pi_q(Y, y_0), \qquad r_*: H_q(Z_f; Z) \to H_q(Y; Z)$$

are isomorphisms for all $q \geqslant 1$. Since $f = r \circ i$, there are factorizations

$$f_*: \pi_q(X, x_0) \xrightarrow{i_*} \pi_q(Z_f, y_0) \xrightarrow{r_*} \pi_q(Y, y_0),$$
$$f_*: H_q(X; Z) \xrightarrow{i_*} H_q(Z_f; Z) \xrightarrow{r_*} H_q(Y; Z).$$

The proof is thrown back on the embedding i, taking into account the fact that when Y is simply connected, the same is true of Z_f.

Care must be taken *not* to jump to the conclusion that two spaces having the same homotopy groups (resp. homology groups) have the same homology groups (resp. homotopy groups). Examples are known to the contrary ([440], p. 420).

§ 6. Homotopy Properties of CW-Complexes

A. Aspherical Spaces

In his fourth Note of 1935–1936 on homotopy groups [256] Hurewicz, from his isomorphism theorem for $(n - 1)$-connected spaces and the Hopf–Hurewicz classification theorem, deduced a consequence that would later lead to unsuspected developments in several directions (chap. V, § 1).

He said that an arcwise-connected space X is *aspherical* if $\pi_j(X) = 0$ for *all integers j at least* 2. If X has a universal covering space $\tilde{X}$, this is equivalent to saying that $\tilde{X}$ is *n-connected for any* $n \geqslant 1$. Hurewicz discovered that for a finite simplicial complex X that is aspherical, *the homotopy type of* X *is determined by its fundamental group.*

First consider an arcwise-connected finite simplicial complex X and an aspherical space Y; for each map $f: (X, x_0) \to (Y, y_0)$ of pointed spaces, the group homomorphism $f_*: \pi_1(X, x_0) \to \pi_1(Y, y_0)$ only depends on the homotopy class $[f] \in [X, x_0; Y, y_0]$. Hurewicz showed* that the map

$$[f] \mapsto f_* \tag{121}$$

is a *bijection* of $[X, x_0; Y, y_0]$ onto $\mathrm{Hom}(\pi_1(X, x_0), \pi_1(Y, y_0))$. To prove injectivity, assume that $f_* = g_*$ and use induction on the dimension n of X. If $n = 1$, use the description of $\pi_1(X, x_0)$ by means of a maximal tree T in X (chap. I, §3,B) and induction on the number of generators $[v_1, v_2]$ (*loc. cit.*). For $n > 1$, the inductive assumption shows that the restrictions $f|X_{n-1}$ and $g|X_{n-1}$ to the $(n-1)$-skeletons are homotopic; since $\pi_n(Y) = 0$, the homotopy can be extended to a homotopy between f and g, by the special case of obstruction theory that Hurewicz had used earlier in his proof of his isomorphism theorem (§4,A).

Surjectivity is also proved by induction on n. Let $\varphi: \pi_1(X, x_0) \to \pi_1(Y, y_0)$ be the given homomorphism; for $n = 1$, again use the description of $\pi_1(X, x_0)$ (chap. I, §3,B). The map f is taken on T equal to the constant y_0; for each generator $[v_1, v_2]$ of the free group $\pi_1(X, x_0)$, f is defined on the 1-simplex $\{v_1, v_2\}$ (identified with the interval $[0, 1]$) as a loop of origin y_0 belonging to the class in $\pi_1(Y, y_0)$ of $\varphi([v_1, v_2])$. For $n = 2$, consider the 1-skeleton X_1 and the composite map

$$\pi_1(X_1, x_0) \to \pi_1(X, x_0) \xrightarrow{\varphi} \pi_1(Y, y_0)$$

and define f on X_1 as before. Then, if σ is a 2-simplex $\{v_1, v_2, v_3\}$ of X, f is extended from its known values on the 1-simplices $\{v_1, v_2\}$, $\{v_2, v_3\}$, and $\{v_3, v_1\}$ to a map of σ into Y, using the fact that $\pi_2(Y, y_0) = 0$. For $n \geqslant 3$, the maps $\pi_1(X_k, x_0) \to \pi_1(X_{k+1}, x_0)$ for the skeletons X_k are bijective for $k \geqslant 2$ (chap. I, §3,B). So when f is extended from X_k to X_{k+1}, using the fact that $\pi_{k+1}(Y, y_0) = 0$, the map $(f|X_{k+1})_*$ is always equal to $\varphi|\pi_1(X_{k+1}, x_0)$.

As a corollary, when both X and Y are aspherical finite simplicial complexes and there exist isomorphisms $\varphi: \pi_1(X, x_0) \xrightarrow{\sim} \pi_1(Y, y_0)$, $\psi: \pi_1(Y, y_0) \xrightarrow{\sim} \pi_1(X, x_0)$ inverse to each other, there exist maps $f: (X, x_0) \to (Y, y_0)$, $g: (Y, y_0) \to (X, x_0)$ such that $f_* = \varphi$ and $g_* = \psi$, hence

$$g_* \circ f_* = 1_{\pi_1(X)}, \qquad f_* \circ g_* = 1_{\pi_1(Y)}.$$

The injectivity of the map (121) implies that $g \circ f \sim 1_X$ and $f \circ g \sim 1_Y$, which

* His formulation was more complicated because he considered the homomorphisms of $\pi_1(X, x_0)$ into an abstract group isomorphic to $\pi_1(Y, y_0)$, and these are only determined up to inner automorphisms of that group.

means X and Y have the *same homotopy type* (and in particular the same homology).

B. The Second Whitehead Theorem

The preceding result of Hurewicz may be expressed by saying that if X and Y are *aspherical* finite-dimensional simplicial complexes, any continuous map $f\colon X \to Y$ such that $f_*\colon \pi_r(X) \to \pi_r(Y)$ is an isomorphism for *all* $r \geqslant 0$, is a *homotopy equivalence.*

In 1939 [494] J.H.C. Whitehead obtained a far reaching generalization of that theorem. A continuous map $f\colon (X, x_0) \to (Y, y_0)$ is called a *weak homotopy equivalence* if it is an n-equivalence (§ 5F) for *every* $n \geqslant 0$: the homomorphism $f_*\colon \pi_r(X, x_0) \to \pi_r(Y, y_0)$ is *bijective* for every n. Whitehead's theorem is that if X and Y are CW-*complexes,* then *any weak homotopy equivalence is in fact a homotopy equivalence.*

Whitehead proved the theorem by means of his theory of simple homotopy type (§ 7). It can be derived from a more general result, showing that a weak homotopy equivalence $f\colon X \to Y$ allows "lifting" a continuous map $h\colon P \to Y$ to a map $h'\colon P \to X$ "up to homotopy," that is, h is homotopic to $f \circ h'$.

It is convenient to work here with *relative CW-complexes* (X, A) (Part 2, chap. V, § 3,C). The proof is done in several steps [459].

1. Suppose X is obtained by attachment of a family (e_α^n) of n-cells to A. Let (Y, B) be a pair such that $\pi_n(Y, B, y_0) = 0$. Then for any map $f\colon (X, A, x_0) \to (Y, B, y_0)$ there is a relative homotopy with respect to A between f and a map $g\colon (X, x_0) \to (B, y_0)$. Suppose the family (e_α^n) consists of a single n-cell e^n, and let $p\colon (\mathbf{D}_n, \mathbf{S}_{n-1}) \to (X, A)$ be the corresponding map with $p(\mathring{\mathbf{D}}_n) = e^n$ and $p(\mathbf{S}_{n-1}) \subset A$. The assumption on Y implies that there exists a continuous map $F\colon \mathbf{D}_n \times [0,1] \to Y$ such that $F(z,t) = f(p(z))$ for $z \in \mathbf{S}_{n-1}$, $F(z,0) = f(p(z))$, and $F(z,1) \in B$ for $z \in \mathbf{D}_n$. This allows us to define a continuous map G: $X \times [0,1] \to Y$ by

$$G(x,t) = f(x) \text{ if } x \in A, \qquad G(p(z),t) = F(z,t) \quad \text{for } z \in \mathbf{D}_n.$$

When the family (e_α^n) is arbitrary, the map $F_\alpha\colon \mathbf{D}_n \times [0,1] \to Y$ for each α is defined in the same way, and then G is defined by

$$G(x,t) = f(x) \text{ if } x \in A, \qquad G(p_\alpha(z),t) = F_\alpha(z,t) \quad \text{for } z \in \mathbf{D}_n.$$

2. If (X, A) has relative dimension n and (Y, B) is n-connected, there is again a relative homotopy with respect to A between $f\colon (X, A) \to (Y, B)$ and a map $g\colon (X, x_0) \to (B, y_0)$, by induction on the relative dimension of (X, A), using the fact that if (X′, A) is a relative subcomplex of (X, A), the injection $X' \to X$ is a cofibration (§ 2,D).

3. The crucial point of the proof is the following *lemma*:

Let $f\colon X \to Y$ be an *n-equivalence* (§ 5,F), and (P, Q) a relative CW-complex of relative dimension $\leqslant n$. Suppose there are continuous maps $h\colon P \to Y$,

$g: Q \to X$ such that $h|Q = f \circ g$. Then there is a continuous map $g': P \to X$ such that $g'|Q = g$, and h is relatively homotopic to $f \circ g'$ with respect to Q.

This is seen by considering the *mapping cylinder* Z_f, the injections $i: X \to Z_f$, $j: Y \to Z_f$ and the retraction $r: Z_f \to Y$, for which Y is a strong deformation retract of Z_f (Part 2, chap. V, § 3,A). From the definition of n-equivalence and the homotopy exact sequence

$$\cdots \to \pi_s(X, x_0) \xrightarrow{i_*} \pi_s(Z_f, y_0) \to \pi_s(Z_f, X, y_0) \xrightarrow{\partial} \pi_{s-1}(X, x_0) \xrightarrow{i_*} \pi_{s-1}(Z_f, y_0) \to \cdots$$

it follows that (Z_f, X) is *n-connected*, because $f = r \circ i$, $f_*: \pi_s(X, x_0) \to \pi_s(Y, y_0)$ is an isomorphism for $s \leqslant n$, and $r_*: \pi_s(Z_f, y_0) \to \pi_s(Y, y_0)$ is the identity for all s.

Since $Q \to P$ is a cofibration (§ 2,D), there is a map $h': P \to Z_f$ such that $h'|Q = i \circ g$ and $r|h'$ is relatively homotopic to $r \circ j \circ h$ with respect to Q. By 2, h' is relatively homotopic to a map $g': P \to X$ with respect to Q. Then $g'|Q = g$, and $f \circ g' = r \circ i \circ g'$ is relatively homotopic to $r \circ h'$, hence to $r \circ j \circ h = h$.

4. From 3 it follows in particular that if P is a *CW-complex of dimension* $\leqslant n$ and $f: X \to Y$ an n-equivalence, then the map

$$f_*: [P; X] \to [P; Y] \tag{122}$$

is *surjective*. If we also suppose that the dimension of P is $\leqslant n - 1$, then f_* is *injective*. Suppose g_0, g_1 are two maps $P \to X$ such that $f \circ g_0$ is homotopic to $f \circ g_1$. This means there exists a map $h: P \times [0, 1] \to Y$ such that $h(z, 0) = f(g_0(z))$ and $h(z, 1) = f(g_1(z))$. Now $(P \times [0, 1], P \times \{0, 1\})$ is a relative CW-complex of relative dimension $\leqslant n$, and the restriction of h to $P \times \{0, 1\}$ is the map g such that $g(z, 0) = g_0(z)$ and $g(z, 1) = g_1(z)$; by 3, there is a map h': $P \times [0, 1] \to X$ such that $h'|(P \times \{0, 1\}) = g$, which means that h' is a homotopy between g_0 and g_1.

5. The Whitehead theorem is a corollary of 4. Let X, Y be two CW-complexes, and suppose $f: X \to Y$ is a weak homotopy equivalence. Then the corresponding maps

$$f_*^{(X)}: [X; X] \to [X; Y], \qquad f_*^{(Y)}: [Y; X] \to [Y; Y] \tag{123}$$

are bijective. Therefore, there is a map $g: Y \to X$ such that $f_*^{(Y)}([g]) = [1_Y]$; and $f_*^{(X)}([g \circ f]) = [f \circ g \circ f] = [1_Y \circ f] = [f \circ 1_X] = f_*^{(X)}([1_X])$, hence $[g \circ f] = [1_X]$ or $g \circ f \sim 1_X$; this proves that f is a homotopy equivalence.

6. When the first (§ 5,F) and second Whitehead theorems are combined, the following useful result is obtained:

If X, Y are two *simply connected* CW-complexes and $f: X \to Y$ is a continuous map such that $f_*: H_\bullet(X; \mathbf{Z}) \to H_\bullet(Y; \mathbf{Z})$ is *bijective*, then f is a homotopy equivalence.

In his work on cobordism (chap. VII, § 1) Thom proved a variant of that result: if the continuous map $f: X \to Y$ for two simply connected CW-complexes X, Y is such that for *every* prime p, $f^*\colon H^r(Y; \mathbf{F}_p) \to H^r(X; \mathbf{F}_p)$ is bijective for $r < k$ and injective for $r = k$, then $f_*\colon \pi_r(X) \to \pi_r(Y)$ is bijective for $r \leqslant k - 1$.

C. Lemmas on Homotopy in Relative CW-Complexes

In 1951–1953 Blakers and Massey [51] made a thorough study of the relative homotopy groups $\pi_r(X, A, x_0)$ when (X, A) is a *relative CW-complex* (Part 2, chap. V, § 3,C); this gave rise to many applications to homotopy theory. Their proofs relied on Eilenberg's obstruction theory (§ 4,C) and can be simplified by using a few preliminary lemmas which make more extensive use of simplicial approximation [459].

Since the elements of relative homotopy groups are classes of maps

$$(\mathbf{D}_n, \mathbf{S}_{n-1}) \to (X, A)$$

it is convenient to consider more general maps

$$(K, L) \to (X, A)$$

where K is a finite simplicial complex and L is a (closed) subcomplex.

The general results rely on the study of the elementary case in which

$$X = A \cup_g e^n$$

is obtained by attachment of a *single* n-cell e^n to a Hausdorff space A by a map $g\colon \mathbf{S}_{n-1} \to A$. Let $p\colon \mathbf{D}_n \to X$ be the corresponding natural map. Then a slight variation of the Alexander process of simplicial approximation (Part 1, chap. II, § 3), which uses induction on the skeletons of K, shows that for any continuous map $f\colon (K, L) \to (X, A)$ there is a map $f_1\colon (K, L) \to (X, A)$, relatively homotopic to f with respect to a neighborhood of $f^{-1}(A)$ in K, with the following property.

(SAL) There exist a triangulation T′ of $\mathbf{D}_n$ and a *simplicial* map $h\colon (K, T) \to (\mathbf{D}_n, T')$ for a sufficiently fine subdivision T of K such that for any simplex σ of T for which $f_1(\sigma) \subset e^n$ and $f_1(\sigma) \cap p(\frac{1}{2}\mathbf{D}_n) \neq \varnothing$, $f_1|\sigma = p \circ (h|\sigma)$.

As a first consequence, if (X, A) is *any* relative CW-complex, then for *any* $n \geqslant 1$,

$$\pi_r(X, (X, A)^n, x_0) = 0 \quad \textit{for all } r \leqslant n; \tag{124}$$

the pair $(X, (X, A)^n)$ is thus *n-connected* for any n.

To prove this, a relative homotopy with respect to $\mathbf{S}_{r-1}$ must be found between f and a map $f'\colon \mathbf{D}_r \to (X, A)^n$, for any map $f\colon (\mathbf{D}_r, \mathbf{S}_{r-1}) \to (X, (X, A)^n)$ with $r \leqslant n$. The compact set $f(\mathbf{D}_r)$ is contained in some skeleton $(X, A)^m$ and only meets a *finite* number of the cells e^j_λ (with $j \leqslant m$) in that skeleton. Suppose that $m > n$ (otherwise there is nothing to prove) and that there is a cell e^m_α; property (SAL) yields a relative homotopy with respect to $\mathbf{S}_{r-1}$ between f and a map $f_1: (\mathbf{D}_r, \mathbf{S}_{r-1}) \to (X, A)$ such that there is a point $x \in e_\alpha^m$ *not* in $f_1(\mathbf{D}_r)$. Since $X - e_\alpha^m$ is a strong deformation retract of $X - \{x\}$, there is another relative homotopy between f_1 and a map f_2 such that $f_2(\mathbf{D}_r) \cap e_\alpha^m = \varnothing$. Repeat the process until, after a convenient relative homotopy, $f(\mathbf{D}_r)$ does not meet any e_λ^j with $j > n$, so that $f(\mathbf{D}_r) \subset (X, A)^n$.

The fact that (X, X^n) is n-connected for an "absolute" CW-complex X implies by the homotopy exact sequence (103) that for the injection $j: X^n \to X$ the homomorphism $j_*: \pi_r(X^n, x_0) \to \pi_r(X, x_0)$ is *bijective for $r < n$ and surjective for $r = n$.*

An important special case of (124) is the one in which for two spheres $(\mathbf{S}_p, *)$ and $(\mathbf{S}_q, *)$,

$$\pi_r(\mathbf{S}_p \times \mathbf{S}_q, \mathbf{S}_p \vee \mathbf{S}_q) = 0 \quad \text{for } r < p + q.$$

This follows from the fact that $\mathbf{S}_p \times \mathbf{S}_q$ is obtained by attachment of $\mathbf{D}_p \times \mathbf{D}_q$ to $\mathbf{S}_p \vee \mathbf{S}_q$.

Another consequence of (SAL) is the possibility of reducing *n-connected* relative CW-complexes, *up to homotopy type*, to the special case in which $(X, A)^n = A$, so that only cells e_α^m of dimension $m > n$ are attached to A.

More precisely, it is possible to attach families (e_α^{n+1}) and (e_α^{n+2}) of cells to X such that if X″ is the union of X and these cells and A′ is the union of A and the cells (e_α^{n+1}), then:

(R) X is a strong deformation retract of X″, A is a strong deformation retract of A′, and the n-skeleton of (X″, A′) is given by

$$(X'', A')^n = A'. \tag{125}$$

The proof is by induction on n. Since (X, A) is $(n-1)$-connected, we may assume that $(X, A)^{n-1} = A$. Let (e_α^n) be the family of n-cells of (X, A) and let

$$f_\alpha: (\mathbf{D}_n, \mathbf{S}_{n-1}, *) \to (X, A, x_0)$$

be the map such that $f_\alpha(\mathbf{D}_n) = \overline{e_\alpha^n}$ and $f_\alpha(\mathring{\mathbf{D}}_n) = e_\alpha^n$. If $\pi_n(X, A, x_0) = 0$, then for each α there is a relative homotopy with respect to $\mathbf{S}_{n-1}$ between f_α and a map $\mathbf{D}_n \to A$. If D_n^+ and D_n^- are the hemispheres of $\mathbf{S}_n$ respectively defined by $\xi_{n+1} \geqslant 0$ and $\xi_{n+1} \leqslant 0$, f_α can be considered a map $(D_n^+, \mathbf{S}_{n-1}) \to (X, A)$ and the homotopy a map $g_\alpha: \mathbf{D}_{n+1} \to X$ such that

$$g_\alpha | D_n^+ = f_\alpha, \qquad g_\alpha(D_n^-) \subset A.$$

We saw above that $(X, (X, A)^{n+1})$ is $(n+1)$-connected, hence we may assume that $g_\alpha(\mathbf{D}_{n+1}) \subset (X, A)^{n+1}$. For each α attach $\mathbf{D}_{n+1}$ to X by the map $g_\alpha | \mathbf{S}_n$: $\mathbf{S}_n \to X$; let $h_\alpha: \mathbf{D}_{n+1} \to X \cup_{g_\alpha} e_\alpha^{n+1}$ be the corresponding map such that

$$h_\alpha(\mathbf{D}_{n+1}) = \overline{e_\alpha^{n+1}}, \qquad h_\alpha(\mathring{\mathbf{D}}_{n+1}) = e_\alpha^{n+1}, \qquad h_\alpha | \mathbf{S}_n = g_\alpha | \mathbf{S}_n.$$

Let X′ be the space obtained by the attachment to X of all the e_α^{n+1}, and

$$A' = A \cup \left(\bigcup_\alpha \overline{e_\alpha^{n+1}} \right);$$

since $e_\alpha^n \subset \overline{e_\alpha^{n+1}}$ for all α, $(X', A')^n = A'$; and A' is obtained by attaching the ball $\mathbf{D}_{n+1}$ to A, for each α, by the map $g_\alpha | D_n^-$. Since D_n^- is a strong deformation retract of $\mathbf{D}_{n+1}$, A is a *strong deformation retract* of A'.

Next, if D_{n+1}^+ and D_{n+1}^- are similarly defined, g_α may be considered a map $D_{n+1}^- \to X'$ and h_α a map $D_{n+1}^+ \to X'$; attach $\mathbf{D}_{n+2}$ to X' by the map $k_\alpha: \mathbf{S}_{n+1} \to X'$ defined by $k_\alpha | D_{n+1}^+ = h_\alpha$, $k_\alpha | D_{n+1}^- = g_\alpha$; if X'' is the space obtained by attaching to X' the cells e_α^{n+2} by the maps k_α, then, similarly, X is a *strong deformation retract of* X'', and relation (125) follows from the definitions.

In the particular case of an n-connected "absolute" CW-complex (X, x_0), A' has the homotopy type of a point: it is contractible. Since the injection $A' \to X''$ is a cofibration, if $X_0 = X''/A'$, the collapsing map $(X'', A') \to (X_0, \bar{x}_0)$ (with $\bar{x}_0 = A'$) is a homotopy equivalence (§ 2,D); the n-skeleton X_0^n is then reduced to the point $\bar{x}_0$.

D. The Homotopy Excision Theorem

Let A, B be two subspaces of a space X, with $x_0 \in A \cap B$ and $X = A \cup B$. The excision axiom of homology theory (Part 1, chap. IV, § 6,B) states that if $A = X - U$ is closed and $\bar{U} \subset \mathring{B}$, then the injection of pairs $j: (A, A \cap B) \to (X, B)$ induces an *isomorphism* of relative singular homology

$$j_*: H_\bullet(A, A \cap B; \mathbf{Z}) \to H_\bullet(X, B; \mathbf{Z}).$$

There is no such general result for homotopy; under the same conditions,

$$j_*: \pi_r(A, A \cap B, x_0) \to \pi_r(X, B, x_0) \tag{126}$$

is not always bijective ([51], pp. 9–10). The homotopy groups of the triad (X, A, B, x_0) (§ 5,C) were invented by Blakers and Massey precisely to "measure" how (126) "deviates" from bijectivity.

Their most important result [51] was the discovery of cases in which these triad homotopy groups vanish. Suppose $(A, A \cap B)$ is an *n-connected relative CW-complex* and $(B, A \cap B)$ an *m-connected relative CW-complex*. Then

$$\pi_r(X, A, B, x_0) = 0 \quad \textit{for } 2 \leqslant r \leqslant m + n \tag{127}$$

or, equivalently, by the exact sequence (112), the map (126) is *bijective for* $1 \leqslant r < m + n$, *surjective for* $r = m + n$.

This can be proved as a consequence of the spectral sequence of a fibration (chap. IV, § 3), but a method of Boardman [459] only uses the elementary lemmas (SAL) and (R) proved above, by a succession of simple steps.

I. The main part of the proof treats the simplest case, in which the space A (resp. B) is obtained from C by attachment of a single n-cell e^n (resp. a single m-cell e^m) to $C = A \cap B$. Suppose $2 \leqslant r \leqslant m + n - 2$. Elements of $\pi_r(X, A, B, x_0)$ are classes of maps $f: \mathbf{D}_{r-1} \times [0, 1] \to X$ such that

$$f(\mathbf{S}_{r-2} \times [0,1]) \subset A, \qquad f(\mathbf{D}_{r-1} \times \{1\}) \subset B, \qquad f(\mathbf{D}_{r-1} \times \{0\}) = \{x_0\} \tag{128}$$

and $f(*, t) = x_0$ for $t \in [0, 1]$. The idea is to show that there exists a point

$p \in e^m$ and a point $q \in e^n$ such that f is relatively homotopic, with respect to $(\mathbf{D}_{r-1} \times \{0\}) \cup (\mathbf{S}_{r-2} \times [0,1])$, to a map f' such that

$$f'(\mathbf{D}_{r-1} \times [0,1]) \subset \mathrm{X} - \{p\} \quad \text{and} \quad f'(\mathbf{D}_{r-1} \times \{1\}) \subset \mathrm{X} - \{p, q\}. \tag{129}$$

Now B is a strong deformation retract of $\mathrm{X} - \{q\}$, so the injection

$$j_1 : (\mathrm{X}, \mathrm{A}, \mathrm{B}, x_0) \to (\mathrm{X}, \mathrm{A}, \mathrm{X} - \{q\}, x_0)$$

yields an isomorphism for all r:

$$j_{1*} : \pi_r(\mathrm{X}, \mathrm{A}, \mathrm{B}, x_0) \cong \pi_r(\mathrm{X}, \mathrm{A}, \mathrm{X} - \{q\}, x_0).$$

Next the injection

$$j_2 : (\mathrm{X} - \{p\}, \mathrm{A}, \mathrm{X} - \{p, q\}, x_0) \to (\mathrm{X}, \mathrm{A}, \mathrm{X} - \{q\}, x_0)$$

defines a homomorphism for all r:

$$j_{2*} : \pi_r(\mathrm{X} - \{p\}, \mathrm{A}, \mathrm{X} - \{p, q\}, x_0) \to \pi_r(\mathrm{X}, \mathrm{A}, \mathrm{X} - \{q\}, x_0)$$

and the homotopy between f and f' implies that

$$j_{1*}([f]) = j_{2*}([f']).$$

But A is a strong deformation retract of $\mathrm{X} - \{p\}$, so by the exact homotopy sequence of a triad, $\pi_r(\mathrm{X} - \{p\}, \mathrm{A}, \mathrm{X} - \{p, q\}, x_0) = 0$ for all r, and therefore $j_{1*}([f]) = 0$, hence $[f] = 0$.

The construction of f' uses (SAL) (§6,C). Let $u : \mathbf{D}_n \to \mathrm{A}$, $v : \mathbf{D}_m \to \mathrm{B}$ be the maps such as $u(\mathring{\mathbf{D}}_n) = e^n$, $u(\mathring{\mathbf{D}}_m) = e^m$; there is a point $p \in v(\frac{1}{2}\mathbf{D}_m)$ in the image $v(\sigma)$ of an m-simplex, and $f^{-1}(p)$ is a subcomplex of $\mathbf{D}_{r-1} \times [0,1]$ of dimension $\leqslant r - m$. Therefore $\mathrm{pr}_1^{-1}(\mathrm{pr}_1(f^{-1}(p)))$ is a subcomplex K of dimension $\leqslant r - m + 1 \leqslant n - 1$, hence $f(\mathrm{K})$ does not contain $u(\frac{1}{2}\mathbf{D}_n)$, and $q \in u(\frac{1}{2}\mathbf{D}_n)$ is taken such that $f^{-1}(q) \cap \mathrm{K} = \varnothing$. It is then easy to find a homotopy between the identity of $\mathbf{D}_{r-1} \times [0,1]$ and a map

$$g : \mathbf{D}_{r-1} \times [0,1] \to (\mathbf{D}_{r-1} \times [0,1]) - f^{-1}(p)$$

such that $h((z,1),t) \notin f^{-1}(q)$ for all $t \in [0,1]$ and h leaves the points of $\mathbf{S}_{r-2} \times [0,1]$ fixed; there is then a homotopy $f \circ h$ between f and the requested $f' = f \circ g$.

The general homotopy excision theorem is then progressively reduced to the preceding case:

II. Using the lemma (R) of §3,C, there is a relative CW-complex $(\mathrm{B}', \mathrm{C}')$ such that $(\mathrm{B}', \mathrm{C}')^m = \mathrm{C}'$ and C (resp. B) is a strong deformation retract of C' (resp. B'); then if $\mathrm{A}' = \mathrm{A} \cup \mathrm{C}'$, $\mathrm{X}' = \mathrm{A} \cup \mathrm{B}'$, A (resp. X) is a strong deformation retract of A' (resp. X'); since $\pi_r(\mathrm{X}, \mathrm{A}, \mathrm{B}, x_0) \cong \pi_r(\mathrm{X}', \mathrm{A}', \mathrm{B}', x_0)$ for all r, it is enough to consider the case in which $(\mathrm{B}, \mathrm{C})^m = \mathrm{C}$. A similar application of (R) allows us also to assume $(\mathrm{A}, \mathrm{C})^n = \mathrm{A}$.

III. Since $f(\mathbf{D}_{r-1} \times [0,1])$ is compact it only meets a finite number of cells

$$e^{n_1}, e^{n_2}, \ldots, e^{n_h} \quad \text{with } n_j > n \text{ for } 1 \leqslant j \leqslant h, \text{ in } (\mathrm{A}, \mathrm{C})$$

$$[\text{resp. } e^{m_1}, e^{m_2}, \ldots, e^{m_k} \quad \text{with } m_i > m \text{ for } 1 \leqslant i \leqslant k, \text{ in } (\mathrm{B}, \mathrm{C})].$$

It may thus be assumed that

$$A = C \cup e^{n_1} \cup e^{n_2} \cup \cdots \cup e^{n_h}, \qquad B = C \cup e^{m_1} \cup e^{m_2} \cup \cdots \cup e^{m_k}. \quad (130)$$

IV. Let $B_0 = C$, $B_i = C \cup e^{m_1} \cup \cdots \cup e^{m_i}$ for $1 \leqslant i \leqslant k$, and $X_i = A \cup B_i$. The injection u: $(A, C) \to (X, B)$ can be factored in

$$(A, C) = (X_0, B_0) \xrightarrow{u_1} (X_1, B_1) \xrightarrow{u_2} (X_2, B_2) \to \cdots \xrightarrow{u_k} (X_k, B_k) = (X, B)$$

so that it is enough to prove the theorem for each u_i.

V. Suppose that $B = C \cup e^{m'}$ with $m' > m$, and define

$$A_0 = C, \qquad A_j = C \cup e^{n_1} \cup \cdots \cup e^{n_j} \quad \text{similarly for } 1 \leqslant j \leqslant h$$

and $X_j = B \cup A_j$. Consider the commutative diagram of homotopy exact sequences of triples

$$\begin{array}{ccccccccc}
\pi_{r+1}(A_j, A_{j-1}, x_0) & \longrightarrow & \pi_r(A_{j-1}, C, x_0) & \longrightarrow & \pi_r(A_j, C, x_0) & \longrightarrow & \pi_r(A_j, A_{j-1}, x_0) & \longrightarrow & \pi_{r-1}(A_{j-1}, C, x_0) \\
\downarrow{\scriptstyle w_j} & & \downarrow{\scriptstyle v_{j-1}} & & \downarrow{\scriptstyle v_j} & & \downarrow{\scriptstyle w_j} & & \downarrow{\scriptstyle v_{j-1}} \\
\pi_{r+1}(X_j, X_{j-1}, x_0) & \longrightarrow & \pi_r(X_{j-1}, B, x_0) & \longrightarrow & \pi_r(X_j, B, x_0) & \longrightarrow & \pi_r(X_j, X_{j-1}, x_0) & \longrightarrow & \pi_{r-1}(X_{j-1}, B, x_0)
\end{array}$$

By I all w_j are bijective for $r < m + n$ and surjective for $r = m + n$. Use induction on j: for $j = 0$, v_0 is trivially the identity; suppose v_{j-1} is bijective for $r < m + n$ and surjective for $r = m + n$; then the 5-lemma (Part 1, chap. IV, § 5,A) shows that v_j has the same properties.

E. The Freudenthal Suspension Theorems

After Hurewicz had defined the homotopy groups, one of the main problems of homotopy theory (still only partially solved today) was to determine the *homotopy groups of spheres* $\pi_m(S_n)$ explicitly for $m \geqslant n$ [175]. This became the main motivation for many of the techniques introduced between 1935 and the present day, each of which provided more knowledge of these groups.

The Hurewicz isomorphism theorem determined $\pi_n(S_n) \simeq \mathbf{Z}$ (§ 4,A), and in his study of the homotopy groups of homogeneous spaces (chap. III, § 2,A), he obtained a very simple proof of the isomorphism $\pi_3(S_2) \simeq \mathbf{Z}$, essentially equivalent to Hopf's theorem of 1930 (§ 1,B).

The next significant advance was made by H. Freudenthal in 1937 [20]. We have mentioned that in that paper he invented the suspension SX of a pointed space X (Part 2, chap. V, § 2,C) and the homotopy suspension

$$E\colon [X, x_0; Y, y_0] \to [SX, x_0; SY, y_0]$$

(§ 3,D). In particular, for $X = S_r$ and $Y = S_n$ this gave him group homomorphisms

$$E\colon \pi_r(S_n) \to \pi_{r+1}(S_{n+1}) \quad \text{for } r \geqslant 1, n \geqslant 1, \qquad (131)$$

and he obtained the striking result that (131) is *bijective for* $1 \leqslant r < 2n - 1$ *and surjective for* $r = 2n - 1$ (which is now often called the "easy" Freudenthal theorem). He also discovered a remarkable relation between suspension and the Hopf invariant $f \mapsto \gamma(f)$ (§§ 1,B and 1,C), considered as a homomorphism

$$\gamma\colon \pi_{2r+1}(\mathbf{S}_{2r}) \to \mathbf{Z}; \tag{132}$$

he showed that the kernel of γ is the image of

$$\mathrm{E}\colon \pi_{2r}(\mathbf{S}_r) \to \pi_{2r+1}(\mathbf{S}_{r+1}). \tag{133}$$

He also obtained partial results on the kernel of (131) for $r = 2n - 1$; for instance, the kernel of $\mathrm{E}\colon \pi_3(\mathbf{S}_2) \to \pi_4(\mathbf{S}_3)$ is the subgroup of classes of maps f with *even* Hopf invariant, which showed that

$$\pi_4(\mathbf{S}_3) \simeq \mathbf{Z}/2\mathbf{Z}. \tag{134}$$

The "easy" theorem showed that

$$\mathrm{E}\colon \pi_{r+k}(\mathbf{S}_{n+k}) \to \pi_{r+k+1}(\mathbf{S}_{n+k+1})$$

is bijective as soon as $r + k < 2(n + k) - 1$, that is, $k > r - 2n + 1$, so that the infinite sequence

$$\pi_r(\mathbf{S}_n) \xrightarrow{\mathrm{E}} \pi_{r+1}(\mathbf{S}_{n+1}) \xrightarrow{\mathrm{E}} \cdots \to \pi_{r+k}(\mathbf{S}_{n+k}) \xrightarrow{\mathrm{E}} \cdots \tag{135}$$

is *stationary*: for each value of $r > 0$, only a *finite* number of homotopy groups $\pi_{r+k}(\mathbf{S}_k)$ may be distinct. The group $\pi_{r+k}(\mathbf{S}_k)$ for $k > r + 1$, which is independent of k (up to isomorphism) is called the *r-stem* or the *r-th stable homotopy group*; this was the first example of "stability" in an infinite sequence of homomorphisms, which later turned up in many circumstances in algebraic topology.

Freudenthal's proofs were patterned after Hopf's methods of geometric constructions for simplicial maps (§ 1), and they used some of Hopf's results. Suppose $f\colon \mathbf{S}_{2r+1} \to \mathbf{S}_{r+1}$ is a simplicial map (for suitable triangulations) such that the Hopf invariant $\gamma(f) = 0$. Freudenthal showed that the class $[f]$ in $\pi_{2r+1}(\mathbf{S}_{r+1})$ is 0 by considering a point $x \in \mathbf{S}_{r+1}$ and its inverse image $f^{-1}(x)$, which is a subcomplex of dimension r; he showed that there is a homotopy between f and a simplicial map f' for which $f'^{-1}(x)$ has only one element. The proof of $[f'] = 0$ is then relatively easy, but the definition of f' is long and difficult.

Freudenthal's theorems attracted much attention from topologists. Simpler proofs using more refined tools were found later, and we shall have the opportunity to describe them as well as their generalizations. One of the earliest generalizations of the "easy" Freudenthal theorem was found by Blakers and Massey as a consequence of their homotopy excision theorem (§ 6,D): if X is any *n-connected CW-complex*, then the suspension

$$\mathrm{E}\colon \pi_r(\mathrm{X}, x_0) \to \pi_{r+1}(\mathrm{SX}, x_0) \tag{136}$$

is *bijective for* $1 \leqslant r \leqslant 2n$ and *surjective for* $r = 2n + 1$.

The proof uses the natural homeomorphism $\mathrm{SX} \simeq \mathrm{CX}/\mathrm{X}$ (Part 2, chap. V, § 2,C). Since CX is contractible, the homotopy exact sequence (103) shows that

$$\partial\colon \pi_{r+1}(\mathrm{CX}, \mathrm{X}, x_0) \to \pi_r(\mathrm{X}, x_0)$$

is an isomorphism for all $r \geqslant 1$. The proof of the theorem is thus reduced to the same statement for the homomorphism

$$p_*\colon \pi_{r+1}(\mathrm{CX}, \mathrm{X}, x_0) \to \pi_{r+1}(\mathrm{CX}/\mathrm{X}, \bar{x}_0)$$

deduced from the collapsing map $p\colon (\mathrm{CX}, x_0) \to (\mathrm{CX}/\mathrm{X}, \bar{x}_0)$. This in turn derives from a more general result concerning a relative CW-complex (X, A): if A is *m-connected* and (X/A) *n-connected*, then

$$p_*\colon \pi_r(\mathrm{X}, \mathrm{A}, x_0) \to \pi_r(\mathrm{X}/\mathrm{A}, \bar{x}_0) \tag{137}$$

is *bijective for* $2 \leqslant r \leqslant m + n$ and *surjective for* $r = m + n + 1$. If $i\colon \mathrm{A} \to \mathrm{X}$ is the natural injection, X/A is naturally homeomorphic to $(\mathrm{X} \cup_i \mathrm{CA})/\mathrm{CA}$ (Part 2, chap. V, § 3,B), but as CA is contractible, the collapsing map

$$(\mathrm{X} \cup_i \mathrm{CA}, \mathrm{CA}) \to ((\mathrm{X} \cup_i \mathrm{CA})/\mathrm{CA}, \bar{x}_0)$$

is a homotopy equivalence (§ 2,D); the proof is finally reduced to showing that the map

$$\pi_r(\mathrm{X}, \mathrm{A}, x_0) \to \pi_r(\mathrm{X} \cup_i \mathrm{CA}, \mathrm{CA}, x_0) \tag{138}$$

is *bijective* for $2 \leqslant r \leqslant m + n$ and *surjective* for $r = m + n + 1$. That is a corollary of the homotopy excision theorem (§ 6,D) applied to the triad

$$(\mathrm{X} \cup_i \mathrm{CA}, \mathrm{X}, \mathrm{CA}, x_0);$$

$\pi_r(\mathrm{CA}, \mathrm{A}, x_0) \simeq \pi_{r-1}(\mathrm{A}, x_0) = 0$ for $1 \leqslant r \leqslant m + 1$ by the homotopy exact sequence, and (X, A) is n-connected.

F. Realizability of Homotopy Groups

We have already seen (chap. I, § 4,B) that Veblen ([474], p. 145) had constructed simplicial complexes having *arbitrary* fundamental groups. The introduction of CW-complexes allowed a considerable extension of that result to all homotopy groups.

I. First, there is a CW-complex X for any $n \geqslant 1$ and any *free group* F (commutative if $n \geqslant 2$) such that

$$\pi_r(\mathrm{X}) = 0 \quad \text{for } r < n, \qquad \pi_n(\mathrm{X}) \simeq \mathrm{F}. \tag{139}$$

This was proved for $n = 1$ in chap. I, § 4,A. Suppose $n \geqslant 2$ and let $\mathrm{F} = \mathbf{Z}^{(\mathrm{I})}$; take for X the "infinite wedge" obtained by attaching a family $(e_\alpha^n)_{\alpha \in \mathrm{I}}$ of n-cells to a single point $\{x_0\}$; then that CW-complex satisfies (139). For $\mathrm{I} = \{1, 2\}$ this is a particular case of the isomorphism

$$\pi_r(\mathbf{S}_m \vee \mathbf{S}_n) \simeq \pi_r(\mathbf{S}_m) \oplus \pi_r(\mathbf{S}_n) \quad \textit{for } r \leqslant m + n; \tag{140}$$

indeed, if $\mathbf{S}_m$ and $\mathbf{S}_n$ are considered CW-complexes (Part 2, chap. V, § 3,C), the cells in $\mathbf{S}_m \times \mathbf{S}_n$ are $\mathbf{D}_m \times \{*\}$, $\{*\} \times \mathbf{D}_n$, and $\mathbf{D}_m \times \mathbf{D}_n$, so $(\mathbf{S}_m \times \mathbf{S}_n)^{m+n-1} = \mathbf{S}_m \vee \mathbf{S}_n$. Now apply (124), which gives

$$\pi_r(\mathrm{S}_m \times \mathrm{S}_n, \mathrm{S}_m \vee \mathrm{S}_n, (*, *)) = 0 \quad \text{for } r \leqslant m + n$$

and (140) follows from (108).

For a finite set I, (139) follows from (140) by induction on Card(I). Finally, if I is arbitrary, it is only necessary to observe that the image of any continuous map $\mathrm{S}_r \to \mathrm{X}$ or of any homotopy $\mathrm{S}_r \times [0, 1] \to \mathrm{X}$ only meets a finite number of n-cells.

II. Next suppose G is *any* group (commutative if $n \geqslant 2$); it can be written F/R, where F is a free group (commutative if $n \geqslant 2$); then R is also a free group (chap. I, §4,A). Let $(x_\alpha)_{\alpha \in \mathrm{I}}$ [resp. $(r_\gamma)_{\gamma \in \mathrm{J}}$] be a family of free generators of F (resp. R), and use the result of I to construct a CW-complex X^n such that

$$\pi_r(\mathrm{X}^n, x_0) \cong \begin{cases} 0 & \text{for } r < n, \\ \mathrm{F} & \text{for } r = n. \end{cases} \tag{141}$$

There is therefore a map h_γ: $(\mathrm{S}_n, *) \to (\mathrm{X}^n, x_0)$ for each $\gamma \in \mathrm{J}$ such that $[h_\gamma] = r_\gamma$. Attach an $(n + 1)$-cell e_γ^{n+1} to X^n by the map h_γ for each $\gamma \in \mathrm{J}$ and let X^{n+1} be the CW-complex thus obtained. Consider the collapsing map p: $\mathrm{X}^{n+1} \to \mathrm{X}^{n+1}/\mathrm{X}^n$; by (137)

$$p_*: \pi_r(\mathrm{X}^{n+1}, \mathrm{X}^n, x_0) \to \pi_r(\mathrm{X}^{n+1}/\mathrm{X}^n, \bar{x}_0) \tag{142}$$

is bijective for $2 \leqslant r \leqslant n - 1$. Since $\mathrm{X}^{n+1}/\mathrm{X}^n$ is obtained by attachment of each $(n + 1)$-cell $p(e_\gamma^{n+1})$ to a single point, by (§6,C)

$$\pi_r(\mathrm{X}^{n+1}/\mathrm{X}^n, \bar{x}_0) = 0 \quad \text{for } r < n + 1, \quad \text{and} \quad \pi_{n+1}(\mathrm{X}^{n+1}/\mathrm{X}^n, \bar{x}_0) \cong \mathrm{R}. \tag{143}$$

In the homotopy exact sequence

$$\pi_{n+1}(\mathrm{X}^{n+1}, \mathrm{X}^n, x_0) \xrightarrow{\partial} \pi_n(\mathrm{X}^n, x_0) \to \pi_n(\mathrm{X}^{n+1}, x_0) \to \pi_n(\mathrm{X}^{n+1}, \mathrm{X}^n, x_0) = 0$$

the image of the homomorphism ∂ is identified with R when $\pi_n(\mathrm{X}^n, x_0)$ is identified with F, so that

$$\pi_r(\mathrm{X}^{n+1}, x_0) = 0 \quad \text{for } r < n \quad \text{and} \quad \pi_n(\mathrm{X}^{n+1}, x_0) \simeq \mathrm{G}$$

since $(\mathrm{X}^{n+1}, \mathrm{X}^n)$ is n-connected by (124).

III. We shall see later (chap. V, §1,D) how Hurewicz's theorem on aspherical spaces (§6,A) led Eilenberg and Mac Lane to consider, more generally, the homology of spaces X for which $\pi_r(\mathrm{X}, x_0) = 0$ *except for one value* n of r, which might be any integer $n \geqslant 1$; but in their first papers on that topic ([179], [181]) they do not seem to have tried to prove the *existence* of such spaces. This was done in 1949 by J.H.C. Whitehead, using CW-complexes [500].

The proof is by induction on $m > n$, the inductive assumption being that there exists a CW-complex X^m such that

$$\pi_r(\mathrm{X}^m, x_0) \simeq \begin{cases} \mathrm{G} & \text{for } r = n, \\ 0 & \text{for } 1 \leqslant r < m \text{ and } r \neq n \end{cases} \tag{144}$$

(for $m = n + 1$, this is the result of II above). A CW-complex X^{m+1} is constructed by attaching $(m + 1)$-cells e_α^{m+1} to X^m in the following way. Consider the group $\pi_m(X^m, x_0)$ and let $[g_\alpha]$ be a set of generators of that group, with $g_\alpha: (S_m, *) \to (X^m, x_0)$. Attach each e_α^{m+1} to X^m by the attaching map g_α; it follows from the definitions that in the homotopy exact sequence

$$\pi_{m+1}(X^{m+1}, X^m, x_0) \xrightarrow{\partial} \pi_m(X^m, x_0) \to \pi_m(X^{m+1}, x_0) \to \pi_m(X^{m+1}, X^m, x_0)$$

the last term is 0 by (124), and $[g_\alpha] = \partial[f_\alpha]$, where $f_\alpha: \mathbf{D}_{m+1} \to X^{m+1}$ is such that $f_\alpha | \mathbf{S}_m = g_\alpha$ and $f_\alpha(\mathring{\mathbf{D}}_{m+1}) = e_\alpha^{m+1}$. So ∂ is surjective, and $\pi_m(X^{m+1}, x_0) = 0$. The process is repeated indefinitely by induction, and X is defined as the union of the X^m, with the fine topology (Part 2, chap. V, § 3,C).

IV. The CW-complex X having all homotopy groups equal to 0 except for $\pi_r(X) = G$ is *unique up to homotopy equivalence*. This is a consequence of the following more general theorem:

Suppose (X, x_0) is an $(n-1)$-connected CW-complex and (Y, y_0) is a pointed space such that $\pi_r(Y, y_0) = 0$ for $r > n$; then, for any homomorphism

$$\varphi: \pi_n(X, x_0) \to \pi_n(Y, y_0)$$

there is a map of pointed spaces $f: (X, x_0) \to (Y, y_0)$ such that $f_* = \varphi$.

Up to homotopy equivalence, the $(n-1)$-skeleton $X^{n-1} = \{x_0\}$ (§ 6,C), so that X^n is obtained by attaching a family (e_α^n) of n-cells to the point x_0. Let $i_\alpha: (\mathbf{S}_n, *) \to (X, x_0)$ be the homeomorphism of $\mathbf{S}_n$ onto $\overline{e_\alpha^n} = e_\alpha^n \cup \{x_0\}$. Then $\varphi([i_\alpha]) \in \pi_n(Y, y_0)$, and there is a map $f_\alpha^n: (\mathbf{S}_n, *) \to (Y, y_0)$ such that $[f_\alpha^n] = \varphi([i_\alpha])$; a map $f_n: (X^n, x_0) \to (Y, y_0)$ is then defined by taking $f_n(i_\alpha(z)) = f_\alpha^n(z)$ for each α and $z \in \mathbf{S}_n$. By obstruction theory (§ 4,C) it is then possible to extend f_n to a map $f: (X, x_0) \to (Y, y_0)$ such that $f_* = \varphi$, and if two maps f, f' are such that $f_* = f'_* = \varphi$, they are homotopic.

The uniqueness of homotopy type of a CW-complex X with $\pi_r(X, x_0) = 0$ for $r \neq n$ and $\pi_n(X, x_0) \simeq G$ is then seen by considering two such CW-complexes X, X′ and applying the preceding theorem to define maps $f: X \to X'$ and $g: X' \to X$, with f_* and g_* the *identity* in G; then, for *all* integers $r \geqslant 1$, $(f \circ g)_* = f_* \circ g_*$ and $(g \circ f)_* = g_* \circ f_*$ are the identity, hence, by the uniqueness of f and g up to homotopy, $f \circ g \sim 1_{X'}$ and $g \circ f \sim 1_X$.

V. An arcwise-connected space X such that $\pi_i(X) = 0$ for $i \neq n$ is called an *Eilenberg–Mac Lane space*. If $\pi_n(X)$ is isomorphic to a group Π, the space is *of type* (Π, n), and often written $K(\Pi, n)$. All Eilenberg–Mac Lane spaces of type (Π, n) which are CW-complexes have the *same homotopy type*.

Interesting Eilenberg–Mac Lane CW-complexes are obtained from the construction of projective spaces. The real projective space $\mathbf{P}_n(\mathbf{R})$ can be defined by attachment of a single n-cell e^n to $\mathbf{P}_{n-1}(\mathbf{R})$: identify the unit disc $\mathbf{D}_n$ with the upper hemisphere D_n^+ of $\mathbf{S}_n$, of which it is the orthogonal projection, and let h_n be the restriction to D_n^+ of the natural map $\mathbf{S}_n \to \mathbf{P}_n(\mathbf{R})$; the restriction $h_n | \mathbf{S}_{n-1}$ is then the natural map $\mathbf{S}_{n-1} \to \mathbf{P}_{n-1}(\mathbf{R})$, and that restriction is an attaching map of D_n^+ to $\mathbf{P}_{n-1}(\mathbf{R})$, so that the n-cell $e^n = h_n(\mathring{D}_n^+)$. The same argument as in II then shows that

$$\pi_r(\mathbf{P}_n(\mathbf{R})) = 0 \quad \text{for } 2 \leqslant r \leqslant n. \tag{145}$$

The *infinite-dimensional real projective space* $\mathbf{P}_\infty(\mathbf{R})$ is defined by considering the infinite sequence (h_n) and taking the fine topology on the CW-complex thus defined (Part 2, chap. V, § 3,C). By (124),

$$\pi_1(\mathbf{P}_\infty(\mathbf{R})) \cong \pi_1(\mathbf{P}_1(\mathbf{R})) \cong \mathbf{Z}/2\mathbf{Z}, \qquad \pi_r(\mathbf{P}_\infty(\mathbf{R})) = 0 \text{ for } r \geqslant 2$$

so that $\mathbf{P}_\infty(\mathbf{R})$ is an Eilenberg–Mac Lane CW-complex of type $(\mathbf{Z}/2\mathbf{Z}, 1)$.

With similar constructions, the *infinite-dimensional complex* (resp. *quaternionic) projective space* $\mathbf{P}_\infty(\mathbf{C})$ [resp. $\mathbf{P}_\infty(\mathbf{H})$] is defined; it is an Eilenberg–Mac Lane space of type $(\mathbf{Z}, 2)$ [resp. $(\mathbf{Z}, 4)$].

VI. It is now easy to conclude, with J.H.C. Whitehead, that *any* sequence $(G_n)_{n\geqslant 1}$ of groups (commutative for $n \geqslant 2$) is "realizable" as the sequence of homotopy groups of a space X: consider for each n an Eilenberg–Mac Lane space X_n of type (G_n, n) and take $X = \prod_{n=1}^{\infty} X_n$.

G. Spaces Having the Homotopy Type of CW-Complexes

All homotopy properties of CW-complexes that do not use the cell structure in their formulaton are still valid for spaces which have the *homotopy type* of a CW-complex, and it is therefore useful to have criteria showing that a space has that homotopy type. In 1949 [499] J.H.C. Whitehead proved that if a space X is *dominated* (§ 2,C) by a CW-complex X′ having at most a countable set of cells, then X has the homotopy type of such a CW-complex. Hanner showed in 1950 that all (metrizable separable) ANR's have that property [219]. Another important result already proved by Kuratowski in 1935 [291] is that if X is an ANR, then for any compact metric space Y the function space $\mathscr{C}(Y; X)$ is also an ANR.

In 1950 J.H.C. Whitehead generalized his result by dropping the restriction on the cardinal of the set of cells of X′; later Milnor [343] proved that $\mathscr{C}(Y; X)$ has the homotopy type of a CW-complex when Y is any compact space and X has the homotopy type of a CW-complex; he also generalized his results to triads and m-ads.

§ 7. Simple Homotopy Type

A. Formal Deformations

In the 1920s M.H.A. Newman [356] and Alexander [14] developed a notion of "combinatorial equivalence" for finite combinatorial complexes (Part 1, chap. II, § 2) based on processes that include the usual barycentric subdivisions but are more general. Let K be an arbitrary finite combinatorial complex and $\sigma = (a_0, a_1, \ldots, a_p)$ be a p-simplex of K. Consider the disjoint union of the set of vertices of K and a single element b; Alexander defined a process (b, σ) called *elementary subdivision*, which deduces a new combinatorial complex K′ from K in the following way:

1. the simplices of K which do not contain σ are also simplices of K′;
2. every $(p + k)$-simplex of K

$$(a_0, a_1, \ldots, a_p, c_1, \ldots, c_k)$$

containing σ is deleted, and is replaced by $p + 1$ new $(p + k)$-simplices

$$(b, a_0, \ldots, \hat{a}_j, \ldots, a_p, c_1, \ldots, c_k) \quad \text{for } 0 \leqslant j \leqslant p$$

in K′, and of course the q-simplices $(q < p + k)$ contained in these $(p + k)$-simplices are also simplices of K′.

The new notion was to introduce also the *inverse process* $(b, \sigma)^{-1}$ deducing K from K′, when it is defined. Two combinatorial complexes K, K′ are then *combinatorially equivalent* if there is a sequence

$$K = K_0, K_1, \ldots, K_m = K' \tag{146}$$

such that K_{j+1} is deduced from K_j, *either* by an elementary subdivision, *or* by the inverse of such a subdivision (when defined). With these definitions Newman and Alexander were able to show that "combinatorial equivalence" is the same as "rectilinear equivalence" for euclidean simplicial complexes, meaning by that the existence of subdivisions* K_1 (resp. K'_1) of K (resp. K′) and of a *simplicial bijection* of K_1 onto K'_1.

In 1938 J.H.C. Whitehead conceived an ambitious and highly original program aiming at treating "combinatorially" in a similar way the *homotopy* theory of simplicial complexes. Impressed by Reidemeister's theory of complexes with automorphisms (Part 2, chap. VI, § 3,A and Part 3, chap. II, § 2,B), he sought to attach some algebraic object to a simplicial complex K in such a way that it would be invariant under "combinatorial" homotopy.

Let K be a finite euclidean simplicial complex and σ be a p-simplex of K that is a face of a *unique* $(p + 1)$-simplex τ of K; let K′ be the complex obtained by deleting both σ and τ from K; Whitehead called the process of passing from K to K′ an *elementary contraction of order* $p + 1$ and the inverse process an *elementary expansion of order* $p + 1$. If there is a sequence (146) such that the passage of each K_j to K_{j+1} is *either* an elementary contraction (of any order) *or* an elementary expansion (of any order), he said K′ is obtained from K (or K from K′) by a *formal deformation* or that K and K′ have the same *nucleus*, a term that he replaced by *simple homotopy type* after 1948.

These are purely combinatorial notions, but their geometric meaning is clear. The box lemma (chap. II, § 2,D), or rather the similar lemma for simplices instead of cubes, shows that if K′ is an elementary contraction of K, there is a retraction r: K → K′ for which K′ is a strong deformation retract of K and the natural injection K′ → K is a homotopy inverse of r. Applying this remark to the sequence (146) defining a formal deformation, the injections *or* retrac-

* The general definition of a subdivision K_1 of a cell complex K is that each cell of K is union of cells of K_1.

tions $K_j \to K_{j+1}$ give by composition a homotopy equivalence $K \to K'$. Such special homotopy equivalences associated to formal deformations are called *simple homotopy equivalences*; the central problem to which Whitehead addressed himself was whether, for two euclidean simplicial complexes K, K′, a given homotopy equivalence $K \to K'$ is *homotopic to a* simple *homotopy equivalence*, and if not, to classify the simple homotopy types within a given homotopy type.

In his first paper on the subject [494] Whitehead did not use homotopy groups in his arguments, but introduced another kind of equivalence between simplicial complexes: he called the *perforation of order n* of such a complex the deleting of one of its n-simplices, and the *filling of order n* the inverse process (when defined); he then said two simplicial complexes K, L have the *same m-group* if one is deduced from the other by a succession of processes that may be elementary contractions, elementary expansions, perforations of order $> m$, or fillings of order $> m$. It transpired in his later work [502] that this is equivalent to saying that the homotopy groups $\pi_r(K)$ and $\pi_r(L)$ are isomorphic for all $r \leqslant m$.

In this first paper Whitehead was still unable to solve the fundamental problems, but he displayed the combinatorial virtuosity that he had acquired under Newman in a series of constructions of simplicial complexes too intricate to be described here in detail. We shall limit ourselves to enumerating his most significant results:

1. The "nucleus" is the same for "combinatorially equivalent" complexes.
2. Suppose K′ is obtained from K by a formal deformation. Then there is a simplicial complex K_0 such that K *and* a simplicial complex K″ deduced from K′ by a succession of "elementary subdivisions" are *both* obtained from K_0 by a succession of elementary *contractions*.
3. Suppose K and L are connected simplicial complexes of dimension $\leqslant n$ that have the same "n-group"; then there exists a *wedge* (Part 2, chap. V, § 2,D) of K and of a finite number of spheres of dimension n that is a formal deformation of a similar wedge of L and a finite number of spheres of dimension n (not necessarily the same number as for K).
4. Finally, from 3 Whitehead was able to deduce the first version of what later became his "second theorem" on homotopy (chap. II, § 6,B): if two connected simplicial complexes have the same "m-group" for *every m*, then they have the same homotopy type.

In a second paper published in 1941 [496] Whitehead tackled the problem of adapting Reidemeister's ideas to his purpose in earnest. He therefore introduced, for a finite euclidean simplicial complex K, its universal covering complex $\tilde{K}$, and considered the p-chains of $\tilde{K}$ as forming a free module over the group algebra $\mathbf{Z}[\pi_1(K)]$. He wanted to find algebraic processes on the "incidence matrices" of K (with elements in $\mathbf{Z}[\pi_1(K)]$) that would correspond to the processes of elementary contractions, elementary expansions, perforations, and fillings described in his first paper. The results are very complicated and cannot be described in simple terms; they were superseded by his later work on torsion (see below); however they enabled him to prove his criterion characterizing the homotopy of lens spaces (chap. II, § 2,B).

B. The Whitehead Torsion

Finally, after the war Whitehead returned to the theory of simple homotopy type and was able to construct an algebraic structure incorporating the Reidemeister theory, with which he could translate into purely algebraic terms the properties of homotopy type and simple homotopy type for CW-complexes. This major achievement was a consequence of the introduction of both new topological and algebraic ideas [502].

I. In his papers of 1939 and 1940 Whitehead, following an earlier definition of Aronszajn [32], had considered attachment of simplicial disks to simplicial complexes by simplicial maps (Part 2, chap. V, § 3). It is thus that he arrived at the definition of CW-complexes (*loc. cit.*) and realized their adequacy for homotopy theory.

He used all properties of CW-complexes described in earlier sections (Part 2, chap. V, § 3,C and Part 3, chap. II, § 6); he also needed a concept that would replace simplicial maps for CW-complexes. If K and L are CW-complexes, K^n and L^n are their n-skeletons, a continuous map $f: K \to L$ is called *cellular* if $f(K^n) \subset L^n$ for every $n \geqslant 0$. The theorem on simplicial approximation (Part 1, chap. II, § 3) is then replaced by the following one:

Any continuous map $f: K \to L$ is *homotopic to a cellular map*; any two cellular maps g_1, g_2 that are homotopic are also homotopic by a *cellular homotopy*, i.e., a cellular map $h: K \times [0,1] \to L$ ($K \times [0,1]$ being considered as a CW-complex).

The first statement is proved by constructing a sequence of maps $f_r: K \to L$ and homotopies h_r between f_{r-1} and f_r, such that $f_0 = f$, $f_r(K^r) \subset L^r$ and h_r does not change the restriction of f_{r-1} to K^{r-1}. This is done by application of the lemma in step 3 of the proof of the second Whitehead theorem (§ 6,B) with $X = L^r$, $Y = L$, $P = K^r$, $Q = K^{r-1}$. The second statement follows from the application of the first to a map $h: K \times [0,1] \to L$.

II. Let K be any CW-complex; we have seen (Part 2, chap. V, § 3,C) that each **Z**-module $H_n(K^n, K^{n-1})$ is *free*, with a basis equipotent to the set of n-cells of K. There is also a **Z**-homomorphism

$$\Delta_n: H_n(K^n, K^{n-1}) \to H_{n-1}(K^{n-1}, K^{n-2}) \tag{147}$$

such that $\Delta_{n-1} \circ \Delta_n = 0$ [Part 1, chap. IV, § 6,B), formula (94)]. This therefore defines a *chain complex*

$$C_\bullet(K): \cdots \longrightarrow H_n(K^n, K^{n-1}) \xrightarrow{\Delta_n} H_{n-1}(K^{n-1}, K^{n-2}) \xrightarrow{\Delta_{n-1}} H_{n-2}(K^{n-2}, K^{n-3}) \longrightarrow \cdots \tag{148}$$

and its fundamental property is that the singular homology $H_\bullet(K)$ *is naturally isomorphic to the homology* $H_\bullet(C_\bullet(K))$ of that complex. This is seen by considering the map

$$H_n(j)\colon H_n(K^n) \to H_n(K^n, K^{n-1}) \tag{149}$$

and using the homology exact sequence [*loc. cit.*, formula (92)] for the pairs (K^{n+1}, K^n), (K^n, K^{n-1}), and (K^{n-1}, K^{n-2}); the image of $H_n(j)$ is the module of n-cycles in $C_n(K)$, and the image by $H_n(j)$ of the kernel of the map $H_n(K^n) \to H_n(K)$ is the module of n-boundaries.

III. These preliminaries allow the application of the Reidemeister method (Part 2, chap. VI, § 3,A) to connected CW-complexes; in this theory extensive use is made of the results on chain homotopies and chain equivalences (Part 1, chap. IV, § 5,F), *generalized* to left modules over *any* ring A instead of **Z**-modules, and to A-homomorphisms instead of **Z**-homomorphisms.

Let K be a connected CW-complex and $\tilde{K}$ be its universal covering space, made into a CW-complex by taking as n-cells of $\tilde{K}$ those that project on the n-cells of K (chap. I, § 3,C); then

$$C_n(\tilde{K}) = H_n(\tilde{K}^n, \tilde{K}^{n-1}) \tag{150}$$

is a *free* $\mathbf{Z}[\pi_1(K)]$-*module*, having as basis a set which may be identified with a set of n-cells of $\tilde{K}$ such that exactly one of them is above each n-cell of K.

If $f\colon K \to L$ is a *cellular* map of CW-complexes, it can be lifted to a cellular map $\tilde{f}\colon \tilde{K} \to \tilde{L}$ (chap. I, § 2,V); passing to homology, $\tilde{f}$ yields a *chain transformation*

$$C_\bullet(\tilde{f})\colon C_\bullet(\tilde{K}) \to C_\bullet(\tilde{L}) \tag{151}$$

and a homomorphism of groups

$$f_*\colon \pi_1(K) \to \pi_1(L). \tag{152}$$

Since a cellular homotopy between two cellular maps f, g of K into L lifts to a cellular homotopy between $\tilde{f}$ and $\tilde{g}$, $C_\bullet(\tilde{f}) = C_\bullet(\tilde{g})$. From I it follows that for *any* continuous map $f\colon K \to L$ it is possible to define $C_\bullet(f)$ equal to $C_\bullet(\tilde{g})$ for any cellular map g homotopic to f. If f_1, f_2 are homotopic maps of K into L, then $C_\bullet(f_1) = C_\bullet(f_2)$. Finally if $f\colon K \to L$ and $g\colon L \to M$ are any two continuous maps of CW-complexes, then $C_\bullet(g \circ f) = C_\bullet(g) \circ C_\bullet(f)$.

Since $C_\bullet(f)$ only depends on the homotopy class of f in $[K; L]$ when $f\colon K \to L$ is a *homotopy equivalence*, $C_\bullet(f)\colon C_\bullet(\tilde{K}) \to C_\bullet(\tilde{L})$ is a *chain equivalence* when $\pi_1(K)$ and $\pi_1(L)$ are identified by f_*. The first important result is the *converse* of that statement: if $C_\bullet(f)$ is a chain equivalence, then f is a homotopy equivalence.

Indeed, from the definition it follows that $f_*\colon \pi_1(K) \to \pi_1(L)$ is an isomorphism, and

$$H_n(C_\bullet(f))\colon H_n(C_\bullet(\tilde{K})) \to H_n(C_\bullet(\tilde{L}))$$

is an isomorphism for every $n \geqslant 1$. But from II it follows that

$$H_n(\tilde{f}): H_n(\tilde{K}) \to H_n(\tilde{L})$$

is also an isomorphism for $n \geqslant 1$; since $\tilde{K}$ and $\tilde{L}$ are simply connected, the first and second Whitehead theorems imply that $\tilde{f}$ is a homotopy equivalence (§ 6,B) and $\tilde{f}_*: \pi_n(\tilde{K}) \to \pi_n(\tilde{L})$ is an isomorphism for $n \geqslant 2$, so the same is true for $f_*: \pi_n(K) \to \pi_n(L)$ (§ 3,B); since $f_*: \pi_1(K) \to \pi_1(L)$ is also an isomorphism, the first Whitehead theorem (§ 5,F) ends the proof.

IV. The problem is now to define among *all* cellular homotopy equivalences $f: K \to L$ those which will be called *simple*. Instead of doing this as he had in his previous papers by using "formal deformations", Whitehead first attached to *every* homotopy equivalence f an element $\tau(f)$ of a commutative group depending only on $\pi_1(K)$ [isomorphic to $\pi_1(L)$]; this is called the *torsion* of f and is an extensive generalization of the Reidemeister–Franz–de Rham torsion. *Simple* homotopy equivalences are then defined as those for which $\tau(f) = 0$, and only *afterward* did Whitehead show that this notion can also be obtained by the use of "formal deformations."

The definition of Whitehead torsion is only reached after a long series of intricate and ingenious algebraic manipulations.

V. The first algebraic step is a streamlined version of the notion of "distinguished bases" (Part 2, chap. VI, § 3,A). Given an arbitrary ring A (commutative or not), all groups GL(n, A) of invertible $n \times n$ matrices with elements in A for variable n may be considered as subgroups of their *direct limit* GL(A) for the direct system of injections GL(n, A) $\to$ GL($n+1$, A) defined by

$$X \mapsto \begin{pmatrix} X & 0 \\ 0 & 1 \end{pmatrix}$$

so that GL(A) can also be assumed to consist of *infinite matrices* $(a_{ij})_{i\geqslant 1, j\geqslant 1}$ of elements of A, for each of which there is an integer n_0 depending on the matrix such that for $i > n_0$ and $j > n_0$, $a_{ij} = 0$ unless $i = j$, when $a_{ii} = 1$.

The "transvection" matrices in GL(n, A)

$$I_n + \lambda E_{ij} \quad (i \neq j, \lambda \in A),$$

where E_{ij} is the matrix whose elements are all 0 except the one at the (i,j)-th place which is equal to 1, generate a subgroup SL(n, A), *normal* in GL(n, A) and contained in the commutator subgroup [GL(n, A), GL(n, A)]; when A is a commutative field with at least three elements this commutator subgroup is *equal* to SL(n, A), but this is not true for all rings A. However, Whitehead showed, by a simple computation of matrices, that

$$[GL(n, A), GL(n, A)] \subset SL(2n, A) \tag{153}$$

so that, if SL(A) is the direct limit of the groups SL(n, A), then

$$SL(A) = [GL(A), GL(A)]. \tag{154}$$

Suppose that G is now an *arbitrary* group and $A = \mathbf{Z}[G]$ is its group algebra over **Z**; Whitehead considered the subgroup E(G) in GL(A) generated by SL(A) and all *diagonal* matrices $(a_{ii}) \in$ GL(A) such that $\pm a_{ii}$ belongs to G.

Then E(G) is a normal subgroup of GL(A), and the quotient

$$\mathrm{Wh(G) = GL(A)/E(G)} \tag{155}$$

is called the *Whitehead group* of G; it is commutative, since $\mathrm{E(G) \supset SL(A)}$; for every $X \in \mathrm{GL(A)}$ the image $\tau(X)$ by the natural surjection

$$\tau\colon \mathrm{GL(A)} \to \mathrm{W}h\mathrm{(G)}$$

is called the *torsion* (or *Whitehead torsion*) of the matrix X.

This construction very soon attracted the attention of algebraists as well as topologists and later came to play an important part in algebraic K-theory and its relations with vector bundles. The computation of Wh(G) is not easy; $\mathrm{Wh}(\mathbf{Z}) = 0$ and $\mathrm{Wh}(\mathbf{Z}/m\mathbf{Z}) = 0$ for $2 \leqslant m \leqslant 4$, but G. Higman showed that $\mathrm{Wh}(\mathbf{Z}/5\mathbf{Z})$ is infinite [222].

If C is now a free left $\mathbf{Z}[\mathrm{G}]$-module that has a *finite* basis, the number of elements of any basis of C is the same ("dimension" of C). A family of "distinguished bases" of C is such that the matrix with elements in $\mathbf{Z}[\mathrm{G}]$ transforming one of these bases into another one always has *zero torsion*. We shall say for short that a free $\mathbf{Z}[\mathrm{G}]$-module equipped with a family of distinguished bases is a *DB-module*. Then, if $f\colon \mathrm{C} \to \mathrm{C}'$ is a $\mathbf{Z}[\mathrm{G}]$-*isomorphism* where C and C′ are DB-modules, the matrix of f with respect to any pair of distinguished bases in C and in C′ has a torsion *independent* of the choice of these bases; it is by definition the *torsion* $\tau(f)$; the isomorphism f is called *simple* if $\tau(f) = 0$.

A *DB-submodule* C′ of a DB-module C is a free $\mathbf{Z}[\mathrm{G}]$-submodule of C having a distinguished basis B′ that is part of a distinguished basis B of C; the $\mathbf{Z}[\mathrm{G}]$-submodule C″ with basis $\mathrm{B'' = B - B'}$ is then a DB-module written $\mathrm{C - C'}$; it is independent of B, up to simple isomorphism, and $\mathrm{C = C' \oplus C''}$.

VI. For short, we shall call *DB-chain complex* a finite chain complex of DB-modules

$$\mathrm{C_\bullet}\colon 0 \to \mathrm{C}_n \xrightarrow{\mathbf{b}_n} \mathrm{C}_{n-1} \to \cdots \to \mathrm{C}_1 \xrightarrow{\mathbf{b}_1} \mathrm{C}_0 \to 0 \tag{156}$$

where the C_j are DB-modules; the largest number r for which $\mathrm{C}_r \neq 0$ will be called the *dimension* of $\mathrm{C_\bullet}$. A DB-chain *subcomplex* $\mathrm{C'_\bullet} = (\mathrm{C}'_j)$ of $\mathrm{C_\bullet}$ is by definition such that each C'_j is a DB-submodule of C_j as defined in V; with the notations of V, the DB-submodules $\mathrm{C}''_j = \mathrm{C}_j - \mathrm{C}'_j$ then form a DB-chain complex when the boundary operator $\mathbf{b}''_j\colon \mathrm{C}''_j \to \mathrm{C}''_{j-1}$ is defined as $p_{j-1} \circ \mathbf{b}_j$, where p_{j-1} is the projection $\mathrm{C}_{j-1} \to \mathrm{C}''_{j-1}$; if $\mathrm{C''_\bullet} = (\mathrm{C}''_j)$, then one writes $\mathrm{C''_\bullet} = \mathrm{C_\bullet} - \mathrm{C'_\bullet}$.

A *simple isomorphism* $f_\bullet\colon \mathrm{C_\bullet} \to \mathrm{C'_\bullet}$ of DB-chain complexes is a chain transformation

$$\begin{array}{ccccccccccccc}
 & 0 & \longrightarrow & \mathrm{C}_n & \xrightarrow{\mathbf{b}_n} & \mathrm{C}_{n-1} & \longrightarrow & \cdots & \longrightarrow & \mathrm{C}_1 & \xrightarrow{\mathbf{b}_1} & \mathrm{C}_0 & \longrightarrow 0 \\
f_\bullet\colon & & & \downarrow{\scriptstyle f_n} & & \downarrow{\scriptstyle f_{n-1}} & & & & \downarrow{\scriptstyle f_1} & & \downarrow{\scriptstyle f_0} & \\
 & 0 & \longrightarrow & \mathrm{C}'_n & \xrightarrow[\mathbf{b}'_n]{} & \mathrm{C}'_{n-1} & \longrightarrow & \cdots & \longrightarrow & \mathrm{C}'_1 & \xrightarrow[\mathbf{b}'_1]{} & \mathrm{C}'_0 & \longrightarrow 0
\end{array} \tag{157}$$

where the f_j are simple isomorphisms for all indices j [i.e., $\tau(f_j) = 0$].

An *elementary trivial DB-chain complex* has only two nonzero homogeneous components

$$\mathrm{T}_{\bullet}: 0 \to \mathrm{T}_r \xrightarrow{\mathbf{b}_r} \mathrm{T}_{r-1} \to 0$$

where $\mathbf{b}_r$ is a *simple isomorphism*. A *trivial* DB-chain complex is a *finite direct sum* of elementary trivial DB-chain complexes. A trivial DB-chain complex can also be defined as follows:

$$\mathrm{T}_{\bullet}: 0 \to \mathrm{A}_n \oplus \mathrm{Z}_n \xrightarrow{\mathbf{b}_n} \mathrm{A}_{n-1} \oplus \mathrm{Z}_{n-1} \to \cdots \to \mathrm{A}_1 \oplus \mathrm{Z}_1 \xrightarrow{\mathbf{b}_n} \mathrm{A}_0 \oplus \mathrm{Z}_0 \to 0$$

where the A_j and Z_j are DB-modules, $\mathbf{b}_j(\mathrm{Z}_j) = 0$ and $\mathbf{b}_j|\mathrm{A}_j$ is a *simple isomorphism* $\mathrm{A}_j \xrightarrow{\sim} \mathrm{Z}_{j-1}$.

Two DB-chain complexes $\mathrm{C}_{\bullet}$, $\mathrm{C}'_{\bullet}$ are *equivalent* if there exist two *trivial* DB-chain complexes $\mathrm{T}_{\bullet}$, $\mathrm{T}'_{\bullet}$, and a *simple isomorphism*

$$\mathrm{C}_{\bullet} \oplus \mathrm{T}_{\bullet} \xrightarrow{\sim} \mathrm{C}'_{\bullet} \oplus \mathrm{T}'_{\bullet}.$$

The main result proved by Whitehead concerning these notions is that it is possible to attach to each *acyclic* DB-chain complex $\mathrm{C}_{\bullet}$ an element $\tau(\mathrm{C}_{\bullet}) \in \mathrm{Wh}(\mathrm{G})$ such that $\tau(\mathrm{C}_{\bullet}) = \tau(\mathrm{C}'_{\bullet})$ *if an only if* $\mathrm{C}_{\bullet}$ *and* $\mathrm{C}'_{\bullet}$ *are equivalent*. He first used the following remark, a special case of the Hopf theorem of Part 1, chap. IV, § 6,F. Since $\mathrm{C}_{\bullet}$ is *both* free and acyclic, the two chain transformations $1_{\mathrm{C}_{\bullet}}$ and $0_{\mathrm{C}_{\bullet}}$ are chain homotopic; in the notation of (157), there is a chain homotopy $h_{\bullet} = (h_j)$ such that

$$\mathbf{b}_{j+1} h_j + h_{j-1} \mathbf{b}_j = 1_{\mathrm{C}_j} \quad \text{for all } j. \tag{158}$$

This implies that $\mathbf{b}'_{\bullet} = (\mathbf{b}'_j)$ with

$$\mathbf{b}'_j = h_j \mathbf{b}_{j+1} h_j \tag{159}$$

is again a chain homotopy, such that

$$\mathbf{b}_{j+1} \mathbf{b}'_j + \mathbf{b}'_{j-1} \mathbf{b}_j = 1_{\mathrm{C}_j} \quad \text{for all } j, \tag{160}$$

and

$$\mathbf{b}'_{j-1} \mathbf{b}'_j = 0 \quad \text{for all } j. \tag{161}$$

Thus

$$\Delta_{\bullet} = \mathbf{b}_{\bullet} + \mathbf{b}'_{\bullet}: \mathrm{C}_{\bullet} \to \mathrm{C}_{\bullet}$$

is a chain transformation such that $\Delta_{\bullet}^2 = 1_{\mathrm{C}_{\bullet}}$. This can be expressed by considering the DB-modules

$$\mathrm{C}_{\mathrm{ev}} = \bigoplus_{i \geqslant 1} \mathrm{C}_{2i}, \qquad \mathrm{C}_{\mathrm{odd}} = \bigoplus_{i \geqslant 0} \mathrm{C}_{2i+1}; \tag{162}$$

$\Delta_{\mathrm{ev}} = (\Delta_{2i})$ is an isomorphism of DB-modules

$$\Delta_{\mathrm{ev}}: \mathrm{C}_{\mathrm{ev}} \xrightarrow{\sim} \mathrm{C}_{\mathrm{odd}} \tag{163}$$

and the *torsion* $\tau(\mathrm{C}_{\bullet})$ of the *acyclic* DB-chain complex $\mathrm{C}_{\bullet}$ is defined by

$$\tau(C_\bullet) = \tau(\Delta_{ev}). \tag{164}$$

Of course it must be shown that this element is independent of the choice of the chain homotopy $h_\bullet$, but this easily follows from the remark that if $h'_\bullet$ is another chain homotopy and Δ'_{ev} the corresponding isomorphism, then $1 + h'_\bullet h_\bullet$ is a simple isomorphism, and

$$\Delta'_{ev}(1 + h'_\bullet h_\bullet) = (1 + h'_\bullet h_\bullet)\Delta_{ev}.$$

It now follows from the definition (164) that, if $C'_\bullet$ and $C''_\bullet$ are two acyclic DB-chain complexes,

$$\tau(C'_\bullet \oplus C''_\bullet) = \tau(C'_\bullet) + \tau(C''_\bullet). \tag{165}$$

Since $\tau(T_\bullet) = 0$ for a trivial DB-chain complex $T_\bullet$, this proves at once that if $C_\bullet$ and $C'_\bullet$ are equivalent acyclic DB-chain complexes, $\tau(C_\bullet) = \tau(C'_\bullet)$.

The proof of the converse is done in two steps. First, show that *any* acyclic DB-chain complex $C_\bullet$ is equivalent to an acyclic DB-chain complex $\bar{C}_\bullet$ with only two homogeneous components,

$$\bar{C}_\bullet: 0 \to \bar{C}_1 \xrightarrow{\mathbf{b}} \bar{C}_0 \to 0, \tag{166}$$

by induction on the "dimension" n of the DB-chain complex (156): compare $C_\bullet$ to a DB-chain complex $C''_\bullet$ of "dimension" $n - 1$. From the acyclicity of $C_\bullet$ it follows that there is an injective $\mathbf{Z}[G]$-homomorphism $\mathbf{b}'_0: C_0 \to C_1$ such that

$$\mathbf{b}_1 + \mathbf{b}'_0: C_2 \oplus C_0 \to C_1$$

is surjective; then $C''_\bullet = (C''_j)$ is such that $C''_j = C_j$ for $j \geqslant 3$ and $\mathbf{b}''_j = \mathbf{b}_j$ for $j \geqslant 4$. Replace the five last terms of (156) by

$$C_3 \xrightarrow{\mathbf{b}_1} C_2 \oplus C_0 \xrightarrow{\mathbf{b}_2 + \mathbf{b}'_0} C_1 \to 0 \to 0.$$

It is clear that $C''_\bullet$ is again acyclic; the fact that $C_\bullet$ and $C''_\bullet$ are DB-equivalent can be proved by constructing a simple isomorphism $f_\bullet$ of the form

$$\begin{array}{ccccccccccccccccc}
 & 0 & \longrightarrow & C_n & \longrightarrow & C_{n-1} & \longrightarrow & \cdots & \longrightarrow & C_3 & \xrightarrow{\mathbf{b}_3} & C_2 \oplus C_0 & \xrightarrow{\mathbf{b}_2 \oplus 1} & C_1 \oplus C_0 & \xrightarrow{\mathbf{b}_1} & C_0 & \longrightarrow 0 \\
f_\bullet: & & & \| & & \| & & & & \| & & \| & & \downarrow{\scriptstyle f_1} & & \downarrow{\scriptstyle -1} & \\
 & 0 & \longrightarrow & C_n & \longrightarrow & C_{n-1} & \longrightarrow & \cdots & \longrightarrow & C_3 & \xrightarrow[\mathbf{b}_3]{} & C_2 \oplus C_0 & \xrightarrow[\mathbf{b}_2 + \mathbf{b}'_0]{} & C_1 \oplus C_0 & \xrightarrow[0+1]{} & C_0 & \longrightarrow 0.
\end{array}$$

The proof is thus reduced to the second step:

$$C_\bullet: 0 \to C_1 \xrightarrow{\mathbf{b}} C_0 \to 0,$$
$$C'_\bullet: 0 \to C'_1 \xrightarrow[\mathbf{b}']{} C'_0 \to 0,$$

with $\tau(\mathbf{b}) = \tau(\mathbf{b}')$. The DB-equivalence is then proved by constructing in a similar way a simple isomorphism

$$
\begin{array}{ccccccccc}
 & 0 & \longrightarrow & C_1 \oplus C'_1 & \xrightarrow{\mathbf{b} \oplus 1} & C_0 \oplus C'_1 & \longrightarrow & 0 \\
g_\bullet: & & & \| & & \downarrow g_0 & & \\
 & 0 & \longrightarrow & C_1 \oplus C'_1 & \xrightarrow[1 \oplus \mathbf{b}']{} & C_1 \oplus C'_0 & \longrightarrow & 0
\end{array}
$$

Whitehead also showed that if $C_\bullet$ and $C'_\bullet$ are acyclic DB-chain complexes, they are simply equivalent if and only if $\tau(C_\bullet) = \tau(C'_\bullet)$, and if $C_\bullet = C'_\bullet \oplus C''_\bullet$, then $C_\bullet$ is simply equivalent to $C'_\bullet$ if and only if $\tau(C''_\bullet) = 0$.

VII. The final algebraic step introduces the *torsion* $\tau(f_\bullet)$ of a *chain equivalence* $f_\bullet: C_\bullet \rightleftarrows C'_\bullet$ of two *arbitrary* DB-chain complexes. Whitehead constructed the algebraic counterpart of the mapping cylinder of a continuous map (Part 2, chap. V, § 3,A), which may be called the *combinatorial mapping cylinder* $Z_\bullet(f_\bullet)$ of a chain equivalence

$$
\begin{array}{ccccccccccccc}
 & 0 & \longrightarrow & C_n & \xrightarrow{\mathbf{b}_n} & C_{n-1} & \longrightarrow & \cdots & \longrightarrow & C_1 & \xrightarrow{\mathbf{b}_1} & C_0 & \longrightarrow 0 \\
f_\bullet: & & & \downarrow f_n & & \downarrow f_{n-1} & & & & \downarrow f_1 & & \downarrow f_0 & \\
 & 0 & \longrightarrow & C'_n & \xrightarrow[\mathbf{b}'_n]{} & C'_{n-1} & \longrightarrow & \cdots & \longrightarrow & C'_1 & \xrightarrow[\mathbf{b}'_1]{} & C'_0 & \longrightarrow 0
\end{array}
$$

It is the DB-chain complex

$$0 \to C'_n \oplus C_n \oplus C_{n-1} \overset{\mathbf{b}''_n}{\to} C'_{n-1} \oplus C_{n-1} \oplus C_{n-2} \to \cdots \to C'_1 \oplus C_1 \oplus C_0$$
$$\overset{\mathbf{b}''_1}{\to} C'_0 \oplus C_0 \to 0$$

where the boundary operator is given by

$$\mathbf{b}''_j(x'_j \oplus x_j \oplus x_{j-1}) = (\mathbf{b}'_j x'_j + f_{j-1}(x_{j-1})) \oplus (\mathbf{b}_j x_j - x_{j-1}) \oplus (-\mathbf{b}_{j-1} x_{j-1}). \quad (167)$$

The first property is that $C''_\bullet = Z_\bullet(f_\bullet) - C'_\bullet$ is *trivial*, hence the chain transformation $r_\bullet: Z_\bullet(f_\bullet) \to C'_\bullet$ a *simple equivalence*. Indeed $C''_j = C_j \oplus C_{j-1}$, and the boundary operator is

$$\bar{\mathbf{b}}''_j(x_j \oplus x_{j-1}) = (\mathbf{b}_j x_j - x_{j-1}) \oplus (-\mathbf{b}_{j-1} x_{j-1}).$$

Consider the DB-chain complex $C^*_\bullet$, having the same homogeneous components as $C''_\bullet$, but with boundary operator

$$\mathbf{b}^*_j(x_j \oplus x_{j-1}) = x_{j-1} \oplus 0.$$

Each map $g_j: C''_j \to C_j$ defined by

$$g_j(x_j \oplus x_{j-1}) = x_j \oplus (\mathbf{b}_j x_j - x_{j-1})$$

is a simple isomorphism, and $g_\bullet = (g_j)$ a chain equivalence. As $C^*_\bullet$ is trivial by VI, so is $C''_\bullet$.

From (167) it follows that $C_\bullet = (C_n)$ is a DB-chain subcomplex of $Z_\bullet(f_\bullet)$; the DB-chain complex

$$M_\bullet(f_\bullet) = Z_\bullet(f_\bullet) - C_\bullet = (C'_n \oplus C_{n-1}) \quad (168)$$

with boundary operator $x'_j \oplus x_{j-1} \to (\mathbf{b}'_j x'_j + f_{j-1}(x_{j-1})) \oplus (-\mathbf{b}_j x_{j-1})$ is called the *cone* of $f_.$. As $f_.: \mathrm{C}_. \xrightarrow{i_.} \mathrm{Z}_.(f_.) \xrightarrow{r_.} \mathrm{C}'_.$ and $r_.$ are chain equivalences, $i_.$ is also a chain equivalence. From the exact sequence

$$0 \to \mathrm{C}_. \xrightarrow{i_.} \mathrm{Z}_.(f_.) \to \mathrm{M}_.(f_.) \to 0$$

follows the homology exact sequence

$$\mathrm{H}_n(\mathrm{C}_.) \xrightarrow{i_*} \mathrm{H}_n(\mathrm{Z}_.(f_.)) \to \mathrm{H}_n(\mathrm{M}_.(f_.)) \xrightarrow{\partial} \mathrm{H}_{n-1}(\mathrm{C}_.) \xrightarrow{i_*} \mathrm{H}_{n-1}(\mathrm{Z}_.(f_.))$$

and since i_* is an isomorphism, $\mathrm{H}_n(\mathrm{M}_.(f_.)) = 0$ for all n, $\mathrm{M}_.(f_.))$ is *acyclic*.

Therefore its *torsion* is defined by VI; by definition

$$\tau(f_.) = \tau(\mathrm{M}_.(f_.)). \tag{169}$$

Whitehead proved the following properties of the torsion $\tau(f_.)$:

(i) If $\mathrm{C}_.$ and $\mathrm{C}'_.$ are acyclic,

$$\tau(f_.) = \tau(\mathrm{C}'_.) - \tau(\mathrm{C}_.).$$

(ii) If $f_.$ is a simple isomorphism (157),

$$\tau(f_.) = 0.$$

(iii) If there is a chain homotopy between the chain equivalences $f_.: \mathrm{C}_. \to \mathrm{C}'_.$ and $g_.: \mathrm{C}_. \to \mathrm{C}'_.$, then $\tau(f_.) = \tau(g_.)$.

(iv) If $f_.: \mathrm{C}_. \to \mathrm{C}'_.$, $g_.: \mathrm{C}'_. \to \mathrm{C}''_.$ are chain equivalences, then

$$\tau(g_. \circ f_.) = \tau(g_.) + \tau(f_.).$$

VIII. Suppose $\mathrm{T}_.$ is a *trivial* DB-chain complex, and consider the chain transformations

$$\mathrm{C}_. \underset{r_.}{\overset{j_.}{\rightleftarrows}} \mathrm{C}_. \oplus \mathrm{T}_. \underset{s_.}{\overset{p_.}{\rightleftarrows}} \mathrm{T}_.. \tag{170}$$

Then $j_.$, $p_.$, $r_.$, $s_.$, are all chain equivalences, and $\tau(r_.) = \tau(j_.) = 0$. To prove that $j_.$ and $r_.$ are chain equivalences, observe that $p_.s_. = 1_{\mathrm{T}_.}$; since $\mathrm{T}_.$ is free and acyclic, it follows from (158) that there exists a chain homotopy $k_.$ such that

$$1_{\mathrm{T}_.} = k_.\mathbf{b}''_. + \mathbf{b}''_.k_. \quad \text{for the boundary operator } \mathbf{b}''_. \text{ of } \mathrm{T}_.; \tag{171}$$

hence

$$s_.p_. = s_.(k_.\mathbf{b}''_. + \mathbf{b}''_.k_.)p_.; \tag{172}$$

but since $s_.\mathbf{b}''_. = \mathbf{b}_.s_.$ and $\mathbf{b}''_.p_. = p_.\mathbf{b}_.$,

$$s_.p_. = (s_.k_.p_.)\mathbf{b}_. + \mathbf{b}_.(s_.k_.p_.), \tag{173}$$

so (171) and (173) show that $p_.$ and $s_.$ are chain equivalences, each one a homotopy inverse to the other. Since

$$1_{\mathrm{C}_.\oplus\mathrm{T}_.} - j_.r_. = s_.p_.$$

and $r_.j_. = 1_{\mathrm{C}_.}$, $j_.$ and $r_.$ are also chain equivalences, homotopy inverses to each other. Finally, from the definition,

$$\mathrm{M}_{\bullet}(j_{\bullet}) = \mathrm{M}_{\bullet}(1_{\mathrm{C}_{\bullet}}) \oplus \mathrm{T}_{\bullet},$$

therefore $\tau(j_{\bullet}) = \tau(1_{\mathrm{C}_{\bullet}}) + \tau(\mathrm{T}_{\bullet}) = 0$, and as $0 = \tau(1_{\mathrm{C}_{\bullet}}) = \tau(r_{\bullet}) + \tau(j_{\bullet})$, $\tau(r_{\bullet}) = 0$.

The main result of the last algebraic part of the proof is that for a chain equivalence $f_{\bullet}: \mathrm{C}_{\bullet} \to \mathrm{C}'_{\bullet}$, the condition $\tau(f_{\bullet}) = 0$ is *necessary and sufficient* for the existence of two trivial DB-chain complexes $\mathrm{T}_{\bullet}$, $\mathrm{T}'_{\bullet}$ and a *simple isomorphism* $h_{\bullet}: \mathrm{C}_{\bullet} \oplus \mathrm{T} \xrightarrow{\sim} \mathrm{C}'_{\bullet} \oplus \mathrm{T}'_{\bullet}$ such that the diagram

$$\begin{array}{ccc} \mathrm{C}_{\bullet} & \xrightarrow{f_{\bullet}} & \mathrm{C}'_{\bullet} \\ {\scriptstyle j_{\bullet}}\downarrow & & \downarrow{\scriptstyle p'_{\bullet}} \\ \mathrm{C}_{\bullet} \oplus \mathrm{T}_{\bullet} & \xrightarrow[h_{\bullet}]{} & \mathrm{C}'_{\bullet} \oplus \mathrm{T}'_{\bullet} \end{array}$$

commutes *up to chain homotopy* (in other words, there is a chain homotopy between $f_{\bullet}$ and $p'_{\bullet} h_{\bullet} j_{\bullet}$).

The sufficiency follows from the above properties of (170), since $\tau(j_{\bullet}) = \tau(p'_{\bullet}) = 0$; hence $\tau(f_{\bullet}) = \tau(h_{\bullet}) = 0$. To prove the necessity, $\tau(\mathrm{M}_{\bullet}(f_{\bullet})) = \tau(f_{\bullet}) = 0$ by definition, and as in VI, this implies that $\mathrm{M}_{\bullet}(f_{\bullet})$ is simply equivalent to 0; therefore $j_{\bullet}: \mathrm{C}_{\bullet} \to \mathrm{Z}_{\bullet}(f_{\bullet})$ is a simple equivalence, and since $r_{\bullet}: \mathrm{Z}_{\bullet}(f_{\bullet}) \to \mathrm{C}'_{\bullet}$ is a simple equivalence by VII, the same is true for $f_{\bullet} = r_{\bullet} j_{\bullet}$.

IX. Now that we have constructed this heavy algebraic machinery, we can at last return to the homotopy of CW-complexes. Let $f: \mathrm{K} \to \mathrm{L}$ be a *cellular* map of CW-complexes; its mapping cylinder Z_f (Part 2, chap. V, § 3,A) is naturally given a structure of CW-complex in the following way. Consider the natural map

$$p: (\mathrm{K} \times [0,1]) \coprod \mathrm{L} \to \mathrm{Z}_f$$

and take for n-cells the images $p(e^n \times \{0\})$, $p(e^{n-1} \times]0,1[)$, and $p(e'^n)$, where e^n (resp. e'^n) are the n-cells of K (resp. L) and e^{n-1} the $(n-1)$- cells of K. The natural maps (chap. II, § 2,D)

$$f: \mathrm{K} \xrightarrow{i} \mathrm{Z}_f \xrightarrow{r} \mathrm{L} \tag{174}$$

are then cellular; they lift to the universal coverings

$$\tilde{f}: \tilde{\mathrm{K}} \xrightarrow{\tilde{i}} \tilde{\mathrm{Z}}_f \xrightarrow{\tilde{r}} \tilde{\mathrm{L}} \tag{175}$$

and $\tilde{\mathrm{L}}$ is again a strong deformation retract of $\tilde{\mathrm{Z}}_f$ for the retraction $\tilde{r}$. By II, from (174) are deduced homomorphisms

$$f_*: \pi_1(\mathrm{K}) \xrightarrow{i_*} \pi_1(\mathrm{Z}_f) \xrightarrow{r_*} \pi_1(\mathrm{L})$$

and from (175) chain transformations

$$\mathrm{C}_{\bullet}(f): \mathrm{C}_{\bullet}(\tilde{\mathrm{K}}) \xrightarrow{\mathrm{C}_{\bullet}(i)} \mathrm{C}_{\bullet}(\tilde{\mathrm{Z}}_f) \xrightarrow{\mathrm{C}_{\bullet}(r)} \mathrm{C}_{\bullet}(\tilde{\mathrm{L}}). \tag{176}$$

Suppose the cellular map f is now a *homotopy equivalence*; then i and r are homotopy equivalences, hence i_* and r_* are isomorphisms and $C_\bullet(i)$ and $C_\bullet(r)$ *chain equivalences* of DB-chain complexes. Whitehead demonstrated that the DB-chain complex $C_\bullet(\tilde{Z}_f)$ with its boundary operator is naturally identified with the "combinatorial mapping cylinder" $Z_\bullet(C_\bullet(f))$ defined in VII, so that $M_\bullet(C_\bullet(f)) = C_\bullet(\tilde{Z}_f) - C_\bullet(\tilde{K})$. The *torsion* of the homotopy equivalence f is then defined as

$$\tau(f) = \tau(C_\bullet(f)). \tag{177}$$

From III it follows that this definition can be extended to *any* homotopy equivalence $f\colon K \to L$ by taking $\tau(f) = \tau(g)$ for any cellular map g homotopic to f.

The CW-complexes K and L are said to have the same *simple homotopy type* if there exists a homotopy equivalence $f\colon K \to L$ with $\tau(f) = 0$, called a *simple homotopy equivalence*.

The first result Whitehead proved is that if the CW-complex K′ is a *subdivision** of the CW-complex K, then the identity map $i\colon K \to K'$ is a simple homotopy equivalence. To show this he only had to observe that the mapping cylinder Z_i of the map i is a subdivision of the mapping cylinder $Z_{1_K} = K \times [0, 1]$ (where only the cells of $K \times \{1\}$ are submitted to subdivision, giving a CW-complex isomorphic to K′). Subdividing one cell of K at a time as in Part 2, chap. VI, § 3,A, he showed that $C_\bullet(\tilde{Z}_i)$ is isomorphic to $C_\bullet(\tilde{K}) \oplus T_\bullet$, where $T_\bullet$ is trivial, so that $\tau(C_\bullet(i)) = 0$.

X. The last part of Whitehead's ambitious undertaking makes the connection between torsion and his notion of "formal deformation" (§ 7,A). He first extended that notion to CW-complexes. The basic definition is the *elementary expansion* of a CW-complex K obtained by attaching in succession an $(n + 1)$-cell e^{n+1} and an $(n + 2)$-cell e^{n+2} in the following way. With the notation of § 6,C, consider a continuous map $f'\colon D_{n+1}^+ \to K$; if $g = f'|S_n$, attach the cell e^{n+1} to K by the map g; there is then a map $f''\colon D_{n+1}^- \to \overline{e^{n+1}}$ coinciding with g in S_n and which is a homeomorphism of the interior of D_{n+1}^- onto e^{n+1}; attach the cell e^{n+2} to $K \cup_g e^{n+1}$ by the map $f\colon S_{n+1} \to K \cup_g e^{n+1}$, equal to f' in D_{n+1}^+ and to f'' in D_{n+1}^-. The passage from K to $K \cup_g e^{n+1} \cup_f e^{n+2} = K_1$ is called an elementary expansion; there is a retraction $r\colon K_1 \to K$ for which K is a strong deformation retract of K_1, and the natural injection $j\colon K \to K_1$ is its homotopy inverse. Elementary contractions and formal deformations are then defined as in § 7,A.

Whitehead's final result is that if K, K′ are two connected CW-complexes that have the *same homotopy type*, a necessary and sufficient condition that they have the *same simple homotopy type* is that there exist a *formal deformation* of K into K′.

The sufficiency of the condition is easy: it is enough to obtain K_1 from K

* The definition of a subdivision K′ of K is the same as for classical cell complexes: each cell of K is a finite union of cells of K′.

by a formal expansion and check that for the injection $j: \mathrm{K} \to \mathrm{K}_1$, the cone $\mathrm{M}_\bullet(\mathrm{C}_\bullet(j))$ is isomorphic to $\mathrm{M}_\bullet(\mathrm{C}_\bullet(1_\mathrm{K})) \oplus \mathrm{T}_\bullet$, where $\mathrm{T}_\bullet$ is an elementary trivial DB-chain complex, hence $\tau(j) = 0$.

The necessity is much harder to prove. Again take the mapping cylinder Z_f of a simple homotopy equivalence $f: \mathrm{K} \to \mathrm{L}$ and the splitting of f:

$$f: \mathrm{K} \xrightarrow{i} \mathrm{Z}_f \xrightarrow{r} \mathrm{L}.$$

The retraction r can be considered a succession of formal contractions: if e^n is a cell of K of maximum dimension, so that $\mathrm{K} = \mathrm{K}_0 \cup e^n$, then $\mathrm{Z}_f = \mathrm{Z}_{f_0} \cup e^n \cup e^{n+1}$, with $f_0 = f|\mathrm{K}_0$ and $e^{n+1} = e^n \times]0,1[$, so that Z_{f_0} is an elementary contraction of Z_f; the result follows by induction on the number of cells of K. From the sufficiency proved above, $\tau(r) = 0$, and if $\tau(f) = 0$, then $\tau(i) = 0$; the hard part of the proof is to show that there is then a formal deformation of Z_f into K.

It is convenient to forget the construction of Z_f and to consider more generally a connected CW-complex P of which K is a CW-subcomplex with the property that the relative homotopy groups $\pi_i(\mathrm{P}, \mathrm{K})$ are all 0 for $i \geqslant 1$.

(i) The strategy is first to treat a *special case*, in which P − K is a disjoint union of cells of only *two* dimensions, e_j^{n+1} $(1 \leqslant j \leqslant t)$ and e_i^{n+2} $(1 \leqslant i \leqslant s)$, with $n > \dim(\mathrm{K})$. From the assumption that $\mathrm{C}_\bullet(\tilde{\mathrm{P}}) - \mathrm{C}_\bullet(\tilde{\mathrm{K}})$ is acyclic and has *zero torsion*, we must prove that there is a formal deformation of P to K. The DB-chain complex $\mathrm{C}_\bullet(\tilde{\mathrm{P}}) - \mathrm{C}_\bullet(\tilde{\mathrm{K}})$ has the form

$$0 \to \mathrm{C}_{n+2} \xrightarrow{\mathbf{b}} \mathrm{C}_{n+1} \to 0 \tag{178}$$

where C_{n+1} (resp. C_{n+2}) is a $\mathbf{Z}[\pi_1(\mathrm{K})]$-module having as basis cells $\tilde{e}_j^{n+1}$ (resp. $\tilde{e}_i^{n+2}$), one above each cell e_j^{n+1} (resp. e_i^{n+2}). Since (178) is acyclic, $s = t$, and if

$$\mathbf{b}\tilde{e}_i^{n+2} = \sum_{j=1}^{s} x_{ij}\tilde{e}_j^{n+1} \quad \text{with } x_{ij} \in \mathbf{Z}[\pi_1(\mathrm{K})], \quad 1 \leqslant i \leqslant s \tag{179}$$

the square matrix $X = (x_{ij})$ has *zero torsion*.

Here we must return to the definition of τ in V. The assumption means that the matrix X can be reduced to the empty matrix by a succession of "elementary operations" of four types:

a. Multiply a row (resp. column) by −1.
b. Multiply a row (resp. column) on the left (resp. right) by $g \in \pi_1(\mathrm{K})$.
c. Change X to the matrix of order increased by 1:

$$\begin{pmatrix} X & 0 \\ 0 & 1 \end{pmatrix}$$

or the inverse operation.
d. Replace the i-th row of X by the sum of the i-th and the k-th row for $i \neq k$.

Whitehead showed that for each of these operations transforming X into X', there is a *formal deformation* of P to P′, leaving K invariant and such that the matrix of the boundary operator of $\mathrm{C}_\bullet(\tilde{\mathrm{P}}') - \mathrm{C}_\bullet(\tilde{\mathrm{K}})$ is X'. It is only for operation (d) that a nontrivial construction is needed. Let T be the union of K and the e_j^{n+1} ($1 \leqslant j \leqslant s$), and let g_j be the attaching map of e_j^{n+2} to T; P′ is defined by considering the union $P'' \subset P$ of T and the $s - 1$ cells e_j^{n+2} for $j \neq i$, and attaching a new cell e^{n+2} to T by a map g taken in the homotopy class $[g_i] + [g_k]$ ($i \neq k$) in $\pi_{n+1}(T)$. Thus $P' = P'' \cup_g e^{n+2}$ is deduced from P by a formal deformation. On the other hand, the relative Hurewicz isomorphism (§5,E) identified in

$$C_.(\tilde{P}') - C_.(\tilde{K})\colon 0 \to C'_{n+2} \xrightarrow{\mathbf{b}'} C'_{n+1} \to 0$$

the module C'_{n+1} with $\pi_{n+1}(T, K)$, the module C'_{n+2} with $\pi_{n+2}(P', T)$ and the boundary operator $\mathbf{b}'$ with the composite homomorphism

$$\pi_{n+2}(P', T) \to \pi_{n+1}(T) \to \pi_{n+1}(T, K).$$

The choice of g then shows that X' can be deduced from X by replacing the i-th row by the sum of the i-th and the k-th row.

(ii) To prepare for the reduction of the general case to the special one treated in (i) Whitehead needed two lemmas on formal deformations.

(ii a) Let K_0, K_1 be two CW-subcomplexes of a CW-complex, with $K = K_0 \cap K_1$ a subcomplex, and $K_i = K \cup_{f_i} e_i^n$ for $i = 0, 1$. Suppose the attaching maps $f_i\colon \mathbf{S}_{n-1} \to K$ are *homotopic* in K. There is then a formal deformation $K_1 \to K_0$ leaving K invariant. Suppose that $e_0^n \cap e_1^n = \varnothing$ and let $K^* = K_0 \cup K_1$. Let $g\colon \mathbf{S}_{n-1} \times [0, 1] \to K$ be the homotopy between f_0 and f_1; there is then a continuous map

$$f\colon (\mathbf{D}_n \times \{0\}) \cup (\mathbf{S}_{n-1} \times [0, 1]) \cup (\mathbf{D}_n \times \{1\}) \to K^*$$

such that $x \mapsto f(x, 0)$ and $x \mapsto f(x, 1)$ are the maps of $\mathbf{D}_n$ onto $\overline{e_0^n}$ and $\overline{e_1^n}$ defining the attachments, and that $f(x, t) = g(x, t)$ in $\mathbf{S}_{n-1} \times [0, 1]$. Take f as attaching map of a cell e^{n+1} to K; if $L = K \cup_f e^{n+1}$, it follows from the definition that there are elementary contractions $L \to K_0$ and $L \to K_1$. If, on the other hand, $e_0^n \cap e_1^n \neq \varnothing$, consider a third CW-complex K′ obtained by attaching a new cell e'^n to K by the *same* attaching map f_0, so that now $e'^n \cap e_0^n = e'^n \cap e_1^n = \varnothing$, which reduces to the preceding case.

(ii b) The second lemma introduces a general process of "continuation" of a formal deformation. Consider CW-complexes $P \supset P_0 \supset K$, and for every $n \geqslant 0$ let $k_n(P - P_0)$ be the number of n-cells in $P - P_0$. Let $P_0 \to Q_0 \supset K$ be a formal deformation leaving K invariant. Then there exists a CW-complex $Q \supset Q_0$ and a formal deformation $P \to Q$ leaving K invariant and such that

$$k_n(P - P_0) = k_n(Q - Q_0) \quad \text{for all } n.$$

By double induction on the number of cells in $P - P_0$ and on the number of elementary deformations in the formal deformation $P_0 \to Q_0$ the proof can be reduced to the case in which $P = P_0 \cup_g e^n$, and $P_0 \to Q_0$ is either an elementary expansion or an elementary contraction.

In the first case $Q_0 = P_0 \cup e^{p-1} \cup e^p$. If $e^n \cap (e^{p-1} \cup e^p) = \varnothing$, take $Q = P \cup e^{p-1} \cup e^p$ with the same attaching maps, and $Q = Q_0 \cup_g e^n$ is clearly an elementary expansion of P. If $e^n \cap (e^{p-1} \cup e^p) \neq \varnothing$, apply lemma (ii a) to P and the CW-complex P′ obtained by attaching to P another n-cell e'^n by the *same* attaching map g, but with $e'^n \cap (e^{p-1} \cup e^p) = \varnothing$.

In the second case, $P_0 = Q_0 \cup e^{p-1} \cup e^p$. Consider the map $f: \mathbf{D}_n \to \overline{e^n}$ corresponding to the attachment of e^n; since Q_0 is a strong deformation retract of P_0, there is a continuous map $f': \mathbf{S}_{n-1} \to Q_0$ homotopic to $g = f|\mathbf{S}_{n-1}$ in P_0. Attach an n-cell e'^n to Q_0 by f', and let $P' = P_0 \cup_{f'} e'^n$. By lemma (ii a), there is a formal deformation $P' \to P$ leaving K invariant, and since $f'(\mathbf{S}_{n-1}) \subset Q_0$, $Q = P' - e^p - e^{p-1}$ is deduced from P′ by an elementary contraction. There is therefore a formal deformation $P \to P' \to Q$ with $k_r(P - P_0) = k_r(Q - Q_0)$ for all $r \geqslant 0$.

(iii) To find a formal deformation of P into a CW-complex of the special kind treated in (i), and leaving K invariant, the method proceeds by double induction on the *smallest* dimension p of the cells whose disjoint union is $P - K$. Let $g: \mathbf{S}_{p-1} \to K^{p-1}$ be the attaching map for one of these cells e^p; $f: \mathbf{D}_p \to \overline{e^p}$ is the map corresponding to this attachment, with $f|\mathbf{S}_{p-1} = g$; it defines an element of the relative homotopy group $\pi_p(P, K)$; but as that group is 0 by assumption, there is a map $h: \mathbf{D}_{p+1} \to P^{p+1}$ such that $h|D_p^- = f$ (after the usual identification of $\mathbf{D}_p$ with the southern hemisphere D_p^- of $\mathbf{S}_p$) and $h(D_p^+) \subset K$. Attach a new cell e^{p+1} to P by $h|\mathbf{S}_p$, and let $u: \mathbf{D}_{p+1} \to \overline{e^{p+1}}$ be the corresponding map; attach to $P \cup e^{p+1}$ a new cell e^{p+2} by $v: \mathbf{S}_{p+1} \to P \cup e^{p+1}$ such that $v|D_{p+1}^+ = h$ and $v|D_{p+1}^- = u$; then

$$P \to P^* = P \cup e^{p+1} \cup e^{p+2}$$

is a formal expansion, and

$$K \to K^* = K \cup_g e^p \cup_{h|\mathbf{S}_p} e^{p+1}$$

is also a formal expansion. By the "continuation" lemma (ii b) there is a formal deformation $P^* \to Q$ leaving K invariant, such that, for every integer $q \geqslant 0$, $Q - K$ has as many q-cells as $P^* - K^*$; therefore

$$k_p(Q - K) = k_p(P - K) - 1,$$

$$k_q(Q - K) = k_q(P - K) \quad \text{for } q \geqslant p + 3.$$

The induction then finally yields a CW-complex of the special type considered in (i), with $n = \dim(P - K)$.

XI. In the 1950s this prodigious piling-up of algebraic and topological original devices excited the admiration of topologists, but they sometimes wondered if Whitehead had not embarked on a quixotic adventure perhaps leading to a blind alley. The situation changed in the early 1960s when the relations of Whitehead's torsion with cobordism, K-theory, and their applications began to appear, and Whitehead's insight was fully vindicated when the extraordinary progress in the theory of C^0-manifolds, both finite dimensional and *infinite dimensional*, finally brought in 1972 the proof of *invariance under homeomorphism of the simple homotopy type of CW-complexes* (see [436]).

CHAPTER III

Fibrations

§ 1. Fibers and Fiber Spaces

A. From Vector Fields to Fiber Spaces

The concept of fibration has been one of the most important mathematical tools in the twentieth century; born in geometry and topology, it has gradually invaded many other parts of mathematics.

Like sheaves (with which they have close relations), fibrations are mathematical expressions of the idea of "functions" that associate to every point x of a space X not a point in a fixed space, but a mathematical object (in this case a space Y_x depending on x, or a point of the *variable* space Y_x) which is not an element of a previously given set. Such "functions" have been around for some time, notably in differential geometry: think of the various tangent lines associated to a point of a surface in $\mathbf{R}^3$. With the global theory of differential equations begun by Poincaré in 1880, emerged the concept of *vector field* (already familiar to physicists). As long as the domain of definition of a vector field is an open subset Ω of an $\mathbf{R}^n$ its definition is obvious: a map $X: \Omega \to \mathbf{R}^n$, or (in the style of the time) a map

$$(x_1, x_2, \ldots, x_n) \mapsto (X_1(x_1, \ldots, x_n), \ldots, X_n(x_1, \ldots, x_n))$$

where the X_j are ordinary (real-valed) functions. This agrees with writing an autonomous differential system in Ω as Poincaré does:

$$\frac{dx_1}{X_1} = \frac{dx_2}{X_2} = \cdots = \frac{dx_n}{X_n}. \tag{1}$$

The situation is not very different when it comes to defining a vector field on a manifold M *embedded* in some $\mathbf{R}^N$. The vector $X(x)$ (or rather, in the classical conceptions, that vector translated to have its origin at the point $x \in M$) has to be tangent to M; the integral curves of (1) are then contained in M. This is how Brouwer and Hopf worked with vector fields on $\mathbf{S}_n$ (Part 2, chap. III, § 3).

This easy reduction is not available when a vector field has to be defined

J. Dieudonné, *A History of Algebraic and Differential Topology, 1900–1960*, Modern Birkhäuser Classics, DOI 10.1007/978-0-8176-4907-4_17,

on a C^1 manifold of dimension *n*, *not* embedded in some $\mathbf{R}^N$. A natural idea is to consider a chart $\varphi\colon U \to \mathbf{R}^n$, defined in an open subset U of M and to take a vector field $X = (X_1, X_2, \ldots, X_n)$ on $\varphi(U)$ in the sense defined above. However, if $\varphi'\colon U \to \mathbf{R}^n$ is another chart and $\psi = \varphi' \circ \varphi^{-1}$ is the C^1 homeomorphism of $\varphi(U)$ onto $\varphi'(U)$, the integral curves of the vector field X', the image of X by ψ, should be the images by ψ of the integral curves of X; this leads to the expression of $X' = (X'_1, X'_2, \ldots, X'_n)$ by the formulas

$$X'_j(\psi(y)) = \sum_{i=1}^{n} X_i(y) \frac{\partial \psi_j}{\partial x_i} \quad (1 \leqslant j \leqslant n). \tag{2}$$

These and similar formulas for the expressions of a differential 1-form in two different systems of local coordinates "explain" in some sense the way "tensors" (or more accurately "tensor fields") were defined in the traditional style by Ricci, Levi-Civita, and their successors until an intrinsic definition of a tangent vector was given in Chevalley's 1946 book [131], probably for the first time.

Another origin of fibrations in differential geometry was the theory of *moving frames*. It had long been known that if M is a smooth curve or surface in $\mathbf{R}^3$, then among the orthonormal systems $(\mathbf{e}_1, \mathbf{e}_2, \mathbf{e}_3)$ of three vectors of origin $x \in M$, some are invariantly attached to M (the "Frenet frame" for curves). Ribaucour and Darboux had the idea of expressing the variations of x and of the $\mathbf{e}_j$ along M, not in a fixed system of coordinates in $\mathbf{R}^3$, but relative to the variable frame $(\mathbf{e}_1, \mathbf{e}_2, \mathbf{e}_3)$, so that

$$dx = \sum_{i=1}^{3} \sigma_i \mathbf{e}_i, \qquad d\mathbf{e}_i = \sum_{j=1}^{3} \omega_{ij} \mathbf{e}_j \quad (i = 1, 2, 3) \tag{3}$$

where the σ_i and ω_{ij} are differential 1-forms on M; for surfaces they satisfy easily written conditions stemming from the relations $d(dx) = 0$ and $d(d\mathbf{e}_j) = 0$ for exterior differentiation. Conversely, a system of 1-forms satisfying these conditions characterizes a surface in $\mathbf{R}^3$ up to an *isometry* of $\mathbf{R}^3$.

The frames $(x, \mathbf{e}_1, \mathbf{e}_2, \mathbf{e}_3)$ may be considered the images of a *fixed* frame $(0, \mathbf{u}_1, \mathbf{u}_2, \mathbf{u}_3)$ of $\mathbf{R}^3$ by a variable isometry of $\mathbf{R}^3$. In 1905 E. Cotton generalized that idea: he replaced the group of isometries of $\mathbf{R}^3$ by any Lie group generated by the translations of $\mathbf{R}^3$ and a subgroup leaving the origin invariant, like the Lobatschefsky group $SO(\Phi)$ for a quadratic form Φ of index 1 [136].

Cotton's ideas were further elaborated in 1910 by E. Cartan [160], who applied his earlier results on Lie groups. After 1920 he became interested in general relativity and riemannian and pseudoriemannian manifolds. It is possible to associate to a point x in a riemannian manifold M of dimension n the set $R(x)$ of orthonormal bases in the tangent space at the point x; the orthogonal group acts in a simply transitive fashion on each $R(x)$. In general there will be no Lie group acting *on* M and transforming each $R(x)$ into an $R(y)$; the Levi-Civita connection only shows that for any path γ joining x to y in M, there is a well-determined linear map $R(x) \to R(y)$ depending in general on the path γ.

Neither for frames nor for tangent spaces at variable points of a manifold did it occur to differential geometers at that time to consider a *space* that would be their disjoint union. This was perhaps due to the traditional concept of tangent vectors to a surface embedded in $\mathbf{R}^3$: the tangent plane formed by these vectors at a point was *also embedded in* $\mathbf{R}^3$!

Before 1935 the closest approach to the necessary "extraction" of tangent vectors from the ambient space was to be found in a paper by Hotelling [252]. He considered a tangent line $\mathrm{D}(x)$ varying continuously with x for each point x of a surface S, and described on the three-dimensional *disjoint union* of the $\mathrm{D}(x)$ a differential structure that is not the product structure on the space $\mathrm{S} \times \mathbf{R}$.

B. The Definition of (Locally Trivial) Fiber Spaces

The words "fiber" ("Faser" in German) and "fiber space" ("gefaserter Raum") probably appeared for the first time in a paper by Seifert in 1932 [420], but his definitions are limited to a very special case and his point of view is rather different from the modern concepts. He is interested in the topological structure of three-dimensional manifolds, particularly those that can be considered spaces of orbits of properly discontinuous groups G of isometries of the sphere $\mathbf{S}_3$, such as the lens spaces or the Poincaré space (chap. I, § 1 and § 4,D). Each circle $\gamma_z\colon t \mapsto ze^{it}$ in $\mathbf{S}_3$ in these examples is transformed by an isometry $s \in \mathrm{G}$ into another circle $\gamma_{s\cdot z}$, so that the images of these circles in $\mathbf{S}_3/\mathrm{G}$ are again homeomorphic to a circle and they form a partition of $\mathbf{S}_3/\mathrm{G}$. When G is reduced to the identity, these "fibers" are all the circles γ_z, also defined as the inverse images of the points of $\mathbf{S}_2$ by the Hopf map (chap. II, § 1); the union of those fibers that meet a small open neighborhood of a point of $\mathbf{S}_3$ is homeomorphic to the three-dimensional "open torus"

$$\mathrm{T} = \mathring{\mathbf{D}}_2 \times \mathbf{S}_1. \tag{4}$$

This led Seifert to study in general what he called a fiber space in his paper: a three-dimensional connected C^0 manifold M, equipped with a partition (Γ_α) into "fibers" consisting of closed Jordan curves, each of which has a neighborhood homeomorphic to the "open torus"; there are in general, however, "exceptional" fibers Γ_α, such that there is no homeomorphism of a neighborhood T_α of Γ_α sending *each* fiber Γ_β contained in T_α onto a circle $\{y\} \times \mathbf{S}_1$ of the "standard torus" (4). The main difference from the present day conception of a fiber space, however, was that for Seifert what is now called the "base space" of a fiber space E was not a part of the structure, but derived from it as a quotient space of E.

The first genuine "fiber space" was only defined by Hassler Whitney in 1935 [505] under the name "sphere-space." In 1940 Whitney changed the name to "sphere-bundle" [513] and gave two expository lectures on that topic (with sketches of proofs) in 1937 [508] and in 1941 [514]; he intended to write a book on the theory, but it never materialized. His interest was mainly focused on the "characteristic classes," which he and Stiefel defined independently (see chap. IV, § 1), and on their application to the topology of differential manifolds, so that for him fiber spaces were primarily a tool; hence his definitions were not spelled out in every detail, but his meaning is quite clear.

The central idea is that a "sphere-space" is a disjoint union of subspaces $\{x\} \times \mathbf{S}_m$, where x is any point of a "base space" K (for Whitney, usually a simplicial complex or a differential manifold) and the dimension m is fixed; the "total space" is "locally" a product of an open subset of K and the sphere $\mathbf{S}_m$. To give a precise meaning to this idea, Whitney, in his 1935 Note, considered an open covering (U_i) of K and, for each index i, the product space $U_i \times \mathbf{S}_m$; these spaces had to be "glued together" in such a way that for any point $x \in U_i \cap U_j$, the "fibers" $\{x\} \times \mathbf{S}_m$ in $U_i \times \mathbf{S}_m$ and in $U_j \times \mathbf{S}_m$ could be identified. He did this by using "charts" in a way similar to the definition of a C^0-manifold. For each pair (i,j) such that $U_i \cap U_j \neq \varnothing$, there is given a "transition" *homeomorphism* φ_{ij} of $(U_i \cap U_j) \times \mathbf{S}_m$ onto itself, which has the special form

$$\varphi_{ij}: (x, y) \mapsto (x, \xi_{ij}(x, y)), \tag{5}$$

where ξ_{ij} is a continuous map of $(U_i \cap U_j) \times \mathbf{S}_m$ into $\mathbf{S}_m$ (Whitney assumed it to be C^1 when K is differentiable); then, for each $x \in U_i \cap U_j$,

$$y \mapsto \xi_{ij}(x, y) \tag{6}$$

is a homeomorphism of $\mathbf{S}_m$ onto itself. There is an additional condition implicit in Whitney's definition: for any three indices i, j, k such that $U_i \cap U_j \cap U_k \neq \varnothing$, the "cocycle condition" is required:

$$\varphi_{ik}(x, y) = \varphi_{ij}(x, \varphi_{jk}(x, y)) \quad \text{for } (x, y) \in (U_i \cap U_j \cap U_k) \times \mathbf{S}_m. \tag{7}$$

Immediately following that definition Whitney introduced a more restrictive class of "sphere-spaces," which he called "regular": those for which the maps (6) are *orthogonal* transformations of $\mathbf{S}_m$. In fact, he only considered that class of sphere-spaces in the remainder of the Note. He mentioned that the study of these "regular" sphere-spaces is equivalent to what was later called vector bundles (see below), and his first examples were the tangent bundle and the normal bundle of a C^2 manifold embedded in an $\mathbf{R}^N$ (see section C, III). In his 1940 Note he was again mainly interested in characteristic classes, but mentioned at the beginning that in his 1935 definition, the sphere $\mathbf{S}_m$ could be replaced by any topological space F, and the orthogonal group by any group of homeomorphisms of the "typical fiber" F; he also observed that when F is a *discrete* space, the fiber space is just a covering space of K.

Meanwhile, Ehresmann was studying E. Cartan's notions of "connections" on his "generalized spaces" and was trying to unify them under a general notion derived from the concept of fiber space. In Notes written partly in collaboration with his pupil Feldbau ([159], [166]), he drew attention to Whitney's definition of a fiber space in the case where the "fiber" is a *topological group* G and the maps (6) of G into itself are *left translations*

$$t \mapsto s_{ij}(x)t \tag{8}$$

of the group G; the "cocycle condition" (7) then boils down to

$$s_{ik}(x) = s_{ij}(x)s_{jk}(x) \quad \text{for } x \in U_i \cap U_j \cap U_k. \tag{9}$$

He called these *principal fiber spaces* with *structural group* G. He observed that in Whitney's 1940 definition, if G is a topological group acting faithfully on F by $(s, y) \mapsto s \,.\, y$, then to *any* fiber space E with typical fiber F and base space B is associated a principal fiber space of structural group G and base B, provided that the maps (6) may be written

$$y \mapsto \xi_{ij}(x, y) = g_{ij}(x) \,.\, y \tag{10}$$

where g_{ij} is a continuous (resp. C^1) map of $U_i \cap U_j$ into G; this is done by taking $s_{ij} = g_{ij}$ in (8).

Conversely, given a principal fiber space P with base space B and structural group G and a topological space F on which G acts by $(s, y) \mapsto s \,.\, y$, the maps

$$\varphi_{ij}(x, y) = (x, s_{ij}(x) \,.\, y) \tag{11}$$

define a fiber space of base space B and typical fiber F, which Ehresmann says is *associated* to P and to the action of G on F. It was by this two-way association that Ehresmann could later give his general definition of "infinitesimal connections" [165], taking as principal fiber spaces the spaces of "moving frames" used by E. Cartan in his conception of "generalized spaces."

Another important example of fiber space was later discovered by Ehresmann: let M and N be two C^∞ manifolds, and $f\colon M \to N$ be a *submersion*; if N is connected and, for every $y \in N$, $f^{-1}(y)$ is compact and connected, then (M, N, f) is a *fiber space* [163]. In particular, the maps $\mathbf{S}_3 \to \mathbf{S}_2$, $\mathbf{S}_7 \to \mathbf{S}_4$, $\mathbf{S}_{15} \to \mathbf{S}_8$ defined by H. Hopf (chap. I, § 1,C) define fiber spaces with respective fibers $\mathbf{S}_1$, $\mathbf{S}_3$, and $\mathbf{S}_7$.

C. Basic Properties of Fiber Spaces

During the period between 1940 and 1950 the usefulness of fiber spaces in many problems concerning homotopy, homology, and differential geometry began to be realized. In order to make them available to larger segments of the mathematical community, efforts were made to systematize the basic notions of the theory; at first some confusion of the different notations and definitions adopted by various mathematicians was probably unavoidable, but at the end of that period things settled down in the 1949–1950 Seminar of H. Cartan [423] and in Steenrod's textbook of 1951 [450].

I. Whitney's definition introduced the "total space" E, the "base space" B, and the "fibers" E_b attached to each point $b \in B$; the latter may be considered the inverse images $\pi^{-1}(b)$ of the points of B by a surjective map $\pi\colon E \to B$. Although Whitney did not explicitly say so, he knew the map π (later often called the "projection") to be continuous. Whitney's definition is at the same time a construction of fiber spaces by choice of an open covering and of transition homeomorphisms (5), but a simpler definition is possible: to each point $b \in B$, there exists an open neighborhood U of b, a space F, and a homeomorphism

$$\varphi: U \times F \xrightarrow{\sim} \pi^{-1}(U) \tag{12}$$

such that

$$\pi(\varphi(y,t)) = y \quad \text{for all } y \in U \text{ and } t \in F. \tag{13}$$

This implies that π is an open map and that the space B can be identified to the quotient space E/R, where R is the relation $\pi(x) = \pi(y)$. All fibers $\pi^{-1}(b)$, also written E_b, for b in the same connected component of B, are homeomorphic; when B is connected, the space F is called the "typical fiber" of the fiber space.

It is convenient to write a fiber space as (E, B, π), (E, B, F, π), or (E, B, F) if all fibers are homeomorphic, but these notations are often simply replaced by E (the total space) when no confusion arises.

Although before 1950 the "categorical" notions (Part 1, chap. IV, § 8) were not yet widespread, "morphisms"

$$(f, g): (E, B, \pi) \to (E', B', \pi') \tag{14}$$

were soon defined; $g: B \to B'$ and $f: E \to E'$ are continuous maps, such that

$$\pi'(f(x)) = g(\pi(x)) \quad \text{for all } x \in E. \tag{15}$$

For each $b \in B$, the restriction $f_b = f|E_b: E_b \to E'_{g(b)}$ is a continuous map; Whitney (and even Seifert) had already considered isomorphisms (f, g), which are morphisms such that both f and g are homeomorphisms. A morphism in which $B' = B$ and g is the identity is called a B-*morphism*.

If $(f, 1_B)$ is a B-morphism $(E, B, \pi) \to (E', B, \pi')$ such that E and E' are locally compact, then, if for each $b \in B$, $f_b: E_b \to E'_b$ is a homeomorphism, $(f, 1_B)$ is an isomorphism.

A fiber space $(B \times F, B, pr_1)$ is called *trivial*; a fiber space (E, B, π) is *trivializable* if it is isomorphic to a trivial one; a B-isomorphism

$$(E, B, \pi) \xrightarrow{\sim} (B \times F, B, pr_1)$$

is called a *trivialization*.

In 1935 Whitney hinted at a definition of the *pullback* (or *inverse image*) of a fiber space $\lambda = (E, B, \pi)$ by a continuous map $g: B' \to B$ as a fiber space (E', B', π') such that for every $b' \in B'$ the fiber $E'_{b'}$ is homeomorphic to $E_{g(b')}$. This can be defined intrinsically by taking as "total space" the *fiber product* $E' = E \times_B B'$, which is the subspace of the product $E \times B'$ consisting of the points (x, b') such that $g(b') = \pi(x)$; if π' is the restriction to E' of the second projection $pr_2: E \times B' \to B'$, the pullback, written $g^*(\lambda)$, is (E', B', π'). If f is the restriction to E' of the first projection $pr_1: E \times B' \to E$, the pair (f, g) is a morphism $g^*(\lambda) \to \lambda$ of fiber spaces such that for any $b' \in B'$, $f_{b'}: E'_{b'} \to E_{g(b')}$ is a homeomorphism; $g^*(\lambda)$ has the "universal property" (Part 1, chap. IV, § 8,C) that any morphism

$$(u, g)\colon (\mathrm{E}'', \mathrm{B}', \pi'') \to (\mathrm{E}, \mathrm{B}, \pi)$$

corresponding to the *same* map g, factorizes uniquely into

$$(\mathrm{E}'', \mathrm{B}', \pi'') \xrightarrow{(v, 1_{\mathrm{B}'})} (\mathrm{E}', \mathrm{B}', \pi') \xrightarrow{(f,g)} (\mathrm{E}, \mathrm{B}, \pi).$$

In particular, if B′ is a subspace of B and j: B′ → B is the natural injection, then $j^*(\lambda)$ is called the *restriction* to B′ of λ; its total space is identified with $\pi^{-1}(\mathrm{B}')$ and its projection with the restriction of π to $\pi^{-1}(\mathrm{B}')$.

In his 1937 address Whitney also introduced (under the unfortunate name "projection") what is now called a *section* (or *cross-section*) of a fiber space $(\mathrm{E}, \mathrm{B}, \pi)$ *over* the base space B, namely, a continuous map s: B → E such that $\pi(s(b)) = b$ for all $b \in \mathrm{B}$. For the tangent bundle of a manifold this is the classical "vector field" defined in the whole manifold; the example of the "sphere-space" over S_2 whose fibers are the circles of radius 1 in the tangent planes shows that sections over the whole "base space" may not always exist. If s is a section over B, it is a homeomorphism of B onto a subspace $s(\mathrm{B})$ of E that is closed if the fibers are Hausdorff spaces. For any pullback $g^*(\lambda) = (\mathrm{E}', \mathrm{B}', \pi')$ of $\lambda = (\mathrm{E}, \mathrm{B}, \pi)$ and any section s of λ over B the map $b' \mapsto s(g(b'))$ is a section of $g^*(\lambda)$ over B′.

For any subspace A of B a section s: A → $\pi^{-1}(\mathrm{A})$ of the restriction of λ to A is also called a *section of* λ (or *of* E) *over* A.

If B is paracompact and the typical fiber is *homeomorphic to some* $\mathbf{R}^{\mathrm{N}}$, any section of E over a closed space A can be extended to a section of E over the whole space B (a generalization of the Tietze–Urysohn theorem); in particular, taking for A a single point, there always exist sections over B.

Finally, if $\lambda' = (\mathrm{E}', \mathrm{B}', \pi')$ and $\lambda'' = (\mathrm{E}'', \mathrm{B}'', \pi'')$ are two fiber spaces, $\lambda' \times \lambda'' = (\mathrm{E}' \times \mathrm{E}'', \mathrm{B}' \times \mathrm{B}'', \pi' \times \pi'')$ is a fiber space. In particular, if $\mathrm{B}' = \mathrm{B}'' = \mathrm{B}$, consider the *pullback* $\lambda = \delta^*(\lambda' \times \lambda'')$ by the diagonal map δ: $b \mapsto (b, b)$ of B; it can be identified with the fiber product $\mathrm{E}' \times_{\mathrm{B}} \mathrm{E}''$, and each fiber is naturally homeomorphic to $\mathrm{E}'_b \times \mathrm{E}''_b$.

II. From the definition by charts of a principal fiber space $(\mathrm{P}, \mathrm{B}, \mathrm{G}, \pi)$ with structural group G it follows that G acts on the right on P; indeed (with the notations of § 1,B) since it acts on each product $\mathrm{U}_i \times \mathrm{G}$ by $(b, t).s = (b, ts)$, it need only be shown that

$$(b, \xi_{ij}(b, ts)) = (b, \xi_{ij}(b, t)s),$$

which is true because $\xi_{ij}(b, t) = s_{ij}(b)t$ in G. The action of G on each fiber $\pi^{-1}(b)$ is *simply transitive*. An intrinsic definition of a principal fiber space may therefore be attempted by considering, as Seifert did, a space P on which a topological group G acts *freely* (without fixed points; $z.s = z$ implies $s = e$). The base space should then be the space of orbits $\mathrm{G}\backslash\mathrm{P}$, quotient space of P. For any point $b \in \mathrm{G}\backslash\mathrm{P}$, there must exist a *local section* of P above a neighborhood of b. However, even if P and G are locally compact, $\mathrm{G}\backslash\mathrm{P}$ is not necessarily Hausdorff (hence there may not exist a local section). A sufficient condition for $G\backslash P$ to be Hausdorff is that the subset R of $P \times P$ consisting of the pairs (u, v) such that u and v belong to the same orbit, be *closed* in $P \times P$. Even when P and G as well as $G\backslash P$ are *compact*, there do not necessarily exist local sections of P. An example is given by the infinite product $P = (SO(3))^{\mathbf{N}}$, on which the subgroup $T = \mathbf{S}_1^{\mathbf{N}}$ acts by right translations ([55], p. 38)

Gleason showed [207] that if P is a regular space and G a *Lie group*, there are local sections for every point of $G\backslash P$. This is always the case when P is a locally compact group and G a closed subgroup of P which is a Lie group acting by right translations.*

A type of principal fiber space that occurs frequently is $(G, G/H, H, \pi)$, where G is a Lie group, H is a closed subgroup, and G/H is the homogeneous space of left cosets tH, H acting on G by right translations.

The definition of *morphisms* for principal fiber spaces is more restrictive than for general fiber spaces: a morphism

$$(f, g)\colon (P, B, G, \pi) \to (P', B', G', \pi')$$

must satisfy (15) and there must exist a continuous *homomorphism* $\rho\colon G \to G'$ such that

$$f(x.s) = f(x).\rho(s) \quad \text{for } x \in P, s \in G. \tag{16}$$

The *trivial* principal fiber space with base B and structural group G is the product $B \times G$ on which G acts by $((b, t), s) \mapsto (b, ts)$. A B-isomorphism $(P, B, G, \pi) \cong (B \times G, B, G, pr_1)$ is called a *trivialization* of (P, B, G, π); when such a B-isomorphism exists, (P, B, G, π) is called *trivializable.* A necessary and sufficient condition of trivializability for a *principal* fiber space with base space B is the existence of *one* section *over* B.

The pullback of a principal fiber space (P, B, G, π) by a continuous map $g\colon B' \to B$ is again a principal fiber space (P', B', G, π'), the action of G on the total space $P' = B' \times_B P$ being given by

$$(b', x).s = (b', x.s).$$

III. The most widely used fiber spaces are the *vector bundles* considered by Whitney in 1935. In such a fiber space $\xi = (E, B, F, \pi)$, also called a *vector fibration,* the typical fiber F is a real (resp. complex) vector space of finite dimension (called the *rank* of the vector bundle), each fiber E_b is a real (resp. complex) vector space of same dimension as F, and each $b \in B$ has an open neighborhood U for which there exists a homeomorphism

$$\varphi\colon U \times F \to \pi^{-1}(U)$$

such that $\pi(\varphi(y, t)) = y$ for all $y \in U$ and $t \in F$, and which has the additional property that $t \mapsto \varphi(y, t)$ is an **R**-*linear* (resp. **C**-*linear*) *bijection* of F on E_y for

* The existence of local sections when P is a Lie group and G a closed subgroup of P had already been mentioned in Ehresmann's thesis of 1934 ([155], p. 398).

all $y \in \mathrm{U}$. If n is the rank of E, an equivalent condition is the existence of a system of n continuous sections s_j ($1 \leqslant j \leqslant n$) above U (also called a *frame* over U) such that

$$(y, \xi_1, \xi_2, \ldots, \xi_n) \mapsto \xi_1 s_1(y) + \cdots + \xi_n s_n(y)$$

is a homeomorphism of $\mathrm{U} \times \mathbf{R}^n$ (resp. $\mathrm{U} \times \mathbf{C}^n$) onto $\pi^{-1}(\mathrm{U})$.

Morphisms of vector bundles are again more restrictive than for general fiber spaces; $(f, g)\colon (\mathrm{E}, \mathrm{B}, \mathrm{F}, \pi) \to (\mathrm{E}', \mathrm{B}', \mathrm{F}', \pi')$ must be such that $f_b\colon \mathrm{E}_b \to \mathrm{E}'_{g(b)}$ is an **R**-linear (resp. **C**-linear) map for every $b \in \mathrm{B}$. The map f is often called a *bundle map*; it defines g uniquely, since if $b = \pi(x)$, then $g(b) = \pi'(f(x))$.

A pullback of a vector bundle $(\mathrm{E}, \mathrm{B}, \mathrm{F}, \pi)$ by a continuous map $g\colon \mathrm{B}' \to \mathrm{B}$ is naturally equipped with a structure of a vector bundle. Since $f_{b'}$ is a bijection of $\mathrm{E}'_{b'}$ onto $\mathrm{E}_{g(b')}$, the structure of a vector space on $\mathrm{E}'_{b'}$ is defined by "transporting" the structure of a vector space of $\mathrm{E}_{g(b')}$ by the map $f_{b'}^{-1}$.

The usual constructions of linear algebra (or, in categorical language, functors that associate a vector space to one or several vector spaces) can be "transferred" to vector bundles. If E', E'' are two vector bundles with the same base space B, their direct *sum* $\mathrm{E}' \oplus \mathrm{E}''$ (also called *Whitney sum*) is a vector bundle E with base space B such that each fiber E_b is the direct sum $\mathrm{E}'_b \oplus \mathrm{E}''_b$; its topological structure is uniquely determined by the condition that if s', s'' are, respectively, sections of E', E'' over an open subset $\mathrm{U} \subset \mathrm{B}$, then $b \mapsto s'(b) \oplus s''(b)$ is a section of E over U.

The definitions of the tensor product $\mathrm{E}' \otimes \mathrm{E}''$ and the bundle of B-morphisms $\mathrm{Hom}(\mathrm{E}', \mathrm{E}'')$, whose fiber at $b \in \mathrm{B}$ is $\mathrm{Hom}(\mathrm{E}'_b, \mathrm{E}''_b)$ are similar; the *dual* E^* of a vector bundle E is $\mathrm{Hom}(\mathrm{B}, \mathrm{E} \times \mathbf{R})$ [resp. $\mathrm{Hom}(\mathrm{E}, \mathrm{B} \times \mathbf{C})$]. The exterior power $\bigwedge^m \mathrm{E}$ and the exterior algebra $\bigwedge \mathrm{E}$ are also vector bundles whose fibers at b are $\bigwedge^m \mathrm{E}_b$ and $\bigwedge \mathrm{E}_b$, respectively.

A *vector subbundle* E' of a vector bundle $(\mathrm{E}, \mathrm{B}, \mathrm{F}, \pi)$ is a subspace of E such that for each $b \in \mathrm{B}$, $\mathrm{E}' \cap \mathrm{E}_b = \mathrm{E}'_b$ is a vector subspace of E_b and for each $b \in \mathrm{B}$ there is an open neighborhood U of b and a homeomorphism

$$\varphi\colon \mathrm{U} \times \mathrm{F} \xrightarrow{\sim} \pi^{-1}(\mathrm{U})$$

as above, such that $\pi^{-1}(\mathrm{U}) \cap \mathrm{E}'$ is the image $\varphi(\mathrm{U} \times \mathrm{F}')$, where F' is a vector subspace of F. If E' is a vector subbundle of E, let E'' be the disjoint union of the quotient vector spaces $\mathrm{E}''_b = \mathrm{E}_b/\mathrm{E}'_b$ for $b \in \mathrm{B}$; then there is on E'' a unique structure of vector bundle of base B such that the map $p\colon \mathrm{E} \to \mathrm{E}''$, which, when restricted to each fiber E_b, is the natural surjection $\mathrm{E}_b \to \mathrm{E}_b/\mathrm{E}'_b$, is a B-morphism. The vector bundle E'' is called the *quotient vector bundle* of E by E' and is written E/E'. If B is a paracompact space, E *splits* into a direct sum $\mathrm{E}' \oplus \mathrm{N}$ such that the restriction of $p\colon \mathrm{E} \to \mathrm{E}''$ to N is a B-*isomorphism*.

If $u\colon \mathrm{E} \to \mathrm{F}$ is a B-morphism of vector bundles with base space B, the disjoint union $u(\mathrm{E})$ of the images $\mathrm{I}_b = u(\mathrm{E}_b) \subset \mathrm{F}_b$ of the fibers of E is *not* necessarily a vector subbundle of F. In general, the dimension of the vector space I_b is only a *lower semicontinuous* function of b; for

$$I = \bigcup_{b \in B} I_b$$

to be a vector subbundle of F, a necessary and sufficient condition is that $b \mapsto \dim(I_b)$ be constant in B; I is then written Im(u). The kernels $N_b = u_b^{-1}(0)$ are then also the fibers of a vector subbundle N of E, written Ker(u), and the quotient bundle F/Im(u) is written Coker(u).

A sequence of B-morphisms of vector bundles of base B

$$E \xrightarrow{u} F \xrightarrow{v} G \tag{17}$$

is called *exact* if for each $b \in B$ the sequence of linear maps

$$E_b \xrightarrow{u_b} F_b \xrightarrow{v_b} G_b$$

is exact; $u(E)$ is then a vector subbundle of F, equal to $v^{-1}(G)$.

Examples. The *tangent bundle* of a C^1 manifold M is the disjoint union T(M) of the tangent vector spaces $T_x(M)$ at each point $x \in M$, which are the fibers of T(M), and the projection maps every tangent vector in $T_x(M)$ to the point x. If N is a submanifold of M, the tangent bundle T(N) is a vector subbundle of T(M); the quotient T(M)/T(N) is called the *normal bundle* of N in M. When M is a riemannian manifold and $T_x(N)^{\perp}$ is the subspace of $T_x(M)$ orthogonal to $T_x(N)$ for each $x \in N$, these vector spaces are the fibers of a vector bundle isomorphic to the normal bundle.

The passage of "regular" sphere bundles to vector bundles, mentioned by Whitney, can simply be done by starting from an open covering (U_i) of the base space B and the transition homeomorphisms (5) for which the maps $y \mapsto \xi_{ij}(x, y)$ are orthogonal transformations of S_m; such a transformation extends uniquely to a map $y \mapsto \eta_{ij}(x, y)$ belonging to the orthogonal group O(m), and the vector bundle is defined by the transition homeomorphisms $(x, y) \mapsto (x, \eta_{ij}(x, y))$. Conversely, if (E, B, π) is a vector bundle and B is paracompact, it is easy to define on the set of pairs (x, x') of points of E such that $\pi(x) = \pi(x')$ a continuous function $d(x, x')$ such that its restriction to $E_b \times E_b$ for each $b \in B$ is a euclidean distance. The corresponding sphere bundle is then the union of the unit spheres in the fibers E_b.

IV. Ehresmann's conception of a fiber space *associated* to a locally compact principal fiber space (P, B, G, π) can be presented in an intrinsic way: suppose there is given an *action* $(s, y) \mapsto s \,.\, y$ of G on a locally compact space F; then G operates (on the right) on the product P × F by

$$(x, y) \,.\, s = (x \,.\, s, s^{-1} \,.\, y). \tag{18}$$

The space of orbits G\(P × F), written $P \times {}^{G}F$, is locally compact; let $p\colon P \times F \to P \times {}^{G}F$ be the natural projection, and write

$$x \,.\, y = p(x, y) \tag{19}$$

so that

$$x \,.\, y = (x \,.\, s) \,.\, (s^{-1} \,.\, y) \quad \text{for } s \in G \tag{20}$$

and the relation $x.y = x'.y'$ means that there exists $s \in G$ for which $x' = x.s$, $y' = s^{-1}.y$. For $z = x.y$, let

$$\pi_F(z) = \pi(x) \in B, \tag{21}$$

which does not depend on $y \in F$. If $U \subset B$ is open and $\sigma: U \to \pi^{-1}(U)$ is a section of P above U, the map

$$(b, y) \mapsto \sigma(b).y$$

is a homeomorphism of $U \times F$ onto $\pi_F^{-1}(U)$ such that

$$\pi_F(\sigma(b).y) = b; \tag{22}$$

this shows that $(P \times {}^G F, B, F, \pi_F)$ is a fiber space with base B and typical fiber F *associated* to the principal fiber space P and to the given action of G on F.

With the same notations the sections of $P \times {}^G F$ above U can be written in a unique way

$$b \mapsto \sigma(b).\psi(b) \tag{23}$$

where ψ is a continuous map of U into F. For every $x \in \pi^{-1}(U)$ and $b = \pi(x)$, $\sigma(b) = x.s$ for a unique $s \in G$, hence $\sigma(b).\psi(b) = x.(s.\psi(b))$, and $s.\psi(b)$ only depends on x. If f is a section of $E = P \times {}^G F$ above the *whole space* B, there is therefore a unique map φ_f: $P \to F$ such that

$$f(\pi(x)) = x.\varphi_f(x); \tag{24}$$

φ_f is continuous and satisfies

$$\varphi_f(x.s) = s^{-1}.\varphi_f(x) \quad \text{for } (s, x) \in G \times P. \tag{25}$$

Conversely, a continuous map φ: $P \to F$ such that

$$\varphi(x.s) = s^{-1}.\varphi(x) \quad \text{for } (s, x) \in G \times P \tag{26}$$

defines a section f of E above B by $f(b) = x.\varphi(x)$ for $x \in \pi^{-1}(b)$.

Suppose (P', B', G', π') is a principal fiber space and that the group G acts on the left on that space in such a way that

$$s.(x'.s') = (s.x').s' \quad \text{for } s \in G, x' \in P', s' \in G'. \tag{27}$$

Then $\pi'(s.x') = \pi'(s.(x'.s'))$, hence that element only depends on s and $\pi'(x')$ and can be written $s.\pi'(x')$; this defines an action of G on B′, easily proved continuous. Then the map

$$p': x.x' \mapsto x.\pi'(x') \tag{28}$$

of $P \times {}^G P'$ into $P \times {}^G B'$ is well defined, and $(P \times {}^G P', P \times {}^G B', G', P')$ is a *principal fiber space*. If G′ now acts on a space F′, G acts on the left on $P' \times {}^{G'} F'$ by $s.(x'.y') = (s.x').y'$; then the map

$$(x.x').y' \mapsto x.(x'.y') \tag{29}$$

is well defined and is a homeomorphism

$$(P \times {}^G P') \times {}^{G'} F' \simeq P \times {}^G (P' \times {}^{G'} F') \tag{30}$$

(note that the left-hand side is a fiber space with base $P \times {}^G B'$, whereas the right-hand side is a fiber space with base B).

The construction of an *associated* fiber space $P \times {}^G F$ is particularly important in the following cases.

(i) Let H be a closed subgroup of G that operates on P on the right by restriction of $(x, s) \mapsto x \, . \, s$ to $P \times H$. Then P is also a principal fiber space with structural group H and base space the space of orbits $H \backslash P$, because $H \backslash P$ is naturally homeomorphic to $P \times {}^G (G/H)$. If $\pi \colon H \backslash P \to G \backslash P$ is the map that to each orbit for H associates the unique orbit for G which contains it, $(H \backslash P, G \backslash P, \pi)$ is a fiber space with typical fiber G/H; if H is normal in G, $H \backslash P$ is a principal fiber space with structural group G/H.

(ii) Let (P, B, H, π) be a principal fiber space, and suppose H is a closed subgroup of a Lie group G. Then H acts freely on G by left translations, and we may form the associated fiber space $Q = P \times {}^H G$; it again has B as a base space and is a *principal* fiber space with structural group G. The action of G on Q is given by

$$((x \, . \, s), s') \mapsto x \, . \, (ss') \quad \text{for } x \in P, \; s, s' \text{ in } G,$$

which is meaningful because for any $t \in H$

$$(x \, . \, t) \, . \, ((t^{-1} s) s') = (x \, . \, t) \, . \, (t^{-1}(ss')) = x \, . \, (ss').$$

The map $x \mapsto x \, . \, e$ is a homeomorphism of P on a closed subspace of Q sending each fiber P_b onto a subspace of Q_b; if $y \in P_b$, so that $P_b = y \, . \, H$, then $Q_b = y \, . \, G$. The principal fiber space (Q, B, G, π_G) is deduced from (P, B, H, π) by *extension of the structural group* H *to* G.

A principal fiber space (Q, B, G, p), however, is not always B-isomorphic to a principal fiber space $P \times {}^H G$. If G is considered as a principal fiber space with base space $\{e\}$, it follows from (30) that if $Q = P \times {}^H G$, then $Q \times {}^G (G/H)$ is B-isomorphic to $P \times {}^H (G \times {}^G (G/H))$, and hence to $P \times {}^H (G/H)$. But this last fiber space has a section above B, because if $\bar{e}$ is the class $e \, . \, H = H$ in G/H, the relation $x \, . \, \bar{e} = x' \, . \, \bar{e}$ in $P \times {}^H (G/H)$ means that there exists an element $t \in H$ such that $x' = x \, . \, t$ and $\bar{e} = t^{-1} \, . \, \bar{e}$, and therefore $\pi(x) = \pi(x')$, so that there is a factorization

$$x \overset{\pi}{\mapsto} \pi(x) \overset{\sigma}{\mapsto} x \, . \, \bar{e}$$

and σ is injective, hence a *section* of $P \times {}^H (G/H)$ over B.

Conversely, if there is a section σ of $Q \times {}^G (G/H)$ over B, the inverse image of $\sigma(B)$ by the map $x \mapsto x \, . \, \bar{e}$ is a principal fiber space with base space B and structural group H, from which Q is deduced by extension of H to G. That principal fiber space is then said to be obtained by *restriction* of the structural group G to its subgroup H.

An important case in which restriction is possible is when G is a connected Lie group and H is a maximal compact subgroup of G, because in that case

G/H is homeomorphic to a space $\mathbf{R}^N$, and there exists a section over B for any fiber space of base B and typical fiber G/H.

(iii) Ehresmann's conception of associating a principal fiber space to a vector bundle (§ 1,B) can be presented intrinsically: suppose (E, B, F, π) is a vector bundle, and consider the vector bundle $\mathrm{Hom}(B \times F, E)$, whose fiber at the point $b \in B$ is $\mathrm{Hom}(F, E_b)$; then GL(F) acts on the right on this vector bundle, the action on a fiber being $(u, s) \mapsto u \circ s$ for $s \in \mathrm{GL}(F)$. Then the subspace $\mathrm{Isom}(B \times F, E)$ may be defined such that, for $b \in B$

$$\mathrm{Isom}(B \times F, E) \cap \mathrm{Hom}(F, E_b) = \mathrm{Isom}(F, E_b),$$

the subspace consisting of the bijective linear maps $F \xrightarrow{\sim} E_b$. This is a dense open subspace of $\mathrm{Hom}(B \times F, E)$; it is stable under the action of GL(F), and the action of that group on each fiber $\mathrm{Isom}(F, E_b)$ is simply transitive, so $R = \mathrm{Isom}(B \times F, E)$ is a *principal fiber space* with base space B and structural group GL(F); it is the *fiber space of frames* for the vector bundle E. The space F may always be identified with an $\mathbf{R}^N$ by choice of a basis; then $\mathrm{Isom}(\mathbf{R}^N, E_b)$ is identified with the set of all *bases* of E_b, each isomorphism $v: \mathbf{R}^N \xrightarrow{\sim} E_b$ being identified with the basis

$$(v(\mathbf{e}_1), \ldots, v(\mathbf{e}_N))$$

image by v of the canonical basis. A *section* s of $\mathrm{Isom}(B \times \mathbf{R}^N, E)$ above an open set $U \subset B$ is therefore a map

$$y \mapsto (s_1(y), \ldots, s_N(y))$$

where each s_j is a continuous map of U into E such that $(s_1(y), \ldots, s_N(y))$ is a basis of the vector space E_y for all $y \in U$; this is what has been called a "frame" for $\pi^{-1}(U)$. The action of a matrix

$$t = (t_{ij}) \in \mathrm{GL}(N, \mathbf{R})$$

on a fiber $\mathrm{Isom}(\mathbf{R}^N, E_b)$ is then

$$(a_j) \mapsto \left(\sum_j t_{ij} a_j\right)$$

Conversely, any *linear representation* $\rho: G \to \mathrm{GL}(F)$ enables us to define, for a principal fiber space R of base B and structural group G, an associated fiber space $R \times {}^G F$, which is a *vector bundle* of base B and typical fiber F. Take $G = \mathrm{GL}(F)$ and for R the fiber space of frames defined above; then there is a natural isomorphism of vector bundles

$$E \xrightarrow{\sim} R \times {}^{\mathrm{GL}(F)} F$$

and similarly one has an isomorphism of $\bigwedge^m E$ onto $R \times {}^{\mathrm{GL}(F)}(\bigwedge^m F)$, etc.

In the principal fiber space of frames of the vector bundle E it is possible to *restrict* the structural group $\mathrm{GL}(N, \mathbf{R})$ to the *orthogonal group* O(N), that is a maximal compact subgroup of $\mathrm{GL}(N, \mathbf{R})$. Every real vector bundle may thus be considered as *associated* to a principal fiber space whose structural group is the orthogonal group.

V. The whole theory described in this section can be developed by replacing topological spaces everywhere by C^∞ manifolds (resp. real analytic manifolds, resp. holomorphic complex manifolds) and continuous maps by C^∞ maps (resp. analytic maps, resp. holomorphic maps); topological groups are replaced by Lie groups and definitions and proofs are transcribed according to that dictionary. The only point that requires some care is the structure of the orbit space X/G, where X is a C^∞ manifold, and G is a Lie group acting on X by a C^∞ action $(s, x) \mapsto s.x$. Then, in order for X/G to have a structure of C^∞ manifold for which the natural projection π: X → X/G is a submersion, a necessary and sufficient condition is that the subspace R of X × X consisting of pairs of points on the same orbit is a closed *submanifold* of X × X.

§ 2. Homotopy Properties of Fibrations

A. Covering Homotopy and Fibrations

In his first Note of 1935 on homotopy groups [256] Hurewicz made a special study of the space $\mathscr{C}(\mathrm{Y}; \mathrm{G})$, where G is a Hausdorff *topological group*. In particular he investigated the relations among that space and the spaces $\mathscr{C}(\mathrm{Y}; \mathrm{H})$ and $\mathscr{C}(\mathrm{Y}; \mathrm{G}/\mathrm{H})$ for a *closed subgroup* H of G. Although he only considered the case in which Y is compact, if Y is Hausdorff and $\mathscr{C}(\mathrm{Y}; \mathrm{G})$ is given the usual compact-open topology (chap. II, § 3,A), then it is a topological group for the obvious law of composition $(f, g) \mapsto fg$, where $(fg)(y) = f(y)g(y)$; the neutral element is the constant map Y → $\{e\}$ and the inverse f^{-1} the map $y \mapsto f(y)^{-1}$.

If H is a closed subgroup of G and $\mathrm{H} \xrightarrow{j} \mathrm{G} \xrightarrow{p} \mathrm{G}/\mathrm{H}$ are the natural maps, they define maps

$$\mathscr{C}(\mathrm{Y}; \mathrm{H}) \xrightarrow{\tilde{j}} \mathscr{C}(\mathrm{Y}; \mathrm{G}) \xrightarrow{\tilde{p}} \mathscr{C}(\mathrm{Y}; \mathrm{G}/\mathrm{H})$$

were $\tilde{j}(f) = j \circ f$ and $\tilde{p}(g) = p \circ g$; $\tilde{j}$ and the map

$$\tilde{\imath}: \mathscr{C}(\mathrm{Y}; \mathrm{G})/\tilde{j}(\mathscr{C}(\mathrm{Y}; \mathrm{H})) \to \mathscr{C}(\mathrm{Y}; \mathrm{G}/\mathrm{H}) \tag{31}$$

deduced from $\tilde{p}$ are *injective*.

The main result Hurewicz stated was that if G is a compact Lie group and Y is a compact space, the image of $\tilde{\imath}$ is *open and closed* in $\mathscr{C}(\mathrm{Y}; \mathrm{G}/\mathrm{H})$. This implies that if $\mathscr{C}(\mathrm{Y}; \mathrm{G}/\mathrm{H})$ is arcwise-connected and $\mathscr{C}(\mathrm{Y}; \mathrm{G})$ locally arcwise-connected, the map $\tilde{\imath}$ is *bijective*, so that every arcwise-connected component of $\mathscr{C}(\mathrm{Y}; \mathrm{G})$ must contain an element of $\tilde{j}(\mathscr{C}(\mathrm{Y}; \mathrm{H}))$. In other words (chap. II, § 3,A), if $p \circ f$ is *homotopic* to the constant map Y → $\{\bar{e}_0\} = p(\mathrm{H})$, then f must be *homotopic* to a map g: Y → H.

Disregarding covering spaces, where such a "lifting" of a homotopy is unique (chap. I, § 2,V), this was the first appearance (in a very special case) of what became known as the *covering homotopy* problem: given a continuous map π: E $\to$ B, a continuous map f: X $\to$ E and a homotopy

$$h: X \times [0,1] \to B$$

such that $h(x,0) = \pi(f(x))$ for $x \in X$, does there exist a homotopy

$$k: X \times [0,1] \to E$$

such that $\pi \circ k = h$ and $k(x,0) = f(x)$ for $x \in X$?

A system (X, E, B, π) has the *covering homotopy property* if the preceding problem has *at least* one solution for *every* continuous map f and *every* homotopy h. When (X, E, B, π) has the covering homotopy property, the same is true for (X, E′, B′, π'), where $E' = E \times_B B'$ is the "fiber product" defined by a continuous map g: B′ $\to$ B, and π': E′ $\to$ B′ is the second projection of the product E′ × B′.

Hurewicz did not publish any proof of his theorem on groups before 1940; it then became a consequence of his joint Note with Steenrod (see B, below)

In the meantime, in 1937 Borsuk published a result on spaces of continuous functions [75] that later was seen to imply a covering homotopy property. Let X be a locally compact ANR (chap. II, §2,B), A be a closed subspace of X which is also an ANR, and Y be an arbitrary space, and let

$$E = \mathscr{C}(X; Y), \qquad B = \mathscr{C}(A; Y) \tag{32}$$

with the usual compact-open topology (chap. II, §3,A). Let π: E $\to$ B be the continuous "restriction map"

$$\pi: f \mapsto f|A.$$

Then for any space Z, (Z, E, B, π) has the covering homotopy property. Using the fundamental property of the compact-open topology (*loc. cit.*), what is to be proved is the following: suppose given two continuous maps

$$u: Z \times [0,1] \times A \to Y, \qquad v: Z \times X \to Y$$

such that $u(z,0,x) = v(z,x)$ in Z × A. Then there exists a continuous map

$$w: Z \times [0,1] \times X \to Y$$

such that

$$w(z,t,x) = u(z,t,x) \text{ for } x \in A, \qquad w(z,0,x) = v(z,x) \quad \text{for } x \in X. \tag{33}$$

Borsuk had proved the existence of a retraction r of the space [0, 1] × X on its subspace

$$T = ([0,1] \times A) \cup (\{0\} \times X)$$

(chap. II, §2,D); then

$$(z,t,x) \mapsto (z, r(t,x))$$

is a retraction of Z × [0, 1] × X on its subspace Z × T, hence

$$w(z,t,x) = u(z, r(t,x))$$

satisfies (33). The result extends to the case in which X is a locally compact space and j: A → X a *cofibration* (chap. II, § 2,D).

The work of Serre on homotopy groups (chap. V, § 5,A) brought to light the most important case of Borsuk's theorem (although the theorem itself is not mentioned in Serre's thesis [429]). Y is an arbitrary arcwise-connected space, $X = \mathbf{D}_1 = [-1, 1]$, and $A = \mathbf{S}_0 = \{-1, 1\}$; $E = \mathscr{C}([-1, 1]; Y)$ is then the space of *all paths* between two points of Y, also written PY*; $B = \mathscr{C}(\{-1, 1\}; Y)$ is naturally identified with the product $Y \times Y$, and the map $\pi: E \to B$ is simply

$$\gamma \mapsto (\gamma(-1), \gamma(1)).$$

For a fixed point $y_0 \in Y$, any subspace M of Y may be identified with the subspace $B' = M \times \{y_0\}$ of B, and $E' = \pi^{-1}(B')$ is *the space* $\Omega(Y, M, y_0)$ *of paths with origin at* y_0 *and extremity in* M [chap. II, § 5,A), formula (95)]. If

$$\pi_M: \Omega(Y, M, y_0) \to M,$$

defined by $\pi_M(\gamma) = \gamma(-1)$, is the restriction of π, $(X, \Omega(Y, M, y_0), M, \pi_M)$ has the covering homotopy property for *every* space X.

After Serre showed that the homotopy and homology of fiber spaces only used the covering homotopy property it became customary to call *fibration* any continuous map $\pi: E \to B$ such that (X, E, B, π) has the covering homotopy property for *every* space X; Serre even noted that he only needed the case in which X is a finite simplicial complex, which does not imply that π is a fibration (it is often called *weak fibration*, or *Serre fibration*). The subspaces $\pi^{-1}(b)$ are still called *fibers*, but they are no longer homeomorphic to a fixed space in general [see below (35)]; if B is arcwise-connected, it can only be said that all fibers have the *same homotopy type*. Often [as in the case of $\Omega(Y, M, y_0)$] there is a privileged point b_0 in the case space B; $F = \pi^{-1}(b_0)$ is then called (by abuse of language) *the* fiber of the fibration, and the fibration is also written

$$F \xrightarrow{j} E \xrightarrow{\pi} B \tag{34}$$

where $j: \pi^{-1}(b_0) \to E$ is the natural injection. For $E = \Omega(Y, M, y_0)$, "the" fiber $F = \Omega(Y, y_0)$ is the *space of loops* of origin y_0 in Y. Borsuk's theorem then justifies the name "cofibration," if the map $\mathscr{C}(X; Y) \to \mathscr{C}(A; Y)$ in (32) is considered as a "transposed" map of the injection A → X.

This kind of "duality" extends to other notions. Let $f: Y \to X$ be a continuous map; as $X \times X \xrightarrow{\mathrm{pr}_2} X$ is a fibration, the same is true of the map

$$p: PX \to X$$

defined by $p(\gamma) = \gamma(-1)$ for every path $\gamma \in PX$. Then let

$$P_f = f^*(PX) = PX \times_X Y$$

---

* To keep general notations (chap. II, § 5,A), paths $\gamma: [-1, 1] \to X$ are considered as having $\gamma(1)$ as origin and $\gamma(-1)$ as extremity.

be the pullback of that fibration, equal by definition to the subspace of PX × Y consisting of the pairs (γ, y) such that

$$\gamma(-1) = f(y).$$

The space P_f is called the *mapping path fibration* of f and may be considered as the "dual" of the notion of mapping cylinder Z_f (Part 2, chap. V, § 3,A).

There are two natural maps

$$p'\colon P_f \to Y, \qquad p''\colon P_f \xrightarrow{j} PX \times Y \xrightarrow{\mathrm{pr}_1} PX \xrightarrow{p} X$$

(j being the natural injection) that are *fibrations*. This is obvious for p', which is the pullback of p. For p'' the proof of the covering homotopy property runs as follows. Consider

1. a continuous map $g\colon W \to P_f$ and
2. a homotopy $G\colon W \times [0,1] \to X$ such that

$$G(w,0) = p''(g(w)).$$

A homotopy $H\colon W \times [0,1] \to P_f$ has to be constructed so that

$$p''(H(w,t)) = G(w,t) \quad \text{for } (w,t) \in W \times [0,1],$$
$$H(w,0) = g(w) \qquad \text{for } w \in W.$$

Write $g(w) = (G''(w), g'(w)) \in PX \times Y$, where $G''(w)$ is a path $u \mapsto G''(w)(u)$ defined in $[-1,1]$ for each $w \in W$, with $G''(w)(-1) = f(g'(w))$. A variant of the "box lemma" (chap. II, § 2,D) defines a retraction $(u,t) \mapsto r(u,t)$ of the rectangle $[-1,1] \times [0,1]$ on the union of its sides $[-1,1] \times \{0\}$ and $\{-1\} \times [0,1]$:

$$r(u,t) = \begin{cases} ((2u-t)/(2-t), 0) & \text{for } u \geqslant t-1, \\ (-1, 2(u-t+1)/(u-1)) & \text{for } u \leqslant t-1. \end{cases}$$

Then take for H the homotopy such that

$$H(w,t)(u) = \begin{cases} (G''(w)((2u-t)/(2-t)), g'(w)) & \text{for } t-1 \leqslant u \leqslant 1, \\ (G(w, 2(u-t+1)/(u-1)), g'(w)) & \text{for } -1 \leqslant u \leqslant t-1. \end{cases}$$

Next, there is a section $s\colon Y \to P_f$ such that $f = p'' \circ s$; it is defined by taking for $s(y)$ the pair $(\omega_{f(y)}, y)$, where $\omega_{f(y)}$ is the constant path $u \mapsto f(y)$; s is a *homotopy equivalence*, and $p' \circ s = 1_Y$. Define $F\colon P_f \times [0,1] \to P_f$ by

$$F((\gamma, y), t) = (\gamma_t, y) \quad \text{with} \quad \gamma_t(u) = \gamma((1-t)u - t);$$

clearly

$$F((\gamma, y), 0) = (\gamma, y) \quad \text{and} \quad F((\gamma, y), 1) = (\omega_{f(y)}, y) = s(y).$$

The "duality" is then that the passage from one to the other of the two diagrams

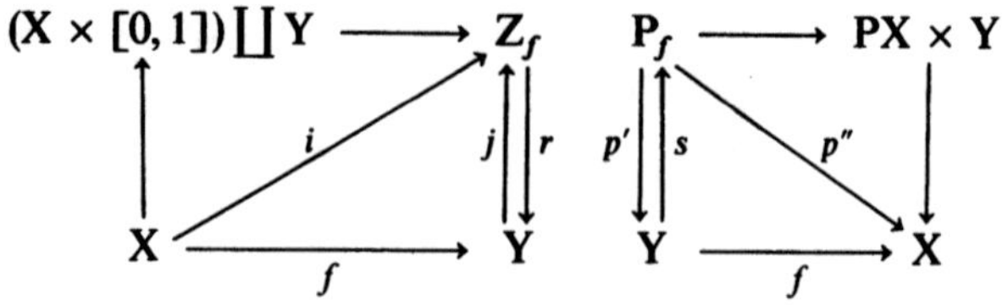

(notations of chap. II, § 2,D for the first diagram) consists in "reversing the arrows" and exchanging "retractions" with "sections" and "cofibrations" with "fibrations." The decomposition

$$f\colon \mathrm{Y} \xrightarrow{s} \mathrm{P}_f \xrightarrow{p''} \mathrm{X}$$

shows that *any* continuous map can be multiplied on the right by a homotopy equivalence such that the product is *homotopic to a fibration.*

For a map $f\colon (\mathrm{Y}, y_0) \to (\mathrm{X}, x_0)$ of pointed spaces, use the subspace $\mathrm{P}_{x_0}\mathrm{X} = \Omega(\mathrm{X}, \mathrm{X}, x_0)$ of paths with origin x_0 instead of PX; the pullback is then P_{f,y_0}, the space of pairs (γ, y) such that $\gamma(1) = x_0 = f(y_0)$ and $\gamma(-1) = f(y)$; the restriction of p'' to P_{f,y_0} is again a fibration. The fibers $p''^{-1}(x_0)$ and $p'^{-1}(y_0)$ are the space of *loops* $\Omega(\mathrm{X}, x_0)$.

Consider also the map $q\colon \mathrm{P}_f \to \mathrm{X}$ such that $q(\gamma, y) = \gamma(1)$; an argument similar to the one proving that p'' is a fibration shows that the same is true for q (the retraction is then on the union of $[-1, 1] \times \{0\}$ and $\{1\} \times [0, 1]$). The fiber $q^{-1}(x_0)$ is then the space P_{f,y_0} considered above.

B. Fiber Spaces and Fibrations

Borsuk's proof of a special case of the covering homotopy property does not seem to have immediately attracted the attention of the topologists who at that time were beginning to investigate fiber spaces (as defined by Whitney). This is not too surprising, since the properties of the "fibers" of a fibration are generally very different from these of the "fibers" of a fiber space.

In 1940, however, Hurewicz and Steenrod published a Note [260] in which they introduced conditions on a map $\pi\colon \mathrm{E} \to \mathrm{B}$ implying the covering homotopy property (given that name for the first time) for $(\mathrm{X}, \mathrm{E}, \mathrm{B}, \pi)$ when X is a *compact* space. They supposed B to be a metric space with distance d and assumed that there exists a number $r > 0$ such that if V_r is the open subset of $\mathrm{E} \times \mathrm{B}$ consisting of the pairs (x, b) for which $d(\pi(x), b) < r$, there exists a continuous map $\varphi\colon \mathrm{V}_r \to \mathrm{E}$ such that

$$\pi(\varphi(x, b)) = b \quad \text{and} \quad \varphi(x, \pi(x)) = x \quad \text{for all } (x, b) \in \mathrm{V}_r. \tag{35}$$

They called φ a "slicing function"; for each $x_0 \in \mathrm{E}$, if $\mathrm{U}_r(x_0)$ is the open ball of center $\pi(x_0)$ and radius r in B, the map $b \mapsto \varphi(x_0, b)$ is a homeomorphism of $\mathrm{U}_r(x_0)$ on its image in E, if E is Hausdorff, and the image of $\pi(x_0)$ by this map is x_0. So these "slices" $b \mapsto \varphi(x_0, b)$ look like the "local sections" of a fiber space for all x_0 in the "fiber" $\pi^{-1}(\pi(x_0))$. The "fibers" are not generally homeomorphic to a fixed space; this is seen in the example given by Hurewicz and Steenrod: E is the subspace of $\mathbf{R}^2$ defined by $0 \leqslant y \leqslant x \leqslant 1$, $B = [0,1]$, and π is the first projection; then

$$\varphi((x,y),x') = \begin{cases} (x',x') & \text{for } x' \leqslant y \\ (x',y) & \text{for } x' > y \end{cases}$$

is a slicing function.

The existence of "slicing functions" makes the proof of the covering homotopy property for compact spaces X almost trivial. Given the homotopy $h\colon X \times [0,1] \to B$, and a map $f\colon X \to E$ such that $\pi(f(x)) = h(x,0)$ for $x \in X$, there is a $\delta > 0$ such that $|t - t'| \leqslant \delta$ implies

$$d(h(x,t),h(x,t')) < r \quad \text{for all } x \in X.$$

Then $[0,1]$ is divided in small intervals $[t_i, t_{i+1}]$ of length $< \delta$ and the covering homotopy $k\colon X \times [0,1] \to E$ is defined inductively by

$$k(x,0) = f(x),$$

$$k(x,t) = \varphi(k(x,t_i),h(x,t)) \quad \text{for } t_i \leqslant t \leqslant t_{i+1}.$$

In their Note Hurewicz and Steenrod proved the existence of slicing functions for the natural map $\pi\colon G \to G/H$, where G is a compact Lie group and H is a closed subgroup; they used the exponential map, as Ehresmann had already done in 1934 ([155], p. 398) to define $\varphi(e,b)$ in a neighborhood of $\bar{e} = \pi(H)$ in G/H, and then took $\varphi(x,b) = x\varphi(e,x^{-1}.b)$ in $G \times (x.U)$.

Although Hurewicz and Steenrod had not yet formulated the complete homotopy sequence of a fibration (see C below), it was for the purpose of computing relative homotopy groups $\pi_i(E,F)$ that they used the covering homotopy property. It was for the explicit purpose of proving the exactness of that sequence (not yet written with arrows) that in 1941 the covering homotopy property was also formulated in two independent papers, one by Ehresmann and Feldbau [166] and the other by Eckmann [148], both without knowledge of the Hurewicz–Steenrod Note, communications having been disrupted by the war. Ehresmann and Feldbau stated the theorem for locally trivial fiber spaces (E,B,F,π) where B is a finite simplicial complex; they merely indicated in a few lines that the proof is by subdivision of the complex B and induction on its dimension.

Eckmann, like Hurewicz and Steenrod, took a more general viewpoint, closer to the case investigated by Borsuk. He considered a compact metric space E and the compact metric space $\mathfrak{R}$ of nonempty closed subsets of E (with the Hausdorff "Abweichung" for distance). For a fixed number $r > 0$ and every set $A \in \mathfrak{R}$, $U(A,r)$ is the open neighborhood of A defined by $d(x,A) < r$. Eckmann called a subspace $\mathfrak{Z} \subset \mathfrak{R}$ *retractible* if there is a *continuous* map

$$Q\colon V_r \to E$$

[where $V_r \subset \mathfrak{Z} \times E$ consists of the pairs (A,x) such that $x \in \overline{U(A,r)}$] having the property of *retracting* $\overline{U(A,r)}$ on A for each $A \in \mathfrak{Z}$, i.e.,

$$Q(A, x) \in A \quad \text{for } x \in \overline{U(A, r)},$$

$$Q(A, x) = x \quad \text{for } x \in A.$$

In this general setup the following property corresponds to the covering homotopy property: for any continuous path

$$\Gamma: [0, 1] \to \mathscr{C}(X; \mathfrak{Z})$$

with $\Gamma(0) = F \in \mathscr{C}(X; \mathfrak{Z})$ and any map $f \in \mathscr{C}(X; E)$ such that

$$f(x) \in F(x) \quad \text{for every } x \in X,$$

there exists at least one continuous path $\gamma: [0, 1] \to \mathscr{C}(X; E)$ such that $\gamma(0) = f$ and $\gamma(t)(x) \in \Gamma(t)(x)$ for $0 \leqslant t \leqslant 1$ and $x \in X$. The proof is again almost trivial, by subdivision of $[0, 1]$ in small intervals $[t_i, t_{i+1}]$ such that the oscillation of Γ in each of them is $< r$, and then taking

$$\gamma(t)(x) = Q(\Gamma(t), \gamma(t_i)) \quad \text{for } t_i \leqslant t \leqslant t_{i+1}.$$

To recover the usual covering homotopy property for a continuous map $\pi: E \to B$ of compact metric spaces, Eckmann took for $\mathfrak{Z}$ the set of "fibers" $\pi^{-1}(b)$ for $b \in B$, and assumed the natural map $\mathfrak{Z} \to B$ to be a homeomorphism. F then becomes a map $X \to B$ and Γ is a homotopy between F and another continuous map $X \to B$, and the relation $f(x) \in F(x)$ for each $x \in X$ means that $F = \pi \circ f$. The proof that $\mathfrak{Z}$ is "retractible" is not given in general, but only when E is a Riemannian manifold and π is a submersion (in fact Eckmann was only interested in the case $E = G$, $B = G/H$, where G is a compact Lie group and H is a closed subgroup).

In 1943 Steenrod [446] gave a proof that the Whitney fiber spaces (E, B, F, π) are fibrations when B is compact; it is relatively simple, since B may be covered by a finite number of open sets over each of which the fiber space is trivializable. The extension to locally compact and paracompact spaces B at first involved a much longer and more intricate proof; Ehresmann gave one in 1944 [161] (for simplicial complexes) and H. Cartan gave another in his 1949–1950 Seminar [423].

After 1950 Serre's work renewed the interest in criteria for the validity of the covering homotopy property. The most general one (implying Cartan's result), which is at the same time the most elegant proof, was given in 1955 by Hurewicz [257], and independently by Huebsch [255]. The space B is any paracompact space, and $\pi: E \to B$ is assumed to be a *local fibration*, meaning that each point $b \in B$ has an open neighborhood U such that the restriction $\pi^{-1}(U) \to U$ is a fibration; the conclusion is that π *is a fibration*. The proof was based on a new idea, namely, that just as in the case of covering spaces (chap. I, § 2,B, V), it should be enough to prove the existence of the lifting of *paths* $\omega: [0, 1] \to B$, provided they "vary continuously" with ω (this is automatic for covering spaces). More precisely:

I. Suppose the covering homotopy property (X, E, B, π) holds for *all* spaces X. Consider the subspace Q of the product $E \times \mathscr{C}([0, 1]; B) \times [0, 1]$ consisting of all triples (x, ω, s) such that $\omega(s) = \pi(x)$. Then there exists a continuous map

$$\Lambda\colon \mathrm{Q} \to \mathscr{C}([0,1]; \mathrm{E})$$

such that

$$\pi(\Lambda(x, \omega, s)(t)) = \omega(t) \quad \text{for } 0 \leqslant t \leqslant 1 \text{ and } \Lambda(x, \omega, s)(s) = x \tag{36}$$

[in other words $\Lambda(x, \omega, s)$ is a path that lifts ω and is equal to x for $t = s$]. This is shown by applying the covering homotopy property to the homotopy $h\colon \mathrm{Q} \times [0,1] \to \mathrm{B}$ such that

$$h((x, \omega, s), t) = \omega(t) \quad \text{for } t \in [0,1] \quad \text{and} \quad h((x, \omega, s), 0) = \pi(x).$$

If $k\colon \mathrm{Q} \times [0,1] \to \mathrm{E}$ is such that $\pi \circ k = h$, take

$$\Lambda(x, \omega, s)(t) = k((x, \omega, s), t).$$

We shall say that a continuous map Λ satisfying (36) is a *general lifting function.*

II. Conversely, consider the subspace $\mathrm{Q}_0 \subset \mathrm{E} \times \mathscr{C}([0,1]; \mathrm{B})$ consisting of the pairs (x, ω) such that $\omega(0) = \pi(x)$. Suppose there exists a continuous map $\lambda\colon \mathrm{Q}_0 \to \mathscr{C}([0,1]; \mathrm{E})$ such that

$$\pi(\lambda(x, \omega)(t)) = \omega(t) \quad \text{for } t \in [0,1] \text{ and } \lambda(x, \omega)(0) = x \tag{37}$$

[so that $\lambda(x, \omega)$ is a path which lifts ω and is equal to x for $t = 0$]. The covering homotopy property $(\mathrm{X}, \mathrm{E}, \mathrm{B}, \pi)$ then holds for *all* spaces X: if $f\colon \mathrm{X} \to \mathrm{E}$ and $h\colon \mathrm{X} \times [0,1] \to \mathrm{B}$ are maps such that $h(x, 0) = \pi(f(x))$ for all $x \in \mathrm{X}$, define $k\colon \mathrm{X} \times [0,1] \to \mathrm{E}$ by

$$k(x, t) = \lambda(f(x), h(x, .))(t).$$

The continuous maps λ satisfying (37) will be called *special lifting functions.*

III. Suppose B now is *paracompact*. There is then a locally finite open covering (U_α) of B such that each $\bar{\mathrm{U}}_\alpha$ is the support of a continuous function $f_\alpha\colon \mathrm{B} \to \mathbf{R}_+$. For each finite sequence $\mu = (\alpha_1, \ldots, \alpha_k)$ of indices let W_μ be the subset of $\mathscr{C}([0,1]; \mathrm{B})$ consisting of the paths ω such that

$$\omega(t) \in \mathrm{U}_{\alpha_j} \quad \text{for } \frac{j-1}{k} \leqslant t \leqslant \frac{j}{k} \text{ and } 1 \leqslant j \leqslant k.$$

The W_μ form a locally finite open covering of the space $\mathscr{C}([0,1]; \mathrm{B})$ (owing to the fact that any compact subset of B has a neighborhood that only meets a finite number of the sets U_α); the function

$$f_\mu(\omega) = \inf_{1 \leqslant j \leqslant k} \left(\inf_{(j-1)/k \leqslant t \leqslant j/k} f_{\alpha_j}(\omega(t)) \right)$$

has support equal to W_μ. Now suppose $\pi\colon \mathrm{E} \to \mathrm{B}$ is a fibration *above each* U_α. Then, by I, there is a general lifting function for every α

$$\Lambda_\alpha\colon \mathrm{Q}_\alpha \to \mathscr{C}([0,1]; \pi^{-1}(\mathrm{U}_\alpha))$$

with

$$Q_\alpha = Q \cap (\pi^{-1}(U_\alpha) \times \mathscr{C}([0,1]; U_\alpha) \times [0,1]).$$

The aim of that part of the proof is to define a general lifting function

$$\Lambda_\mu : Q_\mu \to \mathscr{C}([0,1]; E)$$

with

$$Q_\mu = Q \cap (E \times W_\mu \times [0,1]).$$

Take a point $(x, \omega, s) \in Q_\mu$, and let ω_j be the continuous map $[0,1] \to B$ equal to $\omega(t)$ for $(j-1)/k \leqslant t \leqslant j/k$ and constant in the intervals $[0,(j-1)/k]$ and $[j/k, 1]$. Define $\tau = \Lambda_\mu(x, \omega, s)$ as follows: suppose $(n-1)/k \leqslant s \leqslant n/k$, and take

$$\tau(t) = \Lambda_{\alpha_n}(x, \omega_n | U_{\alpha_n}, s)(t) \quad \text{for } (n-1)/k \leqslant t \leqslant n/k,$$

then extend it left and right of the interval $[(n-1)/k, n/k]$ by

$$\tau(t) = \Lambda_{\alpha_{n-1}}(\tau((n-1)/k), \omega_{n-1} | U_{\alpha_{n-1}}, (n-1)/k)(t) \quad \text{for } \frac{n-2}{k} \leqslant t \leqslant \frac{n-1}{k}$$

$$\cdots\cdots\cdots\cdots$$

$$\tau(t) = \Lambda_{\alpha_{n+1}}(\tau(n/k), \omega_{n+1} | U_{\alpha_{n+1}}, n/k)(t) \quad \text{for } \frac{n}{k} \leqslant t \leqslant \frac{n+1}{k}$$

$$\cdots\cdots\cdots\cdots$$

The verification that Λ_μ is continuous and satisfies (36) is immediate.

IV. To end the proof, it remains to find a special lifting function $\lambda(x, \omega)$ defined in the whole set Q_0, using the general lifting functions Λ_μ. It may be assumed that the set S of all finite sequences μ is totally ordered. Consider an element $(x, \omega) \in Q_0$ and let

$$\mu_1 < \mu_2 < \cdots < \mu_k$$

be the indices μ such that $\omega \in W_\mu$; let

$$q_r(\omega) = \left(\sum_{i=1}^{r} f_{\mu_i}(\omega)\right) \Big/ \left(\sum_{i=1}^{k} f_{\mu_i}(\omega)\right) \quad \text{for } 1 \leqslant r \leqslant k$$

and define the path $\tau = \lambda(x, \omega)$ by

$$\tau(t) = \Lambda_{\mu_1}(x, \omega, 0)(t) \quad \text{for } 0 \leqslant t \leqslant q_1(\omega),$$

$$\tau(t) = \Lambda_{\mu_{i+1}}(\tau(q_i(\omega)), \omega, q_i(\omega))(t) \quad \text{for } q_i(\omega) \leqslant t \leqslant q_{i+1}(\omega).$$

The verification that λ is a special lifting function is again immediate.

C. The Homotopy Exact Sequence of a Fibration

In his first Note of 1935 [256] Hurewicz specialized his general results on the groups $\mathscr{C}(Y; G)$ (see above, section A) to the case $Y = S_n$, in order to obtain relations between the homotopy groups $\pi_n(G)$, $\pi_n(H)$, and $\pi_n(G/H)$. Suppose for simplicity that G is a connected compact Lie group and H is a closed subgroup of G; if H is not connected, $\pi_n(H)$ is, by definition, equal to $\pi_n(H_0)$, where H_0 is the neutral component of G. From the relation (39) of chap. II, §3,A and from chap. I, §4,C it follows that for the group law on

$$\Gamma = \mathscr{C}(S_n, *; G, e)$$

the homotopy class of the product fg in Γ is the product of the homotopy classes of $\tilde{f}$ and $\tilde{g}$ [chap. II, §3, formula (40)] in $\Omega(\mathscr{C}(S_{n-1}, *; G, e), a_n)$. By induction on n the group law on $\pi_n(G, e) = \pi_0(\Gamma) = \Gamma/\Gamma_0$ (Γ_0 being the neutral element of Γ) is the quotient of the group law of Γ. This was the first theorem stated (but not proved) by Hurewicz. Next he stated results that we would now express as the exactness of the sequence

$$\pi_n(H, e) \xrightarrow{j_*} \pi_n(G, e) \xrightarrow{p_*} \pi_n(G/H, \bar{e}_0) \xrightarrow{\partial} \pi_{n-1}(H, e).$$

This was the first appearance of the homotopy exact sequence of a fibration, in a special case. Although in his 1935 Notes Hurewicz did not define in general the homomorphism $f_*: \pi_n(X, x_0) \to \pi_n(Y, y_0)$ for a continuous map $f: (X, x_0) \to (Y, y_0)$ of pointed spaces, it is clear that the natural homomorphism

$$\tilde{j}: \mathscr{C}(S_n, *; H, e) \to \mathscr{C}(S_n, *; G, e)$$

sends each arcwise-connected component of the first group into a well-determined arcwise-connected component of the second, hence the map j_* in (38). The definition of p_* is obvious, and for us the equality $\operatorname{Ker} p_* = \operatorname{Im} j_*$ is a consequence of the covering homotopy property (see below), but the latter was not yet formulated in Hurewicz's Notes.

In their 1940 Note [260] Hurewicz and Steenrod did not explicitly formulate the exact homotopy sequence of a fibration (E, B, p) satisfying their axioms, but they derived from the covering homotopy property the main ingredient of that sequence, namely, the fact that, for a fiber $F = p^{-1}(b_0)$, the map p defines an *isomorphism* $p_*: \pi_n(E, F, x_0) \cong \pi_n(B, b_0)$ for any $x_0 \in F$. To prove injectivity of p_*, observe that if $g: (S_n, *) \to (B, b_0)$ is homotopic to the constant map $S_n \to \{b_0\}$, lifting that homotopy gives a continuous map $f: S_n \to E$ homotopic to a map $f_0: S_n \to F$ and such that $p \circ f = g$. To prove surjectivity, consider a loop γ in

$$\Omega(\mathscr{C}(S_{n-1}, *; B, b_0); a_n)$$

that can be considered a map

$$h: [0, 1] \times S_{n-1} \to B$$

with $h(t, *) = b_0$ for all $t \in [0, 1]$ and $h(0, x) = h(1, x) = b_0$ for $x \in S_{n-1}$. By the covering homotopy property, that map is lifted to a map

$$k: [0, 1] \times S_{n-1} \to E$$

such that $k(t, *) = x_0$ for all $t \in [0, 1]$ and $k(0, x) = x_0 \in F$, and $k(1, x) \in F$ for all $x \in S_{n-1}$. This can be considered a map $k: \mathbf{D}_n \to E$ with $k(S_{n-1}) \subset F$ and

$k(*) = x_0$; its homotopy class $u \in \pi_n(\mathrm{E}, \mathrm{F}, x_0)$ is therefore such that $p_*(u) = z$, the homotopy class of γ.

This result and the homotopy exact sequence [chap. II, § 5,B, formula (103)] give the *homotopy exact sequence of fibrations*

$$\cdots \to \pi_n(\mathrm{F}, x_0) \xrightarrow{j_*} \pi_n(\mathrm{E}, x_0) \xrightarrow{p_*} \pi_n(\mathrm{B}, b_0) \xrightarrow{\partial} \pi_{n-1}(\mathrm{F}, x_0)$$
$$\to \cdots \to \pi_0(\mathrm{F}) \xrightarrow{j_*} \pi_0(\mathrm{E}) \xrightarrow{p_*} \pi_0(\mathrm{B}) \to 0. \tag{38}$$

In the preceding notations $\partial(z)$ is the homotopy class of $k|\mathrm{S}_{n-1}$.

As special cases, if there is a *section* s: $\mathrm{B} \to \mathrm{E}$, it follows from the relation $p_* \circ s_* = \mathrm{Id}$ that

$$\pi_n(\mathrm{E}) \simeq \pi_n(\mathrm{B}) \oplus \pi_n(\mathrm{F}) \quad \text{for } n \geqslant 2.$$

If there is a homotopy in E of j: $\mathrm{F} \to \mathrm{E}$ to the constant map $\mathrm{F} \to \{x_0\}$, then $j_* = 0$ and

$$\pi_n(\mathrm{B}) \simeq \pi_n(\mathrm{E}) \oplus \pi_{n-1}(\mathrm{F}) \quad \text{for } n \geqslant 2. \tag{39}$$

In particular, if the total space E is *contractible*, it follows that

$$\pi_n(\mathrm{B}) \simeq \pi_{n-1}(\mathrm{F}) \quad \text{for } n \geqslant 2. \tag{40}$$

So, if B is an Eilenberg–Mac Lane space $\mathrm{K}(\Pi, n)$ for $n \geqslant 2$, F is then an Eilenberg–Mac Lane space $\mathrm{K}(\Pi, n-1)$.

Under some restrictions, there is also a "covering" theorem for relative homotopy. Let $(\mathrm{E}, \mathrm{B}, \mathrm{F}, \pi)$ be a fibration. Suppose (X, A) is a relative CW-complex of finite dimension; let g: $\mathrm{X} \to \mathrm{B}$ and f: $\mathrm{A} \to \mathrm{E}$ be two continuous maps such that $\pi \circ f$ is the restriction $g|\mathrm{A}$. Then if h: $\mathrm{X} \times [0, 1] \to \mathrm{B}$ is a homotopy relative to A and $h(x, 0) = g(x)$ in X, it can be lifted to a homotopy H: $\mathrm{X} \times [0, 1] \to \mathrm{E}$ relative to A and such that $\pi \circ \mathrm{H} = h$. This is easily seen by "climbing" along the skeletons of (X, A) ([254], p. 64) and applying the covering homotopy property at each step.

Application of this result gives a convenient criterion for the existence of a *section* over the whole base space B of a fibration $(\mathrm{E}, \mathrm{B}, \mathrm{F}, \pi)$. The assumptions are:

(i) The fiber F is $(n-1)$-connected (chap. II, § 4,B).
(ii) There is a closed subspace A of B such that (B, A) is a relative CW-complex of relative dimension $\leqslant n$.

Any section of E over A can then be extended to a section of E over the *whole* space B.

The proof relies first on the lemma in the proof of the second Whitehead theorem (chap. II, § 6,B). The homotopy exact sequence shows that π: $\mathrm{E} \to \mathrm{B}$ is an n-equivalence (chap. II, § 5,F). Let s: $\mathrm{A} \to \mathrm{E}$ be a section of E over A. Then the lemma establishes the existence of a continuous map s': $\mathrm{B} \to \mathrm{X}$ that extends s and is such that $\pi \circ s'$ is relatively homotopic to 1_{B} with respect to A. The preceding "relative" covering homotopy property provides a continuous map s'': $\mathrm{B} \to \mathrm{E}$ that coincides with s in A and is such that $\pi \circ s'' = 1_{\mathrm{B}}$.

D. Applications to Computations of Homotopy Groups

As with practically every new device in algebraic topology, the covering homotopy property of fibrations was harnessed to bring some progress in the persistent and thorny problems of computing homotopy groups, particularly homotopy groups of spheres.

In his first 1935 Note [256] Hurewicz already used his study of the homotopy properties of topological groups to give two simple proofs of the relation $\pi_3(\mathbf{S}_2) \simeq \mathbf{Z}$, which Hopf had obtained with great difficulty in 1930 (chap. II, § 1).

His first argument was based on his study of the group $\mathscr{C}(Y; G)$ for Lie groups G (see above, section A). Suppose G is a connected compact Lie group, and $K \neq G$ is a closed connected subgroup: let j: $K \to G$ be natural injection. Then the arcwise-connected component of 1_G in $\mathscr{C}(G; G)$ does not meet $\tilde{j}(\mathscr{C}(G; K))$. Otherwise, there would exist a continuous map g: $G \to K$ for which $j \circ g$ would be homotopic to 1_G, hence $g^* \circ j^* = 1_{H^{\cdot}(G)}$ in cohomology; but this is absurd, since, if $n = \dim G$, $\dim K < n$, and $H^n(G) \neq 0$ but $H^n(K) = 0$. From his general results on Lie groups Hurewicz concluded that the class $[p]$ of the natural surjection p: $G \to G/K$ cannot be 0. Taking in particular $G = \mathbf{S}_3$ (group of unit quaternions) and $K = \mathbf{S}_1$ yielded Hopf's theorem.

A second argument is even simpler: since $\pi_1(\mathbf{S}_3) = 0$, the natural injection $p^{-1}(x) \to \mathbf{S}_3$ of a fiber for p: $\mathbf{S}_3 \to \mathbf{S}_2$, is homotopic in $\mathbf{S}_3$ to a constant map; hence relation (39) gives

$$\pi_n(\mathbf{S}_2) \simeq \pi_n(\mathbf{S}_3) \oplus \pi_{n-1}(\mathbf{S}_1) \quad \text{for } n \geqslant 2 \tag{41}$$

and in particular

$$\pi_3(\mathbf{S}_2) \simeq \pi_3(\mathbf{S}_3) \simeq \mathbf{Z},$$

and the map p_* is a generator of $\pi_3(\mathbf{S}_2)$. When Freudenthal later proved that $\pi_4(\mathbf{S}_3) \simeq \mathbf{Z}/2\mathbf{Z}$ [chap. II, § 6,E, formula (134)], relation (41) also gave

$$\pi_4(\mathbf{S}_2) \simeq \mathbf{Z}/2\mathbf{Z}. \tag{42}$$

The same argument applies to the two other Hopf fibrations $\mathbf{S}_7 \to \mathbf{S}_4$ and $\mathbf{S}_{15} \to \mathbf{S}_8$ (§ 1,B), the fibers being homeomorphic to $\mathbf{S}_3$ in the first case and to $\mathbf{S}_7$ in the second, so that

$$\pi_n(\mathbf{S}_4) \simeq \pi_n(\mathbf{S}_7) \oplus \pi_{n-1}(\mathbf{S}_3), \tag{43}$$

$$\pi_n(\mathbf{S}_8) \simeq \pi_n(\mathbf{S}_{15}) \oplus \pi_{n-1}(\mathbf{S}_7). \tag{44}$$

Stiefel (and independently Whitney [505]) had defined the Stiefel manifolds $\mathbf{S}_{n,p}$ (Part 2, chap. V, § 4,C) to use their homological properties in their theory of characteristic classes (chap. IV, § 1). However, for his criteria on existence of systems of tangent vector fields (*ibid.*), Stiefel also needed homotopy properties of the $\mathbf{S}_{n,p}$. Using his decomposition $\mathbf{S}_{n,p} = \mathbf{S}'_{n,p} \cup \mathbf{S}''_{n,p}$ (Part 2, chap. V, § 4,C), in order to apply induction on n, he had to investigate the possibility of extending a continuous map $\mathbf{S}_r \to \mathbf{S}_{n,p}$ to the ball $\mathbf{D}_{r+1}$. He could not mention homotopy groups, since Hurewicz's Notes were published at the same time as Stiefel's paper; but the theorems he proved in a direct way amount to the computation of the homotopy groups $\pi_r(\mathbf{S}_{n,p})$; using his determination of the homology groups $\mathrm{H}_r(\mathbf{S}_{n,p})$ for $r \leqslant n - p$ (*loc. cit.*), he showed that for these values of r, $\pi_r(\mathbf{S}_{n,p}) \simeq \mathrm{H}_r(\mathbf{S}_{n,p})$ [of course this amounts to a special case of the absolute Hurewicz isomorphism theorem (chap. II, § 4,A)].

Eckmann [149] and independently J.H.C. Whitehead [498] took up the computation of the groups $\pi_r(\mathbf{S}_{n,p})$ by other means. Whitehead used the structure of CW-complex that he had defined (Part 2, chap. V, § 3,C) while Eckmann applied the fact that by definition $\mathbf{S}_{n,p}(\mathrm{F})$ (for $\mathrm{F} = \mathbf{R}$, $\mathbf{C}$ or $\mathbf{H}$) is a homogeneous space $\mathrm{U}_n(\mathrm{F})/\mathrm{U}_{n-p}(\mathrm{F})$ for the unitary group $\mathrm{U}_n(\mathrm{F})$. From the fact that the grassmannian $\mathrm{G}_{n,p}(\mathrm{F})$ is homeomorphic to the homogeneous space $\mathrm{U}_n(\mathrm{F})/(\mathrm{U}_p(\mathrm{F}) \times \mathrm{U}_{n-p}(\mathrm{F}))$, it follows that $\mathbf{S}_{n,p}(\mathrm{F})$ is a *principal fiber space* with structural group $\mathrm{U}_p(\mathrm{F})$ and base space $\mathrm{G}_{n,p}(\mathrm{F})$. Another fibration stems from the inclusions $\mathrm{U}_n(\mathrm{F}) \supset \mathrm{U}_{n-1}(\mathrm{F}) \supset \mathrm{U}_{n-p}(\mathrm{F})$: applying (§ 1,A, IV a), $\mathbf{S}_{n,p}(\mathrm{F})$ appears as a fiber space with typical fiber $\mathbf{S}_{n-1,p-1}(\mathrm{F})$ and base space $\mathbf{S}_{n-1}$ when $\mathrm{F} = \mathbf{R}$ (resp. $\mathbf{S}_{2n-1}$ when $\mathrm{F} = \mathbf{C}$ and $\mathbf{S}_{4n-1}$ when $\mathrm{F} = \mathbf{H}$).

It is this last fibration that was used by Eckmann for $\mathrm{F} = \mathbf{R}$, *via* the homotopy exact sequence

$$\cdots \to \pi_{i+1}(\mathbf{S}_{n-1}) \to \pi_i(\mathbf{S}_{n-1,p-1}) \to \pi_i(\mathbf{S}_{n,p}) \to \pi_i(\mathbf{S}_{n-1}) \to \cdots. \tag{45}$$

For $p = 1$, $\mathbf{S}_{n,1} = \mathbf{S}_n$, and $\pi_i(\mathbf{S}_n) = 0$ for $2 \leqslant i \leqslant n - 1$. Induction on n then gives back Stiefel's results for $1 \leqslant i \leqslant n - p$; for $i = n - p + 1$, Eckmann used the determination of $\pi_{n+1}(\mathbf{S}_n)$ due to Hopf and Freudenthal (chap. II, § 6,E). Freudenthal had also proved that $\pi_4(\mathbf{S}_2) \simeq \mathbf{Z}/2\mathbf{Z}$, and in 1938 Pontrjagin had announced, without a complete proof, that $\pi_{n+2}(\mathbf{S}_n) = 0$ for $n \geqslant 3$; Eckmann and Whitehead used these values to compute $\pi_{n-p+2}(\mathbf{S}_{n,p})$. Their results had to be corrected when later G. Whitehead [487] and Pontrjagin himself [382] showed that in fact $\pi_{n+2}(\mathbf{S}_n) = \mathbf{Z}/2\mathbf{Z}$ for all values of $n \geqslant 2$.

The first homotopy groups of the classical quasisimple compact Lie groups $\mathrm{SU}(n, \mathrm{F})$ can also be deduced from the fibrations

$$\begin{cases} \mathrm{SO}(n+1)/\mathrm{SO}(n) = \mathbf{S}_n & \text{for } n \geqslant 1, \\ \mathrm{SU}(n+1)/\mathrm{SU}(n) = \mathbf{S}_{2n+1} & \text{for } n \geqslant 1, \\ \mathrm{SU}(n+1, \mathbf{H})/\mathrm{SU}(n, \mathbf{H}) = \mathbf{S}_{4n+3} & \text{for } n \geqslant 0. \end{cases} \tag{46}$$

Use of the homotopy exact sequence shows that

$$\begin{cases} \pi_k(\mathrm{SO}(m)) = \pi_k(\mathrm{SO}(n)) & \text{for } m, n \geqslant k + 2, \\ \pi_k(\mathrm{SU}(m)) = \pi_k(\mathrm{SU}(n)) & \text{for } m, n \geqslant k/2, \\ \pi_k(\mathrm{SU}(m, \mathbf{H})) = \pi_k(\mathrm{SU}(n, \mathbf{H})) & \text{for } m, n \geqslant (k-2)/4. \end{cases} \tag{47}$$

In particular the fundamental groups of $\mathrm{SU}(n)$ and $\mathrm{SU}(n, \mathbf{H})$ are 0 for all $n \geqslant 1$ and those of $\mathrm{SO}(n)$ are equal to $\mathbf{Z}/2\mathbf{Z}$ for $n \geqslant 3$; the second homotopy groups $\pi_2(\mathrm{G})$ are 0 for all these groups [103]. The determination of $\pi_3(\mathrm{G})$,

$\pi_4(G)$, and $\pi_5(G)$, in addition to the homotopy exact sequence, needs special arguments using geometrical properties of spheres; they were developed by Pontrjagin, Eckmann, and J.H.C. Whitehead ([55], p. 426).

Finally, the homotopy exact sequence gave the first (negative) results on the possibility of considering spheres as fiber spaces in cases other than the Hopf fibrations [485].

E. Classifying Spaces I: The Whitney–Steenrod Theorems

In his 1935 Note [505] Whitney had already put in the forefront the problem of classifying the "sphere bundles" with given base space and given typical fiber, up to isomorphism. In his 1937 address [508] he introduced a new idea. Without giving precise definitions he considered "the space $S[\mu, \nu]$ of ν-great spheres in the sphere $\mathbf{S}_\mu$" and sketched a not very clear proof (restricted to the case $\nu = 1$) that for any locally finite simplicial complex B of finite dimension, any sphere bundle (E, B, p) with base space B and typical fiber $\mathbf{S}_\nu$, is B-isomorphic to the *pullback* of a sphere bundle $S[\mu, \nu]$ for $\mu = \nu + \dim B$, by a suitable continuous map of B into the base space of $S[\mu, \nu]$.

In 1943 [446] Steenrod took up the idea, supplied the complete proof of Whitney's assertion, and completed it by a fundamental criterion for the existence of a B-isomorphism between two pullbacks of the same $S[\mu, \nu]$, with base space B. The same results were obtained independently by Pontrjagin [379].

Although Whitney did not give any precise definition in [508], it is clear that he intended $S[\mu, \nu]$ to be a sphere bundle with typical fiber $\mathbf{S}_\nu$ and base space equal to the grassmannian $G_{\mu+1,\nu+1}$, having as elements the vector subspaces of $\mathbf{R}^{\mu+1}$ generated by the "ν-great spheres"; the projection π: $S[\mu, \nu] \to G_{\mu+1,\nu+1}$ maps each such "great sphere" to the vector subspace it generates.

In his paper Steenrod introduced a *principal fiber space*, which he denotes $N_\nu^{\mu-\nu}$; in fact it is identical to the Stiefel manifold $\mathbf{S}_{\mu+1,\nu+1}$, although he makes no mention of Stiefel. Since a sequence of p mutually orthogonal vectors in $\mathbf{S}_n$ uniquely determines a map $\mathbf{S}_p \to \mathbf{S}_n$ preserving orthogonality, it is as the set of such maps that Steenrod defined $\mathbf{S}_{n,p}$. His definition of Whitney's sphere bundle $S[\mu, \nu]$ then amounted to considering it the fiber bundle *associated* (§1,C,IV) to the principal fiber space $\mathbf{S}_{\mu+1,\nu+1}$ of base space $G_{\mu+1,\nu+1}$ and structure group the orthogonal group $O(\nu + 1)$, for the natural action of that group on $\mathbf{S}_\nu$. The space $S[\mu, \nu]$ can be identified with the image of the map

$$\psi: \mathbf{S}_{\mu+1,\nu+1} \times \mathbf{S}_\nu \to G_{\mu+1,\nu+1} \times \mathbf{S}_\mu \tag{48}$$

defined by

$$(w, y) \mapsto (\pi(w), w(y))$$

where $\pi(w)$ is the vector subspace of $\mathbf{R}^{\mu+1}$ generated by $w(\mathbf{S}_\nu)$.

The proof of Whitney's assertion then proceeded as follows. First Steenrod established as a lemma the fact that $\pi_i(\mathbf{S}_{n,m}) = 0$ for $i < n - m$ without mentioning the previous proofs (analyzed above in E). Not even Eckmann's proof in the paper [149] is listed in Steenrod's Bibliography!

A morphism $(f, g)\colon (\mathrm{E}, \mathrm{B}, p) \to (\mathrm{E}', \mathrm{B}', p')$ of "regular sphere bundles" in the sense of Whitney is submitted to the condition that for every $b \in \mathrm{B}$, the map $f_b\colon \mathrm{E}_b \to \mathrm{E}'_{g(b)}$ preserves orthogonality [it corresponds to a morphism of the vector bundles deduced from the given sphere bundles (§ 1,C, III)]. To show that there is a B-isomorphism of $(\mathrm{E}, \mathrm{B}, p)$ onto a pullback $g^*(\mathrm{S}[\mu, \nu])$, it is enough to prove the existence of a continuous map $f\colon \mathrm{E} \to \mathbf{S}_\mu$ such that for every $b \in \mathrm{B}$, the restriction $f_b\colon \mathrm{E}_b \to \mathbf{S}_\mu$ is injective, preserves orthogonality, and maps E_b onto a "ν-great sphere" of $\mathbf{S}_\mu$.

Steenrod's method was to define, for each $b \in \mathrm{B}$, the map f_b as a composite

$$f_b\colon \mathrm{E}_b \xrightarrow{h_b} \mathbf{S}_{\mu+1,\nu+1} \times \mathbf{S}_\nu \xrightarrow{\psi} \mathbf{G}_{\mu+1,\nu+1} \times \mathbf{S}_\mu \xrightarrow{\mathrm{pr}_2} \mathbf{S}_\mu$$

with $h_b(y) = (\omega_b, \rho_b(y))$ for $y \in \mathrm{E}_b$, where ρ_b is a bijection $\mathrm{E}_b \simeq \mathbf{S}_\nu$ preserving orthogonality and ω_b is an element of $\mathbf{S}_{\mu+1,\nu+1}$ independent of y. Each f_b then has the required properties, and the crux of the proof is to show that ω_b and ρ_b may be chosen in such a way that the map h coinciding with h_b on each E_b is *continuous in* E. Then $g\colon \mathrm{B} \to \mathbf{G}_{\mu+1,\nu+1}$ is defined as the map that associates to $b \in \mathrm{B}$ the vector subspace of $\mathbf{R}^{\mu+1}$ generated by $f_b(\mathrm{E}_b)$; g is continuous, and it follows from the "universal" property of a pull-back (§ 1,C, I) that the natural factorization of

$$(\psi \circ h, g)\colon (\mathrm{E}, \mathrm{B}, p) \to (\mathrm{S}[\mu, \nu], \mathbf{G}_{\mu+1,\nu+1}, \pi)$$

defines a B-isomorphism of $(\mathrm{E}, \mathrm{B}, p)$ onto $g^*(\mathrm{S}[\mu, \nu])$.

By suitable subdivision it may be assumed that E is trivializable above a neighborhood U_σ of $\bar{\sigma}$ for each simplex σ of B. There is therefore a homeomorphism for each σ

$$\varphi_\sigma\colon p^{-1}(\bar{\sigma}) \simeq \bar{\sigma} \times \mathbf{S}_\nu$$

with $\varphi_\sigma(p^{-1}(b)) = \{b\} \times \mathbf{S}_\nu$. For every simplex τ contained in $\bar{\sigma} - \sigma$, and any $b \in \tau$, the map

$$\theta_{\sigma\tau}(b)\colon x \mapsto \mathrm{pr}_2(\varphi_\tau(\varphi_\sigma^{-1}(b, x)))$$

is an element of the orthogonal group $\mathrm{O}(\nu + 1)$, and $b \mapsto \theta_{\sigma\tau}(b)$ is continuous in $\mathrm{U}_\sigma \cap \mathrm{U}_\tau$.

For each simplex σ of B Steenrod defined a continuous map $\lambda_\sigma\colon \bar{\sigma} \to \mathbf{S}_{\mu+1,\nu+1}$ by induction on the dimension of σ in such a way that it verifies the condition that if τ is a simplex of B contained in $\bar{\sigma} - \sigma$ and $b \in \tau$, then

$$\lambda_\sigma(b) = \lambda_\tau(b) . \theta_{\sigma\tau}(b) \tag{49}$$

so that $\lambda_\sigma(b)(\mathbf{S}_\nu) = \lambda_\tau(b)(\mathbf{S}_\nu)$. The induction then consists in taking arbitrary λ_v for the vertices v of B; once the λ_τ are defined for all simplices $\tau \subset \bar{\sigma} - \sigma$, condition (49) shows that together they define a continuous map of $\bar{\sigma} - \sigma$ into $\mathbf{S}_{\mu+1,\nu+1}$; if the dimension of B is $< \mu - \nu$, that map can be extended continuously in the interior σ of $\bar{\sigma}$ [although when $b \in \sigma$ tends to a point $b' \in \tau$, $\lambda_\sigma(b)$ does not tend to $\lambda_\tau(b')$ in general].

Once the $\lambda_\sigma^{\bullet}$ have been defined for all simplices in B, ω_b and $\rho_b(y)$ are defined for $b \in \sigma$ and $y \in E_b$ by

$$\omega_b = \lambda_\sigma(b), \qquad \rho_b(y) = \lambda_\sigma(b).\mathrm{pr}_2(\varphi_\sigma(y))$$

[ρ_b determines $\lambda_\sigma(b)$]; h is then defined as the map

$$y \mapsto (\omega_{p(y)}, \rho_{p(y)}(y));$$

its continuity in E is easily verified.

Steenrod added to Whitney's theorem the fundamental fact that if $\mu > \nu + 1 + \dim(B)$ and two continuous maps g', g'' of B into $G_{\mu+1,\nu+1}$ give B-*isomorphic* pullbacks (E', B, p') and (E'', B, p'') of $S[\mu, \nu]$, then g' and g'' are *homotopic*. The proof uses the product $E = E' \times [0,1]$ which may be considered as a sphere-bundle over $B \times [0,1]$, the fiber $E_{(b,t)}$ being the image of E'_b by the homeomorphism $y \mapsto (y,t)$. Steenrod showed that $(E, B \times [0,1], p)$ is isomorphic to a pullback $h^*(S[\mu,\nu])$ by a continuous map $h: B \times [0,1] \to G_{\mu+1,\nu+1}$ such that

$$h(b,0) = g'(b), \qquad h(b,1) = g''(b) \quad \text{for } b \in B, \tag{50}$$

thus establishing the required homotopy.

Let $(1_B, v): (E', B, p') \cong (E'', B, p'')$ be the given B-*isomorphism*. Denote by ρ'_b the map $E'_b \to S_\nu$ corresponding to the pullback of $S[\mu,\nu]$ by g', by ρ''_b the similar map $E''_b \to S_\nu$. Having taken a fine enough subdivision of the complex $B \times [0,1]$ [of dimension $1 + \dim(B)$], such that $B \times \{0\}$ and $B \times \{1\}$ are subcomplexes, Steenrod proceeded as in the previous proof, with the following difference: when a simplex σ of $B \times [0,1]$ is contained in $B \times \{0\}$ or in $B \times \{1\}$, the maps $\rho_{(b,0)}$ and $\rho_{(b,1)}$ are not constructed by induction on the dimension of σ but are already *determined* in σ, namely:

(i) if $\sigma \subset B \times \{0\}$, $\rho_{(b,0)}$ is the map $(y,0) \to \rho'_b(y)$;
(ii) if $\sigma \subset B \times \{1\}$, $\rho_{(b,1)}$ is the composite

$$(y,1) \mapsto \rho''_b(v_b(y)).$$

The remainder of the construction is the same, and the map h thus obtained satisfies (50).

Finally, Steenrod proved a *converse* to the preceding result: if the simplicial complex B is *compact*, and two continuous maps g_0, g_1 of B into $G_{\mu+1,\nu+1}$ are homotopic, then the pullbacks $g_0^*(S[\mu,\nu])$ and $g_1^*(S[\mu,\nu])$ are B-isomorphic. It is enough to show that there is a number $\varepsilon > 0$ such that the conclusion holds as soon as

$$\sup_{b \in B} d(g_0(b), g_1(b)) \leqslant \varepsilon \tag{51}$$

for a distance d on $G_{\mu+1,\nu+1}$. Then, if $h: B \times [0,1] \to G_{\mu+1,\nu+1}$ is a homotopy between two maps g_0, g_1, the compactness of B allows $[0,1]$ to be divided in small intervals $[t_i, t_{i+1}]$ such that (51) holds for any pair $h(.,t_i)$, $h(.,t_{i+1})$.

The result under assumption (51), however, follows from the geometry of the Stiefel manifold $S_{\mu+1,\nu+1}$. More generally, consider a homogeneous space

G/H, where G is a compact Lie group and H is a closed subgroup. The exponential map then shows that, for the principal fiber space $(G, G/H, H, \pi)$, there is a neighborhood V of the point $x_0 = \pi(H)$ of G/H, and a section φ of G over V such that $\varphi(hxh^{-1}) = \varphi(x)$ for $h \in H$ and $x \in V$. Then, for $x_1 = g.x_0$ and $x_2 \in g.V$ for a $g \in G$, define

$$r(x_1, x_2) = g^{-1}\varphi(g.x_2)g.$$

This is independent of the choice of g such that $x_1 = g.x_0$ and arbitrarily close to e when V is taken small enough.

Apply this to $G = O(\mu + 1)$, $H = O(\nu + 1) \times O(\mu - \nu)$; observe that G acts naturally (on the left) on the Stiefel manifold $S_{\mu+1,\nu+1}$, hence also on $S[\mu, \nu]$. Define the map ψ of $B \times S[\mu, \nu]$ into itself by

$$\psi(b, y) = (b, r(g_0(b), g_1(b)).y);$$

Then, if ε is taken small enough, so that $g_1(b)$ belongs to $g_0(b).V$ for all $b \in B$, ψ defines a B-isomorphism of $g_0^*(S[\mu, \nu])$ onto $g_1^*(S[\mu, \nu])$.

These theorems of Steenrod showed that for given $\mu > \nu \geqslant 1$, and *any* simplicial complex B of dimension $\leqslant \mu - 1$, there is a canonical *bijection* of the set of classes of B-isomorphic sphere bundles with base space B and fibers of dimension ν, onto the set $[B; G_{\mu+1,\nu+1}]$ of homotopy classes, associating to the homotopy class of a map $g: B \to G_{\mu+1,\nu+1}$ the B-isomorphism class of the sphere bundle $g^*(S[\mu, \nu])$; $G_{\mu+1,\nu+1}$ is called the *classifying space* for the sphere bundles under consideration and $S[\mu, \nu]$ the corresponding *universal bundle.*

The sphere-bundles with base space B thus depend only on the *homotopy type* of B, up to B-isomorphism; in particular, if B is contractible, every sphere-bundle with base space B is trivializable, a result proved directly in 1939 by Feldbau [196], but already known to Whitney in 1937 ([508], p. 788).

The Whitney–Steenrod theorems immediately extend to the category of pointed spaces; the point associated to $S_{\mu+1,\nu+1}$ (resp. $G_{\mu+1,\nu+1}$) is then the image of the neutral element of $O(\mu + 1)$, and the point associated to $S[\mu, \nu]$ is the image of the pair of points associated to $S_{\mu+1,\nu+1}$ and S_ν, respectively.

F. Classifying Spaces: II. Later Improvements

The preceding results soon began to be studied and applied in several contexts in algebraic and differential topology, and were extended to more general situations, in which the base spaces are not necessarily simplicial complexes.

If P is a principal fiber space with base space B and structural group G, and F is a space on which G operates, there is a natural B′-isomorphism

$$g^*(P \times {}^G F) \simeq g^*(P) \times {}^G F \tag{52}$$

for any continuous map $g: B' \to B$. This reduces the problem of classifying fiber spaces to the *principal* fiber spaces (at least for those bundles that can be associated to a principal fiber space, such as vector bundles or sphere bundles). Most papers therefore were mainly concerned with principal fiber spaces.

For the family of principal fiber spaces (P, B, G, p) with *fixed* structural group G, the search for classifying spaces is part of the more general study of the *existence* of morphisms

$$(f, g): (P, B, G, p) \to (P', B', G, p') \tag{53}$$

for two such fiber spaces, since when such a morphism g exists, P is automatically B-isomorphic to $g^*(P')$.

These morphisms are in one to one correspondence with the *sections* over B of a fiber space associated with P, by the following argument [423]. The group G may be considered as operating on P′ on the *left*, by $(s, x') \mapsto x' . s^{-1}$. Consider the associated fiber space

$$E = P \times {}^G P' \tag{54}$$

with base space B. We have seen (§1,C) that sections φ of E over B are in one-to-one correspondence with continuous maps $f: P \to P'$ such that

$$f(x . s) = s^{-1} . f(x) = f(x) . s$$

for $x \in P$, $s \in G$, and these are exactly the maps f defining a morphism (53), since g is then defined by the relation

$$g(p(x)) = p'(f(x));$$

recall that f is obtained from φ by the relation

$$x . f(x) = \varphi(p(x)) \quad \text{for } x \in P$$

[*loc. cit.*, formula (24)].

We may therefore apply the criterion for existence of a section over B, described in §2,C, to the following situation:

The base space B is a CW-*complex of dimension* $\leqslant n$, and the principal fiber space P′ is *n-connected.*

This yields three general results, of which the Whitney–Steenrod theorems are very special cases:

I. The criterion applied to E defined by (54) shows that there always exist exist morphisms (53), generalizing Whitney's theorem.

II. Let $h: B \times [0, 1] \to B'$ be a homotopy between two maps g_0, g_1 of B into B′. Consider the principal fiber space $g_0^*(P') \times [0, 1]$ with base space $B \times [0, 1]$; the criterion establishes the existence of a morphism

$$(f, h): (g_0^*(P') \times [0, 1], B \times [0, 1], G, (\pi, \mathrm{Id})) \to (P', B', G, p')$$

which *extends* the natural morphism

$$(g_0^*(P'), B, G, \pi) \to (P', B', G, p').$$

Therefore $g_0^*(P') \times [0, 1]$ is $(B \times [0, 1])$-isomorphic to $h^*(P')$. Let $j: B \to B \times [0, 1]$ be defined by $j(b) = (b, 1)$; then $g_1 = h \circ j$, so that

$$g_1^*(P') = j^*(h^*(P')) \cong j^*(g_0^*(P') \times [0,1]).$$

However, the restriction of $g_0^*(P') \times [0,1]$ to $B \times \{1\}$ is isomorphic to $g_0^*(P')$ [more precisely, it can be written $i^*(g_0^*(P'))$, where $i: (b,1) \mapsto b$ is a homeomorphism of $B \times \{1\}$ onto B]; so $g_0^*(P')$ and $g_1^*(P')$ are isomorphic. This is independent of any assumption on the homotopy groups of P′ ([459], p. 200).

III. Conversely, suppose two maps g_0, g_1 of B into B′ are such that $g_0^*(P')$ and $g_1^*(P')$ are B-isomorphic. This can be interpreted as meaning that there is a *section* of the fiber space

$$((g_0^*(P') \times^G P') \times [0,1], B \times [0,1], (\pi, \mathrm{Id}))$$

over the subspace $(B \times \{0\}) \cup (B \times \{1\})$ of $B \times [0,1]$, corresponding to the natural morphism $g_0^*(P') \to P'$ over $B \times \{0\}$ and to the composite of the given B-isomorphism $g_0^*(P') \simeq g_1^*(P')$ and the natural morphism $g_1^*(P') \to P'$ over $B \times \{1\}$.

The criterion on the extension of sections proves that this section can be extended to a section over the whole base space $B \times [0,1]$, hence there is a morphism

$$(f, h): (P \times [0,1], B \times [0,1], (p, \mathrm{Id})) \to (P', B', p')$$

which extends both morphisms corresponding to the maps g_0 and g_1; thus h is a homotopy between g_0 and g_1.

These results show that classifying spaces for a principal fiber space (P, B, G, p) whose base space B is a CW-complex of dimension $\leqslant n$ are found by constructing a principal fiber space (P', B', G, p') for which $\pi_i(P') = 0$ for $1 \leqslant i \leqslant n$; such a fiber space is called *n-universal* and its base space B′ *n-classifying*. When $G = U(k, F)$, with the field $F = \mathbf{R}$, $\mathbf{C}$, or $\mathbf{H}$, the Stiefel manifolds $\mathbf{S}_{n+k+1,k}(F)$ are n-universal and their base spaces $G_{n+k+1,k}(F)$ are n-classifying.

The necessity of limiting the dimension of the base space B for these classifying spaces is often a nuisance in the applications. This was improved by the process of taking the *direct limit* of the n-classifying spaces [a process already used in the definition of the most general CW-complexes (Part 2, chap. V, § 3,C), and in the construction of the Eilenberg–Mac Lane spaces (chap. II, § 6,F)]. In general, let $Y = \bigcup_n Y_n$ be a union of an increasing sequence $Y_1 \subset Y_2 \subset \cdots \subset Y_n \subset \cdots$ of subsets; suppose each Y_n is equipped with a topology $\mathscr{T}_n$, such that $\mathscr{T}_n$ induces $\mathscr{T}_{n-1}$ on Y_{n-1} for each $n \geqslant 2$. Then the *direct limit* $\mathscr{T}$ of the $\mathscr{T}_n$ is the *finest* topology on Y that induces $\mathscr{T}_n$ on each Y_n; its open sets are the sets U such that $U \cap Y_n$ is open in Y_n for each n; the space obtained by taking on Y the topology $\mathscr{T}$ is the *direct limit* of the spaces Y_n, sometimes written $Y = \varinjlim_n Y_n$.

The natural injections

$$\mathbf{R}^{k(n+k)} \to \mathbf{R}^{k(n+k+1)}$$

determine injections

$$G_{n+k,k} \to G_{n+k+1,k} \quad (\text{resp. } \mathbf{S}_{n+k,k} \to \mathbf{S}_{n+k+1,k})$$

and these spaces and their union

$$\mathrm{BO}(k) \quad [\text{resp. } \mathrm{EO}(k)]$$

satisfy the preceding conditions, so that on BO(k) [resp. EO(k)] the direct limit topologies can be taken. It can be proved that EO(k) is a principal fiber space with base space BO(k) and structural group O(k); EO(k) is universal and BO(k) is classifying for *every* principal fiber space with structural group O(k) and base space *any* CW-complex. Using the complex (resp. quaternionic) Stiefel manifolds in a similar way, we obtain universal principal fiber spaces EU(k) [resp. EU(k, **H**)] with base space BU(k) [resp. BU(k, **H**)] and structural group U(k) [resp. U(k, **H**)].

Since any compact Lie group G can be embedded as a subgroup of some U(k), there exists a classifying space B_G and a universal principal fiber space E_G with structural group G and base space B_G; it can be shown that E_G is *contractible.*

G. Classifying Spaces: III. The Milnor Construction

In 1956 [339] Milnor invented a new method for giving a classifying space and a universal principal fiber space for principal fiber spaces (P, B, G, π) where G is *any* topological group, and the only assumption is:

(D) There is a sequence $(u_n)_{n\geqslant 0}$ of continuous maps $B \to [0, 1]$ such that the open sets $U_n = u_n^{-1}(]0, 1])$ form an open covering of B and P is trivializable over each U_n. Such fiber spaces are sometimes called *numerable.*

The universal fiber space E_G is defined as the "infinite join"

$$G * G * \cdots * G * \cdots$$

and may be considered a direct limit: simply define the finite join $A_1 * A_2 * \cdots * A_n$ as for $n = 2$ in Part 2, chap. V, § 2,E, and then by induction on n as $(A_1 * A_2 * \cdots * A_{n-1}) * A_n$. It is just as simple to give direct definitions.

Consider the set $G^{\mathbf{N}} \times \Delta_\infty$ of infinite sequences

$$(\mathsf{x}, \mathsf{t}) = ((x_0, t_0), (x_1, t_1), \ldots, (x_n, t_n), \ldots), \tag{55}$$

where the x_i are *arbitrary* elements of G and $(t_n)_{n\geqslant 0}$ is a sequence of numbers $t_n \geqslant 0$ that have only a *finite* number of terms $t_n \neq 0$ and satisfy $\sum_n t_n = 1$ (so that Δ_∞ may be considered the direct limit of the standard simplices Δ_n). Consider in that set the equivalence relation

$$(\mathsf{x}, \mathsf{t}) \equiv (\mathsf{x}', \mathsf{t}') \tag{56}$$

which means that $\mathsf{t}' = \mathsf{t}$ and $x_i' = x_i$ for all indices *such that* $t_i \neq 0$. The quotient set of $G^{\mathbf{N}} \times \Delta_\infty$ by that relation is written E_G, and $\langle \mathsf{x}, \mathsf{t}\rangle$ denotes the equivalence class of (x, t). The group G *acts* on E_G by

$$\langle \mathsf{x}, \mathsf{t}\rangle . y = \langle \mathsf{x} . y, \mathsf{t}\rangle \quad \text{for } y \in G, \text{ with } \mathsf{x} . y = (x_0 y, x_1 y, \ldots, x_n y, \ldots). \tag{57}$$

To define a topology on E_G, consider, for each $i \geqslant 0$, the maps

$$\tau_i\colon E_G \to [0,1] \quad \text{defined by } \tau_i(\langle x, t\rangle) = t_i; \tag{58}$$

$$\xi_i\colon \tau_i^{-1}(]0,1]) \to G \quad \text{defined by } \xi_i(\langle x, t\rangle) = x_i \tag{59}$$

[by definition of the equivalence relation (56) they make sense]. The topology on E_G is the *coarsest* for which all functions τ_i and ξ_i are continuous. It is clear that for any $a \in E_G$ and any $y \in G$,

$$\tau_i(a\,.\,y) = \tau_i(a) \quad \text{and} \quad \xi_i(a\,.\,y) = \xi_i(a)y, \tag{60}$$

and this shows that the action of G on E_G is continuous; it is without fixed point.

Next consider the space of orbits $B_G = E_G/G$ and the projection $p\colon E_G \to B_G$. First prove that each point of B_G has a neighborhood over which E_G has a section. Consider the sets $\tau_i^{-1}(]0,1])$, that are open in E_G, invariant under G, and cover E_G; then the $V_i = p(\tau_i^{-1}(]0,1]))$ form an open covering of B_G. In the open set $\tau_i^{-1}(]0,1])$, the map

$$s_i'\colon a \mapsto a\,.\,\xi_i(a)^{-1} \tag{61}$$

is continuous, and for every $y \in G$

$$s_i'(a\,.\,y) = (a\,.\,y)\,.\,(\xi_i(a\,.\,y))^{-1} = (a\,.\,y)\,.\,(y^{-1}\xi_i(a)^{-1}) = s_i'(a),$$

so $s_i' = s_i \circ p$, where s_i is a section of E_G over V_i.

Let $R \subset E_G \times E_G$ be the set of pairs (u, v) belonging to the same orbit, so that $v = u\,.\,\rho(u,v)$ with $\rho(u,v) \in G$. In each set $(p^{-1}(V_i) \times p^{-1}(V_i)) \cap R$ the map ρ is continuous, because in that set $\rho(u,v) = \xi_i(u)^{-1}\xi_i(v)$. From this and the existence of the section s_i, E_G is a *locally trivial* fiber space with base space B_G; indeed

$$\varphi_i(z, y) = s_i(z)\,.\,y \tag{62}$$

is a continuous map of $V_i \times G$ onto $p^{-1}(V_i)$, and

$$u \mapsto (p(u), \rho(s_i(p(u)), u)) \tag{63}$$

is the inverse of φ_i.

Finally, let (P, B, G, π) be any numerable principal fiber space with structural group G. By assumption (D) there is a homeomorphism

$$h_n\colon U_n \times G \xrightarrow{\sim} \pi^{-1}(U_n)$$

for each n, such that for $(b, s) \in U_n \times G$,

$$\pi(h_n(b,s)) = b \quad \text{and} \quad h_n(b,s)\,.\,s' = h_n(b, ss') \quad \text{for } s' \in G.$$

A continuous map $f\colon P \to E_G$ is then defined by taking $f(z)$ equal to the equivalence class of

$$((\mathrm{pr}_2(h_0^{-1}(z)), u_0(\pi(z))), \ldots, (\mathrm{pr}_2(h_n^{-1}(z)), u_n(\pi(z))), \ldots)$$

with the convention that, when $\pi(z) \notin U_n$ [so that $h_n^{-1}(z)$ is not defined, but

$u_n(\pi(z)) = 0$] the meaningless element $\mathrm{pr}_2(h_n^{-1}(z))$ is replaced by the neutral element e. It is readily verified that $f(z.s) = f(z).s$ for $s \in G$, so that the map g: $B \to B_G$ can be defined by $g(\pi(z)) = p(f(z))$, and

$$(f, g): (P, B, G, \pi) \to (E_G, B_G, G, p)$$

is a morphism of principal fiber spaces.

H. The Classification of Principal Fiber Spaces with Base Space S_n

As a consequence of his theorem on the triviality of fiber spaces with a contractible base space Feldbau [194] reduced the classification of principal fiber spaces (P, S_n, G, p) with given base space a *sphere* S_n (for $n \geqslant 2$) to a problem in homotopy theory (see [450]).

His starting point was the fact that there is a covering of S_n by two open contractible sets, V_1 defined by $x_{n+1} > -\frac{1}{2}$ and V_2 defined by $x_{n+1} < \frac{1}{2}$. Any principal fiber space (P, S_n, G, p) is thus trivializable over V_1 and V_2, and is therefore defined by "gluing" together two spaces $V_1 \times G$, $V_2 \times G$ by a transition homeomorphism ψ_{12} of $(V_1 \cap V_2) \times G$ onto itself, such that

$$\psi_{12}(x, s) = (x, g_{12}(x)s)$$

where g_{12} is *any* continuous map of $V_1 \cap V_2$ into G [there are *no* "cocycle condition" (9)]. Furthermore two principal fiber spaces over S_n with same structural group G defined by two maps g_{12}, g'_{12} are isomorphic if and only if there are two continuous maps λ_1: $V_1 \to G$, λ_2: $V_2 \to G$ such that

$$g'_{12}(x) = \lambda_1(x)^{-1} g_{12}(x) \lambda_2(x) \quad \text{for } x \in V_1 \cap V_2. \tag{64}$$

A first reduction allows the assumption that for the point $e_1 = *$ of S_{n-1}, $g_{12}(*) = e$: if $g_{12}(*) = a \in G$, then take $\lambda_1(x) = a$ and $\lambda_2(x) = e$ in (64).

A second step allows us to work only with the restriction $T = g_{12}|S_{n-1}$, which is a continuous map of pointed spaces

$$T: (S_{n-1}, *) \to (G, e).$$

This map can be chosen *arbitrarily*, since there is a retraction r: $V_1 \cap V_2 \to S_{n-1}$ mapping x to the point where the great circle through x and e_{n+1} meets S_{n-1}. Then define $g_{12} = T \circ r$. For two principal fiber spaces corresponding to T, T′ to be isomorphic, it is necessary and sufficient that there exist $a \in G$ such that T′ and aTa^{-1} are *homotopic*. The necessity follows from the fact that if $T'(x) = \lambda_1(x)^{-1} T(x) \lambda_2(x)$, then $\lambda_2(*) = \lambda_1(*) = a \in G$. There is a homotopy in G between λ_1 and the constant map $V_1 \to \{a\}$ and another one between λ_2 and the constant map $V_2 \to \{a\}$ (both leaving $*$ fixed); so T′ is homotopic to $a^{-1}Ta$. To prove sufficiency, one may replace T by aTa^{-1} without disturbing the isomorphism and assume that T′ and T are homotopic; therefore $T'T^{-1}$ is homotopic to the constant map $S_{n-1} \to \{e\}$. This implies that $T'T^{-1}$, defined in S_{n-1}, can be extended to a continuous map v of the upper hemisphere D_+ of S_n (defined by $x_{n+1} \geqslant 0$) into G. Let $V_2' \subset V_2$ be defined by $x_{n+1} < 0$, and let f_{12}, f_{12}' be the restrictions of g_{12}, g_{12}' to $V_1 \cap V_2'$. The elementary theory of fiber spaces shows that the principal fiber space defined by f_{12} (resp. f_{12}') is isomorphic to the one defined by g_{12} (resp. g_{12}'). However $f_{12}' f_{12}^{-1}$ can be extended to a continuous map $\mu_1 : V_1 \to G$ equal to v in D_+, so that $f_{12}'(x) = \mu_1(x) f_{12}(x)$ for $x \in V_1 \cap V_2'$; this proves that the principal fiber spaces defined by g_{12} and g_{12}' are isomorphic.

The group G acts on the homotopy group $\pi_{n-1}(G, e)$ by $(a, [F]) \to [aFa^{-1}]$. The conclusion is that the set of *orbits* in $\pi_{n-1}(G, e)$ for that action is applied *bijectively* on the set of *isomorphism classes* of principal fiber spaces (P, S_n, G, p). When G is arcwise-connected, $[F] = [aFa^{-1}]$ for every $a \in G$, and therefore there is a bijection of $\pi_{n-1}(G, e)$ on the set of isomorphism classes of principal fiber spaces (P, S_n, G, p).

CHAPTER IV

Homology of Fibrations

§ 1. Characteristic Classes

A. The Stiefel Classes

In 1935, at the instigation of Hopf, his student Stiefel undertook in his dissertation [457] to extend Hopf's work on vector fields (Part 2, chap. III, § 3). Given an n-dimensional compact C^∞ manifold M, the problem was to investigate whether there exists on M, not only *one* nowhere vanishing vector field, but a system of m vector fields X_j $(1 \leqslant j \leqslant m)$ for some $m \leqslant n$, subject to the condition that at each point $x \in$ M, the m tangent vectors $X_j(x)$ are *linearly independent* (hence $\neq 0$). The case $m = n$ is particularly interesting in differential geometry, because the existence of such systems of n vector fields is equivalent to the existence of a *parallelism* on M.

It was in order to attack that problem that Stiefel first defined and studied what are now called the Stiefel manifolds $\mathbf{S}_{n,m}$ (Part 2, chap. V, § 4,C). He also defined the manifold $\mathrm{V}_{n,m} \subset (\mathbf{R}^n)^m$ consisting of sequences of m vectors of $\mathbf{R}^n$ that are linearly independent; clearly $\mathbf{S}_{n,m} \subset \mathrm{V}_{n,m}$, and the Gram–Schmidt orthonormalization process shows that $\mathbf{S}_{n,m}$ is a *strong deformation retract* of $\mathrm{V}_{n,m}$, hence has the same homotopy type.

Stiefel first considered what he called *m-fields in* $\mathbf{R}^n$: a continuous map $X^{(m)}$: A $\rightarrow \mathrm{V}_{n,m}$ defined in some part A of $\mathbf{R}^n$. For his later arguments he needed the case in which A is homeomorphic to a ball $\mathbf{D}_{r+1}$: if B is the image by that homeomorphism of the frontier $\mathbf{S}_r$ of $\mathbf{D}_{r+1}$, can an m-field $\mathrm{X}^{(m)}$ defined in B be *extended* to A? It was enough to take A $= \mathbf{D}_{r+1}$ and B $= \mathbf{S}_r$; we have seen (chap. III, § 2,D) that in that case Stiefel proved two lemmas. The first one is that such an extension is always possible for $r < n - m$; for $r = n - m$, the condition of possibility for the extension is that the homotopy class $\alpha \in \pi_{n-m}(\mathbf{S}_{n,m})$ of the m-field be 0 [recall that for $m = 1$ or $n - m$ even, $\pi_{n-m}(\mathbf{S}_{n,m})$ is isomorphic to $\mathbf{Z}$, and for $n - m$ odd and $m \neq 1$, $\pi_{n-m}(\mathbf{S}_{n,m})$ is isomorphic to $\mathbf{Z}/2\mathbf{Z}$]. Stiefel called α the "characteristic"* of the m-field $X^{(m)}$.

* The words "characteristic," "regular," and "normal" are unfortunately overworked terms used without restraint by so many mathematicians that it is almost impossible to understand what they mean without an explanatory context.

J. Dieudonné, *A History of Algebraic and Differential Topology, 1900–1960*,
Modern Birkhäuser Classics, DOI 10.1007/978-0-8176-4907-4_18,

The second lemma is that if two m-fields in $\mathbf{D}_{n-m-1}$ have restrictions to $\mathbf{S}_{n-m-2}$ that are homotopic, then they themselves are homotopic.

After these preliminaries Stiefel turned to the central theme of his investigations. He chose a triangulation T on M such that every simplex of T is contained in the domain of definition of a chart. Using his first lemma on m-fields in balls, he showed that on the skeleton T^{n-m} of T it is always possible to define m-fields of vectors tangent to M.

Recall that cohomology was defined precisely at the time Stiefel was writing his paper (Part 1, chap. IV, § 3), and it is unlikely that he had heard of it before the paper was printed. Nevertheless, what he did amounts, in our language, to associating to each m-field $X^{(m)}$ of vectors defined on T^{n-m} a cochain

$$\chi^{(m)} \in \mathrm{C}^{n-m+1}(\mathrm{T};\mathrm{G}) = \mathrm{Hom}(\mathrm{C}_{n-m+1}(\mathrm{T}),\mathrm{G})$$

with values in $\mathrm{G} = \pi_{n-m}(\mathbf{S}_{n,m})$; for each $(n-m+1)$-simplex σ of T, the value of $\chi^{(m)}(\sigma)$ is the "characteristic" of the restriction of the m-field $X^{(m)}$ to the frontier of σ. Next he showed that $\chi^{(m)}$ vanishes on $(n-m+1)$-boundaries, i.e., it is a *cocycle* (in present day language). From his second lemma Stiefel deduced that for another m-field $X'^{(m)}$ on T^{n-m}, the difference of the corresponding cocycles $\chi^{(m)} - \chi'^{(m)}$ is a coboundary. Using Poincaré duality, he finally obtained a *homology class* $\mathrm{F}_{m-1} \in \mathrm{H}_{m-1}(\mathrm{M};\mathrm{G})$, depending only on the differential structure of M.

As an application, Stiefel showed that any compact orientable *three-dimensional* C^∞ manifold is always *parallelizable*; the proof uses a triangulation and is complicated (see section B for a simpler proof). He also showed that the homology class F_1 is $\neq 0$ for the projective spaces $\mathbf{P}_{4k+1}(\mathbf{R})$.

B. Whitney's Work

In the language of fiber spaces, Stiefel had only considered the tangent bundle T(M) of a differential manifold (or rather the corresponding sphere bundle); for any $m \leqslant \dim \mathrm{M}$, he had associated to T(M) a homology class in $\mathrm{H}_m(\mathrm{M};\mathrm{G})$. In his first Note of 1935 on "sphere spaces" Whitney independently applied a very similar strategy to an *arbitrary* sphere bundle $(\mathrm{E},\mathrm{B},p)$ where the base space B is a locally finite simplicial complex. He only used the homology groups of the Stiefel manifolds (with coefficients in $\mathbf{Z}$), and, without proof, stated that $\mathrm{H}_{n-m}(\mathbf{S}_{n,m})$ is isomorphic to $\mathbf{Z}$ for $m = 1$ or $n - m$ even, and to $\mathbf{Z}/2\mathbf{Z} = \mathbf{F}_2$ for $n - m$ odd and $m \neq 1$. Then, again with only a sketch of a proof, he showed that after a sufficiently fine subdivision of B, it is possible to define for $r \leqslant m$ (by induction on r) on the skeleton $\mathbf{B}^r$, a continuous map

$$\varphi \colon \mathbf{B}^r \to \mathbf{S}_{n,m-r+1}.$$

The restriction of φ to the frontier of an r-simplex σ of B defines a singular homology class $\lambda(\sigma,\varphi)$ in $\mathrm{H}_{r-1}(\mathbf{S}_{n,m-r+1}) = \mathrm{G}$. Following the same pattern as Stiefel, he obtained a homomorphism

$$\tilde{w}_r \colon \mathrm{H}_r(\mathrm{B}) \to \mathrm{G}$$

for every $r \leqslant \inf(n, m)$ and stated that this homomorphism only depends on the isomorphism class of the fibration (E, B, p).

In his 1937 address [508] Whitney used the language of the newly defined cohomology to describe the classes $\tilde{w}_r$. In 1940 [513] he returned to these cohomology classes stating without proof that for a continuous map $g\colon B' \to B$, if $E' = g^*(E)$ is the pullback of E, then

$$\tilde{w}_r(E') = g^*(\tilde{w}_r(E)). \tag{1}$$

In this formula G depends on r and is equal to $\mathbf{Z}$ or $\mathbf{Z}/2\mathbf{Z} = \mathbf{F}_2$. Already in his 1935 Note Whitney considered the images $w_r \in H^r(B; \mathbf{F}_2)$ of the classes $\tilde{w}_r$ by the homomorphism

$$H^r(B; G) \to H^r(B: \mathbf{F}_2)$$

induced by the natural homomorphism $G \to \mathbf{Z}/2\mathbf{Z}$; it is these classes that are now called the *Stiefel–Whitney characteristic classes* of the sphere bundle (E, B, p) or of the corresponding vector bundle ξ; they are written $w_r(\xi)$ or $w_r(E)$.

The methods of Stiefel and Whitney may be subsumed under the general theory of obstructions (chap. II, § 4,C). On the pattern of the association of the fiber space of frames to a vector bundle, for a vector bundle $\xi = (E, B, p)$ of rank n and for every $k \leqslant n$, a fiber space $(F_k, B, S_{n,k})$ with base space B, with fibers Stiefel manifolds $S_{n,k}$ is more generally defined. A k-field on B is just a *section* over B of the fiber space F_k. Taking a simplicial complex for B, the construction of Stiefel defines a section of F_k over the $(n-k)$-skeleton of B. For the extension of that section to the $(n-k+1)$-skeleton, there is an obstruction cocycle whose class belongs to the group

$$H^{n-k+1}(B; (G_x))$$

where (G_x) is the local system (Part 1, chap. IV, § 7,A) consisting of the homotopy groups $\pi_{n-k}((F_k)_x)$, isomorphic to $\pi_{n-k}(S_{n,k})$. Using the natural homomorphism of $\pi_{n-k}((F_k)_x)$ onto $\mathbf{F}_2$, we obtain an *obstruction class* $\mathfrak{o}_{n-k+1}(\xi) \in H^{n-k+1}(B; \mathbf{F}_2)$, and one can show that this class is exactly the *Stiefel–Whitney class* $w_{n-k+1}(\xi)$ ([347], pp. 140–141).

The most remarkable result of Whitney's 1940 Note is the formula giving the classes $w_r(E' \oplus E'')$ of the "Whitney sum" of two vector bundles E', E'' over the same base space B:

$$w_r(E' \oplus E'') = \sum_i w_i(E') \smile w_{r-i}(E'') \tag{2}$$

(in the Note the formula is written for sphere-bundles). He said that his proof for $r \geqslant 4$ was "very hard," giving only a few indications of how he proceeded by "climbing" along the skeletons of B, and using deformations as well as modifications of the cup products. In his 1941 lecture [514] he only detailed a proof for E' and E'' of rank 1 ("*line bundles*"), using the special features of that case.

For a vector bundle E of rank r, the Stiefel–Whitney classes $w_j(E)$ are only defined for $1 \leqslant j \leqslant r$; in his 1940 Note Whitney introduced the formal power series with coefficients in $H^\bullet(B; \mathbf{F}_2)$

$$w(E;t) = \sum_{j=0}^{\infty} w_j(E)t^j \tag{3}$$

with $w_j(E) = 0$ by convention for $j > r$, so that (3) is in fact a polynomial. Then formulas (2) take the simple form

$$w(E' \oplus E'';t) = w(E';t) \smile w(E'';t). \tag{4}$$

The first published proofs of Whitney's formula (2) are due to Wu Wen Tsün [521] and Chern [126] (both in the same volume of the *Annals of Mathematics*). We shall deal with Chern's proof in section D. For convenience Wu Wen Tsün worked with vector bundles instead of sphere bundles. His idea was to consider two vector bundles (E_1, B_1, p_1), (E_2, B_2, p_2) with two not necessarily identical base spaces B_1, B_2 (he limited himself to finite simplicial complexes); the product $(E_1 \times E_2, B_1 \times B_2, p_1 \times p_2)$ is then defined as a vector bundle, and he proved the formula

$$w(E_1 \times E_2;t) = w(E_1;t) \times w(E_2;t), \tag{5}$$

for that product, $H^{\bullet}(B_1 \times B_2; F_2)$ being identified to $H^{\bullet}(B_1; F_2) \times H^{\bullet}(B_2; F_2)$ by the Künneth formula. He then derived (4) and (5) by the usual "Lefschetz trick" of identifying $E' \oplus E''$ to the pullback by the diagonal map $\delta: B \to B \times B$ of the product $E' \times E''$ over $B \times B$ as base space.

Wu Wen Tsün's proof of (5) used a construction of Stiefel ([457], p. 237) generalizing the concept of "*m*-field" to "*m*-field with singularities." He considered a vector bundle (E, B, p) of rank ν over a simplicial complex B and for *every* $m \leqslant \nu$, showed that it is possible to define systems

$$\varphi = (\varphi_1, \varphi_2, \ldots, \varphi_m)$$

of "canonical" sections of E over B which have the following properties for *every* skeleton B^r of B:

(i) if $r \leqslant \nu - m$, $\varphi_1, \varphi_2, \ldots, \varphi_m$ are linearly independent over B^r;
(ii) if $r > \nu - m$, there exists an index i such that $0 \leqslant i \leqslant m - (\nu - r)^+$, for which $\varphi_1, \varphi_2, \ldots, \varphi_{i+(\nu-r)^+}$ are linearly independent over B^r, while all other sections $\varphi_{i+(\nu-r)^++1}, \ldots, \varphi_m$ are identically 0 on a proper subcomplex of the first barycentric subdivision of B.

To prove (5), Wu Wen Tsün assumed that the vector bundles have ranks ν_1, ν_2, respectively; for $m \leqslant \nu_1 + \nu_2$, he defined m sections of $E_1 \times E_2$ over $B_1 \times B_2$ by

$$\varphi_i(b_1, b_2) = (\varphi_{i,1}(b_1), \varphi_{m-i+1,2}(b_2)) \quad \text{for } 1 \leqslant i \leqslant m, \tag{6}$$

where the $\varphi_{i,1}$ and $\varphi_{j,2}$ are the "canonical" sections of E_1 and E_2 defined above. The φ_i are linearly independent at each point of $B_1 \times B_2$. There follows a long argument, the purpose of which is to deform the φ_i in such a way that the proof of (5) is reduced to the product formula for the degree of a map $f_1 \times f_2$, for two maps $f_1: B_1 \to E_1$ and $f_2: B_2 \to E_2$ (Part 2, chap. I, § 1).

Applications. From formula (1) follows that if E is a *trivial* vector bundle, then $w_j(\mathrm{E}) = 0$ for all j, since there then exists a map g: $\mathrm{B} \to \{x_0\}$ on a singleton such that E is the pullback of a vector bundle with base space reduced to one point. From (4), if E″ is a trivial bundle, then

$$w(\mathrm{E}' \oplus \mathrm{E}''; t) = w(\mathrm{E}'; t).$$

This result already shows that Stiefel–Whitney classes may give useful information on vector bundles. Suppose E has rank r and has k linearly independent cross sections; they generate a *trivializable* subbundle E′; therefore (chap. III, § 1,C), if B is paracompact, $\mathrm{E} = \mathrm{E}' \oplus \mathrm{E}''$ where E″ has rank $r - k$, and this yields the necessary conditions

$$w_{r-k+1}(\mathrm{E}) = w_{r-k+2}(\mathrm{E}) = \cdots = w_r(\mathrm{E}) = 0.$$

A large part of Whitney's 1940 Note and his 1941 lecture was devoted to the case in which B is a C^1 manifold, and the vector bundles he considered are the tangent bundle $\mathrm{T(B)} = \mathrm{T}$ and the normal bundle $\mathrm{N(B)} = \mathrm{N}$ when B is a submanifold of some larger C^1 manifold. In particular, if B is embedded in some $\mathbf{R}^n$, $\mathrm{T} \oplus \mathrm{N}$ is the restriction to B of the trivial tangent bundle $\mathrm{T}(\mathbf{R}^n)$; hence, for the formal power series $w(\mathrm{T}; t)$ and $w(\mathrm{N}; t)$,

$$1 = w(\mathrm{T} \oplus \mathrm{N}; t) = w(\mathrm{T}; t) \smile w(\mathrm{N}; t).$$

This proves that if $w_i = w_i(\mathrm{T})$, the Whitney classes $\bar{w}_i = w_i(\mathrm{N})$ are given by the formula

$$(1 + w_1 t + \cdots + w_m t^m + \cdots)^{-1} = 1 + \bar{w}_1 t + \cdots + \bar{w}_m t^m + \cdots. \tag{7}$$

Whitney called the $\bar{w}_i$ "dual classes" of the w_i in $\mathrm{H}^\bullet(\mathrm{B}; \mathbf{F}_2)$ and formula (5) the "duality theorem".

Formula (7) gave the first results on the *immersion problem.* For a given smooth manifold M of dimension n, determine the *smallest* k such that there is an immersion of M in $\mathbf{R}^{n+k}$ ($k \leqslant n$ by Whitney's immersion theorem). If there exists such an immersion, the normal bundle N has rank k, so $w_i(\mathrm{N}) = 0$ for $i > k$, and formula (7) implies restrictions on the w_i, since $\bar{w}_i = 0$ for $i > k$. The computation of the Stiefel–Whitney classes for the tangent bundle to $\mathbf{P}_n(\mathbf{R})$ shows that $\mathbf{P}_8(\mathbf{R})$ cannot be immersed in $\mathbf{R}^{14}$; on the other hand, if $\mathbf{P}_{2^r}(\mathbf{R})$ can be immersed in $\mathbf{R}^{2^r+k}$, then $k \geqslant 2^r - 1$, which shows that Whitney's theorem cannot be improved for *general n.*

Another application is to a simple proof of Stiefel's theorem on the parallelizability of any orientable compact three-dimensional manifold M (chap. IV, § 1,A). It is enough to prove the existence of *two* linearly independent vector fields X_1, X_2 on M, since if a third vector $X_3(x)$ is defined for each $x \in \mathrm{M}$ orthogonal to $X_1(x)$ and $X_2(x)$ (for a riemannian structure on M) and such that the sequence $X_1(x)$, $X_2(x)$, $X_3(x)$ is direct (for a given orientation on M), then there is a system of three linearly independent vector fields. Since $\pi_1(\mathbf{S}_{3,2}) \simeq \mathbf{Z}/2\mathbf{Z}$, there is an obstruction to defining a 2-field on the 2-

skeleton of M; but since M is orientable, $w_1(\mathrm{T(M)}) = 0$ (section C), so that obstruction is 0. To extend that 2-field to the whole manifold M, check that $\pi_2(\mathbf{S}_{3,2}) = 0$; but $\mathbf{S}_{3,2}$ is homeomorphic to SO(3), which is itself homeomorphic to $\mathbf{P}_3(\mathbf{R})$, quotient of the unit quaternion group $\mathbf{S}_3$; and since $\pi_2(\mathbf{S}_3) = 0$, $\pi_2(\mathbf{P}_3(\mathbf{R})) = 0$.

C. Pontrjagin Classes

Although he was the first to characterize up to isomorphism the vector bundles (E, B, p) having a simplicial complex B as base space, by the pullback process from Stiefel manifolds (chap. III, § 2,E), Whitney does not seem to have thought of defining the Stiefel–Whitney classes as images of the cohomology classes of Grassmannians with coefficients in $\mathbf{F}_2$. This idea was introduced by Pontrjagin ([379], [381]). In his first Note [379] (later developed in [381]) he only considered orientable C^1 manifolds M and their tangent bundles T(M); he also supposed M embedded in some $\mathbf{R}^n$ [due to Whitney's embedding theorem (Part 1, chap. III, § 1), this does not restrict the generality]. The novelty in his approach was that instead of taking the Grassmannian $\mathrm{G}_{n,m} = \mathrm{G}_{n,m}(\mathbf{R})$ consisting of the vector subspaces of dimension m in $\mathbf{R}^n$, he considered the *special Grassmannian* $\mathrm{G}'_{n,m}$ consisting of the *oriented* vector subspaces of dimension m in $\mathbf{R}^n$ (Part 2, chap. V, § 4,B). He constructed a cellular decomposition of $\mathrm{G}'_{n,m}$ that projects onto the one constructed by Ehresmann for $\mathrm{G}_{n,m}$, using Schubert varieties (*loc. cit.*). That construction was presented in a simpler way by Wu Wen Tsün in his thesis in 1948 [524]. Slightly changing Schubert's notations, Pontrjagin and Wu Wen Tsün considered functions

$$\omega\colon \{1, 2, \ldots, m\} \to \{0, 1, 2, \ldots, n\}$$

such that

$$0 \leqslant \omega(1) \leqslant \omega(2) \leqslant \cdots \leqslant \omega(m) \leqslant n, \tag{8}$$

and they attached to each ω the m-dimensional oriented vector subspace X_ω of $\mathbf{R}^{m+n}$ having as basis the vectors

$$\mathbf{e}_{\omega(1)+1}, \mathbf{e}_{\omega(2)+2}, \ldots, \mathbf{e}_{\omega(m)+m}$$

of the canonical basis of $\mathbf{R}^{m+n}$, that order defining the orientation. They considered the set U_ω^+ (resp. U_ω^+) of oriented vector subspaces $\mathrm{V} \in \mathrm{G}'_{m+n,m}$ for which the canonical projection on X_ω is bijective and preserves (resp. reverses) orientation and for which

$$\dim(\mathrm{V} \cap \mathbf{R}^{\omega(i)+i}) \geqslant i \quad \text{for } 1 \leqslant i \leqslant m. \tag{9}$$

Both U_ω^+ and U_ω^- are homeomorphic to an open ball in $\mathbf{R}^{d(\omega)}$ with

$$d(\omega) = \sum_{i=1}^{m} \omega(i). \tag{10}$$

and when ω varies in the set of functions satisfying (8) they constitute a cellular decomposition of $\mathrm{G}'_{m+n,m}$.

Pontrjagin also determined the boundary operator of that cell complex, and used that result to compute some of the homology groups of $G'_{m+n,m}$; Wu Wen Tsün, using cohomology instead of homology, put Pontrjagin's results in a simpler form.

The cohomology group $H^j(G'_{m+n,m}(\mathbf{R}); \mathbf{Z})$ for $j \leqslant m$ is a direct sum of a free $\mathbf{Z}$-module and a vector space over $\mathbf{F}_2$. Pontrjagin selected chains

$$R_\omega = U_\omega^+ \pm U_\omega^- \tag{11}$$

for suitable choices of the sign $\pm$ (depending on ω): when considered as chains with coefficients in $\mathbf{F}_2$, they are $d(\omega)$-cycles, and when considered as chains with integer coefficients, they are $d(\omega)$-cycles for *some* of the functions ω.

During the period 1935–1950 many topologists identified p-cochains on an n-dimensional combinatorial manifold with the $(n - p)$-chains of a *dual* cellular decomposition, the bilinear form $\langle \zeta, \zeta' \rangle$ being identified to the intersection product. Wu Wen Tsün did this to obtain cocycles from chains of type (11) by a careful study of a dual cell decomposition; in particular he singled out the following functions ω:

(a) $\omega(j) = 0$ for $1 \leqslant j \leqslant m - k$, $\omega(j) = 1$ for $m - k + 1 \leqslant j \leqslant m$, which he wrote ω_k^m, for all $k \leqslant m$; then $d(\omega_k^m) = k$. The corresponding cochains are cocycles in $Z^k(G'_{m+n,m}: \mathbf{F}_2)$, and ω_m^m is even a cocycle in $Z^m(G'_{m+n,m}; \mathbf{Z})$;
(b) $\omega(j) = 0$ for $1 \leqslant j \leqslant m - 2k$, $\omega(j) = 2$ for $m - 2k + 1 \leqslant j \leqslant m$, which he wrote $\omega_{2k,2k}^m$, for all $k \leqslant m/2$; then $d(\omega_{2k,2k}^m) = 4k$. The corresponding cochains are cocycles in $Z^{4k}(G'_{m+n,m}; \mathbf{Z})$.

To each ω_k^m corresponds a cohomology class in $H^k(G'_{m+n,m}; \mathbf{F}_2)$; they are the Stiefel–Whitney classes for the sphere bundle $S'[n + m - 1, m - 1]$, which is the pullback of Whitney's sphere bundle $S[n + m - 1, m - 1]$ (chap. III, §2,E) by the natural projection $G'_{m+n,m} \to G_{m+n,m}$.

However, the cohomology classes $p_k \in H^{4k}(G'_{m+n,m}; \mathbf{Z})$ deduced from the $\omega_{2k,2k}^m$ for $2k \leqslant m$, and $e = e_m \in H^m(G'_{m+n,m}; \mathbf{Z})$ deduced from ω_m^m, had never been considered before Pontrjagin; the p_k are now called the *Pontrjagin classes* and e is the *Euler class* of $S'[n + m - 1, m - 1]$; the relation

$$e_m \smile e_m = p_{m'} \quad \text{if } m = 2m' \tag{12}$$

holds in the cohomology algebra $H^\bullet(G'_{m+n,m}; \mathbf{Z})$.

Pontrjagin also studied the natural involution $\rho: G'_{m+n,m} \to G'_{m+n,m}$ which exchanges the two points of $G'_{m+n,m}$ above the same point of $G_{m+n,m}$; he proved in substance the important results written by Wu Wen Tsün as

$$\rho^*(p_k) = p_k, \qquad \rho^*(e_m) = -e_m, \tag{13}$$

which show that the p_k are natural images of classes in the cohomology algebra $H^\bullet(G_{m+n,m}; \mathbf{Z})$, which are also written p_k.

The spherical fibrations $(S'[n + m, m], G'_{m+n+1,m+1}, \pi')$ are the "universal" ones for what are called *oriented* spherical fibrations $\xi = (E, B, p)$; these can be defined by replacing, in Whitney's definition of "regular" sphere bundles (chap. III, §1,B), the orthogonal group $O(m + 1)$ by the special orthogonal group (or group of rotations) $SO(m + 1)$. The Stiefel manifold $\mathbf{S}_{m+n,m}$ can be written as the homogeneous space $SO(m+n)/SO(n)$ for $m \geqslant 1$, $n \geqslant 1$, and $G'_{m+n,m}$ as the homogeneous space $SO(m+n)/(SO(n) \times SO(m))$, so that (chap. III, § 1,C) $S_{n+m,m}$ is also a principal fiber space with base space $G'_{m+n,m}$ and structural group $SO(m)$. Therefore $S_{n+m+1,m}$ is *n-universal* and its base space $G'_{n+m+1,m}$ is *n-classifying* for principal fiber spaces with structural group $SO(m)$. Using the natural injections $G'_{n+k,k} \to G'_{n+k+1,k}$ and $S_{n+k,k} \to S_{n+k+1,k}$, define the direct limits $BSO(k)$ and $ESO(k)$ of those spaces (chap. III, § 2,F); $ESO(k)$ is universal and $BSO(k)$ is classifying for *every* principal fiber space with structural group $SO(k)$ and base space *any* CW-complex.

The sphere bundle $S'[n+m-1, m-1]$ is then *associated* to the principal fiber space $(S_{n+m,m}, G'_{m+n,m}, SO(m))$ and to the natural action of $SO(m)$ on S_{m-1}. The Whitney–Steenrod therems (chap. III, § 2,E) apply to oriented spherical fibrations, by considering pullbacks of $(S_{n+m,m}, G'_{m+n,m}, SO(m))$ instead of pullbacks of $(S_{n+m,m}, G_{n+m,m}, O(m))$.

In the same way that vector bundles correspond to Whitney's "regular" sphere bundles, *oriented* vector fibrations $\xi = (E, B, F, p)$ correspond to oriented spherical fibrations; the fibers of such a fibration are *oriented* vector spaces over $\mathbf{R}$, and the typical fiber F is also oriented. The homeomorphisms

$$\varphi\colon U \times F \to p^{-1}(U) \tag{14}$$

defining the fiber space structure must be such that, for each $b \in U$, the map $t \mapsto \varphi(b, t)$ of F into E_b is a linear bijection *preserving orientation.* If

$$\varphi'\colon U' \times F \to p^{-1}(U') \tag{15}$$

is another such homeomorphism and $U \cap U' \neq \varnothing$, the transition homeomorphism in $U \cap U'$ can be written

$$(y, t) \mapsto (y, A(y).t) \tag{16}$$

where $A(y) \in GL(m)$ *and* $\det(A(y)) > 0$ for $y \in U \cap U'$.

Given a vector bundle (E, B, F, p) of rank m, it is not always possible to find an open covering (U_α) of B for which there exist homeomorphisms (14) for each U_α satisfying the preceding condition. A necessary and sufficient condition is that for the *m-th exterior power* $\bigwedge^m E$ (which is a line bundle) there exist *a section* w *over* B *for which* $w(b) \neq 0$ *for each* $b \in B$; the vector fibration $\xi = (E, B, F, p)$ is then called *orientable.* Any two sections w_1, w_2 of $\bigwedge^m E$ over B that are $\neq 0$ at every point are such that $w_2 = \lambda w_1$, where $b \mapsto \lambda(b)$ is a continuous map $B \to \mathbf{R}$ that is $\neq 0$ in B; w_1 and w_2 are called *equivalent* if $\lambda(b) > 0$ in B; the equivalence classes for that relation are called the *orientations* of ξ. If w belongs to an orientation of ξ, then the orientation to which $-w$ belongs is *opposite* to the one to which w belongs.

A trivial vector fibration $(B \times \mathbf{R}^m, B, \mathrm{pr}_1)$ is always orientable. To say that a C^∞ manifold M is orientable means that the tangent fibration $\tau = (T(M), M, p)$ is orientable and the orientations of M are in one-to-one correspondence with those of τ. When a vector fibration $\xi = (E, B, F, p)$ has a C^∞ manifold B as base space, the "total space" E is also a C^∞ manifold, and care must be taken not to confuse the property "ξ is orientable" with "E is orientable." If M is any C^∞ manifold, then the "total space" T(M) is always orientable, even if M is not; and the trivial fibration $(M \times \mathbf{R}^m, M, pr_1)$ is orientable even for nonorientable manifolds M.

In more recent work on characteristic classes, the sphere bundle $S[m+n-1, m-1]$ of Whitney was replaced by the corresponding vector bundle $U_{m+n,m}(\mathbf{R}) = U_{m+n,m}$ with base space $G_{m+n,m}$; it can be defined as the subspace of $G_{m+n,m} \times \mathbf{R}^{m+n}$ consisting of the pairs (V, x), where V is an m-dimensional vector subspace of $\mathbf{R}^{m+n}$, and $x \in V$; it is often called the *tautological* bundle with base space $G_{m+n,m}$, and it can also be considered the *associated* vector bundle to the Stiefel principal fiber space $\mathbf{S}_{m+n,m}$, for the natural action of $O(m)$ on $\mathbf{R}^m$. Similarly, $S'[m+n-1, m-1]$ is replaced by the double covering $U'_{m+n,m}$ of $U_{m+n,m}$, defined by replacing, in the preceding definition, m-dimensional vector subspaces by *oriented* m-dimensional vector subspaces. This is again associated to $\mathbf{S}_{m+n,m}$, this time for the natural action of $SO(m)$ on $\mathbf{R}^m$.

For every vector fibration (resp. oriented vector fibration) $\xi = (E, B, p)$, which is a pullback of $(U_{m+n,m}, G_{m+n,m}, \pi)$ [resp. of $(U'_{m+n,m}, G'_{m+n,m}, \pi')$] by a continuous map $g\colon B \to G_{m+n,m}$ (resp. $g\colon B \to G'_{m+n,m}$), the cohomology class

$$p_k(\xi) = g^*(p_k) \in H^{4k}(B; \mathbf{Z}) \quad \text{for } 2k \leqslant m$$

$$[\text{resp. } e(\xi) = g^*(e) \in H^m(B; \mathbf{Z})]$$

is called the *k-th Pontrjagin class* (resp. *Euler class*) of ξ. (The motivation for the name "Euler class" will appear below in section E.) From (13) it follows that if the orientation of an oriented vector bundle is replaced by its opposite, the Euler class changes sign. For an oriented vector bundle ξ of *odd* rank, there is an automorphism of ξ reversing orientation, so $2e(\xi) = 0$. If ξ is an oriented vector bundle of *even* rank $m = 2m'$,

$$e(\xi) \smile e(\xi) = p_{m'}(\xi). \tag{17}$$

The relation between Pontrjagin classes and Stiefel–Whitney classes was later shown to be the following: if $\bar{p}_k$ is the image of p_k by the natural homomorphism $H^{4k}(B; \mathbf{Z}) \to H^{4k}(B; \mathbf{F}_2)$, then

$$\bar{p}_k = w_{2k} \smile w_{2k} \tag{18}$$

in the algebra $H^{\bullet}(B; \mathbf{F}_2)$; similarly, the image $\bar{e}$ of e in $H^m(B; \mathbf{F}_2)$ is the *top Stiefel–Whitney class* $w_m(\xi)$. In his first Note on "sphere spaces" [505] Whitney had shown directly from his definition that the condition $w_1 = 0$ is necessary and sufficient for the restriction to the 1-skeleton B^1 of the vector fibration ξ to be *orientable*. Using the definition of orientable vector bundles as pullbacks of universal bundles with base $G'_{m+n,m}$, Wu Wen Tsün showed that the condition $w_1 = 0$ implies that ξ itself is orientable ([524], p. 45).

The Euler classes $e(\xi)$ can also be related to the obstruction class defined in section B for the case $k = 1$; that class belongs to $H^m(B; (G_x))$, and since ξ is oriented, there is a natural isomorphism of $\pi_{m-1}((F_1)_x)$ onto $\mathbf{Z}$; this gives an *obstruction class* $\mathfrak{o}_m(\xi) \in H^m(B; \mathbf{Z})$ equal to $e(\xi)$ ([347], p. 147).

D. Chern Classes

In 1945 S.S. Chern extended Pontrjagin's idea to *complex* vector bundles (or their corresponding sphere bundles), to which the Whitney–Steenrod theorems generalize immediately; the real Stiefel manifold $S_{n,m}(\mathbf{R})$ [resp. the real Grassmannian $G_{n,m}(\mathbf{R})$] is replaced by the complex Stiefel manifold $S_{n,m}(\mathbf{C})$ [resp. the complex Grassmannian $G_{n,m}(\mathbf{C})$]. At first Chern only considered tangent bundles of complex manifolds and worked from the start with cohomology instead of homology [125]. Here again, Wu Wen Tsün in his thesis [524] generalized Chern's paper to *arbitrary* complex vector bundles over *arbitrary* finite simplicial complexes. Both Chern and Wu Wen Tsün used Ehresmann's cellular decomposition of $G_{m+n,m}(\mathbf{C})$ by Schubert varieties. In Wu Wen Tsün's notation the vector space $X_\omega \subset \mathbf{R}^{m+n}$ of section C is replaced by its complexification $Z_\omega \subset \mathbf{C}^{m+n}$, U_ω^+ and U_ω^- are replaced by the set W_ω of m-dimensional complex vector subspaces of $\mathbf{C}^{m+n}$ that project bijectively on Z_ω; the Schubert varieties are the closures $\overline{W}_\omega$, and the dimension of the cell W_ω is $2d(\omega)$. He then introduced the special functions ω for which

$$\omega(j) = 0 \quad \text{for } 1 \leqslant j \leqslant m-1, \omega(m) = k,$$

which he wrote $\bar{\omega}_k^m$, so that $2d(\bar{\omega}_k^m) = 2k$; the corresponding cochains are in fact cocycles in $Z^\cdot(G_{m+n,m}(\mathbf{C}): \mathbf{Z})$ and the cohomology classes $c_k \in H^{2k}(G_{m+n,m}(\mathbf{C}); \mathbf{Z})$ corresponding to the $\bar{\omega}_k^m$ are by definition the *Chern classes* of the tautological vector bundle $U_{m+n,m}(\mathbf{C})$, defined as $U_{m+n,m}(\mathbf{R})$ by replacement of real vector spaces by complex ones. From this definition, one gets by the usual pullback process the Chern classes of any complex vector bundle.

In 1947 [126] Chern determined the cohomology *algebra* $H^\cdot(G_{m+n,m}(\mathbf{R}); \mathbf{F}_2)$ and proved that the Stiefel–Whitney classes w_k of $U_{m+n,m}(\mathbf{R})$ form a system of *generators* for that algebra. Without going into details, Wu Wen Tsün stated that similarly the Chern classes c_k are generators of the algebra $H^\cdot(G_{m+n,m}(\mathbf{C}); \mathbf{Z})$; this had earlier been proved by Chern, using differential forms (see section E); Wu Wen Tsün also mentioned that similar properties hold for the quaternionic Grassmannian $G_{m+n,m}(\mathbf{H})$.

The fact that all Chern classes belong to even-dimensional cohomology groups implies that $H^{2j+1}(G_{m+n,m}(\mathbf{C}); \mathbf{Z}) = 0$ for $j \geqslant 0$, a property known to Ehresmann for homology ([155], p. 418).

It was later shown [347] that there are no polynomial relations

$$P(c_1, c_2, \ldots, c_m) = 0$$

for $P \neq 0$ and $v_1 + 2v_2 + \cdots + mv_m \leqslant n$ if v_k is the degree of $P(T_1, \ldots, T_m)$ with respect to T_k (the coefficients of P being rational).

In his thesis Wu Wen Tsün simplified the proof he had given earlier of the product formula (4) for Stiefel–Whitney classes; he extended it to Chern classes:

$$c(E' \oplus E''; t) = c(E'; t) \smile c(E''; t) \tag{19}$$

where for a complex vector bundle E, the *total Chern class*

$$c(\mathrm{E}; t) = \sum_i c_i(\mathrm{E})t^i. \tag{20}$$

For a complex vector bundle (E, B, p), he defined the *conjugate* vector bundle $\mathrm{E}^\dagger$ by taking on each fiber E_b the complex structure for which the scalar multiplication is $(\lambda, z) \mapsto \bar{\lambda}z$. Wu Wen Tsün showed ([524], p. 43) that for Chern classes

$$c_k(\mathrm{E}^\dagger) = (-1)^k c_k(\mathrm{E}). \tag{21}$$

If (E, B, p) is a *real* vector bundle, $\mathrm{E}_{(\mathbf{C})} = \mathrm{E} \otimes_{\mathbf{R}} \mathbf{C}$ is a complex vector bundle with base space B, having a complex rank equal to the real rank of E. Consider the Chern classes

$$c_k(\mathrm{E}_{(\mathbf{C})}) \in \mathrm{H}^{2k}(\mathrm{B}; \mathbf{Z}); \tag{22}$$

it is easily seen that $(\mathrm{E}_{(\mathbf{C})})^\dagger$ is isomorphic to $\mathrm{E}_{(\mathbf{C})}$, hence

$$c_k(\mathrm{E}_{(\mathbf{C})}) = (-1)^k c_k(\mathrm{E}_{(\mathbf{C})}) \tag{23}$$

so that $2c_k(\mathrm{E}_{(\mathbf{C})}) = 0$ for *odd k*. Wu Wen Tsün showed that the Pontrjagin classes are given by

$$p_j(\mathrm{E}) = (-1)^j c_{2j}(\mathrm{E}_{(\mathbf{C})}) \tag{24}$$

so that properties of Pontrjagin classes can immediately be deduced from those of Chern classes. In particular, for two real vector bundles E′, E″ with same base space,

$$2p(\mathrm{E}' \oplus \mathrm{E}''; t) = 2p(\mathrm{E}'; t) \smile p(\mathrm{E}''; t). \tag{25}$$

Wu Wen Tsün also proved for total Pontrjagin classes a formula similar to the formula (5) for Stiefel–Whitney classes.

If E is now a *complex* vector bundle of complex rank m, it can be considered a real vector bundle of real rank $2m$ (using the canonical injection $\mathbf{R} \to \mathbf{C}$); it is then written $\mathrm{E}_{\mathbf{R}}$. It has a privileged orientation, and for that orientation, the Euler class

$$e(\mathrm{E}_{\mathbf{R}}) = c_m(\mathrm{E}) \in \mathrm{H}^{2m}(\mathrm{B}; \mathbf{Z}). \tag{26}$$

Furthermore, $(\mathrm{E}_{\mathbf{R}})_{(\mathbf{C})}$ is naturally isomorphic to $\mathrm{E} \oplus \mathrm{E}^\dagger$, hence the relation

$$p(\mathrm{E}_{\mathbf{R}}; t) = c(\mathrm{E}; t) \smile c(\mathrm{E}; -t) \tag{27}$$

or equivalently

$$\begin{aligned} &1 - p_1(\mathrm{E}_{\mathbf{R}}) + p_2(\mathrm{E}_{\mathbf{R}}) + \cdots + (-1)^m p_m(\mathrm{E}_{\mathbf{R}}) \\ &\quad = (1 - c_1(\mathrm{E}) + c_2(\mathrm{E}) - \cdots + (-1)^m c_m(\mathrm{E})) \smile (1 + c_1(\mathrm{E}) + c_2(\mathrm{E}) + \cdots + c_m(\mathrm{E})) \end{aligned} \tag{28}$$

in the algebra $\mathrm{H}^\bullet(\mathrm{B}; \mathbf{Z})$.

E. Later Results

Using the principal fiber spaces with the unitary group $U(m)$ as structure group, Hirzebruch [235] gave a new proof of the product formula (19) for Chern classes. Any complex vector fibration $\xi = (E, B, p)$ of complex rank m can be considered as associated to a principal fiber space $(P, B, U(m), \pi)$ and to the natural action of $U(m)$ on $\mathbf{C}^m$ (chap. III, § 1,C, IV). In $U(m)$, consider a maximal torus $\mathbf{T}^m$; P is also a principal fiber space with base space

$$X = \mathbf{T}^m \backslash P \simeq P \times^{U(m)} (U(m)/\mathbf{T}^m)$$

and structural group $\mathbf{T}^m$ (chap. III, § 1,C, IV); if $\rho: X \to B$ is the projection, then $\rho^*(\xi)$ is a complex vector fibration over the base space X associated to the principal fiber space $(P, X, \mathbf{T}^m, \pi')$, hence a *direct sum of complex line bundles*

$$\lambda_1 \oplus \lambda_2 \oplus \cdots \oplus \lambda_m;$$

this is due to the fact that any unitary representation of $\mathbf{T}^m$ splits into one-dimensional representations. From results of A. Borel on the cohomology of principal fiber spaces with compact Lie groups as structure groups (see § 4) it follows that the cohomology homomorphism

$$\rho^*: H^{\bullet}(B; \mathbf{Z}) \to H^{\bullet}(X; \mathbf{Z}) \tag{29}$$

is *injective*, and the total Chern class

$$c(\rho^*(\xi); t) = \prod_{i=1}^{m} c(\lambda_i; t) \tag{30}$$

in the cohomology algebra $H^{\bullet}(X; \mathbf{Z})$.

If (E', B, p'), (E'', B, p'') are two complex vector bundles of respective ranks p, q over the same base space B, there is a continuous map $g: X \to B$ such that $g^*: H^{\bullet}(B; \mathbf{Z}) \to H^{\bullet}(X; \mathbf{Z})$ is injective and that

$$g^*(E') = \bigoplus_{i=1}^{p} L'_i, \qquad g^*(E'') = \bigoplus_{j=1}^{q} L''_j$$

where the L'_i and L''_j are complex line bundles, so that

$$g^*(E' \oplus E'') = \Big(\bigoplus_i L'_i\Big) \oplus \Big(\bigoplus_j L''_j\Big)$$

and

$$\begin{aligned} c(g^*(E' \oplus E''); t) &= \prod_i c(L'_i; t) \smile \prod_j c(L''_j; t) \\ &= c(g^*(E'); t) \smile c(g^*(E''); t); \end{aligned}$$

using the injectivity of the cohomology homomorphism g^*, the product formula (19) follows. The method yields more relations, for instance,

$$g^*(\mathrm{E}' \otimes \mathrm{E}'') = \bigoplus_{i,j} \mathrm{L}'_i \otimes \mathrm{L}''_j, \tag{31}$$

$$g^*\left(\bigwedge^r \mathrm{E}'\right) = \bigoplus_{i_1 < i_2 < \cdots < i_r} (\mathrm{L}'_{i_1} \otimes \mathrm{L}'_{i_2} \otimes \cdots \otimes \mathrm{L}'_{i_r}) \tag{32}$$

and for two complex *line bundles* L′, L″

$$c_1(\mathrm{L}' \otimes \mathrm{L}'') = c_1(\mathrm{L}') + c_1(\mathrm{L}''). \tag{33}$$

Hence formulas (31) and (32) allow the computation of $c(\mathrm{E}' \otimes \mathrm{E}''; t)$ and $c(\bigwedge^r \mathrm{E}'; t)$ for complex vector bundles.

The *Chern character* of a direct sum $\mathrm{L}_1 \oplus \mathrm{L}_2 \oplus \cdots \oplus \mathrm{L}_p$ of complex *line bundles* is defined as*

$$\exp(c_1(\mathrm{L}_1)) + \exp(c_1(\mathrm{L}_2)) + \cdots + \exp(c_1(\mathrm{L}_p)) \quad \text{in } \mathrm{H}^\bullet(\mathrm{B}; \mathbf{Q}).$$

The Hirzebruch method allows the transfer of that definition to *any* complex vector bundle; writing ch(E) or ch(ξ) for the Chern character of a complex vector fibration $\xi = (\mathrm{E}, \mathrm{B}, p)$; then

$$\mathrm{ch}(\mathrm{E}' \oplus \mathrm{E}'') = \mathrm{ch}(\mathrm{E}') + \mathrm{ch}(\mathrm{E}''), \tag{34}$$

$$\mathrm{ch}(\mathrm{E}' \otimes \mathrm{E}'') = \mathrm{ch}(\mathrm{E}') \smile \mathrm{ch}(\mathrm{E}'') \tag{35}$$

in the algebra $\mathrm{H}^\bullet(\mathrm{B}; \mathbf{Q})$.

Earlier Wu Wen-Tsün had made computations with Stiefel–Whitney classes by using "phantom" indeterminates $t_1, \ldots, t_m$ for a vector bundle ξ of rank m such that its Stiefel–Whitney classes are formally *elementary symmetric polynomials* in $t_1, \ldots, t_m$:

$$w_1 = t_1 + t_2 + \cdots + t_m,$$

$$w_2 = t_1 t_2 + \cdots + t_{m-1} t_m,$$

$$\cdots\cdots$$

$$w_m = t_1 t_2 \cdots t_m.$$

Polynomials in the w_j with integer coefficients can be more easily handled as *symmetric polynomials* in $t_1, \ldots, t_m$. This can also be done for Pontrjagin classes and Chern classes; for the latter, the indeterminates t_j can be interpreted as the Chern classes $c_1(\lambda_j)$ defined above. For instance, it is useful to consider the symmetric polynomial

$$s_k = t_1^k + \cdots + t_m^k$$

which is a polynomial $s_k(c(\xi))$ in the Chern classes $c_j(\xi)$; with that notation the Chern character can be written

* This is meaningful since the Chern classes are nilpotent elements in $\mathrm{H}^\bullet(\mathrm{B}; \mathbf{Q})$ when B has finite dimension.

$$\mathrm{ch}(\xi) = \sum_{k=0}^{\infty} \frac{1}{k!} s_k(c(\xi))$$

(cf. chap. VII, § 2,A).

Other proofs of the product formula (19) are to be found in [347] and [262]; all these proofs also apply to Stiefel–Whitney and Euler classes.

In 1944 Pontrjagin [380], studying the tangent bundle of a Riemannian manifold M, showed that the images of Pontrjagin and Euler classes, in the de Rham real cohomology $H^{\bullet}(M; \mathbf{R})$, could be expressed as cohomology classes of explicitly determined closed differential forms, computed by means of the Riemann–Christoffel tensor; he proved in particular that, for an oriented compact C^{∞} manifold M, the Euler–Poincaré characteristic is given by the formula

$$\langle e(T(M)), [M] \rangle = \chi(M) \tag{36}$$

(hence the name "Euler class" later given to $e(\xi)$).

This method was extended by Chern [125] to the Chern classes of the tangent vector bundle of a complex manifold, using a hermitian metric on the manifold and working with E. Cartan's connexion differential forms instead of the Riemann–Christoffel tensor. Later A. Weil showed ([482], pp. 422–436 and 567–570) that this method of computation can be extended to an *arbitrary* principal fiber space X with base space a C^{∞} manifold B, and a Lie group G as structural group. To a principal connexion (invariant under G)

$$P: T(B) \times_B Y \to T(X) \tag{37}$$

and any p-linear complex form F on $\mathfrak{g}$ (the Lie algebra of G), *invariant* under the adjoint action of G on $\mathfrak{g}$, associate the exterior $2p$-form

$$F(\Omega, \Omega, \ldots, \Omega) \tag{38}$$

where Ω is the curvature 2-form of the connexion P; this form is *closed*, and from the invariance of F it follows that it only depends on the fibers X_b, so that it defines a closed exterior $2p$-form $F_B(\Omega)$ *on* B; the remarkable thing is that the cohomology class of $F_B(\Omega)$ in $H^{\bullet}(B; \mathbf{R})$ is *independent* of the chosen connexion P on X; so that this finally defines the *Weil homomorphism*

$$I(G) \to H^{\bullet}(B; \mathbf{R})$$

of the real algebra of multilinear forms on $\mathfrak{g}$, *invariant* under G, into the cohomology algebra $H^{\bullet}(B; \mathbf{R})$. When G is a complex Lie group, the same process applies to complex multilinear forms and to the algebra $H^{\bullet}(B; \mathbf{C})$. The Chern and Pontrjagin classes are obtained when this process is applied to the corresponding principal fiber spaces to which the vector bundles are associated [347]. For Euler classes it is necessary to restrict the connexion to those compatible with a Riemannian structure on B ([347], pp. 312–314).

Hirzebruch showed that Chern classes can be characterized by a system of axioms (in the style of Eilenberg–Steenrod) including the product formula

([235], pp. 60–63); this later proved very useful in defining similar notions in categories other than the category of topological spaces.

We shall return to the characteristic classes in § 2 for their relations with the Gysin sequence, and in chap. VI for the relations between the Stiefel–Whitney classes and the Steenrod squares.

§ 2. The Gysin Exact Sequence

In his work on mappings of spheres into spheres (chap. II, § 1) Hopf had defined, for two compact, connected, orientable combinatorial manifolds X, Y of *same* dimension n, and for a simplicial map $f: X \rightarrow Y$, a group homomorphism

$$\varphi: H_\bullet(Y) \rightarrow H_\bullet(X)$$

which is also an *algebra* homomorphism for the intersection products on X and Y (Part 1, chap. IV, § 4,A). He did not try to define this "Umkehr-homomorphismus," as he called it, when X and Y have different dimensions; if $n = \dim X > \dim Y = m$, his method would have produced a map φ *increasing the dimension* of homology classes.

Hopf was certainly aware of that possibility, for in his 1931 paper on $\pi_3(S_2)$ he refers to the "Umkehrhomomorphismus" in a footnote ([243], p. 45). Recall that, at the beginning of that paper, he precisely considered a simplicial map $f: X \rightarrow Y$ for $m = 2, n > 2$, and for any $\xi \in Y$ belonging to a two-dimensional simplex, he showed that $f^{-1}(\xi)$ could be considered an $(n - 2)$-cycle. But he also was interested in other maps: when $n = 3$ and $H_1(X) = 0$ (which is the case for $X = S_3$), $f^{-1}(\xi)$, being a 1-cycle, is also a 1-boundary. If K is a 2-chain having $f^{-1}(\xi)$ as boundary, $\tilde{f}(K)$ is a 2-cycle, hence a multiple $c[Y]$ of the fundamental cycle of Y, and it is that integer c that became known as the *Hopf invariant* of f.

In 1941 Gysin, a student of Hopf, investigated in his dissertation [216] whether similar maps could be constructed in the general case $m < n$ mentioned above, which would reduce in the case $m = 2$ to those defined by Hopf for his special purposes. Recall that, for $n = m$, Freudenthal had defined the "Umkehrhomomorphismus"

$$\Phi_\bullet: H_\bullet(Y; \mathbf{Q}) \rightarrow H_\bullet(X; \mathbf{Q})$$

using the cohomology algebra homomorphism

$$f^*: H^\bullet(Y; \mathbf{Q}) \rightarrow H^\bullet(X; \mathbf{Q})$$

and Poincaré duality [Part 1, chap. IV, § 4, formula (26)]. Although Gysin did not mention Freudenthal, he followed exactly the same procedure for $m < n$. Consider the Poincaré isomorphisms

$$j_p: H^p(X; \mathbf{Q}) \xrightarrow{\sim} H_{n-p}(X; \mathbf{Q}), \qquad j'_q: H^q(Y; \mathbf{Q}) \xrightarrow{\sim} H_{m-q}(Y; \mathbf{Q}). \tag{39}$$

Then, for $d = n - m$, Φ_p is defined as the composite

$$H_p(Y;Q) \xrightarrow{j'^{-1}_{m-p}} H^{m-p}(Y;Q) \xrightarrow{f^*} H^{m-p}(X;Q) \xrightarrow{j_{m-p}} \dot{H}_{p+d}(X;Q). \qquad (40)$$

In fact, Gysin (like many topologists of the same period) identified p-cochains on an n-dimensional orientable combinatorial manifold with the $(n-p)$-chains of the *dual* triangulation, the bilinear form $\langle \zeta, \zeta' \rangle$ between p-chains and $(n-p)$-chains being the intersection product (§ 1,C). This allowed him to deduce f^* in (40) from a cochain transformation $\tilde{f}$ assigning to each $(n-p)$-cell σ' of the dual triangulation of Y the sum (with suitable signs) of the $(n-p)$-cells of the dual triangulation of X that are mapped bijectively onto σ' by the simplicial map f. This generalized Hopf's construction of the 0-chain, which Hopf wrote $\varphi_\sigma(\xi)$ (chap. II, § 1), except that, in contrast with Hopf, for whom ξ was variable, Gysin only considered inverse images by f of dual cells of a *fixed* triangulation of Y. The map Φ_p is thus deduced from a homomorphism

$$\varphi_p\colon Z_p(Y;Q) \to Z_{p+d}(X;Q) \qquad (41)$$

at the level of *cycles* in Y.

This method enabled Gysin to follow the argument that led Hopf to his invariant: he considered the homology classes $z \in H_p(Y;Q)$ that belong to the *kernel* of Φ_p. At the level of p-cycles $\zeta \in z$ this means that $\varphi_p(\zeta)$ is the *boundary* of a $(p+d+1)$-chain Γ. Continuing to follow the pattern of Hopf's argument, the boundary of the $(p+d+1)$-chain $\tilde{f}(\Gamma)$ is a *degenerate* $(p+d)$-chain, hence $\tilde{f}(\Gamma)$ is a $(p+d+1)$-*cycle*. Furthermore, if Γ_0, Γ_1 are two $(p+d+1)$-chains that have the same boundary $\varphi_p(\zeta)$, then

$$\tilde{f}(\Gamma_1) - \tilde{f}(\Gamma_0) = \tilde{f}(Z),$$

where Z is a $(p+d+1)$-cycle. Passing to homology, the image of $\tilde{f}(\Gamma)$ in $H_{p+d+1}(Y;Q)$ is thus only determined modulo the image $f_*(H_{p+d+1}(X;Q))$; this defines a homomorphism

$$h_p\colon \operatorname{Ker}\Phi_p \to \operatorname{Coker} f_{*,p+d+1}. \qquad (42)$$

Gysin's chief concern was the case in which X is a sphere-bundle with base space Y and typical fiber S_d; his main theorem was that in this case, h_p is an *isomorphism*. The proof is long and intricate, based on the local study of the complex $f^{-1}(\sigma')$ considered as a product for small enough simplices σ' in Y.

Exact sequences were not yet used in 1941, but Gysin's main theorem can be expressed as the exactness of a homology sequence

$$\cdots \to H_p(X) \xrightarrow{f_*} H_p(Y) \xrightarrow{\Psi_p} H_{p-d-1}(Y) \xrightarrow{\Phi_p} H_{p-1}(X) \to \cdots \qquad (43)$$

where the map Ψ_p is the composite

$$H_p(Y) \to \operatorname{Coker} f_{*p} \xrightarrow{h_p^{-1}} \operatorname{Ker}\Phi_{p-d-1} \to H_{p-d-1}(Y). \qquad (44)$$

After 1947 there was a renewal of interest in Gysin's paper, from several directions. In 1947 Steenrod [448] defined maps in cohomology that, for sphere-bundles (X, Y, p), gave an exact cohomology sequence (which he did not write explicitly)

$$\cdots \to H^p(Y) \to H^p(X) \xrightarrow{\Psi_p} H^{p-d}(Y) \to H^{p+1}(Y) \to \cdots. \tag{45}$$

In 1948 A. Lichnerowicz [326] considered in general a fiber space (E, B, F, p), where E, B, F are C^∞ manifolds of respective dimensions $m+n$, n, m, and p is a C^∞ map. Using de Rham cohomology, he defined homomorphisms

$$H^{j+m}(E) \to H^j(B) \tag{46}$$

that, at the level of cocycles [i.e., closed $(j+m)$-forms on E], are given by the process called *integration along the fiber*. For a trivial fiber space $E = B \times F$, after fixing a point $b \in B$ and a tangent j-vector $z \in \bigwedge^j T_b(B)$, an $(m+j)$-form ω on E naturally defines a tangent m-covector

$$u \mapsto \omega((b, y), z \wedge u) \tag{47}$$

for $u \in \bigwedge^m T_{(b,y)}(E_b)$, where z (resp. u) is identified with its image in $\bigwedge T_{(b,y)}(E)$; these m-covectors are the values of an m-form ω_b on $E_b = \{b\} \times F$, and the integral

$$\Omega\colon b \mapsto \int_{E_b} \omega_b$$

is a j-form on B, "ω integrated along the fibers E_b," sometimes written $\oint_F \omega$. The map $\omega \mapsto \Omega$ commutes with exterior differentiation, and therefore yields the homomorphism (46) in the usual way.

When the slant product was later defined (Part 1, chap. IV, §5,H) it was realized that if (with the same notations) u is the cohomology class of ω, then the cohomology class of Ω is the slant product $u/[F]$.

When F is a sphere S_d, the homomorphism (46) turns out to be the homomorphism Ψ_j of the cohomology exact sequence (45). This was seen independently in 1950 by Thom [462] and Chern and Spanier [128]. The latter two extended the definition of "integration along the fiber" to the more general case in which B is a finite simplicial complex and m is the largest integer for which $H^m(F) \neq 0$; their method used the relative cohomology sequence for the successive skeletons of B and their inverse images in E.

Thom derived the exact sequence (45) for sphere bundles from a general result on fiber spaces that behave like products in cohomology: the "Leray–Hirsch theorem" (see §3,A), applicable to *oriented* vector fibrations $\xi = (E, B, p)$ of rank m (§1,C). This is due to two properties of these fibrations:

1. If E^0 is the open set in E complementary to the zero section (which can be identified with the base space B), then for each fiber E_b the relative cohomology $H^\bullet(E_b, E_b \cap E^0; \mathbf{Z})$ is reduced to the single module

$$H^m(E_b, E_b \cap E^0; \mathbf{Z}) \simeq H^m(\mathbf{R}^m, \mathbf{R}^m - \{0\}; \mathbf{Z}) \simeq \mathbf{Z}.$$

2. The existence of an orientation for ξ implies that there is a unique element u in $H^m(E, E^0; \mathbf{Z})$ such that for all linear bijections $i_b\colon \mathbf{R}^m \to E_b$ preserving orientation for each $b \in B$, $i_b^*(u)$ is the fundamental class of relative cohomo-

logy in $H^m(\mathbf{R}^m, \mathbf{R}^m - \{0\}; \mathbf{Z})$ (Part 2, chap. IV, § 3,A); this class u is called the *orientation class of* ξ (or of E), or the *fundamental class* of relative cohomology in $H^m(E, E^0; \mathbf{Z})$. The proof of the existence of such a class is done by steps, starting with the case of a trivial bundle $B \times \mathbf{R}^m$ and then taking an open covering of B by sets above which E is trivializable; the Mayer–Vietoris sequence concludes the proof.

From that definition, it follows that if $g: B' \to B$ is a continuous map, $\xi' = g^*(\xi) = (E', B', \mathbf{R}^m, p')$ the pullback of ξ by g, and $f: E' \to E$ the corresponding bundle map, then ξ' is orientable and $u' = f^*(u)$ is an orientation class of ξ'.

If $j_b: E_b \to E$ is the natural injection for each $b \in B$, $j_b^*(u)$ is a *generator* of the $\mathbf{Z}$-module $H^m(E_b, E_b \cap E^0; \mathbf{Z})$, and the Leray–Hirsch theorem can be applied: the $\mathbf{Z}$-homomorphism

$$\Phi_u: c \mapsto p^*(c) \smile u \tag{48}$$

of $H^j(B; \mathbf{Z})$ into $H^{j+m}(E, E^0; \mathbf{Z})$ is *bijective* for all $j \in \mathbf{Z}$; Φ_u is called the *Thom isomorphism.* For a pullback $\xi' = g^*(\xi)$, the diagram (with the above notations)

$$\begin{array}{ccc} H^{j+m}(E', E'^0; \mathbf{Z}) & \xrightarrow{f^*} & H^{j+m}(E, E^0; \mathbf{Z}) \\ \uparrow{\scriptstyle \Phi_{u'}} & & \uparrow{\scriptstyle \Phi_u} \\ H^j(B'; \mathbf{Z}) & \xrightarrow[g^*]{} & H^j(B; \mathbf{Z}) \end{array}$$

is commutative.

Now consider the cohomology exact sequence

$$\cdots \to H^r(E, E^0; \mathbf{Z}) \to H^r(E; \mathbf{Z}) \xrightarrow{h^*} H^r(E^0; \mathbf{Z}) \to H^{r+1}(E, E^0; \mathbf{Z}) \to \cdots$$

where $h: E^0 \to E$ is the natural injection. The Thom isomorphism (48) and the fact that

$$p^*: H^\bullet(B; \mathbf{Z}) \to H^\bullet(E; \mathbf{Z})$$

is an isomorphism give rise to the *Gysin cohomology exact sequence*

$$\cdots \to H^{r-m}(B; \mathbf{Z}) \xrightarrow{g} H^r(B; \mathbf{Z}) \xrightarrow{p_0^*} H^r(E^0; \mathbf{Z}) \to H^{r-m+1}(B; \mathbf{Z}) \to \cdots \tag{49}$$

where $p_0: E^0 \to B$ is the restriction of the projection p, and $g: c \mapsto c \smile e_u(\xi)$ is the cup product with the *Euler class* of ξ (§ 2,C).

For an *unoriented* vector fibration $\xi = (E, B, \mathbf{R}^m, p)$ of rank m similar arguments prove the existence of a unique element $u \in H^m(E, E^0; \mathbf{F}_2)$ such that $i_b^*(u)$ is the unique element $\neq 0$ in $H^m(\mathbf{R}^m, \mathbf{R}^m - \{0\}; \mathbf{F}_2)$; u is called the *fundamental class* of relative cohomology in $H^m(E, E^0; \mathbf{F}_2)$; the map Φ_u of (48) is now an isomorphism $H^j(B; \mathbf{F}_2) \simeq H^{j+m}(E, E^0; \mathbf{F}_2)$, again called the *Thom isomorphism.* The Gysin exact sequence is

$$\cdots \to H^{r-m}(B; \mathbf{F}_2) \xrightarrow{g} H^r(B; \mathbf{F}_2) \xrightarrow{p_0^*} H^r(E^0; \mathbf{F}_2) \to H^{r-m+1}(B; \mathbf{F}_2) \to \cdots \tag{50}$$

where $g: c \mapsto c \smile w_m(\xi)$ is now the cup product with the top Stiefel–Whitney class of ξ.

From the exact sequences (49) and (50) the structure of the cohomology algebras of the complex and real Grassmannians may be deduced, and, in consequence, the properties of Chern classes and of Stiefel–Whitney classes [347]. In particular, let $U_{n+1,1}(\mathbf{C})$ be the tautological complex line bundle with base $\mathbf{P}_n(\mathbf{C})$ (defined as above for the real Grassmannians). The cohomology algebra of the projective space $\mathbf{P}_n(\mathbf{C})$ over Z is generated by the Chern class $a = c_1(U_{n+1,1}(\mathbf{C})) \in H^2(\mathbf{P}_n(\mathbf{C}); \mathbf{Z})$. Since $a^m = 0$ for $m > 2n$, $H^{\bullet}(\mathbf{P}_n(\mathbf{C}); \mathbf{Z})$ is isomorphic to $\mathbf{Z}[a] = \mathbf{Z}[T^2]/(T^{2n})$.

There is also a Thom isomorphism in homology for an oriented vector fibration:

$$\Psi_u: z \mapsto p_*(u \frown z) \tag{51}$$

mapping $H_q(E, E^0; \mathbf{Z})$ onto $H_{q-m}(B: \mathbf{Z})$. From this isomorphism we deduce as above the *Gysin homology exact sequence*

$$\cdots \to H_q(E^0; \mathbf{Z}) \xrightarrow{p_{0*}} H_q(B; \mathbf{Z}) \to H_{q-m-1}(B; \mathbf{Z}) \to H_{q-1}(E^0; \mathbf{Z}) \to \cdots \tag{52}$$

for orientable vector bundles, and a similar exact sequence for homology with coefficients in $\mathbf{F}_2$ for unoriented vector bundles.

The Wang Exact Sequence

In 1949 [481] H. Wang considered fiber spaces (E, B, F, π), where the base space B is homeomorphic to a *sphere* S_n and the typical fiber F is a finite simplicial complex such that $H^s(F) = 0$ for $s \geqslant n - 1$. He then obtained an exact homology sequence

$$\cdots \to H_{r-n+1}(F) \to H_r(F) \to H_r(E) \to H_{r-n}(F) \to \cdots \tag{53}$$

by the following very simple argument. Consider a fiber $F_0 = E_{b_0}$ and the exact homology sequence

$$\cdots \to H_{r+1}(E, F_0) \to H_r(F_0) \to H_r(E) \to H_r(E, F_0) \to \cdots. \tag{54}$$

However (Part 1, chap. IV, §5,A), $H_r(E, F_0) \simeq H_r(E - F_0)$, and $E - F_0$ is a fiber space with a contractible base space $B - \{b_0\}$, hence $E - F_0$ is trivializable by Feldbau's theorem, and

$$H_r(E - F_0) \simeq H_{r-n}(F)$$

by Künneth's theorem.

The Wang and Gysin sequences were later obtained as consequences of the spectral sequences of a fibration (§3,A and C).

§3. The Spectral Sequences of a Fibration

A. The Leray Cohomology Spectral Sequence of a Fiber Space

When Leray invented the spectral sequence of a continuous map (Part 1, chap. IV, §7,E), he chiefly had in mind the case of a projection $\pi: X \to B$ of a (locally trivial) fiber space; almost immediately after his 1946 Notes defining sheaf cohomology and spectral sequences he published two Notes ([316], [317]) announcing the results obtained by applying these notions to fiber spaces; he gave the detailed proofs of these results in his 1950 papers ([321], [322]).

The assumptions on the fiber space (X, B, F, π) are that X, B, F are locally compact and arcwise-connected and B is locally arcwise-connected. Let Λ be a principal ideal ring and M is a Λ-module, and consider the *Alexander–Spanier cohomology with compact supports and with coefficients in* M. In the formation of Leray's spectral sequence for π the sheaf $\mathscr{H}^{\bullet}(\pi; M)$ has at each point $b \in B$ a stalk isomorphic to $H^{\bullet}(X_b; M)$ [hence to $H^{\bullet}(F; M)$]; it can therefore be identified to the *local system* $(H^{\bullet}(X_b; M))$ (Part 1, chap. IV, § 7,A). Leray's spectral sequence then has E_2 terms given by

$$E_2^{pq} = H^p(B; (H^q(X_b, M))). \tag{55}$$

When the fundamental group $\pi_1(B)$ acts trivially on $H^{\bullet}(X_b; M)$ for one point $b \in B$ (hence for all $b \in B$), one says the system $(H^{\bullet}(X_b; M))$ is *simple*, and (55) may be replaced by

$$E_2^{pq} = H^p(B; H^q(F; M)). \tag{56}$$

If, in that case, M is also a *field*,

$$E_2^{pq} \simeq H^p(B; M) \otimes H^q(F; M) \tag{57}$$

and similarly, when B or F has no torsion and $M = \mathbf{Z}$,

$$E_2^{pq} \simeq H^p(B; \mathbf{Z}) \otimes H^q(F; \mathbf{Z}). \tag{58}$$

The homomorphisms π^* and i^* in cohomology (where $i: F \to X$ is the natural injection for a fiber F) can be related to the M-modules $E_r^{p,o}$ and $E_r^{o,q}$:

$$E_2^{p,o} \simeq H^p(B; M), \tag{59}$$

and the elements of $E_r^{p,o}$ are d_r-cocycles, so $E_{r+1}^{p,o}$ is a quotient of $E_r^{p,o}$, and π^* can be written as a composite of surjective homomorphisms

$$H^p(B; M) = E_2^{p,o} \to E_3^{p,o} \to \cdots \to E_{p+1}^{p,o} = E_\infty^{p,o} \subset H^p(X; M). \tag{60}$$

Similarly, $E_2^{o,q} = H^q(F; M)^f$, the submodule consisting of the elements of $H^q(F; M)$ *fixed* under the action of $\pi_1(B)$. The elements of $E_r^{o,q}$ cannot be coboundaries for d_r, so $E_{r+1}^{o,q}$ is the submodule of $E_r^{o,q}$ consisting of its d_r-cocycles; there is a succession of inclusions

$$H^q(F; M)^f = E_2^{o,q} \supset E_3^{o,q} \supset \cdots \supset E_{q+2}^{o,q} = E_\infty^{o,q}, \tag{61}$$

and $E_\infty^{o,q}$ is a quotient of $H^q(X; M)$.

An important special case is when the homomorphism

$$i^*: H^{\bullet}(X; M) \to H^{\bullet}(F; M)$$

is *surjective*; then one says F is *totally nonhomologous to* 0. If in addition X is compact and $M = K$ is a field, then $\pi_1(B)$ acts trivially on $H^{\bullet}(F; K)$, all differentials d_r are 0 for $r \geqslant 2$, so that

$$E_2^{pq} = H^p(B; K) \otimes H^q(F; K) \simeq E_\infty^{pq}, \tag{62}$$

and the vector space $H^\cdot(X; K)$ over K is isomorphic to the tensor product $H^\cdot(B; K) \otimes H^\cdot(F; K)$. This result was independently obtained by G. Hirsch in 1948 [228] as a special case of a different approach to the study of $H^\cdot(X; K)$; working at the level of cochains, he defined an increasing sequence of subgroups of $H^\cdot(F; K)$ by induction, and then a boundary operator on the vector space $C^\cdot(B; K) \otimes H^\cdot(F; K)$, such that the corresponding homology is isomorphic to $H^\cdot(X; K)$ (see [231]).

This *Leray–Hirsch theorem* for fibers totally nonhomologous to 0 has since been proved in a more elementary way [262]: for each $p \geqslant 0$, there is in $H^p(X; K)$ a family $(a_{p,1}, a_{p,2}, \ldots, a_{p,r_p})$ such that the elements

$$i^*(a_{p,1}), i^*(a_{p,2}), \ldots, i^*(a_{p,r_p})$$

form a basis of $H^p(F; K)$ over K; it is then proved that for the map

$$(c, a) \mapsto \pi^*(c) \smile a \tag{63}$$

$H^\cdot(X; K)$ is a $H^\cdot(B; K)$-module which is free and has the basis

$$(a_{pk})_{1 \leqslant k \leqslant r_p, 0 \leqslant p \leqslant n}$$

if $H^s(F; K) = 0$ for $s > n$. This is first done for a trivial fiber space $B \times F$ and next for an open covering (U_α) of B such that X is trivializable over each U_α, using induction and the Mayer–Vietoris sequence. The argument works when K is replaced by any ring M, provided $H^\cdot(F; M)$ is free.

The fact that i^* is surjective implies that π^* is injective, but not necessarily the converse. G. Hirsch gave an example in which π^* may be injective and i^* not surjective ([57], p. 56).

In the general case, assuming that $\pi_1(B)$ acts trivially on $H^\cdot(F; M)$, if two of the cohomology modules $H^\cdot(B; M)$, $H^\cdot(F; M)$, $H^\cdot(X; M)$ are finitely generated, the same is true for the third one. If $M = K$ is also a field, then the Betti numbers of X, B, F satisfy inequalities

$$b_k(X) \leqslant b_k(B \times F) \tag{64}$$

(also proved independently by Hirsch), giving the relation between Euler–Poincaré characteristics

$$\chi(X) = \chi(B)\chi(F). \tag{65}$$

Another application of the Leray spectral sequence is the determination of all possible locally trivial fibrations of a space $\mathbf{R}^n$ with connected fibers; A. Borel and Serre showed that the base space and the typical fiber must have the same homology as an $\mathbf{R}^p$ and an $\mathbf{R}^{n-p}$, respectively, for some $p \leqslant n$ [67].

Finally, the cohomology exact sequences of Gysin and Wang (§2) can be deduced from Leray's spectral sequence; it is enough to assume, first that $\pi_1(B)$ acts trivially on $H^\cdot(F; M)$, and second, that $H^\cdot(F; M) = H^\cdot(S_k; M)$ for the Gysin sequence and $H^{\cdot}(B; M) = H^{\cdot}(S_k; M)$ for the Wang sequence. This follows from the fact that the only differential d_r in the spectral sequence which is $\neq 0$ is d_{k+1} for the Gysin sequence and d_k for the Wang sequence. Leray also observed that when M is a principal ideal ring, the homomorphisms

$$\theta: H^i(F; M) \to H^{i-k+1}(F; M) \tag{66}$$

in the Wang sequence are *derivations* if k is odd, *antiderivations* if k is even [for the ring structure of $H^{\cdot}(F; M)$].

B. The Transgression

The first step in the construction outlined in G. Hirsch's 1948 Note [228] is [for a locally trivial fiber space (X, B, F, π)] the definition of a submodule of $H^s(F; M)$, a quotient module of $H^{s+1}(B; M)$, and a homomorphism of the first into the second. At the same time Koszul [286] was working on the cohomology of homogeneous spaces G/U of compact Lie groups for his thesis; but, using E. Cartan's method (Part 2, chap. VI, § 1), he exclusively considered the Lie algebras and the cohomology groups of these algebras as they had been defined a little earlier (independently of the theory of Lie groups) by Chevalley and Eilenberg ([113], [134]). Koszul's study therefore actually belonged to homological algebra, although the inspiration derived from Lie group theory was everywhere apparent.

The algebraic counterpart of the relation between the cohomology groups of G, U, and G/U was the relation between the cohomology of a Lie algebra $\mathfrak{a}$ and a Lie subalgebra $\mathfrak{b}$. It is beyond the scope of this historical survey to analyze Koszul's results in detail; it is enough to mention that, besides the cohomology algebras $H^{\cdot}(\mathfrak{a})$ and $H^{\cdot}(\mathfrak{b})$, he introduced a "relative" cohomology algebra $H^{\cdot}(\mathfrak{a}, \mathfrak{b})$, the counterpart of $H^{\cdot}(G/U)$ in Lie algebra theory. He also defined a spectral sequence, in which the E_2 term is

$$H^{\cdot}(\mathfrak{b}) \otimes H^{\cdot}(\mathfrak{a}, \mathfrak{b})$$

and the abutment is a filtered module associated to $H^{\cdot}(\mathfrak{a})$.

Koszul finally introduced notions that corresponded in Lie algebras to those considered by Hirsch: for each p, he defined a subgroup T^p of $H^p(\mathfrak{b})$, a quotient group of $H^{p+1}(\mathfrak{a}, \mathfrak{b})$, and a homomorphism τ_p of T^p into that quotient and said that the elements of T^p are *transgressive*, calling τ_p the *transgression* (for dimension p). He also observed that τ_p could be identified with a homomorphism

$$E_{p+1}^{o,p} \to E_{p+1}^{p+1,o}$$

in his spectral sequence, namely, the restriction of the differential d_{p+1}.

In 1949 Chevalley, H. Cartan, and A. Weil joined Koszul in the development of his methods, and applied them to the cohomology of principal fiber spaces with compact Lie groups as structure groups. Like Koszul they worked with Lie algebras, and their results only dealt with cohomology with *real* coefficients (see § 4).

In his thesis ([429], p. 434) Serre gave a definition of the transgression for the spectral sequence of a *general* graded differential module $M_{\bullet}$ (Part 1, chap. IV, §7,D), under the single assumption that $M_q = \{0\}$ for $q < 0$, and that if $w(x)$ is the largest integer p such that the homogeneous element $x \in F^p(M)$, then $0 \leqslant w(x) \leqslant \deg(x)$.

Finally A. Borel, in his thesis [58], made extensive use of the transgression to obtain his results on cohomology of the classifying spaces of Lie groups (see §4). He returned to the situation originally considered by Hirsch: for any locally trivial fiber space (X, B, F, π), an element $x \in H^q(F; M)$ is *transgressive* if there exists a cochain $c \in C^q(X; M)$ such that the image $\tilde{\imath}(c)$ (where $i: F \to X$ is the natural injection) is a *cocycle of class* x and $\mathbf{d}c = \tilde{\pi}(b)$, where $b \in C^{q+1}(B; M)$. This implies that b is a cocycle; its class in $H^{q+1}(B; M)$ is only defined modulo a submodule N_{q+1} of $H^{q+1}(B; M)$. It turned out that the submodule $T^q(F; M)$ of transgressive elements in $H^q(F; M)$ is naturally isomorphic to the module $E_{q+1}^{o,q}$ of the cohomology spectral sequence, and its image in $H^{q+1}(B; M)/N_{q+1}$ is naturally isomorphic to $E_{q+1}^{q+1,o}$; the *transgression* $\tau_q: T^q(F; M) \to H^{q+1}(B; M)/N_{q+1}$ is then identified with the restriction to $E_{q+1}^{o,q}$ of the differential d_{q+1}.

C. The Serre Spectral Sequences [429]

We shall see in chap. V that after the Eilenberg–Mac Lane spaces $K(\Pi, n)$ were defined (chap. II, §6,F), the determination of their cohomology became an important problem. Serre observed that the Leray spectral sequence might give information about that problem, provided he could define a fibration (X, B, F), where B is a space $K(\Pi, n)$ and X is *contractible*. He then had the idea of considering, for *any* arcwise-connected space B, the fibration (P, B, Ω), where $P = \Omega(B, B, b_0)$ is the *space of paths* in B with origin b_0 and arbitrary extremity in B; it is easy to see that P is contractible, but (chap. III, §2,A) it is not a locally trivial fiber space with base space B, and it is not locally compact, so that Leray's spectral sequence, using Alexander–Spanier cohomology, is not applicable. To achieve his purpose Serre had to use *singular cohomology* (with an arbitrary group of coefficients G), and to show that for all fibrations there exists a spectral sequence, based on that cohomology, having the same properties as the Leray spectral sequence.

Serre also wanted to have a spectral sequence for *singular homology*. He first had to define the spectral sequence of a differential module (M, d) equipped with an *increasing filtration* (M_p) such that $d(M_p) \subset M_p$; if $Z = \operatorname{Ker} d$, $B = \operatorname{Im} d$, the $Z_p = M_p \cap Z$ (resp. $B_p = M_p \cap B$) are increasing filtrations on Z and B, respectively; Z_p^r is then defined as the submodule of M_p consisting of the elements z such that $dz \in M_{p-r}$, and B_p^r is the submodule $d(Z_{p+r}^r)$. He then introduced the modules

$$E_p^r = Z_p^r/(Z_{p-1}^{r-1} + B_p^{r-1}); \tag{67}$$

d defines in E_p^r a homomorphism

$$d_r\colon \mathrm{E}_p^r \to \mathrm{E}_{p-r}^r \tag{68}$$

such that $d_r \circ d_r = 0$; if $\mathrm{E}_\bullet^r = \bigoplus_p \mathrm{E}_p^r$, $(\mathrm{E}_\bullet^r)$ is the *spectral sequence* defined by the increasing filtration (M_p), and for all p,

$$\mathrm{H}_p(\mathrm{E}_p^r) = \mathrm{E}_p^{r+1}. \tag{69}$$

The homology module H(M) is filtered by the images $\mathrm{Z}_p/\mathrm{B}_p$ of the Z_p, and the corresponding graded module $\mathrm{gr}_\bullet(\mathrm{H}(\mathrm{M}))$ has as homogeneous components the

$$\mathrm{E}_p^\infty = \mathrm{Z}_p/(\mathrm{Z}_{p-1} + \mathrm{B}_p)$$

so that the E_p^r may be considered as approximations of the E_p^∞; $\mathrm{gr}_\bullet(\mathrm{H}(\mathrm{M}))$ is again called the *abutment* of the spectral sequence $(\mathrm{E}_\bullet^r)$.

In most cases the process again starts with a *graded* differential module $\mathrm{M}^\bullet = \bigoplus_q \mathrm{M}^q$, with $d(\mathrm{M}^q) \subset \mathrm{M}^{q-1}$ and a filtration compatible with the grading, so that $\mathrm{M}_p = \bigoplus_q (\mathrm{M}^q \cap \mathrm{M}_p)$. Then, if

$$\mathrm{Z}_{pq}^r = \mathrm{Z}_p^r \cap \mathrm{M}^{p+q}, \qquad \mathrm{B}_{pq}^r = \mathrm{B}_p^r \cap \mathrm{M}^{p+q},$$

the

$$\mathrm{E}_{pq}^r = \mathrm{Z}_{pq}^r/(\mathrm{Z}_{p-1,q+1}^{r-1} + \mathrm{B}_{pq}^{r-1})$$

form a grading of E_p^r.

Serre elected to work with *cubical* singular homology and cohomology (Part 1, chap. IV, § 6,C), which he found more convenient than simplices when dealing with product spaces. As he only considered arcwise-connected spaces, he could restrict the singular cubes $\mathrm{I}^n \to \mathrm{X}$ to those for which the images of *all* the vertices of I^n are always equal to a *fixed* point $x_0 \in \mathrm{X}$; we shall again write $\mathrm{Q}_\bullet(\mathrm{X}; \mathbf{Z})$ the chain complex of these singular cubes, and $\bar{\mathrm{Q}}_\bullet(\mathrm{X}; \mathbf{Z})$ is its quotient by the degenerate cubes (*loc. cit.*).

For a fibration $(\mathrm{X}, \mathrm{B}, \mathrm{F}, \pi)$, the starting point is the definition of an *increasing filtration* (T_p) on $\mathrm{Q}_\bullet(\mathrm{X}; \mathbf{Z})$; Serre's definition is

$$\mathrm{T}_p = \bigoplus_q \mathrm{T}_{pq}$$

where T_{pq} is the subgroup generated by the singular cubes $u\colon \mathrm{I}^{p+q} \to \mathrm{X}$ such that $\pi(u(t_1, \ldots, t_p, t_{p+1}, \ldots, t_{p+q}))$ is *independent* of $t_{p+1}, \ldots, t_{p+q}$. The filtration on the chain complex $\bar{\mathrm{Q}}_\bullet(\mathrm{X}; \mathbf{Z})$, from which the spectral sequence is constructed, then consists of the images A_p of the T_p.

The main technical point Serre had to establish was the expression of the terms E_{pq}^1 of the spectral sequence. Following the pattern of Leray's spectral sequence, he showed that E_{pq}^1 is naturally isomorphic to

$$\bar{\mathrm{Q}}_p(\mathrm{B}; \mathbf{Z}) \otimes \mathrm{H}_q(\mathrm{F}; \mathrm{G}), \tag{70}$$

by first considering the case $\mathrm{G} = \mathbf{Z}$ and showing that there is a *chain equivalence* (Part 1, chap. IV, § 5,F) of the chain complex $\mathrm{E}_p^0 = \mathrm{A}_p/\mathrm{A}_{p-1}$ onto the chain complex

$$J_p = \bar{Q}_p(B; \mathbf{Z}) \otimes \bar{Q}_{\bullet}(F; \mathbf{Z}). \tag{71}$$

In one direction, it is natural to associate to a singular $(p + q)$-cube $u \in T_{pq}$ the pair consisting of:

a. the singular p-cube Bu of B defined by

$$Bu(t_1, \ldots, t_p) = \pi(u(t_1, \ldots, t_p, y_1, \ldots, y_q))$$

where the y_j are fixed arbitrary numbers in I;

b. the singular q-cube Fu of F defined by

$$Fu(t_1, \ldots, t_q) = u(0, \ldots, 0, t_1, \ldots, t_q)$$

(F is identified to $\pi^{-1}(x_0)$).

There is thus a chain transformation

$$\varphi\colon E_p^o \to J_p.$$

More technique is required to define a chain transformation in the opposite direction

$$\psi\colon J_p \to E_p^o$$

such that $\varphi \circ \psi = \mathrm{Id}$, and then to prove the existence of a chain homotopy (Part 1, chap. IV, § 5,F) between $\psi \circ \varphi$ and Id; the details are straightforward but fairly long, using induction on q, and it is there that the covering homotopy property is a crucial ingredient ([429], pp. 459–464). Once the chain equivalence of E_p^o and J_p has been constructed, it is extended to a similar one between the term E_p^o corresponding to the filtration of $\bar{Q}_{\bullet}(X; \mathbf{Z}) \otimes G$, and the chain complex $J_p \otimes G$, for any group of coefficients G; finally the expression (70) is obtained, from which (using the chain transformations φ and ψ)

$$E_{pq}^2 \simeq H_p(B; (H_q(X_b; G))) \tag{72}$$

the coefficients being the local system $(H_q(X_b; G))$ as in the Leray spectral sequence.

The corresponding properties for cubical singular cohomology are easily deduced by duality: on the cochain complex $\mathrm{Hom}(\bar{Q}_{\bullet}(X; \mathbf{Z}), G)$ define a decreasing filtration (M^p) by taking for M^p the subgroup of cochains vanishing on all singular cubes belonging to the A_q for $q \leqslant p - 1$. If $J^p = \mathrm{Hom}(J_p, G)$, the maps $\varphi' = \mathrm{Hom}(\varphi)$ and $\psi' = \mathrm{Hom}(\psi)$ define a chain equivalence, which enabled Serre to show that in the spectral sequence for singular cohomology,

$$E_2^{pq} \simeq H^p(B; (H^q(X_b; G))) \tag{73}$$

as in Leray's spectral sequence. When G is a commutative ring, Serre also showed how to define cup-products in cubical singular cohomology, and proved that this yields a structure of graded algebra on each $E_r^{\bullet}$ ($p + q$ being the degree of E_r^{pq}) that is anticommutative.

Once these foundational preliminaries were out of the way Serre could extend the results established by Leray for locally trivial fiber spaces to his cohomology spectral sequence, and he proved similar ones for the homology spectral sequence.* He also pointed out that when $\pi_1(B)$ acts trivially on $H_\bullet(F)$ and one has $H_i(B) = 0$ for $0 < i < p$ and $H_j(F) = 0$ for $0 < j < q$, then there is a natural exact sequence (often named the *Serre exact sequence*)

$$\begin{aligned} H_{p+q-1}(F) \to H_{p+q-1}(X) \to H_{p+q-1}(B) \xrightarrow{t} H_{p+q-2}(F) \to \\ \cdots \to H_2(B) \xrightarrow{t} H_1(F) \to H_1(X) \to H_1(B) \to 0 \end{aligned} \tag{74}$$

where t is the transgression.

Serre's main applications of his spectral sequences in his thesis were to the *space of paths* $P = \Omega(X, X, x_0)$ of fixed origin x_0 in an arcwise-connected space X and to the fibration (P, X, π), where $\pi: P \to X$ is the map that to each path $\gamma \in P$ assigns its extremity in X; the fiber $\pi^{-1}(x_0)$ is the *space of loops* $\Omega = \Omega(X, x_0)$ of origin x_0 and P is contractible.

As general results on the homology of Ω when X is *simply connected*, Serre showed that if Λ is a principal ideal ring and the $H_i(X; \Lambda)$ are finitely generated Λ-modules, then the same is true for the $H_i(\Omega; \Lambda)$, due to the Leray theorem mentioned in part A, generalized to singular homology. Next, if X is again *simply connected*, k is a field, $H_i(X; k) = 0$ for $i > n \geqslant 2$ and $H_n(X; k) \neq 0$, then for *any* dimension $i \geqslant 0$ such that $H_i(\Omega; k) \neq 0$, there is a j such that $0 < j < n$ and $H_{i+j}(\Omega; k) \neq 0$. The proof is by contradiction. Since

$$E^2_{n-r, i+r-1} \simeq H_{n-r}(X; k) \otimes H_{i+r-1}(\Omega; k),$$

one would have $E^2_{n-r, i+r-1} = 0$ for $2 \leqslant r \leqslant n$, hence $E^r_{n-r, i+r-1} = 0$; this is also true of course if $r > n$, so that $d_r(E^r_{n,i}) = 0$ for *all* $r \geqslant 2$; the elements of $E^r_{n,i}$ cannot be boundaries for d_r, so all $E^r_{n,i}$ are isomorphic, hence $E^\infty_{n,i} \simeq E^r_{n,i}$; however, as $H_i(\Omega; k) \neq 0$, $E^2_{n,i} \simeq H_n(X; k) \otimes H_i(\Omega; k) \neq 0$. This implies $E^\infty_{n,i} \neq 0$, which is absurd since $\tilde{H}_\bullet(P) = 0$.

In particular, under the same assumptions $H_i(\Omega; k) \neq 0$ for *infinitely many* values of i. Since Ω has the same homotopy type as the space of paths $P_{x,y}$ joining two distinct points x, y of X, application of a result of M. Morse gives the result that there are *infinitely many* geodesic arcs $[0, 1] \to X$ joining x and y.†

Serre was especially interested in the space of loops of the spheres S_n ($n \geqslant 2$), in relation with the applications to the homotopy groups of S_n (see chap. V, § 5,A). Using Wang's homology exact sequence, he first gave a much simpler proof of a result of Morse, namely,

$$\begin{cases} H_i(\Omega; Z) \simeq Z & \text{for } i \equiv 0 \,(\text{mod}\,. n-1), \\ H_i(\Omega; Z) = 0 & \text{for the other indices } i. \end{cases} \tag{75}$$

---

* A slight difference with the Leray spectral sequence is that the relation $\chi(X) = \chi(B)\chi(F)$ is not always true when $\pi_1(B)$ does not act trivially on $H_\bullet(X_b; G)$; a counterexample has been given by A. Douady ([425], p. 3-02).

† The example of $X = S_n$ shows that the *images* of these arcs in X may only consist in a finite number of curves; the only exception to that in S_n occurs when x and y are antipodal points.

In addition, using the fact [formula (66)] that in the Wang cohomology sequence, the maps

$$\theta\colon \mathrm{H}^i(\Omega; \mathbf{Z}) \to \mathrm{H}^{i-n+1}(\Omega; \mathbf{Z})$$

are derivations when n is odd and antiderivations when n is even, Serre showed that for each integer $p \geqslant 0$ there is an element $e_p \in \mathrm{H}^{p(n-1)}(\Omega; \mathbf{Z})$ which is a basis for that module and satisfies the multiplication rules

$$e_p \smile e_q = c_{pq} e_{p+q} \tag{76}$$

where

$$\begin{cases} c_{pq} = (p+q)!/p!\,q! & \text{when } n \text{ is odd,} \\ c_{pq} = 0 & \text{if } n \text{ is even, } p \text{ and } q \text{ odd,} \\ c_{pq} = [(p+q)/2]!/[p/2]!\,[q/2]! & \text{in the other cases} \end{cases} \tag{77}$$

($[x]$ integer such that $[x] \leqslant x < [x] + 1$).

A little later Bott and Samelson [83] applied the homology spectral sequence (with a slight change in the filtration) to determine the structure of the ring $\mathrm{H}_\bullet(\Omega; \mathbf{Z})$ (for the Pontrjagin product, see Part 2, chap. VI, § 2,B) when X is a wedge of a finite family of spheres (Part 2, chap. V, § 2,D).

§ 4. Applications to Principal Fiber Spaces

The second of Leray's Notes on fiber spaces in 1946 [317] was concerned with applications of the spectral sequence of fiber spaces to the particular case (G, G/T, T), where G is a *classical* connected quasisimple compact Lie group, and T is a maximal torus in G; Leray was able to determine the cohomology $\mathrm{H}^\bullet(\mathrm{G/T}; \mathbf{R})$ with real coefficients explicitly in these cases. Then in 1949 [320] he considered more generally (for the same group G) a closed connected subgroup U of G such that $\mathrm{T} \subset \mathrm{U} \subset \mathrm{G}$ (so that G and U have the same *rank* $l = \dim \mathrm{T}$), and he studied the relations between the cohomology with real coefficients of G, U and G/U.

G. Hirsch was interested in the same problem at the same time. By Hopf's theorem (Part 2, chap. VI, § 2,A), the Poincaré polynomials of G and U can be written

$$\mathrm{P}(\mathrm{G}; t) = (1 + t^{2m_1 - 1}) \cdots (1 + t^{2m_l - 1}), \tag{78}$$

$$\mathrm{P}(\mathrm{U}; t) = (1 + t^{2n_1 - 1}) \cdots (1 + t^{2n_l - 1}), \tag{79}$$

for two increasing sequences of integers $m_1 \leqslant m_2 \leqslant \cdots \leqslant m_l$, $n_1 \leqslant n_2 \leqslant \cdots \leqslant n_l$; Hirsch conjectured that

$$\mathrm{P}(\mathrm{G/U}; t) = \frac{(1 - t^{2m_1}) \cdots (1 - t^{2m_l})}{(1 - t^{2n_1}) \cdots (1 - t^{2n_l})}. \tag{80}$$

For the classical quasisimple groups G, Leray proved the Hirsch formula

(80) as an application of the spectral sequence of the principal fiber space (G, G/U, U) [320]. At the same time the formula was proved by H. Cartan and Koszul ([109], [287]) in the context of their work (jointly with Chevalley and A. Weil) on reductive Lie algebras over **R**, and using the corresponding transgression (§ 3,B).

It would be too long and difficult to examine in detail the contributions of each of these mathematicians, especially since many of them are stated either without proof or with very sketchy ones.* Fortunately, in his thesis [58] and in several subsequent papers ([59], [61]), A. Borel built up a unifield theory that incorporated as particular cases all the previous results; we shall restrict ourselves to a description of the most salient features of that theory. The main tool is the transgression in the spectral sequence of an *n-universal* principal fiber space (P_G, B_G, G) for a given compact Lie group G, and for large enough n (chap. III, § 2,F).

The central result is a *purely algebraic* investigation of "abstract" spectral sequences (E_r^{pq}) that are not even supposed to be constructed by the Leray–Koszul method from some filtration. They are only assumed to satisfy the following axioms:

1. $(E_r^{\bullet}, d_r)$ is a differential algebra, bigraded by the E_r^{pq}, anticommutative for the total degree $p + q$, such that

$$d_r(E_r^{pq}) \subset E_r^{p+r,q-r+1};$$

2. $E_r^{pq} = 0$ for $p < 0$ or $q < 0$ ("first quadrant sequence");
3. $H^{\bullet}(E_r^{\bullet}) = E_{r+1}^{\bullet}$ for d_r.

By 3, there is a projection k_{r+1}^r of the module of cycles in $E_r^{\bullet}$, onto $E_{r+1}^{\bullet}$. A cycle $x \in E_r^{\bullet}$ is called a *permanent cycle* if it satisfies the infinite sequence of relations

$$d_r x = 0,\ d_{r+1} k_{r+1}^r(x) = 0, \ldots, d_s k_s^{s-1} k_{s-1}^{s-2}, \ldots, k_{r+1}^r(x) = 0, \ldots$$

and $E_{\infty}^{\bullet}$ is defined as the *direct limit* of the modules of permanent cycles, for the maps k_{r+1}^r.

Finally, the transgression is defined for every r as above, being the restriction

$$E_{r+1}^{0,r} \to E_{r+1}^{r+1,0}$$

of the differential d_{r+1}.

With these definitions, the *assumptions* of Borel's main theorem are as follows:

(i) The ring of coefficients is a *field* K.
(ii) One has

$$E_2^{pq} = B^p \otimes (\textstyle\bigwedge (x_1, \ldots, x_j, \ldots)^q), \tag{81}$$

* For a thorough exposition of the work of the "quartet" Cartan–Chevalley–Koszul–Weil, with complete proofs, see [211], vol. III.

where $B^\bullet$ is an anticommutative graded algebra with grading (B^p) and B^o has the unit element of $B^\bullet$ as a basis; $\bigwedge(x_1, \ldots, x_j, \ldots)$ is the exterior algebra of a graded vector space having a basis consisting of homogeneous elements x_j (of odd degree if K has a characteristic $\neq 2$), such that $\deg x_i \leqslant \deg x_j$ if $i \leqslant j$, and only finitely many x_j can have the same degree; then $\bigwedge(x_1, \ldots, x_j, \ldots)$ is a graded algebra where $(\bigwedge(x_1, \ldots, x_j, \ldots))^q$ is the vector subspace having as basis the products $x_{i_1} x_{i_2} \cdots x_{i_m}$ with $i_1 \leqslant i_2 \leqslant \cdots \leqslant i_m$ and

$$\deg x_{i_1} + \cdots + \deg x_{i_m} = q.$$

(iii) $E_\infty^{oo} = K$ and $E_\infty^{pq} = 0$ if $p + q > 0$.

The *conclusion* is that there are transgressive homogeneous elements x'_j with $\deg x'_j = \deg x_j$ generating $\bigwedge(x_1, \ldots, x_j, \ldots)$, and that

$$B^\bullet \simeq K[y_1, \ldots, y_j, \ldots] \tag{82}$$

is an algebra of *polynomials* in the y_j, where y_j is the image of x'_j by *transgression* (so that $\deg y_j = 1 + \deg x_j$).

The proof is long and intricate, a succession of inductive arguments (for other approaches to similar results, see [187] and [426], exp. 7).

The main application of this algebraic method is to the universal principal fiber space (P_G, B_G, G) having as structural group G an arbitrary compact connected Lie group. Elements of $H^\bullet(G; K)$ which are transgressive for that fiber space are called *universally transgressive*.

We now describe Borel's principal results:

I. If $H^\bullet(G; K) = \bigwedge(x_1, \ldots, x_m)$, where the x_j are homogeneous elements forming the basis of a vector subspace of $H^\bullet(G; K)$ and having *odd* degrees [by Hopf's theorem this is the case when K has characteristic 0 (Part 2, chap. VI, §2,A)], then there are *universally transgressive* elements $x'_1, \ldots, x'_m$ such that x_j and x'_j have the same degree and $H^\bullet(G; K) = \bigwedge(x'_1, \ldots, x'_m)$. Furthermore, if y_j is the image of x'_j by transgression (so that $\deg y_j = 1 + \deg x_j$)

$$H^\bullet(B_G; K) = K[y_1, \ldots, y_m] \tag{83}$$

where, in the ring of polynomials $K[y_1, \ldots, y_m]$, $H^q(B_G; K)$ is the vector space generated by the monomials $y_1^{\alpha_1} y_2^{\alpha_2} \cdots y_m^{\alpha_m}$ such that $\sum_{j=1}^m \alpha_j \deg y_j = q$.

The spectral sequence of the principal fiber space (P_G, B_G, G) has for E_2 terms

$$E_2^\bullet \simeq H^\bullet(B_G; K) \otimes H^\bullet(G; K) \simeq K[y_1, \ldots, y_m] \otimes \bigwedge(x'_1, \ldots, x'_m). \tag{84}$$

This is the algebra that Weil had introduced in 1949 ([482], p. 434) and that became the central concept in the "quartet's" work on the Koszul transgression. Borel acknowledges ([58], p. 124) that this work influenced his ideas on the cohomology of principal fiber spaces; relation (84) puts the "Weil algebra" in a context more natural than its rather arbitrary introduction by Weil, and is valid for fields K of arbitrary characteristic.

II. Conversely, if $H^{\bullet}(B_G; K) = K[y_1, \ldots, y_m]$, where the y_j have *even* degrees, then $H^{\bullet}(G; K) = \bigwedge(x_1, \ldots, x_m)$, where the x_j are universally transgressive elements whose images by transgression are the y_j.

The results of I and II extend to cohomology with integer coefficients when G has no torsion.

III. If K has characteristic $p > 0$, a necessary and sufficient condition for G to have no p-torsion is that

$$H^{\bullet}(G; K) \simeq \bigwedge(x_1, \ldots, x_m)$$

where the x_j have odd degrees; then B_G also has no p-torsion.

IV. When K has characteristic 2, there is a more special result. One says $H^{\bullet}(G; K)$ has a *simple system* of generators x_j when the x_j are homogeneous and the products $x_{i_1} x_{i_2} \cdots x_{i_r}$ with $i_1 < i_2 < \cdots < i_r$ form a basis of $H^{\bullet}(G; K)$ (it is not assumed that $x_j^2 = 0$). Then if $H^{\bullet}(G; K)$ has such a simple system of generators, and if these generators are universally transgressive, $H^{\bullet}(B_G; K) = K[y_1, \ldots, y_m]$, where y_j is the image of x_j by transgression.

The assumption on the x_j is often satisfied even when G has 2-torsion, but there are examples of the contrary ([55], pp. 367 and 704).

V. An important part of Borel's thesis was the general study of the relations between the cohomology of B_G and of B_U when U is a closed subgroup of a compact connected Lie group G. U acts naturally on the universal principal fiber space P_G, and B_U can be considered the space of orbits for that action, so that (P_G, B_U, U) is a universal principal fiber space for U; the map $\rho = \rho(U, G)$, that to each orbit in B_U assigns the unique orbit in B_G in which it is contained, defines $(B_U, B_G, G/U, \rho)$ as a fiber space [chap. III, § 1,C, IV, (i)].

When *no* further assumptions are made on U, but the field of coefficients K has *characteristic* 0, Borel gave a topological proof of a theorem first proved by H. Cartan in the context of Lie algebras ([109], p. 65), using the Hirsch construction (see § 3,A) modified by Koszul. $H^{\bullet}(G/U; K)$ is isomorphic to the cohomology of the tensor product

$$H^{\bullet}(B_U; K) \otimes H^{\bullet}(G; K) \tag{85}$$

for a cohomology operator Δ, which is explicitly determined by $\rho(U, G)$ and the transgression in P_G. Cartan's result was later generalized to fields K of arbitrary characteristic.

VI. Borel completed the investigation of the cohomology of G/T begun by Leray when T is a *maximal torus* in G. He considered the principal fiber space (P_G, B_T, T) and the action on $P_T = P_G$ of the Weyl group W of G, defined as follows. Since W is the quotient N(T)/T of the normalizer N(T) of T, an element of N(T) acts on P_T by permuting the cosets xT in a fiber, and the action only depends on the class mod. T of that element. Thus W acts naturally on the orbit space B_T, and on the spectral sequence of P_T. Since $T \simeq (S_1)^l$ has no torsion, the $E_2^{\bullet}$ term of that sequence is given by

$$E_2^{\bullet} \simeq H^{\bullet}(B_T; \mathbf{Z}) \otimes_{\mathbf{Z}} H^{\bullet}(T; \mathbf{Z}).$$

$H^{\cdot}(T; \mathbf{Z})$ is the exterior algebra $\bigwedge(x_1, \ldots, x_l)$ of a free $\mathbf{Z}$-module with l generators x_j of degree 1, which can be taken transgressive; as in the spectral sequence the differential $d_r = 0$ for $r \geqslant 3$, the transgression τ maps $H^1(T; \mathbf{Z})$ isomorphically onto $H^2(B_T; \mathbf{Z})$; so, by I,

$$H^{\cdot}(B_T; \mathbf{Z}) = \mathbf{Z}[v_1, \ldots, v_l]$$

where $v_j = \tau(x_j)$ have degree 2. If I_G is the ring of polynomials belonging to $\mathbf{Z}[v_1, \ldots, v_l]$ and *invariant* under W, it is a direct summand in $H^{\cdot}(B_T; \mathbf{Z})$; so, for a field K, $I_G \otimes_{\mathbf{Z}} K$ is canonically embedded in $H^{\cdot}(B_T; K)$; it is contained in the ring of polynomials in $H^{\cdot}(B_T; K)$ invariant under W, and is equal to it when K has characteristic 0, but may be different from it otherwise.

The ring I_G is related to the cohomology of the classifying space B_G by the following theorem. If $H^{\cdot}(G; K)$ is the exterior algebra of an s-dimensional subspace, having a basis consisting of homogeneous elements of odd degree, then $s = \dim T$ (the rank of G) and $\rho^*(T, G)$ maps $H^{\cdot}(B_G; K)$ isomorphically onto $I_G \otimes K$; the field K can be replaced by $\mathbf{Z}$ when G has no torsion.

The assumptions of that theorem are satisfied by Hopf's theorem when K has characteristic 0; so the ring I_G has l algebraically independent generators, which are polynomials of degrees $2m_1, \ldots, 2m_l$, if

$$H^{\cdot}(G; K) = \bigwedge(x_1, \ldots, x_l)$$

and x_j has dimension $2m_j - 1$. When G is quasisimple, the m_j are the integers such that the complex numbers

$$e^{2\pi i m_j/h} \quad (1 \leqslant j \leqslant l)$$

are the eigenvalues of the Coxeter transformation (of order h). The Weyl group W can then be considered to be acting on the universal covering $\mathbf{R}^l$ of T, and it is generated by reflections (symmetries with respect to hyperplanes of $\mathbf{R}^l$). By purely algebraic methods Chevalley proved that for any finite group Γ generated by reflections in a space $\mathbf{R}^n$, the subring of $K[t_1, \ldots, t_n]$ consisting of polynomials in n variables invariant under Γ is generated by algebraically independent polynomials [133].

VII. Suppose G is a semisimple connected compact Lie group and U is the *centralizer* of a torus S in G (not necessarily a maximal one). Borel found that if G^c is the complexification of G, G/U can be identified with G^c/V, where V is a complex Lie subgroup of G^c which contains a *maximal connected solvable* subgroup B of G^c (the first appearance of what later will be called a *Borel subgroup*). Then the orbits of B acting on G^c/V constitute a decomposition of G^c/V into *cells*, each of which is biregularly equivalent to a complex affine space $\mathbf{C}^k$. This generalizes the cellular decomposition of the complex Grassmannians described by Ehresmann (Part 2, chap. V, §4,B). M. Gôto, independently of Borel, had also considered [210] such a decomposition. This cellular decomposition implies that G/U has no torsion.

In particular, for a maximal torus T (equal to its centralizer) G/T is homeomorphic to G^c/B; in that case the cells of the preceding decomposition are in one-to-one correspondence with the elements of W, so that the Euler–Poincaré characteristic $\chi(G/T) = \text{Card}(W)$, a result first proved by A. Weil in 1935 ([482], p. 111) and found again in 1941 by Hopf and Samelson [251]. The fact that G/T has no torsion also derives from Bott's work on the application of Morse theory to homotopy (chap. V, § 6,B). Another older result of Leray [317] is a consequence of Borel's methods: the natural representation of W into the vector space $H^{\bullet}(G/T; \mathbf{R})$ is equivalent to the regular representation of W.

VIII. Finally, suppose the connected closed subgroup U of G has the same rank as G. By considering the spectral sequence of the fiber space $(G/T, G/U, U/T)$ for a maximal torus T of G such that $T \subset U$, Borel showed that if G and U have no p-torsion and K_p is a field of characteristic p, $H^{\bullet}(G/U; K_p)$ is isomorphic to the quotient of $I_U \otimes K_p$ by the ideal generated by the homogeneous elements of degree >0 in the ring $I_G \otimes K_p$. This proves Hirsch's formula (80) and extends it to fields of coefficients of arbitrary characteristic.

These papers of Borel's were later completed by investigations on the Steenrod reduced powers in Lie groups and homogeneous spaces by himself, Serre, and several other mathematicians (chap. VI, § 4,D).

All these results finally made possible the complete determination of the cohomology ring $H^{\bullet}(G; \mathbf{Z})$ for *all* compact connected Lie groups. Recall that the Betti numbers of the quasisimple classical groups had been determined before 1940 (Part 2, chap. VI, § 1); for the five exceptional groups, the values of the Betti numbers were announced by Chih-Tah Yen in a Note in 1949 [525], with only sketches of proofs, based on the Hirsch formula; complete proofs were later given in a joint paper by Chevalley and Borel [64]. The computation of the torsion coefficients took much longer; their values are explicitly given in [191] and [55], pp. 420–421. Similar methods also have given the cohomology of some homogeneous spaces of compact Lie groups; for noncompact Lie groups, some results on homogeneous spaces are known ([55], p. 429).

CHAPTER V

Sophisticated Relations between Homotopy and Homology

§ 1. Homology and Cohomology of Discrete Groups

A. The Second Homology Group of a Simplicial Complex

The last of the most important papers written by Heinz Hopf on algebraic topology date from the period 1941–1945. By their originality, they exerted the same considerable impact on the theory as his preceding ones on homotopy groups of spheres (chap. II, § 1) and on the homology of H-spaces (Part 2, chap. VI, § 2,A), and opened up entirely new fields of research.

The first paper of that series [248] is devoted to the structure of the second simplicial homology group $H_2(K)$ of an arcwise-connected locally finite simplicial complex K. It was known that in general the image $h_2(\pi_2(K))$ of the second homotopy group by the Hurewicz homomorphism (chap. II, § 4,A) is not the whole group $H_2(K)$. Hopf's remarkable discovery was that the quotient

$$H_2(K)/h_2(\pi_2(K)) \tag{1}$$

only depends on the *fundamental group* $\pi_1(K)$ and can be described explicitly by a purely algebraic construction that applies to *any* group G. This construction is based on the expression of G "by generators and relations," i.e., $G = F/R$, where F is a free group and R is a normal subgroup of F. Hopf considered the commutative group

$$G_1^* = (R \cap [F, F])/[F, R] \tag{2}$$

and proved, first that this group only depends on G and not on its particular expression F/R, and second that for $G = \pi_1(K)$ there is a natural isomorphism of the group (1) on G_1^*. This presentation of his results obviously does not correspond to the way they were conceived; it was by displaying the same wonderful geometric imagination as in his work of 1926–1935 that he discovered the relation between the group (1) and the fundamental group $\pi_1(K)$ and was led to the algebraic construction (2).

Although Hopf was aware of the connections between his procedure and

J. Dieudonné, *A History of Algebraic and Differential Topology, 1900–1960*,
Modern Birkhäuser Classics, DOI 10.1007/978-0-8176-4907-4_19,

the Hurewicz Notes, he used neither the homotopy group $\pi_2(K)$ nor the Hurewicz homomorphism h_2 directly.* He dealt with simplicial homology, and probably had in mind some sort of "map" from the **Z**-module $C_2(K)$ of 2-chains to the space $\Omega(K_1, x_0)$ of "edge-loops" in K (chap. I, § 3,B) with fixed origin at a vertex x_0 of K. There is no clear way of defining such a "map"; Hopf turned the difficulty by considering, instead of $C_2(K)$, the set $\Delta_2(K)$ of *all simplicial maps* $f\colon \mathbf{D}_2 \to K_2$ of the disk $\mathbf{D}_2$ in the 2-skeleton of K, such that $f(*) = x_0$, for *arbitrary* triangulations T of $\mathbf{D}_2$ for which $\mathbf{S}_1$ is a union of 1-simplices and 0-simplices of T (i.e., a subcomplex of $\mathbf{D}_2$). Then $f|\mathbf{S}_1$ is an element of $\Omega(K_1, x_0)$, which Hopf calls a "Homotopie-Rand"; by definition of edge-loops, it has a well-defined image in the fundamental group $F = \pi_1(K_1)$.

We have seen (chap. I, § 4,A) that F is a *free group* and that there is a natural surjective homomorphism $p\colon F \to \pi_1(K)$ (chap. I, § 3,B). Let R be the kernel of p, so that $\pi_1(K) \simeq F/R$. The first key observation made by Hopf was that for $f \in \Delta_2(K)$, the image in F of $f|\mathbf{S}_1$ *belongs to* R, and conversely all elements of $\Omega(K_1, x_0)$ that have images in R are "Homotopie-Ränder."

Let $\Delta_2'(K)$ be the subset of $\Delta_2(K)$ consisting of maps f such that the edge-loop $f|\mathbf{S}_1$ is homotopic to a constant map *in* K_2; Hopf's second result was that the images in $R \subset F$ of these loops belong to the group $[F, R]$ generated by commutators of an element of F and an element of R; conversely any element of $[F, R]$ is such an image.

The third step is the connection with homology. Each simplicial map $f \in \Delta_2(K)$ corresponding to a triangulation T of $\mathbf{D}_2$, gives rise to a homomorphism $\tilde{f}\colon C_2(T) \to C_2(K)$ of 2-chain modules; Hopf showed that each 2-chain in $C_2(K)$ is an image $\tilde{f}(\mathbf{D}_2)$ for a suitable triangulation T of $\mathbf{D}_2$. The elements $f \in \Delta_2'(K)$ are those for which $\tilde{f}(\mathbf{D}_2)$ is homologous to a 2-cycle $\tilde{g}(\mathbf{S}_2)$ for a simplicial map $g\colon \mathbf{S}_2 \to K_2$ relative to some triangulation of $\mathbf{S}_2$; let $\Sigma_2(K) \subset C_2(K)$ be the subgroup consisting of these "spherical" 2-cycles. Finally, $\tilde{f}(\mathbf{D}_2)$ is a *cycle* if and only if the image of $f|\mathbf{S}_1$ in F is in the *commutator subgroup* $[F, F]$.

Hopf carefully proved all these statements using geometrical arguments which occupy seven pages of the paper; putting them together, he defined a surjective homomorphism

$$u\colon C_2(K) \to R/[F, R] \tag{3}$$

whose kernel is $\Sigma_2(K)$; as $\Sigma_2(K)$ is contained in the subgroup $Z_2(K)$ of 2-cycles, there is a surjective homomorphism

$$u|Z_2(K)\colon Z_2(K) \to (R \cap [F, F])/[F, R] \tag{4}$$

hence an *isomorphism*

$$(u|Z_2(K))^{\sim}\colon Z_2(K)/\Sigma_2(K) \xrightarrow{\sim} (R \cap [F, F])/[F, R]. \tag{5}$$

But $\Sigma_2(K)$ contains the subgroup $B_2(K)$ of 2-boundaries; in the homology

* In what follows, we translate Hopf's terminology into modern terms.

group $H_2(K) = Z_2(K)/B_2(K)$, $\Sigma_2(K)/B_2(K)$ is exactly the image $h_2(\pi_2(K))$ of the Hurewicz homomorphism; this finally gives the Hopf isomorphism

$$H_2(K)/h_2(\pi_2(K)) \simeq G_1^*. \tag{6}$$

B. The Homology of Aspherical Simplicial Complexes

In his 1941 paper Hopf naturally wondered if there were theorems similar to his for homology groups $H_n(K)$ with $n \geqslant 3$. He had observed that obvious generalizations certainly do not exist, because he easily found examples of simplicial n-dimensional complexes K with *arbitrary* fundamental group $\pi_1(K)$ and *arbitrary* homology groups $H_j(K)$ for $j \geqslant 3$ [whereas one consequence of Hopf's theorem is that if, for instance, $\pi_1(K)$ is a free commutative group of rank p, the second Betti number must be $\geqslant p(p-1)/2$].

Nevertheless, the quotient groups $H_n(K)/h_n(\pi_n(K))$ still had topological significance for any n. In 1942 Hopf observed that for any $n \geqslant 2$ the set of "spherical" n-cycles, that is, the images $\tilde{g}(S_n)$ for simplicial maps $g: S_n \to K$ (for all triangulations of S_n), form a *subgroup* $\Sigma_n(K)$ of the group $Z_n(K)$ of n-cycles of K *; he showed that the group $Z_n(K)/\Sigma_n(K)$ is isomorphic to

$$Q_n(K) = H_n(K)/h_n(\pi_n(K)). \tag{7}$$

This brought him in contact with Hurewicz's results on aspherical simplicial complexes (chap. II, § 6,A). Hurewicz had observed in a footnote ([256], IV, p. 219) that his characterization of the homotopy classes $[f]$ in $[X; Y]$ still holds if the dimension of X is $\leqslant n$ and if $\pi_i(Y) = 0$ for $2 \leqslant i \leqslant n$ only. Hopf applied this to simplicial maps $f: K_n \to K'_n$ between the n-skeletons of two simplicial complexes K, K′: if $\pi_i(K') = 0$ for $2 \leqslant i \leqslant n-1$, the homomorphism $Q_n(K) \to Q_n(K')$ defined by f is entirely *determined* by the homomorphism $f_*: \pi_1(K) \to \pi_1(K')$. In particular, if $\pi_1(K)$ and $\pi_1(K')$ are isomorphic, and $\pi_i(K) = \pi_i(K') = 0$ for $2 \leqslant i \leqslant n-1$, then $Q_n(K)$ and $Q_n(K')$ are isomorphic. Under the assumption $\pi_i(K) = 0$ for $2 \leqslant i \leqslant n-1$, $Q_n(K)$ is thus *determined by the fundamental group* $\pi_1(K)$, but at that time Hopf could not find a purely algebraic construction deriving $Q_n(K)$ from $\pi_1(K)$ for $n \geqslant 3$.

The breakthrough came 2 years later [249] when Hopf had the idea of considering *free resolutions* (finite or infinite) *of left* P-*modules* (Part 1, chap. IV, § 5,F) for an arbitrary ring of scalars P, and proved their existence and uniqueness up to chain equivalence. His idea probably stemmed from an "iteration" of the presentation of a group as quotient F/R of a free group F, obtained by applying it to R and repeating the process indefinitely. This idea was essentially the same as the one Hilbert had used to build the chain of "syzygies" in invariant theory, but at that time this was not mentioned by Hopf.

* In 1940 [167] Eilenberg had already shown that the homology classes of the elements of $\Sigma_n(K)$ formed a subgroup of $H_n(K)$.

For his purpose, he considered a left P-module J and a two-sided ideal $\mathfrak{a}$ of P; he took a free resolution of J

$$0 \leftarrow J \xleftarrow{\varepsilon} X_0 \xleftarrow{\mathbf{b}_1} X_1 \xleftarrow{\mathbf{b}_2} X_2 \leftarrow \cdots \xleftarrow{\mathbf{b}_n} X_n \leftarrow \cdots \tag{8}$$

and formed the chain complex

$$0 \leftarrow \mathfrak{a}X_0 \xleftarrow{\mathbf{b}'_1} \mathfrak{a}X_1 \xleftarrow{\mathbf{b}'_2} \mathfrak{a}X_2 \leftarrow \cdots \xleftarrow{\mathbf{b}'_n} \mathfrak{a}X_n \leftarrow \cdots \tag{9}$$

where $\mathbf{b}'_n$ is the restriction of $\mathbf{b}_n$ to the submodule $\mathfrak{a}X_n$; he took the homology of that complex, namely, the groups

$$(\mathfrak{a}X_n \cap Z_n)/\mathfrak{a}Z_n \tag{10}$$

where $Z_n = \operatorname{Im} \mathbf{b}_{n+1} = \operatorname{Ker} \mathbf{b}_n$. Since $\mathfrak{a}X_j$ is isomorphic to $\mathfrak{a} \otimes_P X_j$ and $\mathbf{b}'_j$ can be identified with $1 \otimes \mathbf{b}_j$, the general construction of chain homotopies from a free chain complex to an acyclic one (*loc. cit.*) shows that these groups are independent of the choice of the resolution (8) of J; in the later development of homological algebra these groups were written

$$\operatorname{Tor}_n^P(\mathfrak{a}, J). \tag{11}$$

Hopf then specialized the previous construction to the case in which $P = \mathbf{Z}[G]$ is the algebra of an arbitrary group G, and the ideal $\mathfrak{a}$ is the kernel of the homomorphism $S: \mathbf{Z}[G] \to \mathbf{Z}$ such that

$$S\left(\sum_j \lambda_j g_j\right) = \sum_j \lambda_j. \tag{12}$$

In that case he named the group (10) G^n, later to be identified with the homology group $H_n(G; \mathbf{Z})$ (see below, section C).

That specialization of course was dictated by the way Hopf intended to use this algebraic machinery. Inspired (as J.H.C. Whitehead was at the same time, with different purposes) by the Reidemeister device (Part 2, chap. VI, § 3,A), he considered, for any locally finite simplicial complex K, a Galois covering complex $\tilde{K}$ with projection $p: \tilde{K} \to K$ so that the group $C_n(\tilde{K})$ of n-chains in $\tilde{K}$ becomes a *free* $\mathbf{Z}[G]$-module, where $G = \operatorname{Aut}_K(\tilde{K})$. The $\mathbf{Z}$-homomorphism $\tilde{p}: C_n(\tilde{K}) \to C_n(K)$ is surjective and has a kernel equal to $\mathfrak{a}C_n(\tilde{K})$. It is then clear that $\tilde{p}(Z_n(\tilde{K})) \subset Z_n(K)$ for the n-cycles, and the surjectivity of $\tilde{p}$ implies that $\tilde{p}(B_n(\tilde{K})) = B_n(K)$ for the n-boundaries. From this it follows that

$$H_n(K)/p_*(H_n(\tilde{K})) \simeq Z_n(K)/\tilde{p}(Z_n(\tilde{K})). \tag{13}$$

His main result was an *isomorphism*

$$Z_n(K)/\tilde{p}(Z_n(\tilde{K})) \simeq (\mathfrak{a}C_{n-1}(\tilde{K}) \cap B_{n-1}(\tilde{K}))/\mathfrak{a}B_{n-1}(\tilde{K}). \tag{14}$$

Suppose that K is now *aspherical up to dimension* N, that is

$$\pi_i(K) = 0 \quad \text{for } 2 \leqslant i \leqslant N. \tag{15}$$

Then if $G = \pi_1(K)$, it follows from Hurewicz's absolute isomorphism theorem (chap. II, § 4,A) that the chain complex of $\mathbf{Z}[\pi_1(K)]$-modules

$$0 \leftarrow \mathbf{Z} \xleftarrow{\mathbf{b}_0} C_0(\tilde{K}) \xleftarrow{\mathbf{b}_1} C_1(\tilde{K}) \leftarrow \cdots \xleftarrow{\mathbf{b}_N} C_N(\tilde{K}) \tag{16}$$

is free and acyclic. Applying (14), Hopf's determination of the homology (10) of the chain complex (9) yields the result

$$H_n(K)/p_*(H_n(\tilde{K})) \simeq G^n \quad \text{for } 1 \leqslant n \leqslant N \tag{17}$$

and as $H_n(\tilde{K}) = 0$ for $1 \leqslant n \leqslant N - 1$,

$$H_n(K) \simeq G^n \quad \text{for } 1 \leqslant n \leqslant N - 1. \tag{18}$$

C. The Eilenberg Groups

As soon as the first papers of Hopf on the second Betti group were published they attracted the attention of Eilenberg and Mac Lane, who announced in 1943 [179] a solution to the problem of the determination of the homology of an aspherical space one year before the publication of the Hopf paper [249] we analyzed in B. This solution was based on an algebraic construction different from Hopf's, but which later turned out to be equivalent.

Eilenberg and Mac Lane worked with singular homology, which enabled them to deal with arbitrary spaces instead of simplicial complexes. Their fundamental idea was a relation between singular homology and homotopy, published without proof by Eilenberg in 1940 [169], and for which he provided proofs in his foundational paper on singular homology [172] (partly described in Part 1, chap. IV, § 2).

In the chain complex $S_\bullet(X)$ of all singular chains in an arcwise-connected pointed space (X, x_0), he considered a descending sequence of subchain complexes

$$S_\bullet(X) = S_{0\bullet}(X) \supset S_{1\bullet}(X) \supset \cdots \supset S_{n\bullet}(X) \supset \cdots \tag{19}$$

where $S_{np}(X)$ is the subgroup of $S_p(X)$ having as $\mathbf{Z}$-basis the set of singular p-simplices $s\colon \bar{\Delta}_p \to X$ such that for all $q < n$, *all* q-simplices in $\bar{\Delta}_p - \Delta_p$ are mapped by s onto the point x_0. The natural injection

$$\eta_{nm}\colon S_{n\bullet}(X) \to S_{m\bullet}(X) \quad \text{for } m \leqslant n \tag{20}$$

is a chain transformation satisfying

$$\eta_{nm} \circ \eta_{ml} = \eta_{nl} \quad \text{for } l \leqslant m \leqslant n, \tag{21}$$

and for any continuous map $f\colon (X, x_0) \to (Y, y_0)$ of pointed spaces, the chain transformation $\tilde{f}\colon S_\bullet(X) \to S_\bullet(Y)$ maps each $S_{n\bullet}(X)$ into $S_{n\bullet}(Y)$; its restriction to $S_{n\bullet}(X)$ is a chain transformation.

Eilenberg's main result was a "reduction theorem": if $\pi_n(X) = 0$, then the chain transformation

$$\eta_{n+1,n}\colon S_{n+1,\bullet}(X) \to S_{n\bullet}(X) \tag{22}$$

is a *chain equivalence* (for $n = 0$, the assumption is merely that X is arcwise-connected). The problem is to find a chain transformation

$$\zeta\colon S_{n\cdot}(X) \to S_{n+1,\cdot}(X) \tag{23}$$

such that $\zeta \circ \eta_{n+1,n} = 1_{S_{n+1,\cdot}(X)}$ and $\eta_{n+1,n} \circ \zeta$ is chain homotopic to $1_{S_{n\cdot}(X)}$. The solution starts with the consideration of the "prism" $\bar{\Delta}_p \times [0,1]$ (Part 1, chap. II, § 3) for each $p > 0$; preceeding as Lefschetz did [*loc. cit.*, formula (16)], it is enough to define a continuous map $F_s\colon \bar{\Delta}_p \times [0,1] \to X$ for each p and each continuous map $s\colon \bar{\Delta}_p \to X$ such that $s \in S_{np}(X)$, satisfying $F_s(y,0) = s(y)$ and such that $y \mapsto F_s(y,1)$ belongs to $S_{n+1,p}(X)$.

If $s \in S_{n+1,p}(X)$ already, take $F_s(y,t) = s(y)$ for all $t \in [0,1]$. If not (which can only happen when $p \geqslant n$), proceed by induction on p; the crucial step is $p = n$, where $F_s(y,1)$ is taken to be the constant x_0, and the extension of F to the interior of $\bar{\Delta}_n \times [0,1]$ is possible because $\pi_n(X) = 0$. For $p > n$ the inductive assumption enables $F_s(z,t)$ to be defined for z in any face of $\bar{\Delta}_p$ in such a way that $z \mapsto F_s(z,1)$ belongs to $S_{n+1,p-1}(X)$. An application of the box lemma (chap. II, § 2,D) defines F_s in $\bar{\Delta}_p \times [0,1]$.

In particular, if for $q > 1$, $\pi_i(X) = 0$ for all $i < q$, then $\eta_{q,0}$ (simply written η_q) is a chain equivalence; hence for any commutative group G

$$\eta_{q*}\colon H_q(S_{q\cdot}(X); G) \to H_q(X; G) \tag{24}$$

is an *isomorphism*.

The importance of the Eilenberg groups stems from the fact that, by Čech's definition of homotopy groups, the elements of $\pi_q(X, x_0)$ are *the homotopy classes of the elements of* $S_{q,q}(X)$. As two homotopic elements of $S_{q,q}(X)$ are homologous, there is a natural homomorphism

$$v_q\colon \pi_q(X) \to H_q(S_{q\cdot}(X)) \tag{25}$$

and from the definition of the sum in $\pi_q(X)$, it follows that $[\mathbf{b}g] = 0$ for any singular $(q+1)$-simplex $g \in S_{q,q+1}(X)$; this proves that v_q is an *isomorphism* for $q > 1$.

In particular, this leads to a simple proof of Hurewicz's absolute isomorphism theorem: it is an immediate consequence of the fact that v_q is an isomorphism, and (24) is also an isomorphism when $\pi_i(X) = 0$ for all $i < q$.

From the fact that v_q is an isomorphism under the preceding assumptions Eilenberg also deduced (independently of Hopf) that the image $h_q(\pi_q(X))$ in $H_q(X)$ is the subgroup $\Sigma_q(X)$ of the classes of spherical q-cycles.

D. Homology and Cohomology of Groups

In their work on aspherical arcwise-connected spaces, Eilenberg and MacLane used only the chain equivalence, a special case of the chain transformation (22):

$$S_\cdot(X) \simeq S_{1\cdot}(X)$$

so that there is a natural isomorphism

$$H_\cdot(X) \simeq H_\cdot(S_{1\cdot}(X)). \tag{26}$$

All vertices $\mathbf{e}_i$ of $\bar{\Delta}_p$ are now mapped on x_0 by s for a singular p-simplex $s\colon \bar{\Delta}_p \to X$ belonging to $S_{1,p}(X)$, hence every ordered 1-simplex $\{\mathbf{e}_i, \mathbf{e}_j\}$ for $i \neq j$

is mapped onto a *loop*

$$t \mapsto s((1-t)\mathrm{e}_i + t\mathrm{e}_j) \quad \text{for } 0 \leqslant t \leqslant 1$$

belonging to $\Omega(\mathrm{X}, x_0)$; let u_{ij} be its class in $\pi_1(\mathrm{X}, x_0)$. Define $u_{ii} = e$ [neutral element of $\pi_1(\mathrm{X})$]; then the elements u_{ij} for $0 \leqslant i, j \leqslant p$ satisfy the relations

$$u_{ji} = u_{ij}^{-1}, \qquad u_{ij}u_{jk} = u_{ik}. \tag{27}$$

Let $U(s)$ be the $(p+1) \times (p+1)$ matrix $(u_{ij})_{0 \leqslant i,j \leqslant p}$ with elements in $\pi_1(\mathrm{X})$; if $U^{ii}(s)$ is the principal submatrix of $U(s)$ obtained by erasing the ith row and ith column, then for the *boundary* $\mathbf{b}_p s$

$$U(\mathbf{b}_p s) = \sum_{i=0}^{p} (-1)^i U^{ii}(s). \tag{28}$$

This motivated Eilenberg and MacLane to define, for an *arbitrary* group Π, a chain complex

$$\mathrm{K}_{\bullet}(\Pi)\colon 0 \leftarrow \mathrm{K}_0(\Pi) \xleftarrow{\mathbf{b}_1} \mathrm{K}_1(\Pi) \leftarrow \cdots \xleftarrow{\mathbf{b}_n} \mathrm{K}_n(\Pi) \leftarrow \cdots \tag{29}$$

where $\mathrm{K}_n(\Pi)$ is the free $\mathbf{Z}$-module with a basis of all $(n+1) \times (n+1)$ matrices with elements u_{ij} in Π *satisfying* (27), for which the boundary operator is given by

$$\mathbf{b}_n U = \sum_{i=0}^{n} (-1)^i U^{ii}. \tag{30}$$

They also gave simpler equivalent definitions. First they observed that (27) implies

$$u_{ij} = u_{0i}^{-1} u_{0j} \tag{31}$$

so that a matrix $U \in \mathrm{K}_n(\Pi)$ is entirely determined by its first line $(1, x_1, \ldots, x_n)$, where the x_j are arbitrary elements of the group Π. Therefore $\mathrm{K}_n(\Pi)$ can also be considered as the free $\mathbf{Z}$-module having as a basis the product set Π^n, and the boundary operator corresponding to (30) is easily computed to be

$$\mathbf{b}_n(x_1, x_2, \ldots, x_n) = (x_2, \ldots, x_n) + \sum_{i=1}^{n-1} (-1)^i (x_1, \ldots, x_{i-1}, x_i x_{i+1}, x_{i+2}, \ldots, x_n)$$
$$+ (-1)^n (x_1, \ldots, x_{n-1}). \tag{32}$$

This description of $\mathrm{K}_n(\Pi)$ is the one Eilenberg and MacLane called "non-homogeneous." They also used a third "homogeneous" description, in which the basis of $\mathrm{K}_n(\Pi)$ consists of the *equivalence classes* $[x_0, x_1, \ldots, x_n]$ of $(n+1)$-tuples of elements of Π for the equivalence relation in the set Π^{n+1}

$$(x_0, x_1, \ldots, x_n) \equiv (xx_0, xx_1, \ldots, xx_n) \quad \text{for } x \in \Pi; \tag{33}$$

the boundary operator then takes the simple form

$$\mathbf{b}_n[x_0, x_1, \ldots, x_n] = \sum_{j=0}^{n} (-1)^j [x_0, \ldots, x_{j-1}, x_{j+1}, \ldots, x_n] \tag{33}$$

(also written $\sum_{j=0}^{n} (-1)^j [x_0, \ldots, \hat{x}_j, \ldots, x_n]$, the "hat" meaning "omission").

In contrast with Hopf, Eilenberg and MacLane were active propagandists of cohomology; so, having defined the chain complex $K_{\bullet}(\Pi)$, they also introduced the cochain complex

$$K^{\bullet}(\Pi; G) = \mathrm{Hom}(K_n(\Pi), G))_{n \geqslant 0}$$

for any commutative coefficient group G (Part 1, chap. IV, § 3); $K^n(\Pi; G)$ is thus the free $\mathbf{Z}[G]$-module having as basis the maps $f: \Pi^n \to G$, with coboundary given by

$$(\mathbf{d}_n f)(x_1, \ldots, x_{n+1}) = f(x_2, \ldots, x_{n+1}) + \sum_{i=1}^{n-1} (-1)^i f(x_1, \ldots, x_{i-1}, x_i x_{i+1}, x_{i+2}, \ldots, x_{n+1}) + (-1)^{n+1} f(x_1, \ldots, x_n). \quad (34)$$

Of course they also considered the chain complexes $K_{\bullet}(\Pi; G)$ defined by the usual procedure (Part 1, chap. IV, § 2).

With these definitions Eilenberg and MacLane created the concepts of *homology and cohomology groups* $H_n(\Pi; G)$ and $H^n(\Pi; G)$ for an *arbitrary* group Π and an arbitrary commutative coefficient group G; this was the starting point of the new mathematical theory called *homological algebra*, which has played an ever increasing part in the evolution of contemporary mathematics. We shall not try to retrace the history of the tremendous expansion of that discipline and of its numerous ramifications not directly associated with topology; we refer the reader to the books and papers [113], [215], and [330].

We only mention here the enlargement of the concepts of the joint Note of 1943 made by Eilenberg in an announcement of 1944, with detailed proofs published in 1947 [173]. He considered the case in which the group Π *acts linearly* on the commutative group G (written additively). Then, in order to define the cohomology group $H^n(\Pi; G)$, he only considered *equivariant* cochains, namely (for the "homogeneous" description of chains), the maps $f: \Pi^{n+1} \to G$ such that

$$w \,.\, f(w_0, \ldots, w_n) = f(w \,.\, w_0, \ldots, w \,.\, w_n) \quad (35)$$

for all elements w and w_j in Π. In the same paper* Eilenberg directly checked that Hopf's groups written Π^n in his notation coincided, up to isomorphism, with the homology groups $H_n(\Pi; \mathbf{Z})$ (with trivial action of Π on $\mathbf{Z}$).

Eilenberg and Mac Lane determined the homology and cohomology of an aspherical space by a procedure different from Hopf's (which was only published a year later). They used Eilenberg's chain equivalence $S_{\bullet}(X) \rightleftarrows S_{1\bullet}(X)$ and the chain transformation

$$\kappa_{\bullet}: S_{1\bullet}(X) \to K_{\bullet}(\pi_1(X)) \quad (36)$$

* This seems to have been the first paper in which cohomology was written with upper indices H^n; previously, H^n meant homology, and H_n cohomology.

defined by the map $s \mapsto U(s)$ considered above. They showed that in general the group $\Sigma_n(X) = \Sigma_n(S_1.(X))$ of homology classes of *spherical cycles* is contained in the kernel of κ_n, and when X is *aspherical*, they showed that $\kappa_.$ is a *chain equivalence*.

To prove this last property, they defined a chain transformation

$$\bar{\kappa}_.: K.(\pi_1(X)) \to S_1.(X)$$

and showed that $\kappa_. \circ \bar{\kappa}_.$ and $\bar{\kappa}_. \circ \kappa_.$ are chain homotopic to the respective identities.

The maps

$$\bar{\kappa}_n: K_n(\pi_1(X)) \to S_{1,n}(X)$$

are defined by induction on n. For $n = 1$, a 1-simplex of $K_1(\pi_1(X))$ is a matrix

$$U = \begin{pmatrix} 1 & u \\ u^{-1} & 1 \end{pmatrix}$$

with $u \in \pi_1(X)$; $\bar{\kappa}_1(U)$ is a singular 1-simplex (i.e., a loop of origin x_0 in X) whose class in $\pi_1(X)$ is u. Next assume that the maps $\bar{\kappa}_p$ for $p \leqslant n-1$ have been defined in such a way that, for a matrix $U = (u_{ij}) \in K_{n-1}(\pi_1(X))$, if $\sigma_i: \bar{\Delta}_{n-2} \to \bar{\Delta}_{n-1}$ is the standard affine map of $\bar{\Delta}_{n-2}$ on the i-th face of $\bar{\Delta}_{n-1}$, then for each principal minor U^{ii},

$$\bar{\kappa}_{n-1}(U) \circ \sigma_i = \bar{\kappa}_{n-2}(U^{ii}) \quad \text{for } 0 \leqslant i \leqslant n-1. \tag{37}$$

The equation $u_{12}u_{23} = u_{13}$ for $n = 2$ shows that the juxtaposition of the three loops $\bar{\kappa}_1(u_{12})$, $\bar{\kappa}_1(u_{23})$, and $\bar{\kappa}_1(u_{13}^{-1})$ is homotopic to the constant map $[0, 1] \to \{x_0\}$; hence there is a continuous map $s: \bar{\Delta}_2 \to X$ such that $s \circ \sigma_k = \bar{\kappa}_1(u_{ij})$ for $\{i, j, k\} = \{1, 2, 3\}$. For $n \geqslant 2$, using (37) and the assumption $\pi_n(X) = 0$, we can proceed in the same way.

This construction proves that $\bar{\kappa}_.$ is a chain transformation and that $\kappa_. \circ \bar{\kappa}_. = 1_{K.(\pi_1(X))}$. The crux of the proof is to show that $\bar{\kappa}_. \circ \kappa_.$ is chain homotopic to $1_{S_1.(X)}$; the construction of the chain homotopy is quite similar to the one in Eilenberg's "reduction theorem" (see section C above) using "singular prisms." Here, for a singular p-simplex $s \in S_{1,p}(X)$, a continuous map $F_s: \bar{\Delta}_p \times [0, 1] \to X$ must be defined with

$$F_s(y, 0) = s(y), \qquad F_s(y, 1) = (\bar{\kappa}_p(\kappa_p(s)))(y).$$

It is then easy to apply induction on p, the "lateral sides" of a singular prism being themselves singular prisms of lower dimensions, and all vertices being mapped to x_0; the relation $\pi_{p+1}(X) = 0$ must be used to extend F_s from the frontier of $\bar{\Delta}_p \times [0, 1]$ to its interior.

Eilenberg and MacLane observed that their method also works for spaces "aspherical up to dimension r," i.e., such that $\pi_i(X) = 0$ for $1 < i < r$: then

$$\begin{cases} H_n(X; G) \simeq H_n(\pi_1(X); G) \\ H^n(X; G) \simeq H^n(\pi_1(X); G) \end{cases} \quad \text{for } n < r \tag{38}$$

and

$$H_r(X; G)/\Sigma_r(X; G) \simeq H_r(\pi_1(X); G).$$

By using (38) they determined the groups $H_n(\Pi; G)$ for various groups Π. They took a connected graph for X such that $\pi_1(X) = \Pi$ when Π is a free noncommutative group (chap. I, § 4,B) and obtained $H_n(\Pi; G) = 0$ for $n > 1$, and $H_1(\Pi; G) \cong \mathrm{Hom}(\Pi, G)$. If $\Pi = \mathbf{Z}^r$, X is the torus $\mathbf{T}^r$, and $H_n(\Pi; G) = 0$ for $n > r$, $H_n(\Pi; G) = G^{\binom{r}{n}}$ for $n \leqslant r$. If $\Pi = \mathbf{Z}/m\mathbf{Z}$, X is a suitable lens space, and

$$H_{2n}(\Pi; G) = G/mG, \qquad H_{2n+1}(\Pi; G) = {}_mG$$

where ${}_mG$ is the subgroup of elements $g \in G$ such that $mg = 0$.

At the end of their paper, as a kind of afterthought, they gave a sketchy description of the generalization of their results when aspherical spaces are replaced by spaces X for which $\pi_i(X) = 0$ *for* $i > m$. They then considered the chain complex $S_{m.}(X)$ instead of $S_{1.}(X)$, and for any *commutative* group Π, written additively, they replaced the chain complex $K_.(\Pi)$ by the generalization $K_.(\Pi, m)$, defined as follows. The n-simplices are now the maps

$$U: \{0, 1, \ldots, n\}^m \to \Pi$$

written $(a_0, a_1, \ldots, a_m) \mapsto u_{a_0 a_1 \cdots a_m}$ [which for $m = 1$ can be considered as $(n + 1) \times (n + 1$ matrices], subject to the two conditions:

$$u_{a_0 a_1 \cdots a_m} = 0 \quad \text{if two of the arguments } a_j \text{ are equal} \tag{39}$$

(this is of course always the case when $n + 1 < m$);

$$\sum_{i=0}^{m} (-1)^i u_{a_0 \cdots \hat{a}_i \cdots a_m} = 0 \tag{40}$$

[for $m = 1$, this is condition (27) written additively].

The boundary operator is defined by

$$\mathbf{b}_n U: (b_0, b_1, \ldots, b_m) \mapsto \sum_{i=0}^{n} (-1)^i u_{b_0 \cdots b_m} \tag{41}$$

where in the ith term of the sum

$$b_j = a_j \text{ if } i > a_j, \qquad b_j = a_j - 1 \quad \text{if } i \leqslant a_j - 1$$

[by (39) the numbers b_j for which $u_{b_0 \cdots b_m} \neq 0$ take their values in $\{0, 1, \ldots, n - 1\}$, so that $\mathbf{b}_n$ maps $K_n(\Pi, m)$ into $K_{n-1}(\Pi, m)$].

The chain transformation $\kappa_{m.}: S_{m.}(X) \to K_.(\pi_m(X), m)$, which generalizes $\kappa_.(= \kappa_{1.})$, is then defined by associating to a singular p-simplex $s \in S_{m,p}(X)$ the map

$$U(s): \{0, 1, \ldots, p\}^m \to \pi_m(X)$$

defined as follows:

(i) $u_{a_0 a_1 \cdots a_m} = 0$ if two of the a_j in $\{0, 1, \ldots, p\}$ are equal;
(ii) if not (which implies $p \geqslant m$), consider the restriction of s to the simplex

$$\sigma_{a_0 a_1 \cdots a_m} \subset \bar{\Delta}_p - \Delta_p$$

which is the image of $\bar{\Delta}_m$ by the affine map sending each $\mathbf{e}_j$ $(0 \leqslant j \leqslant m)$ to $\mathbf{e}_{a_j}$. As all faces of $\sigma_{a_0 a_1 \cdots a_m}$ are mapped on the point x_0 by s, s can be identified with a continuous map $\mathbf{D}_m \to \mathrm{X}$ sending $\mathbf{S}_{m-1}$ onto x_0; this map therefore has a well determined image in $\pi_m(\mathrm{X})$, and that image by definition is $u_{a_0 a_1 \cdots a_m}$.

Without entering into details, Eilenberg and MacLane stated that their arguments for $m = 1$ could easily be generalized to any m, yielding the following result: *if* $\pi_i(\mathrm{X}) = 0$ *for* $i > m$, then

$$\kappa_{m\bullet}\colon \mathrm{S}_{m\bullet}(\mathrm{X}) \to \mathrm{K}_\bullet(\pi_m(\mathrm{X}), m)$$

is a *chain equivalence*, hence the isomorphisms

$$\begin{cases} \mathrm{H}_n(\mathrm{S}_{m\bullet}(\mathrm{X}); \mathrm{G}) \simeq \mathrm{H}_n(\mathrm{K}_\bullet(\pi_m(\mathrm{X}), m); \mathrm{G}) \\ \mathrm{H}^n(\mathrm{S}_{m\bullet}(\mathrm{X}); \mathrm{G}) \simeq \mathrm{H}^n(\mathrm{K}_\bullet(\pi_m(\mathrm{X}), m); \mathrm{G}) \end{cases} \quad \text{for } all\ n \geqslant 1. \tag{42}$$

Finally, at the very end of their paper the *Eilenberg–MacLane spaces* (chap. II. § 6,F) made their first appearance: if $\pi_n(\mathrm{X}) = 0$ *for all* n *except for* $n = m$, then the homotopy group $\pi_m(\mathrm{X})$ *explicitly determines all homology and cohomology groups of* X *with arbitrary coefficients*: indeed $\mathrm{S}_{m\bullet}(\mathrm{X}) \to \mathrm{S}_\bullet(\mathrm{X})$ is a chain equivalence, so

$$\mathrm{H}_n(\mathrm{X}; \mathrm{G}) \simeq \mathrm{H}_n(\mathrm{K}_m(\pi_m(\mathrm{X}), m); \mathrm{G}),$$

$$\mathrm{H}^n(\mathrm{X}; \mathrm{G}) \simeq \mathrm{H}^n(\mathrm{K}_m(\pi_m(\mathrm{X}), m); \mathrm{G})$$

for all $n \geqslant 1$.

In this chapter and the following ones we shall see how these spaces became the focus of extensive research between 1950 and 1960 that greatly increased our knowledge of homology and homotopy groups.

One of the most important properties of the Eilenberg–McLane spaces is the existence of a natural bijection

$$[\mathrm{X}, \mathrm{K}(\Pi, n)] \simeq \mathrm{H}^n(\mathrm{X}; \Pi) \tag{43}$$

for any commutative group Π and any CW-complex X. This is essentially a special case of Eilenberg's obstruction theory (chap. II, § 4,C, VIII and IX). $\pi_n(\mathrm{K}(\Pi, n))$ is naturally identified with Π, and $\mathrm{H}^n(\mathrm{K}(\Pi, n); \mathrm{G})$ is naturally isomorphic to $\mathrm{Hom}(\Pi, \mathrm{G})$; if ι is the element of $\mathrm{H}^n(\mathrm{K}(\Pi, n); \Pi)$ [called the *canonical cohomology class* of $\mathrm{K}(\Pi, n)$] that corresponds to the identity in $\mathrm{End}(\Pi)$ by that isomorphism, the map (43) is just

$$[f] \mapsto f^*(\iota) \tag{44}$$

by *loc. cit.*.

E. Application to Covering Spaces

Let B be an arcwise-connected space and X be a *Galois covering space* of B (chap. I, § 2,VIII) with automorphism group Π. Then Π acts on the singular complex $\mathrm{S}_\bullet(\mathrm{X})$, hence on the homology $\mathrm{H}_\bullet(\mathrm{X}; \Lambda)$ and cohomology $\mathrm{H}^\bullet(\mathrm{X}; \Lambda)$

with coefficients in any ring Λ. Leray and H. Cartan [114] showed that there is a cohomology spectral sequence with E_2 terms consisting of *cohomology groups of the group* Π

$$E_2^{pq} \simeq H^p(\Pi; H^q(X; \Lambda))$$

and abutment $H^\bullet(B; \Lambda)$ with a suitable filtration. There is a similar homology spectral sequence with

$$E_{pq}^2 \simeq H_p(\Pi; H_q(X; \Lambda))$$

and abutment $H_\bullet(B; \Lambda)$.

If $H^q(X; \Lambda) = 0$ for $0 < q < n$, the spectral sequence gives isomorphisms

$$H^q(B; \Lambda) \simeq H^q(\Pi; \Lambda) \quad \text{for } q < n$$

and an exact sequence

$$H^n(\Pi; \Lambda) \to H^n(B; \Lambda) \to (H^n(X; \Lambda))^\Pi \to H^{n+1}(\Pi; \Lambda) \to H^{n+1}(B; \Lambda) \qquad (45)$$

where $(H^n(X; \Lambda))^\Pi$ is the subgroup of elements of $H^n(X; \Lambda)$ fixed under the action of Π. Similarly, if $H_q(X; \Lambda) = 0$ for $0 < q < n$, there are isomorphisms

$$H_q(B; \Lambda) \simeq H_q(\Pi; \Lambda) \quad \text{for } q < n$$

and an exact sequence

$$H_{n+1}(B; \Lambda) \to H_{n+1}(\Pi; \Lambda) \to (H_n(X; \Lambda))_\Pi \to H_n(B; \Lambda) \to H_n(\Pi; \Lambda) \to 0 \quad (46)$$

where $(H_n(X; \Lambda))_\Pi$ is the quotient of the group $H_n(X; \Lambda)$ by the subgroup generated by the elements $s.a - a$ with $a \in H_n(X; \Lambda)$ and $s \in \Pi$. Particular cases of these properties had already been obtained by Eckmann [150] and Eilenberg and Mac Lane [ ].

The spectral sequences of Leray–Cartan yield exact sequences when $\Pi = \mathbf{Z}$:

$$0 \to (H^{n-1}(X; \Lambda))_\Pi \to H^n(B; \Lambda) \to (H^n(X; \Lambda))^\Pi \to 0,$$

$$0 \to (H_n(X; \Lambda))_\Pi \to H_n(B; \Lambda) \to (H_{n-1}(X; \Lambda))^\Pi \to 0.$$

If X is an n-dimensional manifold such that $H_0(X; \mathbf{Z}) \simeq \mathbf{Z}$, and $H_q(X; \mathbf{Z}) = 0$ for $q > 0$ (for instance $X = \mathbf{R}^n$), then $H_p(B; \mathbf{Z}) \simeq H_p(\Pi; \mathbf{Z})$ and $H^p(B; \mathbf{Z}) \simeq H^n(\Pi; \mathbf{Z})$ for all p, Π acting trivially on $\mathbf{Z}$. Since B is also an n-dimensional manifold, $H^q(B; \mathbf{Z}) = 0$ for $q > n$, which imposes limitations on the action of Π on X; in particular, Π cannot be finite: for instance, $\mathbf{R}^n$ cannot be a Galois covering space with finite group.

Finally, the Leray–Cartan spectral sequences give information on the (necessarily finite) groups Π that can be properly discontinuous groups of homeomorphisms of an odd-dimensional sphere S_n *; one must have

---

* The only automorphisms of an *even*-dimensional sphere S_n that have no fixed point must have degree -1 (Part 2, chap. III, § 1); in a properly discontinuous group of automorphisms of S_n, two elements s, s' other than the neutral element e must be such that $ss' = e$; the only properly discontinuous group of automorphisms of S_n not reduced to e thus consists of e and the symmetry $x \mapsto -x$.

$$\mathrm{H}^n(\Pi;\mathbf{Z}) = 0,$$
$$\mathrm{H}^{n+1}(\Pi;\mathbf{Z}) \simeq \mathbf{Z}/q\mathbf{Z} \qquad \text{if } q = \mathrm{Card}(\Pi),$$
$$\mathrm{H}^{i+n+1}(\Pi;\mathbf{Z}) \simeq \mathrm{H}^i(\Pi;\mathbf{Z}) \quad \text{for } i \geqslant 1. \text{ ([113], p. 358)}$$

This has been the starting point of the determination of *all* such groups Π, which has prompted a large number of papers using all sorts of different theories, and which is not yet complete [337].

§ 2. Postnikov Towers and Eilenberg–MacLane Fibers

A. The Eilenberg–MacLane Invariant

In 1946 Eilenberg and MacLane observed that although by Hopf's theorem the quotient

$$\mathrm{H}_2(\mathrm{X})/h_2(\pi_2(\mathrm{X}))$$

only depends on $\pi_1 = \pi_1(\mathrm{X})$, it is not true that $h_2(\pi_2(\mathrm{X}))$ only depends on π_1, $\pi_2 = \pi_2(\mathrm{X})$ and on the action of π_1 on π_2 (chap. II, § 3,F). They constructed two spaces X, Y with same first and second homotopy groups, the action of π_1 on π_2 being trivial in both cases, but $\mathrm{H}_2(\mathrm{X}) = 0$ and $\mathrm{H}_2(\mathrm{Y}) = \mathbf{Z}/2\mathbf{Z}$ ([183], p. 50). They showed that to determine $\mathrm{H}_2(\mathrm{X})$ completely up to isomorphism, additional information is necessary, namely, an element of the cohomology group $\mathrm{H}^3(\pi_1;\pi_2)$ of the group π_1 acting on π_2. In the same paper they generalized that result to the case in which X is "aspherical up to dimension $q - 1$" [i.e., $\pi_j(\mathrm{X}) = 0$ for $1 < j < q$]; then the additional element belongs to $\mathrm{H}^{q+1}(\pi_1;\pi_q)$ [remember $\mathrm{H}^j(\mathrm{X};\mathrm{G}) \simeq \mathrm{H}^j(\pi_q(\mathrm{X});\mathrm{G})$ for $j < q$ and all commutative coefficient groups G (§ 1,B)]. They used, as had Hopf, the passage to the universal covering space $\tilde{\mathrm{X}}$ in order to reduce the problem to an algebraic one on chain complexes with operators.

The next year [184] they enlarged their investigation further by considering an arcwise-connected space X with

$$\pi_i(\mathrm{X}) = 0 \quad \text{for } i < n \text{ and } n < i < q. \tag{47}$$

They simplified and generalized their method to show that $\mathrm{H}^q(\mathrm{X};\mathrm{G})$ is entirely determined by $\pi_n = \pi_n(\mathrm{X})$, $\pi_q = \pi_q(\mathrm{X})$ (and, for $n = 1$, the action of π_1 on π_q), and a cohomology class

$$\boldsymbol{k}^{q+1} \in \mathrm{H}^{q+1}(\pi_n;\pi_q). \tag{48}$$

Their starting point was the same as in their earlier paper defining the cohomology of groups (§ 1,D): the chain transformation

$$\kappa_{n\bullet}\colon \mathrm{S}_{n\bullet}(\mathrm{X}) \to \mathrm{K}_\bullet(\pi_n, n)$$

defined in that paper. This time it was not a chain equivalence. They constructed a "minimal" chain subcomplex $\mathrm{M}_{n\bullet}$ of $\mathrm{S}_{n\bullet}(\mathrm{X})$ such that

$$\mathrm{H}^\bullet(\mathrm{M}_{n\bullet};\mathrm{G}) \simeq \mathrm{H}^\bullet(\mathrm{S}_{n\bullet}(\mathrm{X});\mathrm{G}) \simeq \mathrm{H}^\bullet(\mathrm{X};\mathrm{G})$$

and showed that there is a simplicial map $\bar{\kappa}_.$ of the q-dimensional skeleton $K_.^q(\pi_n(X), n)$ into $M_{n.}$ such that

$$\kappa_.(\bar{\kappa}_.(\sigma)) = \sigma$$

for all j-simplices σ of $K_.(\pi_n(X), n)$ with $j \leqslant q$. There is then an obstruction to extending $\bar{\kappa}_.$ to each $(q+1)$-simplex σ of $K_.(\pi_n(X), n)$: there is a map $f(\sigma)$ of the frontier $\bar{\Delta}_{q+1} - \Delta_{q+1}$ into X, coinciding with $\bar{\kappa}_.(\sigma_i)$ on each q-simplex σ_i opposite to the vertex $\mathbf{e}_i$ for $0 \leqslant i \leqslant q+1$, but in general that map cannot be extended to the interior Δ_{q+1}, hence it determines an element

$$k_n^{q+1}(\sigma) \in \pi_q.$$

These elements define a cochain k_n^{q+1} in $C^{q+1}(K_.(\pi_n(X), n); \pi_q)$ with values in π_q. Eilenberg and MacLane showed that k_n^{q+1} is in fact a *cocycle* and that its cohomology class $\boldsymbol{k}_n^{q+1}(X)$ is an element of $H^{q+1}(\pi_n; \pi_q)$ independent of the choices made in its definition.

Next, an analysis of the construction of the subcomplex $M_{n.}$ led them to a purely algebraic construction, starting with data which consist of three commutative groups Π, Γ, G, and a cocycle

$$k \in Z^{q+1}(K_.(\Pi, n); \Gamma)$$

where $0 < n < q$. They constructed the group $Z^q(k; G)$ of pairs

$$(\rho, r) \in \mathrm{Hom}(\Gamma, G) \times C^q(K_.(\Pi, n); G)$$

such that $\mathbf{d}r = \rho \circ k$. The group of cocycles $Z^q(K_.(\Pi, n); G)$ is a subgroup of $Z^q(k; G)$, and they considered the quotient

$$E^q(k; G) = Z^q(k; G)/B^q(K_.(\Pi, n); G)$$

by the subgroup of coboundaries; they proved that $E^q(k; G)$ only depends on the cohomology class $\boldsymbol{k}$ of k in $H^{q+1}(K_.(\Pi, n); G)$. Returning to the cohomology of the space X satisfying (47), their main result was that if k is any cocycle in $\boldsymbol{k}_n^{q+1}(X)$, then , taking $\Pi = \pi_n$ and $\Gamma = \pi_q$ in the definition of $E^q(k; G)$,

$$H^q(X; G) \simeq E^q(k; G); \tag{49}$$

this shows that $H^q(X; G)$ is determined by π_n, π_q, the action of π_1 on π_q for $n = 1$, and $\boldsymbol{k}_n^{q+1}(X)$.

B. The Postnikov Invariants

The method used by J.H.C. Whitehead to prove the "realizability" of a sequence of groups by the homotopy groups of some space (chap. II, § 3,F) can immediately be applied to prove the following result:

For each $n \geqslant 0$ and arcwise-connected space X, there exists a continuous injection i: X → Z such that (Z, X) is a relative CW-complex (Part 2, chap. V, § 3,C) having cells only in dimension $n + 1$ and

(a) $\pi_{n+1}(Z) = 0$;
(b) $i_*: \pi_m(X) \to \pi_m(Z)$ is an isomorphism for $1 \leqslant m \leqslant n$.

Attach $(n+1)$-cells e_α^{n+1} to X by maps $g_\alpha: S_n \to X$ such that the homotopy classes $[g_\alpha]$ generate $\pi_{n+1}(X)$; since $\pi_m(Z, X) = 0$ for $m \leqslant n + 1$ [chap. II, § 3,C, formula (124)], the homotopy exact sequence shows that i_* is bijective for $m \leqslant n$ and surjective for $m = n + 1$; but the definition of the g_α shows that $i_*([g_\alpha]) = 0$, hence $\pi_{n+1}(Z) = 0$.

Now iterate the process, obtaining an increasing sequence

$$Z = Z_1 \subset Z_2 \subset \cdots Z_{k-1} \subset Z_k \subset \cdots$$

such that $\pi_{n+k}(Z_k) = 0$ and $\pi_m(Z_k) \simeq \pi_m(Z_{k-1})$ for $m < n + k$. Then the space Y^n, union of all the Z_k with the fine topology (Part 2, chap. V, § 3,C) is such that

(i) $\pi_q(Y^n) = 0$ for $q > n$;
(ii) $j_{n*}: \pi_q(X) \to \pi_q(Y^n)$ is an isomorphism for $1 \leqslant q \leqslant n$, j_n being the composite injection $X \xrightarrow{i} Z \to Y^n$;
(iii) (Y^n, X) is a relative CW-complex with no cell of dimension $\leqslant n$.

If X is a CW-complex and is simple, this construction enables us to define for each $n \geqslant 1$ a continuous map $f_n: Y^n \to Y^{n-1}$ for which the triangle

$$\begin{array}{ccc} Y^n & \xrightarrow{\;f_n\;} & Y^{n-1} \\ {\scriptstyle j_n}\nwarrow & & \nearrow{\scriptstyle j_{n-1}} \\ & X & \end{array}$$

is commutative; all such maps are homotopic with respect to $j_{n-1}(X)$. This follows from the above properties and from obstruction theory (chap. II, § 4,C). Indeed, to extend the natural homeomorphism $h: j_n(X) \to j_{n-1}(X)$ by "climbing" along the skeletons, only cells of dimension $\geqslant n + 1$ must be considered in (Y^n, X), so the possible obstructions are in

$$H^{q+1}(Y^n, j_n(X); \pi_q(Y^{n-1}))$$

which is 0 for $q \geqslant n$ since $\pi_q(Y^{n-1}) = 0$; similarly the fact that any two such maps are homotopic with respect to $j_n(X)$ follows from the fact that $H^q(Y^n, j_n(X); \pi_q(Y^{n-1})) = 0$ for $q \geqslant n$.

Thus, to X is associated a sequence of maps

$$\begin{array}{ccccccccc} \cdots \longrightarrow & Y^n & \xrightarrow{\;f_n\;} & Y^{n-1} & \longrightarrow \cdots \longrightarrow & Y^1 & \xrightarrow{\;f_1\;} & Y^0 \\ & {\scriptstyle j_n}\nwarrow & & \nwarrow{\scriptstyle j_{n-1}} & & \nearrow{\scriptstyle j_1} & & \nearrow{\scriptstyle j_0} \\ & & & & X & & & \end{array}$$

determined up to homotopy; it is called a *homotopy resolution* of X.

A well-determined cohomology class can also be defined for each n, generalizing the Eilenberg–Mac Lane invariant:

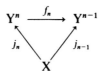

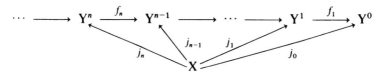

$$k^{n+2} \in H^{n+2}(Y^n; \pi_{n+1}(X)) \tag{50}$$

such that the map f_{n+1} is determined, up to homotopy equivalence, when the pair (Y^n, X), the homotopy group $\pi_{n-1}(X)$ and the invariant k^{n+2} are known. This was first shown by Postnikov [383], and can be presented in the following way [490].

Consider, in the mapping cylinder $Z_{f_{n+1}}$ of $f_{n+1}: Y^{n+1} \to Y^n$, the image of the subspace $j_{n+1}(X) \times [0,1]$ of $Y^{n+1} \times [0,1]$; it is homeomorphic to $X \times [0,1]$ since $f_{n+1} \circ j_{n+1} = j_n$. We define a space $\hat{Y}^n$ by attaching $Z_{f_{n+1}}$ to X along $X \times [0,1]$ by the map $q: (x,t) \mapsto x$ (Part 2, chap. V, § 3); the space Y^n is still a strong deformation retract of $\hat{Y}^n$, so that $\pi_q(\hat{Y}^n) \simeq \pi_q(Y^n)$ for all $q \geqslant 1$. The homotopy exact sequence

$$\cdots \to \pi_{q+1}(Y^{n+1}) \to \pi_{q+1}(\hat{Y}^n) \to \pi_{q+1}(\hat{Y}^n, Y^{n+1}) \xrightarrow{\partial} \pi_q(Y^{n+1}) \to \pi_q(\hat{Y}^n) \to \cdots$$

together with properties (i) and (ii) of the injections j_n, shows that

$$\pi_q(\hat{Y}^n, Y^{n+1}) = 0 \quad \text{if } q \neq n+2, \tag{51}$$

$$\partial: \pi_{n+2}(\hat{Y}^n, Y^{n+1}) \to \pi_{n+1}(Y^{n+1}) \quad \text{is an isomorphism,} \tag{52}$$

and therefore so is the composite

$$\pi_{n+2}(\hat{Y}^n, Y^{n+1}) \xrightarrow{\partial} \pi_{n+1}(Y^{n+1}) \xrightarrow{(j_{n+1})_*^{-1}} \pi_{n+1}(X).$$

From (51) it follows that the relative Hurewicz homomorphism

$$\rho: \pi_{n+2}(\hat{Y}^n, Y^{n+1}) \to H_{n+2}(\hat{Y}^n, Y^{n+1})$$

is bijective, and this yields an isomorphism

$$\kappa_{n+2}: H_{n+2}(\hat{Y}^n, Y^{n+1}) \xrightarrow{\rho^{-1}} \pi_{n+2}(\hat{Y}^n, Y^{n+1}) \xrightarrow{(j_{n+1})_*^{-1} \circ \partial} \pi_{n+1}(X).$$

On the other hand, $H_{n+1}(\hat{Y}^n, Y^{n+1}) = 0$, again by the relative Hurewicz isomorphism; hence [Part 1, chap. IV, § 5,D, formula (66)] the map

$$H^{n+2}(\hat{Y}^n, Y^{n+1}; G) \to \operatorname{Hom}(H_{n+2}(\hat{Y}^n, Y^{n+1}), G)$$

is bijective for any commutative group G. In particular κ_{n+2} is the image of a well-determined cohomology class

$$k_0^{n+2} \in H^{n+2}(\hat{Y}^n, Y^{n+1}; \pi_{n+1}(X))$$

which itself, by the cohomology exact sequence, and the fact that Y^n is a strong deformation retract of $\hat{Y}^n$, is mapped on a well-determined cohomology class

$$k^{n+2} \in H^{n+2}(Y^n; \pi_{n+1}(X));$$

this class is the *Postnikov invariant* of X, of dimension $n+2$.

Given a commutative group Π, a CW-complex Y, and a cohomology class

$$k \in H^{n+2}(Y; \Pi);$$

there is a general process of *amplification* that associates to these data a fibration (W, Y, F, q) with $F = K(\Pi, n+1)$ as fiber, well determined up to Y-isomorphism. Indeed, the fundamental property (43) of Eilenberg–Mac Lane spaces associates to $\boldsymbol{k}$ a continuous map

$$h\colon \mathrm{Y} \to \mathrm{K}(\Pi, n+2)$$

whose homotopy class is well determined. Consider the fibration $(\mathrm{P}_{x_0}\mathrm{K}(\Pi, n+2), \mathrm{K}(\Pi, n+2), \mathrm{F}, p)$ of the space of paths $\mathrm{P}_{x_0}\mathrm{K}(\Pi, n+2)$ in $\mathrm{K}(\Pi, n+2)$ with fixed origin x_0 whose fiber $\mathrm{F} = \Omega(\mathrm{K}(\Pi, n+2); x_0)$ is a space $\mathrm{K}(\Pi, n+1)$ [chap. III, §2,C, formula (40)]. The amplification of $\boldsymbol{k}$ is the pullback $(\mathrm{W}, \mathrm{Y}, \mathrm{F}, q)$ of that fibration by h, and is determined up to Y-isomorphism by Π. Y, and $\boldsymbol{k}$ (chap. III, §2,A). The homotopy exact sequence of that fibration shows that

$$q_*\colon \pi_i(\mathrm{W}) \to \pi_i(\mathrm{Y}) \tag{53}$$

is an isomorphism for $i \neq n+1, n+2$, and there is an exact sequence

$$0 \to \pi_{n+2}(\mathrm{W}) \to \pi_{n+2}(\mathrm{Y}) \to \Pi \to \pi_{n+1}(\mathrm{W}) \to \pi_{n+1}(\mathrm{Y}) \to 0. \tag{54}$$

Furthermore, any continuous map $f\colon \mathrm{Z} \to \mathrm{Y}$ can be "lifted" as

$$f\colon \mathrm{Z} \xrightarrow{g} \mathrm{W} \xrightarrow{q} \mathrm{Y}. \tag{55}$$

Consider the "mapping cone" $\mathrm{CY} \cup_f \mathrm{Z}$ of f (Part 2, chap. V, §3,B): if $j\colon \mathrm{Y} \to \mathrm{CY} \cup_f \mathrm{Z}$ is the injection, the map $j \circ f$ is homotopic to a constant map; h can be chosen as a composite

$$\mathrm{Y} \xrightarrow{j} \mathrm{CY} \cup_f \mathrm{Z} \xrightarrow{h_1} \mathrm{K}(\Pi, n+2).$$

By the covering homotopy property, $h \circ f = h_1 \circ (j \circ f)$ can be lifted to $u\colon \mathrm{Z} \to \mathrm{P}_{x_0}\mathrm{K}(\Pi, n+2)$, so that $p \circ u = h \circ f$. But since W is the fiber product of Y and $\mathrm{P}_{x_0}\mathrm{K}(\Pi, n+2)$ for the maps h and p, there is a map $g\colon \mathrm{Z} \to \mathrm{W}$ such that $q \circ g = f$.

Returning to the Postnikov invariant $\boldsymbol{k}^{n+2}$ of X, consider its amplification $(\mathrm{W}^{n+1}, \mathrm{Y}^n, \mathrm{F}, q_{n+1})$ and apply the preceding "lifting" to the map $f_{n+1}\colon \mathrm{Y}_n^{n+1} \to \mathrm{Y}^n$, so that $f_{n+1} = q_{n+1} \circ g_{n+1}$. Since $\pi_{n+1}(\mathrm{Y}^n) = \pi_{n+2}(\mathrm{Y}^n) = 0$, the exact sequence (54) shows that $\pi_{n+2}(\mathrm{W}^{n+1}) = 0$ and $\pi_{n+1}(\mathrm{W}^{n+1}) \simeq \Pi$; together with (53), this proves that g_{n+1} is a *weak homotopy equivalence*, so that f_{n+1} is determined by Y^n, Π, and $\boldsymbol{k}^{n+2}$ up to weak homotopy equivalence.

C. Fibrations with Eilenberg–Mac Lane Fibers

The amplification $(\mathrm{W}, \mathrm{Y}, \mathrm{F}, q)$ defined in section B is a special case of fibrations with Eilenberg–Mac Lane spaces as fibers. The interesting thing about such fibrations $(\mathrm{Y}, \mathrm{X}, \mathrm{F}, p)$ is that the knowledge of the homology $\mathrm{H}_\bullet(\mathrm{Y})$ gives information on the *Eilenberg groups* $\mathrm{H}_\bullet(\mathrm{S}_{n\bullet}(\mathrm{X}))$ (§1,C), and in particular on the homotopy groups $\pi_n(\mathrm{X})$, due to the isomorphism (25) (*loc. cit.*). This follows from a general lemma:

If a continuous map $f\colon \mathrm{Y} \to \mathrm{X}$ of CW-complexes is such that

$$f_*\colon \pi_i(\mathrm{Y}) \to \pi_i(\mathrm{X})$$

is an isomorphism for $i > n$, then

$$\tilde{f}\colon \mathrm{S}_{n\cdot}(\mathrm{Y}) \to \mathrm{S}_{n\cdot}(\mathrm{X})$$

is a *chain equivalence.*

The proof consists in reducing this to the case in which f is injective, by the consideration of the mapping cylinder Z_f. One then defines a chain transformation $\tilde{g}\colon \mathrm{S}_{n\cdot}(\mathrm{X}) \to \mathrm{S}_{n\cdot}(\mathrm{Y})$ by induction on the skeletons of X of dimension $> n$; the assumptions on the homotopy groups of X and Y makes it possible to construct homotopy operators establishing the necessary chain homotopies, in the same way as in the proof of Eilenberg's "reduction theorem" (§ 1,C).

In 1952 G.W. Whitehead and Cartan and Serre in a joint note [115] independently described a method to "kill" homotopy groups of an arcwise-connected space X: just as a universal covering space $\tilde{\mathrm{X}}$ of a space X is such that $\pi_i(\tilde{\mathrm{X}}) \simeq \pi_i(\mathrm{X})$ for $i \geqslant 2$, and $\pi_1(\tilde{\mathrm{X}}) = 0$ (i.e., the first homotopy group is "killed") (chap. II, § 3,B), they demonstrated the existence of a *fibration* $(\mathrm{W}_{n+1}, \mathrm{X}, p)$ such that $\pi_i(\mathrm{W}_{n+1}) = 0$ for $i \leqslant n$, and $p_*\colon \pi_i(\mathrm{W}_{n+1}) \to \pi_i(\mathrm{X})$ is bijective for $i > n$.

The starting point is the same as in Section B above, namely, the construction of an injection $j_n\colon \mathrm{X} \to \mathrm{Y}^n$ satisfying properties (i), (ii), (iii) in section B. If $x_0 \in \mathrm{X}$ and $y_0 = j_n(x_0)$, consider the fibration $\xi = (\mathrm{W}'_{n+1}, \mathrm{Y}^n, p'')$ defined in chap. III, § 2,A, where W'_{n+1} is the space of paths P_{j_n, y_0}; its restriction $j_n^*(\xi) = (\mathrm{W}_{n+1}, \mathrm{X}, \mathrm{F}, p)$ with base space X has the same fiber equal to the space of loops

$$\mathrm{F} = p^{-1}(x_0) = p''^{-1}(y_0) = \Omega(\mathrm{Y}^n, y_0). \tag{56}$$

From the commutative diagram of homotopy exact sequences

$$\begin{array}{ccccccccccccc}
\cdots \to & \pi_{q+1}(\mathrm{X}) & \overset{\partial}{\to} & \pi_q(\mathrm{F}) & \to & \pi_q(\mathrm{W}_{n+1}) & \overset{p_*}{\to} & \pi_q(\mathrm{X}) & \overset{\partial}{\to} & \pi_{q-1}(\mathrm{F}) & \to & \pi_{q-1}(\mathrm{W}_{n+1}) & \to \cdots \\
 & {\scriptstyle j_{n*}}\downarrow & & \| & & \downarrow & & {\scriptstyle j_{n*}}\downarrow & & \| & & \downarrow & \\
\cdots \to & \pi_{q+1}(\mathrm{Y}^n) & \underset{\partial}{\to} & \pi_q(\mathrm{F}) & \to & \pi_q(\mathrm{W}'_{n+1}) & \underset{p''_*}{\to} & \pi_q(\mathrm{Y}^n) & \underset{\partial}{\to} & \pi_{q-1}(\mathrm{F}) & \to & \pi_{q-1}(\mathrm{W}'_{n+1}) & \to \cdots \\
 & & & & & \| & & & & & & \| & \\
 & & & & & 0 & & & & & & 0 &
\end{array} \tag{57}$$

the following results can be derived:

(a) for $q > n$, $\pi_q(\mathrm{Y}^n) = \pi_{q+1}(\mathrm{Y}^n) = 0$, hence $\pi_q(\mathrm{F}) = 0$, $j_{n*}(\pi_q(\mathrm{X})) = 0$, and $\partial(\pi_q(\mathrm{X})) = 0$, so $p_*\colon \pi_q(\mathrm{W}_{n+1}) \to \pi_q(\mathrm{X})$ is bijective;

(b) for $q \leqslant n$, $j_{n*}\colon \pi_q(\mathrm{X}) \to \pi_q(\mathrm{Y}^n)$ is bijective, and since $\partial\colon \pi_q(\mathrm{Y}^n) \to \pi_{q-1}(\mathrm{F})$ is also bijective by (57), $\partial\colon \pi_q(\mathrm{X}) \to \pi_{q-1}(\mathrm{F})$ is bijective. If $q = n$, as $\pi_{n+1}(\mathrm{Y}^n) = 0$, $\pi_q(\mathrm{F}) = 0$ by (57); as $\pi_n(\mathrm{X}) \to \pi_{n-1}(\mathrm{F})$ is bijective, (57) shows that $\pi_n(\mathrm{W}_{n+1}) = 0$; if $q < n$, $\pi_{q+1}(\mathrm{X}) \to \pi_q(\mathrm{F})$ is bijective, hence again $\pi_q(\mathrm{W}_{n+1}) = 0$.

The conclusion is that $(\mathrm{W}_{n+1}, \mathrm{X}, \mathrm{F}, p)$ is the fibration which "kills" the homotopy groups $\pi_q(\mathrm{X})$ for $q \leqslant n$.

That construction can also be made by steps, applied in succession to

$W_1 = X$, $n = 1$, giving a fibration (W_2, W_1, F_1), then to W_2 and $n = 2$, giving a fibration (W_3, W_2, F_2), and so on. Since $\pi_i(W_{n+1}) = 0$ for $i \leqslant n$ and the map $\pi_i(W_{n+1}) \to \pi_i(W_n)$ deduced from the projection is bijective for $i > n$, the homotopy exact sequence shows that the fiber F_n is such that $\pi_i(F_n) = 0$ for all $i \neq n - 1$, and $\pi_{n-1}(F_n) \simeq \pi_n(X)$; in other words, the fiber F_n is an *Eilenberg–Mac Lane space* $K(\pi_n(X), n - 1)$.

It follows from chap. III, § 2,A that there is also a fibration $(P_{j_n}, K(\pi_n(X), n), W_{n+1})$ with P_{j_n} of same homotopy type as W_n and fiber W_{n+1}.

Another example of Eilenberg–Mac Lane fibers is given by the *fibered Postnikov systems*, which can be associated to any *homotopy resolution* of a space X (section B) with contractible space Y^0. First apply the method of chap. III, § 2,A to $f_1 \colon Y^1 \to Y^0$ to obtain a fibration (Z^1, Y^1, F_1, q_1) with a section $j_1 \colon Y^1 \to Z^1$ that is a homotopy equivalence and a fiber $F_1 \simeq K(\pi_1(X), 1)$; then by induction, construct a sequence (Z^n, Y^n, F_n, q_n) of fibrations and a commutative diagram

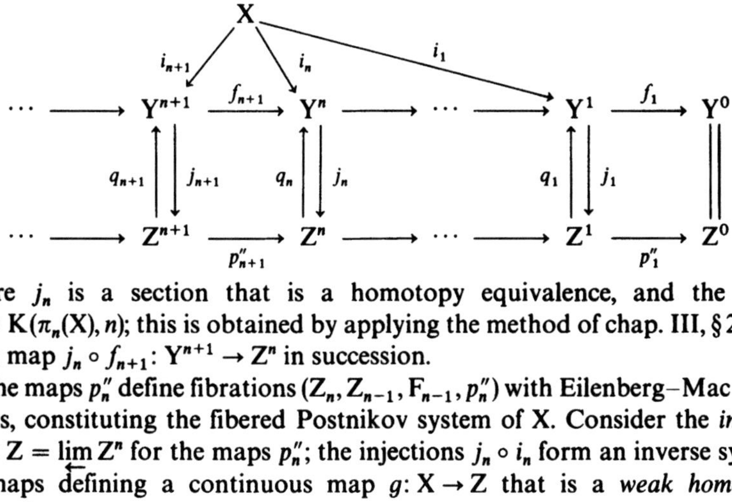

(58)

where j_n is a section that is a homotopy equivalence, and the fiber $F_n \simeq K(\pi_n(X), n)$; this is obtained by applying the method of chap. III, § 2,A to each map $j_n \circ f_{n+1} \colon Y^{n+1} \to Z^n$ in succession.

The maps p''_n define fibrations $(Z_n, Z_{n-1}, F_{n-1}, p''_n)$ with Eilenberg–Mac Lane fibers, constituting the fibered Postnikov system of X. Consider the *inverse limit* $Z = \varprojlim Z^n$ for the maps p''_n; the injections $j_n \circ i_n$ form an inverse system of maps defining a continuous map $g \colon X \to Z$ that is a *weak homotopy equivalence*.

It is possible to use the fibered Postnikov systems to present the question of *extension* of continuous maps and the theory of obstructions in a different way (chap. II, § 4,C). Let us keep the preceding notations and suppose X is n-simple for every n; let $r_j \colon X \to Z^j$ be the composite map defined in (58). Let (W, A) be a relative CW-complex (Part 2, chap. V, § 3,C) and $F \colon A \to X$ be a continuous map. Suppose Z^0 is reduced to a single point, so that $r_0 \circ F \colon A \to Z^0$ has an extension $F_0 \colon W \to Z^0$. Then consider the maps $r_n \circ F \colon A \to Z^n$ and try to extend $r_n \circ F$ to $F_n \colon W \to Z^n$, such that $p''_n \circ F_n = F_{n-1}$; the obstruction to the extension of $r_{n+1} \circ F$ to a lifting of F_n is a cohomology class in $H^{n+2}(W, A; \pi_{n+1}(X))$ (chap. II, § 4,C); let $\mathcal{O}_{n+2}(F)$ be the set of these obstructions.

On the other hand, if W_n is the n-skeleton $(W, A)^n$ and $G \colon W_{n+1} \to X$ is an extension of F to W_{n+1}, there is an obstruction $\beta(G) \in H^{n+2}(W, A; \pi_{n+1}(X))$ to the extension of G to W_{n+2} (*loc. cit.*, IX); let $\mathcal{O}'_{n+2}(F)$ be the set of these obstructions; then one can prove that

$$\mathcal{O}_{n+2}(F) = \mathcal{O}'_{n+2}(F)$$

([490], pp. 450–453; see also [440], pp. 437–452).

D. The Homology Suspension

In their study of the homology of the chain complex $K_{\bullet}(\Pi, n)$ (see below, § 3) Eilenberg and Mac Lane introduced ([185], I) a chain transformation

$$S_{\bullet}: K_{\bullet}(\Pi, n) \to K_{\bullet}(\Pi, n+1) \tag{59}$$

which they considered an "analog" to Freudenthal's "suspension" (chap. II, § 6,E) and to which they therefore gave the same name; they deduced from it homomorphisms in cohomology

$$S^*: H^{n+k}(K_{\bullet}(\Pi, n+1); \Lambda) \to H^{n+k-1}(K_{\bullet}(\Pi, n); \Lambda) \tag{60}$$

and proved that S^* is bijective for $k \leqslant n$ and injective for $k = n + 1$.

Serre greatly generalized that definition in his thesis [429]. He first showed that a "suspension" can be defined in cubical homology for *any* fibration (X, B, F, p) with $F = p^{-1}(b_0)$. Consider the homology exact sequence for the pair (X, F) and the injection $j: F \to X$

$$\cdots \to H_q(F; \Lambda) \to H_q(X; \Lambda) \xrightarrow{j_*} H_q(X, F; \Lambda) \xrightarrow{\partial} H_{q-1}(F; \Lambda) \to \cdots. \tag{61}$$

If $H'_q(X, F; \Lambda) = \operatorname{Coker} j_*$ and $H'_{q-1}(F; \Lambda) = \operatorname{Im} \partial$, then ∂ can be considered an *isomorphism*

$$\partial: H'_q(X, F; \Lambda) \to H'_{q-1}(F; \Lambda).$$

The projection $p: X \to B$ also defines a homomorphism at the level of singular chains

$$\tilde{p}: C_{\bullet}(X, F; \Lambda) \to C_{\bullet}(B; \Lambda)$$

commuting with boundary operators, hence a homomorphism of graded homology modules

$$p_*: H_{\bullet}(X, F; \Lambda) \to H_{\bullet}(B; \Lambda)$$

such that $p_* \circ j_* = 0$. The *suspension* defined by Serre is the composite

$$\sigma: H'_{q-1}(F; \Lambda) \xrightarrow{\partial^{-1}} H'_q(X, F; \Lambda) \xrightarrow{p_*} H_q(B; \Lambda).$$

The most interesting case is when $H_{q-1}(X; \Lambda) = H_q(X; \Lambda) = 0$; then ∂ is an isomorphism in (61) and the suspension becomes a homomorphism in homology

$$\sigma: H_{q-1}(F; \Lambda) \to H_q(B; \Lambda). \tag{62}$$

Furthermore, the spectral sequence of the fibration yields a diagram

$$\begin{array}{ccc} E^q_{q,0} & \xrightarrow{d_q} & E^q_{0,q-1} \\ {\scriptstyle\alpha}\big\downarrow & & \big\uparrow{\scriptstyle\beta} \\ H_q(B;\Lambda) & \xleftarrow[\sigma]{} & H_{q-1}(F;\Lambda) \end{array}$$

where α is injective, $\text{Im}(\alpha) = \text{Im}(\sigma)$, β is surjective, and $\text{Ker}(\beta) = \text{Ker}(\sigma)$. Since d_q is the transgression (chap. IV, § 3,B), the suspension appears as a kind of "inverse" of the transgression.*

In particular, using the Serre exact sequence [chap. IV, § 3,C, formula (74)], if $H_i(X;\Lambda) = 0$ for all $i \geqslant 1$ and $H_i(B;\Lambda) = 0$ for $0 < i < p$, then the suspension (62) is bijective for $1 < q < 2p - 1$, surjective for $i = 2p - 1$, and $H_i(F;\Lambda) = 0$ for $0 < i < p - 1$.

In the particular case of a fibration (P, B, F, π) of a space of paths in B with fixed origin x_0, where $\pi^{-1}(x_0) = F = \Omega(B, x_0)$, the suspension is defined in the whole group $H_{q-1}(F;\Lambda)$ for every $q > 1$; Serre showed that it could be derived from a homomorphism at the level of singular cubical *chains*. He defined a chain transformation

$$k_\bullet\colon Q_\bullet(P) \to Q_\bullet(P)$$

in the following way: if $u\colon [0,1]^n \to P$ is a singular n-cube so that $t \mapsto u(t_1,\ldots,t_n)(t)$ is a path in B with origin x_0, then $(k_n \circ u)(t_1,\ldots,t_{n+1})$ is the path in B

$$t \mapsto u(t_2,\ldots,t_{n+1})(tt_1).$$

It is easily checked that $k_{n-1} \circ (\mathbf{b}u) + \mathbf{b}(k_n \circ u) = u$. From k deduce a chain transformation

$$s_\bullet = \tilde{\pi} \circ k_\bullet\colon Q_\bullet(F) \to Q_\bullet(B)$$

such that

$$(s_n \circ u)(t_1,\ldots,t_{n+1}) = u(t_2,\ldots,t_{n+1})(t_1), \tag{63}$$

and $s_{n-1} \circ (\mathbf{b}u) + \mathbf{b}(s_n \circ u) = 0$ if $u \in Q_n(F)$ with $n > 0$.

Since $s_\bullet$ transforms degenerate cubes in F into degenerate cubes in B, it defines a homomorphism in homology

$$s_*\colon H_i(F) \to H_{i+1}(B) \tag{64}$$

for all dimensions $i > 0$, that coincides with the suspension (62) defined for general fibrations. In this special case of a space of paths the definition (63) shows that the image by $s_n \circ u$ of the cube $[0,1]^{n+1}$ may be considered a continuous image of the *reduced suspension* of the cube $\{\frac{1}{2}\} \times [0,1]^n$ in the

---

* Serre also defined the suspension for the spectral sequence of a filtered differential group ([429], p. 434).

sense of Freudenthal (Part 2, chap. V, § 2,C). This gives a more precise meaning to the "analogy" mentioned by Eilenberg and Mac Lane, since the definition (64) applies in particular to $B = K(\Pi, q+1)$ and $F = K(\Pi, q)$ and coincides with the homomorphism defined by them.

The properties obtained by Eilenberg and Mac Lane for the maps (60) were generalized by Serre to the fibration $(P_{x_0}B, B, F, p)$ of the space of paths $P_{x_0}B$ with fixed origin in *any* arcwise-connected space B. If $H_i(B; \Lambda) = 0$ for $0 < i < q$, the suspension (64) is bijective for $0 < i < 2q - 2$ and surjective for $i = 2q - 2$.

§ 3. The Homology of Eilenberg–Mac Lane Space

Application of the spectral sequences of fibrations to fibrations with Eilenberg–Mac Lane fibers (§ 2,C) immediately called for the determination of the homology of the spaces $K(\Pi, n)$ in order to compute the E^2 terms of these sequences. This thorny problem engaged the efforts of several mathematicians between 1950 and 1954, and was finally completely solved by H. Cartan in 1954.

The abridged notation $H_i(\Pi, n; G)$ is used for $H_i(K(\Pi, n); G)$, and similarly for cohomology. The problem depends very much on the nature of the group G of coefficients.

A. The Topological Approach

In his thesis [429] Serre applied to the problem the homology spectral sequence of the fibration

$$(P_{x_0}K(\Pi, q+1), K(\Pi, q+1), K(\Pi, q), p)$$

of the space of paths in $K(\Pi, q+1)$ with fixed origin; the E^2 terms are given by

$$E^2_{r,s} = H_r(\Pi, q+1; H_s(\Pi, q; G)) \tag{65}$$

and the abutment is 0.

This yields a method of computation of $H_\bullet(\Pi, q; G)$ by induction on q, starting from the $H_n(\Pi, 1; G)$, which, when Π is *commutative*, are the homology groups $H_n(\Pi; G)$ of the *group* Π acting trivially on the group G. As particular cases, this method shows that, if Π is finitely generated so are the $H_i(\Pi, q; \mathbf{Z})$ for all values of i and q; if Π is *finite* and k is a field such that $\Pi \otimes_{\mathbf{Z}} k = 0$, then $H_i(\Pi, q; k) = 0$ for all $q > 0$ and $i > 0$; in particular $H_i(\Pi, q; \mathbf{Z})$ is *finite*. Since $K(\Pi, q)$ has the homotopy type of the space of loops $\Omega K(\Pi, q+1)$ [chap. II, § 3,C, formula (50)], it is a H-*group* (*loc. cit.*); hence its cohomology algebra $H^\bullet(\Pi, q; k)$ is a *Hopf algebra* for any commutative field k, when Π is *commutative* and *finitely generated*. In particular, when k has characteristic 0, Serre showed that when q is even (resp. odd), $H^\bullet(\mathbf{Z}, q; k)$ is a polynomial algebra (resp. an exterior algebra) generated by 1 and a single element of degree q.

In 1952 Serre used the same method, together with properties of the Steenrod squares, to determine the cohomology $H^\bullet(\Pi, q; \mathbf{F}_2)$ for finitely generated commutative groups Π ([431]; see chap. VI. § 3,B).

B. The Bar Construction

Meanwhile, in 1950 Eilenberg and Mac Lane [185] had begun to attack the problem of determination of $H_{\bullet}(\Pi, n; \Lambda)$ for general rings Λ by using algebraic methods applied to the chain complex $K_{\bullet}(\Pi, n)$ which they had defined earlier [181]. Their idea was to substitute for that chain complex another with the same homology, but more easily computable. We can only give here a rough sketch of the intricate algebra used by them and later by H. Cartan.

The simplified presentation given by Cartan starts from an arbitrary associative *differential graded augmented anticommutative* (DGA for short) Λ-algebra A: A is the direct sum of Λ-modules A_k for $k \geqslant 0$, with unit element $1 \in A_0$ and injective map $\lambda \mapsto \lambda . 1$ of Λ into A_0, so that Λ is identified with its image in A_0; the A_k and Λ form an augmented chain complex (Part 1, chap. IV, § 5,F)

$$0 \leftarrow \Lambda \overset{\varepsilon}{\leftarrow} A_0 \overset{d}{\leftarrow} A_1 \overset{d}{\leftarrow} A_2 \leftarrow \cdots \leftarrow A_k \overset{d}{\leftarrow} A_{k+1} \leftarrow \cdots$$

with the conditions

$$d(xy) = (dx)y + (-1)^k x(dy) \quad \text{for } x \in A_k, \tag{66}$$

$$\varepsilon(xy) = \varepsilon(x)\varepsilon(y) \quad \text{for } x, y \text{ in } A_0, \tag{67}$$

$$\varepsilon \circ d = 0 \quad \text{in } A_1. \tag{68}$$

The augmentation ε is extended to the whole algebra A by taking $\varepsilon(x) = 0$ for $x \in A_k$ when $k \geqslant 1$.

Finally,

$$A_h A_k \subset A_{h+k}, \tag{69}$$

$$yx = (-1)^{hk} xy \quad \text{for } x \in A_h, y \in A_k. \tag{70}$$

From relation (66), a Λ-*algebra* structure on the homology graded group $H_{\bullet}(A)$ is deduced.

Let $\bar{A}$ be the graded quotient Λ-module A/Λ, $[a]$ the image in $\bar{A}$ of $a \in A$; define $\bar{A}^{\otimes o} = \Lambda$, $\bar{A}^{\otimes k}$ the k-th tensor power of the Λ-module $\bar{A}$ for $k \geqslant 1$, and

$$\bar{\mathscr{B}}(A) = \bigoplus_{k \geqslant 0} \bar{A}^{\otimes k}; \tag{71}$$

the element $[a_1] \otimes [a_2] \otimes \cdots \otimes [a_k]$ for $k \geqslant 1$ is written $[a_1, a_2, \ldots, a_k]$; it is 0 when one of the a_j is in $\Lambda \subset A_0$; $\bar{\mathscr{B}}(A)$ is *graded* by the graduation

$$\deg[a_1, a_2, \ldots, a_k] = k + \sum_{i=1}^{k} \deg(a_i) \quad \text{for } k \geqslant 1, \tag{72}$$

the elements of $\bar{A}^{\otimes o}$ having degree 0.

Cartan's addition to that definition was to consider also the direct sum of tensor products

$$\mathscr{B}(A) = \bigoplus_{k \geqslant 0} (A \otimes_{\Lambda} \bar{A}^{\otimes k}); \tag{73}$$

it is an A-module and the tensor product

$$a \otimes [a_1] \otimes [a_2] \otimes \cdots \otimes [a_k]$$

is written $a[a_1, a_2, \ldots, a_k]$; $\mathscr{B}(\mathrm{A})$ is graded by

$$\deg(a[a_1, a_2, \ldots, a_k]) = \deg(a) + k + \sum_{i=1}^{k} \deg(a_i) \tag{74}$$

and an augmentation ε is given by

$$\varepsilon(a \otimes 1) = \varepsilon(a), \qquad \varepsilon(a[a_1, a_2, \ldots, a_k]) = 0 \quad \text{if } k \geqslant 1. \tag{75}$$

The differential d in $\mathscr{B}(\mathrm{A})$ is defined in two steps: first an A-endomorphism s: $\mathscr{B}(\mathrm{A}) \to \mathscr{B}(\mathrm{A})$ of degree $+1$ by

$$s(a[a_1, a_2, \ldots, a_k]) = [a, a_1, a_2, \ldots, a_k] \in \bar{\mathscr{B}}_{k+1}(\mathrm{A}); \tag{76}$$

$s(1) = 0$ and $s \circ s = 0$. Then d is an endomorphism of degree -1 which on $\mathscr{B}_0(\mathrm{A})$ is the differential of A, and which satisfies

$$d(ax) = (da)x + (-1)^h a(dx) \quad \text{for } a \in \mathrm{A}_h,\ x \in \mathscr{B}(\mathrm{A}), \tag{77}$$

$$d(s(x)) + s(dx) = x - \varepsilon(x) \qquad \text{for } x \in \mathscr{B}(\mathrm{A}). \tag{78}$$

It has to be checked that this defines d on every $\mathscr{B}_k(\mathrm{A})$ by induction on k, and that $d \circ d = 0$; the main property is that the augmented chain complex $\mathscr{B}(\mathrm{A})$ is *acyclic* (Part 1, chap. IV, § 5,F).

An anticommutative multiplication can be defined on $\mathscr{B}(\mathrm{A})$, for which it is a DGA algebra, in which $\bar{\mathscr{B}}(\mathrm{A})$ is a subalgebra [the multiplication on $\bar{\mathscr{B}}(\mathrm{A})$ had been defined directly by Eilenberg and Mac Lane, using operations which they called "shuffles"]. If I is the two-sided ideal $\mathrm{Ker}\,\varepsilon$ in A, the natural homomorphism $\mathscr{B}(\mathrm{A}) \to \mathscr{B}(\mathrm{A})/\mathrm{I}\mathscr{B}(\mathrm{A})$, restricted to $\bar{\mathscr{B}}(\mathrm{A})$, is an isomorphism of graded Λ-algebras of $\bar{\mathscr{B}}(\mathrm{A})$ onto $\mathscr{B}(\mathrm{A})/\mathrm{I}\mathscr{B}(\mathrm{A})$; this defines by transport of structure a differential $\bar{d}$ on $\bar{\mathscr{B}}(\mathrm{A})$ for which $\bar{\mathscr{B}}(\mathrm{A})$ is a DGA algebra. The original definition of $\bar{d}$ by Eilenberg and Mac Lane was more complicated.

The algebra $\bar{\mathscr{B}}(\mathrm{A})$ was called the *bar construction* by Eilenberg and Mac Lane. Since it is also a DGA algebra, the bar construction $\bar{\mathscr{B}}(\bar{\mathscr{B}}(\mathrm{A}))$ on $\bar{\mathscr{B}}(\mathrm{A})$ may be considered, and more generally the *iterated bar constructions*

$$\bar{\mathscr{B}}^{(n)}(\mathrm{A}) = \bar{\mathscr{B}}(\bar{\mathscr{B}}^{(n-1)}(\mathrm{A})) \tag{79}$$

for any integer $n > 1$. To each one is associated an *acyclic* DGA algebra

$$\mathscr{B}^{(n)}(\mathrm{A}) = \mathrm{A} \otimes_{\Lambda} \bar{\mathscr{B}}^{(n)}(\mathrm{A}). \tag{80}$$

The goal pursued by Eilenberg and Mac Lane in the bar construction stems from their discovery that if $\mathrm{A} = \Lambda[\Pi]$, the group algebra of the commutative group Π with coefficients in Λ in which all elements of A are taken of degree 0 and $d = 0$, there exists an *isomorphism* of graded homology algebras

$$\mathrm{H}_{\bullet}(\bar{\mathscr{B}}^{(n)}(\Lambda[\Pi])) \cong \mathrm{H}_{\bullet}(\mathrm{K}_{\bullet}(\Pi, n); \Lambda). \tag{81}$$

Using this isomorphism they were able to compute the homology groups $\mathrm{H}_{n+q}(\mathbf{Z}, n; \mathbf{Z})$ with $n > q$, for $q \leqslant 10$ [185].

C. The Cartan Constructions

H. Cartan generalized the bar construction in the following way. He called *construction* on A a triple (A, N, M), where A is a DGA algebra over Λ, $N = \bigoplus_{k \geqslant 0} N_k$ a graded anticommutative Λ-algebra, with $N_0 = \Lambda . 1$; finally, on the skew tensor product $M = A \,{}^{g}\!\otimes_\Lambda N$ of A and N over Λ, which is an anticommutative graded Λ-algebra, there must exist a differential d such that, together with the augmentation $\varepsilon(a \otimes 1) = \varepsilon(a)$, M becomes an *acyclic* DGA algebra, and the map $j\colon a \mapsto a \otimes 1$ is an injective homomorphism of DGA algebras. Then the homomorphism $p\colon a \otimes n \mapsto \varepsilon(a)n$ identifies N with a quotient algebra $\overline{M}$ of M, and N is equipped with the image $\bar{d}$ by p of the differential of M; the kernel of p contains $j(A)$.

If there is a sequence of constructions

$$(A, A^{(1)}, B^{(1)}), (A^{(1)}, A^{(2)}, B^{(2)}), \ldots, (A^{(n-1)}, A^{(n)}, B^{(n)})$$

we say $(A^{(n-1)}, A^{(n)}, B^{(n)})$ is an *iterated construction* on A.

The Cartan constructions constitute a kind of algebraic counterpart of a fibration (X, B, F, p) of a *contractible* space X, where M corresponds to the chain complex $C_{\bullet}(X)$, N to $C_{\bullet}(B)$, and A to $C_{\bullet}(F)$. To the classical homology exact sequence derived from

$$0 \to C_{\bullet}(F) \to C_{\bullet}(X) \to C_{\bullet}(X)/C_{\bullet}(F) \to 0$$

corresponds the homology exact sequence derived from

$$0 \to A \xrightarrow{j} M \to M/A \to 0,$$

i.e.,

$$\cdots \to H_{q+1}(M) \to H_{q+1}(M/A) \xrightarrow{\partial} H_q(A) \to H_q(M) \to \cdots, \tag{82}$$

where ∂ is an isomorphism for $q \geqslant 1$ since M is acyclic.

Consider, on the other hand, the homomorphism

$$p_*\colon H_{q+1}(M/A) \to H_{q+1}(N)$$

deduced from p; the *suspension* homomorphisms in the constructions were defined by Cartan as

$$S_q = p_* \circ \partial^{-1}\colon H_q(A) \to H_{q+1}(N) \quad \text{for } q \geqslant 1$$

(cf. §2,D).

Cartan called a construction *special* if there is a Λ-algebra endomorphism k of M [which corresponds to the chain transformation $k_{\bullet}$ defined by Serre for fibrations (§2,D)], such that

(i) $d(kx) + k(dx) = x - \varepsilon x$;
(ii) $k(1) = 0$, $k(M_j) \subset M_{j+1}$, $k \circ k = 0$;
(iii) for $n \in N_j$, $j \geqslant 1$, $1 \otimes n \in k(M)$.

From k [still following Serre's definition of the suspension for spaces of paths (§2,D)] Cartan deduced the composite homomorphism

$$s\colon A \xrightarrow{j} M \xrightarrow{k} M \xrightarrow{p} N;$$

in N, $s(da) + \bar{d}(sa) = 0$ for $a \in A_q$, $q \geqslant 1$, so that s defines a homomorphism in homology

$$s_*: H_\bullet(A) \to H_\bullet(N) \tag{83}$$

which is the *suspension* defined above.

In particular, the bar construction is special and the suspension (83) coincides with the one defined by Eilenberg and Mac Lane.

Cartan's main result was that, for a construction (A, N, M) and a *special* construction (A′, N′, M′), any DGA algebra homomorphism $f: A \to A'$ extends to a homomorphism $g: M \to M'$, such that, if $f_*: H_\bullet(A) \to H_\bullet(A')$ is an *isomorphism*, and if $\bar{g}: N \to N'$ is the homomorphism deduced from g by passage to quotients, then the map $\bar{g}_*: H_\bullet(N) \to H_\bullet(N')$ is also an *isomorphism*. This shows in particular that, for *any* iterated construction $(A^{(n-1)}, A^{(n)}, B^{(n)})$ the homomorphism

$$A^{(n)} \to \bar{\mathscr{B}}^{(n)}(A)$$

deduced from the identity isomorphism of A, yields *isomorphisms*

$$H_\bullet(A^{(n)}) \simeq H_\bullet(\bar{\mathscr{B}}^{(n)}(A)).$$

Taking $A = \Lambda[\Pi]$, it follows from the Eilenberg–Mac Lane isomorphism (81) that there is an isomorphism

$$H_\bullet((\Lambda[\Pi])^{(n)}) \simeq H_\bullet(K_\bullet(\Pi, n); \Lambda) = H_\bullet(\Pi, n; \Lambda)$$

compatible with the suspensions.

This is how, by a clever choice of the construction $(\Lambda[\Pi], N, M)$, Cartan succeeded in determining $H_\bullet(\Pi, n; \Lambda)$; he defined the successive algebras $A^{(n)}$ for $n \geqslant 2$ in his construction by using exterior algebras and polynomial algebras with only one generator as "building blocks" (which was to be expected from Serre's earlier results). The details are very lengthy and intricate [424].

§ 4. Serre's $\mathscr{C}$-Theory

A. Definitions

In his thesis ([429], p. 491) Serre proved results from which could be deduced, for the spaces of class (ULC) (see definition in § 5), a far reaching extension of Hurewicz's absolute isomorphism theorem (chap. II, § 4,A): if such a space X is arcwise-connected and simply connected, and if the group $\pi_i(X)$ is finitely generated (resp. finite) for $2 \leqslant i < n$, then the homology groups $H_i(X)$ are also finitely generated (resp. finite) for $0 < i < n$.

In 1953 [430] he showed that these results could be derived from a general theory based on a concept that covered both concepts of finite commutative groups finitely generated commutative groups, as well as many others, which he called the *$\mathscr{C}$-classes*. In modern language $\mathscr{C}$ can be considered a subcategory of the category ***Ab*** of all commutative groups, that satisfies the following condition:

(I) In an exact sequence $L \to M \to N$ of commutative groups, if L and N are in $\mathscr{C}$, so in M.

In the (abusive) notation $A \in \mathscr{C}$, axiom (I) is equivalent to the statement that $\{0\} \in \mathscr{C}$, that subgroups and quotient groups of groups of $\mathscr{C}$ are in $\mathscr{C}$, and that an extension of two groups of $\mathscr{C}$ is in $\mathscr{C}$.

Examples of $\mathscr{C}$-classes are finitely generated groups, finite groups, and torsion groups (which are finite or infinite direct sums of p-groups for a finite or infinite number of primes p).

In applications of $\mathscr{C}$-theory to homology and homotopy groups, the groups belonging to $\mathscr{C}$ appear most of the time as objects that can be somewhat neglected, just as meager sets in topology or sets of measure 0 in measure theory. This leads to the definitions:

1. a *$\mathscr{C}$-null* group is a group belonging to $\mathscr{C}$;
2. a *$\mathscr{C}$-injective* (resp. *$\mathscr{C}$-surjective*) homomorphism $f\colon A \to B$ of commutative groups is a homomorphism such that $\operatorname{Ker} f$ (resp. $\operatorname{Coker} f$) is $\mathscr{C}$-null;
3. if f is both $\mathscr{C}$-injective and $\mathscr{C}$-surjective, it is called *$\mathscr{C}$-bijective* or a *$\mathscr{C}$-isomorphism.*

Serre used these notions to examine theorems in which the assumption is the *vanishing* of homology or homotopy groups and see what could be said when "zero" is replaced by "$\mathscr{C}$-null."

He first considered fibrations (X, B, F, p), and for later uses he had to extend the machinery of spectral sequences to *relative* homology: when B' is a subspace of B and $X' = p^{-1}(B')$, so that $(X', B', F, p|X')$ is the induced fibration, relate $H_\bullet(X, X'; \Lambda)$ to $H_\bullet(B, B'; \Lambda)$ and to the local system $(H_\bullet(X_b; \Lambda))$. The fact that for any $b \in B'$ and $F = p^{-1}(b)$ the map

$$p_*\colon \pi_i(E, F) \to \pi_i(B, b)$$

is bijective for all $i > 0$ (chap. III, § 2,C) is generalized to the fact that

$$p_*\colon \pi_i(X, X') \to \pi_i(B, B')$$

is also bijective for all $i \geqslant 0$; this follows from the 5-lemma applied to the commutative diagram of exact sequences

$$\begin{array}{ccccccccc}
\pi_i(X', F) & \to & \pi_i(X, F) & \to & \pi_i(X, X') & \to & \pi_{i-1}(X', F) & \to & \pi_{i-1}(X, F) \\
\downarrow & & \downarrow & & \downarrow & & \downarrow & & \downarrow \\
\pi_i(B', b) & \to & \pi_i(B, b) & \to & \pi_i(B, B') & \to & \pi_{i-1}(B', b) & \to & \pi_{i-1}(B, b).
\end{array}$$

The construction of the spectral sequence for relative homology starts by introducing on the chain complexes $C_\bullet(X')$ and $C_\bullet(X, X') = C_\bullet(X)/C_\bullet(X')$ the filtrations deduced from the filtration on $C_\bullet(X)$, and similarly for $C_\bullet(B)$, $C_\bullet(B')$ and $C_\bullet(B, B')$. Using the exact sequences

$$0 \to C_q(B') \otimes C_\bullet(F) \to C_q(B) \otimes C_\bullet(F) \to C_q(B, B') \otimes C_\bullet(F) \to 0$$

the computation of the terms E^1 and E^2 in the spectral sequence of $C_\bullet(X)$ (chap. IV, § 3,C) is easily transferred to the spectral sequence of $C_\bullet(X, X')$ and yields the fundamental isomorphism

$$E^2_{r,s} \simeq H_r(B, B'; (H_s(X_b; \Lambda))) \tag{84}$$

and the fact that $E^\infty_\bullet$ is the graded group associated to a filtration on $H_\bullet(X, X'; \Lambda)$.

The consequences of that method concern properties of the homomorphism

$$p_*: H_\bullet(X, X') \to H_\bullet(B, B') \tag{85}$$

when B, B′ are arcwise-connected spaces and $\pi_1(B)$ acts trivially on $H_\bullet(X_b; \Lambda)$. There are two principal theorems, under different assumptions on $\mathscr{C}$ and B, B′, F.

Theorem A: $\mathscr{C}$ satisfies the additional axiom

($\mathrm{II_A}$) If M and N are in $\mathscr{C}$, so are $M \otimes N$ and Tor(M, N).

The assumptions on B, B′, F are:

(i) $H_1(B, B') = 0$, $H_i(B, B') \in \mathscr{C}$ for $0 \leqslant i \leqslant q$.
(ii) $H_j(F) \in \mathscr{C}$ for $0 < j < r$.

The conclusion is that if $s = \inf(q, r + 1)$, the map (85) is $\mathscr{C}$-bijective for $i \leqslant s$ and $\mathscr{C}$-surjective for $i = s + 1$.
Theorem B: $\mathscr{C}$ satisfies the additional axiom

($\mathrm{II_B}$) If $M \in \mathscr{C}$, then $M \otimes N \in \mathscr{C}$ for *any* commutative group N.

The assumptions on B, B′, F are:

(i) $H_i(B, B') \in \mathscr{C}$ for $0 \leqslant i < q$;
(ii) $H_j(F) \in \mathscr{C}$ for $0 < j < r$.

The conclusion is that, if $s = q + r - 1$, the map (85) is $\mathscr{C}$-bijective for $i \leqslant s$, $\mathscr{C}$-surjective for $i = s + 1$.

The proofs consist in looking at the E^2_{mn} terms (84) for suitable values of m, n, and applying the machinery of the spectral sequence to deduce the needed properties of E^∞_{mn}.

Observe that ($\mathrm{II_A}$) [but not ($\mathrm{II_B}$)] is satisfied when $\mathscr{C}$ is the category of finite commutative groups; ($\mathrm{II_B}$) is satisfied when $\mathscr{C}$ is the category of torsion groups.

Taking B′ as a one point set, these theorems give relations between the homology of X, B, and F in $\mathscr{C}$-theory. A space X is *$\mathscr{C}$-acyclic* if $H_i(X) \in \mathscr{C}$ for all $i > 0$. Axiom ($\mathrm{II_A}$) implies that if *two* of the spaces X, B, F are $\mathscr{C}$-acyclic, so is the third one. Axiom ($\mathrm{II_B}$) implies that if the base space B is $\mathscr{C}$-acyclic, the homomorphism $H_i(F) \to H_i(X)$ is *$\mathscr{C}$-bijective* for all $i \geqslant 0$ ("Feldbau's theorem" modulo $\mathscr{C}$); and if F is $\mathscr{C}$-acyclic, the homomorphism $H_i(X) \to H_i(B)$ is *$\mathscr{C}$-bijective* for all $i \geqslant 0$ ("Vietoris' theorem" modulo $\mathscr{C}$).

Another additional axiom for $\mathscr{C}$ is:

(III) For each group $M \in \mathscr{C}$, the homology groups $H_i(M; \mathbf{Z})$ of the group M (§1) are in $\mathscr{C}$.

This is satisfied by finitely generated groups and by torsion groups.

If $\mathscr{C}$ satisfies (II_A) and (III), then, for any group $\Pi \in \mathscr{C}$, the homology groups $H_i(\Pi, n; \mathbf{Z})$ of Eilenberg–Mac Lane spaces (§3) are in $\mathscr{C}$ for all $n \geqslant 1$, $i \geqslant 1$.

The most useful results in $\mathscr{C}$-theory are analogs of both absolute and relative Hurewicz isomorphism theorems and of the first Whitehead theorem (chap. II, §4,A, §5,E, and §5,F).

B. The Absolute $\mathscr{C}$-Isomorphism Hurewicz Theorem

Axioms (II_A) and (III) are assumed. If X is an arcwise-connected and simply connected space, and if $\pi_i(X) \in \mathscr{C}$ for $i < n$, then $H_i(X) \in \mathscr{C}$ for $0 < i < n$, and the Hurewicz homomorphism $\pi_n(X) \to H_n(X)$ is $\mathscr{C}$-bijective.

There are two proofs. The first one assumes X is a (ULC) space (see §5) and therefore applies in particular to ANR's. Use induction on n; it is only necessary to show $\pi_n(X) \to H_n(X)$ is $\mathscr{C}$-bijective.

Let (P, X, Ω, p) be the fibration of paths in X with origin at x_0 and $\Omega = p^{-1}(x_0)$. The idea of the proof is to consider the commutative diagram

$$\begin{array}{ccccc} \pi_{n-1}(\Omega) & \xleftarrow{\partial} & \pi_n(P, \Omega) & \longrightarrow & \pi_n(X) \\ \downarrow & & \downarrow & & \downarrow \\ H_{n-1}(\Omega) & \xleftarrow[\partial]{} & H_n(P, \Omega) & \longrightarrow & H_n(X). \end{array}$$

If the horizontal arrows are proved $\mathscr{C}$-bijective as well as the Hurewicz homomorphism $\pi_{n-1}(\Omega) \to H_{n-1}(\Omega)$, the result follows.

1. By general properties of fibrations, $\pi_n(P, \Omega) \to \pi_n(X)$ is bijective (chap. III, §2,C).
2. From the fact that P is contractible and the homotopy exact sequence, it follows that $\partial\colon \pi_n(P, \Omega) \to \pi_{n-1}(\Omega)$ is also bijective.
3. By the hypotheses, theorem A above can be applied for $B = X$, $B' = \{x_0\}$, $F = \Omega$, $q = n$, and $r = n - 1$ provided $H_i(\Omega) \in \mathscr{C}$ for $0 < i < n - 1$. It will establish that $H_n(P, \Omega) \to H_n(X)$ is $\mathscr{C}$-bijective.
4. The remaining part of the proof shows at the same time that $H_i(\Omega) \in \mathscr{C}$ for $0 < i < n - 1$ and that $\pi_{n-1}(\Omega) \to H_{n-1}(\Omega)$ is $\mathscr{C}$-bijective. The universal covering space $\tilde{\Omega}$ of Ω exists and is (ULC) since X is (ULC), and $\pi_i(\tilde{\Omega}) = \pi_{i+1}(X)$ for $i \geqslant 2$. The inductive assumption applied to $\tilde{\Omega}$ shows that $H_i(\tilde{\Omega}) \in \mathscr{C}$ for $0 < i < n - 1$ and that $\pi_{n-1}(\tilde{\Omega}) \to H_{n-1}(\tilde{\Omega})$ is $\mathscr{C}$-bijective.

The Cartan–Leray spectral sequence of the covering space $\tilde{\Omega}$ has E^2 terms

$$E^2_{rs} \simeq H_r(\Pi) \otimes H_s(\tilde{\Omega}) \oplus \mathrm{Tor}(H_{r-1}(\Pi), H_s(\tilde{\Omega}))$$

with $\Pi = \pi_1(\tilde{\Omega}) = \pi_2(X)$, because Ω is simple (chap. II, § 3,F). Axiom (II_A) shows that $E^2_{rs} \in \mathscr{C}$ for $r \geqslant 0, 0 < s < n-1$ and for $s = 0, r > 0$; thus $E^2_{0,n-1} \simeq H_{n-1}(\tilde{\Omega})$ is the only term of total degree $n-1$ for which proof that it is in $\mathscr{C}$ is needed. An inspection of the differentials d_m shows that $H_{n-1}(\tilde{\Omega}) \to H_{n-1}(\Omega)$ is $\mathscr{C}$-bijective and $H_i(\Omega) \in \mathscr{C}$ for $0 < i < n-1$. Finally $\pi_i(\tilde{\Omega}) \to \pi_i(\Omega)$ is bijective for $i \geqslant 2$ and $\mathscr{C}$-bijective for $i = 1$, hence $\pi_{n-1}(\Omega) \to H_{n-1}(\Omega)$ is $\mathscr{C}$-bijective.

When $\mathscr{C}$ consists of the one-element groups $\{0\}$, this argument gives a new proof of the original absolute Hurewicz isomorphism theorem (chap. II, § 4,A), and since Ω is then simply connected, it is unnecessary to assume in that proof that X is (ULC).

A second proof of the absolute $\mathscr{C}$-isomorphism Hurewicz theorem uses the sequence (W_n) of spaces "killing" the homotopy groups of X (§ 2,C); here $W_2 = X$ by assumption, and X is not necessarily (ULC). The original Hurewicz theorem shows that $H_n(W_n) \simeq \pi_n(W_n) \simeq \pi_n(X)$. As in the first proof, use induction on n, and show that $\pi_n(X) \to H_n(X)$ is $\mathscr{C}$-bijective. Consider the commutative diagram

$$\begin{array}{ccc} \pi_n(W_n) & \longrightarrow & H_n(W_n) \\ \downarrow & & \downarrow \\ \pi_n(X) & \longrightarrow & H_n(X) \end{array}$$

where the vertical arrows are deduced from the projection $W_n \to X$. The maps $\pi_n(W_n) \to H_n(W_n)$ and $\pi_n(W_n) \to \pi_n(X)$ are bijective; what remains to be seen is that $H_n(W_n) \to H_n(X)$ is $\mathscr{C}$-bijective. $\pi_i(W_j) \in \mathscr{C}$ for $j < n$, $i < n$, hence $H_i(W_j) \in \mathscr{C}$ for these values of i, j by the inductive assumption. The proof is concluded by showing that the map $H_i(W_{j+1}) \to H_i(W_j)$ is $\mathscr{C}$-bijective for $j < n$, $i < n$ and $\mathscr{C}$-surjective for $j < n$, $i = n + 1$. This is a consequence of theorem A applied to the relative fibration consisting in the fibration $(W_{j+1}, W_j, K(\pi_j(X), j-1))$ and one of its fibers, with $q = n$ and $r = \infty$; the assumptions of theorem A are satisfied because from the fact that $\pi_j(X) \in \mathscr{C}$ for $j < n$ it follows that $H_i(K(\pi_j(X), j-1)) \in \mathscr{C}$ for all $i > 0$.

It follows as a corollary, that if $\pi_1(X) = 0$ and $H_i(X) \in \mathscr{C}$ for $0 < j < n$, then $\pi_i(X) \in \mathscr{C}$ for $i < n$; use induction on n.

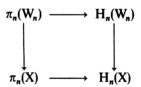

Serre also mentioned without detailed proof an improvement of the absolute $\mathscr{C}$-isomorphism Hurewicz theorem in the following special case: X is a finite CW-complex with $\pi_i(X) = 0$ for $1 \leqslant i \leqslant k-1$, with $k \geqslant 2$. Then if $\mathscr{C}$ is the class of finite commutative groups the Hurewicz homomorphism $\pi_r(X) \to H_r(X; \mathbf{Z})$ is *$\mathscr{C}$-bijective for $r < 2k - 1$*. Serre indicated a possible proof by cohomotopy theory ([430], p. 202) and Milnor sketched a direct proof in ([347], p. 207): using Serre's finiteness theorem for the $\pi_m(S_n)$, it is clear that the result is true for $X = S_n$ with $n \geqslant k$. From the Künneth formula and the relative Hurewicz theorem, the result follows for a wedge of spheres. Finally, using maps $\mathbf{S}_{r_j} \to X$ whose classes are generators of the $\pi_{r_j}(X)$, and combining these maps into a single map

$$f: \mathbf{S}_{r_1} \vee \mathbf{S}_{r_2} \vee \cdots \vee \mathbf{S}_{r_p} \to X$$

the validity of the result for wedges of spheres and the first Whitehead $\mathscr{C}$-theorem (section D) conclude the proof.

C. The Relative $\mathscr{C}$-Isomorphism Hurewicz Theorem

This time it is assumed that axioms (II_B) and (III) are satisfied; X and $A \subset X$ are arcwise-connected and simply connected spaces and the injection $j: A \to X$ is such that $j_*: \pi_2(A) \to \pi_2(X)$ is surjective. Then if $\pi_i(X, A) \in \mathscr{C}$ for $i < n$, $H_i(X, A) \in \mathscr{C}$ for $0 < i < n$, and the Hurewicz map $\pi_n(X, A) \to H_n(X, A)$ is $\mathscr{C}$-bijective.

The strategy again consists in using induction on n and considering the space of paths P in X with fixed origin $x_0 \in A$, and the usual fibration (P, X, F, p) with $F = p^{-1}(x_0) = \Omega(X, x_0)$. Let $P' = p^{-1}(A)$, so that $\pi_i(X, A) \simeq \pi_i(P, P') \simeq \pi_{i-1}(P')$ by the preliminary result on relative fibrations, using the fact that P is contractible. From the homotopy exact sequence for the pair (X, A) and the assumptions it follows that $\pi_1(X, A) = \pi_2(X, A) = 0$, hence $\pi_0(P') = \pi_1(P') = 0$; since by assumption $\pi_i(P') \in \mathscr{C}$ for $i < n - 1$, the absolute $\mathscr{C}$-isomorphism Hurewicz theorem shows that $\pi_{n-1}(P') \to H_{n-1}(P')$ is $\mathscr{C}$-bijective. Consider then the commutative diagram

$$\begin{array}{ccccc} \pi_{n-1}(P') & \longleftarrow & \pi_n(P, P') & \longrightarrow & \pi_n(X, A) \\ \downarrow & & \downarrow & & \downarrow \\ H_{n-1}(P') & \longleftarrow & H_n(P, P') & \longrightarrow & H_n(X, A) \end{array} \tag{86}$$

Application of theorem B with $q = n$, $r = 1$ shows that the map $H_n(P, P') \to H_n(X, A)$ is $\mathscr{C}$-bijective; in (86) all horizontal arrows and one vertical arrow are $\mathscr{C}$-bijective, hence the other vertical arrows are also $\mathscr{C}$-bijective.

It may also be shown that $\pi_{n+1}(X, A) \to H_{n+1}(X, A)$ is $\mathscr{C}$-surjective. In the relative $\mathscr{C}$-isomorphism theorem the axioms (II_B) and (III) may be dropped when $H_j(X)$ and $H_j(A)$ are finitely generated for all j.

D. The First Whitehead $\mathscr{C}$-Theorem

From the relative $\mathscr{C}$-isomorphism Hurewicz theorem the same argument as in the proof of the first Whitehead theorem (chap. II, § 5,F) yields the following generalization in $\mathscr{C}$-theory.

Suppose that $\mathscr{C}$ is a class satisfying (II_B) and (III); X and Y are arcwise-connected and simply connected spaces and $f: X \to Y$ is a continuous map such that $f_*: \pi_2(X) \to \pi_2(Y)$ is surjective. Then the two following properties are equivalent:

(i) f_*: $H_i(X) \to H_i(Y)$ is $\mathscr{C}$-bijective for $i < n$, $\mathscr{C}$-surjective for $i = n$.
(ii) f_*: $\pi_i(X) \to \pi_i(Y)$ is $\mathscr{C}$-bijective for $i < n$, $\mathscr{C}$-surjective for $i = n$.

The method consists of reducing the proof to the case in which A is a subspace of X and f: A $\to$ X the natural injection, by means of the mapping cylinder of f. Then (i) and (ii) are respectively equivalent to:

(i′) $H_i(X, A) \in \mathscr{C}$ for $i \leqslant n$.
(ii′) $\pi_i(X, A) \in \mathscr{C}$ for $i \leqslant n$.

The equivalence of (i′) and (ii′) follows from the relative $\mathscr{C}$-isomorphism Hurewicz theorem.

Serre applied these theorems to homotopy theory mainly in cases in which $\mathscr{C}$ is the class of *finite* groups, or the class of finite groups whose order is only divisible by prime numbers belonging to some fixed set of primes. Then assumption (i) in the Whitehead $\mathscr{C}$-theorem can be replaced by assumptions on the homology groups with coefficients in a *field*.

When $\mathscr{C}$ is the class of finite groups, (i) is equivalent to

(i_0) For a field k of characteristic 0, f_*: $H_i(X; k) \to H_i(Y; k)$ is bijective for $i < n$, surjective for $i = n$.

When $\mathscr{C}$ is the class of finite groups whose order is not divisible by p, (i) can be replaced by

(i_p) For a field k of characteristic $p > 0$, f_*: $H_i(X; k) \to H_i(Y; k)$ is bijective for $i < n$, surjective for $i = n$.

There is also a theorem in $\mathscr{C}$-theory which generalizes the Blakers–Massey homotopy excision theorem (chap. II, § 5,D); see ([5], pp. 102–112).

§ 5. The Computation of Homotopy Groups of Spheres

We have already mentioned several times (chap. II, § 6,E; chap. III, § 2,D) the challenge that polarized the efforts of many topologists, namely, the computation of the homotopy groups of spheres $\pi_m(S_n)$ for $m > n$. After slow and painful beginnings, the veil of mystery shrouding these groups began to be partly lifted in the early 1950s, with *general* theorems going much beyond the determination of individual groups; however, even now the problem is far from being completely solved.

A. Serre's Finiteness Theorem for Odd-Dimensional Spheres

The principle of Serre's method for the determination of the $\pi_m(S_n)$ is to use the original Hurewicz definition of homotopy groups to reduce the computation of homotopy groups of an arcwise-connected space X to the homology groups of *iterated loop spaces* over X (chap. II, § 3,C). The most useful results Serre obtained on the homology of a loop space $\Omega(X) = \Omega(X; x_0')$ assumed that X is *simply connected* (chap. IV, § 3,C); so he had to modify the iteration process slightly by defining two sequences of spaces:

$$\begin{cases} X_0 = X, \\ T_1 = \tilde{X}_0, \quad \text{universal covering space of } X_0 \\ X_1 = \Omega(T_1), \\ T_2 = \tilde{X}_1, \quad \text{universal covering space of } X_1 \\ \cdots\cdots\cdots\cdots \end{cases} \tag{87}$$

In this sequence the Hurewicz definition gives

$$\pi_i(X_n) = \pi_{i+1}(T_n) \quad \text{for } i \geqslant 1;$$

hence, by induction,

$$\pi_1(X_n) = \pi_{n+1}(X), \ldots, \pi_i(X_n) = \pi_{i+n}(X), \ldots \tag{88}$$

and since $\pi_1(X_n)$ is commutative,

$$\pi_{n+1}(X) = H_1(X_n; Z). \tag{89}$$

The formation of the sequence (87) assumes that *each* space X_n does possess a universal covering space (cf. chap. I, § 2,IX). Serre showed that this is implied by a property which he called (ULC), a kind of "uniform" local contractibility*: there exists a neighborhood U of the diagonal Δ in $X \times X$ and a continuous map $F: U \times [0,1] \to X$ such that

1. $F(x, x, t) = x$ for all $t \in [0, 1]$;
2. $F(x, y, 0) = x$, $F(x, y, 1) = y$ for any $(x, y) \in U$.

This is in particular satisfied if X is an ANR.

The study of the homology of the spaces X_n and T_n then proceeds in several steps; in each of them the results are translated by (89) into properties of the homotopy groups of X. The space X is always assumed arcwise-connected, simply connected (so that $T_1 = X_0$), and (ULC).

I. If all the $H_i(X; Z)$ are assumed finitely generated, the same is true for all the $H_i(T_n; Z)$ and $H_i(X_n; Z)$. The proof is by induction on n; to pass from T_n to $X_n = \Omega(T_n)$, it is enough to apply the general result on fibrations that extends Leray's theorem on the relations among the homology of the base space, the fiber, and the total space (chap. IV, § 3,C). To pass from X_{n-1} to $T_n = \tilde{X}_{n-1}$, use the Leray–Cartan spectral sequence of a covering space (§ 1,D); since X_{n-1} is an H-space, $\Pi = \pi_1(X_{n-1}) = H_1(X_{n-1}; Z)$ operates trivially on

* The notation recalls the local property LC of Lefschetz (chap. II, § 2,B), and the q-ulc property of Wilder (Part 2, chap. IV, § 3,B) relative to homology instead of homotopy.

$H_\bullet(X_{n-1})$ (chap. II, § 3,F), so that the E^2 term is given by

$$E^2_{pq} = H_p(\Pi; Z) \otimes_Z H_q(T_n; Z) \oplus \mathrm{Tor}(H_{p-1}(\Pi; Z), H_q(T_n; Z)).$$

Since Π and the $H_i(X_{n-1}; Z)$ are finitely generated, the same Leray–Serre theorem may be applied. In particular the $\pi_{n+1}(X) = H_1(X_n; Z)$ are finitely generated.

II. Along with assumptions in I, assume now that for a field k, $H_i(X; k) = 0$ for $0 < i < n$. Then the conclusion is that:

(i) $H_i(X_j; k) = 0$ for $i > 0$, $i + j < n$;
(ii) $H_i(X_j; k) \simeq H_n(X; k)$ for $i > 0$, $i + j = n$.

The proof is done by induction on j. The passage from X_{j-1} to $T_j = \tilde{X}_{j-1}$ follows from the inductive assumption that for $\Pi = \pi_1(X_{j-1}) = H_1(X_{j-1}; Z)$, $\Pi \otimes k = H_1(X_{j-1}; k) = 0$, and since Π is finitely generated by I, Π is *finite* with order prime to the characteristic of k if it is >0. Using the Leray–Cartan spectral sequence of covering spaces and the values of $H_\bullet(\Pi; k)$ for a finite commutative group (§ 1,D), the isomorphism $H_i(T_j; k) \xrightarrow{\sim} H_i(X_{j-1}; k)$ follows for each $i \geqslant 0$.

The passage from T_j to $X_j = \Omega(T_j)$ then appeals to Serre's generalization to spaces of paths of properties of the homology suspension (§ 2,D) $H_i(X_j; Z) \to H_{i+1}(T_j; Z)$.

Since $\pi_i(X) \otimes k = H_1(X_{i-1}; k)$, the conclusion of II implies that

$$\begin{cases} \pi_i(X) \otimes k = 0 & \text{for } 0 < i < n, \\ \pi_n(X) \otimes k \simeq H_n(X; k). \end{cases} \tag{90}$$

III. The next step comes closer to the loop spaces of spheres. The crux of the matter is the determination of the cohomology algebra of the space $\Omega(S_n)$ [chap. IV, § 3, formulas (76) and (77)]. These formulas show that if n is *odd* and K a field of *characteristic* 0, $H^\bullet(\Omega(S_n); K)$ is isomorphic to an algebra of polynomials $K[e_{n-1}]$, where e_{n-1} has degree $n - 1$. To use the sequence (87) with $X_0 = S_n$, n *odd* $\geqslant 3$, take $T_1 = X_0$, $X_1 = \Omega(S_n)$; since $T_2 = X_1$, evaluate $H^\bullet(\Omega(X_1); K)$. This is settled by the following general lemma:

Let Y be an arcwise-connected and simply connected space such that $H^\bullet(Y; K) \simeq K[u]$, where $u \in H^{2p}(Y; K)$ (here K may have any characteristic). Then $H^\bullet(\Omega(Y); K)$ is an *exterior algebra* $\bigwedge(v)$ with $v \in H^{2p-1}(\Omega(Y); K)$.

The proof first shows that $H^i(\Omega(Y); K) = 0$ for $0 < i < 2p - 1$; this is due to Serre's version of the Eilenberg–MacLane suspension theorem (§ 2,D): $H^i(\Omega(Y); K) \to H^{i+1}(Y; K)$ is an isomorphism for $i < 4p - 2$. What remains to be shown is that $H^i(\Omega(Y); K) = 0$ for $i \geqslant 2p$. Serre considered the cohomology spectral sequence of the fibration $(P, Y, \Omega(Y))$ of the space of paths in Y with fixed origin; its E_2 term is

$$E_2^\bullet \simeq H^\bullet(Y; K) \otimes H^\bullet(\Omega(Y); K).$$

The proof is by contradiction; starting from a homogeneous element w of minimum degree $\geqslant 2p$ among those which are $\neq 0$ in $H^{\bullet}(\Omega(Y); K)$, an examination of the differentials $d_r w$ shows that they are all 0, hence w would define an element $\neq 0$ in $E_{\infty}^{\bullet}$, and this is absurd since $E_{\infty}^{\bullet} = 0$, P being contractible.

IV. With the help of the lemma in III it is now possible to compute the cohomology algebras $H^{\bullet}(X_m; K)$ and $H^{\bullet}(T_m; K)$ when $X = S_n$ (with n *odd*) for $m \leqslant n - 1$. Indeed, $H^{\bullet}(X_2; K)$ is an exterior algebra generated by a single element of degree $n - 2$; in other words, $H^{\bullet}(X_2; K) \simeq H^{\bullet}(S_{n-2}; K)$. Formulas (76) and (77) of chap. IV, §3 are still applicable to a space having the same homology (with coefficients in K) as a sphere (a "homology sphere"), since the Wang sequence only uses the homology of the base space of the fibration. So the argument can be repeated on X_2 instead of X_0; by induction, $H^{\bullet}(X_m; K)$ (with $m \leqslant n - 1$) is an algebra of polynomials, generated by a single element of degree $n - m$ if m is *odd*, and an exterior algebra generated by a single element of degree $n - m$ if m is *even*; this ends with $H^{\bullet}(X_{n-1}; K)$, an exterior algebra generated by a single element of degree 1. As $\pi_1(X_{n-1}) \simeq \pi_n(S_n) \simeq \mathbf{Z}$, the spectral sequence of covering spaces (§ 1,D) may be applied, and this shows that $H_i(T_n; K) = 0$ for $i > 0$. Hence, II proves that $\pi_i(T_n) \otimes_{\mathbf{Z}} K = 0$ for all $i > 0$. Since $\pi_i(T_n) = \pi_{i+n-1}(S_n)$, the conclusion is that

$$\pi_i(S_n) \otimes K = 0 \quad \textit{for } i > n.$$

But as $\pi_i(S_n)$ is finitely generated by I, we finally arrive at the celebrated Serre theorem: *the homotopy groups $\pi_i(S_n)$ for n odd and $i > n$ are finite* [429].

V. Using $\mathscr{C}$-theory, where $\mathscr{C}$ is the class of finitely generated commutative groups, the proof of the preceding theorem can be simplified ([254], p. 317). Consider the fibration (W_{n+1}, S_n, F, π) that "kills" the homotopy groups $\pi_q(S_n)$ for $q \leqslant n$ (§2,C). Since $\pi_*\colon \pi_q(W_{n+1}) \to \pi_q(S_n)$ is bijective for $q > n$, it follows from the homotopy exact sequence that $\pi_q(F) = 0$ for $q \neq n - 1$, and $\pi_{n-1}(F) \simeq \pi_n(S_n) \simeq \mathbf{Z}$; in other words, F is a $K(\mathbf{Z}, n - 1)$. On the other hand, since $\pi_m(S_n)$ is finitely generated, the absolute $\mathscr{C}$-isomorphism Hurewicz theorem shows that $H_m(W_{n+1})$ is also finitely generated for all m.

Since $n - 1$ is even, for any field K of characteristic 0, $H^{\bullet}(F; K)$ is a polynomial algebra $K[u_{n-1}]$ with $u_{n-1} \in H^{n-1}(F; K)$. Consider the Wang cohomology sequence

$$\cdots \to H^m(W_{n+1}; K) \to H^m(F; K) \xrightarrow{\theta} H^{m-n+1}(F; K) \to H^{m+1}(W_{n+1}; K) \to \cdots. \tag{91}$$

Since n is odd, θ is a derivation (chap. IV, §3,A), and since $H^n(W_{n+1}; K) = H^{n-1}(W_{n+1}; K) = 0$, by the absolute Hurewicz isomorphism theorem, θ is bijective for $m = n - 1$, hence $\theta(u_{n-1}) \in H^0(F; K) \simeq K$ is not 0. On the other hand, $H^{p(n-1)}(F; K)$ is a one-dimensonal K-vector space with basis u_{n-1}^p; since θ is a derivation,

$$\theta(u_{n-1}^p) = p u_{n-1}^{p-1} \theta(u_{n-1}) \neq 0,$$

hence $\theta\colon H^{p(n-1)}(F; K) \to H^{(p-1)(n-1)}(F; K)$ is bijective for all p. From the Wang sequence (91) and the fact that $H^m(F; K) = 0$ for all m that are not multiple of $n - 1$ it follows that $H^m(W_{n+1}; K) = 0$ for all $m > 0$. But since $H_m(W_{n+1}; K)$ is finitely generated, and K is a field,

$$0 = H^m(W_{n+1}; K) = \mathrm{Hom}(H_m(W_{n+1}; K), K) \quad \text{(Part 1, chap. IV, § 5D, formula (65))}$$

and this is only possible if $H_m(W_{n+1}; K)$ is 0, hence $H_m(W_{n+1}; \mathbf{Z})$ is *finite* for all $m > 0$; the absolute $\mathscr{C}$-isomorphism Hurewicz theorem then shows that $\pi_m(W_{n+1})$ is finite for all m, and the same is true for $\pi_m(\mathbf{S}_n)$.

A third short proof can be obtained as a special case of a general result stated by H. Cartan and Serre in their second Note of 1952 [116]. It concerns an arcwise-connected and simply connected space X; it is assumed that $H^\bullet(X; \mathbf{Q})$ is a tensor product of an exterior algebra generated by elements of odd degree and a polynomial algebra generated by elements of even degree, the number d_n of generators of degree n being finite for every n; the conclusion is that the rank of every group $\pi_n(X)$ is equal to d_n. The short sketch of a proof uses the fibration $(W_n, W_{n-1}, K(\pi_{n-1}(X), n - 2))$ in order to compute the cohomology algebra $H^\bullet(W_n; \mathbf{Q})$, using the expression of the cohomology $H^\bullet(\Pi, n; \mathbf{Q})$ of the Eilenberg–MacLane spaces.

B. Serre's Finiteness Theorem for Even-Dimensional Spheres

Study of the map $f: \mathbf{S}_{2m-1} \to \mathbf{S}_m$ defined by Hopf for *even m* (chap. II, § 1,C) shows that the group $\pi_{2m-1}(\mathbf{S}_m)$ contains an infinity of elements $p[f]$ for $p \in \mathbf{Z}$; this had to be taken into account in the structure of the group $\pi_{2m-1}(\mathbf{S}_m)$. Serre thought to use the Stiefel manifold $\mathbf{S}_{m+1,2}$ (of dimension $2m - 1$), which can be identified with a sphere bundle having $\mathbf{S}_m$ as base space and $\mathbf{S}_{m-1}$ as fiber: simply identify a pair (u, v) of orthogonal unit vectors in $\mathbf{R}^{m+1}$ with the pair consisting of a point u in $\mathbf{S}_m$ and a unit tangent vector v to $\mathbf{S}_m$ at the point u. There is therefore a homotopy exact sequence

$$\cdots \to \pi_i(\mathbf{S}_{m-1}) \to \pi_i(\mathbf{S}_{m+1,2}) \to \pi_i(\mathbf{S}_m) \to \pi_{i-1}(\mathbf{S}_{m-1}) \to \pi_{i-1}(\mathbf{S}_{m+1,2}) \to \cdots.$$

If $i > m$, the groups $\pi_i(\mathbf{S}_{m-1})$ and $\pi_{i-1}(\mathbf{S}_{m-1})$ are finite by A, hence the group $\pi_i(\mathbf{S}_m)$ will be finite if $\pi_i(\mathbf{S}_{m+1,2})$ is finite, and, more generally, the *rank* of $\pi_i(\mathbf{S}_m)$ will be equal to the rank of $\pi_i(\mathbf{S}_{m+1,2})$.

If $m = 2$, $\mathbf{S}_{m+1,2}$ is homeomorphic to $\mathbf{P}_3(\mathbf{R})$, so $\pi_i(\mathbf{S}_{m+1,2})$ is finite for all $i \geqslant 4$. The information needed on the groups $\pi_i(\mathbf{S}_{m+1,2})$ for $m \geqslant 4$ can be obtained by mimicking the procedure of section A, IV, replacing $\mathbf{S}_m$ (with m odd) by $\mathbf{S}_{m+1,2}$ (with m even) in the formation of the sequence of spaces (87). The basic fact is that the homology of $\mathbf{S}_{m+1,2}$ computed by Stiefel (Part 2, chap. V, § 3,C) is "almost" the same as the homology of a sphere, namely,

$$H_0(\mathbf{S}_{m+1,2}) \simeq \mathbf{Z}, \qquad H_{2m-1}(\mathbf{S}_{m+1,2}) \simeq \mathbf{Z}, \qquad H_{m-1}(\mathbf{S}_{m+1,2}) \simeq \mathbf{Z}/2\mathbf{Z}$$

and all other $H_i(\mathbf{S}_{m+1,2}) = 0$. The homotopy groups of $\mathbf{S}_{m+1,2}$ are finitely generated; for K a field of characteristic 0, $H_i(\mathbf{S}_{m+1,2}; K) = 0$ for $0 < i < 2m - 1$, so that

$$\pi_i(\mathbf{S}_{m+1,2}) \otimes \mathrm{K} = 0 \quad \text{for } i < 2m - 1,$$

$$\pi_{2m-1}(\mathbf{S}_{m+1,2}) \otimes \mathrm{K} \simeq \mathrm{K}.$$

To obtain the other homotopy groups Serre observed that the Wang cohomology sequence with coefficients in K is the same for $\mathbf{S}_{m+1,2}$ as for a sphere. So the arguments of section A, IV can be repeated, and $\pi_i(\mathbf{S}_{m+1,2}) \otimes \mathrm{K} = 0$ for $i > 2m - 1$.

The conclusion is the finiteness theorem for spheres of even dimension: *for n even, the groups $\pi_i(\mathbf{S}_n)$ are finite for $i > n$, with the exception of $\pi_{2n-1}(\mathbf{S}_n)$, which is a direct sum of* $\mathbf{Z}$ *and a finite group.*

C. Wedges of Spheres and Homotopy Operations

Serre's finiteness theorems give no information regarding the *structure* of the homotopy groups $\pi_m(\mathbf{S}_n)$ for $m > n$. Until 1950 the only groups $\pi_{n+k}(\mathbf{S}_n)$ that were known were the $\pi_{n+1}(\mathbf{S}_n)$, isomorphic to $\mathbf{Z}$ for $n = 2$ (Hopf's theorem in 1930) and to $\mathbf{Z}/2\mathbf{Z}$ for $n \geqslant 3$ (Freudenthal's theorems) (chap. II, § 1,B and § 6,E).

After 1946 topologists started renewed attacks, using more refined tools applied to the Freudenthal suspension, the Hopf invariant, and the Whitehead product.

In these papers the wedges $\mathbf{S}_m \vee \mathbf{S}_n$ of two spheres (Part 2, chap. V, § 2,D) were used in several contexts. One of the reasons for their intervention is that the pair consisting of $\mathbf{S}_n \vee \mathbf{S}_m$ and an element $\alpha \in \pi_p(\mathbf{S}_n \vee \mathbf{S}_m)$ *represents a functor* ξ: ***PT*** $\to$ ***Set*** defined by the conditions that*

$$\xi(\mathrm{X}) \in \mathrm{Hom}(\pi_n(\mathrm{X}) \times \pi_m(\mathrm{X}), \pi_p(\mathrm{X})) \tag{92}$$

and that for any continuous map v: X → Y of pointed spaces the diagram

$$\begin{array}{ccc} \pi_n(\mathrm{X}) \times \pi_m(\mathrm{X}) & \xrightarrow{v_* \times v_*} & \pi_n(\mathrm{Y}) \times \pi_m(\mathrm{Y}) \\ \Big\downarrow {\scriptstyle \xi(\mathrm{X})} & & \Big\downarrow {\scriptstyle \xi(\mathrm{Y})} \\ \pi_p(\mathrm{X}) & \xrightarrow[v_*]{} & \pi_p(\mathrm{Y}) \end{array} \tag{93}$$

is commutative (Part 1, chap. IV, § 8,C).

Consider the natural injections

$$i_n: x \mapsto (x, *), \qquad i_m: y \mapsto (*, y)$$

of $\mathbf{S}_n$ and $\mathbf{S}_m$ into $\mathbf{S}_n \vee \mathbf{S}_m$, and their homotopy classes

$$\iota_n \in [\mathbf{S}_n, \mathbf{S}_n \vee \mathbf{S}_m] = \pi_n(\mathbf{S}_n \vee \mathbf{S}_m), \qquad \iota_m \in [\mathbf{S}_m, \mathbf{S}_n \vee \mathbf{S}_m] = \pi_m(\mathbf{S}_n \vee \mathbf{S}_m). \tag{94}$$

Let

* A map between sets is here identified to its graph, hence is an object of the category ***Set***.

$$\alpha = \xi(\mathbf{S}_n \vee \mathbf{S}_m)(\iota_n, \iota_m) \in \pi_p(\mathbf{S}_n \vee \mathbf{S}_m). \tag{95}$$

Consider two arbitrary elements for any pointed space X:

$$\beta \in \pi_n(\mathrm{X}), \qquad \gamma \in \pi_m(\mathrm{X}).$$

They are classes $\beta = [g]$, $\gamma = [h]$ of maps $g\colon \mathbf{S}_n \to \mathrm{X}$, $h\colon \mathbf{S}_m \to \mathrm{X}$. From these maps, define a map

$$v\colon \mathbf{S}_n \vee \mathbf{S}_m \to \mathrm{X}$$

such that $v \circ i_n = g$, $v \circ i_m = h$. Now use the diagram (93) with X replaced by $\mathbf{S}_n \vee \mathbf{S}_m$ and Y by X;

$$\xi(\mathrm{X})(\beta, \gamma) = v_*(\xi(\mathbf{S}_n \vee \mathbf{S}_m)(\iota_n, \iota_m)) = v_*(\alpha)$$

for *any* v; this proves that $(\mathbf{S}_n \vee \mathbf{S}_m, \alpha)$ represents the functor ξ.

The same argument applies to a functor

$$\mathrm{X} \mapsto \xi(\mathrm{X}) \in \mathrm{Hom}(\pi_{n_1}(\mathrm{X}) \times \pi_{n_2}(\mathrm{X}) \times \cdots \times \pi_{n_k}(\mathrm{X}), \pi_p(\mathrm{X}))$$

for an arbitrary integer $k \geqslant 1$. If $k = 1$, the functor is represented by $(\mathbf{S}_{n_1}, \alpha)$ with $\alpha \in \pi_p(\mathbf{S}_{n_1})$; for arbitrary k, by the wedge

$$\mathrm{W} = \mathbf{S}_{n_1} \vee \mathbf{S}_{n_2} \vee \cdots \vee \mathbf{S}_{n_k} \tag{96}$$

and an element $\alpha \in \pi_p(\mathrm{W})$.

The determination of the homotopy groups $\pi_p(\mathrm{W})$ was therefore a problem that was considered by several mathematicians between 1941 and 1955. After partial results had been obtained the general problem was completely solved by P. Hilton in 1955 [226], using iterated Whitehead products (chap. II, § 3,E) of the elements ι_j $(1 \leqslant j \leqslant k)$, which are the homotopy classes of the natural injections

$$i_j\colon \mathbf{S}_{n_j} \to \mathbf{S}_{n_1} \vee \mathbf{S}_{n_2} \vee \cdots \vee \mathbf{S}_{n_k} \quad \text{for } 1 \leqslant j \leqslant k.$$

As a preliminary to the statement and proof of Hilton's theorem it is necessary to select a subset of the set of iterated products; the elements of that subset are called *basic* products, an *order* is defined on the subset, and an integer assigned to each basic product, called its *weight*. These objects are defined inductively in the following way. The ι_{n_j} have weight *one* and are ordered by $\iota_{n_1} < \iota_{n_2} < \cdots < \iota_{n_k}$. When the basic products of weight $< w$ have been defined and ordered, the basic products of weight w are all products $[a, b]$ where a and b are basic products such that: (1) the weights u of a and v of b must be such that $u + v = w$; (2) $a < b$; (3) if b itself has the form $[c, d]$, then $c \leqslant a$. Such restrictions are imposed in order to introduce only linearly independent basic products, owing to the Jacobi identity [chap. II, § 3,E), formula (62)]. If $k = 2$,

$[\iota_1, \iota_2]$ is the only basic product of weight 2;

$[\iota_1, [\iota_1, \iota_2]]$ and $[\iota_2, [\iota_1, \iota_2]]$ are the basic products of weight 3;

$[\iota_1, [\iota_1, [\iota_1, \iota_2]]]$, $[\iota_2, [\iota_1, [\iota_1, \iota_2]]]$, $[\iota_2, [\iota_2, [\iota_1, \iota_2]]]$ the basic products of weight 4.

Once the basic products of weight w have been enumerated they are arbitrarily ordered among themselves and are all taken greater than the basic products of weight $< w$.

Each basic product a belongs to a well-determined homotopy group $\pi_{m(a)}(W)$, where the dimension $m(a)$ is determined inductively by $m(\iota_{n_j}) = n_j$, and for $a = [b, c]$, where b and c are basic products, $m(a) = m(b) + m(c) - 1$.*

Each basic product a is now the homotopy class of a map

$$g_a \colon S_{m(a)} \to W$$

and therefore defines a homomorphism

$$f_a = (g_a)_* \colon \pi_p(S_{m(a)}) \to \pi_p(W)$$

which can be written

$$\beta \mapsto a \circ \beta$$

using the composition product in homotopy (chap. II, §3,B). Finally, define from these homomorphisms

$$f = \bigoplus_a f_a \colon \bigoplus_a \pi_p(S_{m(a)}) \to \pi_p(W) \tag{97}$$

[this is meaningful because only a finite number of integers $m(a)$ are $\leqslant p$]. Hilton's theorem is that *for every p, the map* (97) *is bijective*.

The strategy of the proof is to use the spaces of loops in the following way:

1. The homology algebra $H_\bullet(\Omega(W))$ (for the Pontrjagin product, cf. Part 2, chap. VI, §2,B) is given by the theorem of Bott and Samelson mentioned in chap. IV, §3,C; it is the *free associative algebra* generated by the elements x_1, $x_2, \ldots, x_k$ corresponding by transgression in the space of paths in W (chap. IV, §3,B) to the fundamental classes of the spheres $S_{n_1}, \ldots, S_{n_k}$.

2. For each weight w let T_w be the product of all spaces $\Omega(S_{m(a)})$ for the basic products a of weight $\leqslant w$. They form a direct system of spaces by inclusion; let T be their direct limit. Since $H_\bullet(\Omega(S_{m(a)}))$ is a polynomial algebra generated by a single element y_a of degree $m(a) - 1$ (chap. IV, §3,C), $H_\bullet(T)$ has a basis consisting of the products

$$y_{a_1}^{r_1} \otimes y_{a_2}^{r_2} \otimes \cdots \otimes y_{a_j}^{r_j} \tag{98}$$

for all choices of j, of $(a_i)_{1 \leqslant i \leqslant j}$, and of the exponents $r_j > 0$. The map $g_a \colon S_{m(a)} \to W$ for each basic product a defines a map $u_a \colon \gamma \mapsto g_a \circ \gamma$ of $\Omega(S_{m(a)})$ into $\Omega(W)$. Derive the product map $u_w \colon T_w \to \Omega(W)$ from the maps u_a for each weight w, and by going to the direct limit, a multiplicative map $u \colon T \to \Omega(W)$. Let $z_a = u_*(y_a)$, so that the image of the element (98) by u_* is the product

$$z_{a_1}^{r_1} z_{a_2}^{r_2} \cdots z_{a_j}^{r_j}. \tag{99}$$

3. The crucial point in the proof is to show that the elements (99) form a

* The number $m(a)$ has been explicitly computed by Witt [520].

basis of $H_\bullet(\Omega(W))$. Use is made of a relation given by Samelson [405] between the Whitehead product and the Pontrjagin product in the homology of the H-space $\Omega(W)$ (Part 2, chap. VI, § 2,C) which we write $x.y$: consider the natural maps (for an arcwise-connected space Y)

$$\tau_p\colon \pi_{p+1}(Y) \simeq \pi_p(\Omega(Y)) \xrightarrow{h_p} H_p(\Omega(Y)).$$

Then for $\beta \in \pi_{p+1}(Y)$, $\gamma \in \pi_{q+1}(Y)$,

$$\tau_{p+q}([\beta, \gamma]) = (-1)^p(\tau_p(\beta).\tau_q(\gamma) - (-1)^{pq}\tau_q(\gamma).\tau_p(\beta)). \tag{100}$$

This shows that if the operation $[x, y]$ *in* $H_\bullet(\Omega(W))$ is defined by

$$[x, y] = (-1)^p(x.y - (-1)^{pq}y.x) \tag{101}$$

for $x \in H_p(\Omega(W))$ and $y \in H_q(\Omega(W))$, then each z_a is obtained from the elements $x_1, x_2, \ldots, x_k$ by the same algebraic process as the basic product a is obtained from $\iota_1, \iota_2, \ldots, \iota_k$, the preceding product $[x, y]$ in $H_\bullet(\Omega(W))$ replacing the Whitehead product. The argument Hilton used to prove that the products (99) form a basis of $H_\bullet(\Omega(W))$ was patterned after similar arguments for the free Lie algebras due to P. Hall and W. Magnus.

4. Once this had been done Hilton proved that $u_*\colon H_\bullet(T) \to H_\bullet(\Omega(W))$ is bijective. Next he considered the map $\tilde{u}\colon \tilde{T} \to (\Omega(W))^\sim$ of the universal covering spaces lifted from u (chap. I, § 2,IX) and showed that $\tilde{u}_*\colon H_\bullet(\tilde{T}) \to H_\bullet(\Omega(W)^\sim)$ is also bijective. He first generalized Serre's result on the simplicity of H-spaces (chap. II, § 3,F): if X is a H-space and Y is a covering space of X, then Y is also a H-space and the quotient $\pi_1(X)/\pi_1(Y)$ acts trivially on $H_\bullet(Y)$. The proof of the bijectivity of $\tilde{u}_*$ was patterned after an argument of Serre's thesis ([429], p. 503).

Finally, the relative Hurewicz isomorphism theorem applied to the mapping cylinder of u showed that the map f of formula (97) is bijective for all $p \geqslant 1$.

Of course the Hilton theorem implies that all homotopy operations on k variables are generated by composition of homotopy classes, Whitehead products, and addition.

D. Freudenthal Suspension, Hopf Invariant, and James Exact Sequence

We have seen [chap. II, § 3,D, formula (55)] that Freudenthal's homotopy suspension can be derived from a natural injection of pointed spaces, $s\colon (Y, y_0) \to (\Omega SY, y_1)$.

Beginning in 1950 several mathematicians took up the study of the homotopy suspension by new methods. The exact homotopy sequence of the space ΩSY and its subspace $s(Y)$ can be written

$$\cdots \to \pi_{q+1}(\Omega SY, s(Y)) \xrightarrow{\partial} \pi_q(Y) \xrightarrow{s_*} \pi_q(\Omega SY) \to \pi_q(\Omega SY, s(Y)) \xrightarrow{\partial} \pi_{q-1}(Y) \to \cdots \tag{102}$$

and since $\pi_q(\Omega SY) \simeq \pi_{q+1}(SY)$, s_* is the homotopy suspension E.

Many investigations, which we cannot describe in detail, were concerned with the groups $\pi_q(\Omega SY, s(Y))$ ([491], p. 10). If $\pi_q(Y) = 0$ for $q < n$, $\pi_q(\Omega SY, s(Y)) = 0$ for $q < 2n$, and there is an explicit isomorphism

$$\pi_n(Y) \otimes \pi_n(Y) \rightleftarrows \pi_{2n}(\Omega SY, s(Y)).$$

Hence the exact sequence (102) shows that for $q < n$, the suspension $\pi_q(Y) \to \pi_q(\Omega SY)$ is bijective; furthermore, if

$$W: \pi_n(Y) \otimes \pi_n(Y) \to \pi_{2n-1}(Y)$$

is the Whitehead product map $(\alpha, \beta) \mapsto [\alpha, \beta]$, the sequence

$$\pi_n(Y) \otimes \pi_n(Y) \overset{W}{\to} \pi_{2n-1}(Y) \to \pi_{2n-1}(\Omega SY) \to 0$$

is exact; this gives back Freudenthal's theorems [chap. II, §6,E, formula (136)] with a slight generalization.

For $Y = S_n$, $\Omega S(S_n) \simeq \Omega(S_{n+1})$ (Part 2, chap. V, §2,C), and

$$\pi_q(\Omega(S_{n+1}), S_n) = 0 \quad \text{for } q < 2n,$$

$$\pi_{2n}(\Omega(S_{n+1}), S_n) \simeq \mathbf{Z},$$

so that (102) and the fact that $\pi_q(\Omega(S_{n+1})) = \pi_{q+1}(S_{n+1})$ yield an exact sequence

$$\pi_{2n}(S_n) \overset{E}{\to} \pi_{2n+1}(S_{n+1}) \overset{H}{\to} \mathbf{Z} \to 0$$

and Freudenthal had shown that $H(\alpha)$ is the *Hopf invariant* (chap. II, §1,C) of a map f such that $[f] = \alpha$.

Equivalent definitions of the Hopf invariant have been given by Steenrod [448] and Serre [430]. Steenrod's definition of the "functional cup-product" (Part 1, chap. IV, §4,B) had been inspired by the desire to find a process that would generalize Hopf's construction of his invariant for a map $S_3 \to S_2$ (chap. I, §2), translated from homology to cohomology. Consider a continuous map $f: S_{2n-1} \to S_n$; for any element $u \in H^n(S_n)$, $u \smile u = 0$, $f^*(u) = 0$, and $f^*(H^{2n-1}(S_n)) = 0$. Steenrod's definition of the functional cup-product then showed that $u \smile_f u$ is defined for every element $u \in H^n(S_n)$ and is not a coset but an element of $H^{2n-1}(S_{2n-1})$. Taking $u = s_n$, write $s_n \smile_f s_n = \gamma(f) s_{2n-1}$ with $\gamma(f) \in \mathbf{Z}$. Steenrod showed that $\gamma(f)$ is indeed the Hopf invariant $H(f)$.

An equivalent definition follows from the way Steenrod linked his definition of the functional cup-product to the mapping cylinder Z_f (Part 2, chap. V, §3,A). The relative cohomology algebra $H^\bullet(Z_f, S_{2n-1}; \mathbf{Z})$ is a free $\mathbf{Z}$-module with a basis of two elements $v \in H^n(Z_f, S_{2n-1}; \mathbf{Z})$ and $w \in H^{2n}(Z_f, S_{2n-1}; \mathbf{Z})$; Steenrod showed that

$$v \smile v = H(f) w. \tag{103}$$

Starting from that definition, and using the Hurewicz definition $\pi_{2n-1}(S_n) = \pi_{2n-2}(\Omega(S_n))$, it follows from the structure of $H_\bullet(\Omega(S_n))$ (chap. IV, §3,C) that $H_{2n-2}(\Omega(S_n)) \simeq \mathbf{Z}$, so the Hurewicz homomorphism $h: \pi_{2n-2}(\Omega(S_n)) \to H_{2n-2}(\Omega(S_n))$ can be considered a homomorphism into $\mathbf{Z}$.

Identifying $[f]$ as an element of $\pi_{2n-2}(\Omega(S_n))$, Serre showed that $h([f]) = H(f)$, the Hopf invariant as defined by Steenrod. The proof is quite intricate; it uses the fibration $(P, Z_f, \Omega(Z_f), p)$ of the space of paths with fixed origin in Z_f and the spectral sequence for the "relative" fibration corresponding to the subspace $P' = p^{-1}(S_{2n-1})$ of P (§ 4).

Returning to the groups $\pi_q(\Omega(S_{n+1}), S_n)$, I. James discovered a remarkable isomorphism for *all* values of q when n is *odd* [265]:

$$\pi_q(\Omega(S_{n+1}), S_n) \simeq \pi_q(\Omega(S_{2n+1})) \simeq \pi_{q+1}(S_{2n+1}), \tag{104}$$

giving in that case the *infinite James exact sequence*

$$\cdots \to \pi_q(S_n) \xrightarrow{E} \pi_{q+1}(S_{n+1}) \xrightarrow{H} \pi_{q+1}(S_{2n+1}) \xrightarrow{\partial} \pi_{q-1}(S_n) \to \cdots. \tag{105}$$

James also showed that if n is *even*, the sequence (105) remains exact if the groups are replaced by their 2-*components*. James' proof used a construction (independently found by Toda) of what he called a "reduced product," a CW-complex obtained by successive attachment to S_n of one cell of each dimension $2n, 3n, \ldots, pn, \ldots$ with the same homotopy type as $\Omega(S_{n+1})$. Another proof may be found in [424]; both use spectral sequences of suitable fibrations.

Earlier, G. Whitehead had shown [488] that for *all* n, the sequence (104) is exact *from the term* $\pi_{3n-2}(S_n)$ *on*. His idea was to consider the natural homomorphism $g: S_n \to S_n \vee S_n$ when the wedge $S_n \vee S_n$ is considered to be the space S_n/S_{n-1} obtained by collapsing the equator S_{n-1} to a point (Part 2, chap. V, § 2,A). Whitehead already knew at that time that for $q \leqslant 3n - 2$,

$$\pi_q(S_n \vee S_n) \simeq \pi_q(S_n) \oplus \pi_q(S_n) \oplus \pi_q(S_{2n-1})$$

(a special case of Hilton's theorem of section C); hence there is a natural projection $\pi_q(S_n \vee S_n) \to \pi_q(S_{2n-1})$; when composed with $g_*: \pi_q(S_n) \to \pi_q(S_n \vee S_n)$, it gives a homomorphism $H: \pi_q(S_n) \to \pi_q(S_{2n-1})$ that is the same as the homomorphism H in James' sequence (104). This was called the *generalized Hopf homomorphism* by G. Whitehead.

Hilton generalized that idea, using his theorem to decompose $\pi_p(S_n \vee S_n)$ into a direct sum

$$\pi_p(S_n \vee S_n) \simeq \pi_p(S_n) \oplus \pi_p(S_n) \oplus \pi_p(S_{2n-1}) \oplus \pi_p(S_{3n-2}) \oplus \cdots \tag{106}$$

for all values of p and n; he defined *generalized Hopf homomorphisms* $H_0, H_1, H_2, \ldots$, as the composites of g_* with projections on the components of the direct sum (106). This gives the general expression of $g_*(\alpha)$ for the map g and $\alpha \in \pi_p(S_n)$:

$$g_*(\alpha) = \iota_1 \circ \alpha + \iota_2 \circ \alpha + [\iota_1, \iota_2] \circ H_0(\alpha) + [\iota_1, [\iota_1, \iota_2]] \circ H_1(\alpha) + \cdots.$$

For any space X and two elements β, γ in $\pi_n(X)$, this implies the expression

$$(\beta + \gamma) \circ \alpha = \beta \circ \alpha + \gamma \circ \alpha + [\beta, \gamma] \circ H_0(\alpha) + [\beta, [\beta, \gamma]] \circ H_1(\alpha) + \cdots$$

giving a solution to the problem of "distributivity on the right" for the composition of homotopy classes, which had puzzled topologists for some time.

E. The Localization of Homotopy Groups

In the part of his thesis devoted to homotopy groups Serre introduced a new idea, namely, to evaluate, not the groups $\pi_m(S_n)$ themselves, but their *p-component* for a prime number p. His approach was to compute $\pi_m(S_n) \otimes K$, where K is now a field of *characteristic* p; he reiterated the method he had followed when K has characteristic 0 (section A), namely, the computation of the cohomology $H^{\bullet}(X_m; K)$ for the spaces X_m of the sequence (87). He needed to refine his former results on the cohomology algebra $H^{\bullet}(\Omega(S_n))$ [chap. IV, § 3,C, formula (77)], and more generally to obtain bases for the cohomology spaces $H^q(X; K)$ when K has characteristic p and all that is known on X is that

$$H^i(X; K) \simeq H^i(S_n; K)$$

for some values of i. The technique used as before is the fibration $(P, X, \Omega(X))$ of the space of paths in X with fixed origin. The main results are as follows:

If $m \geqslant 3$ is *odd*, then

$$\pi_i(S_m) \otimes \mathbf{F}_p = 0 \quad \text{for } m < i < m + 2p - 3,$$

$$\pi_i(S_m) \otimes \mathbf{F}_p \simeq \mathbf{F}_p \quad \text{for } i = m + 2p - 3.$$

If $m \geqslant 2$ is *even*, the p-components of $\pi_i(S_m)$ and $\pi_{i-1}(S_{m-1})$ are isomorphic if $2m - 1 < i < 2m + 2p - 4$.

In his paper on $\mathscr{C}$-theory [430] Serre improved on the results of his thesis, also using the Freudenthal suspension and his own definition of the Hopf invariant (section D). He showed that for *odd n* the iterated suspension

$$E^2 \colon \pi_m(S_n) \to \pi_{m+2}(S_{n+2})$$

is a $\mathscr{C}$-isomorphism for $m < p(n + 1) - 3$ when $\mathscr{C}$ is the class of finite commutative groups of order not divisible by the prime p. From this he deduced that if n is *odd*, p is prime, and $m < n + 4p - 6$, the p-components of $\pi_m(S_n)$ and $\pi_{m-n+3}(S_3)$ are isomorphic.

He also gave a different proof for a result which follows from H. Cartan's determination of the homology groups of the spaces $K(\Pi, n)$ (§ 3): if n is *odd* and p is prime and $\neq 2$, then the p-component of $\pi_i(S_n)$ is 0 if $i < n + 2p - 3$ or $n + 2p - 3 < i < n + 4p - 6$ and equal to $\mathbf{Z}/p\mathbf{Z}$ if $i = n + 2p - 3$.

Serre proved that for *even n*, $\pi_m(S_n)$ is $\mathscr{C}$-isomorphic to the direct sum $\pi_m(S_{2n-1}) \oplus \pi_{m-1}(S_{n-1})$, where $\mathscr{C}$ is the class of finite commutative 2-groups. This implies that for $p \neq 2$ the p-components of $\pi_m(S_n)$ and $\pi_{m-1}(S_{n-1})$ are isomorphic if $2n - 1 < m < 2n + 2p - 4$. These results may be considered as constituting the first step in what later became the localization theory of *spaces.*

F. The Explicit Computation of the $\pi_{n+k}(S_n)$ *for* $k > 0$.

The first advance beyond $k = 1$ was independently made in 1950 by Pontrjagin [382] and G.W. Whitehead [487]: the groups $\pi_{n+2}(S_n)$ are $\mathbf{Z}/2\mathbf{Z}$ for all $n \geqslant 2$.

Then in 1951–1952 the group $\pi_6(S_3)$ became the focus of active research by Cartan–Serre [116], Blakers–Massey [51], G. Whitehead [488], and Barratt–Paechter [45]; they first proved that the group has order 12, and then that it contains an element of order 4, so that it must be the cyclic group $\mathbf{Z}/12\mathbf{Z}$. Using the Steenrod squares (chap. VI, § 1), Serre then computed all groups $\pi_{n+k}(S_n)$ for $3 \leqslant k \leqslant 8$ ([431], [432]); they also were computed for $k = 3$ by Rokhlin [399].

After 1953 the structure of the groups $\pi_{n+k}(S_n)$ was systematically investigated by Toda and his school [469]. Tables giving all these groups for $k \leqslant 22$ have been published; by 1983 they had been enlarged to $k \leqslant 30$. No regularity is apparent in these tables, not even for the *stable* groups (chap. II, § 6,E).

Among the tools used by Toda are the general results of Serre described above in sections A, B and E and the James exact sequences (section D); the 2-components require much more work than the p-components for odd primes p. Toda also used what are now called Toda brackets $\langle\alpha,\beta,\gamma\rangle$ (or *secondary homotopy operations*). In general, when there are three homotopy classes $\alpha \in [Y;Z]$, $\beta \in [X;Y]$, and $\gamma \in [W;X]$ satisfying the composition relations $\alpha \circ \beta = 0$ and $\beta \circ \gamma = 0$, $\langle\alpha,\beta,\gamma\rangle$ is a double coset in the group $[SW;Z]$ with respect to the subgroups $\alpha \circ [SW;Y]$ and $[SX;Z] \circ S\gamma$ ([491], p. 17). Special elements of the groups $\pi_m(S_n)$ also play a part, such as $\eta_2 \in \pi_3(S_2)$ and $\nu_4 \in \pi_7(S_4)$ defined by the Hopf fibrations, and the suspensions $\eta_n = E^{n-2}\eta_2$; for example, the element of order 4 in $\pi_6(S_3)$ determined by Barratt–Paechter is $\langle\eta_3, 2\iota_4, \eta_4\rangle$.

Of course, the larger k is, the more intricate are the computations; for $k > 14$, the Toda school also has to use Serre's methods bringing in the Steenrod squares.

§ 6. The Computation of Homotopy Groups of Compact Lie Groups

A. Serre's Method

Before 1950 the only general results on the homotopy groups $\pi_i(G)$ for a compact, connected, semisimple Lie group G were:

1. $\pi_1(G)$ is commutative and finite, a result proved by H. Weyl in 1925, and
2. $\pi_2(G) = 0$, proved by E. Cartan in 1936 [103];

both of these results use the fact that the "singular" elements in a compact Lie group (those belonging to more than one maximal torus) form a set having codimension $\geqslant 3$ (see [103], p. 1314);

3. $\pi_3(G)$ is isomorphic to $\mathbf{Z}$ if G is almost simple, which had been noted by E. Cartan without proof.

Serre's approach [430] to the computation of the $\pi_i(G)$ was through a refinement of the Hopf theorem (Part 2, chap. VI, § 2,A). The latter simply says that if k is a field of characteristic 0, the cohomology algebra $H^{\bullet}(G; k)$ is isomorphic to $H^{\bullet}(X_G; k)$, where

$$X_G = S_{m_1} \times S_{m_2} \times \cdots \times S_{m_l},$$

l being the rank of G and the m_j being odd integers (chap. IV, §4). Serre defined a continuous *map*

$$f: G \to X_G \tag{107}$$

such that $f^*: H^i(X_G; k) \to H^i(G; k)$ is bijective for every $i \geqslant 0$; this follows from preliminary properties of the homology and homotopy of spheres S_n for *n odd*.

I. Using Hopf's construction of a continuous map $(x, y) \mapsto x \, . \, y$ of $S_n \times S_n$ into S_n such that $y \mapsto x \, . \, y$ has degree 1 for all x, and $x \mapsto x \, . \, y$ has degree 2 for all y (chap. II, § 1,C), Serre first proved the existence of a map

$$u: (S_n)^m \to S_n \quad \text{for arbitrary } m \tag{108}$$

such that if j_q is the natural injection of the q-th factor S_n in the product, $u \circ j_q: S_n \to S_n$ has degree 2 for each q. Composing with the diagonal $\delta: S_n \to (S_n)^k$, he concluded that for any continuous map $h: S_n \to S_n$ of degree $2k$ and for any homotopy class $\alpha \in \pi_i(S_n)$ (with arbitrary i) such that $k\alpha = 0$, then $h_*(\alpha) = 0$.

II. His next step was to define, for any finite simplicial complex K and any element $z \in H^n(K; \mathbf{Z})$, where n is *odd*, a continuous map

$$f: K \to S_n$$

such that if s_n is the fundamental cohomology class of S_n, $f^*(s_n) = N \, . \, z$ for some integer $N \neq 0$. The proof is done by "climbing" along the skeletons K_i of K for $i \geqslant n + 1$. Identify $H^n(K)$ to $H^n(K_{n+1})$; it then follows from obstruction theory (chap. II, §4,C, VI and X) that there exists a continuous map $f_n: K_{n+1} \to S_n$ such that $f_n^*(s_n) = z$. Let r_i be the number of elements of the *finite* group $\pi_i(S_n)$ for $i > n$; by I, for a map $g_i: S_n \to S_n$ of degree $2r_i$, $(g_i)_*(\alpha) = 0$ for all $\alpha \in \pi_i(S_n)$. Define $f_i: K_{i+1} \to S_n$ by induction on $i > n$ such that $f_i | K_i = g_i \circ f_{i-1}$. This is possible because if σ is an $(i + 1)$-simplex, the restriction of f_{i-1} to its frontier defines an element $\alpha_\sigma \in \pi_i(S_n)$, and $(g_i)_*(\alpha_\sigma) = 0$. Since $f_i^*(s_n) = f_{i-1}^*(g_i^*(s_n)) = 2r_i f_{i-1}^*(s_n)$, for $m = \dim K$, $f_m^*(s_n) = N \, . \, z$ with

$$N = 2^{m-n-1} \prod_{i=n+1}^{m-1} r_i.$$

This implies that if k is a field of characteristic 0, for each $z \in H^n(K; k)$ with *odd n*, there exists a map $f: K \to S_n$ and an element $u \in H^n(S_n; k) \simeq k$ such that $f^*(u) = z$.

III. Hopf's theorem shows that for a field k of characteristic 0, $H^{\bullet}(G; k)$ is an exterior algebra generated by l elements $z_1, z_2, \ldots, z_l$ of odd degrees m_1, $m_2, \ldots, m_l$ (Part 2, chap. VI, §2,A). By II, there is a map $f_i: G \to S_{m_i}$ and an element $u_i \in H^{m_i}(S_{m_i})$ for each i such that $f_i^*(u_i) = z_i$. Then the map (107) constructed by Serre is simply the product

$$s \mapsto f(s) = (f_1(s), f_2(s), \ldots, f_l(s)).$$

IV. Serre used this result to obtain information on the homotopy groups $\pi_i(G)$; since the universal covering group $\tilde{G}$ is semisimple and compact, $H^\bullet(\tilde{G}; k) = H^\bullet(G; k)$ and $\pi_i(\tilde{G}) = \pi_i(G)$ for $i \geqslant 2$. From the Whitehead $\mathscr{C}$-theorem (§ 4,C), for $\mathscr{C}$ equal to the class of finite commutative groups it follows that $f_*: \pi_i(G) \to \pi_i(X_G)$ is a $\mathscr{C}$-isomorphism for *all* $i > 0$ (§ 4,C). In particular, the *rank* of $\pi_q(G)$ is the number of integers j such that $m_j = q$. If q is *even*, $\pi_q(G)$ is finite; this is obvious for X_G and follows for G by the previous theorem.

V. This raised a further question: for a prime p, is the p-component of $\pi_q(G)$ isomorphic to the direct sum of the p-components of the groups $\pi_q(S_{m_j})$? Serre showed by a similar method that this is indeed the case when G has no p-torsion and $2p - 1 \geqslant m_j$ for all j; in particular it is true if $2p - 1 \geqslant \frac{n}{l} - 1$, where n is the dimension of G.

VI. These results are helpful when trying to reduce the computation of the homotopy groups of compact Lie groups to the computation of the homotopy groups of spheres. For the *classical* groups there are also the usual diffeomorphisms

$$U(m+1)/U(m) \simeq \mathbf{S}_{2m+1}, \tag{109}$$

$$SO(m+1)/SO(m) \simeq \mathbf{S}_m, \tag{110}$$

$$U(m+1, \mathbf{H})/U(m, \mathbf{H}) \simeq \mathbf{S}_{4m+3}. \tag{111}$$

Each one defines a fibration, hence a homotopy exact sequence; for example, (109) gives the exact sequence

$$\cdots \to \pi_i(\mathbf{S}_{2m+1}) \to \pi_{i-1}(U(m)) \to \pi_{i-1}(U(m+1)) \to \pi_{i-1}(\mathbf{S}_{2m+1}) \to \cdots. \tag{112}$$

This shows in particular that the map

$$\pi_{i-1}(U(m)) \to \pi_{i-1}(U(m+1)) \tag{113}$$

is *bijective* as soon as $2m \geqslant i$, and that the map

$$\pi_{2m}(U(m)) \to \pi_{2m}(U(m+1))$$

is *surjective*. All isomorphic $\pi_i(U(m))$ for $2m \geqslant i + 1$ are called the *stable* homotopy groups of the unitary groups. They are written $\pi_i(\mathbf{U})$ and may be considered the homotopy groups of the direct limit $\mathbf{U} = \varinjlim U(n)$. Similar remarks may be made for the groups $SO(m)$ and $U(m+1, \mathbf{H})$.

B. Bott's Periodicity Theorems

Computations of the homotopy groups $\pi_i(G)$ of compact Lie groups G for small values of i were carried out by several mathematicians in the early 1950s, using the preceding methods and the value of homotopy groups of spheres. From the tables thus obtained for the classical groups of low dimension (see for instance [55], p. 433) it appeared that the unitary groups $U(n)$ became very irregular for $i \geqslant 2n$, but exhibited a remarkable regularity for $i < 2n$: for $2 \leqslant i < 2n$ the groups were alternatively 0 and **Z** *; similar periodicities appeared for the groups SO(n) and U(n, **H**) [also written Sp(n) in spite of possible confusion with the symplectic group]. By very ingenious and entirely new methods, Bott showed in 1956 these facts to be valid for *all dimensions* ([79], [80]).

Whereas all preceding results on spheres and Lie groups made no use of the natural riemannian structures on these manifolds, Bott showed that these structures can contribute much more information on the spaces of loops on these manifolds. In particular, since 1953 he had become convinced that Morse theory (Part 2, chap. VII, § 3) could yield more topological results than those derived from the Morse inequalities (Part 2, chap. V, § 5).

We shall give brief descriptions of Bott's main contributions, going into detail only for the unitary groups U(m); we will essentially follow Milnor's exposition [345].

I. The first step is the interpretation of Morse's results on the critical points of a C^∞ function f on a smooth manifold M in terms of CW-complexes, which we gave in Part 2, chap. V, § 5, and which seems to have first been proposed by E. Pitcher [363] and Thom [461]. We only mentioned there the consequences for homology, but in fact we may say more: *if f has only* nondegenerate *critical points in finite number, then* M *has the* same homotopy type *as a finite* CW-*complex having one cell of dimension k for every critical point of index k* ([345], p. 22).

Using Morse's "reduction" theorem (Part 2, chap. VII, § 3), this interpretation is transferred to the metric space $\Omega^c(M; p, q)$ of piecewise smooth paths ω on M with extremities p, q and an energy $E(\omega) < c$. Suppose M is complete and that for some $a < c$, p and q are *not conjugate* along *any* smooth geodesic of length $\leqslant \sqrt{a}$. Recall that there is a subset $B \subset \Omega^c$ consisting of broken geodesics, equipped with a structure of smooth manifold, on which the energy function is C^∞ and the critical points are the smooth geodesics joining p and q of length $< \sqrt{c}$; these critical points are nondegenerate since p and q are not conjugate on any of these geodesics. Then the subset $B^a \subset B$ of broken geodesics of energy $\leqslant a$ is compact and is a deformation retract of Ω^c for c close to a. By the finite-dimensional theorem, Ω^c *has the* homotopy type *of a finite CW complex with one cell of dimension k for every geodesic in Ω^c along which the hessian* E_{**} *has index k*.

II. The next step is to relate the usual space $\Omega(M; p, q)$ of *all* continuous paths from p to q in a riemannian manifold M to the space of piecewise smooth paths from p to q, which we now write $\Omega^\infty(M; p, q)$ instead of $\Omega(M; p, q)$, and which is the union of all spaces $\Omega^c(M; p, q)$ for $c > 0$.

---

* In the tables reproduced by A. Borel the groups π_{10} seemed to deviate from this pattern; it was found by Borel and Hirzebruch [65] that this was due to an error in Toda's computations.

The natural injection $j\colon \Omega^\infty(\mathrm{M}; p, q) \to \Omega(\mathrm{M}; p, q)$ is continuous by definition of the topologies; in fact, it is a *homotopy equivalence*. This can be demonstrated by arguments similar to those used in the proof of Morse's reduction theorem (Part 2, chap. VII, § 3): compare an arbitrary path in $\Omega(\mathrm{M}; p, q)$ to an "inscribed polygon" whose sides are sufficiently small minimal geodesics ([345], p. 94).

On the other hand, the structure of $\Omega^\infty(\mathrm{M}; p, q)$ is obtained by a passage to the limit on the spaces $\Omega^a(\mathrm{M}; p, q)$ when a tends to $+\infty$, by a simple device using Whitehead's second theorem (chap. II § 6,B) ([345], pp. 149–153). The conclusion is that $\Omega(\mathrm{M}; p, q)$ has the homotopy type of a countable CW-complex having one cell of dimension λ for every geodesic from p to q of index λ. The same is true of the loop space $\Omega(\mathrm{M}; p)$, since both $\Omega(\mathrm{M}; p, q)$ and $\Omega(\mathrm{M}; p)$ are fibers of the fibration of the space of paths in M with fixed origin at p.

III. Having thus related homotopy theory of riemannian manifolds to Morse theory, Bott's first observation was that for a *symmetric space* G/K, where G is semisimple and K is a maximal compact subgroup, the index of a geodesic has a very simple expression. From Lie theory it first follows that the geodesics in G/K are the natural images by the projection $\mathrm{G} \to \mathrm{G/K}$ of the *translates of one parameter groups* in G. The differential equation of Jacobi fields along a geodesic γ [Part 2, chap. VII, § 3, formula (47)] has very special properties here. In the same notation as *loc.cit.*, the linear map

$$K_{\mathrm{V}(t)}\colon \mathrm{W} \mapsto \mathrm{R}(\mathrm{V}(t) \wedge \mathrm{W}) . \mathrm{V}(t)$$

in the tangent space $\mathrm{T}_{\gamma(t)}(\mathrm{M})$ is self-adjoint. Consider an orthonormal frame $(\mathrm{U}_1(t), \ldots, \mathrm{U}_n(t))$ along γ moving by parallel translation such that for $t = 0$, the vectors $\mathrm{U}_i(0)$ are *eigenvectors* of $K_{\mathrm{V}(0)}$, so that

$$K_{\mathrm{V}(0)}(\mathrm{U}_i(0)) = e_i \mathrm{U}_i(0) \quad \text{for } 1 \leqslant i \leqslant n,$$

with *real* numbers e_i. Then for *every* t,

$$K_{\mathrm{V}(t)}(\mathrm{U}_i(t)) = e_i \mathrm{U}_i(t) \quad \text{for } 1 \leqslant i \leqslant n.$$

Hence, given a Jacobi field

$$\mathrm{J}(t) = w_1(t)\mathrm{U}_1(t) + \cdots + w_n(t)\mathrm{U}_n(t),$$

the Jacobi equation is equivalent to a system of n second-order linear differential equations with *constant* coefficients

$$\frac{d^2 w_i}{dt^2} + e_i w_i = 0 \quad \text{with } w_i(0) = 0 \text{ for } 1 \leqslant i \leqslant n. \tag{114}$$

If $e_i \leqslant 0$, $w_i(t)$ only vanishes for $t = 0$, but if $e_i > 0$, the zeros of w_i are the integral multiples $k\pi/\sqrt{e_i}$. This shows that the points of γ that are conjugate to p are the $\gamma(k\pi/\sqrt{e_i})$ for $e_i > 0$ and the multiplicity of such a point $\gamma(t)$ is the sum of the integers $\mu_{k,i}$ for all pairs (k, i) such that $k\pi/\sqrt{e_i} = t$, $\mu_{k,i}$ being the multiplicity of the eigenvalue e_i of $K_{\mathrm{V}(0)}$.

IV. Now take as symmetric space a compact, simply connected, semisimple

Lie group G, considered as a space of orbits for G × G. It can then be assumed that the origin p of a geodesic is the neutral element e, so that the tangent space $T_{\gamma(0)}$ is the Lie algebra $\mathfrak{g}$ of G. Lie theory then shows that for $V(0) = X$ in $\mathfrak{g}$,

$$K_X(Y) = \tfrac{1}{4}[[X, Y], X]$$

or equivalently $K_X = -\frac{1}{4}\,\mathrm{ad}(X) \circ \mathrm{ad}(X)$. Now ad(X) is skew symmetric for the natural riemannian metric on $\mathfrak{g}$, i.e.,

$$\langle \mathrm{ad}(X).Y, Z\rangle = -\langle Y, \mathrm{ad}(X).Z\rangle$$

for X, Y, Z in $\mathfrak{g}$. There is therefore an orthonormal basis in $\mathfrak{g}$, such that the matrix of ad(X) for that basis has the form

$$\begin{pmatrix} 0 & a_1 & & & \\ -a_1 & 0 & & & \\ & & 0 & a_2 & \\ & & -a_2 & 0 & \\ & & & & \ddots \end{pmatrix}$$

hence the matrix of K_X is diagonal

$$\frac{1}{4}\begin{pmatrix} a_1^2 & & & & \\ & a_1^2 & & & \\ & & a_2^2 & & \\ & & & a_2^2 & \\ & & & & \ddots \end{pmatrix}$$

and therefore the nonzero eigenvalues of K_X are all >0 and occur in pairs. This implies Bott's first result [78]: *the loop space $\Omega(G)$ has the homotopy type of a CW-complex with no odd-dimensional cells and only finitely many cells of each even dimension.* In particular, this was the first proof that $H_j(\Omega(G)) = 0$ for odd j and that $H_j(\Omega(G))$ is a *free* **Z**-module of finite rank for even j.

By similar methods Bott also gave a proof that for the centralizer U of a torus in G, G/U has no torsion, without having to use the classification of simple Lie groups, and he showed how to compute the Betti numbers of G/U (chap. IV, §4, VII).

V. To approach homotopy problems, Bott introduced a new device, the consideration of the subspace $\Omega_{\min}(M; p, q)$ of $\Omega^\infty(M; p, q)$ consisting of all *minimal* geodesics joining p and q in a complete riemannian manifold M. For simplicity we shall only give details on the arguments for the homotopy groups $\pi_i(SU(m))$ of unitary groups as presented in [345]. Consider the space $\Omega_{\min}(SU(2m); I_{2m}, -I_{2m})$ of minimal geodesics joining the unit matrix I_{2m} to its negative $-I_{2m}$; the program is divided in two steps:

(a) proof that the injection $\Omega_{\min} \to \Omega^\infty$ determines isomorphisms for homotopy groups in dimension $\leqslant 2m$, and
(b) computation of homotopy groups of $\Omega_{\min}(SU(2m); I_{2m}, -I_{2m})$.

VI. The first step is deduced from a more general statement concerning $\Omega_{\min}(M; p, q)$ for a complete riemannian manifold M, using the following (in appearance very restrictive) assumptions:

(i) $\Omega_{\min}(M; p, q)$ is a C^∞ manifold, and
(ii) the *nonminimal* geodesics joining p and q have an index $\geqslant \lambda_0$.

The conclusion is that the relative homotopy groups

$$\pi_i(\Omega^\infty(M; p, q), \Omega_{\min}(M; p, q)) = 0 \quad \text{for } 0 \leqslant i < \lambda_0. \tag{115}$$

Hence, by the homotopy exact sequence

$$\pi_i(\Omega^\infty(M; p, q)) \simeq \pi_i(\Omega_{\min}(M; p, q)) \quad \text{for } 0 \leqslant i \leqslant \lambda_0 - 2. \tag{116}$$

The method is to perform a "reduction" using the construction in the proof of Morse's "reduction theorem" (Part 2, chap. VII, § 3). For c large enough, the space $\Omega_{\min}(M; p, q)$ is contained in the open subspace B of $\Omega^c(M; p, q)$ defined in I, and B is a deformation retract of $\Omega^c(M; p, q)$. As it is enough to prove that

$$\pi_i(\Omega^c(M; p, q), \Omega_{\min}(M; p, q)) = 0 \quad \text{for } i < \lambda_0$$

when c is large enough, we finally need to prove that

$$\pi_i(B, \Omega_{\min}(M; p, q)) = 0 \quad \text{for } i < \lambda_0. \tag{117}$$

Recall that the energy function E on B (*loc.cit.*) has as critical points the smooth geodesics joining p and q; its minimum value d in B is the energy of minimal geodesic arcs from p to q, so that $E(B) = [d, c[$ and $E^{-1}(d) = \Omega_{\min}(M; p, q)$. It is convenient to introduce a diffeomorphism F of $[d, c[$ onto $[0, +\infty[$ and to consider the composite function $f = F \circ E : B \to \mathbf{R}$. The proof is then reduced to establishing the following general property.

Let M be a smooth manifold and $f: M \to \mathbf{R}$ be a C^∞ function defined on M such that for all $c > 0$, $f^{-1}([0, c]) = M^c$ is compact and $M^o = f^{-1}(0)$ is a smooth submanifold of M. Then if each critical point of f in $M - M^o$ has an index $\geqslant \lambda_0$,

$$\pi_i(M, M^o) = 0 \quad \text{for } i < \lambda_0. \tag{118}$$

For the proof, consider a continuous map

$$h: I^r \to M \quad \text{with } I = [0, 1] \tag{119}$$

such that $h(\mathrm{Fr}(I^r)) \subset M^o$; it must be shown that for $r < \lambda_0$, h is homotopic relative to $\mathrm{Fr}(I^r)$ to a continuous map

$$h'': I^r \to M^o. \tag{120}$$

Let U be a tubular neighborhood of M^o in M such that M^o is a strong deformation retract of U (Part 1, chap. III, § 1); let

$$c = \sup_{x \in h(I^r)} f(x) > 0, \quad 3\delta = \inf_{x \in M - U} f(x) > 0. \tag{121}$$

The idea of the proof is to construct a C^∞ function

$$g\colon M^{c+2\delta} \to \mathbf{R}$$

approximating f in the C^2 topology in such a way that:

(i) $|f(x) - g(x)| < \delta$ for $x \in M^{c+2\delta}$;
(ii) g has only nondegenerate critical points;
(iii) the index of every critical point of g that lies in the compact set $f^{-1}([\delta, c + 2\delta])$ is $\geqslant \lambda_0$.

Consider an arbitrary neighborhood W of f in the space of C^2 functions on $M^{c+2\delta}$ for the C^2 topology and cover the set $f^{-1}([\delta, c + 2\delta])$ by finitely many compact sets K_i, each of which lies in the domain U_i of a chart of M. Replacing the functions by their expressions in local coordinates in each U_i, it must be proved that if two C^∞ functions f, g in a neighborhood V of a compact set K in $\mathbf{R}^n$ are such that the differences

$$|f(x) - g(x)|, \qquad \left|\frac{\partial f}{\partial x_j} - \frac{\partial g}{\partial x_j}\right|, \qquad \left|\frac{\partial^2 f}{\partial x_j \partial x_k} - \frac{\partial^2 g}{\partial x_j \partial x_k}\right|$$

are all $\leqslant \varepsilon$ in K, if all critical points of f in K have an index $\geqslant \lambda_0$, and if ε is small enough, then all critical points of g in K also have an index $\geqslant \lambda_0$. To see this, let

$$k_g(x) = \sum_{j=1}^{n} \left|\frac{\partial g}{\partial x_j}\right|$$

and write

$$e_g^1(x) \leqslant e_g^2(x) \leqslant \cdots \leqslant e_g^n(x)$$

the n eigenvalues of the symmetric hessian $(\partial^2 g/\partial x_j \partial x_k)$; finally, let $m_g(x) = \sup(k_g(x) - e_g^{\lambda_0}(x))$. If all critical points of g have index $\geqslant \lambda_0$, then $m_g(x) > 0$ in K. Using the fact that the eigenvalues of a matrix depend continuously on the matrix, it follows that if $m_f(x) > 0$ in K and ε is small enough, then $m_g(x) > 0$ in K. This takes care of conditions (i) and (iii) when W is small enough, and Morse showed that any neighborhood W of f for the C^2 topology contains C^∞ functions with only nondegenerate critical points (Part 2, chap. V, § 5), so (ii) is also satisfied.

Now $h(\mathrm{I}^r)$ is contained in $g^{-1}(]-\infty, c + \delta])$. But by I, $g^{-1}(]-\infty, c + \delta])$ has the homotopy type of the space obtained by attaching cells of dimension $\geqslant \lambda_0$ to $g^{-1}(]-\infty, 2\delta])$; since $r < \lambda_0$, h is homotopic, relative to M^0, within the space $g^{-1}(]-\infty, c + \delta])$, to a map

$$h'\colon \mathrm{I}^r \to g^{-1}(]-\infty, 2\delta])$$

coinciding with h in M^o. Since $g^{-1}(]-\infty, 2\delta]) \subset U$ and M^o is a strong deformation retract of U, h' is homotopic, relative to M^o, to a map (120).

In passing, note that part of the Freudenthal suspension theorem (chap. II, § 6,E) is a consequence of the isomorphism (116) when $M = \mathbf{S}_{n+1}$ and p and q

are antipodal points on $\mathbf{S}_{n+1}$. Then clearly $\Omega_{\min}(\mathrm{M}; p, q)$ is diffeomorphic to $\mathbf{S}_n$ and the nonminimal geodesics in $\Omega(\mathrm{M}; p, q)$ are paths along a great circle winding at least one and a half times around $\mathbf{S}_{n+1}$ and so contain at least two conjugate points to p in their interior, each of multiplicity n; so $\lambda_0 \geqslant 2n$, and since $\pi_i(\Omega(\mathbf{S}_{n+1})) \simeq \pi_{i+1}(\mathbf{S}_{n+1})$, it is thus shown that $\pi_i(\mathbf{S}_n)$ is isomorphic to $\pi_{i+1}(\mathbf{S}_{n+1})$ for $i \leqslant 2n - 2$.

VII. The next step in the study of the homotopy groups of unitary groups is a study of the space $\Omega_{\min}(\mathrm{SU}(2m); I_{2m}, -I_{2m})$. The Lie algebra $\mathfrak{su}(2m)$ consists of $2m \times 2m$ complex skew hermitian matrices A (i.e., $A + A^* = 0$) with trace 0; the geodesics in $\mathrm{SU}(2m)$ issued from I_{2m} are the one parameter subgroups of $\mathrm{SU}(2m)$

$$\gamma\colon t \mapsto \exp(tA) = \sum_{k=0}^{\infty} t^k A^k/k!$$

for $A \in \mathfrak{su}(2m)$; the length from $t = 0$ to $t = 1$ is $(\mathrm{Tr}(AA^*))^{1/2}$. We are only considering the geodesics for which $e^A = -I_{2m}$. By a suitable choice of basis in $\mathfrak{su}(2m)$

$$A = \begin{pmatrix} ia_1 & & & 0 \\ & ia_2 & & \\ & & \ddots & \\ 0 & & & ia_{2m} \end{pmatrix}$$

where the a_j are real and $a_1 + a_2 + \cdots + a_{2m} = 0$. Then $e^A = -I_{2m}$ if and only if

$$A = \begin{pmatrix} k_1 i\pi & & & 0 \\ & k_2 i\pi & & \\ & & \ddots & \\ 0 & & & k_{2m} i\pi \end{pmatrix} \tag{122}$$

where $k_1 \geqslant k_2 \geqslant \cdots \geqslant k_{2m}$ are *odd* rational integers such that $\sum_{j=1}^{2m} k_j = 0$. The length of the corresponding geodesic from I_{2m} to $-I_{2m}$ is $(k_1^2 + k_2^2 + \cdots + k_{2m}^2)^{1/2}$, hence it is *minimal* if and only if $k_j = \pm 1$ for all j and $\sum_{j=1}^{2m} k_j = 0$. This means that A has only *two* distinct eigenvalues $\pm i\pi$, and that the corresponding eigenspaces E_+ and E_- in $\mathbf{C}^{2m}$ are orthogonal for the hermitian scalar product $\sum_{j=1}^{2m} z_j \bar{z}'_j$, and *both* have complex dimension m. Clearly A is thus entirely determined by the subspace E_+, which can be an arbitrary subspace of complex dimension m. The conclusion is that *the space* $\Omega_{\min}(\mathrm{SU}(2m);\ I_{2m}, -I_{2m})$ *is homeomorphic to the complex grassmannian* $\mathrm{G}_{2m,m}(\mathbf{C})$. Thus condition (i) of VI is verified.

To check condition (ii) of VI for a *nonminimal* geodesic of $\mathrm{SU}(2m)$ joining I_{2m} and $-I_{2m}$, use the results of III. For the matrix A given by (122), one must compute the eigenvalues of the self-adjoint operator

$$K_A\colon Y \mapsto -\tfrac{1}{4}[A, [A, Y]] \quad \text{in } \mathfrak{su}(2m).$$

If Y is the matrix (y_{jl}), a short computation gives

$$-\tfrac{1}{4}[A,[A,Y]] = \left(\frac{\pi^2}{4}(k_j - k_l)^2 y_{jl}\right)_{1 \leqslant j,l \leqslant 2m}.$$

This shows that a basis of $\mathfrak{su}(2m)$ consisting of eigenvectors of K_A is the set of matrices:

(i) the matrices E_{jl} with $y_{jl} = 1$, $y_{lj} = -1$ for $j < l$, $y_{rs} = 0$ for other pairs of indices; the corresponding eigenvalues are

$$e_{jl} = \frac{\pi^2}{4}(k_j - k_l)^2; \tag{123}$$

(ii) the matrices iE_{jl}, with the same eigenvalues;
(iii) the matrices forming a basis of the subspace of diagonal matrices belonging to $\mathfrak{su}(2m)$, with eigenvalues all 0.

The nonzero eigenvalues of K_A are thus the numbers (123), for $k_j > k_l$, each counted twice. This corresponds on the geodesic $\gamma\colon t \mapsto e^{tA}$ to the values $r\pi/\sqrt{e_{jl}}$ for integers $r \geqslant 1$, i.e.,

$$\frac{2}{k_j - k_l}, \frac{4}{k_j - k_l}, \ldots, \frac{2r}{k_j - k_l}, \ldots.$$

Their number in the open interval $]0, 1[$ is $\frac{1}{2}(k_j - k_l) - 1$, hence (by III) they give a contribution $(k_j - k_l) - 2$ to the index of γ, and finally

$$\lambda = \sum_{k_j > k_l} (k_j - k_l - 2). \tag{124}$$

If γ is *nonminimal*, there are two possibilities.

a) At least $m + 1$ of the k_j are $\leqslant -1$. Then at least one k_j must be $\geqslant 3$ since $\sum_{j=1}^{2m} k_j = 0$, and therefore

$$\lambda \geqslant \sum_{1}^{m+1} (3 - (-1) - 2) = 2(m + 1).$$

b) There are exactly m of the $k_j \leqslant -1$ and exactly m that are $\geqslant 1$, but not all are ± 1. Therefore, one of the k_j is $\geqslant 3$ and another is $\leqslant -3$; hence

$$\lambda \geqslant \sum_{1}^{m-1} (3 - (-1) - 2) + \sum_{1}^{m-1} (1 - (-3) - 2) + (3 - (-3) - 2)$$
$$= 4(m - 4) + 4 = 4m \geqslant 2(m + 1).$$

This ends the proof of condition (ii) in VI for $\lambda_0 = 2m$. Returning to IV, we therefore have isomorphisms

$$\begin{aligned} \pi_i(\mathbf{G}_{2m,m}(\mathbf{C})) &\simeq \pi_i(\Omega(\mathbf{SU}(2m); I_{2m}, -I_{2m})) \\ &\simeq \pi_i(\Omega(\mathbf{SU}(2m); I_{2m}) \\ &\simeq \pi_{i+1}(\mathbf{SU}(2m)) \end{aligned} \tag{125}$$

for $i \leqslant 2m$.

VIII. Now recall that the complex Stiefel manifolds are homogeneous spaces

$$\mathbf{S}_{2m,m}(\mathbf{C}) = \mathrm{U}(2m)/\mathrm{U}(m)$$

and therefore give rise to a homotopy exact sequence

$$\cdots \to \pi_i(\mathbf{S}_{2m,m}(\mathbf{C})) \xrightarrow{\partial} \pi_{i-1}(\mathrm{U}(m)) \to \pi_{i-1}(\mathrm{U}(2m)) \to \pi_{i-1}(\mathbf{S}_{2m,m}(\mathbf{C}))\cdots. \quad (126)$$

We saw in section A that the map

$$\pi_{i-1}(\mathrm{U}(m)) \to \pi_{i-1}(\mathrm{U}(2m)) \quad (127)$$

is bijective for $i \leqslant 2m$ and surjective for $i = 2m + 1$ [formula (112)]. Hence $\pi_i(\mathbf{S}_{2m,m}(\mathbf{C})) = 0$ for $i \leqslant 2m$.

On the other hand, $\mathrm{G}_{2m,m}(\mathbf{C}) = \mathrm{U}(2m)/(\mathrm{U}(m) \times \mathrm{U}(m))$, and the fibration $(\mathbf{S}_{2m,m}(\mathbf{C}), \mathrm{G}_{2m,m}(\mathbf{C}), \mathrm{U}(m))$ (chap. III, § 2,F) gives another homotopy exact sequence

$$\cdots \to \pi_i(\mathbf{S}_{2m,m}(\mathbf{C})) \to \pi_i(\mathrm{G}_{2m,m}(\mathbf{C})) \xrightarrow{\partial} \pi_{i-1}(\mathrm{U}(m)) \to \pi_{i-1}(\mathbf{S}_{2m,m}(\mathbf{C})) \to \cdots \quad (128)$$

hence the isomorphism

$$\pi_i(\mathrm{G}_{2m,m}(\mathbf{C})) \simeq \pi_{i-1}(\mathrm{U}(m)) \quad \text{for } i \leqslant 2m. \quad (129)$$

Finally the obvious fibration $(\mathrm{U}(n), \mathrm{SU}(n), \mathbf{S}_1)$ similarly gives the isomorphism

$$\pi_j(\mathrm{SU}(n)) \simeq \pi_j(\mathrm{U}(n)) \quad \text{for all } j \geqslant 2. \quad (130)$$

Putting together (125), (127), (129), and (130), we get *Bott's periodicity theorem* for unitary groups

$$\pi_{i-1}(\mathrm{U}(m)) \simeq \pi_{i+1}(\mathrm{U}(m)) \quad \text{for } 1 \leqslant i \leqslant 2m. \quad (131)$$

Introducing the direct limit $\mathbf{U}$ of the $\mathrm{U}(m)$ (chap. III, § 2,F) this can also be written as an isomorphism

$$\pi_{i-1}(\mathbf{U}) \simeq \pi_{i+1}(\mathbf{U}) \quad \textit{for all } i \geqslant 1. \quad (132)$$

Since $\mathrm{U}(1) = \mathbf{S}_1$, $\pi_0(\mathrm{U}(1)) = 0$ and $\pi_1(\mathrm{U}(1)) \cong \mathbf{Z}$; so the groups $\pi_i(\mathrm{U})$ are 0 for *i even*, $\mathbf{Z}$ for *i odd*.

By a more thorough study of $\Omega(\mathrm{U}(m))$ Bott also proved that for all m

$$\pi_{2m}(\mathrm{U}(m)) \simeq \mathbf{Z}/m!\mathbf{Z}. \quad (133)$$

IX. With the partial collaboration of Samelson, Bott was able to use Morse theory to investigate the structure of $\Omega_{\min}(\mathrm{M}; p, q)$, not only for $\mathrm{M} = \mathrm{SU}(2m)$, but for *symmetric spaces* $\mathrm{M} = \mathrm{G}/\mathrm{K}$, where G is *any* compact connected Lie group ([77], [83]). An extension of the finite-dimensional Morse theory (Part 2, chap. V, § 5) is necessary: to consider the case in which the critical points of a C^∞ function f on a smooth manifold M are not isolated, but form a set that is a union of disjoint *connected smooth submanifolds* of M. Such a submanifold V is called *nondegenerate* for f if for each point $x \in \mathrm{V}$, the subspace of the tangent space $\mathrm{T}_x(\mathrm{M})$ in which the hessian $\mathrm{H}_x(f)$ vanishes is exactly the tangent space $T_x(V)$ (when V is reduced to a point, the point is nondegenerate); the index of $H_x(f)$ is then constant in V and is called the *index* of V. Bott showed in [79] that it is possible to extend to that situation the description of M as a relative CW-complex. The usual "reduction" process then transfers that description to the space $\Omega^a(M; p, q)$ when the set of geodesics of M that belong to that space is a union of nondegegerate critical submanifolds for the energy.

To apply that description to $\Omega(G/K; p, q)$, consider the neutral component K_{pq} of the subgroup of G leaving invariant p and q, and for any geodesic γ joining p and q, the subgroup K_γ of K_{pq} leaving invariant each point of γ. Clearly K_{pq} acts on the space $\Omega^a(M; p, q)$, and the orbits of that action are diffeomorphic to K_{pq}/K_γ for the geodesics $\gamma \in \Omega^a(M; p, q)$; in [84] Bott and Samelson showed that the multiplicity of q as conjugate point of p on γ is $\dim(K_{pq}/K_\gamma)$, hence the orbits of K_{pq} in $\Omega^a(M; p, q)$ are nondegenerate critical submanifolds for the energy. When γ is *minimal*, Bott also proved that K_{pq}/K_γ is a symmetric space by returning to the original definition of these spaces by E. Cartan: he observed that the midpoint $r = \gamma(\frac{1}{2})$ of γ is then not conjugate to p on γ, and that this implies $K_{pr} = K_\gamma$; from that fact and the existence in M of an isometric involution leaving invariant r and reversing the geodesics through r, he constructed an involution of K_{pq} whose fixed point set contains K_γ; this defines K_{pq}/K_γ as a *symmetric space*, hence also the connected components of $\Omega_{\min}(M; p, q)$.

Bott then went on to explicitly determine the critical submanifolds and their index for symmetric spaces G/K of a *classical group* G, using the method he had already introduced in [77] and [78]. Let $\mathfrak{g} = \mathfrak{k} \oplus \mathfrak{m}$ be the Cartan decomposition of the Lie algebra $\mathfrak{g}$ of G, $\mathfrak{m}$ being an orthogonal supplement of $\mathfrak{k}$. It is convenient to identify G/K with the subspace $M = \exp(\mathfrak{m})$ of G; the geodesics of M with origin e are then the one parameter subgroups of G contained in M. Let $\mathfrak{h}_\mathfrak{m}$ be a maximal commutative Lie algebra contained in $\mathfrak{m}$, $\mathfrak{h} \supset \mathfrak{h}_\mathfrak{m}$ be a Cartan subalgebra of $\mathfrak{g}$ and Σ be a system of positive roots of G relative to $\mathfrak{h}$. The index of the geodesic $\gamma\colon t \mapsto e^{tX}$ for $X \in \mathfrak{m}$, $0 \leqslant t \leqslant 1$ is computed by considering only the case in which $X \in \mathfrak{h}_\mathfrak{m}$, all other geodesics being conjugate to one of those. Consider then the hyperplanes in $\mathfrak{h}_\mathfrak{m}$ of equation

$$\langle \alpha, Y \rangle = n \quad \text{for } \alpha \in \Sigma \text{ and } n \in \mathbf{Z}; \tag{134}$$

the index of γ is the sum of the numbers of solutions of each equation

$$t\langle \alpha, X \rangle = n$$

in the interval $0 \leqslant t \leqslant 1$ for all $\alpha \in \Sigma$ and $n \in \mathbf{Z}$ [more intuitively, it is the number of intersections of the segment joining 0 and X in $\mathfrak{h}_\mathfrak{m}$ with the hyperplanes (134), each counted with its multiplicity]. In this way Bott obtained expressions for the loop spaces of seven symmetric spaces (the notation $S \cup e_k \ldots$ means that the space is obtained by attaching to S cells of dimension $\geqslant k$):

$$\begin{aligned}
\Omega\mathrm{U}(2n) &\simeq \mathrm{U}(2n)/(\mathrm{U}(n)\times\mathrm{U}(n))\cup e_{2n+2}\cdots\\
\Omega\mathrm{SO}(2n) &\simeq \mathrm{SO}(2n)/\mathrm{U}(n)\cup e_{2n-2}\cdots\\
\Omega\mathrm{Sp}(n) &\simeq \mathrm{Sp}(n)/\mathrm{U}(n)\cup e_{2n+2}\cdots\\
\Omega(\mathrm{Sp}(n)/\mathrm{U}(n)) &\simeq \mathrm{U}(n)/\mathrm{O}(n)\cup e_{n+1}\cdots\\
\Omega(\mathrm{U}(2n)/\mathrm{O}(2n)) &\simeq \mathrm{O}(2n)/(\mathrm{O}(n)\times\mathrm{O}(n))\cup e_{n+1}\cdots\\
\Omega(\mathrm{SO}(4n)/\mathrm{U}(2n)) &\simeq \mathrm{U}(2n)/\mathrm{Sp}(n)\cup e_{4n-2}\cdots\\
\Omega(\mathrm{U}(4n)/\mathrm{Sp}(2n)) &\simeq \mathrm{Sp}(2n)/(\mathrm{Sp}(n)\times\mathrm{Sp}(n))\cup e_{4n+1}\cdots
\end{aligned}\tag{135}$$

Since for all these symmetric spaces depending on an integer n, the homotopy groups are *stable* (for each dimension i, the homotopy group π_i is the same for all n large enough), the preceding relations yield the isomorphisms

$$\pi_{k+1}(\mathbf{U}) \simeq \pi_k(\mathbf{U}/(\mathbf{U}\times\mathbf{U})), \qquad \pi_{k+1}(\mathbf{O}) \simeq \pi_k(\mathbf{O}/\mathbf{U}), \qquad \text{etc.} \tag{136}$$

Finally, the theory of classifying spaces gives isomorphisms

$$\pi_k(\mathbf{U}) \simeq \pi_{k+1}(\mathbf{U}/(\mathbf{U}\times\mathbf{U})), \qquad \pi_k(\mathbf{O}) \simeq \pi_{k+1}(\mathbf{O}/(\mathbf{O}\times\mathbf{O})),$$
$$\pi_k(\mathbf{Sp}) \simeq \pi_{k+1}(\mathbf{Sp}/(\mathbf{Sp}\times\mathbf{Sp}))$$

Putting all these together, Bott established three periodicities:

$$\begin{cases}\pi_k(\mathbf{U}) \simeq \pi_{k+2}(\mathbf{U}),\\ \pi_k(\mathbf{O}) \simeq \pi_{k+4}(\mathbf{Sp}),\\ \pi_k(\mathbf{Sp}) \simeq \pi_{k+4}(\mathbf{O}),\end{cases}\tag{137}$$

so that the periods for $\mathbf{U}$, $\mathbf{O}$, and $\mathbf{Sp}$ are

for $\mathbf{U}$ and $k = 0, 1$, $\quad 0, \mathbf{Z}$,

for $\mathbf{O}$ and $0 \leqslant k \leqslant 7$, $\quad \mathbf{Z}/2\mathbf{Z}, \mathbf{Z}/2\mathbf{Z}, 0, \mathbf{Z}, 0, 0, 0, \mathbf{Z}$,

for $\mathbf{Sp}$ and $0 \leqslant k \leqslant 7$, $\quad 0, 0, 0, \mathbf{Z}, \mathbf{Z}/2\mathbf{Z}, \mathbf{Z}/2\mathbf{Z}, 0, \mathbf{Z}$.

In addition,

$$\begin{cases}\pi_k(\mathbf{Sp}/\mathbf{U}) \simeq \pi_{k+1}(\mathbf{Sp}),\\ \pi_k(\mathbf{U}/\mathbf{O}) \simeq \pi_{k+2}(\mathbf{Sp}),\\ \pi_k(\mathbf{O}/\mathbf{U}) \simeq \pi_{k+1}(\mathbf{O}),\\ \pi_k(\mathbf{U}/\mathbf{Sp}) \simeq \pi_{k+2}(\mathbf{O}).\end{cases}\tag{138}$$

From these results and the homotopy exact sequence of fibrations, results on the classifying spaces can be derived at once; for instance, the spaces $\Omega\mathbf{U}$ and $\mathbf{Z}\times \mathrm{B}_{\mathrm{U}}$ are weakly homotopically equivalent.

C. Later Developments

Bott's periodicity theorems immediately attracted the attention of topologists. We shall see in chap. VII how these theorems became a central ingredient in

K-theory; they led to proofs within that theory that did not use Morse theory at all. Earlier, such a proof had been devised by J.C. Moore and published with full details in H. Cartan's Seminar [426] (see also [468]). He worked directly with the limit groups **U**, **O** and **Sp**; they are defined as groups of linear transformations in a prehilbert space V with denumerable basis. **U** is the group U(V) consisting of the linear transformations leaving invariant the scalar product $(x|y)$ in a complex prehilbert space V, and which are equal to the identity in a subspace (depending on the transformation) of finite codimension. The classifying space B_U (chap. III, § 2,G) is identified with the quotient $U(V)/(U(V^+) \times U(V^-))$, where $V = V^+ \oplus V^-$ is a decomposition of V into a direct sum of two orthogonal subspaces of infinite dimension. Moore's idea was to define explicitly (by very elementary constructions) a map

$$f\colon U(V)/(U(V^+) \times U(V^-)) \to \Omega(SU(V)) \qquad (139)$$

giving an *isomorphism* f_* *in homology*. He had seven such maps in all, corresponding to the seven maps (135); from the isomorphisms in homology he deduced the isomorphisms (137) and (138) by an application of Whitehead's first theorem (chap. II, § 5,F).

CHAPTER VI

Cohomology Operations

§ 1. The Steenrod Squares

A. Mappings of Spheres and Cup-Products

Recall that by 1937 for a finite euclidean simplicial complex K of *dimension n*, the set $[K; S_n]$ of homotopy classes of continuous maps $K \to S_n$ had been completely determined; in the language of cohomology, the Hopf–Hurewicz–Whitney theorem (chap. II, § 4,C) gives a natural *bijection*

$$[K; S_n] \to H^n(K; \pi_n(S_n)) \simeq H^n(K; \mathbf{Z})$$

in which the element $f^*(s_n)$ is associated to the class $[f]$ of a map f, where s_n is the cohomology fundamental class (Part 2, chap. I, § 3,A) of S_n.

Hopf's determination of $\pi_3(S_2)$ (chap. II, § 1,B) followed by the proof of the isomorphism

$$\pi_{n+1}(S_n) \simeq \mathbf{Z}/2\mathbf{Z} \quad \text{for } n \geqslant 3$$

by Freudenthal and Pontrjagin independently (chap. II, § 6,E) naturally led to the problem of computing $[K; S_n]$ when $\dim K = n + 1$. In a 1938 Note [376] Pontrjagin announced without proof a solution of that problem, which later was found to be erroneous for $n \geqslant 3$. In 1941 he published a long paper [378] giving a complete proof of his theorem for $n = 2$.

The problem is to determine when two continuous maps f, g of K into S_2 are homotopic. A first necessary condition is that the restrictions of f and g to the 2-skeleton K^2 be homotopic; by the Hopf–Hurewicz–Whitney theorem this means $f^*(s_2) = g^*(s_2)$. Assuming this, there is a continuous map $g'\colon K \to S_2$ homotopic to g and coinciding with f on K^2. Then the *deviation*

$$\gamma_3(f, g') \in H^3(K; \pi_3(S_2))$$

is defined (chap. II, § 4,C); Pontrjagin found that the condition for f and g' (hence f and g) to be homotopic is the existence of a cohomology 1-class $e_1 \in H^1(K)$ such that

$$\gamma_3(f, g') = 2e_1 \smile f^*(s_2). \tag{1}$$

J. Dieudonné, *A History of Algebraic and Differential Topology, 1900–1960*,
Modern Birkhäuser Classics, DOI 10.1007/978-0-8176-4907-4_20,

Steenrod undertook to generalize Pontrjagin's theorem to $[K; S_n]$ for $\dim K = n + 1$ and arbitrary $n \geqslant 3$. He found that he could express the condition for f and g to be homotopic by the introduction of *new products*

$$(u, v) \mapsto u \smile_i v \tag{2}$$

generalizing the cup-product $u \smile v = u \smile_0 v$ at the level of *cochains* and mapping $C^p(K) \times C^q(K)$ into $C^{p+q-i}(K)$ and that these products could give new maps *in cohomology* for $p = q$ [447].

We shall postpone to § 4,A details of the methods used by Pontrjagin and Steenrod in the determination of $[K; S_n]$ and shall first describe the cohomology operations discovered by Steenrod.

B. The Construction of the Steenrod Squares

Steenrod's initial idea was to generalize the Čech–Whitney definition of the cup-product for a finite simplicial complex K [Part 1, chap. IV, § 4, formula (22)]. Recall that in the latter a $(p + q)$-simplex ζ is decomposed into the join of a p-simplex σ and a q-simplex τ having as their intersection a common vertex; the cup product $u \smile v$ for a p-cochain u and a q-cochain v has the value $\langle u \smile v, \zeta \rangle = \langle u, \sigma \rangle \langle v, \tau \rangle$ in the group of coefficients G. Steenrod considered a $(p + q - i)$-simplex ζ to be the join of a p-simplex σ and a q-simplex τ having $i + 1$ *common vertices*. This may be done in many ways; Steenrod took, as the value of the i-product $u \smile_i v$ of a p-cochain u and a q-cochain v,

$$\langle u \smile_i v, \zeta \rangle = \sum \pm \langle u, \sigma \rangle \langle v, \tau \rangle, \tag{3}$$

where the sum extends over *some* of the splittings of ζ, with suitable signs. To control the unwieldy formulas Steenrod had to impose a total order on the vertices of the simplicial complex K and to define, for that order, the splittings which have to be considered on the right-hand side of (3) and the signs affected to them. The resulting calculations are quite complicated, and it must be shown that the cohomology operations they define are in fact independent of the order chosen on the vertices of K.

In his 1949–1950 Seminar [423] H. Cartan was able to give a simpler presentation of Steenrod's construction. He worked with the Alexander–Spanier cohomology of a space X (Part 1, chap. IV, § 3). When that cohomology has its coefficients in a ring Λ, the cup product $f \smile g$ of *functions* $f \in C^p(X; \Lambda)$ and $g \in C^q(X; \Lambda)$ is the function $f \smile g \in C^{p+q}(X; \Lambda)$ defined by

$$(f \smile g)(x_0, \ldots, x_{p+q}) = f(x_0, \ldots, x_p) g(x_p, \ldots, x_{p+q}) \tag{4}$$

for all points $(x_0, \ldots, x_{p+q}) \in X^{p+q+1}$, the product being taken in the ring Λ. To define the Steenrod squares, it is convenient to consider f and g as taking their values in an additive commutative group G, and $f \smile g$ in another additive commutative group G′; there is also a bilinear map

$$(\alpha, \beta) \mapsto \varphi(\alpha, \beta)$$

of $G \times G$ into G′ that is *symmetric*, i.e., $\varphi(\beta, \alpha) = \varphi(\alpha, \beta)$. Then, instead of (4), for $f \in C^m(X; G)$ and $g \in C^n(X; G)$, a function $p_0(f, g) \in C^{m+n}(X; G')$ (instead of $f \smile g$) is defined by

$$p_0(f, g)(x_0, \ldots, x_{m+n}) = \varphi(f(x_0, \ldots, x_m), g(x_m, \ldots, x_{m+n})). \tag{5}$$

In order to get manageable formulas, Cartan introduced the map T_a: $C^m(X; G) \to C^{m-1}(X; G)$ for every $a \in X$ defined by

$$(T_a f)(x_0, \ldots, x_{m-1}) = f(a, x_0, x_1, \ldots, x_{m-1}) \tag{6}$$

(with $T_a f = 0$ if $m = 0$), and a similar map $C^m(X; G') \to C^{m-1}(X; G')$. He observed that to define a function $f \in C^m(X; G)$ it is enough to define a function $T_a f \in C^{m-1}(X; G)$ for *every* $a \in X$.

Starting from p_0, Cartan defined $p_i(f, g) \in C^{m+n-i}(X; G')$ for $f \in C^m(X; G)$, $g \in C^n(X; G)$ and $i \geqslant 1$ by a double induction on i and m: if $m = 0$, define $p_i(f, g) = 0$ for all $i \geqslant 1$; if $m \geqslant 1$ and $n \geqslant 0$ is arbitrary, define

$$T_a p_i(f, g) = p_i(T_a f, g) + (-1)^{m(n+1)} p_{i-1}(T_a g, T_a f) \quad \text{for any } a \in X. \tag{7}$$

From this it easily follows that if $i > \inf(m, n)$, then $p_i(f, g) = 0$.

The fundamental formula of the theory is the expression of the *coboundary*

$$\begin{aligned} \delta(p_i(f, g)) = p_i(\delta f, g) + (-1)^m p_i(f, \delta g) + (-1)^{m+n+i} p_{i-1}(f, g) \\ + (-1)^{m+n+mn} p_{i-1}(g, f). \end{aligned} \tag{8}$$

For $i = 0$, this is formula (23) of Part 1, chap. IV, § 4. The proof of (8) is again by double induction on i and m. For $i \geqslant 2$ and $m = 0$, both sides of (8) are 0. The case $i = 1$, $m = 0$ has to be considered separately, namely, the formula

$$p_1(\delta f, g) + (-1)^{n+1} p_0(f, g) + (-1)^n p_0(g, f) = 0. \tag{9}$$

This follows from (7) applied for $i = 1$, $m = 0$, giving

$$T_a p_1(\delta f, g) = (-1)^{n+1} p_0(T_a g, T_a f),$$

hence

$$p_1(\delta f, g)(x_0, x_1, \ldots, x_n) = \varphi(f(x_n) - f(x_0), g(x_0, x_1, \ldots, x_n))$$

and that is (9) by definition. For $i > 1$ use the relation

$$\delta T_a f + T_a \delta f = f$$

and the computation is routine.

The next step is to pass from the functions in $C^\bullet(X; G)$ and $C^\bullet(X; G')$ to the Alexander–Spanier cochains, elements of the quotients

$$\bar{C}^\bullet(X; G) = C^\bullet(X; G)/C_0^\bullet(X; G), \qquad \bar{C}^\bullet(X; G') = C^\bullet(X; G')/C_0^\bullet(X; G'):$$

even if only one of the two functions f, g is 0 in a neighborhood of the diagonal, it follows from formula (7) that $p_i(f, g)$ has the same property. Thus are defined the *i-products* $f \smile_i g$ of Alexander–Spanier cochains in $\bar{C}^\bullet(X; G)$ with values in $\bar{C}^\bullet(X; G')$; all preceding formulas are still valid when $p_i(f, g)$ is replaced by $f \smile_i g$.

To get down to cohomology classes f and g must be *cocycles*; making $\delta f = \delta g = 0$ in the formula for $\delta(f \smile_i g)$ deduced from (8) gives

$$\delta(u \smile_{i+1} v) = (-1)^{i+1} u \smile_i v + (-1)^m v \smile_i u \quad \text{for } cocycles\ u, v \tag{10}$$

and if $v = u$,

$$\delta(u \smile_i u) = ((-1)^i + (-1)^m)(u \smile_{i-1} u). \tag{11}$$

So, in general, $u \smile_{i+1} v$ will *not* be a cocycle; however, $u \smile_i u$ is a *cocycle* for $m - i$ *odd* and $u \in Z^m(X; G)$. Suppose $u = \delta w$, where $w \in \bar{C}^{m-1}(X; G)$; then

$$\delta w \smile_i \delta w = \delta(w \smile_{i-1} w + w \smile_i \delta w) + ((-1)^i + (-1)^m)(w \smile_{i-2} w + w \smile_{i-1} \delta w). \tag{12}$$

Again if $m - i$ is *odd* and $u \in Z^m(X; G)$ is a *coboundary*, $u \smile_i u$ is also a coboundary. If both u and v are *m-cocycles* (and $m - i$ is odd)

$$u \smile_i v + v \smile_i u \quad \text{is a coboundary,}$$

and in particular $2u \smile_i u$ is a coboundary.

Replacing u by $u - v$ for two m-cocycles u, v, it follows that if $u - v$ is a coboundary, so is $u \smile_i u - v \smile_i v$. Finally, for two m-cocycles u, v

$$(u + v) \smile_i (u + v) - (u \smile_i u + v \smile_i v) \quad \text{is a coboundary.}$$

These results prove that if $m - i$ is *odd*, the cohomology class of $u \smile_i u$ for an m-cocycle u *only depends on the cohomology class* z of u; if $\mathrm{Sq}_i z$ is the class of $u \smile_i u$, the map

$$z \mapsto \mathrm{Sq}_i z$$

is a *homomorphism*

$$\mathrm{Sq}_i\colon H^m(X; G) \to H^{2m-i}(X; G'). \tag{13}$$

Now, if $m - i$ is *even* and $u \in Z^m(X; G)$, $u \smile_i u$ is not a cocycle in $\bar{C}^{2m-i}(X; G')$ in general, but its *class* in $\bar{C}^{2m-i}(X; G'/2G')$ is again a cocycle by (11); for any m-cochain $w \in \bar{C}^m(X; G)$, the class of $\delta w \smile_i \delta w$ is a coboundary in $\bar{C}^{2m-i}(X; G'/2G')$ by (12); this defines a *homomorphism*

$$H^m(X; G) \to H^{2m-i}(X; G'/2G').$$

Of course, reducing mod. $2G'$ in (13) also gives a homomorphism. It is convenient to write that homomorphism

$$\mathrm{Sq}^i = \mathrm{Sq}_{m-i}$$

so that for *all integers* $m \geqslant 0$, a group homomorphism is defined:

$$\mathrm{Sq}^i\colon H^m(X; G) \to H^{m+i}(X; G'/2G'). \tag{14}$$

This definition is *functorial*: for any continuous map $f\colon X \to Y$,

$$\mathrm{Sq}^i \circ f^* = f^* \circ \mathrm{Sq}^i. \tag{15}$$

Steenrod also showed that if A is a closed subspace of X, it is possible to define similarly the i-products $u \smile_i v$ for *relative* cocycles in $Z^{\bullet}(X, A; G)$ and to obtain homomorphisms

$$Sq^i\colon H^m(X, A; G) \to H^{m+i}(X, A; G'/2G').$$

The diagram of cohomology exact sequences

$$\begin{array}{ccccccccc}
\cdots \to & H^m(X;G) & \to & H^m(A;G) & \xrightarrow{\partial} & H^{m+1}(X,A;G) & \to \cdots \\
& \downarrow{\scriptstyle Sq^i} & & \downarrow{\scriptstyle Sq^i} & & \downarrow{\scriptstyle Sq^i} & \\
\cdots \to & H^{m+i}(X;G'/2G') & \to & H^{m+i}(A;G'/2G') & \xrightarrow[\partial]{} & H^{m+i+1}(X,A;G'/2G') & \to \cdots
\end{array} \tag{16}$$

is commutative.

An important consequence of the commutativity of the diagram (16) is that for the unreduced suspension $\tilde{S}X$ of a space X, there is a commutative diagram

$$\begin{array}{ccc}
H^m(X;G) & \xrightarrow{\tilde{\partial}} & H^{m+1}(\tilde{S}X;G) \\
\downarrow{\scriptstyle Sq^i} & & \downarrow{\scriptstyle Sq^i} \\
H^{m+i}(X;G'/2G') & \xrightarrow{\tilde{\partial}} & H^{m+1+i}(\tilde{S}X;G'/2G')
\end{array} \tag{17}$$

The homomorphisms Sq^i are known as the *Steenrod squares*; they are mostly used when G and G′ are the additive group $\{0, 1\}$ of the field $\mathbf{F}_2$, so that $G'/2G' = G' = G$ and φ is the multiplication in $\mathbf{F}_2$. Then

$$Sq^i(x) = 0 \qquad \text{if } \dim x < i, \tag{18}$$

$$Sq^i(x) = x \smile x \quad \text{if } \dim x = i, \tag{19}$$

$$Sq^o(x) = x; \tag{20}$$

$Sq^1(x)$ coincides with the cohomology Bockstein operator corresponding to the exact sequence

$$0 \to \mathbf{Z}/2\mathbf{Z} \to \mathbf{Z}/4\mathbf{Z} \xrightarrow{.2} \mathbf{Z}/2\mathbf{Z} \to 0$$

(Part 1, chap. IV, § 5,D) which gives the exact cohomology sequence

$$\cdots \to H^n(X, A; \mathbf{Z}/4\mathbf{Z}) \to H^n(X, A; \mathbf{F}_2) \xrightarrow{Sq^1} H^{n+1}(X, A; \mathbf{F}_2) \to H^{n+1}(X, A; \mathbf{Z}/4\mathbf{Z}) \to \cdots.$$

In 1950 [107] H. Cartan showed how i-products could be defined in the cohomology of the product $X \times Y$ of two spaces, starting from the product $p_i(f, g)$ for X and the similar product for Y, written $q_i(f', g')$ to avoid confusion; then he defined

$$r_i(f \otimes f', g \otimes g') = (-1)^{m'n} \sum_j p_{2j}(f, g) \otimes q_{2i-j}(f', g')$$
$$+ (-1)^{m'(n+n')+m'+n'} \sum_j p_{2j+1}(f, g) \otimes q_{i-2j-1}(g', f')$$

for $f \in C^m(X; G)$, $g \in C^n(X; G)$, $f' \in C^{m'}(Y; G)$, $g' \in C^{n'}(Y; G)$. From this construction he deduced two formulas that had been conjectured by Wu Wen-Tsün and Thom for the Steenrod squares

$$\mathrm{Sq}_i(x \times y) = \sum_{j+k=i} \mathrm{Sq}_j(x) \times \mathrm{Sq}_k(y) \tag{21}$$

for a class $x \in H^{\bullet}(X; \mathbf{F}_2)$ and a class $y \in H^{\bullet}(Y; \mathbf{F}_2)$;

$$\mathrm{Sq}_i(x \smile y) = \sum_{j+k=i} \mathrm{Sq}_j(x) \smile \mathrm{Sq}_k(y) \tag{22}$$

for two classes x, y in $H^{\bullet}(X; \mathbf{F}_2)$.

§ 2. The Steenrod Reduced Powers

A. New Definition of the Steenrod Squares

In 1950 [449] Steenrod announced at the International Congress the discovery of new cohomology operations, now called the *Steenrod reduced powers*. He had been led to their definition by a new definition of the operations Sq^i which he connected with the group algebra of the cyclic group $\mathbf{Z}/2\mathbf{Z}$, inspired by the method by which P. Smith and M. Richardson had defined "special" cohomology groups related to that group algebra (Part 2, chap. VI, § 3,C).

Steenrod's point of departure was Lefschetz's definition of the cup-product $a \smile b$ of cohomology classes in $H^{\bullet}(X; \Lambda)$ [Part 1, chap. IV, § 5,H, formula (81)] as the pullback of the cross-product $a \times b$ in $H^{\bullet}(X \times X; \Lambda)$ by the diagonal map

$$\delta_X : X \to X \times X.$$

In singular homology δ_X defines a chain transformation $\tilde{\delta}_X : S_{\bullet}(X) \to S_{\bullet}(X \times X)$ by $\tilde{\delta}_X(s) = \delta_X \circ s$ for any singular simplex s in $S_{\bullet}(X)$; so a cup-product $u \smile v$ of two singular *cochains* in $S^{\bullet}(X; G)$ may be defined by the formula

$$\langle u \smile v, s \rangle = \langle u \times v, \tilde{\delta}_X(s) \rangle \tag{23}$$

for any singular simplex s. If δ_{p+q} is the diagonal map

$$\Delta_{p+q} \to \Delta_{p+q} \times \Delta_{p+q}$$

then for a singular $(p+q)$-simplex s, one has a commutative diagram

$$\begin{array}{ccc} \Delta_{p+q} \times \Delta_{p+q} & \xrightarrow{s \times s} & X \times X \\ \uparrow{\scriptstyle \delta_{p+q}} & & \uparrow{\scriptstyle \delta_X} \\ \Delta_{p+q} & \xrightarrow[s]{} & X \end{array}$$

and therefore

$$\langle u \times v, \tilde{\delta}_X(s) \rangle = \langle u \times v, (s \times s) \circ \delta_{p+q} \rangle$$

for a p-cochain u and a q-cochain v in $S^{\bullet}(X; G)$.

Now consider Δ_{p+q} to be a simplicial complex consisting of all its facets, and consider also a simplicial subdivision of the product $\Delta_{p+q} \times \Delta_{p+q}$. For these simplicial complexes δ_{p+q} is not a simplicial map, but $C_{\bullet}(\Delta_{p+q})$ is free and acyclic, and $C_{\bullet}(\Delta_{p+q} \times \Delta_{p+q})$ is acyclic, so, by the Hopf construction (Part 1, chap. IV, § 5,F) there is a chain transformation

$$d_0: C_{\bullet}(\Delta_{p+q}) \to C_{\bullet}(\Delta_{p+q} \times \Delta_{p+q})$$

coinciding with δ_{p+q} on the vertices of Δ_{p+q}; since Δ_{p+q} is contractible, the map $s \mapsto \langle u \otimes v, D_0 s \rangle$ may be used in the computation of the cup-product *in cohomology* by Lefschetz's method instead of the map

$$s \mapsto \langle u \times v, (s \times s) \circ \delta_X \rangle,$$

where D_0 is the chain transformation $S_{\bullet}(X) \to S_{\bullet}(X) \otimes S_{\bullet}(X)$ defined by

$$s \mapsto D_0 s = \zeta((s \times s) \circ d_0) \tag{24}$$

and $\zeta: S_{\bullet}(X \times X) \to S_{\bullet}(X) \otimes S_{\bullet}(X)$ is the Eilenberg–Zilber chain equivalence (Part 1, chap. IV, § 5,G).

Now let T be the automorphism $s \otimes t \mapsto t \otimes s$ of $S_{\bullet}(X) \otimes S_{\bullet}(X)$ exchanging the factors; then TD_0 is different from D_0, but both chain transformations are chain homotopic, since they coincide in dimension 0; in other words, there is a chain homotopy

$$D_1: S_{\bullet}(X) \to S_{\bullet}(X) \otimes S_{\bullet}(X) \tag{25}$$

such that

$$TD_0 - D_0 = \mathbf{b}D_1 + D_1\mathbf{b} \quad \text{(Part 1, chap. IV, § 5,F).} \tag{26}$$

Steenrod had the idea of applying T to both sides of (26); since T^2 is the identity and T commutes with the boundary operator **b**,

$$\mathbf{b}h + h\mathbf{b} = 0 \quad \text{for } h = TD_1 + D_1. \tag{27}$$

Next he observed that the proof given by Hopf for the existence of chain transformations (Part 1, chap. IV, § 5,F) can be generalized in the following way: let $C_{\bullet} = (C_q)$ and $C'_{\bullet} = (C'_q)$ be two augmented chain complexes such that $C_{\bullet}$ is free and $C'_{\bullet}$ is acyclic. For any $i \geqslant 0$, call a graded homomorphism $h = (h_q): C_{\bullet} \to C'_{\bullet}$ a *homomorphism of degree i* if each h_q is a homomorphism

$$h_q: C_q \to C'_{q+i}.$$

Thus a chain transformation is a particular homomorphism of degree 0 and a chain homotopy a particular homomorphism of degree 1. A homomorphism of degree 0 is a chain transformation if and only if it commutes with the boundary operator **b**. Steenrod considered for a graded homomorphism h of degree $i > 1$ the similar condition

$$\mathbf{b}h + (-1)^{i+1} h\mathbf{b} = 0$$

and by the same argument as Hopf (induction on the degree) he showed that there is a homomorphism k *of degree* $i + 1$ such that

$$h = \mathbf{b}k + (-1)^{i+2} k\mathbf{b}.$$

Returning to the analysis of the chain transformation D_0 defined by (24), Steenrod applied this general lemma to $h = TD_1 + D_1$, and from (27) he deduced the existence of a homomorphism D_2 of degree 2 such that

$$\mathbf{b}D_2 - D_2\mathbf{b} = TD_1 + D_1.$$

The process can obviously be continued, and yields an infinite sequence (D_i) of graded homomorphisms

$$D_i\colon S_{\bullet}(X) \to S_{\bullet}(X) \otimes S_{\bullet}(X)$$

of degree i such that

$$\begin{aligned} \mathbf{b}D_{2i-1} + D_{2i-1}\mathbf{b} &= TD_{2i-2} - D_{2i-2}, \\ \mathbf{b}D_{2i} - D_{2i}\mathbf{b} &= TD_{2i-1} + D_{2i-1}. \end{aligned} \tag{28}$$

Steenrod then checked that the i-products of cochains that he had defined earlier satisfy the relations

$$\langle u \smile_i v, s\rangle = \langle u \times v, D_i s\rangle$$

for any singular simplex

$$s\colon \Delta_{p+q-i} \to X,$$

any p-cochain u and any q-cochain v.

B. The Steenrod Reduced Powers: First Definition

Steenrod observed that the relations (28) could be expressed in a way introducing the group algebra $\mathbf{Z}[\Pi]$ of the cyclic group $\Pi = \mathbf{Z}/2\mathbf{Z}$ of order 2: if $\Pi = \{1, \gamma\}$, where $\gamma^2 = 1$, Π acts on $C'_{\bullet} = S_{\bullet}(X) \otimes S_{\bullet}(X)$ by $\gamma . z = Tz$; hence $C'_{\bullet}$ is a $\mathbf{Z}[\Pi]$-module, and relations (28) can be written

$$\left.\begin{aligned} \mathbf{b}D_{2i-1} + D_{2i-1}\mathbf{b} &= (\gamma - 1) . D_{2i-2} \\ \mathbf{b}D_{2i} - D_{2i}\mathbf{b} &= (\gamma + 1) . D_{2i-1} \end{aligned}\right\} \quad \text{for } i \geqslant 1.$$

This led him [451] to the following generalization. Consider an augmented chain complex $C'_{\bullet}$ and a finite group Π acting on $C'_{\bullet}$ in such a way that for every $g \in \Pi$, $z \mapsto g . z$ is a chain transformation. Then the group algebra $\mathbf{Z}[\Pi]$ acts on $C'_{\bullet}$ by

$$\alpha . z = \sum_i x_i (g_i . z)$$

for any $\alpha = \sum_i x_i g_i$ with $x_i \in \mathbf{Z}$ and $g_i \in \Pi$, and each map $z \mapsto \alpha . z$ is a chain transformation. Let $s(\alpha) = \sum_i x_i \in \mathbf{Z}$. Steenrod called a sequence $(\alpha_i)_{i \geqslant 1}$ of elements of $\mathbf{Z}[\Pi]$ a 0-*sequence* if it satisfies the relations

$$s(\alpha_1) = 0, \qquad \alpha_{i+1}\alpha_i = 0 \quad \text{in } \mathbf{Z}[\Pi] \text{ for } i \geqslant 1. \tag{29}$$

If $\Pi = \mathbf{Z}/2\mathbf{Z}$, then $\alpha_{2j-1} = \gamma - 1$ and $\alpha_{2j} = \gamma + 1$ form a 0-sequence.

Suppose $\mathrm{C}'_\bullet$ is free and acyclic; using as before the Hopf method, Steenrod proved the existence of graded homomorphisms $\mathrm{D}_i\colon \mathrm{C}'_\bullet \to \mathrm{C}'_\bullet$ of degree $i \geqslant 0$ satisfying

$$\mathbf{b}\mathrm{D}_i + (-1)^{i+1}\mathrm{D}_i\mathbf{b} = \alpha_i \,.\, \mathrm{D}_{i-1} \quad \text{for all } i \geqslant 1. \tag{30}$$

The usual functorial argument shows that this construction can be extended simply by assuming the identity functor in $\mathrm{C}'_\bullet$ to be free, and to have acyclic models (Part 1, chap. IV, § 5,G).

Steenrod then generalized the Sq_i in the following way. Let $\mathrm{C}_\bullet = \mathrm{S}_\bullet(\mathrm{X})$ be the singular complex of a space X, $\mathrm{C}'_\bullet = (\mathrm{S}_\bullet(\mathrm{X}))^{\otimes p}$ its p-th tensor power for any integer $p \geqslant 2$. The symmetric group $\mathfrak{S}_p$ operates on $\mathrm{C}'_\bullet$ by a slight modification of the usual action; the transposition $\tau = (i \;\; i+1)$ acts on a tensor product $s_1 \otimes s_2 \otimes \cdots \otimes s_p$ by

$$\tau\,.\,(s_1 \otimes s_2 \otimes \cdots \otimes s_p) = (-1)^{q_i q_{i+1}+1} s_1 \otimes \cdots \otimes s_{i-1} \otimes s_{i+1} \otimes s_i \otimes \cdots \otimes s_p, \tag{31}$$

if q_i is the dimension of the singular simplex s_i. Clearly, $z \mapsto \tau\,.\,z$ is a chain transformation, and $\mathrm{C}'_\bullet$ is free and has acyclic models (Part 1, chap. IV, § 5,G); hence, for any subgroup Π of $\mathfrak{S}_p$ and any 0-sequence $(\alpha_i)_{i \geqslant 1}$ for Π there exists a sequence $(\mathrm{D}_i)_{i \geqslant 0}$ of graded homomorphisms satisfying (30), By duality, Π acts on the cochain complex $\mathrm{C}'^\bullet = \mathrm{Hom}(\mathrm{C}'_\bullet, \mathbf{Z})$ in a natural way, and by transposition, D_i gives a graded homomorphism of $\mathrm{C}'^\bullet$, of degree $-i$, which Steenrod also writes D_i. Then, for an *r-cocycle* w of $\mathrm{C}'^\bullet$, the coboundary of $\mathrm{D}_i w$ is given by

$$\mathbf{d}(\mathrm{D}_i w) = (-1)^{i+1}\mathrm{D}_{i-1}(\alpha_i \,.\, w). \tag{32}$$

Now consider two commutative groups G, G′ and a p-linear map $\varphi\colon \mathrm{G}^p \to \mathrm{G}'$; for *cocycles* $u_1, u_2, \ldots, u_p$ in $\mathrm{Z}^q(\mathrm{X}; \mathrm{G})$, define a "cross-product" $u_1 u_2 \cdots u_p$ as the pq-cocycle in $\mathrm{Z}^{pq}(\mathrm{X}^p; \mathrm{G}')$ such that

$$\langle u_1 u_2 \cdots u_p, z_1 \times z_2 \times \cdots \times z_p \rangle = \varphi(\langle u_1, z_1 \rangle, \ldots, \langle u_p, z_p \rangle) \tag{33}$$

for any cycles $z_j \in \mathrm{Z}_q(\mathrm{X}; \mathrm{G})$. For a q-cocycle $u \in \mathrm{Z}^q(\mathrm{X}; \mathrm{G})$, u^p is defined by (33) with all u_j equal to u; Steenrod obtained the formulas (for a 0-sequence (α_i) in $\mathbf{Z}[\mathrm{G}']$)

$$\mathbf{d}(\mathrm{D}_i u^p) = \begin{cases} (-1)^{i+1} s(\alpha_i)\mathrm{D}_{i-1}u^p & \text{if } q \text{ is even,} \\ (-1)^{i+1} t(\alpha_i)\mathrm{D}_{i-1}u^p & \text{if } q \text{ is odd,} \end{cases} \tag{34}$$

where $s(\alpha)$ is defined above, and $t\colon \mathbf{Z}[\mathrm{G}'] \to \mathbf{Z}$ is the homomorphism such that $t(g) = 1$ if g is an even permutation, $t(g) = -1$ if g is an odd permutation.

Put $n = n(q, \alpha_i)$ equal to $s(\alpha_i)$ if q is even and to $t(\alpha_i)$ if q is odd and let

$$G'_n = G'/nG'; \tag{35}$$

then

$$D_i u^p \in Z^{pq-i}(X; G'_n).$$

Steenrod showed that for a given 0-sequence (α_i), the *cohomology class* of $D_i u^p$ in $H^{pq-i}(X; G'_n)$ only depends on the class of u in $H^q(X; G)$, not on the choice of the sequence (D_i) of graded homomorphisms satisfying (30). Thus to each element v of $H^q(X; G)$ he attached an element of $H^{pq-i}(X; G'_n)$ that he wrote $\mathscr{P}_i^v v$ and called the *i-fold reduction of the p-th power of* v; he observed that the map

$$\mathscr{P}_i^p\colon H^p(X; G) \to H^{pq-i}(X; G'_n)$$

is *not* necessarily a homomorphism, although it is functorial in X. He singled out explicitly the 0-sequence (inspired by the work of Smith and Richardson)

$$\alpha_{2j-1} = \gamma - 1, \qquad \alpha_{2j} = 1 + \gamma + \gamma^2 + \cdots + \gamma^{p-1}$$

in the symmetric group $\mathfrak{S}_p$, where γ is the circular permutation $(1\ 2\ \ldots\ p)$ and said that $\mathscr{P}_i^p$ is a *cyclic reduced power*.

C. The Steenrod Reduced Powers: Second Definition

A little later [453] Steenrod realized he could give another, simpler definition for the reduced powers. Inspired by the method J. Adem used for the proof of the relations between the Sq^i (see § 3,C), he connected this new definition with the cohomology of groups and the definition of the slant product (Part 1, chap. IV, § 5,H), which he introduced for that purpose.

The definition of the slant product given *loc. cit.* has to be slightly modified to account for the presence of operators acting on chain complexes. There are two cell complexes (finite or infinite) W and X, and a finite group Π operates *freely* on W, in such a way that its action on the chain complex $C_\bullet(W)$ is a chain transformation. The group Π is made to operate on $W \times X$ by the condition

$$\alpha.(e \times \sigma) = (\alpha.e) \times \sigma$$

for all cells $e \in W$ and $\sigma \in X$. The group Π also operates on two additive commutative groups A and G; it then operates on the tensor product $A \otimes G$ by $\alpha.(a \otimes g) = (\alpha.a) \otimes (\alpha.g)$; if $R(A \otimes G)$ is the subgroup of $A \otimes G$ generated by the elements $\alpha.s - s$ for all $s \in A \otimes G$ and all $\alpha \in \Pi$, then one defines $A \otimes_\Pi G$ as the quotient

$$(A \otimes G)/R(A \otimes G).$$

An *equivariant* r-cochain of $W \times X$ with values in G is a cochain v such that

$$\langle v, \alpha.c\rangle = \alpha.\langle v, c\rangle$$

for any $\alpha \in \Pi$ and any chain $c \in C_r(W \times X)$.

Write $a \otimes e$ instead of ae for $a \in A$ and any cell e of W and $A \otimes C_\bullet(W)$ instead of $C_\bullet(W; A)$, the chain complex of chains of W with coefficients in A; the group Π operates on $A \otimes C_\bullet(W)$ by

$$\alpha.(a \otimes e) = (\alpha.a) \otimes (\alpha.e).$$

The chains in $A \otimes C_\bullet(W)$ that are sums of chains of the form $\alpha.c - c$ for $c \in A \otimes C_\bullet(W)$ and $\alpha \in \Pi$ generate a chain subcomplex $R(A \otimes C_\bullet(W))$ of $A \otimes C_\bullet(W)$; the quotient chain complex is written $A \otimes_\Pi C_\bullet(W)$. Eilenberg showed [173] that if W is *acyclic* (and Π operates freely on W), the homology of $A \otimes_\Pi C_\bullet(W)$ only depends on the group A and the action of Π on A, and is in fact the *homology* $H_\bullet(\Pi; A)$ *of the group* Π with coefficients in A (chap. V, § 1,D).

Now, for an i-chain z in $A \otimes C_i(W)$, an $(r-i)$-cell σ in X, and an *equivariant* r-cochain v of $W \times X$, by definition of the slant product

$$\langle (v/(\alpha.z - z)), \sigma \rangle = \alpha.\langle v, z \times \sigma \rangle - \langle v, z \times \sigma \rangle$$

so that the $(r-i)$-cochain $v/(\alpha.z - z)$ of X has its value in the subgroup $R(A \otimes G)$. This also means that, for all elements $\alpha \in \Pi$ the image of $v/(\alpha.z)$ in $\mathrm{Hom}(C_\bullet(X), A \otimes_\Pi G)$ is the same; in other words, that cochain only depends on the image $\bar{z}$ of z in $A \otimes_\Pi C_\bullet(W)$; it is written again as the *slant product* $v/\bar{z}$ of the equivariant r-cochain v of $W \times X$ and the i-chain $\bar{z}$ in $A \otimes_\Pi C_i(W)$, and it is an element of $C^{r-i}(X; A \otimes_\Pi G)$. If v is a cocycle (resp. a coboundary) and z is a cycle (resp. a boundary), $v/\bar{z}$ is a cocycle (resp. a coboundary).

To define the reduced powers, the preceding notions are specialized. Let B be any commutative group and $u \in C^q(X; B)$ be a q-cochain; for any integer $n \geqslant 1$, an n-th power u^n is defined by

$$\langle u^n, \sigma_1 \times \sigma_2 \times \cdots \times \sigma_n \rangle = \langle u, \sigma_1 \rangle \otimes \langle u, \sigma_2 \rangle \otimes \cdots \otimes \langle u, \sigma_n \rangle \qquad (36)$$

for arbitrary q-cells $\sigma_1, \ldots, \sigma_n$ in X; it is an element of

$$C^{nq}(X^n; B^{\otimes n}),$$

i.e., an nq-cochain taking its values in the n-th tensor power $B^{\otimes n}$ of the group B.

The group Π is specialized to a subgroup of the symmetric group $\mathfrak{S}_n$ and made to operate on the chain complex $(C_\bullet(X))^{\otimes n}$ by defining for every transposition $(i \;\; i+1)$ its action on a tensor product by formula (31). We also have to define the action of Π on $B^{\otimes n}$ in such a way that u^n becomes an *equivariant* cochain; for that purpose, it is enough to take

$$\alpha.(b_1 \otimes b_2 \otimes \cdots \otimes b_n) = \varepsilon b_{\alpha^{-1}(1)} \otimes \cdots \otimes b_{\alpha^{-1}(n)}, \qquad (37)$$

where $\varepsilon = 1$ if q is even and ε is the signature of the permutation α when q is odd.

Now take an *acyclic* cell complex W on which Π operates freely and let Π operate (also freely) on $W \times X$ by

$$\alpha.(e \times \sigma) = (\alpha.e) \times \sigma.$$

As the chain complex $(C_\bullet(X))^{\otimes n}$ [equivalent to $C_\bullet(X^n)$ by the Eilenberg–Zilber theorem] has acyclic models, it is possible to define a chain transformation

$$\varphi_{\bullet}\colon C_{\bullet}(W \times X) \to C_{\bullet}(X^n)$$

by the usual induction on the dimension, with a slight change: once $\varphi_{\bullet}(e \times \sigma)$ has been defined for all cells σ of a given dimension and *one* cell e of an equivalence class $\{\alpha . e | \alpha \in \Pi\}$ then $\varphi_{\bullet}(\alpha . (e \times \sigma))$ is taken equal to $\alpha . (\varphi_{\bullet}(e \times \sigma))$ for all $\alpha \in \Pi$; this ensures that

$$\varphi_{\bullet} \circ \alpha = \alpha \circ \varphi_{\bullet} \quad \text{for all } \alpha \in \Pi. \tag{38}$$

By duality, $\varphi_{\bullet}$ gives a cochain transformation

$$\varphi^{\bullet}\colon C^{\bullet}(X^n; B^{\otimes n}) \to C^{\bullet}(W \times X; B^{\otimes n}),$$

and for each equivariant q-cochain $u \in C^q(X^n; B^{\otimes n})$, $\varphi^{\bullet}u^n$ is an equivariant nq-cochain in $C^{nq}(W \times X; B^{\otimes n})$ (recall the action of Π on $B^{\otimes n}$ depends on the parity of q).

The slant product for an i-chain $\bar{z} \in A \otimes_{\Pi} C_i(W)$,

$$\varphi^{\bullet}u^n/\bar{z} \in C^{nq-i}(X; G_{n,q}),$$

is therefore defined, $G_{n,q}$ being the group $A \otimes_{\Pi} B^{\otimes n}$ for the actions of Π on A and $B^{\otimes n}$. Steenrod states that when u is a cocycle and $\bar{z}$ is a cycle, the cohomology class of $\varphi^{\bullet}u^n/\bar{z}$ only depends on the cohomology class of u in $H^q(X; B)$ and the homology class of $\bar{z}$ in $H_i(\Pi; A)$. He thus defined a homomorphism

$$H^q(X; B) \to H^{nq-i}(X; G_{n,q})$$

which he wrote

$$v \mapsto v^n/c$$

for any $c \in H_i(\Pi; A)$; he said that v^n/c is the power v^n "reduced by c." That definition is functorial: for any continuous map $f\colon X \to Y$, $f^*(v^n/c) = (f^*v)^n/c$.

If $\Pi \subset \Pi'$ are two subgroups of $\mathfrak{S}_n$, then

$$u^n/\lambda_* c = \lambda'(u^n/c)$$

where $\lambda\colon \Pi \to \Pi'$ is the natural injection, and λ' is the homomorphism of the cohomology groups deduced from the natural map $G_{n,q} \to G'_{n,q}$, where $G'_{n,q}$ is the group $A \otimes_{\Pi'} B^{\otimes n}$ for the actions of Π' on A and on $B^{\otimes n}$.

Steenrod connected this new definition with the former one by starting from a 0-sequence $(\alpha_i)_{i \geqslant 1}$ in the algebra $\mathbf{Z}[\Pi]$. If z_0 is a vertex in W, it is possible to construct a sequence $(z_i)_{i \geqslant 0}$, where z_i is an i-chain in $C_i(W)$ and $\mathbf{b}z_i = \alpha_i . z_{i-1}$; $\varphi^{\bullet}u^n/\bar{z}_i$ is then identified with the $D_i u^n$ of the first definition.

As he had done for the first definition, Steenrod specialized the group Π to the subgroup of $\mathfrak{S}_n$ generated by the neutral element and the circular permutation $\gamma = (1\ 2\ \dots\ n)$; the coefficient groups A and B are the same both equal to $\mathbf{Z}$ or to $\mathbf{Z}/n\mathbf{Z}$. Let α_n and α^n be the Bockstein homomorphisms in homology and in cohomology corresponding to the exact sequence

$$0 \to \mathbf{Z} \xrightarrow{.n} \mathbf{Z} \to \mathbf{Z}/n\mathbf{Z} \to 0 \tag{39}$$

He then showed that for $c \in H_i(\Pi; \mathbf{Z}/n\mathbf{Z})$ and $v \in H^q(X; \mathbf{Z})$,

$$v^n/\alpha_n c = (-1)^{i+1}\alpha^n(v^n/c) \tag{40}$$

(for *any* subgroup $\Pi \subset \mathfrak{S}_n$). As in the first definition, he introduced the two elements of $\mathbf{Z}[\Pi]$

$$\tau = \gamma - 1, \qquad \sigma = 1 + \gamma + \gamma^2 + \cdots + \gamma^{n-1}$$

and used them to determine the complex W with which he computed the homology of the cyclic group Π: for each dimension $i \geqslant 0$, W has n cells $\sigma^k . e_i$ $(0 \leqslant k \leqslant n-1)$ and the boundary operator is given by

$$\mathbf{b}e_{2i+1} = \tau . e_{2i}, \qquad \mathbf{b}e_{2i+2} = \sigma . e_{2i+1}.$$

This defines W as a free and acyclic complex, and easily yields the homology (for trivial actions of Π on $\mathbf{Z}$ or $\mathbf{Z}/n\mathbf{Z}$)

$$\mathrm{H}_{2i+1}(\Pi; \mathbf{Z}) = \mathbf{Z}/n\mathbf{Z}, \qquad \mathrm{H}_{2i+2}(\Pi; \mathbf{Z}) = 0 \quad \text{for } i \geqslant 0;$$
$$\mathrm{H}_j(\Pi; \mathbf{Z}/n\mathbf{Z}) = \mathbf{Z}/n\mathbf{Z} \quad \text{for } j \geqslant 0.$$

In the cases H_j is not 0, it is generated by the class $\bar{e}_j$ of the cycle, image of $1 \otimes e_j$ in $\mathbf{Z} \otimes_\Pi \mathrm{C}_.(\mathrm{W})$ [resp. $(\mathbf{Z}/n\mathbf{Z}) \otimes_\Pi \mathrm{C}_.(\mathrm{W})$]. The Bockstein operator corresponding to (39) is such that

$$\alpha_n \bar{e}_{2i} = \bar{e}_{2i-1} \quad \text{in } \mathrm{H}_{2i-1}(\Pi; \mathbf{Z})$$

and $\alpha_n \bar{e}_{2i}$ is mapped onto $\bar{e}_{2i-1}$ by the map

$$\mathrm{H}_{2i-1}(\Pi; \mathbf{Z}) \to \mathrm{H}_{2i-1}(\Pi; \mathbf{Z}/n\mathbf{Z}).$$

Formula (40) then shows that the cyclic reduced powers v^n/c, for c in $\mathrm{H}_{2i-1}(\Pi; \mathbf{Z})$ or in $\mathrm{H}_{2i-1}(\Pi; \mathbf{Z}/n\mathbf{Z})$ are entirely determined by the reduced powers $v^n/\bar{e}_{2i}$ for $\bar{e}_{2i} \in \mathrm{H}_{2i}(\Pi; \mathbf{Z}/n\mathbf{Z})$.

Steenrod then showed that the Cartan formulas for the Sq^i [§ 1,B, formulas (21) and (22)] extend to the cyclic reduced powers for the cohomology with coefficients in $\mathbf{Z}/n\mathbf{Z}$ when n is any odd number: for the cross product

$$(u \times v)^n/\bar{e}_{2i} = \pm \sum_{j=0}^{i} (u^n/\bar{e}_{2j}) \times (u^n/\bar{e}_{2i-2j}) \tag{41}$$

where the sign depends on the degrees of the classes u and v. He realized that the formula is in fact a consequence of properties of the homology of the *group* Π with coefficients in $\mathbf{Z}/n\mathbf{Z}$, namely, the expressions

$$\delta_* \bar{e}_{2i} = \sum_{j=0}^{i} \bar{e}_{2j} \times \bar{e}_{2i-2j},$$

$$\delta_* \bar{e}_{2i+1} = \sum_{s=0}^{2i+1} \bar{e}_s \times \bar{e}_{2i+1-s},$$

for the diagonal map $\delta: \Pi \to \Pi \times \Pi$.

Already in that paper [453] the special case in which n is an *odd prime p* appears; its importance will be made clearer later (§ 3). Acting upon a suggestion of Serre (who had done the same thing for the Steenrod squares), Steenrod revised the notation he had introduced in [451] for the cyclic reduced powers of *odd prime* order p. He wrote

$$P_p^k\colon \mathrm{H}^q(\mathrm{X};\mathbf{F}_p) \to \mathrm{H}^{q+2k(p-1)}(\mathrm{X};\mathbf{F}_p) \quad (\text{for all } q \geqslant 0) \tag{42}$$

the cyclic reduced powers

$$u \mapsto (-1)^r \left(\left(\frac{p-1}{2}\right)!\right)^{2k-q} u^p/\bar{e}_{(q-2k)(p-1)}$$

with $r = (p-1)(k + \frac{1}{2}q(q-1))$. Its main properties are quite similar to the properties (18), (19), (20) of the squares:

$$\mathrm{P}_p^k u = 0 \qquad \text{if } \dim u < 2k, \tag{43}$$

$$\mathrm{P}_p^k u = u^{\smile p} \quad \text{if } \dim u = 2k, \tag{44}$$

$$\mathrm{P}_p^o u = u, \tag{45}$$

$$\mathrm{P}_p^k(u \smile v) = \sum_{h=0}^{k} \mathrm{P}_p^h u \smile \mathrm{P}_p^{k-h} v. \tag{46}$$

Furthermore, the reduced powers can be defined in the same way for *relative* cohomology $\mathrm{H}^{\bullet}(\mathrm{X},\mathrm{A};\mathbf{F}_p)$, and there is a commutative diagram of exact sequences similar to (17).

§ 3. Cohomology Operations

A. Cohomology Operations and Eilenberg–Mac Lane Spaces

Historically, the definition of homotopy operations (chap. V, § 5,C) had been preceded by a similar definition, this time for cohomology, introduced by Serre [431] and independently by Eilenberg and Mac Lane ([185], IV). The data, here, are two integers $q > 0$, $n > 0$ and two commutative groups A, B; we write as ***CW*** the category of CW-complexes. A *cohomology operation of type* $(q, n, \mathrm{A}, \mathrm{B})$ can be considered a functor

$$\xi = \xi_{q,n,\mathrm{A},\mathrm{B}}\colon \boldsymbol{CW} \to \boldsymbol{Set}$$

defined by the conditions that for any CW-complex X

$$\xi(\mathrm{X}) \in \mathrm{Map}(\mathrm{H}^q(\mathrm{X};\mathrm{A}), \mathrm{H}^n(\mathrm{X};\mathrm{B}))$$

(set of all maps, not necessarily homomorphisms), and that for any continuous map $v\colon \mathrm{Y} \to \mathrm{X}$ of CW-complexes, the diagram

$$\begin{array}{ccc} \mathrm{H}^q(\mathrm{X};\mathrm{A}) & \xrightarrow{v^*} & \mathrm{H}^q(\mathrm{Y};\mathrm{A}) \\ \Big\downarrow{\scriptstyle \xi(\mathrm{X})} & & \Big\downarrow{\scriptstyle \xi(\mathrm{Y})} \\ \mathrm{H}^n(\mathrm{X};\mathrm{B}) & \xrightarrow[v^*]{} & \mathrm{H}^n(\mathrm{Y};\mathrm{B}) \end{array} \tag{47}$$

commutes.

Serre's result was that there is an element

$$\alpha \in H^n(K(A, q); B) \tag{48}$$

such that the pair $(K(A, q), \alpha)$ *represents* the functor ξ (Part 1, chap. IV, § 8,C). To define α, recall that there is a natural bijection of $H^q(K(A, q); A)$ onto the set of homotopy classes $[K(A, q); K(A, q)]$ [chap. V, § 1,D, formula (43)]. Let ι be the class in $H^q(K(A, q); A)$ corresponding by that bijection to the homotopy class of the *identity*. For any CW-complex X, and any class $x \in H^q(X; A)$, there is a continuous map g_x: $X \to K(A, q)$, determined up to homotopy, such that

$$g_x^*(\iota) = x \quad (\textit{loc. cit.}). \tag{49}$$

The class α is then defined by

$$\alpha = \xi(K(A, q))(\iota) \in H^n(K(A, q); B) = H^n(A, q; B). \tag{50}$$

Indeed, the commutative diagram (47) applied to g_x gives

$$\begin{array}{ccc} H^q(K(A, q); A) & \xrightarrow{g_x^*} & H^q(X; A) \\ {\scriptstyle \xi(K(A, q))}\downarrow & & \downarrow{\scriptstyle \xi(X) = C} \\ H^n(K(A, q); B) & \xrightarrow[g_x^*]{} & H^n(X; B) \end{array} \tag{51}$$

hence

$$C(x) = g_x^*(\alpha) \tag{52}$$

by (49). Conversely, *any* $\alpha \in H^n(K(A, q); B)$ defines a cohomology operation of type (q, n, A, B) by formula (52).

Examples

I. For $0 < n < q$, necessarily $C = 0$, since $H^n(A, q; B) = 0$ by Hurewicz's theorem.

II. For $n = q$, $H^q(A, q; B) = \mathrm{Hom}(A, B)$ (chap. V, § 1,D); to each homomorphism φ: $A \to B$ is thus associated the cohomology operation

$$C_{q,\varphi}: H^q(X; A) \to H^q(X; B)$$

(Part 1, chap. IV, § 3).

III. For $n = q + 1$, suppose L is an extension of B by A, so that we have an exact sequence of commutative groups

$$0 \to B \to L \to A \to 0.$$

Then, in the corresponding cohomology exact sequence

$$\cdots \to H^q(X; L) \to H^q(X; A) \xrightarrow{\beta} H^{q+1}(X; B) \to H^{q+1}(X; L) \to \cdots$$

the Bockstein homomorphism β (Part 1, chap. IV, § 5,D) is a cohomology operation.

A cohomology operation defined by an element $\alpha \in H^n(A, q; B)$ does not necessarily define *homomorphisms* $H^q(X; A) \to H^n(X; B)$; conditions for this "additivity" of a cohomological operation were given by Eilenberg and Mac Lane; the simplest sufficient condition is that α be in the image of the Eilenberg –Mac Lane cohomology suspension $H^{n+1}(A, q+1; B) \to H^n(A, q; B)$ (chap. V, § 2,D) (cf. [424], pp. 14–10).

B. The Cohomology Operations of Type $(q, n, \Pi, \mathbf{F}_2)$

Serre was particularly interested in the case $B = \mathbf{F}_2$, so he had to determine the cohomology algebra $H^{\bullet}(\Pi, q; \mathbf{F}_2)$ for commutative groups Π (he only considered finitely generated groups). His starting point was the observation made in his thesis ([429], p. 457) that for a fibration (X, B, F) the Steenrod squares commute with the transgression (cf. chap. IV, § 3,B)

$$\tau \colon H^n(F; \mathbf{F}_2) \to H^{n+1}(B; \mathbf{F}_2)$$

as a consequence of the commutativity of the diagram (16) in § 1,B). He then used A. Borel's algebraic theorem on the determination of $H^{\bullet}(B; \mathbf{F}_2)$ when $H^{\bullet}(F; \mathbf{F}_2)$ is an algebra of polynomials over $\mathbf{F}_2$ in a family of transgressive elements z_i of degrees n_i (chap. IV, § 4); recall that $H^{\bullet}(B; \mathbf{F}_2)$ is then the polynomial algebra

$$\mathbf{F}_2[(t_i)]$$

where the $t_i = \tau(z_i)$ are the images of the transgressive elements z_i.

For any finite sequence $I = (i_1, i_2, \ldots, i_r)$ of integers $i_h \geqslant 1$, Serre wrote

$$\mathrm{Sq}^{I} = \mathrm{Sq}^{i_1} \circ \mathrm{Sq}^{i_2} \circ \cdots \circ \mathrm{Sq}^{i_r} \tag{53}$$

the composite which sends each $H^n(X; \mathbf{F}_2)$ into

$$H^{n+i_1+i_2+\cdots+i_r}(X; \mathbf{F}_2).$$

From § 1,B, formula (19), it follows that for $x \in H^a(X; \mathbf{F}_2)$ (for the cup-product multiplication)

$$\mathrm{Sq}^a(x) = x^2,\ \mathrm{Sq}^{2a}(x^2) = x^4, \ldots, \mathrm{Sq}^{2^{r-1}a}(x^{2^{r-1}}) = x^{2^r}$$

so that

$$x^{2^r} = \mathrm{Sq}^{L(a,r)} \tag{54}$$

with

$$L(a, r) = (2^{r-1}a, 2^{r-2}a, \ldots, 2a, a). \tag{55}$$

Apply this to $H^{\bullet}(B; \mathbf{F}_2)$ under the above assumption on F; the z_i^{2r} constitute a *simple* system of generators of $H^{\bullet}(F; \mathbf{F}_2)$ (chap. IV, § 4,IV); since $\mathrm{Sq}^a \circ \tau = \tau \circ \mathrm{Sq}^a$,

$$H^{\bullet}(B; \mathbf{F}_2) = \mathbf{F}_2[(\mathrm{Sq}^{L(n_i,r)}(t_i))] \tag{56}$$

where i is any integer $\geqslant 1$, r any integer $\geqslant 0$.

The first application is to the determination of the cohomology $H^\bullet(\mathbf{Z}/2\mathbf{Z}, q; \mathbf{F}_2)$; by definition of the Eilenberg–Mac Lane spaces, for $0 < i < q$, $H^i(\mathbf{Z}/2\mathbf{Z}, q; \mathbf{F}_2) = 0$ and $H^q(\mathbf{Z}/2\mathbf{Z}, q; \mathbf{F}_2) \cong \mathbf{Z}/2Z$; write u_q the unique element $\neq 0$ of $H^q(\mathbf{Z}/2\mathbf{Z}, q; \mathbf{F}_2)$. Serre determined the other groups $H^n(\mathbf{Z}/2\mathbf{Z}, q; \mathbf{F}_2)$ by induction on q. He started from the fact that $K(\mathbf{Z}/2\mathbf{Z}, 1)$ is the infinite-dimensional real projection space $\mathbf{P}_\infty(\mathbf{R})$ (chap. II, § 6,F), and therefore $H^\bullet(\mathbf{Z}/2\mathbf{Z}, 1; \mathbf{F}_2) = \mathbf{F}_2[u_1]$. The Borel theorem recalled above determines $H^\bullet(\mathbf{Z}/2\mathbf{Z}, q; \mathbf{F}_2)$ from $H^\bullet(\mathbf{Z}/2\mathbf{Z}, q-1; \mathbf{F}_2)$ by consideration of the space of paths of $K(\mathbf{Z}/2\mathbf{Z}, q)$ with fixed origin, that has $K(\mathbf{Z}/2\mathbf{Z}, q-1)$ as a fiber. To express the result, call a sequence of integers $\geqslant 1$

$$I = (i_1, i_2, \dots, i_r)$$

admissible if it satisfies the inequalities

$$i_1 \geqslant 2i_2, i_2 \geqslant 2i_3, \dots, i_{r-1} \geqslant 2i_r;$$

call $i_1 + i_2 + \cdots + i_r$ the *degree* of I and

$$\begin{aligned} e(I) &= (i_1 - 2i_2) + (i_2 - 2i_3) + \cdots + (i_{r-1} - 2i_r) + i_r \\ &= 2i_1 - (i_1 + i_2 + \cdots + i_r) \end{aligned}$$

its *excess*. Then

$$H^\bullet(\mathbf{Z}/2\mathbf{Z}, q; \mathbf{F}_2) = \mathbf{F}_2[(\mathrm{Sq}^I u_q)] \tag{57}$$

where I takes as values *all admissible sequences of excess* $e(I) < q$.

Next Serre applied a similar method to $H^\bullet(\mathbf{Z}, q; \mathbf{F}_2)$, this time starting from the fact that $K(\mathbf{Z}, 1)$ has the homotopy type of $\mathbf{S}_1$ and $K(\mathbf{Z}, 2)$ is the infinite-dimensional complex projective space $\mathbf{P}_\infty(\mathbf{C})$ (chap. II, § 6,F). The result was that

$$H^\bullet(\mathbf{Z}, q; \mathbf{F}_2) = \mathbf{F}_2[(\mathrm{Sq}^I v_q)] \tag{58}$$

where v_q is the unique element $\neq 0$ of $H^q(\mathbf{Z}, q; \mathbf{F}_2) = \mathbf{Z}/2\mathbf{Z}$, and I is any admissible sequence such that $e(I) < q$ *and* $i_r > 1$.

For odd m, $H^n(\mathbf{Z}/m\mathbf{Z}, q; \mathbf{F}_2) = 0$ for all $n > 0$, since $(\mathbf{Z}/m\mathbf{Z}) \otimes \mathbf{F}_2 = 0$ (chap. V, § 5,A). On the other hand, $K(\Pi, q) \times K(\Pi', q)$ has the homotopy type of $K(\Pi \oplus \Pi', q)$, so the determination of $H^\bullet(\Pi, q; \mathbf{F}_2)$ is complete when it is known also for the groups $\Pi = \mathbf{Z}/2^h\mathbf{Z}$ with $h \geqslant 2$. The method is again the same, but here the initial algebra $H^\bullet(\mathbf{Z}/2^h\mathbf{Z}, 1; \mathbf{F}_2)$ is a little different from the case $h = 1$: it is the tensor product

$$\mathbf{F}_2[v_2] \otimes \bigwedge(u_1)$$

where v_2 is the image of the canonical generator of $H^1(\mathbf{Z}/2^h\mathbf{Z}, 1; \mathbf{Z}/2^h\mathbf{Z})$ by the Bockstein operator

$$\partial_{h,1}\colon H^1(\mathbf{Z}/2^h\mathbf{Z}, 1; \mathbf{Z}/2^h\mathbf{Z}) \to H^2(\mathbf{Z}/2^h\mathbf{Z}, 1; \mathbf{F}_2)$$

coming from the exact sequence

$$0 \to \mathbf{F}_2 \to \mathbf{Z}/2^{h+1}\mathbf{Z} \xrightarrow{\cdot 2} \mathbf{Z}/2^h\mathbf{Z} \to 0.$$

Similarly, if

$$\partial_{h,q}\colon \mathrm{H}^q(\mathbf{Z}/2^h\mathbf{Z}, q; \mathbf{Z}/2^h\mathbf{Z}) \to \mathrm{H}^{q+1}(\mathbf{Z}/2^h\mathbf{Z}, q; \mathbf{F}_2)$$

is again the Bockstein operator coming from the same exact sequence and v_q is the image by $\partial_{h,q}$ of the canonical generator $\mathrm{H}^q(\mathbf{Z}/2^h\mathbf{Z}, q; \mathbf{Z}/2^h\mathbf{Z})$, then the element written $\mathrm{Sq}_q^{\mathrm{I}}(u_q)$ is defined as equal to $\mathrm{Sq}^{\mathrm{I}}(u_q)$ if $i_r > 1$ and to

$$\mathrm{Sq}^{i_1}\,\mathrm{Sq}^{i_2}\cdots\mathrm{Sq}^{i_{r-1}}(v_q)$$

if $i_r = 1$. Then the final result is that

$$\mathrm{H}^{\bullet}(\mathbf{Z}/2^h\mathbf{Z}, q; \mathbf{F}_2) = \mathbf{F}_2[(\mathrm{Sq}_h^{\mathrm{I}}(u_q))]$$

for admissible sequences I such that $e(\mathrm{I}) < q$.

The Eilenberg–Mac Lane theorem on homology suspension (chap. V, § 2,D) shows that for $q > n$ all cohomology groups $\mathrm{H}^{n+q}(\Pi, q; \mathrm{G})$ are naturally isomorphic; they are called the *stable* cohomology groups and the cohomological operations corresponding to their elements (§ 3,A) the *stable* cohomology operations. Write $\mathrm{A}_n^n(\Pi, \mathrm{G})$ for the stable group $\mathrm{H}^{n+q}(\Pi, q; \mathrm{G})$ for $q = n + 1$. For $\mathrm{G} = \mathbf{F}_2$ and Π equal to one of the groups $\mathbf{Z}$, $\mathbf{Z}/2^h\mathbf{Z}$, the stable groups are explicitly determined by the preceding computations.

C. The Relations between the Steenrod Squares

In the special case $\Pi = \mathbf{Z}/2\mathbf{Z}$ the determination of the cohomology algebra $\mathrm{H}^{\bullet}(\mathbf{Z}/2\mathbf{Z}, q; \mathbf{F}_2)$ implies, by § 3,A above, the determination of *all* cohomology operations of type $(q, n; \mathbf{F}_2, \mathbf{F}_2)$: every such operation

$$\mathrm{C}\colon \mathrm{H}^q(\mathrm{X}; \mathbf{F}_2) \to \mathrm{H}^n(\mathrm{X}; \mathbf{F}_2)$$

for fixed q and arbitrary n, can be written

$$x \mapsto \mathrm{C}(x) = \mathrm{P}(\mathrm{Sq}^{\mathrm{I}_1}(x), \ldots, \mathrm{Sq}^{\mathrm{I}_k}(x)) \tag{59}$$

where P is a polynomial (for the cup-product) and the $\mathrm{Sq}^{\mathrm{I}_1}, \ldots, \mathrm{Sq}^{\mathrm{I}_k}$ correspond to admissible sequences of excess $< q$. There are corresponding results for the cases $\Pi = \mathbf{Z}$ and $\Pi = \mathbf{Z}/2^h\mathbf{Z}$ with $h \geqslant 2$. Since only admissible sequences I occur in (59), there must be formulas expressing the Sq^{I} for *nonadmissible* sequences by linear combinations of the Sq^{I} for admissible sequences. It is enough to express in that way all products $\mathrm{Sq}^a\,\mathrm{Sq}^b$ for $a < 2b$; the precise formula giving that product was conjectured by Wu Wen Tsün and first proved in 1952 by Adem [6]:

$$\mathrm{Sq}^a\,\mathrm{Sq}^b = \sum_{0 \leqslant c \leqslant a/2} \binom{b-c-1}{a-2c} \mathrm{Sq}^{a+b-c}\,\mathrm{Sq}^c. \tag{60}$$

Adem's proof was an adaptation of Steenrod's definition of the reduced powers by means of the operators D_i (§ 2,A); he considered a group Π of order 4 contained in a dihedral subgroup of $\mathfrak{S}_4$ of order 8; this gave him a cohomology operation

$$[{}^i_k]\colon \mathrm{H}^q(\mathrm{X}) \to \mathrm{H}^{4q-1}(\mathrm{X})$$

depending on an additional parameter k, and he proved it by direct computation to be a sum of iterated squares $\sum \mathrm{Sq}^r \mathrm{Sq}^s$. He also showed that $[{}^{i}_{k}] = [{}^{i}_{i-k}]$, and this gave him relation (60).

Serre described ([431], p. 224) a simpler method suggested by Wu Wen Tsün's study of the Stiefel–Whitney characteristic classes (§ 4,B); he considered the real projective space of infinite dimension $\mathbf{P}_\infty(\mathbf{R}) = \mathrm{Y}$ (chap. II, § 6,F), which has the homotopy type of $\mathrm{K}(\mathbf{Z}/2\mathbf{Z}, 1)$, and the product $\mathrm{X} = \mathrm{Y}^q$; then $\mathrm{H}^\bullet(\mathrm{X}; \mathbf{F}_2)$ is a polynomial algebra $\mathbf{F}_2[y_1, y_2, \ldots, y_q]$, where the y_j are independent elements of $\mathrm{H}^1(\mathrm{X}; \mathbf{F}_2)$. It is easy to compute the Sq^i in $\mathrm{H}^\bullet(\mathrm{X}; \mathbf{F}_2)$ ([463], p. 38). From their expression is deduced the following lemma: if w_q is the product $y_1 \smile y_2 \smile \cdots \smile y_q$, the $\mathrm{Sq}^\mathrm{I}(w_q)$ for all admissible sequences $\mathrm{I} = (i_1, i_2, \ldots, i_k)$ with $i_1 + i_2 + \cdots + i_k \leqslant q$ are linearly independent. Therefore, if C is a sum of operators Sq^I with *arbitrary* sequences I of degree $\leqslant q$, the relation $\mathrm{C}(w_q) = 0$ implies $\mathrm{C} = 0$.

The general result from which the Adem relations can be deduced is then: if C is any sum of operators Sq^I such that for every space T the relation $\mathrm{C}(y) = 0$ for an element $y \in \mathrm{H}^\bullet(\mathrm{T}; \mathbf{F}_2)$ implies $\mathrm{C}(x \smile y) = 0$ for every element $x \in \mathrm{H}^1(\mathrm{T}; \mathbf{F}_2)$, then $\mathrm{C} = 0$. This is proved by taking $\mathrm{T} = \mathrm{Y}^q$ for q large enough, because $\mathrm{C}(1) = 0$, hence $\mathrm{C}(y_1 \smile y_2 \smile \cdots \smile y_i) = 0$ for every i, and therefore in particular $\mathrm{C}(w_q) = 0$, which implies $\mathrm{C} = 0$ by the preceding lemma. Adem's formula is then proved by induction on $a + b$ and repeated application of Cartan's formula (22).

D. The Relations between the Steenrod Reduced Powers, and the Steenrod Algebra

In 1953 J. Adem [7] generalized to the Steenrod reduced powers the method he had used to prove (60) for the Steenrod squares by considering the homology of the Sylow subgroups of the symmetric group $\mathfrak{S}_{p^2}$. In 1954 H. Cartan [112] obtained the Adem relations by another method, extending the one used by Serre and Thom for the Sq^i, described in section C above.

For an odd prime p, Cartan used the notation

$$\beta_p\colon \mathrm{H}^q(\mathrm{X}; \mathbf{F}_p) \to \mathrm{H}^{q+1}(\mathrm{X}; \mathbf{F}_p)$$

for the Bockstein operator corresponding to the exact sequence

$$0 \to \mathbf{Z}/p\mathbf{Z} \to \mathbf{Z}/p^2\mathbf{Z} \xrightarrow{\cdot p} \mathbf{Z}/p\mathbf{Z} \to 0$$

already considered by Steenrod (§ 2,C). For any integer $a > 0$ such that

$$a \equiv 0 \ (\mathrm{mod}.\, 2p-2) \quad \text{or} \quad a \equiv 1 \ (\mathrm{mod}.\, 2p-2) \tag{61}$$

he wrote

$$\mathrm{St}_p^a\colon \mathrm{H}^q(\mathrm{X}; \mathbf{F}_p) \to \mathrm{H}^{q+a}(\mathrm{X}; \mathbf{F}_p) \tag{62}$$

for the operator defined as follows: if $a = 2k(p-1)$, then $\mathrm{St}_p^a = \mathrm{P}_p^k$; if $a = 2k(p-1) + 1$, $\mathrm{St}_p^a = \beta_p \circ \mathrm{P}_p^k$ (one writes β and St^a when no confusion can arise).

For each sequence $I = (a_1, a_2, \ldots, a_r)$ where the a_j are congruent to 0 or 1 mod. $(2p - 2)$, let

$$\mathrm{St}_p^{\mathrm{I}} = \mathrm{St}_p^{a_1} \circ \mathrm{St}_p^{a_2} \circ \cdots \circ \mathrm{St}_p^{a_r}; \tag{63}$$

the *degree* of the sequence I is the number $d = a_1 + a_2 + \cdots + a_r$; $\mathrm{St}_p^{\mathrm{I}}$ is a graded endomorphism of degree d of the vector space $\mathrm{H}^{\bullet}(\mathrm{X}; \mathbf{F}_p)$.

The *Steenrod algebra* $\mathscr{A}_p$ is introduced in the following way. Consider the vector space E over $\mathbf{F}_p$ with denumerable basis $(\mathrm{T}_j)_{j \geq 1}$. The *tensor algebra* $\mathscr{T}(\mathrm{E})$ over $\mathbf{F}_p$ has as basis the unit element 1 of $\mathbf{F}_p$ and the tensor products

$$\mathrm{T}_{j_1} \otimes \mathrm{T}_{j_2} \otimes \cdots \otimes \mathrm{T}_{j_r}$$

for arbitrary finite sequences $(j_h)_{1 \leq h \leq r}$ of integers ≥ 1 (it is also called the *free associative algebra* over $\mathbf{F}_p$ with generators T_j). Consider the increasing sequence of all integers $a \geq 1$ satisfying (61), and write $a(m)$ the m-th term of that sequence; then for any space X the tensor algebra $\mathscr{T}(\mathrm{E})$ acts on the cohomology $\mathrm{H}^{\bullet}(\mathrm{X}; \mathbf{F}_p)$ by

$$(\mathrm{T}_{j_1} \otimes \mathrm{T}_{j_2} \otimes \cdots \otimes \mathrm{T}_{j_r}) . u = (\mathrm{St}^{a(j_1)} \circ \mathrm{St}^{a(j_2)} \circ \cdots \circ \mathrm{St}^{a(j_r)}) . u. \tag{64}$$

Let $\mathscr{I}$ be the subset of $\mathscr{T}(\mathrm{E})$ consisting of all elements ξ such that $\xi . \mathrm{H}^{\bullet}(\mathrm{X}; \mathbf{F}_p) = 0$ for *all* spaces X. It is clear that $\mathscr{I}$ is a *two-sided ideal* of the algebra $\mathscr{T}(\mathrm{E})$. By definition, the Steenrod algebra $\mathscr{A}_p$ is the quotient algebra $\mathscr{T}(\mathrm{E})/\mathscr{I}$. Since every element $\xi \in \mathscr{T}(\mathrm{E})$ is acting on $\mathrm{H}^{\bullet}(\mathrm{X}; \mathbf{F}_p)$ as a cohomology operation, the characterization of these operators (section B) shows that the necessary and sufficient condition for ξ to belong to $\mathscr{I}$ is that

$$\xi(\mathrm{K}(\mathbf{Z}/p\mathbf{Z}, n))(\iota_n) = 0 \tag{65}$$

for *every* $n \geq 1$, where ι_n is the canonical cohomology class in $\mathrm{H}^n(\mathrm{K}(\mathbf{Z}/p\mathbf{Z}, n); \mathbf{F}_p)$ (chap. V, § 1,D). The $\mathrm{St}_p^{\mathrm{I}}$, however, commute with the Eilenberg–Mac Lane cohomology suspension [chap. V, § 2,D, formula (60)]

$$\mathrm{H}^{r+1}(\mathrm{K}(\mathbf{Z}/p\mathbf{Z}, n + 1); \mathbf{F}_p) \to \mathrm{H}^r(\mathrm{K}(\mathbf{Z}/p\mathbf{Z}, n); \mathbf{F}_p).$$

Therefore, from the definition (64) it follows that if for a given ξ relation (65) is satisfied for *large enough* integers n, this is enough to conclude that it is satisfied for *all* $n \geq 1$. However, H. Cartan, in his determination of the cohomology of Eilenberg–Mac Lane spaces (chap. V, § 3,C), had proved that for $n > q$ the $\mathbf{F}_p$-vector space $\mathrm{H}^{n+q}(\mathrm{K}(\mathbf{Z}/p\mathbf{Z}, n); \mathbf{F}_p)$ has a basis consisting of the $\mathrm{St}_p^{\mathrm{I}}(\iota_n)$, where $\mathrm{I} = (a_1, a_2, \ldots, a_r)$ is any sequence such that

$$a_1 + a_2 + \cdots + a_r = q \quad \text{and} \quad a_j \geq p a_{j+1} \quad \text{for } 1 \leq i \leq r - 1. \tag{66}$$

This generalized Serre's result for the cohomology of $\mathrm{K}(\mathbf{Z}/2\mathbf{Z}, q)$ with coefficients in $\mathbf{F}_2$ (section B). Cartan similarly said that sequences satisfying (66) are *admissible* for any q; from the argument given above, it follows that the $\mathrm{St}_p^{\mathrm{I}}$ for *all* admissible sequences form a *basis* of the $\mathbf{F}_p$-vector space $\mathscr{A}_p$ (Adem had shown earlier that these operators *generate* the vector space $\mathscr{A}_p$).

To complete the determination of the *algebra* structure of $\mathscr{A}_p$, Cartan had to compute the products $P^k P^h$ and $P^k \beta P^h$ for $k < ph$. He showed that

$$P^k P^h = \sum_{t \geqslant 0} c^t_{k,h} P^{k+h-t} P^t \quad \text{for } k < ph, \tag{67}$$

$$P^k \beta P^h = \sum_{t \geqslant 0} c'^t_{k,h} P^{k+h-t} \beta P^t + \sum_{t \geqslant 0} c''^t_{k,h} \beta P^{k+h-t} P^t \quad \text{for } k \leqslant ph, \tag{68}$$

where the values of the coefficients $c^t_{k,h}$, $c'^t_{k,h}$, $c''^t_{k,h}$ in $\mathbf{F}_p$ are explicitly given, by adapting Serre's procedure described in section C to a space $K(\mathbf{Z}/p\mathbf{Z}, 1)^{n+n'}$ (whose cohomology is known) for large values of n, n'.

In 1957 [341] Milnor took a different approach to the structure of $\mathscr{A}_p$. He showed that there is a comultiplication

$$\Delta: \mathscr{A}_p \to \mathscr{A}_p \, {}^g\!\otimes \mathscr{A}_p$$

for which $\mathscr{A}_p$ becomes a *Hopf algebra* (Part 2, chap. VI, § 2,B) which is *coassociative* [*loc. cit.*, formula (18)] and *co-anticommutative.* This means that the diagram

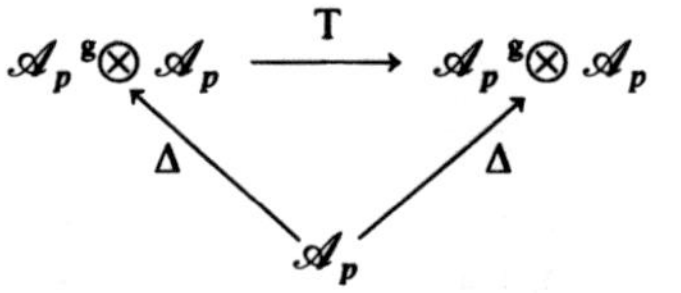

is commutative, where T is the bijection such that

$$T(x_k \otimes y_h) = (-1)^{kh} y_h \otimes x_k$$

when k is the degree of x and h the degree of y.

The existence of Δ is another way of expressing the Cartan–Steenrod formula for $P^k(u \smile v)$:

$$\Delta(P^k) = P^k \otimes 1 + P^{k-1} \otimes P^1 + \cdots + P^1 \otimes P^{k-1} + 1 \otimes P^k$$

and there is a similar formula for the Bockstein operator

$$\Delta(\beta) = \beta \otimes 1 + 1 \otimes \beta.$$

Let $\mathscr{A}_p^*$ be the graded dual of the vector space $\mathscr{A}_p$; the transposed map

$${}^t\Delta: \mathscr{A}_p^* \, {}^g\!\otimes \mathscr{A}_p^* \to \mathscr{A}_p^*$$

defines on $\mathscr{A}_p^*$ a structure of *associative* and *anticommutative algebra*; and if $m: \mathscr{A}_p \, {}^g\!\otimes \mathscr{A}_p \to \mathscr{A}_p$ is the linear map defining the algebra structure of $\mathscr{A}_p$, its transposed map

$${}^t m: \mathscr{A}_p^* \to \mathscr{A}_p^* \, {}^g\!\otimes \mathscr{A}_p^*$$

together with ${}^t\Delta$, defines on $\mathscr{A}_p^*$ a structure of *coassociative Hopf algebra.*

Milnor has determined the *algebra* structure of $\mathscr{A}_p^*$; his results can be presented in the following way. For an odd prime p, the cohomology $H^{\bullet}(\mathbf{Z}/p\mathbf{Z}, 1; \mathbf{F}_p)$ is the tensor product of an *exterior algebra* $\bigwedge(x)$ with one generator of degree 1:

$$x \in \mathrm{H}^1(\mathbf{Z}/p\mathbf{Z}, 1; \mathbf{F}_p), \tag{69}$$

and an algebra of *polynomials* $\mathbf{F}_p[y]$ with one generator of degree 2:

$$y \in \mathrm{H}^2(\mathbf{Z}/p\mathbf{Z}, 1; \mathbf{F}_p), \quad \text{and } y = \beta x \tag{70}$$

([110], p. 1306). The action of the Steenrod algebra $\mathscr{A}_p$ on $\mathrm{H}^{\bullet}(\mathbf{Z}/p\mathbf{Z}, 1; \mathbf{F}_p)$ is given by the rules

$$\mathrm{St}^{\mathrm{I}} x = 0$$

unless $\mathrm{St}^{\mathrm{I}} = \beta$, when $\beta x = y$, or if $\mathrm{St}^{\mathrm{I}} = \mathrm{P}^{p^k}\mathrm{P}^{p^{k-1}} \cdots \mathrm{P}^p\mathrm{P}^1$, when $\mathrm{St}^{\mathrm{I}} x = y^{p^{k+1}}$;

$$\mathrm{St}^{\mathrm{I}} y = 0$$

unless $\mathrm{St}^{\mathrm{I}} = \mathrm{P}^{p^k}\mathrm{P}^{p^{k-1}} \cdots \mathrm{P}^p\mathrm{P}^1$, when $\mathrm{St}^{\mathrm{I}} y = y^{p^{k+1}}$.

These rules entirely determine the $\mathscr{A}_p$-module structure of $\mathrm{H}^{\bullet}(\mathbf{Z}/p\mathbf{Z}, 1; \mathbf{F}_p)$; in particular, for each $s \in \mathscr{A}_p$ and for the generators x, y defined in (69) and (70),

$$s.y = \sum_{i \geqslant 0} a_i(s) y^{p^i}, \qquad s.x = a_0(s)x + \sum_{i \geqslant 0} b_i(s) y^{p^i}$$

where $\xi_i\colon s \mapsto a_i(s)$ and $\tau_i\colon s \mapsto b_i(s)$ are *linear forms* on the $\mathbf{F}_p$-vector space $\mathscr{A}_p$, or in other words, elements of the *dual space* $\mathscr{A}_p^*$, of respective degrees $2p^i - 2$ and $2p^i - 1$.

Milnor's theorem is that the *algebra* $\mathscr{A}_p^*$ is isomorphic to the skew tensor product

$$\bigwedge(\tau_0, \tau_1, \ldots, \tau_i, \ldots) \,{}^{g}\!\otimes\, \mathbf{F}_p[\xi_0, \xi_1, \ldots, \xi_i, \ldots] \tag{71}$$

of the exterior algebra over $\mathbf{F}_p$ with generators 1 and the τ_i and of the algebra of polynomials with generators 1 and the ξ_i; note that this agrees with the general theorem of Leray and Borel on the structure of infinite-dimensional anticommutative Hopf algebras (Part 2, chap. VI, § 2,A).

The comultiplication ${}^t m$ in $\mathscr{A}_p^*$ is given by

$${}^t m(\xi_k) = \sum_{0 \leqslant i \leqslant k} (\xi_{k-i})^{p^i} \otimes \xi_i,$$

$${}^t m(\tau_k) = \tau_k \otimes 1 + \sum_{0 \leqslant i \leqslant k} (\xi_{k-i})^{p^i} \otimes \tau_i.$$

Finally, the general theory of Hopf algebras (or of bigebras) [346] establishes that if A is a Hopf algebra over $\mathbf{F}_p$ such that $\mathrm{A}^0 = \mathbf{F}_p.1$ (so-called "connected" Hopf algebras), there is a unique *antipodism*, defined as a graded linear map of degree 0,

$$a\colon \mathrm{A} \to \mathrm{A}$$

such that $a(1) = 1$, and for an $x \in \mathrm{A}$, if

$$\Delta(x) = \sum_i x_i' \otimes x_i''$$

then

$$\sum_i x_i' a(x_i'') = 0.$$

The map a is an *antiautomorphism*, a linear bijection such that

$$a(x_p x_q) = (-1)^{pq} a(x_q) a(x_p)$$

when x_p has degree p and x_q has degree q. The origin of that map is to be sought in the particular case $A = H_\bullet(G)$, the homology Hopf algebra of a Lie group G, where a is just the map in homology

$$i_*: H_\bullet(G) \to H_\bullet(G)$$

corresponding to the inverse map $i: x \mapsto x^{-1}$ in G.

For the Steenrod algebra $\mathscr{A}_p$, the antipodism a was introduced by Thom ([463], p. 60). Milnor then gave the expression of $a(x)$ for any element x of a Milnor basis.

There are corresponding results for the Steenrod algebra $\mathscr{A}_2$ over $\mathbf{F}_2$, generated by the Sq^i ([425], exps. 10 and 11).

E. The Pontrjagin p-th Powers

In a Note of 1942 [378a] Pontrjagin used an operation that transforms cohomology classes with coefficients in $\mathbf{Z}/2\mathbf{Z}$ into cohomology classes with coefficients in $\mathbf{Z}/4\mathbf{Z}$. It was called the *Pontrjagin square* and was generalized later by Steenrod and E. Thomas ([454], [465]). For each prime p they defined cohomology operations

$$\mathfrak{P}_p: H^i(X; \mathbf{Z}/p^h\mathbf{Z}) \to H^{pi}(X; \mathbf{Z}/p^{h+1}\mathbf{Z})$$

which have the properties:

1. If $\eta: \mathbf{Z}/p^{h+1}\mathbf{Z} \to \mathbf{Z}/p^h\mathbf{Z}$ is the natural homomorphism, η_* the corresponding homomorphism of cohomology rings, then
$$\eta_*(\mathfrak{P}_p x) = x^p.$$
2. If $\rho: \mathbf{Z}/p^h\mathbf{Z} \to \mathbf{Z}/p^{h+1}\mathbf{Z}$ is the natural injection, and ρ_* is the corresponding homomorphism of cohomology rings, then
$$\mathfrak{P}_p(x + y) = \mathfrak{P}_p x + \mathfrak{P}_p y + \sum_{i=1}^{p-1} \left(\frac{1}{p}\binom{p}{i}\right) \rho_*(x^i \smile y^{p-i}).$$
3. $\mathfrak{P}_p(x \smile y) = \mathfrak{P}_p x \smile \mathfrak{P}_p y$.
4. If p is odd and x has odd degree, then $\mathfrak{P}_p x = 0$.

It follows from H. Cartan's general description of the cohomology $H^\bullet(K(\Pi, n); G)$ for all finitely generated commutative groups Π, G (chap. V, § 3,C) that all cohomology operations of type (q, n, A, B) for finitely generated commutative groups A, B can be obtained by composition and addition from the special cohomology operations described in this chapter.

§4. Applications of Steenrod's Squares and Reduced Powers

A. The Steenrod Extension Theorem

Pontrjagin's proof of formula (1) of §1, classifying up to homotopy the continuous maps of a three-dimensional simplicial complex into $\mathbf{S}_2$, was long and intricate; it independently developed parts of the obstruction theory published in 1940 by Eilenberg (chap. II, §4,C), and used the technique of inverse images inaugurated by Hopf in his 1930 paper on $\pi_3(\mathbf{S}_2)$ (chap. II, §1,B). Steenrod took another approach, linking the problem to the computation of what he called a "secondary obstruction." In the Eilenberg theory of obstructions (*loc. cit.*) applied to the special case in which $\mathrm{Y} = \mathbf{S}_n$, consider a continuous map $f\colon \mathrm{X}_{q-1} \to \mathbf{S}_n$ of the $(q-1)$-skeleton X_{q-1} of X, and suppose it can be extended to a continuous map $f'\colon \mathrm{X}_q \to \mathbf{S}_n$, so that the obstruction q-cocycle $b_q(f)$ in $\mathrm{Z}^q(\mathrm{X};\pi_q(\mathbf{S}_n))$ is 0. The obstruction $(q+1)$-cocycle $b_{q+1}(f')$ in $\mathrm{Z}^{q+1}(\mathrm{X};\pi_q(\mathbf{S}_n))$ is then defined, and its cohomology class in $\mathrm{H}^{q+1}(\mathrm{X};\pi_q(\mathbf{S}_n))$ is independent of the particular extension f' of f to X_q; it can therefore be written $z_{q+1}(f)$ and called the *secondary obstruction* to the extension of f to X [it is of course only defined when the *primary* obstruction $b_q(f) = 0$].

Consider now a CW-complex X of dimension $n+2$ and suppose that there exists a continuous map $f\colon \mathrm{X}_n \to \mathbf{S}_n$ that can be extended to a continuous map $f'\colon \mathrm{X}_{n+1} \to \mathbf{S}_n$. Steenrod's theorem is that the secondary obstruction $z_{n+2}(f')$ to the extension of f' only depends on f (and not on the particular extension f') and is given by the $(n-2)$-*product* (§1,B)

$$z_{n+2}(f') = f^*(s_n) \smile_{n-2} f^*(s_n). \tag{72}$$

The proof is ingenious and done in several steps [447].

I. The first and second steps are devoted to what look like very special cases of formula (72). First Steenrod considered the four-dimensional complex projective plane $\mathbf{P}_2(\mathbf{C})$ and its standard definition as a CW-complex (chap. II, §6,F). He extended the natural map of $\mathbf{S}_3 \subset \mathbf{C}^2$ onto the line at infinity of $\mathbf{P}_2(\mathbf{C})$ to a continuous map ψ_4 of the ball $\mathbf{D}_4$:

$$|z_1|^2 + |z_2|^2 \leqslant 1$$

onto $\mathbf{P}_2(\mathbf{C})$, such that $\psi_4(z_1, z_2)$ is the point of $\mathbf{P}_2(\mathbf{C})$ having homogeneous coordinates $(z_1, z_2, (1 - |z_1|^2 - |z_2|^2)^{1/2})$; he then defined the three cells e^0, e^2, e^4 of $\mathbf{P}_2(\mathbf{C})$ as follows:

i. $\overline{e^0}$ is the point $\psi_4(1,0)$;

ii. $\overline{e^2}$ is the image by ψ_4 of the disk $\mathbf{D}_2 \subset \mathbf{C}$ defined by $|z_1| \leqslant 1$, $z_2 = 0$, the intersection of $\mathbf{D}_4$ with the projective line $z_2 = 0$; $\overline{e^2}$ is identified with the sphere $\mathbf{S}_2$; ψ_4, restricted to the interior $\mathring{\mathbf{D}}_2$ of the disk, is a homeomorphism onto $e^2 = \overline{e^2} - e_0$ and maps $\mathbf{S}_1\colon |z_1| = 1$ on the point e^0;

iii. $\overline{e^4} = \mathbf{P}_2(\mathbf{C})$; the restriction of ψ_4 to the interior $\mathring{\mathbf{D}}_4$ of the ball is a homeomorphism onto $e^4 = \overline{e^4} - \overline{e^2}$; and the restriction of ψ_4 to the frontier $\mathbf{S}_3: |z_1|^2 + |z_2|^2 = 1$ is the Hopf map $\mathbf{S}_3 \to \mathbf{S}_2$ (chap. II, § 1,B).

In this part of the proof he considered the extension of the identity $f = \mathrm{Id}_2: \overline{e^2} \to \mathbf{S}_2$ to a map $\mathbf{P}_2(\mathbf{C}) \to \mathbf{S}_2$. Since the 3-skeleton of $\mathbf{P}_2(\mathbf{C})$ coincides with the 2-skeleton $\overline{e^2}$, the obstruction 4-cocycle is defined by

$$\langle b_4(\mathrm{Id}_2), [\overline{e^4}]\rangle = \eta_2,$$

the homotopy class of the Hopf map in $\pi_3(\mathbf{S}_2)$, of which it is a generator; in other words, in the cohomology group $\mathrm{H}^4(\mathbf{P}_2(\mathbf{C}); \pi_3(\mathbf{S}_2))$ the obstruction class $\beta_4(\mathrm{Id}_2)$ can be identified with the fundamental cohomology class s_4 of $\mathbf{P}_2(\mathbf{C})$ when $\pi_3(\mathbf{S}_2)$ is identified with $\mathbf{Z}$.

On the other hand, the computation of the cohomology of a CW-complex (Part 2, chap. V, § 3,C) shows that the natural injection

$$j: \mathbf{S}_2 = \mathbf{P}_1(\mathbf{C}) \to \mathbf{P}_2(\mathbf{C})$$

gives in cohomology an isomorphism

$$j^*: \mathrm{H}^2(\mathbf{P}_2(\mathbf{C}); \mathbf{Z}) \to \mathrm{H}^2(\mathbf{P}_1(\mathbf{C}); \mathbf{Z})$$

by the cohomology exact sequence. The fundamental cohomology class s_2 of $\mathbf{P}_1(\mathbf{C})$ (or $\mathbf{S}_2$) is thus the image of a well-determined element of $\mathrm{H}^2(\mathbf{P}_2(\mathbf{C}); \mathbf{Z})$ that we again write s_2 [up to sign, the Euler class of the tautological vector bundle over $\mathbf{P}_2(\mathbf{C})$ (chap. IV, § 1,D)]; furthermore,

$$s_2 \smile s_2 = \pm s_4 \tag{73}$$

in the cohomology algebra $\mathrm{H}^{\bullet}(\mathbf{P}_2(\mathbf{C}); \mathbf{Z})$.

As a basis for the induction in step II it is convenient to consider s_4 to be taking its values in $\pi_3(\mathbf{S}_2)$ and s_2 in $\pi_2(\mathbf{S}_2)$; the equation (73) can then be written in the notation of § 1,B:

$$s_2 \smile_0 s_2 = s_4 \tag{74}$$

for a suitable bilinear symmetric map $\varphi: \pi_2(\mathbf{S}_2) \times \pi_2(\mathbf{S}_2) \to \pi_3(\mathbf{S}_2)$; the preceding argument then shows that for $f = f' = \mathrm{Id}_2$,

$$z_4(f) = s_2 \smile_0 s_2 \tag{75}$$

which means that relation (72) is satisfied.

II. Steenrod then defined a sequence of CW-complexes P_n of dimension n by induction on n, starting with $\mathrm{P}_4 = \mathbf{P}_2(\mathbf{C})$ and its cells e^0, e^2, e^4 defined in I. The space P_{n+1} is taken as the *unreduced suspension* $\tilde{\mathrm{S}}\mathrm{P}_n$ (Part 2, chap. V, § 2,C); it is convenient to consider this the *join* of P_n and two points A'_n, B'_n (*loc. cit.*). The idea is to repeatedly "suspend" all the features of the situation in I.

First define $\mathrm{N}_4 = \mathbf{D}_4$; N_{n+1} is the unreduced suspension $\tilde{\mathrm{S}}\mathrm{N}_n$ identified with the join of N_n and two points A_n, B_n, so that N_n is homeomorphic to the closed ball $\mathbf{D}_n$. Starting with the map ψ_4, define by "suspension" a sequence of continuous maps $\psi_n: \mathrm{N}_n \to \mathrm{P}_n$: for each $y \in \mathrm{N}_n$, ψ_{n+1} sends the segment of extremities A_n, y (resp. B_n, y) linearly onto the segment of extremities A'_n, $\psi_n(y)$ [resp. B'_n, $\psi_n(y)$].

The subspace M_{n-1} of N_n, homeomorphic to S_{n-1}, is inductively defined by $M_3 = S_3$, $M_n = \tilde{S}M_{n-1}$; similarly, the subspace Q_{n-2} of P_n, homeomorphic to S_{n-2}, is defined by $Q_2 = S_2$, $Q_{n-1} = \tilde{S}Q_{n-2}$. The subspace L_{n-2} of N_n, homeomorphic to $\mathbf{D}_{n-2}$, is defined by $L_2 = \mathbf{D}_2$, $L_{n-1} = \tilde{S}L_{n-2}$, and the subspace K_{n-3} of L_{n-2}, homeomorphic to S_{n-3}, by $K_1 = S_1$ and $K_{n-2} = \tilde{S}K_{n-3}$.

The structure of the CW-complex of P_n is then defined by three cells e^0, e^{n-2}, e^n. The point e^0 is the same for all P_n; $\overline{e^{n-2}} = Q_{n-2}$ is the image $\psi_n(L_{n-2})$, and ψ_n, restricted to the cell $L_{n-2} - K_{n-3}$, is a homeomorphism on e^{n-2}; finally ψ_n maps M_{n-1} onto Q_{n-2}, and its restriction to the cell $N_n - M_{n-1}$ is a homeomorphism on e^n.

The investigation in that part of the proof concerns the extension of the identity map $f = \mathrm{Id}_n \colon Q_n \to Q_n (= S_n)$ to a continuous map $P_{n+2} \to Q_n$. Again, the $(n+1)$-skeleton and the n-skeleton of P_n are both $\overline{e^n} = Q_n$, so that the obstruction $(n+2)$-cocycle $b_{n+2}(\mathrm{Id}_n)$ is defined by

$$\langle b_{n+2}(\mathrm{Id}_n), [\overline{e^{n+2}}] \rangle = \eta_n,$$

the homotopy class of the restriction $\psi_{n+2} | M_{n+1}$ which can be identified with an element of $\pi_{n+1}(S_n)$. It follows at once from the definitions that

$$\eta_n = \mathrm{E}(\eta_{n-1})$$

where E is the Freudenthal homotopy suspension (chap. II, §6,E). In the infinite sequence

$$\pi_3(S_2) \xrightarrow{\mathrm{E}} \pi_4(S_3) \xrightarrow{\mathrm{E}} \pi_5(S_4) \to \cdots$$

the first map is surjective and the other ones are bijective by Freudenthal's theorems (*loc. cit.*). Since η_2 is a generator of $\pi_3(S_2) \simeq \mathbf{Z}$, each of the other classes η_n is the unique element $\neq 0$ in $\pi_{n+1}(S_n) \simeq \mathbf{Z}/2\mathbf{Z}$. Define a bilinear map

$$\varphi \colon \pi_n(S_n) \times \pi_n(S_n) \to \pi_{n+2}(S_{n+1})$$

by $\varphi(\iota_n, \iota_n) = \eta_{n+1}$, where ι_n is the homotopy class of Id_n.

Now consider the fundamental class $s_n \in H^n(S_n; \mathbf{Z})$ taking its values in $\pi_n(S_n)$. Since the natural isomorphism $H^n(S_n) \xrightarrow{\sim} H^{n+1}(S_{n+1})$ defined by the suspension (Part 2, chap. V, §2,C) maps s_n onto s_{n+1}, and the similar isomorphism $H^{n+2}(P_{n+2}) \xrightarrow{\sim} H^{n+3}(P_{n+3})$ maps $b_{n+2}(\mathrm{Id}_n)$ onto $b_{n+3}(\mathrm{Id}_{n+1})$, it follows by induction from (74) and from the commutativity of the diagram (17) in §1,B) that

$$z_{n+2}(\mathrm{Id}_n) = s_n \smile_{n-2} s_n \tag{76}$$

so relation (72) is satisfied.

III. Now consider the general case of a continuous map $f \colon X_n \to S_n$ under the assumption that there is a continuous extension $f' \colon X_{n+1} \to S_n$ of f; (72) can be proved by establishing that when S_n is identified with the subspace Q_n of P_{n+2} defined in II, there exists a continuous extension $f'' \colon X_{n+2} \to P_{n+2}$ of f' such that the obstruction $(n+2)$-cocycle $b_{n+2}(f')$ is given by

$$b_{n+2}(f') = \tilde{f}''(b_{n+2}(\mathrm{Id}_n)) \in \mathrm{Z}^{n+2}(\mathrm{X}; \pi_{n+1}(\mathbf{S}_n)). \tag{77}$$

Then, from the functoriality of i-products and relation (76), derive

$$z_{n+2}(f') = f''^*(z_{n+2}(\mathrm{Id}_n)) = f''^*(s_n \smile_{n-2} s_n)$$
$$= f''^*(s_n) \smile_{n-2} f''^*(s_n) = f^*(s_n) \smile_{n-2} f^*(s_n)$$

proving (72).

To define f'' satisfying (77), it can first be supposed that $f(\mathrm{X}_{n-1})$ is reduced to the point $*$ in $\mathbf{S}_n$; this follows from the argument detailed in the definition of the Eilenberg groups (chap. V, § 1,C), originally due to Whitney [509]. Because of the different natures of the homotopy groups $\pi_{n+1}(\mathbf{S}_n)$ for $n = 2$ and $n > 2$, both cases must be considered separately.

First suppose $n > 2$ and consider any $(n+2)$-cell σ in X; if the obstruction $(n+2)$-cochain $b_{n+2}(f')$ is such that $\langle b_{n+2}(f'), \sigma \rangle = 0$, then f', by definition, can be extended continuously from $\mathrm{Fr}(\sigma) \subset \mathrm{X}_{n+1}$ to σ. If $\langle b_{n+2}(f'), \sigma \rangle \neq 0$, its value is the unique element $\eta_n \neq 0$ in $\pi_{n+1}(\mathbf{S}_n)$. Choose a map $g\colon \mathrm{Fr}(\sigma) \to \mathrm{M}_{n+1}$ of degree 1; since by **II** the homotopy class of $\psi_{n+2}|\mathrm{M}_{n+1}$ is also η_n, the restrictions of f' and $\psi_{n+2} \circ g$ to $\mathrm{Fr}(\sigma)$ are homotopic. Since g can obviously be extended to a map g' of $\bar{\sigma}$ onto N_{n+2}, $\psi_{n+2} \circ g'$ extends to $\psi_{n+2} \circ g'\colon \bar{\sigma} \to \mathrm{P}_{n+2}$, and therefore $f'|\mathrm{Fr}(\sigma)$ can be extended to a map $f''_\sigma\colon \bar{\sigma} \to \mathrm{P}_{n+2}$ of degree 1.

Now suppose $n = 2$; then $\langle b_4(f'), \sigma \rangle = m\eta_2$ for some integer m; the same construction can be made, with the sole difference that g has degree m.

In both cases f'' is thus defined as coinciding with f''_σ on each $\bar{\sigma} \subset \mathrm{X}_{n+2}$. Using the construction of f'' and the Brouwer–Hopf theorem, Steenrod showed that

$$\tilde{f}''(b_{n+2}(\mathrm{Id}_n)) = b_{n+2}(\mathrm{Id}_n \circ f') = b_{n+2}(f'),$$

and this ends the proof of (72).

Having done this, Steenrod applied standard obstruction theory and recovered Pontrjagin's result (1) for $n = 2$. Next he obtained its generalization for all $n \geqslant 3$: if f_0, f_1 are two continuous maps of X_{n+1} into $\mathbf{S}_n$ that coincide in X_n, a necessary and sufficient condition for f_0 and f_1 to be homotopic with respect to X_n is the existence of a cohomology $(n-1)$-class e_{n-1} in $\mathrm{H}^{n-1}(\mathrm{X}; \pi_n(\mathbf{S}_n))$ such that, in the group $\mathrm{H}^{n+1}(\mathrm{X}; \pi_{n+1}(\mathbf{S}_n))$, the deviation of f_0 and f_1 satisfies

$$\gamma_{n+1}(f_0, f_1) = e_{n-1} \smile_{n-3} e_{n-1}. \tag{78}$$

In the notation Sq^i, this result can be expressed by saying that for a given class $c \in \mathrm{H}^n(\mathrm{X}; \mathbf{Z})$, the homotopy classes of all continuous maps $f\colon \mathrm{X}_{n+1} \to \mathbf{S}_n$ such that $f^*(s_n) = c$ are in one-to-one correspondence with the group

$$\mathrm{H}^{n+1}(\mathrm{X}; \mathbf{F}_2)/\mathrm{Sq}^2(\mathrm{H}^{n-1}(\mathrm{X}; \mathbf{Z})) \tag{79}$$

where Sq^2 is considered as mapping cohomology with coefficients in $\mathbf{Z}$ into cohomology with coefficients in $\mathbf{F}_2$ [§ 1,B, formula (14)].

Another proof can be given using the Postnikov factorization ([440], p. 460).

B. Steenrod Squares and Stiefel–Whitney Characteristic Classes

In 1950 Thom discovered a remarkable relation between the Steenrod squares and the Stiefel–Whitney characteristic classes of an unoriented vector fibration $\xi = (E, B, F, p)$ [462]. Suppose the typical fiber F is a real vector space of dimension m, and consider the Thom isomorphism

$$\Phi\colon H^0(B; \mathbf{F}_2) \xrightarrow{\sim} H^m(E, E^0; \mathbf{F}_2)$$

[chap. IV, § 2, formula (48)]. Then the Stiefel–Whitney classes are given by the formula

$$w_i(\xi) = \Phi^{-1}(\mathrm{Sq}^i(\Phi(1))) \tag{80}$$

where the Steenrod squares are operating on the *relative* cohomology $H^{\bullet}(E, E^0; \mathbf{F}_2)$.

Thom's proof used a triangulation of B and the definition of the Stiefel–Whitney classes as obstructions to the extensions of sections of E above the skeletons of B. A simpler proof [347] can be based on an axiomatic definition of the Stiefel–Whitney classes, patterned after Hirzebruch's similar axioms for Chern classes (chap. IV, § 1,B). There are two essential axioms, the stipulation of the functoriality of the maps $\xi \mapsto w_i(\xi)$ and the Whitney product formula [chap. IV, § 1,B, formula (2)]; two other axioms "normalize" the classes by the conditions $w_0(\xi) = 1$, $w_i(\xi) = 0$ if $m = \text{rank}\ \xi < i$, and finally $w_1(U_{2,1}) \neq 0$ for the tautological bundle $U_{2,1}$ with base space $\mathbf{P}_1(\mathbf{R})$ [chap. IV, § 1,C]. It is then an easy matter to check that these axioms uniquely characterize the Stiefel–Whitney classes and that the right-hand sides of (80) verify these axioms ([347], pp. 86 and 92).

Thom was particularly interested in the Stiefel–Whitney classes of the tangent bundle of a smooth manifold and the normal bundle of a submanifold; these would play a fundamental part in his theory of cobordism, conceived in 1953 (chap. VII, § 1,G).

One of Thom's first results was that the Stiefel–Whitney classes of the tangent bundle of a smooth compact manifold X only depend on the topology of X, not on its differential structure. This can be deduced at once from a useful formula found by Wu Wen Tsün ([522], [523]) for the computation of these classes. Its proof can be presented as follows [347].

I. First, Wu Wen Tsün introduced cohomology classes related to the Sq^i for any space X satisfying the following assumptions:

1. $H^i(X; \mathbf{F}_2) = 0$ for $i > n$;
2. $H^n(X; \mathbf{F}_2)$ has dimension 1 over $\mathbf{F}_2$, and has therefore a unique element $u \neq 0$;
3. for $0 \leqslant k \leqslant n$, $H^k(X; \mathbf{F}_2)$ is put in duality with $H^{n-k}(X; \mathbf{F}_2)$ by the formula

$$x_k \smile y_{n-k} = \langle x_k, y_{n-k} \rangle u \tag{81}$$

for $x_k \in \mathrm{H}^k(\mathrm{X}; \mathbf{F}_2)$, $y_{n-k} \in \mathrm{H}^{n-k}(\mathrm{X}; \mathbf{F}_2)$.

It is clear, by Poincaré duality (Part 2, chap. IV, § 3,A), that these assumptions are verified by any smooth compact connected manifold X.

Under these assumptions, for each k such that $0 \leqslant k \leqslant n$ there is a unique cohomology class $v_k \in \mathrm{H}^k(\mathrm{X}; \mathbf{F}_2)$ such that

$$v_k \smile x = \mathrm{Sq}^k(x) \in \mathrm{H}^n(\mathrm{X}; \mathbf{F}_2) \tag{82}$$

for every $x \in \mathrm{H}^{n-k}(\mathrm{X}; \mathbf{F}_2)$; the v_k are now called the *Wu classes* of X, and

$$v = v_0 + v_1 + \cdots + v_n \in \mathrm{H}^{\bullet}(\mathrm{X}; \mathbf{F}_2) \tag{83}$$

the *total Wu class.*

In this notation Wu Wen Tsün's formulas for the Stiefel–Whitney classes of the tangent bundle T(X) of a smooth compact manifold X are

$$w_k(\mathrm{T}(\mathrm{X})) = \sum_{i+j=k} \mathrm{Sq}^i(v_j) \tag{84}$$

or, if the total Wu class v and the "total Steenrod square"

$$\mathrm{Sq} = \mathrm{Sq}^0 + \mathrm{Sq}^1 + \cdots + \mathrm{Sq}^n \tag{85}$$

are introduced, formula (84) can be written

$$w(\mathrm{T}(\mathrm{X})) = \mathrm{Sq}(v). \tag{86}$$

II. The proof uses properties of the normal bundle of a submanifold (chap. III, § 1,C). Let Y be a smooth manifold and X be a compact submanifold of Y of codimension k. Assuming that Y is given a riemannian structure, let $\mathrm{N}(\varepsilon)$ be the open subset of the normal bundle N of X in Y consisting of the points of N at a distance $<\varepsilon$ from X. Then the "exponential map" of riemannian geometry is defined in $\mathrm{N}(\varepsilon)$ for ε small enough and is a diffeomorphism of $\mathrm{N}(\varepsilon)$ on a tubular neighborhood of X in Y (Part 1, chap. III, § 1). By excision, this gives the existence of a natural isomorphism of relative cohomology algebras

$$\psi: \mathrm{H}^{\bullet}(\mathrm{Y}, \mathrm{Y} - \mathrm{X}; \Lambda) \simeq \mathrm{H}^{\bullet}(\mathrm{N}, \mathrm{N}^0; \Lambda) \tag{87}$$

where N^0 is the complement $\mathrm{N} - \mathrm{X}$.

III. This leads to an expression for the top Stiefel–Whitney class $w_k(\mathrm{N})$ for the normal bundle N. The Thom isomorphism

$$\Phi: \mathrm{H}^0(\mathrm{X}; \mathbf{F}_2) \simeq \mathrm{H}^k(\mathrm{N}, \mathrm{N}^0; \mathbf{F}_2)$$

defines in $\mathrm{H}^k(\mathrm{N}, \mathrm{N}^0 \colon \mathbf{F}_2)$ the fundamental cohomology class $u = \Phi(1)$ (chap. IV, § 2); hence it defines a cohomology class

$$u' = \psi^{-1}(u) \in \mathrm{H}^k(\mathrm{Y}, \mathrm{Y} - \mathrm{X}; \mathbf{F}_2)$$

called the *dual cohomology class of* X *in* Y. That name is justified by the fact ([347], p. 136) that the image of u' in the cohomology exact sequence

$$\mathrm{H}^k(\mathrm{Y}, \mathrm{Y} - \mathrm{X}; \mathbf{F}_2) \xrightarrow{a} \mathrm{H}^k(\mathrm{Y}; \mathbf{F}_2)$$

is the *Poincaré dual* of the homology class $j_*([\mathrm{X}])$, where $j\colon \mathrm{X} \to \mathrm{Y}$ is the natural injection and $[\mathrm{X}] \in \mathrm{H}_{n-k}(\mathrm{X}; \mathbf{F}_2)$ the fundamental class of the manifold X (Part 2, chap. I, § 3,A).

Now consider the commutative diagram

$$\begin{array}{ccc} \mathrm{H}^k(\mathrm{N}, \mathrm{N}^0; \mathbf{F}_2) & \xrightarrow{\quad a \quad} & \mathrm{H}^k(\mathrm{N}; \mathbf{F}_2) \\ \Phi \Big\uparrow & & s^* \Big\downarrow \Big\uparrow p^* \\ \mathrm{H}^0(\mathrm{X}; \mathbf{F}_2) & \xrightarrow[\quad g \quad]{} & \mathrm{H}^k(\mathrm{X}; \mathbf{F}_2) \end{array}$$

leading to the Gysin sequence for the fibration $(\mathrm{N}, \mathrm{X}, \mathbf{R}^k, p)$ [chap. IV, § 2, formula (50)], where a is the homomorphism in the cohomology exact sequence for $(\mathrm{N}, \mathrm{N}^o)$, and $s\colon \mathrm{X} \to \mathrm{N}$ the zero section. Since $g = s^* \circ a \circ \Phi$, $s^*(a(u)) = g(1) = w_k(\mathrm{N})$. Now using the commutative diagram

$$\begin{array}{ccc} \mathrm{H}^k(\mathrm{Y}, \mathrm{Y} - \mathrm{X}; \mathbf{F}_2) & \xrightarrow{\quad a \quad} & \mathrm{H}^k(\mathrm{Y}; \mathbf{F}_2) \\ \psi \Big\downarrow & & \Big\downarrow j^* \\ \mathrm{H}^k(\mathrm{N}, \mathrm{N}^0; \mathbf{F}_2) & \xrightarrow[\quad s^* \circ a \quad]{} & \mathrm{H}^k(\mathrm{X}; \mathbf{F}_2) \end{array}$$

we obtain $j^*(a(u')) = w_k(\mathrm{N})$, where $j\colon \mathrm{X} \to \mathrm{Y}$ is the natural injection.

IV. These general properties of normal bundles are now specialized to the case of the diagonal embedding

$$\delta\colon \mathrm{X} \to \mathrm{X} \times \mathrm{X}.$$

There is then a natural isomorphism of the tangent bundle T(X) over X onto the normal bundle N over the submanifold $\delta(\mathrm{X})$ in $\mathrm{X} \times \mathrm{X}$: to the tangent vector $v \in \mathrm{T}_x(\mathrm{X})$ it associates the normal vector $(-v, v) \in \mathrm{T}_{\delta(x)}(\mathrm{X} \times \mathrm{X})$. If u' is the dual cohomology class in $\mathrm{H}^n(\mathrm{X} \times \mathrm{X}, \mathrm{X} \times \mathrm{X} - \delta(\mathrm{X}); \mathbf{F}_2)$ defined in III, its image u'' by the map

$$a\colon \mathrm{H}^n(\mathrm{X} \times \mathrm{X}, \mathrm{X} \times \mathrm{X} - \delta(\mathrm{X}); \mathbf{F}_2) \to \mathrm{H}^n(\mathrm{X} \times \mathrm{X}; \mathbf{F}_2)$$

is called the *diagonal cohomology class* in $\mathrm{H}^n(\mathrm{X} \times \mathrm{X})$. From the fact that the two projections pr_1, pr_2 of $\mathrm{X} \times \mathrm{X}$ coincide on $\delta(\mathrm{X})$ it is easily deduced that for any class $c \in \mathrm{H}^r(\mathrm{X}; \mathbf{F}_2)$,

$$(c \times 1) \smile u'' = (1 \times c) \smile u''. \tag{88}$$

Suppose X is compact and connected; it then follows from the characterization of the slant product (Part 1, chap. IV, § 5,H)

$$\mathrm{H}^\bullet(\mathrm{X} \times \mathrm{X}) \otimes \mathrm{H}_\bullet(\mathrm{X}) \to \mathrm{H}^\bullet(\mathrm{X})$$

that

$$u''/[X] = 1 \in H^0(X; \mathbf{F}_2). \tag{89}$$

By (88) and (89),

$$\mathrm{Sq}^i(u'')/[X] = w_i(T(X)). \tag{90}$$

Take a basis $(b_i)_{1 \leqslant i \leqslant r}$ of the $\mathbf{F}_2$-vector space $H^{\bullet}(X; \mathbf{F}_2)$ and a "dual basis" $(b'_i)_{1 \leqslant i \leqslant r}$ of the same space satisfying

$$\langle b_i \smile b'_j, [X] \rangle = \delta_{ij} \tag{91}$$

the existence of which follows from Poincaré duality (Part 2, chap. IV, § 3,A); using the Künneth formula for $H^{\bullet}(X \times X; \mathbf{F}_2)$,

$$u'' = \sum_i (-1)^{\dim b_i} b_i \times b'_i. \tag{92}$$

On the other hand, Thom's formula (80) gives

$$\mathrm{Sq}^i(u'') = (w_i(T(X)) \times 1) \smile u''$$

and since $((w_i(T(X)) \times 1) \smile u'')/[X] = w_i(T(X)) \smile (u''/[X]) = w_i(T(X))$ by (88), this yields (90).

Wu Wen Tsün formula (86) is now easily proved; for any $x \in H^{\bullet}(X; \mathbf{F}_2)$,

$$x = \sum_i b_i \langle x \smile b'_i, [X] \rangle$$

and in particular, for the total Wu class

$$v = \sum_i b_i \langle v \smile b'_i, [X] \rangle = \sum_i b_i \langle \mathrm{Sq}(b'_i), [X] \rangle$$

hence

$$\begin{aligned} \mathrm{Sq}(v) &= \sum_i \mathrm{Sq}(b_i) \langle \mathrm{Sq}(b'_i), [X] \rangle \\ &= \sum_i (\mathrm{Sq}(b_i) \times \mathrm{Sq}(b'_i))/[X] \end{aligned}$$

and finally, by Cartan's formula

$$\mathrm{Sq}(v) = \mathrm{Sq}(u'')/[X] = w(T(X)).$$

Application. As an example of application of (86), consider a compact connected manifold X for which the cohomology algebra $H^{\bullet}(X; \mathbf{F}_2)$ is generated by a single element $a \in H^k(X; \mathbf{F}_2)$ for a $k \geqslant 1$, so that $H^j(X; \mathbf{F}_2) = 0$ when j is not divisible by k and $H^{km}(X; \mathbf{F}_2) = \mathbf{F}_2 . a^m$; the dimension of X is thus a multiple kq of k. Then for the total Steenrod square

$$\mathrm{Sq}(a) = a + a^2$$

and, by Cartan's formula

$$\mathrm{Sq}(a^m) = a^m(1 + a)^m$$

so that the total Wu class is

$$v = \sum_{m=0}^{n} \binom{n-m}{m} a^m.$$

Wu's formula (86) gives therefore for the total Stiefel–Whitney class

$$w(\mathrm{T}(\mathrm{X})) = (1 + a)^{q+1} = 1 + \binom{q+1}{1} a + \cdots + \binom{q+1}{q} a^q \tag{93}$$

where the binomial coefficients are reduced modulo 2. Among such spaces are the spheres $\mathbf{S}_n$ with $k = n$, the projective spaces $\mathbf{P}_n(\mathbf{R})$ with $k = 1$, $\mathbf{P}_n(\mathbf{C})$ with $k = 2$ and $\mathbf{P}_n(\mathbf{H})$ with $k = 4$. In addition G. Hirsch also discovered in 1947 another example, of dimension 16 with $k = 8$; he used octonions to give its definition [227], and that manifold is now called the *Cayley plane**; A. Borel later showed that it is diffeomorphic to the homogeneous space $\mathrm{F}_4/\mathrm{Spin}(9)$ [56].

In 1958 J.F. Adams closed that list by showing that there are no other compact CW-complexes whose cohomology algebra over $\mathbf{F}_2$ is generated by a single element (cf. § 5,D).

From the consideration of the cohomology of the product of two Grassmannians, Wu Wen Tsün obtained a general formula for a vector fibration ξ

$$\mathrm{Sq}^k(w_m) = w_k \smile w_m + \binom{k-m}{1} w_{k-1} \smile w_{m+1} + \cdots + \binom{k-m}{k} w_0 \smile w_{m+k} \tag{94}$$

where $w_j = w_j(\xi)$ and $m < k$ [347].

Finally, a little later Wu Wen Tsün defined other characteristic classes for cohomology with coefficients in $\mathbf{F}_p$, p is an odd prime. He extended Thom's formula (80) to the Steenrod reduced powers and defined for a vector fibration ξ

$$q_n(\xi) = \Phi^{-1}(\mathrm{P}_p^n(\Phi(1))). \tag{95}$$

They played an important part in the further developments of the theory of fibrations.

C. Application to Homotopy Groups

Serre's determination of the cohomology algebra $\mathrm{H}^{\bullet}(\Pi, q; \mathbf{F}_2)$ for a finitely generated commutative group Π (§ 3,B) allowed him to prove an interesting general result on homotopy groups. Let X be an arcwise-connected and simply connected space satisfying the following conditions:

1. $\mathrm{H}_i(\mathrm{X}; \mathbf{Z})$ is finitely generated for every $i \geqslant 0$;
2. $\mathrm{H}_i(\mathrm{X}; \mathbf{F}_2) = 0$ for large enough i;
3. $\mathrm{H}_i(\mathrm{X}; \mathbf{F}_2) \neq 0$ for at least one $i > 0$.

These conditions are satisfied by a finite simplicial complex.

---

* From a strictly algebrogeometric point of view, the Cayley plane had been considered by R. Moufang, *Hamburg. Abhandl.*, **9** (1933), 207–222.

The theorem is that for *infinitely* many dimensions i, the homotopy group $\pi_i(X)$ contains a subgroup isomorphic to $\mathbf{Z}$ or to $\mathbf{Z}/2\mathbf{Z}$; this is equivalent to saying

$$\pi_i(X) \otimes \mathbf{F}_2 \neq 0 \tag{96}$$

for infinitely many values of i.

The proof is by contradiction, assuming that there is a *largest* integer q such that

$$\pi_q(X) \otimes \mathbf{F}_2 \neq 0. \tag{97}$$

Consider the sequence (W_n) of spaces "killing" the homotopy groups of X (chap. V, § 2,C). The assumption (97) and the definition of W_{q+1} imply that

$$\pi_r(W_{q+1}) \otimes \mathbf{F}_2 = 0 \quad \text{for } all \text{ values } r > 0;$$

hence (chap. V, § 4)

$$H_r(W_{q+1}; \mathbf{F}_2) = 0 \quad \text{for } all\ r > 0. \tag{98}$$

Next, W_q has the homotopy type of the space P of a fibration $(P, K(\pi_q(X), q), W_{q+1})$, and by Serre's exact sequence [chap. IV, § 3,C, formula (74)]

$$H^i(W_q; \mathbf{F}_2) = H^i(P; \mathbf{F}_2) = H^i(\pi_q(X), q; \mathbf{F}_2) \quad \text{for all } i > 0. \tag{99}$$

Finally, for $2 \leqslant j \leqslant q$ there is a fibration

$$(W_j, W_{j-1}, K(\pi_{j-1}(X), j-2)). \tag{100}$$

Serre used these facts to obtain majorations of the (positive) coefficients of the *Poincaré series*

$$P_Y(t) = \sum_{n=0}^{\infty} \dim H^n(Y; \mathbf{F}_2) . t^n \tag{101}$$

when Y is one of the spaces W_j for $2 \leqslant j \leqslant q$.

Write

$$\theta(\Pi, q; t) = P_{K(\Pi, q)}(t). \tag{102}$$

Then by (99),

$$P_{W_q}(t) = \theta(\pi_q(X), q; t) \tag{103}$$

and for $3 \leqslant i \leqslant q$,

$$P_{W_i}(t) \prec P_{W_{i-1}}(t) . \theta(\pi_{i-1}(X), i-2; t), \tag{104}$$

where $A(t) \prec B(t)$ for power series with positive coefficients is the Cauchy "majorant" inequality, meaning that each coefficient in $A(t)$ is at most equal to the corresponding one in $B(t)$. The relations (104) follow from the existence of the fibrations (100) and the Leray–Hirsch inequalities for the dimensions of the homology spaces [chap. IV, § 3,A, formula (64)]. Finally, from (103) and (104)

$$\theta(\pi_q(X), q; t) \prec P_X(t) . \prod_{1 \leqslant i \leqslant q} \theta(\pi_i(X), i - 1; t). \tag{105}$$

To use this inequality, Serre applied his determination of the cohomology $H^{\bullet}(\Pi, q; F_2)$ to a study of the Poincaré series $\theta(\Pi, q; t)$. He showed that the series is convergent for $|t| < 1$. If

$$\varphi(\Pi, q; x) = \log_2 \theta(\Pi, q; 1 - 2^{-x}) \quad \text{for } 0 \leqslant x < +\infty, \tag{106}$$

he could obtain asymptotic evaluations of these functions when x tends to $+\infty$: let $r \geqslant 0$ (resp. $s \geqslant 0$) be the number of groups isomorphic to a 2-group $\mathbf{Z}/2^h\mathbf{Z}$ (resp. to $\mathbf{Z}$) in the decomposition of Π into cyclic groups of prime power order or infinite. Then:

1. if $r \geqslant 1$, $\varphi(\Pi, q; x) \sim rx^q/q!$;
2. if $r = 0$ and $s \geqslant 1$, $\varphi(\Pi, q; x) \sim s . x^{q-1}/(q - 1)!$;
3. if $r = s = 0$, $\varphi(\Pi, q; x) = 0$.

Now, since $P_X(t)$ is by assumption a polynomial, the Cauchy majoration (105) implies, for $|t| < 1$,

$$\theta(\pi_q(X), q; t) \leqslant C. \prod_{1 \leqslant i \leqslant q} \theta(\pi_i(X), i - 1; t) \tag{107}$$

for a constant $C > 0$, or, equivalently

$$\varphi(\pi_q(X), q; x) \leqslant \log_2 C + \sum_{1 \leqslant i \leqslant q} \varphi(\pi_i(X), i - 1; x). \tag{108}$$

Since $\pi_q(X) \otimes \mathbf{F}_2 \neq 0$, one of the evaluations 1 or 2 above may be used for $\varphi(\pi_q(X), q; x)$, but by the same evaluations, the right-hand side of (107) is majorized by Ax^{q-2} for some constant A, and thus the desired contradiction is reached.

In the last part of his paper [431] Serre also showed how the knowledge of the cohomology algebra $H^{\bullet}(\Pi, q; \mathbf{F}_2)$ could yield information on the homotopy groups $\pi_n(S_3)$ [remember that up to 1951 the only homotopy groups $\pi_m(S_n)$ explicitly known for $m > n$ were $\pi_{n+1}(S_n)$ and $\pi_{n+2}(S_n)$]. Serre again used the sequence (W_q) of spaces that "kill" the homotopy groups $\pi_i(S_n)$, so that $\pi_q(S_3) = H_q(W_q; \mathbf{Z})$; thus

$$H^q(W_q; \mathbf{F}_2) = \mathrm{Hom}(H_q(W_q; \mathbf{Z}), \mathbf{F}_2) = \mathrm{Hom}(\pi_q(S_3), \mathbf{F}_2)$$

gives information on the 2-component of $\pi_q(S_3)$. Serre carried out computations for the cases $q = 4, 5, 6, 7$; he used the spectral sequence of the Serre–Cartan fibrations defining the spaces W_q (chap. V, § 2,C), which can be done given the cohomology of the Eilenberg–Mac Lane spaces with coefficients in $\mathbf{F}_2$. In this manner, and by using previously incomplete information on $\pi_6(S_3)$ and $\pi_7(S_3)$, he could show that $\pi_6(S_3) = \mathbf{Z}/12\mathbf{Z}$ and $\pi_7(S_3) = \mathbf{Z}/2\mathbf{Z}$. Other results on the Freudenthal suspension then allowed him to determine the groups $\pi_{n+3}(S_n)$ and $\pi_{n+4}(S_n)$ completely; later he also determined $\pi_{n+k}(S_n)$ for $5 \leqslant k \leqslant 8$ [432].

D. Nonexistence Theorems

The discovery of unsuspected cohomology operations provided new ways of proving that two spaces do not necessarily have the same homotopy type even if their cohomology algebras and their fundamental groups are isomorphic. Let ξ be a cohomology operation of type $(q, n, \mathrm{A}, \mathrm{B})$; consider a continuous map $f\colon \mathrm{Y} \to \mathrm{X}$ and the corresponding commutative diagram (§ 3,A)

$$\begin{array}{ccc} \mathrm{H}^q(\mathrm{X};\mathrm{A}) & \xrightarrow{f^*} & \mathrm{H}^q(\mathrm{Y};\mathrm{A}) \\ \Big\downarrow{\scriptstyle \xi(\mathrm{X})} & & \Big\downarrow{\scriptstyle \xi(\mathrm{Y})} \\ \mathrm{H}^n(\mathrm{X};\mathrm{B}) & \xrightarrow[f^*]{} & \mathrm{H}^n(\mathrm{Y};\mathrm{B}) \end{array}$$

If there is $u \neq 0$ in $\mathrm{H}^q(\mathrm{X};\mathrm{A})$ such that $\xi(\mathrm{X})(u) = 0$, but $\xi(\mathrm{Y})(v) \neq 0$ for $v \neq 0$ in $\mathrm{H}^q(\mathrm{Y};\mathrm{A})$, then f^* cannot be injective.

Thom gave such an example ([462], p. 142): he considered the product $\mathbf{S}_2 \times \mathbf{S}_k$ for $k > 2$ and the (unique) nontrivial fiber space E with base space $\mathbf{S}_2$ and typical fiber $\mathbf{S}_k$; their cohomology algebras are isomorphic, but $\mathrm{Sq}^2(1 \times s_k) \neq 0$ in $\mathrm{H}^\bullet(\mathrm{E};\mathbf{F}_2)$, whereas $\mathrm{Sq}^2(1 \times s_k) = 0$ in $\mathrm{H}^\bullet(\mathbf{S}_2 \times \mathbf{S}_k;\mathbf{F}_2)$.

Another example was given by Steenrod: consider the real projective spaces

$$\mathbf{P}_5(\mathbf{R}) \supset \mathbf{P}_4(\mathbf{R}) \supset \mathbf{P}_2(\mathbf{R}).$$

Let $\mathrm{X} = \mathbf{P}_5(\mathbf{R})/\mathbf{P}_2(\mathbf{R})$ be the "truncated (or "stunted") projective space" obtained by collapsing $\mathbf{P}_2(\mathbf{R})$ to a point (Part 2, chap. V, § 2,A), and let $f\colon \mathbf{P}_5(\mathbf{R}) \to \mathrm{X}$ be the collapsing map and A be the image $f(\mathbf{P}_4(\mathbf{R}))$ in X, homeomorphic to $\mathbf{P}_4(\mathbf{R})/\mathbf{P}_2(\mathbf{R})$. Then there is no retraction of X onto A, although there are algebra homomorphisms

$$\varphi\colon \mathrm{H}^\bullet(\mathrm{A};\mathbf{F}_2) \to \mathrm{H}^\bullet(\mathrm{X};\mathbf{F}_2), \qquad \psi\colon \mathrm{H}^\bullet(\mathrm{X};\mathbf{F}_2) \to \mathrm{H}^\bullet(\mathrm{A};\mathbf{F}_2)$$

such that $\psi \circ \varphi = \mathrm{Id}$. This is because if $u \neq 0$ in $\mathrm{H}^3(\mathrm{X};\mathbf{F}_2)$, one has $\mathrm{Sq}^2(u) \neq 0$, whereas $\mathrm{H}^5(\mathrm{A};\mathbf{F}_2) = 0$.

We postpone to § 5,B other kinds of nonexistence theorems, which Adem was first to deduce from his relations between Steenrod squares, and which a little later were greatly generalized by J.F. Adams in his work on the maps of spheres with Hopf invariant 1. A. Borel and Serre also obtained interesting results by computing the Steenrod reduced powers for the classical compact Lie groups [68]; this is easily done, using the known description of the cohomology algebras of these groups and the Cartan–Steenrod formulas.

The most interesting consequence of these computations is the fact that on a sphere $\mathbf{S}_{2n}$ with $n \geqslant 4$ *there does not exist a structure of almost complex manifold*, which means that on the tangent bundle $\mathrm{T}(\mathbf{S}_{2n})$ there is no structure of *complex* vector bundle of rank n for which the underlying structure of *real* vector bundle would be the usual one. The proof is by contradiction: there would exist Chern classes $c_i \in H^{2i}(S_{2n}; \mathbf{Z})$, which would be images of the Chern classes $C_i \in H^{2i}(B_{U(n)}; \mathbf{Z})$ by f^*, for a continuous map $f: S_{2n} \to B_{U(n)}$. Using the structure of the cohomology of $B_{U(n)}$ determined by Borel (chap. IV, § 4) and the fact that $H^{2i}(S_{2n}; \mathbf{Z}) = 0$ for $0 < i < n$, they showed that the Chern class c_n should be divisible by an odd prime p, owing to the computations of the reduced Steenrod p-powers in $H^\bullet(B_{U(n)}; \mathbf{F}_p)$. However, c_n would be the Euler class $e(T(S_{2n}))$ [chap. IV, § 3,D, formula (26)] and if u is the fundamental cohomology class of S_{2n}, $e(T(S_{2n})) = \chi(S_{2n})u = 2u$ [chap. IV, § 3,E, formula (36)], giving the required contradiction.

Another result concerns the existence of sections of a fibration (E, S_r, F, π) with base S_r. Let β be the fundamental cohomology class in $H^r(S_r; \mathbf{F}_p)$; suppose $\gamma = \pi^*(\beta) \in H^\bullet(E: \mathbf{F}_p)$ is equal to a sum of cup-products of reduced p-powers of elements of $H^\bullet(E; \mathbf{F}_p)$ of dimension $< r$; then E cannot have a section $s: S_r \to E$. This follows from the fact that $s^* \circ \pi^* = \text{Id.}$, so $\beta = s^*(\pi^*(\beta)) = s^*(\gamma)$ would similarly be a sum of cup-products of reduced p-powers of elements of $H^\bullet(S_r; \mathbf{F}_p)$ of dimension $< r$ that are all 0, hence one would have the contradiction $\beta = 0$. In this way Borel and Serre proved that the fibration $(SU(n), S_{2n-1}, SU(n-1))$ has no section for $n \geqslant 3$ and similar results for the groups $SO(n)$ and $U(n; \mathbf{H})$.

§ 5. Secondary Cohomology Operations

A. The Notion of Secondary Cohomology Operation

The idea that led to secondary cohomology operations was first formulated by Adem in his work on the Steenrod squares [6]. One of his results in that Note was a continuation of Steenrod's investigations on the cohomological characterization of the set $[X; S_n]$ of homotopy classes when $\dim X = n + 1$ (§§ 1 and 4,A); Adem tackled the next case, when $\dim X = n + 2$. Suppose that a continuous map $f: X_n \to S_n$ defined in the n-skeleton of X can be extended to a continuous map $f': X_{n+2} \to S_n$; then the secondary obstruction

$$Sq^2(f^*(s_n)) = 0$$

by Steenrod's result, and the obstruction $(n + 3)$-cocycle $b_{n+3}(f')$ can be defined. Here (contrary to Steenrod's case) it depends on the particular extension f' chosen, but the cohomology classes of all these cocycles form a *coset* in the quotient space

$$H^{n+3}(X; \mathbf{F}_2)/Sq^2(H^{n+1}(X; \mathbf{F}_2))$$

that only depends on f and is called the *tertiary obstruction* to the extension of f. Adem found that the map that sends $u = f^*(s_n)$ to the tertiary obstruction can be defined for *all* cohomology classes u in $H^n(X; \mathbf{F}_2)$ for which $Sq^2 u = 0$ (without reference to a map f), and he showed that there is a natural homomorphism Φ from the *kernel* of

$$\mathrm{Sq}^2 \colon \mathrm{H}^n(\mathrm{X}; \mathbf{F}_2) \to \mathrm{H}^{n+2}(\mathrm{X}; \mathbf{F}_2)$$

to the *cokernel* of

$$\mathrm{Sq}^2 \colon \mathrm{H}^{n+1}(\mathrm{X}; \mathbf{F}_2) \to \mathrm{H}^{n+3}(\mathrm{X}; \mathbf{F}_2);$$

he called that homomorphism a *secondary cohomology operation.*

Adem obtained the existence and uniqueness of Φ by the use of the relations he had found between the Sq^i. This method was greatly generalized by J.F. Adams in his work of 1958 on the Hopf invariant ([1], [2]).

B. General Constructions

Adams starts from the graded Steenrod algebra $\mathscr{A}_2$ over $\mathbf{F}_2$ (§ 3,D). Consider a pair of free left graded $\mathscr{A}_2$-modules $\mathrm{C}_0, \mathrm{C}_1$ and a graded $\mathscr{A}_2$-homomorphism

$$d \colon \mathrm{C}_1 \to \mathrm{C}_0.$$

Let $(c_{0,\lambda}), (c_{1,\mu})$ be bases of C_0 and C_1, respectively, consisting of homogeneous elements, with $\deg c_{0,\lambda} = l_\lambda$, $\deg c_{1,\mu} = m_\mu$; write

$$d(c_{1,\mu}) = \sum_\lambda a_{\mu\lambda} c_{0,\lambda} \quad \text{with } a_{\mu\lambda} \in \mathscr{A}_2.$$

Finally, let $z = \sum_\mu b_\mu c_{1,\mu}$ be an element of $\mathrm{Ker}(d)$, with $b_\mu \in \mathscr{A}_2$.

Let $\mathrm{H}^+(\mathrm{X}) = \bigoplus_{i \geqslant 1} \mathrm{H}^i(\mathrm{X}; \mathbf{F}_2)$ for any space X, so that $\mathrm{H}^+(\mathrm{X})$ is a graded left $\mathscr{A}_2$-module. Define two $\mathscr{A}_2$-modules $\mathrm{D}^n(d, \mathrm{X})$, $\mathrm{Q}^n(z, \mathrm{X})$ for each integer $n \geqslant 0$:

$\mathrm{D}^n(d, \mathrm{X})$ is the module consisting of all graded $\mathscr{A}_2$-homomorphisms $e \colon \mathrm{C}_0 \to \mathrm{H}^+(\mathrm{X})$ of degree n such that $e \circ d = 0$. If

$$e(c_{0,\lambda}) = x_\lambda \in \mathrm{H}^{n+l_\lambda}(\mathrm{X}; \mathbf{F}_2),$$

$\mathrm{D}^n(d, \mathrm{X})$ may be identified with the submodule of $\bigoplus_\lambda \mathrm{H}^{n+l_\lambda}(\mathrm{X}; \mathbf{F}_2)$ consisting of the families (x_λ) such that

$$\sum_\mu a_{\mu\lambda} \cdot x_\lambda = 0.$$

$\mathrm{Q}^n(z, \mathrm{H}) \subset \mathrm{H}^+(\mathrm{X})$ consists of the elements $\xi(z)$ where $\xi \colon \mathrm{C}_1 \to \mathrm{H}^+(\mathrm{X})$ is any graded $\mathscr{A}_2$-homomorphism of degree $n - 1$; therefore

$$\mathrm{Q}^n(z, \mathrm{X}) = \sum_\mu b_\mu \,.\, \mathrm{H}^{n+m_\mu-1}(\mathrm{X}; \mathbf{F}_2).$$

Adams defined a family $(\Phi_{n,\mathrm{X}})_{n \geqslant 0}$ of *stable secondary cohomology operations* associated to d and z as a family of maps

$$\Phi_{n,\mathrm{X}} \colon \mathrm{D}^n(d, \mathrm{X}) \to \mathrm{H}^+(\mathrm{X})/\mathrm{Q}^n(z, \mathrm{X})$$

satisfying three conditions:

I. *Functoriality*: for any continuous map $f \colon \mathrm{X} \to \mathrm{Y}$, the diagram

$$\begin{array}{ccc} D^n(d, X) & \xrightarrow{\Phi_{n,X}} & H^+(X)/Q^n(z, X) \\ {\scriptstyle f^*}\Big\uparrow & & \Big\uparrow{\scriptstyle \bar{f}^*} \\ D^n(d, Y) & \xrightarrow[\Phi_{n,Y}]{} & H^+(Y)/Q^n(z, Y) \end{array}$$

is commutative ($\bar{f}^*$ being the map deduced from f^* by passage to quotients).

II. *Commutation with the suspension*: the diagram

$$\begin{array}{ccc} D^n(d, X) & \xrightarrow{\Phi_{n,X}} & H^+(X)/Q^n(z, X) \\ \Big\uparrow{\scriptstyle \sigma} & & \Big\uparrow{\scriptstyle \bar{\sigma}} \\ D^{n+1}(d, SX) & \xrightarrow[\Phi_{n+1,SX}]{} & H^+(SX)/Q^{n+1}(z, SX) \end{array}$$

is commutative [σ and $\bar{\sigma}$ are induced by the natural isomorphisms $H^{m+1}(SX; \mathbf{F}_2) \to H^m(X; \mathbf{F}_2)$].

III. For *any* continuous injective map i: $Y \to X$ such that $i^* \circ e = 0$ and *any* pair of graded homomorphisms

$$\eta: C_0 \to H^+(X, Y) \quad \text{of degree } n$$

$$\zeta: C_1 \to H^+(Y) \quad \text{of degree } n - 1$$

such that the diagram

$$\begin{array}{ccccccccc} H^+(X) & \xrightarrow{i^*} & H^+(Y) & \xrightarrow{\partial} & H^+(X, Y) & \xrightarrow{j^*} & H^+(X) & \xrightarrow{i^*} & H^+(Y) \\ & & \Big\uparrow{\scriptstyle \zeta} & & \Big\uparrow{\scriptstyle \eta} & \overset{e}{\nearrow} & & & \\ & & C_1 & \xrightarrow[d]{} & C_0 & & & & \end{array}$$

is commutative,

$$\bar{i}^*(\Phi_{n,X}(e(c_{0,\lambda}))) = \zeta(c_{0,\lambda}) + i^*(Q^n(z, X)).$$

The existence of such operations $\Phi_{n,X}$ is proved by using "universal spaces" as in the determination of "primary" cohomology operations (§ 3,A); instead of Eilenberg–Mac Lane spaces, here it is necessary to use fibrations with base space and typical fiber equal to suitable products of Eilenberg–Mac Lane spaces ([2], and [425], pp. 13-07 to 13-29).

Adams showed that if $(\Phi_{n,X})$ and $(\Phi'_{n,X})$ are two families of secondary cohomology operations corresponding to the same data (C_0, C_1, d, z) and if $\psi = \Phi_{n,X} - \Phi'_{n,X}$, there is a homomorphism

$$\varphi: D^n(d, X) \to H^+(X)$$

such that each $\psi((x_\lambda))$ is in the class of $\varphi((x_\lambda))$ modulo $Q^n(z, X)$.

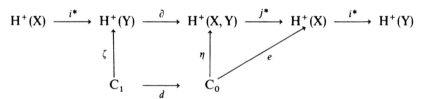

C. Special Secondary Cohomology Operations

Adams applied his general construction to define a special kind of secondary cohomology operations, in order to use them in his solution of the Hopf invariant problem (see Section D). He considered $\mathbf{F}_2$ an $\mathscr{A}_2$-module, the operation $(a, 1) \mapsto a.1$ being such that its value is 0 if $\deg a > 0$, and 1 if $a = 1$. He then considers a limited free resolution of that $\mathscr{A}_2$-module (Part 1, chap. IV, § 5,F)

$$0 \leftarrow \mathbf{F}_2 \overset{\varepsilon}{\leftarrow} \mathrm{C}_0 \overset{d_1}{\longleftarrow} \mathrm{C}_1 \overset{d_2}{\longleftarrow} \mathrm{C}_2 \tag{109}$$

constructed in the following way. First $\mathrm{C}_0 = \mathscr{A}_2$, and the augmentation ε is defined by $\varepsilon(1) = 1$. Next C_1 is defined as a free graded $\mathscr{A}_2$-module with one basis element c_i for each degree $i \geqslant 0$, and d_1 is defined by

$$d_1 c_i = \mathrm{Sq}^{2^i}.$$

Finally C_2 is a free graded $\mathscr{A}_2$-module with homogeneous basis elements c_{ij} for all pairs of indices such that $0 \leqslant i \leqslant j$ and $j \neq i + 1$; d_2 is defined by

$$d_2 c_{ij} = \mathrm{Sq}^{2^i}.c_j + \sum_{0 \leqslant k < j} b_k . c_k$$

where the elements $b_k \in \mathscr{A}_2$ are elements that appear in one Adem relation (§ 3,C) of the form

$$\mathrm{Sq}^{2^i}\mathrm{Sq}^{2^j} + \sum_{0 \leqslant k < j} b_k \mathrm{Sq}^{2^k} = 0 \quad \text{for } 0 \leqslant i \leqslant j, j \neq i + 1;$$

there may be several such relations for given i and j, just pick one; the b_k are homogeneous, and

$$\deg b_k \leqslant 2^i + 2^j - 1$$

so that $d_1(d_2(c_{ij})) = 0$.

For each $j \geqslant 0$, let $\mathrm{C}_1(j)$ be the submodule of C_1 generated by the c_k with $0 \leqslant k \leqslant j$ and $d_1(j)$ be the restriction of d_1 to $\mathrm{C}_1(j)$. The general construction of section B is applied by specializing the C_1 of that construction to $\mathrm{C}_1(j)$, the map d to $d_1(j)$, and the element z to $z_{ij} = d_2 c_{ij}$.

The secondary cohomology operation $\Phi_{ij,\mathrm{X}}$ obtained in this fashion is defined on the subspace $\mathrm{D}^n_{ij}(\mathrm{X})$ that consists of the elements $x \in \mathrm{H}^n(\mathrm{X}; \mathbf{F}_2)$ such that $\mathrm{Sq}^{2^k} x = 0$ for $0 \leqslant k \leqslant j$; it takes its value in the quotient

$$\mathrm{H}^{n+2^i+2^j-1}(\mathrm{X}; \mathbf{F}_2)/\mathrm{Q}^n_{ij}(\mathrm{X})$$

where

$$\mathrm{Q}^n_{ij}(\mathrm{X}) = \mathrm{Sq}^{2^i}.\mathrm{H}^{n+2^j-1}(\mathrm{X}; \mathbf{F}_2) + \sum_{0 \leqslant k < j} b_k.\mathrm{H}^{n+2^k-1}(\mathrm{X}; \mathbf{F}_2).$$

Note that the operation Φ defined by Adem [6] is equal to $\Phi_{1,1}$ in the preceding notation.

D. The Hopf Invariant Problem

Recall that when Hopf defined the Hopf invariant $H(f)$ for a continuous map $f: \mathbf{S}_{2n-1} \to \mathbf{S}_n$ for $n \geqslant 1$ (chap. II, § 1,C) he asked for conditions on n that would imply the possibility of having $H(f) = \pm 1$ and showed by explicit examples that the values $n = 2$, 4, and 8 have that property.

The problem attracted several mathematicians. In 1950 G.W. Whitehead [486], using his study of the homotopy of the wedge $\mathbf{S}_n \vee \mathbf{S}_n$, proved that a necessary condition for the existence of an f with $H(f) = \pm 1$ is that $n = 2$ or $n \equiv 0 \pmod{4}$.

The use of Steenrod squares brought a decisive turn in the problem. Steenrod had given a cohomological definition of the Hopf invariant (chap. V, § 5,D): using the mapping cylinder Z_f, it followed from that definition [*loc. cit.*, formula (103)] that the relation $H(f) = \pm 1$ implies that the map

$$H^n(Z_f; \mathbf{F}_2) \to H^{2n}(Z_f; \mathbf{F}_2) \tag{110}$$

defined by $u \mapsto u \smile u$ is $\neq 0$. But that map is exactly Sq^n [§ 1,B, formula (19)]. Therefore, the condition $H(f) = \pm 1$ *cannot be satisfied if the map*

$$Sq^n\colon H^n(X; \mathbf{F}_2) \to H^{2n}(X; \mathbf{F}_2) \tag{111}$$

is 0 *for all spaces* X *such that* $H^1(X; \mathbf{F}_2) = 0$ *for* $n < i < 2n$ (which is the case for Z_f).

Adem had the original idea of using the relations he had found between the Steenrod squares to find values of n for which the map (111) is *always zero* for the spaces under consideration. He showed that the only exceptions would necessarily be *powers of* 2 [7]. As an example, the proof that the map (111) is 0 for $n = 6$ relies on the relation

$$Sq^6 = Sq^2\, Sq^4 + Sq^5\, Sq^1.$$

Since $H^i(X; \mathbf{F}_2) = 0$ for $i = n + 4$ and $i = n + 1$, $Sq^6(H^n(X; \mathbf{F}_2)) = 0$.

The argument does not apply if n is a power of 2 and Adams undertook to generalize Adem's use of secondary cohomology operations in that case to obtain an expression for Sq^{2^k} in terms of these operations. His result was that if a space X is such that

$$H^i(X; \mathbf{F}_2) = 0 \quad \text{for } m < i < m + n, \tag{112}$$

then the map

$$Sq^n\colon H^m(X; \mathbf{F}_2) \to H^{m+n}(X; \mathbf{F}_2) \tag{113}$$

is 0 *provided that n is not* 2, 4, *or* 8; this establishes in particular that the *only* values of n for which $H(f) = \pm 1$ are 2, 4, or 8.

Adams' proof is extremely long and intricate; it can be divided into three parts.

The first part establishes a relation valid for *every* space X: there is a $\lambda \in \mathbf{F}_2$

(thus equal to 0 or 1) and elements $a_{ijk} \in \mathscr{A}_2$, such that if $u \in \mathrm{H}^m(\mathrm{X}; \mathbf{F}_2)$ satisfies the relations

$$\mathrm{Sq}^{2^r} u = 0 \quad \text{for } 0 \leqslant r \leqslant k, \tag{114}$$

then

$$\lambda[\mathrm{Sq}^{2^{k+1}} u] = \sum_{\substack{0 \leqslant i \leqslant j \leqslant k \\ j \neq i+1}} a_{ijk} \cdot \Phi_{ij,\mathrm{X}}(u), \quad \textit{provided } k \geqslant 3, \tag{115}$$

where $[\mathrm{Sq}^{2^{k+1}} u]$ is the class modulo $\sum_{i,j,k} a_{ijk} \cdot \mathrm{Q}_{ij}^m(\mathrm{X})$.

The proof needs some heavy machinery from homological algebra. The sequence (109) is the beginning of a free resolution of the $\mathscr{A}_2$-module $\mathbf{F}_2$. Because of the construction of the $\Phi_{ij,\mathrm{X}}$, relations such as (115) are obtained by a detailed investigation of the *cohomology algebra* $\mathrm{H}^{\cdot}(\mathscr{A}_2)$ as defined by Hochschild and Cartan–Eilenberg [113]; in particular it is necessary to have precise information on $\mathrm{H}^i(\mathscr{A}_2)$ for $i = 1, 2, 3$. This is obtained by using the results of Milnor on the structure of Hopf algebra of $\mathscr{A}_2$ (§ 3,D). It is also necessary to consider an increasing sequence (B'_n) of sub-Hopf algebras of the dual Hopf algebra $\mathscr{A}_2^*$ that is a direct limit of the B'_n; $\mathrm{H}^{\cdot}(\mathscr{A}_2)$ can be studied by considering the algebras $\mathrm{H}^{\cdot}(\mathrm{B}_n)$, where B_n is the Hopf algebra dual to B'_n; the passage from $\mathrm{H}^{\cdot}(\mathrm{B}_{n-1})$ to $\mathrm{H}^{\cdot}(\mathrm{B}_n)$ is obtained by defining an appropriate filtration and using the corresponding spectral sequence; the latter is similar to the one Adams had introduced earlier for the study of stable homotopy (see chap. VII, § 5,D).

The second part of the proof consists in showing that in (115) the coefficient λ is *equal to* 1 by choosing a particular space X for which both sides of (115) may be computed for well chosen elements u. Adams takes for X the infinite-dimensional complex projective space $\mathbf{P}_\infty(\mathbf{C})$ (chap. II, § 6,F) that has the homotopy type of $\mathrm{K}(\mathbf{Z}, 2)$. The cohomology algebra $\mathrm{H}^{\cdot}(\mathrm{X}; \mathbf{F}_2)$ is then a polynomial algebra $\mathbf{F}_2[y]$ with $y \in \mathrm{H}^2(\mathrm{X}; \mathbf{F}_2)$, and $u = y^{2^j}$, at first for any integer $j \geqslant 0$. Preliminary computations show that $\Phi_{ij}(y^t)$ is defined if and only if $t \equiv 0 \,(\mathrm{mod}\,.\, 2^j)$, that it is a class consisting of a single element, and that element is 0 unless $i = 0$ and $j \geqslant 2$. A double induction on j and t then gives

$$\Phi_{0,j}(y^{2^j t}) = t y^{2^j t + 2^j - 1}.$$

For $k \geqslant 3$, there is then a relation between *elements* of $\mathrm{H}^{\cdot}(\mathrm{X}; \mathbf{F}_2)$, not merely between classes modulo a subgroup:

$$\lambda\, \mathrm{Sq}^{2^{k+1}}(y^{2^k}) = a_{0,k,k} \cdot \Phi_{0,k}(y^{2^k}) = a_{0,k,k} \cdot y^{3 \cdot 2^{k-1}}.$$

Finally the right-hand side is shown to be equal to $y^{2^{k+1}}$ and the left-hand side to $\lambda y^{2^{k+1}}$, hence $\lambda = 1$.

The third part of the proof now follows the pattern of Adem's proof. If n is not a power of 2, Sq^n is decomposable as a sum or products of Sq^j with $j < n$; from the assumption on the groups $\mathrm{H}^{m+j}(\mathrm{X}; \mathbf{F}_2)$ for $j < n$, it follows that Sq^n is 0 in $\mathrm{H}^m(\mathrm{X}; \mathbf{F}_2)$. If $n = 2^{k+1}$ with $k \geqslant 3$, $\mathrm{Sq}^{2^s}(u) = 0$ for $u \in \mathrm{H}^m(\mathrm{X}; \mathbf{F}_2)$ and $0 \leqslant s \leqslant k$, so (115) can be applied; but the $\Phi_{ij,\mathrm{X}}(u)$ belong to a quotient of

$$\bigoplus_{i,j} \mathrm{H}^{m-1+2^i+2^j}(\mathrm{X}; \mathbf{F}_2)$$

with $2^i + 2^j - 1 < 2^{k+1}$ since $i \leqslant j \leqslant k$ and $i + 1 \neq j$. So again the right-hand side of (115) is 0, and this ends the proof of the vanishing of the map (113).

In 1966 Adams and Atiyah found a very simple proof of the vanishing of the map (111), based on K-theory (chap. VII, § 5,A).

E. Consequences of Adams' Theorem

It had earlier been recognized that the existence of continuous maps f: $\mathbf{S}_{2n+1} \to \mathbf{S}_{n+1}$ with $\mathrm{H}(f) = \pm 1$ is equivalent to other properties of $\mathbf{S}_n$:

1. If ι_n is the class of the identity in $\pi_n(\mathbf{S}_n)$, then the Whitehead product $[\iota_n, \iota_n]$ (chap. II, § 3,E) is 0.
2. There is a structure of H-space (Part 2, chap. VI, § 2,A) on $\mathbf{S}_n$.
3. The Freudenthal suspension E: $\pi_{2n-1}(\mathbf{S}_n) \to \pi_{2n}(\mathbf{S}_{n+1})$ is bijective.

Indeed, G.W. Whitehead [488] proved that in the exact sequence

$$\pi_{2n+1}(\mathbf{S}_{n+1}) \xrightarrow{\mathrm{H}} \underset{\substack{\wr\Vert \\ \mathbf{Z}}}{\pi_{2n+1}(\mathbf{S}_{2n+1})} \xrightarrow{\partial} \pi_{2n-1}(\mathbf{S}_n) \xrightarrow{\mathrm{E}} \pi_{2n}(\mathbf{S}_{n+1})$$

[chap. V, § 5,D), formula (105)], the image of ∂, equal to the kernel of E, is the subgroup generated by $[\iota_n, \iota_n]$ in $\pi_{2n-1}(\mathbf{S}_n)$. Hence $[\iota_n, \iota_n] = 0$, or equivalently E is injective [hence bijective by Freudenthal's theorem (chap. II, § 6,E)] only if H is *surjective*, i.e., $n = 1$, 3, or 7 by Adams' theorem.

If $\mathbf{S}_n$ has a structure of H-space, the maps f_1: $x \mapsto xe$ and f_2: $x \mapsto ex$ have degree 1 and $f_1 \vee f_2$: $\mathbf{S}_n \vee \mathbf{S}_n \to \mathbf{S}_n$ has a continuous extension $\mathbf{S}_n \times \mathbf{S}_n \to \mathbf{S}_n$ by definition. The definition of the Whitehead product then implies that $[f_1, f_2] = 0$, hence $[\iota_n, \iota_n] = 0$, so $n = 1$, 3, or 7.

Finally, it was proved by Dold that if $\mathbf{S}_n$ is given a differential structure compatible with its topology (it may be the usual ones or "exotic" ones, see chap. VII, § 2,B), then $\mathbf{S}_n$ is *parallelizable* for that structure if and only if it has a structure of H-space, so again $n = 1$, 3, or 7. Milnor, Bott, and Kervaire obtained proofs for that result using Bott's work on homotopy of compact Lie groups ([82], [272]).

§ 6. Cohomotopy Groups

Homotopy groups arise from the consideration of the *sets* of homotopy classes

$$[\mathbf{S}_n, *; \mathrm{X}, x_0]$$

for pointed spaces (X, x_0) and the possibility on defining naturally a group structure on such a set. Similarly, in 1936 Borsuk [74] considered the *sets* of homotopy classes

$$[\mathrm{X}, x_0; \mathbf{S}, *] \tag{116}$$

and showed that for *some* values of n they could naturally be given a group structure; his results were later extended by Spanier [439].

A. Cohomotopy Sets

Let (X, A) be a pair of pointed spaces; recall that the set

$$[X, A; S_n, *] \tag{117}$$

is the set of homotopy classes $[f]$ of continuous maps $f: X \to S_n$ such that $f(A) = \{*\}$; for a homotopy $(x, t) \mapsto F(x, t)$ between two such maps $F(x, t) = \{*\}$ for all $x \in A$ and $t \in [0, 1]$. The set (117) is called the *n-th cohomotopy set* of the pair (X, A) and written $\pi^n(X, A)$; it must be considered a pointed set with a privileged element, the homotopy class of the constant map $X \to \{*\}$, usually written 0.

It is clear that $(X, A) \mapsto \pi^n(X, A)$ is a *contravariant functor*: for any map $f: (X, A) \to (Y, B)$ of pairs of spaces,

$$f^*: \pi^n(Y, B) \to \pi^n(X, A)$$

is defined by the relation

$$f^*([u]) = [u \circ f]$$

for every map $u: (Y, B) \to (S_n, *)$, and $f^*(0) = 0$. When X is compact, Y is paracompact, A is closed in X, B is closed in Y, and the restriction $f|(X - A)$: $X - A \to Y - B$ is a *homeomorphism*, it can be shown that f^* is *bijective* ([254], p. 207).

Write

$$\pi^n(X) = \pi^n(X, \varnothing). \tag{118}$$

When X is *paracompact* and A is closed in X, it is possible to define a map of pointed sets

$$\partial: \pi^n(A) \to \pi^{n+1}(X, A) \tag{119}$$

such that in the infinite sequence

$$\begin{aligned}\pi^1(X, A) \xrightarrow{j^*} \pi^1(X) \xrightarrow{i^*} \pi^1(A) \xrightarrow{\partial} \pi^2(X, A) \to \cdots \pi^m(A) \xrightarrow{\partial} \pi^{m+1}(X, A)\\ \xrightarrow{j^*} \pi^{m+1}(X) \xrightarrow{i^*} \pi^{m+1}(A) \to \cdots\end{aligned} \tag{120}$$

the composite of two successive maps has image $\{0\}$ [the image of each map is contained in the "kernel" (chap. II, § 5,D) of the next one]; $i: A \to X$ is the usual injection and $j: X \to (X, A)$ is the identity.

To define the map (119), consider the hemispheres D_{n+1}^+ and D_{n+1}^- of S_{n+1}, defined, respectively, by $x_{n+2} \geqslant 0$ and $x_{n+2} \leqslant 0$. Take $\partial = \gamma^{-1} \circ \beta \circ \alpha^{-1}$, where α and γ are two natural bijections

$$\alpha: [X, A; D_{n+1}^+, S_n] \to [A; S_n],$$

$$\gamma: [X, A; S_{n+1}, *] \to [X, A; S_{n+1}, D_{n+1}^-],$$

and

$$\beta\colon [X, A; D_{n+1}^+, S_n] \to [X, A; S_{n+1}, D_{n+1}^-]$$

is the map associating to the class $[f]$ of a map $f\colon (X, A) \to (D_{n+1}^+, S_n)$ the class $[h \circ f]$, where

$$h\colon (D_{n+1}^+, S_n) \to (S_{n+1}, D_{n+1}^-)$$

is the natural injection. The fact that α is surjective is a consequence of the Tietze–Urysohn extension theorem, since D_{n+1}^+ is homeomorphic to $[0, 1]^{n+1}$; the fact that it is injective is a consequence of the homotopy extension property proved by Borsuk, since D_{n+1}^+ is an ANR (chap. II, § 2,D). The bijectivity of γ follows at once from the fact that D_{n+1}^- is contractible to the point $*$ in S_{n+1}. The verification of the relations $\partial \circ i^* = 0$ and $j^* \circ \partial = 0$ are straightforward, and $i^*([g]) = 0$ if and only if $[g] = j^*([f])$ for some $f\colon (X, A) \to (S_n, *)$. In general the sequence (120) of pointed sets is not necessarily exact at $\pi^m(A)$ and $\pi^{m+1}(X, A)$ in the sense of the category ***PSet*** (chap. II, § 5,D) ([254], p. 227).

B. Cohomotopy Groups

For simplicity's sake, we shall now assume that (X, A) is a compact *relative* CW-*complex* of finite dimension. Then *for* $2m - 1 > \dim(X, A)$ (Part 2, chap. V, § 3,C), there is a natural structure of *commutative group* on the cohomotopy set $\pi^m(X, A)$. Consider the product space $S_m \times S_m$ a pointed space with $(*, *)$ as a privileged point and its closed suspace, the wedge

$$S_m \vee S_m = (S_m \times \{*\}) \cup (\{*\} \times S_m).$$

Let (α, β) be any pair of elements in $\pi^m(X, A) \times \pi^m(X, A)$, with $\alpha = [f]$, $\beta = [g]$. Consider the map $(f, g)\colon x \mapsto (f(x), g(x))$ of X into $S_m \times S_m$; if $x \in A$, $(f(x), g(x)) = (*, *)$. Now the injection $i\colon (S_m \vee S_m, (*, *)) \to (S_m \times S_m, (*, *))$ defines a *bijection* $i_*\colon [X, A; S_m \vee S_m, (*, *)] \to [X, A; S_m \times S_m, (*, *)]$ associating to the class $[u]$ of $u\colon (X, A) \to (S_m \vee S_m, (*, *))$ the class of $[i \circ u]$. This follows from the fact that $\pi_r(S_m \times S_m, S_m \vee S_m) = 0$ for $r < 2m$ (chap. II, § 4,C), (X, A) is a relative CW-complex of dimension $2m - 2$, and $(X \times I, A \times I)$ is a relative CW-complex of dimension $< 2m - 1$; it is therefore possible, by "climbing" on the skeletons, to extend a continuous map $(X, A) \to (S_m \vee S_m, (*, *))$ [resp. $(X \times I, A \times I) \to (S_m \vee S_m, (*, *))$] to a continuous map $(X, A) \to (S_m \times S_m, (*, *))$ [resp. $(X \times I, A \times I) \to (S_m \times S_m, (*, *))$] (chap. II, § 3,C).

Next consider the natural map $j\colon S_m \vee S_m \to S_m$ equal to $(x, *) \mapsto x$ in $S_m \times \{*\}$ and to $(*, x) \mapsto x$ in $\{*\} \times S_m$, and let j_* be the map which associates the class $[j \circ u]$ to the class of $u\colon (X, A) \to (S_m \vee S_m, (*, *))$. Finally, let

$$h\colon \pi^m(X, A) \times \pi^m(X, A) \to [X, A; S_m \vee S_m, (*, *)]$$

be the map which, to every pair (α, β) of elements $\alpha = [f]$, $\beta = [g]$ of $\pi^m(X, A)$ associates the class $[(f, g)]$, which only depends on α and β. Then define a map $s\colon \pi^m(X, A) \times \pi^m(X, A) \to \pi^m(X, A)$ by

$$s = j_* \circ i_*^{-1} \circ h; \tag{121}$$

s straightforwardly defines a structure of commutative group on $\pi^m(X, A)$, written $s(\alpha, \beta) = \alpha + \beta$; the element $0 \in \pi^m(X, A)$ defined in section A is the neutral element of that group. For $\alpha \in \pi^m(X, A)$, the element $-\alpha$ is defined in the following way: if $\alpha = [f]$, $-\alpha = [r \circ f]$, where $r: \mathbf{S}_m \to \mathbf{S}_m$ is a continuous map of degree -1 ([254], pp. 211–212).

Suppose X is now a finite CW-complex of dimension $< 2n - 1$ and A is a CW-subcomplex of X. Then $\pi^m(X)$ and $\pi^m(A)$ are also commutative groups for $m \geqslant n$, the maps in the sequence

$$\pi^n(X, A) \xrightarrow{j^*} \pi^n(X) \to \cdots \to \pi^m(A) \xrightarrow{\partial} \pi^{m+1}(X, A) \xrightarrow{j^*} \pi^{m+1}(X) \xrightarrow{i^*} \cdots \tag{122}$$

are homomorphisms of groups, and the sequence is *exact*. Spanier, under the same assumptions, defined two mappings between cohomotopy and cohomology groups. First, for $m \geqslant n$ there is a homomorphism

$$\Phi: \pi^m(X, A) \to H^m(X, A; \pi_n(\mathbf{S}_n)). \tag{123}$$

It is defined in the following way: if $\alpha \in \pi^m(X, A)$ is the class $[f]$ of a map $f: (X, A) \to (\mathbf{S}_m, *)$, the element $f^*(s_m)$, the image of the fundamental cohomology class $s_m \in H^m(\mathbf{S}_m; \mathbf{Z})$, only depends on the class α, and $\Phi(\alpha) = f^*(s_m)$; it is convenient to identify $\mathbf{Z}$ and $\pi_n(\mathbf{S}_n)$.

The second map goes instead from cohomology to cohomotopy:

$$\Lambda: H^{n+2}(X, A; \pi_{n+2}(\mathbf{S}_{n+1})) \to \pi^{n+1}(X, A) \tag{124}$$

and is only defined for $\dim(X, A) \leqslant n + 2$. Let u be a cohomology class in $H^{n+2}(X, A; \pi_{n+2}(\mathbf{S}_{n+1}))$, and c be a cocycle in the class u; therefore, for any $(n + 2)$-cell σ in X, $c(\sigma) \in \pi_{n+2}(\mathbf{S}_{n+1})$ and $c(\sigma) = 0$ if $\sigma \subset A$. Let X_{n+1} be the $(n + 1)$-skeleton of X; by definition, there is a continuous map $\varphi_\sigma: \mathbf{D}_{n+2} \to X$ such that $\varphi_\sigma(\mathbf{S}_{n+1}) \subset X_{n+1}$, and the restriction of φ_σ to the interior $\mathring{\mathbf{D}}_{n+2}$ is a homeomorphism onto σ. Let $g_\sigma: (\mathbf{S}_{n+2}, *) \to (\mathbf{S}_{n+1}, *)$ be a continuous map whose homotopy class is $c(\sigma)$; g_σ can also be considered a continuous map $\mathbf{D}_{n+2} \to \mathbf{S}_{n+1}$ such that $g_\sigma(\mathbf{S}_{n+1}) = \{*\}$. Define

$$f_\sigma: (X_{n+1} \cup \sigma, X_{n+1}) \to (\mathbf{S}_{n+1}, *)$$

by $f_\sigma(x) = *$ if $x \in X_{n+1}$ and $f_\sigma(x) = g_\sigma(z)$ if $x = \varphi_\sigma(z)$ and $z \in \mathring{\mathbf{D}}_{n+2}$. There is a continuous map $f: (X, X_{n+1}) \to (\mathbf{S}_{n+1}, *)$ equal to f_σ in each $(n + 2)$-cell σ of X, since there are no m-cells in X for $m > n + 2$; by definition, $\Lambda(u) = [f]$. Under the condition $\dim(X, A) \leqslant n + 2$, Spanier proved that the sequence

$$\begin{aligned} \pi^n(X, A) &\xrightarrow{\Phi} H^n(X, A; \pi_n(\mathbf{S}_n)) \xrightarrow{Sq^2} H^{n+2}(X, A; \pi_{n+2}(\mathbf{S}_{n+1})) \\ &\xrightarrow{\Lambda} \pi^{n+1}(X, A) \xrightarrow{\Phi} H^{n+1}(X, A; \pi_{n+1}(\mathbf{S}_{n+1})) \end{aligned} \tag{125}$$

is *exact*. He showed that this result gives another proof of Steenrod's results on the extension problems described in § 4,A.

A little later F. Peterson, in his thesis [360], obtained similar results for the kernel and cokernel of the map (123) for larger values of m, by considering only the *p-components* of the groups for any prime $p > 2$ and using the corresponding Steenrod reduced powers.

CHAPTER VII

Generalized Homology and Cohomology

The ten years between 1942 and 1952 had seen an explosive development of algebraic topology, with a totally unexpected wealth of new methods and new results. The next decade was just as fruitful, and just as unpredictable.

The two main novelties were cobordism and K-theory. The former inaugurated a revival of differential topology, which has kept its momentum to this day; the latter emerged in algebraic geometry, but very soon spread over all topology and all algebra. A large part of their rapid success was due to the availability of the new techniques acquired in algebraic topology during the preceding decade.

A little later it was realized that both theories, as well as all previous definitions of homology and cohomology, could be encompassed in the very general concepts of *generalized* (or *extraordinary*) *homology and cohomology*, which have dominated algebraic topology and differential topology ever since.

§ 1. Cobordism

A. The Work of Pontrjagin

In 1950, in order to compute the homotopy groups $\pi_{n+1}(\mathbf{S}_n)$ and $\pi_{n+2}(\mathbf{S}_n)$, Pontrjagin [382] described a new approach to the computation of $\pi_{n+k}(\mathbf{S}_n)$ for any $k \geqslant 1$. His central concept was that of a *framed manifold* M of dimension k, embedded in a space $\mathbf{R}^{n+k}$: M is a smooth manifold whose normal bundle in $\mathbf{R}^{n+k}$ is *trivializable*; equivalently, there exist n continuous maps $x \mapsto u_j(x)$ of M into $\mathbf{R}^{n+k}$ $(1 \leqslant j \leqslant n)$ such that the $u_j(x)$ form an orthonormal basis of the normal subspace N_x to M at the point x for every $x \in \mathrm{M}$. Not every smooth submanifold of $\mathbf{R}^{n+k}$ is framed, for instance, no nonorientable submanifold may have that property.

However, it is easy to give examples of framed manifolds: consider a C^∞ map $f: \mathrm{X} \to \mathrm{Y}$ of an $(n + k)$-dimensional manifold X into an n-dimensional manifold Y; then, if $y \in \mathrm{Y}$ does not belong to the image $f(\mathrm{E})$ of the set E of critical points of f (Part 1, chap. III, § 1), $f^{-1}(y)$ is a k-dimensional submanifold

J. Dieudonné, *A History of Algebraic and Differential Topology, 1900–1960*,
Modern Birkhäuser Classics, DOI 10.1007/978-0-8176-4907-4_21,

of X whose normal bundle in X is trivializable. The proof is an immediate consequence of the implicit function theorem.

It is that property that Pontrjagin used to study the maps $f\colon \mathbf{S}_{n+k} \to \mathbf{S}_n$; since any such continuous map is homotopic to a C^∞ map (Part 1, chap. III, § 1), it may be assumed that f is C^∞. Then by Sard's theorem (*loc. cit.*), there is a point $z \in \mathbf{S}_n$ that does not belong to the image of the set of critical points, and (barring the uninteresting case in which f is not surjective) $f^{-1}(z)$ is a compact k-dimensional framed submanifold of $\mathbf{S}_{n+k}$, and since it cannot be $\mathbf{S}_{n+k}$ itself, it can be considered as a submanifold of $\mathbf{R}^{n+k}$. Conversely, a framed k-dimensional submanifold M of $\mathbf{R}^{n+k}$ has a tubular neighborhood N that is diffeomorphic to the product $\mathrm{M} \times \mathbf{R}^n$, and there is therefore a C^∞ map $g\colon \mathrm{N} \to \mathbf{R}^n$ that tends to infinity on the frontier of N; if $\mathbf{R}^{n+k}$ and $\mathbf{R}^n$ are considered as open subsets of $\mathbf{S}_{n+k}$ and $\mathbf{S}_n$, respectively, g can be continued to a map $f\colon \mathbf{S}_{n+k} \to \mathbf{S}_n$ that sends $\mathbf{S}_{n+k} - \mathrm{N}$ to the point at infinity of $\mathbf{S}_n$, and $\mathrm{M} = f^{-1}(0)$.

To study homotopy classes of maps $\mathbf{S}_{n+k} \to \mathbf{S}_n$ by means of framed submanifolds, Pontrjagin introduced an equivalence relation between framed k-dimensional submanifolds of $\mathbf{R}^{n+k}$, that his student Rokhlin called "intrinsic homology": two framed submanifolds M_0, M_1 of $\mathbf{R}^{n+k}$ are "homologous" if there is a (noncompact) $(k+1)$-dimensional framed submanifold W of $\mathbf{R}^{n+k} \times \mathbf{R}$ such that $\mathrm{M}_0 \times \{0\}$ and $\mathrm{M}_1 \times \{1\}$ are the intersections of W with the hyperplanes $x_{n+k+1} = 0$ and $x_{n+k+1} = 1$, and the normal bundles of $\mathrm{M}_0 \times \{0\}$ and $\mathrm{M}_1 \times \{1\}$ are intersections of these hyperplanes with the normal bundle of W.

Pontrjagin's main result was then that two smooth maps f, g of $\mathbf{S}_{n+k}$ into $\mathbf{S}_n$ are homotopic if and only if the two framed submanifolds $f^{-1}(z)$, $g^{-1}(z)$ (for a point z which is not in the images of the critical points of f and g) are "homologous" in his sense. This enabled him to determine the groups $\pi_{n+1}(\mathbf{S}_n)$ and $\pi_{n+2}(\mathbf{S}_n)$, and Rokhlin also determined the groups $\pi_{n+3}(\mathbf{S}_n)$ in that way [399], but the study of "intrinsic homology" for $k \geqslant 4$ became too complicated to compete with the other methods of computation of the homotopy groups of spheres (chap. V, § 5,F).

This work of Pontrjagin may be considered the germ of the much more extensive theory of *cobordism* inaugurated by Thom in 1953 [463].

B. Transversality

Let X, Y be two smooth manifolds* and $f\colon \mathrm{X} \to \mathrm{Y}$ be a C^∞ map. A question that recurs in many problems is whether the inverse image $f^{-1}(\mathrm{Z})$ of a smooth submanifold Z of Y is also a submanifold of X. When $\mathrm{Z} = \{z\}$ is reduced to a point, a sufficient condition is that $f^{-1}(z)$ does not contain any critical point of f (section A). This generalizes to any submanifold Z: if $f^{-1}(\mathrm{Z})$ does not contain any critical point of f, $f^{-1}(\mathrm{Z})$ is a submanifold, but the condition is very restrictive.

---

* Manifolds are always assumed to be locally compact, metrizable, and separable.

Thom found a much less restrictive condition, which turned out to be a very flexible and useful tool in all differential topology. The map f is said to be *transverse over* Z *at the point* $x \in f^{-1}(Z)$ if the tangent space $T_{f(x)}(Y)$ is the *sum* (not necessarily a direct sum)

$$T_{f(x)}(Z) + T_x(f)(T_x(X));$$

it is *transverse over* Z if that condition is satisfied at *all* points $x \in f^{-1}(Z)$. The set of points of $f^{-1}(Z)$ at which f is transverse over Z is open in $f^{-1}(Z)$.

As an example, let X be the plane $x = 0$ in $\mathbf{R}^3$, Y be the plane $z = 0$, Z be the line $y = z = 0$, and f be the projection $(y, z) \mapsto y$ of X into Y; then f is transverse over Z, but the tangent mapping $T_x(f)\colon T_x(X) \to T_{f(x)}(Y)$ is never surjective.

If f is transverse over Z, $f^{-1}(Z)$ is a submanifold of codimension in X equal to the codimension of Z in Y; the normal bundle of $f^{-1}(Z)$ in X is the *pullback* $f^*(N)$ of the normal bundle N of Z in Y. The question being a local one, the proof of these statements is immediately reduced to the case in which X, Y, and Z are vector spaces, and is then a consequence of the implicit function theorem.

The main result concerning transverse mappings is Thom's *transversality theorem*:

Let d be a distance defining the topology of Y and $f\colon X \to Y$ be an *arbitrary* smooth mapping. Then for any $\varepsilon > 0$ there is a smooth map $g\colon X \to Y$ such that $d(f(x), g(x)) \leqslant \varepsilon$ in X and g is *transverse over* Z. If A is a closed set of $f^{-1}(Z)$ such that f is already transverse over Z at the points $x \in A$, then it may be assumed that $f|A = g|A$.

Thom's proof ([463], pp. 22–26) made use of the group $\mathscr{H}$ of diffeomorphisms of Y onto itself that leave fixed the points of Y − T, where T is an open tubular neighborhood of Z; his idea was to take $g = A \circ f$, where $A \in \mathscr{H}$ is close enough to the identity for a suitable topology. Milnor provided a more elementary proof, which can be broken up as follows ([347], p. 212).

I. There is an open neighborhood U of A in X such that f is still transverse over Z at every point of $U \cap f^{-1}(Z)$. Take an open denumerable covering $(Y_i)_{i \geqslant 0}$ of Y such that $Y_0 = Y - Z$, Y_i is the domain of a chart of Y for $i \geqslant 1$, and $Z \subset \bigcup_{i \geqslant 1} Y_i$. Take an open denumerable covering $(V_j)_{j \geqslant 0}$ of X by domains of charts of X, finer than the covering consisting of X − A and U, and also finer than the covering consisting of the $f^{-1}(Y_i)$. Finally, let $(W_k)_{k \geqslant 0}$ be an open locally finite covering such that $(\overline{W}_k)$ is finer than (V_j).

II. The strategy of the proof is to define by induction a sequence (f_i) of smooth maps of X into Y, having the following properties:

(a) $f_0 = f$, $d(f_i(x), f_{i-1}(x)) \leqslant \varepsilon/2^i$ for all $x \in X$;
(b) f_i coincides with f_{i-1} in the complement of $\overline{W}_i$;
(c) f_i is transverse over Z at every point of $f_i^{-1}(Z) \cap \bigcup_{j \leqslant i} \overline{W}_j$.

Then the sequence (f_i) converges uniformly in X to a smooth map $g: X \to Y$, which satisfies the conclusions of the theorem.

III. By assumption, there is an index $j(i)$ such that $f_{i-1}(V_i)$ is contained in an open set $Y_{j(i)}$. Using the charts with domains V_i and $Y_{j(i)}$, the problem is reduced to the following:

Let V be an open set in $\mathbf{R}^n$, W be an open set such that $\overline{W} \subset V$, Y be an open set in $\mathbf{R}^p$, $Z = \mathbf{R}^q \cap Y$ for a value $q < p$, K be a compact subset of $\mathbf{R}^n$, and $f: V \to Y$ be a smooth map that is transverse over Z at every point of $K \cap V$. For a given $\delta > 0$ define a smooth map $g: V \to Y$ that coincides with f outside a closed neighborhood of $\overline{W}$ in V, such that $|g(x) - f(x)| \leqslant \delta$ for all $x \in V$, and that g is transverse over Z.

To define g, first consider a smooth map $\lambda: V \to [0, 1]$, equal to 1 in $\overline{W}$ and whose compact support T is contained in V. Let ρ be the projection of $\mathbf{R}^p$ on the supplementary space $\mathbf{R}^{p-q}$ of $\mathbf{R}^q$; $Z = Y \cap \rho^{-1}(0)$. There is, by Sard's theorem, a point $y \in \mathbf{R}^{p-q}$ arbitrarily close to 0, which is not an image of a critical point of $\rho \circ f$. Take

$$g(x) = f(x) - \lambda(x)y.$$

If y is close enough to 0, it is possible to assume $|g(x) - f(x)| \leqslant \delta$ in V, and that the derivative $Dg(x) = Df(x) - (D\lambda(x))y$ is arbitrarily close to $Df(x)$ in V. Since $\rho(Df(x))$ has rank $p - q$ in $f^{-1}(\rho^{-1}(0)) \cap K$, the same is true of $\rho(Dg(x))$ when y is close enough to 0. On the other hand, $D\lambda(x) = 0$ in $\overline{W}$, so $Dg(x) = Df(x)$; the points of $\overline{W}$ such that $\rho(g(x)) = 0$ are those for which $\rho(f(x)) = y$, and therefore at all points $x \in \overline{W} \cap g^{-1}(\rho^{-1}(0))$, $\rho(Dg(x))$ has rank $p - q$, which means that g is transverse over $Z = Y \cap \rho^{-1}(0)$. Since $g(x) = f(x)$ outside T, g has the required properties.

C. Thom's Basic Construction

There are two basic ideas in Thom's fundamental paper [463].

I. The first one is the association, to each vector fibration $\xi = (E, B, \mathbf{R}^k, p)$, of the *mapping cone* of the map $p: E \to B$ (Part 2, chap. V, § 3,B); it is now called the *Thom space* $T(\xi)$ of the fibration. When B is compact, hence E is locally compact, $T(\xi)$ can be identified to the compact space obtained by adjoining to E a "point at infinity" t_0. When B is paracompact, there exists a distance d on E that has as a restriction to each fiber E_b a euclidean distance on that vector space. Recall that for the sphere bundle S corresponding to E each fiber S_b is the sphere of radius 1 and center b in E_b (chap. III, § 1,C). Let $E(\alpha)$ be the open subset of E such that $E(\alpha) \cap E_b$ is the open ball of radius α in E_b. Then $T(\xi)$ can also be identified to the space $\overline{E(1)}/S$ obtained by collapsing the sphere bundle S to a single point t_0 in the closure $\overline{E(1)}$, or equivalently to the space $E/(E - E(1))$.

II. Thom used the spaces $T(\xi)$ in an original way to establish, by means of the transversality theorem (section B), a two-way correspondence between *submanifolds* of a smooth manifold X and *continuous maps* $X \to T(\xi)$ into Thom spaces.

Let Z be a submanifold of X, of codimension k; there is then an open tubular neighborhood N of Z. If $\xi = (\mathrm{E}, \mathrm{Z}, \mathbf{R}^k)$, there is an isomorphism of vector fibrations g: $\mathrm{N} \to \mathrm{E}(1)$, which is extended to a continuous map $\bar{\mathrm{N}} \to \overline{\mathrm{E}(1)}$, sending the frontier of N in X onto the sphere bundle $\overline{\mathrm{E}(1)} - \mathrm{E}(1)$. If φ: $\mathrm{E} \to \mathrm{T}(\xi) = \mathrm{E}/(\mathrm{E} - \mathrm{E}(1))$ is the collapsing map, $\varphi \circ g$ is extended to a *continuous map* f: $\mathrm{X} \to \mathrm{T}(\xi)$, constant and equal to the "point at infinity" t_0 in X − N, and coinciding with the identity in Z.

Conversely, suppose X is a *compact* smooth manifold and let f: $\mathrm{X} \to \mathrm{T}(\xi)$ be a continuous map for a fibration $\xi = (\mathrm{E}, \mathrm{B}, \mathbf{R}^k, p)$ with base space a smooth manifold B; let d be a distance on E such that its restriction to each fiber E_b is a euclidean distance. Consider the open set $f^{-1}(\mathrm{E})$ in X; there is a smooth map $g: f^{-1}(\mathrm{E}) \to \mathrm{E}$ such that $d(f(x), g(x)) \leqslant \varepsilon$ for $x \in f^{-1}(\mathrm{E})$ and a homotopy $\mathrm{F}: f^{-1}(\mathrm{E}) \times [0, 1] \to \mathrm{E}$ such that $d(\mathrm{F}(x, s), f(x)) \leqslant 2\varepsilon$ for all (x, s), and $\mathrm{F}(x, 0) = f(x)$, $\mathrm{F}(x, 1) = g(x)$. The choice of d implies that F can be extended by continuity to $\mathrm{X} \times [0, 1]$ by taking $\mathrm{F}(x, s) = t_0$ for $x \notin f^{-1}(\mathrm{E})$. Now use the transversality theorem to define a smooth map $h: f^{-1}(\mathrm{E}) \to \mathrm{E}$ that is *transverse over* B, and a homotopy $\mathrm{H}: f^{-1}(\mathrm{E}) \times [0, 1] \to \mathrm{E}$ such that $d(\mathrm{H}(x, s), g(x)) \leqslant \varepsilon$, $\mathrm{H}(x, 0) = g(x)$, $\mathrm{H}(x, 1) = h(x)$; it is extended by continuity to $\mathrm{X} \times [0, 1]$. The map h is thus *homotopic to* f, and $h^{-1}(\mathrm{B})$ is a *smooth submanifold* of X.

D. Homology and Homotopy of Thom Spaces

Suppose B is a finite CW-complex; then for any vector fibration $\xi = (\mathrm{E}, \mathrm{B}, \mathbf{R}^k, p)$ of rank k with base space B, the Thom space $\mathrm{T}(\xi)$ is a $(k - 1)$-*connected* CW-complex. Indeed, if e_α is any n-cell of B, $p^{-1}(e_\alpha) = e'_\alpha$ is a $(k + n)$-cell by Feldbau's theorem, and $\mathrm{T}(\xi)$ is the disjoint union of the cells e'_α and the 0-cell $\{t_0\}$ (chap. II, §6,C).

The homology of $\mathrm{T} = \mathrm{T}(\xi)$ is easily computed. If $\mathrm{T}^0 = \mathrm{T} - \mathrm{B}$, T^0 is contractible, and the homology exact sequence of the triple $(\mathrm{T}, \mathrm{T}^0, \{t_0\})$ gives natural isomorphisms $\mathrm{H}_i(\mathrm{T}, \{t_0\}; \mathbf{F}_2) \simeq \mathrm{H}_i(\mathrm{T}, \mathrm{T}^0; \mathbf{F}_2)$. There is then an excision isomorphism $\mathrm{H}_i(\mathrm{T}, \mathrm{T}^0; \mathbf{F}_2) \simeq \mathrm{H}_i(\mathrm{E}, \mathrm{E}^0; \mathbf{F}_2)$, and finally the Thom isomorphism $\mathrm{H}_i(\mathrm{E}, \mathrm{E}^0; \mathbf{F}_2) \simeq \mathrm{H}_{i-k}(\mathrm{B}; \mathbf{F}_2)$ [chap. IV, §2, formula (51)]. By composition, there is an isomorphism $\mathrm{H}_i(\mathrm{T}, \{t_0\}; \mathbf{F}_2) \simeq \mathrm{H}_{i-k}(\mathrm{B}; \mathbf{F}_2)$; from the homology exact sequence, $\mathrm{H}_i(\mathrm{T}; \mathbf{F}_2) \simeq \mathrm{H}_i(\mathrm{T}, \{t_0\}; \mathbf{F}_2)$ for $i \geqslant 2$, and since T is connected, this holds for $i = 1$; finally, there is a natural isomorphism

$$\mathrm{H}_i(\mathrm{T}(\xi); \mathbf{F}_2) \simeq \mathrm{H}_{i-k}(\mathrm{B}; \mathbf{F}_2) \quad \text{for all } i \geqslant 1. \tag{1}$$

Since the homotopy groups $\pi_i(\mathrm{T}(\xi))$ are 0 for $1 \leqslant i \leqslant k - 1$, it follows from the improvement of the $\mathscr{C}$-absolute Hurewicz theorem mentioned in chap. V, §4,A that if $\mathscr{C}$ is the class of *finite* commutative groups, there is a $\mathscr{C}$-*isomorphism*

$$\pi_{n+k}(\mathrm{T}(\xi)) \to \mathrm{H}_n(\mathrm{B}; \mathbf{Z}) \tag{2}$$

for all $n < k - 1$.

For the applications he had in mind (see below sections E, F, G) Thom considered the vector bundles of rank k (or the corresponding sphere bundles) associated to the principal fiber spaces

$$(\mathrm{EO}(k), \mathrm{BO}(k), \mathrm{O}(k)) \quad \text{and} \quad (\mathrm{ESO}(k), \mathrm{BSO}(k), \mathrm{SO}(k))$$

(chap. III, § 2,E). It is more convenient to work with locally compact bundles that have these as their direct limits, namely, the vector fibrations (for large N)

$$\gamma_{\mathrm{N},k} = (\mathrm{U}_{\mathrm{N},k}, \mathrm{G}_{\mathrm{N},k}, \mathbf{R}^k) \quad \text{and} \quad \gamma'_{\mathrm{N},k} = (\mathrm{U}'_{\mathrm{N},k}, \mathrm{G}'_{\mathrm{N},k}, \mathbf{R}^k), \tag{3}$$

where $\mathrm{U}_{\mathrm{N},k}$ is the tautological bundle with base the real Grassmannian $\mathrm{G}_{\mathrm{N},k}$, and $\mathrm{U}'_{\mathrm{N},k}$ is its double covering, a vector bundle over the special Grassmannian $\mathrm{G}'_{\mathrm{N},k}$ (chap. IV, § 1,C). Thom wrote the corresponding Thom spaces $\mathrm{MO}(k)$ and $\mathrm{MSO}(k)$, and he made in [463] a deeper study of their homology and homotopy.

Recall that the cohomology algebra $\mathrm{H}^\bullet(\mathrm{G}_{\mathrm{N},k}; \mathbf{F}_2)$ is generated by the Stiefel–Whitney classes $w_1, w_2, \ldots, w_k$ of $\gamma_{\mathrm{N},k}$ (chap. IV, § 1,C). Using (1), Thom showed that the cohomology algebra $\mathrm{H}^\bullet(\mathrm{MO}(k); \mathbf{F}_2)$ can be identified with *the ideal of* $\mathrm{H}^\bullet(\mathrm{G}_{\mathrm{N},k}; \mathbf{F}_2)$ *generated by the top class* w_k. He proved in the same way that for odd primes p, $\mathrm{H}^\bullet(\mathrm{MO}(k); \mathbf{F}_p)$ is the ideal in $\mathrm{H}^\bullet(\mathrm{G}_{\mathrm{N},k}; \mathbf{F}_p)$ generated by the cup-square $e \smile e$ of the Euler class $e = e(\gamma_{\mathrm{N},k})$. There are similar results for $\mathrm{MSO}(k)$.

Concerning homotopy groups, the Freudenthal theorem on homotopy suspension (chap. II, § 6,E), together with the variant of the first Whitehead theorem which he proved for that purpose (chap. II, § 5,F), enabled Thom to prove that for $i < k - 1$ there is a natural isomorphism

$$\pi_{k-1+i}(\mathrm{MO}(k)) \simeq \pi_{k+i}(\mathrm{MO}(k)) \tag{4}$$

and there is a similar isomorphism for $\mathrm{MSO}(k)$. For given i the groups $\pi_{k+i}(\mathrm{MO}(k))$ [resp. $\pi_{k+i}(\mathrm{MSO}(k))$] are all isomorphic as soon as $k > i + 1$, the same phenomenon of *stability* as for the homotopy groups of spheres (chap. II, § 6,E).

Inspired by the results on the cohomology of the Eilenberg–Mac Lane spaces $\mathrm{K}(\mathbf{Z}/2\mathbf{Z}, n)$ obtained by Serre (chap. VI, § 3,E), and with some assistance from the latter, Thom was able to prove similar results for the cohomology of the spaces $\mathrm{MO}(k)$. He first showed that the classes $\mathrm{Sq}^{\mathrm{I}}(w_k)$, for admissible sequences I of degree k (*loc. cit.*) are linearly independent in $\mathrm{H}^\bullet(\mathrm{G}_{\mathrm{N},k}; \mathbf{F}_2)$ for large N. Expressing $w_1, w_2, \ldots, w_k$ as elementary symmetric polynomials in "phantom" indeterminates $t_1, t_2, \ldots, t_k$ [cf. Hirzebruch's method for Chern classes, chap. IV, § 1,E formula (30)], Thom wrote a basis over $\mathbf{F}_2$ of

$$\mathrm{H}^{k+h}(\mathrm{MO}(k); \mathbf{F}_2) \quad \text{for } h \leqslant k$$

as symmetrized polynomials* of monomials in the t_j:

* The symmetric group $\mathfrak{S}_n$ acts in an obvious way on any polynomial in $t_1, t_2, \ldots, t_n$. The *symmetrized* polynomial s(P) of a polynomial P is the sum of the distinct polynomials in the *orbit* of P under the action of $\mathfrak{S}_n$.

$$X_\omega^h = s((t_1)^{a_1+1}(t_2)^{a_2+1}\cdots(t_r)^{a_r+1}t_{r+1}\cdots t_k), \tag{5}$$

where $\omega = \{a_1, a_2, \ldots, a_r\}$ is an *arbitrary partition* of the number h. A study of the behavior of the elements $\mathrm{Sq}^1(X_\omega^h)$ then enabled him to define a continuous map

$$F\colon MO(k) \to Y = \prod_{h=0}^{k} (K(\mathbf{Z}/2\mathbf{Z}, k+h))^{d(h)} \tag{6}$$

where $d(h)$ is the *number of partitions* ω of h into integers a_j such that $a_j + 1$ *is not a power of* 2 (*nondyadic* partitions). The crucial property of that map is that if Y_{2k} is the $2k$-skeleton of Y, there exists a continuous map $g\colon Y_{2k} \to MO(k)$ such that $g \circ F$ is *the identity on the* $(2k-1)$*-skeleton of* $MO(k)$. This shows in particular that for $h < k$, $\pi_{k+h}(MO(k))$ is an $\mathbf{F}_2$-vector space of *dimension* $d(h)$.

Thom could also describe the vector spaces $H^m(MO(k); \mathbf{F}_2)$ explicitly for small values of m and k.

The description of the "stable" homotopy of $MSO(k)$ proved more difficult, and Thom could only obtain partial results and compute explicitly the stable groups $\pi_{k+i}(MSO(k))$ for $i \leqslant 7$. This was because in order to prove the existence of a map similar to the map (6), the product of Eilenberg–Mac Lane spaces on the right-hand side of (6) must be replaced by a tower of nontrivial fibrations with Eilenberg–Mac Lane fibers. Thom only considered a fibration with base space $K(\mathbf{Z}, k)$ and fiber $K(\mathbf{Z}, k+4)$, defined by an Eilenberg–Mac Lane invariant (chap. V, §2,A), which turns out to be a lifting to the cohomology group $H^{k+5}(\mathbf{Z}, k; \mathbf{Z})$ of the Steenrod reduced power $\mathrm{St}_3^5(\iota)$ of the canonical class ι of $K(\mathbf{Z}, k)$ (chap. V, §1,D).

E. The Realization Problem

The first problem to which Thom applied his basic construction in his fundamental paper [463] was: given a smooth compact manifold X of dimension n and a homology class $z \in H_{n-k}(X; \mathbf{F}_2)$, is there a smooth closed *submanifold* W of X, of codimension k, such that $z = i_*([W])$ for the natural injection $i\colon W \to X$? If such is the case, the class z is said to be *realizable* by the submanifold W.

Thom's construction allowed him to replace that problem by an equivalent one concerning the cohomology class $z^* \in H^k(X; \mathbf{F}_2)$ *dual* to z by Poincaré duality (Part 2, chap. I, §3).

A *canonical cohomology class* $u_{T(\xi)} \in H^k(T(\xi); \mathbf{F}_2)$ is defined for every vector fibration $\xi = (E, B, \mathbf{R}^k)$ of rank k; it is the image of the fundamental class u_{E,E^o} of relative cohomology in $H^k(E, E^o; \mathbf{F}_2)$ (chap. IV, §2) by the map

$$H^k(E, E^o; \mathbf{F}_2) \simeq H^k(T, \{t_0\}; \mathbf{F}_2) \simeq H^k(T; \mathbf{F}_2)$$

(section D). Thom's theorem is then:

A necessary and sufficient condition for $z \in H_{n-k}(X; \mathbf{F}_2)$ *to be realizable is that the corresponding cohomology class* $z^* \in H^k(X; \mathbf{F}_2)$ *be such that* $z^* = f^*(u_k)$, *where* u_k *is the canonical cohomology class of* $T(\gamma_{m,k})$ *with large* m, *and* f *a continuous map* $f: X \to T(\gamma_{m,k})$.

Necessity. Let W be a smooth submanifold of X of codimension k, and consider an open tubular neighborhood N of W, identified with the total space of the normal bundle $\nu = (N, W, \mathbf{R}^k)$ of W in X. There is a continuous map $g: W \to G_{m,k}$ for a large integer m, such that ν is the pullback $g^*(\gamma_{m,k})$ (chap. III, § 2,E). Let $G: N \to U_{m,k}$ be the corresponding bundle map, which extends continuously to a continuous map $f: X \to T(\gamma_{m,k})$ by taking $f(x) = t_0$ for $x \in X - N$. This gives the commutative diagram

$$\begin{array}{ccc} H^k(T(\gamma_{m,k}); \mathbf{F}_2) & \xrightarrow{\;f^*\;} & H^k(X; \mathbf{F}_2) \\ j^* \uparrow \wr & & \uparrow j^* \\ H^k(T(\gamma_{m,k}), \{t_0\}; \mathbf{F}_2) & \xrightarrow[G^*]{} & H^k(X, X - W; \mathbf{F}_2) \\ e \uparrow \wr & & \uparrow \wr\, e \\ H^k(U_{m,k}, U^0_{m,k}; \mathbf{F}_2) & \xrightarrow[G^*]{} & H^k(N, N^0; \mathbf{F}_2) \end{array} \qquad (7)$$

where the arrows e are isomorphisms defined in section B and j^* comes from the exact cohomology sequence. The left vertical arrows are isomorphisms, and $G^*(u_{U_{m,k}, U^0_{m,k}} = u_{N,N^0}$ (chap. IV, § 2); therefore

$$f^*(u_k) = j^*(e(u_{N,N^0})).$$

But $j^*(e(u_{N,N^0}))$ is the cohomology class *dual* to $i_*([W])$ (chap. VI, § 4,B).

Sufficiency. Suppose $f: X \to T(\gamma_{m,k})$ is such that $f^*(u_k) = z^*$. Then Thom's construction (section C) yields a map $h: X \to T(\gamma_{m,k})$ homotopic to f such that h is C^∞ in $h^{-1}(U_{m,k})$, h is transverse over $G_{m,k}$, and therefore $W = h^{-1}(G_{m,k})$ is a smooth submanifold and the pullback $h^*(\gamma_{m,k})$ is isomorphic to the normal bundle of W in X. Then $h^*(u_k) = z^*$, and the first part of the proof shows that z^* is the dual cohomology class of $i_*([W])$, where $i: W \to X$ is the natural injection.

The results obtained by Thom on the cohomology of MO(k) (section D) allowed him to apply the preceding theorem to the realization problem for homology modulo 2. From the properties of the map F [formula (6)] follows the existence of a continuous map g of the $2k$-skeleton $(K(\mathbf{Z}/2\mathbf{Z}, k))_{2k}$ into $T(\gamma_{m,k})$ such that

$$g^*(u_k) = e_k^*$$

the unique cohomology class $\neq 0$ in $H^k(K(\mathbf{Z}/2\mathbf{Z}, k); \mathbf{F}_2)$. On the other hand, *if* X *has dimension* $\leqslant 2k$, for every class $z^* \in H^k(X; \mathbf{F}_2)$, there is a continuous map $h: X \to (K(\mathbf{Z}/2\mathbf{Z}, k))_{2k}$ such that $z^* = h^*(e_k^*)$ (chap. V, § 1,D). Thus for

$f = g \circ h\colon X \to T(\gamma_{m,k})$, $z^* = f^*(u_k)$, hence *every homology class in* $H^k(X; \mathbf{F}_2)$ *is realizable if* X *is a smooth manifold of dimension* $\leqslant 2k$. Serre pointed out that $H^\bullet(K(\mathbf{Z}/2\mathbf{Z}, k); \mathbf{F}_2)$ has infinite dimension, whereas $H^\bullet(T(\gamma_{m,k}); \mathbf{F}_2)$ is finite dimensional; for n large enough, there cannot exist a continuous map g: $(K(\mathbf{Z}/2\mathbf{Z}, k))_n \to T(\gamma_{m,k})$ that would be a homotopy inverse (chap. II, §2,C) to the continuous map $T(\gamma_{m,k}) \to (K(\mathbf{Z}/2\mathbf{Z}, k))_n$. Hence, for large enough n, there are homology classes in $H_n(X; \mathbf{F}_2)$ which are *not* realizable if $\dim X > n$.

Nevertheless, a more detailed study of the cohomology mod. 2 of MO(k) shows that for a manifold X of dimension n, all classes in $H_{n-1}(X; \mathbf{F}_2)$ are realizable for any n, all classes in $H_{n-2}(X; \mathbf{F}_2)$ if $n < 6$, all classes in $H_{n-3}(X; \mathbf{F}_2)$ if $n < 8$, and finally, all classes in $H_i(X; \mathbf{F}_2)$ if $i \leqslant n/2$, for every n.

Thom also considered the realization problem for classes in $H_\bullet(X; \mathbf{Z})$ when X is an *orientable* smooth manifold and the submanifolds W that must "realize" homology classes are also restricted to *orientable* ones. The normal bundle of W in X is then also orientable (chap. IV, §1,C), and is therefore associated to a principal fiber space with structural group SO(k) (*loc. cit.*). If X is orientable, a homology class in $H_\bullet(X; \mathbf{Z})$ is *realizable for the rotation group* SO(k) if it is realizable by an orientable submanifold of codimension k. Thom's main theorem reducing the problem to a problem on cohomology classes can then be extended by replacing $T(\gamma_{m,k})$ by $T(\gamma'_{m,k})$. Using his results on the cohomology of MSO(k) (section D), he could show that for any integer $n \geqslant 1$ in an n-dimensional compact connected orientable manifold X, all homology classes in $H_{n-1}(X; \mathbf{Z})$, $H_{n-2}(X; \mathbf{Z})$, and $H_i(X; \mathbf{Z})$ for $i \leqslant 5$ are realizable for the rotation group. He also showed that for given n and k, there is an integer $N > 0$, depending only on n and k, such that for *every* class $z \in H_k(X; \mathbf{Z})$, $N . z$ is realizable for the rotation group.

F. Smooth Classes in Simplicial Complexes

Thom also applied his method to a related problem raised by Steenrod [175]: if $K \subset \mathbf{R}^N$ is a finite euclidean simplicial complex, and $z \in H_r(K; \Lambda)$, does there exist a *smooth* compact connected manifold M of dimension r and a continuous map $f\colon M \to K$ such that $f_*([M]) = z$, [M] being the fundamental class of M in $H_r(M; \Lambda)$? For $\Lambda = \mathbf{Z}$ or $\Lambda = \mathbf{F}_2$ Thom reduced the problem to the realization problem.

Since K is an ANR (chap. II, §2,B), it is possible to find an open neighborhood $U \supset K$ with a *smooth* frontier T in $\mathbf{R}^N$, for which there is a continuous retraction $r\colon \bar{U} \to K$. Take an open neighborhood V of K such that K is a retract of V; by a theorem of Whitney there is then a continuous map $h\colon V \to [0, +\infty[$ such that $h^{-1}(0) = K$ and which is C^∞ in $V - K$; by Sard's theorem there exists some $c > 0$ such that $U = h^{-1}([0, c[)$ has the required properties. Now let X be the *double* of $\bar{U}$ (Part 2, chap. V, §2,F), and consider the natural projection $p\colon X \to \bar{U}$; it is clear that $r' = r \circ p\colon X \to K$ is a retraction, and if $i\colon K \to X$ is the natural injection, $r' \circ i$ is the identity in K, so $r'^*\colon H^\bullet(K; \Lambda) \to H^\bullet(X; \Lambda)$ is *injective*.

Using this construction, Steenrod's question is answered by the following theorem: for $z \in H_r(K; \Lambda)$ to be equal to $f_*([M])$ for a continuous map $f: M \to K$ with M a smooth manifold, it is necessary and sufficient that, after embedding K in an $\mathbf{R}^N$ with sufficiently large N and considering a "double" smooth manifold X in $\mathbf{R}^N$ containing K, the class $i_*(z) \in H_r(X; \Lambda)$ be realizable.

The sufficiency is evident, since if $i_*(z) = [W]$ for a submanifold W of X, $r'_*([W]) = r'_*(i_*(z)) = z$.

To prove necessity, consider a continuous map $f: M \to K$ such that $f_*([M]) = z$. It is possible to suppose that the origin 0 of $\mathbf{R}^N$ is in K; on the other hand, there is an N′ such that there is an embedding $g: M \to \mathbf{R}^{N'}$. Consider K as embedded in $\mathbf{R}^{N+N'}$, and let X be a corresponding smooth manifold that is the "double" of a closed neighborhood $\bar{U}$ of K in $\mathbf{R}^{N+N'}$. It is enough to show that there is a homotopy $F: M \times [0,1] \to U$ between the map f and a map $h: M \to U$ such that $h(M) \subset U$ is a smooth manifold diffeomorphic to M. Such a homotopy can be defined by

$$F(x,t) = ((1-t)i(f(x)), atg(x))$$

where $i: K \to U$ is the natural injection and $a > 0$ is taken small enough so that the values of $F(x,t)$ are all in U. Then $x \mapsto h(x) = F(x,1) = ag(x)$ maps M homeomorphically on a submanifold of $U \cap \mathbf{R}^{N'} \subset X$, so that $h_*([M]) = i_*(f_*([M])) = i_*(z)$.

Since in that theorem N can be taken arbitrarily large, the results obtained for the realization problem give a complete solution to Steenrod's problem for $\Lambda = \mathbf{F}_2$: *any* homology class $z \in H_r(K; \mathbf{F}_2)$ can be written $f_*([M])$ for a continuous map $f: M \to K$ where M is a smooth manifold.

The results are much less complete for $\Lambda = \mathbf{Z}$. If $r \leqslant 5$, then all classes $z \in H_r(K; \mathbf{Z})$ are again images of fundamental classes of smooth manifolds. But a deeper study, using the antipodism of the Steenrod algebra (chap. VI, § 3,D) enabled Thom to prove that the same statement is false for all $r \geqslant 7$. However, for any $r > 0$, there is an integer $N > 0$ depending only on r, such that $N.z$ is the image of the fundamental class of a smooth manifold for every finite simplicial complex K and every $z \in H_r(K; \mathbf{Z})$.

G. Unoriented Cobordism

The most remarkable part of Thom's fundamental paper [463] in his creation of *cobordism*. He described two theories: *unoriented cobordism* dealing with smooth manifolds, which may be orientable or not; and *oriented cobordism*, where only oriented manifolds are considered.

The basic notion is that of a smooth n-dimensional *manifold-with-boundary*. This is a paracompact Hausdorff space X, equipped with a family of *charts* $\varphi_\alpha: U_\alpha \to H_n$, where (U_α) is an open covering of X, and H_n is the *closed half-space* of $\mathbf{R}^n$ consisting of points $(x_1, x_2, \ldots, x_n)$ such that $x_1 \geqslant 0$; each φ_α is a homeomorphism of U_α onto an open subset *of* H_n; if $U_\alpha \cap U_\beta \neq \emptyset$, $\varphi_\beta \circ \varphi_\alpha^{-1}$:

$\varphi_\alpha(U_\alpha \cap U_\beta) \to \varphi_\beta(U_\alpha \cap U_\beta)$ is a *diffeomorphism.** The points $x \in X$ for which there exists a chart φ_α: $U_\alpha \to H_n$ for a domain U_α containing x such that $\varphi_\alpha(x)$ belongs to the hyperplane $x_1 = 0$, frontier of H_n in $\mathbf{R}^n$, are called the *boundary points* of X. The set ∂X of boundary points of X is called the *boundary* of X, a smooth $(n-1)$-dimensional manifold (without boundary, connected or not). A smooth manifold-with-boundary X is triangulable in such a way that ∂X becomes a simplicial subcomplex of X for the triangulation.† At a boundary point $x \in \partial X$, the tangent space $T_x(\partial X)$ is a hyperplane in the tangent space $T_x(X)$. There exists a diffeomorphism h of $\partial X \times [0, 1[$ onto an open subset of X that maps $\partial X \times \{0\}$ onto ∂X; for any $x \in \partial X$ the tangent vectors v in $T_x(X)$ which are the images by $T_x(h)$ of the vectors (u, t) in $T_x(\partial X) \times [0, 1[$ with $t > 0$ form a half-space of $T_x(X)$ and are said to "point inside X" ($-v$ is then said to "point outside X"). The normal bundle of $T(X)/T(\partial X)$ is a trivial line bundle.

The following problem was raised by Steenrod [175]: find necessary and sufficient conditions for a smooth n-dimensional compact manifold M (without boundary, but not necessarily connected) to be diffeomorphic to the boundary ∂X of an $(n+1)$-dimensional smooth manifold-with-boundary X. Pontrjagin [379] had found necessary conditions on the cohomology of the tangent bundle T(M) that we shall describe a little later. In his 1954 paper [463] Thom introduced a new method using his basic construction (section C) that allowed him to completely solve Steenrod's problem.

His idea was to introduce among n-dimensional smooth manifolds (without boundary) an *equivalence relation* R_n: two such manifolds M, M′ are *cobordant* (or cobordant mod. 2) if their *disjoint union* $M \coprod M'$ is diffeomorphic to the boundary of a smooth $(n+1)$-dimensional manifold-with-boundary. The relation is obviously *symmetric*; to prove that it is *reflexive* it is only necessary to observe that if M is any smooth n-dimensional manifold, the union of $M \times \{0\}$ and $M \times \{1\}$ in $M \times \mathbf{R}$ is the boundary of the manifold-with-boundary $M \times [0, 1]$. Finally, to show that R_n is an equivalence relation, it only remains to prove it *transitive*. Suppose there is a diffeomorphism h_1: $M_1 \coprod M_2 \simeq \partial W_1$ and a diffeomorphism h_2: $M_2 \coprod M_3 \simeq \partial W_2$. Form a space W by identifying in $W_1 \coprod W_2$ the points $h_1(x)$ and $h_2(x)$ for every $x \in M_2$, and let q: $W_1 \coprod W_2 \to W$ be the natural surjection. To define a differential structure on W, it is enough to define charts only for the points $y = q(h_1(x)) = q(h_2(x))$ for $x \in M_2$, the other ones being obvious. Let φ_1: $U_1 \to H_{n+1}$, φ_2: $U_2 \to H_{n+1}$ be charts for neighborhoods U_1 of $h_1(x)$ in W_1 and U_2 of $h_2(x)$ in W_2; it may be assumed that $\varphi_1(U_1)$ and $\varphi_2(U_2)$ are both the half ball $z_1 \geqslant 0, |z| < 1$ in $\mathbf{R}^{n+1}$

---

* A C^∞ map u of an open set $V \subset H_n$ into H_n must be such that for every point $z \in V$, there is a neighborhood W of z *in* $\mathbf{R}^n$ such that $W \cap H_n \subset V$ and $u|(W \cap H_n)$ is the restriction of a C^∞ map of W in $\mathbf{R}^n$. A diffeomorphism of $V \subset H_n$ onto $V' \subset H_n$ must be such that it is bijective and both it and its inverse are C^∞.

† For such a triangulation X is a pseudomanifold-with-boundary (Part 2, chap. I, § 3).

and that if $q(h_1(x')) = q(h_2(x'))$ for $x' \in M_2$, then $\varphi_1(h_1(x')) = \varphi_2(h_2(x')) \in \mathbf{R}^n$, with $\varphi_1(h_1(x)) = \varphi_2(h_2(x)) = 0$. Then take as neighborhood of y in W the union $U = q(U_1) \cup q(U_2)$; the chart φ: $U \to \mathbf{R}^{n+1}$ is defined as follows: if $y' = q(h_1(x')) \in q(U_1)$, then $\varphi(y') = \varphi_1(h_1(x'))$; if $y'' = q(h_2(x'')) \in q(U_2)$, then $\varphi(y'') = -\varphi_2(h_2(x''))$. When $y' = y''$, $\varphi(y') = \varphi(y'')$.

Let $\mathfrak{N}_n$ be the set* of equivalence classes for the relation R_n, and write cl(M) the equivalence class in $\mathfrak{N}_n$ of the smooth compact n-dimensional manifold M. It is possible to define in $\mathfrak{N}_n$ a law of composition

$$(\mathrm{cl}(M), \mathrm{cl}(M')) \mapsto \mathrm{cl}(M \coprod M'),$$

written cl(M) + cl(M'). It is necessary to verify that if $\mathrm{cl}(M_1) = \mathrm{cl}(M_2)$, then $\mathrm{cl}(M_1 \coprod M') = \mathrm{cl}(M_2 \coprod M')$: if $M_1 \coprod M_2$ is the boundary of W, we already know that $M' \coprod M'$ is the boundary of a manifold-with-boundary W′, so

$$(M_1 \coprod M') \coprod (M_2 \coprod M') = (M_1 \coprod M_2) \coprod (M' \coprod M')$$

is the boundary of $W \coprod W'$.

The law thus defined on $\mathfrak{N}_n$ is obviously *associative* and *commutative*; it has a *neutral element*, written 0, which is the class of all manifolds M that are diffeomorphic to boundaries of $(n + 1)$-dimensional manifolds-with-boundary: these manifolds are equivalent for R_n, because if $M_1 = \partial Q_1$ and $M_2 = \partial Q_2$, $M_1 \coprod M_2 = \partial(Q_1 \coprod Q_2)$. If M′ is any smooth compact n-manifold, $(M_1 \coprod M') \coprod M' = M_1 \coprod (M' \coprod M')$ is the boundary of $Q_1 \coprod W'$, so $\mathrm{cl}(M_1 \coprod M') = \mathrm{cl}(M')$.

We have already seen that $\mathrm{cl}(M \coprod M) = 0$, so each element of $\mathfrak{N}_n$ is its own "inverse"; in other words, $\mathfrak{N}_n$ is a *commutative group each element of which has order* 2, or equivalently a *vector space over* $\mathbf{F}_2$.

In addition, if P is any smooth compact n-dimensional manifold, $\mathrm{cl}(M \times P) = \mathrm{cl}(M' \times P)$ in $\mathfrak{N}_n$ when $\mathrm{cl}(M) = \mathrm{cl}(M')$: indeed, if $M \coprod M' = \partial Q$, $(M \times P) \coprod (M' \times P) = (M \coprod M') \times P$ is the boundary of $Q \times P$. There is therefore a natural map

$$(\mathrm{cl}(M), \mathrm{cl}(P)) \mapsto \mathrm{cl}(M \times P)$$

and it can immediately be verified that it is *bilinear*. The direct sum

$$\mathfrak{N}_\bullet = \bigoplus_{n \geqslant 0} \mathfrak{N}_n \tag{8}$$

(with $\mathfrak{N}_0 = \mathbf{F}_2$) is thus a *graded commutative algebra over* $\mathbf{F}_2$.

Thom was able to completely determine the structure of that algebra: it is a *graded algebra of polynomials*

---

* Each compact finite-dimensional smooth manifold (or manifold-with-boundary) is diffeomorphic to a manifold (or manifold-with-boundary) that is a subspace of the infinite product space $\mathbf{R}^{\mathbf{N}}$. We can therefore speak of the *set* of equivalence classes of smooth manifolds that are subspaces of $\mathbf{R}^{\mathbf{N}}$, and it is that set which is taken as the union of the $\mathfrak{N}_n$.

$$\mathbf{F}_2[\mathsf{v}_2, \mathsf{v}_4, \mathsf{v}_5, \mathsf{v}_6, \mathsf{v}_8, \mathsf{v}_9, \ldots] \tag{9}$$

with exactly one generator v_k *of degree* k, *for each integer* k *such that* $k + 1$ *is not a power of* 2.

The proof of this remarkable theorem relies on the basic construction of section C and the study of the homology and homotopy of the Thom space MO(k) (section D).

I. First he defined, for each integer $m > k$ and every vector fibration $\xi = (\mathrm{E}, \mathrm{B}, \mathbf{R}^k)$, a natural homomorphism

$$\lambda_\xi\colon \pi_m(\mathrm{T}(\xi)) \to \mathfrak{N}_{m-k}. \tag{10}$$

By transversality (section B), each continuous map $f\colon \mathbf{S}_m \to \mathrm{T}(\xi)$ is homotopic to a continuous map $g\colon \mathbf{S}_m \to \mathrm{T}(\xi)$ which maps the point $*$ of $\mathbf{S}_m$ to t_0, is C^∞ in $g^{-1}(\mathrm{E})$, and is transverse over B, so that $g^{-1}(\mathrm{B})$ is an $(m - k)$-dimensional compact submanifold of $\mathbf{S}_m$. He then had to show that if g_0, g_1 are two such maps that are homotopic, they are *cobordant*. Starting with a homotopy

$$\mathrm{F}\colon \mathbf{S}_m \times [0, 1] \to \mathrm{T}(\xi)$$

such that $\mathrm{F}(x, 0) = g_0(x)$, $\mathrm{F}(x, 1) = g_1(x)$, assume that $\mathrm{F}(x, s) = g_0(x)$ for $0 \leqslant s \leqslant 1/3$ and $\mathrm{F}(x, s) = g_1(x)$ for $2/3 \leqslant s \leqslant 1$. Let U be the open subset in $\mathbf{S}_m \times [0, 1]$

$$\mathrm{U} = \mathrm{F}^{-1}(\mathrm{E}) \cap (\mathbf{S}_m \times]0, 1[).$$

By transversality it is possible to approximate $\mathrm{F}|\mathrm{U}$ by a smooth map $\mathrm{G}\colon \mathrm{U} \to \mathrm{E}$, transverse over B and such that $\mathrm{G}(x, s) = \mathrm{F}(x, s)$ for $(x, s) \in \mathrm{U}$ and $s \notin [1/4, 3/4]$. That map is extended as usual by continuity to a map G_1: $\mathbf{S}_m \times [0, 1] \to \mathrm{T}(\xi)$ such that $\mathrm{G}_1(x, s) = t_0$ for $(x, s) \in (\mathbf{S}_m \times]0, 1[) - \mathrm{U}$ and $\mathrm{G}_1(x, s) = \mathrm{F}(x, s)$ for $s = 0$ and $s = 1$. Then $(\mathrm{F}|\mathrm{U})^{-1}(\mathrm{B})$ is a smooth $(m + 1)$-dimensional submanifold of $\mathbf{S}_m \times]0, 1[$ and its intersection with $\mathbf{S}_m \times \{s\}$ is $g_0^{-1}(\mathrm{B}) \times \{s\}$ for $0 < s \leqslant 1/4$ and $g_1^{-1}(\mathrm{B}) \times \{s\}$ for $3/4 \leqslant s < 1$; this proves our assertion, and λ_ξ is well defined by

$$\lambda_\xi([f]) = \mathrm{cl}(g^{-1}(\mathrm{B})). \tag{11}$$

The fact that it is a homomorphism follows from the definition of the sum $[g_1] + [g_2]$ in $\pi_m(\mathrm{T}(\xi))$, since by that definition (chap. II, §3D) $g_1^{-1}(\mathrm{B})$ and $g_2^{-1}(\mathrm{B})$ must be disjoint in $\mathbf{S}_m$.

II. Thom next specialized (10) to the case

$$\lambda_{\gamma_{m,k}}\colon \pi_{n+k}(\mathrm{T}(\gamma_{m,k})) \to \mathfrak{N}_n \tag{12}$$

for large m and k and proved that the map (12) is *bijective*.

(a) He first proved (12) *surjective*. For $k > n + 1$, any smooth n-dimensional compact manifold M can be embedded in $\mathbf{S}_{n+k}$ by Whitney's theorem (Part 1, chap. III, §1). Consider an open tubular neighborhood N of M in $\mathbf{S}_{n+k}$, identified with the normal bundle of M in $\mathbf{S}_{n+k}$. Since $\mathbf{R}^{n+k}$ is diffeomorphic to $\mathbf{S}_{n+k} - \{*\}$, assume that $\mathrm{N} \subset \mathbf{R}^{n+k}$; there is then a smooth map $g\colon \mathrm{M} \to \mathrm{G}_{n+k,k} = \mathrm{G}_{n+k,k}(\mathbf{R})$ that to every $x \in \mathrm{M}$ associates the k-dimensional vector subspace of $\mathbf{R}^{n+k}$ orthogonal to the tangent vector space $T_x(M)$ (the so-called "Gauss map"). Since the ranks of N and $U_{n+k,k}$ are equal, N can be identified with the pullback $g^*(U_{n+k,k})$, and it follows from the definition that the bundle map $F: N \to U_{n+k,k}$ is *transverse over* $G_{n+k,k}$. Composing F with the natural bundle map $U_{n+k,k} \to U_{m,k}$ for arbitrary $m > n+1$, one has a bundle map $F_1: N \to U_{m,k}$ that is still transverse over $G_{m,k}$, and $M = F_1^{-1}(G_{m,k})$. As usual F_1 extends by continuity to a map $f: \mathbf{S}_{n+k} \to T(\gamma_{m,k})$ equal to t_0 in $\mathbf{S}_{n+k} - N$; cl(M) is clearly the image by the homomorphism (12) of the homotopy class $[f]$.

(b) The proof of the *injectivity* of (12) is more involved. Start with a continuous map $g: \mathbf{S}_{n+k} \to T(\gamma_{m,k})$, which is C^∞ in $g^{-1}(U_{m,k})$ and transverse over $G_{m,k}$, so that $M = g^{-1}(G_{m,k})$ is a smooth submanifold of dimension n of $\mathbf{S}_{n+k}$. Assume that M is diffeomorphic to the boundary ∂Q of an $(n+1)$-dimensional manifold-with-boundary Q; we must show that g is homotopic to a *constant map*. The proof (somewhat simplified by Milnor) can be divided in two parts; the first treats a very special case and the second reduces the general case to the special one.

In the special case $\mathbf{S}_{n+k}$ is identified to $\mathbf{R}^{n+k}$ with the point $*$ at infinity; it is assumed that Q is contained in $\mathbf{R}^{n+k} \times [0, \frac{1}{2}]$ and that the intersection of Q with $\mathbf{R}^{n+k} \times [0, 1/4]$ is the "cylinder" $M \times [0, 1/4]$. If V, defined by $d(x, Q) < \varepsilon$ in $\mathbf{R}^{n+k} \times [0, 1]$, is a sufficiently small tubular neighborhood of Q in $\mathbf{R}^{n+k+1}$, then $U = V \cap (\mathbf{R}^{n+k} \times \{0\})$ is the tubular neighborhood of $M \times \{0\}$ in $\mathbf{R}^{n+k} \times \{0\}$ defined by $d(x, M \times \{0\}) < \varepsilon$; when V is identified with the normal bundle of Q in $\mathbf{R}^{n+k} \times [0, 1]$, U is identified with the normal bundle of M in $\mathbf{R}^{n+k}$.

Let h be a bundle map of the vector bundle $(U, M, \mathbf{R}^k, p)$ into $(U_{m,k}, G_{m,k}, \mathbf{R}^k)$; it is possible to extend it to a bundle map F of $(V, Q, \mathbf{R}^k)$ into $(U_{m,k}, G_{m,k}, \mathbf{R}^k)$ by Steenrod's technique (chap. III, § 2) after having triangulated Q in such a way that M becomes a subcomplex of Q. If h is the restriction to U of a continuous map $f: \mathbf{S}_{n+k} \to T(\gamma_{m,k})$ with value t_0 in $\mathbf{S}_{n+k} - U$, then F can be extended to a continuous map $G: \mathbf{S}_{n+k} \times [0, 1] \to T(\gamma_{m,k})$ with value t_0 in the complement of V. Since $G(x, 0) = f(x)$ and $G(x, 1) = t_0$ for $x \in \mathbf{S}_{n+k}$, f is homotopic to a constant map.

The problem in this special case is to obtain a map f homotopic to the given g having a restriction h to U that is a bundle map. Consider first the restriction $g_1 = g|U$; for any $x \in U$ in the normal vector space to M at the point $p(x)$, sx, for $0 \leqslant s \leqslant 1$, means the image of x by the homotethy $z \mapsto sz$ in that vector space. Then $H(x, s) = g_1(sx)/s$ is defined for $0 \leqslant s \leqslant 1$, has a limit $H(x, 0) = \lim_{s \to 0} H(x, s)$ that is not 0 for $x \neq p(x)$, and gives a homotopy H of $g_1(x)$ to $h(x) = H(x, 0)$, which is easily seen to be a bundle map. However, if g is taken equal to t_0 in $\mathbf{S}_{n+k} - U$ (as can always be done by homotopy), H does not map $\operatorname{Fr}(U) \times [0, 1]$ to t_0. It can be modified in the following way. There is a $\delta > 0$ such that $d(x, M) \leqslant \frac{1}{2}\varepsilon$ implies $|H(x, s)| \geqslant \delta$ (for a euclidean distance on $U_{m,k}$) when $x \in U$. Then take

$$H_1(x, s) = \varphi(|H(x, s|/\delta) . H(x, s)$$

where $\zeta \mapsto \varphi(\zeta)$ is a C^∞ function defined in $[0, +\infty[$, increasing in that interval, equal to 1 in a neighborhood of 0 and tending to $+\infty$ with ζ. $H_1(x, 0) = h_1(x)$ restricted to U is again a bundle map, and $H_1(x, 1)$ is homotopic to $g(x)$ by a homotopy sending $(S_{n+k} - U) \times [0, 1]$ to t_0.

To treat the general case, start from a diffeomorphism h of $M \times [0, 1[$ onto a neighborhood of ∂Q in Q that maps M on ∂Q. It is then possible to define a *smooth map* h_1 of Q onto a "hut" the union of the "cylinder" $M \times [0, 1/4]$ and a "roof", the "cone" union of the line segments in $\mathbf{R}^{n+k+1}$ joining a point $p = (x_0, \frac{1}{2})$ to the points of $M \times [0, 1/4]$. This is done as follows:

(a) for $y = h(x, s)$ with $x \in M$, $0 \leqslant s \leqslant \frac{1}{2}$, take $h_1(y) = (x, \frac{1}{2}s)$;
(b) for $y \notin h(M \times [0, 1[)$, take $h_1(y) = p$;
(c) for $y = h(x, s)$ with $x \in M$, $\frac{1}{2} \leqslant s < 1$, take

$$h_1(y) = (1 - \beta(s)).(x, 1/4) + \beta(s).p$$

where β is an increasing C^∞ function in $[\frac{1}{2}, 1]$, equal to 0 in a neighborhood of $\frac{1}{2}$, to 1 in a neighborhood of 1.

Standard techniques of approximation of smooth maps ([232], p. 53) then yield an embedding $h_2: Q \to \mathbf{R}^{n+k+1}$ arbitrarily close to h_1 (for k large enough) and equal to h_1 in $h(M \times [0, \frac{1}{2}])$; the manifold-with-boundary $Q_2 = h_2(Q)$ is of the special type considered in the first part of the proof.

This already proves that *the vector space $\mathfrak{N}_n$ has dimension $d(n)$ over the field* $\mathbf{F}_2$.

III. Before dealing with multiplication in $\mathfrak{N}_\bullet$ Thom derived, from what he had already proved, *the solution to Steenrod's problem*. This uses the concept of Stiefel–Whitney numbers of a compact n-dimensional smooth manifold M. Let $w_1(\tau), \ldots, w_n(\tau)$ be the Stiefel–Whitney classes of the tangent bundle $\tau = (T(M), M, \mathbf{R}^n)$. For each sequence $(r_1, r_2, \ldots, r_n)$ of integers $\geqslant 0$ such that $r_1 + 2r_2 + \cdots + nr_n = n$, the cup-product

$$w_1(\tau)^{r_1} w_2(\tau)^{r_2} \cdots w_n(\tau)^{r_n}$$

is a cohomology class in $H^n(M; \mathbf{F}_2)$; the elements

$$\langle w_1(\tau)^{r_1} \cdots w_n(\tau)^{r_n}, [M] \rangle \in \mathbf{F}_2 \qquad (13)$$

are called the *Stiefel–Whitney numbers* of M.

In his work (section A) Pontrjagin found a *necessary* condition for M to be diffeomorphic to a boundary ∂Q: *all Stiefel–Whitney numbers of* M *must be* 0.

The proof is an application of the homology and cohomology exact sequences of the pair (Q, M) when $M = \partial Q$. The homomorphism

$$\partial_{n+1}: H_{n+1}(Q, M; \mathbf{F}_2) \to H_n(M; \mathbf{F}_2)$$

maps the fundamental class $\mu_{Q,M}$ of the pair (Q, M) on the fundamental class [M] ([440], p. 304). By duality between homology and cohomology with coefficients in a field (Part 1, chap. IV, § 5,D) we have, for every cohomology class $z \in H^n(M; \mathbf{F}_2)$,

$$\langle z, [\mathrm{M}]\rangle = \langle z, \partial_{n+1}\mu_{\mathrm{Q,M}}\rangle = \langle {}^t\partial_{n+1} z, \mu_{\mathrm{Q,M}}\rangle. \tag{14}$$

Now the normal bundle of M in Q is a *trivializable* line bundle since it has a section nowhere equal to 0; for the natural injection i: $\mathrm{M} \to \mathrm{Q}$, we therefore have for the Stiefel–Whitney classes

$$i^*(w_j(\mathrm{T(Q)})) = w_j(i^*(\mathrm{T(Q)})) = w_j(\tau) \quad \text{for } 1 \leqslant j \leqslant n. \tag{15}$$

From the exact cohomology sequence

$$\mathrm{H}^n(\mathrm{Q}; \mathbf{F}_2) \xrightarrow{i^*} \mathrm{H}^n(\mathrm{M}; \mathbf{F}_2) \xrightarrow{{}^t\partial_{n+1}} \mathrm{H}^{n+1}(\mathrm{Q}, \mathrm{M}; \mathbf{F}_2)$$

it follows that ${}^t\partial_{n+1}(i^*(z)) = 0$ for all $z \in \mathrm{H}^n(\mathrm{Q}; \mathbf{F}_2)$, so by (14), $\langle i^*(z), [\mathrm{M}]\rangle = 0$; in particular, for $r_1 + 2r_2 + \cdots + nr_n = n$,

$$\langle w_1(\tau)^{r_1} \cdots w_n(\tau)^{r_n}, [\mathrm{M}]\rangle = \langle i^*(w_1(\mathrm{T(Q)})^{r_1} \cdots w_n(\mathrm{T(Q)})^{r_n}), [\mathrm{M}]\rangle = 0.$$

The elucidation of the structure of the unoriented cobordism group $\mathfrak{N}_n$ enabled Thom to prove the *converse*: if all Stiefel–Whitney numbers of a smooth manifold M are 0, then M *is diffeomorphic to a boundary* ∂Q.

The proof consists in showing that if $\mathrm{cl(M)} \neq 0$ in $\mathfrak{N}_n$, then at least one of the Stiefel–Whitney numbers of M is not 0. From II, M can be identified with a submanifold $f^{-1}(\mathrm{G}_{m,k}) \subset \mathbf{S}_{n+k}$ for large m and k, where f: $\mathbf{S}_{n+k} \to \mathrm{T}(\gamma_{m,k})$ is C^∞ in $f^{-1}(\mathrm{U}_{m,k})$, transverse over $\mathrm{G}_{m,k}$, and *the class* $[f]$ *in* $\pi_{n+k}(\mathrm{T}(\gamma_{m,k}))$ *is* $\neq 0$. For a tubular neighborhood N of M in $\mathbf{S}_{n+k}$, N may be identified with the normal bundle of M in $\mathbf{S}_{n+k}$ and is the pullback $f^*(\mathrm{U}_{m,k})$. Consider the commutative diagram deduced from (7) for $\mathrm{X} = \mathbf{S}_{n+k}$, where we have suppressed the field of coefficients $\mathbf{F}_2$,

$$\begin{array}{ccc}
\mathrm{H}^{n+k}(\mathrm{T}(\gamma_{m,k})) & \xrightarrow{f^*} & \mathrm{H}^{n+k}(\mathbf{S}_{n+k}) \\
{\scriptstyle j^*}\uparrow\wr & & \uparrow{\scriptstyle j^*} \\
\mathrm{H}^{n+k}(\mathrm{T}(\gamma_{m,k}), \{t_0\}) & \xrightarrow{f^*} & \mathrm{H}^{n+k}(\mathbf{S}_{n+k}, \mathbf{S}_{n+k} - \mathrm{M}) \\
{\scriptstyle e}\uparrow\wr & & \wr\uparrow{\scriptstyle e} \\
\mathrm{H}^{n+k}(\mathrm{U}_{m,k}, \mathrm{U}^0_{m,k}) & \xrightarrow{f^*} & \mathrm{H}^{n+k}(\mathrm{N}, \mathrm{N}^0) \\
{\scriptstyle \Phi}\uparrow\wr & & \wr\uparrow{\scriptstyle \Phi} \\
\mathrm{H}^n(\mathrm{G}_{m,k}) & \xrightarrow{g^*} & \mathrm{H}^n(\mathrm{M})
\end{array} \tag{16}$$

where g: $\mathrm{M} \to \mathrm{G}_{m,k}$ is the restriction of f, and Φ is the Thom isomorphism. The elements of $\mathrm{H}^n(\mathrm{G}_{m,k})$ are polynomials of weight n in the generators w_1, $w_2, \ldots, w_k$ of the algebra $\mathrm{H}^\bullet(\mathrm{G}_{m,k})$ (chap. IV, § 1,D); their images in $\mathrm{H}^n(\mathrm{M})$ are polynomials of weight n (with coefficients in $\mathbf{F}_2$) in the Stiefel–Whitney classes $w_1(\mathrm{N}), \ldots, w_k(\mathrm{N})$ of the normal bundle N; since the sum $\mathrm{T(M)} \oplus \mathrm{N}$ is the trivial tangent bundle of $\mathbf{R}^{n+k}$, the $g^*(w_j)$ are also polynomials of weight n in the Stiefel–Whitney classes $w_1(\tau), \ldots, w_n(\tau)$.

For any vector fibration ξ of rank k and any partition $\omega = \{a_1, a_2, \ldots, a_k\}$ of an integer n let

$$s_\omega(w(\xi)) = s_\omega(w_1(\xi), \ldots, w_k(\xi)) = \mathrm{s}(t_1^{a_1} t_2^{a_2} \cdots t_k^{a_k}) \tag{17}$$

when the Stiefel–Whitney classes are expressed as elementary symmetric functions

$$w_j(\xi) = \sigma_j(t_1, \ldots, t_k) \tag{18}$$

in "phantom" indeterminates $t_1, \ldots, t_k$. For large m the $s_\omega(w_1, \ldots, w_k)$ form a *basis* of $\mathrm{H}^n(\mathrm{G}_{m,k})$ when ω runs through *all* partitions of the integer n. Suppose that for the given map f such that $[f] \neq 0$ there exists a partition ω of n such that for the corresponding element X_ω^n in $\mathrm{H}^{n+k}(\mathrm{T}(\gamma_{m,k}))$ defined by (5), $f^*(\mathrm{X}_\omega^n) \neq 0$ in $\mathrm{H}^{n+k}(\mathrm{S}_{n+k})$. Write $\varphi_{m,k}$ (resp. φ) the composite of the left (resp. right) vertical arrows in (16); by (5) and the definition of the Thom isomorphism (chap. IV, §2) $\mathrm{X}_\omega^n = \varphi_{m,k}(s_\omega(w_1, \ldots, w_k))$; the commutativity of the diagram (16) then gives

$$f^*(\mathrm{X}_\omega^n) = \varphi(g^*(s_\omega(w_1, \ldots, w_k)). \tag{19}$$

Let $z_\omega = g^*(s_\omega(w_1, \ldots, w_k))$; from the properties of the fundamental classes relative to the homology exact sequences and relative to the Thom isomorphism (chap. IV, §2), it follows easily that

$$\langle f^*(\mathrm{X}_\omega^n), [\mathrm{S}_{n+k}] \rangle = \langle z_\omega, [\mathrm{M}] \rangle. \tag{20}$$

Since $f^*(\mathrm{X}_\omega^n) \neq 0$ in $\mathrm{H}^n(\mathrm{S}_{n+k})$, the element $\langle z_\omega, [\mathrm{M}] \rangle$ is not 0 in $\mathbf{F}_2$, and since z_ω is a polynomial in the Stiefel–Whitney classes $w_j(\tau)$, not all Stiefel–Whitney numbers of M can be 0.

To prove the existence of a partition ω such that $f^*(\mathrm{X}_\omega^n) \neq 0$, note that from the properties of the map F in (6), by composition of that map with the projection on $(\mathrm{K}(\mathbf{Z}/2\mathbf{Z}, n+k))^{d(n)}$, we obtain a map

$$\mathrm{F}_n\colon \mathrm{T}(\gamma_{m,k}) \to \mathrm{K}(\mathbf{Z}/2\mathbf{Z}, n+k)^{d(n)}$$

which, in *homotopy*, yields an *isomorphism*

$$\mathrm{F}_{n,k}\colon \pi_{n+k}(\mathrm{T}(\gamma_{m,k})) \xrightarrow{\sim} (\pi_{n+k}(\mathrm{K}(\mathbf{Z}/2\mathbf{Z}, n+k))^{d(n)} \cong \mathbf{F}_2^{d(n)}.$$

Now consider the commutative diagram

$$\begin{array}{ccc}
\pi_{n+k}(\mathrm{T}(\gamma_{m,k})) & \xrightarrow{\mathrm{F}_{n,k}} & (\pi_{n+k}(\mathrm{K}(\mathbf{Z}/2\mathbf{Z}, n+k))^{d(n)} \\
\Big\downarrow{\scriptstyle h} & & \Big\downarrow{\scriptstyle h'^{d(n)}} \\
\mathrm{H}_{n+k}(\mathrm{T}(\gamma_{m,k})) & \xrightarrow[\mathrm{F}_{n*}]{} & (\mathrm{H}_{n+k}(\mathrm{K}(\mathbf{Z}/2\mathbf{Z}, n+k))^{d(n)}
\end{array}$$

(coefficients of homology in $\mathbf{F}_2$), where the vertical arrows are the Hurewicz homomorphisms. By definition of the Eilenberg–Mac Lane spaces $h'^{d(n)}$ is bijective, so $\mathrm{F}_{n*} \circ h$ is bijective; this implies that h is injective and its image $h(\pi_{n+k}(\mathrm{T}(\gamma_{m,k})))$ is a vector subspace supplementary to the kernel $\mathrm{F}_{n*}^{-1}(0)$ and has dimension $d(n)$ over $\mathbf{F}_2$; the assumption on f thus implies $h([f]) \neq 0$. Passing to cohomology, since F_n^* is transposed of F_{n*}, the map

$$F_n^*: (H^{n+k}(K(\mathbf{Z}/2\mathbf{Z}, n+k))^{d(n)} \to H^{n+k}(T(\gamma_{m,k}))$$

is injective, and its image is naturally identified with the *dual* space of $h(\pi_{n+k}(T(\gamma_{m,k})))$. But by definition (section D) this image of F_n^* has a basis over $\mathbf{F}_2$ consisting of the X_ω^n for all *nondyadic* partitions of n. Since $h([f]) \neq 0$, there is therefore at least one of these nondyadic partitions ω for which

$$\langle X_\omega^n, h([f])\rangle \neq 0$$

and by definition of the Hurewicz homomorphism [chap. II, § 4,A, formula (67)], this means that $f^*(X_\omega^n) \neq 0$.

IV. To determine the multiplication in $\mathfrak{N}_\bullet$ Thom used special n-dimensional manifolds M_ω, indexed by the nondyadic partitions ω of n, such that the classes $\mathrm{cl}(M_\omega)$ form a *basis* for $\mathfrak{N}_n$. From the arguments of III it follows that there is in $\pi_{n+k}(T(\gamma_{m,k}))$ (for large m and k) a basis $([f_\omega])$ such that $(h([f_\omega]))$ is the *dual basis* of the basis (X_ω^n) of the image of F_n^*. It may be assumed that f_ω is transverse over $G_{m,k}$, and $M_\omega = f_\omega^{-1}(G_{m,k})$ in S_{n+k}. By definition $f_\omega^*(X_{\omega'}^n) = 0$ for two different nondyadic partitions ω, ω' of n; if N_ω is the normal bundle of M_ω in S_{n+k}, it follows from (19) and (20) that if $z_{\omega'} = s_{\omega'}(w_1(N_\omega), \ldots, w_k(N_\omega))$,

$$\langle z_{\omega'}, [M_\omega]\rangle = 0 \quad \text{for } \omega \neq \omega'. \tag{21}$$

It is enough to show that if $n = r + s$, ω_1 (resp. ω_2) is a nondyadic partition of r (resp. s), and $\omega = \omega_1 \coprod \omega_2$, then in the algebra $\mathfrak{N}_\bullet$,

$$\mathrm{cl}(M_\omega) = \mathrm{cl}(M_{\omega_1} \times M_{\omega_2}) = \mathrm{cl}(M_{\omega_1})\mathrm{cl}(M_{\omega_2}). \tag{22}$$

Indeed, if $v_a = \mathrm{cl}(M_{\{a\}})$ for every integer a which is not of the form $2^m - 1$, it follows from (22) by induction on n that for every nondyadic partition $\omega = \{a_1, a_2, \ldots, a_n\}$ of n,

$$\mathrm{cl}(M_\omega) = v_{a_1} v_{a_2} \cdots v_{a_n}, \tag{23}$$

and the structure theorem follows.

Relation (22) means that M_ω and $M_{\omega_1} \times M_{\omega_2}$ are *cobordant*. This follows from the Pontrjagin–Thom criterion proved in III if M_ω and $M_{\omega_1} \times M_{\omega_2}$ have *the same Stiefel–Whitney numbers*.

Let N_ω be the normal bundle of M_ω in $\mathbf{R}^{n+k}$ and N_{ω_1} (resp. N_{ω_2}) be the normal bundle of M_{ω_1} in $\mathbf{R}^{r+k'}$ (resp. the normal bundle of M_{ω_2} in $\mathbf{R}^{s+k''}$) with $k' + k'' = k$ (k' and k'' large); then the normal bundle of $M_{\omega_1} \times M_{\omega_2}$ can be written as a Whitney sum

$$N_{\omega_1, \omega_2} = \mathrm{pr}_1^* N_{\omega_1} \oplus \mathrm{pr}_2^* N_{\omega_2}. \tag{24}$$

For clarity, we write as $z_{\omega'}(N_\omega)$ the element of $H^n(M_\omega)$ written $z_{\omega'}$ above, and similarly $z_{\omega_1'}(N_{\omega_1})$, $z_{\omega_2'}(N_{\omega_2})$ are the elements of $H^r(M_{\omega_1})$ and $H^s(M_{\omega_2})$ defined in the same manner, ω_1' (resp. ω_2') running through all nondyadic partitions of r (resp. s). If

$$z_{\omega'}(\mathrm{N}_{\omega_1,\omega_2}) \in \mathrm{H}^n(\mathrm{M}_{\omega_1} \times \mathrm{M}_{\omega_2})$$

is also defined in the same way for every nondyadic partition of n, we must prove that

$$\langle z_{\omega'}(\mathrm{N}_\omega), [\mathrm{M}_\omega]\rangle = \langle z_{\omega'}(\mathrm{N}_{\omega_1,\omega_2}), [\mathrm{M}_{\omega_1} \times \mathrm{M}_{\omega_2}]\rangle \tag{25}$$

for all such partitions ω'; this will imply that all Stiefel–Whitney numbers of M_ω are equal to the corresponding Stiefel–Whitney numbers of $\mathrm{M}_{\omega_1} \times \mathrm{M}_{\omega_2}$, because the Stiefel–Whitney classes of a tangent bundle of a submanifold of an $\mathbf{R}^\mathrm{N}$ are expressed as polynomials in the Stiefel–Whitney classes of the normal bundle of that submanifold [chap. IV, § 1,B, formula (7)].

The computation of the classes $z_{\omega'}(\mathrm{N}_{\omega_1,\omega_2})$ is, by (24), a special case of the computation of $s_\omega(w(\xi \oplus \xi'))$ for two vector fibrations ξ, ξ' with base space a smooth manifold of dimension n and a partition ω of n, the ranks of ξ and ξ' being $>n$. The general formula Thom proved is

$$s_\omega(w(\xi \oplus \xi')) = \sum_{\omega_1 \coprod \omega_2 = \omega} s_{\omega_1}(w(\xi)) \smile s_{\omega_2}(w(\xi')) \tag{26}$$

the sum being over *all* disjoint unions $\omega_1 \coprod \omega_2 = \omega$. Using "phantom" indeterminates $t_1, \dots, t_k, t_{k+1}, \dots, t_{2k}$ with $k > n$ to express the Stiefel–Whitney classes of ξ and ξ', the verification of (26) is easily reduced to the purely algebraic formula

$$s_\omega(\sigma_1, \dots, \sigma_n) = \sum_{\omega_1 \coprod \omega_2 = \omega} s_{\omega_1}(\sigma'_1, \sigma'_2, \dots) s_{\omega_2}(\sigma''_1, \sigma''_2, \dots) \tag{27}$$

where σ_j (resp. σ'_j, σ''_j) is the j-th elementary function of $t_1, \dots, t_{2k}$ (resp. $t_1, \dots, t_k$, resp. $t_{k+1}, \dots, t_{2k}$); the verification of that formula is straightforward.

Applying (26) to the expression (24) of the normal bundle $\mathrm{N}_{\omega_1,\omega_2}$, we get

$$\begin{aligned}&\langle s_{\omega'}(w(\mathrm{N}_{\omega_1,\omega_2})), [\mathrm{M}_{\omega_1} \times \mathrm{M}_{\omega_2}]\rangle \\ &\quad = \sum_{\omega'_1 \coprod \omega'_2 = \omega'} \langle \mathrm{pr}_1^*(s_{\omega'_1}(w(\mathrm{N}_{\omega_1})) \smile \mathrm{pr}_2^*(s_{\omega'_2}(w(\mathrm{N}_{\omega_2})), [\mathrm{M}_{\omega_1}] \times [\mathrm{M}_{\omega_2}]\rangle \end{aligned} \tag{28}$$

the sum being over *all* disjoint unions $\omega'_1 \coprod \omega'_2 = \omega'$; but only the terms of that sum for which ω'_1 is a partition of r and ω'_2 is a partition of s can be $\neq 0$, and the sum is then [Part 1, chap. IV, § 5,H, formula (81)]

$$\sum_{\omega'_1 \coprod \omega'_2 = \omega'} \langle s_{\omega'_1}(w(\mathrm{N}_{\omega_1})), [\mathrm{M}_{\omega_1}]\rangle . \langle s_{\omega'_2}(w(\mathrm{N}_{\omega_2})), [\mathrm{M}_{\omega_2}]\rangle$$

restricted to those partitions of r and s. Finally, by using (21) we see that $\langle z_{\omega'}(\mathrm{N}_{\omega_1,\omega_2}), [\mathrm{M}_{\omega_1} \times \mathrm{M}_{\omega_2}]\rangle = 0$ except when $\omega' = \omega = \omega_1 \coprod \omega_2$, ending the proof.

V. The manifolds $\mathrm{M}_{\{a\}}$ for integers a other than the numbers $2^m - 1$ were not all explicitly described by Thom; he only showed that for *even* values of a, $\mathrm{cl}(\mathbf{P}_a(\mathbf{R})) = \mathsf{v}_a$, and he gave examples of other manifolds whose classes in $\mathfrak{N}_\bullet$ are the v_a for $a \leqslant 8$. The description of the generators v_a for *odd* a was completed by Dold [141]. Let $\mathrm{P}(m,n)$ be the quotient space of $\mathbf{S}_m \times \mathbf{P}_n(\mathbf{C})$ by the equivalence relation

$$(x, z) \equiv (-x, \bar{z})$$

so that $\mathbf{S}_m \times \mathbf{P}_n(\mathbf{C})$ is the universal covering space of $\mathrm{P}(m, n)$. Then for *odd* $j = 2^r(2s + 1) - 1$ with $r \geqslant 1$, $s \geqslant 1$, v_j is the class of the manifold $\mathrm{P}(2^r - 1, s \,.\, 2^r)$; Dold's proof consists in explicitly computing the Stiefel–Whitney numbers of $\mathrm{P}(m, n)$ and checking that they are not all 0 for $\mathrm{P}(2^r - 1, s \,.\, 2^r)$.

H. Oriented Cobordism

The theory of *oriented cobordism* introduces an equivalence relation R'_n between *oriented* n-dimensional smooth compact manifolds (without boundary). To say that an $(n + 1)$-dimensional smooth manifold-with-boundary W is *oriented* means that its tangent bundle is oriented (chap. IV, § 1,C). Since the normal bundle of $\partial \mathrm{W}$ in W is trivializable, the tangent bundle to $\partial \mathrm{W}$ is *orientable*: it is given the orientation *induced* by the orientation of W by taking in each tangent space $\mathrm{T}_x(\partial \mathrm{W})$ as oriented basis $(u_2, \ldots, u_{n+1})$ such that if u_1 is a tangent vector in $\mathrm{T}_x(\mathrm{W})$ which "points out" of W, $(u_1, u_2, \ldots, u_{n+1})$ is an oriented basis of $\mathrm{T}_x(\mathrm{W})$.

For an oriented manifold X (with or without boundary), write $-\mathrm{X}$ the same manifold with the opposite orientation. Then two oriented n-dimensional smooth compact manifolds M, M′ are equivalent for R'_n (or *cobordant* if no confusion is possible) if there exists an $(n + 1)$-dimensional smooth manifold-with-boundary W such that $\partial \mathrm{W}$, with the orientation induced by the orientation of W, is diffeomorphic* to $\mathrm{M} \coprod (-\mathrm{M}')$. The fact that R'_n is an equivalence relation is easily verified in the same way as in unoriented cobordism; reflexivity here is due to the fact that $\mathrm{M} \coprod (-\mathrm{M})$ is the *oriented* boundary of $\mathrm{M} \times [0, 1]$. Thom wrote Ω_n for the set of equivalence classes for R'_n. If the equivalence class of M for R'_n is again written cl(M), a law of composition in Ω_n is defined by

$$(\mathrm{cl}(\mathrm{M}), \mathrm{cl}(\mathrm{M}')) \mapsto \mathrm{cl}(\mathrm{M} \coprod \mathrm{M}')$$

written $\mathrm{cl}(\mathrm{M}) + \mathrm{cl}(\mathrm{M}')$. It is again easily verified that this definition is meaningful, and gives a structure of a *commutative group* on Ω_n, the neutral element 0 being the class of n-dimensional oriented manifolds that are boundaries of oriented $(n + 1)$-dimensional manifolds-with-boundary; here Ω_n may have elements of finite or infinite order.

The *graded algebra*

$$\Omega_{\bullet} = \bigoplus_{n \geqslant 0} \Omega_n \tag{29}$$

(with $\Omega_0 = \mathbf{Z}$) is defined in the same way as $\mathfrak{N}_{\bullet}$, but it is now *anticommutative*, since in an exterior algebra $z_p \wedge z_q = (-1)^{pq} z_q \wedge z_p$ when z_p has degree p and z_q has degree q.

---

* We say that two oriented manifolds are diffeomorphic if there exists a diffeomorphism between them that *preserves orientation*.

The notion of transversality and the transversality theorem (section B) are easily extended to manifolds-with-boundary. The argument leading to the isomorphism theorem (12) may then be repeated with slight modifications when $\gamma_{m,k}$, $G_{m,k}$, $U_{m,k}$ are respectively replaced by $\gamma'_{m,k}$, $G'_{m,k}$, and $U'_{m,k}$; they establish an *isomorphism* of commutative groups

$$\lambda_{\gamma'_{m,k}}: \pi_{n+k}(T(\gamma'_{m,k})) \xrightarrow{\sim} \Omega_n \tag{30}$$

for large m and k.

As Thom's results for the cohomology of MSO(k) were much less complete than for MO(k), he could only explicitly determine the groups Ω_n for $n \leqslant 7$:

$$\Omega_1 = \Omega_2 = \Omega_3 = 0,^* \qquad \Omega_4 = \mathbf{Z}, \qquad \Omega_5 = \mathbf{Z}/2\mathbf{Z}, \qquad \Omega_6 = \Omega_7 = 0. \tag{31}$$

The $\mathscr{C}$-isomorphism (for the class $\mathscr{C}$ of commutative finite groups)

$$\pi_r(T(\gamma'_{m,k})) \to H_r(G'_{m,k}; \mathbf{Z}) \quad \text{for } r < 2k - 1$$

[section D, formula (2)] also enabled Thom to show that Ω_n *is finite for* $n \not\equiv 0$ (mod. 4), *and* $\Omega_{4r} \otimes_{\mathbf{Z}} \mathbf{Q}$ *has dimension* $c(r)$ *over* $\mathbf{Q}$, *where* $c(r)$ *is the number of all partitions of* r.

For a smooth, oriented manifold M of dimension $4r$, the *Pontrjagin numbers* are defined in the same way as Stiefel–Whitney numbers: for any partition $\{a_1, a_2, \ldots, a_r\}$ of r,

$$\langle p_1^{a_1} p_2^{a_2} \cdots p_r^{a_r}, [M] \rangle \in \mathbf{Z}$$

where $p_j = p_j(T(M)) \in H^{4j}(M; \mathbf{Z})$ is the j-th Pontrjagin class of the tangent bundle T(M) (chap. IV, § 1,C). By the same argument as in section G Pontrjagin showed that when M is the boundary of an $(n + 1)$-dimensional manifold-with-boundary *all Pontrjagin numbers of* M *are* 0.

Using results of A. Borel and Serre on the generation of the cohomology of special grassmannians by Pontrjagin classes [68], Thom repeated his arguments of section G for $\Omega_{4r} \otimes_{\mathbf{Z}} \mathbf{Q}$. For each partition ω of r there is a $4r$-dimensional oriented manifold W_ω with all its Pontrjagin numbers equal to 0 except the one corresponding to the partition ω. He showed that in $\Omega_{4r} \otimes \mathbf{Q}$, $\mathrm{cl}(W_\omega) = \mathrm{cl}(W_{\omega_1}) . \mathrm{cl}(W_{\omega_2})$ if $\omega = \omega_1 \coprod \omega_2$. It is easily verified that for each integer $a \geqslant 1$, $W_{\{a\}}$ is cobordant to $\mathbf{P}_{2a}(\mathbf{C})$, and therefore the algebra $\Omega_{\bullet} \otimes_{\mathbf{Z}} \mathbf{Q}$ is *isomorphic to the algebra of polynomials* $\mathbf{Q}[\mathrm{w}_4, \mathrm{w}_8, \ldots]$ with $\mathrm{w}_{4n} = \mathrm{cl}(\mathbf{P}_{2n}(\mathbf{C}))$. Finally, Thom proved that if an oriented manifold M has all its Pontrjagin numbers equal to 0, then cl(M) has *finite order* in the group $\Omega_{\bullet}$.

I. Later Developments

Thom's paper immediately attracted much attention among topologists. The first challenge was to complete the determination of the structure of the ring $\Omega_{\bullet}$. This was done between 1956 and 1959 by the combined contributions of

* The relation $\Omega_3 = 0$ (in another formulation) had been announced earlier by Rokhlin [397], without a complete proof.

several mathematicians. Thom had left open the question of the structure of the *torsion subgroup* for every group Ω_n with $n \geqslant 8$. In 1959 Milnor [344] and Averbuch [36] independently obtained an important result in that problem: there is *no odd torsion*; the order of an element of finite order in a group Ω_n is always *a power of* 2. Milnor's proof used a deep study of the Steenrod algebra $\mathscr{A}_p$ for odd p and the Adams spectral sequence (see below, § 5,D).

The complete determination of the torsion of the groups Ω_n and the algebra structure of $\Omega_{\bullet}$ was accomplished in 1959 by C.T.C. Wall [479]. He proved the following theorems.

I. Every element $\neq 0$ and of finite order in a group Ω_n has order 2.

II. For each $q \geqslant 1$, there is a subgroup $\mathfrak{W}_q$ of $\mathfrak{N}_q$ and a homomorphism $d\colon \mathfrak{W}_q \to \Omega_{q-1}$ such that the infinite sequence

$$\cdots \to \Omega_q \xrightarrow{.2} \Omega_q \xrightarrow{r} \mathfrak{W}_q \xrightarrow{d} \Omega_{q-1} \xrightarrow{.2} \Omega_{q-1} \to \cdots \tag{32}$$

is *exact*, where .2 is multiplication by 2, and r the "forgetful" map that to the class of an oriented manifold in Ω_q associates the class of the same manifold in $\mathfrak{N}_q$ when its orientation is disregarded.

III. The direct sum $\mathfrak{W}_{\bullet} = \bigoplus_{q \geqslant 0} \mathfrak{W}_q$ (with $\mathfrak{W}_0 = \mathbf{F}_2$) is a subalgebra of $\mathfrak{N}_{\bullet}$ and the composite maps

$$d'_q\colon \mathfrak{W}_q \xrightarrow{d} \Omega_{q-1} \xrightarrow{r} \mathfrak{W}_{q-1} \tag{33}$$

define a *derivation* d' of the algebra $\mathfrak{W}_{\bullet}$.

IV. $\mathfrak{W}_{\bullet}$ is a graded algebra of polynomials

$$\mathbf{F}_2[\mathrm{z}_2, \mathrm{z}_4, \mathrm{z}_5, \mathrm{z}_6, \mathrm{z}_8, \mathrm{z}_9, \ldots] \tag{34}$$

with exactly one generator z_k of degree k for each integer k such that $k+1$ is not a power of 2:

$$\mathrm{z}_{2q-1} = \mathrm{v}_{2q-1} \quad \text{for } q \text{ not a power of 2}, \qquad \mathrm{z}_{2^j+1} = \mathrm{v}_{2^j}^2 \quad \text{for } j \geqslant 1, \tag{35}$$

and

$$d'(\mathrm{z}_{2q}) = \mathrm{z}_{2q-1} \quad \text{for } q \text{ not a power of 2}, \qquad d'(\mathrm{z}_{2^j+1}) = 0, \quad d'(\mathrm{z}_{2q-1}) = 0. \tag{36}$$

V. The ideal Θ of torsion elements in $\Omega_{\bullet}$ is an algebra over $\mathbf{F}_2$ isomorphic to the subalgebra $\mathrm{Im}(d')$ of $\mathfrak{W}_{\bullet}$. The quotient algebra $\Omega_{\bullet}/\Theta$ is isomorphic to the algebra of polynomials

$$\mathbf{Z}[\mathrm{w}_4, \mathrm{w}_8, \mathrm{w}_{12}, \ldots] \tag{37}$$

and the multiplication in $\Omega_{\bullet}$ is such that for $\mathrm{U} \in \Theta = \mathrm{Im}(d')$, $\mathrm{w}_{4j}\mathrm{U} = \mathrm{z}_{2i}^2\mathrm{U} \in \mathrm{Im}(d') = \Theta$.

VI. In order that an oriented manifold be diffeomorphic to the boundary of an oriented manifold-with-boundary, it is necessary and sufficient that its Stiefel–Whitney numbers and its Pontrjagin numbers all be 0 (the solution of Steenrod's problem for oriented manifolds conjectured by Thom).

Wall used the Milnor–Averbuch result and the fact that the sequence $\Omega_q \xrightarrow{.2} \Omega_q \xrightarrow{r} \mathfrak{N}_q$ is exact, announced earlier by Rokhlin [390] and for which

Dold had given a proof [142]. The center of the proof is a construction of $\mathfrak{W}_q$ and of the map d inspired by a similar construction by Rokhlin [399], but Wall did not use any of Rokhlin's results. For clarity it is convenient to write $\mathrm{cl}_\Omega(\mathrm{M})$ and $\mathrm{cl}_\mathfrak{N}(\mathrm{M})$ the classes in $\Omega_\bullet$ and $\mathfrak{N}_\bullet$.

The elements of $\mathfrak{W}_q$ are the classes $\mathrm{cl}_\mathfrak{N}(\mathrm{M})$ of manifolds M of dimension q for which the first Stiefel–Whitney class $w_1(\mathrm{T}(\mathrm{M}))$ is the *image* of a cohomology class $u \in \mathrm{H}^1(\mathrm{M}; \mathbf{Z})$ by the natural homomorphism

$$\mathrm{H}^1(\mathrm{M}; \mathbf{Z}) \to \mathrm{H}^1(\mathrm{M}; \mathbf{F}_2).$$

Since there is a natural bijection $\mathrm{H}^1(\mathrm{M}; \mathbf{Z}) \simeq [\mathrm{M}; \mathrm{K}(\mathbf{Z}, 1)]$ (chap. V, § 1,D) and $\mathrm{K}(\mathbf{Z}, 1)$ has the homotopy type of $\mathbf{S}_1$, there is a continuous map $f: \mathrm{M} \to \mathbf{S}_1$ such that $u = f^*(s_1)$, where s_1 is a generator of $\mathrm{H}^1(\mathbf{S}_1; \mathbf{Z}) \simeq \mathbf{Z}$, so $w_1(\mathrm{T}(\mathrm{M})) = f^*(\bar{s}_1)$, where $\bar{s}_1$ is the unique element $\neq 0$ in $\mathrm{H}^1(\mathbf{S}_1; \mathbf{F}_2)$. If $w_1(\mathrm{T}(\mathrm{M})) = 0$, take $d(\mathrm{cl}_\mathfrak{N}(\mathrm{M}))) = 0$; we may therefore suppose $w_1(\mathrm{T}(\mathrm{M})) \neq 0$. Since $\bar{s}_1$ is the Stiefel–Whitney class of the nontrivial double sheet covering space S'_1 of $\mathbf{S}_1$, $w_1(\mathrm{T}(\mathrm{M}))$ is the Stiefel–Whitney class of a nontrivial double sheet covering space $f^*(\mathrm{S}'_1)$ of M. By the transversality theorem, it may be assumed that $f^{-1}(*) = \mathrm{V}$ is a $(q-1)$-dimensional submanifold of M. Wall proved that V is orientable, that $\mathrm{cl}_\Omega(\mathrm{V})$ only depends on $\mathrm{cl}_\mathfrak{N}(\mathrm{M})$, and that $2 . \mathrm{cl}_\Omega(\mathrm{V}) = 0$. It is convenient to identify $\mathbf{S}_1$ with the space obtained by identification of the points 0, 1 in the interval $[0, 1]$ of $\mathbf{R}$. Since f is transverse over a small interval $[0, \delta]$ and the covering space S'_1 is trivial above that interval, $f^*(\mathrm{S}'_1)$ is trivial over $f^{-1}([0, \delta])$, which is therefore an orientable manifold-with-boundary; that boundary consists of V and $\mathrm{V}_1 = f^{-1}(\delta)$, and is therefore orientable. By transversality, the normal bundle of V in M is trivial, so there is a diffeomorphism of V onto V_1 preserving orientation. The same argument may be repeated for $f^{-1}([1-\delta, 1])$ and for its boundary, the disjoint union of V and $\mathrm{V}_2 = f^{-1}(1-\delta)$. The covering space S'_1 is also trivial above the open interval $]0, 1[$, so $f^{-1}([\delta, 1-\delta])$ is an orientable manifold-with-boundary, its boundary being the disjoint union $\mathrm{V}_1 \coprod \mathrm{V}_2$. Finally, the orientations induced on V_1 and V_2 give, by the diffeomorphisms mentioned above, the *same* orientation on V, since otherwise the covering space $f^*(\mathrm{S}'_1)$ would be trivial; this proves that $\mathrm{cl}_\Omega(\mathrm{V} \coprod \mathrm{V}) = 0$.

If $\mathrm{cl}_\mathfrak{N}(\mathrm{M}) = \mathrm{cl}_\mathfrak{N}(\mathrm{M}')$, so that there is a manifold-with-boundary B such that $\partial \mathrm{B} = \mathrm{M} \coprod \mathrm{M}'$, and if we have two maps $f: \mathrm{M} \to \mathbf{S}_1$, $f': \mathrm{M}' \to \mathbf{S}_1$ with $f^{-1}(*) = \mathrm{V}$, $f'^{-1}(*) = \mathrm{V}'$, there is a map $\mathrm{F}: \mathrm{B} \to \mathbf{S}_1$ extending f and f' and transverse over the point $*$, so that $\mathrm{F}^{-1}(*) = \mathrm{C}$ is a q-dimensional orientable manifold-with-boundary; its boundary is $(\pm\mathrm{V}) \coprod (\pm\mathrm{V}')$, and therefore $\mathrm{cl}_\Omega(\mathrm{V}) = \pm\mathrm{cl}_\Omega(\mathrm{V}')$; but since $2 . \mathrm{cl}_\Omega(\mathrm{V}) = 0$, this means $\mathrm{cl}_\Omega(\mathrm{V}) = \mathrm{cl}_\Omega(\mathrm{V}')$. The map $d: \mathfrak{W}_q \to \Omega_{q-1}$ is thus well defined, and easily seen to be a homomorphism.

Next Wall proved that the sequence $\mathfrak{W}_q \xrightarrow{d} \Omega_{q-1} \xrightarrow{.2} \Omega_{q-1}$ is exact. Suppose V is an oriented $(q-1)$-dimensional manifold such that $2 . \mathrm{cl}_\Omega(\mathrm{V}) = 0$; there is therefore an oriented q-dimensional manifold-with-boundary M' such that $\partial \mathrm{M}' = \mathrm{V}_1 \coprod \mathrm{V}_2$, where V_1, V_2 are oriented manifolds diffeomorphic to V. Let ρ be a smooth distance on M' such that $\rho(\mathrm{V}_1, \mathrm{V}_2) > 1$; consider a map $g: \mathrm{M}' \to [0, 1]$ such that

$$\begin{aligned} g(x) &= \rho(x, V_1) && \text{if } \rho(x, V_1) < \tfrac{1}{2}, \\ g(x) &= 1 - \rho(x, V_2) && \text{if } \rho(x, V_2) < \tfrac{1}{2}, \\ g(x) &= \tfrac{1}{2} && \text{at other points.} \end{aligned}$$

If $h: V_1 \to V_2$ is a diffeomorphism preserving orientation, let M be the manifold (without boundary) obtained by identifying x and $h(x)$ for every $x \in V_1$, and let V be the submanifold of M image of $V_1 \coprod V_2$ by that identification; since $g(x) = 0$ for $x \in V_1$ and $g(x) = 1$ for $x \in V_2$, there is a continuous map $f: M \to S_1$ such that $f^{-1}(*) = V$ and $f(x) = g(x)$ for $x \notin V$. Then one shows that $\text{cl}_{\mathfrak{N}}(M)$ belongs to $\mathfrak{W}_q$, and by the construction, $d(\text{cl}_{\mathfrak{N}}(M)) = \text{cl}_{\Omega} V$.

Wall postponed the proof of the exactness of the sequence $\Omega_q \xrightarrow{r} \mathfrak{W}_q \xrightarrow{d} \Omega_{q-1}$ to the end of his paper, listing it as a consequence of the other theorems on $\Omega_.$. It may be derived directly from the definitions by a construction of Dold [143], inspired by another construction of Rokhlin. If $\text{cl}_{\mathfrak{N}}(M) \in \mathfrak{W}_q$, and the submanifold V such that $\text{cl}_{\Omega}(V) = d(\text{cl}_{\mathfrak{N}}(M))$ verifies $\text{cl}_{\Omega}(V) = 0$, there is an oriented q-dimensional manifold L such that $\partial L = V$. There is a tubular neighborhood T of V in M, and a diffeomorphism $(x, t) \mapsto h(x, t)$ of $V \times [-1, 1]$ onto $\bar{T}$. In the disjoint union

$$(M \times [-1, 1]) \coprod (L \times [-1, 1])$$

identify each point $(x, t) \in V \times [-1, 1] \subset L \times [-1, 1]$ to the point $(h(x, t), 1) \in M \times \{1\}$. This gives (after "smoothing the corners" in a standard way [232]) a manifold-with-boundary P whose boundary is the disjoint union of $M \times \{-1\}$ and of the oriented manifold $((M - T) \times \{1\}) \cup (L \times \{-1\}) \cup (L \times \{1\}) = M'$; therefore

$$\text{cl}_{\mathfrak{N}}(M) = r(\text{cl}_{\Omega}(M')).$$

The fact that $\mathfrak{W}_.$ is an algebra is an immediate consequence of the definition and of the Whitney–Wu Wen Tsün formula

$$w_1(\xi \times \xi') = w_1(\xi) \times 1 + 1 \times w_1(\xi')$$

(chap. IV, § 1,B). In Wall's construction $w_k(T(\dot{V})) = j^*(w_k(T(M)))$ for the injection $j: V \to M$, since the normal bundle of V in M is trivial; the homology class $j_*([V]) = w_1(T(M)) \frown [M]$ by Poincaré duality and the definition of $\mathfrak{W}_q$; for any monomial $z_\omega = w_1(T(V))^{a_1} \cdots w_r(T(V))^{a_r}$ of weight $q - 1$, $z_\omega = j^*(z'_\omega)$ with $z'_\omega = w_1(T(M))^{a_1} \cdots w_r(T(M))^{a_r}$, hence the relations

$$\begin{aligned} \langle z_\omega, [V] \rangle = \langle z'_\omega, j_*([V]) \rangle &= \langle z'_\omega, w_1(T(M)) \frown [M] \rangle \\ &= \langle z'_\omega \smile w_1(T(M)), [M] \rangle \end{aligned}$$

express the Stiefel–Whitney numbers of V in terms of those of M. Together with the Whitney–Wu Wen Tsün formula for the total Stiefel–Whitney class $w(T(M \times M'))$, this shows that

$$d'(\text{cl}_{\mathfrak{N}}(M)\,\text{cl}_{\mathfrak{N}}(M')) = d'(\text{cl}_{\mathfrak{N}}(M)) . \text{cl}_{\mathfrak{N}}(M') + \text{cl}_{\mathfrak{N}}(M) . d'(\text{cl}_{\mathfrak{N}}(M'))$$

because the manifolds representing the cobordism classes of both sides have the same Stiefel–Whitney numbers.

The greater part of Wall's paper is devoted to the structure of the algebra $\mathfrak{W}_\bullet$ and the formulas (36). In addition to the generators v_{2q-1} and v_{2^j} of the algebra $\mathfrak{N}_\bullet$ (section G, V), new generators z_{2q} had to be introduced. In the space $P(m,n)$ defined by Dold (*loc. cit.*), define an involution h which associates the image of $(s\,.\,x, z)$ to the image of (x, z) in $P(m,n)$, where s is the symmetry in $\mathbf{R}^{m+1}$ with respect to the hyperplane $x_m = 0$. Then consider the space $Q(m,n)$ obtained by identifying in $P(m,n) \times [0,1]$ the points $(y,0)$ and $(h(y),1)$ for $y \in P(m,n)$; for $2q = 2^r(2s+1)$ with $s > 0$, z_{2q} is the class in $\mathfrak{N}_\bullet$ of the space $Q(2^r - 1, 2^r s)$. The most difficult part of the proof of IV giving the structure of $\mathfrak{W}_\bullet$ is to show that the subalgebra generated by the z_k is equal to $\mathfrak{W}_\bullet$; it requires a delicate study of $H^\bullet(MO(n); F_2)$ considered as a module over the Steenrod algebra $\mathscr{A}_2$. Then the derivation d' defines $\mathfrak{W}_\bullet$ as a graded differential algebra, and a study of its homology of that algebra enabled Wall to prove that:

$$\mathrm{Ker}(d')/\mathrm{Im}(d') = \mathbf{F}_2[z_2^2, z_4^2, \ldots, z_{2q}^2, \ldots]; \tag{38}$$

$$\mathrm{Im}(d') = r(\Theta). \tag{39}$$

From (39) it follows that no element of Θ may have an order 2^n with $n \geqslant 2$, because for such an element $x \in \Omega_q$ (with *maximal* n), $r(x) = d'(x') = r(d(x'))$, so $r(x - d(x')) = 0$, and by exactness $x - d(x') = 2x''$; if $n > 1$, then $0 = 2^n x'' = 2^{n-1}(x - d(x'))$ would hold, and since $2d(x') = 0$, this would give $2^{n-1}x = 0$, contradicting the maximality of n.

The characterization VI of boundaries of oriented manifolds-with-boundary follows: if the Stiefel–Whitney numbers of M are 0, then $\mathrm{cl}_{\mathfrak{N}}(M) = 0$ by Thom's theorem (section G,III); if the Pontrjagin numbers are also 0, then $x = \mathrm{cl}_\Omega(M) \in \Theta$ (section H); but by exactness $x = 2y$ with $y \in \Omega_\bullet$, and since there is no odd torsion, $2y = 0$, so $x = 0$.

Finally, $\mathrm{Ker}(d) = \mathrm{Ker}(d')$: obviously $\mathrm{Ker}(d) \subset \mathrm{Ker}(d')$; if $x \in \mathrm{Ker}(d')$, $r(d(x)) = 0$, so $d(x) = 2y$, and since $2d(x) = 0$, $4y = 0$, so $2y = 0$ and $x \in \mathrm{Ker}(d)$. The same argument shows that the restriction of r to Θ is injective, and (39) shows that is isomorphic to $\mathrm{Im}(d')$; since $(\Omega_\bullet/\Theta) \otimes \mathbf{F}_2 = \mathrm{Ker}(d')/\mathrm{Im}(d')$, the knowledge of the structure of the algebra $\mathfrak{W}_\bullet$ implies the results on the structure of $\Omega_\bullet$. In particular

$$\Omega_8 \cong \mathbf{Z} \oplus \mathbf{Z}, \qquad \Omega_9 \cong (\mathbf{Z}/2\mathbf{Z}) \oplus (\mathbf{Z}/2\mathbf{Z}), \qquad \Omega_{10} \cong \mathbf{Z}/2\mathbf{Z}, \qquad \Omega_{11} \cong \mathbf{Z}/2\mathbf{Z},$$

and all groups Ω_n with $n \geqslant 8$ are $\neq 0$.

In his 1954 paper Thom had already remarked that his basic construction (section C) could apply not only to the orthogonal groups $O(k)$ and $SO(k)$, but to any closed subgroup G of $O(k)$; it was only necessary to replace the principal fiber space $(EO(k), BO(k), O(k))$ by (E_G, B_G, G), where B_G is a classifying space for G and E_G the corresponding universal fiber space (chap. III, § 2,F). In 1960 Milnor and Novikov were the first to investigate such a theory for a group G other than O(k) and SO(k), namely, the *unitary group* U(k) [344]. The pattern of the study of the homology and homotopy of the corresponding Thom space MU(k) follows Thom's methods in outline, but with the injection of a strong dose of more sophisticated tools, such as the Adams spectral sequence and concepts derived from the idea of "spectrum" of spaces (see below § 5,C).

There is a corresponding "cobordism" theory called *complex cobordism.* It deals with "stably almost complex manifolds," with or without boundary: an n-dimensional manifold M defined by the property that it can be embedded in a sphere S_{n+2m} of high dimension for which the normal bundle of M in S_{n+2m} is associated with a principal fiber space having U(m) as structural group [344].

After 1960 all other "classical groups" were likewise investigated from the point of view of cobordism [458].

§ 2. First Applications of Cobordism

A. The Riemann–Roch–Hirzebruch Theorem

We saw earlier (Part 2, chap. VII, § 1) how during the first half of the twentieth century the concepts of algebraic topology (cycles, simplicial homology, de Rham cohomology, homology of currents and Hodge theory) were harnessed to prove new results in algebraic geometry. After 1950 the new topological tools (fiber spaces, characteristic classes, sheaf cohomology) brought even more spectacular progress in the study of algebraic varieties, and, by an unforeseen backlash, this study was the origin of an entirely new chapter of topology and algebra, the K-*theory.* I think it is worthwhile to give a brief account of these events, which provide a magnificent example of how different parts of mathematics react with another.

I. The Arithmetic Genus

During the third part of the nineteenth century the study of algebraic surfaces brought to light an unsuspected phenomenon. Whereas *one* integer was attached to a complex algebraic curve, invariant under birational equivalence, namely, the Riemann *genus*, for a smooth complex algebraic surface embedded into some $\mathbf{P}_n(\mathbf{C})$, there were *two* such invariants, written p_g and p_a, called, respectively, the "geometric" and the "arithmetic" genus, and in general having different values.

The geometric genus p_g was easily defined, being the dimension over $\mathbf{C}$ of the vector space of *holomorphic complex differential* 2-*forms* on the surface (a four-dimensional manifold from the point of view of algebraic topology). But the definition of p_a turned out to be much more involved; it was found that $p_a \leqslant p_g$ always, and for nonsingular surfaces that can be *embedded in* $\mathbf{P}_3(\mathbf{C})$, $p_a = p_g$; the number $q = p_g - p_a$ was in consequence called the *irregularity* of the surface. It was soon realized that the number q is linked to the one-dimensional homology of the surface. Picard had already observed that $q = 0$ when the first Betti number $R_1 = 0$. In 1905 Castelnuovo and Severi published a proof that in general $q = \frac{1}{2}R_1$ and that q is the dimension over $\mathbf{C}$ of the *holomorphic complex differential* 1-*forms*. That proof was based on an earlier theorem of Enriques, but the proof of that theorem was later found defective and it was only in 1910 that Poincare provided a correct proof for both the Enriques and the Castelnuovo–Severi theorem.

When in 1905 Severi started a program aimed at extending the theory of algebraic curves and surfaces to algebraic varieties of arbitrary dimension n, the definition of the geometric genus g_n was naturally [for a smooth algebraic manifold embedded in some $\mathbf{P}_N(\mathbf{C})$] the dimension of the vector space of holomorphic complex differential n-forms on the manifold. But when he tried to define a general expression that would be the "arithmetic genus" Severi was confronted with three possibilities (see below, in A,V) he thought should give the same number, a conjecture he was not able to prove. The most striking of these tentative definitions was the alternating sum

$$p_a = g_n - g_{n-1} + g_{n-2} - \cdots + (-1)^{n-1} g_1 \tag{40}$$

where g_j is the dimension over $\mathbf{C}$ of the vector space of *holomorphic differential* j-*forms*; it obviously generalizes the expression $p_a = p_g - q$ for surfaces by the interpretation of q.

II. The Todd Genus

The work of the Italian geometers on algebraic surfaces had been centered on the study of what they called "systems" ("linear" or "continuous" ones) of algebraic curves on a surface. As Picard, Poincaré, and later Lefschetz pointed out, from the point of view of algebraic topology these curves are special 2-*cycles* on the surface (Part 2, chap. VII, § 1); but the Italians defined more restrictive equivalence relations between these cycles than mere "homology." These concepts were extended by Severi and Lefschetz to algebraic cycles on complex algebraic varieties of arbitrary dimension. In 1937 J.A. Todd [470] and independently M. Eger [153], following research by Severi on "equivalence" of algebraic cycles on a smooth complex algebraic manifold M of dimension n, introduced "canonical classes", special homology classes in the groups $H_{2(n-j)}(M; \mathbf{Z})$ for $0 \leqslant j \leqslant n-1$; Todd conjectured that the expression (40) proposed by Severi could be expressed by *universal polynomials* with rational coefficients (independent of the manifold) in these classes. It was later discovered that the Todd–Eger classes correspond by Poincaré duality to the *Chern classes* c_j $(1 \leqslant j \leqslant n)$ of the tangent bundle T(M), and Hirzebruch wrote the formula Todd conjectured in the form

$$(-1)^n p_a + 1 = \langle T_n(c_1, c_2, \ldots, c_n), [M] \rangle \tag{41}$$

where the T_n are called the *Todd polynomials* [233]. They can be obtained in the following way, essentially the one used by Todd [470]: if $c_1, c_2, \ldots, c_n$ are expressed as elementary symmetric functions of "phantom" indeterminates t_1,

$t_2, \ldots, t_n$, then the power series in z

$$\prod_{j=1}^{n} \frac{t_j z}{1 - e^{-t_j z}} = \sum_{k=0}^{\infty} \mathrm{T}_k(c_1, c_2, \ldots, c_n) z^k; \tag{42}$$

the first Todd polynomials are

$$\mathrm{T}_1(c_1) = \frac{1}{2} c_1, \qquad \mathrm{T}_2(c_1, c_2) = \frac{1}{12}(c_1^2 + c_2), \qquad \mathrm{T}_3(c_1, c_2, c_3) = \frac{1}{24} c_1 c_2,$$

$$\mathrm{T}_4(c_1, c_2, c_3, c_4) = \frac{1}{720}(-c_4 + c_1 c_3 + 3c_2^2 + 4c_2 c_1^2 - c_1^4). \tag{43}$$

Todd published a proof of (41), but it was based on a theorem of Severi of which the proof was later found insufficient.

III. Divisors and Line Bundles

In 1949 A. Weil found a new way to apply algebraic topology to algebraic geometry. He showed that on a smooth complex analytic manifold the classical concept of *divisor* on algebraic manifolds could be generalized in such a way that it becomes equivalent to the notion of *holomorphic line bundle.*

These ideas can best be explained in the simplest case of smooth algebraic curves, where the concept of divisor was first introduced to algebraically express the results obtained by Riemann in his "transcendental" theory of algebraic functions and their integrals. On a smooth algebraic curve $\Gamma \subset \mathbf{P}_n(\mathbf{C})$ (or equivalently a compact Riemann surface), a *divisor* is simply a 0-*cycle*

$$\mathrm{D} = \sum_{x \in \Gamma} \alpha_x . x$$

where the $\alpha_x \in \mathbf{Z}$; its *degree* $\deg(\mathrm{D}) = \sum_{x \in \Gamma} \alpha_x$ (the value of the "augmentation"), and D is called *positive* if $\alpha_x \geqslant 0$ for all x. Algebraic geometry enters the picture with the notion of *principal divisor.* If a meromorphic function f on Γ is not identically 0, it has only a finite number of zeros and of poles; to f is then associated its "divisor of zeros" $(f)_0$, which is the sum $\sum_{x \in \Gamma} m_x . x$, with m_x equal to the multiplicity of x as a zero of f; the "divisor of poles" $(f)_\infty$ is defined as $(1/f)_0$, and $(f) = (f)_0 - (f)_\infty$ is the principal divisor of f (the divisor of the function 0 is taken 0 by convention). Now $(fg) = (f) + (g)$, $\deg(f) = 0$, and the relation $(f) = (g)$ is equivalent to $f = c . g$ for a constant $c \neq 0$ in $\mathbf{C}$. Whereas for 0-cycles "homology" is a trivial equivalence relation (any two points are "homologous" if Γ is connected), the important relation between divisors is *linear equivalence*, meaning

$$\mathrm{D} - \mathrm{D}' = (f) \quad \text{for some meromorphic } f. \tag{44}$$

Ever since Riemann first mentioned it, a central problem in the theory of algebraic curves has been to find the *positive* divisors linearly equivalent to a given divisor D, i.e. to find the *dimension* of the vector space L(D) of all meromorphic functions f such that $(f) + \mathrm{D} \geqslant 0$. The relation $(f) + \mathrm{D} =$

$(g) + D$ for two elements f, g of L(D) means that f/g is a constant, so the set |D| of positive divisors linearly equivalent to D can be identified to the projective space **P**(L(D)) of dimension dim L(D) − 1.

It is easy to see in this simple case how the determination of L(D) can be translated into a problem of the theory of line bundles. Take an open covering (U_λ) of Γ such that U_λ is the domain of a chart $\varphi_\lambda: U_\lambda \to \mathbf{C}$ of the analytic manifold Γ. For each λ let h_λ be the meromorphic function in U_λ that has the expression

$$z \mapsto \prod_{x \in U_\lambda} (z - x)^{\alpha_x} \tag{45}$$

in local coordinates. Clearly, if $U_\lambda \cap U_\mu \neq \emptyset$, the function h_μ/h_λ in $U_\lambda \cap U_\mu$ is *holomorphic and nowhere* 0. A *holomorphic line bundle* $(B(D), \Gamma, \mathbf{C}, \pi)$ can then be defined by charts

$$\psi_\lambda: \pi^{-1}(U_\lambda) \xrightarrow{\sim} U_\lambda \times \mathbf{C} \tag{46}$$

with transition functions $\psi_\mu \circ \psi_\lambda^{-1}: (U_\lambda \cap U_\mu) \times \mathbf{C} \to (U_\lambda \cap U_\mu) \times \mathbf{C}$ given by

$$(x, z) \mapsto \left(x, \frac{h_\mu(x)}{h_\lambda(x)} z\right). \tag{47}$$

Now suppose that $s: \Gamma \to B(D)$ is a *holomorphic section*; for each λ the restriction $s|U_\lambda: U_\lambda \to \pi^{-1}(U_\lambda)$ is such that

$$\chi_\lambda((s|U_\lambda)(x)) = (x, s_\lambda(x))$$

where s_λ is holomorphic in U_λ; if $U_\lambda \cap U_\mu \neq \emptyset$, then for $x \in U_\lambda \cap U_\mu$,

$$s_\mu(x) = (h_\mu(x)/h_\lambda(x))s_\lambda(x)$$

or equivalently

$$s_\lambda(x)/h_\lambda(x) = s_\mu(x)/h_\mu(x)$$

(the values being in $\bar{\mathbf{C}} = \mathbf{C} \cup \{\infty\}$). This means there is a *meromorphic function* $f: \Gamma \to \bar{\mathbf{C}}$ such that $f|U_\lambda = s_\lambda/h_\lambda$ for every λ; by definition of D this is equivalent to $(f) + D \geqslant 0$.

It easily follows from the definitions that

$$B(D + D') = B(D) \otimes B(D'), \tag{48}$$

$$B(-D) = B(D)^* \tag{49}$$

[the *dual* line bundle $\mathrm{Hom}(B(D), \Gamma \times \mathbf{C})$].

These definitions are easily generalized to an *arbitrary* complex manifold M (algebraic or not, projective or not) of arbitrary (complex) dimension n. For a covering (U_λ) of M by domains of charts $U_\lambda \to \mathbf{C}^n$, call (f_λ) a *defining family* if each f_λ is an elementary meromorphic function* and if, when $U_\lambda \cap U_\mu \neq \emptyset$,

* This means a quotient p_λ/q_λ of two holomorphic functions defined in U_λ.

the function f_μ/f_λ (defined in $U_\lambda \cap U_\mu$) is *holomorphic and nowhere* 0. Two defining families (f_λ) and (f'_ρ), relative to respective coverings (U_λ) and (U'_ρ), define the *same divisor* D if whenever $U_\lambda \cap U'_\rho \neq \varnothing$, f_λ/f'_ρ is holomorphic and nowhere 0 in $U_\lambda \cap U'_\rho$. When a divisor D′ has a defining family (g_λ) for the covering (U_λ), D + D′ is the divisor defined by the family $(f_\lambda g_\lambda)$. The relation $D \geqslant 0$ means that all f_λ are holomorphic.

When there is a complex function f defined in an open and dense subset of M such that for each λ, $f|U_\lambda$ is an elementary meromorphic function, then the functions $f|U_\lambda$ constitute a defining family; f is a meromorphic function in M and the divisor defined by $(f|U_\lambda)$ is the *principal divisor* (f). The line bundle B(D) is defined as above, and relations (48) and (49) are valid; the vector space L(D) of meromorphic functions in M such that $(f) + D \geqslant 0$ is identified with the space of holomorphic sections of B(D) above M. Two line bundles B(D), B(D′) are isomorphic if and only if D − D′ is a principal divisor ("linear equivalence" of divisors).

The definitions show that when M is a compact algebraic manifold embedded in a $\mathbf{P}_N(\mathbf{C})$ ("projective" algebraic manifold) of (complex) dimension n, a divisor D may be identified with a $(2n-2)$-*cycle*, a linear combination $\sum_j \alpha_j S_j$ with integer coefficients α_j of cycles which are intersections of M with algebraic hypersurfaces S_j in $\mathbf{P}_N(\mathbf{C})$ not containing M; this is a direct generalization of divisors on an algebraic curve defined above. An important role is played by the divisor H, the cycle intersection of M and a hyperplane in $\mathbf{P}_N(\mathbf{C})$ (or any linearly equivalent divisor). For algebraic projective surfaces, the intersection number (D . D′) of two arbitrary divisors is therefore defined (Part 2, chap. VII, § 1) and invariant under linear equivalence; similarly, for three-dimensional M, the intersection number (D . D′ . D″), etc.

A *meromorphic differential n-form* ω on an *arbitrary* complex manifold M of (complex) dimension n is by definition an n-form defined in an open dense subset of M such that for a covering (U_λ) by domains of charts, each restriction $\omega|U_\lambda$ has a local expression

$$f_\lambda dz_1 \wedge dz_2 \wedge \cdots \wedge dz_n$$

where f_λ is an elementary meromorphic function in U_λ. Such forms need not always exist, but they do if M is a smooth algebraic projective manifold. For such a form (f_λ) is a defining family of a *divisor* $\Delta(\omega)$; any two divisors $\Delta(\omega)$, $\Delta(\omega')$ are linearly equivalent; they are called the *canonical divisors* on M. A canonical divisor K is not always positive; if it is, L(K) has dimension g_n, the geometric genus.

IV. The Riemann–Roch Problem

In his theory of abelian integrals Riemann obtained, for the dimension of L(D) for a divisor D on an algebraic curve, the inequality

$$\dim L(D) \geqslant \deg(D) + 1 - p_g; \tag{50}$$

his student Roch completed that result by giving an expression for the difference between the two sides of (50), using the canonical divisor K; the equation

$$\dim \mathrm{L}(\mathrm{D}) - \dim(\mathrm{K} - \mathrm{D}) = \deg(\mathrm{D}) + 1 - p_g \tag{51}$$

is called the *Riemann–Roch theorem for curves*.

The Italian geometers investigated the possibility of proving a similar relation for algebraic surfaces, but they only could prove an inequality

$$\dim \mathrm{L}(\mathrm{D}) + \dim \mathrm{L}(\mathrm{K} - \mathrm{D}) \geqslant p_a + (\mathrm{D}\,.\,\mathrm{D}) - \pi(\mathrm{D}) + 1 \tag{52}$$

where $\pi(\mathrm{D})$ is what they called the "virtual genus" of the divisor D (see below in A,VI).

The first progress beyond the inequality (52) was accomplished by Kodaira in 1950 [277]. He worked on compact complex kählerian manifolds (Part 2, chap. VII, § 1), which of course include as special cases smooth projective algebraic surfaces; in (52) he replaced p_a by $p_g - \frac{1}{2}\mathrm{R}_1$, which made sense for nonalgebraic surfaces. He used Hodge theory, the homology of currents (Part 1, chap. III, § 3), and the interpretation of divisors by complex line bundles; by intricate computations he obtained an expression of the difference between the left- and right-hand sides of (52) (which he called the "superabundance" of D), in which the dimension of subspaces of the vector space of holomorphic differential 1-*forms* on M is noticeable.

In 1952 Kodaira extended his methods to three-dimensional compact kählerian manifolds; as could be expected, his expression for $\dim \mathrm{L}(\mathrm{D})$ was even more complicated, having no less than nine terms [278]. Clearly he could not push much farther in that direction.

V. Virtual Genus and Arithmetic Genus

At the same time Zariski was investigating and putting on secure algebraic foundations the concepts of arithmetic genus and virtual genus of a divisor which had been introduced by Severi [526]. His starting point was the *Hilbert polynomial* $\chi(\mathrm{V})$ of an irreducible algebraic variety of dimension n embedded in $\mathbf{P}_\mathrm{N}(\mathbf{C})$. Let $\mathbf{C}[\mathrm{V}]$ be the graded ring of the restrictions to V of homogeneous polynomials in the $\mathrm{N} + 1$ coordinates of a point of $\mathbf{P}_\mathrm{N}(\mathbf{C})$, and R_m be the complex vector space of the elements of $\mathbf{C}[\mathrm{V}]$ of degree m. Then the dimension $\chi(\mathrm{V}, m)$ of R_m is equal, for large enough integers m, to a polynomial in m, the Hilbert polynomial of V

$$\chi(\mathrm{V}, m) = a_0\binom{m}{n} + a_1\binom{m}{n-1} + \cdots + a_{n-1}\binom{m}{1} + a_n \tag{53}$$

where the a_j are integers. For large m,

$$\chi(\mathrm{V}, m) = 1 + \dim|\mathrm{C}_m| \tag{54}$$

where $|\mathrm{C}_m|$ is the projective space of the positive divisors linearly equivalent to the intersection C_m of V and a hypersurface of degree m in $\mathbf{P}_\mathrm{N}(\mathbf{C})$. The first definition Severi proposed for the arithmetic genus of V was

$$p_a(\mathrm{V}) = (-1)^n(a_n - 1). \tag{55}$$

Zariski proved rigorously that to a divisor D on V is attached an integer $p_a(\mathrm{D})$, its *virtual arithmetic genus*, invariant by linear equivalence and such that

$$p_a(\mathrm{D} + \mathrm{C}_m) = p_a(\mathrm{D}) + p_a(\mathrm{C}_m) + p_a(\mathrm{D}.\mathrm{C}_m) \tag{56}$$

a relation which gives a definition of $p_a(\mathrm{D})$ by induction on the dimension n. He also showed that

$$\dim(\mathrm{D} + \mathrm{C}_m) = (-1)^n(p_a(\mathrm{V}) + p_a(-\mathrm{D} - \mathrm{C}_m)) - 1 \tag{57}$$

and that the second definition of the arithmetic genus proposed by Severi could be written

$$\mathrm{P}_a(\mathrm{V}) = (-1)^n(p_a(\mathrm{V}) + p_a(-\mathrm{K})) - 1 \tag{58}$$

where K is a canonical divisor; the equality $p_a(\mathrm{V}) = \mathrm{P}_a(\mathrm{V})$ conjectured by Severi is thus equivalent to

$$p_a(-\mathrm{K}) = \begin{cases} 1 & \text{if } n \text{ is even} \\ 1 - 2p_a(\mathrm{V}) & \text{if } n \text{ is odd.} \end{cases} \tag{59}$$

Kodaira proved this in [275] by the same methods as in his two previous papers on the Riemann–Roch theorem, applied to special types of divisors.

VI. The Introduction of Sheaves

Meanwhile, in 1950–1951 H. Cartan had found in the notion of sheaf (Part 1, chap. IV, § 7,B) a particularly useful tool for the expression of the results obtained by Thullen, Oka, K. Stein, and himself during the preceding 20 years in analytic geometry (at that time still called "the theory of analytic functions of several complex variables"). Acting on a suggestion of Serre, he also showed that sheaf cohomology could lead to generalizations and simplifications of these results. It has indeed become the frame within which have been conceived all subsequent works in that theory [235].

On a complex manifold M of arbitrary dimension the fundamental object is the *structure sheaf* $\mathcal{O}_\mathrm{M}$, whose stalks consist in "germs of holomorphic functions"; it can be defined as the presheaf $\mathrm{U} \mapsto \mathcal{O}(\mathrm{U})$ (Part 1, chap. IV, § 7,B), where $\mathcal{O}(\mathrm{U})$ is the vector space over **C** of all functions holomorphic in the open set U (it is obvious that this presheaf is indeed a sheaf). The sheaves used in analytic geometry are mostly $\mathcal{O}_\mathrm{M}$-*Modules*; the most useful ones are the *coherent* $\mathcal{O}_\mathrm{M}$-Modules. Such a sheaf $\mathscr{F}$ is characterized by the property that each point of M has an open neighborhood V for which there is an exact sequence of sheaves

$$\mathcal{O}_\mathrm{M}^p|\mathrm{V} \to \mathcal{O}_\mathrm{M}^q|\mathrm{V} \to \mathscr{F}|\mathrm{V} \to 0$$

for some integers p, q.

To any complex holomorphic vector bundle E with base space M, is associated the sheaf $\mathscr{F} = \mathcal{O}(\mathrm{E})$ of "germs of holomorphic sections" of E, defined as above as the presheaf $U \mapsto \mathscr{F}(U)$, where $\mathscr{F}(U)$ is the vector space of holomorphic sections of E over U. The sheaves $\mathcal{O}(E)$ are *locally free*: every point of M has an open neighborhood V such that $\mathcal{O}(E)|V$ is isomorphic to $\mathcal{O}_M^r|V$, where r is the (complex) rank of E. Conversely, every locally free sheaf over M has the form $\mathcal{O}(E)$ with E a holomorphic vector bundle.

In particular, the sheaf corresponding to the vector bundle $\bigwedge^p T(M)^*$ of the cotangent p-vectors of M is written Ω_M^p or Ω^p, so that $H^0(M; \Omega_M^p)$ is the vector space of holomorphic differential p-forms on M.

For a divisor D on M, the sheaf $\mathcal{O}(B(D))$ corresponding to the line bundle B(D) is written $\mathcal{O}_M(D)$; it turns out that computation with those sheaves is more flexible than that with line bundles. In particular, $\mathcal{O}_M(D + D') = \mathcal{O}_M(D) \otimes \mathcal{O}_M(D')$ for two divisors D, D′, and $\mathcal{O}_M(-D) = \mathcal{O}_M(D)^\vee$, the *dual* $\mathscr{H}om(\mathcal{O}_M(D), \mathcal{O}_M)$ of the sheaf $\mathcal{O}_M(D)$. The divisor D, D′ are linearly equivalent if and only if $\mathcal{O}_M(D)$ and $\mathcal{O}_M(D')$ are isomorphic.

The main applications of sheaf cohomology given by H. Cartan and Serre were, relative to domains of holomorphy and their generalizations, the Stein manifolds; for these manifolds, Cartan's "Theorem B" states that $H^q(M; \mathscr{F}) = 0$ for all $q \geqslant 1$ and all coherent sheaves $\mathscr{F}$ on M.

VII. The Sprint

In the year 1953 the pace of research on the Riemann–Roch problem suddenly accelerated, so that the problem was solved at the end of the year.

It all started in January, when P. Dolbeault published a *Comptes Rendus* Note [140] that acted as a catalyst for all subsequent papers of the year. His note dealt with C^∞ (not necessarily holomorphic) differential complex forms of type (p, q) (Part 2, chap. VII, § 1) on any complex manifold M; they are C^∞ sections over M of a sheaf $\mathscr{A}^{p,q}$ with $\Omega^p \subset \mathscr{A}^{p,0}$. The local expression of a form ω of type (p, q) shows that $d\omega = d'\omega + d''\omega$, where $d'\omega$ (resp. $d''\omega$) is a well-determined form of type $(p + 1, q)$ [resp. $(p, q + 1)$], and it is clear that $d'' \circ d'' = 0$. Localizing above any open subset of M, there is a cochain complex of sheaves

$$0 \to \Omega^p \to \mathscr{A}^{p,0} \xrightarrow{d''} \mathscr{A}^{p,1} \xrightarrow{d''} \cdots \mathscr{A}^{p,n} \to 0. \qquad (60)$$

Grothendieck had recently proved for the operator d'' a lemma corresponding to the "Poincaré lemma" for d [214]: if $d''\omega = 0$ for a form ω of type (p, q), then $\omega = d''\alpha$ *locally* on M, where α (only defined locally) is a form of type $(p, q - 1)$. From this result it follows that the sequence (60) is *exact* (in other words, a *resolution* of Ω^p); on the other hand the sheaves $\mathscr{A}^{p,q}$ are *fine*. The sheaf-theoretic proof of the de Rham theorem (Part 1, chap. IV, § 7,E) can therefore be extended to (60): the q-th cohomology group $H^{p,q}(M)$ of the cochain complex

$$0 \to H^0(M; \Omega^p) \to H^0(M; \mathscr{A}^{p,0}) \to \cdots \to H^0(M; \mathscr{A}^{p,q}) \to \cdots \qquad (61)$$

is *isomorphic to* $H^q(M; \Omega^p)$. If M is a compact *kählerian* manifold,

$$\mathrm{H}^q(\mathrm{M};\Omega^p) \simeq \mathrm{H}^{p,q}(\mathrm{M})$$

the vector space of complex *harmonic forms* of type (p, q) (Part 2, chap. VII, § 1).

Since for a divisor D on a complex compact manifold M the vector space L(D) can be written $\mathrm{H}^0(\mathrm{M};\mathcal{O}_\mathrm{M}(\mathrm{D}))$, Serre, led by Cartan's results and his own for Stein manifolds, undertook to apply also sheaf cohomology to the Riemann–Roch problem. He was immediately confronted with the need to show that the vector spaces $\mathrm{H}^q(\mathrm{M};\mathscr{F})$ have *finite* dimension for the $\mathcal{O}_\mathrm{M}$-Modules $\mathscr{F}$ he had to consider; in collaboration with Cartan, he proved that this is true for *arbitrary* compact complex manifolds M and any *coherent* sheaf $\mathscr{F}$ [117]. The main idea of the proof is that for sufficiently small open sets U in M the space of sections $\mathrm{H}^0(\mathrm{U};\mathscr{F})$ with the compact open topology is a *Fréchet space*; the known properties of these spaces enabled L. Schwartz to provide tools that finished the proof [419].

A few months earlier Serre had used this technique of Fréchet spaces to prove probably his most original contribution to that part of the theory, the *duality theorem.* He showed that the simultaneous appearance in all "Riemann–Roch" equalities or inequalities known until that time of a divisor D and the divisor K − D for the canonical divisor K was not fortuitous, but a natural consequence of a very general phenomenon, now called *Serre duality*. This is valid on an *arbitrary* compact complex manifold M of dimension n; for any holomorphic vector bundle E of base M the finite-dimensional vector spaces

$$\mathrm{H}^q(\mathrm{M};\Omega^p \otimes \mathcal{O}(\mathrm{E})) \quad \text{and} \quad \mathrm{H}^{n-q}(\mathrm{M};\Omega^{n-p} \otimes \mathcal{O}(\mathrm{E}^*))$$

are *dual* to each other in a natural way; in particular, for every divisor D on M

$$\mathrm{H}^q(\mathrm{M};\mathcal{O}_\mathrm{M}(\mathrm{D})) \quad \text{and} \quad \mathrm{H}^{n-q}(\mathrm{M};\mathcal{O}_\mathrm{M}(\mathrm{K}-\mathrm{D}))$$

are in duality.

The proof starts with an extension of Dolbeault's theorem to differential forms with values in a vector bundle on an arbitrary complex manifold M of dimension n (compact or not): the group

$$\mathrm{H}^q(\mathrm{M};\Omega^p \otimes \mathcal{O}(\mathrm{E}))$$

is isomorphic to the q-th cohomology group of the cochain complex

$$0 \to \mathrm{H}^0(\mathrm{M};\Omega^p \otimes \mathcal{O}(\mathrm{E})) \to \mathrm{A}^{p,0}(\mathrm{E}) \xrightarrow{d''} \cdots \xrightarrow{d''} \mathrm{A}^{p,q}(\mathrm{E}) \to \cdots$$

where

$$\mathrm{A}^{p,q}(\mathrm{E}) = \mathrm{H}^0(\mathrm{M};\mathscr{A}^{p,q} \otimes \mathcal{O}(\mathrm{E})).$$

Then the spaces $\mathrm{A}^{p,q}(\mathrm{E})$ are given topologies of *Fréchet spaces* in a natural way. Using the duality of such spaces and its application to the theory of currents (Part 1, chap. III, § 3), Serre proved that the *topological dual* of

$$\mathrm{H}^q(\mathrm{M};\Omega^p \otimes \mathcal{O}(\mathrm{E}))$$

is isomorphic (as a vector space without topology) to

$$H_c^{n-q}(M; \Omega^{n-p} \otimes \mathcal{O}(E^*))$$

provided that both maps

$$A^{p,q-1}(E) \xrightarrow{d''} A^{p,q}(E) \xrightarrow{d''} A^{p,q+1}(E) \tag{62}$$

are *strict homomorphisms.** This condition is always satisfied when the spaces $H^q(M; \Omega^p \otimes \mathcal{O}(E))$ have *finite dimension*; such is always the case when M is compact because $\Omega^p \otimes \mathcal{O}(E)$ is a coherent sheaf ([428], p. 305).

When sheaf cohomology is used, the expression (40) proposed by Severi for the arithmetic genus is written

$$\dim H^0(\Omega^n) - \dim H^0(\Omega^{n-1}) + \cdots + (-1)^{n-1} \dim H^0(\Omega^1)$$

so that for complex compact manifolds this is equal to

$$\dim H^n(\mathcal{O}_M) - \dim H^{n-1}(\mathcal{O}_M) + \cdots + (-1)^{n-1} \dim H^1(\mathcal{O}_M)$$

by the duality theorem.

This led Serre to introduce, for *any* coherent sheaf $\mathcal{F}$ on a compact complex manifold M of dimension n, what he called the *Euler–Poincaré characteristic of* $\mathcal{F}$

$$\chi(\mathcal{F}) = \dim H^0(\mathcal{F}) - \dim H^1(\mathcal{F}) + \cdots + (-1)^n \dim H^n(\mathcal{F}) \tag{63}$$

also written $\chi(D)$ when $\mathcal{F} = \mathcal{O}_M(D)$.

He then endeavored to express the Riemann–Roch relation for algebraic curves and surfaces in the language of sheaves. For curves, he showed that the classical Riemann–Roch theorem could be written

$$\chi(D) = \langle f + \tfrac{1}{2}c_1, [M] \rangle \tag{64}$$

where c_1 is the unique Chern class of the tangent bundle T(M) and f is the Chern class of the line bundle B(D). For surfaces, he interpreted Kodaira's formula as

$$\chi(D) = \left\langle \frac{1}{2} f(f + c_1) + \frac{1}{12}(c_2 + c_1^2), [M] \right\rangle \tag{65}$$

where f has the same meaning as above and c_1, c_2 are the two Chern classes of T(M). This led him to conjecture that a general Riemann–Roch relation for projective algebraic manifolds of dimension n should have the form

$$\chi(D) = \langle P(f, c_1, c_2, \ldots, c_n), [M] \rangle \tag{66}$$

for some polynomial P in f and the Chern classes $c_1, c_2, \ldots, c_n$ of T(M) with rational coefficients.

---

* A linear continuous map u: E → F, where E and F are topological vector spaces, is a strict homomorphism if in the canonical factorization

$$u: E \xrightarrow{v} E/\mathrm{Ker}(u) \xrightarrow{w} u(E) \to F$$

w is an isomorphism of topological vector spaces.

Around the beginning of 1953 D. Spencer (then professor at Princeton University), whose work until that time had chiefly been concerned with analytic functions of one complex variable, became convinced of the usefulness of sheaf cohomology in the theory of compact complex manifolds; independently of Serre he introduced the Euler–Poincaré characteristic (63) of a sheaf. He teamed with Kodaira (then at the Institute for Advanced Study) to apply sheaf cohomology to the problems on which the latter had already been working for several years. Through A. Borel (then also in Princeton) Serre was kept informed of their work; in April he sent Borel a long letter [433] describing his own results, which Borel communicated to Kodaira and Spencer.

The focus of their research in early 1953 was to settle the Severi conjectures on the arithmetic genus. To avoid confusions, write $a(\mathrm{M})$ the right-hand side of (40) for a compact n-dimensional kählerian manifold M. For any divisor D on M and a sufficiently large integer m, there is a positive divisor $\mathrm{S}_m \in |\mathrm{D} + \mathrm{C}_m|$ (in the notation of IV) which is a nonsingular hypersurface; Kodaira and Spencer showed that $a(\mathrm{S}_m)$ for large m is equal to a polynomial in m which they wrote $a(m; \mathrm{D}, \mathrm{M})$, and they defined

$$a_{\mathrm{M}}(\mathrm{D}) = a(0; \mathrm{D}, \mathrm{M}) \tag{67}$$

[so that $a_{\mathrm{M}}(\mathrm{D}) = a(\mathrm{D})$ when D is itself a nonsingular hypersurface]. In his earlier paper [275], Kodaira had shown that the number $p_a(\mathrm{M})$ defined in (55) satisfies the relation

$$p_a(\mathrm{M}) = a(\mathrm{M}) - a_{\mathrm{M}}(-\mathrm{K}) - (-1)^n. \tag{68}$$

Since he had also shown that $\mathrm{P}_a(\mathrm{M}) = p_a(\mathrm{M})$, the point which remained to be settled was the relation

$$a_{\mathrm{M}}(-\mathrm{K}) = (1 - (-1)^n)a(\mathrm{M}). \tag{69}$$

In the Note Kodaira and Spencer published in May [279] this was done by an interpretation of $a_{\mathrm{M}}(\mathrm{D})$ in terms of sheaf cohomology, which also identified $a_{\mathrm{M}}(\mathrm{D})$ to the virtual arithmetic genus $p_a(\mathrm{D})$ defined by Zariski when M is a projective algebraic variety.

Their method was to define for each integer $r \leqslant n$ a subsheaf

$$\Omega_{\mathrm{D}}^r \subset \Omega^r \quad [\text{also written } \Omega_{\mathrm{D}}^r(\mathrm{M})]$$

which they called the "sheaf of germs of meromorphic r-forms multiples of $-$D." It is enough to define it when D is a nonsingular hypersurface S; then for a point $x \in \mathrm{S}$, there are local coordinates $z_1, z_2, \ldots, z_n$ on M in a neighborhood U of x such that $\mathrm{U} \cap \mathrm{S}$ is defined by the equation $z_1 = 0$. The meromorphic r-form τ in U is then "multiple of $-$S" if the form $z_1 . \tau$ is holomorphic in U.

They then wrote

$$\chi_{\mathrm{M}}^r(\mathrm{D}) = \chi(\Omega_{\mathrm{D}}^r) \quad \text{for each } r \leqslant n$$

and in particular $\chi_M(D) = \chi_M^n(D)$, and proved the formula

$$a_M(D) = (-1)^{n-1}(\chi_M(D) - \chi(M)). \tag{70}$$

Like Zariski's (see IV), their proof proceeded by induction on the dimension n of M, but using the exact sequence of sheaves for a generic hypersurface S,

$$0 \to \Omega_D^n \to \Omega_{D+S}^n \to \Omega_{D.S}^{n-1}(S) \to 0.$$

Such uses of exact sequence of sheaves henceforth became a standard method in all works on algebraic geometry, both classical (i.e., over the complex field) and "abstract." We will not follow these developments, which are beyond the scope of this book. We only mention here that that is the way Serre found that a "weak" form of the Riemann–Roch relation expressing the Euler–Poincaré characteristic $\chi(D)$ of a divisor

$$\chi(D) = \chi(M) - \chi_M(-D) \tag{71}$$

where $\chi_M(-D)$ is his notation for the virtual arithmetic genus of $-D$.

VIII. The Grand Finale

In the Spring of 1953 Thom published four *Comptes Rendus* Notes in which he summarized the results of the 1954 paper he had just completed [463]. When these Notes reached Princeton, Hirzebruch, who was spending a year at the Institute for Advanced Study, became convinced from their contents that they would enable him to prove the Todd conjecture (see II) on the arithmetic genus.

Thom's Notes did not mention the Todd genus, but in his 1952 thesis [462] he had studied a purely topological invariant, the *signature* (or *index*) of a CW-complex M (connected or not). If M has even dimension $2m$, the map

$$(x, y) \mapsto \langle x \smile y, [M] \rangle \tag{72}$$

is a *bilinear form* on the vector space $H^m(M; \mathbf{Q})$; it is symmetric if m is even, antisymmetric if m is odd. For $m = 2k$ the signature of the symmetric form (72) is therefore an invariant of the homotopy type* of M and is now called the *signature* of M, which we write $\sigma(M)$. By convention $\sigma(M) = 0$ when the dimension of M is not a multiple of 4.

In his thesis Thom showed that if an oriented compact smooth manifold M is the boundary of an oriented $(4k + 1)$-dimensional manifold-with-boundary W, then $\sigma(M) = 0$. The proof relies on the commutative diagram of exact sequences

* Recall that the signature of a quadratic form with real coefficients is the difference between the number of plus signs and the number of minus signs in the decomposition of the form as combinations of squares of independent linear forms with coefficients ± 1.

$$
\begin{array}{ccccccccc}
\cdots \longrightarrow & H^{2k}(W) & \xrightarrow{j^*} & H^{2k}(M) & \xrightarrow{\partial} & H^{2k+1}(W,M) & \longrightarrow \cdots \\
& \downarrow u & & \downarrow v & & \downarrow w & \\
\cdots \longrightarrow & H_{2k+1}(W,M) & \xrightarrow[\partial]{} & H_{2k}(M) & \xrightarrow[j_*]{} & H_{2k}(W) & \longrightarrow \cdots
\end{array}
\tag{73}
$$

(homology and cohomology with coefficients in $\mathbf{Q}$; $j\colon M \to W$ being the natural injection). Here u, v, w are bijective. If $A = \operatorname{Im} j^*$ and $B = \operatorname{Ker} j^*$, $\dim A = \dim B$; A is also the dual vector space of $H_{2k}(M)/B$. This implies that for the bilinear form (72) A is a maximal isotropic subspace of dimension equal to one half the Betti number b_{2k}, the dimension of $H_{2k}(M)$; therefore $\sigma(M) = 0$. After introducing cobordism Thom added to that result the observation that $\sigma(-M) = -\sigma(M)$ and $\sigma(M \coprod M') = \sigma(M) + \sigma(M')$ for two oriented manifolds of same dimension; finally, for two oriented manifolds of any dimension M, N, the Künneth formula implies

$$\sigma(M \times N) = \sigma(M)\sigma(N). \tag{74}$$

The signature $\sigma(M)$ therefore only depends on the *class* $\mathrm{cl}_\Omega(M)$ and the map

$$\mathrm{cl}_\Omega(M) \to \sigma(M)$$

thus defined is a *ring homomorphism* of $\Omega_{\bullet}$ into $\mathbf{Z}$.

Recall that Pontrjagin had shown that two oriented cobordant manifolds have the same Pontrjagin numbers (§ 1,H); it was therefore natural to expect relations between signature and Pontrjagin numbers. Using the isomorphism $\Omega_4 \simeq \mathbf{Z}$ that he had discovered, Thom showed that for four-dimensional compact manifolds

$$\sigma(M) = \tfrac{1}{3}\langle p_1, [M]\rangle. \tag{75}$$

This was generalized by Hirzebruch, who obtained an explicit expression of $\sigma(M)$ for *any* oriented compact manifold M as a linear combination of Pontrjagin numbers with rational coefficients.

In general, suppose that for any oriented compact smooth manifold M of dimension $4r$ (where r is any integer $\geqslant 1$), a rational number $\psi(M)$ is defined with the properties:

(i) $\psi(-M) = -\psi(M)$ and $\psi(M \coprod M') = \psi(M) + \psi(M')$ for two oriented manifolds M, M′ of same dimension $4r$;
(ii) $\psi(M \times N) = \psi(M)\psi(N)$ for two oriented manifolds of dimensions multiples of 4;
(iii) $\psi(M) = 0$ if M is the boundary of a $(4r + 1)$-dimensional oriented manifold-with-boundary.

Thom's theory showed that $\psi(M)$ only depends on $\mathrm{cl}_\Omega(M)$, and $\mathrm{cl}_\Omega(M) \to \psi(M)$ is a ring homomorphism of $\Omega_{\bullet} \otimes \mathbf{Q}$ into $\mathbf{Q}$.

Hirzebruch showed that there exists a unique sequence of polynomials

$(K_r(T_1, T_2, \ldots, T_r))_{r \geqslant 1}$ of weight $4r$ in the indeterminates T_j of weight $4j$ such that for any oriented manifold M of dimension $4r$

$$\psi(M) = \langle K_r(p_1, p_2, \ldots, p_r), [M] \rangle \tag{76}$$

where the p_j $(1 \leqslant j \leqslant r)$ are the Pontrjagin classes of T(M).

His method was a generalization of Todd's procedure for the definition of his polynomials. He started from a formal power series with constant coefficient 1 in $\mathbf{Q}((x))$

$$K(x) = 1 + a_1 x + a_2 x^2 + \cdots. \tag{77}$$

Write the Pontrjagin classes p_j $(1 \leqslant j \leqslant r)$ of T(M) for a compact oriented manifold M of dimension $4r$ as elementary symmetric polynomials in the "phantom" indeterminates $t_1, t_2, \ldots, t_r$ (chap. IV, § 1,D), and consider the product

$$K(t_1 x)K(t_2 x) \cdots K(t_r x) = \sum_{m=0}^{\infty} K_m(p_1, \ldots, p_r)x^m. \tag{78}$$

Write

$$K(M) = \langle K_r(p_1, \ldots, p_r), [M] \rangle \in \mathbf{Q}. \tag{79}$$

Then:

1. if M is the boundary of an oriented $(4r + 1)$-dimensional manifold-with-boundary, all Pontrjagin numbers of M are 0 (§ 1,H), hence $K(M) = 0$;
2. $K(-M) = -K(M)$ since the Pontrjagin classes of M and $-M$ are the same;
3. if M, M′ are two oriented $4r$-dimensional compact manifolds, $K(M \coprod M') = K(M) + K(M')$;
4. from the Wu Wen Tsün formula for Pontrjagin classes of the tangent bundle of a product (chap. IV, § 1,D), it follows that

$$K(M \times N) = K(M)K(N).$$

Thus K(M) only depends on the class $\text{cl}_\Omega(M)$; in the same way as ψ, it defines a **Q**-algebra homomorphism

$$\text{cl}_\Omega M \to K(M)$$

of $\Omega_\bullet \otimes \mathbf{Q}$ into **Q**. To prove that $K(M) = \psi(M)$, it is only necessary to show that

$$K(\mathbf{P}_{2k}(\mathbf{C})) = \psi(\mathbf{P}_{2k}(\mathbf{C}))$$

since the classes $\text{cl}_\Omega(\mathbf{P}_{2k}(\mathbf{C}))$ generate the **Q**-algebra $\Omega_\bullet \otimes \mathbf{Q}$ (§ 1,H).

In the case of the signature $\sigma(M)$, if a is the first Chern class of $T(\mathbf{P}_{2k}(\mathbf{C}))$, the vector space $H^{2k}(\mathbf{P}_{2k}(\mathbf{C}); \mathbf{Q})$ has a basis consisting of the unique element a^k, and

$$\langle a^k \smile a^k, [\mathbf{P}_{2k}(\mathbf{C})] \rangle = 1 \quad \text{(chap. IV, § 2)}$$

hence $\sigma(\mathbf{P}_{2k}(\mathbf{C})) = 1$. On the other hand (*loc. cit.*), the total Pontrjagin class of

$T(P_{2k}(C))$ is equal to $(1 + a^2)^{2k+1}$; in other words, for the Pontrjagin classes p_j

$$\sum_{j=0}^{\infty} p_j x^{4j} = (1 + a^2 x^4)^{2k+1}.$$

The formal power series corresponding to σ, written $L(x)$, is therefore determined by the condition that the coefficient of x^k in $(L(x))^{2k+1}$ must be equal to 1 for all $k \geqslant 1$. By an elementary computation of residues Hirzebruch proved that

$$L(x) = \frac{\sqrt{x}}{\operatorname{th}\sqrt{x}}. \tag{80}$$

This gives as values of the first polynomials L_j:

$$L_1(p_1) = \frac{1}{3}p_1, \qquad L_2(p_1, p_2) = \frac{1}{45}(7p_2 - p_1^2),$$

$$L_3(p_1, p_2, p_3) = \frac{1}{945}(62p_3 - 13p_1p_2 + 2p_1^3). \tag{81}$$

These results were published by Hirzebruch in July [233]; but, although he saw that the Todd polynomials are obtained by the same method applied to the series $x/(1 - e^{-x})$, he could not prove at that time that, for a projective algebraic manifold of dimension n

$$\chi(M) = \tau(M) = \langle T_n(c_1, c_2, \ldots, c_n), [M] \rangle. \tag{82}$$

He mentioned in that Note that Kodaira could prove (82) for complete intersections, and he began to investigate the relations between the Todd polynomials of an algebraic manifold M and a fiber space E with base M.

Hirzebruch pushed this idea to a successful conclusion in December of the same year [234]. The proof is intricate and clever, using results of Kodaira and Spencer just published [281] and other results of Kodaira still unpublished. The main idea is to associate to a projective algebraic nonsingular variety M another variety M^*. Let $P(M)$ be the principal bundle with base space M and structure group $GL(n, C)$ to which the tangent bundle $T(M)$ is associated; M^* is the total space of the bundle with base space M associated to $P(M)$ and to the action of $GL(n, C)$ on the homogeneous space $GL(n, C)/\Delta(n, C)$, where $\Delta(n, C)$ is the subgroup of lower triangular matrices; M^* has dimension $m = n + \frac{1}{2}n(n-1)$, and it is easily seen that $\chi(M^*) = \chi(M)$. From the results Hirzebruch had proved in his first Note it follows that $\tau(M^*) = \tau(M)$; the proof of (82) for M is thus reduced to establishing the same equation for M^*.

In the construction of M^* it is possible to define an increasing sequence of holomorphic vector bundles W_j, with base space M^* and complex rank j, such that $W_m = T(M^*)$ and it is possible to express $\tau(M^*)$ by a combination of signatures of the line bundles W_j/W_{j-1}. On the other hand, using a paper of

Kodaira and Spencer [276], Hirzebruch defined a "virtual index" χ_1 for algebraic varieties, and showed that $\chi(M^*)$ can be expressed as linear combination of values of χ_1 for Whitney sums of line bundles W_j/W_{j-1}. The final step of the proof appealed to a previous theorem of Hodge: for any compact kählerian manifold V, namely $\chi_1(V) = \tau(V)$.*

From the formula (82) the passage to the general Riemann–Roch formula was easier, as had been forecast by Serre; the final result obtained by Hirzebruch can be written, for any holomorphic vector bundle E over the projective algebraic smooth manifold M

$$\chi(\mathcal{O}(E)) = \kappa_{2n}(\mathrm{ch}(E) \smile \mathrm{td}_M) \tag{83}$$

where ch(E) is the Chern character of E (chap. IV, § 1,E) and td_M is the sum $\sum_k T_k(c_1, c_2, \ldots, c_k)$ of the Todd polynomials; if u is the sum of the terms in the cup-product which belong to $H^{2n}(M; \mathbf{Q})$, the right-hand side of (83) is $\langle u, [M] \rangle$.

B. Exotic Spheres

In 1956 Milnor startled the mathematical community with a completely unexpected theorem. He showed that on the sphere S_7 it is possible to define a structure of C^∞ manifold *not diffeomorphic* to the usual one (induced by the standard differential structure of $\mathbf{R}^8$) [340].

The method can be presented in two steps:

I. definition of the "exotic" differential structure on S_7 and
II. proof that this structure is not diffeomorphic to the usual one.

I. The sphere S_7 (with its usual structure) can be considered the total space of the Hopf fibration (S_7, S_4, S_3, π) (chap. III, § 1,C). The complements U_+, U_- of the points $-e_5$, e_5, respectively, in $S_4 \subset \mathbf{R}^5$ are the domains of two charts

$$s_+ : U_+ \to \mathbf{R}^4, \qquad s_- : U_- \to \mathbf{R}^4$$

which are the stereographic projections with poles at $-e_5$ and e_5, respectively. The sphere S_7 can be defined as the set of pairs of quaternions $(z_1, z_2) \in \mathbf{H}^2$ (considered as left vector space over $\mathbf{H}$) satisfying the equation

$$|z_1|^2 + |z_2|^2 = 1.$$

For $z_2 \neq 0$, if $\sigma_+ = s_+ \circ \pi$,

$$\sigma_+(z_1, z_2) = z_2^{-1} z_1 \in \mathbf{H} \qquad (= \mathbf{R}^4).$$

For any quaternion u, $\sigma_+^{-1}(u)$ is therefore the fiber consisting of the pairs of

* In his proof of the Riemann–Roch theorem (65) for surfaces Serre had also used Hodge's result.

quaternions

$$\left(\frac{qu}{\sqrt{1+|u|^2}}, \frac{q}{\sqrt{1+|u|^2}}\right) \quad \text{for } |q| = 1 \text{ (i.e., } q \in \mathbf{S}_3\text{).}$$

Similarly, for $z_1 \neq 0$ and $\sigma_- = s_- \circ \pi$,

$$\sigma_-(z_1, z_2) = \bar{z}_2 \bar{z}_1^{-1},$$

and $\sigma_-^{-1}(u)$ is the fiber consisting of the pairs

$$\left(\frac{q}{\sqrt{1+|u|^2}}, \frac{uq}{\sqrt{1+|u|^2}}\right) \quad \text{for } |q| = 1.$$

The diffeomorphism of transition $\sigma_- \circ \sigma_+^{-1}$ of $(\mathbf{H} - \{0\}) \times \mathbf{S}_3$ onto itself is therefore

$$\psi : (u, q) \mapsto \left(\frac{u}{|u|^2}, \frac{qu}{|u|}\right). \tag{84}$$

The "exotic" differential structures are defined on $\mathbf{S}_7$ by taking the same charts σ_+ and σ_-, but for a diffeomorphism of transition

$$\psi_h : (u, q) \mapsto \left(\frac{u}{|u|^2}, \frac{(u^h q u^{-h})u}{|u|}\right) \quad \text{for } h \in \mathbf{Z}, \tag{85}$$

taking into account that for $u \neq 0$, $q \mapsto u^h q u^{-h}$ is an element of the rotation group SO(4) acting on $\mathbf{R}^4$ (with that notation, $\psi = \psi_0$). Call $\mathbf{M}_h$ the $\mathbf{C}^\infty$ manifold defined by ψ_h.

II. To show that for some integers h, $\mathbf{M}_h$ is not diffeomorphic to $\mathbf{S}_7$, Milnor attached to every oriented seven-dimensional compact $\mathbf{C}^\infty$ manifold M for which

$$\mathrm{H}^3(\mathrm{M}; \mathbf{Z}) = \mathrm{H}^4(\mathrm{M}; \mathbf{Z}) = 0, \tag{86}$$

an element $\lambda(\mathrm{M}) \in \mathbf{Z}/7\mathbf{Z}$, invariant under diffeomorphism. Since Thom proved that $\Omega_7 = 0$ (§ 1,H), M is the boundary of an oriented eight-dimensional manifold-with-boundary B. From the exact cohomology sequence

$$\cdots \to \mathrm{H}^3(\mathrm{M}; \mathbf{Z}) \to \mathrm{H}^4(\mathrm{B}, \mathrm{M}; \mathbf{Z}) \xrightarrow{j^*} \mathrm{H}^4(\mathrm{B}; \mathbf{Z}) \to \mathrm{H}^4(\mathrm{M}; \mathbf{Z}) \to \cdots$$

and assumptions (86) it follows that j^* is bijective. Define on $\mathrm{H}^4(\mathrm{B}, \mathrm{M}; \mathbf{Q})$ a symmetric bilinear form

$$(x, y) \mapsto \langle x \smile y, \mu_{\mathrm{B},\mathrm{M}} \rangle$$

where $\mu_{\mathrm{B},\mathrm{M}}$ is the fundamental homology class in $\mathrm{H}^8(\mathrm{B}, \mathrm{M} : \mathbf{Z})$ (Part 2, chap. IV, § 3,A); let $\sigma(\mathrm{B})$ be the signature of that form. Now let $p_1(\mathrm{B})$ be the first Pontrjagin class of T(B), and write

$$q(\mathrm{B}) = \langle j^{*-1}(p_1(\mathrm{B}))^2, \mu_{\mathrm{B},\mathrm{M}} \rangle.$$

Milnor showed that the class of

$$2q(\mathrm{B}) - \sigma(\mathrm{B})$$

in $\mathbf{Z}/7\mathbf{Z}$ is *independent* of the choice of the eight-dimensional manifold-with-boundary B such that $\partial B = M$. The proof consists in considering two such manifolds-with-boundary B_1, B_2 and the eight-dimensional oriented manifold C (without boundary) obtained by "gluing" together B_1 and $-B_2$ (cf. § 1,G). Milnor showed that $\sigma(C) = \sigma(B_1) - \sigma(B_2)$ and that if $p_1(C)$ is the first Pontrjagin class of T(C) and $q(C) = \langle p_1(C)^2, [C] \rangle$, then

$$q(C) = q(B_1) - q(B_2).$$

Hirzebruch's explicit determination of the signature as a function of Pontrjagin numbers [relations (81)] gave

$$\sigma(C) = \frac{1}{45}\langle (7p_2(C) - p_1(C)^2), [C] \rangle$$

hence

$$45\sigma(C) + q(C) = 7\langle p_2(C), [C] \rangle \equiv 0 \pmod{7}$$

so that $2q(C) - \sigma(C) \equiv 0 \pmod{7}$, and therefore

$$2q(B_1) - \sigma(B_1) \equiv 2q(B_2) - \sigma(B_2) \pmod{7}.$$

Milnor computed the invariant $\lambda(M_h)$ for each integer h. There is a canonical manifold-with-boundary B such that $\partial B = M_h$, namely, the fiber space with base S_4, each fiber of which is the closed 4-ball with frontier $\pi^{-1}(x) \simeq S_3$ at a point x. For that space B, $\sigma(B) = 1$ and $q(B) = 4k^2$ with $k = 2h - 1$. So $2q(B) - \sigma(B) = 8k^2 - 1 \equiv k^2 - 1 \pmod{7}$, and for $k \not\equiv \pm 1 \pmod{7}$, $\lambda(M_h) \neq 0$, whereas $\lambda(S_7) = 0$.

A little later Thom made a complete study of the oriented vector fibrations $\xi = (E, S_4, \mathbf{R}^4, \pi)$ with base space S_4 and rank 4 ([347], pp. 243–245); he showed that for every pair of integers k, l such that $k \equiv 2l \pmod{4}$, there exist such fibrations for which the Pontragin and Euler classes are given by

$$p_1(\xi) = ks_4, \qquad e(\xi) = ls_4$$

where s_4 is the fundamental cohomology class of S_4. Taking $l = 1$, $k \equiv 2 \pmod{4}$, he showed that the total space M of the corresponding sphere bundle (set of vectors of E of length 1) is homeomorphic to S_7, but is not diffeomorphic to S_7 for $k \not\equiv \pm 2 \pmod{7}$. The corresponding Thom space $T(\xi)$ is a C^0 manifold which admits a triangulation, but there is no C^∞ differential structure on $T(\xi)$ for which that triangulation is smooth.

Still later Kervaire and Milnor were able to define on the set of diffeomorphism classes of C^∞ structures on S_n for $n \neq 3$ a structure of *finite group*; for $n = 7$ that group has order 28, and for $n = 11$, order 992.

Finally, exotic spheres appeared in a very natural way in classical differential geometry: E. Brieskorn showed that the intersection of the sphere S_{4m+1}: $\sum_{j=0}^{2m} z_j \bar{z}_j = 1$ in $\mathbf{C}^{2m+1}$ with the algebraic variety $z_0^3 + z_1^{6k-1} + z_2^2 + \cdots z_{2m}^2 = 0$ is an "exotic" sphere for the differential structure induced by the usual structure of $\mathbf{R}^{4m+2}$.

§ 3. The Beginnings of K-Theory

A. The Grothendieck Groups

Barely was the ink dry on Hirzebruch's Note when new problems arose about the Riemann–Roch relation. One was part of the general trend toward an exclusive use of algebraic tools in algebraic geometry, which of course made possible the replacement of the complex field **C** by any algebraically closed field with arbitrary characteristic. We mentioned earlier (Part 2, chap. VII, § 1) this new "abstract" algebraic geometry, which, in 1950, had already reached a high level of sophistication in the works of van der Waerden, A. Weil, and Zariski. It received a new impetus when Serre, in 1954, discovered the possibility of transplanting to that general situation the use of sheaves and sheaf cohomology. He showed that his methods and those of Kodaira based on various tools from functional analysis could be replaced by purely algebraic ones, yielding similar results valid in the "abstract" case, such as his duality theorem (§ 2,A, VII) and the "weak" form (71) of the Riemann–Roch relation. A little later the precise form (83) Hirzebruch gave to that relation was also generalized independently by Washnitzer and Grothendieck ([55], pp. 659–698).

The latter's method was based on extremely original ideas, which were to have unexpected consequences reaching far beyond algebraic geometry. We can only examine those consequences that relate to algebraic and differential topology, and shall therefore, in sketching Grothendieck's work on the Riemann–Roch relation, restrict the exposition to the case of *smooth complex projective algebraic manifolds*, although his explicit purpose was to deal with algebraic varieties over arbitrary algebraically closed fields.

In the late 1950s the growing usefulness of "categorical" notions gradually convinced mathematicians that "morphisms" rather than "objects" had to be emphasized in many situations. It was that trend that led Grothendieck to believe that the Riemann–Roch–Hirzebruch formula (83) is only a special case of a "relative" Riemann–Roch relation dealing with a morphism $f: X \to Y$ of smooth projective algebraic varieties; the relation (83) would then be the case in which Y is reduced to a single point. The problem was thus to replace both sides of (83) by meaningful generalizations when X, Y and f are arbitrary.

For the right-hand side of (83) the "integration" map $u \mapsto \langle u, [M] \rangle$ sending $H^{\bullet}(X; \mathbf{Q})$ into **Q** has to be replaced by a map $H^{\bullet}(X; \mathbf{Q}) \to H^{\bullet}(Y; \mathbf{Q})$ that goes "against" the fact that cohomology is a *contravariant* functor. Because X and Y are oriented manifolds, this can be done by using Poincaré duality in a way "dual" to the way it had been used by Gysin (chap. IV, § 2). Let m and n be the complex dimensions of X and Y, and consider the Poincaré isomorphisms

$$j_p: H^p(X; \mathbf{Q}) \xrightarrow{\sim} H_{2m-p}(X; \mathbf{Q}), \qquad j'_q: H^q(Y; \mathbf{Q}) \xrightarrow{\sim} H_{2n-q}(Y; \mathbf{Q}).$$

Then a homomorphism

$$f_*: H^p(X; \mathbf{Q}) \to H^{p+d}(Y; \mathbf{Q})$$

with $d = 2(n - m)$ is defined as the composite

$$\mathrm{H}^p(\mathrm{X};\mathbf{Q}) \xrightarrow{j_p} \mathrm{H}_{2m-p}(\mathrm{X};\mathbf{Q}) \xrightarrow{f_*} \mathrm{H}_{2m-p}(\mathrm{Y};\mathbf{Q}) \xrightarrow{j'^{-1}_{p+d}} \mathrm{H}^{p+d}(\mathrm{Y};\mathbf{Q})$$

(for $n = 0$, $p = 2m$, this is the "integration" $\mathrm{H}^{2m}(\mathrm{X};\mathbf{Q}) \to \mathbf{Q}$).

On the other hand, there was no obvious way for Grothendieck to define an expression that would be the generalization of the left-hand side of (83)

$$\chi(\mathscr{F}) = \sum_k (-1)^k \dim \mathrm{H}^k(\mathscr{F}) \tag{87}$$

for a coherent sheaf $\mathscr{F}$ on X. Such an expression should belong to the cohomology ring $\mathrm{H}^\bullet(\mathrm{Y};\mathbf{Q})$, but how could he define "alternating sums" of elements of that ring which would take the place of dimensions in (87)? It was here that Grothendieck broke entirely new ground by his concept of what are now called *Grothendieck groups* and by the imaginative use he made of them.

Let C(X) be the set of *isomorphism classes* of coherent sheaves on X, and for each coherent sheaf $\mathscr{F}$ let $[\mathscr{F}]$ be its class in C(X). Consider the free **Z**-module $\mathbf{Z}^{(\mathrm{C(X)})}$ with basis $(e_{[\mathscr{F}]})$ indexed by C(X). Let E(X) be the submodule of that module generated by the elements

$$e_{[\mathscr{F}]} - e_{[\mathscr{F}']} - e_{[\mathscr{F}'']} \tag{88}$$

for all triples $(\mathscr{F}, \mathscr{F}', \mathscr{F}'')$ of coherent sheaves on X for which there exists an exact sequence of homomorphisms

$$0 \to \mathscr{F}' \to \mathscr{F} \to \mathscr{F}'' \to 0. \tag{89}$$

The quotient $\mathrm{K(X)} = \mathbf{Z}^{(\mathrm{C(X)})}/\mathrm{E(X)}$ is the Grothendieck group of coherent sheaves on X. It has the "universal" property that any map

$$\varphi\colon \mathbf{Z}^{(\mathrm{C(X)})} \to \mathrm{G}$$

into a commutative group G such that

$$\varphi([\mathscr{F}]) = \varphi([\mathscr{F}']) + \varphi([\mathscr{F}'']) \tag{90}$$

for all exact sequences (89), factorizes into

$$\varphi\colon \mathbf{Z}^{(\mathrm{C(X)})} \xrightarrow{\gamma} \mathrm{K(X)} \xrightarrow{\psi} \mathrm{G} \tag{91}$$

where γ (also written γ_{X}) is the natural map and ψ is a homomorphism of *groups.*

From the fact that X is smooth and compact it follows that any coherent sheaf $\mathscr{F}$ on X has a *finite* resolution by locally free sheaves

$$0 \leftarrow \mathscr{F} \leftarrow \mathscr{L}_0 \leftarrow \mathscr{L}_1 \leftarrow \cdots \leftarrow \mathscr{L}_m \leftarrow 0. \tag{92}$$

This enables us to define the *total Chern class* $c(\mathscr{F}) \in \mathrm{H}^\bullet(\mathrm{X};\mathbf{Q})$ of a coherent sheaf: for a locally free sheaf $\mathscr{L}$, the Chern classes are by definition the Chern classes of the corresponding vector bundles on X. Then the total Chern class of $\mathscr{F}$ is defined by considering a finite locally free resolution (92) of $\mathscr{F}$ and taking

$$c(\mathscr{F}) = c(\mathscr{L}_0)c(\mathscr{L}_1)^{-1}c(\mathscr{L}_2)\cdots c(\mathscr{L}_m)^{(-1)^m}. \tag{93}$$

This definition does not depend on the choice of the locally free resolution (92) of $\mathscr{F}$. The *Chern character* $\mathrm{ch}(\mathscr{F})$ [also written $\mathrm{ch}_X(\mathscr{F})$] is then defined by the general formula (chap. IV, § 1,E); it satisfies the relation

$$\mathrm{ch}(\mathscr{F}) = \mathrm{ch}(\mathscr{F}') + \mathrm{ch}(\mathscr{F}'') \tag{94}$$

for every exact sequence (89). Since it is obviously invariant under isomorphism, it factorizes as $\mathscr{F} \mapsto \gamma(\mathscr{F}) \mapsto \mathrm{ch}(\gamma(\mathscr{F}))$ for a homomorphism $\mathrm{ch}\colon \mathrm{K}(\mathrm{X}) \to \mathrm{H}^{\bullet}(\mathrm{X}; \mathbf{Q})$.

Now consider a morphism $f\colon \mathrm{X} \to \mathrm{Y}$ of smooth compact algebraic manifolds. The "relative" objects corresponding to the "absolute" cohomology groups $\mathrm{H}^q(\mathscr{F})$ are the Leray "higher direct images" by f of a coherent sheaf $\mathscr{F}$ on X, which we have written $\mathscr{H}^q(f, \mathscr{F})$ (Part 1, chap. IV, § 7,E), and which were written $\mathrm{R}^q f_!(\mathscr{F})$ by Grothendieck [$\mathrm{R}^0 f_!(\mathscr{F}) = f_!(\mathscr{F})$ is the direct image of $\mathscr{F}$, written also $f_*(\mathscr{F})$]. He proved that the $\mathrm{R}^q f_!(\mathscr{F})$ are *coherent* sheaves on Y and that $\mathrm{R}^q f_!(\mathscr{F}) = 0$ when q is large enough. The finite sum

$$\sum_q (-1)^q \gamma_{\mathrm{Y}}(\mathrm{R}^q f_!(\mathscr{F})) = \varphi(\mathscr{F}) \tag{95}$$

is therefore defined in the group K(Y). It obviously only depends on the isomorphism class of $\mathscr{F}$. The crucial fact is that it only depends on the class $\gamma_{\mathrm{X}}(\mathscr{F})$ in K(X), and if we write it as

$$f_!(\gamma_{\mathrm{X}}(\mathscr{F}))$$

we define a *homomorphism of groups* $f_!\colon \mathrm{K}(\mathrm{X}) \to \mathrm{K}(\mathrm{Y})$. This amounts to saying that

$$\varphi(\mathscr{F}) = \varphi(\mathscr{F}') + \varphi(\mathscr{F}'') \tag{96}$$

in K(Y), for every exact sequence (89) of coherent sheaves. From that sequence an exact cohomology sequence can be derived:

$$\cdots \to \mathrm{R}^q f_!(\mathscr{F}') \to \mathrm{R}^q f_!(\mathscr{F}) \to \mathrm{R}^q f_!(\mathscr{F}'') \to \mathrm{R}^{q+1} f_!(\mathscr{F}') \to \cdots;$$

and since this sequence is finite, the alternating sum of the values of γ_{Y} for the sheaves of the sequence is 0, which proves (96).

With these definitions the Riemann–Roch–Grothendieck theorem may be formulated:

$$\mathrm{ch}_{\mathrm{Y}}(f_!(x)) \smile \mathrm{td}_{\mathrm{Y}} = f_*(\mathrm{ch}_{\mathrm{X}}(x) \smile \mathrm{td}_{\mathrm{X}}) \tag{97}$$

in $\mathrm{H}^{\bullet}(\mathrm{Y}; \mathbf{Q})$, for every $x \in \mathrm{K}(\mathrm{X})$. It is easy to check that it reduces to equation (83) when Y is reduced to a point.

The strategy of the proof imagined by Grothendieck was completely different from Hirzebruch's method: its principle was to decompose $f\colon \mathrm{X} \to \mathrm{Y}$ as the immersion $h\colon \mathrm{X} \to \mathrm{X} \times \mathrm{Y}$ sending X onto the graph of f, followed by the projection $g\colon \mathrm{X} \times \mathrm{Y} \to \mathrm{Y}$; of course he had to prove that $(g \circ h)_! = g_! \circ h_!$. The case of the projection g is fairly easy, but the immersion h requires intricate arguments.

B. Riemann–Roch Theorems for Differentiable Manifolds

Since the Riemann–Roch–Hirzebruch theorem dealt with vector bundles, it would have seemed more natural to generalize it for locally free sheaves instead of going to the larger category of coherent sheaves. Unfortunately, the higher direct images of locally free sheaves are not locally free in general, which explains Grothendieck's approach. However, if a "Grothendieck group" $K_1(X)$ is defined by replacing coherent sheaves by locally free ones in the definition, there is a natural homomorphism $K_1(X) \to K(X)$, and Grothendieck proved that (for smooth projective algebraic manifolds) it is *bijective*.

In so doing, he used the language of abelian categories (Part 1, chap. IV, §8) with which he had already been familiar for some time [215]; for such a category $\boldsymbol{C}$, the group $K(\boldsymbol{C})$ can be defined in the same way as in the particular case of coherent sheaves, since the concept of exact sequence applies to abelian categories.

In particular, the abelian category of complex vector bundles on X may be considered instead of the category of their associated locally free sheaves, and then X may be *any space*. Furthermore, instead of exact sequences to define the elements (88), the Whitney sum of vector bundles may be used here; consider the subgroup generated by the elements

$$e_{[E \oplus F]} - e_{[E]} - e_{[F]}$$

for all pairs (E, F) of vector bundles over X. When X is paracompact, both definitions are in fact equivalent, since the existence of an exact sequence $0 \to E' \to E \to E'' \to 0$ of vector bundles implies that E is isomorphic to $E' \oplus E''$.

In 1959 [34] Atiyah and Hirzebruch investigated the group K(X) [also written $K_C(X)$] for the category of complex vector bundles over a finite-dimensional *CW-complex* X. They noted (as Grothendieck had already done) that K(X) has the natural structure of a *commutative ring* stemming from the tensor product of vector bundles: $\gamma(E)\gamma(F)$ is defined as equal to $\gamma(E \otimes F)$, and there is a unit element, the class of trivial line bundles on X. The contravariant functor $K: X \mapsto K(X)$ is defined by associating to any continuous map $f: X \to Y$ the map $\gamma(F) \mapsto \gamma(f^*(F))$ of K(Y) into K(X); $K(\{x_0\})$ is naturally isomorphic to the ring **Z** for any point $x_0 \in X$, and the embedding $\{x_0\} \to X$ defines a surjective ring homomorphism $K(X) \to K(\{x_0\})$; its kernel $\tilde{K}(X)$ is an ideal in K(X).

The classification theorem of vector bundles (chap. III, §2) shows that $\tilde{K}(X)$ may be identified to the group $[X, B_U]$ of homotopy classes, where **U** is the "infinite" unitary group. With Bott's periodicity, which establishes a weak homotopy equivalence between B_U and $\Omega^2(B_U)$ (chap; V, §2,B), this enabled Atiyah and Hirzebruch to establish the central result of their investigations: the natural map

$$\beta: K(X) \otimes K(S_2) \to K(X \times S_2)$$

deduced from the product of vector bundles on X and S_2 is an *isomorphism* of rings, and the diagram

$$\begin{array}{ccc} K(X) \otimes K(S_2) & \xrightarrow{\beta} & K(X \times S_2) \\ \Big\downarrow {\scriptstyle ch \otimes ch} & & \Big\downarrow {\scriptstyle ch} \\ H^{\cdot}(X; \mathbf{Q}) \otimes H^{\cdot}(S_2; \mathbf{Q}) & \xrightarrow[\alpha]{} & H^{\cdot}(X \times S_2 : \mathbf{Q}) \end{array} \tag{98}$$

is commutative (α being the cup-product).

This result was a base for their proof of a Riemann–Roch relation analogous to Grothendieck's for *compact* C^∞ *manifolds* X, Y and a continuous map $f: X \to Y$. Some restrictions on these data have to be made: the difference of the dimensions of X and Y must be *even*, and the relation

$$w_2(Y) = f^*(w_2(X)) \tag{99}$$

between the second Stiefel–Whitney classes of the tangent bundles of X and Y must be satisfied. The Atiyah–Hirzebruch theorem then asserts the existence of a homomorphism of groups

$$g: K(X) \to K(Y)$$

such that

$$ch(g(x)) \smile A(T(Y)) = f_*(ch(x) \smile A(T(X)) \tag{100}$$

for all $x \in K(X)$. Here, for any vector fibration ξ, $A(\xi)$ is a polynomial in the Pontrjagin classes of ξ defined by the Hirzebruch method (§2,A, VIII): if the Pontrjagin classes are considered elementary symmetric polynomials in the "phantom" indeterminates t_j, then $A(\xi) = K(1)$ for the power series

$$\dot{K}(x) = \prod_j a(t_j x)$$

where

$$a(x) = \tfrac{1}{2}\sqrt{x}/\mathrm{sh}(\tfrac{1}{2}\sqrt{x}). \tag{101}$$

The proof follows a pattern similar to Grothendieck's: f is assumed to be C^∞, and there is a smooth embedding $g: X \to S_{2n}$ for n large enough. Then f is factored into $v \circ u$, where

$$u = (f, g): X \to Y \times S_{2n}$$

and v is the first projection $Y \times S_{2n} \to Y$. Then (100) is proved in succession for v and for u; for v the fundamental diagram (98) is used.

Shortly afterward [35] Atiyah and Hirzebruch developed their study of K(X) into a "generalized cohomology" (see §5,A). After 1960 the concepts of K-theory invaded many parts of mathematics, particularly algebra and number theory (see [33]).

§4. S-Duality

The general idea of duality has been one of the guiding concepts in algebraic topology since Poincaré, and we saw in the preceding chapters the various forms it has taken, particularly after the introduction of cohomology. The properties of cohomotopy groups (chap. VI, §6) pointed toward a possible duality theory for homotopy, but these groups $\pi^n(X)$ are not defined for arbitrary values of $n > 0$ and X, in contrast with the homotopy groups $\pi_n(X)$. In 1953 E. Spanier and J.H.C. Whitehead endeavored to define in homotopy theory a genuine duality [441]. Given two pointed spaces X, Y, they wrote SX, SY for their unreduced suspensions (Part 2, chap. V, §2C) and they considered the infinite sequence of sets

$$[X;Y] \xrightarrow{S} [SX;SY] \xrightarrow{S} [S^2X;S^2Y] \to \cdots \xrightarrow{S} [S^nX;S^nY] \to \cdots$$

where the map $S: [f] \to [Sf]$ is the Freudenthal suspension. Recall that for $k \geqslant 2$, $[S^kX;S^kY]$ is a commutative group and S is a homomorphism of groups (chap. II, §3D); then they define the direct limit commutative group

$$\{X;Y\} = \varinjlim_{k} [S^kX;S^kY]. \tag{102}$$

The elements of $\{X;Y\}$ are called S-*homotopy classes* or S-*maps*; two continuous maps $g: S^pX \to S^pY$, $h: S^qX \to S^qY$ belong to the same S-map if and only if for some $n > \text{Max}(p,q)$, $S^{n-p}g$ and $S^{n-q}h$ are homotopic. Note that for some finite CW-complexes the map $S: [S^nX;S^nY] \to [S^{n+1}X;S^{n+1}Y]$ is bijective for large n (chap. II, §6,E).

These definitions allow us in particular to define *generalized homotopy and cohomotopy groups* for every dimension $p \geqslant 1$:

$$\Sigma_p(X) = \{S_p;X\}, \qquad \Sigma^p(X) = \{X;S_p\}. \tag{103}$$

S-maps may be composed: if the maps $f: S^kX \to S^kY$, $g: S^kY \to S^kZ$ are in the respective classes $\alpha \in \{X;Y\}$, $\beta \in \{Y;Z\}$ for some large k, $\beta \circ \alpha \in \{X;Z\}$ is by definition the class of $g \circ f: S^kX \to S^kZ$. This defines the S-*category*, in which the objects are the finite CW-complexes, and the morphisms are the S-maps. The suspension

$$S: \{X;Y\} \to \{SX;SY\} \tag{104}$$

is defined in an obvious way. If $f: S^kX \to S^kY$ is in the class $\alpha \in \{X;Y\}$, then $S\alpha$ is the class of $Sf: S^{k+1}X \to S^{k+1}Y$; the map (104) is obviously bijective.

Since the suspension defines isomorphisms (Part 2, chap. V, §2,C)

$$H_q(X) \simeq H_{q+1}(SX), \qquad H^{q+1}(SX) \simeq H^q(X)$$

in homology and cohomology, any S-map $\alpha \in \{X;Y\}$ defines in a natural way homomorphisms in homology and cohomology

$$\alpha_*: H_q(X) \to H_q(Y), \qquad \alpha^*: H^q(Y) \to H^q(X): \tag{105}$$

if $f: S^kX \to S^kY$ is in the class α, α_* is the unique homomorphism for which the diagram

$$\begin{array}{ccc} H_q(X) & \xrightarrow{\alpha_*} & H_q(Y) \\ \downarrow{S^k} & & \downarrow{S^k} \\ H_{q+k}(S^kX) & \xrightarrow[f_*]{} & H_{q+k}(S^kY) \end{array}$$

is commutative, and similarly for cohomology.

An S-*equivalence* for the S-category is an element $\alpha \in \{X;Y\}$ that has an inverse $\alpha^{-1} \in \{Y;X\}$. This is the case if and only if $\alpha_*: H_q(X) \to H_q(Y)$ is bijective for every $q > 0$: if that condition is satisfied, since S^kX and S^kY are simply connected CW-complexes for $k \geqslant 3$, a map $f: S^kX \to S^kY$ which belongs to α is a homotopy equivalence by the second Whitehead theorem (chap. II, § 6,B).

The technical tool at the basis of the S-duality theory of Spanier and Whitehead is the notion of S-*deformation retract* for a CW-subcomplex A of the CW-complex X: the class $\iota \in \{A;X\}$ of the natural injection $A \to X$ is an S-equivalence; there is therefore an inverse $\iota^{-1} \in \{X;A\}$, called an S-*retraction*.

S-duality is concerned with finite euclidean simplicial complexes (which we call *polyhedra* for short) contained in a *fixed* sphere S_n; its purpose is to extend Alexander duality. Suppose $X \subset S_n$ is a polyhedron; then an *n-dual* of X is a polyhedron $D_nX \subset S_n - X$ that is an S-deformation retract of $S_n - X$. Such a space is not uniquely determined, but the S-homotopy type is the same for all of them, and X is an n-dual of D_nX.

The main result of Spanier and Whitehead was the definition of a homomorphism of groups

$$D_n: \{X;Y\} \to \{D_nY;D_nX\} \tag{105}$$

for all pairs of polyhedra X, Y contained in S_n; it has the properties expected from a "duality":

1. if $X \subset Y$ and $\iota \in \{X;Y\}$ is the class of the natural injection $X \to Y$, then $D_nY \subset D_nX$ and $D_n\iota = \iota'$, the class in $\{D_nY;D_nX\}$ of the natural injection $D_nY \to D_nX$;
2. if X, Y, Z are three polyhedra contained in S_n, and $\alpha \in \{X;Y\}$, $\beta \in \{Y;Z\}$, then
$$D_n(\beta \circ \alpha) = D_n\alpha \circ D_n\beta;$$
3. for all $\alpha \in \{X;Y\}$, $D_n(D_n\alpha) = \alpha$;
4. for $\alpha \in \{X;Y\}$ and $p < n$, the diagram

$$\begin{array}{ccc} H_p(X) & \xrightarrow{\alpha_*} & H_p(Y) \\ \phi \downarrow \wr & & \wr \downarrow \phi \\ H^{n-p-1}(D_nX) & \xrightarrow[(D_n\alpha)^*]{} & H^{n-p-1}(D_nY) \end{array}$$

commutes, where the vertical arrows are the Alexander duality isomorphisms;

5. for $X = S_p$ with $p < n$, one may take $D_n X = S_{n-p-1}$, and D_n can be considered a duality in homotopy theory

$$\theta\colon \Sigma_p(Y) \to \Sigma^{n-p-1}(D_n Y) \tag{106}$$

such that the diagram

$$\begin{array}{ccc} \Sigma_p(Y) & \xrightarrow{\tau} & H_p(Y) \\ \theta\big\downarrow & & \big\downarrow\phi \\ \Sigma^{n-p-1}(D_n Y) & \xrightarrow[\tau^*]{} & H^{n-p-1}(D_n Y) \end{array}$$

commutes; τ is the Hurewicz homomorphism, and τ^* is defined by

$$\tau^*(\beta) = \beta^*(s_{n-p-1})$$

for

$$\beta \in \Sigma^{n-p-1}(D_n Y) = \{D_n Y; S_{n-p-1}\}.$$

The proofs are broken down into a long series of lemmas, most of which use the straightforward combinatorial methods of simplicial homology with which Whitehead was familiar from his work on combinatorial homotopy. The major part of the proof is the definition of the map D_n in (105); it turns out that this is fairly easily done when a map

$$\Delta_n\colon [X; Y] \to \{D_n Y; D_n X\}$$

can be defined in a reasonable way for two polyhedra X, Y contained in S_n such that $X \cap Y = \varnothing$. Let $f\colon X \to Y$ be a simplicial map and W be a "polyhedral mapping cylinder" of f having the same homotopy type as the usual mapping cylinder Z_f; it may be supposed that W (containing X and Y as subpolyhedra) is contained in a sphere S_q for some $q > n$. There are q-duals X^*, Y^*, and W^* of X, Y, W in S_q, such that $W^* \subset X^* \cup Y^*$ and

$$D_q X = S^{q-n} D_n X \subset X^*, \qquad D_q Y = S^{q-n} D_n Y \subset Y^*.$$

Then S-maps

$$D_q Y \xrightarrow{i} Y^* \xrightarrow{r^*} W^* \xrightarrow{i^*} X^* \xrightarrow{r} D_q X$$

can be defined, where i, i^* are S-inclusions and r, r^* S-retractions. The element $\Delta_n([f]) \in \{D_n Y; D_n X\}$ is then defined by

$$\Delta_n([f]) = S^{n-q}(r \circ i^* \circ r^* \circ i),$$

using the fact that S has an inverse S^{-1} in $\{D_q Y; D_q X\}$. Showing that the element $\Delta_n([f])$ is well defined, independently of all the choices made, constitutes the bulk of the proof.

The notion of n-dual can be extended to any finite CW-complex X; X has the homotopy type of a polyhedron, and any polyhedron is homeomorphic to a subpolyhedron of some S_p. Two finite CW-complexes X, X* will then be called *weakly p-dual* if there are S-equivalences $\xi: X \to X_1$, $\xi^*: X^* \to X_1^*$ such that X_1 and X_1^* are p-dual subpolyhedra of S_p. It is then possible to define

$$D_p: \{X; Y\} \to \{Y^*; X^*\}$$

in an obvious way when X*, Y* are respectively weak p-duals of X, Y.

§ 5. Spectra and Theories of Generalized Homology and Cohomology

Around 1959 several mathematicians, working in different directions, were led to consider systems of covariant functors h_n: $\boldsymbol{C} \to \boldsymbol{Ab}$ (resp. contravariant functors h^n: $\boldsymbol{C}^0 \to \boldsymbol{Ab}$), where n takes all values in **N** or **Z**, and $\boldsymbol{C}$ is a subcategory of the category $\boldsymbol{T}$ of topological spaces. These functors verified all Eilenberg–Steenrod axioms for homology (resp. cohomology) *with the exception of the dimension axiom*: the groups h_n(pt.) [resp. h^n(pt.)] might be $\neq 0$ for $n \neq 0$. The theory of such functors became known as *generalized* (or *extraordinary*) *homology* (resp. *cohomology*).

A. K-Theory as Generalized Cohomology

The first of these theories to appear in print was Atiyah and Hirzebruch's [35]. They restricted the contravariant functor K that they had defined after Grothendieck (§ 3,B) to pointed spaces X equipped with a structure of finite CW-complex and pairs (X, Y), where Y is a sub-CW-complex of the finite CW-complex X. At first, for $n \geqslant 0$ and for such a pair (X, Y), they defined, by means of the iterated reduced suspension $S^n X = S_n \wedge X$ (Part 2, chap. V, § 2,C and D),

$$K^{-n}(X, Y) = K(S^n(X/Y)), \tag{107}$$

$$K^{-n}(X) = K^{-n}(X, \varnothing) \tag{108}$$

so that $K^0(X) = K(X)$, $K^0(X, Y) = K(X, Y)$. Using the Puppe exact sequence (chap. II, § 5,D), they obtained a long exact sequence

$$\cdots \to K^{-n-1}(X) \xrightarrow{\partial} K^{-n}(X, Y) \to K^{-n}(X) \to K^{-n}(Y) \to \cdots \to K^0(X) \to K^0(Y) \tag{109}$$

and a similar one for the "reduced" functor $\tilde{K}$. Applying the latter to the pair $(X \times Y, X \vee Y)$, they got the split exact sequences

$$0 \to \tilde{K}^{-n}(X \wedge Y) \to \tilde{K}^{-n}(X \times Y) \to \tilde{K}^{-n}(X \vee Y) \to 0.$$

From the Bott isomorphism (§ 3,B) they derived an explicit isomorphism

$$K^{-n}(X, Y) \simeq K^{-n-2}(X, Y)$$

so that the exact sequence (109) could be extended to a doubly infinite exact sequence

$$\cdots \to K^n(X,Y) \to K^n(X) \to K^n(Y) \xrightarrow{\partial} K^{n+1}(X,Y) \to \cdots \quad (n \in \mathbf{Z}).$$

This led them to consider that the K^n and the homomorphism ∂ (for $n \in \mathbf{Z}$) defined a "generalized" cohomology, for which they could also prove the other Eilenberg–Steenrod axioms, except the dimension axiom, since $K^n(\text{pt.}) \cong \mathbf{Z}$ for n even, $K^n(\text{pt.}) = 0$ for n odd.

They also filtered each $K^n(X)$ by the subgroups

$$K_p^n(X) = \text{Ker}(K^n(X) \to K^n(X^{p-1})) \tag{110}$$

where X^p is the p-th skeleton of X. This gave them a spectral sequence linking ordinary cohomology and K-theory; they found

$$E_2^{pq} \simeq H^p(X; K^q(\text{pt.})), \tag{111}$$

$$E_\infty^{pq} \simeq K_p^{p+q}(X)/K_{p+1}^{p+q}(X). \tag{112}$$

B. Spectra

In 1958 E. Lima wrote a thesis in Chicago under the guidance of Spanier [327]. He noted that S-duality, as defined by Spanier and Whitehead, did not immediately extend to subspaces of $\mathbf{S}_n$ more general than finite CW-complexes. For instance, in $\mathbf{R}^2$, if X is a circle and Y is the Alexandroff subspace defined in Part 1, chap. IV, § 2, it is easily verified that $\{X;Y\} = 0$ but $\{\mathbf{S}_2 - Y; \mathbf{S}_2 - X\} \cong \mathbf{Z}$. Lima undertook to generalize S-duality by introducing what he called *spectra.* Instead of considering only the successive suspensions

$$X \xrightarrow{S} SX \xrightarrow{S} S^2X \to \cdots \xrightarrow{S} S^nX \to \cdots$$

of a space X, he took a more general sequence (X_n) of spaces together with a sequence of continuous maps

$$\rho_n \colon SX_n \to X_{n+1}. \tag{113}$$

He showed that it is possible to define morphisms of spectra in a natural way, so that spectra become a category. Using this category (in which he called the objects "direct spectra") and a similar one with "reversed arrows," called "inverse spectra," he was able to generalize the notion of "n-dual" in such a way that an arbitrary closed subset X in $\mathbf{S}_n$ has $\mathbf{S}_n - X$ as an n-dual, and for a second closed subset Y of $\mathbf{S}_n$ there is a natural isomorphism

$$\{X;Y\}_s \xrightarrow{\sim} \{\mathbf{S}_n - Y; \mathbf{S}_n - X\}_c$$

where both sides are generalizations of the Spanier–Whitehead group $\{X;Y\}$, but the left-hand side is based on "direct spectra" and the right-hand side on "inverse spectra."

A little later Spanier described another generalization of S-duality based on a special kind of spectrum: for a pointed space M,

$$X_n = \mathscr{C}(M; S_n)$$

with the usual compact-open topology (chap. II, § 2A); ρ_n is the natural map

$$\rho_n: S\mathscr{C}(M; S_n) \to \mathscr{C}(M; S_{n+1})$$

that associates to Su: $SM \to S_{n+1}$ the composite map

$$M \xrightarrow{u} S_n \xrightarrow{S} S_{n+1}.$$

C. Spectra and Generalized Cohomology

Recall the fundamental relation between Eilenberg–Mac Lane spaces and cohomology [chap. V, § 1,D, formula (43)]

$$[X; K(\Pi, n)] \simeq H^n(X; \Pi) \tag{114}$$

which can be extended to

$$[X/Y; K(\Pi, n)] \simeq H^n(X, Y; \Pi). \tag{115}$$

Recall also the existence of a weak homotopy equivalence

$$K(\Pi, n) \xrightarrow{\sim} \Omega K(\Pi, n+1)$$

(chap. IV, § 2,C), which can, by the consideration of adjoint functors, be written as a weak homotopy equivalence

$$\rho_n: SK(\Pi, n) \to K(\Pi, n+1). \tag{116}$$

Since the suspension defines isomorphisms

$$H^q(X; \Pi) \xrightarrow{\sim} H^{q+1}(SX; \Pi),$$

the iteration of the Freudenthal suspension homomorphism defines isomorphisms of groups for $q < n$,

$$[X; K(\Pi, q)] \xrightarrow{\sim} [S^{n-q}X; K(\Pi, n)]$$

so that relation (114) may be written

$$H^q(X; \Pi) \simeq \varinjlim_n [S^n X; K(\Pi, q+n)]. \tag{117}$$

The maps ρ_n define the sequence $(K(\Pi, n))$ as a spectrum, the *Eilenberg–Mac Lane spectrum*, written $K(\Pi)$. The relation (117) led to the consideration, for *any spectrum*

$$E: E_1 \to E_2 \to \cdots \to E_n \to \cdots$$

of pointed CW-complexes, and, for a finite pointed CW-complex X, of a *generalized cohomology group*

$$H_E^q(X) = \varinjlim_n [S^n X; E_{q+n}]. \tag{118}$$

For a CW-subcomplex Y of X, formula (115) similarly leads to the definition of relative cohomology groups

$$H^q_E(X, Y) = H^q_E(X/Y). \tag{119}$$

It can then be checked that these generalized cohomology groups satisfy the Eilenberg–Steenrod axioms, with the exception of the dimension axiom, which is replaced by

$$H^q_E(pt.) \simeq \varinjlim_n \pi_n(E_{n+q}).$$

The Atiyah–Hirzebruch spectral sequence (111), (112) generalizes to any cohomology theory defined by a spectrum: simply replace K^q by H^q_E in the definition.

In 1959 [93] E.H. Brown characterized contravariant functors

$$F: \boldsymbol{CW}^0 \to \boldsymbol{Set}$$

defined in the category of pointed CW-complexes, which can be written

$$X \mapsto [X; Y]$$

for a CW-complex Y. It is necessary and sufficient that they satisfy the two following conditions.

1. For any wedge $\bigvee_\alpha X_\alpha$ of pointed CW-complexes (with arbitrary set of indices), the map defined by the injections $i_\beta: X_\beta \to \bigvee_\alpha X_\alpha$

$$F((i_\alpha)): F\left(\bigvee_\alpha X_\alpha\right) \to \prod_\alpha F(X_\alpha)$$

is bijective.

2. For any pair (A_1, A_2) of CW-subcomplexes of a CW-complex X, such that $X = A_1 \cup A_2$, let

$$i_1: A_1 \to X, \qquad i_2: A_2 \to X, \qquad i_{21}: A_1 \cap A_2 \to A_1, \qquad i_{12}: A_1 \cap A_2 \to A_2$$

be the natural injections. Then for any pair of elements

$$x \in F(A_1), \qquad x_2 \in F(A_2)$$

satisfying the condition $F(i_{21})(x_1) = F(i_{12})(x_2)$, there is a $y \in F(X)$ such that $F(i_1)(y) = x_1$ and $F(i_2)(y) = x_2$.

When F satisfies these two conditions the construction of Y is done in the following way. By induction on n, an increasing sequence

$$\{y_0\} \subset Y_1 \subset Y_2 \subset \cdots \subset Y_n \subset \cdots$$

and a sequence of elements $u_n \in F(Y_n)$ are defined simultaneously to verify the two conditions:

(i) $F(i)(u_n) = u_{n-1}$ for the natural injection $i: Y_{n-1} \to Y_n$.
(ii) Let $T_{u_n}: \pi_q(Y_n) \to F(S_q)$ be defined by

$$T_{u_n}([f]) = F(f)(u_n);$$

then T_{u_n} is bijective for $q < n$, surjective for $q = n$.

That construction is done by attaching n-cells to Y_{n-1} in a suitable way; the techniques are similar to those used in the proof of the second Whitehead theorem (chap. II, § 6,B).

From the Brown theorem it easily follows that *any* generalized cohomology defined for CW-complexes can be defined by a spectrum $E = (E_n)$ such that ΩE_{n+1} is weakly homotopically equivalent to E_n.

D. Generalized Homology and Stable Homotopy

In 1954 G.W. Whitehead studied the homotopy of the smash product $X \wedge Y$ of two pointed CW-complexes X, Y. He had the idea of considering the filtration $(X^p \wedge Y)$ of $X \wedge Y$, where X^p is the p-skeleton of X. This gave him a spectral sequence with E^2 terms which, for p, q satisfying some restrictions, are

$$E^2_{pq} \simeq H_p(X; \pi_q(Y))$$

and the terms E^∞ are given by a suitable filtration on

$$\pi_\bullet(X \wedge Y) = \bigoplus_n \pi_n(X \wedge Y).$$

When $Y = K(\Pi, n)$, it follows from that spectral sequence that for given p and large n,

$$\pi_{p+n}(X \wedge K(\Pi, n)) \simeq H_p(X; \Pi). \tag{120}$$

In particular, when $X = K(G, m)$ with large m, the homology of Eilenberg–Mac Lane spaces is

$$H_{p+m}(G, m; \Pi) \simeq \pi_{p+m+n}(X \wedge Y) \simeq \pi_{p+n+m}(Y \wedge X) \simeq H_{p+n}(\Pi, n; G)$$

a "symmetry" which had been noted by H. Cartan in his earlier work on that homology [110].

Relation (120) can be written

$$H_q(X; \Pi) \simeq \varinjlim_n \pi_{n+q}(X \wedge K(\Pi, n)). \tag{121}$$

When Lima introduced the idea of spectrum, G. Whitehead, by analogy with (121), considered, for an *arbitrary* spectrum $E = (E_n)$, the direct limit

$$H^E_q(X) = \varinjlim_n \pi_{n+q}(X \wedge E_n) \tag{122}$$

for the system of homomorphisms

$$\pi_{n+q}(X \wedge E_n) \xrightarrow{S_*} \pi_{n+1+q}(X \wedge SE_n) \xrightarrow{(1 \wedge \rho_n)_*} \pi_{n+1+q}(X \wedge E_{n+1}).$$

He called (122) the *generalized homology groups* for the spectrum E.

In a paper which laid the foundations for both the generalized homology defined by (122) and the generalized cohomology defined by (118), he showed that definition (122) satisfies the Eilenberg–Steenrod axioms for homology (except of course the dimension axiom), and that Alexander and Poincaré duality can be generalized in that context [489]. There is again a spectral sequence going from the E^2 terms

$$E^2_{pq} \simeq H_p(X; H^E_p(\text{pt.}))$$

to the graded module associated to a filtration on $H^E_{p+q}(X)$.

The "coefficient groups" $H^E_p(\text{pt.})$ are here the *stable homotopy groups*

$$\sigma_n(E) = \varinjlim_k \pi_{n+k}(E_k). \tag{123}$$

It is therefore important to be able to obtain information on these groups for applications of generalized homology.

In the case $E_{k+1} = SE_k$ for all k, J.F. Adams discovered in 1958 [1] a remarkable spectral sequence that links the cohomology of the Steenrod algebra $\mathscr{A}_p$ for each prime p to the p-components of the stable homotopy groups $\varinjlim_k \pi_{n+k}(S^k X)$.

The construction of that spectral sequence is very involved and impossible to describe here in detail. The main idea is to build a decreasing sequence of space for each n

$$S^n X = Y_0 \supset Y_1 \supset Y_2 \supset \cdots \supset Y_s \supset \cdots$$

such that $\bigoplus_s H^\bullet(Y_s, Y_{s+1}; F_p)$ is a free resolution of $H^\bullet(X; F_p)$ considered as a left module over the Steenrod algebra $\mathscr{A}_p$. In the spectral sequence the groups E_r^{st} and E_∞^{st} are quotients of subgroups of the relative homotopy groups

$$\pi_{n+t-s}(Y_s, Y_{s+1});$$

then E_∞^{st} is one of the quotients of a suitable filtration of $\pi_{t-s}(X)$, and

$$E_2^{st} \simeq \text{Ext}^{s,t}_{\mathscr{A}_p}(H^\bullet(X; F_p), Z/pZ). \tag{124}$$

Here $\text{Ext}^s_{\mathscr{A}_p}(H^\bullet(X; F_p), Z/pZ)$ is the Ext functor of Cartan–Eilenberg (Part 1, chap. IV, §8,A); it is given a grading $(E^{s,t}_{\mathscr{A}_p}(H^\bullet(X; F_p), F_p)_t)$ arising from the gradings of $H^\bullet(X; F_p)$ and of the Steenrod algebra $\mathscr{A}_p$; the result (124) is derived from the consideration of the Hurewicz homomorphisms.

Later Adams' spectral sequence was generalized to any spectrum [459]. He had chiefly used it in his solution of the Hopf invariant problem (chap. VI, §5,D) but it was very useful in the various cobordism.theories developed after 1960 [458].

In 1960 the spectra used in algebraic topology were the Eilenberg–Mac Lane spectrum, giving the "ordinary" homology and cohomology theories, and the spectrum BU obtained by taking $E_{2n} = B_U$ and $E_{2n+1} = U$, with the corresponding homomorphisms $U \to \Omega B_U$ and $\Omega U \to Z \times B_U$ (the second being the Bott isomorphism); K-theory is the corresponding cohomology theory, and similar spectra were later defined for all classical groups. Finally, the Thom spaces MO(k) and MSO(k) also define spectra, and similar ones were also defined for all classical groups.

Generalized homology and cohomology have provided a host of new tools, built on the model of the classical theories, and that have shown their value in the uninterrupted progress made by algebraic and differential topology since 1960 [459].

Bibliography

[1] J.F. Adams: On the structure and applications of the Steenrod algebra, *Comment. Math. Helv.*, **32** (1958), 80–214.

[2] J.F. Adams: On the non-existence of elements of Hopf-invariant one, *Ann. of Math.*, **72** (1960), 20–104.

[3] J.F. Adams: Stable homotopy theory, *Lect. Notes in Maths.*, **3** (1964).

[4] J.F. Adams: Lectures on generalized cohomology, *Lect. Notes in Math.*, **99** (1966).

[5] J.F. Adams: *Algebraic Topology: A Student's Guide*, Lond. Math. Soc. Lect. Notes Series 4, Cambridge U.P., 1972.

[6] J. Adem: The iteration of the Steenrod squares in algebraic topology, *Proc. Nat. Acad. Sci. USA*, **38** (1952), 720–724.

[7] J. Adem: The relations in Steenrod powers of cohomology classes, *Alg. Geom. and Topology, Symposium in Honor of S. Lefschetz*, Princeton Univ. Press, 1957.

[8] J.W. Alexander: Sur les cycles des surfaces algébriques et sur une définition topologique de l'invariant de Zeuthen-Segre, *Rendic. dei Lincei*, (2), **23** (1914), 55–62.

[9] J.W. Alexander: A proof of the invariance of certain constants of analysis situs, *Trans. Amer. Math. Soc.*, **16** (1915), 148–154.

[10] J.W. Alexander: Note on two three-dimensional manifolds with the same group, *Trans. Amer. Math. Soc.*, **20** (1919), 330–342.

[11] J.W. Alexander: A proof and extension of the Jordan–Brouwer separation theorem, *Trans. Amer. Math. Soc.*, **23** (1922), 333–349.

[12] J.W. Alexander: An example of a simply connected surface bounding a region which is not simply connected, *Proc. Nat. Acad. Sci. USA*, **10** (1924), 8–10.

[13] J.W. Alexander: New topological invariants expressible as tensors, *Proc. Nat. Acad. Sci. USA*, **10** (1924), 99–101.

[14] J.W. Alexander: Combinatorial Analysis Situs, *Trans. Amer. Math. Soc.*, **28** (1926), 301–329.

[15] J.W. Alexander: Topological invariants of knots and links, *Trans. Amer. Math. Soc.*, **30** (1928), 275–306.

[16] J.W. Alexander: The combinatorial theory of complexes, *Ann. of Math.* **31** (1930), 294–322.

[17] J.W. Alexander: On the chains of a complex and their duals, *Proc. Nat. Acad. Sci. USA*, **21** (1935), 509–511.

[18] J.W. Alexander: On the ring of a compact metric space, *Proc. Nat. Acad. Sci. USA*, **21** (1935), 511–512.

[19] J.W. Alexander: On the connectivity ring of an abstract space, *Ann. of Math.*, **37** (1936), 698–708.

[20] J.W. Alexander: A theory of connectivity in terms of gratings, *Ann. of Math.*, **39** (1938), 883–912.

[21] J.W. Alexander and O. Veblen: Manifolds of N dimensions, *Ann. of Math.*, **14** (1913), 163–178.

[22] P. Alexandroff: Über kombinatorische Eigenschaften allgemeiner Kurven, *Math. Ann.*, **96** (1926), 512–554.

[23] P. Alexandroff: Über die Dualität zwischen den Zusammenhang einer abgeschlossenen Menge und des zu ihr komplementären Raumes, *Nachr. Ges. Wiss. Göttingen*, 1927, 323–329.

[24] P. Alexandroff: Über den allgemeinen Dimensionsbegriff und seine Beziehung zur elementaren geometrischen Anschauung, *Math. Ann.*, **98** (1928), 617–636.

[25] P. Alexandroff: Une définition des nombres de Betti pour un ensemble fermé quelconque, *C. R. Acad. Sci. Paris*, **184** (1927), 317–319.

[26] P. Alexandroff: Sur la décomposition de l'espace par des ensembles fermés, *C. R. Acad. Sci. Paris*, **184** (1927), 425–428.

[27] P. Alexandroff: Untersuchung über Gestalt und Lage abgeschlossener Mengen beliebiger Dimension, *Ann. of Math.*, **30** (1928), 101–187.

[28] P. Alexandroff: On local properties of closed sets, *Ann. of Math.*, **36** (1935), 1–35.

[29] P. Alexandroff: Die Topologie in und um Holland in den Jahren 1920–1930, *Nieuw Arch. voor Wisk.*, (3), **17** (1969), 109–127.

[30] P. Alexandroff and H. Hopf: *Topologie* I, Berlin, Springer, 1935 (Die Grundlehren der math. Wiss., Bd. 45).

[31] P. Alexandroff and L. Pontrjagin: Les variétés à n dimensions généralisées, *C. R. Acad. Sci. Paris*, **202** (1936), 1327–1329.

[32] N. Aronszajn: Sur les lacunes d'un polyèdre et leurs relations avec les groupes de Betti, *Proc. Akad. Wetensch. Amsterdam*, **40** (1937), 67–69.

[33] M. Atiyah: *K-theory*, Benjamin, New York, 1967.

[34] M. Atiyah and F. Hirzebruch: Riemann-Roch theorems for differentiable manifolds, *Bull. Amer. Math. Soc.*, **65** (1959), 276–281.

[35] M. Atiyah and F. Hirzebruch: Vector bundles and homogeneous spaces, *Proc. Symp. Pure Math., III, Differential Geometry*, Amer. Math. Soc. 1961, pp. 7–38.

[36] S. Averbukh: The algebraic structure of the intrinsic homology groups, *Dokl. Akad. Nauk USSR*, **125** (1959), 11–14.

[37] R. Baer: Erweiterung von Gruppen und ihren Isomorphismen, *Math. Zeitschr.*, **35** (1934), 375–416.

[38] R. Baer: Automorphismen von Erweiterungsgruppen, *Act. Scient. Ind.*, **205**, Paris, Hermann, 1935.

[39] R. Baer: Abelian groups that are direct summands of every containing abelian group, *Bull. Amer. Math. Soc.*, **46** (1940), 800–806.

[40] R. Baire: Sur la non-applicabilité de deux continus à n et $n + p$ dimensions, *C. R. Acad. Sci. Paris*, **144** (1907), 318–321.

[41] R. Baire: Sur la non-applicabilité de deux continus à n et $n + p$ dimensions, *Bull. Sci. Math.*, **31** (1907), 94–99.

[42] S. Banach: Sur les opérations dans les ensembles abstraits et leur application aux équations intégrales, *Fund. Math.*, **3** (1923), 133–181.

[43] M. Barratt: Track groups I, II, *Proc. Lond. Math. Soc.*, **5** (1955), 71–106 and 285–329.

[44] M. Barratt and J. Milnor: An example of anomalous singular homology, *Proc.*

Amer. Math. Soc., **13** (1962), 293–297.

[45] M. Barratt and G. Paechter: A note on $\pi_r(V_{n,m})$, *Proc. Nat. Acad. Sci. USA*, **38** (1952), 119–121.

[46] E. Begle: Locally connected spaces and generalized manifolds. *Amer. J. Math.*, **64** (1942), 553–574.

[47] E. Begle: The Vietoris mapping theorem for bicompact spaces, *Ann. of Math.*, **51** (1950), 534–543.

[48] G. D. Birkhoff: *Collected Mathematical Papers*, vol. II, Amer. Math. Soc., New York, 1950.

[49] G.D. Birkhoff: Dynamical systems with two degrees of freedom, *Trans. Amer. Math. Soc.*, **18** (1917), 199–300 (also in [48], pp. 1–102).

[50] G.D. Birkhoff and O. Kellogg: Invariant points in function space, *Trans. Amer. Math. Soc.*, **23** (1922), 96–115.

[51] A. Blakers and W. Massey: The homotopy groups of a triad, *Ann. of Math.*, I, II, III, **53** (1951), 161–205; **55** (1952), 192–201; **58** (1953), 401–417.

[52] S. Bochner: Remark on the theorem of Green, *Duke Math. J.*, **3** (1937), 334–338.

[53] M. Bokstein: Universal systems of ∇-homology rings, *Dokl. Akad. Nauk USSR*, **37** (1942), 243–245.

[54] M. Bokstein: Homology invariants of topological spaces, *Trudy Mosk. Mat. Obsc.* **5** (1956), 3–80.

[55] A. Borel: *Oeuvres*, vol. 1, Springer, Berlin–Heidelberg–New York–Tokyo, 1983.

[56] A. Borel: Le plan projectif des octaves et les sphères comme espaces homogènes, *C. R. Acad. Sci. Paris*, **230** (1950), 1378–1380 (also in [55], pp. 39–41).

[57] A. Borel: Cohomologie des espaces localement compacts, d'après J. Leray, Sém. de Top. alg., ETH, *Lect. Notes* 2, 1964, 3ᵉ éd.

[58] A. Borel: Sur la cohomologie des espaces fibrés principaux et des espaces homogènes des groupes de Lie compacts, *Ann. of Math.*, **57** (1953), 115–207 (also in [55], pp. 121–216).

[59] A. Borel: Sur l'homologie et la cohomologie des groupes de Lie compacts connexes, *Amer. J. Math.*, **76** (1954), 273–342 (also in [55], pp. 322–391).

[60] A. Borel: Kählerian coset spaces of semi-simple Lie groups, *Proc. Nat. Acad. Sci. USA*, **40** (1954), 1147–1151 (also in [55], 397–401).

[61] A. Borel: Sur la torsion des groupes de Lie, *J. Math. Pures Appl.*, (9), **35** (1955), 127–139 (also in [55], pp. 477–489).

[62] A. Borel: The Poincaré duality in generalized manifolds, *Mich. Math. J.*, **4** (1957), 227– 239 (also in [55], pp. 565–577).

[63] A. Borel: *Seminar on transformation groups*, Princeton Univ. Press, 1960 (Ann. of Math. Studies No. 46).

[64] A. Borel and C. Chevalley: The Betti numbers of the exceptional groups, *Memoirs Amer. Math. Soc.*, **14** (1955), 1–9 (also in [55], pp. 451–459).

[65] A. Borel and F. Hirzebruch: Characteristic classes and homogeneous spaces, I, *Amer. J. Math.*, **80** (1958), 458–538 (also in [55], pp. 578–648).

[66] A. Borel and J.C. Moore: Homology theory for locally compact spaces, *Mich. Math. J.*, **7** (1960), 137–159.

[67] A. Borel and J-P. Serre: Impossibilité de fibrer un espace euclidien par des fibres compactes, *C. R. Acad. Sci. Paris*, **230** (1950), 2258–2260 (also in [428], pp. 3–4).

[68] A. Borel and J-P. Serre: Groupes de Lie et puissances réduites de Steenrod, *Amer. J. Math.*, **73** (1953), 409–448 (also in [55], pp. 262–301).

[69] K. Borsuk: *Collected papers*, vol. I, PWN, Warszawa, 1983.

[69a] K. Borsuk, *Theory of retracts*, PWN, Warszawa, 1967.

[70] K. Borsuk: Sur les rétractes, *Fund. Math.*, **17** (1931) (also in [69], pp. 2–20).
[71] K. Borsuk: Über eine Klasse von lokal zusammenhängende Räume, *Fund. Math.*, **19** (1932), 220–240 (also in [69], pp. 102–124).
[72] K. Borsuk: Zur kombinatorischen Eigenschaften des Retraktes, *Fund. Math.*, **21** (1933), 91–98 (also in [69], pp. 167–174).
[73] K. Borsuk: Über den Lusternik-Schnirelmann Begriff der Kategorie, *Fund. Math.*, **26** (1936), 123–136 (also in [69], pp. 279–292).
[74] K. Borsuk: Sur les groupes des classes de transformations continues, *C. R. Acad. Sci. Paris*, **202** (1936), 1400–1403 (also in [69], pp. 296–298).
[75] K. Borsuk: Sur les prolongements des transformations continues, *Fund. Math.*, **28** (1937), 99–130 (also in [69], pp. 348–359).
[76] R. Bott: On torsion in Lie groups, *Proc. Nat. Acad. Sci. USA*, **40** (1954), 586–588.
[77] R. Bott: An application of the Morse theory to the topology of Lie groups, *Bull. Soc. Math. France*, **84** (1956), 251–281.
[78] R. Bott: The space of loops on a Lie group, *Mich. Math. J.*, **5** (1958), 35–61.
[79] R. Bott: The stable homotopy of the classical groups, *Ann. of Math.*, **70** (1959), 313–337.
[80] R. Bott: Quelques remarques sur les théorèmes de périodicité, *Bull. Soc. Math. France*, **87** (1959), 293–310.
[81] R. Bott: A report on the unitary group, *Proc. Symp. Pure Math.* vol. III, pp. 1–6, Amer. Math. Soc., Providence, R. I., 1961.
[82] R. Bott and J. Milnor: On the parallelizability of spheres, *Bull. Amer. Math. Soc.*, **64** (1958), 87–89.
[83] R. Bott and H. Samelson: On the cohomology ring of G/T, *Proc. Nat. Acad. Sci. USA*, **41** (1955), 490–493.
[84] R. Bott and H. Samelson: Applications of the theory of Morse to symmetric spaces, *Amer. J. Math.*, **80** (1958), 964–1029.
[85] N. Bourbaki: *Topologie Générale*, chap. I, § 3, Eléments de Mathématique, nouv. éd., 1971, Hermann, Paris, 1971.
[86] R. Brauer: Sur les invariants intégraux des variétés des groupes de Lie simples clos, *C. R. Acad. Sci. Paris*, **201** (1935), 419–421.
[87] G. Bredon: *Sheaf Theory*, McGraw-Hill, New York, 1967.
[88] L.E.J. Brouwer: *Collected Works*, vol. I, North Holland, Amsterdam, 1975.
[89] L.E.J. Brouwer: *Collected Works*, vol. II, North Holland, Amsterdam, 1976.
[90] A.B. Brown: Functional dependence, *Trans. Amer. Math. Soc.*, **38** (1935), 379–394.
[91] A.B. Brown and B. Koopman: On the covering of analytic loci by complexes, *Trans. Amer. Math. Soc.*, **34** (1932), 231–251.
[92] E.H. Brown: Finite computability of Postnikov complexes, *Ann. of Math.*, **65** (1957), 1–20.
[93] E.H. Brown: Cohomology theories, *Ann. of Math.*, **75** (1962), 467–484 and corr., **78** (1963), 201.
[94] M. Brown: Locally flat imbeddings of topological manifolds, *Ann. of Math.*, **75** (1962), 331–341.
[94a] R.F. Brown, *The Lefschetz fixed point theorem*, Glenview, Illinois, 1971.
[95] D. Buchsbaum: Exact categories and duality, *Trans. Amer. Math. Soc.*, **80** (1955), 1–34.
[96] S. Cairns: The cellular division and approximation of regular spreads, *Proc. Nat. Acad. Sci. USA*, **16** (1930), 488–490.
[97] S. Cairns: On the triangulation of regular loci, *Ann. of Math.*, **35** (1934), 579–587.

[98] E. Cartan: *Oeuvres Complètes*, vol. I_2, Gauthier-Villars, Paris, 1952.

[99] E. Cartan: *Oeuvres Complètes*, vol. III_1, Gauthier-Villars, Paris, 1955.

[100] E. Cartan: La structure des groupes de transformations continus et la théorie du trièdre mobile, *Bull. Sci. Math.*, **34** (1910), 1–34 (also in [99], pp. 145–178).

[101] E. Cartan: Sur les nombres de Betti des espaces de groupes clos, *C. R. Acad. Sci. Paris*, **187** (1928), 196–198 (also in [98], pp. 999–1001).

[102] E. Cartan: Sur les invariants intégraux de certains espaces homogènes clos et les propriétés topologiques de ces espaces, *Ann. Soc. Pol. Math.*, **8** (1929), 181–225 (also in [98], pp. 1081–1125).

[103] E. Cartan: La topologie des espaces représentatifs des groupe de Lie, *Act. Sci. Ind.*, No. 358, Hermann, Paris, 1936 (also in [98], pp. 1307–1330).

[104] E. Cartan: *Leçons sur les Invariants Intégraux*, Hermann, Paris 1922.

[105] H. Cartan: *Oeuvres*, vol. III, Springer, Berlin–Heidelberg–New York, 1979.

[106] H. Cartan: Méthodes modernes en Topologie algébrique, *Comment. Math. Helv.*, **18** (1945), 1–15 (also in [105], pp. 1164–1178).

[107] H. Cartan: Une théorie axiomatique des carrés de Steenrod, *C. R. Acad. Sci. Paris*, **230** (1950), 425–427 (also in [105], pp. 1252–1254).

[108] H. Cartan: Notions d'algèbre différentielle; applications aux groupes de Lie et aux variétes où opère un groupe de Lie, *Coll. de Topologie (espaces fibrés), Bruxelles 1950*, C.R.B.M., Liége et Paris, 1951, 15–27 (also in [105], pp. 1255–1267).

[109] H: Cartan: La transgression dans un groupe de Lie et dans unespace fibré principal, *Coll. de Topologie (espaces fibrés), Bruxelles 1950*, C.R.B.M., Liége et Paris 1951, 57–71 (also in [105], pp. 1268–1282).

[110] H. Cartan: Sur les groupes d'Eilenberg-Mac Lane, I, II, *Proc. Nat. Acad. Sci. USA*, **40** (1954), 467–471 and 704–707 (also in [105] pp. 1300–1308).

[111] H. Cartan: Algèbres d'Eilenberg-Mac Lane, *Sém. H. Cartan, ENS, 1954–55*, exp. 2 to 11, 2nd ed., 1956 (also in [105], pp. 1309–1394).

[112] H. Cartan: Sur l'itération des opérations de Steenrod, *Comment. Math. Helv.*, **29** (1955), 40–58 (also in [105], pp. 1395–1413).

[113] H. Cartan and S. Eilenberg: *Homological Algebra*, Princeton Univ. Press, 1956.

[114] H. Cartan and J. Leray: Relations entre anneaux de cohomologie et groupe de Poincaré, *Coll. Top. alg. C. N. R. S. Paris* (1947), 83–85.

[115] H. Cartan and J-P. Serre: Espaces fibrés et groupes d'homotopie. I. Constructions générales, *C. R. Acad. Sci. Paris*, **234** (1952), 288–290 (also in [105], pp. 1294–1296 and [428], pp. 105–107).

[116] H. Cartan and J-P. Serre: Espaces fibrés et groupes d'homotopie. II. Applications, *C. R. Acad. Sci. Paris*, **234** (1952), 393–395 (also in [105], pp. 1297–1299 and [428], pp. 108–110).

[117] H. Cartan and J-P. Serre: Un théoreme de finitude concernant les varietés analytiques compactes, *C. R. Acad. Sci. Paris*, **237** (1953), 128–130 (also in [428], pp. 271–273).

[118] E. Čech: *Topological papers*, Prague, 1968, Akademia.

[119] E. Čech: Théorie générale de l'homologie dans un espace quelconque, *Fund. Math.*, **19** (1932), 149–183 (also in [118], pp. 90–117).

[120] E. Čech: Théorie générale des variétés et de leurs théorèmes de dualité, *Ann. of Math.*, **34** (1933), 621–730 (also in [118], pp. 183–286).

[121] E. Čech: Höherdimensionalen Homotopiegruppen, *Verhandl. des intern. Math. Kongresses, Zürich, 1932*, Bd. 2, 203.

[122] E. Čech: Sur les nombres de Betti locaux, *Ann. of Math.*, **35** (1934), 678–701 (also in [118], pp. 336–359).
[123] E. Čech: Multiplications on a complex, *Ann. of Math.*, **37** (1936), 681–697 (also in [118], pp. 417–433).
[124] S.S. Chern: *Selected Papers*, Springer, Berlin–Heidelberg–New York, 1978.
[125] S.S. Chern: Characteristic classes of hermitian manifolds, *Ann. of Math.*, **47** (1946), 85–121 (also in [124], pp. 101–137).
[126] S.S. Chern: On the multiplication in the characteristic ring of a sphere bundle, *Ann. of Math.*, **49** (1948), 362–372 (also in [124], pp. 148–158).
[127] S.S. Chern: On the characteristic classes of complex sphere bundles and algebraic varieties, *Amer. J. Math.*, **75** (1953), 565–597 (also in [124], pp. 165–198).
[128] S.S. Chern and E. Spanier: The homology structure of fibre bundles, *Proc. Nat. Acad. Sci. USA*, **36** (1950), 248–255.
[129] S.S. Chern, F. Hirzebruch and J-P. Serre: On the index of a fibered manifold, *Proc. Amer. Math. Soc.*, **8** (1957), 587–596 (also in [124], pp. 259–268).
[130] C. Chevalley: Sur la définition des groupes de Betti des ensembles fermés, *C. R. Acad. Sci. Paris*, **200** (1935), 1005–1007.
[131] C. Chevalley: *Theory of Lie Groups I*, Princeton Univ. Press, 1946.
[132] C. Chevalley: The Betti numbers of the exceptional Lie groups, *Proc. Int. Math. Congress of Math., Cambridge (Mass.), 1950*, Amer. Math. Soc., Providence, R. I. 1952, vol. 2, pp. 21–24.
[133] C. Chevalley: Invariants of finite groups generated by reflections, *Amer. J. Math.*, **77** (1955), 778–782.
[134] C. Chevalley and S. Eilenberg: Cohomology theory of Lie groups and Lie algebras, *Trans. Amer. Math. Soc.*, **63** (1948), 85–124.
[135] C. Chevalley and J. Herbrand: Groupes topologiques, groupes fuchsiens, groupes libres, *C. R. Acad. Sci. Paris*, **192** (1931), 724–726.
[136] E. Cotton: Généralisation de la théorie du trièdre mobile, *Bull. Soc. Math. France*, **33** (1905), 1–23.
[137] M. Dehn: Die beiden Kleeblattschlingen, *Math. Ann.*, **75** (1914), 402–413.
[138] M. Dehn and P. Heegaard: Analysis Situs, *Enzykl. der math. Wiss.*, III 1 AB 3, Teubner, Leipzig, 1907.
[139] J. Dieudonné: *History of Algebraic Geometry*, Wadsworth, Monterey, CA, 1985.
[140] P. Dolbeault: Sur la cohomologie des variétés analytiques complexes, *C. R. Acad. Sci. Paris*, **236** (1953), 175–177.
[141] A. Dold: Erzeugende der Thomschen Algebra $\mathfrak{N}$, *Math. Zeitschr.*, **65** (1956), 25–35.
[142] A. Dold: Démonstration élémentaire de deux résultats du cobordisme, *Sém. Ehresmann*, 1959.
[143] A. Dold: Structure de l'anneau de cobordisme $\Omega^{\bullet}$ d'après les travaux de V.A. Rokhlin et de C.T.C. Wall, *Sém. Bourbaki*, No. 188, 1959.
[144] C. Dowker: Mapping theorems for non compact spaces, *Amer. J. Math.* **69** (1947), 200–242.
[145] C. Dowker: Čech cohomology theory and the axioms, *Ann. of Math.*, **51** (1950), 278–292.
[146] C. Dowker: Homology groups of relations, *Ann. of Math.*, **56** (1952), 84–95.
[147] W.v. Dyck: Beiträge zur Analysis Situs II: Mannigfaltigkeiten von *n* Dimensionen, *Math. Ann.*, **37** (1890), 275–316.
[148] B. Eckmann: Zur Homotopietheorie gefaserter Räume, *Comment. Math. Helv.*, **14** (1941–1942), 141–192.

[149] B. Eckmann: Über die Homotopiegruppen von Gruppenräume, *Comment. Math. Helv.*, **14** (1941–42), 234–256.

[150] B. Eckmann: Coverings and Betti numbers, *Bull. Amer. Math. Soc.*, **55** (1949), 95–101.

[151] B. Eckmann: Espaces fibrés et homotopie, *Coll. de Topologie (espaces fibrés) Bruxelles, 1950*, C.R.B.M., Liége et Paris, 1951, 83–99.

[152] B. Eckmann: Homotopy and cohomology theory, *Proc. Int. Congress of Math., Stockholm 1962*, Inst. Mittag-Leffler, 1963, 59–73.

[153] M. Eger: Les systèmes canoniques d'une variété algébrique à plusieurs dimensions, *Ann. Ec. Norm. Sup.*, **60** (1943), 143–172.

[154] C. Ehresmann: *Oeuvres Complètes et Commentées*, parties 1-1 et 1-2, Amiens, 1984 [suppl. 1 et 2 au vol. XXIV (1983) des *Cahiers de Top. et Géom. diff.*].

[155] C. Ehresmann: Sur la topologie de certains espaces homogènes, *Ann. of Math.*, **35** (1934), 396–443 (also in [154], pp. 3–54).

[156] C. Ehresmann: Sur la topologie de certaines variétés algébriques réelles, *J. Math. Pures Appl.*, (9), **16** (1937), 69–100 (also in [154], pp. 55–86).

[157] C. Ehresmann: Sur la variété des génératrices planes d'une quadrique réelle et sur la topologie du groupe orthogonal à n variables, *C. R. Acad. Sci. Paris*, **208** (1939), 321–323 (also in [154], pp. 304–306).

[158] C. Ehresmann: Sur la topologie des groupes simples, *C. R. Acad. Sci. Paris*, **208** (1939), 1263–1265 (also in [154], pp. 307–309).

[159] C. Ehresmann: Espaces fibrés associés, *C. R. Acad. Sci. Paris*, **213** (1941), 762–764 (also in [154], pp. 313–315).

[160] C. Ehresmann: Espaces fibrés de structures comparables, *C. R. Acad. Sci. Paris*, **214** (1942), 144–147 (also in [154], pp. 316–318).

[161] C. Ehresmann: Sur les applications continues d'un espace dans un espace fibré ou dans un revêtement, *Bull. Soc. Math. France*, **72** (1944), 37–54 (also in [154], pp. 105–132).

[162] C. Ehresmann: Sur la théorie des espaces fibrés, *Coll. Top. alg. Paris 1947*, C.N.R.S., 3–15 (also in [154], pp. 133–146).

[163] C. Ehresmann: Sur les espaces fibrés différentiables, *C. R. Acad. Sci. Paris*, **224** (1947), 1611–1612 (also in [154], pp. 326–328).

[164] C. Ehresmann: Sur les variétés presque complexes, *Proc. Int. Congress of Math. Cambridge 1950*, Amer. Math. Soc., Providence, R.I, 1952, vol. 2, pp. 412–419 (also in [154], pp. 147–152).

[165] C. Ehresmann: Les connexions infinitésimales dans un espace fibré différentiable, *Coll. de Topologie (espaces fibrés), Bruxelles 1950*, C.R.B.M., Liége et Paris, 1951, pp. 29–55 (also in [154], pp. 179–206).

[166] C. Ehresmann and J. Feldbau: Sur les propriétés d'homotopie des espaces fibrés, *C. R. Acad. Sci. Paris*, **212** (1941), 945–948 (also in [154], pp. 310–312).

[167] S. Eilenberg: On the relation between the fundamental group and the higher homotopy groups, *Fund. Math.*, **32** (1939), 167–175).

[168] S. Eilenberg: Cohomology and continuous mappings, *Ann. of Math.*, **41** (1940), 231–260.

[169] S. Eilenberg: On homotopy groups, *Proc. Nat. Acad. Sci. USA*, **26** (1940), 563–565.

[170] S. Eilenberg: On spherical cycles, *Bull. Amer. Math. Soc.*, **47** (1941), 432–434.

[171] S. Eilenberg: Extension and classification of continuous mappings, *Lectures in Topology, Conf. at Univ. of Michigan, 1940*, U. of Michigan Press, 1941, pp. 57–99.

[172] S. Eilenberg: Singular homology, *Ann. of Math.*, **45** (1944), 407–447.
[173] S. Eilenberg: Homology of spaces with operators, I, *Trans. Amer. Math. Soc.*, **61** (1947), 378–417.
[174] S. Eilenberg: Singular homology in differential manifolds, *Ann. of Math.*, **48** (1947), 670–681.
[175] S. Eilenberg: On the problems of topology, *Ann. of Math.*, **50** (1949), 247–260.
[176] S. Eilenberg and S. Mac Lane: Infinite cycles and homology, *Proc. Nat. Acad. Sci. USA*, **27** (1941), 535–539.
[177] S. Eilenberg and S. Mac Lane: Group extensions and homology, *Ann. of Math.*, **43** (1942), 758–831.
[178] S. Eilenberg and S. Mac Lane: Natural isomorphisms in group theory, *Proc. Nat. Acad. Sci. USA*, **28** (1942), 537–543.
[179] S. Eilenberg and S. Mac Lane: Relations between homology and homotopy groups, *Proc. Nat. Acad. Sci. USA*, **29**, (1943), 155–158.
[180] S. Eilenberg and S. Mac Lane: General theory of natural equivalences, *Trans. Amer. Math. Soc.*, **58** (1945), 231–294.
[181] S. Eilenberg and S. Mac Lane: Relations between homology and homotopy groups of spaces, I, *Ann. of Math.*, **46** (1945), 480–509.
[182] S. Eilenberg and S. Mac Lane: Determination of the second homology and cohomology groups of a space by means of homotopy invariants, *Proc. Nat. Acad. Sci. USA*, **32** (1946), 277–280.
[183] S. Eilenberg and S. Mac Lane: Homology of spaces with operators, II, *Trans. Amer. Math. Soc.*, **65** (1949), 49–99.
[184] S. Eilenberg and S. Mac Lane: Relations between homology and homotopy groups of spaces, II, *Ann. of Math.*, **51** (1950), 514–533.
[185] S. Eilenberg and S. Mac Lane: Cohomology theory of abelian groups and homotopy theory, *Proc. Nat. Acad. Sci. USA*; I, **36** (1950), 443–447; II, **36** (1950), 657–663; III, **37** (1951), 307–310; IV, **38** (1952), 325–329.
[186] S. Eilenberg and S. Mac Lane: Acyclic models, *Amer. J. Math.*, **79** (1953), 189–199.
[187] S. Eilenberg and J.C. Moore: Homology and fibrations, I: *Comment. Math. Helv.*, **40** (1966), 201–236.
[188] S. Eilenberg and N. Steenrod: Axiomatic approach to homology theory, *Proc. Nat. Acad. Sci. USA*, **31** (1945), 177–180.
[189] S. Eilenberg and N. Steenrod: *Foundations of Algebraic Topology*, Princeton Univ. Press, 1952.
[190] S. Eilenberg and J. Zilber: On products of complexes, *Amer. J. Math.*, **75** (1953), 200–204.
[191] *Encyclopaedic Dictionary of Mathematics*, 2 vol., 2nd ed., Math. Soc. of Japan, 1968, transl. by Math. Soc. of Japan and Amer. Math. Soc., Mass. Institute of Technology, 1977. Articles on algebraic topology:
1: Topology, 409. 2: Complexes, 73. 3: Manifolds, 259. 4: Homology groups, 203. 5: Cohomology rings, 68. 6: Cohomology operations, 67. 7: Hopf algebras, 207. 8: Homotopy, 204. 9: Fundamental Group, 175. 10: Covering spaces, 93. 11: Knot theory, 234. 12: Degree of mapping, 102. 13: Fixed-point theorems, 163. 14: Obstructions, 300. 15: Homotopy groups, 205. 16: Homotopy operations, 206. 17: Eilenberg–Mac Lane complexes, 137. 18: Topology of Lie groups and homogeneous spaces, 411. 19: Fiber spaces, 156. 20: Fiber bundles, 155. 21: Characteristic classes, 58. 22: K-theory, 236. 23: Differential topology, 117. 24:

Topology of differentiable manifolds, 410. 25: Immersions and embeddings, 211. Tables 6 and 7 in Appendix A.

[192] I. Fary: Sur une nouvelle démonstration de l'unicité de l'algèbre de cohomologie à supports compacts d'un espace localement compact, *C. R. Acad. Sci. Paris*, **237** (1953), 552–554.

[193] J. Feldbau: Sur la classification des espaces fibrés, *C. R. Acad. Sci. Paris*, **208** (1939), 1621–1623.

[194] J. Feldbau: (under the name J. Laboureur) Les structures fibrées sur la sphère et le probleme du parallélisme, *Bull. Soc. Math. France*, **70** (1942), 181–183.

[195] E. Floyd: Periodic maps via Smith theory, in [63], pp. 35–47.

[196] R. Fox: On the Lusternik-Schnirelmann category, *Ann. of Math.*, **42** (1941), 333–370.

[197] R. Fox: On fibre spaces, I, II, *Bull. Amer. Math. Soc.*, **49** (1943), 553–557 and 733–735.

[198] R. Fox: On homotopy type and deformation retracts, *Ann. of Math.*, **44** (1943), 40–50.

[199] W. Franz: Über die Torsion einer Überdeckung, *J. für reine u. angew. Math.*, **173** (1935), 245–254.

[200] W. Franz: Abbildungsklassen und Fixpunktklassen dreidimensionaler Linsenräume, *J. für reine u. angew. Math.*, **185** (1943), 65–77.

[201] H. Freudenthal: Über die Klassen von Sphärenabbildungen, *Comp. Math.*, **5** (1937), 299–314.

[202] H. Freudenthal: Alexanderscher und Gordonscher Ring und ihrer Isomorphie, *Ann. of Math.*, **38** (1937), 647–655.

[203] H. Freudenthal: Zum Hopfschen Umkehrhomomorphismus, *Ann. of Math.*, **38** (1937), 847–853.

[204] H. Freudenthal: Die Triangulation der differenziehbaren Mannigfaltigkeiten, *Proc. Akad. Wetensch. Amsterdam*, **42** (1939), 880–901.

[205] T. Ganea: Lusternik-Schnirelmann category and cocategory, *Proc. Lond. Math. Soc.*, **10** (1960), 623–629.

[206] C. F. Gauss: Zur Elektrodynamik, *Werke*, Bd. 5, 605.

[207] A. Gleason: Spaces with a compact Lie group of transformations, *Proc. Amer. Math. Soc.*, **1** (1950), 35–43.

[208] R. Godement: *Topologie algébrique et théorie des faisceaux*, Publ. de l'Inst. math. de Strasbourg, XII, Hermann, Paris, 1958.

[209] I. Gordon: On intersection invariants of a complex and its complementary spaces, *Ann. of Math.*, **37** (1936), 519–525.

[210] M. Gôto: On algebraic homogeneous spaces, *Amer. J. Math.*, **76** (1954), 811–818.

[211] W. Graeub, S. Halperin, and R. Vanstone, *Connections, Curvature and Cohomology*, vol. III, Academic Press, New York, 1976.

[212] M. Greenberg: *Lectures on Algebraic Topology*, Benjamin, New York, 1967.

[213] H.B. Griffiths: The fundamental group of two spaces with a common point, *Quart. J. Math.*, **5** (1954), 175–190.

[214] A. Grothendieck: See [140], p. 176.

[215] A. Grothendieck: Sur quelques points d'algèbre homologique, *Tohoku Math. J.*, (2), **9** (1957), 119–221.

[216] W. Gysin: Zur Homologietheorie der Abbildungen und Faserungen der Mannigfaltigkeiten , *Comment. Math. Helv.*, **14** (1941), 61–122.

[217] J. Hadamard: Sur quelques applications de l'indice de Kronecker, Appendix

to J. Tannery, *Théorie des fonctions*, Paris, 1910 (also in *Oeuvres*, vol. II, pp. 875–915).

[218] A. Haefliger: Knotted $(4k-1)$-spheres in $6k$-space, *Math. Ann.*, **75** (1962), 452–466.

[219] O. Hanner: Some theorems on absolute neighborhood retracts, *Ark. Math.*, **1** (1950), 389–408.

[220] F. Hausdorff: *Mengenlehre*, de Gruyter, Berlin, 1927.

[221] P. Heegaard: Sur l'Analysis Situs, *Bull. Soc. Math. France*, **44** (1916), 161–242.

[222] G. Higman: The units of group rings, *Proc. Lond. Math. Soc.*, (2), **46** (1940), 231–249.

[223] D. Hilbert: *Gesammelte Abhandlungen*, 3 vol., Springer, Berlin, 1932–1935.

[224] P. Hilton: Suspension theorems and generalized Hopf invariants, *Proc. Lond. Math. Soc.*,**1** (1951), 462–493.

[225] P. Hilton: The Hopf invariant and homotopy groups of spheres, *Proc. Camb. Phil. Soc.*, **48** (1952), 547–554.

[226] P. Hilton: On the homotopy groups of the union of spheres, *Journ. Lond. Math. Soc.*, **30** (1955), 154–172.

[227] G. Hirsch: La géométrie projective et la topologie des espaces fibrés, *Coll. de Top. Algébrique*, Paris, 1947, C.N.R.S., 35–42.

[228] G. Hirsch: Un isomorphisme attaché aux structures fibrées, *C. R. Acad. Sci. Paris*, **227** (1948), 1328–1330.

[229] G. Hirsch: L'anneau de cohomologie d'un espace fibré et les classes caractéristiques, *C. R. Acad. Sci. Paris*, **229** (1949), 1297–1299.

[230] G. Hirsch: Quelques relations entre l'homologie dans les espaces fibrés et les classes caractéristiques relatives à un groupe de structure, *Coll. de Topologie (espaces fibrés) Bruxelles 1950*, C.R.B.M., Liége et Paris, 1951, pp. 123–136.

[231] G. Hirsch: Sur les groupes d'homologie des espaces fibrés, *Bull. Soc. Math. Belgique*, **6** (1954), 79–96.

[232] M. Hirsch: *Differential Topology*, Springer, Berlin–Heidelberg–New York, 1976.

[233] F. Hirzebruch: On Steenrod's reduced powers, the index of inertia and the Todd genus, *Proc. Nat. Acad. Sci. USA*, **39** (1953), 951–956.

[234] F. Hirzebruch: Arithmetic genera and the theorem of Riemann-Roch for algebraic varieties, *Proc. Nat. Acad. Sci. USA*, **40** (1954), 110–114.

[235] F. Hirzebruch: *Neue topologische Methoden in der algebraischen Geometrie*, Springer, Berlin, 1956 (Erg. der Math., neue Folge, Heft 9).

[236] J. Hocking and G. Young: *Topology*, Addison-Wesley, Reading, 1961.

[237] W.V.D. Hodge: *The Theory and Applications of Harmonic Integrals*, Cambridge Univ. Press, 1941.

[238] H. Hopf: *Selecta*, Springer, Berlin–Göttingen–Heidelberg–New York, 1964.

[239] H. Hopf: Abbildungsklassen n-dimensionaler Mannigfaltigkeiten, *Math. Ann.*, **96** (1926), 209–224.

[240] H. Hopf: Vektorfelder in n-dimensionalen Mannigfaltigkeiten *Math. Ann.*, **96** (1926), 225–250.

[241] H. Hopf: Eine Verallgemeinerung der Euler-Poincaréschen Formel, *Nachr. Ges. Wiss. Göttingen*, 1928, 127–136 (also in [238], pp. 5–13).

[241a] H. Hopf: A new proof of the Lefschetz formula on invariant points, *Proc. Nat. Acad. Sci. USA*, **14** (1928), 149–153.

[241b] H. Hopf: Ueber die algebraische Anzahl von Fixpunkten, *Math. Zeitschr.* **29** (1929), 493–524.

[242] H. Hopf: Zur Algebra der Abbildungen von Mannigfaltigkeiten, *J. für reine u.*

angew. Math., **105** (1930), 71–88 (also in [238], pp. 14–37).

[243] H. Hopf: Über die Abbildungen der dreidimensionalen Sphäre auf die Kugelfläche, *Math. Ann.*, **104** (1931), 637–665 (also in [238], pp. 38–63).

[244] H. Hopf: Die Klassen der Abbildungen der n-dimensionalen Polyeder auf die n-dimensionalen Sphäre, *Comment. Math. Helv.*, **5** (1933), 39–54 (also in [238], pp. 80–94).

[245] H. Hopf: Über die Abbildungen von Sphären auf Sphären von niedriger Dimension, *Fund. Math.*, **25** (1935), 427–440 (also in [238], pp. 95–106).

[246] H. Hopf: Über die Topologie der Gruppenmannigfaltigkeiten und ihre Verallgemeinerungen, *Ann. of Math.*, **42** (1941), 22–52 (also in [238], pp. 119–151).

[247] H. Hopf: Über den Rang geschlossener Liescher Gruppen, *Comment. Math. Helv.*, **13** (1940/41), 119–143 (also in [238], pp. 152–174).

[248] H. Hopf: Fundamentalgruppe und zweite Bettische Gruppe, *Comment. Math. Helv.*, **14** (1942), 257–309 (also in [238], pp. 186–206).

[249] H. Hopf: Über die Bettischen Gruppen die einer beliebigen Gruppen gehören, *Comment. Math. Helv.*, **17** (1944/45), 39–79 (also in [238], pp. 211–234).

[250] H. Hopf: Einige persönliche Erinnerungen aus der Vorgeschichte der heutigen Topologie, *Colloque de Topologie Bruxelles, 1964*, C.R.B.M., Louvain et Paris, 1966, pp. 9–20.

[251] H. Hopf and H. Samelson: Ein Satz über die Wirkungsräume geschlossener Liescher Gruppen, *Comment. Math. Helv.*, **13** (1940–41), 241–251.

[252] H. Hotelling: Three dimensional manifolds of states of motions, *Trans. Amer. Math. Soc.*, **27** (1925), 329–344.

[253] S.T. Hu: An exposition of the relative homotopy theory, *Duke Math. J.*, **14** (1947), 991–1033.

[254] S.T. Hu: *Homotopy Theory*, Academic Press, New York and London, 1959.

[255] W. Huebsch: On the covering homotopy theorem, *Ann. of Math.*, **61** (1955), 555–563.

[256] W. Hurewicz: Beiträge zur Topologie der Deformationen, *Proc. Akad. Wetensch. Amsterdam*; I: Höherdimensionalen Homotopiegruppen, **38** (1935), 112–119; II: Homotopie- und Homologiegruppen, **38** (1935), 521–528; III: Klassen und Homologietypen von Abbildungen, **39** (1936), 117–126; IV: Asphärische Räume, **39** (1936), 215–224.

[257] W. Hurewicz: On duality theorems, *Bull. Amer. Math. Soc.*, **47** (1941), 562–563.

[258] W. Hurewicz: On the concept of fibre spaces, *Proc. Nat. Acad. Sci. USA*, **41** (1955), 60–64.

[259] W. Hurewicz, J. Dugundji, and C. Dowker: Connectivity groups in terms of limit groups, *Ann. of Math.*, **49** (1948), 391–406.

[260] W. Hurewicz and N. Steenrod: Homotopy relations in fibre spaces, *Proc. Nat. Acad. Sci. USA*, **27** (1941), 60–64.

[261] W. Hurewicz and H. Wallman: *Dimension Theory*, Princeton Univ. Press, 1941.

[262] D. Husemoller: *Fibre bundles*, McGraw-Hill, New York, 1966.

[263] I. James: Reduced product spaces, *Ann. of Math.*, **62** (1955), 170–197.

[264] I. James: On the suspension triad, *Ann. of Math.*, **63** (1956), 191–247.

[265] I. James: The suspension triad of a sphere, *Ann. of Math.*, **63** (1956), 407–429.

[266] I. James and J.H.C. Whitehead: The homotopy of sphere bundles over spheres, *Proc. Lond. Math. Soc.*, **4** (1954), 196–218.

[267] Z. Janiszewski: *Oeuvres Choisies*, Warszawa, 1962.

[268] C. Jordan: Recherches sur les polyèdres, *J. für reine u. angew. Math.*, **66** (1866),

22–85 (also in *Oeuvres*, vol. IV, Gauthier-Villars, Paris, 1964, pp. 15–78).

[269] V.G. Kac: Torsion in cohomology of compact Lie groups, *Math. Sci. Res. Inst. Berkeley*, 1984.

[270] D. Kan: Adjoint functors, *Trans. Amer. Math. Soc.*, **87** (1958), 294–329.

[271] J. Kelley and E. Pitcher: Exact homomorphisms sequences in homology theory, *Ann. of Math.*, **48** (1947), 682–709.

[272] M. Kervaire: Nonparallelizability of the n-sphere for $n > 7$, *Proc. Nat. Acad. Sci. USA*, **44** (1958), 280–283.

[273] F. Klein: *Gesammelte mathematische Abhandlungen*, 3 vol., Springer, Berlin, 1921–1923.

[274] H. Kneser: Die Topologie der Mannigfaltigkeiten, *Jahresber. der DMV*, **34** (1925), 1–14.

[275] K. Kodaira: *Collected Works*, vol. I, Princeton Univ. Press, 1975.

[276] K. Kodaira: *Collected Works*, vol. II, Princeton Univ. Press, 1975.

[277] K. Kodaira: The theorem of Riemann-Roch on compact analytic surfaces, *Amer. J. Math.*, **73** (1951), 813–875 (also in [275], pp. 339–401).

[278] K. Kodaira: The theorem of Riemann-Roch for adjoint systems on 3-dimensional algebraic varieties, *Ann. of Math.*, **56** (1952), 298–342 (also in [275], pp. 423–467).

[279] K. Kodaira and D. Spencer: On arithmetic genera of algebraic varieties, *Proc. Nat. Acad. Sci. USA*, **39** (1953), 641–649 (also in [276], pp. 648–656).

[280] K. Kodaira and D. Spencer: Divisor class groups on algebraic varieties, *Proc. Nat. Acad. Sci. USA*, **39** (1953), 872–877 (also in [276], pp. 665–670).

[281] K. Kodaira and D. Spencer: On a theorem of Lefschetz and the lemma of Enriques-Severi-Zariski, *Proc. Nat. Acad. Sci. USA*, **39** (1953), 1273–1278 (also in [276], pp. 677–682).

[282] A. Kolmogoroff: Über die Dualität im Aufbau der kombinatorischen Topologie, *Mat. Sborn.*, **1** (1936), 97–102.

[283] A. Kolmogoroff: Homologiering des Komplexes und des lokal-bicompakten Räumes, *Mat. Sborn.*, **1** (1936), 701–705.

[284] J-L. Koszul: Sur les opérateurs de dérivation dans un anneau, *C. R. Acad. Sci. Paris*, **224** (1947), 217–219.

[285] J-L. Koszul: Sur l'homologie des espaces homogènes, *C. R. Acad. Sci. Paris*, **224** (1947), 477–479.

[286] J-L. Koszul: Homologie et cohomologie des algèbres de Lie, *Bull. Soc. Math. France*, **78** (1950), 65–127.

[287] J-L. Koszul: Sur un type d'algèbres différentielles en rapport avec la transgression, *Coll. de Topologie (espaces fibrés) Bruxelles 1950*, CRBM, Liége et Paris, 1951, 73–81.

[288] L. Kronecker: Über Systeme von Funktionen mehrerer Variabeln, *Monatsh. Berl. Akad. Wiss.* (1869), 159–193 and 688–698 (also in *Werke*, vol. I, Teubner, Leipzig, 1895, pp. 175–226).

[289] H. Künneth: Über die Bettischen Zahlen einer Produktmannigfaltigkeit, *Math. Ann.*, **90** (1923), 65–85.

[290] H. Künneth: Über die Torsionzahlen von Produktmannigfaltigkeiten, *Math. Ann.*, **91** (1924), 125–134.

[291] C. Kuratowski: *Topologie I: Espaces métrisables, espaces complets*, Warszawa, 1933.

[292] H. Lebesgue: *Oeuvres Scientifiques*, vol. IV, L'Enseignement math. Genève, 1973.

[293] H. Lebesgue: Sur la non-applicabilité de deux domaines appartenant respectivement à des espaces à n et $n + p$ dimensions (extrait d'une lettre à M.O. Blumenthal), *Math. Ann.*, **70** (1911), 166–168 (also in [292], pp. 170–172).

[294] H. Lebesgue: Sur l'invariance du nombre de dimensions d'un espace et sur le théorème de M. Jordan relatif aux variétés fermées, *C. R. Acad. Sci. Paris*, **152** (1911), 841–844 (also in [292], pp. 173–175).

[295] H. Lebesgue: Sur les correspondances entre les points de deux espaces, *Fund. Math.*, **2** (1921), 3–32 (also in [292], pp. 177–206).

[296] S. Lefschetz: *Selected papers*, Chelsea, New York, 1971.

[297] S. Lefschetz: Algebraic surfaces, their cycles and integrals, *Ann. of Math.*, **21** (1920), 225–258, and **23** (1922), 33.

[298] S. Lefschetz: Continuous transformations of manifolds, *Proc. Nat. Acad. Sci. USA*, **9** (1923), 90–93.

[299] S. Lefschetz: *L'Analysis Situs et la géométrie algébrique*, Gauthier-Villars, Paris, 1924 (also in [296], pp. 283–442).

[300] S. Lefschetz: Intersections and transformations of complexes and manifolds, *Trans. Amer. Math. Soc.*, **28** (1926), 1–49 (also in [296], pp. 199–247)

[301] S. Lefschetz: Manifolds with a boundary and their transformations, *Trans. Amer. Math. Soc.*, **29** (1927), 429–462 (also in [296], pp. 248–281).

[302] S. Lefschetz: Closed point sets on a manifold, *Ann. of Math.*, **29** (1928), 232–254 (also in [296], pp. 545–568).

[303] S. Lefschetz: Duality relations in topology, *Proc. Nat. Acad. Sci. USA*, **15** (1929), 367–369.

[304] S. Lefschetz: *Topology*, Amer. Math. Soc. Coll. Publ. No. 12, Providence, RI, 1930.

[305] S. Lefschetz: On singular chains and cycles, *Bull. Amer. Math. Soc.*, **39** (1933), 124–129 (also in [296], pp. 479–484).

[306] S. Lefschetz: On generalized manifolds, *Amer. J. Math.*, **55** (1933), 469–504 (also in [296], pp. 487–524).

[307] S. Lefschetz: On locally connected and related sets, *Ann. of Math.*, **35** (1934), 118–129 (also in [296], pp. 610–622).

[308] S. Lefschetz: *Algebraic Topology*, Amer. Math. Soc. Coll. Publ. No. 27, Providence, R. I., 1942.

[309] S. Lefschetz: *Topics in Topology*, Princeton Univ. Press, 1942 (Ann. of Math. Studies, No. 10).

[310] S. Lefschetz: A page of mathematical autobiography, *Bull. Amer. Math. Soc.*, 74 (1968), 854–879 (also in [296], pp. 13–40).

[311] S. Lefschetz and J.H.C. Whitehead: On analytical complexes, *Trans. Amer. Math. Soc.*, **35** (1933), 510–517.

[312] J. Leray: Topologie des espaces de Banach, *C. R. Acad. Sci. Paris*, **200** (1935), 1082–1084.

[313] J. Leray: Sur la forme des espaces topologiques et sur les points fixes des représentations, *J. Math. Pures Appl.*, (9), **24** (1945), 95–248.

[314] J. Leray: L'anneau d'homologie d'une représentation, *C. R. Acad. Sci. Paris*, **222** (1946), 1366–1368.

[315] J. Leray: Structure de l'anneau d'homologie d'une représentation, *C. R. Acad. Sci. Paris*, **222** (1946), 1419–1422.

[316] J. Leray: Propriétés de l'anneau d'homologie de la projection d'un espace fibré sur sa base, *C. R. Acad. Sci. Paris*, **223** (1946), 395–397.

[317] J. Leray: Sur l'anneau d'homologie de l'espace homogène, quotient d'un groupe clos par un sous-groupe abélien, connexe, maximum, *C. R. Acad. Sci. Paris*, **223** (1946), 412–415.
[318] J. Leray: L'homologie filtrée, *Coll. Top. alg. C. N. R. S. Paris* (1947), 61–82.
[319] J. Leray: Applications continues commutant avec les éléments d'un groupe de Lie, *C. R. Acad. Sci. Paris*, **228** (1949), 1784–1786.
[320] J. Leray: Détermination, dans les cas non exceptionnels, de l'anneau de cohomologie de l'espace homogène quotient d'un groupe de Lie compact par un sous-groupe de même rang, *C. R. Acad. Sci. Paris*, **228** (1949), 1902–1904.
[321] J. Leray: L'anneau spectral et l'anneau filtré d'homologie d'un espace localement compact et d'une application continue, *J. Math. Pures Appl.*, (9), **29** (1950), 1–139.
[322] J. Leray: L'homologie d'un espace fibré dont la fibre est connexe, *J. Math. Pures Appl.*, (9), **29** (1950), 169–213.
[323] J. Leray: Sur l'homologie des groupes de Lie, des espaces homogènes et des espaces fibrés principaux, *Coll. de Topologie (espaces fibrés), Bruxelles 1950*, CRBM, Liége et Paris, 1951, pp. 101–115.
[324] J. Leray: La théorie des points fixes et ses applications en Analyse, *Proc. Int. Congress Math. Cambridge 1950*, vol. 2, pp. 202–208.
[325] J. Leray and J. Schauder: Topologie et équations fonctionnelles, *Ann. Ec. Norm. Sup.*, **51** (1934), 43–78.
[326] A. Lichnerowicz: Un théorème sur l'homologie dans les espaces fibrés, *C. R. Acad. Sci. Paris*, **227** (1948), 711–712.
[327] E. Lima: The Spanier-Whitehead duality in new homotopy categories, *Summa Brasil. Math.*, **4** (1959), 91–148.
[328] N. Lloyd: *Degree theory*, Cambridge tracts No. 73, Cambridge Univ. Press, 1978.
[329] L. Lusternik and L. Schnirelmann: Méthodes topologiques dans les problèmes variationnels, *Act. Scient. Ind. No. 188*, Hermann, Paris, 1934.
[330] S. Mac Lane: *Homology*, Springer, Berlin–Heidelberg–New York, 1963 (Die Grundlehren der math. Wiss., Bd. 114).
[331] S. Mac Lane: Duality for groups, *Bull. Amer. Math. Soc.*, **56** (1950), 485–516.
[332] A.A. Markov: The insolubility of the problem of homeomorphy, *Dokl. Akad. Nauk USSR*, **121** (1958), 218–220.
[333] A.A. Markov: The problem of homeomorphy, *Proc. Int. Congress Math. Edinburgh 1958*, 300–306.
[335] W. Massey: Exact couples in algebraic topology, *Ann. of Math.*, **56** (1952), 363–396; **57** (1953), 248–286.
[336] W. Mayer: Über abstrakte Topologie, *Monatsh. für Math. u. Phys.*, **36** (1929), 1–42 and 219–258.
[337] R. Milgram and J. Davis: A survey of the spherical form problem, *Math. Reports*, **2**, Part 2, 1984.
[338] C. Miller: The topology of rotation groups, *Ann. of Math.*, **57** (1953), 95–110.
[339] J. Milnor: Construction of universal bundles, I, II, *Ann. of Math.*, **63** (1956), 272–284 and 430–436.
[340] J. Milnor: On manifolds homeomorphic to the 7-sphere, *Ann. of Math.*, **64** (1956), 399–405.
[341] J. Milnor: The Steenrod algebra and its dual, *Ann. of Math.*, **67** (1958), 150–171.

[342] J. Milnor: Some consequences of a theorem of Bott, *Ann. of Math.*, **68** (1958), 444–449.

[343] J. Milnor: On spaces having the homotopy type of a CW-complex, *Trans. Amer. Math. Soc.*, **90** (1959), 272–280.

[344] J. Milnor: On the cobordian ring Ω^* and a complex analogue, *Amer. J. Math.*, **82** (1960), 505–521.

[345] J. Milnor: *Morse Theory*, Princeton Univ. Press, 1963 (Ann. of Math. Studies No. 51).

[346] J. Milnor and J.C. Moore: On the structure of Hopf algebras, *Ann. of Math.*, **81** (1965), 211–264.

[347] J. Milnor and J. Stasheff: *Characteristic classes*, Princeton Univ. Press, 1974 (Ann. of Math. Studies No. 76).

[348] H. Miyazaki: Paracompactness of CW-complexes, *Tohoku Math. J.*, (2), **4** (1952), 309–313.

[349] E. Moise: Affine structure in 3-manifolds, V, *Ann. of Math.*, **56** (1952), 96–114.

[350] J.C. Moore: Semi-simplicial complexes and Postnikov systems, *Symp. intern. de Topologia algebrica, Mexico City 1958*, Univ. nacional autonoma de Mexico and UNESCO, 1958, pp. 232–247.

[351] R.L. Moore: Concerning upper semi-continuous collections of continua which do not separate a given continuum, *Proc. Nat. Acad. Sci. USA*, **10** (1924), 356–360.

[352] A. Morse: The behavior of a function on its critical set, *Ann. of Math.*, **40** (1939), 62–70.

[353] M. Morse: Relations between the critical points of a real function of *n* independent variables, *Trans. Amer. Math. Soc.*, **27** (1925), 345–396.

[354] M. Morse: *The Calculus of Variations in the Large*, Amer. Math. Soc. Coll. Publ. No. 18, Providence, RI, 1934.

[355] M. Nakaoka and H. Toda: On Jacobi identity for Whitehead products, *J. Inst. Polytechn. Osaka City Univ.*, Ser. A, **5** (1956), 1–13.

[356] M.H.A. Newman: On the foundations of combinatory Analysis Situs, *Proc. Akad. Wetensch. Amsterdam*, **29** (1926), 611–641 and **30** (1927), 670–673.

[357] J. Nordon: Les éléments d'homologie des quadriques et des hyperquadriques, *Bull. Soc. Math. France*, **74** (1946), 116–129.

[358] P. Olum: Obstructions to extensions and homotopies, *Ann. of Math.*, **52** (1950), 1–50.

[359] P. Painlevé: Observation an sujet de la Communication précédente, *C.R. Acad. Sci. Paris*, **148** (1909), 1156–1157.

[360] F. Peterson: Some results on cohomotopy groups, *Amer. J. Math.*, **78** (1956), 243–258.

[361] L. Phragmén: Über die Begrenzung von Continua, *Acta math.* **7** (1885), 43–48.

[362] E. Picard and G. Simart: *Théorie des fonctions algébriques de deux variables indépendantes*, Gauthier-Villars, Paris, vol. I, 1897; vol. II, 1906.

[363] E. Pitcher: Homotopy groups of the space of curves, with applications to spheres, *Proc. Int. Congress of Math. Cambridge, 1950*, American Math. Soc., Providence, RI, 1952, vol. I, p. 528.

[364] H. Poincaré: Analyse de ses travaux scientifiques, *Acta math.*, **38** (1921), 3–135.

[365] H. Poincaré: *Oeuvres*, vol. I, Gauthier-Villars, Paris, 1928.

[366] H. Poincaré: *Oeuvres*, vol. II, Gauthier-Villars, Paris, 1916.

[367] H. Poincaré: *Oeuvres*, vol. III, Gauthier-Villars, Paris, 1934.

[368] H. Poincaré: *Oeuvres*, vol. IV, Gauthier-Villars, Paris, 1950.

[369] H. Poincaré: *Oeuvres*, vol. VI, Gauthier-Villars, Paris, 1953.

[370] H. Poincaré: *Les méthodes nouvelles de la mécanique céleste*, 3 vol., Gauthier-Villars, Paris, 1893–1899.

[371] H. Poincaré: *La valeur de la science*, Flammarion, Paris, 1905.

[372] H. Poincaré: *Dernières pensées*, Flammarion, Paris, 1913.

[373] J.C. Pont: *La topologie algébrique des origines à Poincaré*, Presses Univ. de France, Paris, 1974.

[374] L. Pontrjagin: Über den algebraischen Inhalt topologische Dualitätssätze, *Math. Ann.*, **105** (1931), 165–205.

[375] L. Pontrjagin: The general topological theorem of duality for closed sets, *Ann. of Math.*, **35** (1934), 904–914.

[376] L. Pontrjagin: A classification of continuous transformations of a complex into a sphere, *Dokl. Akad. Nauk USSR*, **19** (1938), 147–149 and 361–363.

[377] L. Pontrjagin: Homologies in compact Lie groups, *Math. Sborn.*, **6** (1939), 389–422.

[378] L. Pontrjagin: A classification of the mappings of the 3-dimensional complex into the 2-dimensional sphere, *Math. Sborn.*, **9** (1941), 331–363.

[378a] L. Pontrjagin: Mappings of a 3-dimensional sphere into an n-dimensional complex, *Dokl. Akad. Nauk USSR*, **34** (1942), 35–37.

[379] L. Pontrjagin: Characteristic cycles on manifolds, *Dokl. Akad. Nauk USSR*, **35** (1942), 34–37.

[380] L. Pontrjagin: On some topologic invariants of Riemannian manifolds, *Dokl. Akad. Nauk USSR*, **43** (1944), 91–94.

[381] L. Pontrjagin: Characteristic classes of differential manifolds, *Math. Sborn.*, **21** (1947), 233–284 [also in *Amer. Math. Soc. Transl.*, (2), **32** (1950)].

[382] L. Pontrjagin: Homotopy classification of the mappings of an $(n + 2)$-dimensional sphere on an n-dimensional, *Dokl. Akad. Nauk USSR*, **70** (1950), 957–959.

[383] M. Postnikov: Investigations in homotopy theory of continuous mappings, *Amer. Math. Soc. Transl*, (2): **7** (1957), 1–134; **11** (1959), 115–153.

[384] D. Puppe: Homotopiemengen und ihre induzierten Abbildungen, *Math. Zeitschr.*, **69** (1958), 299–344.

[385] F. Raymond and W.D. Neumann: Seifert manifolds, plumbing, μ-invariant and orientation reversing maps, *Algebraic and geometric topology, Proc. of a Conference at Santa Barbara, 1977*, pp. 163–196, Lect. Notes in Math., No. 664, 1978.

[386] K. Reidemeister: Fundamentalgruppe und Überlagerungsräume, *Nachr. Ges. Wiss. Göttingen*, 1928, 69–76.

[387] K. Reidemeister: Homotopieringe und Linsenräume, *Hamburg. Abhandl.*, **11** (1935), 102–109.

[388] G. de Rham: *Oeuvres mathématiques*, L'Enseignement math., Genève, 1981.

[389] G. de Rham: Sur l'Analysis Situs des variétés à n dimensions, *J. Math. Pures Appl.*, (9), **10** (1931), 115–200 (also in [388], pp. 23–113).

[390] G. de Rham: Relations entre la Topologie et la théorie des intégrales multiples, *L'Enseignement math.* **4** (1936), 213–228 (also in [388], pp. 125–140).

[391] G. de Rham: Sur les complexes avec automorphismes, *Comment. Math. Helv.*, **12** (1939–40), 191–211 (also in [388], pp. 174–194).

[392] G. de Rham: Complexes à automorphismes et homéomorphie différentiable, *Ann. Inst. Fourier*, 2 (1950), 51–67 (also in [388], 347–363)

[393] G. de Rham: *Variétés différentiables. Formes, courants formes harmoniques*, Hermann, Paris, 1955.

[394] M. Richardson: Special homology groups, *Proc. Nat. Acad. Sci. USA*, **24** (1938), 21–23.

[395] M. Richardson and P. Smith: Periodic transformations of complexes, *Ann. of Math.*, **39** (1938), 611–633.

[396] F. Riesz: *Oeuvres complètes*, vol. I, Gauthier-Villars, Paris, 1960.

[397] V. Rokhlin: A 3-dimensional manifold is the boundary of a 4-dimensional manifold, *Dokl. Akad. Nauk USSR*, **81** (1951), 355.

[398] V. Rokhlin: New results in the theory of 4-dimensional manifolds, *Dokl. Akad. Nauk USSR*, **84** (1952), 221–224.

[399] V. Rokhlin: The theory of intrinsic homologies, *Uspehi Math. Nauk*, **14** (1959), No. 4, 3–20.

[400] D. Rolfson: *Knots and links*, Publish of Perish, Berkeley, CA, 1976.

[401] M. Rueff: Beiträge zur Untersuchung der Abbildungen von Mannigfaltigkeiten, *Comp. Math.*, **6** (1938), 161–202.

[402] H. Samelson: Beiträge zur Topologie der Gruppenmannigfaltigkeiten, *Ann. of Math.*, **42** (1941), 1091–1137.

[403] H. Samelson: Remark on a paper by R. Fox, *Ann. of Math.*, **45** (1944), 448–449.

[404] H. Samelson: Topology of Lie groups, *Bull. Amer. Math. Soc.*, **58** (1952), 2–37.

[405] H. Samelson: A connection between the Whitehead and the Pontrjagin product, *Amer. J. Math.*, **75** (1953), 744–752.

[406] H. Samelson: Groups and spaces of loops, *Comment. Math. Helv.*, **28** (1954), 278–286.

[407] A. Sard: The measure of the critical values of differentiable maps, *Bull. Amer. Math. Soc.*, **48** (1942), 883–890.

[408] J. Schauder: Zur Theorie stetiger Abbildungen in Funktionalräumen, *Math. Zeitschr.*, **26** (1927), 47–65 and 417–431.

[409] J. Schauder: Der Fixpunktsatz in Funktionalräumen, *Studia Math.*, **2** (1930), 171–180.

[410] J. Schauder: Über lineare, vollstetige Operationen, *Studia Math.*, **2** (1930), 183–196.

[411] J. Schauder: Über den Zusammenhang zwischen der Eindeutigkeit und Lösbarkeit partieller Differentialgleichungen zweiter Ordnung ellptischen Typus, *Math. Ann.*, **106** (1932), 661–721.

[412] J. Schauder: Das Anfangswertproblem einer quasilinearen hyperbolischen Differentialgleichung zweiter Ordnung in beliebiger Anzahl unabhängigen Veränderlichen, *Fund. Math.*, **24** (1935), 213–246.

[413] L. Schläfli: *Gesammelte mathematische Abhandlungen*, 3 vol. Birkhäuser, Basel, 1950–1956.

[414] L. Schnirelmann: Über eine neue komninatorische Invariante, *Monatsh. für Math. u. Physik*, **37** (1930), 131–134.

[415] A. Schoenflies: Die Entwicklung der Lehre von der Punktmannigfaltigkeiten, II, *Jahresber. der DMV*, Ergänzungsband II, Leipzig, Teubner, 1908.

[416] O. Schreier: Die Untergruppen der freien Gruppen, *Hamburg. Abhandl.*, **5** (1927), 161–183.

[417] H. Schubert: *Kalkül der abzählende Geometrie*, Teubner, Leipzig, 1879.

[418] I. Schur: Über die Darstellung der endlichen Gruppen durch gebrochene lineare Substitutionen, *J. für reine u. angew. Math.*, **132** (1907), 85–137.

[419] L. Schwartz: Homomorphismes et applications complètement continues, *C. R. Acad. Sci. Paris*, **236** (1953), 2472–2473.

[420] H. Seifert: Topologie dreidimensionaler geschlossener Räume, *Acta math.*, **60** (1932), 147–238.

[421] H. Seifert and W. Threlfall: *Lehrbuch der Topologie*, Teubner, Leipzig-Berlin, 1934.

[422] H. Seifert and W. Threlfall: *Variationsrechnung im Grossen, (Theorie von Morse)*, Teubner, Leipzig–Berlin, 1938.

[423] *Séminaire H. Cartan de l'ENS, 1940–50: Homotopie, espaces fibrés*, Secr. math, 11, R.P. Curie, Paris.

[424] *Séminaire H. Cartan de l'ENS, 1954–55: Algèbres d'Eilenberg-Mac Lane et homotopie*, Secr. math., 11, R.P. Curie, Paris, 1956.

[425] *Séminaire H. Cartan de l'ENS, 1958–59: Invariant de Hopf et opérations cohomologiques secondaires*, Secr. math., 11, R.P. Curie, Paris, 1959.

[426] *Séminaire H. Cartan de l'ENS, 1959–60: Périodi cité des groupes d'homotopie stables des groupes classiques, d'après Bott*, Secr. math., 11, R.P. Curie, Paris, 1961.

[427] *Séminaire G. de Rham, Univ. de Lausanne, 1963–64: Torsion et type simple d'homotopie*, Lect. Notes No. 48, Springer, 1967.

[428] J-P. Serre: *Oeuvres*, vol. I, Springer, Berlin–Heidelberg–New York–Tokyo, 1986.

[429] J-P. Serre: Homologie singulière des espaces fibrés. Applications, *Ann. of Math.*, **54** (1951), 425–505 (also in [428], pp. 24–104).

[430] J-P. Serre: Groupes d'homotopie et classes de groupes abéliens, *Ann. of Math.*, **58** (1953), 258–294 (also in [428], pp. 171–207).

[431] J-P. Serre: Cohomologie modulo 2 des complexes d'Eilenberg-Mac Lane, *Comment. Math. Helv.*, **27** (1953), 198–232 (also in [428], pp. 208–242).

[432] J-P. Serre: Quelques calculs de groupes d'homotopie, *C. R. Acad. Sci. Paris*, **236** (1953), 2475–2477 (also in [428], pp. 256–258).

[433] J-P. Serre: Lettre à Armand Borel, [428], pp. 243–250.

[434] F. Severi: Sulla topologia e sui fondamenti dell'analisi generale, *Rend. Semin. mat. Roma*, (2), **7** (1931), 5–37.

[435] F. Severi: Über die Grundlagen der algebraischen Geometrie, *Hamburg. Abhandl.*, **9** (1933), 335–364.

[436] L. Siebenmann: L'invariance topologique du type simple d'homotopie (d'après T. Chapman et R.D. Edwards), *Sém. Bourbaki*, No. 428, 1972.

[437] P. Smith: The topology of transformation groups, *Bull. Amer. Math. Soc.*, **44** (1938), 497–514.

[438] E. Spanier: Cohomology theory for general spaces, *Ann. of Math.*, **49** (1948), 407–427.

[439] E. Spanier: Borsuk's cohomotopy groups, *Ann. of Math.*, **50** (1949), 203–245.

[440] E. Spanier: *Algebraic Topology*, McGraw-Hill, New York, 1966.

[441] E. Spanier and J.H.C. Whitehead: Duality in homotopy theory, *Mathematika*, **2** (1955), 56–80.

[442] E. Sperner: Neuer Beweis für die Invarianz der Dimensionzahl und des Gebietes, *Hamburg. Abhandl.*, **6** (1928), 265–272.

[443] N. Steenrod: Universal homology groups, *Amer. J. Math.*, **58** (1936), 661–701.

[444] N. Steenrod: Regular cycles on compact metric spaces, *Ann. of Math.*, **41** (1940), 833–851.

[445] N. Steenrod: Homology with local coefficients, *Ann. of Math.*, **44** (1943), 610–627.

[446] N. Steenrod: The classification of sphere bundles, *Ann. of Math.*, **45** (1944), 295–311.

[447] N. Steenrod: Products of cocycles and extensions of mappings, *Ann. of Math.*, **48** (1947), 290–320.
[448] N. Steenrod: Cohomology invariants of mappings, *Ann. of Math.*, **50** (1949), 954–968.
[449] N. Steenrod: Reduced powers of cocycles, *Proc. Intern. Congress Math. Cambridge 1950*, vol. I, p. 530.
[450] N. Steenrod: *The topology of fibre bundles*, Princeton Univ. Press, 1951.
[451] N. Steenrod: Reduced powers of cohomology classes, *Ann. of Math.*, **56** (1952), 47–67.
[452] N. Steenrod: Homology groups of symmetric groups and reduced power operations, *Proc. Nat. Acad. Sci. USA*, **39** (1953), 213–217.
[453] N. Steenrod: Cyclic reduced powers of cohomology classes, *Proc. Nat. Acad. Sci. USA*, **39** (1953), 217–223.
[454] N. Steenrod: Cohomology operations derived from the symmetric group, *Comment. Math. Helv.*, **31** (1956/57), 195–218.
[455] N. Steenrod and D. Epstein, *Cohomology Operations*, Princeton Univ. Press, 1962 (Ann. of Math. Studies, No. 50).
[456] E. Steinitz: Beiträge zur Analysis Situs, *Sitzungsber. Berlin math. Gesellschaft*, **7** (1908), 29–49.
[457] E. Stiefel: Richtungsfelder und Fernparallelismus in Mannigfaltigkeiten, *Comment. Math. Helv.*, **8** (1936). 3–51.
[458] R. Stong: *Notes on Cobordism theory*, Princeton Univ. Press, 1958.
[459] R. Switzer: *Algebraic Topology: Homotopy and Homology*, Springer, Berlin–Heidelberg–New York–Tokyo, 1975.
[460] P.G. Tait: On knots, *Trans. Roy. Soc. Edinburgh*, **28** (1870), 145–190.
[461] R. Thom: Sur une partition en cellules associée à une fonction sur une variété, *C. R. Acad. Sci. Paris*, **228** (1949), 973–975.
[462] R. Thom: Espaces fibrés en sphères et carrés de Steenrod, *Ann. Ec. Norm. Sup.*, **69** (1952), 109–181.
[463] R. Thom: Quelques propriétés globales des variétés différentiables, *Comment. Math. Helv.*, **28** (1954), 17–86.
[465] E. Thomas: The generalized Pontrjagin cohomology operations and rings with divided powers, *Mem. Amer. Math. Soc.*, **27** (1957).
[466] H. Tietze: Über die topologischen Invarianten mehrdimensionaler Mannigfaltigkeiten, *Monatsh. für Math. u. Phys.*, **19** (1908), 1–118.
[467] H. Toda: Calcul de groupes d'homotopie des sphères, *C. R. Acad. Sci. Paris*, **240** (1955), 147–149.
[468] H. Toda: A topological proof of theorems of Bott and Borel-Hirzebruch for homotopy groups of unitary groups, *Mem. Coll. Sci. Univ. Kyoto, Ser. A, Math.*, **32** (1962), 103–119.
[469] H. Toda: *Composition methods in homotopy groups of spheres*, Princeton Univ. Press, 1962 (Ann. of Math. Studies, No. 49).
[470] J.A. Todd: The arithmetical invariants of algebraic loci, *Proc. Lond. Math. Soc.*, (2), **43** (1937), 190–225.
[471] A. Tucker: Degenerate cycles bound, *Math. Sborn.*, **3** (1938), 287–288.
[472] A. Tychonoff: Ein Fixpunktsatz, *Math. Ann.*, **111** (1935), 767–776.
[473] E. van Kampen: On the connection between the fundamental groups of some related spaces, *Amer. J. Math.*, **55** (1933), 255–260.
[474] O. Veblen: *Analysis Situs*, Amer. Math. Soc. Coll. Publ. No. 5 II, New York, 1921.

[475] L. Vietoris: Über den höheren Zusammenhang kompakter Räume und eine Klasse von zusammenhangstreue Abbildungen, *Math. Ann.*, **97** (1927), 454–472.

[476] L. Vietoris: Über die Homologiegruppen der Vereinigung zweier Komplexe, *Monatsh. für Math. u. Phys.*, **37** (1930), 159–162.

[477] B.L. van der Waerden: Kombinatorische Topologie, *Jahresber. der DMV*, **39** (1929), 121–139.

[478] B.L. van der Waerden: Topologische Begründung des Kalküls der abzählende Geometrie, *Math. Ann.*, **102** (1930), 337–362.

[479] C.T.C. Wall: Determination of the cobordism ring, *Ann. of Math.*, **72** (1960), 292–311.

[480] A. Wallace: *Homology theory of algebraic varieties*, Pergamon Press, 1958.

[481] H. Wang: The homology groups of the fiber bundles over a sphere, *Duke Math. J.*, **16** (1949), 33–38.

[482] A. Weil: *Oeuvres scientifiques*, vol. I, Springer, Heidelberg–Berlin–New York, 1979.

[483] H. Weyl: *Die Idee der Riemannschen Fläche*, Teubner, Leipzig, 1913.

[484] H. Weyl: Analysis Situs Combinatorio, *Rev. math. Hisp. amer.* **5** (1923), 209–218, 241–248, 278–279; **6** (1924), 33–41.

[485] G.W. Whitehead: On the homotopy groups of spheres and rotation groups, *Ann. of Math.*, **43** (1942), 634–640.

[486] G.W. Whitehead: A generalization of the Hopf invariant, *Ann. of Math.*, **51** (1950), 192–237.

[487] G.W. Whitehead: The $(n+2)$-nd homotopy of the n-sphere, *Ann. of Math.*, **52** (1950), 245–248.

[488] G.W. Whitehead: On the Freudenthal theorems, *Ann. of Math.*, **57** (1953), 209–228.

[489] G.W. Whitehead: Generalized homology theories, *Trans. Amer. Math. Soc.*, **102** (1962), 227–283.

[490] G.W. Whitehead: *Elements of homotopy theory*, Springer, Berlin–Heidelberg–New York, 1978.

[491] G.W. Whitehead: 50 years of homotopy theory, *Bull. Amer. Math. Soc.*, (N.S.), **8** (1983), 1–29.

[492] J.H.C. Whitehead: *The Mathematical Works of J.H.C. Whitehead*, vol. II: Complexes and Manifolds, Pergamon press London–New York, 1962.

[493] J.H.C. Whitehead: *The Mathematical Works of J.H.C. Whitehead, vol. III: Homotopy Theory*, Pergamon Press, London–New York, 1962.

[494] J.H.C. Whitehead: Simplicial spaces, nuclei and m-groups, *Proc. Lond. Math. Soc.*, (2), **45** (1939), 243–327 (also in [492], pp. 99–184).

[495] J.H.C. Whitehead: On adding relations to homotopy groups, *Ann. of Math.*, **42** (1941), 409–428 (also in [492], pp. 235–258).

[496] J.H.C. Whitehead: On incidence matrices, nuclei and homotopy types, *Ann. of Math.*, **42** (1941), 1197–1239 (also in [492], pp. 259–302).

[497] J.H.C. Whitehead: On C^1-complexes, *Ann. of Math.*, **41** (1940), 809–829. (also in [492], pp. 207–222).

[498] J.H.C. Whitehead: On the groups $\pi_r(V_{n,m})$ and sphere bundles, *Proc. Lond. Math. Soc.*, (2), **48** (1944), 243–291 and (2), **49** (1947), 478–481 (also in [492], pp. 303–356).

[499] J.H.C. Whitehead: Combinatorial homotopy, I, II, *Bull. Amer. Math. Soc.*, **55** (1949), 213–245 and 453–496 (also in [493], pp. 85–162).

[500] J.H.C. Whitehead: On the realizability of homotopy groups, *Ann. of Math.*, **50** (1949), 261–263 (also in [493], pp. 221–224).

[501] J.H.C. Whitehead: A certain exact sequence, *Ann. of Math.*, **52** (1950), 51–110 (also in [493], pp. 261–320).

[502] J.H.C. Whitehead: Simple homotopy types, *Amer. J. Math.*, **72** (1950), 1–57 (also in [493], pp. 163–220).

[503] J.H.C. Whitehead: On the theory of obstructions, *Ann. of Math.*, **54** (1951), 66–84 (also in [493], pp. 321–377).

[504] H. Whitney: Differentiable manifolds in euclidean space, *Proc. Nat. Acad. Sci. USA*, **21** (1935), 462–463.

[505] H. Whitney: Sphere spaces, *Proc. Nat. Acad. Sci. USA*, **21** (1935), 464–468.

[506] H. Whitney: Differentiable manifolds, *Ann. of Math.*, **37** (1936), 645–680.

[507] H. Whitney: The imbedding of manifolds in families of analytic manifolds, *Ann. of Math.*, **37** (1936), 865–878.

[508] H. Whitney: Topological properties of differentiable manifolds, *Bull. Amer. Math. Soc.*, **43** (1937), 785–805.

[509] H. Whitney: The maps of an *n*-complex into an *n*-sphere, *Duke Math. J.*, **3** (1937), 51–55.

[510] H. Whitney: On products in a complex, *Ann. of Math.*, **39** (1938), 397–432.

[511] H. Whitney: Tensor products of abelian groups, *Duke Math. J.*, **4** (1938), 495–528.

[512] H. Whitney: Some combinatorial properties of complexes, *Proc. Nat. Acad. Sci. USA*, **26** (1940), 143–148.

[513] H. Whitney: On the theory of sphere bundles, *Proc. Nat. Acad. Sci. USA*, **26** (1940), 148–153.

[514] H. Whitney: On the topology of differentiable manifolds, *Lectures in topology, Conference at Univ. of Michigan, 1940*, Univ. of Michigan Press, 1941, pp. 101–141.

[515] H. Whitney: The self-intersections of a smooth *n*-manifold in 2*n*-space, *Ann. of Math.*, **45** (1944), 220–246.

[516] H. Whitney: *Geometric Integration Theory*, Princeton, U. P., Princeton, NJ, 1957.

[517] G. Whyburn: *Analytic topology*, Amer. Math. Soc. Coll. Publ. No. 28, 1942.

[518] R. Wilder: *Topology of manifolds*, Amer. Math. Soc. Coll. Publ. No. 32.

[519] W. Wilson: Representation of manifolds, *Math. Ann.*, **100** (1928), 552–578.

[520] E. Witt: Treue Darstellung Liescher Ringe, *J. für reine u. angew. Math.*, **177** (1937), 152–161.

[521] Wu Wen-Tsün: On the product of sphere bundles and the duality theorem modulo two, *Ann. of Math.*, **49** (1948), 641–653.

[522] Wu Wen-Tsün: Classes caractéristiques et *i*-carrés d'une variéte, *C. R. Acad. Sci. Paris*, **230** (1950), 508–509.

[523] Wu Wen-Tsün: Les *i*-carrés dans une variété grassmannienne, *C. R. Acad. Sci. Paris*, **230** (1950), 918–920.

[524] Wu Wen-Tsün: *Sur les classes caractéristiques des structures fibrées sphériques*, Publ. de l'Inst. Math. de l'Univ. de Strasbourg, XI, Paris, Hermann, 1952.

[525] Yen Chih-Tah: Sur les polynômes de Poincaré des groupes de Lie exceptionnels, *C. R. Acad. Sci. Paris*, **228** (1949), 628–630.

[526] O. Zariski: Complete linear systems on normal varieties and a generalization of a lemma of Enriques–Severi, *Ann. of Math.*, **55** (1952), 552–592.

Index of Cited Names

Subject Index

CPSIA information can be obtained at www.ICGtesting.com
Printed in the USA
LVOW072157021011
248798LV00002B/2/P

9 780817 649067